General and Comparative Endocrinology

General and Comparative Endocrinology takes a holistic approach to endocrinology, introducing students to the diverse facets of this interdisciplinary science ranging from the medical to comparative domains, while also exploring evolutionary, environmental, and conservation specializations within the field. The textbook is founded on the principle that students interested in the health sciences will benefit from understanding how proficiency in endocrine function among a diversity of organisms contributes to advances in modern medicine. Likewise, students intrigued by comparative physiology will benefit from the wealth of knowledge derived from medical/clinical endocrinology, the historical bedrock of the field. This textbook represents the modern field of endocrinology in its totality by addressing topics and recent advances not currently discussed in other introductory endocrinology textbooks.

Key Features

- Introduces the broad and interdisciplinary scope of endocrinology
- Provides clear chapter objectives and key concepts
- Includes summary and synthesis questions for each chapter (and answers for instructors) that are suitable for exams and quizzes
- Includes a chapter devoted to endocrine-disrupting chemicals
- Describes the roles played by the endocrine system in important health challenges related to appetite regulation, obesity, diabetes, and other diseases stemming from 'mismatches to modernity'
- Integrates evolutionary and comparative approaches to hormones and health

General and Comparative Endocrinology

An Integrative Approach

A.M. Schreiber

CRC Press
Taylor & Francis Group
Boca Raton London New York

CRC Press is an imprint of the
Taylor & Francis Group, an **informa** business

Cover image art: A.M. Schreiber

First edition published 2024
by CRC Press
6000 Broken Sound Parkway NW, Suite 300, Boca Raton, FL 33487–2742

and by CRC Press
4 Park Square, Milton Park, Abingdon, Oxon, OX14 4RN

CRC Press is an imprint of Taylor & Francis Group, LLC

ISBN: 9781032416861 (hbk)
ISBN: 9781032416847 (pbk)
ISBN: 9781003359241 (ebk)

DOI: 10.1201/9781003359241

Typeset in Sabon
by Apex CoVantage, LLC

Access the Instructor Resources: www.routledge.com/9781032416847

For my students, mentors, and family—past, present, and yet to be. . .

—AMS

Contents

Unit Overview Unit VII: Reproduction

Unit Overview Unit VIII: Endocrine-Disrupting Chemicals

APPENDICES

Foreword

"About half of what you are going to learn in this endocrinology course is established fact, and half is wild speculation. The only problem is that I can't always be certain of which is which!" This was the opening statement, indelibly etched in my memory, that my PhD mentor, Professor Jennifer Specker, gave to her students taking their first endocrinology class in 1994 at the University of Rhode Island. Endocrinology is still a young and rapidly growing science that only originated in the late 19th and early 20th centuries. In their seminal 1962 book, *A Textbook of Comparative Endocrinology*, Aubrey Gorbman and Howard Bern acknowledged this juvenescence by stating at the start of their book, "Nowadays, it might be said that a student comes to a course in endocrinology with a reasonably good understanding of what it is, but that his [sic] instructor ordinarily might not share this enlightenment".[1] Fast-forward to the first half of the 21st century and we find that the field has grown and developed so substantially that most undergraduates now enter endocrinology courses with only vague preconceptions derived primarily from the field's medical applications. We, as endocrinology instructors, certainly have our work cut out for us.

General and Comparative Endocrinology: An Integrated Approach aims to expose students to the field in its totality, integrating essential concepts and contributions from both biomedical (general) and comparative endocrinology. This textbook introduces students to diverse modern facets of this deeply interdisciplinary science ranging from the medical to comparative domains, while also exposing students to relevant evolutionary, environmental, conservation, and toxicology specializations within the field. The book is founded on the principle that students interested in the health sciences will benefit from understanding how proficiency in endocrine function among a diversity of organisms contributes to advances in modern medicine. Likewise, students intrigued by comparative physiology will benefit from the wealth of knowledge derived from medical/clinical endocrinology, the historical bedrock of the field. Indeed, the historical discoveries made by researchers at the field's forefront are sometimes only surpassed by the remarkable life stories and personalities of the researchers themselves, many of whom were/are iconoclasts, risk takers, rule breakers, and even barroom brawlers. If this textbook can be thought of as a recipe for making "endocrine soup", with a tasty assortment of facts and information on a variety of endocrine organs, tissues, hormones, and other delicacies, then this particular ductless gland jambalaya is spiced with more than a dash of the field's vibrant history and colorful characters. A recipe for a savory meal, however, is only as good as the chef's skill and ingenuity in preparing the dish. You, the course instructor, are that innovative chef, and I encourage you to adapt this recipe to your unique palate through creative improvisation, supplementation, and substitution of ingredients and spices. After all, you are preparing this dish for the most discerning of all customers—your students. Bon appétit!

LITERATURE CITED

1. Gorbman A, Bern HA. *A Textbook of Comparative Endocrinology* (1st ed.). Wiley; 1962.

Acknowledgments

Writers love writing, but the same level of enthusiasm for the writer's project is understandably often not shared by their significant others, children, and even pets who compete for the limited resources of time and attention. Although the aforementioned would likely be better off without my having invested time into this project, the completion of this book would never have been possible without their stalwart support, love, understanding, and good humor. It is only fitting, then, that my first acknowledgments go to those who felt the brunt of the writing the most: Max, Sam, Jack, Ben, Maddie, Adrian, Cassidy, and Stormy, and certainly to my Mom, Mona Espy Schreiber (also to Snowball, Seuss, Bjorn, Doom Kitty, and the Big Black Cat).

I wish to express special gratitude to Professor Duncan MacKenzie (Texas A&M University) who went well above and beyond the call of duty as a reviewer of this textbook. His vast experience and excellence in both research and teaching endocrinology has profoundly influenced the final shape of this book. Professor MacKenzie is that critical transcription factor that ensures all the right genes are turned on at the correct time and place during ontogeny. I also want to thank the students from the Fall 2020 Biology 405 Comparative Endocrinology class at Texas A&M University, as well as dozens of students from my BIOL 370 Hormones Disease and Development classes at St. Lawrence University who test-drove many book chapters. Their constructive comments and unique perspectives played an essential role in making the book content as interesting, accessible, and student-friendly as possible.

I am indebted to several amazing St. Lawrence University research librarians who assisted me in obtaining countless interlibrary loans and hard-to-find historical resources. These are Melissa Burchard, Julia Courtney, and Gwendolyn Cunningham. My appreciation is also directed to the many anonymous folks at the Copyright Clearance Center who tirelessly and efficiently processed hundreds of copyright permissions requests. I would also like to acknowledge St. Lawrence University for providing me with two sabbatical leaves dedicated to the writing of this book, as well as grant funding to cover a variety of the book's publication costs.

The first edition of any textbook is particularly challenging to assemble, and I am especially beholden to several folks at the Taylor & Francis Group who were wonderful midwives to this book's birth. The first is Kara Roberts, Editorial Assistant for Life Sciences, who was instrumental in mentoring me through diverse aspects of the complex landscape of digital publishing. The second is Senior Acquisitions Editor for Life Sciences Dr. Chuck Crumly. I knew that I had found a nurturing home for this book project upon learning that in a previous life Chuck was a senior editor for none other than the journal *General and Comparative Endocrinology* and was a contemporary of and influenced by Howard Bern, to boot! Thirdly, I am grateful to Medical Specialist Project Editor Kyle Meyer, as well as to Project Manager Balaji Karuppanan and their teams for copyediting the text and figures and bringing the book to life.

Lastly, the quality of any textbook is only as good as the efforts placed by expert reviewers in evaluating the accuracy, appropriateness, and accessibility of the content. I am in great debt to these "guardians of the academic gates".

Reviewers of the 1st Edition

Lisa Abrams, *Rowan University*
J.P. Advis, *Rutgers University*
Maryam Bamshad, *Lehman College, CUNY*
Morgan Benowitz-Fredericks, *Bucknell University*
Juan Bernal, *Instituto de Investigaciones Biomédicas, Spain*
Don Brown, *Carnegie Institution of Science*
Dan Buccholz, *University of Cincinnati*
Heather Caldwell, *Kent State University*
Russell W. Chesney *University of Tennessee Medical Center*
Brian D. Cohen, *Union College*
Rachel Cohen, *Minnesota State University, Mankato*
Robert V. Considine, *Indiana University School of Medicine*
Cynthia Corbitt, *University of Louisville*
Shannon Davis, *University of South Carolina*
Gregory Demas, *Indiana University*
Bob Denver, *University of Michigan*
Pierre Deviche, *Arizona State University*
Robert Dores, *University of Denver*
Ryan Earley, *University of Alabama*
Lia Edmunds, *University of Pittsburgh*
Buffy S. Ellsworth, *Southern Illinois University*
Michael Ferkin, *University of Memphis*
James Gelsleichter, *University of North Florida*
Dalia Giedrimiene, *University of Saint Joseph*
Andreas Heyland, *University of Guelph*
Mary Holder, *Georgia Institute of Technology*
Robert Hurd, *Xavier University*
Jerry Husak, *University of St. Thomas*
Mark Jackson, *Central Connecticut State University*

Orin James, *University of Pittsburgh-Bradford*
Marlo Jeffries, *Texas Christian University*
Amy E. Jetton, *Middle Tennessee State University*
David Kabelik, *Rhodes College*
Jesse S. Krause, *University of Nevada*
Dianne Figlewicz Lattemann, *VA Puget Sound Health Care System*
Vincent Laudet, *Institut de Génomique Fonctionnelle de Lyon*
Sean Lema, *California Polytechnic State University*
Christopher Loretz, *SUNY Buffalo*
Duncan MacKenzie, *Texas A&M University*
Richard Manzon, *University of Regina*
Steve McCormick, *University of Massachusetts*
Elizabeth McCullagh, *Oklahoma State University*
Christina McKittrick, *Drew University*
Lee Meserve, *Bowling Green State University, Ohio*
Gergana G. Nestorova, *Louisiana Tech University*
Abigail Neyer, *University of North Georgia*
Justicia Opoku-Edusei, *University of Maryland*
Frank V. Paladino, *Purdue University*
Alejandro Relling, *The Ohio State University*
Erin Rhinehart, *Susquehanna University*
Lauren Riters, *University of Wisconsin, Madison*
Aurea Orozco Rivas, *Universidad Nacional Autónoma de México*
Edmund Rodgers, *Georgia State University*
Christopher Rose, *James Madison University*
David C. Rostal, *Georgia Southern University*
David Rubin, *Illinois State University*
Laurent Sachs, *French National Centre for Scientific Research Institute of Biological Sciences*
Wendy Saltzman, *University of California, Riverside*
Haifei Shi, *Miami University, Ohio*
Warner S. Simonides, *VU University Medical Center, Amsterdam*
Mitch T. Sitnick, *Montclair State University*
Wendy Smith, *Northeastern University*
Stacia Sower, *University of New Hampshire*
James Squires, *University of Guelph*
Jacqueline Stephens, *Louisiana State University*
Jianjun Sun, *University of Connecticut*
Alan Vajda, *University of Colorado, Denver*
Brian G. Walker, *Fairfield University*
Michael Wasserman, *Indiana University*
Allison Kaese Wilson, *Benedictine University*
Tristram Wyatt, *University of Oxford*
Zhi Zhang, *University of Michigan-Dearborn*
Tom Zoeller, *University of Massachusetts, Amherst*
Susan Zup, *University of Massachusetts, Boston*

The author is also grateful for constructive comments provided by several anonymous reviewers.

Author Biography

A.M. Schreiber is a comparative endocrinologist at St. Lawrence University where he holds the R. Sheldon '68 and Virginia H. Johnson Professorship in the Sciences. He teaches classes in endocrinology, cell biology, and physiology. His research addresses vertebrate metamorphosis, focusing on the influences of thyroid hormones and glucocorticoid stress hormones on amphibian development. He received a B.A. in biology from the University of Colorado, Boulder. After serving in the U.S. Peace Corps in Kenya, he went on to earn an M.S. in biology from Eastern Washington University studying the osmoregulatory physiology of migrating salmon under the tutelage of Dr. Ronald J. White. He then earned a Ph.D. in zoology at the University of Rhode Island studying the endocrinology of flatfish metamorphosis with Dr. Jennifer L. Specker. He pursued postdoctoral research at the Carnegie Institution's Department of Embryology, studying the molecular biology of amphibian metamorphosis in Dr. Donald Brown's laboratory. His summers are spent in East Africa where he teaches classes in wildlife biology and high altitude physiology.

Abbreviations

1R	whole genome duplication event 1
2R	whole genome duplication event 2
3R	whole genome duplication event 3
3β-HSD	3β-hydroxysteroid dehydrogenase
A	aldosterone
A4	androstenedione
AANAT	N-acetyltransferase
ABP	androgen-binding protein
ACE	angiotensin-converting enzyme
ACTH	adrenocorticotropic hormone
ADH	antidiuretic hormone
ADHD	attention deficit hyperactivity disorder
AdipoQ	adiponectin
AGP	adrenogonadal primordium
AgRP	agouti-related peptide
AGT	angiotensinogen
AhR	aryl hydrocarbon receptors
AMH	anti-Müllerian hormone
AncCR	ancestral corticoid receptor
ANP	atrial natriuretic peptide
ANR	anterior neural ridge
AR	adrenergic receptor
AR	androgen receptor
Arc	arcuate nucleus
ATP	adenosine triphosphate
AVP	arginine vasopressin
AVPV	anteroventral periventricular nucleus
AVT	arginine vasotocin
B	corticosterone
BAT	brown adipose tissue
BeAT	beige adipose tissue
BMAL1	brain and muscle aryl-hydrocarbon receptor nuclear translocator-like 1
BMI	body mass index
BPA	bisphenol A
BRCA	breast cancer gene
brite	brown in white adipocytes
BST	bovine somatotropin
CAH	congenital adrenal hyperplasia
CAIS	complete androgen insensitivity syndrome
cAMP	cyclic AMP
CaMs	calmodulins
CART	cocaine-and amphetamine-related transcript
CaSR	calcium-sensing receptors
CCAC	cervicovaginal clear-cell adenocarcinoma
CCK	cholecystokinin
CFTR	cystic fibrosis transmembrane regulator
CG	chorionic gonadotropin
cGMP	cyclic GMP
CHH	crustacean hyperglycemic hormone
Clk	clock
CLOCK	circadian locomotor output cycles kaput
CNS	central nervous system
COCP	combined oral contraceptive pill
COX	cyclooxygenases
CREBs	cAMP response element-binding proteins
CREs	cAMP response elements
CRH	corticotropin-releasing hormone
CRH-BP	CRH binding protein
CRP	C-reactive protein
CRY	cryptochrome
CS	chorionic somatomammotropin
CT	calcitonin
CTBP	cytosolic T_3-binding proteins
CTX	cholera toxin
cyc	cycle
CYP	cytochrome P-450 monooxidases
D1	type 1 deiodinase
D2	type 2 deiodinase
D3	type 3 deiodinase
DAG	diacylglycerol
DAX1	dosage-sensitive sex reversal, adrenal hypoplasia critical region, on chromosome X, gene 1
DBD	DNA-binding domain
dbt	double-time
DDT	dichloro-diphenyl-trichloroethane
DEHAL1	iodotyrosine dehalogenase 1
DES	diethylstilbestrol
DHEA	dehydroepiandrosterone
DHEA-S	dehydroepiandrosterone-sulfate
DHN	dorsomedial hypothalamic nucleus
DHT	dihydrotestosterone
DIT	diiodotyrosine
DMRT1	doublesex and mab-3-related transcription factor 1
DMV	dorsal motor nucleus of the vagus
DNH/VMH	dorso/ventromedial hypothalamus
DNMTs	DNA methyltransferases
DOC	deoxycorticosterone
DOHaD	developmental origins of health and disease
DSD	disorders of sex development
DSD	differences of sex development
Dyn	dynorphin A
E1	estrone
E2	estradiol
E3	estriol
EcR	ecdysone receptor
EcRE	ecdysone response element
EDC	endocrine-disrupting chemical
EGF	epidermal growth factor
ELISA	enzyme-linked immunosorbent assay

EMT	extraneuronal monoamine transporter
ENaC	epithelial sodium channel
EPO	erythropoietin
EPSP	excitatory post-synaptic potential
ER	endoplasmic reticulum
ER	estrogen receptor
ESD	environment-dependent sex determination
ETH	ecdysis triggering hormone
F	cortisol
FDA	Food and Drug Administration
FFAs	free fatty acids
FGF23	fibroblast growth factor 23
FOXL2	forkhead box protein L2
FSH	follicle-stimulating hormone
G6P	glucose-6-phosphate
G6Pase	glucose-6-phosphatase
GABA	gamma-aminobutyric acid
GDP	guanosine diphosphate
GFR	glomerular filtration rate
GH	growth hormone
GHRH	growth hormone-releasing hormone
GIP	gastric inhibitory peptide
GIP	glucose-dependent insulinotropic polypeptide
GLP	glucagon-like peptide
GLUT2	glucose transporter 2
GnRH	gonadotropin-releasing hormone
GPCR	G protein-coupled receptor
GPER	G protein-coupled estrogen receptor
GR	glucocorticoid receptor
GRE	glucocorticoid response elements
GRKs	G protein receptor kinases
GRP	gastrin-releasing peptide
GSD	genotypic sex determination
GSK3	glycogen synthase kinase 3
GTP	guanosine triphosphate
$G\alpha_i$	inhibitory Gα
$G\alpha_q$	phospholipase-associated Gα
$G\alpha_s$	stimulatory Gα
$G\alpha_t$	transducin Gα
HATs	histone acetyltransferases
hCG	human chorionic gonadotropin
HDAC	histone deacetylase
HDL	high-density lipoprotein
HIOMT	N-acetylserotonin O-methyltransferase
HMTs	histone methyltransferases
HRE	hormone response element
HSD	hydroxysteroid dehydrogenase
HSL	hormone-sensitive lipase
HSPs	heat shock proteins
IGF-I	insulin-like growth factor I
IL6	interleukin-6
INSL3	insulin-like peptide 3
IP_3	inositol triphosphate
ipRGC	intrinsically photosensitive retinal ganglion cells
IPSP	inhibitory post-synaptic potential
IVF	in vitro fertilization
JAK	Janus kinases
JH	juvenile hormone
K_d	equilibrium dissociation constant
KNDy	kisspeptin-NKB-Dyn
LBD	ligand-binding domain
LDL	low-density lipoprotein
LEN	light exposure at night
LEP	leptin
LH	luteinizing hormone
LHA	lateral hypothalamic area
LOX	lipoxygenases
LPH	lipotropic peptide hormone
M-CSF	macrophage colony-stimulating factor
mAChRs	muscarinic acetylcholine receptors
MAP	mitogen-activated protein
MAPK	MAP kinase
MCH	melanin-concentrating hormone
MCR	melanocortin receptor
MCTs	monocarboxylate transporters
MF	methylfarnesoate
MIH	molt-inhibiting hormone
MIT	monoiodotyrosine
MMPs	matrix metalloproteinases
MOIH	mandibular organ-inhibiting hormone
MPOA	medial preoptic area
MR	mineralcorticoid receptor
MRAPs	melanocortin receptor accessory proteins
MSH	melanocyte-stimulating hormone
MT	melatonin
NAFLD	non-alcoholic fatty liver disease
NaPiIIb	sodium-dependent phosphate cotransporter IIb
NCC	sodium chloride cotransporter
NCX	Na^+/Ca^{++} exchangers
NGF	nerve growth factor
NGFI-A	nerve growth factor-inducible protein A
NIS	sodium iodide symporter
NKB	neurokinin B
NPY	neuropeptide Y
NRF	nuclear respiratory factors
NSAIDs	non-steroidal anti-inflammatory drugs
NTS	nucleus of the solitary tract
OATPs	organic anion transporter proteins
OPG	osteoprotegerin
OT	oxytocin
OVLT	organum vasculosum of the lateral terminalis
p,p′-DDE	dichloro-diphenyl-dichloroethylene
P4	progesterone
$P450_{aldo}$	aldosterone synthetase
$P450_{C11b}$	11b-hydroxylase
PAI-1	plasminogen activator inhibitor-1
PAIS	partial androgen insensitivity syndrome
PC	prohormone convertases
PC	pyruvate carboxylase
PCB	polychlorinated biphenyl
PEPCK	carboxykinase
PGC-1α	peroxisome proliferator-activated receptor gamma coactivator 1 alpha
PGCs	primordial germ cells

PI3K phosphoinositide-3-kinase
PIH prolactin-inhibiting hormone
PIP2 phosphatidylinositol-4,5-bisphosphate
PKA protein kinase A
PKC protein kinase C
PKG protein kinase G
PL placental lactogen
PLA2 phospholipase A_2
PLC phospholipase C
PMCA plasma membrane Ca^{++} ATPase
PMDS persistent Müllerian duct syndrome
PNMT phenylethanolamine N-methyltransferase
Polrmt mitochondrial DNA-directed RNA polymerase
POMC proopiomelanocortin
PP pancreatic polypeptide
PPAR peroxisome proliferator-activated receptor
PRH prolactin-releasing hormone
PRL prolactin
PRTH pre-ecdysis triggering hormone
PTH parathyroid hormone
PTHrP parathyroid hormone-related protein
PTSD post-traumatic stress disorder
PTTH prothoracicotropic hormone
PTX pertussis toxin
PVN paraventricular nucleus
PYY peptide YY
R_0 total receptor number
RAAS renin-angiotensin-aldosterone system
RANKL receptor activator of nuclear factor kappa-B ligand
RAR retinoic acid receptor
RBF renal blood flow
RER rough endoplasmic reticulum
RHT retinohypothalamic tract
RIA radioimmunoassay
RNA ribonucleic acid
ROR retinoic acid-related orphan receptor
ROS reactive oxygen species
RSPO1 R-spondin-1
RTKs receptor tyrosine kinases
RXR retinoid X receptor
SAD seasonal affective disorder
SCAT subcutaneous adipose tissue
SCN suprachiasmatic nucleus
SER smooth endoplasmic reticulum
SERMs selective estrogen receptor modulators
SFO subfornical organ
SHBG sex hormone-binding globulin
SHOX short stature homeobox
SL somatolactin
SOCS suppressor of cytokine signaling
SON supraoptic nucleus
SOX9 SRY-box transcription factor 9
SRY sex-determining region on the Y chromosome
SST somatostatin
StAR steroidogenic acute regulatory protein
STAT signal transducers and activators of transcription
T1DM type 1 diabetes mellitus
T2 testosterone
T2DM type 2 diabetes mellitus
T_3 triiodothyronine
T_4 thyroxine
TAG triacylglycerol
TBG thyroxine-binding globulin
TDS testicular dysgenesis syndrome
Tfam mitochondrial transcription factor A
TIDA tuberoinfundibular dopaminergic neurons
tim timeless
TNFa tumor necrosis factor a
TPO thyroid peroxidase
TR thyroid hormone receptor
TRE thyroid hormone response element
TRH thyrotropin-releasing hormone
TRPV transient receptor potential cation channel subfamily V
TSD temperature-dependent sex determination
TSH thyroid-stimulating hormone
TSI thyroid-stimulating immunoglobulins
TTR transthyretin
TTX tetrodotoxin
TZDs thiazolidinediones
UCP1 uncoupling protein 1
UDPGTs uridine diphosphate glucuronyl transferases
USP ultraspiracle protein
UVB ultraviolet B radiation
VAT visceral adipose tissue
VDR vitamin D receptor
VEGFA vascular endothelial growth factor A
VIP vasoactive intestinal peptide
VLDL very low-density lipoprotein
VTG vitellogenin
WAT white adipose tissue
WNT4 wingless-type MMTV integration site family member 4
ZES Zollinger–Ellison syndrome
αGSU alpha glycoprotein subunit

Unit Overview
Unit I

Introduction to Endocrinology

An "endocrine bouquet", illustrating how diverse animal taxa draw life from a common pool of evolutionarily conserved hormones.

DOI: 10.1201/9781003359241-1

OPENING QUOTATION:

> "It seems to me fairly evident that physiological factors, especially our endocrines, control our destiny."
> —Albert Einstein (1929), From "What Life Means to Einstein: An Interview by George Sylvester Viereck", *The Saturday Evening Post*, Oct. 26

Welcome to the world of endocrinology! Endocrinology is a branch of physiology that studies a specific class of blood-borne chemical signals called *hormones*, as well as the cells that generate hormones and their actions on target cells and tissues. Unit I serves as a broad introduction to the topic. Chapter 1 defines the scope of study of endocrinology, describes the extensive influences that hormones have over diverse physiological and behavioral processes among animals, and discusses the relevance of endocrinology to society. The chapter also provides a historical overview of the field's growth and development, from its ancestral origins to maturation into its two modern incarnations: general (biomedical) and comparative endocrinology. Chapter 2 introduces the reader to some of the fundamental features of hormones and endocrine signaling systems. This includes familiarization with the chemical diversity of hormones, their methods of transport through the blood, interactions with target-specific receptors and intracellular signaling pathways, as well as the higher-order notion of feedback regulation for the maintenance of homeostasis. Chapter 3 describes how an understanding of the mechanisms of evolution and the results of natural selection contribute to advancements in both human and comparative endocrinology.

Chapter 1: **The Scope and Growth of Endocrinology**
Chapter 2: **Fundamental Features of Endocrine Signaling**
Chapter 3: **Evolution of Endocrine Signaling**

CHAPTER

1

The Scope and Growth of Endocrinology

CHAPTER LEARNING OBJECTIVES:

- Define "endocrinology", and describe how the concept of chemical communication is integral to it.
- Describe the relevance and benefits of endocrinology to society, as well as some of the challenges associated with certain hormone technologies.
- Define the scope of inquiry of modern endocrinology, and describe how the fields of "general endocrinology" and "comparative endocrinology" interact to advance endocrine science as a whole.
- Describe key advances in physiological knowledge prior to the 19th century that laid the foundation for the emergence of endocrinology, and discuss seminal endocrine experiments during the late 19th and early 20th centuries that established the field.
- Describe some paradigm-shifting concepts and technologies that emerged during the late 20th and early 21st centuries that transformed endocrinology into its modern incarnations.

OPENING QUOTATIONS:

> "The history of the discovery of hormones abounds with bizarre experiments, remarkable characters, wrong turns, opportunism and quackery. However, there are elements of genius and amazing tales of survival."
>
> —John Wass, Professor of Endocrinology at Oxford University[1]

> "We recognize that each tissue and, more generally, each cell of the organism secretes . . . special products or ferments into the blood which thereby influence all the other cells thus integrated with each other by a mechanism other than the nervous system."
>
> —Charles Édouard Brown-Séquard and J. d'Arsonval (1891), article in *Comptes Rendus de la Société de Biologie*

KEY CONCEPTS:

- Endocrinology is a branch of physiology that studies blood-borne chemical signals called "hormones", the cells that produce them, and their actions on target tissues.
- Hormones influence virtually all aspects of physiology, from mediating growth, development, and reproduction to metabolism and homeostasis.
- Endocrine signaling systems consist of cells that transmit hormone signals, binding proteins that may help transport hormones through the blood, receptors on/in target cells that receive the signals, and intracellular chemical pathways that transduce and amplify signals into responses.
- Disruption of endocrine signaling anywhere along the hormone transmission-reception-transduction pathway may lead to disease.

Endocrinology: A Science of Chemical Communication

LEARNING OBJECTIVE Define "endocrinology" and describe how the concept of chemical communication is integral to it.

Welcome to the exciting world of endocrinology! Endocrinology is a branch of **physiology**, the discipline in the life sciences that studies the forms, functions, and mechanisms of action of living systems, ranging from biomolecules and cells to tissues, organs, and organisms. Before we tackle the specific question of "what is endocrinology?", it is useful to first take a big step back and ponder a much more fundamental question: "what is life?"

What Is Life?

From art to religion, and from philosophy to physics, there are numerous ways of defining life. For our purpose, perhaps the most useful definition of life is that put forth by Arnold De Loof, a comparative endocrinologist and evolutionary biologist who elegantly described life using the principle "Life is a *verb* and its very essence refers to communication".[2] Using simple mathematical language, De Loof defined life as:

$$L = \sum C$$

DOI: 10.1201/9781003359241-2

where "life" (L) is an activity denoted by the total sum (Σ) of all acts of "communication" (C). More specifically, life can be expressed in terms of mathematical integration as

$$L = \sum_{i=1}^{N} C_i$$

where acts of communication are executed at all levels of **compartmental organization** (i), from the lowest level of compartmentalization (1 = an organelle within a cell) to the highest level of compartmentalization in question (N = a cell, tissue, organ, organism, population, community, ecosystem, etc.) Indeed, the notion of compartmental organization is integral to life, and the term "organism", which is an example of a highly structured, self-organizing system, reflects this.

Building on the notion of compartmental organization, a *communication system* consists of a *transmitter* located in one compartment that emits messages, such as radio signals, sound waves, and electrical or chemical signals, that are written in coded form. A *receiver* located in another compartment decodes the signal, which is then *transduced*, or converted from one form of signal to another. Finally, the signal is *amplified*, or its output becomes proportionately greater compared to the original input, into a response that entails some form of work by the receiving compartment. Within an organism an example of a biochemical transmitter could be a cell that emits a chemical signal that is transported by the circulatory system; a receiver may be a protein receptor located in the plasma membrane of a target cell; various different proteins and enzymes within the cell may subsequently transduce and amplify the signal into a response that could manifest as an increase in transcription of a certain gene. An analogy between a radio and a cellular chemical signaling pathway is shown in Figure 1.1. As proper communication is essential to a healthy life, it follows that errors in signaling interactions may result in disease. Signal disruptions may be caused by abnormally high or low levels of transmission, reception, transduction, or amplification. Signals may be transmitted at the wrong time or to the incorrect target. Furthermore, communication may be interfered with through the presence of disruptive signals or excessive background noise. Some examples of human diseases associated with cellular dysfunction in one component of a chemical signaling pathway are described in Table 1.1.

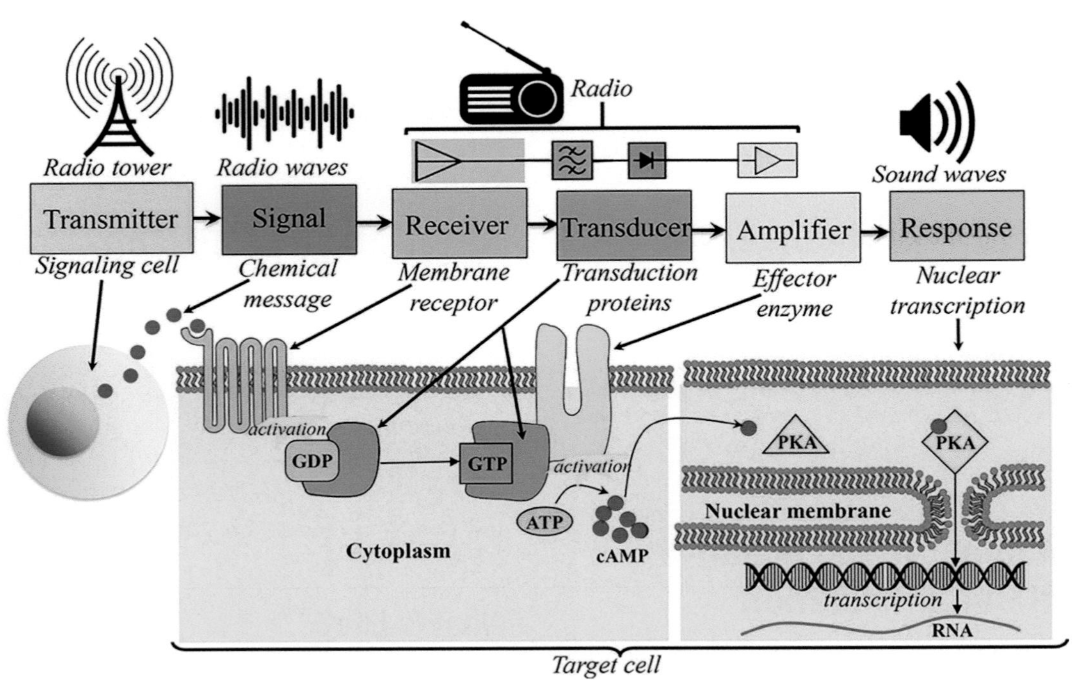

Figure 1.1 Conceptual similarities between radio and cellular signaling pathways. A radio tower transmits signals in the form of radio waves (electromagnetic radiation). A radio contains a receiver in the form of an antenna, as well as transducers that convert radio waves into electrical signals and amplifiers that boost the signal strength. Ultimately a speaker responds to the electrical signals by transducing them into sound waves. A signaling cell secretes a signal in the form of a chemical messenger into the extracellular fluid or blood. The signal then binds to a receptor on the target cell. Depicted here is a common type of plasma membrane-localized receptor called a G protein-coupled receptor (GPCR). Intracellular transducer proteins in this pathway, called G proteins, interact with the ligand-bound receptor and transduce the signal via a series of conformational changes mediated by the hydrolysis of GTP that subsequently activate an effector enzyme. The enzyme amplifies the signal by synthesizing large amounts of a new chemical called a "second messenger". The effector in this example is called "adenylyl cyclase", an enzyme that synthesizes the second messenger, cAMP, from ATP. In this pathway the amplified signal interacts with other downstream proteins that ultimately promote a response in the form of gene transcription.

Table 1.1 Some Human Diseases Associated with Cellular Dysfunction in One Component of a Chemical Signaling Pathway

	SIGNALING PATHWAY			
DISEASE	**Transmission**	**Reception**	**Transduction/ amplification**	**EFFECT**
Type I diabetes	Inhibited ability of pancreatic beta cells to synthesize insulin	Normal insulin receptor function	Normal transduction of signal within target tissues	High blood glucose
Hyperthyroidism	Excessive synthesis of thyroid hormones by the thyroid gland	Normal thyroid receptor function	Normal transduction of signal within target tissues	Increased metabolic rate
Androgen insensitivity syndrome	Normal ability of testes to synthesize androgens	Androgen receptor has a loss-of-function mutation	Transduction machinery intact	Genetic XY individuals develop with an androgen-resistant female phenotype
Human epidermal growth factor receptor 2 (HER2)-related cancers	Normal synthesis of epidermal growth factors	Excessive receptor activation due to high levels of constitutively active HER2 receptor	Normal transduction machinery	Excessive cell proliferation
Type II diabetes	Normal ability of pancreatic beta cells to synthesize insulin	Normal insulin receptor function	Inhibited transduction of signal within target tissues	High blood glucose
Ras-associated cancers	Normal synthesis of growth factors	Normal growth factor receptor function	Excessive signal transduction activity, even in the absence of growth factor signal. Caused by a gain of function mutation in the "Ras" transducer protein	Excessive cell proliferation

Note: The darkly shaded region indicates the dysfunctional component of the signaling pathway.

What Is Endocrinology?

So, what exactly is endocrinology, and how is the notion of communication so integral to the field? Broadly defined, **endocrinology** is a branch of physiology that studies a specific class of blood-borne chemical signals called **hormones**, as well as the cells that generate hormones, and the resulting hormone actions on target cells and tissues. The word "endocrinology", derived from the Greek roots *endon*, meaning "within," and *krinein*, meaning "to release", literally denotes the study of substances that are released "internally", or directly into the blood. "Hormones", derived from the Greek word *hormôn*, to excite or arouse, are so named due to the excitatory effects they impart on their target tissues. Hence, endocrinology is, both literally and figuratively, a very *exciting* field of study!

Hormones are secreted by diverse organs and tissues (Figure 1.2) into the surrounding extracellular fluids, cerebral spinal fluids, and blood of all animals possessing these body substances. After being transported varying distances though these media, hormones bind to specific receptors on or inside target cells and thereby facilitate communication among the various "compartments" that constitute living organisms. As such, hormones profoundly influence countless aspects of normal development, physiology, and behavior. Some examples of hormone-mediated processes, shown in Figure 1.3, include the following:

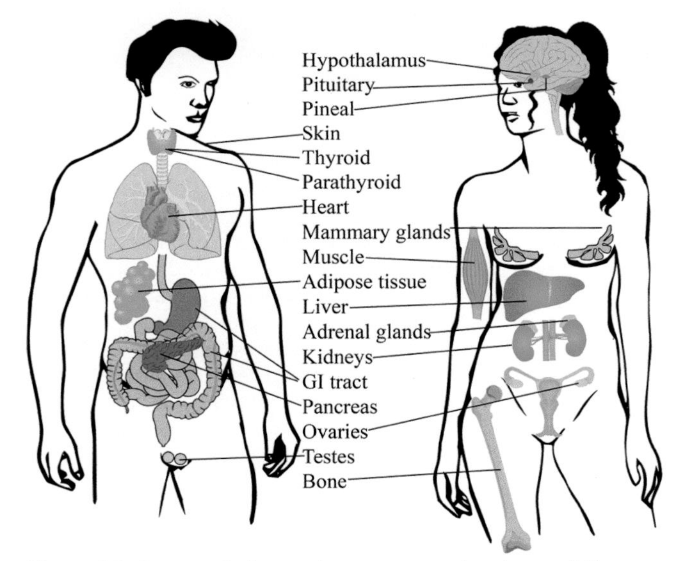

Figure 1.2 Some of the major organs, glands, and tissues that possess endocrine functions in the human body. In addition to classical endocrine glands, like the thyroid, pancreas, ovaries, and testes, we now know that other organs, like bone, adipose tissue, cardiac, and skeletal muscle also release hormones. One important transient endocrine organ that is not depicted is the placenta, which is attached to the uterus and is only present in the pregnant female.

Source of images: Modified from Flaws, J., et al. 2020. Plastics, EDCs and Health. Washington DC: Endocrine Society.

- Coordination of early developmental events like organogenesis. Indeed, in mammals the fetal endocrine system

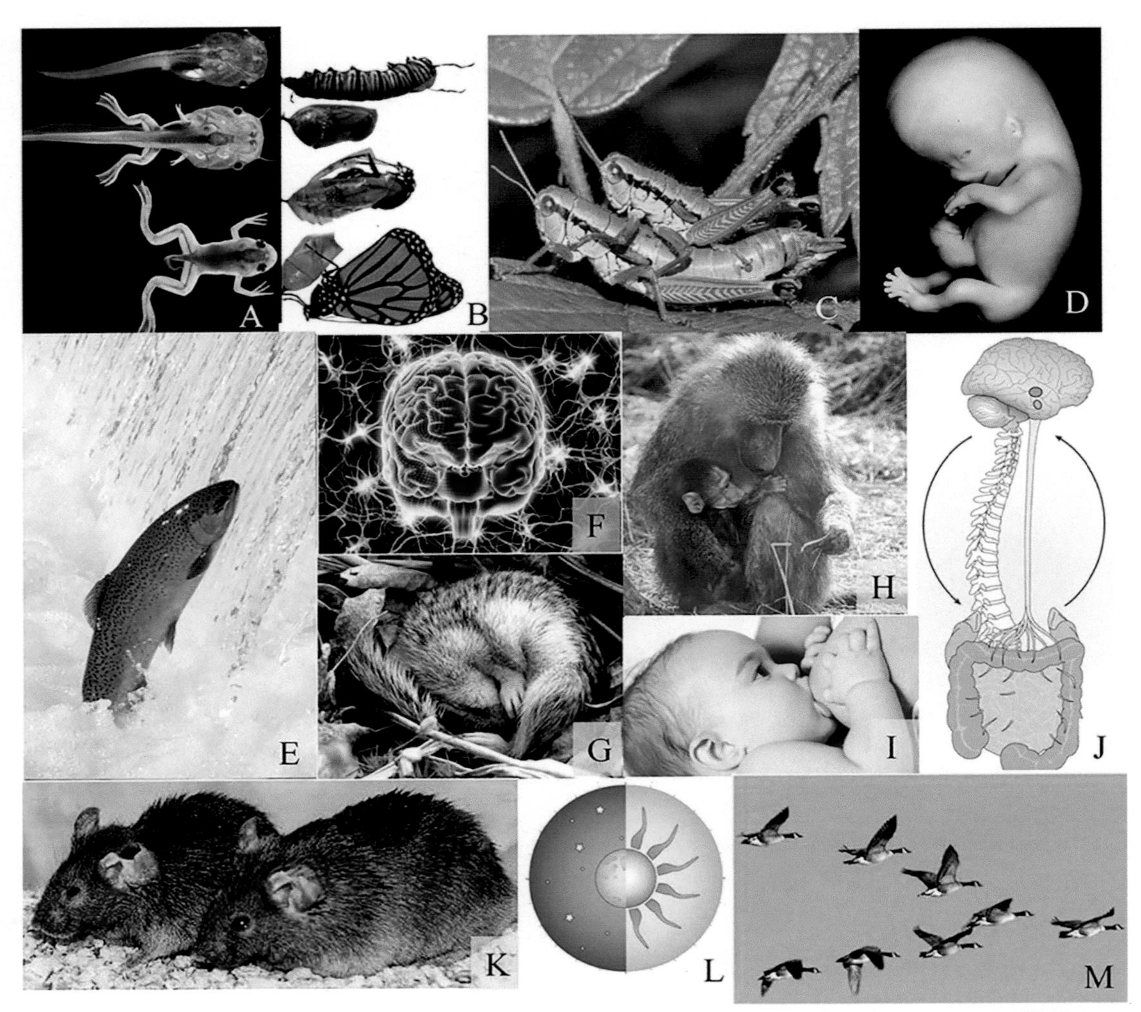

Figure 1.3 The broad reach of hormones: some examples of normal hormone-mediated processes. **(A)** Metamorphosis in amphibians is driven primarily by thyroid hormones and glucocorticoids, whereas in insects **(B)** this transition is mediated by changes in ecdysone and juvenile hormone. **(C)** Mating physiology and behavior in all animals are regulated by complex endocrine interactions between the brain and gonads. **(D)** During the fetal period the growth and development of all organs and tissues are influenced by diverse hormones of both maternal and fetal origin. **(E)** Breeding-associated salmon migration is also accompanied by dramatic changes in water salinity that are mediated by hormones with osmoregulatory function such as prolactin and glucocorticoids. **(F)** The brain is a complex neuroendocrine organ, and its study forms the basis of the field of neuroscience. **(G)** The actions of various metabolic hormones promote seasonal ground squirrel hibernation. **(H)** Bonding behaviors among mammals is mediated in part by the pituitary hormones oxytocin and vasopressin. **(I)** Lactation in mammals is facilitated by the synergistic actions of prolactin and oxytocin on mammary tissue. **(J)** The diverse hormones of the brain-gut axis mediate eating behaviors, digestion, and metabolism. **(K)** Senescence reflects the actions of a broad spectrum of hormones. An aging mouse treated to clear senescent cells, seen on the right, has a healthier appearance than the untreated littermate on the left. **(L)** Circadian and other biological rhythms are under control by hormones such as melatonin. **(M)** Seasonal bird migration is driven by reproductive and metabolic hormones.

Source of images: (B) Courtesy of Ba Rea (barea@basrelief.org); (C) Lemonnier-Darcemont, M. and Darcemont, C., 2021. J. Sci. Env. Technol. 3, pp. 150-160; (D) Moore, K.L., Persaud, T.V.N., et al., 2018. The developing human-e-book: clinically oriented embryology; used with permission (E) U.S. Fish and Wildlife Service. (F) Baker, D.J. et al. 2016. Nature, 530, pp. 184-189; used with permission (G) Feng, N.Y. et al., 2019. Curr. Biol. 29, pp. 3053-3058; used with permission (H) Boulinguez-Ambroise, G., et al. 2020. Sci. Rep. 10(1), pp. 1-8; (I) National Health Service, UK. (J) Morais, L.H. et al., 2021. Nat. Rev. Microbiol, 19(4), pp. 241-255; used with permission (K) Baker, D.J. et al., 2016. Nature, 530(7589), pp. 184-189; used with permission (L) Gamble, K.L. et al. 2014. Nature Rev. Endocrinol., 10(8), pp. 466-475; used with permission (M) Mirzaeinia, A., et al., 2020. Swarm Intelligence, 14, pp. 117-141; used with permission.

is the first physiological system to begin to develop, and it functions throughout pregnancy.[3]

- Growth and development of the brain.
- Control of appetite, growth, and metabolism.
- Maintenance of body homeostasis, such as concentrations of blood electrolytes, pH, glucose, metabolites, and temperature.
- Modulation of immune function.

- Mediation of major life history transitions, like metamorphosis, puberty, and senescence.
- Facilitation of organismal responses to environmental changes, such as fluctuations in temperature, salinity, photoperiod, predation, and food availability.
- Regulation of the timing of reproduction, migration, and hibernation.
- Mediation of many aspects of gamete production, sexual differentiation, and development of sexual behaviors.
- Promotion of parenting, bonding, and nurturing behaviors.

Considering the broad importance of hormones in mediating complex communications among virtually every organ system in the body, a deficiency, disruption, or overproduction of any given hormone can lead to abnormal development, disease, or death (Figure 1.4). In this chapter you will learn that from basic biomedical research to clinical

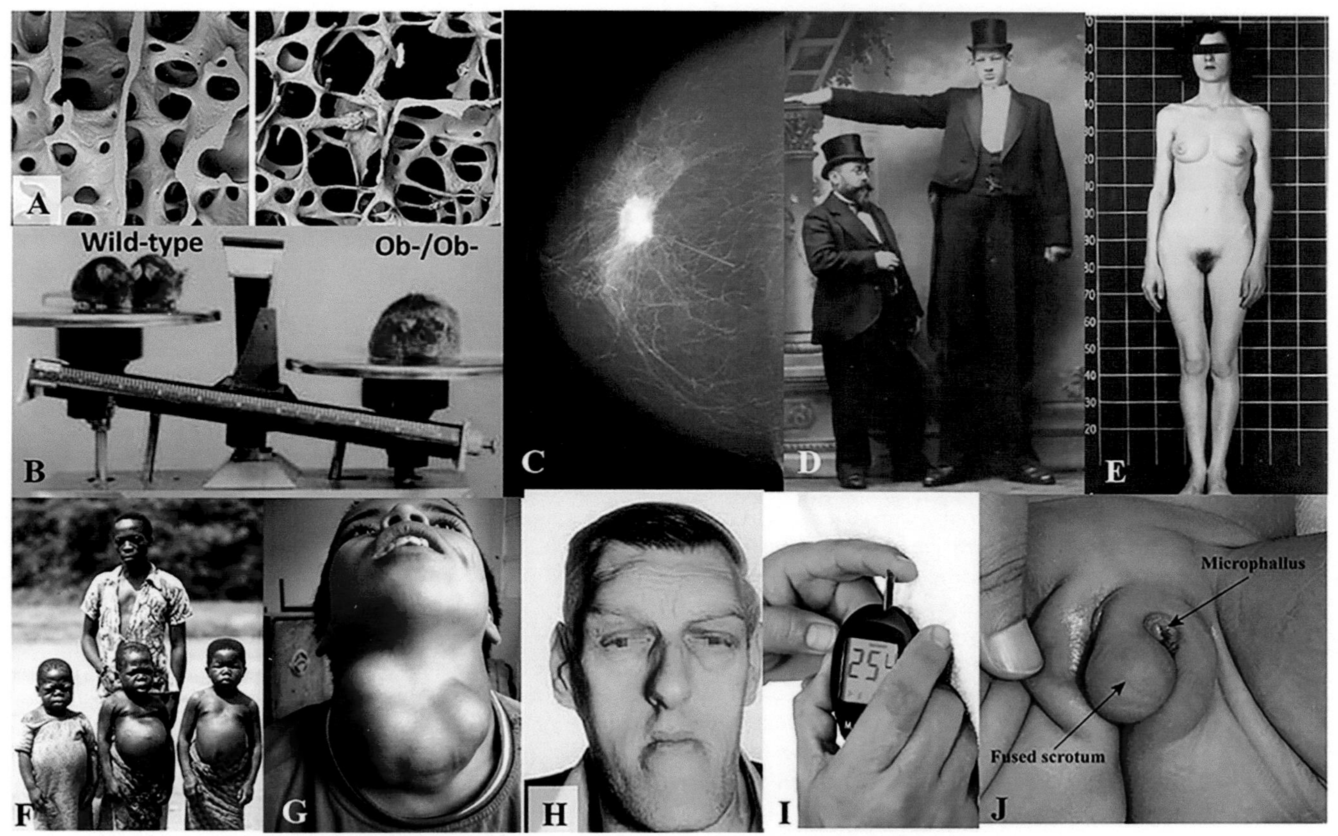

Figure 1.4 Some examples of human endocrine pathology. (A) Osteoporosis can be caused by calcium and estrogen insufficiency. The left shows healthy bone under high magnification, and the right denotes osteoporotic bone. **(B)** Obesity is a disorder of appetite and metabolism. The obese mouse on the right is lacking the gene for the hormone, leptin, which promotes satiety. **(C)** An estrogen-dependent breast cancer tumor. **(D)** Julius Koch, who measured up to 259 cm in height, suffered from gigantism, a condition caused by excess growth hormone synthesis starting in childhood. **(E)** Androgen resistance occurs in people with a male XY genotype but who display a typical female XX phenotype due to a nonfunctional androgen receptor. **(F)** Modern cretinism (hypothyroidism with onset in the perinatal period) in the Democratic Republic of Congo, a region with severe iodine deficiency. Iodine is an essential component of thyroid hormone. Four inhabitants aged 18 to 20 years are depicted: a normal male and three cretinous females with severe hypothyroidism with dwarfism, retarded sexual development, puffy features, dry skin, and severe mental retardation. **(G)** Large nodular thyroid goiter in a 14-year-old boy in an area of severe iodine-deficiency in northern Morocco. **(H)** A patient with acromegaly (adult-onset gigantism) exhibiting the classic facial appearance, including an elongated lower jaw. **(I)** Diabetes types I and II are characterized by elevated blood glucose levels due to an insulin synthesis or signaling dysfunction, respectively. **(J)** Ambiguous genitalia in a newborn infant. Such a phenotype typically manifests from disrupted gonadal steroid hormone function during fetal development.

Source of images: (A) Boyde, A., 2002. Endocrine, 17(1), pp. 5-14; used with permission (B) Courtesy of Dr. Jeffrey M. Friedman (C) Sturesdotter, L., et al. 2020. Sci. Rep., 10(1), pp. 1-10. (D) Beckers, A., et al. 2018. Nature Rev. Endocrinol., 14(12), pp. 705-720; used with permission (E) Moore, K.L., et al. 2018. The developing human-e-book: clinically oriented embryology. Elsevier Health Sciences; used with permission (F) Eastman CJ, Zimmermann MB. Endotext.; used with permission (G) Zimmermann, M.B., et al. 2008. Lancet, 372(9645), pp. 1251-1262; used with permission (H) Greenwood, M. and Meechan, J.G., 2003. Br. Dent. J., 195(3), pp. 129-133; used with permission (I) National Institute of Diabetes and Digestive and Kidney Diseases; (J) Mehta, P. and Rajender, S., 2021. Andrologia, 53(2), p. e13937; used with permission.

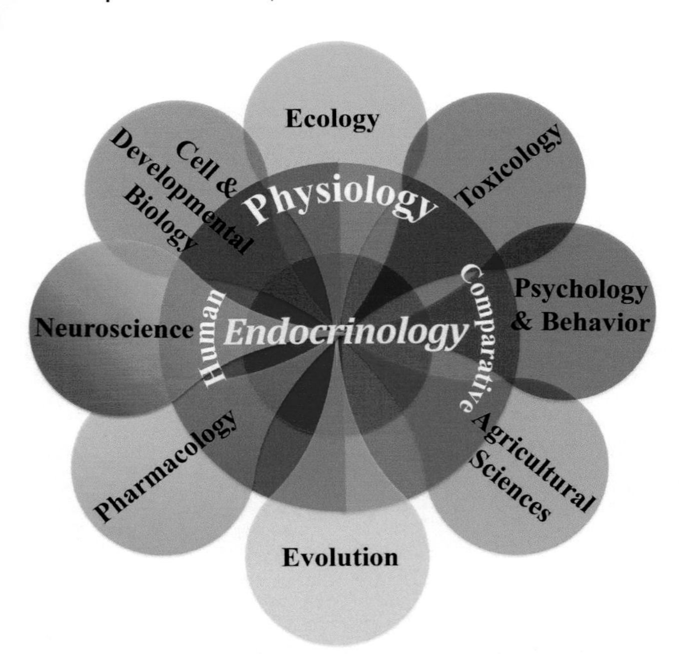

Figure 1.5 Endocrinology is a highly integrated discipline. An endocrine-centric, "spinning pinwheel" view of the biological sciences.

and veterinary applications, from neuroscience to cancer research, and from comparative animal physiology to ecotoxicology an understanding of endocrinology is profoundly relevant to many aspects of today's world. Endocrinology's scope of inquiry is particularly broad and borrows from and contributes to many other disciplines, making endocrinology one of the most integrative of all life sciences (Figure 1.5). Indeed, in the title for this textbook, *General and Comparative Endocrinology: An Integrative Approach*, the adjective "integrative" was chosen specifically to reflect the profoundly interdisciplinary nature of the field, in addition to emphasizing the key notions that virtually all physiological systems within an organism are interconnected via hormone communication and that the endocrine systems among all organisms share a common ancestry.

SUMMARY AND SYNTHESIS QUESTIONS

1. How does the science of endocrinology play a prominent role in Arnold De Loof's definition of life: $L = \sum_{i=1}^{N} C_i$?
2. Provide endocrine analogies for the following terms borrowed from the world of telecommunication: transmitter, signal, receiver, transducer, amplifier, response.

The Relevance of Endocrinology to Society

LEARNING OBJECTIVE Describe the relevance and benefits of endocrinology to society, as well as some of the challenges associated with certain hormone technologies.

KEY CONCEPTS:

- An understanding of endocrinology is essential not just to the medical and veterinary sciences, but also to the fields of neuroscience, psychology, pharmacology, and toxicology.
- Endocrine-based drug discoveries and technological breakthroughs have profoundly influenced human welfare, animal husbandry, and agriculture science through the synthesis of diverse compounds designed to mimic hormones or block their actions.
- Challenges associated with some hormone technologies include unintended side effects of hormone analogs, hormone abuse, and toxic hormone waste emanating from city sewage and concentrated animal husbandry facilities.
- Synthetic endocrine-disrupting chemicals (EDCs) are a broad category of ubiquitous endocrine-toxic pollutants that may be detrimental to the health of both human and wildlife populations alike.

The study of hormones has produced enormous benefits to human health, social welfare, and economic development. Many of the most important advances in basic and medical science of the past century have been attributed to endocrinologists. Indeed, since 1903, at least 35 Nobel Prizes in "Physiology or Medicine" and "Chemistry" have been awarded to scientists conducting endocrine or endocrine-related research (Appendix 1). Such breakthroughs include the discovery, isolation, and characterization of diverse hormones, including insulin, sex and adrenal steroid hormones, prostaglandins, neurohormones, and growth factors. Other discoveries have elucidated mechanisms of hormone biosynthesis, intracellular signaling pathways, the development of assays for measuring hormones, and advancements that led to the treatment of major endocrine-related diseases such as diabetes, breast and prostate cancers, kidney disease, and cardiovascular disorders.

Beyond the basic medical and veterinary sciences, endocrinology is fundamental to understand *neuroscience*, a dynamic field that examines the development, structure, function and evolution of the brain and nervous system. This allows neuroscientists to better understand behavior, cognition, and diseases of the nervous system, such as Alzheimer's disease, multiple sclerosis, post-traumatic stress disorder (PTSD), and behavioral disorders like attention deficit hyperactivity disorder (ADHD). Not only do hormones from diverse tissues profoundly influence brain development, but the brain itself is the source of many hormones that exert acute effects on virtually every tissue in the body. An understanding of endocrinology is also essential to the field of *toxicology*, the study of the adverse effects of synthetic and naturally occurring chemicals, many of which disrupt endocrine signaling pathways. Another closely allied discipline is *pharmacology*, which is concerned with the synthesis of drugs and the study of their actions, many of which affect hormone signaling. The trillion dollar global *pharmaceutical industry* develops new drugs and medical devices to treat or cure diseases. Some notable endocrine-based drug discoveries and technological breakthroughs that have revolutionized medicine, animal husbandry, and many aspects of modern society (Figure 1.6) include advances in the following:

1. Reproduction and fertility:
   - The development of female oral contraception ("the pill"), a feat that has revolutionized societies with access to these drugs, empowering 50% of the population with greater control over their own fertility.
   - Assays for home pregnancy testing.
   - In vitro fertilization (IVF) and other treatments for infertility.
   - Oral pharmaceuticals for the early and safe termination of pregnancy.

Figure 1.6 Some notable endocrine breakthroughs that have revolutionized health, agriculture, and society. (A) Synthetic estrogens and progesterone-based birth control pills. **(B)** Pregnancy test kits measure the presence of human chorionic gonadotropin in the urine. **(C)** Sildenafil citrate (sold under the Pfizer brand name, Viagra) is a nitric oxide synthase stimulator used to treat erectile dysfunction. **(D)** Inhalers use epinephrine and glucocorticoid hormones to treat asthma attacks. **(E)** Recombinant insulin, shown administered in the form of an injection, is used to treat diabetes. **(F)** Intravenous infusions of estrogen receptor blockers, such as tamoxifen, are used to treat some breast and ovarian cancers. **(G)** Melatonin is the only "over the counter" hormone, and it is used to treat jet lag and insomnia. **(H)** Drugs that inhibit androgen synthesis or androgen receptor function are used to treat pattern baldness. **(I)** Insecticide sprays inhibit hormones that mediate metamorphosis and reproduction. **(J)** Aspirin and many other non-steroidal anti-inflammatory drugs that are used to treat pain and inflammation function by inhibiting prostaglandin synthesis. **(K)** AquaBounty's transgenic AquAdvantage salmon expresses an extra growth hormone gene, growing much more rapidly and to a larger size than a conventional Atlantic salmon. Both fish are 18 months old. **(L)** Injections of bovine somatotropin are used to increase milk production in dairy cows.

Source of images: (A-J) and (L) Shutterstock. (K) Marris, E., 2010. Nature, 467(7313), pp. 259-259; used with permission.

2. Large-scale industrial synthesis of steroid hormones, recombinant protein hormones, and other analogs for hormone replacement and treatment therapies:
   - Insulin for the treatment of diabetes.
   - Erythropoietin for enhancing red blood cell numbers in diseases like leukemia.
   - Growth hormone for the treatment of growth disorders.
   - Glucocorticoids for treatment of inflammation, asthma, autoimmune disorders and immune suppression for organ transplants.
   - Enrichment of milk and other products and supplements with the hormone vitamin D to promote healthy blood calcium balance and the proper development of bone and other tissues.
   - Sex steroid hormone analogs for alleviating some of the symptoms associated with menopause and aging, such as osteoporosis and high blood pressure. Sex steroid analogs are also used for the treatment of developmental disorders such as congenital ambiguous genitalia and delayed puberty, as well as for transgender hormone therapy.
   - Synthetic thyroid hormone for the treatment of hypothyroidism.
   - Synthetic melatonin for treatment of sleep and circadian disorders.
3. Hormone receptor antagonists ("blockers") and chemical inhibitors of hormone biosynthesis and signaling pathways:
   - Sex steroid hormone receptor antagonists for treatment of sex hormone-dependent cancers (e.g. some breast and prostate cancers).
   - Adrenergic receptor antagonists ("beta blockers") for treatment of hypertension and some cardiovascular disorders.
   - Non-steroidal anti-inflammatory drugs (NSAIDs, prostaglandin synthesis inhibitors) for treatment of pain, inflammation, excessive blood clotting, and fever.
   - Inhibitors of the renin-angiotensin-aldosterone signaling pathway for the treatment of high blood pressure.
   - Thyroid hormone synthesis inhibitors for the treatment of hyperthyroidism.
4. Agriculture and animal husbandry biotechnology:
   - Insecticides that specifically disrupt hormone-mediated components of insect reproduction or larval development.
   - Hormone supplementation in livestock to increase protein and to decrease fat content (sex steroids), and to enhance milk production (recombinant bovine growth hormone, rBGH, also called bovine somatotropin, or BST).
   - Genetically modified animals for meat. Salmon that are genetically modified to express an additional copy of the growth hormone gene are the first US Food and Drug Administration (FDA)-approved meat to be allowed on the market.

Endocrine Diseases Associated with "Modernity"

Modern humans originated approximately 200,000 years ago in Africa. As humans multiplied and colonized the planet, new and diverse selective pressures were generated by the thousands of local variations in disease and diet that were encountered, contributing to genome divergence among local populations.[4] Today, culture and technology in many societies is changing at an unprecedented rate, bringing new challenges to human health and quality of life that our species as a whole has not experienced before. In effect, the rate of human "cultural evolution" now greatly exceeds the pace of physiological evolution.[5] Some resulting "mismatches to modernity" that are particularly relevant to endocrinology include (1) overnutrition and an increasingly sedentary lifestyle, which contribute to obesity, type 2 diabetes, and poor cardiovascular health; (2) longer lives have introduced menopause as a norm to the human life cycle, as well as many new associated health dysfunctions, such as dementia, osteoporosis, hypertension, and cancers; and (3) exposures to toxic endocrine-disrupting compounds, light at night, and other chronic stressors, which are associated with diverse maladies described throughout this textbook. These new and rapidly growing health threats can only be effectively addressed by understanding the endocrine bases of these mismatches to modernity.

Some Challenging Aspects of Hormone Technologies

Technology is often a two-edged sword, and there certainly are societal and environmental costs associated with some of these endocrine technologies. However, research by endocrinologists can contribute in substantive ways to provide the information necessary to understand the nature and scope of potentially negative impacts and design solutions to mitigate them.

Unintended Side Effects of Hormone Analogs

Some pharmaceutical hormone analogs that were originally designed to be therapeutic may have unintended side effects. For example, diethylstilbestrol (DES) was the first synthetic compound to be specifically designed and marketed as a medically prescribed estrogen mimic for human use (Figure 1.7). Beginning in 1938, DES was initially prescribed to prevent miscarriage and ultimately was advertised and used as a pregnancy supplement for producing "stronger babies".[6] Tragically, not only was DES ineffective at preventing miscarriage (indeed, it was ultimately found to increase the chance of miscarriage), but the compound was found to cause extensive reproductive harm to many of the developing male and female fetuses, damage that in most cases would not manifest until years or decades after birth. One of the first observed consequences of fetal exposure to DES was that girls born to mothers who were prescribed DES during pregnancy were much more likely to develop an extremely rare form of cervical cancer.[7,8]

Figure 1.7 Pharmaceutical company advertisement espousing the virtues of DES. Advertisement for the Grant Chemical Company, Brooklyn, NY, printed in the *American Journal of Obstetrics & Gynecology* in 1957.

Source of images: DES Daughter.

Another example of unintended consequences was the first generation of estrogen replacement therapy drugs for treating side effects associated with menopause and aging in women, but were also found to increase the risks of developing breast cancer, dementia, stroke, and cardiovascular disease in some women. Importantly, the development of new selective estrogen receptor modulators (SERMs) shows great promise. SERMs selectively block the negative actions of estrogen on breast tissue, for example, while simultaneously promoting estrogen's beneficial effects on bone and liver.

Beyond humans, hormone supplementation in livestock can also exert negative physiological effects on the animals such as increased mastitis (inflammation of the udder), lameness, and infertility following the treatment of cattle with growth hormone.[9] Furthermore, hormone augmentation in animals can chemically modify the composition of food destined for human consumption, such as elevating insulin-like growth factor I (IGF-I) levels in milk and the sex steroid hormone content of meat, variables that may impact human health.[10]

Hormone Abuse

A second prominent challenge associated with hormone technology consists of hormone abuse, specifically in the use of performance-enhancing drugs ("doping") in competitive sports, a problem that even extends to the world of horse racing and other competitive animal sports. Commonly abused hormones include testosterone and growth hormone (both anabolic hormones used to build muscle mass and strength), as well as erythropoietin (EPO), which is used to elevate red blood cell counts and enhance blood oxygen-carrying capacity in endurance sports. Not only are these drugs illegal for use by athletes in competition, but they are dangerous and can result in sterility, cardiovascular dysfunction, stroke, and death.

Toxic Hormone Waste

A third category of costs is associated with the massive quantities of hormone waste that pours into the world's waterways every day from city sewage and also seeps into groundwater reservoirs from garbage dumps and landfills. One example of such waste arrives in the form of synthetic estrogens used for human contraception. Importantly, these and other "environmental estrogens" are well documented to influence the health of fish and other species that live amongst this waste by "feminizing" males and disrupting reproduction.[11] In addition, concentrated animal-feeding operations associated with raising animals for meat, milk, or fur can release high levels of natural and synthetic hormones into surrounding soils and waterways, potentially affecting human and wildlife health.[12] Regarding the husbandry of genetically modified animals, such as the previously mentioned growth hormone-enhanced salmon, there are concerns that their inadvertent release into the wild could result in their interaction with wild animals, including breeding with them.[13]

Endocrine-Disrupting Compounds

Beyond some of the negative effects associated with the use and abuse of synthetic hormones described earlier, a comparatively much larger global threat to human and wildlife health stems not from chemicals purposefully designed to act as hormones, but rather from the endocrine-toxic effects of certain chemicals produced by industries in vast quantities for other purposes. These include many plastics, flame retardants, cosmetics, fragrances, industrial lubricants, ingredients of fracking fluids, electrical insulators, solvents, agricultural pesticides, and heavy metals. Such chemicals, called **endocrine-disrupting compounds** (EDCs), can disturb endocrine signaling by mimicking hormones, blocking hormone–receptor interactions, or altering signaling pathways, rates of hormone synthesis, or degradation. Indeed, of the approximately 80,000 known synthetic chemicals in our environment today, about 1,500 have been classified as EDCs,[14] a small fraction that will certainly increase as thousands of new chemicals (most of which are developed with little to no toxicological testing) are manufactured every year. Alarmingly, EDCs have been found to be present in virtually every human and animal tested, from the Arctic to the Antarctic, and everywhere in between.[15,16] EDCs may significantly impact the health of both human and wildlife populations alike, though their impacts on wildlife and laboratory animals are typically studied and observed much more easily compared with humans (Figure 1.8). The potential negative impacts of EDCs on human and wildlife health are so profound and worrisome that, after climate change, EDCs have been described as "the second greatest environmental challenge of our time".[17] The final unit of this textbook, Unit VIII: Endocrine-Disrupting Chemicals, is devoted to understanding the potential impacts of EDCs on human and wildlife endocrine systems.

Beyond synthetic chemical compounds, another artificial "product" with significant endocrine-disrupting capacity to humans and wildlife is artificial light. Specifically, **light exposure at night** (LEN), or "light pollution" caused by indoor and outdoor lighting, is known to disrupt the normal circadian patterns and fluctuations of daily hormone secretions. Disturbance of circadian hormone production in humans through night-shift work and LEN has been associated with increased incidence of obesity, diabetes, cardiovascular disease, rheumatoid arthritis, peptic ulcers, cancer, depression, and other disorders.[18–23] Indeed, in 2019 the World Health Organization's International Agency for Research on Cancer (IARC) officially classified night-shift work as a "Group 2A carcinogen" that, based on strong mechanistic evidence in experimental animals, is "probably carcinogenic to humans".[24] Today, 99% of the human population in the United States and Europe, and 62% of

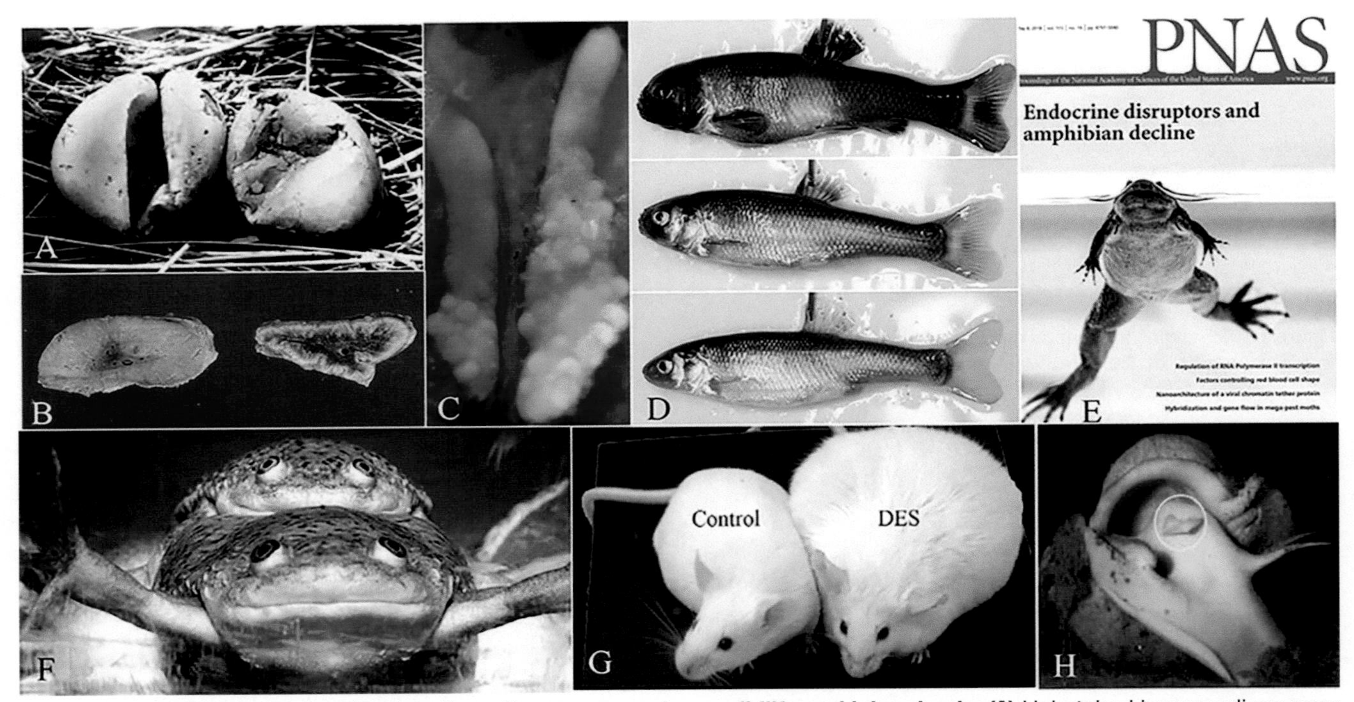

Figure 1.8 Some impacts of endocrine-disrupting compounds on wildlife and lab animals. (A) Unhatched brown pelican eggs damaged because of DDT pesticide poisoning that weakened its eggshell. **(B)** Transected adrenal glands from a Baltic gray seal (left) and a gray seal from Svalbard (right). The adrenal gland from the Baltic seal shows massive hyperplasia of the adrenal cortex linked to high body burdens of PCBs, DDT, and their persistent methyl sulphone metabolites. **(C)** Gonads from a male leopard frog (*Rana pipiens*) exposed to the herbicide atrazine (an estrogen-inducing compound) has numerous oocytes protruding through the posterior surface of the testes. **(D)** Feminization of a male fathead minnow in response to estrogens in the water from human prescription drugs. The top fish is a normal male fathead minnow. The bottom fish is a normal female fathead minnow. The middle fish is a male that was exposed to female human hormones from water effluent containing birth control drugs and that looks more like a female than a male. **(E)** A *Proceedings of the National Academy of Sciences* (PNAS) cover story from a featured article titled "Unexpected metabolic disorders induced by endocrine disruptors in *Xenopus tropicalis* provide new lead for understanding amphibian decline". **(F)** A genetic male African clawed frog (bottom), feminized by exposure to the herbicide atrazine is mating with a normal unexposed male sibling (top). The mating produced viable larvae that survived to metamorphosis and adulthood. **(G)** Estrogenic substances can exert "obesogenic" effects. Mice treated with a low dose of the estrogenic compound diethylstilbestrol (DES) on days 1–5 of neonatal life did not affect body size during treatment, but the mice became obese as adults. **(H)** A female whelk from the North Sea showing the development of a penis (circled region), a condition called imposex, caused by exposure to tributyltin, an antifouling agent from ship-hull paints.

Source of images: (A) Jehl, J.R., 1973. Condor, 75(1), pp. 69-79; used with permission. (B and H) Vos, J.G. et al. 1999. Scientific Committee on Toxicity, Ecotoxicity, and the Environment; used with permission. (C) Hayes, T.B., 2004. Bioscience, 54(12), pp. 1138-1149; used with permission (D) Courtesy of Adam Scheindt. (E) Regnault, C., et al. 2018. Proc. Nat. Acad. Sci. USA, 115(19), pp. E4416-E4425; used with permission (F) Hayes, T.B., et al. 2010. PNAS, 107(10), pp. 4612-4617; used with permission. (G) Newbold RR, Padilla-Banks E, Jefferson WN. Mol Cell Endocrinol. 2009;304:84–9; used with permission.

the world's remaining population, are exposed to LEN.[25] Additionally, seasonal and daily natural fluctuations in photoperiod and light intensity are used as environmental cues by many animals to synchronize time-and energy-consuming behavioral and physiological processes (e.g. seasonal reproduction and migration) with the environment in order to optimize energy utilization, reproduction, and survival. LEN has been shown to alter reproductive behavior in sea turtles near illuminated beaches,[26] promote misorientation/disorientation of migratory birds near urban sky glow,[27–29] cause daily desynchronization of biological rhythms in a nocturnal primate,[30] and may contribute to the extinction of some firefly species.[31]

SUMMARY AND SYNTHESIS QUESTIONS

1. In most physiology textbooks, chapters that address endocrinology are typically among the last in the book. Why do you think this is the case?
2. Evolutionary biologist Stephen Stearns has stated, "Mismatches to modernity produce diseases that result from the inability of our biology to keep pace with cultural change."[4] What is meant by "mismatches to modernity", in an endocrine context?

General and Comparative Endocrinology: Two Synergistic Sides of the Same Coin

LEARNING OBJECTIVE Define the scope of inquiry of modern endocrinology, and describe how the fields of "general endocrinology" and "comparative endocrinology" interact to advance endocrine science as a whole.

KEY CONCEPTS:

- Modern endocrinology emphasizes two areas of study: general (clinical) endocrinology and comparative endocrinology.
- Since physiological systems are conserved among animals, the study of endocrine systems among diverse taxa promotes advances in both basic and clinical human endocrinology.
- Because more is known about human endocrinology compared with that of any other organism, knowledge gained by studying human hormone signaling helps researchers understand how the endocrine systems of other animals function.
- Comparative endocrinologists study the differences and similarities of hormone systems among diverse species (including humans) and also try to understand endocrine function in the contexts of ecology and evolution.

Endocrinology is a relatively young field within physiology, with its first professional society founded in 1916 by a group of physicians dedicated to the study of "internal secretions", as hormones were then also known. The newly formed Association for the Study of Internal Secretions would change its name to the "Endocrine Society" in 1952. Modern endocrinology emphasizes two areas of study: general and comparative endocrinology.

General Endocrinology

Originally established as a field of internal medicine, **general endocrinology** now has two major emphases: (1) *clinical endocrinology*, which studies the origins and treatments of endocrine disorders, and (2) *basic endocrinology*, which researches fundamental principles of endocrine signaling without the bias of application (Figure 1.9). *Translational research* (sometimes called "bench to bedside" medicine) describes the effort to transform basic laboratory findings into new clinical applications and therapies.

Comparative Endocrinology

Comparative endocrinology studies the differences and similarities of hormone systems among diverse vertebrate and invertebrate species, including humans. Research in comparative endocrinology grew rapidly in the beginning of the 20th century following the discovery and isolation of the first hormones, such as thyroid hormones and sex steroid hormones, from animals.[32] The discipline of

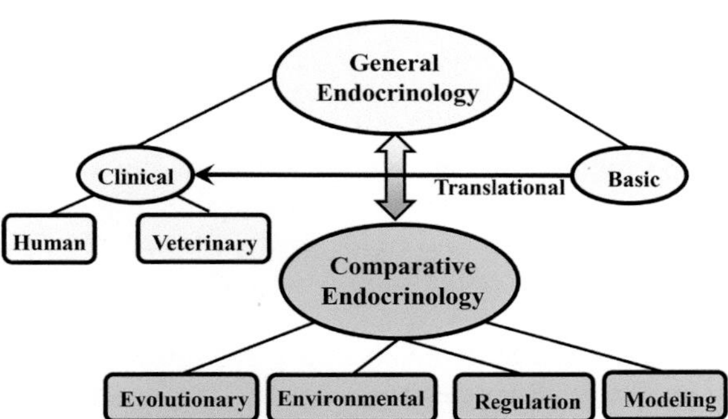

Figure 1.9 Modern fields of inquiry within endocrinology. General endocrinology consists of applied human and veterinary clinical practice, as well as basic research into fundamental aspects of endocrine function. Translational research describes the effort to adapt basic laboratory findings into new clinical applications and therapies. Comparative endocrinology studies the mechanisms of endocrine system function and evolution among diverse animal taxa, including humans. Applications in comparative endocrinology are often in the form of environmental research (endocrine disruption and conservation biology) and the development of model systems for studying endocrinology. Although general and comparative endocrinologists address distinct questions, the two fields often interact and inform one another.

Source of images: Norris, D.O., 2018. Integr. Comp. Biol., 58(6), pp. 1033-1042; used with permission.

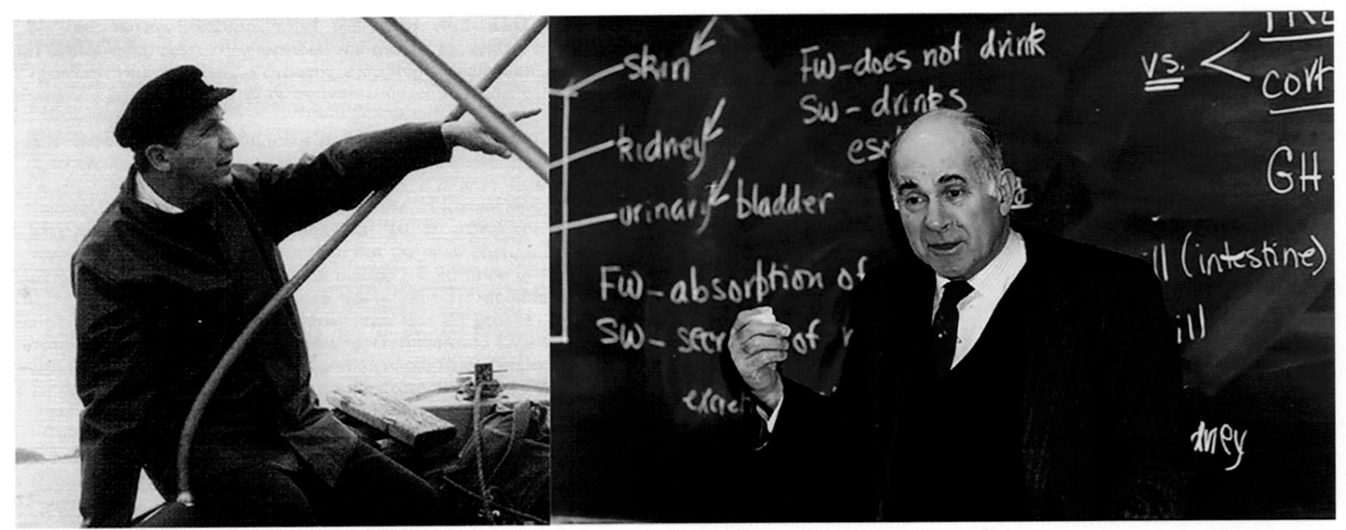

Figure 1.10 The field of comparative endocrinology was founded by Dr. Aubrey Gorbman (left) and Dr. Howard Bern (right), who together in 1962 co-authored the definitive text in the field, *A Textbook of Comparative Endocrinology*, which set the platform for research in this biological discipline. Dr. Gorbman founded *General and Comparative Endocrinology*, the journal for the field of comparative endocrinology, in the early 1960s, where he served as the editor in chief for over three decades. Dr. Bern founded the field of research on "endocrine disruptors", which studies chemicals that influence hormonal systems.

Source of images: (left) Courtesy of Stacia Sower; (right) Courtesy of Allan Bern.

comparative endocrinology was officially founded in the early 1960s when Aubrey Gorbman and Howard Bern (Figure 1.10) co-authored the definitive text in the field, *A Textbook of Comparative Endocrinology*,[33] and Aubrey Gorbman founded the journal *General and Comparative Endocrinology*. Historically, the field of comparative endocrinology has emphasized four areas of study (summarized in Figure 1.9).

1. *Regulation of hormone actions among diverse animal species*. In comparing and contrasting hormone regulation among species, two of the most fundamentally important discoveries by researchers in this area are that (1) hormones and endocrine tissues are structurally similar among evolutionarily distant vertebrates, and (2) although vertebrate endocrine systems are structurally and often functionally conserved, hormone actions may vary with species and target tissue (Figure 1.11).
2. *Evolution of endocrine systems*. Evolutionary endocrinologists study how endocrine glands, hormones, and signaling pathways have evolved from vertebrate and invertebrate ancestors to control diverse developmental, physiological, and behavioral processes. Researchers in this area also address how evolving hormone systems facilitated the development of new traits that promoted or constrained species adaptation to different environments.
3. *Development of "model" systems to study endocrinology*. Of the currently estimated 7.7 million animal species living on earth,[34] we have relatively detailed physiological understandings for only a tiny fraction. Unsurprisingly, the most well-studied animal on earth is *Homo sapiens*. Next on the list are animals that humans have high affinities to, such as those bred for work, food, pets, hobbyists, animal sport, and scientific research. In the world of physiological research, **model organisms** are animals that have been very broadly studied, usually because they are relatively easy to breed in a laboratory environment, have a rapid generation time (time

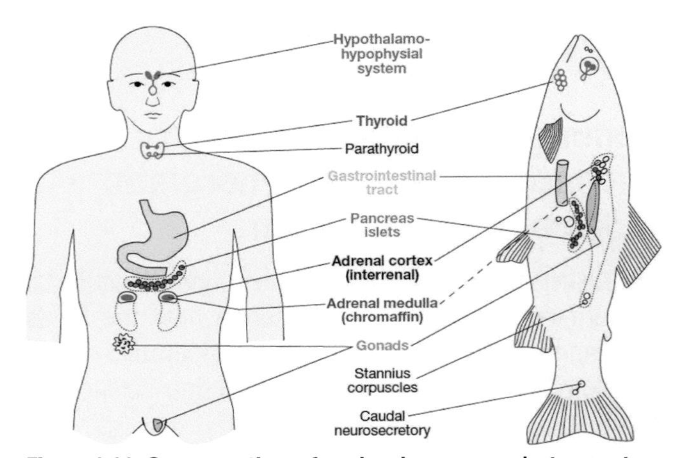

Figure 1.11 Conservation of endocrine organs between two evolutionarily distant vertebrates, a human and a trout. Although most endocrine tissues present in humans have homologs in fish, their actions may vary with species and target tissue. Two endocrine organs not shared between mammals and fish are the parathyroid glands (present only in tetrapods) and Stannius corpuscles (present only in teleost fish). Both endocrine organs are involved in calcium regulation and were gained and lost, respectively, following the ocean-to-land evolutionary transition.

Source of images: Tata, J.R., 2005. EMBO Rep., 6(6), pp. 490-496.; Used with permission.

between two consecutive generations), are cost-effective to raise, and can be genetically manipulated. In addition, each model possesses specialized attributes useful for the design and analysis of specific experiments, such as the appropriateness, complexity, and simplicity of the physiological system in question. Commonly used model organisms for studying basic endocrine function include mammals (typically rodents, like mice and rats), amphibians (e.g. the African clawed frog, *Xenopus laevis*), fish (e.g. zebrafish, *Danio rerio*), insects (e.g. fruit flies, *Drosophila melanogaster*), and worms (e.g. *Caenorhabditis elegans*) (Figure 1.12). Importantly, due to the aforementioned conserved nature of many endocrine structures and functions among animals, these and other models have contributed tremendously to basic and clinical endocrine research. For example, dogs were the animal model of choice that led to the discoveries of both secretin and insulin; the former was the first hormone to be discovered, and the latter hormone would yield the first effective treatment for type I diabetes. Due to their relative simplicity and tractability, non-mammalian models are often used to study basic endocrine principles that are too difficult or unethical to pursue in humans or other mammals. Indeed, the use of non-mammalian models has contributed to numerous seminal advances in the field of basic endocrinology, ranging from the first reported endocrine experiment in 1849 using chickens to the discovery of neuroendocrine signaling using insects and fish. Also, many new hormones that were originally discovered in non-mammalian vertebrates and invertebrate species are now known to have structural and functional counterparts in mammals and also have important functions in human physiology and disease.[32] Some notable examples of endocrine breakthroughs using non-mammalian model organisms are listed in Appendix 2.

Curiously Enough . . . The world's first accurate and widely used human pregnancy test performed in hospitals from the 1940s–1970s was the injection of women's urine into female *Xenopus laevis* frogs, which would respond by laying eggs if the women were pregnant due to the presence of the embryo-stimulated hormone, chorionic gonadotropin.[35]

Although the model organismal approach has provided valuable insights to many basic endocrine processes, this approach is limited to studying only a small subset of animals. Considering the tremendous diversity in animal physiology, behavior, life history strategies, and responses to the environment, non-model organisms harbor a vast wealth of new information, and their study by comparative endocrinologists is also yielding new insights and applications to basic and translational research. Critically, resident wildlife are also essential for studying the impacts of endocrine-disrupting compounds on ecological health.

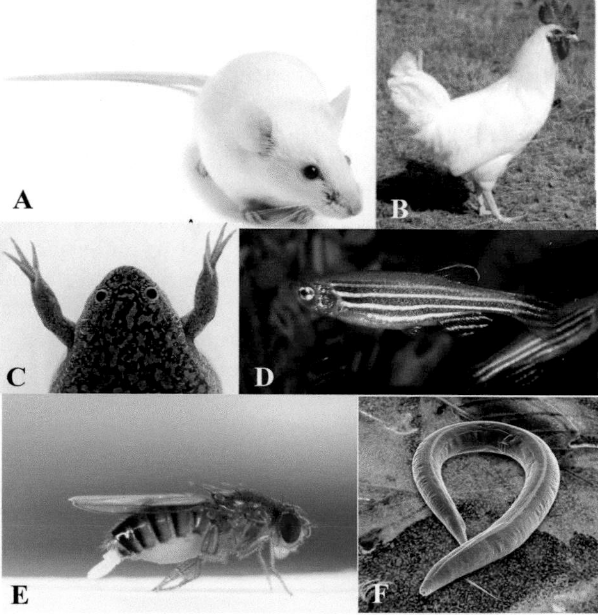

Figure 1.12 Some model organisms commonly used to study endocrine function. (A) The house mouse, *Mus musculus*, is the most commonly used mammal in laboratory research. **(B)** The chicken, *Gallus gallus*, is an ideal system for the study of vertebrate development, particularly due to easy access to the chicken embryo using incubated eggs and the ease of embryo manipulation. **(C)** The western clawed frog, *Xenopus tropicalis*, is amenable to transgenic gene expression and for studying the genes associated with both embryogenesis and metamorphosis, a form of post-embryonic development that is under endocrine control. **(D)** The zebrafish, *Danio rerio*, has embryos that develop rapidly outside the mother, are optically clear, are genetically tractable, and are easily accessible for experimentation and observation. **(E)** The fruit fly, *Drosophila melanogaster*, is a complex organism, but its rapid generation time and excellent genetic tools have made it indispensable for basic genetic and developmental research. This organism also undergoes a highly tractable hormone-mediated metamorphic transition from a larva to a juvenile. **(F)** Although the microscopic, soil-dwelling nematode worm, *Caenorhabditis elegans*, is a relatively simple organism, many of the molecular signals controlling its development are also found in more complex organisms, like vertebrates. They also have a short life cycle of only two weeks and produce over 1,000 eggs every day.

Source of images: (A) Enríquez, J.A. Nat Metab 1, 5–7 (2019); used with permission. (B) Lázár, B., et al. 2021. Poult. Sci. 100(8), p. 101207; used with permission. (C) Brod, S., et al. 2019. Lab. Anim., 48(1), pp. 16-18; used with permission. (D) Valavanidis, A., et al. 2006. Ecotoxicol. Environ. Saf., 64(2), pp. 178-189; used with permission. (E) Kutzer, M.A. and Armitage, S.A., 2016. Ecol. Evol., 6(13), pp. 4229-4242. (F) Jorgensen, E.M. and Mango, S.E., 2002. Nat. Rev. Genet., 3(5), pp. 356-369; used with permission.

4. *Environmental endocrinology*. One of the greatest scientific challenges of our time is understanding how anthropogenic changes to earth's environment impact the quality of life and survival of many species on the planet. In addition to EDCs, other major environmental impacts include climate change and related weather

disasters caused by the burning of fossil fuels, which is perhaps the greatest environmental challenge that we face this century. Other challenges include light pollution (the brightening of the night sky caused by artificial lights, or LEN), noise pollution, habitat partitioning (e.g. roads, trails, fences, and dams), habitat destruction (e.g. deforestation), the introduction of invasive species, and ecotourism (Figure 1.13). While it is self-evident that endocrinology is critical to understanding environmental dangers imposed by EDCs, the relevance of endocrinology to these other environmental threats may be less obvious. Because hormones often transduce environmental signals into appropriate physiological and behavioral responses, a complete understanding of an organism's capacity to respond to environmental change, as well as the resilience of the organism challenged by environmental changes, is necessary for understanding the impact of extreme anthropogenic change on the viability of populations. Researchers in the field of **environmental endocrinology** study how animals transduce signals received from their natural biotic and abiotic environment into endocrine responses that mediate diverse adaptive behavioral and physiological changes.[36] In addition to contributing to the basic understandings of endocrine function, environmental endocrinologists also study how anthropogenic environmental disturbances affect key physiological and behavioral parameters in wildlife, such as stress, the timing of important life history transitions, and the ability to reproduce[37] (Figure 1.14). The health of a wild population can be evaluated in part by collection and analysis of *endocrine biomarkers*, or hormones that indicate an animal's level of stress, metabolic health, or reproductive status. Such hormones are collected directly from saliva or blood, or indirectly and noninvasively from urine or feces samples from animals in the field. Changes in environmental conditions may lead to inappropriate timing of hormone-controlled life history

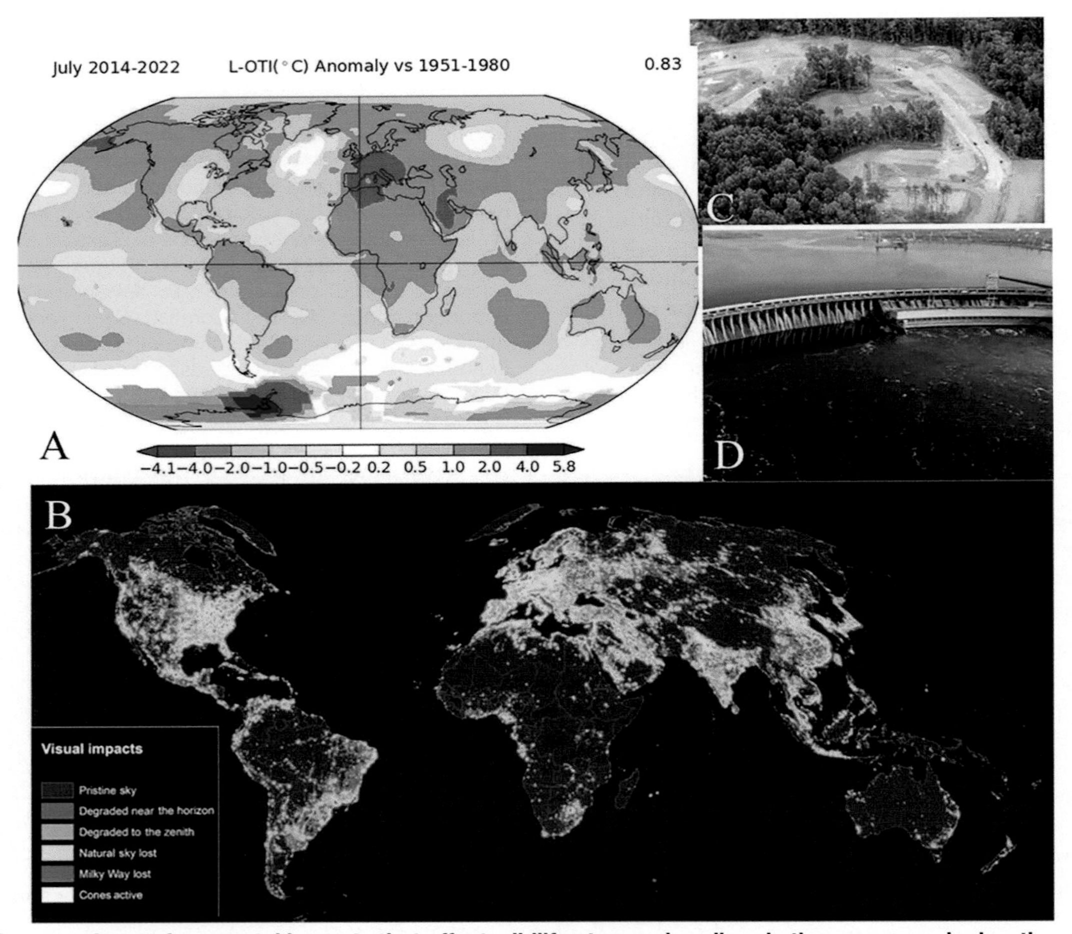

Figure 1.13 Some major environmental impacts that affect wildlife stress, circadian rhythms, seasonal migrations, and reproduction. (A) Climate change: average global temperatures from 2014 to 2022 compared to a baseline average from 1951 to 1980, according to NASA's Goddard Institute for Space Studies. **(B)** Light exposure at night: map of light pollution's visual impact on the night sky. **(C)** Habitat destruction: deforestation in Indonesia. **(D)** Habitat partitioning: hydroelectric dam in China.

Source of images: (A) courtesy of NASA. (B) Falchi, F., et al. 2016. Sci. Adv., 2(6), p. e1600377; used with permission. (C) Scanes, C.G., 2018. Human activity and habitat loss: destruction, fragmentation, and degradation. In Animals and human society (pp. 451-482). Academic Press; used with permission. (D) Han, R., et al. 2020. Sustainability, 12(9), p. 3609.

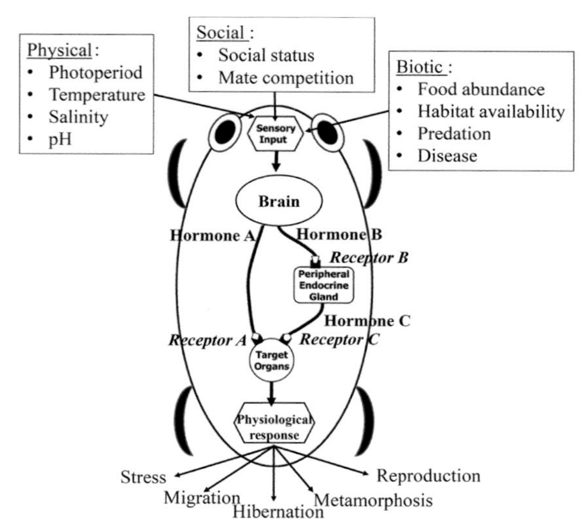

Figure 1.14 Neural integration of environmental information by a hypothetical animal, a "bugphibian". External changes are perceived by the brain, which results in the release of neuroendocrine factors (hormones A and B) that travel in the blood and bind to receptors on either target organs or as peripheral endocrine organs that function as an intermediate along the signaling pathway. Ultimately, the activation of downstream target organs and cells results in coordinated and adaptive physiological responses.

Source of images: McCormick, S.D. and Romero, L.M., 2017. BioScience, 67(5), pp. 429-442; used with permission.

events, which will manifest as altered patterns of circulating hormones.[38] Endocrine biomarkers may also be useful in detecting the presence and effects of EDCs and other harmful toxins. As such, environmental endocrinologists not only are concerned with the health and welfare of wildlife, but also study the effects of pollutants and other environmental disturbances on wildlife to understand their potential impacts on human health.

SUMMARY AND SYNTHESIS QUESTIONS

1. How is comparative endocrinology relevant to human medical endocrinology?
2. How is human medical endocrinology relevant to comparative endocrinology?
3. Describe some advantages and disadvantages of using model systems for basic endocrine research.
4. How is studying the effects of EDCs on wildlife relevant to human health?

Ancestral Origins of a Young Science

LEARNING OBJECTIVE Describe key advances in physiological knowledge prior to the 19th century that laid the foundation for the emergence of endocrinology, and discuss seminal endocrine experiments during the late 19th and early 20th centuries that established the field.

KEY CONCEPTS:

- Endocrinology is a relatively young science that originated in the late 19th and early 20th centuries.
- Two of the most ancient practices that make use of fundamental endocrine principles include organotherapy and castration.
- The foundations of endocrinology rest on knowledge assembled from the 16th–18th centuries for the presence of "ductless" glands and an accurate understanding of circulatory system function.
- The first sophisticated endocrine experiment took place in 1849 when Arnold Berthold transplanted testes into castrated male chicks to show that the organs communicated with target tissues via the bloodstream.
- Claiming that he had found a way to reverse the aging process, Charles-Édouard Brown-Séquard revived the ancient practice of organotherapy during the late 19th century.
- Applying 19th-century approaches to the ancient principles of organotherapy, George Murray successfully treated patients with hypothyroidism with thyroid extracts, the first example of a human endocrine disease to be successfully treated with hormone replacement therapy.
- In 1902 the first hormone to be purified and its actions characterized was "secretin", discovered by Ernest Henry Starling and William Maddock Bayliss.
- In 1905 Starling coined the word "hormone".

Endocrine dysfunction in humans has been documented since antiquity. For example, historical references to goiter (enlarged thyroid gland) in Chinese texts date back as far as 2700 BC, and Chinese physicians recorded using burnt sponge and seaweed to successfully treat it beginning around 1600 BC.[39] Gigantism, a condition caused by excess pituitary growth hormone production beginning in childhood, was well known, with Goliath, the most famous giant in the Bible, described in the 11th century BC as "a champion out of the camp of the Philistines, whose height was six cubits and a span [equal to 9 feet 9 inches or 2.97 meters]" (Samuel 17:4).[40] Additionally, the symptoms of diabetes were accurately reported in the 5th century BC by the Indian surgeon Sushruta, who used the term "madhumeha" (honey-like urine) and pointed out its ability to attract ants.[41] At the time the causes of these and other hormone disorders were, of course, unknown. This section describes the origins of some key concepts and discoveries that ultimately led to the founding of the field of endocrinology in the early 20th century and its subsequent maturation into a modern science that gave physicians the tools to understand and treat these types of diseases.

Ancient Practices

Animism and Humorism

Whereas the science of endocrinology is relatively new, some of the fundamental principles it rests upon are ancient. Among the earliest beliefs about the human body was the view that within each organ resides some special "virtue" that is transferable to a person who consumes that organ. In antiquity one method of augmenting one's own virtue was via the cannibalistic act of eating organs from respected deceased relatives or from brave opponents defeated in battle who possessed strong virtue.[42–44] However, such a belief system (called totemism or animism by anthropologists) did not recognize a physical basis for the existence of these organ-specific virtues, but rather that an animal's "vital force" derives from an immaterial source, the "anima" or soul. One of the first attempts to provide a physical explanation for the integration of body functions was put forth by ancient Greek and Roman physicians in the form of "humorism". This was the notion that health arises from a balance among four physical bodily fluids, termed *humors*: phlegm (a water-based humor), "melancholy" (a so-called black bile thought to be secreted by the spleen and kidneys), "choler" (yellow bile secreted by the liver), and blood. An imbalance among the humors was thought to promote disease and also contribute to a person's temperament. For example, consider the meanings and etymological derivations of the adjectives melancholic, phlegmatic, choleric, and sanguine and the term "good humor".

Organotherapy

Organotherapy, or the practice of eating a specific organ from an animal for medicinal purposes in order to treat a disease or deficiency in a person's corresponding malfunctioning organ, was also widely practiced in antiquity. Ancient Chinese, Egyptian, Greek, and Roman medical texts describe the use of wolf's liver for hepatic disorders, heart preparations for cardiac disease, pig's kidneys for kidney disease, toad's skin for eczema, chicken's gizzard for stomach disorders, rabbit's uterus for female infertility, and even whale's semen for male impotence.[42] Greek and Roman medicine prescribed the consumption of specific organs as a mechanism to balance the humors.

Castration

The notion that specific organs influence the functions of other tissues and organs in the body, as well as behaviors, was well established in the form of animal and human **castration**, or the removal of the testes. Castration of immature male chickens to generate "capons", as well as of livestock to produce "steer" and "oxen", was and continues to be a common practice to render the males more docile and their meat more tender and palatable. The castration of young male chickens produces not only changes in behavior upon reaching adulthood, such as reduced crowing, aggression, and sexual interest in hens, but also distinct changes in morphology that are more similar to those of hens, such as smaller combs and wattles (Figure 1.15). Observations of such changes date back to at least the time of Aristotle (384–322 BC). Furthermore, the castration of humans (such individuals are called *eunuchs*) has also been practiced since antiquity, primarily for three purposes. First, to provide guards with reduced sex drive for working in harems and for escorting female royalty in the Middle East. Second, to provide a source of less aggressive (and therefore, less threatening or ambitious) political advisors to Chinese emperors and European kings. Third, eunuchs were employed as choir boys and adult singers in Asia and Europe. These latter eunuchs, known as *Castrati*, possessed the singing range of a soprano, but with larger and more powerful lungs. Compared with eunuchs castrated as young men (e.g. harem guards and political advisors), Castrati were castrated as prepubescent boys and therefore developed distinct morphological characteristics, tending to be taller than average, with long slender arms, unusually wide hips, and a lack of facial hair.[45] In Europe, the use of Castrati was popular from the Renaissance (14th–17th centuries) through the 18th century.

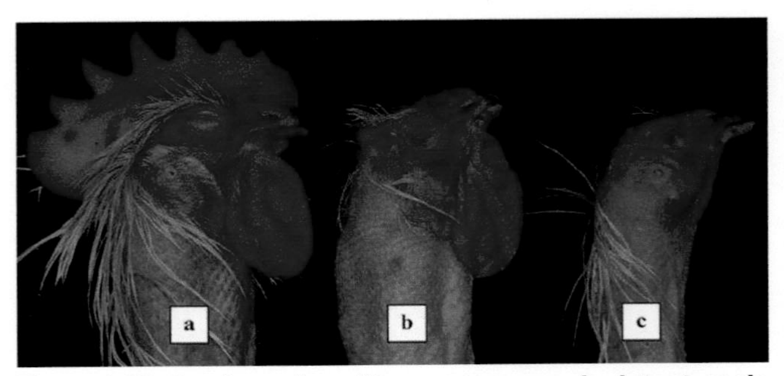

Figure 1.15 Comb and wattle appearance in intact male (a, "cock"), partially caponized male (b, "slip"), and completely caponized male (c, "capon") birds at slaughter.

Source of images: Sirri, F., et al. 2009. Poult. Sci., 88(7), pp. 1466-1473; used with permission.

1543–1778 Foundations of Endocrinology: Ductless Glands and the Circulatory System

Among the physiological sciences, endocrinology's historical focus has been on blood-borne chemical messengers that are secreted by diverse tissues and organs. As such, the required foundations upon which to build endocrine theory included an accurate knowledge of (1) the anatomy and physiology of the circulatory system needed to transport secreted chemicals, (2) the locations of the body's major organs and glands, and (3) the notion that glands and organs can secrete chemicals directly into the blood. The first accurate descriptions for the anatomy and physiology of the circulatory system, which provided

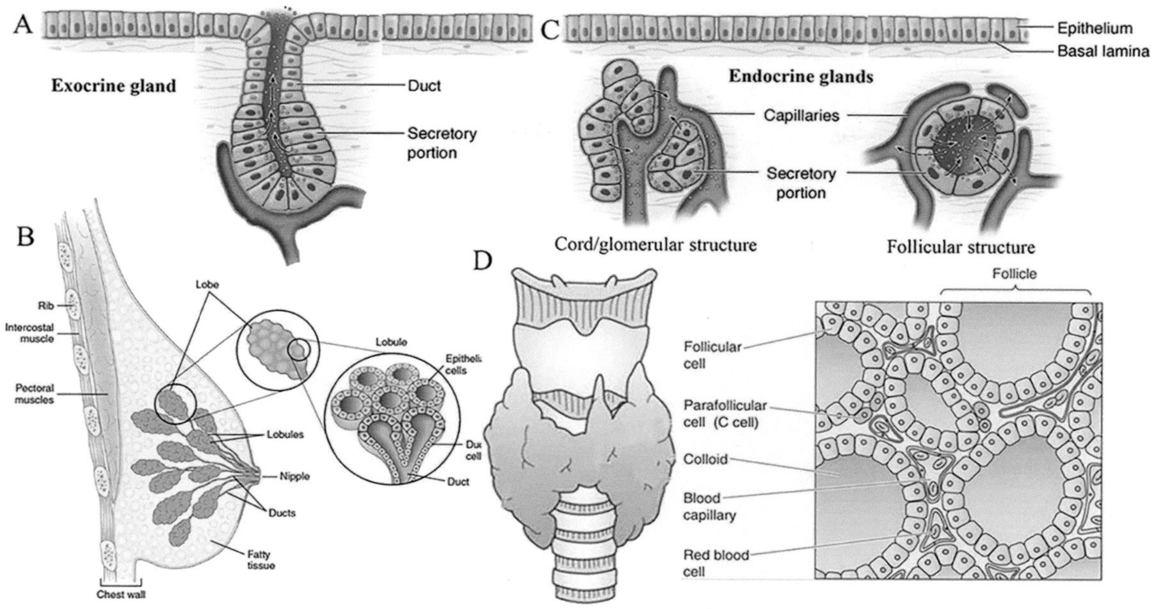

Figure 1.16 Examples of ducted and ductless glands. (A) Glands with ducts (also called exocrine glands) convert substrates obtained from blood into products that are secreted into the lumen and are transferred onto an epithelial surface via a duct. **(B)** The mammary gland is an example of an exocrine gland whose secretory cells (mammary lobules) release milk into mammary ducts that lead to the nipple. Milk substrates, like proteins and lipids, are transferred from surrounding capillaries into the lumen of a mammary lobule. **(C)** Ductless glands (also called endocrine glands) secrete substances into the extracellular spaces and blood. Endocrine cells may be arranged as cords or glomeruli juxtaposed to a dense capillary network, or as follicles that temporarily store products in a central extracellular compartment formed by the secretory cells, before secreting the activated compounds into capillaries surrounding the follicles. **(D)** The thyroid gland is an example of an endocrine organ with a follicular structure. The precursor substrate to thyroid hormone, thyroglobulin, is synthesized by follicle cells and stored extracellularly as colloid within thyroid follicles. Thyroglobulin is subsequently converted into thyroid hormone by the follicular cells and is then secreted into surrounding blood vessels.

Source of images: (A, C) Junqueira, L.C. and Mescher, A.L., 2013. Junqueira's basic histology: text & atlas; McGraw-Hill Education; used with permission. (B) Norman, A.W. and Henry, H.L., 2022. Hormones. Academic Press; used with permission. (D) Boron, W.F. and Boulpaep, E.L., 2005. Medical Physiology: A Cellular and Molecular Approach. Elsevier Health Sciences; used with permission.

the critical mechanism to facilitate the transport of the proposed (but as of yet unproven) internal secretions, had been published in 1543 by Andreas Vesalius in his monumental anatomical text, *De Humani Corporis Fabrica Libri Septem*, and in 1628 by William Harvey in his equally towering book, *On the Circulation of the Blood*. In 1656 Thomas Wharton published *Adenographia*, the first book devoted entirely to describing the anatomy of the body's various glands, consolidating into one book centuries of glandular anatomical knowledge. Two categories of glands had been described: glands with *ducts* (e.g. pancreatic and bile ducts, salivary ducts, mammary ducts, sweat ducts) that secrete substances onto internal or external epithelial surfaces, and *"ductless" glands* (e.g. the thyroid, thymus, spleen, gonads, and adrenals) whose functions were entirely unknown at the time (Figure 1.16). Another century would pass by before the French physician Théophile de Bordeu[46] proposed the notion that ductless glands release "emanations" into the blood, substances that would later be called **internal secretions**. By 1778 the stage was set, but the world would have to wait almost another century for the new field of endocrinology to emerge.

1849 The First Endocrine Experiment? Arnold Berthold and His Castrated Chickens

Prior to the discovery of hormones in the late 19th and early 20th centuries, physiologists assumed that all long-distance communication among organs, such as between the brain and the gonads, and between the gonads and its target tissues, was mediated exclusively by nerves that physically interconnect the communicating organs. On February 8, 1849, the German physiologist Arnold Berthold (see Figure 1.17) spoke at a meeting of the Royal Scientific Society in Gottingen, telling the audience of a remarkable experiment he had performed using chickens that contradicted this dogma.[47] Berthold's experimental design was conceptually simple (Figure 1.18) but required great surgical skill to perform successfully. After removing the testes from six sexually immature male chickens, he then created three different experimental groups consisting of two chickens each, as well one positive control group where the testes of two more chickens remained intact throughout the experiment. The first experimental group (the negative control group) was raised without testes over

Figure 1.17 Giants of the 19th and early 20th centuries whose contributions generated the field of endocrinology. (A) Arnold Adolph Berthold (1803–1861), a German scientist. **(B)** Claude Bernard (1813–1878), a French physiologist. **(C)** Thomas Addison (1793–1860), an English physician. **(D)** Charles-Édouard Brown-Séquard (1817–1894), a French-Mauritian physiologist. **(E)** Ernest Henry Starling (1866–1927), an English physiologist.

Source of images: (A-C and E) National Library of Medicine. (D) Gagliano-Jucá, T., et al. 2022. Rev. Endocr. Metab. Disord., pp. 1-10; used with permission.

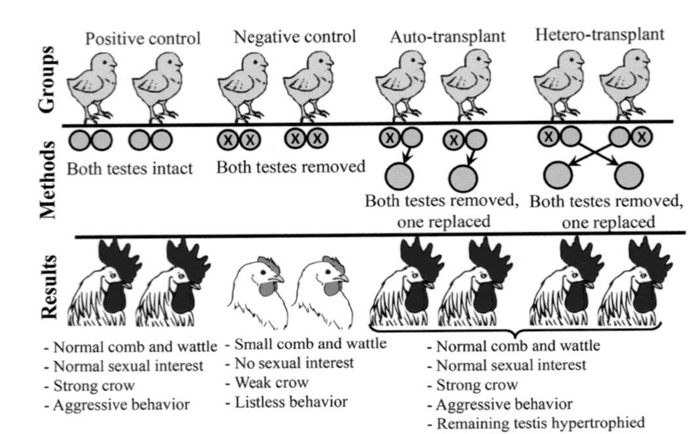

Figure 1.18 Arnold Berthold's classic experiment with young roosters demonstrating that a transplanted testis devoid of neural connections is sufficient to promote the development of adult rooster morphology and behaviors. Berthold's findings suggested that the testes act to somehow modify the blood. Berthold's elegant experiment is often considered to be history's first sophisticated experiment demonstrating what we now call endocrine secretion.

Source of images: Modified from Mac E. Hadley and Jon E. Levine, 2007. Endocrinology. 6th edition. Pearson Prentice Hall.

the next several months, upon which the chickens subsequently developed into "capons" (described earlier), or adult male chickens devoid of aggression or sexual interest in hens, that crow weakly, and possess underdeveloped wattles and combs resembling those of female hens. In the second experimental group, immediately following castration Berthold replaced one of each chickens' own testes back into its abdominal cavity (an *auto-transplant* group) and also raised them over several months. In the third experimental group, Berthold again transplanted one testis into the abdominal cavity, but this time into the *other* animal (a *hetero-transplant* group). Strikingly, unlike the control group that developed into capons, the chickens in the auto- and hetero-transplant groups all grew into normal-behaving and normal-looking roosters (Figure 1.18). After the experiment, the transplanted testes from the chickens were examined and were found to not only be living and revascularized, but also became *hypertrophied* (grew larger in size) to compensate for the absence of a second testis. Notably, the testes were not re-enervated.

While it had been known for centuries prior that the testes were necessary for the development of many adult male rooster characteristics, the results of Berthold's experiment provided critical new information. Specifically, Berthold knew that one effect of the transplants was to sever all nerve connections to the testes, and nerves were known to lack the capacity to regenerate. Therefore, any effect that the transplanted testes exerted on the recipient chickens' behavior and morphology in these experiments must have resulted from non-neural interactions stemming from the testes that somehow affected target organ functions. Berthold concluded, rather vaguely, that the normal development in both groups of transplanted chickens must originate from the testes "through their action on the blood, and then through the suitable ensuing action of the blood on the organism as a whole".[47] That is, the testes somehow modified the blood, and the altered blood then somehow interacted with the other tissues and/or nerves. However, in keeping with the physiological thinking of the time, blood modification did not necessarily signify the secretion of substances into the blood, but could also reflect the removal of inhibitory factors from the blood or the transformation/activation of preexisting blood materials.

Although it is not clear from Berthold's conclusions if his original intent was to actually demonstrate endocrine secretion by the testes, his elegant experimental design, methods, and findings can certainly be interpreted (even if just retrospectively) as history's first sophisticated endocrine experiment.

1855–1856 Internal Secretions Come into Vogue

Claude Bernard (1813–1878) (Figure 1.18B) is considered one of the most important physiologists of the 19th century. He formed the concept of **homeostasis**, or the constancy of the internal environment that is actively maintained by organisms, as well as the notion of eliminating experimental bias by performing "blind" experiments, among many others. Possibly his most important discovery, in 1855

Bernard demonstrated that the liver produced not one but two different categories of secretions, and he coined two new words to reflect this. The first secretion, bile, was already known to be released through the bile duct into the lumen of the intestine. Bernard termed this form of secretion an **external secretion** (*sécretion externe*), or a product secreted external to the blood such as onto an epithelial surface associated with either a lumen of an organ or the skin. The second secretion, which he discovered and termed an **internal secretion** (*sécretion interne*), or a product secreted directly into the blood, was glucose that was released from liver glycogen stores into the circulatory system.[48] Although glucose is a metabolic fuel and not a classical chemical messenger, it was the first internally secreted compound to be discovered. Importantly, Bernard also suspected that other ductless glands, like the adrenal, thyroid, spleen, and thymus glands, were also sources of as of yet unidentified internal secretions.

Also in 1855, Thomas Addison (1793–1860) (Figure 1.18C) of Guy's Hospital London reported that patients whose adrenal glands were destroyed by tuberculous infection died. This observation suggested that the adrenals were essential for life, albeit via an unknown mechanism, and Addison inaugurated the study of diseases of the "ductless glands".[49] Intrigued by Addison's findings, a year later the French-Mauritian physiologist, Charles-Édouard Brown-Séquard (Figure 1.18D) reported that adrenalectomy in various animals (dogs, rabbits, guinea pigs, and mice) resulted in death within a day, and he equated their condition to that of patients dying from Addison's disease.[50] Critically, Brown-Séquard's research also showed that adrenalectomized animals survived if they were transfused with blood from intact animals, strongly suggesting that the adrenals secreted some substance into the blood that was vital for life.

1889–1891 Charles-Édouard Brown-Séquard: An Unconventional Midwife for the Birth of Endocrinology

Charles-Édouard Brown-Séquard was a remarkably accomplished scientist of his time, having published over 500 papers, serving as the editor of many prestigious journals, and being particularly well-regarded in the field of neurophysiology. However, he was a highly unconventional and eccentric researcher who was prone to self-experimentation. Some experiments included ingesting the vomit of cholera patients to study its treatment and transfusing his own blood into guillotined criminals to demonstrate the continued viability of the tissues.[51] Such was the setting when on June 1, 1889, the 72-year-old Brown-Séquard published the findings of a most extraordinary claim, that he had reversed some of the effects of aging by injecting himself over a 3-week period with extracts derived from the testicles of guinea pigs and dogs.[52] Brown-Séquard proposed that some of the effects of aging, as well as the symptoms of some diseases, were caused by a deficiency of internal secretions, and that these could be treated by administering extracts from animal glands to compensate for their deficiency. Although this was a testable hypothesis, Brown-Séquard only reported subjective examples of his own rejuvenation, and these included enhanced mental concentration, physical strength, endurance, increased sexual potency, as well as more forceful urinary streams and bowel movements. He did not test his extracts on patients, preferring to let others pursue such clinical experiments.

The testes extracts used by Brown-Séquard most likely did not contain any active hormone since, although the testes synthesize testosterone, they cannot store it. The subjective "rejuvenative" effects were, therefore, almost certainly caused by the placebo effect, and his claims were ridiculed by most scientists. However, his widely publicized work renewed public interest in the ancient use of organotherapy as a method for treating many maladies, and charlatans made fortunes by selling Brown-Séquard-inspired extracts labeled "Elixir of Life" and "Sequarine" for treating virtually every malady imaginable, from tuberculosis to leprosy (Figure 1.19A). However, Brown-Séquard did not profit from his method, and he provided his extract free of charge to medical practitioners.[53] Notably, in 1889 the Pittsburgh Major League Baseball and Hall of Fame pitcher, Jim "Pud" Galvin, became the first player in baseball history to use a supposed performance-enhancing drug (a "Brown-Séquard Elixir") to try to improve his pitching performance in a game against Boston (Figure 1.19B). Although Galvin pitched a two-hit shutout and was uncharacteristically successful at the plate,[51] this was most likely in spite of, and not by virtue of, his consumption of the elixir.

Ultimately, the conclusions that Brown-Séquard drew from his testes extract self-experimentations in 1889 were clearly premature and devoid of experimental validation. However, as will be seen next, his ideas were nonetheless prescient and highly influential, and he played a significant role in ushering in the new science of endocrinology.[51]

Curiously Enough . . . Brown-Séquard was author Robert Louis Stevenson's neighbor. Stevenson was the author of several well-known books, including *Treasure Island* and *Strange Case of Dr Jekyll and Mr Hyde*. Brown-Séquard was reportedly the inspiration for Stevenson's famous potion-imbibing characters of Dr Jekyll and Mr. Hyde[54] (see Figure 1.19C).

1891–1892 Organotherapy Is Successfully Used to Treat Hypothyroid Patients

Only two years after Brown-Séquard's publication, physician George Murray successfully treated a patient with injected extracts of thyroid gland for the treatment of symptoms associated with hypothyroidism,[55] and the following

Figure 1.19 The first "performance-enhancing" drug. (A) An advertisement for organotherapeutical remedy claiming to be based on Sequard's method. **(B)** In 1889 the Hall of Fame Pittsburgh Major League Baseball pitcher, Jim "Pud" Galvin, became the first player in baseball history to use a performance-enhancing drug (a "Brown-Séquard Elixir") to try to improve his pitching performance in a game against Boston. Though almost certainly a result of the placebo effect, in that game Galvin pitched a two-hit shutout and was uncharacteristically successful at the plate. **(C)** Brown-Séquard, who was author Robert Louis Stevenson's neighbor, was reportedly the inspiration for Stevenson's famous characters of Dr. Jekyll and Mr. Hyde.

Source of images: (A) Lindholm, J. and Laurberg, P., 2011. J. of Thyroid Res., 2011. (B-C) U.S. Library of Congress.

year an oral treatment with sheep's thyroid was also found to be effective.[56] Thus, hypothyroidism became the first human disease to be successfully treated with endocrine organ extracts and the first clinical vindication of Brown-Séquard's notions. A transition from organotherapy to what would eventually be called "hormone replacement therapy" in the 20th century was underway, and continues to this day.

Early–Mid 20th Century: The Isolation and Characterization of Hormones

1902 marked the year that the first hormone was purified and its actions characterized. This hormone, termed "secretin", was discovered by Ernest Henry Starling and William Maddock Bayliss[57] (see Figure 1.18E). Using dogs, they demonstrated that the secretion of secretin by the duodenum, the anterior-most region of the small intestine, into the bloodstream triggers the release of pancreatic digestive fluids into the duodenal lumen. Specifically, they showed that when they injected an extract made from the mucosal lining of the duodenum from one dog into the circulatory system of another dog whose duodenal nerve supply had been severed, this induced pancreatic secretion. It was also Starling who in 1905 coined the term "hormone" to describe chemical messengers that "have to be carried from the organ where they are produced to the organ which they affect by means of the blood stream".[58] Soon after, extracts from many different endocrine organs were shown to exert biological effects, including the use of posterior pituitary extracts to successfully treat diabetes insipidus, a condition of excess urine production caused by a deficiency in antidiuretic hormone, in human patients (Table 1.2).[59,60] However, it was the isolation of insulin from dog pancreatic extracts by Frederick Banting and colleagues in 1922 that marked the most significant achievement yet in the fledgling world of endocrinology, providing for the first time an effective clinical treatment for the devastating disease

Table 1.2 Isolation and Characterization of Hormones

Year	Accomplishment and Discoverers
1900	Purification of epinephrine. J. Takamine, T.B. Aldrich
1915	Crystallization of thyroxine. E.C. Kendall
1923	Purification of insulin. F. Banting, C. Best, J.B. Collip
1925	Purification of parathyroid hormone. J.B. Collip
1927	Synthesis of thyroxine. C.R. Harrington
1928	Isolation of pituitary neurohypophysis hormones. O. Kamm et al.
1923–1934	Isolation and characterization of ovarian and testicular hormones. E.A. Allen, W.M. Allen, S. Aschhein, A.F.J. Butenandt, G.W. Corner, E.A. Doisy, G.F. Marrian, L. Ruzicka, O.P. Wintersteiner
1927–1934	Isolation and characterization of the pituitary and chorionic gonadotropin. S. Aschheim, H.M. Evans, H.L. Fevold, F.L. Hisaw, E. Laquerer, G.F. Marrian, B. Zondek
1933	Isolation of pituitary prolactin. O. Riddle
1936–1938	Isolation and characterization of adrenal cortical hormones. T. Reichstein, J. von Euw, E.C. Kendall, A. Grollman
1944	Isolation of growth hormone. C.H. Li, H.M. Evans
1953–1954	Synthesis of oxytocin and vasopressin. V. du Vigneaud et al.

Source: After Fisher, D.A., 2004. A short history of pediatric endocrinology in North America. *Pediatric Research*, 55(4), p. 716.

Table 1.3 The Heroic Age of Steroid Hormone Discovery, 1924–1954

Hormone	Starting Material
Estrone	Pools of female urine
Estradiol	4 tons hog ovaries
Progesterone	20 kg hog corpora lutea
Vitamin D2	8 g irradiated ergosterol
Androsterone	25,000 L male urine
Testosterone	"tons" of bull testes
Corticosterone/DOCA	1,000 kg beef adrenal
Cortisone	150 tons beef adrenal
Aldosterone	500 kg beef adrenal
Ecdysone	500 kg silk moth *Bombyx mori* pupae

Source: After Wilson, J.D., 2005. The evolution of endocrinology: Plenary lecture at the 12th International Congress of Endocrinology, Lisbon, Portugal, 31 August 2004. *Clinical Endocrinology*, 62(4), pp. 389–396.

type 1 diabetes mellitus, a condition of chronically elevated blood glucose caused by insulin insufficiency.[61] By the mid-1950s not only had the chemical structures of thyroxine, epinephrine, insulin, oxytocin, and vasopressin been determined, but so had the structures of most steroid hormones. Indeed, 1929–1954 is often referred to as the "heroic" age of steroid hormone discovery, as massive quantities of animal tissue and human urine were required for their extraction and purification (Table 1.3).[62]

SUMMARY AND SYNTHESIS QUESTIONS

1. Which one of the following statements is true? Explain your reasoning. Berthold's experiment with chickens conclusively demonstrated that:
    a. The testes release a compound into the blood that communicates with target tissues directly
    b. The testes remove a compound from the blood whose absence affects target tissues directly
    c. The testes modify a preexisting compound from the blood that affects target tissues directly
    d. The testes communicate with target tissues via the blood independent of neuronal action
2. Under what condition can glucose be considered an "internal secretion"?
3. By 1922, three human maladies could be effectively treated with endocrine gland extracts. What were they?
4. Regarding the discovery of secretin, Bayliss and Starling showed that when they injected an extract made from the mucosal lining of the duodenum from one dog into the circulatory system of another dog whose duodenal nerve supply had been severed, this induced pancreatic secretion. Why was it important that they severed the nerves in this experiment?

Modern Endocrinology (Late 20th Century to Present): Integration of Reductionist and Systems Thinking

LEARNING OBJECTIVE Describe some paradigm-shifting concepts and technologies that emerged during the late 20th and early 21st centuries that transformed endocrinology into its modern incarnations.

KEY CONCEPTS:

- In the 1970s Rosalyn Yalow and Solomon A. Berson developed the "radioimmunoassay", allowing miniscule levels of specific hormones in the blood or another body fluid to be accurately quantified for the first time.
- The emergence of molecular biology in the late 20th century provided an arsenal of powerful new tools for elucidating the kinetics of intracellular signaling pathways activated by hormones in their target tissues.
- Whereas the 20th century was characterized by a reductionist approach to endocrinology, the 21st marked the emergence of "systems" approaches for understanding endocrinology.

The second half of the 20th century featured the advent of paradigm-shifting concepts in endocrinology. Among these was a growing appreciation for the stunning complexity of endocrine systems. Broad examples of such endocrine complexity include:

- **Neurosecretion**: some hormones are secreted by neurons into the blood, blurring the distinction between classical nervous and endocrine tissues. Additionally, some classical hormones were discovered to also function as neurotransmitters, even further muddying the water. While the concept of neurosecretion began to be explored in the early 20th century, it only became widely accepted by the late 20th century.
- Hormone **pleiotropy**: one hormone can produce different effects on different tissues, indicating an association with multiple receptor types and/or intracellular signaling pathways.
- Short-distance signaling: in addition to exerting long-distance effects via circulatory system transport, hormones can also exert local effects on adjacent cells without entering the blood.
- *Feedback control*: variable concentrations of circulating hormone influence their own production via both negative-and positive-feedback mechanisms.
- In addition to inducing rapid but short-term physiological responses in target cells by activating intracellular enzymes, many hormones also modulate slower but longer-lasting responses by altering gene expression.

The second half of the 20th century also gave rise to technologies and tools that would dramatically accelerate the pace of discovery in endocrinology. These included advances in organic chemistry and biochemistry, light and electron microscopy, cell biology, neuroscience, and genetics. Perhaps no single technology had as profound an effect on the field of endocrinology in the 20th century as did the development of the radioimmunoassay by Rosalyn Yalow and Solomon A. Berson in the 1970s. **Radioimmunoassay** (RIA) is a method that uses radioactively labeled antibodies to specifically recognize and bind to miniscule levels of a hormone of interest in the blood or another body fluid and accurately quantify its concentration. In 1977 Rosalyn Yalow (Figure 1.20) received the *Nobel Prize in Physiology or Medicine* for her and Berson's development of the RIA. Since recipients of the Nobel Prize must be living, Yalow was awarded the prize without Berson, who died in 1972. The prize was shared with Roger Guillemin and Andrew Schally, who were honored for their own work on hormone production in the brain. Yalow was the second woman to be awarded the Nobel Prize in this field.

Figure 1.20 Rosalyn Sussman Yalow (1921–2011) was an American medical physicist and a co-winner of the 1977 Nobel Prize in Physiology or Medicine (together with Roger Guillemin and Andrew Schally) for development of the radioimmunoassay (RIA) technique.

Source of images: Glick, S., 2011. Rosalyn sussman yalow (1921–2011). Nature, 474(7353), pp. 580-580; used with permission.

The emergence of the field of molecular biology in the late 20th century provided an arsenal of powerful new tools for elucidating not just cellular pathways associated with hormone biosynthesis, but also the tissue-specific expressions of receptors and the kinetics of intracellular signaling pathways activated by hormones in their target tissues. As such, the late 20th century marked a shift from the characterization of endocrine tissues and hormones themselves to the downstream actions of receptors, intracellular second messenger pathways, and the enzymatic and transcriptional responses that mediate the effects of hormones in target cells. Indeed, it is now recognized that pathologies associated with resistance to hormone action, such as type II diabetes mellitus, due to altered receptor and/or intracellular signaling pathways may be more common than diseases caused by states of deficiency or excess of the hormones themselves.[63]

Much of the approach to endocrine research in the 20th century can be described as *reductionist*, or the analysis of complex signaling pathways by breaking them down into simpler, discreet components. These include studying the actions of specific genes, proteins, enzymes, hormones, receptors, and linear pathways. While reductionist methods are powerful and continue to drive much of endocrine research today, dividing a problem into its parts can sometimes lead to the loss of important information about the whole. The transition to the 21st century marked the emergence of "systems" approaches for understanding endocrinology. *Systems biology* makes use of large amounts of data, often in the forms of gene sequences, RNA transcripts, and protein activity, to computationally model complex physiological systems and predict the outcome of groups of dynamically interacting components. To achieve this, researchers make use of technologies such as genomics, transcriptomics, proteomics, metabolomics, bioinformatics, and mathematical modeling to predict the potential effects of perturbations to a system and to design further experiments to study the system. Systems approaches provide a more holistic account of complex networks of interacting genes, hormones, and signaling pathways and are emerging as critical tools for understanding the etiologies of complex endocrine diseases such as obesity, type II diabetes mellitus, and other metabolic disorders, as well as for predicting the effects of exposure to endocrine-disrupting chemicals on fetal development and the onset of adult diseases.

SUMMARY AND SYNTHESIS QUESTIONS

1. "Metabolic syndrome" is a cluster of risk factors that occur together, increasing the risk of developing type 2 diabetes, heart disease, and stroke. Contrast reductionist vs. systems approaches to studying and treating this syndrome.

Summary of Chapter Learning Objectives and Key Concepts

LEARNING OBJECTIVE **Define** "endocrinology", and describe how the concept of chemical communication is integral to it.

- Endocrinology is a branch of physiology that studies blood-borne chemical signals called "hormones", the cells that produce them, and their actions on target tissues.
- Hormones influence virtually all aspects of physiology, from mediating growth, development, and reproduction to metabolism and homeostasis.
- Endocrine signaling systems consist of cells that transmit hormone signals, receptors on/in target cells that receive the signals, and intracellular chemical pathways that transduce and amplify signals into responses.

- Disruption of endocrine signaling anywhere along the hormone transmission-reception-transduction pathway may lead to disease.

LEARNING OBJECTIVE Describe the relevance and benefits of endocrinology to society, as well as some of the challenges associated with certain hormone technologies.

- An understanding of endocrinology is essential not just to the medical and veterinary sciences, but also to the fields of neuroscience, psychology, pharmacology, and toxicology.
- Endocrine-based drug discoveries and technological breakthroughs have profoundly influenced human welfare, animal husbandry, and agriculture science through the synthesis of diverse compounds designed to mimic hormones or block their actions.
- Challenges associated with some hormone technologies include unintended side effects of hormone analogs, hormone abuse, and toxic hormone waste emanating from city sewage and concentrated animal husbandry facilities.
- Synthetic endocrine-disrupting chemicals (EDCs) are a broad category of ubiquitous endocrine-toxic pollutants that may be detrimental to the health of both human and wildlife populations alike.

LEARNING OBJECTIVE Define the scope of inquiry of modern endocrinology, and describe how the fields of "general endocrinology" and "comparative endocrinology" interact to advance endocrine science as a whole.

- Modern endocrinology emphasizes two areas of study: general (clinical) endocrinology and comparative endocrinology.
- Since physiological systems are conserved among animals, the study of endocrine systems in diverse taxa promotes advances in both basic and clinical human endocrinology.
- Because more is known about human endocrinology compared with that of any other organism, knowledge gained by studying human hormone signaling helps researchers understand how the endocrine systems of other animals function.
- Comparative endocrinologists study the differences and similarities of hormone systems among diverse species (including humans), and also try to understand endocrine function in the contexts of ecology and evolution.

LEARNING OBJECTIVE Describe key advances in physiological knowledge prior to the 19th century that laid the foundation for the emergence of endocrinology, and discuss seminal endocrine experiments during the late 19th and early 20th centuries that established the field.

- Endocrinology is a relatively young science that originated in the late 19th and early 20th centuries.
- Two of the most ancient practices that make use of fundamental endocrine principles include organotherapy and castration.
- The foundations of endocrinology rest on knowledge assembled from the 16th–18th centuries for the presence of "ductless" glands and an accurate understanding of circulatory system function.
- The first sophisticated endocrine experiment took place in 1849 when Arnold Berthold transplanted testes into castrated male chicks to show that the organs communicated with target tissues via the bloodstream.
- Claiming that he had found a way to reverse the aging process, Charles-Édouard Brown-Séquard revived the ancient practice of organotherapy during the late 19th century.
- Applying 19th-century approaches to the ancient principles of organotherapy, George Murray successfully treated patients with hypothyroidism with thyroid extracts, the first example of a human endocrine disease to be successfully treated with hormone replacement therapy.
- In 1902 the first hormone to be purified and its actions characterized was "secretin", discovered by Ernest Henry Starling and William Maddock Bayliss.
- In 1905 Starling coined the word "hormone".

LEARNING OBJECTIVE Describe some paradigm-shifting concepts and technologies that emerged during the late 20th and early 21st centuries that transformed endocrinology into its modern incarnations.

- In the 1970s Rosalyn Yalow and Solomon A. Berson developed the "radioimmunoassay", allowing miniscule levels of specific hormones in the blood or another body fluid to be accurately quantified for the first time.
- The emergence of molecular biology in the late 20th century provided an arsenal of powerful new tools for elucidating the kinetics of intracellular signaling pathways activated by hormones in their target tissues.
- Whereas the 20th century was characterized by a reductionist approach to endocrinology, the 21st marked the emergence of "systems" approaches for understanding endocrinology.

LITERATURE CITED

1. Wass J. The fantastical world of hormones. Paper presented at: 19th European Congress of Endocrinology; 2017.
2. De Loof A. How to deduce and teach the logical and unambiguous answer, namely L=Σ C, to "What is Life?" using the principles of communication? *Commun Integr Biol.* 2015;8(5): e1059977.
3. Pasqualini JR, Chetrite GS. The formation and transformation of hormones in maternal, placental and fetal compartments: biological implications. *Horm Mol Biol Clin Investig.* 2016;27(1):11–28.
4. Stearns SC. Evolutionary medicine: its scope, interest and potential. *Proc Royal Soc B Biol Sci.* 2012;279(1746):4305–4321.
5. Lieberman D. *The Story of the Human Body: Evolution, Health, and Disease.* Vintage; 2014.
6. Patisaul HB, Adewale HB. Long-term effects of environmental endocrine

disruptors on reproductive physiology and behavior. *Front Behav Neurosci.* 2009;3:10.

7. Herbst AL, Green TH, Ulfelder H. Primary carcinoma of the vagina: an analysis of 68 cases. *Am J Obstet Gynecol.* 1970;106(2):210–218.
8. Herbst AL, Ulfelder H, Poskanzer DC. Adenocarcinoma of the vagina: association of maternal stilbestrol therapy with tumor appearance in young women. *N Engl J Med.* 1971;284(16):878–881.
9. Dohoo IR, DesCôteaux L, Leslie K, et al. A meta-analysis review of the effects of recombinant bovine somatotropin: 2. Effects on animal health, reproductive performance, and culling. *Can J Vet Res.* 2003;67(4):252.
10. Denver RJ, Hopkins PM, McCormick SD, et al. Comparative endocrinology in the 21st century. *Integr Comp Biol.* 2009;49(4):339–348.
11. Sumpter JP. Feminized responses in fish to environmental estrogens. *Toxicol Lett.* 1995;82:737–742.
12. Jensen KM, Makynen EA, Kahl MD, Ankley GT. Effects of the feedlot contaminant 17α-trenbolone on reproductive endocrinology of the fathead minnow. *Environ Sci Technol.* 2006;40(9):3112–3117.
13. Muir WM, Howard RD. Assessment of possible ecological risks and hazards of transgenic fish with implications for other sexually reproducing organisms. *Transgenic Res.* 2002;11(2):101–114.
14. TEDX. The Endocrine Disruptor Exchange. *TEDX List of Potential Endocrine Disruptors.* www.endocrinedisruption.org/. Published 2018. Accessed December 2018.
15. Bergman Å, Heindel JJ, Kasten T, et al. The impact of endocrine disruption: a consensus statement on the state of the science. *Environ Heal Perspect.* 2013;121(4):a104.
16. Kortenkamp A, Martin O, Faust M, et al. State of the art assessment of endocrine disrupters. *Final Report.* 2011;23.
17. Trasande L. *Sicker, Fatter, Poorer: The Urgent Threat of Hormone-Disrupting Chemicals on Our Health and Future . . . and What We Can Do about It.* Houghton Mifflin; 2019.
18. Marciano DP, Chang MR, Corzo CA, et al. The therapeutic potential of nuclear receptor modulators for treatment of metabolic disorders: PPARgamma, RORs, and Rev-erbs. *Cell Metab.* 2014;19(2):193–208.
19. Altman BJ. Cancer clocks out for lunch: Disruption of circadian rhythm and metabolic oscillation in cancer. *Front Cell Dev Biol.* 2016;4:62.
20. Golombek DA, Casiraghi LP, Agostino PV, et al. The times they're a-changing: effects of circadian desynchronization on physiology and disease. *J Physiol-Paris.* 2013;107(4):310–322.
21. Fonken LK, Nelson RJ. The effects of light at night on circadian clocks and metabolism. *Endocr Rev.* 2014;35(4):648–670.
22. Paksarian D, Rudolph KE, Stapp EK, et al. Association of outdoor artificial light at night with mental disorders and sleep patterns among US adolescents. *JAMA Psychiat.* 2020;77(12):1266–1275.
23. Cho Y, Ryu S-H, Lee BR, Kim KH, Lee E, Choi J. Effects of artificial light at night on human health: a literature review of observational and experimental studies applied to exposure assessment. *Chronobiol Int.* 2015;32(9):1294–1310.
24. Ward EM, Germolec D, Kogevinas M, et al. Carcinogenicity of night shift work. *Lancet Oncol.* 2019;20(8):1058–1059.
25. Navara KJ, Nelson RJ. The dark side of light at night: physiological, epidemiological, and ecological consequences. *J Pineal Res.* 2007;43(3):215–224.
26. Salmon M. Protecting sea turtles from artificial night lighting at Florida's oceanic beaches. In: Rich C, Longcore T, eds. *Ecological Consequences of Artificial Night Lighting.* Island Press; 2006:141–168.
27. Gauthreaux SA, Belser CG. Effects of artificial night lighting on migrating birds. In: Rich C, Longcore T, eds. *Ecological Consequences of Artificial Night Lighting.* Island Press; 2006:67–93.
28. Montevecchi WA. Influences of artificial light on marine birds. In: Rich C, Longcore T, eds. *Ecological Consequences of Artificial Night Lighting.* Island Press; 2006:94–113.
29. Becker DJ, Singh D, Pan Q, et al. Artificial light at night amplifies seasonal relapse of haemosporidian parasites in a widespread songbird. *Proc Royal Soc B Biol Sci.* 2020;287(1935):20201831.
30. Le Tallec T, Perret M, Thery M. Light pollution modifies the expression of daily rhythms and behavior patterns in a nocturnal primate. *PLoS One.* 2013;8(11):e79250.
31. Lewis SM, Wong CH, Owens A, et al. A global perspective on firefly extinction threats. *BioScience.* 2020.
32. Norris DO. Comparative endocrinology: Past, present, and future. *Integr Comp Biol.* 2018;58(6):1033–1042.
33. Gorbman A, Bern HA. Textbook of comparative endocrinology. John Wiley & Sons, Inc.; 1962.
34. Mora C, Tittensor DP, Adl S, Simpson AG, Worm B. How many species are there on Earth and in the ocean? *PLoS Biol.* 2011;9(8):e1001127.
35. Green SL. *The Laboratory Xenopus sp.* CRC Press; 2009.
36. Bradshaw D. Environmental endocrinology. *Gen Comp Endocrinol.* 2007;152(2–3):125–141.
37. McCormick SD, Romero LM. Conservation endocrinology. *Bioscience.* 2017;67(5):429–442.
38. Wingfield JC. Coping with change: a framework for environmental signals and how neuroendocrine pathways might respond. *Front Neuroendocrinol.* 2015;37:89–96.
39. Medvei VC. *The History of Clinical Endocrinology: A Comprehensive Account of Endocrinology from Earliest Times to the Present Day.* CRC Press; 1993.
40. Donnelly DE, Morrison PJ. Hereditary gigantism-the biblical giant Goliath and his brothers. *Ulster Med J.* 2014;83(2):86.
41. Karamanou M, Protogerou A, Tsoucalas G, Androutsos G, Poulakou-Rebelakou E. Milestones in the history of diabetes mellitus: the main contributors. *World J Diabetes.* 2016;7(1):1.
42. Eknoyan G. Emergence of the concept of endocrine function and endocrinology. *Adv Chronic Kidney Dis.* 2004;11(4):371–376.
43. Rolleston HD. *The Endocrine Organs in Health and Disease: With an Historical Review.* Oxford University Press; 1936.
44. Medvei VC. *The History of Clinical Endocrinology: A Comprehensive Account of Endocrinology from Earliest Times to the Present Day.* CRC Press; 1993.
45. Jenkins JS. The voice of the castrato. *Lancet.* 1998;351(9119):1877–1880.
46. de Bordeu T. Analyse medicinale du sang. *Oeuvres (Paris, 1818).* 1776;2:930.
47. Berthold AA. Transplantation der hoden. *Arch Anat Physiol.* 1849:42–46.
48. Bernard C. *Sur le mécanisme de la formation du sucre dans le foie.* Mallet-Bachelier; 1855.
49. Addison T. *On the Constitutional and Local Effects of Disease of the Supurarenal Capsules.* Highley; 1855.
50. Brown-Séquard C-E. *Recherches expe (rimentales sur la physiologie et la pathologie des capsules surre (nales).* Imprimerie de Mallet-Bachelier; 1856.
51. Rengachary SS, Colen C, Guthikonda M. Charles-Edouard Brown-Séquard: An eccentric genius. *Neurosurgery.* 2008;62(4):954–964.
52. Brown SC. Note on the effects produced on man by subcutaneous injections of a liquid obtained from the testicles of animals. *The Lancet.* 1889;134(3438):105–107.
53. Borell M. Organotherapy, British physiology, and discovery of the internal secretions. *J Hist Biol.* 1976;9(2):235–268.
54. Edouard L. Commentary: Brown-Séquard, father of endocrinology. *Afr J Reprod Heal.* 2019;23(4):16–18.

55. Murray GR. Note on the treatment of myxoedema by hypodermic injections of an extract of the thyroid gland of a sheep. *Brit Med J.* 1891;2(1606):796.
56. Fox E. A case of myxoedema treated by taking extract of thyroid by the mouth. *Brit Med J.* 1892;2(1661):941.
57. Bayliss WM, Starling EH. The mechanism of pancreatic secretion. *J Physiol.* 1902;28(5):325–353.
58. Starling EH. The Croonian lectures. *Lancet.* 1905;26:579–583.
59. Farini F. Diabete insipido ed opoterapia. *Gazz Osped Clin.* 1913;34:1135–1139.
60. Von den Velden R. Die nierenwirkung von hypophysenextrakten beim menschen. *Berlin Klin Wochenshr.* 1913;45:2083–2087.
61. Banting FG, Best C, Collip J, et al. The effect produced on diabetes by extracts of pancreas. *Trans Assoc Am Phys.* 1922;37:337–347.
62. Wilson JD. The evolution of endocrinology: plenary lecture at the 12th international congress of endocrinology, Lisbon, Portugal, 31 August 2004. *Clin Endocrinol.* 2005;62(4):389–396.
63. Wilson JD. Endocrinology: survival as a discipline in the 21st century? *Ann Rev Physiol.* 2000;62(1):947–950.

CHAPTER

2

Fundamental Features of Endocrine Signaling

CHAPTER LEARNING OBJECTIVES:

- Compare and contrast the classical and modern definitions of "hormone" and "endocrinology".
- Describe the fundamental features of endocrine signaling:
  - Hormone diversity
  - Concentration in circulation
  - Influences on hormone half-life
  - Receptor specificity
  - Intracellular signaling pathways
  - Pleiotropy
  - Feedback control
  - Hormone rhythmicity and pulsatility
  - Complexity of interactions with other hormones
  - Dynamic states of hormone function

OPENING QUOTATIONS:

"These numerous and exceptional qualifying phenomena demonstrate the futility of clinging blindly to interpretations of endocrine and hormonal characteristics within the rigid framework of the Bayliss and Starling definitions. On the other hand, since developments in the endocrine field are continuing to accumulate and to alter the meanings of the basic terms, it is also futile to attempt to fix their definitions at this time."

—Gorbman and Bern (1962)[1]

"I'm a professor of endocrinology at the Harvard Medical School. I'm an attending physician at the Peter Bent Brigham Hospital! I'm a contributing editor to the *American Journal of Endocrinology* and I am a fellow and vice-president of the Eastern Association of Endocrinologists and president of a Journal Club! And I'm not going to listen to any more of your kabbalistic, quantum, friggin' dumb limbo mumbo jumbo!"

—Dr. Mason Parrish, character in Ken Russell's movie *Altered States* (1980)

Classical vs. Modern Endocrinology

LEARNING OBJECTIVE Compare and contrast the classical and modern definitions of "hormone" and "endocrinology".

KEY CONCEPTS:

- Compared to the classical views of hormone signaling established by the early 20th century, modern endocrinology has expanded to include the notions that hormones can be secreted by neurons, and that in addition to long-distance blood-borne transport, hormones also can diffuse over short distances without entering the blood to bind to target receptors.
- Homeostasis arises from the interactions of three classes of "bioregulators": hormones of the endocrine system, neurotransmitters of the nervous system, and cytokines of the immune system.

Our knowledge of endocrinology has skyrocketed since the field's emergence during the late 19th and early 20th centuries. As our understanding of hormone signaling increases, the definitions of "hormone" and "endocrinology" continue to evolve. Some key characteristics of classical hormones, as originally defined by Bayliss and Starling in 1905, include:

1. Hormones are secreted by ductless endocrine glands into the bloodstream.
2. Hormones are transported by the blood some distance away from the gland of origin.
3. Hormones exert their effects on distinct target organs.

In addition to these classical notions of **endocrine signaling**, prior to the 1930s it was also widely assumed that endocrine and neural signaling were two distinct long-distance intercellular communication processes. However, by the mid- to late 20th century many exceptions to the

DOI: 10.1201/9781003359241-3

classical endocrine paradigm began to emerge, including observations that:

1. Hormones can also be secreted into the blood by some neurons, introducing the concept of *neuroendocrine signaling* (Figure 2.1). Furthermore, classical hormones, like epinephrine, can also function as neurotransmitters.
2. In addition to secretion into the blood, hormones can also exert local effects on adjacent cells in a process called **paracrine signaling,** or even target the cell of origin through **autocrine signaling** (Figure 2.1).
3. Hormone synthesis is not confined to a limited range of glands, but rather they can be made by virtually all cell types, including neurons, heart muscle, skeletal muscle, bone, and adipose tissue.
4. The same hormone can be generated by multiple organs, tissues, and even scattered cells.

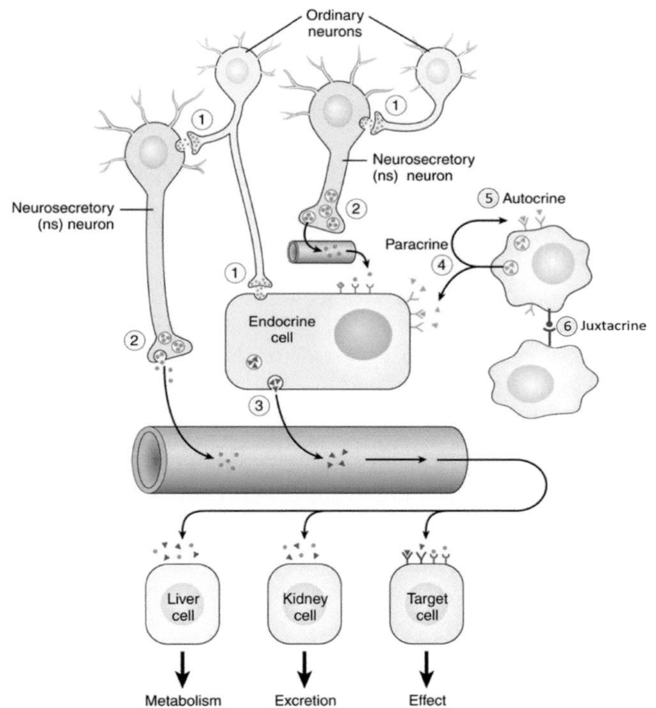

Figure 2.1 The diverse mechanisms of cell-cell signaling. (1) Classical neural signaling occurs when a neuron releases neurotransmitter into a synapse, an extracellular space between the secretory neural cell and the post-synaptic target cell that has receptors for the neurotransmitter. (2) Neuroendocrine secretion occurs when neurons release signaling compounds into the bloodstream. These compounds now behave as classical hormones and are transported to downstream target cells with ligand-specific receptors. (3) Classical endocrine signaling occurs when a non-neuronal cell secretes a signaling compound into circulation, where it binds to target cells possessing ligand-specific receptors. (4) Paracrine signaling occurs when a signaling compound released by a cell travels a short distance within the extracellular fluid to bind to receptors on adjacent target cells. (5) Autocrine signaling occurs when a cell secretes a chemical messenger that binds to receptors on that same cell. (6) Juxtacrine, or contact-dependent signaling, occurs when a plasma membrane-localized signal on one cell activates a plasma membrane-localized receptor on another target cell.

Source of images: Norris, D.O. and Carr, J.A., 2020. Vertebrate endocrinology. Academic Press; used with permission.

In 1977 the Nobel Prize–winning neuroendocrinologist, Roger Guillemin, proposed that the modern endocrine definition of hormone be revised to broadly include "any substance released by a cell and which acts on another cell near or far, regardless of the singularity or ubiquity of the source and regardless of the means of conveyance, blood stream, axoplasmic flow, immediate intracellular space".[2] Typically, hormones are operationally distinguished from two other classes of signaling molecules: **neurotransmitters**, which are released at synapses to promote electrical signaling among neurons, and **cytokines**, which are chemicals involved in immune cell signaling. Despite the somewhat artificial distinctions among these three classes of signaling molecules, the endocrine, immune, and nervous systems communicate with each other to maintain organismal homeostasis. Together these three interacting chemical signaling classes have been referred to as **bioregulators.**[3] As such, the scope of modern endocrinology has exploded from its origins as a specialty within the discipline of physiology that focused solely on endocrine glands and their internal secretions to a discipline addressing the entire spectrum of chemical communication in animals ranging from the molecular and cellular to the organismal and population levels. Some important differences between classical and modern endocrine thinking are summarized in Appendix 3.

SUMMARY AND SYNTHESIS QUESTIONS

1. Compare and contrast the classical and modern definitions of "hormone" and "endocrinology".

Fundamental Features of Endocrine Signaling

LEARNING OBJECTIVE Describe the fundamental features of endocrine signaling.

KEY CONCEPTS:

- Whereas hormones are chemically diverse, ranging from modified amino acids to proteins, lipids, and other chemicals, all of their known receptors are proteins.
- Hormones are active at extremely low concentrations and bind to specific receptors located at the surface and inside of target cells.
- Many hormones bind to blood transport proteins, which increases their longevity and solubility in circulation.
- One hormone can exert distinct effects on different cells and tissues by interacting with different receptors.
- Intracellular signaling pathways transduce and amplify hormone signals.

- Hormones mediate homeostasis via feedback control.
- Hormones can exert "pleiotropic" effects.
- In order to adapt to changing conditions, organisms must be able to change the homeostatic set points for a regulated parameter, a concept called "allostasis".
- Hormones are often secreted in rhythmic and pulsatile manners.
- Hormones exist in a dynamic state and may be present in the blood in inactive forms or activated forms, and they can also be transactivated into hormones with other functions.

Hormones Are Chemically Diverse

Hormone classes are structurally heterogeneous, and include (1) amino acid-derived members, (2) proteins, including glycoproteins and short peptides, (3) lipids, consisting of cholesterol and fatty acid derivatives, (4) terpenes, a type of volatile hydrocarbon found in insects, plants, and other organisms, as well as (5) gases, like nitric oxide (Figure 2.2). Importantly, endogenous metabolites, such as free fatty acids, glucose, and bile acids, as well as ions, like calcium, and even photons can also behave as hormone-like signals, conveying information on metabolic, seasonal, and ion status to target tissues. Hormones range across the solubility spectrum, from hydrophilic to hydrophobic, parameters that greatly influence their transportation in the blood and interactions with target tissues. For example, whereas many hydrophobic hormones, like steroid hormones, can freely cross the plasma membranes of target cells, hydrophilic hormones (e.g. proteins and peptides) cannot.

Hormones Are Active at Extremely Low Concentrations

In the bloodstream and extracellular fluid are diverse molecules that are structurally similar to hormones, including cholesterol, fatty acids, amino acids, peptides, and proteins. Whereas these molecules generally circulate at concentrations in the micro- to millimolar (10^{-6} to 10^{-3} mol/L) range, hormones are present at *much* lower concentrations, typically in the pico- to nanomolar (10^{-12} to 10^{-9} mol/L) range. Table 2.1 shows reference values for some human hormones and other blood molecules. Because hormones are physiologically active in such

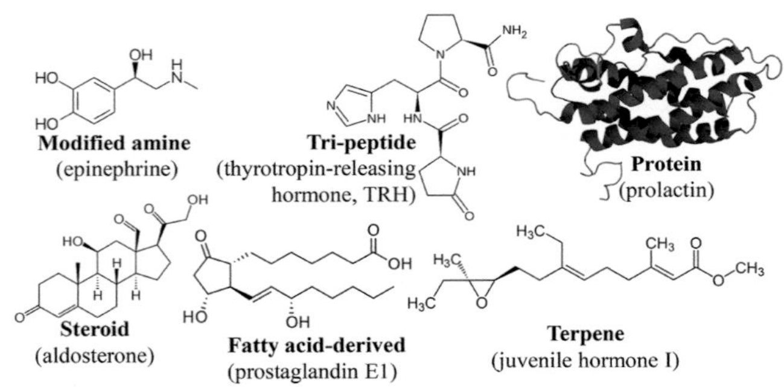

Figure 2.2 Some examples of the broad structural diversity of hormones.

Table 2.1 Comparison of Normal Concentrations of Some Organic Substances and Hormones in Human Blood

Substance	Concentration (mM)
Glucose	5
Cholesterol	5
Albumin	0.7
Antibodies	0.09
High-density lipoprotein	0.013
Thyroxine	0.00009
Testosterone	0.00002
Insulin	0.00000005

Source: After Altman, P.L. In (1961) *Blood and Other Body Fluids* D.S. Dittmer (ed.) *Fed. Am. Soc. Exp. Biol.* 540 pp. and (1979) *Methods of Hormone Radioimmunoassay*, 2nd ed. B.M. Jaffe and H.R. Behrman, eds. Academic, New York, pp. 1005–1014

minute quantities, sensitive receptors and systemic feedback mechanisms have evolved in target tissues to sense and amplify these weak signals.

Many Hormones Bind to Blood Transport Proteins to Increase Their Longevity and Solubility in Circulation

A hormone's **half-life** is the amount of time required for a hormone to fall to half of its initial concentration in the blood. The half-lives of hormones in circulation vary, with amino acid-derived epinephrine from the adrenal medulla ranging in the order of seconds, protein and peptide hormones spanning minutes, and hours for lipophilic steroid and thyroid hormones. The half-lives of some hormones are greatly extended if they bind to *blood transport proteins* (Figure 2.3), which also increase the solubility of lipophilic hormones in the blood. Although not all hormones associate with blood transport proteins,

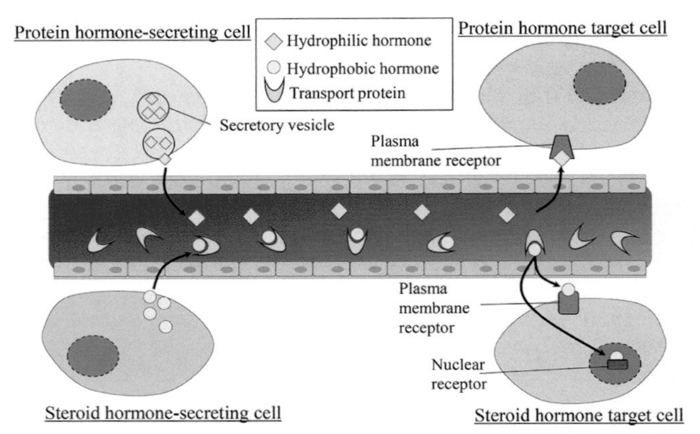

Figure 2.3 Hormone secretion, blood transport, and selectivity by hormone receptors. Unlike most hydrophilic hormones, all lipophilic hormones are transported in the blood attached to transport proteins. Hydrophilic hormones target plasma membrane-localized receptors, whereas lipophilic hormones can either interact with cell-surface receptors or diffuse across the plasma membrane to interact with intracellular and nuclear receptors.

such proteins exist for every hormone class. Blood transport proteins for protein and peptide hormones inhibit hormone destruction by proteases present in the blood plasma. Blood transport proteins for hydrophobic steroid and thyroid hormones allow these molecules to exist in the blood at concentrations several hundred-fold greater than their solubility in water would otherwise permit. In addition, blood transport proteins for small amino acid-derived hormones prolong their circulating half-life by preventing their filtration through the renal glomerulus.

Hormones Bind to Specific Receptors at a Target Cell's Surface and/or Interior

Because hormones circulate at extremely low concentrations, target cells must distinguish not only among different hormones present in miniscule quantities, but also between any given hormone and the 10^6- to 10^9-fold excess of other similar molecules. This extraordinary degree of discrimination is provided by the presence of specialized signal recognition proteins called *receptors* (Figure 2.4). Indeed, a "target cell" is defined by its ability to selectively bind a specific hormone to its cognate receptor.[4]

Curiously Enough . . . In contrast with the broad chemical heterogeneity of hormones themselves, all known hormone receptors are proteins.

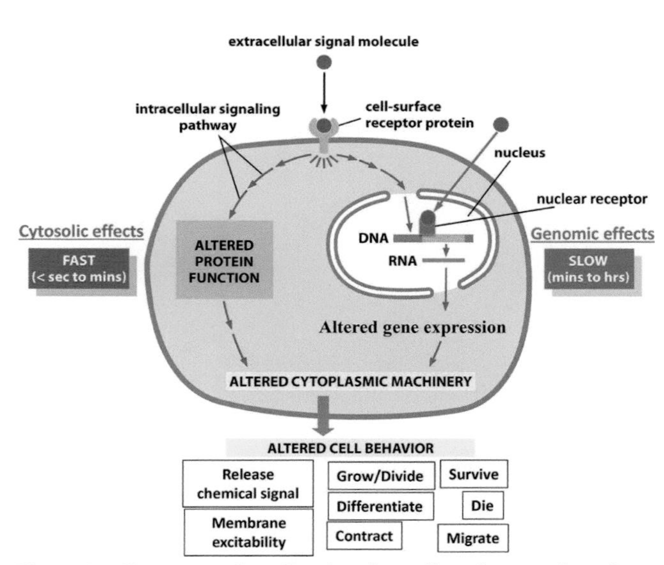

Figure 2.4 Receptors localized to the cell surface and nucleus function by initiating intracellular signaling responses that may result rapidly in the form of altered cytosolic protein function or slowly via the modulation of gene expression. The cell continuously integrates these signals into vital decisions about its function, ultimately yielding a change in cell physiology or behavior.

Source of images: W.W. Norton & Company, Inc., Essential Cell Biology, Fifth Edition by Bruce Alberts, et al. 2019; used with permission.

Hormone receptors may be either localized to the plasma membrane or located intracellularly (e.g. cytoplasm, nucleus, endoplasmic reticulum membrane, or mitochondria) (Figure 2.4). Whereas plasma membrane-localized receptors may bind to both hydrophilic and hydrophobic hormones, intracellular receptors only associate with hydrophobic hormones that can cross the plasma membrane. As such, in order for a plasma membrane receptor-bound hormone to exert its effects on the cell interior, signals must be transmitted from the receptor to the cytoplasm, nucleus, or target organelles. In general, the binding of a ligand to a cell-surface receptor can result in either (1) a "*cytosolic response*", which can include changes to plasma membrane ion permeability, cytosolic protein functions, and/or (2) a "*genomic response*" that is characterized by altered gene expression (Figure 2.4). Because genomic responses require the mobilization of transcriptional and translational machinery to manifest, they take place relatively slowly over the course of hours. In contrast, cytosolic protein responses and alterations to membrane ion permeability following receptor binding occur rapidly, over the course of seconds to minutes. Whereas hydrophilic hormones must act via plasma membrane-localized receptors, many hydrophobic hormones can both bind to plasma membrane-localized receptors and/or cross the plasma membrane and immediately associate with cytosolic or nuclear-localized receptors that function as transcription factors and induce direct genomic responses (Figure 2.4). Importantly, signal transduction pathways emanating from plasma membrane receptors do not always proceed as simple linear chains, but often as highly complex, branching pathways that involve a diversity of signaling mediators (Figure 2.5). A signaling pathway may result in both cytosolic and genomic responses.

The total number of functional receptors for a particular hormone in a target cell helps to determine the sensitivity of that cell to the hormone, with relatively high concentrations of receptors promoting higher sensitivity, and lower receptor numbers reducing sensitivity. Indeed, cells are constantly adjusting their sensitivity to hormones by either **upregulating** (increasing) or **downregulating** (decreasing) functional receptor count. Whereas a low concentration of hormone tends to promote the upregulation of its receptors on target cells, higher concentrations typically downregulate target cell receptors (Figure 2.6). The downregulation of receptors located on the plasma membrane is mediated by their endocytosis following hormone binding and sequestration into small storage and recycling vesicles within the cytoplasm. These stored receptors may then be upregulated via their recycling to the cell surface.

Intracellular Signaling Pathways Transduce and Amplify Hormone Signals

In order for a target cell to respond to a hormone, the signal must be biochemically transduced into a form capable of exerting physiological change to the cell. Reflecting the increasingly common analogies between computer science and biology at the time, Nobel laureate Martin Rodbell (he

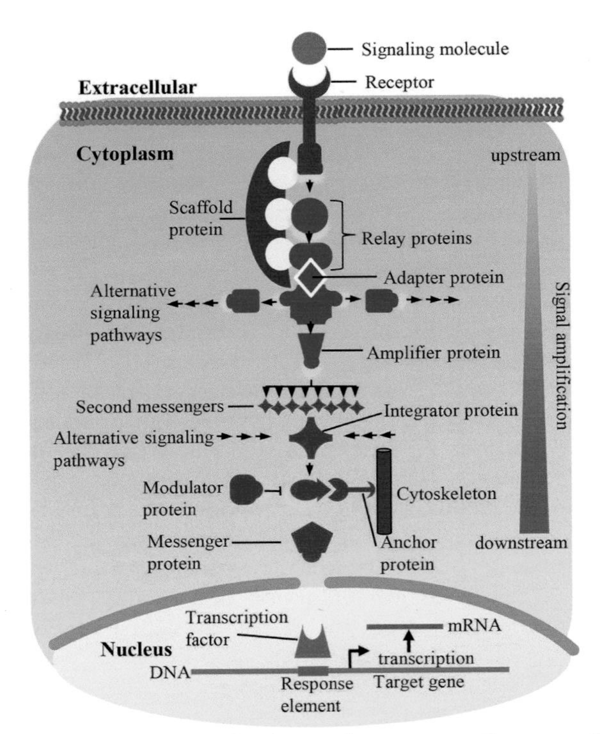

Figure 2.5 Signal transduction pathways are often complex and branching, resulting in both cytosolic and genomic responses. Signaling proteins may include scaffold proteins, relay proteins, bifurcation proteins, adapter proteins, amplifier and transducer proteins, integrator proteins, modulator proteins, messenger proteins, transcription factors, and target proteins.

https://www.open.edu/openlearn/science-maths-technology/cell-signalling/content-section-1.5

shared the *1994 Nobel Prize in Physiology or Medicine* with Alfred G. Gilman for "their discovery of G-proteins and the role of these proteins in signal transduction in cells") proposed that individual cells were analogous to "cybernetic systems" made up of three distinct molecular components: "discriminators", "transducers", and "amplifiers".[5] Specifically, receptors discriminate among hundreds of possible ligands. Other proteins, of which G proteins are one category, transduce the extracellular signal into an intracellular message. Another category of proteins termed "effectors" enzymatically amplify messages by producing second messenger molecules. Such reception, processing, and interpretation of extracellular signals by the intracellular machinery typically consist of complex, highly integrated and ordered branching networks of consecutive protein—protein interactions referred to as **signal transduction pathways.**

A rapid response stimulated by the binding of a hormone to a plasma membrane-localized receptor is often initiated through a "phosphorylation cascade" (Figure 2.7), whereby the ligated receptor first activates a dormant *effector protein*, which is typically an enzyme that regulates the activity of another protein. A common category of plasma membrane effectors are **kinases**, enzymes that catalyze the transfer of the terminal phosphate group from ATP molecules onto specific amino acids within proteins. An activated kinase often phosphorylates another kinase, causing a conformation change in the enzyme, which activates it. The newly activated kinase can go on to phosphorylate and activate yet another group of kinases, and so on, until the phosphorylation of a final *non*-kinase target molecule ultimately results in the alteration of the cell's physiology or behavior in some way. Importantly, these kinase relays

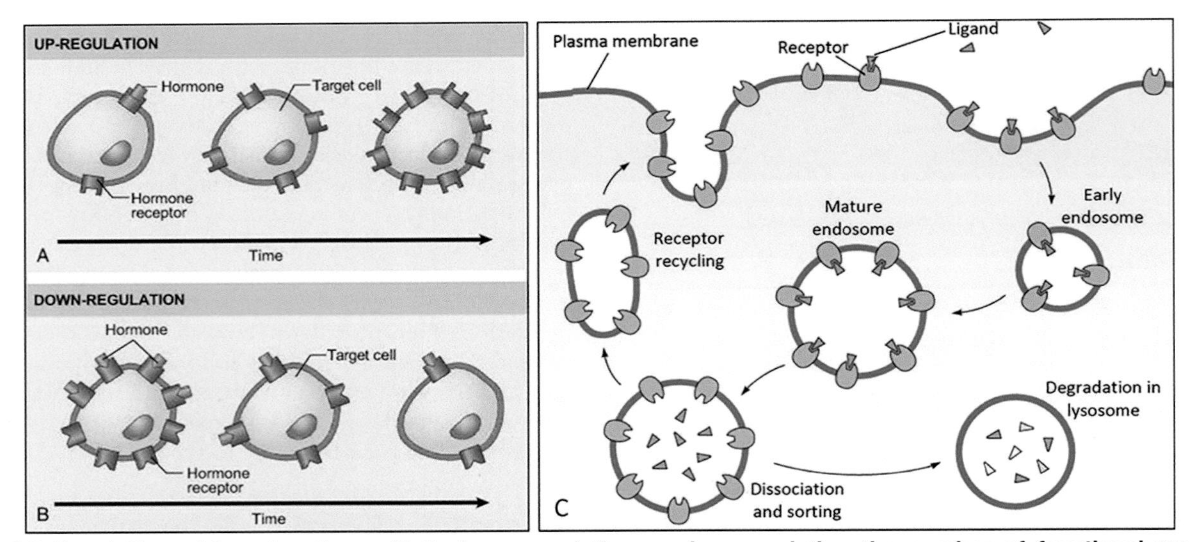

Figure 2.6 Regulation of target cell sensitivity by upregulating or downregulating the number of functional receptors on the plasma membrane. (A) Low levels of circulating hormone typically upregulates (increases the number of) the receptor. **(B)** By contrast, relatively high hormone concentrations downregulate (reduce) the number of receptors. **(C)** Receptor-mediated endocytosis. After a peptide hormone binds to a target cell receptor, it is endocytosed as a small vesicle. The ligand–receptor complex is dissociated, the hormone is degraded, and the receptor may be recycled back to the cell surface.

Source of images: (A) Moini, J., et al. 2021. Epidemiology of Endocrine Tumors. Elsevier; used with permission. (B) Jones, R.E. and Lopez, K.H., 2013. Human reproductive biology. Academic Press; used with permission.

not only occur rapidly, but also dramatically amplify the initial signal. Thus, the binding of a single hormone to one receptor can potentially activate tens of thousands of target molecules, producing a very robust response. Although the notion of signal transduction pathways was originally developed to explain rapid response systems associated with plasma membrane receptors, the concept also applies to the mediation of slower genomic response systems associated with intracellular receptors that function as nuclear transcription factors. In this case, the liganded receptors associate with transcriptional machinery (i.e. coactivators and RNA polymerases) that function as transducers and amplifiers to generate many RNA transcripts.

Hormones Can Exert "Pleiotropic" Effects

In the context of endocrinology, **pleiotropy** occurs when one hormone exerts different effects on different cells and tissues. For example, whereas epinephrine induces the vasodilation of blood vessels associated with skeletal muscle, the same hormone promotes vasoconstriction of intestinal blood vessels (Figure 2.8). In this example, pleiotropy manifests by virtue of the fact that vascular smooth muscle can express one of two different types **adrenergic receptors**, or receptors that bind to epinephrine and/or norepinephrine: β-adrenergic receptors in skeletal muscle vascular tissue and α-adrenergic receptors in intestinal vascular tissue. These receptors each interact with distinct intracellular transducer and effector proteins that produce different outcomes on cell function. Indeed, the adrenergic receptor family is large, with different tissues enriched with different receptor types, which are responsible for mediating a diversity of effects (Table 2.2). Hormones can even simultaneously exert different effects within the same cell. For example, the steroid hormone, estradiol, can bind to several different types of receptors located in different regions of a cell. Among these are (1) cytoplasm and nuclear-localized receptors that function as hormone-activated transcription factors that exert

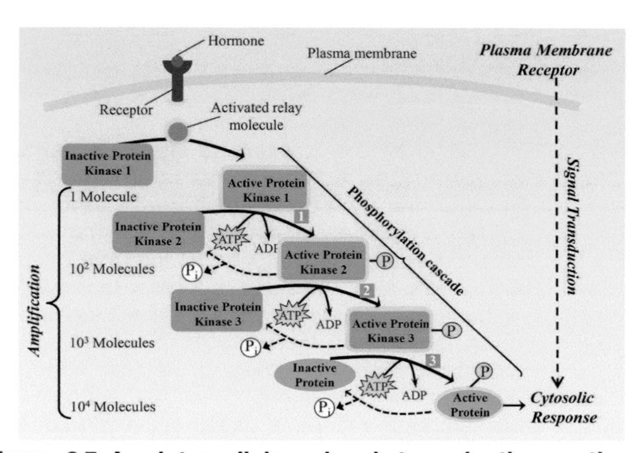

Figure 2.7 An intracellular signal transduction pathway mediated by a phosphorylation cascade. This type of kinase relay not only transduces the signal from the receptor rapidly, but also dramatically amplifies the original signal. The binding of a single hormone to one receptor can potentially activate tens of thousands of target proteins.

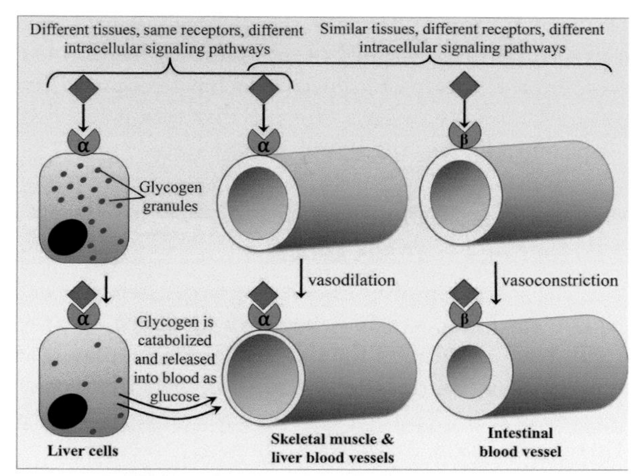

Figure 2.8 Epinephrine exerts pleiotropic effects on different tissues. Tissues may respond differentially to epinephrine by expressing the same receptor types, but by using different intracellular signaling pathways downstream of the receptors. Alternatively, different tissues may respond discretely to epinephrine by expressing different classes of adrenergic receptors (α or β forms), resulting in distinct intracellular signaling pathways.

Table 2.2 Some Adrenergic Receptor Tissue Distributions and Effects

Receptor	Major Target Tissues	Major Effects
α_1	• Vascular smooth muscle (tissues other than liver and skeletal muscle) • Digestive system sphincters • Liver	• Vasoconstriction • Constriction • Glycogenolysis
α_2	• Adipose tissue • Pancreas	• Inhibition of lipolysis • Inhibition of insulin secretion
β_1	• Cardiac muscle • Kidney	• Increase contraction, heart rate • Increase renin
β_2	• Bronchiolar smooth muscle • Liver & skeletal muscle blood vessels • Liver • Skeletal muscle • Pancreas	• Bronchodilation • Vasodilation • Gluconeogenesis and glycogenolysis • Glycogenolysis • Insulin secretion
β_3	• Adipose tissue	• Lipolysis

slow-acting genomic responses, and (2) plasma membrane and endoplasmic reticulum transmembrane receptors (called G protein-coupled estrogen receptors, or GPERs) that facilitate rapid response cytosolic effects on target cells, as well as transcriptional responses[6,7] (Figure 2.9). Importantly, GPERs are known to play key roles in the progression of some breast cancers, an active area of research.[8–11]

Hormones Mediate Homeostasis via Feedback Control

Homeostasis

The concept that the *milieu intérieur* (internal environment) operates optimally when physiological variables such as blood pressure, blood glucose, body temperature, and energy balance are stabilized within a specific range was originally conceived by Claude Bernard in the 19th century.[12] The term "homeostasis" was later developed by Walter Cannon to describe the physiological processes that maintain the constancy of the internal environment.[13] Since then, the notion of homeostasis has become a central tenet of physiology, with a chronic disruption of homeostasis potentially leading to a diseased state.

Homeostasis Is Mediated by Neural and Endocrine Feedback Loops

One of the most common underlying processes that drives physiological homeostasis is **negative feedback**, a process that can been defined as the detection of a regulated variable away from its optimal value, followed by corrective responses that return the perturbed variable back to preperturbation levels.[14] As such, negative feedback helps facilitate long-term homeostasis under conditions of environmental or internal change. A useful way to understand the concepts of homeostasis and negative feedback is to consider how engineers describe the mechanical design of a laboratory incubator that maintains a constant air temperature via a thermostat-regulated heating or cooling system (Figure 2.10). In such a system, the magnitude of the

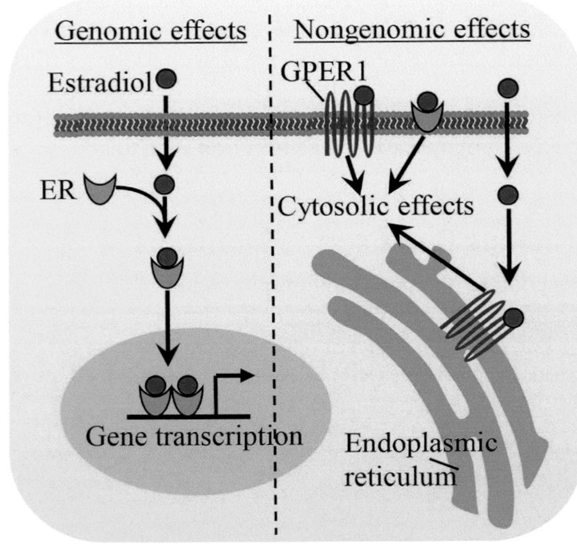

Figure 2.9 Estradiol is known to exert its effects by binding to different receptor types localized to variable regions of a target cell. Estradiol can exert genomic effects by associating with classic cytosolic estrogen receptors (ER) that translocate to the nucleus and become transcription factors that mediate gene transcription. However, the same receptors that mediate the genomic response are also known to sometimes function as plasma membrane-associated receptors that mediate nongenomic, rapid response signaling. Other nongenomic effects of estradiol are mediated by plasma membrane and endoplasmic reticulum transmembrane receptors called G protein-coupled estrogen receptors (GPERs).

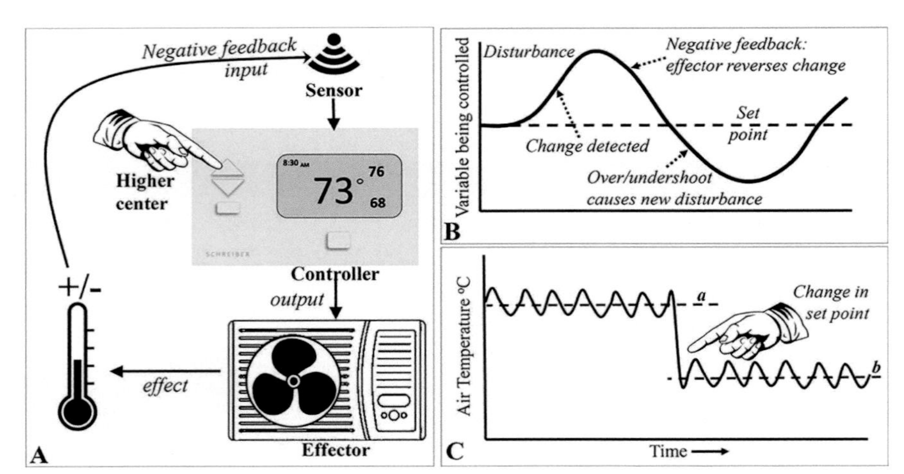

Figure 2.10 The language of homeostasis. (A) The components of a thermostat-regulated heating/cooling system. The sensor measures the magnitude of a parameter (in this case temperature). The controller quantifies the deviation of a parameter from a pre-established set point value. A higher center represents additional circuits that can reset the set point value. The effector changes the magnitude of the parameter. The degree of parameter change is re-evaluated by the sensor via negative feedback. **(B)** The process of attempting to achieve homeostasis typically manifests as a series of overshoots and undershoots past the set point value, producing a sinusoidal curve that tracks the homeostatic set point. **(C)** A change in set point designated by a "higher center". The dashed lines corresponding with "*a*" and "*b*" denote two different set points.

parameter being measured (in this case air temperature) is constantly monitored by a *sensor* (a thermometer in this example). The sensor conveys the parameter's magnitude information to an integrating *controller* (here a thermostat computer). The information received by the controller, called *input*, is compared to a predetermined value called a *set point*. If the magnitude of the input value deviates from the set point, the controller transmits an *output* signal to an *effector* (in this case an air heating/cooling unit), a device that will attempt to correct the value of the parameter being measured and drive it closer to the homeostatic set point. If the corrective change is successful, the controller will be notified through a new signal emanating from the sensor via a *negative feedback pathway*, and the effector will stop working to alter the parameter. By contrast, if the magnitude of change is either insufficient or excessive, the controller will alter its output accordingly to further attempt to minimize the deviation from set point. This process of attempting to achieve homeostasis typically manifests as a series of overshoots and undershoots past the set point value, producing a sinusoidal curve that tracks the homeostatic set point (Figure 2.10). Importantly, other circuits or pathways called *higher centers* can alter the activity of the controller by modifying its set point. In this example of a thermostat-regulated heating/cooling system, a higher center could be a human researcher in the lab who wants to alter the set point of an incubator to conduct a new experiment at a different stable temperature.

Endocrine negative feedback control systems resemble engineering systems in that the concentrations of a hormone in the blood, or a parameter that is modulated by that hormone, regulates the output of the target controller cell. However, in endocrine systems the sensor and controller elements typically both reside within the same cell and can be difficult to distinguish. In a hormone-secreting cell, the sensor is usually a protein receptor, and the controller manifests as a complex chemical signal transduction cascade that activates successive intracellular second messengers and modulates output (secretion) of the hormone in question. An example of a relatively simple endocrine negative feedback circuit is the maintenance of blood calcium homeostasis by *parathyroid hormone* (PTH), a hormone produced by the parathyroid glands that are juxtaposed to the thyroid gland in mammals (Figure 2.11). PTH cells function as both calcium receptor and controller. If blood calcium levels fall below a designated set point (2–2.5 mmol/l in humans), calcium-sensing receptors on the PTH cell plasma membrane transmit input via an intracellular signaling cascade that promotes the fusion of PTH-containing secretory vesicles to the plasma membrane, resulting in the release (output) of PTH into circulation. PTH subsequently binds to receptors on target tissues that function as effectors, which in turn increase blood calcium levels. Two of these effectors are bone, which responds to PTH by releasing calcium from the bone matrix, and the kidneys, which respond to PTH by reabsorbing calcium from the tubule filtrate back into the blood. If blood calcium levels have been successfully increased to the set point value, this information will be conveyed by negative feedback to the calcium-sensing receptors on the parathyroid cells, which will cease input, resulting in reduced PTH secretion. Although the homeostatic set point for calcium-regulated PTH release by the parathyroid glands is not known to normally be under control by higher center circuits, the pathological condition "familial hypocalciuric hypercalcemia" (FHH) manifests as a change in calcium-regulated PTH homeostatic set point that results in abnormally high blood calcium concentrations.[15] The endocrine control of calcium homeostasis will be explored further in Chapter 12: Calcium/Phosphate Homeostasis, Skeletal Remodeling, and Growth.

An example of a much more complex neural-endocrine negative feedback circuit is the mammalian stress response (Figure 2.12), which consists of multiple levels of negative feedback. When such an integration of multiple feedback loops is initiated by the central nervous system, this is often referred to as a *neuroendocrine reflex*. The primary endocrine portion of the stress response axis consists of cells located in the hypothalamus, pituitary gland, and adrenal cortex (known as the "H-P-A axis"), all of which intercommunicate via distinct hormones. At the top of the axis are hypothalamic cells that release *corticotropin-releasing*

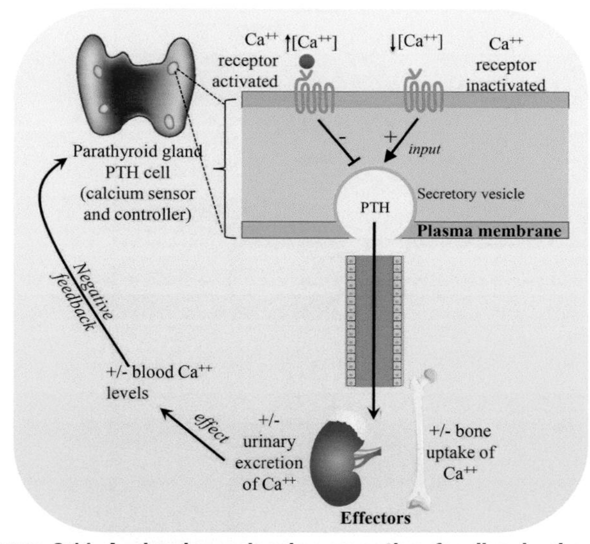

Figure 2.11 A simple endocrine negative feedback circuit: the maintenance of blood calcium homeostasis by parathyroid hormone (PTH). Blood calcium levels are sensed by calcium-sensing receptors on the plasma membrane of PTH cells, located on the thyroid gland. When circulating calcium levels are low, the calcium sensor initiates an intracellular signaling pathway that ultimately results in the fusion of PTH-containing vesicles to the basal (blood vessel-facing) side of the plasma membrane, promoting PTH levels in the blood to rise. Two target tissues of PTH that act as effectors are bone (which responds to PTH by releasing calcium from the bone matrix) and the kidneys (which respond to PTH by reabsorbing calcium from the tubule filtrate back into the blood). If blood calcium levels have been successfully increased to the set point value, this information will be conveyed by negative feedback to the calcium-sensing receptors on the parathyroid cells, which will cease input, resulting in reduced PTH secretion.

hormone (CRH), which targets specific pituitary cells that release *adrenocorticotropic hormone* (ACTH), which in turn induces the release of adrenal *glucocorticoids*, which are steroid hormones that target peripheral tissues and mediate aspects of the stress response. Note that the H-P-A endocrine axis is under higher center nervous control by regions of the brain that respond to real or perceived stress, such as the cerebral cortex, amygdala, and hippocampus. The nuances of this complex feedback system will be explored in detail in Chapter 14: Adrenal Hormones and the Stress Response.

Whereas negative feedback maintains hormone concentrations close to their homeostatic set point values, another form of feedback, called **positive feedback**, temporarily drives hormones away from their set points in a manner that transiently exacerbates the magnitude and direction of an initially small perturbation. For example, hormone X produces more of hormone Y, which in turn promotes more of X in a vicious cycle. An important example of positive feedback between two hormones occurs during the mammalian ovarian cycle, characterized in part by interactions between the ovarian steroid hormone *estradiol* and the pituitary hormone *luteinizing hormone* (LH). Note how the human ovarian cycle (Figure 2.13) is characterized by two distinct feedback occurrences: positive feedback (approximately days 0–16) and negative feedback (approximately days 16–28). Positive feedback manifests as rising levels of estradiol induce LH to rise especially during days 9–14. Indeed, it is the rapidly rising levels of LH that induce ovulation, as will be discussed in Chapter 22: Female Reproductive System. Importantly, note that positive feedback also takes place during days 14–16 as the abrupt decrease in estradiol is accompanied by an equally abrupt decrease in LH levels. As such, positive feedback can describe concomitant increases in two or more variables, as well as concomitant decreases in the same variables. Note that in the human ovarian cycle, the positive feedback that drives ovulation is followed by a distinct period of negative feedback where rising levels of estradiol are accompanied by decreasing levels of LH that are clearly apparent during days 16–22.

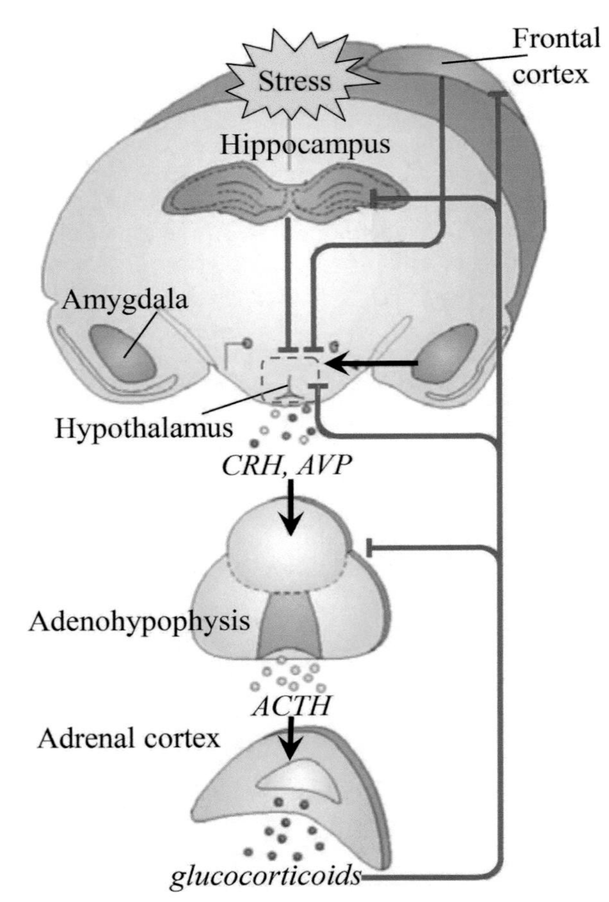

Figure 2.12 A complex endocrine negative feedback circuit: the mammalian stress response pathway. Components of the pathway include higher centers of neural stimulatory (amygdala) and inhibitory (hippocampus and frontal cortex) input by the brain to the endocrine hypothalamic-pituitary (adenohypophysis)-adrenal cortex (HPA) axis. In response to stress stimuli from higher centers, the hypothalamus releases corticotropin-releasing hormone (CRH) and arginine vasopressin (AVP), which stimulates adrenocorticotropic hormone (ACTH) by the adenohypophysis, which in turn stimulates glucocorticoid (GC) synthesis by the adrenal cortex. GCs exert negative feedback on the HPA axis, as well as on some higher control centers of the brain.

Source of images: Holgate, J.Y. and Bartlett, S.E., 2015. Brain Sci., 5(3), pp. 258-274.

Allostasis

Although homeostasis can be viewed as a self-regulating negative feedback system that maintains constancy of the internal environment, the regulated parameters are not always invariant, and homeostatic set points may change to

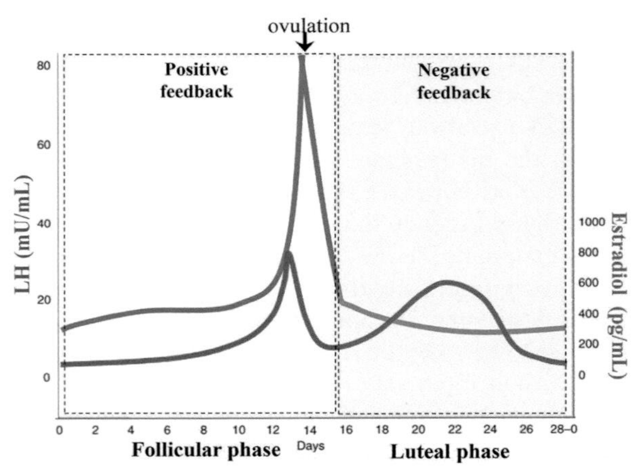

Figure 2.13 Examples of positive and negative feedback between pituitary luteinizing hormone (LH) and ovarian estradiol in an idealized drawing of the human ovarian cycle. Positive feedback between two hormones occurs when increasing levels of one hormone induce the other hormone to also rise, or falling levels for one hormone induce the other hormone to also fall. Whereas positive feedback occurs primarily during the follicular phase, negative feedback is prominent in the luteal phase.

optimize an organism's ability to cope and survive demands presented by a new environment. This is especially true for animals that experience long-term behavioral and physiological changes associated with seasonal hibernation, migration, and/or breeding. For example, consider the tegu lizard (*Salvator merianae*), a reptile that is endemic to South America and exhibits annual cycles of reproduction and high activity during the austral (southern hemisphere) spring and summer and hibernation during the austral winter (Figure 2.14). Glucocorticoids are hormones that influence aspects of energy mobilization and the stress response. In these and in most other seasonally breeding vertebrates, the baseline levels of adrenal glucocorticoid hormones fluctuate seasonally in a predictable manner, presumably to help the animals cope with the energy-demanding processes of mating, nest building, and egg incubation[16,17]

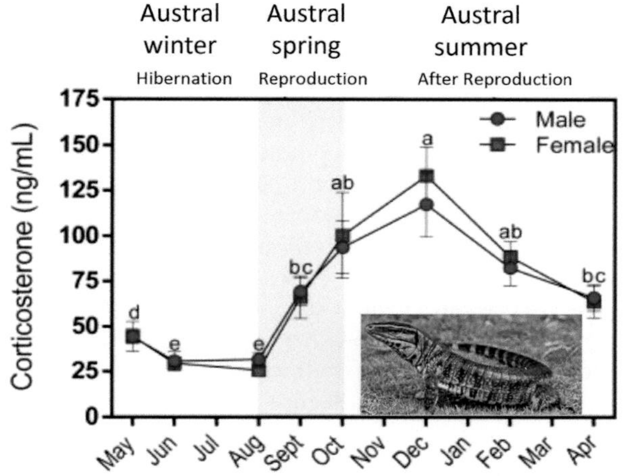

Figure 2.14 Changes in mean monthly plasma corticosterone concentrations with season in tegu lizards, *Salvator merianae*. Corticosterone concentrations were sampled from adult males (n = 9–10) and females (n = 10) raised in an outdoor enclosure in southeastern Brazil. Data are presented as means ± SEM. Different letters denote statistically significant differences. The term "austral" refers to the southern hemisphere.

Source of images: Zena et al. (2019), used with permission. Tegu lizard photograph: Murphy, J.C., et al. 2016. PLoS One, 11(8), p. e0158542.

(Figure 2.14). Compared with the breeding state, the hibernating state demands relatively low energy expenditure, and glucocorticoid levels are relatively low.

In order to adapt to changing conditions, organisms must be able to change the homeostatic set point for a regulated parameter, an important concept called **allostasis**, or literally "remaining stable by being variable", from the word's Greek roots.[18] An *allostatic state* has been defined as "a state of chronic deviation of the regulatory system from its normal (homeostatic) operating level".[19] The breeding season for the tegu lizard and all other seasonally reproducing animals is an example of an allostatic state that is part of an animal's normal life history. The concept of **allostatic load** describes the cumulative energetic costs and wear and tear associated with maintenance of the allostatic state[20] (Figure 2.15). Importantly, if the cumulative energetic costs exceed the ability of the organism to maintain an allostatic state, this places the animal in **allostatic overload**, a pathological state that imparts damage to the body and predisposes the organism to disease[21] (Figure 2.15). The topic of allostasis is particularly important to the field of stress physiology and is discussed further in Chapter 14: Adrenal Hormones and the Stress Response.

Curiously Enough . . . The concepts of allostasis and allostatic load are useful not just to biologists, but also to medicine. Allostatic overload may result in chronic human illnesses, such as atherosclerosis, hypertension, allergies, asthma, diabetes, myocardial infarction, obesity, autoimmune disorders, memory loss, and depression.

Hormones Are Often Secreted in Rhythmic and Pulsatile Manners

The concentrations of most hormones in the circulatory system rarely remain static over the course of a day but instead fluctuate in predictable manners (Figure 2.16). Such physiological fluctuations over a 24-hour period are called *circadian rhythms*, and an understanding of a

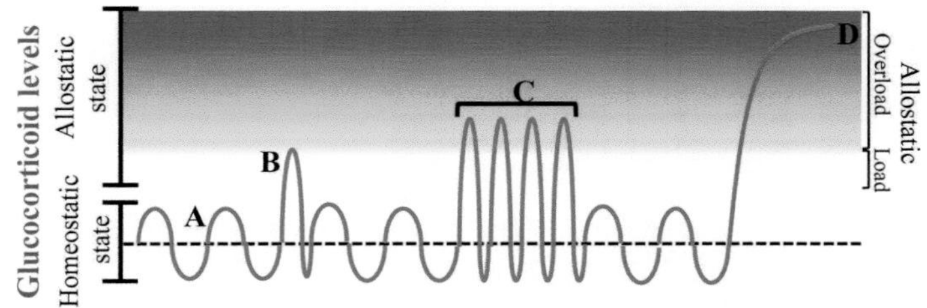

Figure 2.15 Circulating glucocorticoid (GC) levels reflect varying degrees of allostatic load. (A) Under normal conditions, GCs fluctuate in a circadian manner around a homeostatic set point (dashed line). In seasonally migrating or breeding animals, this set point may change with time of year (not shown). **(B)** In order to cope with an unpredictable stressful event (e.g. injury associated with disease or predation), GCs rise transiently beyond homeostatic levels and into the allostatic range to facilitate the mobilization of resources needed to repair damage induced by the stressor. The energy and wear and tear required to maintain the allostatic state is referred to as allostatic load. If the cumulative energetic costs associated with a high frequency of allostatic loading **(C)** or the magnitude of a prolonged allostatic load **(D)** exceeds the ability of the organism to maintain an allostatic state, this places the animal in allostatic overload, a pathological state that imparts damage to the body and predisposes the organism to disease.

hormone-specific circadian cycle is critical for an accurate assessment of that hormone's function. Additionally, superimposed onto the circadian cycle there may exist pulsatile *ultradian rhythms*, or recurring bursts of hormone secretion every few minutes to hours. This is particularly prominent for gonadotropin-releasing hormone (GnRH) and luteinizing hormone (LH), reproductive hormones that in ovariectomized sheep pulse at a rate of approximately once every hour (Figure 2.16). The distinct periods, amplitudes, and frequencies of hormone pulses are necessary for normal responses of target tissues to these signals. The basis and importance of circadian and other endocrine rhythms is discussed in Chapter 9: Central Control of Biological Rhythms.

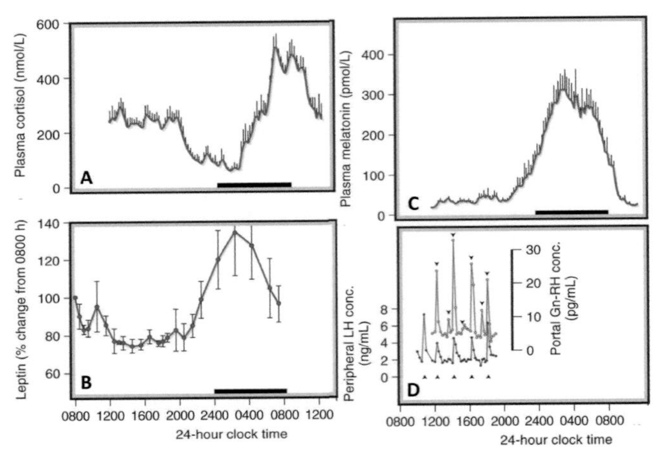

Figure 2.16 Mammalian examples of the circadian and pulsatile secretion of some hormones. (A) Human cortisol. **(B)** Human leptin. **(C)** Human melatonin. **(D)** Pulsatile secretion of gonadotropin-releasing hormone (GnRH; open circles) and luteinizing hormone (LH; filled circles) in sheep.

Source of images: Melmed, S., Polonsky, K.S., Larsen, P.R. and Kronenberg, H.M., 2011. Williams textbook of endocrinology; Elsevier; used with permission.

Multiple Hormones Can Interact in Complex Ways

The combined effects of multiple hormones can influence a physiological response in several different ways (summarized in Figure 2.17).[22] If the hormones each induce the same response and the combined effect equals the sum of their separate actions, this is known as an **additive response.** By contrast, if the combined effect of multiple hormones is less than the sum of their actions, this is referred to as a **nonadditive response.** The effects of multiple hormones are *synergistic* if the result is an amplification that totals to more than the sum of the individual interactions alone. Finally, a hormone can exert a **permissive action** if it elicits no effect alone but its presence is required for another hormone to elicit an effect. In such a case, a permissive hormone may function by either augmenting the number of receptors or intracellular second messengers for the second hormone to act upon. A good example demonstrating additive and synergistic effects of several hormones on blood glucose is shown in Figure 2.17.

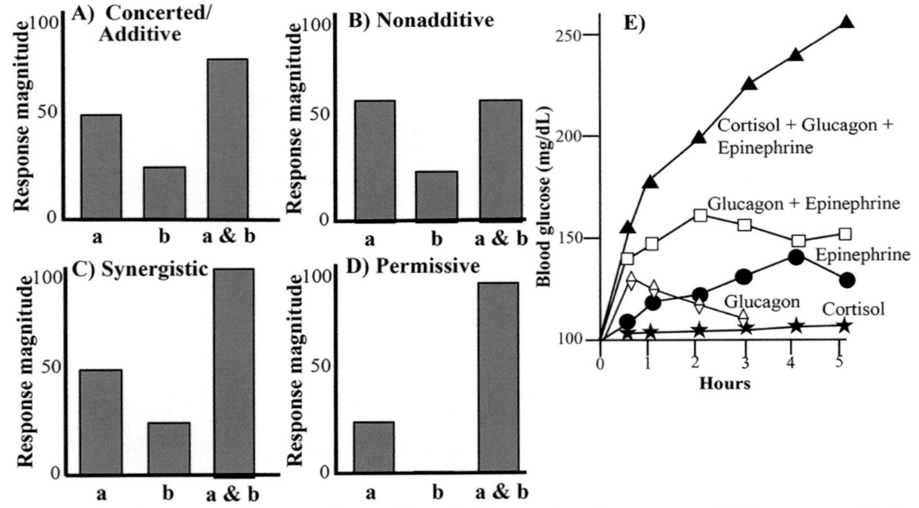

Figure 2.17 Different idealized categories of interactive actions between two hormones. (A) In a concerted/additive response, the combined hormone response is equal to the sum of their individual responses. **(B)** In a nonadditive response, the combined response is less than the sum of the individual responses. **(C)** In a synergistic response, the combined response is greater than the sum of individual responses. **(D)** A permissive response occurs when one hormone has no affect individually, but potentiates a greater response by a second hormone when combined. **(E)** Effects of different hormones, administered alone or in combination, on blood glucose levels of dogs. Injections of epinephrine and glucagon alone each produced an increase in blood glucose, but when administered together their effects were additive. Although cortisol treatment alone exerted very little effect, when administered in conjunction with epinephrine and glucagon it produced a synergistic rise in glucose (the increase in glucose exceeded the sum of the increases by each individual hormone).

Source of images: (A-D) Modified from Squires, E.J., 2010. Applied animal endocrinology. CAB International. (E) Modified from Eigler N, et al. J Clin Invest 63: 114, 1979.

Hormones Exist in Dynamic States of Activity

Some hormones are secreted in an inactive or less active form, sometimes called a **prohormone**. In this form the molecule has no or very low affinity to a receptor and requires *activation* by enzymes to become functional. For example, the prohormone angiotensin I is activated in the blood to angiotensin II, which targets the adrenal cortex and other tissues to elevate blood pressure. The prohormone dehydroepiandrosterone can be activated by peripheral tissues into testosterone. Some already active hormones, like testosterone and thyroxine, can become *potentiated*, or further converted into a form with an even higher receptor affinity, like dihydrotestosterone and triiodothyronine, respectively. In some cases, an active hormone with affinity to one receptor category can be *transactivated* into another hormone that has affinity to an entirely different receptor. For example, testosterone, which has high affinity to the androgen receptor, can be transactivated to estradiol, which binds to the estrogen receptor. Hormones do not remain in an active state for long and are continuously converted into *inactivated* forms by enzymes in target tissues and in the liver. In some instances, hormones can be *interconverted* from an inactive/less active form into an active/more active form and vice versa. An example of this is the cyclical interconversion between the glucocorticoids cortisone (less active form) and cortisol (more active) by activating enzymes in the liver and inactivating enzymes in the kidney. The ultimate fate of all hormones is *catabolism*, or permanent degradation, by target tissues and the liver, and *excretion* of the soluble metabolites by the kidneys into urine or the liver into bile acids that are eliminated via the feces. Some examples of hormones in different dynamic states are listed in Table 2.3, and many of the notions of hormone transport, feedback, and excretion are summarized in Figure 2.18.

Table 2.3 The Possible Chemical Fates of Hormones

Hormone Regulation	Examples
Activation	• Angiotensin I→ Angiotensin II • Dehydroepiandrosterone sulfate (DHEAS)→ Testosterone
Potentiation	• Testosterone→ Dihydrotestosterone (DHT) • Thyroxine (T4)→Triiodothyronine (T3)
Interconversion	• Cortisone ←→ Cortisol
Transactivation	• Testosterone→ Estradiol
Inactivation	• T3→ Diiodothyronine (T2) • Steroid hormones→ Glucuronidated metabolites

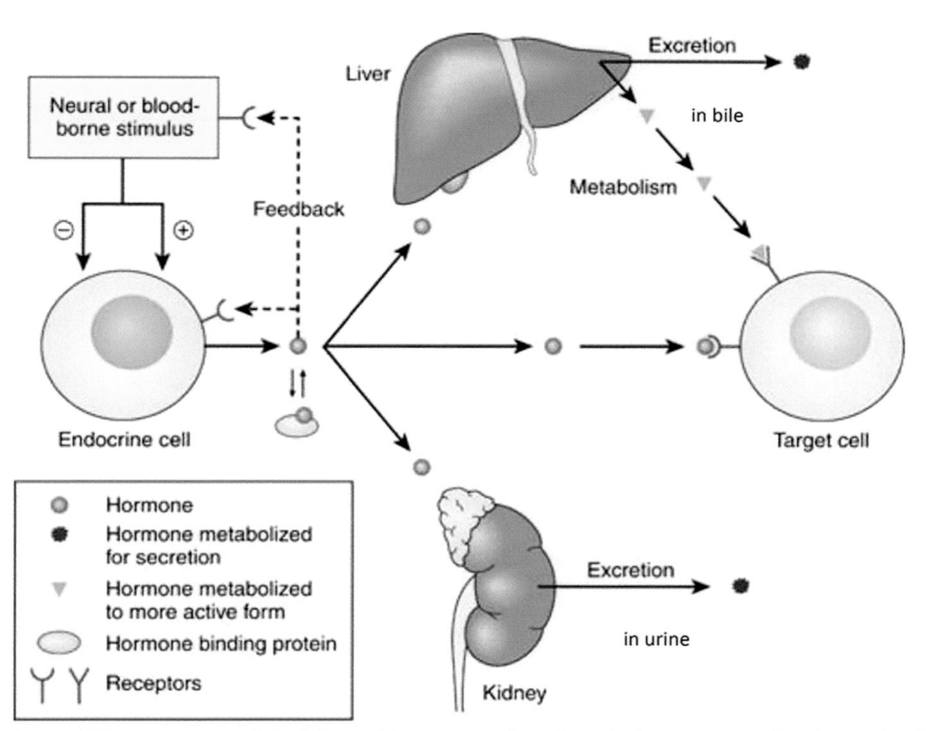

Figure 2.18 An overview of the many possible fates of hormone signaling. Following secretion by endocrine cells, hormones often bind to transport proteins that carry them in the blood to target tissues bearing hormone-specific receptors. In some instances, hormones may first be converted by local converting enzymes at local target tissues to more active or inactive forms. Ultimately, all hormones are catabolized and excreted from the body by the liver and kidneys.

Source of images: Norris, D.O. and Carr, J.A., 2020. Vertebrate endocrinology. Academic Press; used with permission.

SUMMARY AND SYNTHESIS QUESTIONS

1. Contrast the chemical diversity of hormones vs. their receptors.
2. Describe how blood transport proteins increase the half-life of different hormones.
3. In contrast with rapid (seconds) intracellular signaling cascades mediated by plasma membrane-localized receptors, why do genomic responses typically take hours to manifest?
4. How can cells adjust their sensitivity to circulating hormones?
5. The binding of a single hormone to one receptor can potentially activate tens of thousands of target proteins. How is this possible?
6. The hormone vasopressin stimulates blood vessels to vasoconstrict and simultaneously promotes the reabsorption of water and salts by the kidneys. How is it possible for one hormone to exert different effects on different tissues?
7. Refer to Figure 2.12 that describes the mammalian stress response axis. If the numbers of glucocorticoid receptors located in the hypothalamus was pathologically downregulated, what effect would this have on circulating levels of glucocorticoids? Explain your reasoning.
8. Use the topic of blood pressure to define and provide examples of homeostasis, allostasis, allostatic load, and allostatic overload.
9. A novice undergraduate researcher designs her own experiment testing the effects of playing "soothing" classical music on baseline blood glucocorticoid stress hormone levels in mice. Before playing classical music to the mice, she collected baseline glucocorticoid values at 9:00 am. After playing classical music to the mice for 4 hours, she collected blood samples again at 1:00 pm. She repeated this procedure with the same mice every day for a week and demonstrated that the glucocorticoid levels of the mice were statistically lower after 4 hours of listening to classical music, concluding that the music must have reduced mice stress levels. Name one major flaw in the experimental design.
10. Explain how the concentration of a hormone in circulation is not necessarily indicative of its activity.
11. Describe the "life history" of an idealized hormone. Where is it synthesized, how is it transported, and where are its targets located? What types of modifications might the chemical messenger experience, and what is its ultimate fate?

Summary of Chapter Learning Objectives and Key Concepts

LEARNING OBJECTIVE Compare and contrast the classical and modern definitions of "hormone" and "endocrinology".

- Compared to the classical views of hormone signaling established by the early 20th century, modern endocrinology has expanded to include the notions that hormones can be secreted by neurons, and that in addition to long-distance blood-borne transport, hormones also can diffuse short distances without entering the blood to bind to target receptors.
- Homeostasis arises from the interactions of three classes of bioregulators: hormones of the endocrine system, neurotransmitters of the nervous system, and cytokines of the immune system.

LEARNING OBJECTIVE Describe the fundamental features of endocrine signaling.

- Whereas hormones are chemically diverse, ranging from modified amino acids to proteins, lipids, and other chemicals, all of their known receptors are proteins.
- Hormones are active at extremely low concentrations and bind to specific receptors located at the surface and inside of target cells.
- Many hormones bind to blood transport proteins, which increases their longevity and solubility in circulation.
- One hormone can exert distinct effects on different cells and tissues by interacting with different receptors.
- Intracellular signaling pathways transduce and amplify hormone signals.
- Hormones mediate homeostasis via feedback control.
- Hormones can exert "pleiotropic" effects.
- In order to adapt to changing conditions, organisms must be able to change the homeostatic set points for a regulated parameter, a concept called "allostasis".
- Hormones are often secreted in rhythmic and pulsatile manners.
- Hormones exist in a dynamic state and may be present in the blood in inactive forms or activated forms, and they can also be transactivated into hormones with other functions.

LITERATURE CITED

1. Gorbman A, Bern HA. *A Textbook of Comparative Endocrinology* (1st ed.). Wiley; 1962.
2. Guillemin R. The expanding significance of hypothalamic peptides, or, is endocrinology a branch of neuroendocrinology? Paper presented at: Proceedings of the 1976 Laurentian Hormone Conference; 1977.

3. Norris DO, Carr JA. *Vertebrate Endocrinology* (5th ed.). Academic Press; 2013.
4. Weil PA. The diversity of the endocrine system. *Harper's Illust Biochem*. 2015;28.
5. Rodbell M. Signal transduction: evolution of an idea. *Environ Heal Perspect*. 1995;103(4):338–345.
6. Revankar CM, Cimino DF, Sklar LA, Arterburn JB, Prossnitz ER. A transmembrane intracellular estrogen receptor mediates rapid cell signaling. *Science*. 2005;307(5715):1625–1630.
7. Prossnitz ER, Arterburn JB, Smith HO, Oprea TI, Sklar LA, Hathaway HJ. Estrogen signaling through the transmembrane G protein-coupled receptor GPR30. *Ann Rev Physiol*. 2008;70:165–190.
8. Scaling AL, Prossnitz ER, Hathaway HJ. GPER mediates estrogen-induced signaling and proliferation in human breast epithelial cells and normal and malignant breast. *Horm Cancer*. 2014;5:146–160.
9. Lappano R, Pisano A, Maggiolini M. GPER function in breast cancer: an overview. *Front Endocrinol*. 2014;5:1–6.
10. Prossnitz ER, Barton M. The G-protein-coupled estrogen receptor GPER in health and disease. *Nat Rev Endocrinol*. 2011;7(12):715–726.
11. Martin SG, Lebot MN, Sukkarn B, et al. Low expression of G protein-coupled oestrogen receptor 1 (GPER) is associated with adverse survival of breast cancer patients. *Oncotarget*. 2018;9(40):25946.
12. Bernard C. Leç ons sur les phénomenes de la vie communs aux animaux et aux végétaux. Paris: J. *B Baillière*. 1878.
13. Cannon WB. *The Wisdom of the Body* (Revised ed.). W.W. Norton and Co; 1939.
14. Ramsay DS, Woods SC. Clarifying the roles of homeostasis and allostasis in physiological regulation. *Psychol Rev*. 2014;121(2):225.
15. Riccardi D, Brown EM. Physiology and pathophysiology of the calcium-sensing receptor in the kidney. *Am J Physiol Renal Physiol*. 2009;298(3):F485–F499.
16. Zena LA, Dillon D, Hunt KE, Navas CA, Bícego KC, Buck CL. Seasonal changes in plasma concentrations of the thyroid, glucocorticoid and reproductive hormones in the tegu lizard Salvator merianae. *Gen Comp Endocr*. 2019;273:134–143.
17. Romero LM. Seasonal changes in plasma glucocorticoid concentrations in free-living vertebrates. *Gen Comp Endocrinol*. 2002;128(1):1–24.
18. Sterling P. Allostasis: a new paradigm to explain arousal pathology. In: *Handbook of Life Stress, Cognition and Health*. Edited by S. Fisher and J. Reason. John Wiley & Sons; 1988.
19. Koob GF. The role of CRF and CRF-related peptides in the dark side of addiction. *Brain Res*. 2010;1314:3–14.
20. McEwen BS, Wingfield JC. The concept of allostasis in biology and biomedicine. *Horm Behav*. 2003;43(1):2–15.
21. McEwen BS. Stressed or stressed out: what is the difference? *J Psychiat Neurosci*. 2005;30(5):315.
22. Squires EJ. *Applied Animal Endocrinology*. Cabi; 2010.

CHAPTER

3

Evolution of Endocrine Signaling

CHAPTER LEARNING OBJECTIVES:

- Explain how an understanding of the mechanisms of evolution and the results of natural selection contribute to advancements in endocrinology.
- Explain the molecular mechanisms by which endocrine signaling diversity evolved, and why the evolution of hormones was necessary.

OPENING QUOTATION:

> "The comparative biologist generally has one key word to guide him [sic] in assessing his [sic] creative contributions to our field, and that word is *evolution*."
>
> —Howard Bern[1]

The Power of an Evolutionary Perspective on Endocrine Signaling

LEARNING OBJECTIVE Explain how an understanding of the mechanisms of evolution and the results of natural selection contribute to advancements in endocrinology.

KEY CONCEPTS:

- Evolutionary theory not only gives biologists the ability to explain the origins of the diversity of endocrine systems, but also provides them with the tools to predict how chemical signaling systems may change in response to new selective pressures.
- Natural selection acts not only on genes that code for the biosynthesis of hormones and receptors, but also for hormone-binding proteins and components of intracellular signal transduction pathways.

Evolutionary theory constitutes a body of knowledge and a method of analysis that empowers biologists with the ability not only to explain the origins of the diversity of living systems, but also to understand how and why those systems changed in the past. Critically, such knowledge also provides scientists with the tools to predict how living systems may change in the future in response to new selective pressures. Much as an understanding of how Newton's theory of the motion of large objects allows physicists to accurately predict the future locations of Mars and the Moon in space and time, allowing astronauts to travel there, knowledge of the mechanisms that mediate the evolution of endocrine systems among diverse taxa provides scientists with tools to predict how organisms will respond to changing environments, such as to climate change and to endocrine-disrupting chemicals, as well as to diseases such as infertility and metabolic disorders associated with over-or malnutrition. Such knowledge also facilitates the design of appropriate tools to treat these disruptions and pathologies.

There are numerous examples where naturally occurring variation in hormone signaling has been described among individuals, populations, and species.[2,3] For example, populations of mammals (including humans) and birds inhabiting northern latitudes typically have higher circulating levels of thyroid hormones, which are critical for regulating metabolism and body temperature, compared with conspecific populations in more southern latitudes.[4] Furthermore, male birds from polygamous species display different levels of androgen hormones in response to aggression by competitive males compared with those from monogamous species.[5,6] Natural selection acts not only on genes that code for the biosynthesis of hormones and receptors, but also for hormone-binding proteins, plasma membrane transporters, and components of intracellular signal transduction pathways. Therefore, functional variation in the structures of these genes, as well as changes in their magnitude and spatiotemporal expression patterns, are major driving forces for the evolution of animal physiology, life history, and species diversity.[7] In particular, variable expression and function of endocrine components within and among species may profoundly influence evolutionary trajectory by affecting a key aspect of development called **plasticity**, or the adaptability of an organism to changes in its environment. In addition, evolution not only explains

DOI: 10.1201/9781003359241-4

physiological differences among species, but also variable responses to hormones among individuals within the same species. In endocrinology, such an understanding of intraspecific genetic variability linked to human evolution could, for instance, help to optimize the drug dosage and timing associated with hormone replacement therapies.[8–11]

Curiously Enough . . . Genetic variability among humans is the basis of *pharmacogenetics*, a branch of pharmacology that studies genetic influences on drug response in order to reduce drug side effects, enhance drug efficacy, and improve individualized patient care.[12]

SUMMARY AND SYNTHESIS QUESTIONS

1. How is an understanding of the mechanisms and results of natural selection critical to advancements in endocrinology?

How and Why Did Endocrine Signaling Evolve?

LEARNING OBJECTIVE Explain the molecular mechanisms by which endocrine signaling diversity evolved, and why the evolution of hormones was necessary.

KEY CONCEPTS:

- As cell numbers increased and organisms grew larger, circulatory systems became necessary to transport both nutrients and chemical signals across longer distances.
- Endocrine signaling is hypothesized to have evolved from ancestral paracrine systems.
- Some hormone–receptor pairs likely co-evolved via the process of "molecular exploitation", or the recruitment of an older ligand molecule into a new functional complex with a receptor via incremental change.
- Gene families consist of genes derived from a common ancestral gene that underwent one or more rounds of gene duplication and divergence.
- The ancestors of all vertebrates underwent two rounds of whole genome duplication, and the teleost fish ancestors experienced three rounds.
- Whole genome duplication, local duplication, gene divergence, and gene loss together constitute the primary mechanisms that led to the evolution of endocrine diversity.

Why Are Hormones Necessary? The Paracrine-to-Endocrine Transition

Multicellular animals evolved 750–800 million years ago in the late Proterozoic, 1.4 billion years after the evolution of the first single-celled eukaryotes.[13] All single cell prokaryotes (bacteria and archaea) and eukaryotes (like yeast) possess plasma membrane-localized receptors that are capable of not only responding to environmental chemical stimuli, such as the presence of nutrients, but also facilitating a type of cell-cell signaling called *quorum sensing*. This form of signaling is mediated by ancient plasma membrane-localized receptors that coordinate various cell behaviors within a microbial population in a cell-density-dependent manner.[14] These include morphological differentiation, secondary metabolite production, genetic exchange, pathogenesis, and biofilm formation. However, in contrast with single-celled organisms that can obtain all nutritional and quorum sensing signals via passive diffusion from the external environment, the advent of **metazoans**, or multicellular organisms with differentiated tissues, introduced a new problem: how can chemical signals detected by the external layer of cells be communicated to the internal layers? A solution was the development of cell-cell contact-dependent signaling, as well as **paracrine** communication, or the release of signaling molecules into the extracellular fluid with receptors on nearby cells (Figure 3.1). As cell numbers increased and organisms grew larger, circulatory

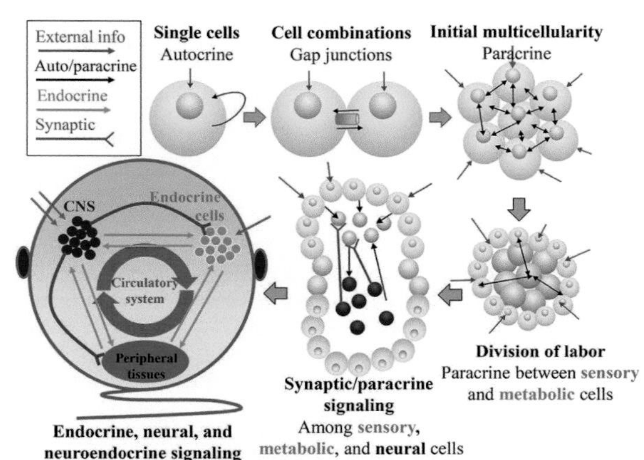

Figure 3.1 Hypothetical stages of evolution of cell signaling with the transition from unicellular to complex multicellular organisms requiring division of labor and long-distance communication among cells. Unicellular cells self-signal via autocrine mechanisms. Combinations of cells may interact via intercellular gap junctions, and cells within a small multicellular organism can communicate via paracrine signaling. As multicellular organisms grew larger and more complex, divisions of labor evolved whereby external sensory cells and internal metabolic cells communicate via paracrine mechanisms. The advent of neurons introduced synaptic signaling among distant cells. Larger organisms with circulatory systems incorporated endocrine and neuroendocrine signaling systems into the signaling repertoire.

Source of images: Modified from Lovejoy, D.A., et al. 2019. Front. Endocrinol., 10, p. 730.

systems became necessary to transport both nutrients and chemical signals across longer distances, and endocrine signaling is therefore hypothesized to have evolved from ancestral paracrine systems.[15] Three important observations support this hypothesis:

1. For any given organ in humans, paracrine regulators greatly outnumber endocrine regulators.[16,17]
2. For almost all endocrine ligand–receptor pairs, a related autocrine or paracrine pair exists within the same receptor family.[16,17]
3. Whereas all the primary endocrine intracellular signaling pathways are present in paracrine systems, several paracrine pathways appear to be absent in endocrine systems.[18]

Because paracrine and endocrine intracellular transduction pathways share some similar signaling machinery, it is therefore not surprising to note that hormones can sometimes behave as paracrine factors, and vice versa. However, because hormones must diffuse throughout the entire body without becoming overly diluted, this means that hormones must be synthesized at relatively higher concentrations compared with paracrine factors. Such a paracrine-to-endocrine transition not only is apparent among taxa of different cellular complexity, but also is evident during vertebrate embryonic and fetal development, with the endocrine system developing only after the embryo has differentiated and grown to a large enough size that necessitates long-distance signaling via the circulatory system. Hormones that signal over long distances likely evolved from modified paracrine ligands. Because organisms as evolutionarily distant as yeast, fruit flies, nematode worms, and mammals each possess many of the receptors and intracellular signaling pathways, this suggests that ancient multicellular organisms must have already established most of the signaling systems that are present in vertebrate endocrine systems today.[18]

Evolution of the Hormone–Receptor Pair

There are several hypotheses for the evolution of the specificity of the hormone–receptor pair, ranging from the separate evolution of each component via chance trial and error to the coevolution of the hormone–receptor complex.[19] Members of the "nuclear receptor superfamily", whose ligands include all steroid hormones, vitamin D, and thyroid hormones, are proposed to have initially arisen devoid of any hormone, with the hormone–receptor complex forming later in evolution.[20] As the receptor evolved, so did the concomitant process for the relay of information.

Throughout animal evolution, several conserved ligand–receptor signaling pathways arose that were sufficient to give rise to all cellular and morphological diversity, as well as to all endocrine signaling pathways. Some of the receptors for these pathways that are critical to endocrine signaling include the nuclear receptors, as well as many plasma membrane-localized receptors including G protein-coupled receptors (GPCRs), receptors with intrinsic enzyme activity, enzyme-associate receptors, and ion channel receptors. These receptors and signaling pathways are discussed throughout this textbook and are detailed in Chapter 5: Receptors.

Developments & Directions: Resurrecting an ancestral gene to reconstruct the evolution of aldosterone-mineralocorticoid receptor interactions

A fascinating hypothesis for the coevolution of the 500 million year-old nuclear receptor superfamily with its various hormone ligands comes in the form of "molecular exploitation", or the recruitment of an older ligand molecule, previously used in a different role, into a new functional complex with a receptor via incremental evolutionary change.[21] Joseph Thornton's group[21,22] have findings that suggest the ancestral adrenal corticosteroid hormone/receptor complex changed following gene duplication and divergence, in tandem with the extension of a novel biosynthetic steroidogenic pathway that produced novel hormones and receptors with slight structural and functional modifications from their ancestral forms.

The glucocorticoid cortisol activates both the glucocorticoid receptor (GR) and the mineralocorticoid receptor (MR) and is present in teleost fish, amphibians, reptiles, birds, and mammals. The mineralocorticoid aldosterone, however, is known to be an evolutionarily recent novelty that is specific to tetrapods. Although the most basal living vertebrates (the jawless fishes, lamprey, and hagfish) synthesize neither aldosterone nor cortisol, *deoxycorticosterone* (DOC), which is an intermediate in the synthesis of corticosteroids in all jawed vertebrates, functions as both a glucocorticoid and a mineralocorticoid in the lamprey.[23,24] Interestingly, although aldosterone is not synthesized by the jawless vertebrates, the hagfish and lamprey corticoid receptor is still activated by the hormone, as measured by cultured cells using a reporter-gene transcription assay.[21] The GR and its closest family member, the MR, descended from duplication of the *ancestral corticoid receptor* (AncCR) gene, approximately 450 million years ago.[21] This AncCR, which is present in lamprey, appears to be the receptor for the ancient vertebrate hormone DOC.

In order to reconstruct the evolution of the MR's interaction with aldosterone, Joseph Thornton's group[21] reconstructed the phylogenetically inferred sequence of the AncCR. A gene sequence construct coding for the ancient receptor was then synthesized and expressed in cultured cells. Using a reporter-gene transcription assay, they made the surprising observation that the resurrected AncCR was activated not only by DOC, but also by aldosterone and cortisol (Figure 3.2). This observation raised the fascinating question, why was the ancestral receptor sensitive to both aldosterone and cortisol millions of years before these hormones evolved? The group proposed that aldosterone is similarly shaped to the more ancient DOC, explaining the receptor's preexisting ability to bind with it. To test this idea, Ortlund et al.[25] determined the crystal structures of AncCR in complex with DOC and with aldosterone. As predicted, the ancestral receptor binds aldosterone in precisely the same way that it binds DOC, indicating that the ancestral DOC receptor was structurally pre-adapted for aldosterone activation. The fact that the AncCR structurally resembles the mineralocorticoid receptor more than the glucocorticoid receptor suggests that the glucocorticoid receptor is the more derived (evolutionarily recent) and the mineralocorticoid is the more ancestral of the two. Thornton's group also identified two mutations that enhanced the specificity of the modern-day receptor for cortisol. These findings are summarized in Figure 3.3.

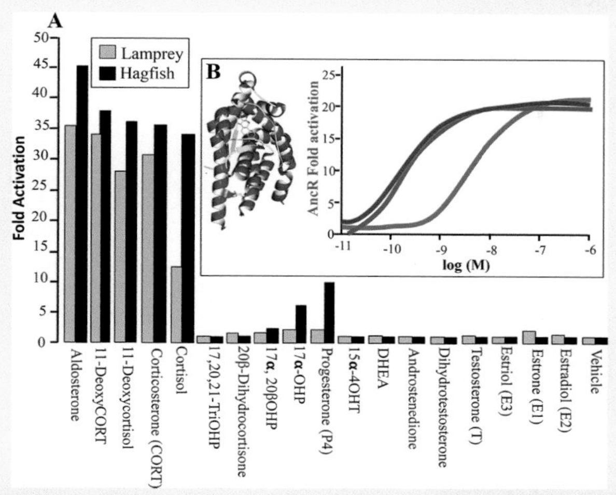

Figure 3.2 Corticoid receptors (CRs) from extant basal vertebrate, as well as the resurrected ancient corticoid receptor (AncCR) are activated by aldosterone. (A) Activation of a luciferase reporter gene by CR ligand-binding domains (LBDs) of hagfish (black) and lamprey (gray) with 100 nM hormone. Fold-activation indicates reporter activity compared with treatment with vehicle only. **(B)** The resurrected AncCR (pink and yellow protein structure) in complex with aldosterone (gray steroid molecule). Hormone-induced activation of a luciferase reporter by the resurrected AncCR-LBD is shown for aldosterone (red line), cortisol (green line), and 11-deoxycorticosterone (blue line).

Source of images: (A) Modified from Bridgham, J.T., et al. 2006. Science, 312(5770), pp. 97-101. (B) PDB ID: 2Q1H DOI: Ortlund, E.A.,Bridgham, J.T., et al. (2007) doi: 10.2210/pdb2q1h/pdb

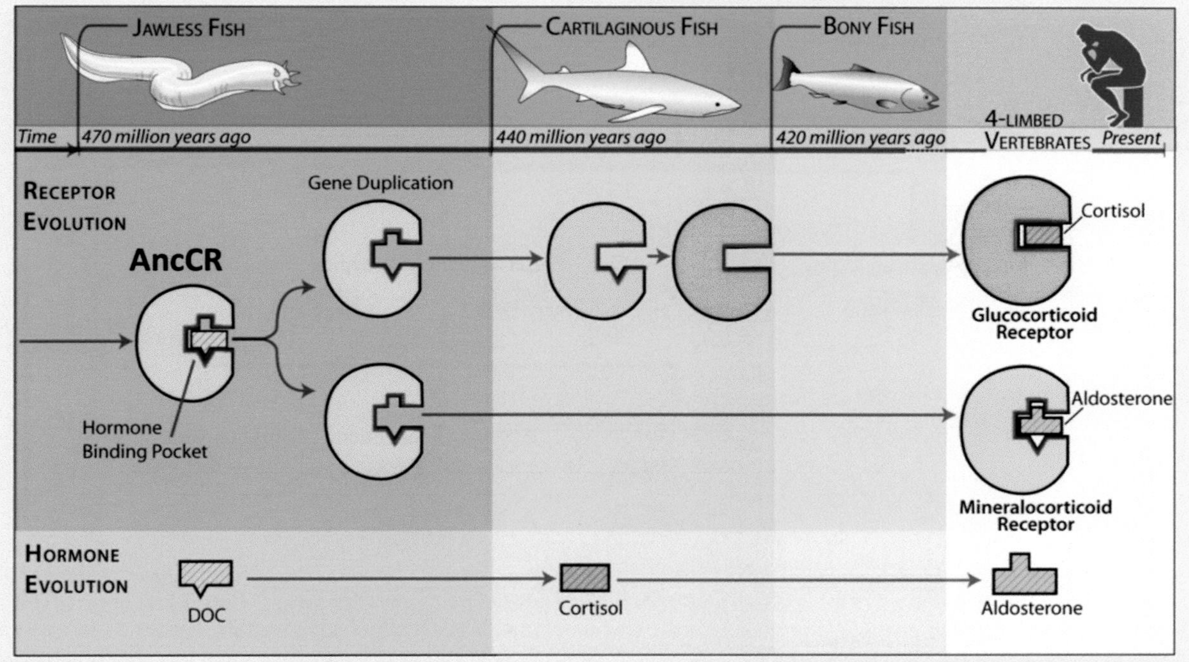

Figure 3.3 Evolution of glucocorticoid and mineralcorticoid receptors from an ancient corticoid receptor (AncCR). Although aldosterone synthesis did not exist at the time of the AncCR, the receptor was already structurally predisposed to binding aldosterone. Following a gene duplication event, two subsequent mutations in the glucocorticoid lineage increased its binding specificity to cortisol.

Source of images: The National Science Foundation.

Foundations & Fundamentals: Gene Duplications Generate Genetic Material for Evolutionary Novelty

Whereas individual genes contain the information necessary for regulating the expression of particular RNAs, *genomes* constitute the entirety of an organism's hereditary information. Genomes can be thought of as documents that record the life history strategies of a species, and their structures and content are continually being edited and rewritten throughout evolutionary history. Major evolutionary mechanisms that have been crucial to structuring genomes and generating differences in gene numbers among species include individual gene duplication, whole genome duplication, and gene loss. This section describes the importance of gene duplication and divergence as mechanisms for generating new genetic material that mutation and natural selection subsequently act upon to produce new gene functions from which evolutionary novelties can originate.

Gene duplication, whether at the level of individual genes or entire genomes, is the most important mechanism for facilitating the evolution of biodiversity and imparting the capability of organisms to adapt to changing environments.[26] Gene duplication events generally arise as errors of DNA repair or replication mishaps that result in either a **regional duplication** (the generation of a second copy of a portion of a gene, a single gene, a portion of a chromosome, or an entire chromosome), or a *whole genome duplication* (Figure 3.4).[27] Following a gene duplication event, both daughter genes typically accumulate sequence changes by mutation unevenly, with one copy radically diverging from the other over evolutionary time, potentially generating new genes with novel functions.[28] Of the two duplications, whole genome duplication is generally considered to play a more important role in evolution, as it allows for the duplication of entire regulatory systems, including hormones, receptors, and intracellular signaling pathways. By contrast, regional duplications typically facilitate the doubling of only one component of a signaling system that may lead to either local optimization or disruption of function, promoting the evolution of lineage-specific variability.[26,27,29]

Some important terms associated with gene duplication and the evolution of novelty are defined later (also refer to Figure 3.5).

Gene divergence: Following a gene duplication event, the asymmetric accumulation of mutations among the daughters of the duplicated genes over evolutionary time promotes functional disparity in the duplicated gene lineages.
Homolog: Genes with shared ancestry.
Paralog: Related genes within the same genome that arose via gene duplication and divergence.
Ortholog: Related genes in different genomes (i.e. different species) that arose via gene duplication and divergence.
Gene loss: The loss of redundant gene duplicates from an organism's genome.
Pseudogene: Genes that remain in the genome, but have lost some or all functionality.
Gene family: Genes that derive from a common ancestral gene that underwent one or more rounds of gene duplication and divergence.

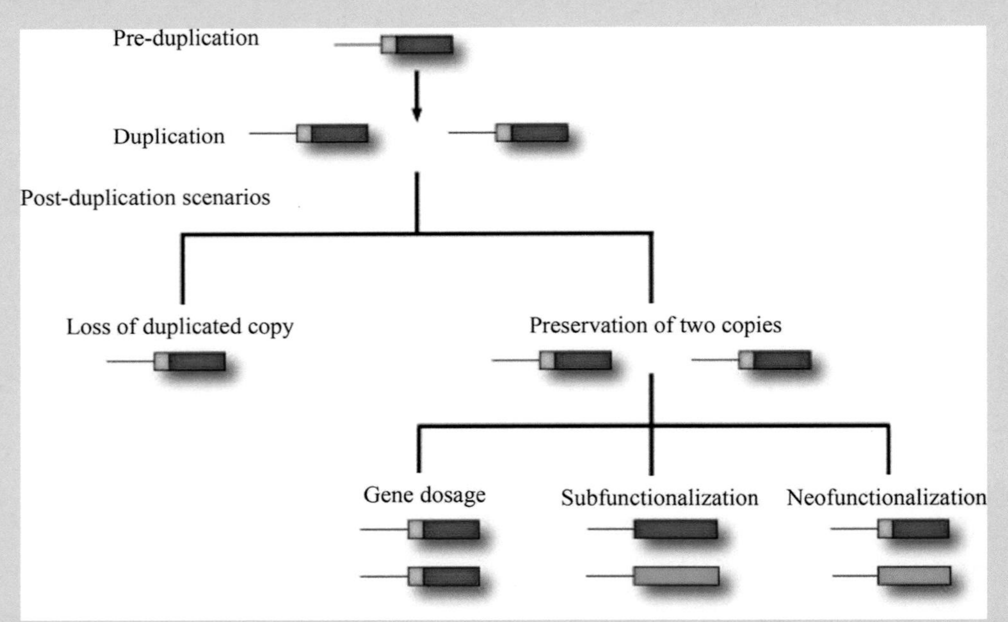

Figure 3.4 Evolutionary fates of duplicated genes. The most common outcome of gene duplication is loss of the duplicated copy from the genome. If the two copies are retained, there may be three different outcomes. Different functions of the genes are denoted with red and blue colors. *Gene dosage* occurs when the two copies continue to perform the same function as the pre-duplication, ancestral copy, and therefore introduce redundancy and/or increased activity of the gene. *Subfunctionalization* occurs if the ancestral gene was multifunctional, and the different functions can be divided over the post-duplication copies. Optimization of the different functions in the different copies is also possible. *Neofunctionalization* occurs when one copy still performs the function(s) of the ancestral gene, and the other copy acquires mutations that can generate a novel function (as indicated by green color).

Source of images: Voordeckers, K. and Verstrepen, K.J., 2015. Curr. Opin. Microbiol., 28, pp. 1-9; used with permission.

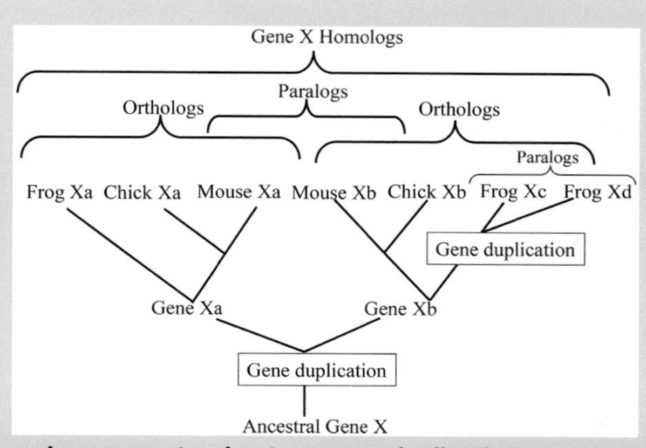

Figure 3.5 Homologous genes can share ancestry due to a gene duplication event within a species (paralog) or due to a speciation event (ortholog). The concept of orthology concerns only the genetic family tree and does not necessarily imply functional equivalence.

Source of images: Cruciani, V. and Mikalsen, S.O., 2006. Cell. Mol. Life Sci., 63(10), pp. 1125-1140; used with permission.

Animal Phylogeny and the Role of Genome Duplication in Chordate Endocrine Evolution

In order to understand the importance of genome duplication on the evolution of endocrine signaling, it is first necessary to provide a brief overview of animal phylogeny[30] (Figure 3.6). The Metazoa comprise animals possessing bilateral symmetry (**bilaterians**) and animals with radial symmetry (**non-bilaterians**). Whereas some Metazoa like the Cnidaria (jellyfishes) possess nervous systems, others such as the Porifera (sponges) lack them. The bilaterians comprise two super-phyla whose constituents possess distinct features of embryogenesis. The first are the **protostomes**, so named because the early embryonic structure, the blastopore, develops into the

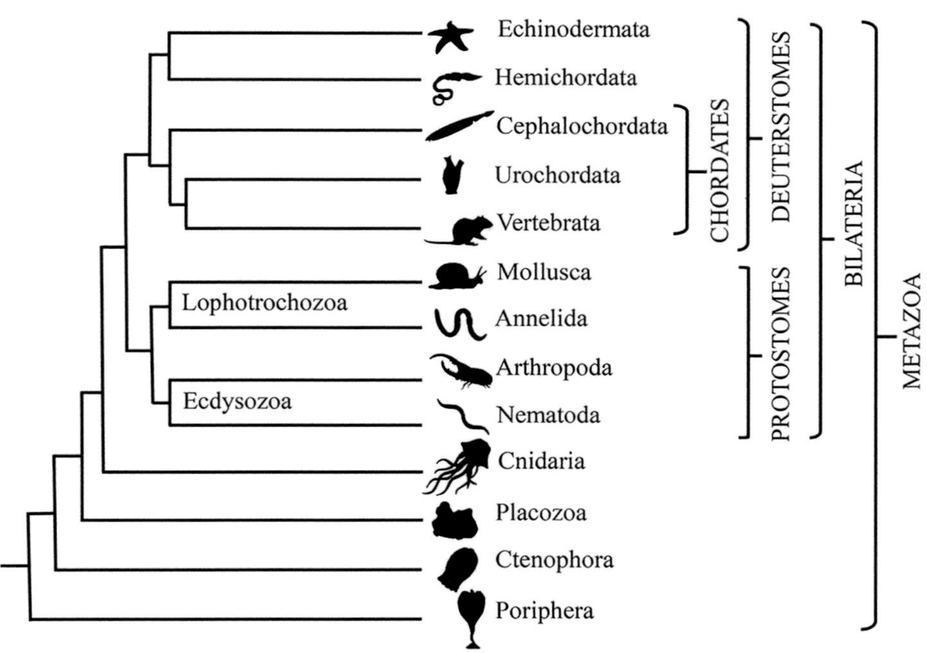

Figure 3.6 Animal phylogeny. Phylogenetic tree showing relationships of selected animal phyla. The Metazoa comprise non-bilaterian phyla and bilaterian phyla. The non-bilaterians include phyla that lack nervous systems (Porifera and Placozoa) and phyla that have nervous systems (Ctenophora and Cnidaria). The bilaterians comprise two super-phyla: the deuterostomes, which include vertebrates, and the protostomes, which include lophotrochozoans (e.g. the mollusc *Aplysia californica*) and ecdysozoans (e.g. the arthropod *Drosophila melanogaster* and the nematode *Caenorhabditis elegans*). Note that the branch lengths in the tree are arbitrary.

Source of images: Elphick, M.R et al. 2018. J. Exp. Biol. 221(3).

mouth. Protostomes include the mollusks (e.g. oysters, squid, and snails), arthropods (e.g. crustaceans and insects), annelids (segmented worms), and nematodes (non-segmented worms related to the arthropods). The second are the **deuterostomes**, so named because the blastopore develops into the anus, and these include the vertebrates, chordates (e.g. tunicates and amphioxus), and protochordates (e.g. echinoderms and hemichordates). This section focuses on the evolution of the **chordates**, a phylum of animals that possess a rigid internal structure called a *notochord* during development (the location of the notochord can be seen in Figure 10.6 of Chapter 10: Thyroid Hormones: Development and Growth). The chordates include the subphyla Vertebrata (vertebrates), Cephalochordata (amphioxus), and Urochordata (tunicates) (Figure 3.6).

Beginning approximately 350 million years ago, vertebrate evolution followed two main trajectories. One group of fishes exploited a *benthic* niche (one associated with the substrates on the ocean bottoms), and the other group exploited a *pelagic* (water column-dwelling) niche. The benthic group were selected for limb-like fins and developed into the lobe-finned teleost fishes (Sarcopterygii) and ultimately into land-dwelling animals (Figure 3.7). The pelagic lineage developed into the ray-finned teleost fishes (Actinopterygii). The coelacanths and several species of lungfish are the only living members of the lobe-finned fishes. Today the vertebrates are subdivided into the jawless vertebrates (the cyclostomes: lampreys and hagfish) and the jawed vertebrates (gnathostomes). The jawed vertebrates are further subdivided into chondrichthyes (cartilaginous vertebrates) and osteichthyes (bony vertebrates). Finally, the bony vertebrates are subdivided into the Actinopterygii (ray-finned fish) and Sarcopterygii, which includes the lobe-finned fish (lungfish and coelacanth) and tetrapods (amphibians, reptiles, birds, and mammals).

Evidence from large-scale genomic studies suggests that early in its evolution, the vertebrate lineage underwent two rounds of whole genome duplication (referred to as *1R* and *2R*),[31–33] followed by a third round of duplication (*3R*) near the start of teleost fish evolution[34–36] (Figure 3.7). The two consecutive rounds of genome duplication that influenced the evolution of all chordates (1R and 2R) resulted in two *tetraploidization* events, where every gene locus in the diploid genome was duplicated. However, due to extensive gene loss, the total number of genes following 2R has not quadrupled.[37] Importantly, many families of hormone receptors, as well as peptide and protein hormones described throughout this book, expanded during 2R. Paralogs of polypeptide hormones such as for gonadotropin-releasing hormone (GnRH) (Figure 3.8), the nonapeptides of the posterior pituitary (Figure 3.9) and hormone receptors, such as the nuclear receptors superfamily, have been well documented. Virtually all mammalian polypeptide hormones and receptors have orthologs among the more ancestral vertebrates. Genes orthologous to mammalian peptide hormones have even been described in the direct ancestors to vertebrates, the protochordates. For example, the most ancient member of the cholecystokinin-gastrin family of peptides, which regulates digestion in vertebrates, originated approximately 500 million years ago in ancestors of the sea squirt (*Ciona intestinalis*).[38] Remarkably, signaling molecules with sequences similar

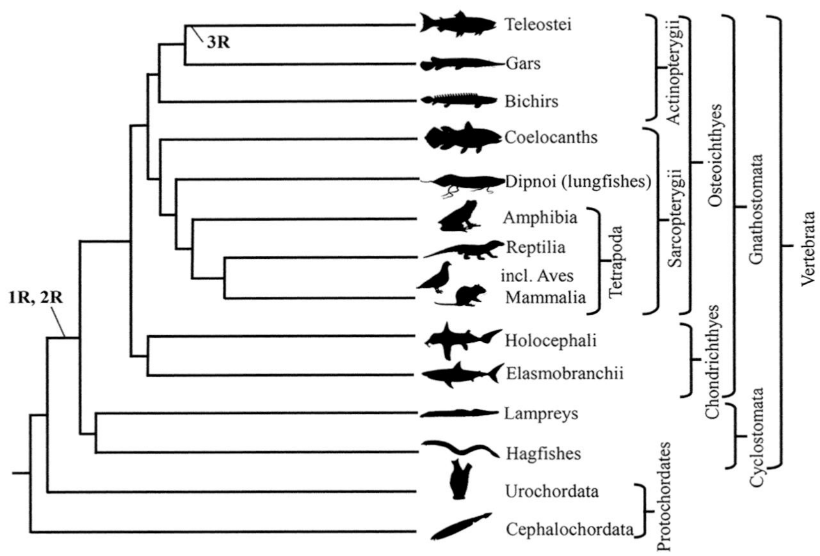

Figure 3.7 Chordate phylogeny. Phylogenetic tree showing relationships of the major chordate lineages. The phylum Chordata comprises three subphyla: Cephalochordata, Urochordata, and Vertebrata. The vertebrates are subdivided into the jawless vertebrates (cyclostomes; lampreys and hagfish) and the jawed vertebrates, which are further subdivided into chondrichthyes (cartilaginous fish) and osteichthyes (bony vertebrates). The bony vertebrates are subdivided into the Actinopterygii (ray-finned fish) and Sarcopterygii (lobe-finned fish and tetrapods). 1R, 2R and 3R denote first, second, and third rounds of genome duplication, respectively.

Source of images: Elphick, M.R et al. 2018. J. Exp. Biol. 221(3).

	1	2	3	4	5	6	7	8	9	10	
Human	pGlu-	His-	Trp-	Ser-	Tyr-	Gly-	Leu-	Arg-	Pro-	Gly-	NH 2
Guinea pig	pGlu-	Tyr-	Trp-	Ser-	Tyr-	Gly-	Val-	Arg-	Pro-	Gly-	NH2
Chicken-I	pGlu-	His-	Trp-	Ser-	Tyr-	Gly-	Leu-	Gln-	Pro-	Gly-	NH2
Chicken-II	pGlu-	His-	Trp-	Ser-	His-	Gly-	Trp-	Tyr-	Pro-	Gly-	NH2
Salmon	pGlu-	His-	Trp-	Ser-	Tyr-	Gly-	Trp-	Leu-	Pro-	Gly-	NH 2
Dogfish	pGlu-	His-	Trp-	Ser-	His-	Gly-	Trp-	Leu-	Pro-	Gly-	NH2
Catfish	pGlu-	His-	Trp-	Ser-	His-	Gly-	Leu -	Gln-	Pro-	Gly-	NH2
Herring	pGlu	-His-	Trp-	Ser-	His-	Gly-	Leu-	Ser-	Pro-	Gly-	NH2
Medaka	pGlu-	His-	Trp-	Ser-	His-	Gly-	Leu-	Ser-	Pro-	Gly-	NH2
Lamprey-I	pGlu-	His-	Trp-	Ser-	Leu-	Glu-	Trp-	Lys-	Pro-	Gly-	NH2
Lamprey-II	pGlu-	His-	Trp-	Ser-	His-	Gly-	Trp-	Phe-	Pro-	Gly-	NH2
Lamprey-II	pGlu-	His-	Trp-	Ser-	His-	Asp-	Trp-	Lys-	Pro-	Gly-	NH2
Frog	pGlu-	His-	Trp-	Ser-	Tyr-	Gly-	Lcu-	Trp-	Pro-	Gly-	NH2
Seabream	pGlu-	His-	Trp-	Ser-	Tyr-	Gly-	Met-	Ser-	Pro-	Gly-	NH2
Whitefish	pGlu-	His-	Trp-	Ser-	Tyr-	Gly-	Met-	Asn-	Pro-	Gly-	NH2

Figure 3.8 The primary structures of the 15 known molecular forms of GnRH in vertebrates. All GnRH forms are decapeptides (composed of ten amino acids). GnRH forms are traditionally named after the species from which they were first identified. Amino acids with the highest positional identity among species are highlighted. I, II, and III indicate the presence of multiple gene copies in the organisms' genome due to gene duplication.

Source of images: Amano, Masafumi. "Reproductive Biology of Salmoniform and Pleuronectiform Fishes with Special Reference to Gonadotropin-Releasing Hormone (GnRH)." Aqua-bioscience Monographs 3 (2010): 39-72.

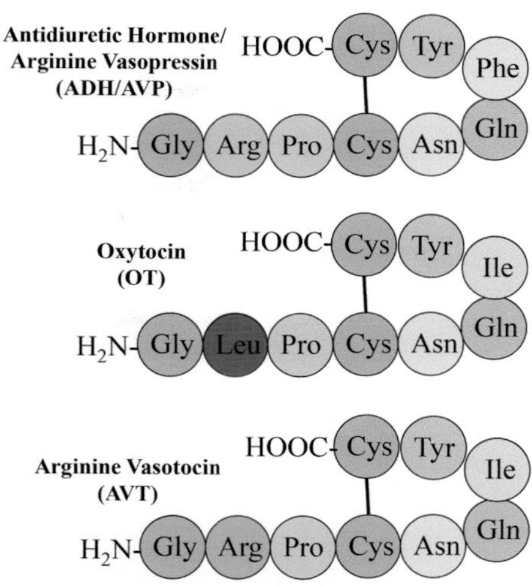

Figure 3.9 Homology in the structures of the three nonapeptides of the neurohypophysis. Whereas ADH and oxytocin are generally present in mammals, vasotocin is found in non-mammalian vertebrates, and also in mammals during fetal development. The three closely related genes arose via gene duplication followed by divergence.

Source of images: White circles denote conserved amino acids, and gray circles denote differences among the hormones.

to mammalian peptides have also been found in insects, yeasts, and bacteria,[39] suggesting that ancestral polypeptide hormones arose early in the evolution of life. Insulin, for example, has remained essentially unchanged in organisms ranging from invertebrates to mammals.[40,41] Discussed next are two case studies of endocrine evolution: the first describes an example of receptor evolution primarily by whole genome duplication, and the second discusses hormone evolution by local duplication. Many other endocrine components evolved via a much more complex combination of the two processes and are described throughout this textbook.

Example: Neuropeptide Y (NPY) Receptor Evolution

NPY and its related peptides (peptide YY, or PYY, and pancreatic polypeptide, PP) comprise a family of related hormones and neurotransmitters that are involved in the regulation of diverse functions, including appetite/satiety, gut motility, cardiovascular activity, and many others.[42] The NPY and PYY signaling peptides arose by a local duplication of an ancestral peptide prior to the radiation of the vertebrates, and PP is a local duplicate of PYY.[43] Mammals typically possess four or five functional receptors for the NPY-family peptides (subtypes Y1, Y2, Y4, Y5, and Y6) (Figure 3.10). The human NPY system, consisting of three peptides and four receptors, displays a degree of complexity that resembles many other vertebrate peptide-receptor systems. The evolution of the ancestral NPY receptor into its modern forms in humans and coelacanth can be explained by two rounds of local duplication of the ancestral gene, followed by two rounds of whole genome duplication that were accompanied by several gene loss events (Figure 3.10).

Example: Evolution of the Vasopressin and Oxytocin Family of Nonapeptides

In mammals the pituitary *nonapeptide* (meaning "9-amino acids") hormones vasopressin and oxytocin have roles in the regulation of osmoregulation and smooth muscle contraction, respectively, and both also play roles in mediating parenting and other behaviors such as monogamy and maternal aggression. Remarkably, all gnathostomes possess at least one homolog each of vasopressin and oxytocin, whereas the jawless vertebrates contain a single member of the nonapeptide hormone family called vasotocin (Figures 3.9 and 3.11). In non-mammalian vertebrates vasotocin also serves as the vasopressin homolog. Mesotocin serves as the oxytocin homolog in non-placental tetrapods and lungfish, and isotocin is the ray-finned fish oxytocin homolog. Interestingly, these pituitary nonapeptide hormones did not arise from a common ancestral peptide as a result of a 1R or 2R whole genome duplication, but instead

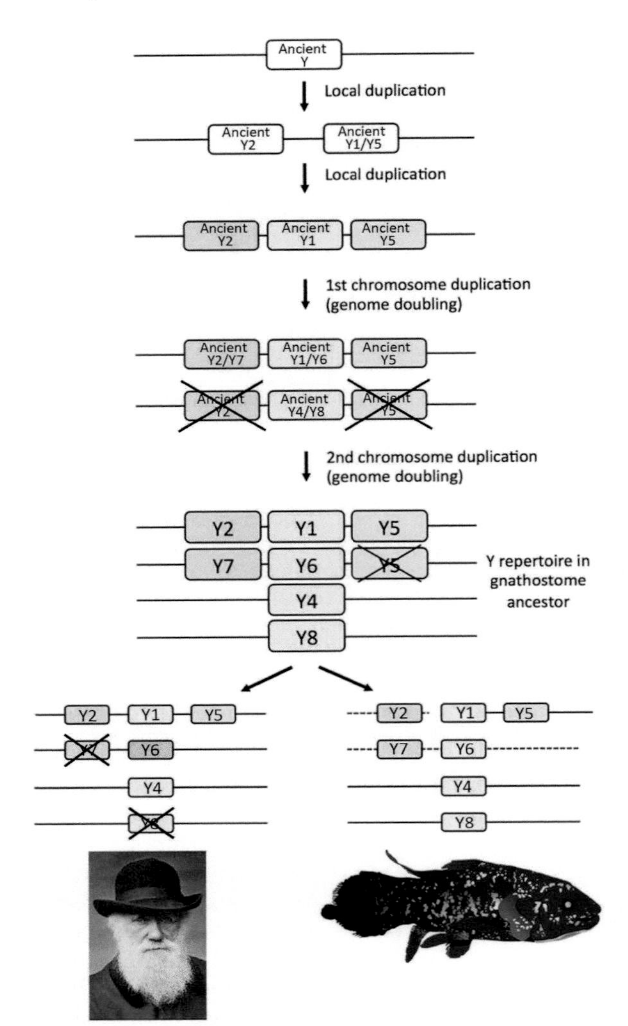

Figure 3.10 Gene duplication events for the NPY receptor family in early vertebrate evolution including the two basal tetraploidizations. Crosses mark gene losses. The human Y6 gene is a pseudogene, as indicated by orange color. The dashed lines for the coelacanth indicate that it is not known if the genes are syntenic (exist on the same chromosome).

Source of images: Larhammar, D. and Bergqvist, C.A., 2013. Front. Neurosci., 7, p. 27.

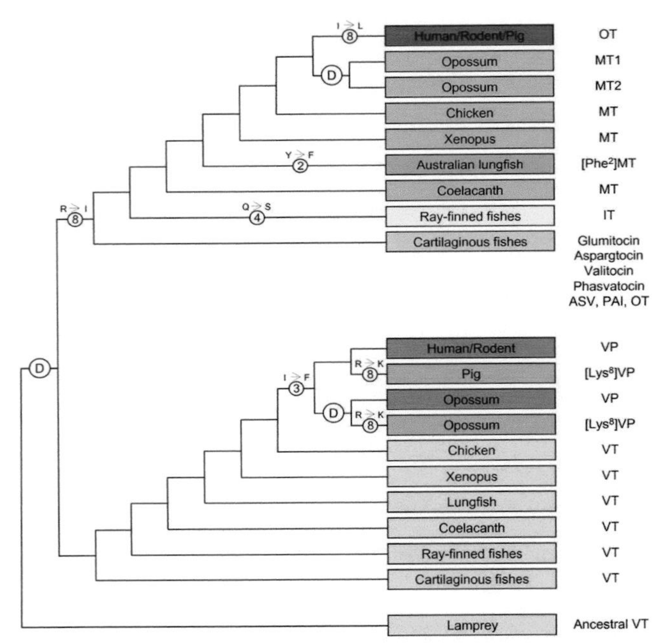

Figure 3.11 Evolution of the vasopressin and oxytocin family of nonapeptides. The letter "D" within the circle represents a gene duplication event. The numbers within the circles denote the position of the amino acid that has been substituted (shown above the circle). OT, oxytocin; MT, mesotocin; IT, isotocin; VP, vasopressin; [Lys8]VP, lysipressin, VT, vasotocin; ASV, asvatocin; PAI, phasitocin.

Source of images: Gwee, P.C., et al. 2008. BMC Evol. Biol., 8(1), p. 93.

one duplication occurred in a gnathostome ancestor following divergence from the lineage leading to the jawless fishes, and a second occurred in the mammalian ancestor[30,44] (Figure 3.11). The result is the generation of a family of nonapeptides whose members differ by only one to two amino acids.

A Pharmacological Cornucopia: How Many Hormones and Receptors Are There?

So, what has 3.5 billion years of gene duplication, divergence, gene loss, and natural selection produced for endocrine diversity? Excluding the estimated several hundred genes and proteins of chemokine and cytokine ligands known to be involved in mammalian immune signaling alone,[45] well over 130 major hormones, growth factors, and paracrine bioregulators have been described for the endocrine system of the most thoroughly studied vertebrate species, *Homo sapiens*, and more continue to be discovered.[46] By contrast, the number of receptors far exceeds the number of known hormones, with plasma membrane-localized receptors being the most abundant type of receptor. Of the 1,352 human plasma membrane-localized receptor genes currently identified, approximately 20% of these receptors bind hormones, growth factors, cytokines, and other substances of endocrine/paracrine origin.[47] Compared with the approximately 280 known plasma membrane-localized endocrine/paracrine receptors, only about 48 nuclear receptor genes have been identified in humans. Many of these hormones are listed in Appendix 4, and their receptors are described in Chapter 5: Receptors. From a human health perspective, the importance of these receptors is clearly evident by the fact that over 50% of the targets of the current 500 major pharmaceutical drugs function by modulating receptor signaling pathways.[48] As new components of endocrine signaling pathways continue to be discovered and understood, the number of pharmaceutical opportunities for disease intervention will surely continue to rise.

Developments & Directions: Steps in the evolution of the endocrine and neuroendocrine systems

The hypothetical stages in the evolution of the endocrine and neuroendocrine systems have been elegantly described by Volker Hartenstein.[49] Cell communication through secreted, diffusible signals is phylogenetically older than either neural transmission or neurosecretion, since cells releasing endocrine signals are found in extant metazoa that lack nerve cells, such as the sponges. These specialized epithelial cells, which are the progenitors of both neurons and neurosecretory cells, likely first colonized surfaces of the epidermis (Figure 3.12[A]). The cells probably responded to chemical or physical stimuli by secreting metabolites that diffused throughout the body, promoting adaptive responses by other tissues. After undergoing further specializations during the course of evolution, these ancestral endocrine cells separated from their epithelial surfaces and formed distinct sensory cell lineages, many becoming neurons associated with a nerve net, similar to the ones still found in present day cnidarians (Figure 3.12[B]). In bilaterians, these cells began to form the central and peripheral nervous systems, as well as specialized endocrine glands, integrating multimodal sensory input (Figure 3.12[C]). During later stages of evolution, neurosecretory neurons and endocrine cells in general underwent three important changes: (1) a loss of sensory function; (2) delamination from the epidermis, pharynx, and intestinal epithelium; and (3) condensation and morphogenesis into distinct classical endocrine glands, such as the thyroid/parathyroid (from the pharyngeal endoderm) and pancreatic islets (from the midgut) (Figure 3.12[D]).

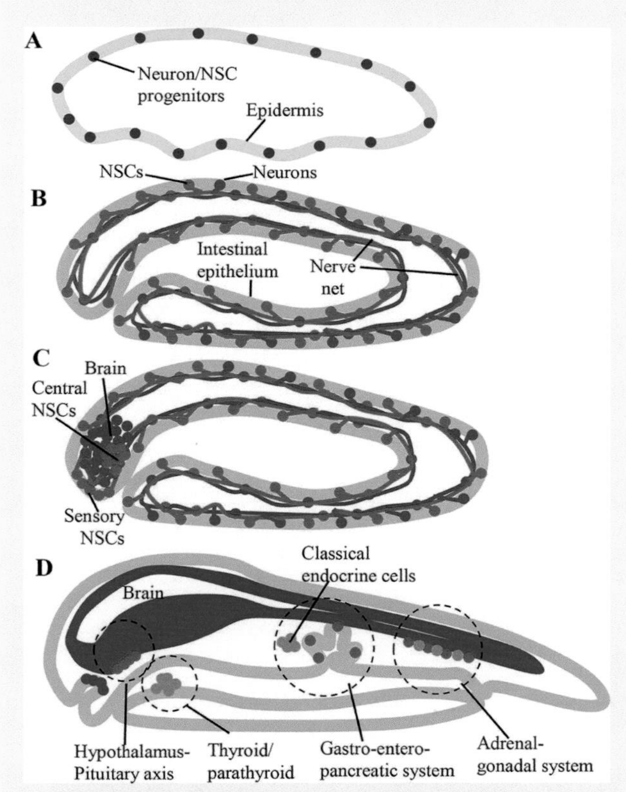

Figure 3.12 Hypothetical stages in the evolution of the endocrine and neuroendocrine systems. See text for details.

Source of images: Hartenstein V. J. Endocrinology. 2006; 190(3):555-570.

SUMMARY AND SYNTHESIS QUESTIONS

1. Describe two observations that support the hypothesis that the evolution of paracrine signaling preceded endocrine signaling.
2. Describe the concept of "molecular exploitation" and how it can be used to understand the coevolution of some hormone–receptor complexes.
3. How is gene duplication essential to the evolution of endocrine novelties?
4. Whereas mammals have about 48 nuclear receptors, teleost fish have about 70. What might explain the difference?
5. The large numbers of hormones, receptors, and intracellular signaling pathways that have evolved in humans can be bewildering. How can this complexity be considered a gift to the pharmaceutical industry?
6. How are hormone–receptor pairs thought to have evolved?
7. Explain the importance of gene and genome duplication in the evolution of endocrine complexity.

Summary of Chapter Learning Objectives and Key Concepts

LEARNING OBJECTIVE Explain how an understanding of the mechanisms of evolution and the results of natural selection contribute to advancements in endocrinology.

- Evolutionary theory not only gives biologists the ability to explain the origins of the diversity of endocrine systems, but also provides them with the tools to predict how chemical signaling systems may change in response to new selective pressures.
- Natural selection acts not only on genes that code for the biosynthesis of hormones and receptors, but also for hormone-binding proteins and components of intracellular signal transduction pathways.

LEARNING OBJECTIVE Explain the molecular mechanisms by which endocrine signaling diversity evolved, and why the evolution of hormones was necessary.

- As cell numbers increased and organisms grew larger, circulatory systems became necessary to transport both nutrients and chemical signals across longer distances.
- Endocrine signaling is hypothesized to have evolved from ancestral paracrine systems.
- Some hormone–receptor pairs likely co-evolved via the process of "molecular exploitation", or the recruitment of an older ligand molecule into a new functional complex with a receptor via incremental change.
- Gene families consist of genes derived from a common ancestral gene that underwent one or more rounds of gene duplication and divergence.

- The ancestors of all vertebrates underwent two rounds of whole genome duplication, and the teleost fish ancestors experienced three rounds.
- Whole genome duplication, local duplication, gene divergence, and gene loss together constitute the primary mechanisms that led to the evolution of endocrine diversity.

LITERATURE CITED

1. Bern HA. Comparative endocrinology—the state of the field and the art. *Gen Comp Endocrinol*. 1972;3:751–761.
2. Adkins-Regan E. *Hormones and Animal Social Behavior*. Princeton University Press; 2005.
3. Bradshaw D. Environmental endocrinology. *Gen Comp Endocrinol*. 2007;152(2–3):125–141.
4. Ishikawa A, Kitano J. Ecological genetics of thyroid hormone physiology in humans and wild animals. *Thyroid Horm*. 2012:37.
5. Wingfield JC, Hegner RE, Dufty Jr AM, Ball GF. The "challenge hypothesis": theoretical implications for patterns of testosterone secretion, mating systems, and breeding strategies. *Am Nat*. 1990;136(6):829–846.
6. Hirschenhauser K, Oliveira RF. Social modulation of androgens in male vertebrates: meta-analyses of the challenge hypothesis. *Anim Behav*. 2006;71(2):265–277.
7. Kitano J, Ishikawa A, Lema SC. Integrated genomics approaches in evolutionary and ecological endocrinology. In: *Ecological Genomics*. Springer; 2014:299–319.
8. Zafón C. Evolutionary endocrinology: A pending matter. *Endocrinol Nutr (English Edition)*. 2012;59(1):62–68.
9. Romijn J, Smit J, Lamberts S. Intrinsic imperfections of endocrine replacement therapy. *Eur J Endocrinol*. 2003;149(2):91–97.
10. Ho KK. Endocrinology: the next 60 years. *J Endocrinol*. 2006;190(1):3–6.
11. Lamberts S, Romijn J, Wiersinga W. The future endocrine patient. Reflections on the future of clinical endocrinology. *Eur J Endocrinol*. 2003;149(3):169–175.
12. Hockett RD, Kirkwood SC, Mitlak BH, Dere WH. Pharmacogenomics in endocrinology. *J Clin Endocrinol Metab*. 2002;87(6):2495–2499.
13. Erwin DH. Early metazoan life: divergence, environment and ecology. *Philosophical Trans Royal Soc B Biol Sci*. 2015;370(1684):20150036.
14. Barriuso J, Hogan DA, Keshavarz T, Martínez MJ. Role of quorum sensing and chemical communication in fungal biotechnology and pathogenesis. *FEMS Microbiol Rev*. 2018;42(5):627–638.
15. Ben-Shlomo I, Hsu SY, Rauch R, Kowalski HW, Hsueh AJ. Signaling receptome: a genomic and evolutionary perspective of plasma membrane receptors involved in signal transduction. *Sci STKE*. 2003;2003(187):re9-re9.
16. Ben-Shlomo I, Yu Hsu S, Rauch R, Kowalski HW, Hsueh AJ. Signaling receptome: a genomic and evolutionary perspective of plasma membrane receptors involved in signal transduction. *Science's STKE: Sig Transduc Knowled Environ*. 2003;2003(187):RE9.
17. Rubin GM, Yandell, MD, Wortman, JR et al. Comparative genomics of the eukaryotes. *Science*. 2000;287(5461):2204–2215.
18. Kronenberg HM, Melmed S, Larsen PR, Polonsky KS. Principles of endocrinology. In: *Williams Textbook of Endocrinology*. Elsevier; 2016:1–11.
19. Nair A, Chauhan P, Saha B, Kubatzky KF. Conceptual evolution of cell signaling. *Int J Mol Sci*. 2019;20(13):3292.
20. Laudet V. Evolution of the nuclear receptor superfamily: early diversification from an ancestral orphan receptor. *J Mol Endocrinol*. 1997;19(3):207–226.
21. Bridgham JT, Carroll SM, Thornton JW. Evolution of hormone-receptor complexity by molecular exploitation. *Science*. 2006;312(5770):97–101.
22. Carroll SM, Bridgham JT, Thornton JW. Evolution of hormone signaling in elasmobranchs by exploitation of promiscuous receptors. *Mol Biol Evol*. 2008;25(12):2643–2652.
23. Weisbart M, Youson JH. In vivo formation of steroids from [1,2,6,7–3H]-progesterone by the sea lamprey, Petromyzon marinus L. *J Steroid Biochem*. 1977;8(12):1249–1252.
24. Close DA, Yun SS, McCormick SD, Wildbill AJ, Li W. 11-deoxycortisol is a corticosteroid hormone in the lamprey. *Proc Natl Acad Sci U S A*. 2010;107(31):13942–13947.
25. Ortlund EA, Bridgham JT, Redinbo MR, Thornton JW. Crystal structure of an ancient protein: evolution by conformational epistasis. *Science*. 2007;317(5844):1544–1548.
26. Ohno S. *Evolution by Gene Duplication*. Springer; 1970.
27. He J, Irwin DM, Zhang Y-P. The power of an evolutionary perspective in studies of endocrinology. In: *Contemporary Aspects of Endocrinology*. Intech; 2011:1–22.
28. Holland PW, Marlétaz F, Maeso I, Dunwell TL, Paps J. New genes from old: asymmetric divergence of gene duplicates and the evolution of development. *Philos Trans Royal Soc B Biol Sci*. 2017;372(1713):20150480.
29. Kuraku S, Meyer A, Kuratani S. Timing of genome duplications relative to the origin of the vertebrates: did cyclostomes diverge before or after? *Mol Biol Evol*. 2008;26(1):47–59.
30. Elphick MR, Mirabeau O, Larhammar D. Evolution of neuropeptide signalling systems. *J Exp Biol*. 2018;221(3).
31. Dehal P, Boore JL. Two rounds of whole genome duplication in the ancestral vertebrate. *PLoS Biol*. 2005;3(10):e314.
32. Nakatani Y, Takeda H, Kohara Y, Morishita S. Reconstruction of the vertebrate ancestral genome reveals dynamic genome reorganization in early vertebrates. *Genome Res*. 2007;17(9):1254–1265.
33. Putnam NH, Butts T, Ferrier DE, et al. The amphioxus genome and the evolution of the chordate karyotype. *Nature*. 2008;453(7198):1064.
34. Jaillon O, Aury J-M, Brunet F, et al. Genome duplication in the teleost fish Tetraodon nigroviridis reveals the early vertebrate proto-karyotype. *Nature*. 2004;431(7011):946.
35. Meyer A, Van de Peer Y. From 2R to 3R: evidence for a fish-specific genome duplication (FSGD). *Bioessays*. 2005;27(9):937–945.
36. Kasahara M, Naruse K, Sasaki S, et al. The medaka draft genome and insights into vertebrate genome evolution. *Nature*. 2007;447(7145):714.
37. Holland PW. More genes in vertebrates? In: *Genome Evolution*. Springer; 2003:75–84.
38. Johnsen AH. Phylogeny of the cholecystokinin/gastrin family. *Front Neuroendocrinol*. 1998;19(2):73–99.
39. Roth J, LeRoith D, Shiloach J, Rosenzweig JL, Lesniak MA, Havrankova J. The evolutionary origins of hormones, neurotransmitters, and other extracellular chemical messengers: implications for mammalian biology. *N Engl J Med*. 1982;306(9):523–527.
40. Ebberink RHM, Smit AB, Van Minnen J. The insulin family: evolution of structure and function in vertebrates and invertebrates. *Biol Bull*. 1989;177:176–182.
41. Smit AB, van Marle A, van Elk R, Bogerd J, van Heerikhuizen H, Geraerts WP. Evolutionary conservation of the insulin gene structure in invertebrates: cloning of the gene encoding molluscan insulin-related peptide III from Lymnaea stagnalis. *J Mol Endocrinol*. 1993;11(1):103–113.
42. Larhammar D, Bergqvist CA. Ancient grandeur of the vertebrate neuropeptide Y system shown by the coelacanth

Latimeria chalumnae. *Front Neurosci.* 2013;7:27.

43. Larhammar D, Söderberg C, Lundell I. Evolution of the neuropeptide Y family and its receptors A. *Ann N Y Acad Sci.* 1998;839(1):35–40.
44. Gwee P-C, Tay B-H, Brenner S, Venkatesh B. Characterization of the neurohypophysial hormone gene loci in elephant shark and the Japanese lamprey: origin of the vertebrate neurohypophysial hormone genes. *BMC Evol Biol.* 2009;9(1):47.
45. Vilcek J, Feldmann M. Historical review: cytokines as therapeutics and targets of therapeutics. *Trends Pharmacol Sci.* 2004;25(4):201–209.
46. Berridge MJ. *Cell Signalling Biology.* Portland Press; 2007. doi:10.1042/csb0001001.
47. Almen MS, Nordstrom KJ, Fredriksson R, Schioth HB. Mapping the human membrane proteome: a majority of the human membrane proteins can be classified according to function and evolutionary origin. *BMC Biol.* 2009;7:50.
48. Drews J. Drug discovery: a historical perspective. *Science.* 2000;287(5460):1960–1964.
49. Hartenstein V. The neuroendocrine system of invertebrates: a developmental and evolutionary perspective. *J Endocrinol.* 2006;190(3):555–570.

Unit Overview
Unit II

Mechanisms of Hormone Action

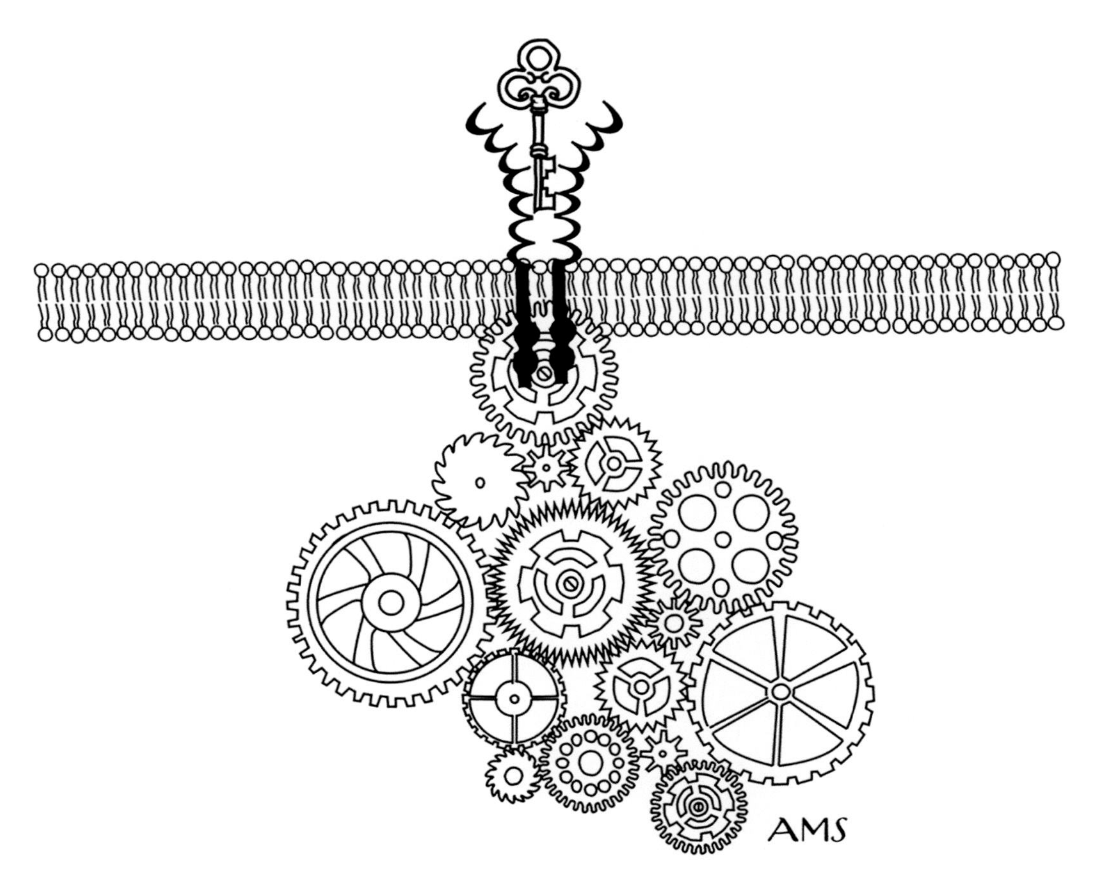

"Signal Transduction", by A.M. Schreiber.

DOI: 10.1201/9781003359241-5

OPENING QUOTATION:

"Finally, one may then ask if we shall ever fully understand how hormones act at the cellular level. The answer has to be a qualified "no," simply because hormone research will continue to reflect our knowledge of regulatory mechanisms at any given time. Conversely, many important advances in biochemistry and in cell and molecular biology have been, and will continue to be, the outcome of work on the mechanism of hormone action."

—Jamshed Tata (2007). A hormone for all seasons. *Perspectives in Biology and Medicine*. 50(1):89–103.

Unit II takes a detailed look at some key mechanisms of hormone biosynthesis and responses resulting from hormone–receptor interactions. Chapter 4 addresses how hormones are categorized based on biochemical and structural criteria and how the major hormone classes are synthesized from distinct biochemical categories of precursor molecules. Chapter 5 concerns itself with the two broad classes of receptors that mediate the effects of hormones. The first class consists of receptors that localize to the cell's plasma membrane and whose actions are conveyed to the cell's interior via complex signal transduction pathways. The second class are nuclear receptors that function as ligand-activated gene transcription factors. Chapter 6 describes how the magnitude of a cellular response to a hormone depends not only on variable hormone and receptor concentrations, but also on a hormone's specificity and affinity to its receptor.

***Chapter 4*: Hormone Classes and Biosynthesis**
***Chapter 5*: Receptors**
***Chapter 6*: Receptor Binding Kinetics**

CHAPTER

4

Hormone Classes and Biosynthesis

CHAPTER LEARNING OBJECTIVES:

- Describe the different methods by which hormones are categorized.
- Identify how different chemical classes of hormones are transported by the blood.
- Describe how hormones are inactivated and excreted.
- List the amine-derived hormones, their precursor substrates, and the organs that synthesize them.
- Describe how polypeptide hormones are categorized and synthesized.
- Provide an overview for the biosynthesis of the two most common categories of lipid hormones (steroid hormones and eicosanoids) and describe their most important functions.

OPENING QUOTATIONS:

"Steroids are the creams that stop skin from peeling. Or make skin more appealing. They are drugs that help a woman get pregnant. Or stop a woman from getting pregnant. Steroids can build hard muscles. Or tenderize meat. They are the vitamins that grow strong bones. Or drugs that grow hair. They help you absorb your food. Or absorb the sun. They are drugs that prolong your life. Or prolong your love life."

—Thom Rooke *The Quest for Cortisone* (2012)

"It was like we had fallen into that magical tunnel described by Lewis Carroll depositing us at the entrance to Wonderland . . . an entrance that was sealed and could be unlocked only by asking the right question."

—Albert Lasker Basic Medical Research Award Acceptance Remarks by Ronald Evans (2004)

Hormone Systematics

LEARNING OBJECTIVE Describe the different methods by which hormones are categorized.

KEY CONCEPTS:

- Hormones can be sorted into three broad categories, based upon their chemical structures: (1) lipids, (2) polypeptides, and (3) modified amino acids.
- Hormones can be divided into three groups based on their solubility and receptor location: (1) hydrophilic hormones with cell-surface receptors, (2) hydrophobic hormones with intracellular receptors, and (3) hydrophobic hormones with cell-surface receptors.

Over 130 major hormones, growth factors, and paracrine bioregulators have been described for vertebrates alone,[1] a conservative estimate that excludes the hundreds of known cytokines and chemokines of the immune system that can also fall into the broad definition of "hormone". Categorizing hormones into both cohesive and meaningful functional groups is no elementary task. The simplest classification scheme partitions hormones based upon their organs of origin. However, considering that the same hormone can be produced by different organs, and one organ typically produces multiple hormones, the limitations of this classification scheme become quickly apparent. Somatostatin, for instance, is made by the hypothalamus, pancreas, and stomach, and the pituitary gland, an organ the size of a pea, produces a dozen different hormones. Another approach classifies hormones according to their general physiological functions: "sex" hormones, "stress" hormones, "growth" hormones, and so on. This method, too, has constraints considering that the functions of many hormones change throughout an animal's life history. Thyroid hormones, for instance, exert profound developmental effects on mammalian fetal and neonatal organ ontogeny, whereas in adults they primarily influence metabolism and potentiate growth. Furthermore, the same hormone may have very different functions among animal taxa. Whereas in mammals prolactin plays a key role in milk production, in fish the same hormone is involved in osmoregulation and adaptation to freshwater. Here, hormones will be classified based upon their immutable physical and chemical characteristics and those of their receptors. These hormone attributes provide information that relates directly to their functions, which will be addressed in detail in subsequent chapters. Note that Appendix 4 lists some of the most common vertebrate and arthropod hormones according to their organ of origin, chemical class, target tissue, solubility, and general actions.

DOI: 10.1201/9781003359241-6

Hormone Classification Based on Chemical Structure

Perhaps the most straightforward method for categorizing hormones is based simply upon their chemical structures. Most hormones fall into one of three broad categories under this paradigm. The first group, the *amine-derived* hormones, consists of small molecules, such as catecholamines (epinephrine and norepinephrine), indolamines (melatonin), and iodothyronines (thyroid hormones) that are all amino acid derivatives (Figure 4.1). The second group constitutes *polypeptide hormones* that range in size from peptides just a few amino acids in length, such as the three amino acid-long thyrotropin-releasing hormone, to proteins and glycoproteins almost 200 amino acids in length, like the pituitary hormone prolactin. The third group consists of *lipid-derived hormones*, such as cholesterol-derived *steroid hormones*, like cortisol, estrogens, and aldosterone in vertebrates and ecdysone in arthropods. Other lipid-derived hormones include **eicosanoids**, like prostaglandins and leukotrienes, and *endocannabinoids*, both of which derive from arachidonic acid. Other classes of hormones include those derived from acetyl CoA, such as juvenile hormone in insects, and retinol, like retinoic acid in vertebrates.

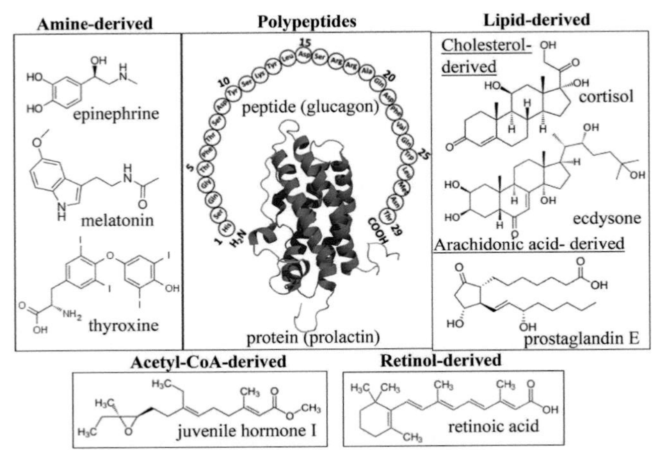

Figure 4.1 Examples of different hormone classes based on chemical structure.

Hormone Classification Based on Solubility and Receptor Location

Hormones can be categorized based upon their solubility and the locations of their receptors within target cells. The mechanism of receptor function varies with its location. Plasma membrane-localized receptors typically initiate an intracellular second messenger signaling cascade that alters cell metabolism or gene transcription after binding the hormone. By contrast, intracellular receptors commonly form hormone–receptor complexes that interact directly with DNA elements to affect gene transcription. Using these criteria, hormones fall into three major categories (Table 4.1):

1. **Hydrophilic hormones with plasma membrane receptors.** Hydrophilic signaling molecules are typically either large or electrically charged, features that prevent them from diffusing freely across the plasma membrane; they instead bind exclusively to cell-surface receptors. This large class consists of small peptide, larger protein, and glycoprotein hormones, as well as the amine-derived catecholamines (epinephrine and norepinephrine). Many hydrophilic hormones induce rapid modifications in the activities of enzymes present in the target cell that typically persist for only a short duration. However, such signals can also give rise to changes in gene transcription that may persist for hours or days.
2. **Hydrophobic hormones with intracellular receptors.** Many lipid-soluble hormones, like steroid hormones,

Table 4.1 General Classifications of Hormones Based upon Solubility and Receptor Location

	Water-Soluble Hormones	Lipid-Soluble Hormones	
Receptor location	Cell surface	Intracellular	Cell surface
Categories	Catecholamines, indolamines (*exception*: melatonin is lipophilic), peptides, proteins, glycoproteins	Steroids, iodothyronines, calcitriol, retinoids some eicosanoids	Eicosanoids melatonin many steroids, and iodothyronines
Action by hormone–receptor complex	Activate 2nd messenger signaling	Alter gene expression	Activate 2nd messenger signaling
Blood carrier proteins	No (*exceptions*: GHBP, IGFBP, CRHBP)	Yes	
Blood serum half-life	Seconds (catecholamines); Minutes-hours (others)	Hours to days	
Secretion	Exocytosis	Simple diffusion across membrane (*exception*: TH move via facilitated diffusion across plasma membrane transporters)	
Intracellular storage time	Catecholamines: hours Peptides/proteins/ glycoproteins: days	Steroids, eicosanoids, melatonin: none Iodothyronines: weeks	

can cross the plasma membrane freely by passive diffusion and interact with receptors in the cytosol or nucleus. The resulting intracellular ligand–receptor complexes formed by most hydrophobic hormones often bind to transcription-control regions of the DNA, thereby directly modulating expression of specific genes.

3. **Hydrophobic hormones with plasma membrane receptors.** The primary lipid-soluble hormones that can bind to cell-surface receptors are the eicosanoids and steroid hormones. Note that in the case of steroid hormones, the same hormone type can associate with both plasma membrane and intracellular receptors. The amine-derived hormone, melatonin, is also highly hydrophobic and binds to cell-surface receptors. As with the cell plasma membrane receptors that bind hydrophilic hormones, the hydrophobic ligands of cell-surface receptors can also induce rapid, nongenomic cellular responses.

SUMMARY AND SYNTHESIS QUESTIONS

1. What are some common ways that hormones are classified?
2. The binding of a hormone to its receptor initiates some type of intracellular signaling pathway. How do plasma membrane-localized receptors generally differ from intracellular receptors?

Transport in the Blood

LEARNING OBJECTIVE Identify how different chemical classes of hormones are transported by the blood.

KEY CONCEPTS:

- Most hydrophobic and some hydrophilic hormones are transported in the blood by transport proteins.
- Transport proteins extend the half-life of a hormone in the blood.
- The free fraction of hormone (non-transport protein-bound) in the blood is considered the most biologically active fraction.

Hormone solubility not only affects the nature of receptor-ligand interactions, but also influences the half-lives of hormones in the blood serum (Table 4.1). The **half-life** of a hormone is the amount of time required for a hormone to fall to half of its concentration as measured at the beginning of the time period. A hydrophilic hormone's half-life tends to be relatively short (seconds to minutes) because these hormones are rapidly catabolized by serum proteases or are filtered from the blood via the kidney. Some hydrophilic polypeptide hormones, such as growth hormone, insulin-like growth factors, and corticotropin-releasing hormone, have longer half-lives due to their association with *blood transport proteins* (Table 4.2) that protect them from degradation, filtration by the kidney, and diffusion into cells. Since hydrophobic hormones are not highly soluble in an aqueous environment, most are also carried in the blood by transport proteins (Table 4.2). These transport proteins allow hydrophobic hormones to exist in the blood plasma at concentrations several hundred-fold greater than their solubility in water would otherwise permit. Indeed, over 90% of hydrophobic steroid and thyroid hormones associate with blood transport proteins, and hydrophobic hormones generally tend to have comparatively longer half-lives than hydrophilic hormones not associated with serum transport proteins (Table 4.1).

Hormones that associate with serum transport proteins can exist in the blood in one of two states: the vast majority constitutes the **bound fraction**, and usually less than 3% of the total hormone comprises the **unbound** or **"free" fraction**. Because the unbound hormone fraction is immediately available for action on target organs, feedback control, and clearance by cellular uptake and metabolism, it is considered to be the most biologically active fraction and is an important diagnostic tool for medicine. The bound hormone represents a "reservoir" of hormone that

Table 4.2 Some Blood Plasma Hormone Transport Proteins and Their Characteristics

Blood Protein	Hormone(s) Bound	Binding Affinity	Binding Capacity
Albumin	All steroid and thyroid hormones	Low	High
α_1-Acid glycoprotein	Progesterone	Low	High
Sex hormone-binding globulin (SHBG)	Dihydrotestosterone, testosterone, estradiol	High	Low
Corticosteroid-binding globulin (CBG)	Corticosteroids, progesterone, 17α-hydroxyprogesterone	High	Low
T_4-binding globulin (TBG)	Thyroid hormones	High	Low
Transthyretin	Thyroid hormones, retinoic acid	High	Low
Retinol binding protein (RBP)	Retinoic acid	High	Low
GHBP	Growth hormone	High	High
IGFBP	Insulin-like growth factor I	High	High
CRHBP	Corticotropin-releasing hormone	High	High

can be transferred to the free fraction and, as such, serves to buffer acute changes in hormone secretion.

The concentrations of free hormone [H], serum transport protein [P], and serum transport protein-bound hormone [HP] in the blood plasma are in equilibrium. Therefore, if free hormone concentrations drop, hormone will be released from the transport proteins to restore equilibrium, and the opposite is true if free hormone concentrations rise. This relationship may be expressed as:

$$[H] \times [P] \rightleftharpoons [HP] \text{ or } K \rightleftharpoons [H] \times [P]/[HP]$$

where [H], [P], and [HP] denote the concentrations of the hormone, transport protein, and hormone-transport protein complex, respectively. K is the dissociation constant, and the double arrows denote a state of chemical equilibrium. Importantly, when evaluating hormonal status it is sometimes necessary to distinguish free hormone concentrations from total hormone concentrations. This is particularly important considering that hormone transport proteins themselves are modified by disease or altered endocrine states.

Hydrophobic hormones can associate with multiple types of blood serum transport proteins, some of which are hormone-specific, like sex hormone-binding globulin (SHBG) and thyroxine-binding globulin (TBG), and bind to hormones with high affinity (Table 4.2). Other transport proteins, like albumin, are less specific and bind to hormones with lower affinity (Table 4.2). Although albumin is a generalist that can bind to different hydrophobic hormones with lower affinity than other more specialized transport proteins, because albumin is the most abundant individual blood protein it often has a higher total hormone-binding capacity than other specialized transport proteins present at lower concentrations. Because of albumin's high abundance and low binding affinity, it plays an important role in maintaining equilibrium with the free fraction of that hormone. That is, if free hormone concentrations drop due to uptake by target tissues or are excreted by the kidney, hormones will be released most readily from albumin first in order to maintain free hormone homeostasis, and from specialized transport proteins second. Therefore, the "*bioavailability*" of different hormone fractions in the blood can be ranked as follows: hormones bound tightly to a specialized serum transport protein are generally unavailable for immediate use; hormones bound weakly to albumin are "bioavailable" and directly supply the unbound free fraction; and the unbound free fraction itself is the most "biologically active".

The body can modulate the amount of the biologically active free fraction in several different ways by altering the following:

1. Rate of hormone synthesized and released into circulation
2. Rate of hormone cleared by tissues and the kidneys
3. Magnitude and relative proportions of albumin and specialized transport proteins in the blood

For example, compared with women, men typically produce more testosterone and also have a much higher serum albumin-to-SHBG ratio, which together ensure a relatively high free fraction of testosterone available to the target cells. Women typically release much lower amounts of testosterone into their blood and also tend to have higher SHBG concentrations and lower albumin than men, which together ensure the maintenance of relatively low free fractions of testosterone (Figure 4.2).

A more complex but illustrative example of the influence of hormone transport protein concentration on hormone availability is the effect of pregnancy on maternal thyroxine in humans (Figure 4.3). The developing fetus relies on maternal TH for much of its development. During pregnancy, increased estrogen synthesis induces the liver to double its synthesis of thyroxine-binding globulin (TBG), an important blood transport protein with high affinity to thyroxine (T_4). Increased TBG concentrations initially lead to lowered maternal blood free T_4 concentrations, but this results in elevated pituitary thyroid-stimulating hormone (TSH) secretion due to negative feedback by lower T_4, ultimately enhancing TH synthesis. In addition, elevated secretion of human chorionic gonadotropin (hCG), a hormone structurally similar to TSH, by the placenta also

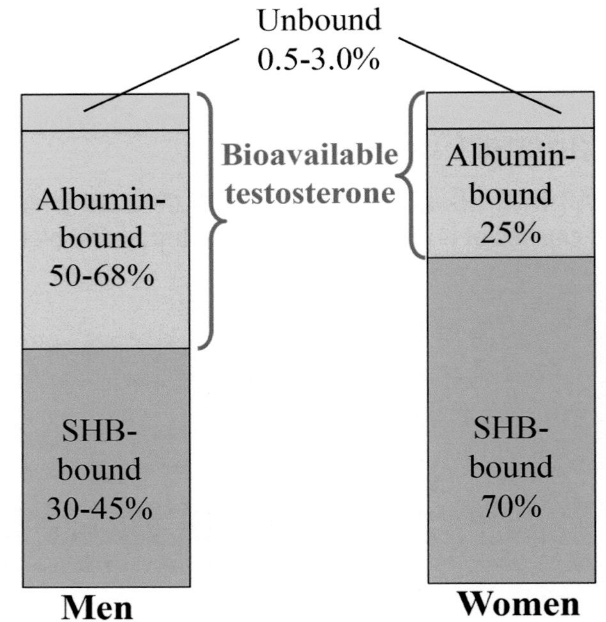

Figure 4.2 Testosterone fractions in the blood serum of human males and females. Blood testosterone is present in three states: bound to a high-affinity carrier protein (sex hormone-binding globulin, SHBG), bound to a low affinity carrier (albumin), or unbound (free fraction). Of the three states, the free and albumin-bound partitions are considered to be bioavailable and may interact with receptors in cells. Whereas the free fraction of testosterone for a typical male ranges from 9 to 30 ng/dL, in typical females it is much lower, ranging from 0.3 to 1.9 ng/dL. This lower bioavailability of testosterone in women is achieved, in part, by maintaining higher SHBG levels and lower albumin compared with men.

Source of images: Lepage, R., 2006. Clin. Biochem., 39(2), pp. 97-108.

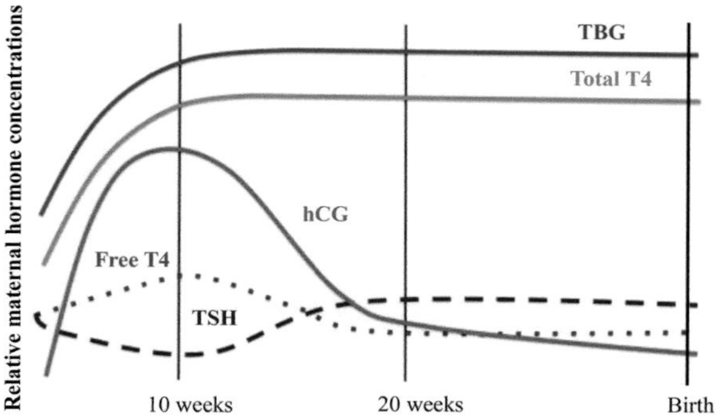

Figure 4.3 Some key changes in maternal blood thyroid parameters during human pregnancy. These changes include rising levels of thyroxine-binding globulin (TBG) and total and free T_4 during the first few critical weeks of fetal life when the fetus is dependent upon maternal T_4 for development. Human chorionic gonadotropin (hCG); thyroid-stimulating hormone (TSH).

Source of images: Salmeen, K.E. and Block-Kurbisch, I.J., 2020. Thyroid physiology during pregnancy, postpartum, and lactation. In Maternal-Fetal and Neonatal Endocrinology (pp. 53-60). Academic Press; used with permission.

stimulates increased maternal TH production by trans-activating the TSH receptor. The effect of elevated TBG and TH concentrations are to establish a new equilibrium between free and bound TH that ultimately increases the bioavailable free fraction during the first critical weeks (weeks 10–20) to support increased demand for TH by the fetus during that time.

SUMMARY AND SYNTHESIS QUESTIONS

1. Why do steroid hormones typically have longer half-lives in the blood compared with most polypeptide hormones?
2. At any given time, why do blood-borne polypeptide hormones generally have a greater "bioavailable" fraction (more hormone able to bind receptors) compared with steroid hormones?
3. You measure concentrations of blood albumin, sex hormone-binding protein (SHBP), and total blood estrogen from two patients. You find that both patients have identical concentrations of total blood estrogen, but whereas patient A has a high proportion of blood albumin compared with SHBP, patient B has a higher proportion of SHBP compared with albumin. Which patient has the lower fraction of bioavailable estrogen, and why?
4. A child is diagnosed with a genetic disorder where the gene coding for sex hormone-binding globulin (SHBG) has been mutated, resulting in very low levels of circulating SHBG protein. Is this child likely to experience delayed puberty, precocious (early) puberty, or no significant effect on the timing of puberty? Explain your reasoning.

Hormone Inactivation and Excretion

LEARNING OBJECTIVE Describe how hormones are inactivated and excreted.

KEY CONCEPTS:

- Hormone inactivation is necessary to ensure steady-state levels of plasma hormones.
- The liver and kidneys are the primary organs involved in the inactivation and excretion of hormones.
- The inactivation of hydrophobic hormones typically requires their conversion into more hydrophilic compounds to ensure their solubility at high concentrations in biological fluids prior to their excretion in the urine.

Hormone inactivation, or the metabolic conversion of an active hormone into an inactive metabolite, is necessary to ensure steady-state levels of plasma hormones. Hormone inactivation can occur at many stages of a hormone's life, including during its transportation in the blood and before or after a hormone's interaction with target cell receptors (Figure 4.4). In many cases hormones are inactivated by

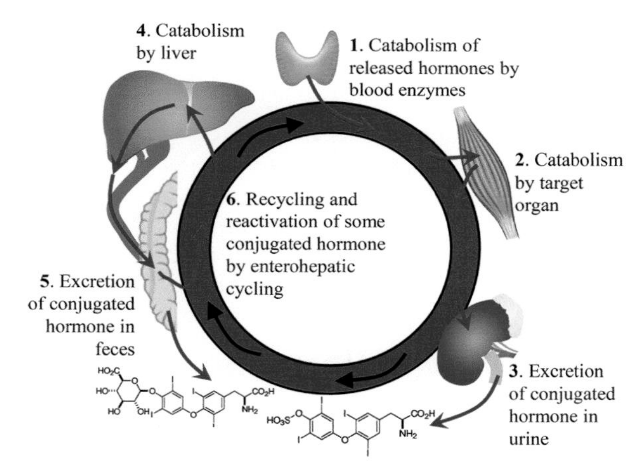

Figure 4.4 The many potential fates of a hormone. (1) Following release of hormones into circulation, some (particularly peptide hormones) may be degraded by proteases in the blood. (2) Hormones that enter into somatic target cells (e.g. thyroid hormones and steroid hormones) may be chemically conjugated and inactivated by the target tissues. (3) Inactivated hormone byproducts are excreted by the kidney into urine. (4) The liver is the primary organ for the inactivation of hormones, particularly of amine and steroid hormones. (5) Following conjugation by the liver, hormone metabolites may be incorporated into bile and excreted in feces, or may be reabsorbed into circulation and excreted in urine. (6) Liver-conjugated hormones can also sometimes be deconjugated and reactivated by the intestine and reabsorbed into the bloodstream to be reused, a process called enterohepatic cycling. Chemical diagrams at bottom denote conjugated/inactivated glucuronide (left) and sulphate (right) catabolic derivatives of thyroxine.

Source of images: A. Gorbman, et al. Comparative Endocrinology, ©1983, John Wiley and Sons, New York.

enzymes inside the target tissues themselves. The primary organ involved in the inactivation of circulating hormones that have not been metabolized by target tissues is the liver, though the kidneys also possess some catabolic activity (Figure 4.4). The inactivation of hydrophobic hormones, such as steroid hormones and thyroid hormones, typically requires their conversion into more hydrophilic compounds to ensure their solubility at high concentrations in biological fluids prior to their excretion in the urine. This process is called **conjugation.** Two major pathways of conjugation include the attachment of hydrophilic sulphate and/or glucuronide derivatives to the hydrophobic hormones (Figure 4.4).[2] Following conjugation, the metabolized hormones are primarily excreted by the kidney in urine. Hormone metabolites processed by the liver can also be incorporated into bile, transferred to the intestine, and excreted in feces. However, there is some evidence that liver-conjugated hormones can also sometimes be deconjugated and reactivated by the intestine and reabsorbed into the bloodstream to be reused, a process called *enterohepatic cycling* (Figure 4.4).[3–8]

SUMMARY AND SYNTHESIS QUESTIONS

1. Why can hormone concentrations in the blood and extracellular fluid vary in a spatial-temporal manner?
2. Describe two ways in which the liver contributes to a hormone's half-life.
3. Some toxic compounds, like phthalates, enhance the liver's ability to conjugate thyroid hormones to sulfates and glucuronidates. How would this affect blood thyroid hormone levels?

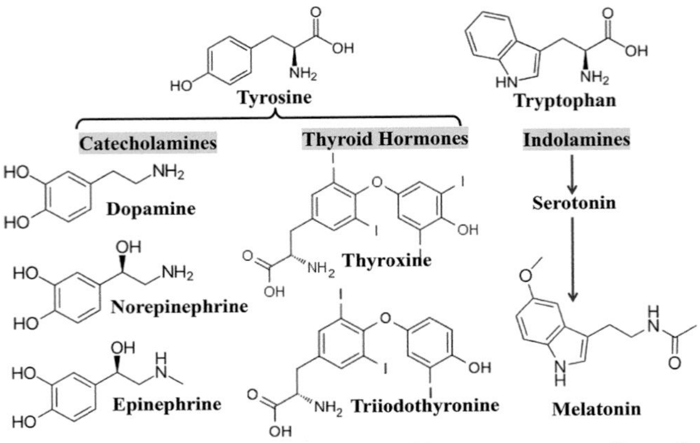

Amine-Derived Hormones

LEARNING OBJECTIVE List the amine-derived hormones, their precursor substrates, and the organs that synthesize them.

KEY CONCEPTS:

- All amine-derived hormones are modifications of the aromatic amino acids tyrosine or tryptophan.
- The primary site of circulating amine-derived hormone inactivation is the liver.

The **catecholamines** (Figure 4.5), which include epinephrine, norepinephrine, and dopamine, are modified from tyrosine by cytoplasmic enzymes and are then packaged into secretory vesicles for export. Whereas catecholamines function as classical neurotransmitters of the sympathetic nervous system when synthesized by neurons, they also function as neurohormones when released into the bloodstream from chromaffin tissue of the adrenal medulla. The **iodothyronines,** such as thyroxine and triiodothyronine, are iodinated derivatives of tyrosine residues. The hormone melatonin is synthesized by the pineal gland from the precursor serotonin, both of which are derivatives of the amino acid tryptophan and belong to a family of molecules called the *indolamines*.

Figure 4.5 Vertebrate amine-derived hormones are all modifications of either tyrosine or tryptophan.

SUMMARY AND SYNTHESIS QUESTIONS

1. Catecholamines and thyroid hormones are both derivatives of the amino acid tyrosine. Name two structural differences between the two categories.

Polypeptide Hormones

LEARNING OBJECTIVE Describe how polypeptide hormones are categorized and synthesized.

KEY CONCEPTS:

- Polypeptides are characterized by unique primary sequences of amino acids that are encoded by the genome.
- Families of genes coding for closely related polypeptides within a species (paralogs) arose by gene duplication followed by gene mutation.
- Virtually all mammalian polypeptide hormones and receptors have orthologs (same gene in a different species) among the more ancestral vertebrates.
- Functional variability in polypeptide hormone function is produced via alternative splicing and alternative posttranslational processing.
- Receptors for all hormones, irrespective of chemical class, are polypeptides.

Polypeptide Hormone Classes

Polypeptide hormones range in length from 3 amino acids (thyrotropin-releasing hormone) to 192 amino acids (growth hormone). They are characterized by unique primary sequences of amino acids that are encoded by the genome and also may undergo extensive post-transcriptional and posttranslational modifications. Although no clear consensus exists, in general, polypeptides consisting of about 2–20 amino acids are loosely referred to as "**oligopeptides**" (*oligo* is Greek for "few") (Figure 4.6), those between 21 and 50 amino acids are called "*peptides*", and those greater than that are "*proteins*". "**Glycoproteins**" are polypeptides that have carbohydrate moieties covalently attached to them (Figure 4.7). Whereas most protein hormones exhibit complex secondary and tertiary three-dimensional structures, the short lengths of oligopeptide hormones limit the formation of complex higher-order conformations.

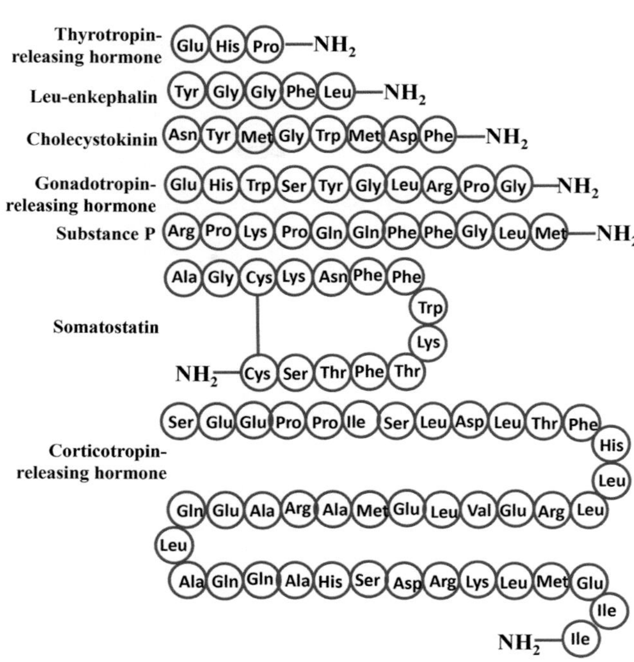

Figure 4.6 Some oligopeptide hormones.

Polypeptide Hormone Synthesis

Transcription

As with all proteins, polypeptide hormones are encoded by genes located within chromosomes contained in the

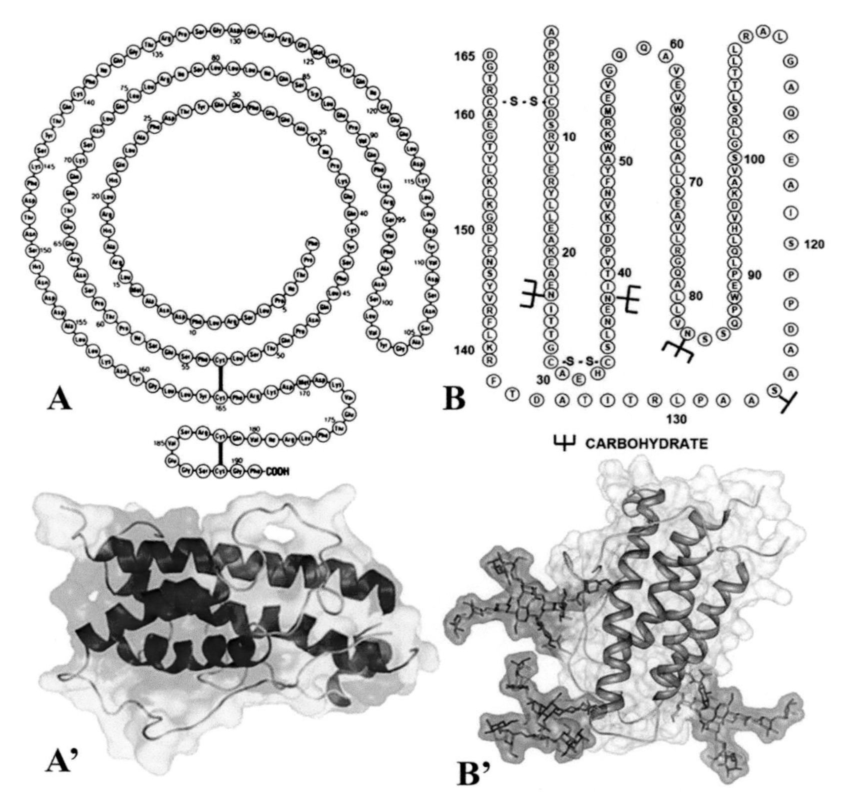

Figure 4.7 Primary sequence and higher-order structures of a protein and glycoprotein hormone. Human growth hormone primary **(A)** and tertiary **(A')** structures. Human erythropoietin primary **(B)** and tertiary **(B')** structures. In addition to its complex 3-D shape, the, erythropoietin glycoprotein **(B')** has several carbohydrate moieties (purple regions) covalently attached to specific asparagine and serine residues. "S-S" denotes disulfide bonds **(A)**.

Source of images: (A) Herman-Bonert, V.S. and Melmed, S., 2022. Growth hormone. In The pituitary (pp. 91-129). Academic Press; used with permission (A') Courtesy of Karl Harrison 3DChem.com. (B) Zoltán Kiss, et al. 2010. Eur. J. Clin. Pharmacol., Volume 66, Issue 4, pp 331-340; used with permission. (B') Jelkmann, W., 2013. Recombinant human erythropoietin and its analogues. In Introduction to Biological and Small Molecule Drug Research and Development (pp. 307-326); used with permission.

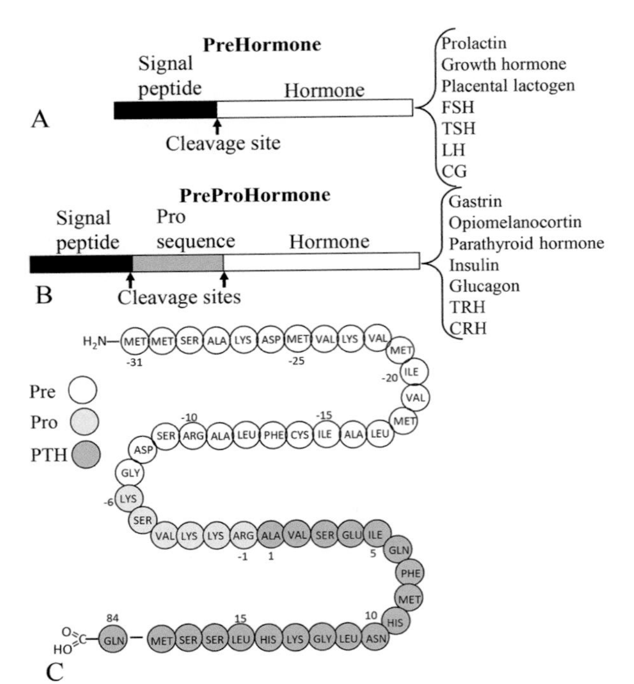

Figure 4.8 Nascent polypeptide hormones are translated into inactive precursors called prehormones and preprohormones. Whereas prehormones develop into mature hormones in the RER after the signal peptide is cleaved **(A)**, preprohormones require further cleavage from their prohormone form before the mature hormone is produced **(B)**. **(C)** Amino acid sequence of bovine preproparathyroid hormone.

Source of images: (B) modified from Habener, J.F. and Kronenberg, H.M., 1978. In Fed. Proc. (Vol. 37, No. 12, pp. 2561-2566).

nucleus. Following transcription, **introns** (*in*tervening sequences that are not expressed in the mature mRNA) are excised from the primary transcript, and **exons** (sequences *ex*pressed in the mature mRNA) are spliced together. **Alternative splicing**, or the generation of multiple mRNA variants through the selective exclusion of specific exons, is a contributor to the sequence diversity of many types of proteins. However, it is not a widely reported feature of polypeptide hormone processing, although this has been described for some hormones, such as insulin-like growth factor-1 (IGF-1).[9] By contrast, alternative splicing can play a prominent role in the isoform variability of many hormone *receptors*, discussed in Chapter 5: Receptors.

Translation and Processing through the Secretory Pathway

Polypeptide hormones are synthesized as precursor proteins that are extended at their N-terminus by hydrophobic sequences of 15 to 30 amino acids, called *signal sequences*. This sequence directs the nascent polypeptide, along with the ribosome it is attached to, to the rough endoplasmic reticulum (RER) where its translation into an inactive precursor is completed. The inactive nascent polypeptide can either be in the form of a **prehormone**, which requires cleavage of the signal sequence by a signal peptidase in the RER for direct conversion into the mature hormone, or a **preprohormone** that requires cleavage of both the "pre" signal sequences. After cleavage, the polypeptide is called a **prohormone** and requires further excision of "pro" sequences within the prohormone to reach full maturation (Figures 4.8–4.9). The cleavage of the "pro" sequence takes place either in the Golgi apparatus or in secretory vesicles before the mature hormone is produced. Whereas most polypeptide hormone mRNAs are translated into preprohormones, hormones with final products that are 100 amino acids long or longer (e.g. GH, prolactin, and the α- and β-subunits of the pituitary and placental glycoprotein hormones) are typically translated into prehormones and have no prohormone intermediate.

Worth noting here is a separate use of the term "prohormone" that is frequently encountered in the literature, and most typically in association with the broad concept of a precursor molecule for *non*-peptide hormones. For example, the steroid dehydroepiandrosterone (DHEA) is often referred to as a prohormone due to its very low affinity to the androgen receptor. DHEA requires conversion to more active hormones like testosterone and androstenedione that have higher affinities to the androgen receptor.[10] In insects, the steroid ecdysone is also generally considered a prohormone and is converted into the active 20-hydroxyecdsyone that has a high affinity to the ecdysone receptor. The thyroid hormone, thyroxine (T_4), is also sometimes called a prohormone due to its conversion into the more biologically active hormone, triiodothyronine (T_3), in target tissues. However, considering that T_4 is also capable of binding to the thyroid receptor and exerts biological activity independent of its conversion to T_3, it is not clear if the term prohormone is accurate.[11]

Polypeptides translated in the RER lumen undergo several forms of posttranslational processing. These include removal of the "pre" signal peptide, folding by *chaperone proteins* into a protein's functional three-dimensional conformation (the "native state"), and in some proteins, *glycosylation* (addition of carbohydrate moieties), and the formation of covalent *disulfide bonds* (refer to Figure 4.7 and Figure 4.9). Polypeptide hormone precursors are subsequently exported to the Golgi for further processing. An important function of the Golgi is the modification of glycosylated residues of proteins fated to become glycoproteins. Vertebrate glycoprotein hormones include erythropoietin, the gonadal polypeptides activin and inhibin, the prohormone proopiomelanocortin (POMC), and a group of four closely related glycoproteins: follicle-stimulating hormone (FSH), luteinizing hormone (LH), thyroid-stimulating hormone (TSH), and chorionic gonadotropin (CG). In general, carbohydrate moieties on glycoprotein hormones are thought to influence clearance rates from circulation and facilitate signal transduction in association with receptors. Interestingly, Golgi hypoglycosylation dysfunctions may result in anemia due to a failure to properly glycosylate erythropoietin, a hormone that induces red blood cell differentiation,[12] as well as ovarian dysfunction due to improper FSH glycosylation.[13]

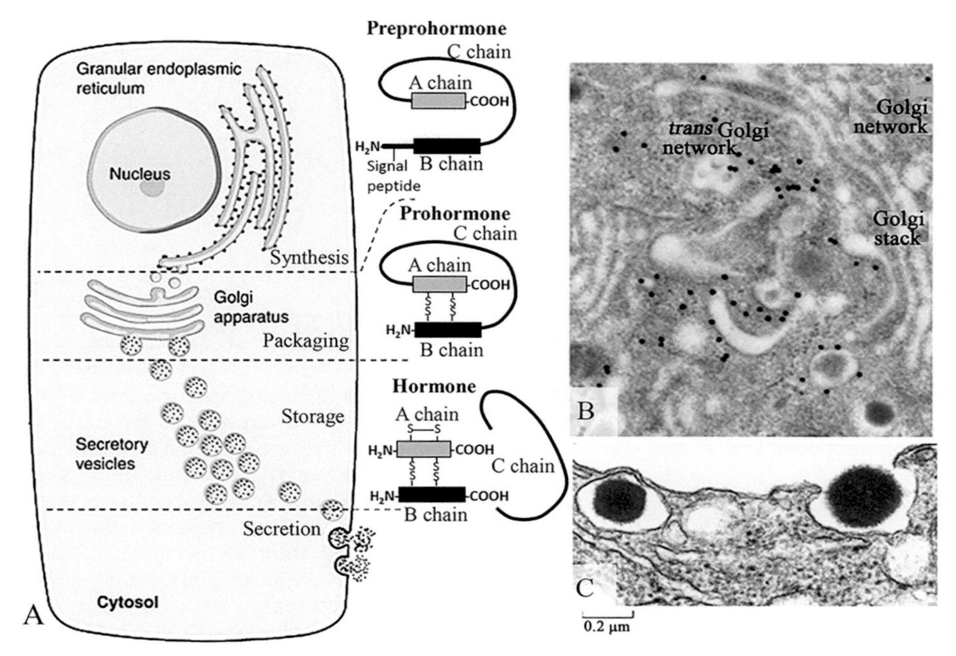

Figure 4.9 Synthesis of preproinsulin and its processing into mature insulin via the regulated secretory pathway in a pancreatic beta cell. (A) The conversion of the nascent preproinsulin molecule into proinsulin in the RER is accompanied by removal of the signal peptide and formation of disulfide bonds. After anterograde sorting through the Golgi is completed, proinsulin is converted into mature insulin within secretory vesicles through the sequential actions of the prohormone convertases (PCs), PC1 and PC2. Mature insulin is then retained in the secretory granule until it, along with the inactive C chain, are secreted via exocytosis. **(B)** Electron micrograph showing immature secretory vesicles containing proinsulin (black dots) within a β cell. **(C)** Electron micrograph showing insulin being released from a β cell. The secretory vesicle on the left is about to fuse with the β cell membrane, while the granule on the right is being released into the extracellular space. Each secretory vesicle holds roughly 800,000 insulin molecules on average.

Source of images: (A) Luesch, H. and Paavilainen, V.O., 2020. Nat. Prod. Rep.; used with permission. (B-C) Handorf, A.M., et al. 2015. J. Diabetes Mellit., 5(04), p. 295.

Activation, Secretion, and Degradation

Another important function of the Golgi is the sorting of proteins into *secretory vesicles* (also called secretory "granules") that bud off of the trans Golgi for export to the plasma membrane, the fate of all polypeptide hormones (Figure 4.9). Prior to their export, the prohormones are cleaved by proteolytic enzymes called **prohormone convertases** contained in the vesicles, resulting in the conversion of the prohormone into its active form (Figure 4.9). Prohormones are stored in their secretory vesicles until the cell receives a signal, possibly from another regulatory hormone, that induces the vesicle to fuse with the plasma membrane and release its contents into the extracellular environment and circulatory system via the process of *exocytosis* (Figure 4.9 and Figure 4.10). The capillary vasculature associated with peptide-secreting endocrine tissues has a fenestrated ("perforated") endothelial layer that facilitates the uptake of these large molecules into the bloodstream for dispersal (Figure 4.11).

Importantly, some prohormones can undergo **alternative posttranslational processing**, yielding variable products depending upon cell type and what prohormone convertases the cell's vesicles contain. For example, whereas proglucagon is processed by the pancreas primarily into

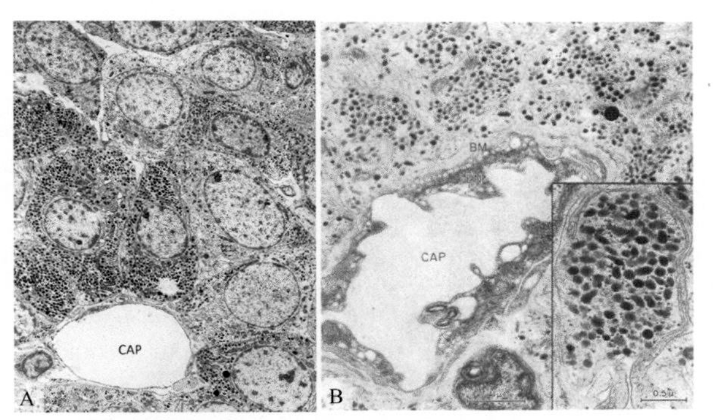

Figure 4.10 Polypeptide hormone-secreting cells are characterized by the presence of protein-rich, electron-dense secretory vesicles on the basal end of cells juxtaposed to blood vessels. (A) An electron micrograph of a mammalian adenohypophysis depicting a cluster of secretory cells surrounding a capillary (CAP). **(B)** Electron micrograph of the goby fish (*Gillichthys*) showing a capillary associated with a neurosecretory organ called the urohypophysis. Note that the surrounding axon profiles contain neurosecretory granules. Insert at right is higher magnification of one axon showing details of granules. BM, basement membrane; EN, endothelium.

Source of images: (A) Courtesy of Dr. Peter Takizawa, Yale University; (B) Fridberg, G. and Bern, H.A., 1968. Biol. Rev., 43(2), pp. 175-199; used with permission.

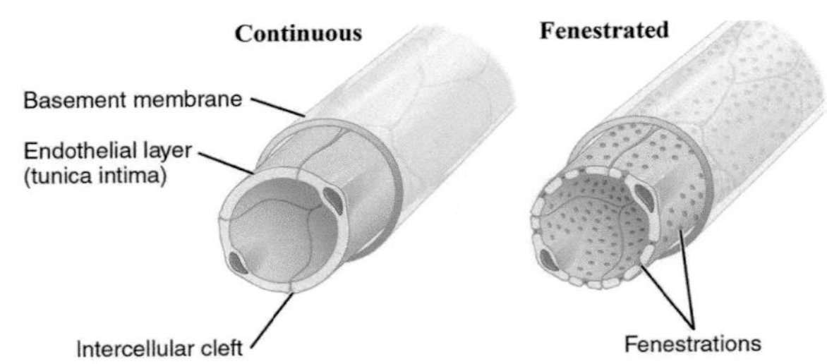

Figure 4.11 Compared with a typical continuous endothelial layer capillary, the capillary vasculature associated with peptide-secreting endocrine tissues has a fenestrated endothelial layer that facilitates the diffusion of these large molecules into the bloodstream.

Source of images: From openstax.org.

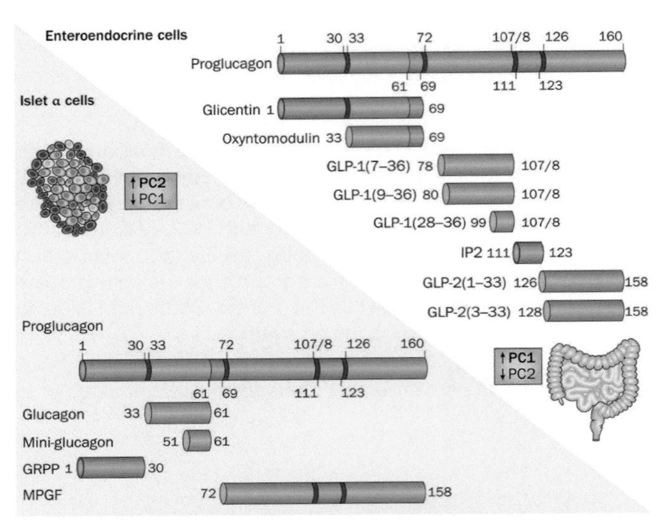

Figure 4.12 Alternative posttranslational processing of proglucagon in the pancreas and intestines. A single mammalian gene gives rise to an identical precursor polypeptide, proglucagon, in enteroendocrine, brain, and pancreatic α cells. This prohormone undergoes differential cell-specific posttranslational processing to yield different peptides, which themselves can undergo further enzymatic cleavage as shown. Abbreviations: GLP-1, glucagon-like peptide 1; GLP-2, glucagon-like peptide 2; GRPP, glicentin-related polypeptide; IP, intervening peptide; MPGF, major proglucagon fragment; PC1, prohormone convertase 1; PC2, prohormone convertase 2.

Source of images: Campbell, J.E. and Drucker, D.J., 2015. Nat. Rev. Endocrinol., 11(6), p. 329; used with permission.

glucagon, the intestine and brain process the same prohormone primarily into two isoforms of glucagon-like peptide (GLP), GLP-1 and GLP-2[14] (Figure 4.12). A prohormone capable of even more complex alternative posttranslational processing is proopiomelanocortin (POMC), which, depending on what pituitary cell type expresses it, can be converted into half a dozen different active hormones (Figure 4.13).

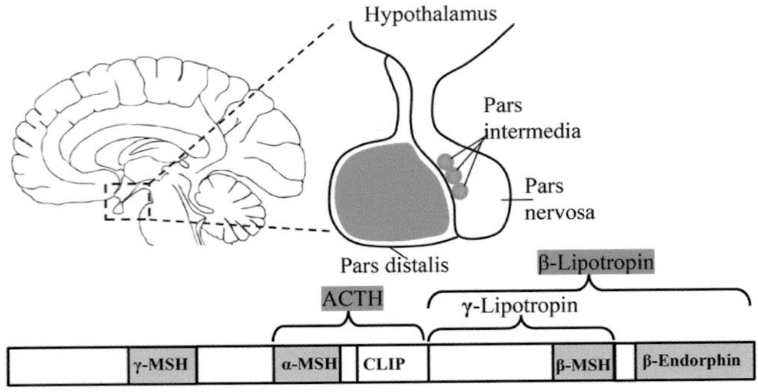

Figure 4.13 Proopiomelanocortin (POMC) is a large precursor peptide expressed by cells in the pars distalis and pars intermedia lobes of the pituitary, as well as other regions of the brain, such as the hypothalamus. This 241 amino acid prohormone is cleaved differentially into several mature hormones by prohormone convertase enzymes differentially expressed in different cells. For example, ACTH and β-lipotropin are products generated in the corticotropic cells of the pars distalis. β-endorphins and melanotropins (α-, β- and γ-MSH) are generated in the pars intermedia and the hypothalamic arcuate nucleus. Corticotropin-like intermediate peptide (CLIP) is a cleavage product of ACTH.

SUMMARY AND SYNTHESIS QUESTIONS

1. The hormones chorionic gonadotropin, TSH, LH, and FSH are all glycoproteins and are considered to be paralogs. The hormone erythropoietin is also a rare example of a glycoprotein hormone, yet it is not considered a paralog of the other four hormones. Why not?
2. Although there is only one gene that codes for the glucocorticoid receptor (GR) in the human genome, there are many different isoforms that exist for the GR protein. How is this possible?
3. Mutations in one or more RER glycosylation enzymes can result in Golgi hypoglycosylation dysfunctions, where carbohydrate groups on the proteins cannot be modified properly. Explain why hypoglycosylation dysfunctions may result in both anemia and infertility.
4. All polypeptide hormones must be translated via ribosomes associated with the RER, as opposed to ribosomes that are free in the cytosol. Why?
5. The large pro-hormone, POMC, is expressed in diverse organs, including the different cell types of the pituitary gland. When expressed in anterior pituitary corticotrope cells it is used to make the adrenocorticotropic hormone (ACTH); when expressed in melanotrope cells of the intermediate lobe of the pituitary it makes melanocyte-stimulating hormone (MSH). How can different cell types convert the same POMC pro-hormone into completely different active hormones?

6. Obesity due to prohormone convertase 1/3 deficiency is a rare genetic disorder characterized by severe diarrhea and intestinal malabsorption, followed by the early onset of obesity and hormonal deficiencies resulting in hypocortisolism (reduced pituitary ACTH synthesis), hypothyroidism (reduced pituitary thyroid-stimulating hormone), diabetes insipidus (reduced pituitary antidiuretic hormone), hypogonadism (reduced pituitary gonadotropins), and diabetes mellitus (reduced insulin).[15] Describe why this deficiency results in reduced synthesis of the aforementioned hormones.
7. Protein hormones, like insulin, can be injected into the body to treat diseases like diabetes. Though it would be much easier to just eat insulin, compared with intravenous delivery, oral consumption would not be an effective way to augment the body's insulin levels. Why not?

Lipid Hormones

LEARNING OBJECTIVE Provide an overview for the biosynthesis of the two most common categories of lipid hormones (steroid hormones and eicosanoids) and describe their most important functions.

KEY CONCEPTS:

- Lipid hormones derive primarily from one of two substrates: cholesterol or arachidonic acid.
- Cholesterol-derived hormones can be divided into seven categories defined broadly by their physiological functions: glucocorticoids, mineralocorticoids, androgens, estrogens, progestogens, the vitamin D family, and in arthropods ecdysteroids.
- The enzymes that mediate steroidogenesis are located primarily in the mitochondria and smooth ER.
- In contrast with steroids, eicosanoids are generally not released into circulation and function as paracrine signaling agents.
- Aspirin-like drugs suppress inflammation by inhibiting prostaglandin synthesis.

Steroid Hormones

Steroid and secosteroid hormones are both derivatives of cholesterol, a hydrocarbon molecule with a characteristic fused four-ring core structure (Figure 4.14). Thom Rooke's opening quotation at the start of this chapter colorfully describes the great extent to which steroid hormones, both natural and synthetic, influence diverse aspects of human physiology. Vertebrate steroid hormones are synthesized predominantly by the adrenal gland and the gonads. A newly described class of steroids, the *neurosteroids*, are synthesized within the brain and modulate neuronal excitability primarily by interaction with neuronal membrane receptors and ion channels.[16–18] These steroids appear to be modified by glial cells from circulating steroid hormone precursors originating from the adrenal gland or gonads and are released locally by the hippocampus and other brain structures. **Secosteroids**, cholesterol-derived relatives of steroids such as vitamin D, are synthesized by the skin in response to UV light and modified by the liver and kidney into more active forms.

Curiously Enough . . . With 30 of the 300 top-selling drugs being either steroids or drugs that affect endogenous steroid hormone synthesis, from a commercial perspective, steroid hormones and their synthetic analogs are among the most important pharmaceutical classes of compounds of all time.[19,20]

Structure and Classification

Functional Classifications

In vertebrates, cholesterol-derived hormones can be divided into six categories defined broadly by their first-described physiological functions and the receptors they bind to: *glucocorticoids* maintain glucose homeostasis; *mineralocorticoids* maintain mineral homeostasis and osmotic pressure; *cholecalciferol* (also known as vitamin D3) maintains calcium homeostasis; *androgens* stimulate development of male sexual characteristics; *estrogens* play roles in female estrus and menstrual cycles; and *progestogens* maintain pregnancy (Figure 4.14). Arthropods (insects and crustaceans) also possess steroid hormones

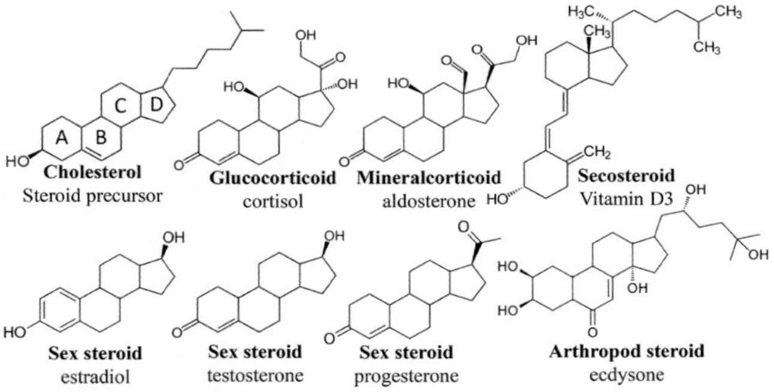

Figure 4.14 Some vertebrate and arthropod steroid hormones. Cholesterol is a hydrocarbon with a unique structure consisting of four fused hydrocarbon rings (A, B, C, D) that form the steroid nucleus. Among vertebrates, cholesterol-derived hormones are divided into six categories defined by their physiological functions. The six categories of vertebrate steroid hormones are glucocorticoids that maintain glucose homeostasis; mineralocorticoids that maintain mineral homeostasis and osmotic pressure; cholecalciferol (a secosteroid also known as vitamin D3) that maintains calcium homeostasis; androgens that stimulate development of male sexual characteristics; estrogens that play roles in female estrus and menstrual cycles; and progestogens that maintain pregnancy. Among arthropods (insects and crustaceans), ecdysone mediates ecdysis (molting), among other developmental processes.

Table 4.3 Some Common Steroid Hormones

Family	# Carbons	Example	Primary Site(s) of Synthesis	Primary Receptor
Progestogens	21	Progesterone	Adrenal cortex	Progesterone Receptor (PR)
			Male: testis	
			Female: ovary, adipose tissue	
Glucocorticoids	21	Cortisol	Adrenal cortex: Zona fasciculata	Glucocorticoid Receptor (GR)
		Corticosterone		
Mineralcorticoids	21	Aldosterone	Adrenal cortex: Zona glomerulosa	Mineralcorticoid Receptor (MR)
Androgens	19	Testosterone	Male: testis	
			Female: ovary, adrenal cortex	Androgen Receptor (AR)
		Dihydroepiandr-osterone	Adrenal cortex: Zona reticularis	
Estrogens	18	Estradiol	Ovary, testis, placenta	Estrogen Receptor (ER)
Vitamin D	25	Cholecalciferol	Skin (in response to UV light)	Vitamin D Receptor (VDR)
		25-Hydroxycholecalciferol	Liver (cholecalciferol)	
		1,25-Dihydroxycholecalciferol	Kidney (25-Hydroxycholecalciferol)	
Neurosteroids	21	Allopregnanolone	Central and peripheral nervous system, especially in myelinating glial cells. Made from circulating progesterone	GABA-A Receptor
		Tetrahydrodeoxycorticosterone	Made from circulating adrenal deoxycortisone	GABA-A Receptor
Ecdysteroids (arthropods only)	25	Ecdysone	Prothoracic gland	Ecdysone Receptor

called *ecdysteroids*, which are used to regulate ecdysis (molting), metamorphosis, and reproduction. Some common steroids, their structural features, sites of synthesis, and functions are summarized in Table 4.3. With regards to the so-called sex hormones, androgens, estrogens, and progesterone, it is important to note that these hormones are each present and play essential roles in both male and female vertebrates, and as such none of these hormones have exclusively male or female functions. Indeed, all estrogens are derived from androgen precursors, and in addition to functioning as a hormone in its own right progesterone is a substrate from which virtually all other vertebrate steroid hormones are synthesized. Furthermore, as will be seen in Unit VII: Reproduction, estrogen regulates male reproductive system functions important to the maturation of sperm and plays key roles in both sexes in mediating bone development and cardiovascular health.

Curiously Enough . . . In developing mice it is estradiol, and not testosterone, that is directly responsible for the development of brain circuits that guide aggressive and territorial behaviors in males.[21]

Cholesterol Synthesis

Although adequate cholesterol levels are critically important for the maintenance of proper cell membrane fluidity and steroid hormone synthesis, high blood cholesterol concentrations may lead to cardiovascular disease and other disorders. Even with a complete abstinence from dietary cholesterol intake, some individuals are genetically predisposed to having high blood cholesterol levels. The reason for this is that most of the cholesterol in the body is synthesized endogenously from hydrocarbon substrates by the liver and packaged into lipoproteins for export to cells in the rest of the body via the bloodstream. Steroid-producing endocrine cells, such as the ovaries, testes, and adrenal cortex, also have the ability to synthesize cholesterol but generally utilize liver-derived cholesterol obtained from the blood as the raw material needed to synthesize their hormones. Following synthesis by the liver, cholesterol may be converted into bile salts and stored in the gall bladder, or cholesterol can be packaged in the form of less soluble cholesteryl esters into **low-density lipoproteins** (LDLs) that are exported to the bloodstream. Target cells possessing LDL receptors take up the lipoprotein via endocytosis, the LDL is digested by lysosomes, and free cholesterol is subsequently used for steroid hormone or membrane synthesis or stored in lipid droplets (Figure 4.15). Excess cholesterol from cells may be returned to the liver via **high-density lipoproteins** (HDLs).

Steroid Hormone Synthesis

Most enzymes that mediate *steroidogenesis* are located in the mitochondria and the smooth ER, and the shuttling of steroid substrates and products between these organelles are features of this process. As such, in addition to containing prominent lipid droplet stores of cholesterol, steroid hormone synthesizing cells are also characterized

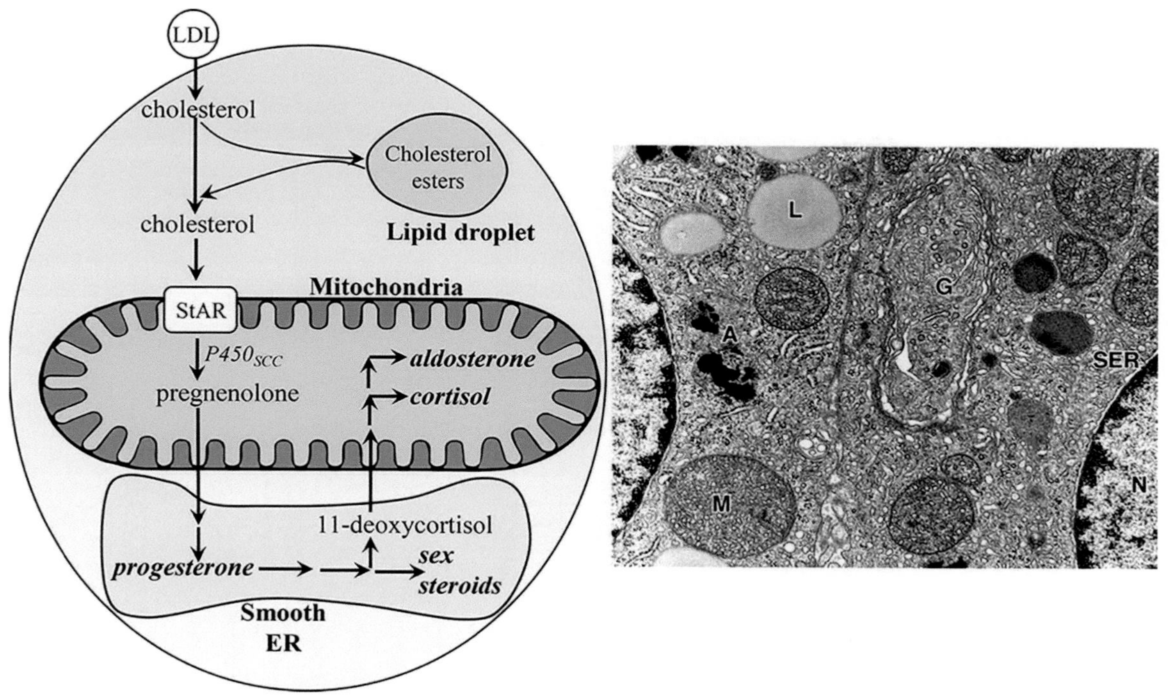

Figure 4.15 Pathways of steroid hormone synthesis from cholesterol. After uptake of cholesterol and cholesteryl esters from circulating low-density lipoproteins (LDLs), cholesteryl esters may be stored in lipid droplets and hydrolyzed to cholesterol as needed by the enzymes cholesterol esterase and hormone-sensitive lipase. Following import into mitochondria via transmembrane steroidogenic acute regulatory (StAR) transporters, cholesterol is converted into pregnenalone. Pregnenalone is converted to progesterone in the smooth ER. The smooth ER of sex steroid producing tissues, such as the testes and ovaries, convert progesterone into sex steroids, which diffuse into the bloodstream. Corticosteroids, such as aldosterone and cortisol, require further processing by the mitochondria prior to release by adrenal cortex cells. The micrograph depicts the ultrastructure of a steroidogenic adrenalocyte. Note the abundance of lipid droplets, smooth endoplasmic reticulum, and mitochondria with tubular cristae in this steroidogenic cell.

Source of images: (right) Junqueira, L.C. and Mescher, A.L., 2013. Junqueira's basic histology: text & atlas/ McGraw Hill LLC.; used with permission.

by the presence of well-developed mitochondria that are juxtaposed with an extensive smooth ER (Figure 4.15). In order to achieve the high surface area necessary for steroid hormone synthesis, the folds of the mitochondria inner membrane, called cristae, are organized into complex and highly involuted tubules, compared with the relatively flat folds that characterize most mitochondria.

Prior to the biochemical conversion of cholesterol into steroid hormones, cholesteryl esters stored in lipid droplets are hydrolyzed and released into the cytoplasm. They are then imported into mitochondria via transmembrane **steroidogenic acute regulatory** (StAR) protein-mediated transport from the outer to inner mitochondrial membrane (Figure 4.15). The differential processing of cholesterol by diverse enzymes located in both the mitochondria and the smooth endoplasmic reticulum determine the type of steroid hormone that is ultimately produced. In some cases the final product of one class of steroid hormone may serve as the starting substrate for another class (Figure 4.15). Also important to note is that unlike cholesterol, which can be stored in lipid droplets as cholesteryl esters, steroid hormones have no storage form and once synthesized diffuse freely across cell membranes and into circulation.

Steroidogenic enzymes are categorized into two superfamilies:

1. **Cytochrome P-450 monooxidases** (CYPs): these enzymes are localized to the inner mitochondria matrix or smooth endoplasmic reticulum where they facilitate electron transfer, functioning as oxidases, aromatases, hydroxylases, and lyases.
2. **Hydroxysteroid dehydrogenase** (HSD): localized to the smooth endoplasmic reticulum, these enzymes can activate or deactivate steroid hormones by catalyzing the dehydrogenation of hydroxysteroids.

The first step of synthesis common to the adrenal and gonadal steroids is the conversion of cholesterol to *pregnenolone*, a C_{21} intermediate, by P450scc, a *side-chain cleaving* enzyme located in the mitochondrial inner membrane (Figure 4.15 and Figure 4.16). Pregnenolone is then exported out of the mitochondria where it is converted into *progesterone* by 3β-hydroxysteroid dehydrogenase (3β-HSD), a membrane-bound enzyme associated with the smooth endoplasmic reticulum (ER). Progesterone is a hormone produced by the corpus luteum of the ovary, and its effects are mediated by progesterone receptors. Additionally, progesterone serves as

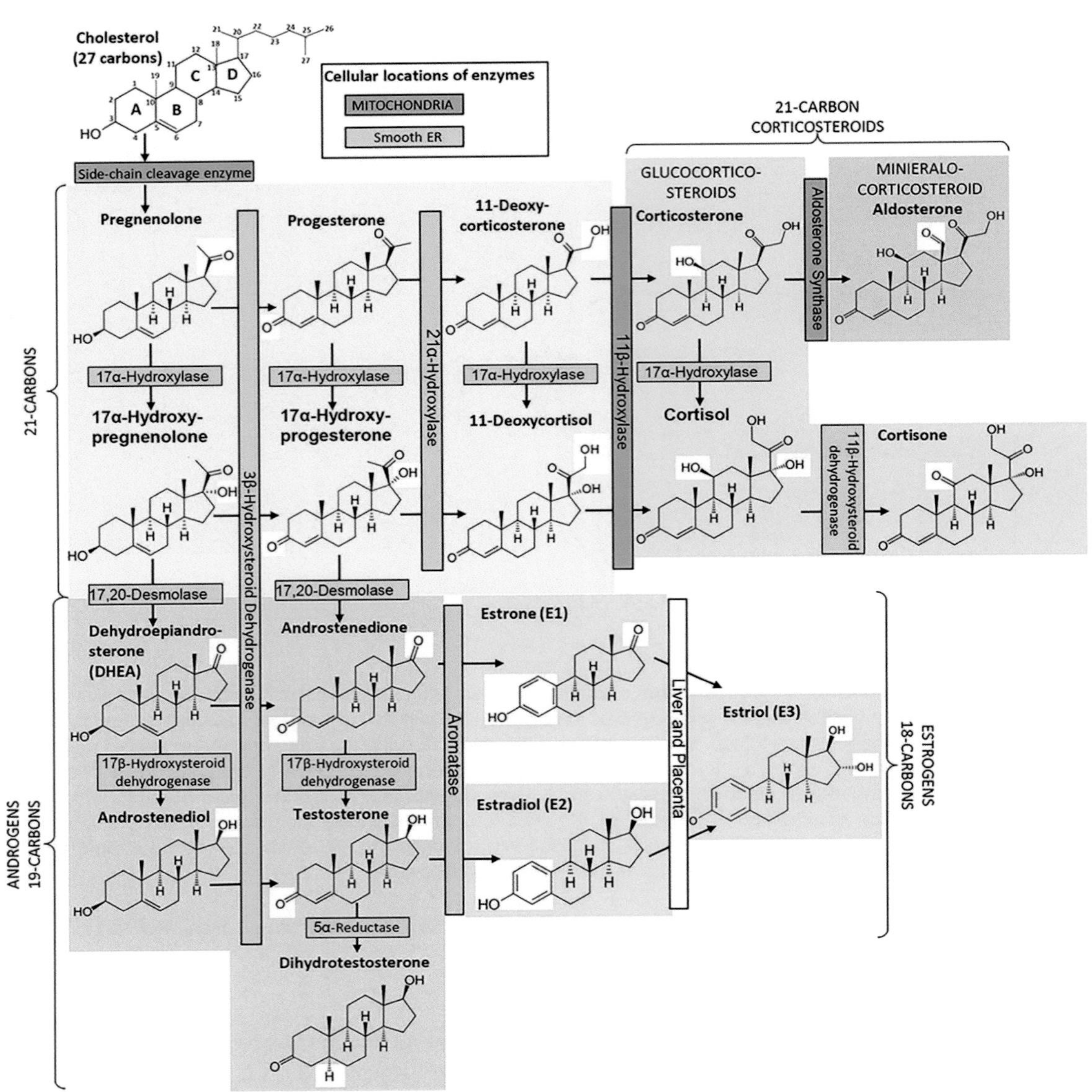

Figure 4.16 Synthesis and metabolism of the major vertebrate steroid hormone classes from cholesterol. The major transformative enzymes are in the horizontal and vertical boxes; they are located in either the smooth endoplasmic reticulum (SER) or the mitochondria. The chemical groups modified by each enzyme are highlighted in the reaction product.

Source of images: Boron, W.F. and Boulpaep, E.L., 2012. Medical physiology, 2e updated edition e-book: with student consult online access. Elsevier health sciences.

the precursor substrate for the synthesis of all other gonadal and adrenal steroids. **Cyanoketone**, a drug that inhibits the activity of 3β-HSD, is therefore an effective inhibitor of both adrenal and gonadal steroid hormone synthesis.

Adrenal Cortical Hormone Synthesis

In the adrenal cortex, progesterone is converted to 11-deoxycortisol in the smooth ER, which is then exported to the mitochondria where in humans and other large mammals it is converted to the glucocorticoid, *cortisol*, by the enzyme *11β-hydroxylase* ($P450_{C11}\beta$) (Figure 4.15 and Figure 4.16). The drug **metyrapone** is a selective inhibitor of 11β-hydroxylase and thus effectively blocks cortisol synthesis by adrenocortical cells. An additional enzyme present in the mitochondria of mineralocorticoid-secreting cells, aldosterone synthetase ($P450_{aldo}$), converts corticosterone to *aldosterone*, a potent mineralocorticoid.

Curiously Enough . . . The enzyme aldosterone synthetase and the mineralcorticoid aldosterone are present only in tetrapods (vertebrates with four limbs) and absent in teleost fishes. As such, the evolution of aldosterone is considered essential for the colonization of land by vertebrates.[22]

Clinical Considerations: The quest for cortisone, a "miracle cure" for autoimmune disorders

In the early 1930s, Edward Kendall at the Mayo Clinic focused his efforts on purifying the steroid hormones of the adrenal cortex.[23] For five years the Kendall Laboratory processed a remarkable 900 pounds of bovine adrenals *each week*,[24] preparing the "cortin" extract used to treat patients with Addison's disease, a form of adrenal insufficiency characterized by reduced cortisol and, often, aldosterone production. Kendall succeeded in isolating six adrenocortical hormones, and identified each by a letter, A through F. Several other groups joined the quest, including one led by Tadius Reichstein in Zurich. Working in intense competition with one another they quickly isolated, purified, and determined chemical structures for 29 different steroids, including what would eventually be characterized as the most biologically active glucocorticoid in humans, *hydrocortisone* (Kendall's "compound F", also called cortisol) and its inactive precursor, *cortisone* ("compound E"), as well as various androgens.[25–29] However, what were the actual physiological functions of these hormones? The researchers of the 1930s knew that adrenal steroids were important, but it took clinical investigations to show that steroids had incredible potential as therapeutic agents.

The first evidence to suggest that an adrenocortical hormone can exert immunosuppressive effects was put forth by Hans Selye, who noted that the exposure of rats to a variety of stress-inducing substances and environments elicited both hyperplasia (increased cell number) of the adrenal cortex and the regression of the *thymus gland*, a key component of the immune system.[30] At about the same time, Philip Hench, a rheumatologist at the Mayo Clinic, noted the symptomatic remission of *chronic rheumatoid arthritis*, a profoundly debilitating autoimmune disorder, in patients who had liver failure caused by hepatitis and jaundice.[31] Hench postulated that the effect was due to an unknown innate "substance X",[32] and in collaboration with Kendall concluded that substance X was in fact "compound E", cortisone. We now know that one critical liver function is to degrade steroid hormones, and that liver failure can therefore indirectly yield elevated blood cortisone. Hench obtained several grams of cortisone and began treating a 29-year-old, wheelchair-bound arthritic woman. Incredibly, after receiving 100 mg daily for four days, she was able to walk out of the hospital to enjoy a three-hour shopping spree.[19] Hench treated another 15 patients with similar spectacular results, this time filming them before and after treatment. Fifty years after watching the movie, one audience member still emotionally recalled the event, "It was like God had touched them".[19] The results were announced in 1949,[33] the world lauded the "miracle cure", and Kendall, Hench, and Reichstein received the *Nobel Prize in Physiology or Medicine* for work with adrenal steroids in 1950.

Gonadal Steroid Hormone Synthesis

In both the testes and ovaries, progesterone is processed into *androstenedione* (Figure 4.16), the precursor to *testosterone*, which is the principal androgen of vertebrates. One of two critical enzymes next acts on testosterone, each producing very different results from the other. The first enzyme, **5α-reductase**, converts testosterone into *dihydrotestoterone* (DHT), a critical enzyme in the development of male genitalia that also plays a prominent role in the development of male pattern baldness. The second enzyme, **CYP19 aromatase** ($P450_{aro}$), converts testosterone into *estradiol*, a C_{18} estrogen. Indeed, the presence or absence of these enzymes during key stages of fetal and adult ontogeny greatly influences the development of diverse primary and secondary sexual characteristics among vertebrates.

Curiously Enough . . . The first androgen to be discovered, androsterone, was purified from 15,000 liters of policemen's urine, and the first estrogen, estrone, was isolated from hundreds of liters of urine from pregnant women.

The names of some major vertebrate steroidogenic enzymes and some of their pharmacological inhibitors are listed in Table 4.4. Circulating steroid hormones are

Table 4.4 Some Cholesterol Synthesis and Steroidogenic Enzymes and Their Inhibitors

Common Name	"Old" Name	Current Name	Subcellular Location	Reaction Catalyzed	Pharmacological Inhibitor
HMG-CoA reductase	HMGCR	HMGCR	Endoplasmic reticulum	HMG-CoA→Mevalonate cholesterol synthesis	Statins
Side-chain cleavage enzyme; desmolase	P450SCC	CYP11A1	Mitochondria	Cholesterol→Pregnenalone	Methoxychlor (pesticide)
3-hydroxysteroid dehydrogenase (3-HSD)	3-HSD	3-HSD	Endoplasmic reticulum	Pregnenalone→Progesterone	Cyanoketone
11-hydroxylase	P450C11	CYP11B1	Mitochondria	11-deoxycortisol→Cortisol	Metyrapone
Aldosterone synthetase	P450C11AS	CYP11B2	Mitochondria	Corticosterone→Aldosterone	LCI699
5-reductase	5-reductase	5-reductase	Nuclear membrane	Testosterone→ Dihydrotestoterone (DHT)	Finasteride (Propecia)
Aromatase	P450aro	CYP19	Endoplasmic reticulum	Testosterone→Estradiol	Exemestane (Aromasin)

inactivated mainly by the liver and kidney, where they are hydroxylated and then conjugated to glucuronides or sulfate to increase their water solubility. Degraded steroid hormones are excreted primarily in the urine.

Developments & Directions: Steroids from soybeans: Percy Julian and the industrial synthesis of steroid hormones

Percy Julian (1899–1975) (Figure 4.17) was one of the great scientists of the 20th century. In a career in chemistry spanning four decades, he made many valuable discoveries, for which he was awarded 130 patents, 18 honorary degrees, and membership to the prestigious National Academy of Sciences—only the second African American bestowed such an honor at the time.

Figure 4.17 Percy Julian in the Minshall Laboratory at DePauw University during his tenure as a research fellow.

Source of images: Figure provided courtesy of the Oak Park River Forest Museum.

Julian is perhaps best known for his synthesis of the vertebrate steroid hormone, cortisone, from soybean sterols. Cortisone, often referred to as one of science's "miracle drugs", is effective in the treatment of rheumatoid arthritis and other inflammatory diseases. Although it is now widely available, in the 1940s the only way to obtain cortisone was to extract it from the bile of slaughtered oxen, and the treatment of a single patient for one year required the bile from nearly 15,000 animals. Because of the limited supply, the cost of cortisone was very high and out of reach of the majority of patients. In 1939 Julian developed a biochemical procedure to isolate *stigmasterol* from soybean oil and convert it into cortisone. For the first time this steroid hormone could be produced synthetically in large quantities, significantly reducing the price of cortisone. By 1940 he was able to produce in bulk the hormone progesterone, allowing the efficient production of other synthetic steroid hormones, such as testosterone. Julian was inducted into the National Inventors Hall of Fame in 1990 for his "Preparation of Cortisone".

Percy Julian was a remarkable chemist, all the more so considering that he grew up in a time when racial tensions against African Americans ran exceptionally high. The grandson of Alabama slaves, Percy Julian met with every possible barrier during his successful career in a deeply segregated America, including numerous acts of violence against himself and his family. "I have had one goal in my life", Julian was quoted as saying, "that of playing some role in making life a little easier for the persons who come after me" (*Ebony*, March 1975, pp. 94–96). His spirit lives on in dozens of lifesaving discoveries, as well as in the halls of Percy L. Julian Junior High School in Oak Park, Illinois, which, in 1985, was renamed in honor of the community's most famous native son.

Eicosanoids

Discovery of Eicosanoids

The **eicosanoids** are a family of hydrophobic hormones derived from the C20 fatty acid arachidonic acid (Figure 4.18). Although they are diverse in structure, many

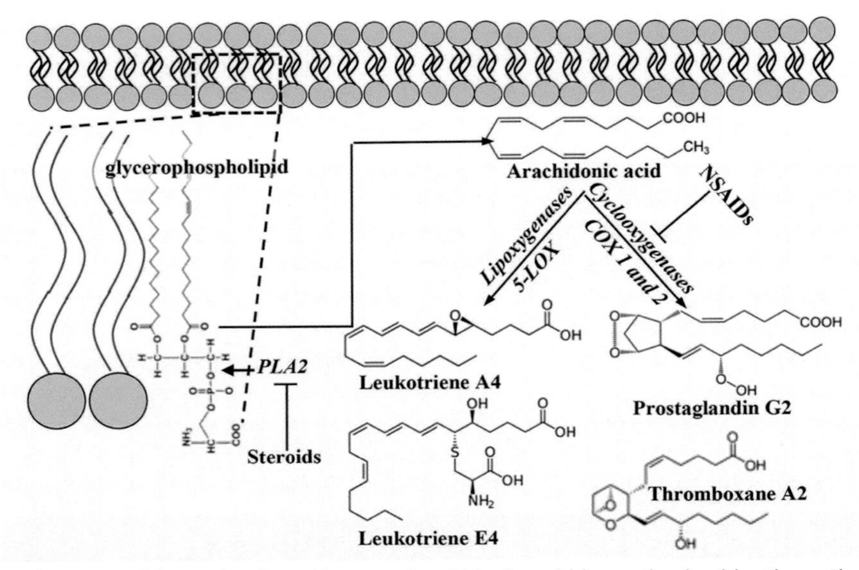

Figure 4.18 Eicosanoid hormone biosynthesis pathways. Arachidonic acid is synthesized by the action of cytosolic phospholipase A2 (PLA2) on smooth ER and plasma membrane phospholipids. Lipoxygenases (LOX) convert arachidonic acid into leukotrienes, and cyclooxygenases (COX) use arachidonic acid to generate prostaglandins, thromboxanes, and prostacyclins. Whereas anti-inflammatory steroids inhibit the synthesis of all eicosanoids by inhibiting PLA2 activity, non-steroidal anti-inflammatory drugs (NSAIDS) specifically inhibit the synthesis of eicosanid products generated by COX enzymes.

Table 4.5 Some Eicosanoids and Their Functions

Physiological Function	Eicosanoids Involved
Vasodilation	PGE_2, PGF, PGI_2, TXA_1
Vasoconstriction	TXA_2
Promote platelet aggregation	TXA_2
Inhibit platelet aggregation	PGI_2, TXA_1
Bronchodilation	PGE_2, PGI_2, LTC_4
Bronchoconstriction	PGF_2, TXA_2, PGD_2
Protect gastric mucosa (↓ gastric acid secretion, ↑ gastric mucus secretion)	PGE_2
Leukocyte recruitment	LTB_4, LTC_4
Contract uterine smooth muscle	PGE_2, PGF_2
Relax uterine smooth muscle	PGI_2
Increase renal blood flow	PGE_2, PGI_2
Sleep/wake cycle regulation	PGD_2

Note: PG (prostaglandin), TX (thromboxane), LT (leukotriene).

eicosanoids have roles in inflammation and smooth muscle contraction, including the regulation of vasodilation, vascular permeability, pain, and recruitment of leukocytes. Many important aspects of immunity, such as cytokine production, antibody formation, cell differentiation, cell proliferation, migration and antigen presentation, are regulated by eicosanoids. In contrast with classical hormones, eicosanoids are typically not released into circulation and are generally considered to function as short-lived paracrine signaling agents. Importantly, synthesis and release of this class of signaling molecules is often coupled to classical endocrine signaling pathways, and these molecules are also known to stimulate the synthesis of classical hormones such as testosterone and corticosteroids.

The history of *prostaglandins*, the first class of eicosanoids to be discovered, began in 1930 when two American gynecologists, Kurzrok and Lieb, found that human semen caused the myometrium (smooth muscle) of the uterine wall to contract and relax.[34] These findings were extended when prostaglandins were isolated from the lipid-soluble fraction of seminal fluid in 1935.[35,36] Prostaglandins were so named because it was originally thought they were produced by the prostate gland, though it is now known that they are in fact produced by the seminal vesicles. There are at least 16 different prostaglandins in nine different chemical classes, designated PGA–PGI, and these molecules are produced in diverse tissues in both males and females. The names of other categories of eicosanoids are also derived from their tissue of synthesis: for example, *thromboxanes* are derived from platelets (thrombocytes), and *leukotrienes* from leukocytes. Eicosanoids have diverse functions, such as promoting uterine smooth muscle contraction and parturition during childbirth, blood clotting, induction of inflammation and fever, as well as vasodilation, vasoconstriction, and bone remodeling (summarized in Table 4.5). In 1982, biochemists Sune K. Bergström, Bengt I. Samuelsson, and John R. Vane jointly received the *Nobel Prize in Physiology or Medicine* for their research on "prostaglandins and related biologically active substances".[37]

Table 4.6 Relative Expression Levels of COX Isozymes in Selected Normal Human Organs and Tissues

COX 1 > COX 2	COX 2 > COX 1
Brain glia	Brain neurons
Stomach	Thyroid
Small intestine	Lung
Colon	Spleen
Heart	Adipose tissue

Source: Data assimilated from Zidar et al., 2009

Eicosanoid Biosynthesis

The first step in eicosanoid synthesis generally involves the cleavage of arachidonic acid from phospholipids in the smooth endoplasmic reticulum and plasma membranes by cytosolic phospholipase A2 (PLA_2) (Figure 4.18). Anti-inflammatory steroids, such as glucocorticoids, are known to inhibit PLA_2 and thus suppress the synthesis of inflammatory prostaglandins and leukotrienes. Following its synthesis, arachidonic acid is then modified by one of two enzyme classes: **lipoxygenases** (LOX) convert the substrate to leukotriene B_4, and **cyclooxygenases** (COX 1 and 2) convert arachidonic acid to prostaglandin H2, products that can be further processed into additional varieties of leukotrienes and prostaglandins.

Importantly, in humans the two COX isoforms (COX 1 and 2) are normally expressed differentially among many tissues[38] (Table 4.6). For example, COX 1 is present in the digestive system at higher levels than COX 2, and therefore COX 1 is primarily responsible for producing prostaglandins involved in protecting the stomach lining from the acidic environment (e.g. prostaglandin E2 both inhibits gastric acid secretion and enhances gastric mucus secretion). In contrast, COX 2 expression is higher than that of COX 1 in tissues such as the lung and brain neurons. In some organs, such as the liver, both isoforms are expressed at equal levels.[38] Notably, aspirin-like drugs (known as **NSAIDs**, an acronym for "**non-steroidal anti-inflammatory drugs**") suppress prostaglandin synthesis by inhibiting both COX 1 and 2, and are hence potent inhibitors of the inflammatory response. However, because of their dual COX suppression characteristics, aspirin-like drugs can also produce gastrointestinal problems such as bleeding and ulcers. As such, new classes of COX 2-selective NSAIDs, such as celecoxib (marketed under the brand name Celebrex), are now marketed as ways to inhibit inflammation without adverse gastrointestinal effects.

SUMMARY AND SYNTHESIS QUESTIONS

1. Like steroid hormones, cholecalciferol (vitamin D3) is derived from cholesterol and binds to receptors that are members of the nuclear receptor superfamily. However, cholecalciferol is technically not a steroid hormone. Why not?

2. Congenital lipoid adrenal hyperplasia is a disease characterized in part by the presence of abnormally large and abundant lipid droplets in adrenal steroidogenic cells. A *deficiency* in which of the following would be the most likely cause of this condition? Explain your reasoning.

   a. LDL
   b. LDL receptor
   c. lysosome enzyme
   d. HMG CoA reductase
   e. Steroidogenic acute regulatory protein
   f. Cycloxygenase
   g. Lipoxygenase

3. Compare and contrast the ultrastructures (cell morphology as viewed by transmission electron microscopy) of a steroid hormone versus a peptide hormone producing cell. What organelles are most developed, and why?
4. Less enlightened students who have not taken an endocrinology course often make the mistake of considering testosterone to be a "male" hormone and estrogen a "female" hormone. Provide examples that demonstrate how this is incorrect.
5. In Chapter 1: The Scope and Growth of Endocrinology, you learned how organotherapy (the practice of eating a specific organ from an animal for medicinal purposes in order to treat a disease or deficiency in a person's corresponding malfunctioning organ) in the form of eating thyroid glands from sheep was an effective treatment for hypothyroidism. In the 19th century, Brown-Séquard made the extraordinary (and highly doubtful) claim that he had reversed some of the effects of aging by injecting himself over a 3-week period with extracts derived from the testicles of guinea pigs and dogs to replace his declining levels of the hormone we now know to be testosterone. Whereas the use of thyroid organotherapy does indeed reverse the effects of hypothyroidism, the organotherapeutic consumption of testicles (whole or in extract) cannot successfully augment circulating levels of testosterone. Why not?
6. Contrast the mechanisms by which glucocorticoid and NSAID drugs function as anti-inflammatory agents.
7. Cyclooxygenases (COX) 1 and 2 catalyze the same reactions, converting arachidonic acid into prostaglandins. The fact that these two enzymes mediate identical reactions may seem biologically redundant, but this redundancy has been exploited by pharmacologists to treat diseases. Explain how this is true.
8. Suppose that you are a doctor treating a patient who suffers from both chronic pain and gastric ulcers. Which of the following anti-inflammatory pain drugs, aspirin, ibuprofen, or Celebrex, would be most effective for this patient, and why?
9. Name two attributes that eicosanoids and endocannabinoids share in common.

Summary of Chapter Learning Objectives and Key Concepts

LEARNING OBJECTIVE Describe the different methods by which hormones are categorized.

- Hormones can be sorted into three broad categories, based upon their chemical structures: (1) lipids, (2) polypeptides, and (3) modified amino acids.
- Hormones can be divided into three groups based on their solubility and receptor location: (1) hydrophilic hormones with cell-surface receptors, (2) hydrophobic hormones with intracellular receptors, and (3) hydrophobic hormones with cell-surface receptors.

LEARNING OBJECTIVE Identify how different chemical classes of hormones are transported by the blood.

- Most hydrophobic and some hydrophilic hormones are transported in the blood by transport proteins.
- Transport proteins extend the half-life of a hormone in the blood.
- The free fraction of hormone (non-transport protein-bound) in the blood is considered the most biologically active fraction.

LEARNING OBJECTIVE Describe how hormones are inactivated and excreted.

- Hormone inactivation is necessary to ensure steady-state levels of plasma hormones.
- The liver and kidneys are the primary organs involved in the inactivation and excretion of hormones.
- The inactivation of hydrophobic hormones typically requires their conversion into more hydrophilic compounds to ensure their solubility at high concentrations in biological fluids prior to their excretion in the urine.

LEARNING OBJECTIVE List the amine-derived hormones, their precursor substrates, and the organs that synthesize them.

- All amine-derived hormones are modifications of the aromatic amino acids tyrosine or tryptophan.
- The primary site of circulating amine-derived hormone inactivation is the liver.

LEARNING OBJECTIVE Describe how polypeptide hormones are categorized and synthesized.

- Polypeptides are characterized by unique primary sequences of amino acids that are encoded by the genome.
- Families of genes coding for closely related polypeptides within a species (paralogs) arose by gene duplication followed by gene mutation.
- Virtually all mammalian polypeptide hormones and receptors have orthologs (same gene in a different species) among the more ancestral vertebrates.
- Functional variability in polypeptide hormone function is produced via alternative splicing and alternative post-translational processing.
- Receptors for all hormones, irrespective of chemical class, are polypeptides.

LEARNING OBJECTIVE Provide an overview for the biosynthesis of the two most common categories of lipid hormones (steroid hormones and eicosanoids) and describe their most important functions.

- Lipid hormones derive primarily from one of two substrates: cholesterol or arachidonic acid.
- Cholesterol-derived hormones can be divided into seven categories defined broadly by their physiological functions: glucocorticoids, mineralocorticoids, androgens, estrogens, progestogens, the vitamin D family, and in arthropods ecdysteroids.
- The enzymes that mediate steroidogenesis are located primarily in the mitochondria and smooth ER.
- In contrast with steroids, eicosanoids are generally not released into circulation and function as paracrine signaling agents.
- Aspirin-like drugs suppress inflammation by inhibiting prostaglandin synthesis.

LITERATURE CITED

1. Berridge MJ. *Cell Signalling Biology*. Portland Press; 2007. doi:10.1042/csb0001001.
2. Gower D. *Steroid Hormones*. Year Book Medical Pub; 1979.
3. Adlercreutz H, Martin F, Järvenpää P, Fotsis T. Steroid absorption and enterohepatic recycling. *Contraception*. 1979;20(3):201–223.
4. Azezli AD, Bayraktaroglu T, Orhan Y. The use of konjac glucomannan to lower serum thyroid hormones in hyperthyroidism. *J Am Coll Nutr*. 2007;26(6):663–668.
5. Clements M, Chalmers T, Fraser D. Enterohepatic circulation of vitamin D: a reappraisal of the hypothesis. *Lancet*. 1984;323(8391):1376–1379.
6. Steinberg SE, Campbell CL, Hillman RS. Kinetics of the normal folate enterohepatic cycle. *J Clin Investig*. 1979;64(1):83–88.
7. Malik MY, Jaiswal S, Sharma A, Shukla M, Lal J. Role of enterohepatic recirculation in drug disposition: cooperation and complications. *Drug Metab Rev*. 2016;48(2):281–327.
8. Roberts MS, Magnusson BM, Burczynski FJ, Weiss M. Enterohepatic circulation. *Clin Pharmacokinet*. 2002;41(10):751–790.
9. Tahimic CG, Wang Y, Bikle DD. Anabolic effects of IGF-1 signaling on the skeleton. *Front Endocrinol (Lausanne)*. 2013;4:6.
10. Brown GA, Vukovich M, King DS. Testosterone prohormone supplements. *Med Sci Sports Exerc*. 2006;38(8):1451–1461.
11. Galton VA. The ups and downs of the thyroxine pro-hormone hypothesis. *Mol Cell Endocrinol*. 2017;458:105–111.
12. Fukuda MN, Sasaki H, Lopez L, Fukuda M. Survival of recombinant erythropoietin in the circulation: the role of carbohydrates. *Blood*. 1989;73(1):84–89.
13. Sanders RD, Spencer JB, Epstein MP, et al. Biomarkers of ovarian function in girls and women with classic galactosemia. *Fertil Steril*. 2009;92(1):344–351.
14. Mojsov S, Heinrich G, Wilson IB, Ravazzola M, Orci L, Habener JF. Preproglucagon gene expression in pancreas and intestine diversifies at the level of post-translational processing. *J Biol Chem*. 1986;261(25):11880–11889.
15. Bandsma R, Sokollik C, Chami R, et al. From diarrhea to obesity in prohormone convertase 1/3 deficiency—Age dependent clinical, pathological and enteroendocrine characteristics. *J Clin Gastroenterol*. 2013;47(10).
16. Lambert JJ, Belelli D, Peden DR, Vardy AW, Peters JA. Neurosteroid modulation of GABAA receptors. *Prog Neurobiol*. 2003;71(1):67–80.
17. Akk G, Covey DF, Evers AS, Steinbach JH, Zorumski CF, Mennerick S. The influence of the membrane on neurosteroid actions at GABA(A) receptors. *Psychoneuroendocrinol*. 2009;34(Suppl 1):S59–S66.
18. Reddy DS. Neurosteroids: endogenous role in the human brain and therapeutic potentials. *Prog Brain Res*. 2010;186:113–137.
19. Rooke T. *The Quest for Cortisone*. Michigan State University Press; 2012.
20. Bai C, Schmidt A, Freedman LP. Steroid hormone receptors and drug discovery: therapeutic opportunities and assay designs. *Assay Drug Dev Technol*. 2003;1(6):843–852.
21. Wu MV, Manoli DS, Fraser EJ, et al. Estrogen masculinizes neural pathways and sex-specific behaviors. *Cell*. 2009;139(1):61–72.
22. Colombo L, Dalla Valle L, Fiore C, Armanini D, Belvedere P. Aldosterone and the conquest of land. *J Endocrinol Invest*. 2006;29(4):373–379.
23. Kendall EC, Mason HL, McKenzie BF, Myers CS, Koelsche GA. Isolation in crystalline form of the hormone essential to life from the supranetal cortex: its chemical nature and physiologic properties. *Trans Assoc Am Physicians*. 1934;48:147–152.
24. Ingle DJ. Biographical memoir of Edward C. Kendall. *Nat Acad Sci*. 1975;47.
25. Wintersteiner O, Pfiffner JJ. Chemical studies of the adrenal cortex. III. Isolation of two new physiologically inactive compounds. *J Biol Chem*. 1936;116:291–305.
26. Mason HL, Meyers CS, Kendall EC. Chemical studies of the suprarenal cortex. II. The identification of a substance which possesses the qualitative action of cortin; its conversion into a diketone closely related to androstenedione. *J Biol Chem*. 1936;116:267–276.
27. Reichstein T. "Adrenosteron". Über die Bestandteile der Nebennierenrinde II (vorläufige Mitteilung). *Helv Chim Acta*. 1936;19:223–225.
28. Reichstein T. Über die Bestandteile der Nebennierenrinde IV. *Helv Chim Acta*. 1936;19:402–412.
29. Steiger M, Reichstein T. Desoxy-corticosteron (21-oxyprogesterone) aus Δ5–3 -xoy-atio-cholensaure. *Helv Chim Acta*. 1937;20:1164–1179.
30. Selye H. A syndrome produced by diverse nocuous agents. *Nature*. 1936:32.
31. Hench PS. Analgesia accompanying hepatitis and jaundice in cases of chronic arthritis, fibrositis and sciatic pain. *Proc Staff Meet Mayo Clin*. 1933;8:430–437.
32. Hench PS. Effect of spontaneous jaundice on rheumatoid arthritis. Attempts to reproduce the phenomenon. *Brit Med J*. 1938;20:394–398.
33. Hench PS, Kendall EC, Slocumb CH, Polley HF. The effect of a hormone of the adrenal cortex (17-hydroxy-11-dehydrocorticosterone [compound E]) and of the pituitary adrenocorticotropic hormone on rheumatoid arthritis. *Proc Staff Meet Mayo Clin*. 1949;24:181–197.
34. Kurzrok R, Lieb C. Biochemical studies of human semen: II. The action of semen on the human uterus. *Proc Soc Exp Biol Med*. 1930;28:268–272.
35. Euler USv. Über die Spezifische Blutdrucksenkende Substanz des Menschlichen Prostata- und Samenblasensekretes. *Klinische Wochenschrift*. 1935;14(33):1182–1183.
36. Goldblatt MW. Properties of human seminal plasma. *J Physiol*. 1935;84(2):208–218.
37. Oates JA. The 1982 Nobel Prize in Physiology or Medicine. *Science*. 1982;218(4574):765–768.
38. Zidar N, Odar K, Glavac D, Jerse M, Zupanc T, Stajer D. Cyclooxygenase in normal human tissues—is COX-1 really a constitutive isoform, and COX-2 an inducible isoform? *J Cell Mol Med*. 2009;13(9B):3753–3763.

CHAPTER

5

Receptors

CHAPTER LEARNING OBJECTIVES:

- Describe the different effectors and signal transduction pathways associated with each major plasma membrane receptor class: Ionotropic receptors
- Describe the different effectors and signal transduction pathways associated with each major plasma membrane receptor class: G protein-coupled receptors (GPCRs)
- Describe the different effectors and signal transduction pathways associated with each major plasma membrane receptor class: Receptors with intrinsic enzymatic activity (receptor tyrosine kinases)
- Describe the different effectors and signal transduction pathways associated with each major plasma membrane receptor class: Receptors with enzyme-associated activity (recruiter receptors/cytokine receptors)
- Address the following aspects of nuclear receptors: Describe some of the earliest experiments that hinted at the role of nuclear receptors in regulating gene expression.
- Address the following aspects of nuclear receptors: Describe the mechanisms of action of the three classes of nuclear receptors.
- Address the following aspects of nuclear receptors: Describe structural and conformational changes associated with nuclear receptor-ligand interaction.
- Address the following aspects of nuclear receptors: Describe epigenetic modifications of chromatin by nuclear receptors.

OPENING QUOTATIONS:

"*Corpora non agunt nisi fixate*" (translation from Latin: "bodies do not act if they are not bound [to receptors]").[1]

—Paul Ehrlich, in proposing the existence of "chemoreceptors" for drugs

"[H]ormones originally thought to act monotheistically actually are pleiotropic agents; i.e., they can do many different things by separate routes. Certainly in my mind, endocrinology was no longer just a science; it was imbued with existentialism!"[2]

—Martin Rodbell, in autobiographical comments associated with his 1994 Nobel Prize

Introduction: Discovery and Importance of Receptors

In the early 1900s, while investigating the actions of the toxins nicotine and curare on nerve function, the physiologist John Newport Langley postulated the existence of "receptive substances" capable of "receiving and transmitting stimuli" in response to an interaction with an "excitable substance".[3] Almost simultaneously and independently, the immunologist Paul Ehrlich proposed the existence of "chemoreceptors" for drugs; his research on this topic contributed to his sharing the *Nobel Prize in Physiology or Medicine* in 1908 with Ilya Ilyich Mechnikov "in recognition of their work on immunity".

Sixty years after Paul Ehrlich's proposal for the existence of chemoreceptors for drugs, Pastan and colleagues[4] demonstrated that the biological effects of thyroid-stimulating hormone (TSH) on cultured thyroid slices, as well as the effects of insulin on cultured muscle tissue, were reversed after the tissues were rinsed in the presence of antibodies against the hormones, which inactivated both the hormone and any protein it was bound to. Since neither antibodies nor large, water-soluble peptide hormones like TSH and insulin can cross the plasma membrane, the researchers concluded, "the initial interaction of polypeptide hormones with target tissue is a rapid firm binding to a superficial cell site, presumably on the external cell membrane".

These initial observations led to the modern concept of *cell-surface receptors*: proteins localized to the plasma membrane that recognize and bind to specific chemical compounds such as hormones, neurotransmitters, toxins, and drugs, and subsequently initiate intracellular signaling pathways culminating in biological responses. Indeed, plasma membrane-localized receptors are the most abundant type of receptor, with 1,352 human genes currently identified. These plasma membrane receptors constitute an estimated 6% of the entire human *proteome*, the total number of proteins derived from the human genome.[5] Approximately 20% of these receptors bind to hormones, growth factors, cytokines, and other substances of endocrine/paracrine origin (Table 5.1). A second category of cell

DOI: 10.1201/9781003359241-7

Table 5.1 Categories and Numbers of Endocrine/Paracrine Cell-Surface Receptors in the Human Genome with Both Known and Unknown Functions

Category	Number
Total number of receptors	**~284**
GPCRs with endogenous ligands	133[#]
Receptor-type kinases	72[*]
Cytokines	51[*]
Tumor necrosis and nerve growth factors (TNF/NGF)	28[*]

Sources: [#]After Vassilatis, D.K., Hohmann, J.G., Zeng, H., Li, F., Ranchalis, J.E., Mortrud, M.T., Brown, A., Rodriguez, S.S., Weller, J.R., Wright, A.C. and Bergmann, J.E., 2003. The G protein-coupled receptor repertoires of human and mouse. *Proceedings of the National Academy of Sciences*, *100*(8), pp. 4903–4908. [*]After Almén, M.S., Nordström, K.J., Fredriksson, R. and Schiöth, H.B., 2009. Mapping the human membrane proteome: a majority of the human membrane proteins can be classified according to function and evolutionary origin. *BMC biology*, *7*(1), pp. 1–14.

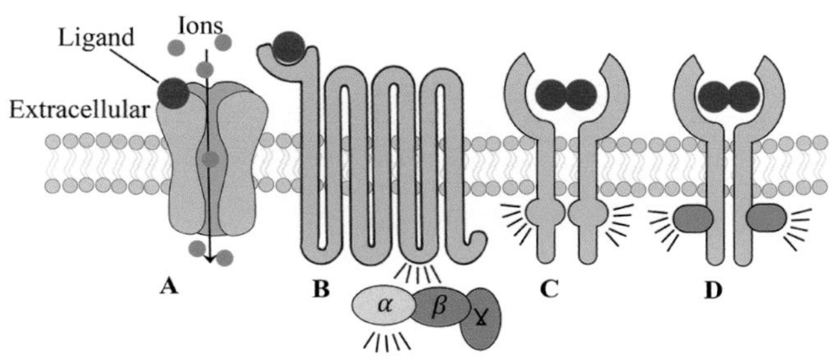

Figure 5.1 Plasma membrane receptors that bind hormones can be divided into four broad categories. (A) ionotropic receptors are ligand-activated ion channels. **(B)** G protein-coupled receptors (GPCRs), also known as 7-transmembrane (7TM) receptors, initiate intracellular signaling cascades mediated by intracellular proteins with GTPase activity ("G proteins"). **(C)** Receptors with intrinsic enzyme activity, where the receptor itself is a ligand-activated enzyme. **(D)** Enzyme-associated receptors, where the receptor itself is not an enzyme, but upon binding a ligand recruits other proteins that are enzymes.

receptors, **nuclear receptors**, are used by lipid-soluble hormones that cross the plasma membrane. A comparatively smaller number of these receptors are present in humans, with only about 48 genes identified.

Despite diverse modes of chemical signal action, a genome-wide analysis has unified the large number of plasma membrane receptors into several distinct families (Figure 5.1): (1) **ionotropic receptors**, which function as ligand-activated ion channels, (2) **G protein-coupled receptors** (GPCRs), which initiate an intracellular signaling cascade mediated by proteins possessing GTPase activity, (3) *receptors with intrinsic enzyme activity*, where the receptor itself is a ligand-activated enzyme, and (4) *enzyme-associated receptors* (recruiter receptors), where the receptor itself is not an enzyme but recruits other proteins that are enzymes. From an evolutionary perspective ionotropic receptors are the most ancient, appearing in archaea and bacteria, as well as in plants[6] (Figure 5.2). GPCRs arose in eukaryotes and are absent in bacteria and archaea.[7] Receptors with intrinsic enzyme activity and enzyme-associated receptors appeared with the metazoans. Some enzyme-associated receptors, such as cytokine type 1 (mediates immune responses) are present only among the chordates, whereas cytokine type 2 receptors (T cell immune receptors) are unique to the vertebrates.

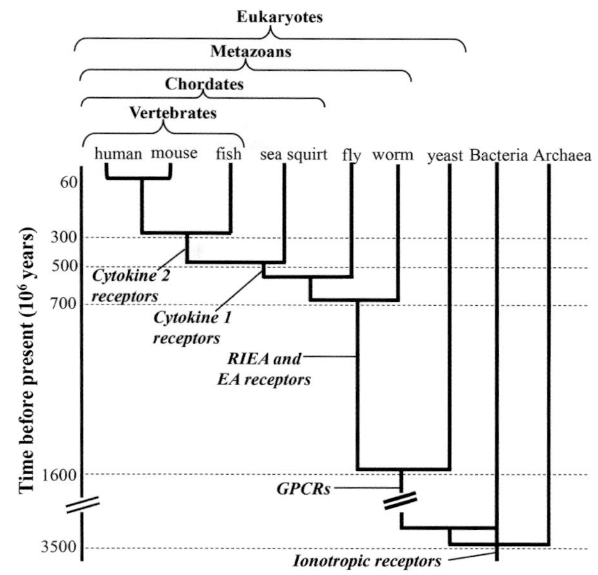

Figure 5.2 Evolutionary origins of the plasma membrane signaling receptor families. Receptor families are listed in the order of their presumed appearance during evolution, based upon living organisms in which the entire genomes have been sequenced. Ionotropic receptors are the most ancient, appearing in archaea and bacteria. G protein-coupled receptors (GPCRs) arose in eukaryotes. Receptors with intrinsic enzyme activity (RIEA) and enzyme-associated receptors (EA) are unique to the metazoans. Some EA receptors, such as cytokine type 1 (which mediate immune responses) are present only among the chordates, whereas cytokine type 2 receptors (T cell immune receptors) are unique to the vertebrates.

Plasma Membrane Receptor Categories and Mechanisms of Action

Ionotropic Receptors

LEARNING OBJECTIVE Describe the different effectors and signal transduction pathways associated with each major plasma membrane receptor class: Ionotropic receptors.

KEY CONCEPTS:

- Ionotropic receptors are ligand-gated ion channels that help modulate a cell's transmembrane potential (voltage).
- Nicotinic cholinergic receptors are acetylcholine-gated cation (Na^+ and K^+) channels.
- GABA receptors are ligand-gated chloride (Cl^-) channels.

The simplest class of plasma membrane receptors are **ionotropic receptors**, multi-pass transmembrane proteins that function as ligand-gated ion channels that alter ion permeability and transmembrane potential. These receptors do not associate with effectors to generate second messengers. The best-characterized receptor of this class is the *nicotinic cholinergic receptor*, which is an acetylcholine-gated cation (Na^+ and K^+) channel. Although ionotropic receptors are best known for rapid chemical synaptic neurotransmission between electrically excitable cells (such as neurons and muscle), some also have roles as modulators of endocrine, paracrine, and autocrine actions. Examples of non-neuronal tissues that use acetylcholine and its receptors for paracrine signaling include bronchial epithelial cells,[8] pancreatic beta cells,[9] the pituitary gland,[10] and vascular endothelial cells.[11] Receptors for the neurotransmitter, *GABA* (gamma-aminobutyric acid), the chief inhibitory neurotransmitter in the vertebrate central nervous system, function as ligand-gated chloride channels. Interestingly, the GABA receptor is also modulated by *neurosteroids*, paracrine signals that are released locally from neurons or glia, which can enhance synaptic inhibition.

SUMMARY AND SYNTHESIS QUESTIONS

1. Considering the receptors that they use, do neurosteroids function as stimulants or depressants?

GPCRs/7TM Receptors

LEARNING OBJECTIVE Describe the different effectors and signal transduction pathways associated with each major plasma membrane receptor class: GPCRs.

KEY CONCEPTS:

- GPCRs constitute the largest superfamily of receptors and are the targets of over half of all current pharmaceutical drugs.
- All GPCRs share seven transmembrane (7TM)-spanning helices.
- Most GPCRs transmit signals that interact with intermediary guanine nucleotide-binding proteins called G proteins.
- G proteins modulate effector enzymes that amplify signals and produce second messenger molecules.
- Although hydrophilic hormones are the primary ligands of GPCRs, some lipid-soluble hormones (e.g. eicosanoids and steroids) can also bind to plasma membrane-localized GPCRs.

The **G protein-coupled receptors** (GPCRs) are not only among the most ancient of plasma membrane receptors, first appearing in unicellular eukaryotic organisms over 1.6 billion years ago, but with over 900 members they also constitute the single largest superfamily of receptors, representing 2% of the human proteome.[12,13] With more receptors than there are known ligands that bind them, this "superfamily" is divided into several families based on their structures.[14] The primary functions of GPCRs are to transduce diverse extracellular stimuli including light, odorants, nucleotides, lipids, steroids, modified amino acids, peptides, and glycoprotein hormones into intracellular signals. Malfunctions of GPCR signaling pathways are associated with many diseases, such as diabetes, blindness, allergies, depression, cardiovascular defects, and certain forms of cancer (refer to Appendix 5 for specific examples).

Curiously Enough . . . Because GPCRs are ubiquitous and involved in virtually all aspects of physiology, they are also the most common target of therapeutic drugs, representing an astounding 50%–60% of all current pharmaceutical targets.[15]

There are two important features of GPCRs that help define them structurally and functionally as a group:

1. All members of the GPCR superfamily share seven membrane-spanning helices (Figure 5.3). As such, GPCRs are also commonly called "seven-transmembrane (*7TM*)" receptors. Despite the structural similarity among GPCR families, the external ligand-binding domains and modes of action vary considerably, facilitating interactions with a particularly diverse array of ligands.
2. Most GPCRs transmit signals to intracellular targets via the intermediary action of guanine nucleotide-binding proteins called *G proteins* (Figure 5.4). In the late 1960s and early 1970s, Martin Rodbell and colleagues showed that signal transduction by the plasma membrane in response to the hormone glucagon required not only the presence of a plasma membrane receptor, but also a separate intracellular "transducer" protein that uses GTP as an energy source to relay the receptor's signal.[16] Alfred G. Gilman and colleagues later characterized the nature of the transducer, discovering and purifying "G proteins".[17] For their discovery of G proteins and the role of these proteins in signal transduction in cells, the *Nobel Prize in Physiology or Medicine* was awarded jointly to Alfred G. Gilman and Martin Rodbell in 1994.

G Proteins

Most GPCRs interact with a 34-member family of "heterotrimeric" G proteins, so named because in their inactive state they form a complex of three distinct interacting subunits (α, β, and γ) (Figure 5.4). As their name implies, G proteins bind to guanine nucleotides, specifically to guanosine triphosphate (GTP) and guanosine diphosphate (GDP). G proteins function as "molecular

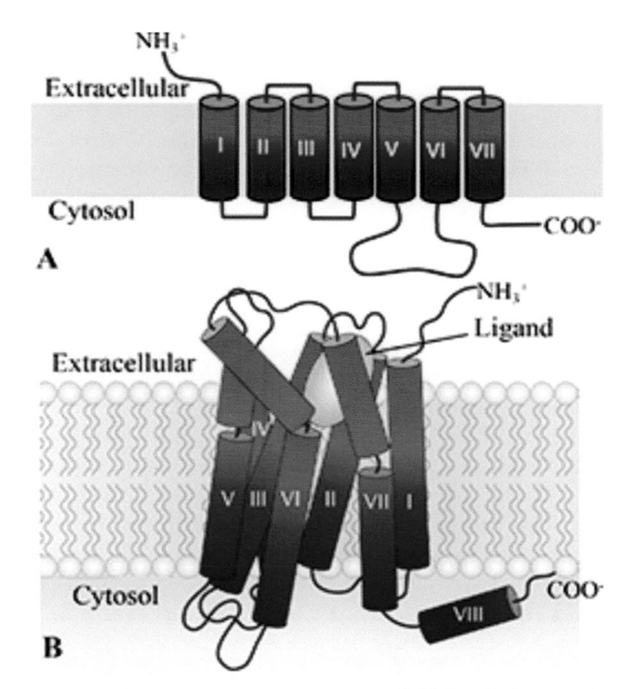

Figure 5.3 The general structure of a GPCR. (A) All members of the GPCR superfamily share seven membrane-spanning helices, depicted in this hypothetical linear arrangement. **(B)** The physiologically active three-dimensional arrangement resembles a barrel shape, with the extracellular "pocket" forming the ligand-binding domain. Compared with the intracellular domain that is relatively conserved among different GPCRs, the extracellular domain is highly variable, reflecting the diversity of different ligands that GPCRs bind.

Source of images: (B) George, S.R., et al. 2002. Nat. Rev. Drug Discov., 1(10), pp. 808-820; used with permission.

switches"—when they bind GTP, they are "on", and when they bind GDP, they are "off". G proteins belong to a larger group of enzymes called "GTPases", which are proteins that hydrolyze GTP to GDP. As such, the "on" and "off" states of G proteins are regulated by factors that control their ability to bind to and hydrolyze GTP to GDP.

Robert Lefkowitz and colleagues made a seminal contribution to the study of GPCRs when they cloned and sequenced the first GPCR receptor for epinephrine, the β-adrenergic receptor (βAR).[18] Brian Kobilka and his colleagues first revealed the three-dimensional structure of a fully functional βAR in complex with both **agonist** and G protein.[19] These key discoveries have allowed researchers to identify the precise regions of GPCRs that facilitate interactions with their ligands and associated G proteins. Such knowledge explains the basis of pathologies where specific interactions are thwarted and permits the development of drugs that may restore normal signaling. Based on this work, the *Nobel Prize in Chemistry* 2012 was awarded jointly to Lefkowitz and Kobilka "for studies of G-protein-coupled receptors". Indeed, the field of GPCR research has certainly produced its fair share of Nobel laureates.

The GPCR Signal Transduction Pathway

Figure 5.4 describes how the binding of a hormone to a GPCR initiates a signal transduction pathway mediated by G proteins. In the unliganded state, GPCRs do not interact with the heterotrimeric G protein complex, whose alpha subunit is bound to GDP. However, the binding of a ligand to the GPCR's extracellular domain induces a conformational change in the intracellular domain, allowing the Gα subunit of the heterotrimer to interact with the GPCR. This interaction produces a conformational change in the Gα subunit that initiates two events: (1) the GDP molecule dissociates from Gα and is replaced by a GTP molecule, and (2) the Gα subunit dissociates from the Gβ/γ complex. These events result in the "activation" of the Gα subunit, such that it is in the "on" state and now has a high binding affinity to a specific target protein (usually a transmembrane enzyme) known as an *effector*. The binding of the activated Gα subunit to the effector activates this target enzyme, which will catalytically convert a specific substrate (e.g. ATP) into *second messengers* (e.g. cAMP), intracellular signaling molecules that activate target enzymes (e.g. kinases) (Figure 5.5). These target enzymes ultimately alter some aspect of cell

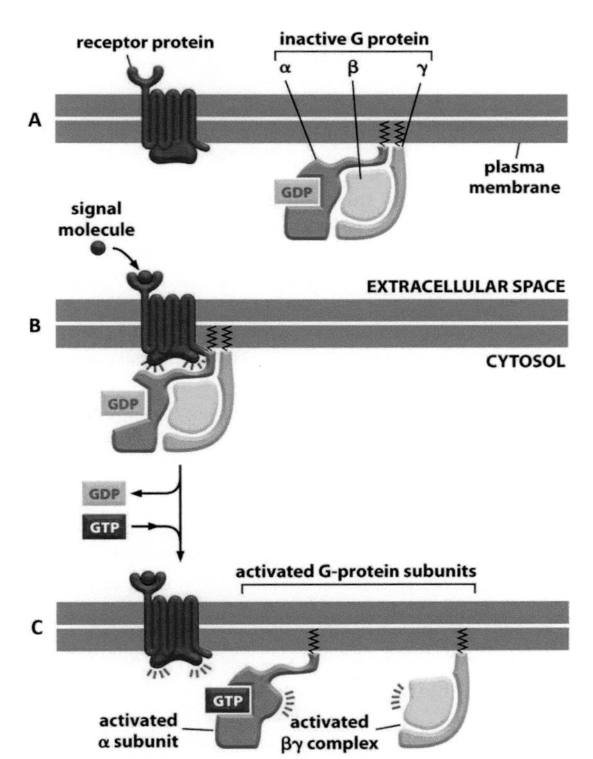

Figure 5.4 The binding of a hormone to its GPCR activates the G protein complex. (A) In the absence of hormone, the heterotrimeric G protein complex does not interact with the GPCR. **(B)** The binding of a hormone induces conformation changes in the GPCR that allow it to interact with and activate the G protein complex. **(C)** The activated Gα has exchanged GDP for GTP, and separated from the Gβγ subunits.

Source of images: W. W. Norton & Company, Inc., Essential Cell Biology, Fifth Edition by Bruce Alberts, et al. 2019; used with permission.

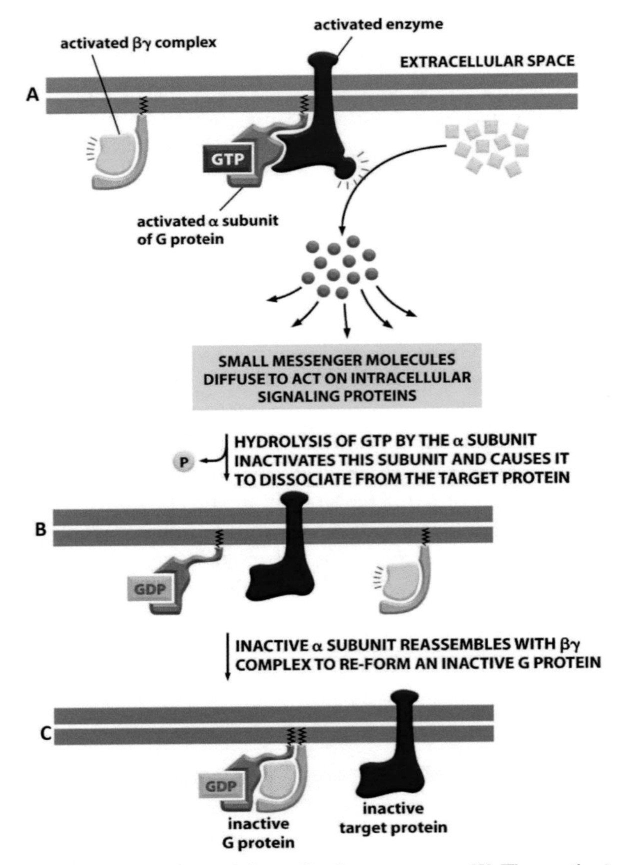

Figure 5.5 Activation of the effector enzyme. (A) The activated Gα-GTP complex binds to and activates an effector enzyme, which catalyzes the synthesis of second messenger molecules that in turn activate intracellular signaling proteins that mediate a biological response. **(B)** The eventual hydrolysis of GTP to GDP by the Gα inactivates the effector enzyme, and **(C)** the system returns to the resting state when the heterotrimeric complex reforms.

Source of images: W. W. Norton & Company, Inc., Essential Cell Biology, Fifth Edition by Bruce Alberts, et al. 2019; used with permission.

metabolism or gene expression. As long as GTP remains bound to the Gα subunit, both the subunit and the effector will remain activated, perpetuating the generation of second messenger molecules. However, the eventual hydrolysis of GTP to GDP by the Gα subunit switches the subunit into an "off" state, causing it to dissociate from the effector, shutting down the production of second messenger molecules. The deactivated Gα subunit now reassociates with the G beta/gamma complex, reforming the heterotrimer, and returning the G proteins to an inactive state.

It is important to note that while there are many stimulatory Gα proteins that activate effectors in the manner just described, there also exist inhibitory Gα proteins that inhibit activated effectors[20] (Figure 5.6). Adenylate cyclase and ion channel effectors can be under dual control by both stimulatory ($G\alpha_s$) and inhibitory ($G\alpha_i$) Gα proteins. Other Gα proteins associated with different receptor-effector pathways have their own unique designations (e.g. $G\alpha_q$, phospholipase C effector; $G\alpha_0$, endorphin effector; $G\alpha_t$, or "transducin", the light effector) (Table 5.2).

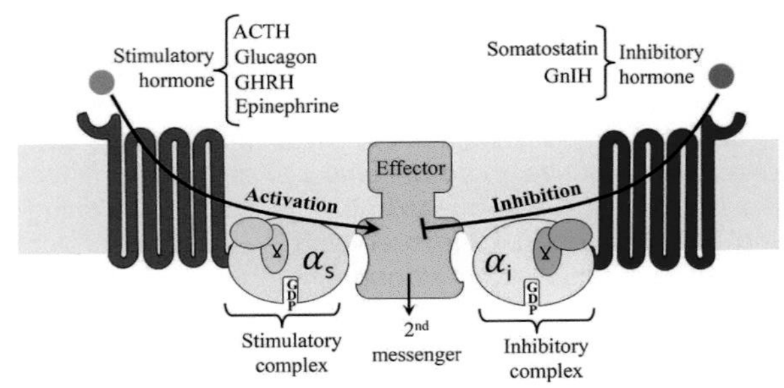

Figure 5.6 Some effector enzymes are under dual antagonistic control by stimulatory and inhibitory G protein complexes. For example, whereas the receptor for growth hormone-releasing hormone (GHRH) in the pituitary gland interacts with stimulatory Gα subunits ($G\alpha_s$) to activate an effector enzyme (in this case, adenylate cyclase) and synthesize second messengers (cAMP), in the same cell the receptor for somatostatin (SST) interacts with inhibitory Gα subunits ($G\alpha_i$) that inhibit the same effector.

Table 5.2 Classes and Functions of Select G Proteins

G Protein	Location	Stimulus	Effector	Effect
$G\alpha_s$	Liver	Epinephrine, glucagon	Adenylyl cyclase	Glycogenolysis
$G\alpha_s$	Adipose tissue	Epinephrine, glucagon	Adenylyl cyclase	Lipolysis
$G\alpha_s$	Kidney	Antidiuretic hormone	Adenylyl cyclase	Water conservation
$G\alpha_s$	Ovarian follicle	Luteinizing hormone	Adenylyl cyclase	Estrogen/progesterone synthesis
$G\alpha_i$	Heart muscle	Acetylcholine	Potassium channel	Reduced heart rate and force of contraction
$G\alpha_{i/O}$	Brain neurons	Enkephalins, endorphins, opioids	Adenylyl cyclase, potassium channels, calcium channels	Altered electrical activity
$G\alpha_q$	Vascular smooth muscle	Angiotensin	Phospholipase C	Vasoconstriction, increased blood pressure
$G\alpha_{olf}$	Nasal neuroepithelial cells	Oderant molecules	Adenylyl cyclase	Oderant detection
$G\alpha_t$	Retinal rod and cone cells	Light	cGMP phosphodiesterase	Light detection

Source: Adapted from Hepler, J., and Gilman, A., 1992. G proteins. *Trends in Biochemical Sciences* 17:383–387

The Adenylate Cyclase Effector

The first GPCR effector to be characterized was **adenylate cyclase**, an enzyme found in all eukaryotic cells and a common effector used by multiple hormone signaling pathways. Adenylate cyclase is a transmembrane enzyme with cytosolic catalytic domains that are activated upon interaction with a GTP-bound G protein alpha subunit. The activated enzyme converts ATP into the potent second messenger, cAMP (cyclic AMP), which binds to and activates protein kinase A, initiating a phosphorylation cascade that ultimately induces a change in cell physiology. One example of a rapid cytosolic response mediated by adenylate cyclase is the release of glucose from glycogen stores in liver and skeletal muscle in response to epinephrine (Figure 5.7). Within 5–8 seconds of hormonal stimulation, this signaling cascade results in the maximal activation of *glycogen phosphorylase*, the enzyme that ultimately depolymerizes glucose from glycogen.[21] cAMP is inactivated by the cytosolic enzyme, **phosphodiesterase**, which converts cAMP into AMP by breaking its phosphodiester bond. Based on his discovery that cAMP functions as an intracellular second messenger of epinephrine and glucagon,[22] Earl Sutherland was awarded the *Nobel Prize in Physiology or Medicine* in 1971 "for his discoveries concerning the mechanisms of the action of hormones".

Curiously Enough . . . *Methylxanthines*, such as caffeine, are secondary plant compounds that inhibit cAMP-dependent phosphodiesterase activity, and are thus potent stimulants of cyclic nucleotide second messenger activity.

Cytosolic vs. Genomic Responses

The activation of adenylate cyclase by GPCRs not only produces a rapid cytosolic response, but also is capable of generating a slower, but longer-lasting, genomic response through the phosphorylation of *cAMP response element-binding proteins* (CREBs) (Figure 5.8). Upon phosphorylation by protein kinase A, CREBs translocate from the cytoplasm to the nucleus, where they function as transcription factors that bind to *cAMP response elements* (CREs) in the promoters of target genes whose expression will be modulated. Whereas epinephrine and glucagon can stimulate glycogenolysis through their respective GPCRs within seconds via the previously described rapid cytosolic pathway, these hormones can also concurrently stimulate the *de novo* synthesis of glucose by inducing the expression of genes that mediate **gluconeogenesis**, the synthesis of

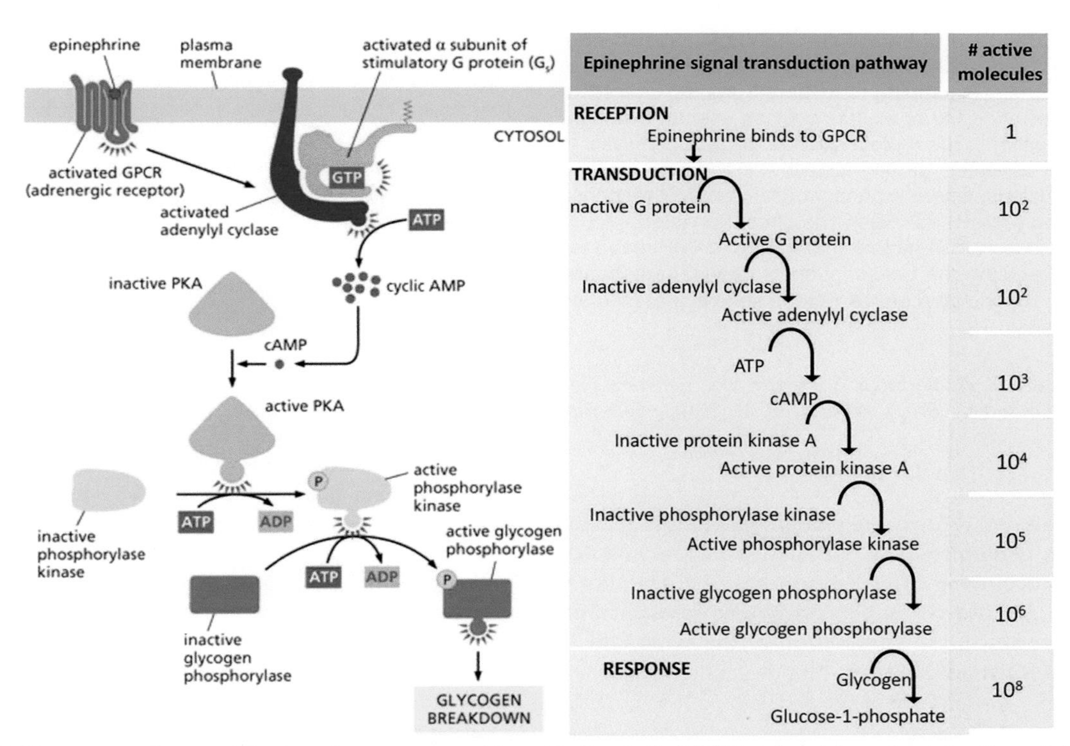

Figure 5.7 GPCRs can mediate rapid cytosolic responses. Upon binding to its GPCR, epinephrine stimulates the catabolism of glycogen into free glucose via a cAMP-driven kinase cascade. The initial hormonal signal is amplified about 1,000-fold at each enzymatic step. This signal amplification ultimately promotes the activation of glycogen phosphorylase (the enzyme that catabolizes glycogen) within 5 to 8 seconds of hormonal stimulation.

Source of images: W. W. Norton & Company, Inc., Essential Cell Biology, Fifth Edition by Bruce Alberts, et al. 2019; used with permission.

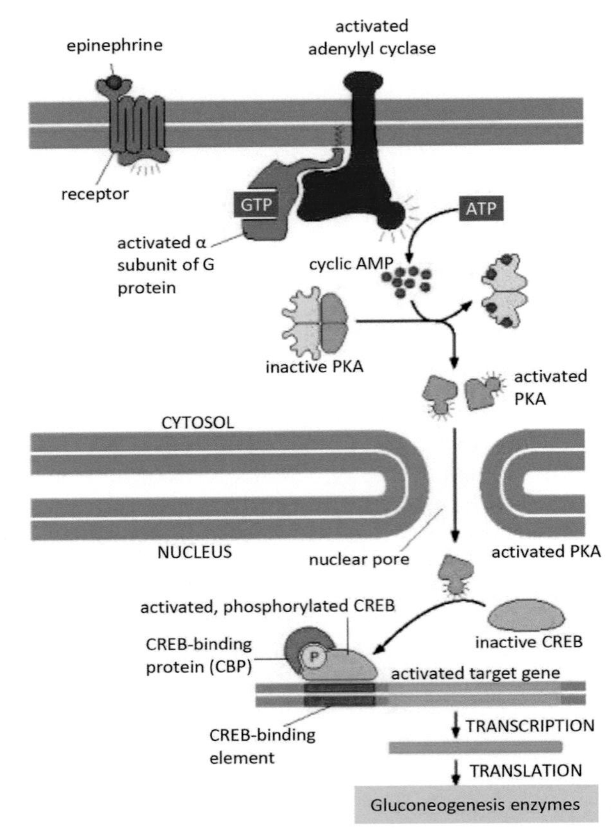

Figure 5.8 GPCRs can modulate genomic responses. Depicted here is the mechanism by which epinephrine elevates blood glucose levels by binding to its beta-adrenergic receptor in liver cells activating the adenylate cyclase effector pathway. In addition to initiating the rapid induction of glycogenolysis in the cytoplasm, activated protein kinase A (PKA) also translocates to the nucleus, where it phosphorylates cAMP response element-binding proteins (CREBs). The phosphorylated CREBs then become transcription factors that bind to cAMP response elements that regulate the transcription of genes. In response to epinephrine, upregulated genes include several that mediate gluconeogenesis.

Source of images: W. W. Norton & Company, Inc., Essential Cell Biology, Fifth Edition by Bruce Alberts, et al. 2019; used with permission.

glucose from non-carbohydrate precursors. This is accomplished via the genomic CREB pathway, with peak rates of transcription occurring after 30 minutes.[23,24] Therefore, in the case of glucose synthesis, concurrent cytosolic and genomic responses to the stimulation of the same GPCR result in a synergistic increase in blood glucose levels.

Phospholipase C Effector

Phospholipases are enzymes that hydrolyze membrane phospholipids into fatty acids and other substances. Phospholipase C (PLC) is the most common GPCR effector of the phospholipase class, and its activation by cell-surface receptors is known to trigger a wide variety of cellular responses that include egg fertilization, immune cell activation, synaptic transmission, and hormone transduction. Upon activation by GTP-bound $G\alpha_q$, the transmembrane PLC cleaves phosphatidylinositol-4,5-bisphosphate (PIP2), an inner leaf plasma membrane phospholipid, to yield two products that function as second messengers: diacylglycerol (DAG), a lipid-soluble molecule that remains in the plasma membrane, and inositol trisphosphate (IP_3), a water-soluble molecule that translocates to the cytosol (Figure 5.9). The binding of IP_3 to IP_3-gated Ca^{++} channels on the smooth endoplasmic reticulum (ER) releases Ca^{++} into the cytoplasm. Whereas Ca^{++} concentrations in the ER lumen and the extracellular fluid are high (approximately 10^{-3} M), cytosolic Ca^{++} is normally kept very low (approximately 10^{-7} M). When a signal transiently opens Ca^{++} channels in either of these membranes, Ca^{++} rushes into the cytosol, increasing the local Ca^{++} concentration by ten- to 20-fold and triggering Ca^{++}-responsive proteins in the cell. This gradient allows Ca^{++} to also function as a potent second messenger with diverse functions, two of which we will consider here.

First, the binding of Ca^{++} to protein kinase C (PKC) causes the enzyme to associate with DAG in the plasma membrane, which activates PKC (Figure 5.9). PKC then phosphorylates proteins that will modulate various physiological responses by the cell. Second, Ca^{++} also commonly binds to a broad class of proteins called *calmodulins* (CaMs), which function as protein switches that activate diverse enzymes, including those that modulate smooth muscle contraction in blood vessels, bronchioles, and intestine. The frequency of Ca^{++} oscillations influences a cell's response. These Ca^{++} oscillations (Figure 5.10) can persist for as long as receptors are activated at the cell surface. Both the waves and the oscillations are thought to depend, in part, on a combination of positive and negative feedback by Ca^{++} on the IP_3-gated Ca^{++}-release channels: the released Ca^{++} initially stimulates more Ca^{++} release, a process known as Ca^{++}-induced Ca^{++} release. But then, as its concentration rises high enough, Ca^{++} inhibits further release.

It was previously described how epinephrine binds to beta-adrenergic GPCR receptors to activate the adenylate cyclase effector, a process that induces glycogenolysis, smooth muscle relaxation, bronchodilation, and heart muscle contraction. Importantly, when the same hormone binds to alpha-1 adrenergic GPCR receptors that activate a PLC effector, this produces different effects, such as smooth muscle contraction and vasoconstriction. This is an example of **pleiotropy**, where a single hormone produces different effects on different tissues, in this case mediated by different receptors and effectors. For example, whereas the heart and lungs are rich in beta-adrenergic receptors with adenylate cyclase effectors, blood vessels are rich in alpha adrenergic receptors with PLC effectors.

Muscarinic Acetylcholine GPCR Effectors

Recall from earlier that nicotinic acetylcholine receptors (nAChRs) are ionotropic receptors that also have a high binding affinity and responsivity to nicotine as an agonist. A second family of acetylcholine receptors is the *muscarinic acetylcholine receptors* (mAChRs), so named

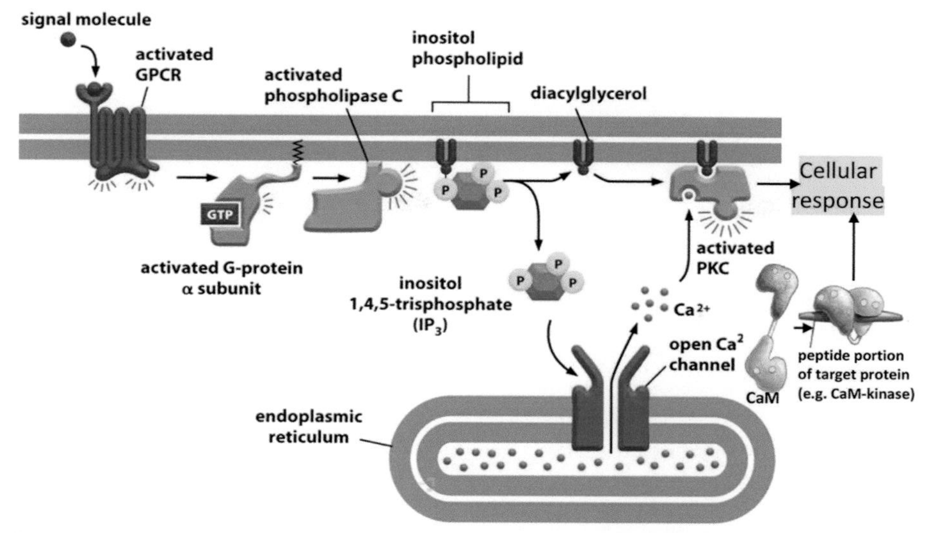

Figure 5.9 The phospholipase C (PLC) effector pathway results in increased cytosolic Ca^{++}. The activated PLC cleaves the membrane phospholipid, PIP2, into two second messenger products: diacylglycerol (DAG) is hydrophobic and remains in the plasma membrane, and inositol trisphosphate (IP_3) is water soluble and translocates to the cytosol. IP_3 binds to ligand-gated Ca^{++} channels on the smooth endoplasmic reticulum, causing Ca^{++} to be released into the cytoplasm. Ca^{++} is also a potent second messenger that modulates a cellular response in two ways. First, the association of Ca^{++} with the cytosolic "switch" protein, calmodulin (CaM), allows CaM to associate with target regions of other proteins and activate them. Second, the binding of Ca^{++} to protein kinase C (PKC) allows it to associate with DAG in the plasma membrane, which activates PKC, allowing it to phosphorylate downstream proteins that facilitate the cell response.

Source of images: W. W. Norton & Company, Inc., Essential Cell Biology, Fifth Edition by Bruce Alberts, et al. 2019; used with permission.

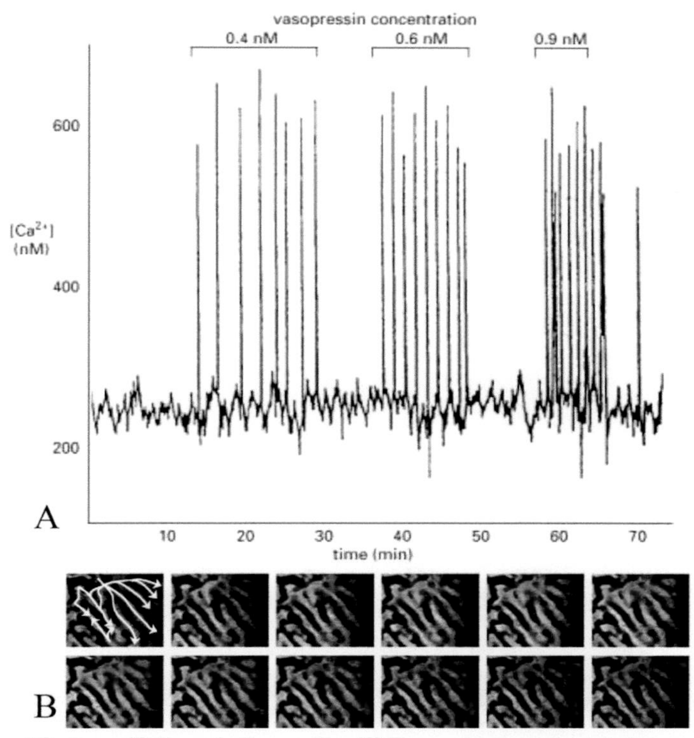

Figure 5.10 Vasopressin-induced Ca^{++} oscillations in liver cells. (A) These readings depict changes in cytosolic Ca^{++} from a single liver cell loaded with the Ca^{++}-sensitive fluorescent indicator aequorin and then exposed to increasing concentrations of vasopressin. Note that the frequency of the Ca^{++} spikes increases with an increasing concentration of vasopressin, but that the amplitude of the spikes is not affected. **(B)** The confocal image time course of intact liver loaded with the Ca^{++}-sensitive indicator fluo-3 depicts a vasopressin-stimulated Ca^{++} wave propagating throughout cells in a liver lobule. Images were acquired every 8 s in response to vasopressin. The arrows indicate the paths of intercellular Ca^{++} waves.

Source of images: (A) N.M. Woods, et al. Nature 319:600–602, 1986; used with permission. (B) Patel, S., et al. 1999. Nature Cell Biol., 1(8), pp. 467-471; used with permission.

because they have a high responsivity to muscarine, a toxin from the mushroom *Amanita muscaria*. In contrast to the ionotropic nAChRs, mAChRs are GPCRs. *M3 muscarinic receptors* are coupled to the Gq class of G proteins, which upregulate phospholipase C, generating inositol trisphosphate and intracellular calcium as second messengers. When this pathway is stimulated in the endothelial cells of blood vessels, calmodulin is induced by Ca^{++} to activate the enzyme *nitric oxide synthase*, which stimulates the production of the gas nitric oxide (NO), which functions as another second messenger (Figure 5.11). NO diffuses from the endothelial cells into adjacent smooth muscle cells, activating *guanylyl cyclase* to generate yet another second messenger, cyclic GMP (cGMP). cGMP activates protein kinase G (PKG), which phosphorylates several muscle proteins to ultimately induce smooth muscle relaxation. In 1999 the *Nobel Prize for Physiology and Medicine* was awarded to R.F. Furchgott, L.J. Ignarro, and F. Murad for their discovery in the 1980s that NO was a hormone of the cardiovascular system.

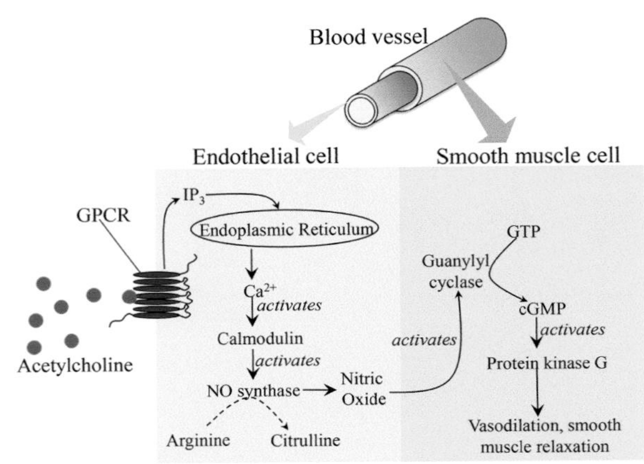

Figure 5.11 The M3 class muscarinic acetylcholine receptor signal transduction pathway induces vascular smooth muscle relaxation. These G protein-coupled receptors (GPCRs) are localized to the plasma membranes of blood vessel endothelial cells. Acetylcholine activates the phospholipase C effector pathway, resulting in the Ca^{++}/calmodulin-induced activation of the enzyme, nitric oxide synthase, which stimulates the production of the gas nitric oxide (NO) using arginine as a substrate. NO then diffuses from the endothelial cells into adjacent smooth muscle cells and activates guanylyl cyclase to generate cyclic GMP (cGMP). The activation of protein kinase G (PKG) by cGMP causes several muscle proteins to become phosphorylated, ultimately inducing smooth muscle relaxation.

Curiously Enough . . . "UK-92,480", the original name of a drug developed in the 1990s for the treatment of hypertension, functions as a potent inhibitor of cGMP-specific phosphodiesterase, the enzyme that inactivates cGMP by converting it to GMP. By prolonging the half-life of cGMP in vascular smooth muscle, the drug promotes vasodilation. However, in testing the drug an unexpected side effect was noted in males, namely the induction of penile erection. Thus was born one of the world's top selling pharmaceutical products in history, Viagra (sildenafil), which is now used to treat male erectile dysfunction, a condition characterized by reduced of blood flow to the penis, preventing erection.

Lipid-Soluble Hormones with Plasma Membrane GPCRs

As will be discussed in Section 5.3: Nuclear Receptors of this chapter, the most well-described effects of steroid and thyroid hormones are mediated by nuclear receptors that function as transcription factors. However, in addition to these "classical" actions of steroid-hormone–receptor complexes, accumulating evidence is showing that steroid and thyroid hormones also exhibit important "rapid response", nongenomic pathways mediated by plasma membrane receptors.[25] For example, steroid hormones can initiate rapid responses in the absence of a functional nucleus,[26,27] showing that a classical nuclear receptor need not necessarily be involved. Additionally, a rapid, nongenomic response of the lung to therapeutic glucocorticoid administration plays a critically important role in treating acute asthma (see Developments & Directions box).

It is now well established that virtually all classes of steroid and thyroid hormones have been shown to be able to induce rapid responses,[25] some by binding to membrane GPCRs. One such receptor, called G protein-coupled estrogen receptor (GPER), is present in both the plasma membrane and the smooth endoplasmic reticulum membrane (see Figure 2.9 of Chapter 2: Fundamental Features of Endocrine Signaling) and has been shown to mediate both estrogen-dependent kinase activation as well as transcriptional responses.[31,32] Importantly, upregulation and activation of GPER by a wide number of natural and synthetic compounds, including estrogens and antiestrogens, is associated with breast cancer progression and the development of resistance to the drug, tamoxifen, a nuclear estrogen receptor **antagonist.**[33–35] Furthermore, both estrogen receptor α (ERα) and estrogen receptor β (ERβ), which commonly function as nuclear-localized receptors, have been shown to also localize to the plasma membrane where they bind to estrogen and induce rapid response intracellular signal transduction pathways.[36] In combination with nuclear receptors, these membrane-localized estrogen receptors contribute to the overall, integrated effects of estrogen signaling to generate biological responses.

Developments & Directions: A rapid plasma membrane response to corticosteroids in the treatment of asthma

Asthma, one of the most common chronic illnesses worldwide, is a condition characterized by the narrowing of the lung's bronchial airways due to (1) *hyperperfusion*, or increased blood flow to the lungs, (2) inflammation and thickening of the bronchiole wall due to increased numbers of immune cells, (3) mucus hypersecretion into the airway, and (4) bronchiole smooth muscle contraction. Inhaled corticosteroids, the gold standard in asthma therapy, is known to suppress the development of inflammation through the activation or repression of various target genes, a genomic response that requires hours to days to manifest. However, the asthmatic lung also exhibits a robust nongenomic response to corticosteroids, namely a rapid decrease in hyperperfusion, which takes place within seconds to minutes.[28–30] This rapid response, which is particularly important in the treatment of acute asthmatic attacks, is thought to be related to the binding of inhaled corticosteroids to as of yet uncharacterized plasma membrane receptors. These activated receptors inhibit the *extraneuronal monoamine transporter* (EMT) on airway vascular smooth muscle cells, thereby increasing norepinephrine concentrations at alpha-1 adrenergic receptors and causing airway vasoconstriction and a decrease in airway blood flow. Therefore, the rapid-response effect of glucocorticoid treatment operates synergistically with endogenous norepinephrine and epinephrine to mitigate the hyperperfusion effect of asthma.

SUMMARY AND SYNTHESIS QUESTIONS

1. In intestinal epithelial cells, cAMP stimulates a transmembrane protein to pump chloride ions into the lumen of the gut, which helps maintain proper ion/water balance. Cholera toxin (CTX) specifically targets the G alpha subunit of G proteins, inhibiting GTPase activity. Considering this, how would the presence of CTX in these cells contribute to the effects of cholera, namely profuse diarrhea and dehydration?
2. Pertussis toxin (PTX) inhibits the G alpha subunit's ability to switch GDP for GTP. Considering the mechanism of action of PTX, is whooping cough (which is caused by the bacterial toxin) a disease of excessive or inhibited G protein signaling?
3. The adenylate cyclase effector of growth hormone producing cells in the anterior pituitary gland are under "dual control" by hypothalamic hormones—it can be activated by GHRH, but inhibited by somatostatin. Explain the mechanisms that make this dual control possible.
4. Methylxanthines (such as caffeine) are stimulants that inhibit cAMP-dependent phosphodiesterase activity. How does the inhibition of this enzyme produce a stimulatory response?
5. Glucagon elevates blood sugar levels by causing the liver to release glucose. The time course of glucose release is bimodal, with some being released within seconds of glucagon binding to its receptor, and more being released over the course of hours. Explain the molecular basis of this bimodal response.
6. List the names and functions of all second messengers in the PLC signal transduction pathway.
7. Epinephrine induces bronchiolar smooth muscle relaxation (bronchodilation) in the lungs, but at the same time causes vascular smooth muscle contraction (vasoconstriction). How can this single hormone produce opposite effects in smooth muscle from different tissues?
8. Compare and contrast the mechanisms of action of nicotinic versus muscarinic acetylcholine receptors.
9. A patient takes a *nitroglycerin* tablet sublingually for chest pain. Chest pain can be caused by high blood pressure that produces too much blood flow to the heart, which is too weak to handle the volume. How would nitroglycerin (which can function as a substrate for the synthesis of the molecule, nitric oxide [NO]) reduce chest pain?

Receptors with Intrinsic Enzymatic Activity (Enzyme-Linked Receptors)

LEARNING OBJECTIVE Describe the different effectors and signal transduction pathways associated with each major plasma membrane receptor class: Receptors with intrinsic enzymatic activity (receptor tyrosine kinases).

KEY CONCEPTS:

- Many members of the enzyme-linked receptor family are "growth factors", which are mis-expressed in some cancers.
- Members of the "receptor tyrosine kinase" (RTK) family have intracellular domains that become auto-phosphorylated, allowing interactions with effector proteins.

In 1952, Rita Levi-Montalcini discovered that a substance derived from mouse tumor extracts, now known as *nerve growth factor* (NGF), promoted neurite outgrowth when injected into chicken embryos.[37] After Levi-Montalcini and Stanley Cohen purified NGF from snake venom and mouse salivary-gland extracts, Cohen isolated and characterized another novel salivary-gland protein termed *epidermal growth factor* (EGF) that induced precocious eyelid opening and tooth eruption when injected into newborn mice.[38,39] For their important discovery of *growth factors*, the *Nobel Prize in Physiology or Medicine* was awarded jointly to Stanley Cohen and Rita Levi-Montalcini in 1986.

The receptors for many growth factors constitute a second major class of cell-surface receptors, those with intrinsic enzymatic activity, and are sometimes called *enzyme-linked receptors*. Since cell growth, proliferation, differentiation, and survival of cells in animal tissues are regulated by growth factors, it is not surprising that disorders of enzyme-linked receptors play major roles in cancer, as well as in type 2 diabetes and atherosclerosis.

The most common category of enzyme-linked receptor are the *receptor tyrosine kinases* (RTKs), a diverse family that includes receptors for hormones and paracrine factors such as insulin, epidermal growth factor (EGF), and nerve growth factor (NGF). All RTKs possess an extracellular domain containing the ligand-binding site, a single transmembrane domain, and an intracellular domain containing the tyrosine kinase domain (Figure 5.12). In the absence of ligand, RTK monomers typically do not interact with each other. However, in the presence of a ligand, RTKs form homodimers whose intracellular kinase domains phosphorylate tyrosine residues on each receptor (Figure 5.12). These phosphorylated regions serve as "docking sites" that allow the receptor to interact with *scaffold/adapter proteins*, many of which contain *SH2* (Src homology 2) domains that allow them to dock to the phosphorylated tyrosine residues. These adapter proteins then interact with other proteins that ultimately modulate effector proteins, such as phospholipase Cγ (PLCγ), inositol trisphosphate kinase (PI3K), and Ras/MAPK that aid in the signal transduction of receptor tyrosine kinase pathways. These transduction pathways can produce cytosolic and/or genomic changes.

The Ras/MAPK Pathway

Ras (an abbreviation of "*Ra*t *s*arcoma", derived from the genes of the first members discovered) is a family of small GTPase proteins closely related to the α subunit of trimeric G proteins. As such, the activation of Ras via proteins associated with the RTK scaffold allow Ras to activate a three-kinase downstream signaling effector module, beginning with an enzyme with the somewhat cumbersome (but apt) name "*MAP kinase kinase kinase*" (Figure 5.13). *MAP* is an acronym for "mitogen-activated protein", and

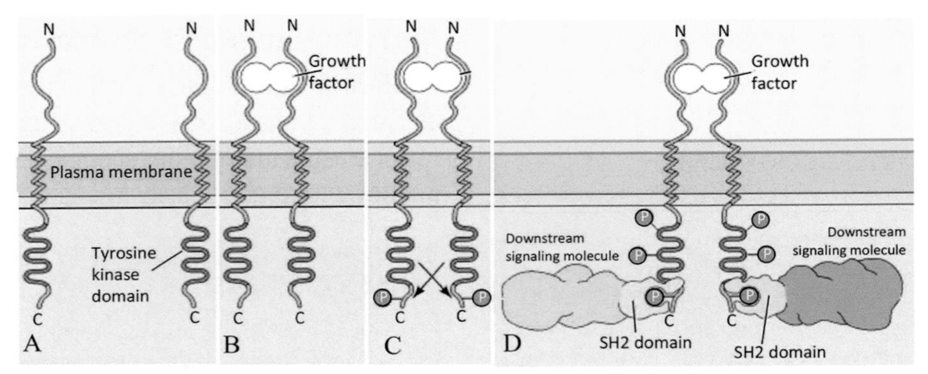

Figure 5.12 Assembly of a typical receptor tyrosine kinase (RTK). (A) In the absence of ligand, RTK monomers do not interact with each other. In the presence of a ligand, RTKs form homodimers **(B)** whose intracellular kinase domains phosphorylate each other **(C)**. **(D)** These phosphorylated regions serve as "docking sites" that allow the receptor to interact with adapter proteins, which facilitate interactions with effector proteins.

Source of images: Cooper, G.M., et al. 2007. The cell: a molecular approach (Vol. 4, pp. 649-656). Washington, DC: ASM press; used with permission.

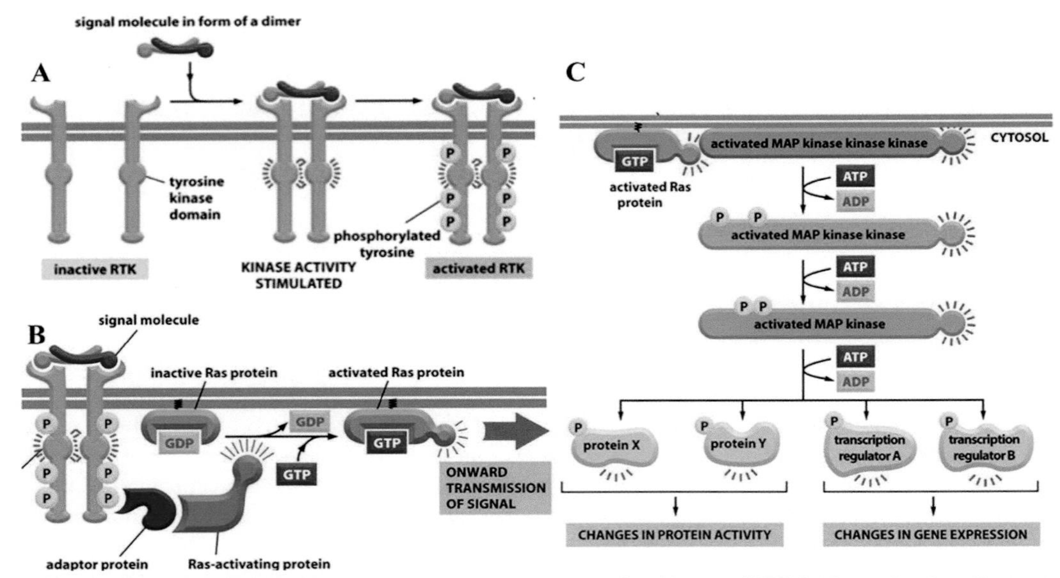

Figure 5.13 The Ras/MAPK signaling pathway. Activated receptor tyrosine kinases (RTKs) often stimulate Ras proteins indirectly via adapter proteins. Once activated, Ras activates a MAP kinase relay that ultimately initiates changes in gene expression and/or cytosolic activity that promote mitogenic activity.

Source of images: W. W. Norton & Company, Inc., Essential Cell Biology, Fifth Edition by Bruce Alberts, et al. 2019; used with permission.

a *mitogen* is a substance that plays an important role in initiating mitosis, or cell proliferation. MAP kinase kinase kinase phosphorylates and activates the second enzyme in the module, called MAP kinase kinase, which, as you may have guessed, phosphorylates and activates the last enzyme in the series, called MAP kinase (MAPK). The activation of MAPK ultimately initiates changes in gene expression and/or cytosolic activity that promote mitogenic activity. Importantly, because these signals result in cell growth and division, constitutively active Ras signaling mutations that fail to hydrolyze GTP to GDP and therefore cannot "turn off" may ultimately promote cancer.[40]

Curiously Enough . . . Ras genes are the most common oncogenes in human cancer, and mutations in the Ras GTPase domain that permanently activate Ras are found in 20–25% of all human tumors and up to 90% in pancreatic cancer.[41]

SUMMARY AND SYNTHESIS QUESTIONS

1. "Wild-type" Ret (a type of RTK receptor) has intramolecular disulfide bonds formed by two cysteine residues within the *same* receptor monomer. In a mutant version of Ret, one of the two cysteine residues instead forms a disulfide bond with a cysteine residue on *another* receptor monomer, forming an RTK dimer. Would the mutant Ret result in reduced signaling, increased signaling, or no effect?
2. Compare and contrast heterotrimeric G proteins that associate with GPCRs with Ras G proteins that associate with RTKs.
3. A single amino acid change in Ras (i.e. a point mutation) eliminates its ability to hydrolyze GTP. Roughly 30% of all human cancers have this mutation in Ras. You have just identified a small molecule inhibitor that prevents dimerization of a receptor tyrosine kinase that signals via Ras. Would you expect this small molecule inhibitor to be useful as a drug to treat cancers that express this common, mutant form of Ras? Why or why not?

Receptors with Enzyme-Associated Activity (Recruiter Receptors/Cytokine Receptors)

LEARNING OBJECTIVE Describe the different effectors and signal transduction pathways associated with each major plasma membrane receptor class: Receptors with enzyme-associated activity (recruiter receptors/cytokine receptors).

KEY CONCEPTS:

- Most recruiter receptor ligands are cytokines that are involved in blood cell development and immune system function.
- Several receptors in this family bind to classical hormones (e.g. growth hormone, prolactin, leptin, and erythropoietin).
- An important category of enzymes that associate with recruiter receptors are Janus kinases (JAK), which activate STAT proteins that modulate gene transcription.

Like the previously described enzyme-linked receptors, receptors with enzyme-associated activity are single transmembrane-domain polypeptides that form dimers. Unlike enzyme-linked receptors, these receptors do not possess intrinsic enzymatic activity, but instead recruit separate enzymes that associate with the receptor and are thus sometimes called **recruiter receptors**. The ligands of recruiter receptors are often broadly referred to as **cytokines**, structurally diverse signaling peptides and proteins best known for their many roles in blood cell development and immune system function. As such, this category of receptors is often called the *cytokine receptors*. Cytokine receptors help regulate hematopoiesis, apoptosis, cell migration, and cell proliferation of particular cell populations and are involved in virtually all aspects of both innate and adaptive immune responses. Leukocytes ("white blood cells") are a primary source of cytokines, although they are produced by many other cell types as well. Several receptors in this family bind to classical hormones, like growth hormone, prolactin, leptin, and erythropoietin.[42] There are currently several hundred known genes and proteins of cytokine ligands, with new ones still being discovered.[43] In contrast, fewer than 100 cytokine receptors have been identified,[5] suggesting that many cytokines bind to the same receptors.

The JAK-STAT Signaling Pathway

The most widely studied enzymes recruited by cytokine receptors are called *Janus kinases* (JAK), so named because when two JAK proteins are bound to their respective homodimerized receptors, their appearance resembles (albeit, with some imagination) the faces of Janus, the two-faced Roman god of gates, doorways, and transitions (one face looks to the past, the other to the future—the month, January, is a tribute to Janus) (Figure 5.14). Cytokine receptors also differ from enzyme-linked receptors in that even in the absence of a ligand they form homodimers and associate with inactive JAK. The binding of a ligand to the receptor induces a conformation change in the receptor's cytosolic domain, which activates JAK. Subsequently, JAK phosphorylates itself, as well as tyrosine residues on the receptor that will function as docking sites for other proteins. Like RTKs, the phosphorylated cytokine receptor can

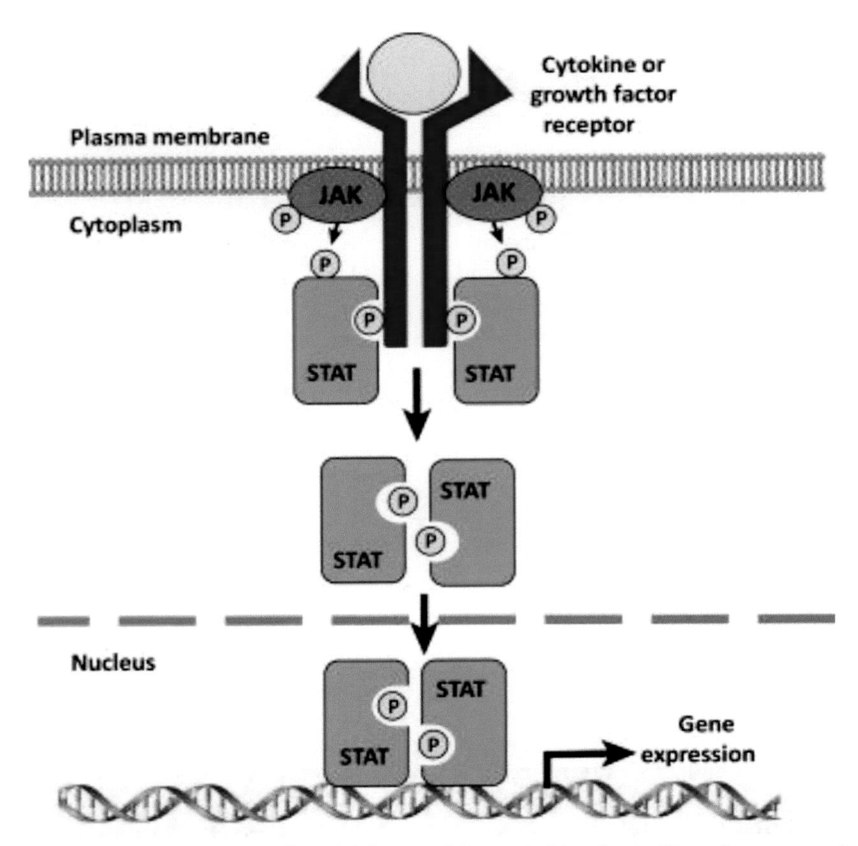

Figure 5.14 The JAK-STAT signaling pathway is widely used by cytokine/recruiter-class receptors. In the absence of ligand, cytokine receptors form homodimers and associate with inactive JAK. However, the binding of a ligand to the receptor activates JAK, which phosphorylates itself, as well as tyrosine residues on the receptor that will function as docking sites for other proteins. STAT proteins bind to the phosphorylated receptor's docking domain, allowing STAT's SH2 domain to becomes phosphorylated by JAK, causing it to dissociate from the receptor. Dissociated STAT molecules form homodimers in the cytosol, then translocate to the nucleus where they function as transcription factors that bind to specific DNA elements in the promoters of STAT-responsive genes.

Source of images: Dodington, D.W., et al. 2018. Trends Endocrinol. Metab., 29(1), pp. 55-65; used with permission.

associate with adapter proteins that activate the Ras/MAPK signaling pathway (not shown). In addition, a common signaling pathway initiated by cytokine receptors is known as the JAK-STAT pathway (Figure 5.14). In this pathway, a group of proteins called signal transducers and activators of transcription (STATs) first bind to the phosphorylated receptor's docking domain. Next, the STAT's SH2 domain becomes phosphorylated by JAK, causing it to dissociate from the receptor, and it forms a homodimer with another STAT in the cytosol. Finally, the homodimerized STATs translocate to the nucleus where they function as transcription factors that bind to specific DNA elements in the promoters of STAT-responsive genes. Examples of genes upregulated via the JAK-STAT pathway in response to growth hormone include insulin-like growth factor-1 (IGF-1) and suppressor of cytokine signaling (SOCS), and genes upregulated in response to the hormone prolactin include the milk proteins casein, whey acidic protein, and beta lactoglobulin.

Receptor Crosstalk

As we have seen in this chapter, different receptor classes can transduce their signals using components of the same pathways. For example, both GPCRs and RTKs can generate IP_3, DAG, and Ca^{++} as second messengers, and both RTKs and cytokine receptors can initiate Ras/MAPK signaling pathways. Because different receptors located on the same plasma membrane can use similar kinase cascades or second messengers, this can produce a phenomenon called *receptor crosstalk*, whereby the components of one pathway affect other parallel pathways. Crosstalk can result in cooperative signaling, whereby the release of Ca^{++} from one pathway can promote enzymatic activity in another parallel pathway. Crosstalk can also produce the opposite effect, where one pathway attenuates another. For example, if one GPCR that initiates "pathway A" produces free $G\beta\gamma$ subunits, the ability of $G\beta\gamma$ to associate with $G\alpha$ subunits derived from "pathway B" could inactivate "pathway B". As such, receptor crosstalk is a variable

Figure 5.15 Overlapping intracellular signal transduction pathways can be a source of headaches for students and professors alike.

Source of images: Tony Bramley, Trends Biochem. Sci. 19:469, 1994. Used with permission.

that profoundly increases the complexity of intracellular signaling pathways, much to the bane of students and professors alike (Figure 5.15).

SUMMARY AND SYNTHESIS QUESTIONS

1. Mutations of Janus kinase3 (Jak3) disrupt the actions of many cytokines involved in immune system development, leading to a condition called severe combined immunodeficiency (SCID), a group of rare, inherited disorders in primary immunity and results in major defects in host defense. Why would a Jak3 mutation be so severe to cytokines?
2. Compare and contrast the four major categories of plasma membrane receptors: ionotrophic, GPCR, RTK, and cytokine.
3. For any given family of plasma membrane receptor (e.g. GPCR, RTK), which domain (intracellular or extracellular) would you expect to exhibit the highest structural variability? Why?

Nuclear Receptors

Importance and Discovery

LEARNING OBJECTIVE Describe some of the earliest experiments that hinted at the role of nuclear receptors in regulating gene expression.

KEY CONCEPTS:

- There are 48 known nuclear receptors in humans.
- Dysfunctions of nuclear receptors and their ligands are associated with major diseases, such as cancer, osteoporosis, and diabetes.
- Some of the earliest evidence that steroid hormone receptors are involved in gene transcription came from observations of "puffs" in fruit fly polytene chromosomes treated with ecdysone.
- The cloning of the first nuclear receptor led to the discovery of a large superfamily of receptors that includes "orphans" with no known ligands.

Nuclear receptors are a "superfamily" of ligand-activated transcription factors that regulate genetic networks controlling diverse aspects of physiology, such as development, homeostasis, reproduction, immunity, and metabolism.[44] Nuclear receptor dysfunctions are also associated with some important diseases (see Appendix 6). As their name implies, the active forms for most of these receptors exert their effects in the nucleus. However, in addition to these "classical" actions of nuclear receptors, accumulating evidence is showing that many members of the nuclear receptor superfamily can also exert their effects via nongenomic, "rapid response" pathways mediated by receptors that localize to the plasma membrane (Figure 5.16). Additionally, there are several other categories of plasma membrane receptors whose signaling pathways affect gene transcription in the nucleus. This section will focus exclusively on the actions of classical nuclear receptors as direct modulators of gene transcription.

In humans, there are 48 known nuclear receptors divided into several subfamilies (Appendix 7). There are 49 nuclear receptors in mice, 21 in flies, and a remarkable 270 in the worm *Caenorhabditis elegans*. Although the receptors themselves are highly conserved, their ligands are chemically diverse. These ligands all share two important properties: they are small, lipophilic molecules that can—with the notable exception of thyroid hormone—move across the plasma and nuclear membranes by simple diffusion. Dysfunctions of nuclear receptors and their ligands are associated with major diseases, such as cancer, osteoporosis, and diabetes, and are thus attractive targets for pharmacological intervention. Their successes as drug targets are highlighted by the common use of retinoic acid receptor agonists for the treatment of acute promyelocytic leukemia, the synthetic estrogen receptor antagonist tamoxifen for treatment of breast cancer, the synthetic glucocorticoid receptor agonist dexamethasone for inflammatory diseases, and synthetic peroxisome proliferator-activated receptor agonists (thiazolidinediones) for type II diabetes. Other notable pharmacological uses of these ligands and their nuclear receptors include steroidal contraceptives, hormone replacement therapies, and lipid-lowering agents.

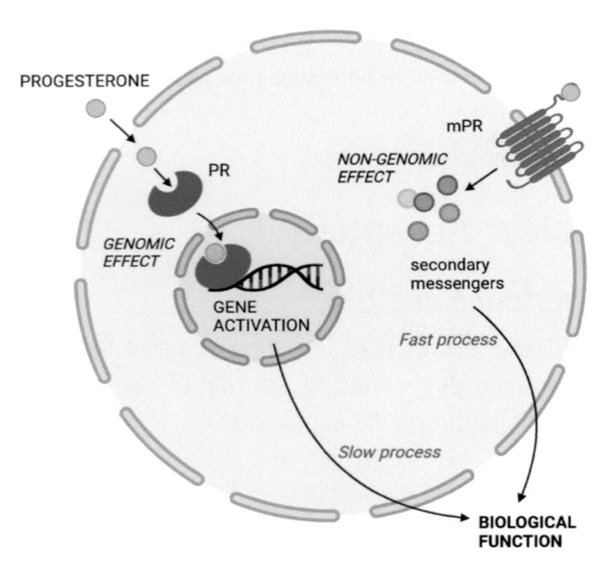

Figure 5.16 Steroid hormone actions are mediated by both genomic (nuclear receptor) and nongenomic (membrane-associated receptor) pathways. The genomic actions of steroid hormones, such as progesterone, are facilitated classically by their passing through the plasma membrane to engage receptors that function as transcription factors that modulate gene transcription. In their nongenomic action, the steroid hormone associates with cell-surface receptors to activate rapid response second messenger-mediated signaling pathways.

Source of images: Kolatorova, L et al. 2022. Int. J. Mol. Sci., 23(14), p. 7989. Used with permission.

> **Curiously Enough . . .** Nuclear receptors are the molecular target of approximately 10%–15% of drugs currently approved by the U.S. Food and Drug Administration, highlighting their tractability for therapeutic intervention.[45]

Hints of Nuclear Functions

The 1950s was the great era of enzymology, and conventional wisdom held that the mechanism of action for steroid hormones was to bind directly to cytosolic enzymes and alter their activity via allosteric changes. However, landmark experiments performed in the late 1950s and early 1960s, primarily in the laboratories of Gerald Mueller and Elwood Jensen, set the stage for the development of the modern concept of steroid hormone action. Jensen's laboratory showed that when tritiated (labeled with radioactive ^{3}H) estradiol was injected into immature rats, most tissues—skeletal muscle, kidneys, and liver, for example—started expelling it within 15 minutes. By contrast, tissues of the reproductive tract known to respond to the hormone held onto it with high affinity, indicating the presence of an "estrogen-binding component" or "estrophilin", later termed "estrogen receptor" by Jensen.[46,47] This was the first time that target tissue specificity was ever observed for a natural hormone. During this period, Mueller's laboratory also reported the key observation that estrogen treatment induced RNA and protein synthesis,[48] opening the real possibility that the estrogen receptor was somehow involved in modulating these processes.

Long Lost Relatives, Orphans, and Adopted Orphans: Discovery of a Superfamily of Nuclear Receptors

With the advent of gene cloning technologies in the 1980s, the first nuclear receptor gene was cloned, the human glucocorticoid receptor, by Ron Evans and his colleagues in 1985. This initial cloning led to the subsequent cloning of other family members by the laboratories of Ron Evans, Pierre Chambon, Björn Vennström, Bert O'Malley, and

Developments & Directions: Gene activation in polytene chromosomes of flies by ecdysone

Giant *polytene chromosomes* found in certain cells of larval flies are chromosomes that have undergone repeated rounds of DNA replication without any cell division, producing a "banded" appearance. Figure 5.17(A) depicts one polytene chromosome from the fruit fly (*Drosophila*) after it has undergone ten rounds of DNA replication, and the maternal and paternal homologs—as well as all their duplicates—are aligned exactly side by side with each other. This polytene chromosome consists of a thick cable containing 2,048 identical strands of DNA. Having multiple copies of genes permits a high level of gene expression, and these chromosomes are located in highly metabolically active tissues, such as the salivary glands. These chromosomes are so large that they can be seen during interphase, even with a low-power light microscope. Clever and Karlson[49] found that the molting steroid hormone *ecdysone* injected into larvae of *Chironomus* (another genus of fly) induced "puffs" at specific sites on the salivary gland polytene chromosomes. These puffs represented regions of decondensed chromatin that we now know are associated with transcription. Importantly, Clever[50] further showed that the appearance of the earliest puffs did not require protein synthesis but, by contrast, the formation of later puffs was dependent on protein synthesis.

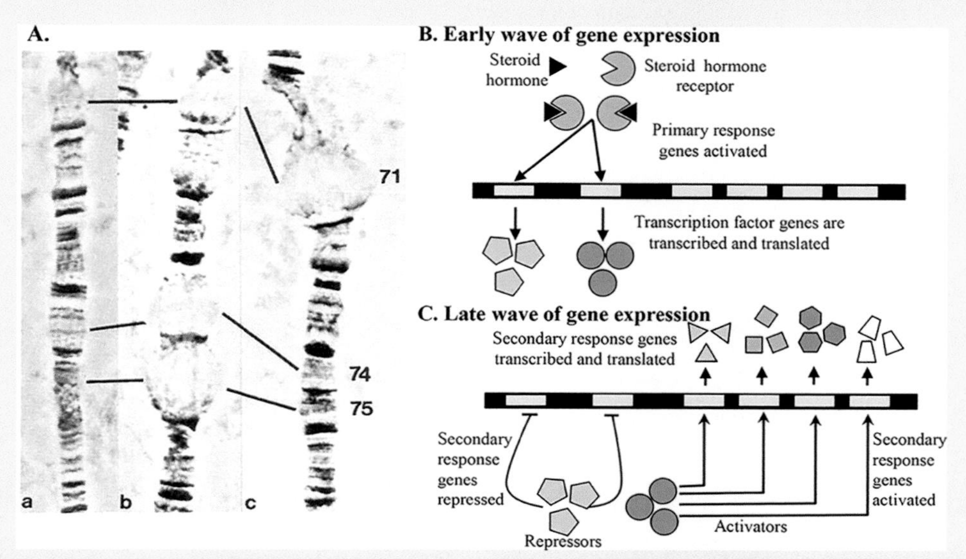

Figure 5.17 Early and late waves of gene transcription following steroid hormone treatment. (A) Ecdysone-induced formation and regression of puffs on chromosome arm 3L of the fly *Drosophila melanogaster* reflect the initiation and cessation of gene transcription events. (a) Before induction. (b) 5 h after induction, showing large early puffs at locations 74EF and 75B. (c) 10 h after induction, when the early puffs have regressed and a large late puff has appeared at 71E. Ashburner and colleagues (1974) postulated a model to explain the control of the puffing sequence, proposing that the early ecdysone responsive genes **(B)** consist of two different categories of transcription factors. **(C)** Whereas transcriptional activators induce the expression of secondary response genes, transcriptional repressors repress the primary response genes.

Source of images: (A) Ashburner, M., 1990. Cell, 61(1), pp. 1-3. Used with permission.

In the late 1960s and early 1970s, Michael Ashburner at Cambridge University extended this analysis of ecdysone-induced chromosome puffing.[51] His observations of *Drosophila* salivary gland chromosomes established distinct puffing stages and showed that the complement of puffing responses could be reproduced in isolated salivary glands cultured *in vitro* on addition of ecdysone to the culture medium (Figure 5.17). Ashburner made several key observations:

1. Prior to the addition of ecdysone (a), the chromosomes exhibit no puffs.
2. Minutes after adding ecdysone (b), six "early" puffs form in the absence of protein synthesis (in agreement with Clever, 1964), indicating that these early puffs are a primary response to the steroid hormone.
3. Following the appearance of the early puffs, a second set of more than 100 "late" ecdysone-induced puffs begins to appear several hours after the start of ecdysone treatment (c).
4. In addition to protein synthesis being required for the formation of the late, but not early, puffs (previously shown by Clever), protein synthesis is also required for the *regression* of the early puffs.

Taken together, these results led to the formulation of the *Ashburner model*[51] for the genetic control of polytene chromosome puffing by ecdysone, depicted in Figures 5.17(B-C). This model proposed that ecdysone acts through its nuclear receptor to directly exert two regulatory functions: to activate the early genes (B) and to suppress premature activation of the late genes (C). Many of the protein products encoded by the early genes are in fact transcription factors that induce late gene expression in a precisely timed sequential manner. The early gene products also repress their own expression, thus establishing the duration of their regulatory function. The late gene products modulate developmental and physiological changes during the early stages of metamorphosis. The Ashburner model provided the first detailed insights into how a developmental pathway for a cell can be initiated by a relatively small number of master regulatory genes and subsequently unfold in a complex, but precisely programmed, network of gene activity.

Developments & Directions: Alternative receptor splicing and translation initiation

Irrespective of the chemical class of ligand that they associate with, the receptors for all hormones are proteins. As with all proteins, receptors are coded for by genes located within chromosomes contained in the nucleus. Following transcription, **introns** (*in*tervening sequences that are not expressed in the mature mRNA) are excised from the primary transcript, and **exons** (sequences expressed in the mature mRNA) are spliced together. **Alternative splicing** of exons yields multiple mRNA variants and is a process known to account for the large diversity of hormone receptor isoforms, such as the glucocorticoid receptor (Figure 5.18).

In addition to alternative splicing of the RNA primary transcript, another level of post-transcriptional processing complexity can be the use of *alternative translation initiation* sites, which potentially yield multiple isoforms of a polypeptide hormone or receptor. For example, in humans there exists only one gene coding for the glucocorticoid receptor (GR), and the RNA transcript for that gene is known to be alternatively spliced into two main isoforms: GRα and GRβ. However, in different tissues the GRα variant mRNA alone is translated from eight different start codon initiation sites into multiple GRα isoforms (Figure 5.18), and these GR molecules are differentially expressed in multiple tissues.[52,53] Therefore, one glucocorticoid receptor gene can potentially generate 16 different receptor isoforms.

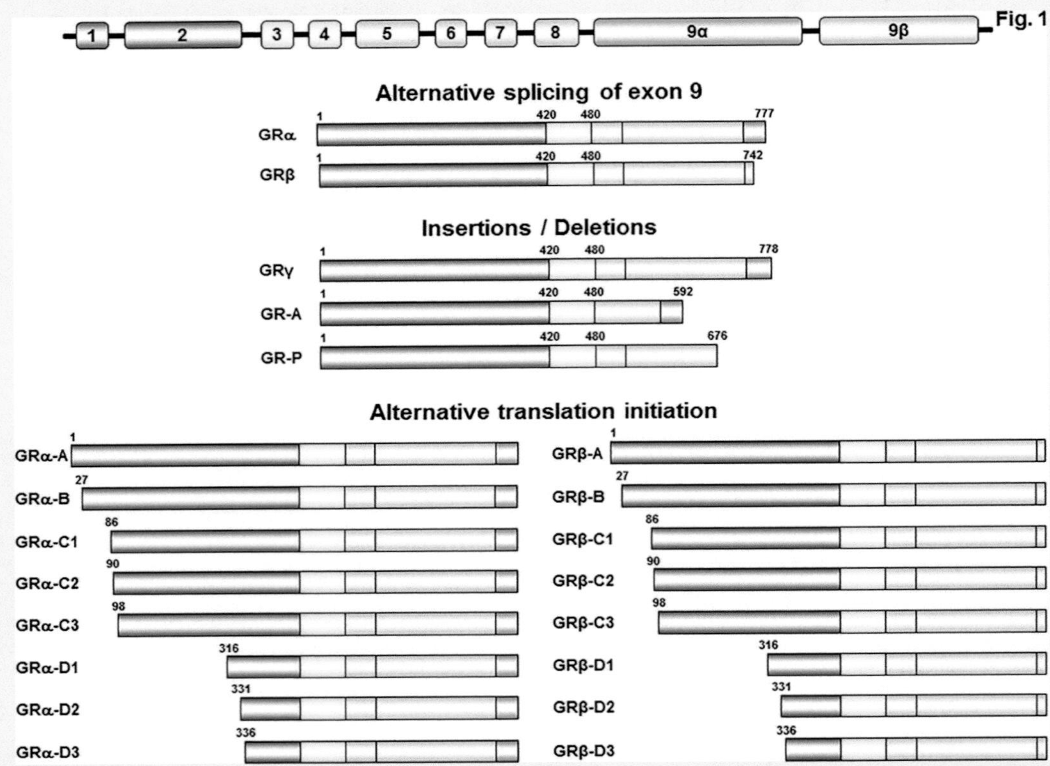

Figure 5.18 Genomic DNA and protein isoforms of the human glucocorticoid receptor (hGR) created through alternative splicing and alternative translation initiation. The hGR gene consists of ten exons. Alternative splicing of exon 9 generates the two main protein isoforms, hGRα and hGRβ. The initiation of hGR mRNA translation from different alternative sites gives rise to eight different protein isoforms.

Source of images: Nicolaides, N.C. and Charmandari, E., 2017. Hormones, 16(2), pp. 124-138. Used with permission.

others, and ultimately led to the surprising discovery of a large superfamily of 48 human nuclear receptors. The endogenous ligands for most of these newly discovered receptors were initially not known, and these receptors were termed "*orphan receptors*". Although ligands have since been identified for over a dozen of these orphan receptors (now termed "*adopted orphans*"), half of the members of this superfamily still remain orphans, and it is postulated that at least some of these may exert their important biological functions independent of any endogenous ligands.

SUMMARY AND SYNTHESIS QUESTIONS

1. Mueller's laboratory in 1961 reported that estrogen treatment induced RNA and protein synthesis. What implications did this have to the mechanism of estrogen receptor action?
2. Studying the phenomenon of "gene puffing" in polytene chromosomes in larval salivary glands of insects, Ulrich Clever's lab found that injection of

ecdysone induced puffing rapidly within 2 hours. What is it about the 2-hour time course that is so interesting?

3. The Ashburner model describing the control of the puffing sequence of insect polytene chromosomes in response to ecdysone proposes two waves of gene expression: early and delayed. If ecdysone was added to cultured salivary glands in the presence of the translational inhibitor, cyclohexamide, what effect would this have on each wave of gene expression?

General Mechanisms of Action

LEARNING OBJECTIVE Describe the mechanisms of action of the three classes of nuclear receptors.

KEY CONCEPTS:

- Class I nuclear receptors function as homodimers and bind all vertebrate steroid hormones.
- Class II nuclear receptors function as heterodimers with RXR.
- Class III nuclear receptors (the orphans) are diverse, with members that function as monomers, homodimers, and heterodimers.

The mechanisms of action of nuclear receptors fall into three major categories: classes I, II, and III receptors.

Class I Nuclear Receptors: The Homodimers

Class I nuclear receptors bind all vertebrate steroid hormones: the sex hormones, glucocorticoids, and mineralocorticoids. Current dogma states that in the absence of ligand, the receptors are inactive and are thought to be sequestered as monomers in the cytoplasm, bound to a class of chaperone proteins called *heat shock proteins* (HSPs) (Figure 5.19). Upon binding hormone, the complex undergoes a conformational change that results in disassociation of the HSPs and release of the monomeric receptor, which translocates to the nucleus. In the nucleus, the activated receptor then homodimerizes with another identical activated receptor, and the homodimer binds to a specific DNA "hormone response element" (HRE) (short DNA sequences, described in detail later) located within the regulatory regions of specific genes. The HRE-bound homodimer recruits various proteins, collectively termed "coactivators", which ultimately promote the transcriptional activation of hormone-response genes. In the case of the glucocorticoid receptor, this model appears to be consistent. However, the cellular localization of the inactive receptor complex is somewhat controversial for other steroid receptors, with cytoplasmic and/or nuclear localization observed depending on the cell type and the conditions examined.[54,55]

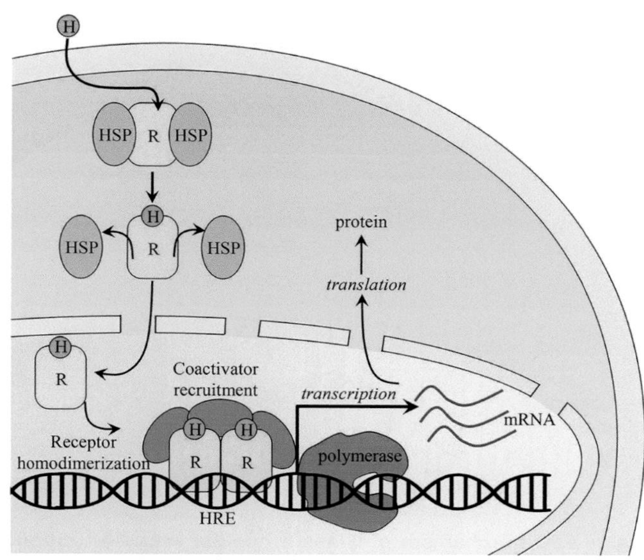

Figure 5.19 Mechanism of Class I nuclear receptor action. Receptors in this class (which include all vertebrate steroid hormone receptors) associate with cytosolic heat shock proteins (HSPs) in the absence of ligand. Hormone binding to the receptor triggers the dissociation of HSPs, translocation to the nucleus, and homodimerization. The ligated homodimer complex binds to a specific sequence of DNA known as a hormone response element (HRE). The nuclear receptor DNA complex in turn recruits coactivator proteins that facilitate transcription of downstream DNA into mRNA, which is eventually translated into protein, which results in a change in cell function.

Class II Nuclear Receptors: The Heterodimers

In vertebrates, the **Class II nuclear receptors** include thyroid hormone receptors (TR), retinoic acid receptor (RAR), and the vitamin D receptor (VDR), and in insects is represented by the ecdysone receptor (EcR). This class also includes adopted orphan receptors, such as peroxisome proliferator-activated receptors (PPARs), retinoid X receptor (RXR), and several others. Many of these adopted orphan receptors form a group sometimes referred to as "*nutrient sensors*", as members bind retinoic acid, fatty acids, bile acids, and cholesterol derivatives that influence metabolism. For example, the adopted PPAR and LXR subfamilies have been shown to orchestrate gene expression programs involved in the control of glucose homeostasis, lipid metabolism, inflammation, and proliferation in the vascular wall. The understanding of their ability to sense and translate metabolic signals into specific gene expression programs has considerably expanded our knowledge on the pathophysiology of metabolic disorders, such as type II diabetes and atherosclerosis.

In marked contrast with all Class I receptors that bind to HREs only in association with a ligand, Class II nuclear receptors display more complex kinetics with several notable differences (Figure 5.20):

1. They do not interact with cytosolic HSPs.
2. They usually form heterodimer partners specifically with RXR for high-affinity DNA binding.

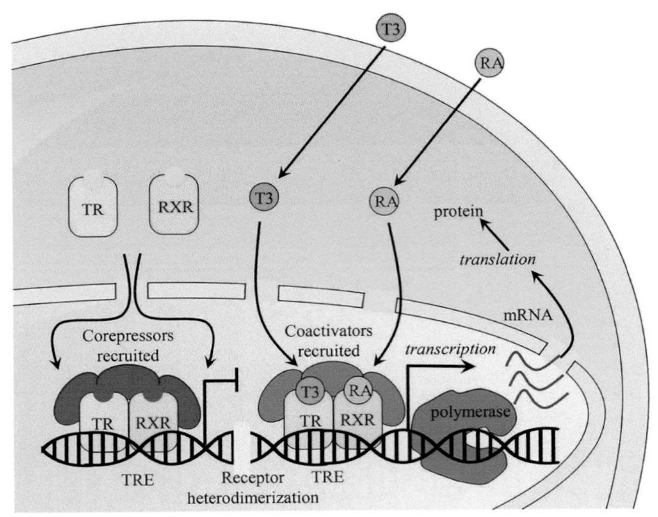

Figure 5.20 Mechanism of Class II nuclear receptor action. Receptors in this class all form heterodimers with the retinoid X receptor (RXR). Regardless of ligand-binding status, the heterodimer complex is located in the nucleus bound to its DNA hormone response element (HRE). The heterodimer serves a dual function as a transcriptional repressor (in the absence of a ligand) and a transcriptional activator (in the presence of ligand). For the purpose of illustration, the nuclear receptor shown here is the thyroid hormone receptor (TR), bound to a thyroid hormone response element (TRE) and heterodimerized to the RXR. In the absence of ligand, the TR recruits corepressor proteins. Ligand binding to TR causes a dissociation of corepressors and recruitment of coactivator proteins, which, in turn, recruits additional proteins such as RNA polymerase that facilitate transcription of downstream DNA into RNA and eventually protein, which results in a change in cell function.

3. In the absence of ligands, the heterodimer still binds to an HRE but recruits corepressor proteins that function to inhibit gene transcription.
4. When bound to ligand, the heterodimer recruits coactivator proteins that induce gene transcription.

Therefore, in contrast with Class I nuclear receptors that only function as transcriptional activators, Class II nuclear receptors exhibit *dual functions*: in the absence of ligand the receptors are transcriptional repressors, whereas in the presence of hormone they are transcriptional activators.

Class III Nuclear Receptors: The Orphans

Of the 48 members of the human nuclear receptor superfamily, almost half are currently considered to be orphan nuclear receptors because they have no known endogenous ligands.[56] For some of these, the search for a natural ligand continues. Others are believed to function naturally as unliganded receptors, the activities of which are regulated by receptor abundance, posttranslational modifications, and/or cofactor recruitment.[57] Although the ligands for some orphan receptors, if any, remain unknown, considerable progress has been made in identifying their regulated target genes and characterizing their physiological functions, which include regulating energy and general metabolism, immunity, growth and reproduction, and many others variables.

SUMMARY AND SYNTHESIS QUESTIONS

1. In the absence of ligand, the glucocorticoid receptor localizes predominantly to the cytoplasm. What prevents its translocation to the nucleus?
2. True or false?: in the absence of ligand, the estrogen receptor binds to DNA.
3. In contrast with Class I receptors, which only function as transcriptional activators, Class II receptors exhibit "dual functions". What is meant by this?
4. The drug tamoxifen is a potent antagonist of classical nuclear estrogen receptors, and is often used as a first line of defense in treating breast cancer. However, up to 50% of breast cancer patients develop resistance to tamoxifen—that is, although tamoxifen still antagonizes the nuclear receptor, some cells still exhibit a pathologically high sensitivity to estrogen. How is this possible?

Structure and Conformational Changes

LEARNING OBJECTIVE Describe structural and conformational changes associated with nuclear receptor-ligand interaction.

KEY CONCEPTS:

- The binding of hormone to a nuclear receptor's ligand-binding domain (LBD) induces conformational changes that activate its DNA-binding domain (DBD).
- The DBD is folded into zinc finger motifs that insert into DNA.
- Nuclear receptors recognize specific DNA sequences called hormone response elements (HREs).
- HREs are hexameric sequences that can be arranged as direct repeats, palindromes, or inverted palindromes.

Structural Domains and Conformational Change

Nuclear receptors share a common overall three-dimensional structure, with several conserved features, including DNA-binding and ligand-binding domains (Figure 5.21). Most nuclear receptors form homo- or heterodimers with another nuclear receptor. The region of greatest homology is the conserved *DNA-binding domain* (DBD) that facilitates its binding to a specific hormone response element with high affinity. The DBD is composed of 66–68 amino acids containing nine perfectly conserved cysteine residues. This domain is folded into two *zinc finger* motifs (Figure 5.22) that are each comprised of four cysteine residues that associate with one zinc ion. The association with zinc permits the tight folding of the protein domain into a finger-like structure. These fingers have the ability to insert into a half-turn of DNA and are structural features of virtually all nuclear receptors.

The *ligand-binding domain* (LBD) selectively binds hormones or metabolites and specifies the biological response. The LBD also contains domains that modulate receptor dimerization, interactions with coactivator or corepressor proteins, and in the case of steroid hormone receptors the binding of heat shock proteins (Figure 5.21). A variable *hinge region* separates the DBD from the LBD, providing flexibility so the receptor can assume various conformations, and also has roles in facilitating dimer formation. For Class I nuclear receptors in the unliganded state, association of the receptor with heat shock and/or other inhibitory proteins is thought to cause the receptor to assume a shape that prevents the DBD from interacting with DNA elements. The binding of a ligand to the LBD, however, induces a conformational change resulting in (1) the dissociation of inhibitory proteins, (2) the binding of the DBD to the DNA hormone response element, and (3) the recruitment of coregulator proteins that modulate gene transcription.

Hormone Response Elements

Hormone response elements (HREs) are short DNA sequences localized predominantly to the regulatory 5′-flanking regions of target genes, typically within several hundred base pairs upstream of the transcription start site. Although some nuclear receptors bind these elements as monomers, most bind as homodimers or heterodimers to their HREs, which usually consist of two hexameric half-sites, of which there are two canonical sequences:

1. **5'-AGAACA-3'** is recognized by glucocorticoid, mineralocorticoid, progesterone, and androgen receptors.
2. **5'-AGGTCA-3'** is recognized by virtually all other nuclear receptors.

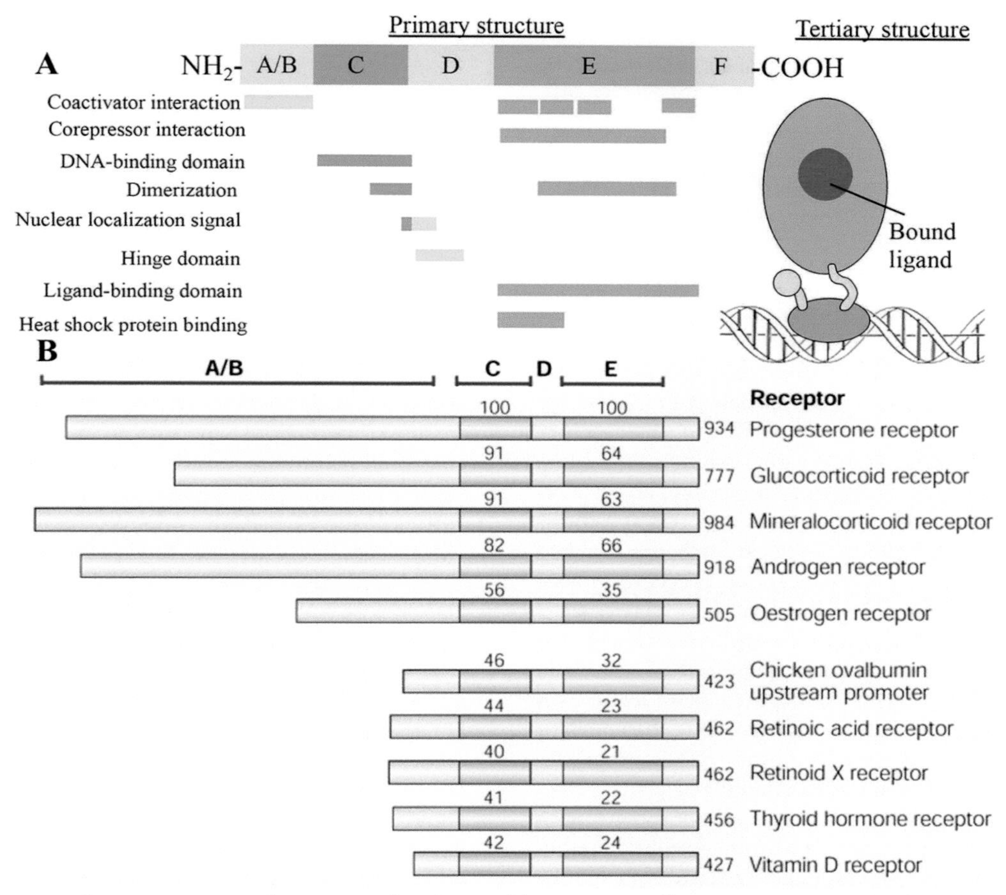

Figure 5.21 The protein structures and conserved amino acid sequence alignments across the human nuclear receptor superfamily. (A) The primary (linear amino acid sequence) and tertiary (3-D) structure of a typical nuclear receptor. **(B)** Amino acid sequence alignments reveal significant conservation of the DNA-binding domain and ligand-binding domain across the human nuclear receptor superfamily. The modular structure is represented by three main domains, A/B, C (ligand-binding domain), and E (DNA-binding domain). The numbers above the bars indicate the percentage amino acid homology of the consensus regions of the DNA- and ligand-binding domains.

Source of images: (B) Tata, J.R., 2002. Nat. Rev. Mol. Cell Biol, 3(9), pp. 702-710. Used with permission.

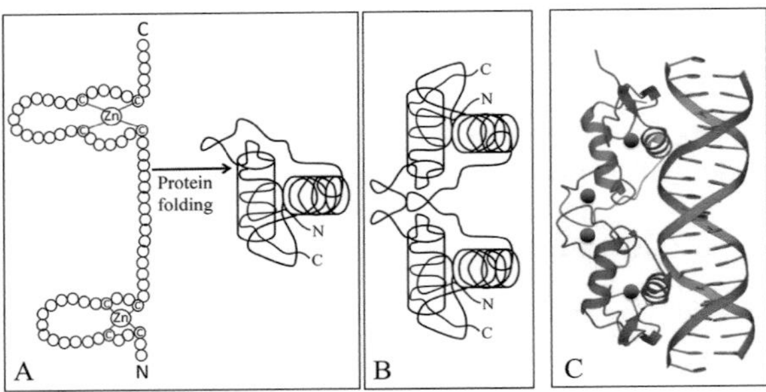

Figure 5.22 Zinc fingers in the steroid hormone receptor DNA-binding domain allow the receptor to bind to specific DNA response elements. (A) DNA-binding domain (DBD) of one receptor. Within the DBD, the association of one zinc atom with four cysteines causes the formation of a "zinc finger" loop. Each of the two zinc fingers folds into an alpha helix barrel shape, and they associate at right angles to each other. **(B)** DNA-binding domains of two dimerized receptors. **(C)** Crystallographic structure of the human estrogen receptor DNA-binding domain dimer (magenta and purple) complexed with a hormone DNA response element (orange and green). Zinc atoms are depicted as gray spheres. Each zinc finger pair interacts with one major groove of the DNA double helix.

Source of images: (A+B) Modified from Aranda, A. and Pascual, A., 2001. Physiol. Rev., 81(3), pp. 1269-1304; (C) Schwabe, J.W., et al. 1993. Cell, 75(3), pp. 567-578.

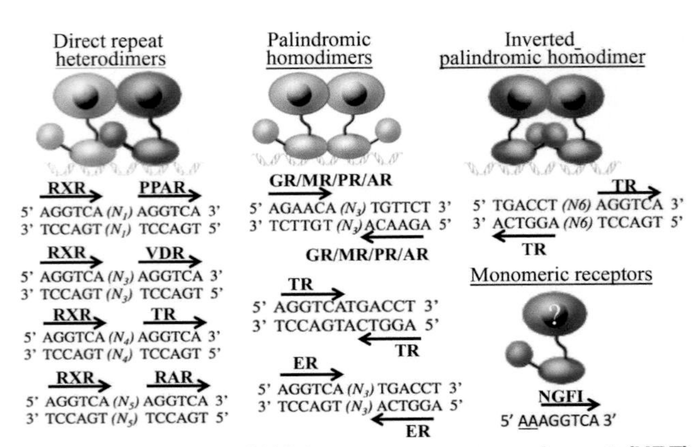

Figure 5.23 Different DNA hormone response element (HRE) structural motifs influence the orientation and dimerization of their nuclear receptors. Direct repeat HREs typically associate with nuclear receptors that form heterodimers with RXR. Palindrome and inverted palindrome HREs are generally bound by nuclear receptors that form homodimers. Some receptors, including many (but not all) orphan receptors, bind to HREs as monomers. Note that thyroid hormone receptors (TR) are unusual in that they can bind to response elements that are arranged as direct repeats (most commonly), palindromes, or inverted palindromes.

Source of images: Glass, C.K., 2006. J. Clin. Investig., 116(3), pp. 556-560.

Analogous to different sentence structures, HREs assume different motifs, which directly influence the orientations of the receptors that bind them (Figure 5.23):

3. *Direct repeats*: Nuclear receptors that form heterodimers with RXR typically bind to HREs arranged as direct repeats of two identical hexameric "half-site" sequences separated by a spacer of one to five nucleotides. Together, these half-site and spacer sequences determine the receptor binding specificity. In general, HREs in this category bind to heterodimeric receptors and exhibit "dual function" modulation of transcription: in the absence of ligand, the **aporeceptors** (unbound receptors) repress transcription below baseline levels, and when associated with a ligand the **holoreceptors** (bound receptors) activate transcription above baseline levels.
4. *Palindromes*: A palindrome is a word or sentence that reads the same forward or backwards, such as "Madam, I'm Adam". In the present context, a palindrome is a nucleotide sequence that reads the same forward or backwards. Glucocorticoid receptors (GR), progesterone receptors (PR), mineralocorticoid receptors (MR), and androgen receptors (AR) bind as homodimers, each to identical HREs consisting of a double-stranded palindrome (an inverted repeat sequence) of two 6-nucleotide half sites separated by a 3-nucleotide spacer. The fact that multiple receptors recognize the same HRE suggests that receptor specificity is provided by additional interactions with nucleotides that lie outside of the core HRE, as well as interactions with different transcriptional coregulators. In general, nuclear receptors that form homodimers bind to palindromic HREs and induce transcription only in the presence of ligand.
5. *Inverted palindromes*: Receptors binding these HREs assume orientations opposite to those of palindromes. Another minor homodimeric form of the thyroid hormone receptor is known to assume this orientation, as are minor forms of the VDR-RXR and several other heterodimers.
6. *Monomeric orphan receptors*: A fourth group of receptors, consisting mostly of orphan receptors, binds to DNA with high affinity as monomers. Members of this group all recognize the canonical AGGTCA half site, but the group can be further subdivided by recognition of a 5'-flanking extension of two to six base pairs outside of the core HRE.[58,59] Importantly, monomeric orphan receptors generally bind constitutively to HREs to induce transcription. As a class, not all orphan receptors bind to HREs exclusively as monomers, and several are known to associate with direct repeat HREs as homodimers, as well as heterodimers partnered with RXR.[60]

Clinical Considerations: Selective estrogen receptor modulators (SERMs)

First synthesized in the early 1960s, the compound tamoxifen was originally classified as an "antiestrogen" due to its estrogen-blocking effects in breast tissue. This pioneering medicine continues to be widely used for the treatment of breast cancer in pre- and postmenopausal women with estrogen receptor-positive tumors, and is also the first drug to be approved by the U.S. Food and Drug Administration for use as a treatment for the prevention of breast cancer in high-risk women.[61] Tamoxifen and another similar estrogen receptor antagonist, raloxifene, exert their antagonistic effects by targeting the estrogen receptor's (ER) alpha helix 12. Compared with the binding of estradiol to the LBD of its receptor, notice how the orientation of alpha helix 12 of the estrogen receptor changes radically in response to the binding of raloxifene to the LBD, which causes the alpha helix to change its conformation, inhibiting activation (Figure 5.24). In contrast to estradiol, ER antagonists, such as raloxifene, contain extra tails that protrude out of the LBD's binding pocket, preventing helix 12 from capping the pocket and therefore inactivating the receptor (Figure 5.25).

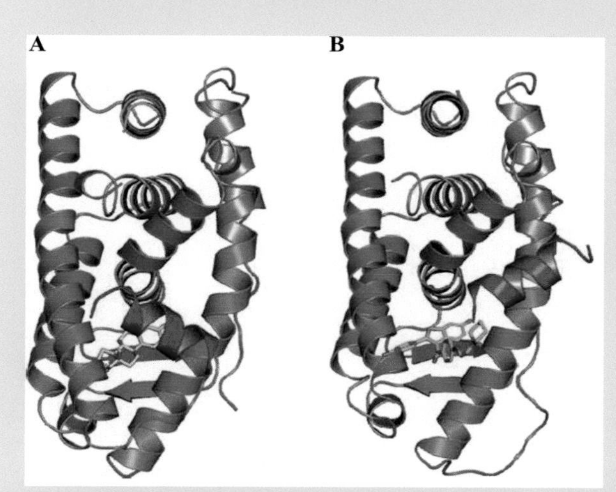

Figure 5.24 Conformational changes in the estrogen receptor (ER) ligand-binding domain (LBD) bound to an agonist or antagonist. The crystal structures of wild-type ERα LBD bound with **(A)** estradiol (an ER agonist) and **(B)** raloxifene (an ER antagonist). Estradiol is shown in yellow, raloxifene is shown in green. Note that when raloxifene binds to the ERα LBD, helix 12 (red) changes its conformation, inhibiting activation.

Source of images: Nuclear Receptors: current Concepts and Future Challenges" (2010) Eds. Bunce and Campbell. Fig 6.5 from: chapter 6: Androgen Receptor. By Dalton and Gao. Used with permission.

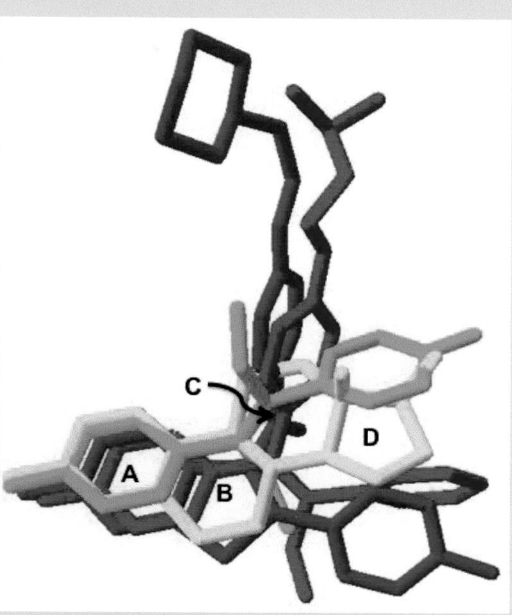

Figure 5.25 A ligand's structure determines the nature of its interaction. The size and shape of the synthetic estrogen agonist diethylstilbestrol (DES, green) is similar to that of estradiol (yellow). Although the structures of synthetic estrogen antagonists tamoxifen (red) and raloxifene (blue) are similar enough to estradiol to bind to the estrogen receptor's ligand-binding domain (LBD), they also contain extra tails that protrude out of the LBD's binding pocket, blocking conformational change by the receptor, resulting in its inactivation.

Source of images: Courtesy of Cresset Group.

However, despite its effectiveness in blocking estrogen action in the breast, by the late 1990s tamoxifen was also found to paradoxically stimulate uterine endometrial cell proliferation, increasing the risk of patients developing endometrial cancer,[62] and also other estrogen-induced cardiovascular maladies such as thromboembolisms. Soon thereafter it was determined that tamoxifen and other similar compounds displayed an unusual tissue-selective pharmacology, functioning as estrogen antagonists in some tissues (e.g. breast and brain), agonists in other tissues (e.g. bone, cardiovascular system, and liver), and mixed agonists/antagonists in the endometrium.[63] Considering these observations, these and other similar pharmacological compounds are now more accurately termed **selective estrogen receptor modulators** (SERMs), compounds that when bound to the estrogen receptor produce a spectrum of responses in a tissue-specific manner, ranging from estrogenic agonism to an antiestrogenic activity. The finding that some SERMs have estrogen-like actions in bone but antiestrogenic effects in the breast has introduced the possibility that both osteoporosis and breast cancer risk can be simultaneously reduced. Indeed, raloxifene was the first SERM approved for the prevention and treatment of osteoporosis that also reduces the incidence of breast cancer.[64] Efforts are currently under way to optimize SERM tissue selectivity for the prevention and treatment of breast cancer and also for alleviating the symptoms of menopause, such as hot flashes and osteoporosis without dangerous side effects.

What provides SERMs with their tissue-specific activity? Many SERM actions can be explained through three synergistic mechanisms[65] (summarized in Figure 5.26):

1. *Differential estrogen-receptor type expression among target tissues*. Estrogen receptors may function as homodimers of estrogen receptor α (ERα), homodimers of estrogen receptor β (ERβ), or as ERα/ERβ heterodimers, with the relative levels of expression of these two receptor isoforms varying among cell types. Whereas ERα homodimers usually function as activators of transcription, ERβ can inhibit the actions of ERα when it heterodimerizes with it. Indeed, whereas tamoxifen and raloxifene appear to function as pure antagonists when acting through ERβ, they operate as partial agonists when acting through estrogen ERα.[66]
2. *Differential estrogen-receptor conformations in association with different ligands*. The binding of different ligands to ERs produces a unique ER conformation for each ligand, ranging from a pure ER agonist-bound conformation to a pure ER antagonist-bound form. SERM-bound ERs may assume a continuum of intermediate conformations.

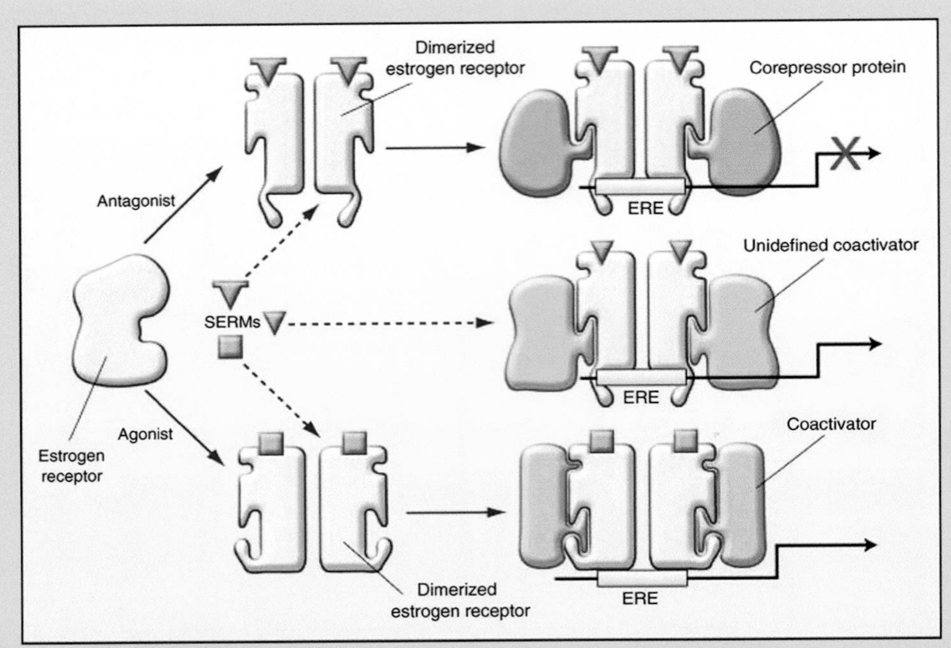

Figure 5.26 Mechanism of SERM action. After binding a SERM (orange shapes), the ER undergoes a conformational change that promotes dimerization with another ER, facilitating the interaction of the dimer with estrogen response elements (ERE) located within the promoter/enhancer regions of target genes. The ligand-bound ER complex allows the binding of various coregulator proteins that promote or inhibit gene transcription. Some estrogen-receptor–SERM complexes favor the recruitment of corepressors (CoR) that, in a given target cell, augment its antagonist activity, whereas others favor the recruitment of coactivator (CoA) that enhance its agonist activity. Some SERMs may allow the ER to interact with as of yet unidentified coactivators (CoAX) with which pure agonists (e.g. estrogens) or pure antagonists would not normally associate, causing the ER to adopt a structure that is intermediate between that observed following the binding of agonists or antagonists. This model implies that SERM activity is influenced by the relative levels of expression of corepressors and coactivators in different target cells.

Source of images: Pickar, J.H., et al. 2010. Maturitas, 67(2), pp. 129-138. Used with permission.

3. *Differential expression of coregulator proteins*. Diverse *coregulator* proteins have been identified that bind to steroid hormone receptors. These coregulators, which may be general or receptor specific, function as either positive transcriptional regulators (*coactivators*) or negative transcriptional regulators (*corepressors*). The relative levels of coregulator protein expression vary among estrogen target cells, and depending on which receptor conformation is induced by the ligand, different coregulator protein combinations will interact with the ER to modulate function. For example, in mammary cells both tamoxifen and raloxifene act as ER antagonists by recruiting corepressors to ERs that target the promoters of estrogen-responsive genes.[67] By contrast, in endometrial cells tamoxifen acts as an ER agonist by recruiting coactivators to ER target promoters, whereas raloxifene does not recruit coactivators and exerts a neutral effect on the endometrium.

   Ultimately, unique ligand-specific ER conformations in combination with variable local concentrations of different ER isoforms and coregulator proteins appear to play key roles in modulating the tissue-selectivity of SERMs.

SUMMARY AND SYNTHESIS QUESTIONS

1. Direct repeats of the HRE sequence 5'-AGGTCA-3' are recognized by many different heterodimer receptors. How is it that these receptors bind to their specific HREs if these sequences are identical?
2. If monomeric orphan receptors don't have ligands, how can they influence gene expression?
3. You have a mutant cell line "A" in which all genes coding for thyroid hormone receptors have been knocked out (deleted) from the genome. You also have a second mutant cell line "B" that has thyroid hormone receptors, but they are all missing their ligand-binding domains. Describe the relative level of expressions of thyroid hormone-responsive genes in these cells under the following conditions: (1) "A" without TH vs. "A" plus TH; (2) "A" without TH vs. "B" without TH; "B" without TH vs. "B" plus TH.

Epigenetic Modifications of Chromatin by Nuclear Receptors

LEARNING OBJECTIVE Describe epigenetic modifications of chromatin by nuclear receptors.

Foundations & Fundamentals: Epigenetic Mechanisms: Histone Tail Modifications

Genes are contained within a complex of nuclear DNA and protein called *chromatin*. About half of the weight of chromatin consists of basic (positively charged) proteins called *histones* that assemble into octamers called *nucleosomes* (Figure 5.27). Nucleosomes are the basic unit of chromatin structure that the negatively charged DNA double helix wraps around. When chromatin is in a relatively transcriptionally inactive state, nucleosomes are thought to be wound into tightly packed structures ranging from 30 nm thick solenoids to 200 nm chromatin fibers that are inaccessible to transcriptional machinery, such as RNA polymerase. In contrast, when chromatin is in a transcriptionally active state it is thought to be arranged in a more loosely packed 11 nm thick "beads on a string" structure, where the beads are individual nucleosomes and the string is DNA. The degree of chromatin condensation influences the accessibility of transcriptional machinery to DNA, and it changes with a cell's life cycle and significantly influences gene expression. For example, the aforementioned gene "puffs" associated with regions of high transcriptional activity in polytene chromosomes of flies are regions of decondensed chromatin (refer to **Developments & Directions** Figure 5.27(A)).

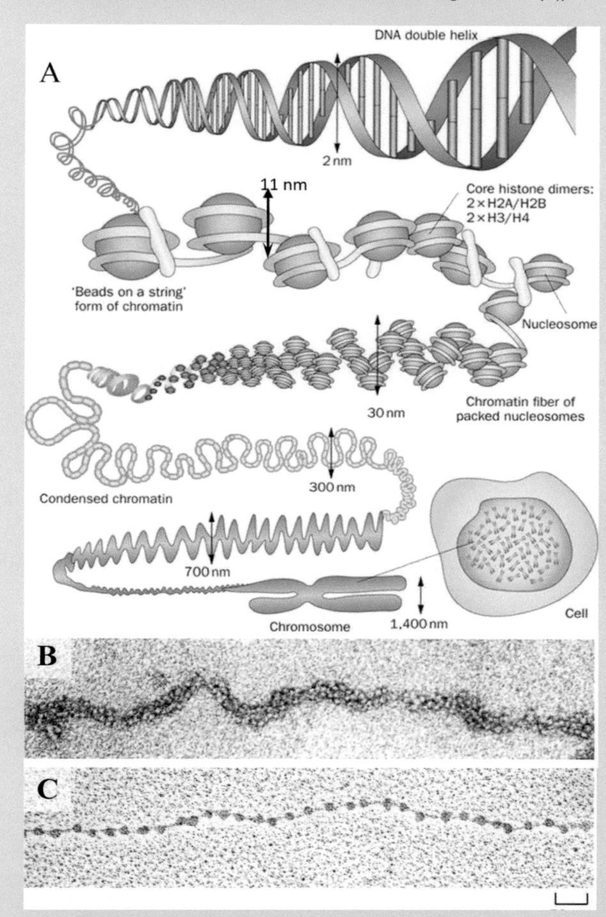

Figure 5.27 The degree of chromatin condensation influences gene transcription. (A) Chromatin is DNA that is complexed to histone and nonhistone nuclear proteins and condenses to form a chromosome. The condensation of chromatin into fibers thicker than the 30 nm "solenoid" appearance promotes a relatively transcriptionally silent state. In contrast, decondensed chromatin of 11 nm thickness (the "beads on a string" appearance) allows greater access of transcriptional machinery to the DNA, and represents a more transcriptionally active state. Electron micrographs depict chromatin in the **(B)** condensed (30 nm thick) and **(C)** decondensed (11 nm thick) states.

Source of images: (A) Tonna, S., et al. 2010. Nat. Rev. Nephrol., 6(6), pp. 332-341; used with permission. (B-C) W. W. Norton & Company, Inc., Essential Cell Biology, Fifth Edition by Bruce Alberts, et al. 2019; used with permission.

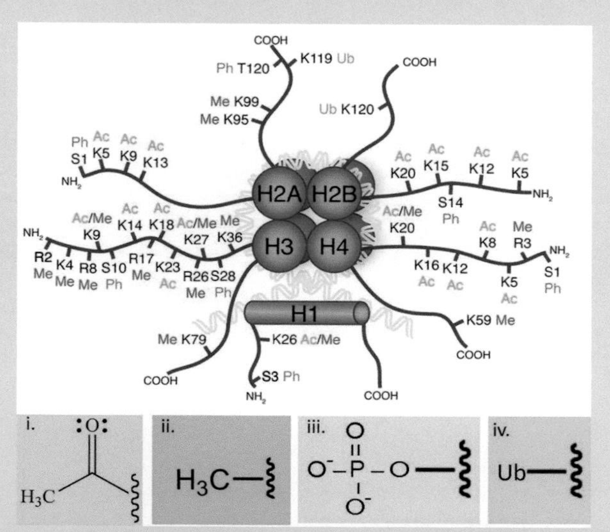

Figure 5.28 Residues in the C- and N-terminus tails of histone proteins that are potentially affected by various covalent modifications. These modifications include (i) acetylation (Ac), (ii) methylation (Me), (iii) phosphorylation (P), and (iv) ubiquitination (Ub) (ubiquitin is a large 76-amino acid protein).

Source of images: (top) Zhao, Y.Q., et al. 2013. Neurotherapeutics, 10(4), pp. 647-663; used with permission.

Changes in chromatin condensation/decondensation and accompanying repression/activation of gene expression are, to a large extent, controlled "epigenetically" by reversible posttranslational modifications of the C and N termini of histone "tails". The term **epigenetic** refers to heritable changes in gene expression that do not result from changes in the underlying DNA sequence. Examples of epigenetic modifications to histone tails include acetylation, methylation, phosphorylation, and ubiquitination (Figure 5.28).

Note that there are additional mechanisms of epigenetic modifications that do not involve histone tails, such as DNA methylation (described in Chapter 14: Adrenal Hormones and the Stress Response) and RNA interference. The two most common categories of histone tail modifications mediated by nuclear receptors are:

1. *Histone acetylation*: the addition of negatively charged acetyl groups (Figure 5.29) by enzymes called **histone acetyltransferases** (HATs) neutralizes the positively charged lysine residues on histone tails, promoting dissociation of the tails from the negatively charged DNA and an overall decondensation of the chromatin, which promotes transcription. In contrast, **histone deacetylase** (HDAC) enzymes promote nucleosome condensation and inhibit transcription.

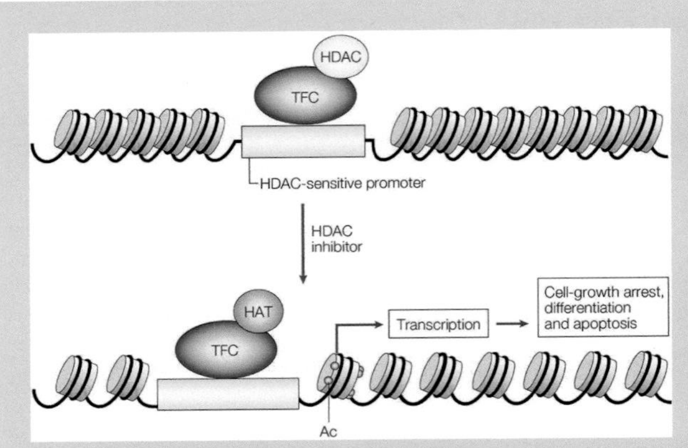

Figure 5.29 Histone acetylases (HATs) promote chromatin decondensation/transcriptional activation, and histone deacetylases (HDACs) promote chromatin condensation/transcriptional silencing. Acetylation neutralizes the positively charged lysine residues on histone tails, promoting dissociation of the tails from the negatively charged DNA and an overall decondensation of the chromatin, which promotes transcription.

Source of images: Marks, P.A., et al. 2001. Nat. Rev. Cancer, 1(3), pp. 194-202. Used with permission.

2. *Histone methylation*: the addition of methyl groups to histones (Figure 5.28) by **histone methyltransferases** (HMTs) can either activate or further repress gene transcription depending on the histone tail and specific residue being methylated and the presence of other histone tail modifications in the vicinity. Methylated amino acids can be recognized by enzymes that influence transcription, which bind the residues.

Less common histone tail modifications associated with nuclear receptors include histone phosphorylation (addition of a negatively charged phosphate group) and ubiquitination (addition of a bulky 76-amino acid protein) (Figure 5.28).

KEY CONCEPTS:

- Epigenetic changes in chromatin condensation influence the local status of gene transcription on a chromosome.
- Histone tail modifications by acetylation and methylation are the most common categories of chromatin remodeling.
- Coregulator proteins recruited by nuclear receptors modify chromatin acetylation and methylation status, serving as transcriptional repressors or activators.

In order to modulate gene transcription, HRE-bound nuclear receptors recruit classes of coregulator proteins collectively termed *corepressors*, which suppress transcription, or *coactivators*, which induce transcription. Table 5.3 lists common nuclear receptor coregulators and their functions. Nuclear receptors heterodimerized with RXR typically exhibit dual functions in transcriptional regulation. In the absence of ligand, aporeceptor heterodimers (such as TR-RXR) bind to their target HREs and recruit corepressor proteins such as the *n*uclear *co*repressor, "*NCoR*",

Table 5.3 Some Nuclear Receptor Coregulator Proteins

Coregulator Class	Function	Name
Coactivator	Histone Acetyltransferases (HATs)	SRC, steroid receptor coactivator NCoA-1, Nuclear receptor coactivator 1 GRIP1, glucocorticoid receptor-interacting protein 1 CBP, CREB-binding protein TRAM-1, thyroid hormone receptor activator molecule 1 P/CAF, P300/CBP-associated factor
	Methylases	CARM1, coactivator-associated arginine methyltransferase 1
		PRMT1, protein arginine methyltransferase 1
	Ubiquitin ligase	E6-AP
Corepressor	Histone deacetylases (HDACs)	NCoR, Nuclear receptor corepressor SMRT, Silencing mediator for RXR and TR
Mediator proteins	Direct contact with basal transcription factors	TRAPs, thyroid hormone receptor-associated proteins DRIPs, Vitamin D receptor-interacting proteins ARC, activator-recruited cofactor

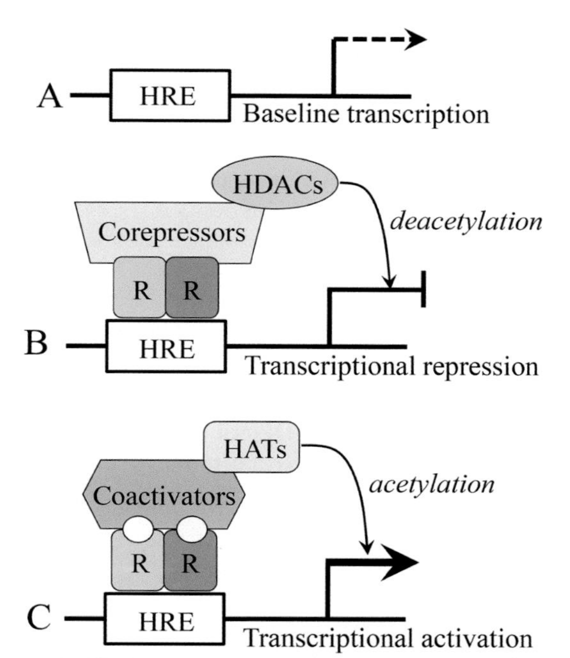

Figure 5.30 Nuclear receptors heterodimerized with RXR function as transcriptional repressors in the absence of ligand, and transcriptional activators in the presence of ligand. (A) In the absence of hormone or receptor, baseline levels of transcription occur. **(B)** In the absence of hormone, Class II receptors heterodimerize, bind to their hormone response element (HRE), and recruit corepressor proteins with histone deacetylase (HDAC) activity, repressing transcription below baseline levels. **(C)** In the presence of ligand, corepressors dissociate and coactivator proteins are recruited. Coactivators typically possess histone acetyltransferase (HAT) activity, which decondenses chromatin and promotes transcription.

and *s*ilencing *m*ediator for *R*AR and *T*R, "*SMRT*"), which recruit other proteins to the complex, such as Class I and II *HDACs* (histone deacetylases). A major activity of this complex is to inhibit transcription from the promoter by histone deacetylation of lysine residues of specific histone tails, placing chromatin in a "closed" state of transcriptional inhibition. The presence of ligand causes the release of the corepressor complex and the recruitment of coactivator complexes, such as "*steroid receptor coactivators*" (SRCs) and *cAMP-response element binding* (CREB) protein. Importantly, these coactivator proteins all possess **histone** *acetyltransferase* (HAT) activity that acetylates specific histones, placing chromatin in an "open" state of transcriptional activation (Figure 5.30).

SUMMARY AND SYNTHESIS QUESTIONS

1. The drug "panobostat" is an inhibitor of the enzyme "*histone deacetylase*". Based on what you know about histone acetylation and its role in histone/DNA interactions, what effect do you predict this drug to have on steroid hormone-induced gene expression (increase, decrease, no effect)?

Summary of Chapter Learning Objectives and Key Concepts

LEARNING OBJECTIVE Describe the different effectors and signal transduction pathways associated with each major plasma membrane receptor class: Ionotropic receptors.

- Ionotropic receptors are ligand-gated ion channels that help modulate a cell's transmembrane potential (voltage).
- Nicotinic cholinergic receptors are acetylcholine-gated cation (Na^+ and K^+) channels.
- GABA receptors are ligand-gated chloride (Cl^-) channels.

LEARNING OBJECTIVE Describe the different effectors and signal transduction pathways associated with each major plasma membrane receptor class: GPCRs.

- GPCRs constitute the largest superfamily of receptors and are the targets of over half of all current pharmaceutical drugs.
- All GPCRs share seven transmembrane (7TM)-spanning helices.
- Most GPCRs transmit signals that interact with intermediary guanine nucleotide-binding proteins called G proteins.
- G proteins modulate effector enzymes that amplify signals and produce second messenger molecules.
- Although hydrophilic hormones are the primary ligands of GPCRs, some lipid-soluble hormones (e.g. eicosanoids and steroids) can also bind to plasma membrane-localized GPCRs.

LEARNING OBJECTIVE Describe the different effectors and signal transduction pathways associated with each major plasma membrane receptor class: Receptors with intrinsic enzymatic activity (receptor tyrosine kinases).

- Many members of the enzyme-linked receptor family are "growth factors", which are mis-expressed in some cancers.
- Members of the "receptor tyrosine kinase" (RTK) family have intracellular domains that become autophosphorylated, allowing interactions with effector proteins.

LEARNING OBJECTIVE Describe the different effectors and signal transduction pathways associated with each major plasma membrane receptor class: Receptors with enzyme-associated activity (recruiter receptors/cytokine receptors).

- Most recruiter receptor ligands are cytokines that are involved in blood cell development and immune system function.
- Several receptors in this family bind to classical hormones (e.g. growth hormone, prolactin, leptin, and erythropoietin).
- An important category of enzymes that associate with recruiter receptors are Janus kinases (JAK), which activate STAT proteins that modulate gene transcription.

LEARNING OBJECTIVE Describe some of the earliest experiments that hinted at the role of nuclear receptors in regulating gene expression.

- There are 48 known nuclear receptors in humans.
- Dysfunctions of nuclear receptors and their ligands are associated with major diseases, such as cancer, osteoporosis, and diabetes.
- Some of the earliest evidence that steroid hormone receptors are involved in gene transcription came from observations of "puffs" in fruit fly polytene chromosomes treated with ecdysone.
- The cloning of the first nuclear receptor led to the discovery of a large superfamily of receptors that includes "orphans" with no known ligands.

LEARNING OBJECTIVE Describe the mechanisms of action of the three classes of nuclear receptors.

- Class I nuclear receptors function as homodimers and bind all vertebrate steroid hormones.
- Class II nuclear receptors function as heterodimers with RXR.
- Class III nuclear receptors (the orphans) are diverse, with members that function as monomers, homodimers, and heterodimers.

LEARNING OBJECTIVE Describe structural and conformational changes associated with nuclear receptor-ligand interaction.

- The binding of hormone to a nuclear receptor's ligand-binding domain (LBD) induces conformational changes that activate its DNA-binding domain (DBD).
- The DBD is folded into zinc finger motifs that insert into DNA.
- Nuclear receptors recognize specific DNA sequences called hormone response elements (HREs).
- HREs are hexameric sequences that can be arranged as direct repeats, palindromes, or inverted palindromes.

LEARNING OBJECTIVE Describe epigenetic modifications of chromatin by nuclear receptors.

- Epigenetic changes in chromatin condensation influence the local status of gene transcription on a chromosome.
- Histone tail modifications by acetylation and methylation are the most common categories of chromatin remodeling.
- Coregulator proteins recruited by nuclear receptors modify chromatin acetylation and methylation status, serving as transcriptional repressors or activators.

LITERATURE CITED

1. Ehrlich P. Chemotherapeutics: scientific principles, methods and results. *Lancet* 1913;182:445–451.
2. Les_Prix_Nobel. Martin Rodbell—Biographical. *Nobel Media*. Les Prix Nobel Website. www.nobelprize.org/nobel_organizations/nobelfoundation/publications/lesprix.html. Published 2014. Accessed.
3. Langley JN. On nerve endings and on special excitable substances. *Proc R Soc Lond*. 1906;78:170–194.
4. Pastan I, Roth J, Macchia V. Binding of hormone to tissue: the first step in polypeptide hormone action. *Proc Natl Acad Sci U S A*. 1966;56(6):1802–1809.
5. Almen MS, Nordstrom KJ, Fredriksson R, Schioth HB. Mapping the human membrane proteome: a majority of the human membrane proteins can be classified according to function and evolutionary origin. *BMC Biol*. 2009;7:50.
6. Croset V, Rytz R, Cummins SF, et al. Ancient protostome origin of chemosensory ionotropic glutamate receptors and the evolution of insect taste and olfaction. *PLoS Genet*. 2010;6(8):e1001064.
7. Ben-Shlomo I, Hsu SY, Rauch R, Kowalski HW, Hsueh AJ. Signaling receptome: a genomic and evolutionary perspective of plasma membrane receptors involved in signal transduction. *Sci STKE*. 2003;2003(187):re9.
8. Proskocil BJ, Sekhon HS, Jia Y, et al. Acetylcholine is an autocrine or paracrine hormone synthesized and secreted by airway bronchial epithelial cells. *Endocrinology*. 2004;145(5):2498–2506.
9. Rodriguez-Diaz R, Dando R, Jacques-Silva MC, et al. Alpha cells secrete acetylcholine as a non-neuronal paracrine signal priming beta cell function in humans. *Nat Med*. 2011;17(7):888–892.
10. Denef C. Paracrinicity: the story of 30 years of cellular pituitary crosstalk. *J Neuroendocrinol*. 2008;20(1):1–70.
11. Cooke JP, Ghebremariam YT. Endothelial nicotinic acetylcholine receptors and angiogenesis. *Trends Cardiovasc Med*. 2008;18(7):247–253.
12. Ben-Shlomo I, Yu Hsu S, Rauch R, Kowalski HW, Hsueh AJ. Signaling receptome: a genomic and evolutionary perspective of plasma membrane receptors involved in signal transduction. *Sci STKE*. 2003;2003(187):RE9.
13. Lander ES, and, many, others. Initial sequencing and analysis of the human genome. *Nature*. 2001;409(6822):860–921.
14. Stevens RC, Cherezov V, Katritch V, et al. The GPCR network: a large-scale collaboration to determine human GPCR structure and function. *Nat Rev Drug Discov*. 2013;12(1):25–34.
15. Lundstrom K. An overview on GPCRs and drug discovery: structure-based drug design and structural biology on GPCRs. *Methods Mol Biol*. 2009;552:51–66.
16. Rodbell M. The role of hormone receptors and GTP-regulatory proteins in membrane transduction. *Nature*. 1980;284(5751):17–22.
17. Northup JK, Sternweis PC, Smigel MD, Schleifer LS, Ross EM, Gilman AG. Purification of the regulatory component of adenylate cyclase. *Proc Natl Acad Sci U S A*. 1980;77(11):6516–6520.
18. Dixon RA, Kobilka BK, Strader DJ, et al. Cloning of the gene and cDNA for mammalian beta-adrenergic receptor and homology with rhodopsin. *Nature*. 1986;321(6065):75–79.
19. Rasmussen SG, DeVree BT, Zou Y, et al. Crystal structure of the beta2 adrenergic receptor-Gs protein complex. *Nature*. 2011;477(7366):549–555.
20. Thompson EB. Comment: single receptors, dual second messengers. *Mol Endocrinol*. 1992;6(4):501.
21. Cohen P. Protein phosphorylation and the control of glycogen metabolism in skeletal muscle. *Philos Trans R Soc Lond B Biol Sci*. 1983;302(1108):13–25.
22. Sutherland EW. Studies on the mechanism of hormone action. *Science*. 1972;177(4047):401–408.
23. Herzig S, Long F, Jhala US, et al. CREB regulates hepatic gluconeogenesis through the coactivator PGC-1. *Nature*. 2001;413(6852):179–183.
24. Hagiwara M, Brindle P, Harootunian A, et al. Coupling of hormonal stimulation and transcription via the cyclic AMP-responsive factor CREB is rate limited by nuclear entry

of protein kinase A. *Mol Cell Biol.* 1993;13(8):4852–4859.

25. Norman AW, Mizwicki MT, Norman DP. Steroid-hormone rapid actions, membrane receptors and a conformational ensemble model. *Nat Rev Drug Discov.* 2004;3(1):27–41.
26. Meizel S, Turner KO, Nuccitelli R. Progesterone triggers a wave of increased free calcium during the human sperm acrosome reaction. *Dev Biol.* 1997;182(1):67–75.
27. Schwartz Z, Sylvia VL, Larsson D, et al. 1alpha,25(OH)2D3 regulates chondrocyte matrix vesicle protein kinase C (PKC) directly via G-protein-dependent mechanisms and indirectly via incorporation of PKC during matrix vesicle biogenesis. *J Biol Chem.* 2002;277(14): 11828–11837.
28. Horvath G, Vasas S, Wanner A. Inhaled corticosteroids reduce asthma-associated airway hyperperfusion through genomic and nongenomic mechanisms. *Pulm Pharmacol Ther.* 2007;20(2):157–162.
29. Rodrigo GJ. Inhaled corticosteroids in the treatment of asthma exacerbations: essential concepts. *Arch Bronconeumol.* 2006;42(10):533–540.
30. Wanner A, Horvath G, Brieva JL, Kumar SD, Mendes ES. Nongenomic actions of glucocorticosteroids on the airway vasculature in asthma. *Proc Am Thorac Soc.* 2004;1(3):235–238.
31. Revankar CM, Cimino DF, Sklar LA, Arterburn JB, Prossnitz ER. A transmembrane intracellular estrogen receptor mediates rapid cell signaling. *Science.* 2005;307(5715):1625–1630.
32. Prossnitz ER, Arterburn JB, Smith HO, Oprea TI, Sklar LA, Hathaway HJ. Estrogen signaling through the transmembrane G protein-coupled receptor GPR30. *Annu Rev Physiol.* 2008;70:165–190.
33. Scaling AL, Prossnitz ER, Hathaway HJ. GPER Mediates estrogen-induced signaling and proliferation in human breast epithelial cells and normal and malignant breast. *Horm Cancer.* 2014.
34. Lappano R, Pisano A, Maggiolini M. GPER function in breast cancer: an overview. *Front Endocrinol.* 2014;5:1–6.
35. Prossnitz ER, Barton M. The G-protein-coupled estrogen receptor GPER in health and disease. *Nat Rev Endocrinol.* 2011;7(12):715–726.
36. Levin ER. Plasma membrane estrogen receptors. *Trends Endocrinol Metab.* 2009;20(10):477–482.
37. Levi-Montalcini R. Effects of mouse tumor transplantation on the nervous system. *Ann N Y Acad Sci.* 1952;55(2):330–344.
38. Cohen S. Isolation of a mouse submaxillary gland protein accelerating incisor eruption and eyelid opening in the new-born animal. *J Biol Chem.* 1962;237:1555–1562.
39. Cohen S. The stimulation of epidermal proliferation by a specific protein (EGF). *Dev Biol.* 1965;12(3):394–407.
40. Goodsell DS. The molecular perspective: the ras oncogene. *Stem Cells.* 1999;17(4):235–236.
41. Downward J. Targeting RAS signalling pathways in cancer therapy. *Nat Rev Cancer.* 2003;3(1):11–22.
42. Baker SJ, Rane SG, Reddy EP. Hematopoietic cytokine receptor signaling. *Oncogene.* 2007;26(47):6724–6737.
43. Vilcek J, Feldmann M. Historical review: Cytokines as therapeutics and targets of therapeutics. *Trends Pharmacol Sci.* 2004;25(4):201–209.
44. Germain P, Staels B, Dacquet C, Spedding M, Laudet V. Overview of nomenclature of nuclear receptors. *Pharmacol Rev.* 2006;58(4):685–704.
45. Overington JP, Al-Lazikani B, Hopkins AL. How many drug targets are there? *Nat Rev Drug Discov.* 2006;5(12): 993–996.
46. Jensen EV, Jacobson HI. Fate of steroid estrogens in target tissue. In: Pincus G, Vollmer EP, eds. *Biological Activities of Steroids in Relation to Cancer.* Academic Press; 1960:161–178.
47. Jensen EV. On the mechanism of estrogen action. *Perspect Biol Med.* 1962;6:47–59.
48. Mueller GC, Gorski J, Aizawa Y. The role of protein synthesis in early estrogen action. *Proc Natl Acad Sci U S A.* 1961;47:164–169.
49. Clever U, Karlson P. Induktion von puff-vera¨nderungen in den speicheldru¨ senchromosomen von Chironomus tentans durch ecdyson. *Exp Cell Res.* 1960;20:623–626.
50. Clever U. Actinomycin and puromycin: effects on sequential gene activation by ecdysone. *Science.* 1964;146(3645): 794–795.
51. Ashburner M, Chihara C, Meltzer P, Richards G. Temporal control of puffing activity in polytene chromosomes. *Cold Spring Harb Symp Quant Biol.* 1974;38:655–662.
52. Lu NZ, Cidlowski JA. Translational regulatory mechanisms generate N-terminal glucocorticoid receptor isoforms with unique transcriptional target genes. *Mol Cell.* 2005;18(3):331–342.
53. Oakley RH, Cidlowski JA. The glucocorticoid receptor. In: Bunce CM, Campbell MJ, eds. *Nuclear Receptors: Current Concepts and Future Challenges,* Vol 8. Springer; 2010:63–90.
54. Gorski J, Hansen JC, Welshons WV. Estrogen receptors as nuclear proteins. *Adv Exp Med Biol.* 1987;230:13–29.
55. Klinge CM, Rao CV. The steroid hormone receptors. In: Sciarra JJ, ed. *Gynecology and Obstetrics CD-ROM,* Vol 5. Lippincott Williams and Wilkins; 2004.
56. Riggins RB, Mazzotta MM, Maniya OZ, Clarke R. Orphan nuclear receptors in breast cancer pathogenesis and therapeutic response. *Endocr Relat Cancer.* 2010;17(3):R213–231.
57. Hummasti S, Tontonoz P. Adopting new orphans into the family of metabolic regulators. *Mol Endocrinol.* 2008;22(8):1743–1753.
58. Harding HP, Lazar MA. The monomer-binding orphan receptor Rev-Erb represses transcription as a dimer on a novel direct repeat. *Mol Cell Biol.* 1995;15(9):4791–4802.
59. Meinke G, Sigler PB. DNA-binding mechanism of the monomeric orphan nuclear receptor NGFI-B. *Nat Struct Biol.* 1999;6(5):471–477.
60. Mangelsdorf DJ, Evans RM. The RXR heterodimers and orphan receptors. *Cell.* 1995;83(6):841–850.
61. Jordan VC. Chemoprevention of breast cancer with selective oestrogen-receptor modulators. *Nat Rev Cancer.* 2007;7(1):46–53.
62. Barakat RR. Tamoxifen and endometrial cancer. *Cancer Control.* 1996;3(2): 107–112.
63. Katzenellenbogen BS, Katzenellenbogen JA. Biomedicine. Defining the "S" in SERMs. *Science.* 2002;295(5564): 2380–2381.
64. Cummings SR, Eckert S, Krueger KA, et al. The effect of raloxifene on risk of breast cancer in postmenopausal women: results from the MORE randomized trial. Multiple Outcomes of Raloxifene Evaluation. *JAMA.* 1999;281(23):2189–2197.
65. Riggs BL, Hartmann LC. Selective estrogen-receptor modulators—mechanisms of action and application to clinical practice. *N Engl J Med.* 2003;348(7): 618–629.
66. Hall JM, McDonnell DP. The estrogen receptor beta-isoform (ERbeta) of the human estrogen receptor modulates ERalpha transcriptional activity and is a key regulator of the cellular response to estrogens and antiestrogens. *Endocrinology.* 1999;140(12):5566–5578.
67. Shang Y, Brown M. Molecular determinants for the tissue specificity of SERMs. *Science.* 2002;295(5564):2465–2468.

CHAPTER 6

Receptor Binding Kinetics

CHAPTER LEARNING OBJECTIVES:

- Describe important variables that influence the strength of the response of a tissue to a hormone.
- Discuss the mechanisms that regulate receptor sensitivity.
- Explain how a ligand–receptor's equilibrium dissociation constant (K_d) influences the strength of the response of a tissue to a hormone.

OPENING QUOTATION:

"The secret of life is molecular recognition; the ability of one molecule to 'recognize' another through weak bonding interactions."

—Linus Pauling (1984) 25th anniversary of the Institute of Molecular Biology at the University of Oregon

Receptors Distinguish Distinctly

LEARNING OBJECTIVE Describe important variables that influence the strength of the response of a tissue to a hormone.

KEY CONCEPTS:

- The strength of the response of a tissue to a hormone depends on
  1. Hormone concentration
  2. Receptor number
  3. The specificity of a hormone to its receptor
  4. The affinity of the receptor for the hormone
- The binding of hormones to receptors is a saturable process.
- Hormone and receptor concentrations vary in a spatial-temporal manner.

In proposing the existence of "chemoreceptors" for drugs, immunologist and Nobel laureate Paul Ehrlich described the specific binding of a ligand to its receptor as resembling a "lock and key" mechanism.[1] In the bloodstream and extracellular fluid there exists a myriad of different molecules that are structurally similar to hormones, including sterols, amino acids, peptides, and proteins. Whereas these molecules generally circulate at concentrations in the micro- to millimolar (10^{-6} to 10^{-3} mol/L) range, hormones are present at *much* lower concentrations, typically in the pico- to nanomolar (10^{-12} to 10^{-9} mol/L) range (refer to Table 2.1 of Chapter 2: Fundamental Features of Endocrine Signaling). Target cells, therefore, must distinguish not only among different hormones present in miniscule quantities, but also between any given hormone and the 10^6 - to 10^9-fold excess of other similar molecules. This extraordinary degree of discrimination is provided by plasma membrane- and intracellular-localized receptors.

We have come a long way from the simple "lock and key" model of hormone–receptor interaction and now know that the strength of the response of a tissue to a hormone depends on many variables, four of which will be addressed in this chapter: (1) hormone concentration, (2) receptor number, (3) the specificity of a hormone to its receptor, and (4) the affinity of the receptor for the hormone.

Hormone Concentration

Hormone concentrations vary in a spatial-temporal manner since some tissues not only degrade hormones at faster rates than others, but rates of hormone synthesis and degradation also change with time of day, month, season, or year. At any given moment in time, there are a finite number of functional receptors for a hormone on or in a target cell. The greater the concentration of hormone is, the greater the number of bound receptors becomes until all the receptors are occupied. Therefore, the binding of hormones to receptors is a "saturable" process (Figure 6.1). The biological response of a cell to hormone is related to the degree of receptor occupancy, and dose-response curves typically increase in a logarithmic manner until the point of receptor saturation is reached. Importantly, note how a small change in hormone concentration at the low end and middle of the curve will have a much greater effect on the response than a similar change in hormone concentration of similar magnitude at the high end of the curve.

DOI: 10.1201/9781003359241-8

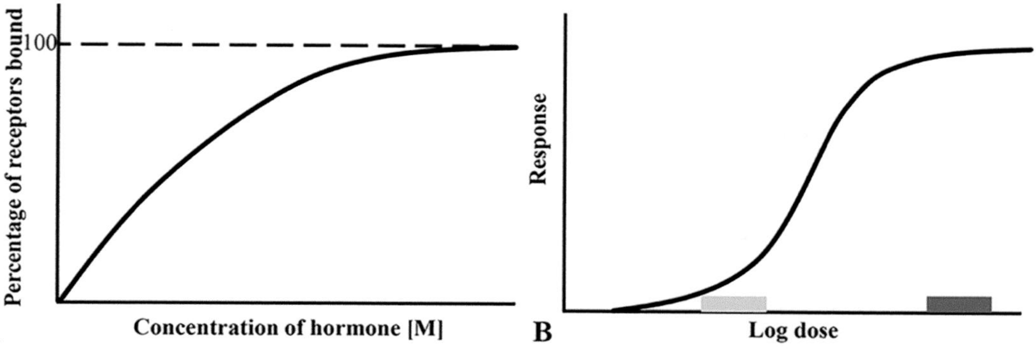

Figure 6.1 The binding of hormone to receptor is a saturable process. (A) The greater the concentration of the messenger the greater the response. This relationship holds until all the receptors become bound to messenger and are occupied. At this point the receptors are saturated. **(B)** As the dose of hormone increases, the response increases in a logarithmic manner until the point of saturation. Note that a small change in hormone concentration at the low end of the curve (gray box) will have much greater effects on the response than a similar change in hormone concentration at the high end of the curve (black box).

Source of Images: (B) Bergman, Å., et al. and World Health Organization, 2013. State of the science of endocrine disrupting chemicals 2012. World Health Organization. Used with permission.

Table 6.1 Estrogen Receptor Content in Target and Nontarget Tissues

Tissue	Receptor Content (pmol/100 mg wet wt.)
Uterus	5.92
Vagina	2.15
Pituitary	1.43
Hypothalamus	0.5–0.10
Kidney	0.20
Diaphragm	0.06
Spleen	0.02

Source: After J.H. Clark and E.J. Peck 1977 Steroid hormone receptors: basic principles and measurement. In: *Receptors and Hormone Action* I Ed. B.W. O'Malley and Lutz Birnbaumer pp. 383–410. Academic Press.

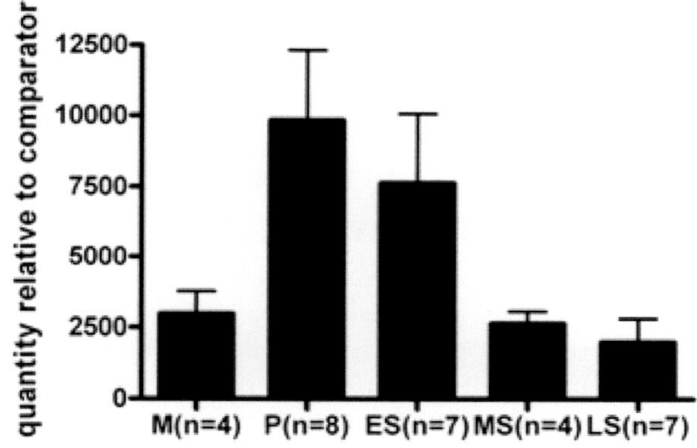

Figure 6.2 Changes in human endometrial estrogen receptor α (ERα) mRNA during different stages of the uterine cycle. M, menstrual; P, proliferative; ES, early secretory; MS, mid-secretory; LS, late secretory.

Source of Images: Bombail, V., et al. 2008. Hum. Reprod., 23(12), pp. 2782-2790; used with permission.

Receptor Number

Like hormone concentration, receptor number also varies in a spatial-temporal manner. For example, although estrogen receptors are present in many tissues, their concentrations are orders of magnitude higher in target tissues, such as the uterus, compared with non-target tissues (Table 6.1). In addition, the expression levels of estrogen receptors in the uterus fluctuate dramatically over time with the changing uterine cycle (Figure 6.2).

SUMMARY AND SYNTHESIS QUESTIONS

1. The concentrations of some hormones are identical throughout the bloodstream. Considering this, how can these hormones have "target tissues" that respond more strongly to the hormones than other tissues?
2. When hormone concentrations are very low, a doubling of hormone concentration will produce a large biological response. When hormone concentrations are already high, a doubling of hormone concentration will produce a comparatively much lower magnitude of change in biological response. Explain the differential response to increasing hormone concentrations at low versus high doses.
3. Why does a small change in hormone concentration at the low end and middle of the curve produce a much greater effect on the response than a similar change in hormone concentration of similar magnitude at the high end of the curve?

Receptor Desensitization, Sequestration, and Up-/ Downregulation

LEARNING OBJECTIVE Discuss the mechanisms that regulate receptor sensitivity.

KEY CONCEPTS:

- The number of receptors present in cells is not fixed but is dynamically up- and downregulated.
- The continuous binding of a hormone to its receptor may promote receptor desensitization, sequestration, or downregulation, all of which reduce cell sensitivity to hormone exposure.
- The absence of hormone–receptor interactions promotes receptor upregulation, which increases cell sensitivity to subsequent hormone exposure.
- "Spare" receptors increase the sensitivity of a cell to low levels of hormone and to small changes in hormone concentration.

Modifying Sensitivity to Hormones by Receptor Up- and Downregulation

The biological response to a given dose of hormone depends on receptor concentration, and the number of receptors present in a given cell is not fixed, but is dynamically regulated. Figure 6.3 portrays data showing how as receptor concentration decreases, a maximum biological response becomes unachievable, irrespective of how high the dose of hormone becomes. Furthermore, as the receptor concentration increases, *sensitivity* to the hormone also increases; that is, it takes significantly less hormone to produce the same biological response. The level of sensitivity to a hormone by a cell can be modulated over time by increasing or decreasing functional receptor numbers (Figure 6.4). Whereas the addition of more functional

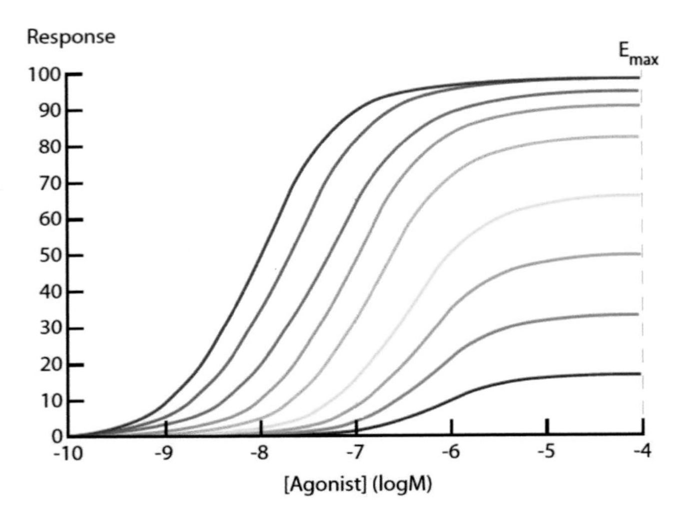

Figure 6.3 The dose response to the hormone depends on receptor concentration. As the receptor concentration increases, it takes significantly less hormone to produce the same response. At low hormone receptor levels, the maximum response is reduced.

Source of Images: Bergman, Å., et al. 2013. State of the science of endocrine disrupting chemicals 2012. World Health Organization. Used with permission.

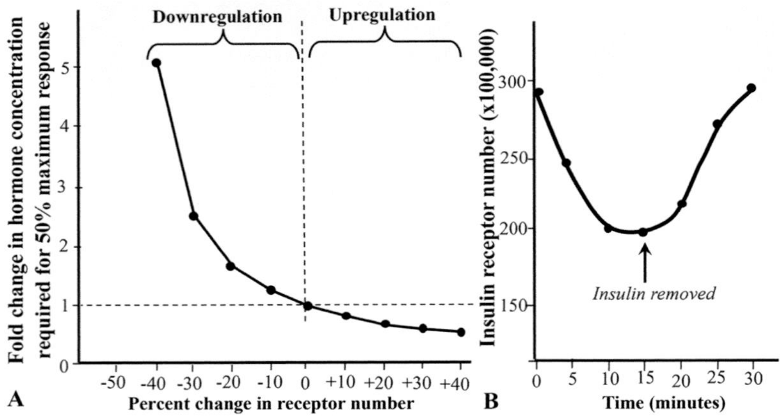

Figure 6.4 A cell's sensitivity to hormones can be modulated via upregulation and downregulation of receptors. (A) Generalized effects of up- or downregulation of receptor number on sensitivity to hormonal stimulation. Note that the abscissa is plotted on a logarithmic scale. **(B)** Effect of insulin on the number of cell-surface insulin receptors. Adipocyte cells were incubated in the presence of 100 ng/ml of insulin from 0 to 15 minutes, and then in the absence of insulin from 15 to 30 minutes. At the indicated times adipocytes were washed to remove extracellular and receptor-bound ligand, and then incubated briefly with radiolabeled insulin to measure surface receptor numbers.

Source of Images: (B) Modified from Marshall, S., 1985. J. Biol. Chem., 260(7), pp. 4136-4144.

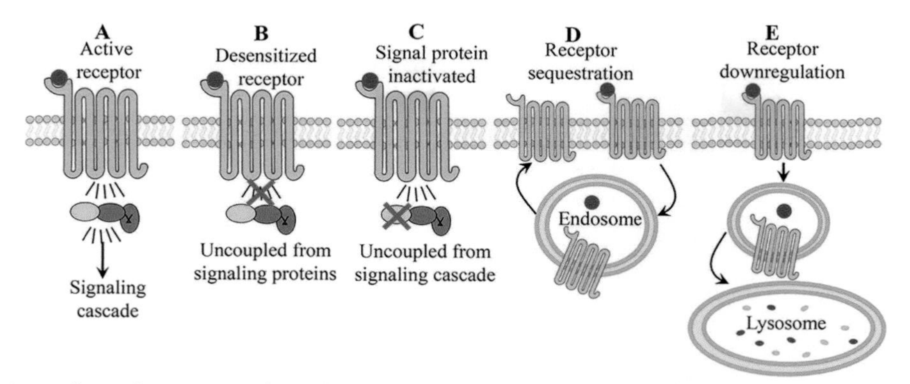

Figure 6.5 Four categories of receptor deactivation. (A) A typical active GPCR signaling pathway. **(B)** Receptor desensitization caused by an uncoupling from its signaling proteins. **(C)** Receptor desensitization caused by inactivation of the signaling proteins. **(D)** Receptor sequestration occurs following removal of the receptor from the plasma membrane and storage in recycling endosomes. **(E)** Receptor downregulation occurs following removal of the receptor from the plasma membrane, followed by its catabolism by the lysosome.

receptors increases the sensitivity of the organism to low hormone levels, reducing functional receptor numbers is an important homeostatic mechanism that protects the organism from the potentially detrimental effects of hormone excess. The attenuation of signaling can be accomplished either via ligand-induced receptor *desensitization* (an uncoupling of the receptor from its downstream signaling pathway), by temporary *sequestration* of the receptor in a recycling endosome following endocytosis, or by the permanent *downregulation* of receptors via lysosomal degradation[2] (Figure 6.5). For example, in the continuing presence of ligand, many GPCRs become desensitized by proteins called *G protein receptor kinases* (GRKs) that phosphorylate the C-terminus of the GPCR, increasing the GPCR's affinity for a protein called *arrestin* that binds to the receptor and prevents it from associating with trimeric G proteins. Arrestin can further target the receptor for endocytosis, leading to receptor downregulation (Figure 6.5). The phenomenon of downregulation in response to a continuous presence of ligand has practical relevance. For example, in order to obtain the optimal therapeutic benefit of hormone therapy, the intervals of hormone administration should be spaced to avoid the downregulation period. Two well-studied examples of receptor downregulation are receptors for vasopressin and for gonadotropin-releasing hormone (GnRH) (Figure 6.6).

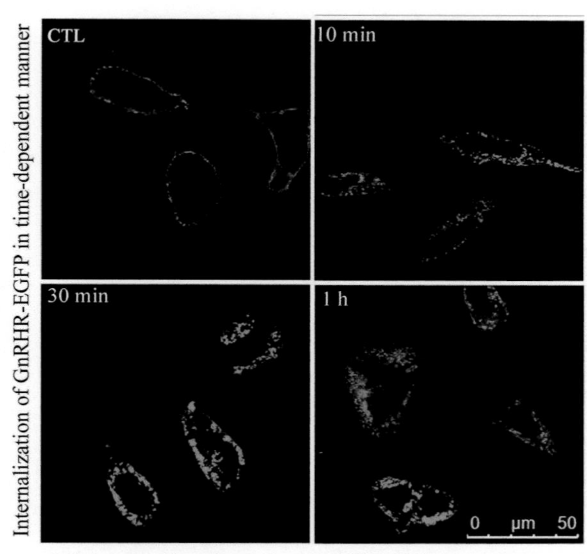

Figure 6.6 Internalization of a GnRHR-EGFP fusion protein by cells occurs in a dose-dependent manner. Cultured cells transfected with GnRHR-EGFP plasmid were activated by treatment with GnRH ligand for different durations of time. In the absence of GnRH (control, CTL), receptors are localized primarily in the plasma membrane. In the presence of GnRH receptor agonist, many of the receptors have become increasingly internalized in recycling endosome vesicles over time.

Source of Images: Yang, J., et al. 2017. JoVE (122), p. e55514. Used with permission.

Whereas the continuous binding of a hormone to its receptor generally promotes receptor downregulation, the absence of hormone–receptor interactions typically promotes the reverse effect—receptor *upregulation*, or an increase of receptor numbers on the cell surface, making the cell more responsive to subsequent exposure to the hormone. Some hormones, such as prolactin and GnRH, induce the transcription, translation, and upregulation of their own receptors, acting as positive regulators.[3–5] More often, hormones positively regulate receptors for other hormones. Well-known examples of *heterologous positive regulation* are the stimulation of luteinizing hormone (LH) receptors by follicle-stimulating hormone (FSH) in the ovarian follicle[6] and the stimulation of beta-adrenergic receptors by thyroid hormone (TH) in the heart muscle.[7]

Spare Receptors

In theory, the biological response of a target cell to a hormone is directly proportional to the number of receptor sites occupied, with the maximum biological response achieved when 100% of the receptor sites are occupied. However, in most cases the maximum biological response

Developments & Directions: Insulin receptor downregulation in response to hyperinsulinemia in lean and obese mice

In humans, obesity often predisposes individuals to developing type 2 diabetes, a condition characterized in part by elevated blood glucose levels. Because blood glucose is high in an obese individual, the β cells of the pancreatic islets of Langerhans must release more insulin than normal to meet the demand for glucose uptake and return the blood to homeostatic levels. In cell culture models, high plasma insulin concentrations have been shown to downregulate insulin receptors on the plasma membrane by increasing the rate of ligand-induced internalization.[8–10] Although the causes of insulin resistance in obese type 2 diabetes patients are particularly complex, the downregulation of insulin receptors in response to a near-constant increase in blood insulin levels is thought to contribute, at least in part, to insulin resistance.

In human and rodent models of obesity, functional insulin receptor levels are reduced.[11–16] Interestingly, Gletsu and colleagues[17] have shown that following a glucose meal, obese mice downregulate their insulin receptors at a significantly greater magnitude compared with lean mice (Figure 6.7), supporting the notion that insulin receptor downregulation contributes to insulin resistance under conditions of obesity.

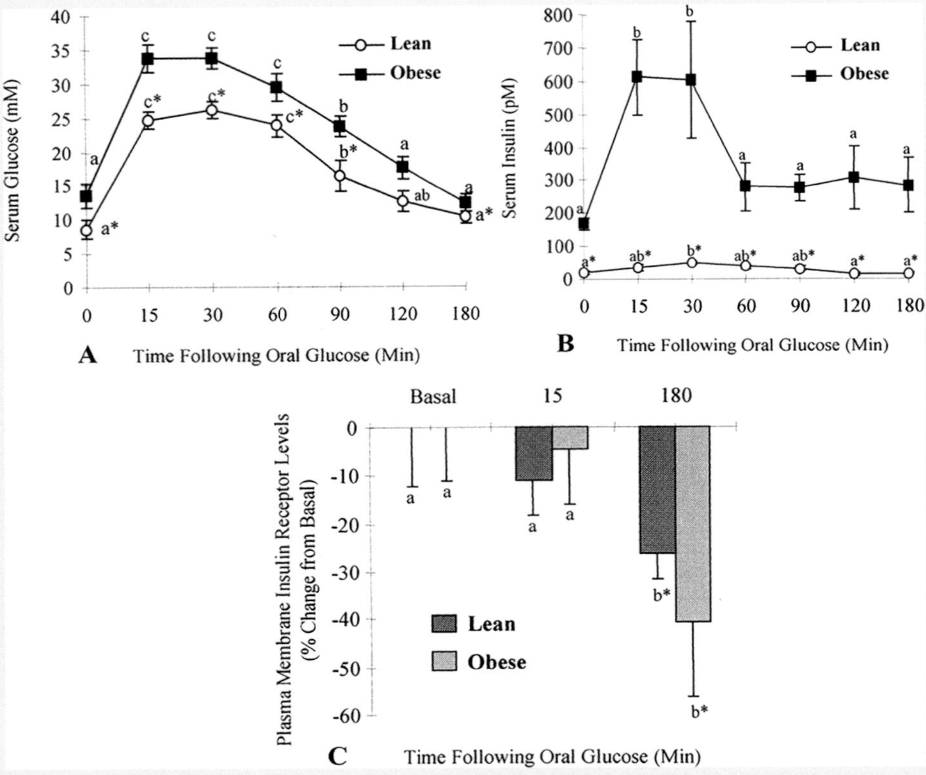

Figure 6.7 Insulin receptor downregulation in response to hyperinsulinemia in lean and obese mice. Compared with lean mice, obese mice display elevated serum glucose **(A)** and insulin **(B)** responses to a glucose meal. **(C)** In response to hyperinsulemia, obese mice display a significantly greater downregulation of insulin receptors from the cell surface, compared with lean mice.

Source of Images: Gletsu, N.A., et al. 1999. Biochim Biophys Acta Mol Basis Dis., 1454(3), pp. 251-260. Used with permission.

is actually achieved at concentrations of hormone that are significantly lower than those required to occupy all of the receptors (Figure 6.8). For example, in cultured adipocytes only a 2%–3% receptor occupancy by insulin is required to achieve the maximum stimulation of glucose oxidation, and in Leydig cells of the testes only a 1% receptor occupancy by gonadotropin is needed for maximum testosterone production.[18] In these examples, 97% or more of the receptors are referred to as **spare receptors**, or the receptors that exist in excess of the minimum number required for a full biological response (Figure 6.8). Importantly, the term "spare receptors" does not mean that the receptors are not being used, but rather that the maximum biological response manifests when all of the receptors on a target cell are occupied on average about 3% of the time.[18] Spare receptors appear to serve two primary functions. First, they increase the sensitivity of a cell to low levels of hormone and to small changes in hormone concentration. Second, considering that occupied plasma membrane receptors typically become internalized, potentially resulting in receptor desensitization or downregulation, the presence of spare receptors may prolong the duration of the biological response in the presence of slowly declining hormone concentrations.

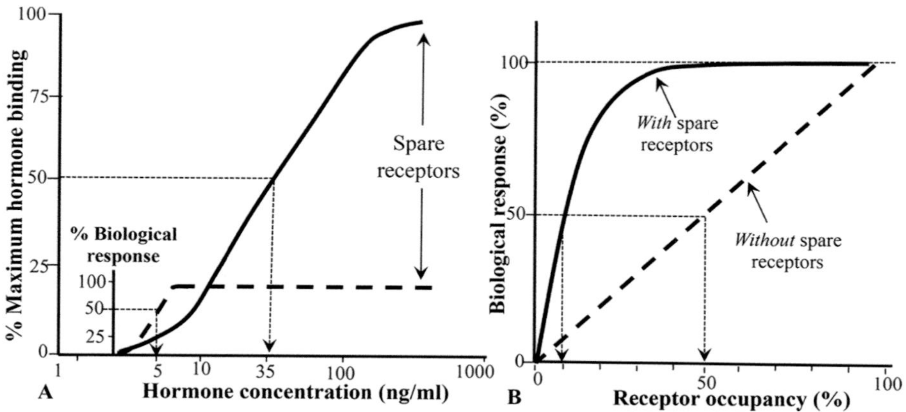

Figure 6.8 "Spare" receptors are those that exist in excess of the minimum number required for a full biological response. **(A)** In this example, the minimum concentration of hormone needed for a half maximal biological response (5 ng/ml) is considerably lower than that needed to occupy half of the receptors (35 ng/ml). **(B)** In this example, without spare receptors a 50% biological response corresponds with a 50% occupancy; in contrast, with spare receptors a 50% response is achieved at about 10% receptor occupancy.

SUMMARY AND SYNTHESIS QUESTIONS

1. When treating a patient using hormone therapy, the timing of treatment is critical. If the space between consecutive administration of the hormone is too brief, this may desensitize the patient's tissues to the hormone. Explain why this is true.
2. Your friend has characterized a certain cell line and found that, on average, each cell has 1000 receptors for ligand X. This friend has further determined that the cell line exhibits a maximum biological response to ligand X under 10% receptor occupancy. Based on these observations, your friend hypothesizes that these cells have more receptors for ligand X than they need. He further predicts that if he alters the cell line such that it now only has 100 receptors for ligand X, that it should still be able to exhibit a maximal biological response. Why is your friend's reasoning incorrect?

Hormone–Receptor Specificity and Affinity

LEARNING OBJECTIVE Explain how a ligand–receptor's equilibrium dissociation constant (K_d) influences the strength of the response of a tissue to a hormone.

KEY CONCEPTS:

- The equilibrium dissociation constant (K_d) is a measure of a receptor's affinity to a ligand.
- Saturation binding assays measure both the number of receptors and their affinity for a hormone.
- Competitive binding assays are used to measure a receptor's binding specificity to different ligands of variable structural resemblance.

A receptor's attraction to a ligand is a function of the ligand's structure and the extent of non-covalent interactions with amino acid residues in the hormone-binding "pocket" of the receptor (Figure 6.9). Hydrophobic interactions generally provide the driving force for the binding reaction, and electrostatic interactions (hydrogen bonds, van der Waals interactions, ionic bonds) provide both the specificity and affinity.[18] For example, although epinephrine and norepinephrine are structurally very similar molecules, epinephrine has a closer association and a much higher affinity to the β2-adrenergic receptor compared to norepinephrine. Pharmaceutical companies take advantage of receptor specificity to manufacture or purify ligand analogs that are structurally similar to the original hormones. If an endogenously produced or exogenously derived natural or synthetic hormone analog occupies the receptor and mimics the effects of the endogenous hormone, potentially with much higher potency, it is referred to as a receptor *agonist*. Hormone analogs that occupy the receptor but inhibit activation by the hormone are referred to as receptor *antagonists*.

Curiously Enough . . . Opium, a substance that is derived from the poppy plant (*Papaver somniferum*), is the source of many narcotics, like morphine, heroine, and codeine. Opium binds to human GPCR "opioid receptors" and mimics the endogenously produced peptide hormones, enkephalins and endorphins.

Hormone–Receptor Affinity

Although many molecules in the extracellular fluid display weak molecular interactions with plasma membrane proteins, most do not exhibit physiologically relevant ligand–receptor interactions. In addition to possessing a high specificity to a hormone, a receptor must also possess a very high affinity to its ligand that allows binding to take place at extremely low concentrations. The greater the affinity of the receptor for the hormone is, the greater the percentage of receptor

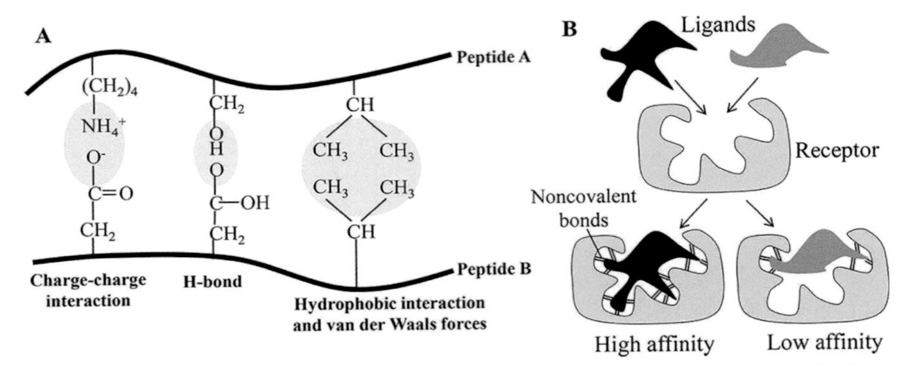

Figure 6.9 Non-covalent interactions provide specificity and affinity of ligands to receptors. (A) Examples of non-covalent bonds include ionic bonds (attractions between opposite charges), hydrogen bonds (hydrogen shared between electronegative atoms), van der Waals forces (fluctuations in electron clouds around molecules that oppositely polarize neighboring atoms), and hydrophobic interactions (hydrophobic amino acids interact with each other to exclude water molecules). **(B)** Ligand–receptor pairs with more non-covalent interactions have higher affinities than those with fewer interactions.

bound and the biological response will be (Figure 6.10). The binding of a hormone, H, to its receptor, R, is a reversible reaction with kinetics that can be described as:

$$[H]+[R] \underset{k_{off}}{\overset{k_{on}}{\rightleftarrows}} [HR] \quad \text{or} \quad [H]\bullet[R]\,k_{on} = [HR]\,k_{off}$$

where k_{on} is the rate of association (number of binding events/time), k_{off} is the rate of dissociation (number of dissociation events/time), [H] and [R] are the concentrations of free (unbound) hormone and receptor (respectively), and [HR] is the concentration of bound hormone–receptor complex.

The affinity of a hormone for its receptor is commonly defined quantitatively in terms of the **equilibrium dissociation constant (K_d)**, where K_d is the concentration of hormone [H] at equilibrium that is required for binding of 50% of its receptor [R] sites:

$$K_d = k_{off} / k_{on} = \frac{[H]\bullet[R]}{[HR]}$$

Since K_d is equal to the ratio of k_{off}/k_{on}, a receptor with a high affinity to a hormone would be expected to have a relatively low K_d value, since the hormone's rate of association (k_{on}) is greater than its rate of dissociation (k_{off}). Therefore, K_d is an inverse measure of receptor affinity, and the higher the affinity of a receptor is for a hormone, the lower is the K_d value. In fact, typical hormone receptors bind their ligands with extremely low K_d values ranging from 10^{-11} to 10^{-8} M.[19]

Importantly, many congenital endocrine disorders caused by mutations in either hormone structure or in a receptor's ligand-binding domain are characterized by significantly altered K_d (Table 6.2), which interferes with normal binding affinity. In most cases, such mutations result in "loss-of-function" phenotypes, where K_d values are significantly increased and binding affinity is lower. However, in some cases a "gain of function" mutation can result in a *higher* affinity of the ligand for its receptor (lower K_d), activating the receptor even when concentrations of hormone are low.

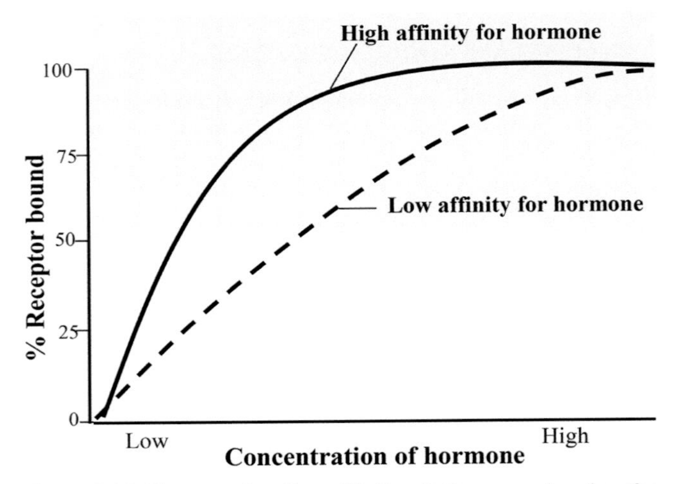

Figure 6.10 The greater the affinity of the receptor for the hormone, the greater the response.

Radioligand Binding Studies

The defining features of hormone receptors described in the previous section can be analyzed by radioligand binding assays. A *radioligand* is a radioactively labeled hormone or drug that binds to a specific receptor. Common radioactive isotopes that are covalently attached to ligands include tritium (^3H) and iodine (^{125}I). The amount of radioactive label in a sample can be measured very precisely using equipment such as a gamma or scintillation counter. Two main uses of radioligands are for conducting (1) saturation binding and (2) competitive binding experiments.

Saturation Binding Assays

Saturation binding assays measure both the number of receptors and their affinity for a hormone. These experiments typically obtain receptors from intact cells, partially purified plasma membranes, or cell extracts of target cells. In these procedures the receptors are incubated in increasing concentrations of radioligand, which will bind in a

Table 6.2 Normal and Mutant K_d Values for Diseases Where Receptors Are Mutated but Ligands Are Normal

Receptor Class	Disease	Mutated Receptor (specific location)	Normal Receptor-Ligand K_d	Diseased Receptor-Ligand K_d
GPCR	Hypogonadism	GnRH receptor (A1290)	1.6 nM	No binding detectable
GPCR	Hereditary nephrogenic diabetes insipidus	Vasopressin type 2 receptor (^{205}Tyr◊Cys)	1.8 nM	19.8 nM
GPCR	Hyperfunctioning thyroid nodule (gain of function)	TSH receptor (L629F)	2.2 mU/ml	0.8 nM
RTK	Insulin resistance	Insulin receptor (R118C)	30.71 nM	125.8 nM
Cytokine	Growth hormone insensitivity	Growth hormone receptor (eD152H)	1.01 nM	1.70 nM
Nuclear	Androgen insensitivity	Androgen receptor (^{855}Arg◊His)	0.59 nM	3.03 nM

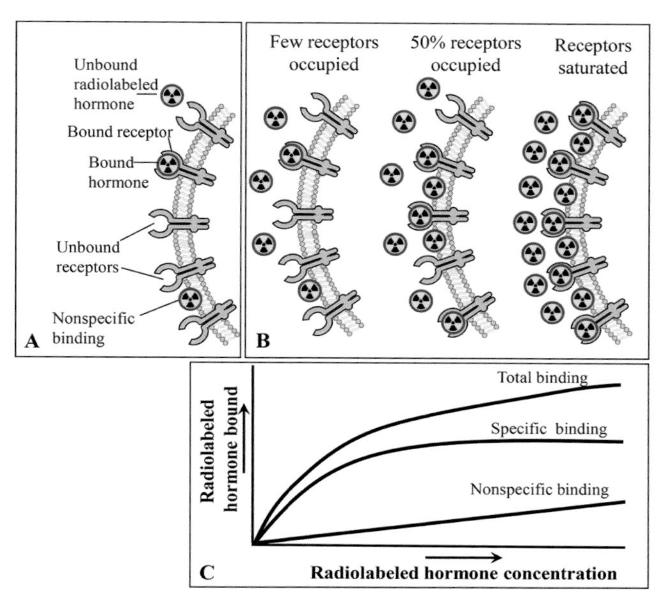

Figure 6.11 Saturation binding assays measure the number of receptors as well as their affinity for a hormone. (A) Radiolabeled hormone binds specifically to its plasma membrane receptors, but also nonspecifically to other parts of the cell membrane. **(B)** Increasing concentrations of radiolabeled hormone corresponds with increased receptor binding, but also to increased nonspecific binding. **(C)** Saturation receptor binding curves in a typical saturation binding assay showing the hypothetical total binding and nonspecific binding for increasing levels of radiolabeled hormone added to assay tubes. Whereas nonspecific binding tends to increase linearly with concentration, specific binding increases logarithmically. Specific binding at each level is determined by subtracting nonspecific binding from total binding.

Source of Images: Hadley, M. E., Levine, J. E., 2007. Endocrinology. Pearson Prentice Hall.

concentration-dependent manner that varies with their affinity (Figure 6.11). After binding has taken place at each concentration, the bound and unbound radioligand fractions are separated from each other, typically by centrifugation or filtration, and the total amount of bound (insoluble) and unbound (soluble) radioligand is measured.

Importantly, there are two categories of bound fraction that must be distinguished: *specifically bound*, where ligand is bound to its receptor, and *nonspecifically bound*, where ligand is bound to protein and/or membrane that is not its receptor. These can be distinguished by repeating the aforementioned incubation with radioligand at increasing concentration, except this time in the presence of 100- to 1,000-fold excess *un*labeled (i.e. non-radioactive) ligand. Whereas the specifically bound fraction will display typical asymptotic maximal saturation binding kinetics in the presence of the excess unlabeled ligand, the nonspecifically bound fraction will continue to increase linearly and indefinitely, since the nonspecific binding sites are so plentiful they are not subject to displacement by the excess unlabeled hormone (Figure 6.11). Therefore, for any given concentration, the specific binding can be determined by simply subtracting the nonspecific binding curve from the total binding curve.

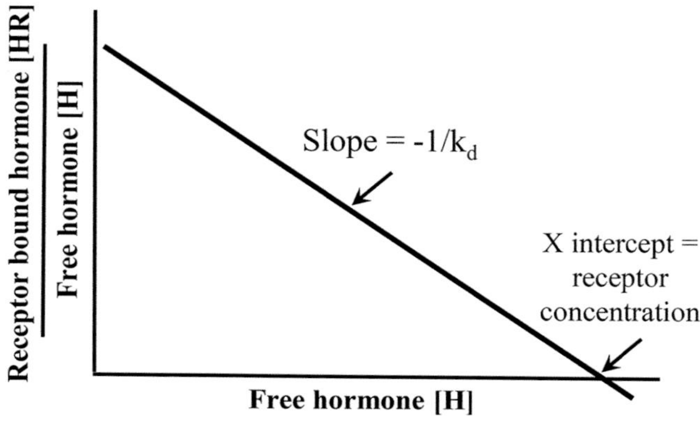

Figure 6.12 A Scatchard plot of hypothetical data derived from a saturation binding assay. Two important types of information are derived from a Scatchard plot: the receptor concentration and the dissociation constant (K_d). [HR] denotes the concentration of hormone bound to receptor. [H] denotes free, or unbound, hormone.

A particularly useful approach to measuring receptor affinity is to derive a *Scatchard plot*, a plot of the ratio of concentrations of bound to unbound ligand versus the bound ligand concentration (Figure 6.12). Conveniently, Scatchard analysis yields a straight line from which two important variables can be derived: (1) the ligand–receptor's *dissociation constant* (K_d) (where $1/K_d$ is equal to the slope of the line), as well as (2) the *total receptor number* (R_0) (which is equal to the x-intercept).

Competitive Binding Assays

Whereas receptor number and affinity can be derived from saturation binding assays and Scatchard plot analysis, a receptor's binding *specificity* to different ligands of variable structural resemblance is typically evaluated using a **competitive binding assay.** In these assays, increasing amounts of a test compound (hormone or drug) are incubated with cells expressing the receptor in question, in the presence of a fixed amount of radiolabeled hormone already known to bind to the receptor. If the test compound displays any specific binding affinity to the receptor, then increasing concentrations of that test compound should competitively displace the radioligand (Figure 6.13). A popular alternative to radiolabeled competitive binding assays is called an *enzyme-linked immunosorbent assay* (ELISA) (Figure 6.14).

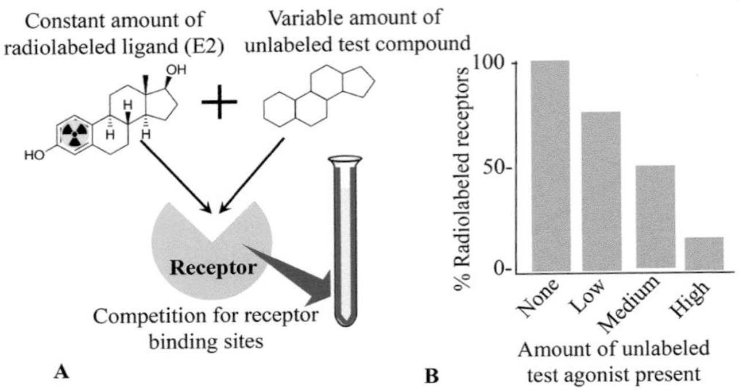

Figure 6.13 Competitive binding assays measure a receptor's binding specificity to a test compound in the presence of a known radiolabeled agonist. (A) In this example, radiolabeled estradiol (E2) is competing with an unlabeled test compound for binding to estrogen receptors that have been coated onto the sides of a test tube. **(B)** If the test compound competitively displaces the known radioligand binding, then the test compound has a measurable affinity to the receptor.

Studies using these assays have provided valuable insight into understanding the binding specificity of receptors to hormones that belong to large families of similarly structured members, such as steroid hormones. For example, estradiol, progesterone, and testosterone have similar overall structures, yet of the three only estradiol displays high specificity to the estrogen receptor in competitive binding assays (Figure 6.15).[20] Interestingly, the synthetic compound diethylstilbesterol (DES) is not a steroid but has a high specificity and affinity to the estrogen receptor (Figure 6.15). DES was mistakenly believed to reduce the risk of pregnancy complications in the 1940s and 1950s, and its use during pregnancy resulted in the occurrence of rare cancers upon reaching adulthood. Competitive receptor binding assays are not only useful for determining if intentionally designed pharmaceutical compounds display specificity and affinity to particular receptors, but they can also determine the degree to which other synthetic compounds released into the environment that are designed for non-pharmacological purposes (such as pesticides, fire-retardants, and plasticizers) display specificity and affinity to hormone receptors, and subsequently assess their potential to "disrupt" endogenous endocrine signaling in humans and wildlife.

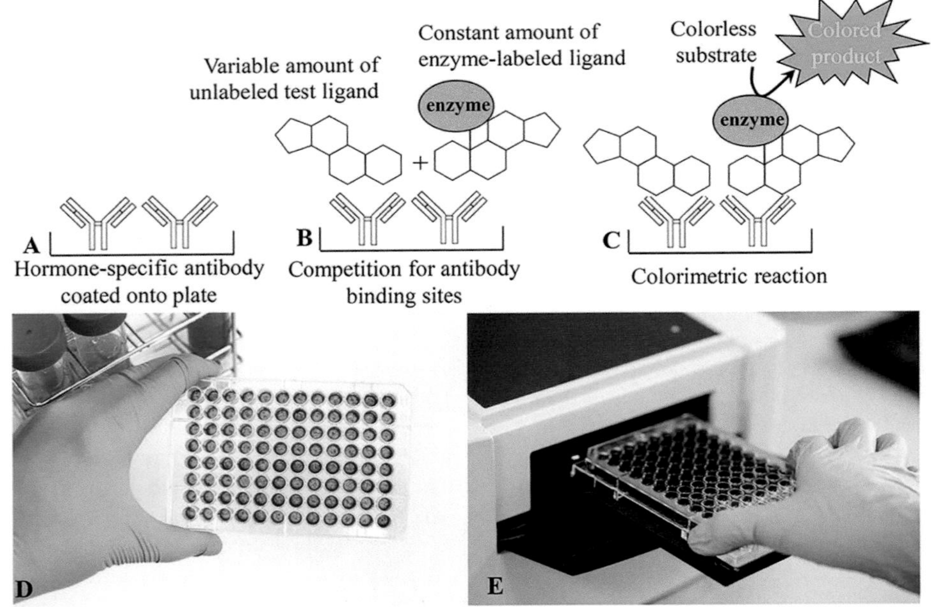

Figure 6.14 An enzyme-linked immunosorbent assay (ELISA). (A) Plastic 96-well plates are coated with a hormone-specific antibody. **(B)** A fixed amount of hormone covalently linked to an enzyme, such as horseradish peroxidase (HRP), competes for antibody binding sites on the side of a microtiter plate with the sample containing an unknown concentration of unlabeled hormone. **(C)** After the competitive binding has taken place, the amount of antibody-bound HRP-linked hormone is measured by adding a colorless substrate, such as tetramethylbenzidine (TMB) that reacts with HRP to generate a color change, in this case to blue. **(D)** The variable amount color change among wells reflects different concentrations of hormone in the samples. **(E)** The intensities of the colorimetric changes are quantifiable using a microplate reader.

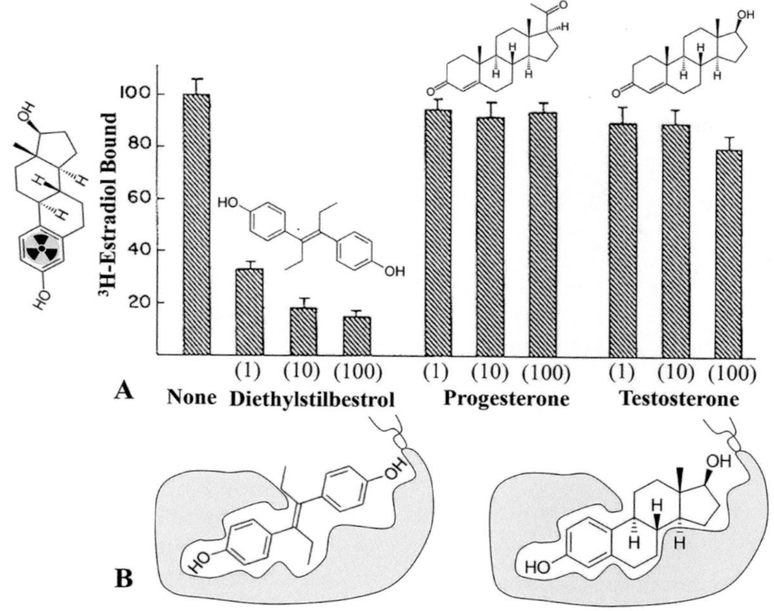

Figure 6.15 Steroid-binding specificity of the estrogen receptor. (A) Effect of various synthetic and natural hormones on ^{3}H-estradiol binding in nuclear fractions of rat uterus. Unlabeled hormone was mixed with ^{3}H-estradiol at the ratios in parentheses (unlabeled hormone/[3H]-estradiol). 3H-estradiol bound in the absence of unlabeled ligand was taken as 100% or total binding. Despite the similar steroid hormone structures, progesterone and testosterone don't bind to the estrogen receptor, whereas diethylstilbesterol (DES) has a high affinity to the receptor. **(B)** Both estradiol and DES have similarly spaced hydroxyl (–OH) groups that facilitate their specific binding to the estrogen receptor (orange). In contrast, neither progesterone nor testosterone possess both hydroxyl groups required for binding specificity.

Source of Images: (A) Clark JH, Peck EJ. Steroid hormone receptors: basic principles and measurement. In: O'Malley BW, Birnbaumer L, eds. Receptors and Hormone Action I Academic Press; 1977:383-410; used with permission.

SUMMARY AND SYNTHESIS QUESTIONS

1. Beta endorphin is a 31 amino acid-long peptide hormone with a molecular weight of 3465 g/mol. Morphine is a small opium-derived alkaloid (molecular weight = 285 g/mol). Considering that beta endorphin is greater than an order of magnitude larger in size than morphine, how is it possible that morphine functions as a potent agonist to the beta endorphin receptor?
2. There are two ligands that have varying affinity to a particular receptor. The K_d for ligand A is 2 nM, and the K_d for ligand B is 1.5 nM. Which ligand has the greater affinity for the receptor? Explain your reasoning.
3. In a saturation binding assay, how is it possible to differentiate receptor specific binding from non-specific binding?
4. This figure (**Summary and Synthesis Figure 6.16**) portrays a Scatchard plot of radiolabeled [^{125}I] bombesin binding to a membrane fraction isolated from H-128 SCLC tumors.[21] Estimate (a) the total receptor concentration in pM and (b) the K_d in *nM*.

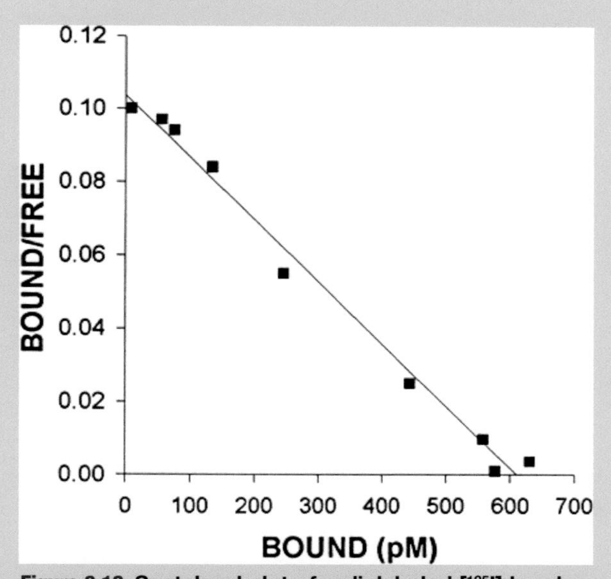

Figure 6.16 Scatchard plot of radiolabeled [^{125}I] bombesin binding to a membrane fraction isolated from H-128 SCLC tumors.

Source of Images: Halmos G, Schally AV. Proc Natl Acad Sci USA. 1997;94(3):956-960. Used with permission.

Summary of Chapter Learning Objectives and Key Concepts

LEARNING OBJECTIVE Describe important variables that influence the strength of the response of a tissue to a hormone.

- The strength of the response of a tissue to a hormone depends on
  1. Hormone concentration
  2. Receptor number
  3. The specificity of a hormone to its receptor
  4. The affinity of the receptor for the hormone
- The binding of hormones to receptors is a saturable process.
- Hormone and receptor concentrations vary in a spatial-temporal manner.

LEARNING OBJECTIVE Discuss the mechanisms that regulate receptor sensitivity.

- The number of receptors present in cells is not fixed but is dynamically up- and downregulated.
- The continuous binding of a hormone to its receptor may promote receptor desensitization, sequestration, or downregulation, all of which reduce cell sensitivity to hormone exposure.
- The absence of hormone–receptor interactions promotes receptor upregulation, which increases cell sensitivity to subsequent hormone exposure.
- "Spare" receptors increase the sensitivity of a cell to low levels of hormone and to small changes in hormone concentration.

LEARNING OBJECTIVE Explain how a ligand–receptor's equilibrium dissociation constant (K_d) influences the strength of the response of a tissue to a hormone.

- The equilibrium dissociation constant (K_d) is a measure of a receptor's affinity to a ligand.
- Saturation binding assays measure both the number of receptors and their affinity for a hormone.
- Competitive binding assays are used to measure a receptor's binding specificity to different ligands of variable structural resemblance.

LITERATURE CITED

1. Prull CR. Part of a scientific master plan? Paul Ehrlich and the origins of his receptor concept. *Med Hist.* 2003;47(3):332–356.
2. Shankaran H, Wiley HS, Resat H. Receptor downregulation and desensitization enhance the information processing ability of signalling receptors. *BMC Syst Biol.* 2007;1:48.
3. Posner BI, Kelly PA, Friesen HG. Prolactin receptors in rat liver: possible induction by prolactin. *Science.* 1975;188(4183):57–59.
4. Katt JA, Duncan JA, Herbon L, Barkan A, Marshall JC. The frequency of gonadotropin-releasing hormone stimulation determines the number of pituitary gonadotropin-releasing hormone receptors. *Endocrinology.* 1985;116(5):2113–2115.
5. Brandebourg TD, Bown JL, Ben-Jonathan N. Prolactin upregulates its receptors and inhibits lipolysis and leptin release in male rat adipose tissue. *Biochem Biophys Res Commun.* 2007;357(2):408–413.
6. Nimrod A, Tsafriri A, Lindner HR. In vitro induction of binding sites for hCG in rat granulosa cells by FSH. *Nature.* 1977;267(5612):632–633.
7. Williams LT, Lefkowitz RJ, Watanabe AM, Hathaway DR, Besch HR, Jr. Thyroid hormone regulation of beta-adrenergic receptor number. *J Biol Chem.* 1977;252(8):2787–2789.
8. Olefsky JM, Reaven GM. Insulin binding in diabetes. Relationships with plasma insulin levels and insulin sensitivity. *Diabetes.* 1977;26(7):680–688.
9. Gavin JR, 3rd, Roth J, Neville DM, Jr., de Meyts P, Buell DN. Insulin-dependent regulation of insulin receptor concentrations: a direct demonstration in cell culture. *Proc Natl Acad Sci U S A.* 1974;71(1):84–88.
10. Kobayashi M, Olefsky JM. Effect of experimental hyperinsulinemia on insulin binding and glucose transport in isolated rat adipocytes. *Am J Physiol.* 1978;235(1):E53–62.
11. Adamo M, Shemer J, Aridor M, et al. Liver insulin receptor tyrosine kinase activity in a rat model of type II diabetes mellitus and obesity. *J Nutr.* 1989;119(3):484–489.
12. Meier DA, Hennes MM, McCune SA, Kissebah AH. Effects of obesity and gender on insulin receptor expression in liver of SHHF/Mcc-FAcp rats. *Obesity Res.* 1995;3(5):465–470.
13. Poole GP, Pogson CI, O'Connor KJ, Lazarus NR. The metabolism of 125I-labelled insulin by isolated Zucker rat hepatocytes. *Biosci Rep.* 1981;1(12):903–910.
14. Svedberg J, Bjorntorp P, Smith U, Lonnroth P. Effect of free fatty acids on insulin receptor binding and tyrosine kinase activity in hepatocytes isolated from lean and obese rats. *Diabetes.* 1992;41(3):294–298.
15. Arner P, Einarsson K, Backman L, Nilsell K, Lerea KM, Livingston JN. Studies of liver insulin receptors in non-obese and obese human subjects. *J Clin Invest.* 1983;72(5):1729–1736.
16. Wigand JP, Blackard WG. Downregulation of insulin receptors in obese man. *Diabetes.* 1979;28(4):287–291.
17. Gletsu NA, Field CJ, Clandinin MT. Obese mice have higher insulin receptor levels in the hepatocyte cell nucleus following insulin stimulation in vivo with an oral glucose meal. *Biochim Biophys Acta.* 1999;1454(3):251–260.
18. Hammes SR, Mendelson CR. Mechanisms of hormone action. In: Kovacs WJ, Ojeda SR, eds. *Textbook of Endocrine Physiology* (6th ed.). Oxford University; 2012:58–98.
19. Hadley ME, Levine JE. *Endocrinology* (6th ed.). Pearson Prentice Hall; 2007.
20. Clark JH, Peck EJ. Steroid hormone receptors: basic principles and measurement. In: O'Malley BW, Birnbaumer L, eds. *Receptors and Hormone Action I.* Academic Press; 1977:383–410.
21. Halmos G, Schally AV. Reduction in receptors for bombesin and epidermal growth factor in xenografts of human small-cell lung cancer after treatment with bombesin antagonist RC-3095. *Proc Natl Acad Sci U S A.* 1997; 94(3): 956–960.

Unit Overview
Unit III

Neuroendocrinology

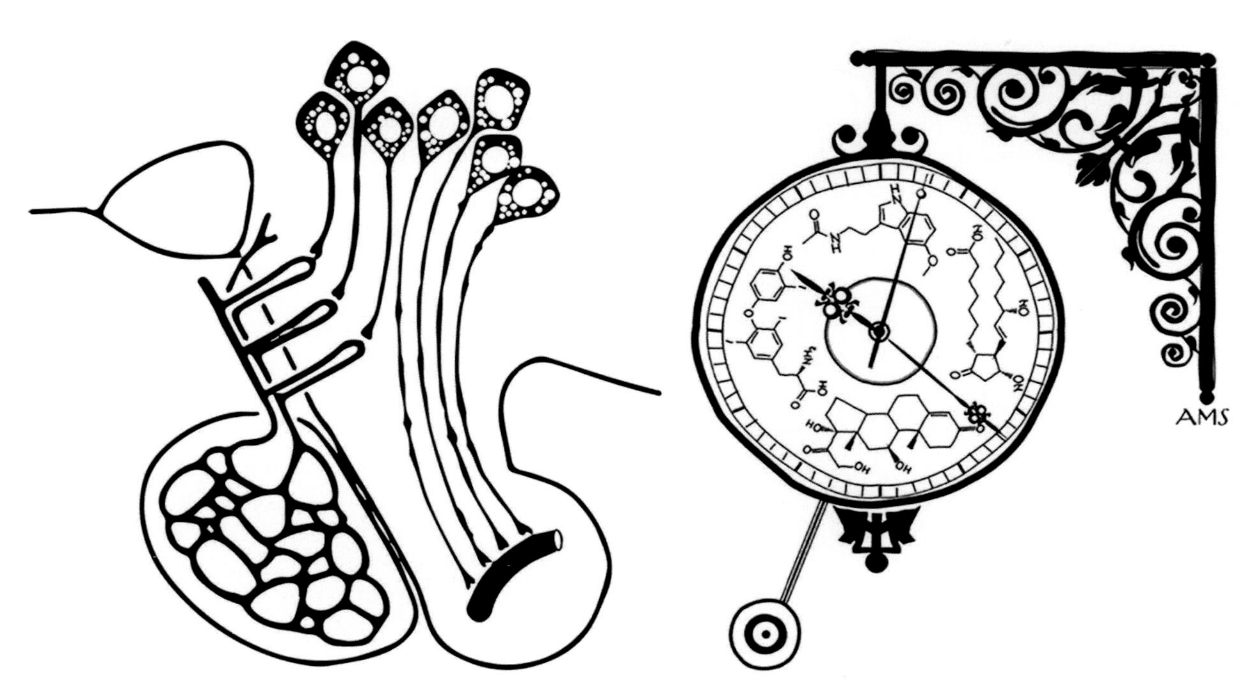

Left: Ernst and Berta Scharrer's classic portrayal of neurosecretion by the hypothalamus-pituitary axis, redrawn by A.M. Schreiber.
Right: "Circadian Time", by A.M. Schreiber.

DOI: 10.1201/9781003359241-9

OPENING QUOTATION:

> "Once interpreted as a small, possibly aberrant minority among a well-established group of neuroregulators, neuropeptides have now moved to center stage. Their importance in the control over a variety of fundamental biological processes can hardly be overestimated. Their versatility seems to account for much of the complexity and subtlety in neurochemical signaling that could not be accomplished by classical synaptic transmission alone."
>
> —Berta Scharrer, 1987

As its name implies, "neuroendocrinology" denotes a fusion of the notions of classical neural with classical endocrine signaling. Unit III addresses the prominent regulatory roles that these chemical signals, which are released into circulation by specialized subsets of neurons, play in endocrine physiology. Chapter 7 begins by discussing the history and significance of the concept of neurosecretion to endocrinology, providing examples of notable neurosecretory organs in invertebrates and vertebrates. The chapter goes on to describe the gross anatomy and function of the hypothalamus and pituitary gland, contrasting mechanisms of hypothalamic control over the different pituitary lobes. An overview of hypothalamic-pituitary-effector organ (HPE) signaling axes in regulating diverse aspects of vertebrate physiology is described, and knowledge of how feedback mechanisms among HPE organs can be used to diagnose endocrine dysfunctions is also discussed. Chapter 8 outlines the development of the pituitary gland, distinctive features of pituitary anatomy among vertebrate taxa, and their evolutionary origins. The chapter also describes the distinguishing biochemical features of each of the major families of pituitary hormones and their physiological functions both within and among vertebrate species. Chapter 9 addresses how the commensurate rhythmic actions of diverse neurohormones promotes the synchronization of time- and energy-consuming metabolic, physiological, and behavioral processes to the environment for the purposes of optimizing energy utilization, reproduction, and survival. The chapter describes the molecular and cellular basis of circadian clock function and the endocrine mechanisms that promote the synchronizations of peripheral clocks with environmental stimuli, such as light. The chapter also describes how circadian disruption (e.g. light exposure at night) can alter seasonal breeding and migrations of wildlife and also influence the progression of diseases in humans such as cancers and cardiovascular and metabolic disorders.

CHAPTER 7

Neurosecretion and Hypothalamic Control of the Pituitary

CHAPTER LEARNING OBJECTIVES:

- Compare and contrast the concept of neurosecretion with that of classical endocrine and neural signaling.
- Describe contributions of insect and fish research to the discovery of neuroendocrine neurons.
- List the locations and functions of the major vertebrate neuroendocrine systems.
- Describe the six basic physiological aspects of homeostasis regulated by the hypothalamus, and discuss the general importance and gross anatomy of the hypothalamus and pituitary gland.
- Contrast mechanisms of hypothalamic control over the pituitary adenohypophysis and neurohypophysis, and list the names and functions of hypothalamic hormones that are involved in pituitary signaling.
- Outline the components of generic hypothalamic-pituitary-end organ (HPE) feedback loops, and explain how endocrine dysfunctions along an HPE axis can be diagnosed.

OPENING QUOTATIONS:

"Here in this well-concealed spot, almost to be covered with a thumbnail, lies the very main spring of primitive existence—vegetative, emotional, reproductive—on which with more or less success, man [sic] has come to superimpose a cortex of inhibitions."

—Harvey Cushing (1932) Papers relating to the pituitary body, hypothalamus, and parasympathetic nervous system[1]

"Today these disciplines tend to converge towards the point where studying the regulation of the neural and hormonal systems becomes increasingly integrated into a science of synthesis: Neuroendocrinology."

—Roussy G, Mosinger M. *Traite de Neuro-Endocrinologie*. Paris: Masson et Cie; 1946[2]

KEY CONCEPTS:

- Classical neurons signal by releasing neurotransmitters into a synapse, which bind to receptors on the postsynaptic cell.
- Neurosecretory neurons are specialized for the synthesis, storage, and release of hormones into the circulatory system.
- Neurohemal organs consist of neurosecretory axonal termini and associated blood vessels.
- Neuroendocrine systems can transduce external environmental information (e.g. photoperiod, temperature) into chemical messages that target diverse physiological systems.

The Concept of Neuroendocrine Secretion

LEARNING OBJECTIVE Compare and contrast the concept of neurosecretion with that of classical endocrine and neural signaling.

Classical Neural Signaling

Prior to the 1950s, the nervous and endocrine systems were considered functionally distinct in their roles of intercellular and interorgan communication. Endocrine cells were thought to be exclusively epithelial-derived glandular cells that released hormones into circulation, allowing the chemical message to be carried to distant target organs. By contrast, nerves were thought to communicate exclusively via rapid electrical signaling along their axons and transduce the electrical signals into chemical signals in the form of neurotransmitters at the synapse to stimulate downstream neurons or other excitatory cells (Figure 7.1). In this classical model of neuron signaling, the cell body, as well as *dendrites* projecting from the cell body, possess two categories of transmembrane receptors: (1) those for *excitatory inputs*, such as from the neurotransmitter acetylcholine, which depolarizes the plasma membrane potential by promoting the influx of Na^+ cations through the ligand-gated receptor channel, and (2) those for *inhibitory inputs*, such as from the neurotransmitter gamma-aminobutyric acid (GABA) that hyperpolarizes the plasma membrane potential by promoting the influx of Cl^- anions and efflux of K^+ cations through the receptor channel. The

DOI: 10.1201/9781003359241-10

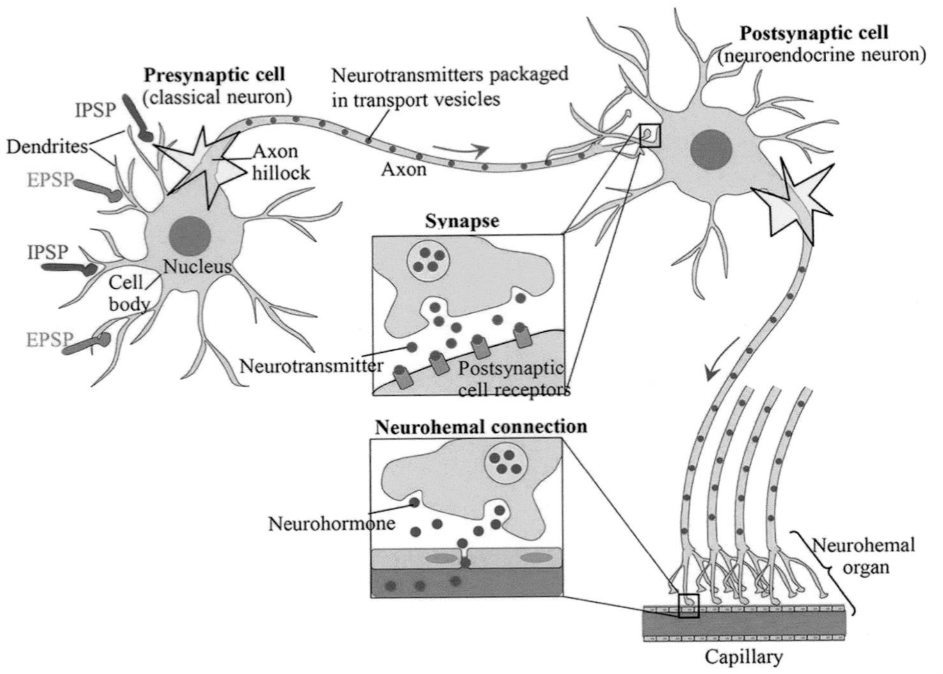

Figure 7.1 Structures and functions of classical synaptic neurotransmitters and neuroendocrine neurons. Both classical synaptic and neurosecretory neurons require electrical stimulation and axon potential generation, and both use small molecules and peptides for intercellular chemical signaling. Electrical stimulation of a neuron arrives from innervation by upstream neurons in the form of excitatory postsynaptic potentials (EPSP) that stimulate the generation of action potential at the axon hillock, and inhibitory postsynaptic potentials (IPSP) that inhibit the generation of action potential by the neuron. The primary difference between the two types of neurons is the distance and mode of transport of the chemical messages to their receptors. In classical neurotransmission, the signals are transmitted across extremely short distances called synapses, whereas neuroendocrine signals are released into the circulatory system and transported across potentially very large distances. The neuroendocrine neuron cell bodies are clustered into nuclei located in the central nervous system and receive classical neuronal input from presynaptic neurons. Neuroendocrine neuron axons project to peripheral release sites associated with capillary beds, and these are called "neurohemal organs".

resting membrane potential of a typical neuron is about −70 mv. At any given instant, the results of summation for all excitatory and inhibitory input occurs in a specialized region of the neuronal cell body called the *axon hillock*, and if a specific voltage threshold potential, typically between −40 and −55 mv, is crossed, action potentials are generated and propagate through the rest of the *axon* mediated by the opening of *voltage-gated sodium channels*. At the *axon terminus*, an influx of Ca^{++} mediated by the opening of *voltage-gated Ca^{++} channels* induces neurotransmitter-containing vesicles to fuse to the plasma membrane, releasing neurotransmitter into the *synaptic cleft*. Receptors for the neurotransmitter are located on the plasma membrane of the adjacent excitatory cell producing either an excitatory post-synaptic potential (EPSP) or an inhibitory post-synaptic potential (IPSP), depending on the nature of the neurotransmitter.

Neurotransmitters can be placed into one of three broad categories: peptides (e.g. opioids, oxytocin, substance P), small molecules (e.g. GABA and amino acid-derived molecules such as epinephrine), and gases (such as nitric oxide). All peptide neurotransmitters are encoded by genes and transcribed into mRNA in the nucleus. Following translation into peptides in the RER and processing by the Golgi, these peptide neurotransmitters are packaged into vesicles that are conveyed by *anterograde* (from the cell body to the axon termini) transport along microtubules down the axon where they are stored in the axonal terminus (Figure 7.1). At the axon terminus, an influx of Ca^{++} caused by membrane depolarization induces exocytosis of the stored peptides into the adjoining extracellular fluid and bloodstream. The enzymes that synthesize small molecule and gaseous neurotransmitters from precursor molecules are also transported down the axon. Whereas the small molecules are packaged locally in the axonal terminus into vesicles for export, gaseous neurotransmitters are not packaged but diffuse freely into the cytoplasm and out of the cell. Used synaptic vesicles may be returned to the cell body via *retrograde* axonal transport from the axon termini to the cell body.

Neuroendocrine Signaling

The differences between classical neurotransmission and classical endocrine signaling became less distinct with the discovery of **neurosecretory neurons**, which are a category of neurons that specialize in the synthesis, storage, and release of hormones into the circulatory system

(Figure 7.1). Like conventional neurons, neurosecretory neurons also integrate information at the axon hillock and conduct electrical impulses along their axons. However, in contrast with conventional neurons whose neurotransmitters are active only within the synapse, the *neurohormones* released by neurosecretory neurons have much more widespread targets. Like classical neuron signaling, neurosecretory neurons also synthesize and release both small molecules and peptides; unlike classical neuron signaling, these enter circulation and are therefore considered hormones. Whereas the cell bodies of neurosecretory neurons localize to specific neurosecretory centers, such as within the vertebrate hypothalamus, their axonal projections associate with blood vessels (Figure 7.1), whose juxtaposition is referred to as a **neurohemal organ**.

Importantly, neuroendocrine and classical neural signaling pathways are often highly integrated, with classical neurons forming synapses with neuroendocrine neurons, and with classical neurons also possessing receptors for neuroendocrine hormones. As such, neuroendocrine systems can function to transduce external environmental signals, such as photoperiod, temperature, salinity, food availability, mate availability, or threats to wellbeing, into hormone messages. These messages subsequently target relevant physiological systems, such as osmoregulatory, stress response, immune, and reproductive systems, to elicit appropriate responses. As a discipline, *neuroendocrinology* studies hormone production by neurons, the responses of target neurons to hormones, as well as the dynamic, bidirectional interactions between neurons and endocrine glands. Although today the concept of neurosecretion is one of the most important and well-established disciplines within endocrinology, it was once revolutionary to consider that nervous tissue was capable of producing hormones.

SUMMARY AND SYNTHESIS QUESTIONS

1. Compare and contrast classical neural signaling with neuroendocrine signaling.

History and Contributions from Fish and Insect Research

LEARNING OBJECTIVE Describe contributions of insect and fish research to the discovery of neuroendocrine neurons.

KEY CONCEPTS:

- In the 1920s, Stefan Kopeć demonstrated that secretions emanating from the brain are required for metamorphosis to take place in gypsy moth caterpillars, the first evidence of neurosecretion.
- Carl Caskey Speidel demonstrated that secretory cells are present in the spinal cords of fish, suggesting the presence of neurosecretory systems in verterbrates.
- Ernst and Berta Scharrer founded the modern field of neuroendocrinology by studying neurosecretion in mammals and insects.

Early 20th Century

The first reports suggesting a role for neurosecretion in insects and fish were published in the second decade of the 20th century, a time when the concept that neurons were capable of producing hormones was difficult to accept. The Polish scientist Stefan Kopec´ (Figure 7.2) suspected that the larval insect brain was the source of the substance that induced metamorphosis. Kopec´ used two simple yet ingenious techniques to demonstrate for the first time that the head and, more specifically, the brain were required for metamorphosis to commence in gypsy moth caterpillars[3,4] (Figure 7.2). The first was head *ligation*, or the use of a loop of fine string or hair pulled tight, like a tourniquet, to interrupt blood flow between the head and body compartments. The second technique was brain removal and retransplantation. He concluded that the influence of the brain was chemical and that it should be considered as a gland of internal secretion. Tragically, in 1941, just when his revolutionary hypothesis was gaining recognition, Kopec´ and his son were executed by the Nazis.[5]

During the same time that Stefan Kopec´ was conducting his insect research, in the United States, Carl Caskey Speidel published a study called "Gland cells of internal secretion in the spinal cord of skates",[6] where he described a group of large cells in the spinal cord of the skate, a cartilaginous fish, that contained prominent secretory vesicles (Figure 7.3). Speidel called these cells "Dahlgren cells", named after his graduate advisor who originally identified these cells. Much later it would be confirmed that these large cells are indeed neurosecretory and project to the urophysis—a neurohemal organ located ventral to the caudal spinal cord in fish—and may be involved with osmoregulation.

Ernst and Berta Scharrer: Neuroendocrinology's Original Power Couple

After each receiving their doctoral degrees from the University of Munich, Ernst and Berta Scharrer (Figure 7.4) worked together for their entire professional careers and became known as the founders of the concept of neurosecretion. They advanced this notion by adopting a powerful comparative approach: Ernst pursued neurosecretion in vertebrates and Berta in invertebrates. Ernst Scharrer first formulated the concept of neurosecretion in vertebrates when he reported the existence of nerve cells that he hypothesized secreted hormones in the brains of fish, describing them as "nerve-gland cells" (Figure 7.4).[7] He boldly proposed that their role may be endocrine and

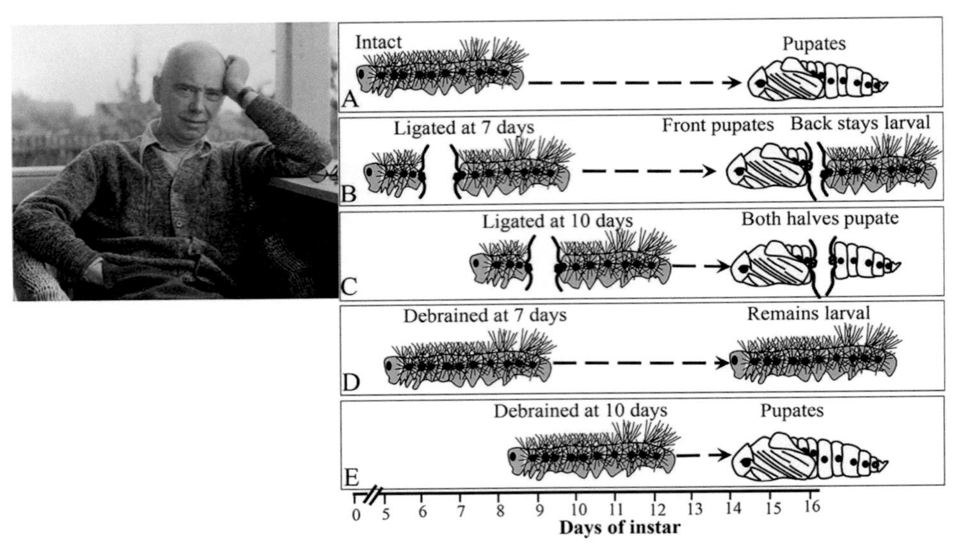

Figure 7.2 Polish biologist Stefan Kopec´ (1888–1941) and his ingenious experiment. Diagram of Stefan Kopec´'s demonstration in 1922 that blood-borne hormones from the brain are necessary for metamorphosis in the gypsy moth. **(A)** Intact caterpillars undergoing normal pupation. **(B)** When anterior and posterior regions of the caterpillar are ligated (tied off with thread and separated) at day 7 of the final larval instar, a week later the front half pupates, but the back half remains larval. This suggests that a pupation initiation substance is produced by the anterior half of the caterpillar and transmitted to the posterior half. **(C)** When the caterpillar is ligated at day 10, both halves pupate, showing that the pupation initiation substance has already been dispersed to both halves by day 10. **(D)** Caterpillars that were debrained (brains removed) before 7 days remained larval, but caterpillars debrained at 10 days **(E)** pupated normally. To eliminate the possibility that the brain might affect pupation by nervous pathways, Kopec´ left the brain intact but removed the subesophageal ganglion, the first in the chain of ventral ganglia that connects the brain with the rest of the nervous system. These caterpillars pupated normally even when the ganglion was absent and nervous transmission from the brain was interrupted. Kopec´ concluded that the influence of the brain was chemical and that it should be considered as a gland of internal secretion.

Source of Images: (A) Courtesy of The Alan Greenwood archive, University of Edinburgh.

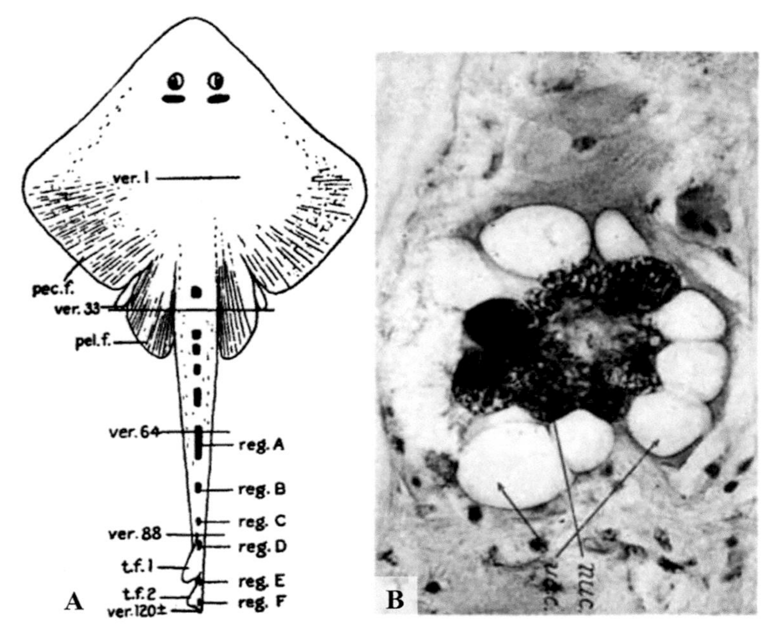

Figure 7.3 Speidel (1919) described groups of large cells in the spinal cord of the skate (*Raia punctata*) that contained prominent secretory vesicles with multilobulated nuclei. (A) The original legend reads: "Diagram of the skate to show the region of the tail in which the large cells of the spinal cord are found". **(B)** A single cell from a longitudinal section of the spinal cord of *Raia punctata*, showing a centrally located multi-lobed nucleus (dark coloration) surrounded by about nine large secretory vesicles (light-colored) in the periphery.

Source of Images: Watts, A.G., 201. Brain Res. Rev., 66(1-2), pp. 174-204. Used with permission.

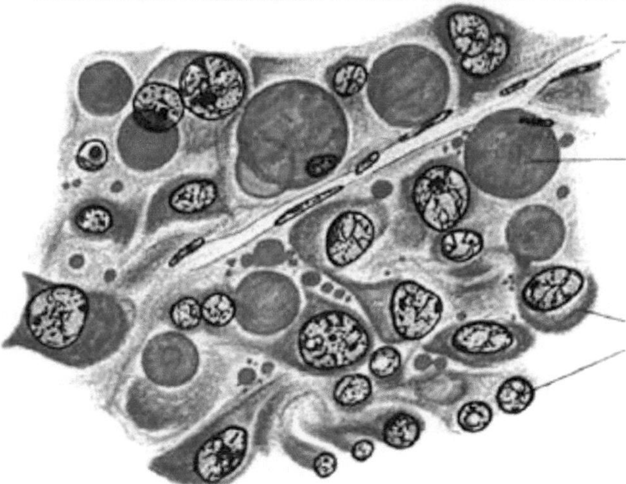

Figure 7.4 Ernst and Berta Scharrer, founders of the field of neuroendocrinology. The illustration by Ernst Scharrer depicts a group of cells in the magnocellular preoptic nucleus of the hypothalamus that appears to contain and be surrounded by secretory vesicles (dark blue depicts nuclei of cells, red represents secretory vesicles, pink denotes cytoplasm). The original legend reads "Histology of the magnocellular preoptic nucleus". From Scharrer (1928).

Source of Images: Photographs of the Scharrers from Bettendorf, G., 1995. Scharrer, Ernst Albert und Scharrer, Berta. J. für Reproduktionsmedizin und Endokrinologie (pp. 482-484); used with permission. Scharrer painting from Watts, A.G., 2011. Brain Res. Rev., 66(1-2), pp. 174-204; used with permission.

suggested a functional relationship with the pituitary gland. The very first description of neuroendocrine activity in an invertebrate was reported by Berta for neural cells in the sea slug, *Aplysia*,[8] again, at a time when the conventional wisdom was that invertebrates did not possess *any* hormones, let alone neurohormones. This was followed in 1937 by a report identifying secretory cells in the nervous systems of a variety of invertebrates, including insects.[9]

After leaving the intolerable working conditions of Nazi Germany for the United States where Ernst accepted a research position at the University of Chicago, Berta, who did not have a paid position at the time, turned to insects as cheap and convenient laboratory subjects. In 1941, she published a detailed description of insect neurosecretory cells from a region of the brain called the corpora cardiaca.[10]

Curiously Enough . . . With the help of a custodian, Berta Scharrer collected cockroaches from the basement of her building, finding that their tolerance to surgical manipulations made them excellent models to study neurosecretion.[11]

SUMMARY AND SYNTHESIS QUESTIONS

1. What intracellular structures, unusual for classical neurons, led Speidel and the Scharrers to hypothesize that "Dahlgren cells" and "nerve-gland cells" in the brains of fish, respectively, were secretory neurons?
2. How does it make sense that the discovery of neurosecretory organs in insects and fish inevitably led to similar discoveries in mammals?

Vertebrate Neuroendocrine Systems

LEARNING OBJECTIVE List the locations and functions of the major vertebrate neuroendocrine systems.

KEY CONCEPTS:

- The primary peripheral vertebrate neurosecretory sites are chromaffin cells of the adrenal medulla and the caudal neurosecretory system of fish.
- The primary neurosecretory centers within the vertebrate brain are the peptide-secreting cells of the hypothalamus and neurohypophysis and the melatonin-secreting cells of the pineal gland.

It is now known that vertebrates possess several neuroendocrine systems that can be separated into categories based upon their anatomical locations: those that stem from *ganglia* (groups of nerve cell bodies located in the peripheral nervous system), those that emanate from cell bodies located in the spinal cord, and those found in the brain.

Chromaffin Cells

Chromaffin cells are modified postganglionic sympathetic neurons located in the *adrenal medulla* of *adrenal glands* (called "interrenal glands" in fish and amphibians) (Figure 7.5). While chromaffin cells are technically not neurons, they are developmentally and functionally very closely related and are therefore generally considered neuroendocrine cells. The primary function of these cells is to secrete epinephrine and norepinephrine into the bloodstream. These cells play key roles in elevating blood pressure and blood glucose levels, particularly during the vertebrate "stress response".

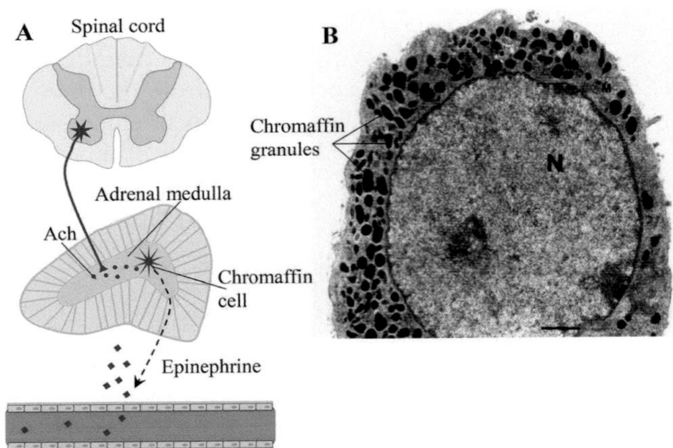

Figure 7.5 The control pathway of chromaffin cell secretion. **(A)** Presynaptic neurons from the spinal cord deliver acetylcholine (Ach) neurotransmitter to the adrenal medulla. Chromaffin tissue, which consists of modified sympathetic cells, respond to Ach by releasing epinephrine and norepinephrine into the blood. **(B)** Electron micrograph showing abundant the secretory granules (dark-stained vesicles) in a bovine adrenal chromaffin cell.

Source of Images: (B) Koval, L.M., et al. 2000. Neuroscience, 96(3), pp. 639-649. Used with permission.

The Caudal Neurosecretory System of Fish

An intriguing neuroendocrine system unique to teleosts (bony fish) and elasmobranchs (cartilaginous fish, like skates and sharks) is termed the **caudal neurosecretory system**. As the name implies, this system is located in the tails of fish, and specifically consists of neuroendocrine cells located in the terminal segments of the spinal cord that project to a neurohemal organ, the ***urophysis***, from which neuropeptides are released (Figure 7.6).[12] These are in fact the same neurosecretory cells identified by Dahlgren and Speidel in the early 20th century, described earlier. The caudal neurosecretory system may be involved in osmoregulation[13] and the stress response.[14] Although the caudal neurosecretory system itself is unique to fish, all of its neurosecretory products are now known to be present in humans and other vertebrates.

Pinealocytes

One group of neurosecretory neurons located in the brain are the **pinealocytes**, which are found in the *pineal gland*

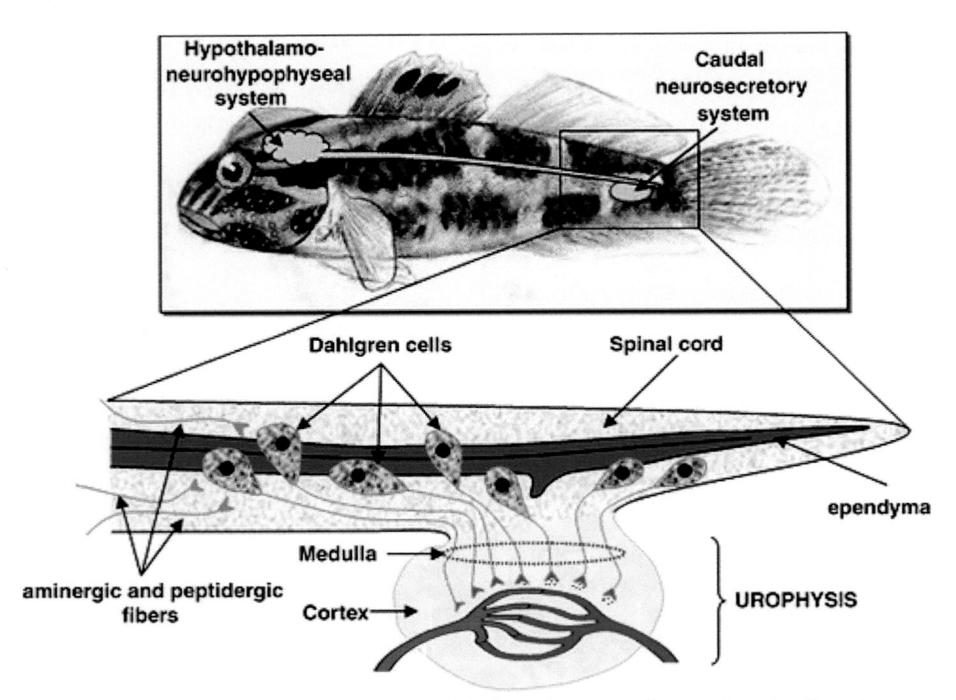

Figure 7.6 The caudal neurosecretory cells in the spinal cord of fish (Dahlgren cells) send projections to a neurohemal gland (the urophysis) that release neurohormones into circulation.

Source of Images: Vaudry, H., et al. 2010. Ann. N. Y. Acad. Sci., 1200(1), pp. 53-66. Used with permission.

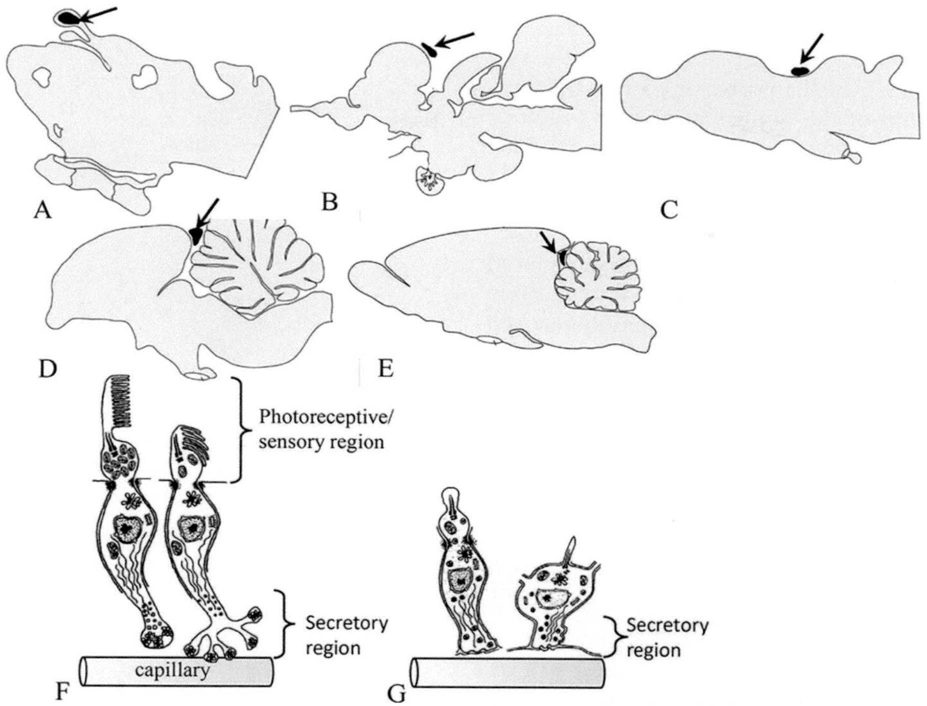

Figure 7.7 Diversity of pineal gland location and pinealocytye morphology among vertebrates. Arrows denote the location of the pineal gland. **(A)** Jawless fishes. **(B)** Teleost fishes. **(C)** Amphibians. **(D)** Birds. **(E)** Mammals. Pinealocytes are developmentally related to neurons and photoreceptors. Their processes, which release the hormone melatonin, terminate in perivascular spaces surrounding capillaries. In contrast to the pinealocytes of amphibians that are photoreceptive **(F)**, mammalian pinealocytes **(G)** have lost their photoreceptive function, and are exclusively secretory cells.

Source of Images: (A-E) Ogawa, S. and Parhar, I.S., 2014. Front. Endocrinol., 5, p. 177; used with permission. (F-G) Ekström, P. and Meissl, H., 2003. Philos. Trans. Royal Soc. B: Biol. Sci., 358(1438), pp. 1679-1700; used with permission.

and whose major function is the synthesis of *melatonin*, a hormone that modulates circadian rhythms (Figure 7.7). Although the pineal gland is part of the brain, pinealocytes themselves are not neurons. Indeed, they are developmentally more closely related to photoreceptors and are therefore technically not *neuro*endocrine cells. However, because they are derived from embryonic neural epithelium and share many neural characteristics, the pineal gland is still typically described as a neuroendocrine organ. This fascinating organ, considered by 17th-century philosopher and mathematician René Descartes to be the "seat of the soul", is discussed further in Chapter 9: Central Control of Biological Rhythms.

The last category of neurosecretory neurons found in the vertebrate brain consist of peptide-secreting cells of the hypothalamus and neurohypophysis, and these are described in the next section of this chapter.

SUMMARY AND SYNTHESIS QUESTIONS

1. List the four major neurosecretory centers of vertebrates.

The Hypothalamic-Pituitary Axis

LEARNING OBJECTIVE Describe the six basic physiological aspects of homeostasis regulated by the hypothalamus, and discuss the general importance and gross anatomy of the hypothalamus and pituitary gland.

KEY CONCEPTS:

- The hypothalamus integrates sensory information received from diverse receptors throughout the body to monitor most aspects of homeostasis.
- The hypothalamus maintains homeostasis through endocrine, autonomic, and behavioral outputs.
- The pituitary gland is a bi-lobed structure consisting of the adenohypophysis and neurohypophysis.
- Magnocellular hypothalamic nuclei project their axons directly into the neurohypophysis, a pituitary lobe that stores neuropeptides hormones produced by the hypothalamus.
- In mammals, the hypothalamus controls adenohypophysis function indirectly, with parvocellular nuclei releasing tropic hormones into portal vessels that carry the signals to the adenohypophysis.

As suggested by its Greek derivation, the **hypothalamus** (*hypo* = below, *thalamus* = bed) is that portion of the vertebrate "diencephalon", or the posterior part of the forebrain that sits on top of the brainstem, that lies inferior to the *thalamus*, a region involved in sensory signal relay (Figures 7.8–7.9). The *pituitary gland*, which lies beneath the hypothalamus, is also called the *hypophysis*, whose Greek derivation describes an "outgrowth from below" the hypothalamus. In this chapter you will be introduced to a system of neuroendocrine interactions that takes place between these two structures, a system commonly referred to as the *hypothalamic-pituitary axis* (H-P axis). The H-P axis represents the core of the vertebrate neuroendocrine system. Its interactions profoundly affect the functions of diverse target organs throughout the body, many of which are endocrine themselves. H-P axis dynamics are constantly influenced by endocrine feedback received from

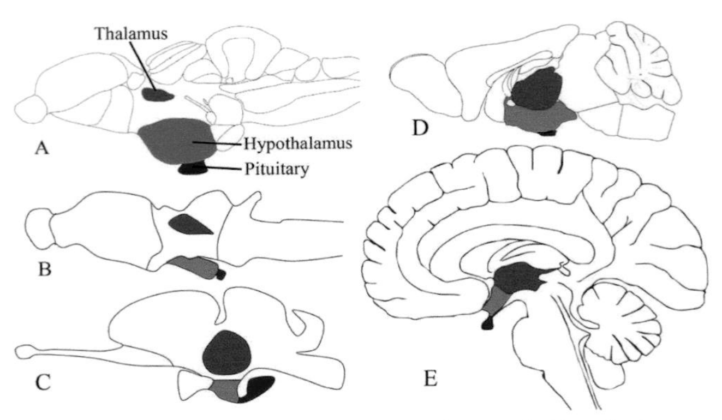

Figure 7.8 The locations and functions of the thalamus (green), hypothalamus (orange), and pituitary gland (purple) are conserved among vertebrates. All images depict medial sagittal sections of the brain. **(A)** Teleost fish. **(B)** Amphibian. **(C)** Reptile. **(D)** Mouse. **(E)** Human.

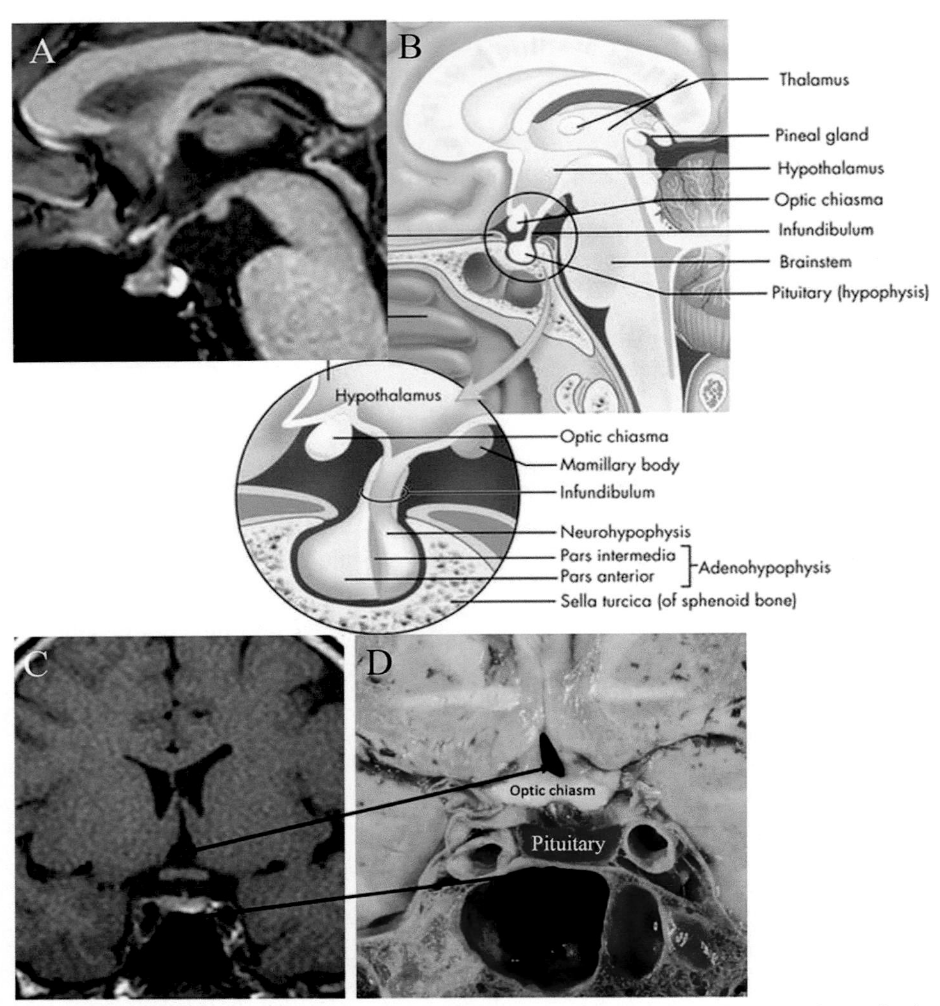

Figure 7.9 Sagittal and coronal views of the human thalamus, hypothalamus, and pituitary. (A) The neurohypophysis displays a characteristically high signal compared with the adenohypophysis in a sagittal MRI scan. **(B)** The diagram depicts the gross anatomy of the region. Coronal slices **(C–D)** through the human brain at the level of the pituitary gland (C, MRI; D, post-mortem *in situ* appearance).

Source of Images: (A) Yamada, K., et al. 2021. Biomarkers Neuropsychiatry, 5, p. 100039; used with permission. (B) Moini, J., et al. 2021. Epidemiology of Endocrine Tumors; Elsevier; used with permission. (C-D) Larkin, S. and Ansorge, O., 2017. Development and microscopic anatomy of the pituitary gland. In: Feingold KR, Anawalt B, Boyce A, et al, editors. Endotext. Used with permission.

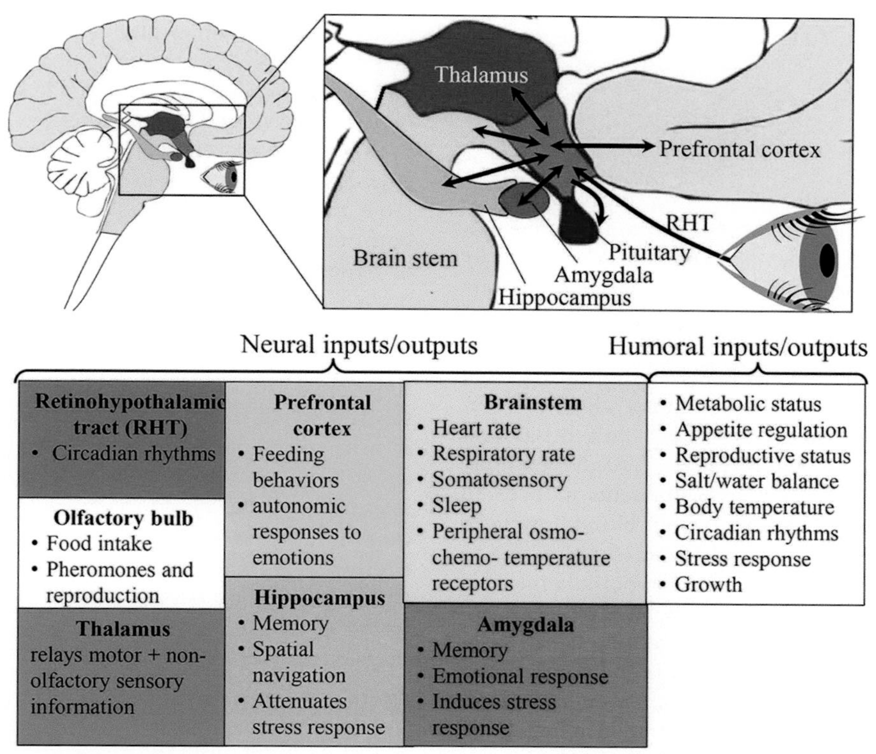

Figure 7.10 Major hypothalamic afferent and efferent pathways. Juxtaposed to the thalamus, amygdala, hippocampus, brainstem, and prefrontal cortex, the hypothalamus receives diverse signals from and transmits its own signals to these and other regions of the brain and body. The retinohypothalamic tract (RHT) transmits nonvisual light information from the retina to the circadian regulatory centers of the hypothalamus. The hypothalamus funnels these diverse sources of inputs primarily to the brainstem and pituitary for the regulation of visceral functions.

multiple organs, as well as neural input from other regions of the nervous system. The emphasis of this chapter will be on the hypothalamic control over pituitary hormone release. The details of the pituitary hormone functions themselves will be addressed in Chapter 8: The Pituitary Gland and Its Hormones.

History and Importance

Historically, the hypothalamus and pituitary gland have been elusive objects of study. The hypothalamus has no clear anatomical demarcation in adult vertebrates, and it was not even recognized as a distinct region until 1893 when Swiss anatomist Wilhelm His coined the term "hypothalamus" after describing its first anatomical subdivision based on fetal development of the human brain. The pituitary, by contrast, has been recognized since the 2nd century AD when Galen of Pergamon first described it. However, due to its small size (in humans the pituitary is about the size of a pea) and the fact that it is largely encased in the bony sella turcica (Figure 7.9), making it difficult to remove without damage, the pituitary has been a tricky gland to study. Galen considered the pituitary to be simply a draining route and receptacle for what was thought to be mucus passing from the brain's cerebrospinal fluid-filled *ventricles* to the nasopharynx—not a particularly glamorous function.[15] In fact, the word "pituitary" is derived from the Latin word *pituita*, which means phlegm, one of the four humors of Hippocratic medicine. The ventricles were believed, at the time, to house *animal spirit*, a substance thought to facilitate all sensory, motor, and mental activities.

> **Curiously Enough . . .** The third ventricle, whose inferior-most region lies juxtaposed to the hypothalamus, was believed to be the site of input and elaboration of the *sensus communis*, Latin for peripheral physical sensations. This is where the modern colloquialism "common sense" is derived.

Today, we know that a primary function of the hypothalamus is to integrate sensory information received from diverse inputs throughout the body (Figure 7.10) in order to monitor and maintain most aspects of **homeostasis**, or the constancy of the internal environment. Six basic physiological aspects of homeostasis regulated by the hypothalamus are:

1. Blood pressure and electrolyte composition
2. Body temperature

3. Energy metabolism (including growth and feeding behavior)
4. Reproduction
5. Stress response
6. Circadian rhythms

Indeed, the hypothalamus is amongst the most evolutionarily conserved regions of the vertebrate brain, and animals cannot survive without it. The hypothalamus continuously compares sensory information with biological set points and adjusts output to maintain homeostasis. For example, using thermoreceptors it compares local temperature to the set point of 37°C in adult humans. If a deviation from the set point is detected, the hypothalamus adjusts a coordinated array of endocrine, autonomic, and behavioral responses to restore homeostasis (Figure 7.10). When body temperature exceeds 37°C, the hypothalamus induces an *autonomic response*, vasodilation, to shift blood flow from deep to cutaneous vascular beds and increases sweating to increase heat loss through the skin. The hypothalamus also induces a concurrent *endocrine response*: it increases vasopressin (an antidiuretic hormone) secretion, which conserves water for use in sweating. Additionally, the hypothalamus activates complementary *behavioral responses*, such as seeking a cooler environment to reduce heat gain. There are set points for diverse physiological measurements, including blood glucose, sodium, pH, osmotic pressure, and levels of virtually all hormones.

This section focuses specifically on the hypothalamic control of pituitary endocrine function, primarily using the well-studied mammalian system as a reference. In response to an altered homeostatic variable, hypothalamic output is relayed to the pituitary gland, which broadcasts this information to diverse organs by releasing target-specific hormones into general circulation. These target organs, many of which have endocrine functions themselves, respond by adjusting aspects of their metabolism to accommodate homeostasis. Because the pituitary controls such a large diversity of vital physiological functions, it has often been referred to as the "master gland". In fact, the effects of **hypophysectomy**, or surgical removal of the pituitary, proved to be so dramatic, particularly on gonad development and growth, that for a short period in the 1930s most scientists in the field thought that the pituitary gland functioned autonomously. However, studies examining the impact of environmental information on physiology in fish,[16] amphibians,[17,18] and birds,[19–21] together with others showing that brain lesions affected pituitary hormone secretion, led to the concept that the pituitary adenohypophysis must be under central nervous control. Therefore, it would now in fact be more accurate to consider the hypothalamus as the primary controller and the pituitary as responsive to it.

Gross Anatomy

Hypothalamus

Among vertebrates the hypothalamus comprises the walls of the inferior portion of the brain's third ventricle and contains a diversity of neurons divided into functional clusters called *nuclei*. In humans, the hypothalamus can be divided into three broad regions: anterior, middle (also called the "tuberal" region), and posterior (Figure 7.11

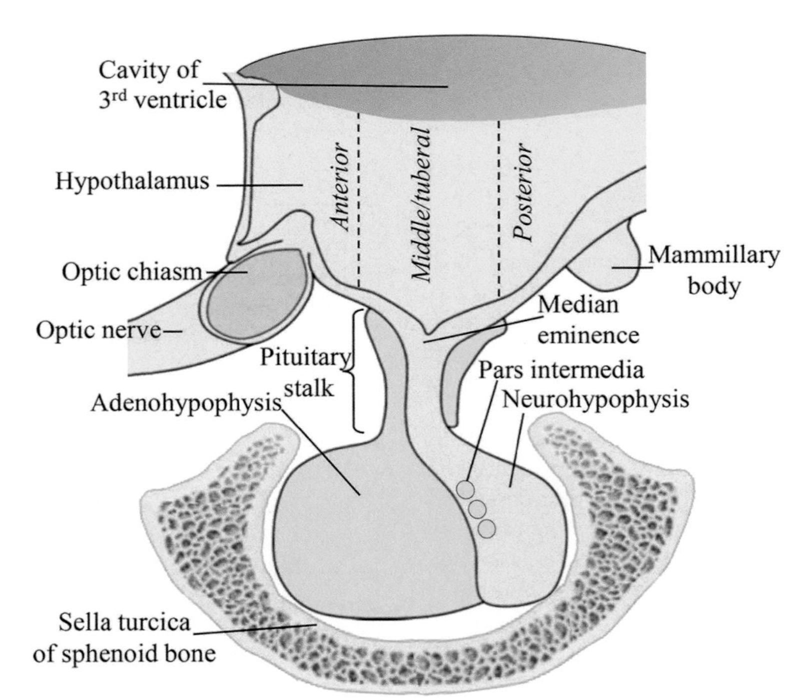

Figure 7.11 Gross anatomical subdivisions of the human hypothalamus and pituitary gland. The pars intermedia is not normally present in human adults but is present in the adults of many non-primate mammals.

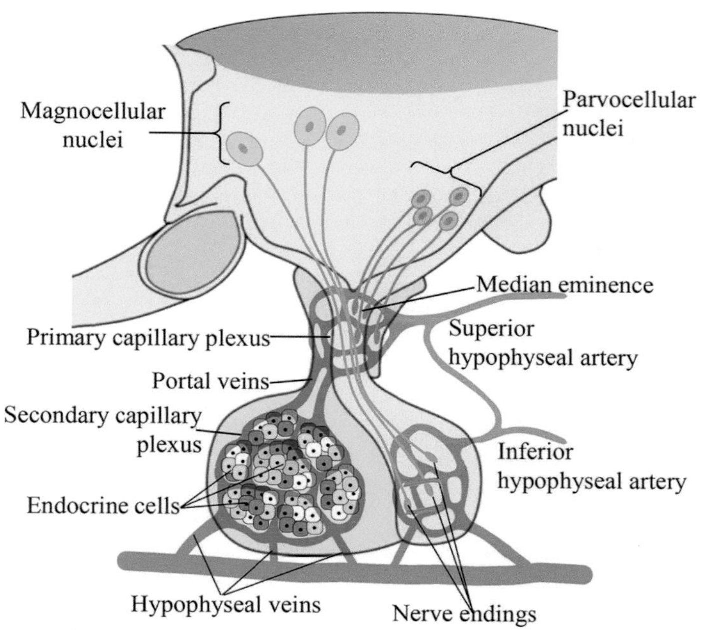

Figure 7.12 Distribution of some hypothalamic nuclei and their nerve endings, as well as endocrine cells of the adenohypophysis and pituitary vasculature.

and Figure 7.12). The middle third of the hypothalamus overlies the pituitary stalk and contains diverse *magnocellular* (large cell body) and *parvocellular* (small cell body) neuroendocrine neurons that communicate with the pituitary gland, as well as other regions of the brain.[22] The most anterior part of the hypothalamus overlays the *optic chiasm* and is known as the preoptic area. This area contains nuclei that are mainly concerned with integration of different kinds of sensory information needed to evaluate deviations from physiological set points, such as for blood pressure and composition, body temperature, and regulation of reproductive and circadian cycles of activity. Importantly, nuclei from the anterior hypothalamic region communicate with neuroendocrine cells in the middle region and, therefore, indirectly modulate neurosecretion. The posterior third of the hypothalamus, which includes the mammillary body and the overlying posterior hypothalamic area, are involved in organizing autonomic responses and behaviors that include wakefulness and arousal. Although hypothalamic architecture is generally similar in females and males, interesting examples of hypothalamic sexual dimorphism do exist, and these will be addressed in Chapter 20: Sexual Determination and Differentiation.

Pituitary Gland

Located beneath the middle region of the hypothalamus, and stemming almost exactly from its inferior midline tip, lies the pituitary gland. The pituitary is a bi-lobed structure, consisting of the **adenohypophysis** (in primates it is also called the "anterior" lobe) and the *neurohypophysis* (the "posterior" lobe in primates) (Figure 7.11 and Figure 7.12). The pituitary is further subdivided into other specialized regions that will be described in the next chapter. Importantly, the two lobes of the pituitary are histologically distinct, with the adenohypophysis composed primarily of classical endocrine glandular tissue, and the neurohypophysis comprised of magnocellular neuronal axon and terminal processes that originate from regions of the hypothalamus, such as the paraventricular and supraoptic nuclei.

The hypothalamus controls pituitary function via both direct and indirect means. Whereas hypothalamic neurons extend directly into the neurohypophysis, in mammals the adenohypophysis does not house neurons of hypothalamic origin. Instead, the nerve endings of the parvocellular hypothalamic nuclei that communicate with the adenohypophysis terminate at a region within the hypothalamus called the **median eminence** (Figure 7.11 and Figure 7.12). The adenohypophysis receives neurohormones from the hypothalamus indirectly via a vascular link, called the **hypophyseal portal vessels**, which drain blood from capillary loops originating in the median eminence down to the adenohypophysis. This notion of direct and indirect control of pituitary function by hypothalamic nuclei forms the very foundation of our modern understanding of hypothalamic-pituitary axis communication.

Curiously Enough . . . Unlike other vertebrate taxa, teleost fish neither possess a median eminence nor a hypophyseal portal system, and the adenohypophysis is innervated directly by the hypothalamus.

SUMMARY AND SYNTHESIS QUESTIONS

1. The hypothalamic-releasing hormone, GHRH, was originally cloned from *pancreatic* tissue. How does this make sense?
2. What are the three different categories of hypothalamic output? Describe an example for how each category responds to reduced body temperature in a mammal.
3. Describe the neurovascular hypothesis.
4. Describe the etymology (origin of the meaning of the word) of each of the following components of the H-P axis: (a) hypothalamus, (b) hypophysis, (c) pituitary, (d) adenohypophysis, (e) neurohypophysis.

A Neuroendocrine Signaling Hierarchy

LEARNING OBJECTIVE Contrast mechanisms of hypothalamic control over the pituitary adenohypophysis and neurohypophysis, and list the names and functions of hypothalamic hormones that are involved in pituitary signaling.

KEY CONCEPTS:

- Neurohypophysial neurons of hypothalamic origin release the neuropeptide hormones vasopressin and oxytocin from nerve endings located in the neurohypophysis.
- Hypophysiotropic neurons of hypothalamic origin project their axons to the median eminence, a neurohemal organ where diverse neuropeptides are released and carried by portal blood vessels to the adenohypophysis.
- There are two categories of hypophysiotropic hormones: releasing hormones promote the release of adenohypophysis hormones and release-inhibiting hormones inhibit the release of adenohypophysis hormones.

Hypothalamic Neurons

The hypothalamus is a very small structure, in humans about the size of a walnut, occupying only about 0.3% of adult human brain weight. However, despite its small size, it is loaded with a very complex array of nuclei. Due to the myriad of neurotransmitters and neuropeptides that it contains, the hypothalamus has been described as a "pharmacological museum".[23] While the study of this large number of signals may be a source of headaches for students of endocrinology who are tasked with understanding them all, their diversity offers hope for pharmacological intervention in numerous medical problems that involve the hypothalamus. In this section, nuclei with specific neuroendocrine functions will be highlighted and the neurons that constitute these nuclei described. Maps showing the locations of these nuclei in mammals are depicted in Figure 7.13, and their neuroendocrine secretions and functions are summarized in Table 7.1. Initial studies suggested that individual neuroendocrine cells were each responsible for producing a single hormone, known as the "one-cell, one-hormone" hypothesis. However, more recent research has shown that many of these cells are multi-messenger systems and can synthesize more than one hormone. Broadly, the hypothalamic neurons that constitute these nuclei can be divided into two categories: neurohypophysial and hypophysiotropic neurons.

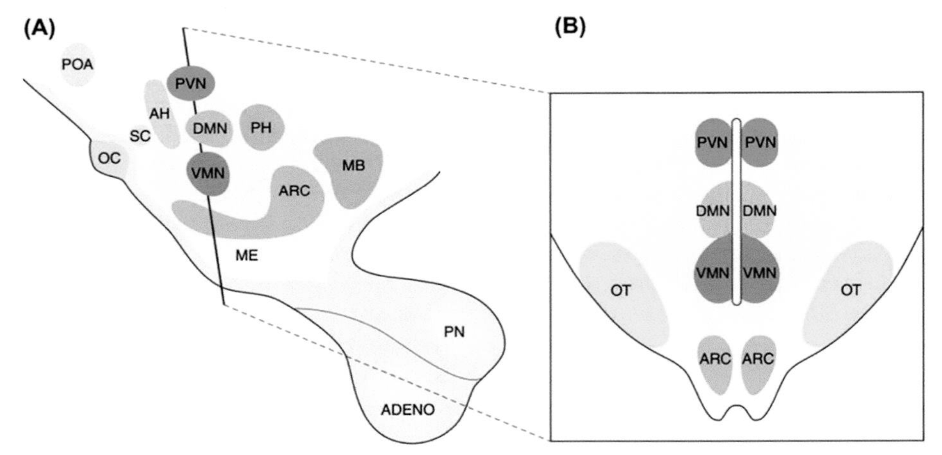

Figure 7.13 The locations of some hypothalamic nuclei in a generalized mammalian hypothalamus-pituitary axis. (A) Sagittal brain section showing nuclei for one side. **(B)** Brain cross section showing paired nuclei on both sides. Nuclei abbreviations: AH, anterior hypothalamic; ARC, arcuate; DMN, dorsomedial; PH, posterior hypothalamic; POA, preopctic area; PVN, paraventricular; SC, suprachiasmatic; VMN, ventromedial. Other abbreviations: ADENO, adenohypophysis; MB, mammillary body; ME, median eminence; OC, optic chiasm; OT, optic tract; PN, pars nervosa.

Source of Images: Norris, D.O. and Carr, J.A., 2020. Vertebrate endocrinology. Academic Press. Used with permission.

Table 7.1 Hypothalamic Nuclei and Some Functions in Mammals

Nucleus	Zone	Cell size	Peptides Synthesized	Functions
Anterior (AN)	Anterior	Parvo		Thermoregulation (cooling) and thirst
Suprachiasmatic (SCN)	Anterior	Magno	AVP, VIP	Circadian rhythms
Supraoptic (SON)	Anterior	Magno	OT, AVP	Electrolyte/water balance, blood pressure, milk ejection, uterine contractility
Paraventricular (PVN)	Anterior, Tuberal	Magno/parvo	CRH, OT, AVP, TRH	*Magnocellular*: Electrolyte/water balance, blood pressure, milk ejection, uterine contractility; *Parvocellular*: stress response, control of metabolism
Dorsomedial (DMN)	Tuberal	Parvo	NPY	Focal point of information processing; receives input from VMN, output to PVN; emotion (rage); regulation of body composition, feeding behavior
Ventromedial (VMN)	Tuberal	Parvo	GHRH, SST	GH release and inhibition; satiety center
Arcuate (ARC)	Tuberal	Parvo	GnRH, GHRH, DA, AgRP, NPY, POMC, kisspeptin	Reproduction, food intake, energy expenditure, control of growth, regulation of HPG
Periventricular	Tuberal	Parvo	TRH, SST, Leptin, GnRH	Inhibit growth
Lateral complex	Tuberal	Parvo	MCH	Appetite control
Posterior	Posterior	Parvo	?	Thermoregulation (heating)
Mammillary	Posterior	Parvo	?	Emotion and short-term memory
*Preoptic area (POA)	Anterior	Parvo	GnRH, TRH	Regulation of HPG and HPT axes, feeding, thermostat

Note: *There is some disagreement on whether the POA should be considered a part of the hypothalamus or a separate telencephalic structure.

Neurohypophysial Neurons

Neurohypophysial neurons are magnocellular neurons that originate from the *paraventricular* (PVN) and *supraoptic* (SON) nuclei (Figure 7.14). Their axons traverse the hypothalamic-pituitary stalk, and they release the neuropeptide hormones *arginine vasopressin* (AVP, also called antidiuretic hormone, or ADH) and *oxytocin* from nerve endings in the neurohypophysis. These neurons typically synthesize either one hormone or the other, but under some conditions can produce both. In addition to direct projections into the neurohypophysis, these neurons are also known to project to a variety of other brain regions where they are thought to elicit a variety of actions.[24] The functions of vasopressin and oxytocin, which include controlling osmoregulation and smooth muscle contraction, respectively, will be covered in detail in Chapter 8: The Pituitary Gland and Its Hormones.

Hypophysiotropic Neurons

In contrast with neurohypophysial neurons, which project directly into the neurohypophysis from only two major groups of magnocellular nuclei, *hypophysiotropic neurons* emanate from a multitude of parvocellular nuclei (Table 7.1) and project their axons to the median eminence (Figure 7.14). The median eminence receives its blood supply from the superior hypophyseal artery, which branches into a rich capillary bed. The capillary loops extend into the median eminence and coalesce to form the hypophyseal portal vessels that traverse the pituitary stalk and end in the adenohypophysis (Figure 7.15). These hypophysiotropic neuropeptides are secreted into the hypophyseal portal vessels. Importantly, the vasculature in most regions of the brain cannot accommodate the transfer of neuropeptides into the bloodstream due to the presence of the *blood-brain barrier*, a highly selective barrier formed by capillary endothelial cells connected by tight junctions that separate the brain's extracellular fluid from circulating blood. However, hypophyseal portal vessels lack this barrier and are instead composed of *fenestrated* ("leaky") capillaries that facilitate the transfer of hypothalamic neuropeptides and their transport to the adenohypophysis. These hypophysiotropic neuropeptides control the secretion of hormones by the adenohypophysis. Most of the blood flow is anterograde, or forward moving from the brain to the pituitary gland. However, retrograde, or reverse, flow from the adenohypophysis to the hypothalamus has also been documented,[25] suggesting a two-way communication between nervous and endocrine systems.

The Hypophysiotropic Hormones

The term *tropic hormone* generally describes a hormone released from one gland that targets a second endocrine gland, which subsequently responds by modulating the production and release into general circulation of a hormone of its own. Hypophysiotropic hormones are tropic hormones of hypothalamic origin that specifically target the

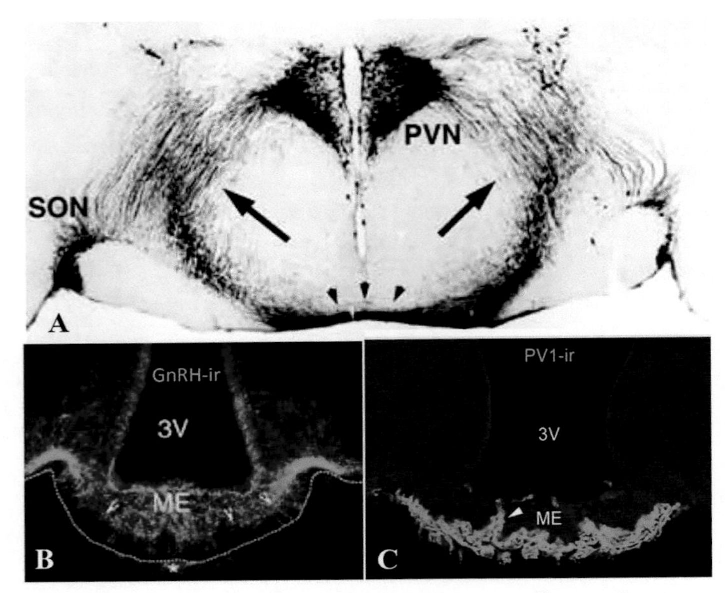

Figure 7.14 Organization of the hypothalamic neurohypophysial tract, median eminence, and associated vasculature. (A) Hypothalamic neurohypophysial tract (arrows). Note arching fibers emanating from magnocellular neurons in the paraventricular nucleus (PVN) as they descend toward and join fibers emanating from the supraoptic nucleus (SON). The fiber tract converges in the midline at the base of the hypothalamus in the retrochiasmatic area (arrowheads) before entering the internal zone of the median eminence. **(B)** GnRH-immunoreactive fibers descending from the periventricular nuclei to the median eminence. **(C)** Intrainfundibular capillary loops (arrowhead) containing PV1 (plasmalemmal vesicle associated protein-1), which specifically labels fenestrated endothelial cells of the primary capillary plexus vasculature.

Source of Images: (A) Lechan, R.M. and Toni, R., 2016. Functional anatomy of the hypothalamus and pituitary. In: Feingold KR, Anawalt B, Boyce A, et al, editors. Endotext; used by permission (B) Parkash, J., et al. 2015. Nat. Comm., 6(1), pp. 1-17; (C) Giacobini, P., et al. 2014.. PLoS biology, 12(3), p. e1001808.

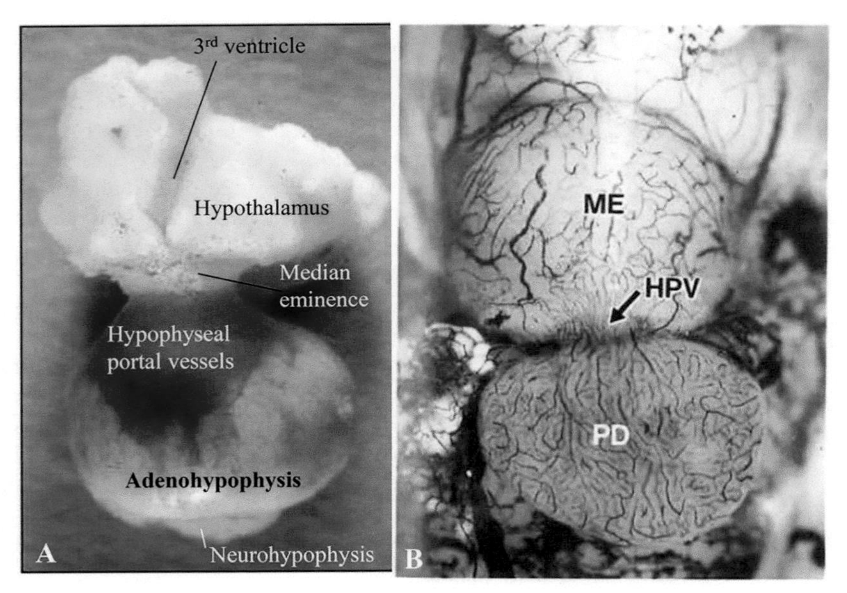

Figure 7.15 Anatomy of the hypophyseal portal vessels. (A) In a human, the hypophyseal portal vessels connect the median eminence to the adenohypophysis, facilitating the transfer of release and release-inhibiting factors from hypothalamic parvocellular nuclei to cells of the anterior pituitary. **(B)** The hypophyseal portal circulation of a frog, ventral view. The median eminence (ME) houses the primary capillary plexus (dark-stained vasculature) that branch off of the infundibular arteries arising from the carotid arteries. The hypophyseal portal vessels (HPV) extend from the primary capillary plexus and extend into the secondary capillary plexus that associates with the pars distalis (PD, the main component of the adenohypophysis) of the pituitary.

Source of Images: (A) Courtesy of Dr. Richard Bowen, Colorado State University; (B) Matsumoto, A. and Ishii, S., 1992. Atlas of endocrine organs: vertebrates and invertebrates. Springer. Used by permission.

adenohypophysis, whose cells possess receptors for these peptides (summarized in Figure 7.16). Hypophysiotropic hormones are generally termed *releasing hormones* if they promote the release of adenohypophysis hormones and *release-inhibiting hormones* if they inhibit the release of adenohypophysis hormones. The classical view of hypothalamic control of adenohypophysis hormone secretion is that each hypothalamic hormone targets a single cell type in the adenohypophysis, which responds by releasing a single hormone of its own. However, hypothalamic control of pituitary function now appears more complex than previously thought, with each of the major adenohypophysis cell types now known to contain discrete subpopulations of *multi-responsive cells* that are able to respond to several hypothalamic-releasing hormones.[26,27] Interestingly, multifunctional cells have been found to be more abundant in females than in males, suggesting that the female pituitary is more plastic compared with that of males.[28] The sizes for some of these neuropeptides are shown in Table 7.2.

Table 7.2 The Principal Human Hypothalamic Neuropeptides and Their Sizes

Name	Size (number of amino acids)
Vasopressin	9
Oxytocin	9
Thyrotropin-releasing hormone	3
Gonadotropin-releasing hormone	10
Corticotropin-releasing hormone	41
Growth hormone-releasing hormone	37–44
Somatostatin	14
Somatostatin-28	28
Vasoactive intestinal peptide	28
Ghrelin	28

The hypophysiotropic system is quite complex, and you will soon discover that endocrinology is a field where exceptions to the rule are often the rule. However, some of the inherent intricacies of hypophysiotropic hormone actions can be tempered in advance by being alerted to examples of some of their eccentricities:

- *Not all adenohypophysis hormones are under dual control by both releasing and release-inhibiting hormones.* The adenohypophysis hormone, adrenocorticotropic hormone (ACTH), for example, is positively controlled by corticotropin-releasing hormone (CRH), but no specific hypothalamic ACTH-inhibiting hormone has been described. For all pituitary hormones, and particularly those not under dual control by the hypothalamus, their regulation via negative feedback control from downstream peripheral hormones is especially important.
- *Non-hypophysiotropic hormones can also modulate the release of adenohypophysis hormones.* For example, growth hormone secretion by the adenohypophysis is stimulated not just by hypothalamic growth hormone-releasing hormone (GHRH), but also via ghrelin, a peptide hormone secreted by the stomach and brain.
- *Hypophysiotropic hormones can target multiple adenohypophysis cell types.* This can be true, despite the specific names attributed to them. For example, growth hormone-inhibiting hormone (GHIH), also known as somatostatin, can also inhibit thyrotropin release, and thyrotropin-releasing hormone (TRH) can also stimulate prolactin synthesis (Figure 7.16).
- *Hypophysiotropic hormones are often secreted in a pulsatile manner, with pulse amplitude and frequency varying with time of day, month, or season.* Rhythmic variability, due to changing input from the central nervous system, produces differential (sometimes opposing) physiological responses.

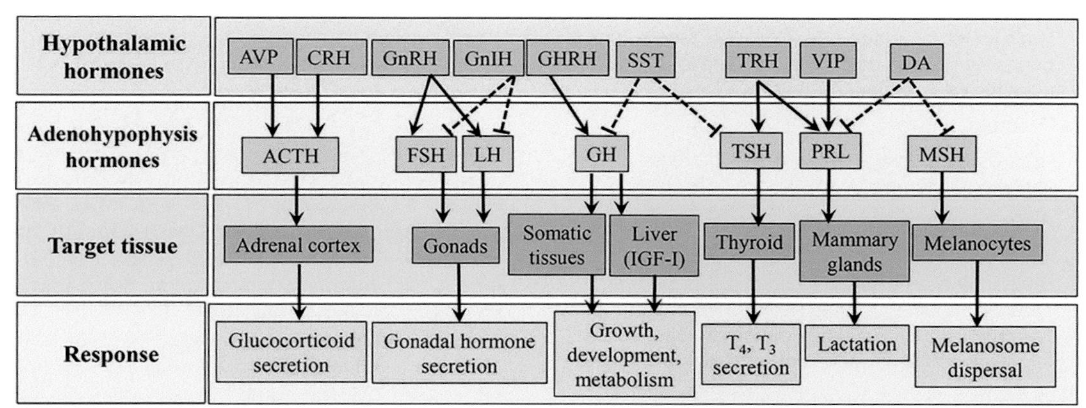

Figure 7.16 Summary of major hypothalamic tropic hormones and some of their influences over adenohypophysis function in mammals. Arginine vasopressin (AVP), corticotropin-releasing hormone (CRH), gonadotropin-releasing hormone, (GnRH), gonadotropin-inhibiting hormone (GnIH), growth hormone-releasing hormone (GHRH), somatostatin (SST), thyroid-releasing hormone (TRH), vasointestinal peptide (VIP), dopamine (DA), adrenocorticotropic hormone (ACTH), insulin-like growth factor-I (IGF-I), thyroxine (T_4), triiodothyronine (T_3).

- *Many hypophysiotropic hormones are also expressed by non-neural cells of different endocrine tissues.* This is particularly true in the gastrointestinal tract and placenta, where they may have significant physiologic functions (see Table 7.3). All the hypophysiotropic hormones are also present in extrahypothalamic brain tissue where they function as neurotransmitters.

Thyrotropin-Releasing Hormone

Thyrotropin-releasing hormone (TRH) is a potent stimulator of the release of pituitary **thyrotropin** (also known as *thyroid-stimulating hormone*, TSH), the hormone that induces the thyroid gland to produce and secrete thyroid hormone (Figure 7.16). TRH was the first hypophysiotropic hormone identified, discovered and purified by Roger Guillemin[29] and Andrew Schally.[30] This tripeptide (pyro-Glu-His-Pro-NH2) is the smallest peptide hormone. It has been estimated that a single molecule of TRH induces, via its stimulation of TSH, the release of more than 100,000 molecules of thyroxine from the thyroid gland. At pathologically high levels, TRH is also implicated in the release of *prolactin* (PRL). TRH is derived from a large, 242 amino acid precursor protein that in humans contains six repeating copies of the sequence, Gln-His-Pro-Gly, which generates mature TRH following excision. Interestingly, this repeating sequence is dispersed throughout the precursor and conserved between mammalian and amphibian prohormones,[31] suggesting that the ability of the prohormone to generate multiple bioactive peptides may be an important mechanism of amplification for production of this hormone production. Guillemin and Schally were awarded the *Nobel Prize in Physiology or Medicine* in 1977 for their purification of TRH.

Table 7.3 Extrahypothalamic Sites of Hypothalamic Peptide Secretion

Peptide	Extrahypothalamic Site of Synthesis
TRH	Brain (cerebral cortex, olfactory lobe, cerebellum, brainstem), spinal cord, fetal pancreatic islet cells, intestine neuroendocrine cells
GHRH	Pancreas
GnRH	Brain (olfactory lobe, limbic system), breast (lactation), placenta
CRH	Brain (cerebral cortex, limbic system, brainstem), spinal cord, placenta, immune system
SST	Brain (cerebral cortex, brainstem, olfactory lobe, cerebellum, pineal gland), spinal cord, retina, peripheral nervous system, pancreatic islet cells, intestinal neuroendocrine cells, thyroid, placenta
Dopamine	Brain, gastrointestinal tract
Ghrelin	Stomach
AVP	Brain (amygdala, hippocampus, diagonal band of Broca, and the choroid plexus)
Oxytocin	Brain (amygdala, hippocampus)

Developments & Directions: Hypogonadism in GnRH mouse mutants

The most effective demonstration that GnRH is an essential component in controlling the brain-pituitary-gonadal axis would be to disrupt the gene encoding this hormone using gene knockout technology in mice, and then describe the resulting physiological consequences. Conveniently, the naturally occurring hypogonadal mouse mutant (hpg) has a massive 33.5 kb deletion in its GnRH gene.[32] George Fink's lab[33] described the resulting phenotype, showing that in *homozygotes* of a population, or individuals that have inherited both alleles for the mutated gene, the hypothalamus and median eminence are both devoid of immunoreactive GnRH.[34] The resulting depletion of pituitary LH and FSH production produces adult male and female mice with extremely small and atrophied gonads and reproductive tracts (Figure 7.17), and these homozygote mutants are unable to reproduce. Remarkably, normal reproductive functions in these homozygous hpg mutants can be completely restored following injections of GnRH, which stimulate pituitary LH and FSH synthesis and secretion and activation of the gonads.[35] Another powerful demonstration of the hypothalamic control of the HPG axis by GnRH was to restore reproductive function to adult hpg mice by transplanting hypothalamic tissue from the preoptic area (POA; the region of the brain rich in GnRH cell bodies) from late fetal/early neonatal donors into the third ventricle of the adult hpg mice.[34,36]

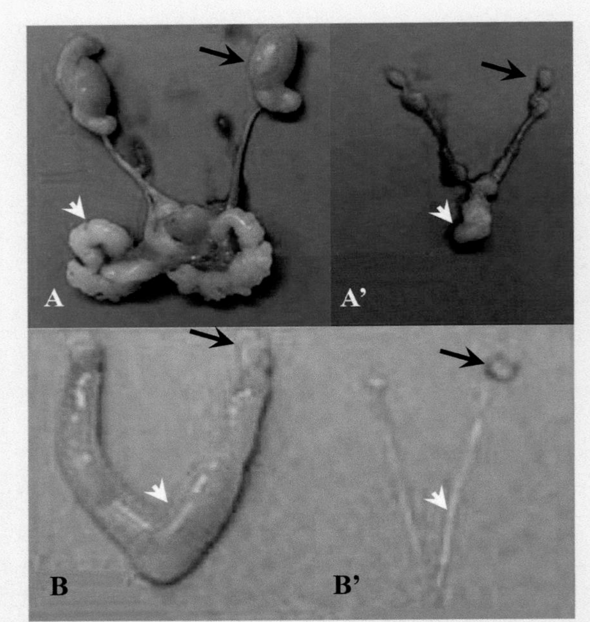

Figure 7.17 Photographs of the gonads and accessory sexual tissue of normal and GnRH deficient mice. (A) Testes (black arrow) and seminal vesicles (white arrowhead) from a wild-type (WT) male mouse compared with **(A')** the comparatively small structures from a mouse genetically altered to inhibit GnRH synthesis. **(B)** Ovaries (black arrow) and uterus (white arrowhead) from a WT female mouse compared with **(B')** the comparatively small structures from a mouse genetically altered to inhibit GnRH synthesis.

Source of Images: Hoffmann, H.M., et al. 2016. J. Neurosci., 36(12), pp. 3506-3518.

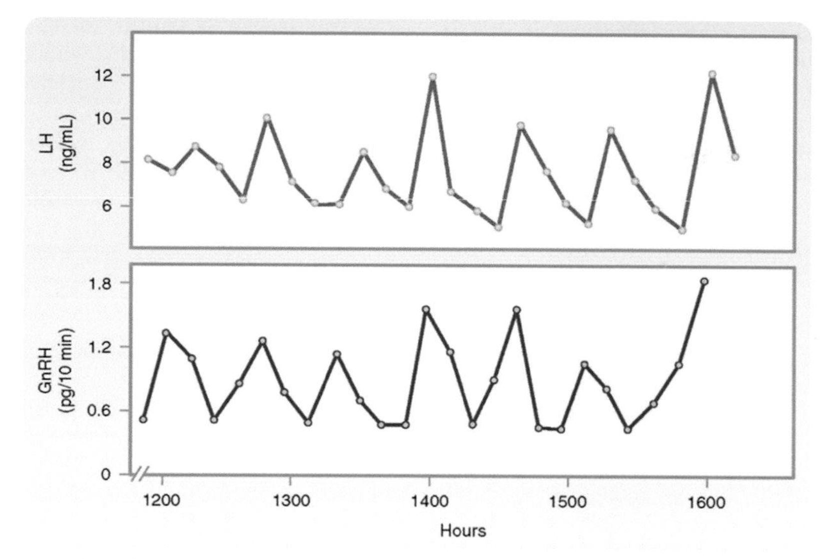

Figure 7.18 Fluctuation of peripheral vein plasma LH levels and portal vein plasma GnRH levels in unanesthetized, ovariectomized female sheep. Each pulse of LH is coordinated with a pulse of GnRH. This supports the view that pulsatility of LH release is dependent on pulsatile stimulation of the pituitary by GnRH.

Source of Images: Levine, J.E., et al. 1982. Endocrinology, 111(5), pp. 1449-1455. Used by permission.

Gonadotropin-Releasing and Inhibiting Hormones

In vertebrates, *gonadotropin-releasing hormone* (GnRH) represents the primary neuroendocrine link between the brain and the reproductive axis. GnRH stimulates the release of two different adenohypophysis "gonadotropins", or peptide hormones that target the gonads: *luteinizing hormone* (LH) and *follicle-stimulating hormone* (FSH). It was initially assumed that LH and FSH were regulated by different hypothalamic factors, and GnRH was originally called LHRH for its greater potency in stimulating LH versus FSH. However, a separate FSH-releasing factor has not been clearly identified, and most endocrinologists consider both LH and FSH to be regulated by GnRH. Interestingly, GnRH is secreted in a pulsatile fashion (Figure 7.18), and changes in GnRH pulsatility are essential for the onset of puberty and for normal fertility.

Among vertebrates all active GnRH forms, regardless of the vertebrate species they derive from, are highly conserved "decapeptides" (composed of ten amino acids) (refer to Figure 3.8 of Chapter 3: Evolution of Endocrine Signaling). Two genes located on separate chromosomes encode GnRHs in mammals: GnRH-I and GnRH-II. GnRH-II is also expressed in a variety of non-mammalian vertebrates.[37] A third form of GnRH, GnRH-III, has been reported in several fish species. Indeed, the GnRH gene is ancient, predating the protostome/deuterostome split, and has undergone at least two gene duplications in chordates.[38,39] Despite its small size, the peptide has accumulated amino acid changes resulting in at least 15 paralogous structural variants, many of which may primarily function within the central nervous system.

Gonadotropin-inhibitory hormone (GnIH) was first discovered in the brain of the Japanese quail, *Coturnix japonica*, and is now known to be a major regulator of reproductive timing in birds. GnIH also possesses mammalian orthologs,[40] one of which can inhibit GnRH activity.[41] Although GnIH is present in mammals, its role in regulating the HPG axis remains unclear and is a topic of current study.

> **Curiously Enough . . .** GnIH peptides were first isolated from various invertebrates, including cnidarians, nematodes, annelids, mollusks, and arthropods,[42,43] where they appear to exert diverse functions beyond the regulation of reproduction.

Corticotropin-Releasing Hormone

Although TRH was the first hypophysiotropic hormone to be purified, *corticotropin-releasing hormone* (CRH) was actually the first hypothalamic-releasing hormone to be recognized. CRH is a 41-amino acid peptide that regulates the production of a large pituitary prohormone, **proopiomelanocortin** (POMC), which is cleaved to produce several different hormones, including ACTH (an important stress hormone), β-endorphin, β-lipotropin, melanocyte-stimulating hormone (MSH), and other peptides.

Somatostatin and Growth Hormone-Releasing Hormone

While searching for the hypophysiotropic hormone responsible for causing the adenohypophysis to release growth hormone (GH), another hypothalamic factor was first isolated and characterized by Guillemin's group.[44] However, this factor was shown to actually inhibit GH secretion. This was a 14-amino acid peptide named *somatostatin* (SST). Interestingly, in humans SST is also a potent inhibitor of TSH, prolactin, glucagon, and insulin release.[45] This peptide has since been found to be quite versatile and is widely expressed in the gastrointestinal tract, the delta cells of the pancreas, and the thyroid-parafollicular cells, as well as the hypothalamus, limbic system, septum, hippocampus, cortex, and medulla.[46]

Growth hormone-releasing hormone (GHRH) is the most recent hypothalamic-releasing factor to be completely characterized. As previously mentioned, many hypophysiotropic peptides have been localized outside the central nervous system. Therefore, it was not too surprising that GHRH was first isolated from patients with pancreatic tumors[47] that stimulated the pituitary to produce excess growth hormone, inducing acromegaly, a condition characterized by excessive growth of bones in the hands, feet, and face in adults.

Dopamine: The Release-Inhibiting Factor for Prolactin and Melanocyte-Stimulating Hormone

Prolactin (PRL) is a hormone released by the adenohypophysis, and in mammals plays an essential role in milk production. The major physiological PRL inhibitory factor (PIF) is not a peptide hormone, but, rather, appears to be the catecholamine *dopamine* (Figure 7.19).[48,49] The primary PRL-regulating dopamine neurons are the *tuberoinfundibular dopaminergic cells*, which have their cell bodies in the hypothalamic arcuate nucleus, and they release dopamine into the median eminence. The blockade of endogenous dopamine receptors by a variety of drugs causes a rise in prolactin levels. Furthermore, patients with lesions in the median eminence or pituitary stalk transects have elevated levels of prolactin due to the inability of dopamine to be transported to the adenohypophysis. Interestingly, the prolactin-producing cells in the pituitaries of dopamine-deficient mouse mutants are significantly hypertrophied (Figure 7.19), indicative of elevated prolactin production.[50] Such findings are not restricted to mammals, but have also been observed in fish, such as tilapia (*Oreochromis mossambicus*).[51] Therefore, in contrast to what is seen with all the other previously described pituitary hormones, the hypothalamus generally suppresses prolactin secretion from the pituitary. Natural decreases in hypothalamic dopamine release, such as during lactation, are likely responsible for mediating the rise in prolactin synthesis.

Melanocyte-stimulating hormone (α-MSH, also called melanotropin) is produced by the pituitary gland and other tissues and is associated with the regulation of skin pigmentation, as well as appetite control. Like ACTH, α-MSH is a derivative of the large prohormone, POMC. Unlike ACTH, however, α-MSH synthesis is not regulated by hypothalamic CRH. Similar to the regulation of prolactin release in mammals, α-MSH release appears to also be predominantly under negative control by dopamine, which functions as a melanotropin release-inhibiting factor. When pituitary α-MSH-producing cells are cultured *in vitro* in the absence of dopamine, they also produce more α-MSH.

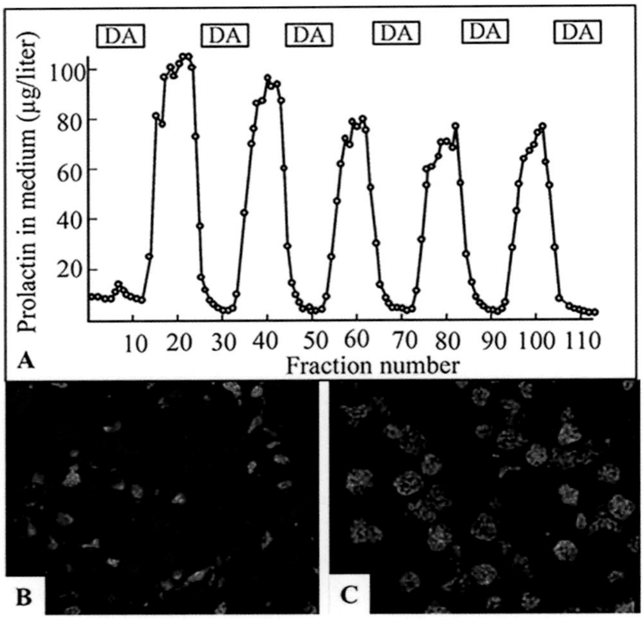

Figure 7.19 Dopamine inhibits prolactin secretion. (A) Response of cultured rat pituitary lactotropes to dopamine. The addition of dopamine (DA) immediately suppresses the secretion of prolactin in culture. PRL immunofluorescence on pituitary sections revealing lactotropes of dopamine-deficient mice **(C)** are hypertrophied compared to control **(B)** mice.

Source of Images: (A) Martini, L. and Besser, G.M. "Clinical Neuroendocrinology". Academic Press, New York, 1977; used by permission. (B) Hnasko, T.S., et al. 2007. Neuroendocrinology, 86(1), pp. 48-57. Used with permission.

SUMMARY AND SYNTHESIS QUESTIONS

1. In mammals, the normal release of what pituitary hormones would be unaffected by damage to the median eminence? Why?
2. Imagine that the portal vessels of the pituitary gland are damaged such that blood no longer flows through them. How would this affect the release of the hormones vasopressin and oxytocin into general circulation?
3. Many "hypothalamic" hormones, such as TRH, CRH, and SST are also produced by other regions of the brain. However, these neuropeptides produced by the other brain regions are thought to signal in an autocrine/paracrine manner, and don't enter the bloodstream. Why don't these particular neuropeptides enter the circulatory system?
4. One molecule of TRH prohormone can generate five molecules of TRH peptides. Explain.
5. The naturally occurring hypogonadal mouse mutant (hpg), has deletion in its GnRH gene, and in homozygotes the resulting phenotype of these mutants the hypothalamus and median eminence are both

devoid of immunoreactive GnRH. Considering that these homozygotes are sterile, how does this mouse population continue to survive in the wild?

6. Many antipsychotic drugs for treating diseases like schizophrenia and bipolar disorder function by blocking dopamine receptors in the brain. Interestingly, two side effects of these drugs may include gynecomastia (enlargement of breast tissue in male patients) and hyperpigmentation (development of patches of dark pigment on the skin). Propose an endocrine explanation for these side effects.
7. Explain why *hypo*thyroidism can result in *hyper*prolactemia.
8. Why can individuals with pituitary tumors often also have vision problems?

Hypothalamic-Pituitary-End Organ Feedback Loops and Diagnoses of Endocrine Dysfunction

LEARNING OBJECTIVE Outline the components of generic hypothalamic-pituitary-end organ (HPE) feedback loops, and explain how endocrine dysfunctions along an HPE axis can be diagnosed.

KEY CONCEPTS:

- Hypothalamic-pituitary-end organ feedback loops may be regulated through both negative and positive feedback.
- There are two broad categories of endocrine dysfunctions: endocrine hyperfunction, or overproduction of a hormone, and endocrine hypofunction, or underproduction of a hormone.
- Suppression tests, which are useful in evaluating endocrine hyperfunction, measure the ability of administered hormone to provide feedback inhibition along the HP axis.
- Stimulation tests, which are useful in evaluating endocrine hypofunction, measure the ability of an administered hormone to increase target gland hormone synthesis and of upstream glands along the HP axis to respond to feedback inhibition.

Historically, it has been well established that the removal of major pituitary target glands, such as the adrenal,[52] the gonads,[53] and thyroid,[54] each results in the *hypertrophy*, or enlargement, of specific cell populations within the adenohypophysis. By contrast, **hypophysectomy**, or the removal of the pituitary gland, induces all of these target glands to degenerate. These types of experiments demonstrated that the secretion of pituitary ACTH, gonadotropins, and thyrotropin are controlled by **negative feedback** exerted by what we now know are the adrenal corticosteroids, gonadal steroids, and thyroid hormones, respectively. With some important exceptions described in other chapters, most hormones produced by target glands feedback onto both the pituitary and the hypothalamus to inhibit their secretion.

Hypothalamic-Pituitary-End Organ Feedback Loops

In this section, the concept of feedback loops that regulate several major hypothalamic-pituitary (adenohypophysis)-end organ (HPE) axes will be described in broad terms. Importantly, not all HPE axes use all three categories of loops. It is also essential to keep in mind that these feedback regulatory systems are superimposed upon hormonal rhythms used for adaptation to the environment. Seasonal changes, daily light-dark cycles, and stress are just a few of many environmental events that have major impacts on the secretion of pituitary hormones. Detailed descriptions for the regulatory feedback loops associated with each endocrine axis will be found in other chapters that are devoted to the respective systems. In general, hypothalamic-pituitary-end organ axes feedback loops fall into one of three categories (Figure 7.20):

Ultra-short loop: a hormone produced by a gland autoinhibits the gland of origin when hormone levels are high.

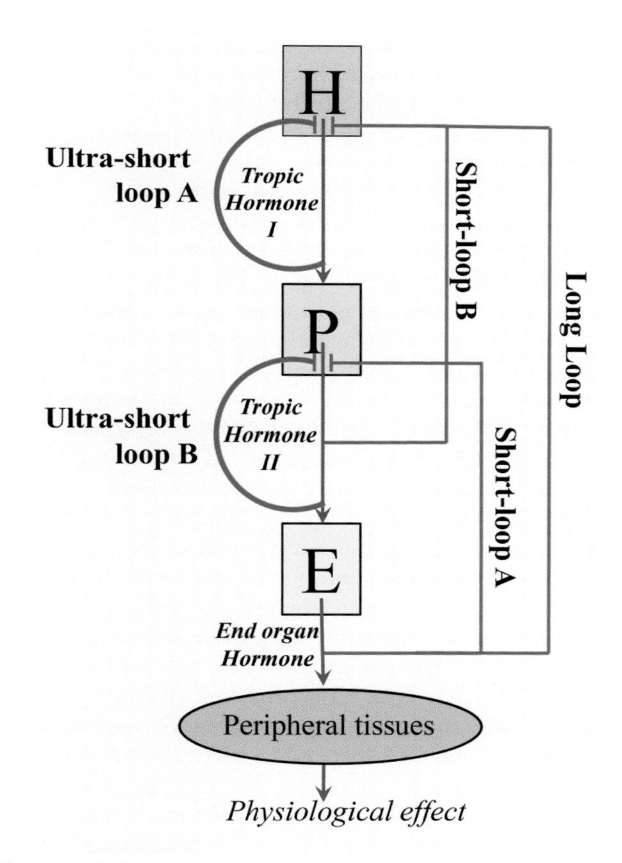

Figure 7.20 Negative feedback loops of the hypothalamic-pituitary-end organ (HPE) axis. Stimulatory and inhibitory actions are denoted by green arrows and red lines, respectively.

For example, an alteration in the release of a hypothalamic hypophysiotropic peptide may be induced by the peptide itself acting on the hypothalamus. Ultra-short loops are, in effect, mediated by autocrine and paracrine means.

Short loop: a second hormone produced downstream by a second gland in response to a hormone from the first gland inhibits the first gland. For example, thyroid hormones and sex steroids may feedback to the pituitary gland.

Long loop: a final hormone produced downstream by a third gland (the "end organ") inhibits all or most stages upstream of the pathway. Hormones mediating this feedback do so via entry into general circulation.

Diagnoses of Endocrine Dysfunction

Endocrine dysfunctions generally fall into two broad categories: *endocrine hyperfunction*, or overproduction of a hormone most typically caused by the presence of a hormone-secreting tumor (see Table 7.4), and *endocrine hypofunction*, or underproduction of a hormone most commonly due to a "lesion", or damage to the organ. Endocrine hypofunction may also manifest as target organ resistance or the absence of functional tropic hormone receptors. Within hypothalamic-pituitary-end organ axes, in particular, these two types of dysfunctions are further subcategorized as follows (Figure 7.21):

1. *Primary disorder*: the functioning of the end organ itself is directly altered, producing abnormally high or low levels of end organ hormone. For example, hypothyroidism may be caused by a primary hypofunction, whereby the thyroid gland itself is damaged and fails to synthesize thyroid hormone.
2. *Secondary disorder*: the pituitary gland is damaged, producing abnormally high/low levels of a pituitary tropic hormone, resulting in increased/decreased end organ hormone synthesis, respectively. For example, gigantism caused by elevated growth rates during childhood can stem from a secondary hyperfunction of pituitary somatotropes due to a tumor in these GH-producing cells.
3. *Tertiary disorder*: the hypothalamus is damaged, producing abnormally high/low levels of hypothalamic tropic hormones, resulting in elevated/reduced pituitary tropic hormone synthesis and subsequent increased/decreased end organ hormone synthesis. For example, the best described form of tertiary disorder is Kallmann syndrome, a genetic condition characterized by the failure of GnRH neurons to migrate properly to the hypothalamus, resulting in hypogonadism and infertility.
4. *Ectopic hyperfunction*: a tumor *ectopic* (lies outside of) to the HPE axis produces elevated levels of hypophysiotropic-releasing hormone, elevated pituitary tropic hormone, or increased end organ hormone. All three instances result in elevated end organ hormone levels. For example, Cushing's disease, characterized by overproduction of adrenal glucocorticoids, can be caused by ectopic ACTH or CRH synthesis by tumors located in the lung, thymus gland, pancreatic islet cells, and even medullary carcinoma of the thyroid.

Table 7.4 Classification of Pituitary Adenomas

Adenoma Cell Origin	Hormone Product	Clinical Syndrome
Lactotrope	PRL	Hypogonadism, galactorrhea
Gonadotrope	FSH, LH	Hypogonadism (compression of LH cells)
Somatotrope	GH	Acromegaly/gigantism
Corticotrope	ACTH	Cushing's disease
Mixed GH and PRL cell	GH, PRL	Acromegaly, hypogonadism, galactorrhea
Acidophil stem cell	GH, PRL	Acromegaly, hypogonadism, galactorrhea
Mammosomatotrope	GH, PRL	Acromegaly, hypogonadism, galactorrhea
Thyrotrope	TSH	Thyrotoxicosis

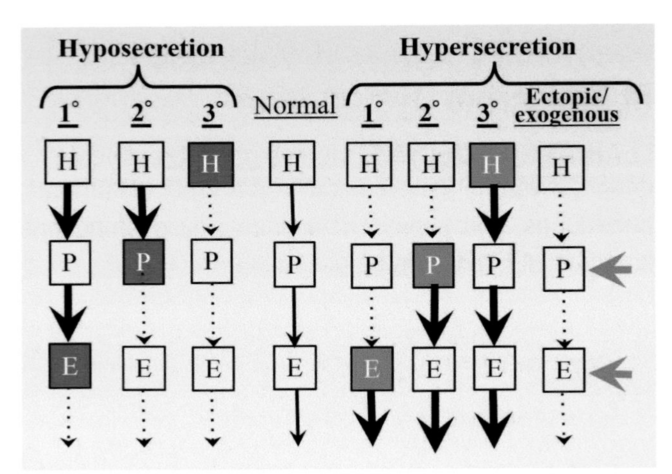

Figure 7.21 Primary, secondary, tertiary, and ectopic disorders of hypothalamic-pituitary-end organ axes.

5. *Exogenous hyperfunction*: the symptoms of abnormally high end organ hormone production can also be derived from disturbances originating from outside the body, often in the form of a drug or medication. For example, when anabolic steroids, such as testosterone, are ingested as a muscle-building supplement, this feeds back negatively onto the hypothalamic-pituitary-gonad axis, causing the gonads to shrink in size. Dexamethasone, a synthetic glucocorticoid commonly prescribed to suppress inflammation, causes the adrenal cortex to degenerate following prolonged use.

Stimulation and Suppression Tests

Two critical features to note regarding hyper/hypofunction disorders of the HPE axes described earlier are that

1. Hypersecretion from an endocrine organ results in increased hormone production downstream, but also decreased hormone production upstream due to negative feedback.

2. Hyposecretion from an endocrine organ results in decreased hormone production downstream, but also increased hormone production upstream also due to negative feedback.

Therefore, both the category (hyperfunction or hypofunction) and the nature of the subcategory (primary, secondary, tertiary, or ectopic/exogenous) can, in theory, be diagnosed by measuring relative concentrations of three hormones along any given axis and comparing them to normal reference values. For example, abnormally high levels of blood cortisol place the disorder firmly in the hyperfunction category. If follow-up blood tests further show that ACTH levels are abnormally elevated and CRH levels are very low, this could suggest the presence of a secondary (pituitary) disorder, such as a hypersecreting tumor of the corticotrope cells. Unfortunately, in practice, the concentrations of hypothalamic hormones, such as CRH, in general circulation are typically too low to measure accurately and do not correlate well with those in the hypothalamic-hypophyseal portal blood plasma. Furthermore, it is important to recall that elevated ACTH can be of ectopic origin, and a simple blood test cannot distinguish between secondary and ectopic ACTH. Therefore, other approaches must be used to help differentiate among secondary, tertiary, and ectopic dysfunctions. For example, if a secondary hyperfunction disorder is present, the pituitary gland might be expected to be abnormally large, possibly due to the presence of an ACTH-producing tumor in the pituitary corticotropes. Pituitary size can be measured using imaging techniques, such as CAT scan or MRI, to assess this possibility.

Additional valuable diagnostic information can be gained by perturbing the feedback system through the administration of hormones using invasive stimulation and suppression tests. *Stimulation tests*, where hypothalamic and/or pituitary tropic hormones are administered and the ability of the end organ to respond is evaluated by measuring its product, are most useful in evaluating endocrine hypofunction. This approach estimates the ability of the target gland to synthesize hormone and of upstream glands along the H-P axis to respond to feedback inhibition. A common example is the TRH stimulation test for hypothyroidism, in which levels of pituitary TSH are measured in response to administering the hypothalamic-releasing hormone. In the case of secondary hypothyroidism, serum TSH fails to increase in response to a standard intravenous injection of TRH. In the case of primary hypothyroidism, in which feedback inhibition by thyroid hormone is very low, the pituitary responds to administered TRH by producing very high levels of TSH. In the case of primary hyperthyroidism, there is minimal or no increases in TSH due to the high feedback inhibition that is already present.[55]

Suppression tests, which are useful in evaluating endocrine hyperfunction, measure the ability of administered hormone to provide feedback inhibition. For example, to test for cortisol hypersecretion in the aforementioned Cushing's syndrome, the cause of cortisol excess can be determined using suppression with dexamethasone, a synthetic glucocorticoid analog. Pituitary tumors that produce excess ACTH frequently retain a susceptibility to feedback inhibition. These tumors are resistant to doses of dexamethasone that suppress normal corticotrope ACTH production but are inhibited by higher doses of dexamethasone. In contrast, adrenal gland tumors and tumors that ectopically produce ACTH are typically resistant to even high doses of dexamethasone.[55]

SUMMARY AND SYNTHESIS QUESTIONS

1. Describe the relative hormone concentration profile (compared to a normal person) along the H-P axis for patients with the following conditions: Patient (A) secondary hyperthyroidism; Patient (B) ectopic hyperprolactinemia; Patient (C) tertiary hypogonadism; Patient (D) extensive treatment with dexamethasone (a synthetic glucocorticoid) to combat a persistent inflammation.
2. What would happen to pituitary prolactin production if the pituitary is removed from the animal and cultured *in vitro* (increase, decrease, remain the same)? What about GH? Why?

Summary of Chapter Learning Objectives and Key Concepts

LEARNING OBJECTIVE Compare and contrast the concept of neurosecretion with that of classical endocrine and neural signaling.

- Classical neurons signal by releasing neurotransmitters into a synapse, which bid to receptors on the postsynaptic cell.
- Neurosecretory neurons are specialized for the synthesis, storage, and release of hormones into the circulatory system.
- Neurohemal organs consist of neurosecretory axonal termini and associated blood vessels.
- Neuroendocrine systems can transduce external environmental information (e.g. photoperiod, temperature) into chemical messages that target diverse physiological systems.

LEARNING OBJECTIVE Describe contributions of insect and fish research to the discovery of neuroendocrine neurons.

- In the 1920s, Stefan Kopeć demonstrated that secretions emanating from the brain are required for metamorphosis to take place in gypsy moth caterpillars, the first evidence of neurosecretion.
- Carl Caskey Speidel demonstrated that secretory cells are present in the spinal cords of fish, suggesting the presence of neurosecretory systems in vetebrates.

- Ernst and Berta Scharrer founded the modern field of neuroendocrinology by studying neurosecretion in mammals and insects.

LEARNING OBJECTIVE List the locations and functions of the major vertebrate neuroendocrine systems.

- The primary peripheral vertebrate neurosecretory sites are chromaffin cells of the adrenal medulla and the caudal neurosecretory system of fish.
- The primary neurosecretory centers within the vertebrate brain are the peptide-secreting cells of the hypothalamus and neurohypophysis and the melatonin-secreting cells of the pineal gland.

LEARNING OBJECTIVE Describe the six basic physiological aspects of homeostasis regulated by the hypothalamus, and discuss the general importance and gross anatomy of the hypothalamus and pituitary gland.

- The hypothalamus integrates sensory information received from diverse receptors throughout the body to monitor most aspects of homeostasis.
- The hypothalamus maintains homeostasis through endocrine, autonomic, and behavioral outputs.
- The pituitary gland is a bi-lobed structure consisting of the adenohypophysis and neurohypophysis.
- Magnocellular hypothalamic nuclei project their axons directly into the neurohypophysis, a pituitary lobe that stores neuropeptides hormones produced by the hypothalamus.
- In mammals, the hypothalamus controls adenohypophysis function indirectly, with parvocellular nuclei releasing tropic hormones into portal vessels that carry the signals to the adenohypophysis.

LEARNING OBJECTIVE Contrast mechanisms of hypothalamic control over the pituitary adenohypophysis and neurohypophysis, and list the names and functions of hypothalamic hormones that are involved in pituitary signaling.

- Neurohypophyseal neurons of hypothalamic origin release the neuropeptide hormones vasopressin and oxytocin from nerve endings located in the neurohypophysis.
- Hypophysiotropic neurons of hypothalamic origin project their axons to the median eminence, a neurohemal organ where diverse neuropeptides are released and carried by portal blood vessels to the adenohypophysis.
- There are two categories of hypophysiotropic hormones: releasing hormones promote the release of adenohypophysis hormones and release-inhibiting hormones inhibit the release of adenohypophysis hormones.

LEARNING OBJECTIVE Outline the components of generic hypothalamic-pituitary-end organ (HPE) feedback loops, and explain how endocrine dysfunctions along an HPE axis can be diagnosed.

- Hypothalamic-pituitary-end organ feedback loops may be regulated through both negative and positive feedback.
- There are two broad categories of endocrine dysfunctions: endocrine hyperfunction, or overproduction of a hormone, and endocrine hypofunction, or underproduction of a hormone.
- Suppression tests, which are useful in evaluating endocrine hyperfunction, measure the ability of administered hormone to provide feedback inhibition along the HP axis.
- Stimulation tests, which are useful in evaluating endocrine hypofunction, measure the ability of an administered hormone to increase target gland hormone synthesis and of upstream glands along the HP axis to respond to feedback inhibition.

LITERATURE CITED

1. Cushing H. *Papers Relating to the Pituitary Body, Hypothalamus and Parasympathetic Nervous System*. Thomas; 1932.
2. Roussy G, Mosinger M. *Traite de Neuro-Endocrinologie*. Paris: Masson et Cie; 1946.
3. Kopeć S. Experiments on metamorphosis of insects. *Bull Int Acad Sci Cracovie*. 1917:57–60.
4. Kopeć S. Studies on the necessity of the brain for the inception of insect metamorphosis. *Biol Bull*. 1922;42 323–342.
5. Prof. S. Kopeć. *Nature*. 1941;148(3761): 655–655.
6. Speidel CC. Gland cells of internal secretion in the spinal cord of skates. *Carn Inst Washingt Pub*. 1919;13:1–31.
7. Scharrer E. Die Lichtempfindlichkeit blinder Elritzen (Untersuchungen fiber das Zwischenhirn der Fische). *Z Vergl Physiol*. 1928;7:1–38.
8. Scharrer B. Uber das Hanstromsche Organ X bei Opisthobranchiern. *Pubbl Stn Zool Napoli*. 1935;15:132–142.
9. Scharrer B. Uber sekretorisch tatige Nervenzellen bei wirbellosen *Tieren Naturwissenschaften*. 1937;25: 131–138.
10. Scharrer B. Neurosecretion. II. Neurosecretory cells in the central nervous system of cockroaches. *J Comp Neurol*. 1941;74:93–108.
11. Klowden MJ. Contributions of insect research toward our understanding of neurosecretion. *Arch Insect Biochem Physiol*. 2003;53(3):101–114.
12. McCrohan CR, Lu W, Brierley MJ, Dow L, Balment RJ. Fish caudal neurosecretory system: a model for the study of neuroendocrine secretion. *Gen Comp Endocrinol*. 2007;153(1–3):243–250.
13. Bern HA. The Elusive Urophysis—Twenty-five years in pursuit of caudal neurohormones. *Amer Zool*. 1985;25:763–769.
14. Lu W, Dow L, Gumusgoz S, et al. Coexpression of corticotropin-releasing hormone and urotensin i precursor genes in the caudal neurosecretory system of the euryhaline flounder (Platichthys flesus): a possible shared role in peripheral regulation. *Endocrinology*. 2004;145(12):5786–5797.
15. Kaplan SA. The pituitary gland: a brief history. *Pituitary*. 2007;10(4):323–325.
16. von Frisch K. Beitrage zur Physiologie der Pigmentzellen in der Fischhaut. *Pfluger's Archges Physiol*. 1911;138:319–387.
17. Hogben LT. The pigmentary effector system IV.—A further contribution to the role of pituitary secretion in amphibian colour response. *Brit J Exp Biol*. 1924;1:249–270.
18. Hogben LT. Chromatic behaviour. *Proc R Soc Lond B*. 1942;131:111–136.

19. Rowan W. On photoperiodism, reproductive periodicity and the annual migration of birds and certain fishes. *Proc Boston Soc Nat Hist.* 1926;38:147–189.
20. Marshall FHA. Sexual periodicity and the causes which determine it. *The Croonian Lecture Proc R Soc Lond B.* 1936;226:423–456.
21. Benoit J. Facteurs externes et internes de l'activite sexuelle. I. Stimulation par la lumiere de l'activite sexuelle chez le Canard et la Cane domestique. *Bull Biol France Belg.* 1936;70:487–533.
22. Kandel ER, Schwartz JH, Jessell TM, Siegelbaum SA, Hudspeth AJ. *Principles of Neural Science* (5th ed.). McGraw-Hill Professional; 2012.
23. Goodman LS, Gilman A. *The Pharmacological Basis of Therapeutics* (2nd ed.). Macmillan; 1955.
24. Kovacs WJ, Ojeda SR. *Textbook of Endocrine Physiology* (6th ed.). Oxford University Press; 2011.
25. Dorsa DM, de Kloet ER, Mezey E, de Wied D. Pituitary-brain transport of neurotensin: functional significance of retrograde transport. *Endocrinology.* 1979;104(6):1663–1666.
26. Villalobos C, Nunez L, Frawley LS, Garcia-Sancho J, Sanchez A. Multi-responsiveness of single anterior pituitary cells to hypothalamic-releasing hormones: a cellular basis for paradoxical secretion. *Proc Natl Acad Sci U S A.* 1997;94(25):14132–14137.
27. Villalobos C, Nunez L, Garcia-Sancho J. Anterior pituitary thyrotropes are multifunctional cells. *Am J Physiol Endocrinol Metab.* 2004;287(6): E1166–1170.
28. Nunez L, Villalobos C, Senovilla L, Garcia-Sancho J. Multifunctional cells of mouse anterior pituitary reveal a striking sexual dimorphism. *J Physiol.* 2003;549(Pt 3):835–843.
29. Boler J, Enzmann F, Folkers K, Bowers CY, Schally AV. The identity of chemical and hormonal properties of the thyrotropin releasing hormone and pyroglutamyl-histidyl-proline amide. *Biochem Biophys Res Commun.* 1969;37(4):705–710.
30. Schally AV, Redding TW, Bowers CY, Barrett JF. Isolation and properties of porcine thyrotropin-releasing hormone. *J Biol Chem.* 1969;244(15):4077–4088.
31. Lechan RM, Wu P, Jackson IM, et al. Thyrotropin-releasing hormone precursor: characterization in rat brain. *Science.* 1986;231(4734):159–161.
32. Mason AJ, Hayflick JS, Zoeller RT, et al. A deletion truncating the gonadotropin-releasing hormone gene is responsible for hypogonadism in the hpg mouse. *Science.* 1986;234(4782):1366–1371.
33. Cattanach BM, Iddon CA, Charlton HM, Chiappa SA, Fink G. Gonadotrophin-releasing hormone deficiency in a mutant mouse with hypogonadism. *Nature.* 1977;269(5626):338–340.
34. Charlton H. Neural transplantation in hypogonadal (hpg) mice—physiology and neurobiology. *Reproduction.* 2004;127(1):3–12.
35. Charlton HM, Halpin DM, Iddon C, et al. The effects of daily administration of single and multiple injections of gonadotropin-releasing hormone on pituitary and gonadal function in the hypogonadal (hpg) mouse. *Endocrinology.* 1983;113(2):535–544.
36. Livne I, Gibson MJ, Silverman AJ. Brain grafts of migratory GnRH cells induce gonadal recovery in hypogonadal (hpg) mice. *Brain Res Develop Brain Res.* 1992;69(1):117–123.
37. King JA, Millar RP. Evolutionary aspects of gonadotropin-releasing hormone and its receptor. *Cell Mol Neurobiol.* 1995;15(1):5–23.
38. Sower SA, Freamat M, Kavanaugh SI. The origins of the vertebrate hypothalamic-pituitary-gonadal (HPG) and hypothalamic-pituitary-thyroid (HPT) endocrine systems: new insights from lampreys. *Gen Comp Endocrinol.* 2009;161(1):20–29.
39. Uchida K, Moriyama S, Chiba H, et al. Evolutionary origin of a functional gonadotropin in the pituitary of the most primitive vertebrate, hagfish. *Proc Natl Acad Sci U S A.* 2010;107(36):15832–15837.
40. Ubuka T, Son YL, Bentley GE, Millar RP, Tsutsui K. Gonadotropin-inhibitory hormone (GnIH), GnIH receptor and cell signaling. *Gen Comp Endocrinol.* 2013;190:10–17.
41. Ducret E, Anderson GM, Herbison AE. RFamide-related peptide-3, a mammalian gonadotropin-inhibitory hormone ortholog, regulates gonadotropin-releasing hormone neuron firing in the mouse. *Endocrinology.* 2009;150(6):2799–2804.
42. Price DA, Greenberg MJ. Structure of a molluscan cardioexcitatory neuropeptide. *Science.* 1977;197(4304):670–671.
43. Elphick MR, Mirabeau O. The evolution and variety of RFamide-type neuropeptides: insights from deuterostomian invertebrates. *Front Endocrinol (Lausanne).* 2014;5:93.
44. Brazeau P, Vale W, Burgus R, et al. Hypothalamic polypeptide that inhibits the secretion of immunoreactive pituitary growth hormone. *Science.* 1973;179(4068):77–79.
45. Yen SS, Lasley BL, Wang CF, Leblanc H, Siler TM. The operating characteristics of the hypothalamic-pituitary system during the menstrual cycle and observations of biological action of somatostatin. *Recent Prog Horm Res.* 1975;31:321–363.
46. Liu J, Mershon J. Neurosecretory peptides. *Glob Libr Women's Med.* 2008. doi:10.3843/GLOWM.10287/Update due. (ISSN: 1756-2228).
47. Thorner MO, Perryman RL, Cronin MJ, et al. Somatotroph hyperplasia. Successful treatment of acromegaly by removal of a pancreatic islet tumor secreting a growth hormone-releasing factor. *J Clin Invest.* 1982;70(5):965–977.
48. Lamberts SW, Macleod RM. Regulation of prolactin secretion at the level of the lactotroph. *Physiol Rev.* 1990;70(2): 279–318.
49. MacLeod RM, Lamberts SW. The biphasic regulation of prolactin secretion by dopamine agonist-antagonists. *Endocrinology.* 1978;103(1):200–203.
50. Hnasko TS, Hnasko RM, Sotak BN, Kapur RP, Palmiter RD. Genetic disruption of dopamine production results in pituitary adenomas and severe prolactinemia. *Neuroendocrinol.* 2007;86(1):48–57.
51. Grau EG, Helms LM. The tilapia prolactin cell: a model for stimulus-secretion coupling. *Fish Physiology and Biochemistry.* 1989;7(1–6):11–19.
52. Schreiber V, Kmentova V. Hypertrophy of the rat pituitary after adrenalectomy, castration, thyroidectomy and combinations of them. *C R Hebd Seances Acad Sci.* 1964;258:4151–4153.
53. Andersen DH, Kennedy HS. The effect of gonadectomy on the adrenal, thyroid, and pituitary glands. *J Physiol.* 1933;79(1):1–30 32.
54. Bryant AR. The effect of total thyroidectomy on the structure of the pituitary gland in the rabbit. *Anat Rec.* 1930;47(2):131–145.
55. Goldman L, Ausiello DA, Arend W, et al. eds. *Cecil Medicine* (23rd ed.). Saunders; 2007.

CHAPTER

8

The Pituitary Gland and Its Hormones

CHAPTER LEARNING OBJECTIVES:

- Summarize the effects of hypophysectomy on peripheral tissues in early experiments with tadpoles and rodents, and discuss their implications.
- Describe the gross anatomy, microanatomy and cytology of the various regions of the pituitary gland.
- Describe the developmental origins of the pituitary gland.
- Describe the distinguishing biochemical features, functions, and evolutionary origins of the principal families of adenohypophyseal hormones.
- Describe the distinguishing biochemical features, functions, and evolutionary origins of the neurohypophyseal family of hormones.
- Describe distinctive features of pituitary anatomy among vertebrate taxa and their evolutionary origins from presumed homologous structures in ancestral protochordates.

OPENING QUOTATION:

"The Lewis Carroll of to-day would have Alice nibble from a pituitary mushroom in her left hand and a lutein [portion of an ovary] in her right hand and presto! She is any height desired!"

—Pituitary surgeon, Harvey Cushing, as quoted by Robert Harrower[1]

Introduction and Historical Perspective

LEARNING OBJECTIVE Summarize the effects of hypophysectomy on peripheral tissues in early experiments with tadpoles and rodents, and discuss their implications.

KEY CONCEPTS:

- Hypophysectomized tadpoles and rats display atrophy of the thyroid gland, reduced skin pigmentation, adrenal cortex atrophy, reduced growth rate, underdeveloped brains, and degenerated reproductive tissues.
- These phenotypes of hypophysectomy could all be reversed following injections of pituitary extracts or pituitary implants, demonstrating that the pituitary released mysterious substances into the bloodstream that stimulated diverse organs.
- It is now well established that the mammalian pituitary gland produces nine major hormones.

The Importance of the Pituitary

In a physiology textbook published in 1895, the state of knowledge about **pituitary gland** function was summarized succinctly: "Concerning the processes which take place in these alveoli [of the pituitary body], and the purposes of the organ as a whole, we know absolutely nothing".[2] The ancient Galenic belief that the pituitary merely collected waste products accumulating in the cerebral ventricles and passed them out to the nasal cavity was not dispelled until the early 20th century, when research demonstrating that the pituitary was a gland of internal secretion was conducted. The first three decades of the 20th century were accompanied by an explosion of understanding of pituitary physiology, with experiments using diverse vertebrate species demonstrating that **hypophysectomy** (ablation of the pituitary gland) alters the functions of many other organs. So profound were the effects of hypophysectomy on other organs that by 1931 the pituitary was dubbed "the leader in the endocrine orchestra"[3] and was thought to function autonomously of the brain (the notion of hypothalamic control over the pituitary had not yet been established).

Mainstream medical thought regarding pituitary function during the early 20th century was led by Harvey Williams Cushing, the distinguished American neurosurgeon who is often referred to as the father of neurosurgery (Figure 8.1). Due to his remarkable technical proficiency, Cushing was one of the first surgeons to be able to access the pituitary both quickly and safely. In 1912, he published "The Pituitary Body and Its Disorders",[4] which served as the foundation for research into the pituitary gland and its functions. Cushing used results from 30 years of experimental and clinical surgery to underscore the importance of the pituitary gland in endocrine physiology.[5,6] In the field of endocrinology, Cushing's name is most commonly associated with his most

DOI: 10.1201/9781003359241-11

Figure 8.1 Dr. Harvey Williams Cushing (1869–1939). The distinguished American neurosurgeon who is often referred to as the father of neurosurgery, Cushing used results from 30 years of experimental and clinical surgery to underscore the importance of the pituitary gland in endocrine physiology.

Source of Images: Ruggieri, M., et al. 2018. Amer. J. of Med. Genet. A, 176(3), pp. 515-550. Used with permission.

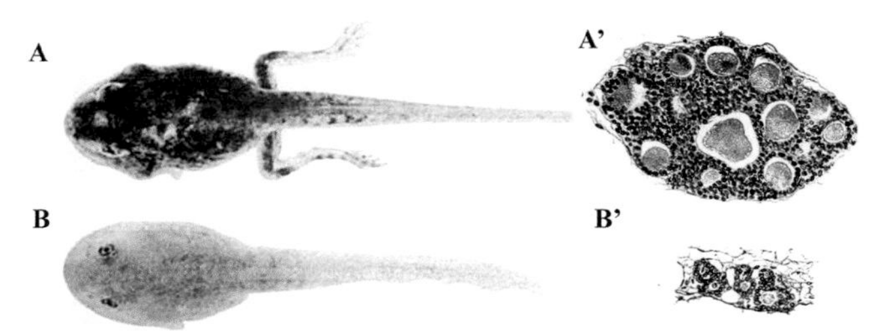

Figure 8.2 Studies in the early 20th century by Philip E. Smith demonstrated that hypophysectomized tadpoles fail to metamorphose and develop with reduced pigmentation and atrophied thyroid glands, as well as numerous other developmental disorders. A normal tadpole **(A)** and a hypophysectomized albino tadpole **(B)**. Note that unlike the normal tadpole, which has grown hind legs, the hypophysectomized tadpole displays no legs or other features of metamorphosis. Camera lucida drawings of the cross section of normal tadpole thyroid gland **(A')** and from a hypophysectomized tadpole **(B')**.

Source of Images: Philip E. Smith (1920) The Pigmentary, Growth and Endocrine Disturbances Induced in the Anuran Tadpole by the Early Ablation of the Pars Buccalis of the Hypophysis. Wistar Institute of Anatomy and Biology. 112 pages. Used with permission.

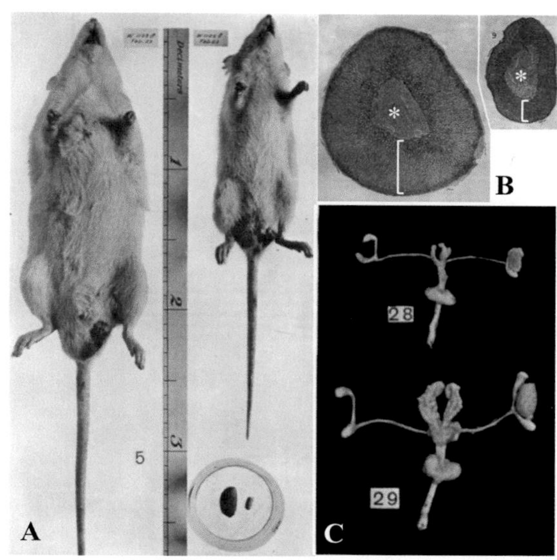

Figure 8.3 Early experiments with hypophysectomized rats showed the pituitary gland is required for normal growth and development of the adrenal gland and gonads. (A) A control rat (left) and its hypophysectomized sibling (right). A testis has been removed from each rat (in petri dish). Note the small size of the testis from the hypophysectomized rat (right) compared with the control (left). **(B)** Cross sections of adrenal glands from a control (left) and hypophysectomized (right) rat. Note the thin adrenal cortex (white brackets) in the hypophysectomized rat's gland compared with the control. By contrast, the size of the adrenal medulla (white asterisk) has not changed. **(C)** An undeveloped reproductive tract from a hypophysectomized female (top). The second female reproductive tract (bottom) is also from a hypophysectomized rat, but it received daily injections of pituitary extract and developed normally.

Source of Images: Philip E. Smith (1930) Hypophysectomy and a replacement therapy in the rat. American Journal of Anatomy Volume 45, Issue 2, pages 205–273, March 1930, plate 2 p 261, 5 p 267, plate 8 p 273. Used with permission.

famous discovery, **Cushing's disease**, caused by a tumor in the pituitary gland that produces large amounts of adrenocorticotropic hormone (ACTH), causing the adrenal glands to produce elevated levels of cortisol.

Experiments addressing the effects of hypophysectomy in tadpoles by Philip Smith[7–9] and Bennet Allen[10–12] provided the first conclusive evidence for any vertebrate of the pituitary's profound role in modulating various aspects of growth and development. Observations following hypophysectomy included atrophy of the thyroid gland, failure to undergo metamorphosis, reduced skin pigmentation, adrenal cortex atrophy, reduced growth rate, and small, underdeveloped brains (Figure 8.2). Remarkably, not only could all of these features be rescued by injections of bovine pituitary extract, but such injections could actually accelerate the onset of metamorphosis, as well as increase growth rates. Smith extended his work on tadpoles to mammals and obtained similarly striking results with a report that early ablation of the adenohypophysis in rats resulted in atrophy of thyroidal, adrenal cortical, testicular, and ovarian tissues, abolishing estrus and stunting growth (Figure 8.3).[13] Importantly, these features of hypophysectomized rodents were all reversed following pituitary implants,[14,15] clearly demonstrating that the pituitary released mysterious substances into the bloodstream that stimulated diverse organs.

Hormones of the Pituitary

It is now well established that the mammalian pituitary gland produces nine major hormones (Figure 8.4), and homologs to these hormones are present in all vertebrates and even some invertebrates. Today, due to advances in molecular biology and gene cloning technologies, large

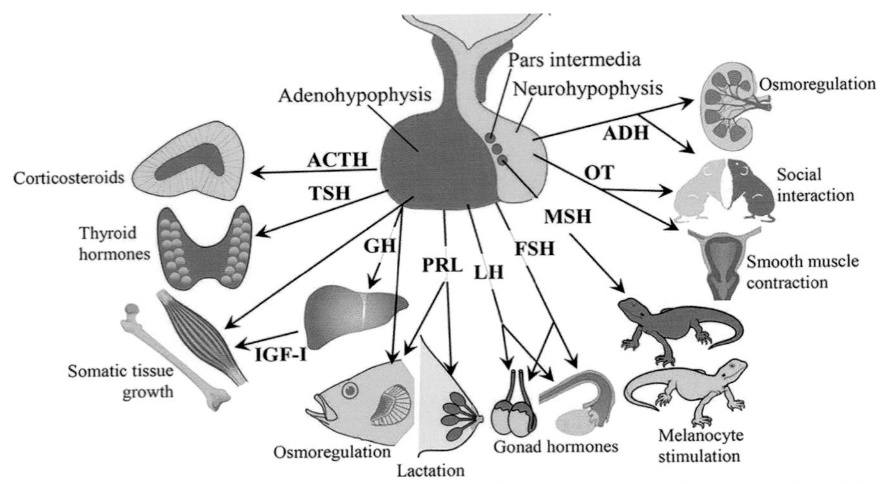

Figure 8.4 Overview of the nine major hormones produced by the pituitary gland and some of their targets. Six hormones are released by the adenohypophysis of the pituitary: adrenocorticotropic hormone (ACTH), thyroid-stimulating hormone (TSH), growth hormone (GH), prolactin (PRL), follicle-stimulating hormone (FSH), and luteinizing hormone (LH). Insulin-like growth factor I (IGF-I), produced by the liver in response to GH, is a potent mediator of somatic growth. Melanocyte-stimulating hormone (MSH) is secreted by the pars intermedia. The two major hormones released by the neurohypophysis are vasopressin/antidiuretic hormone (ADH) and oxytocin (OT).

quantities of highly purified pituitary hormones can be synthesized. The identification and purification of these hormones has led to significant advances in both basic and medical endocrinology, such as the development of stimulation/suppression tests to diagnose pituitary disorders, the creation of effective hormone replacement therapy for patients deficient in the production of one or more pituitary hormones, and generation of the basic resources required for studying hormone actions and hormone evolution. This chapter discusses the gross and fine anatomy of the pituitary gland, its development, the categories and functions of its hormones, and its evolutionary history.

SUMMARY AND SYNTHESIS QUESTIONS

1. Hypophysectomy experiments in tadpoles, frogs, and mice produce profound effects on various organs. For each of the following observations, state which missing pituitary hormone accounts for the effect: atrophy of the thyroid gland, reduced growth rate, reduced testes size and spermatogenic activity, reduced adrenal cortex size, pseudo-albino coloration.
2. Figure 8.3 depicts, in part, the results of hypophysectomy on mouse adrenal gland size. Although the size of the adrenal cortex is dramatically reduced, the size of the adrenal medulla remains unchanged with hypophysectomy. How can this be explained?

Anatomy

LEARNING OBJECTIVE Describe the gross anatomy, microanatomy and cytology of the various regions of the pituitary gland.

KEY CONCEPTS:

- The adenohypophysis of the pituitary consists of classical endocrine glandular cells, while the neurohypophysis is made of neurosecretory axonal termini that extend from hypothalamic nuclei.
- Adenohypophysis cell types can be functionally differentiated based upon the immunohistochemical labeling of specific neuropeptides.

Gross Anatomy

In all vertebrates the pituitary gland consists of two major regions, the adenohypophysis and the neurohypophysis, each of which can be subdivided into multiple morphologically and functionally distinct areas. The generalized tetrapod pituitary morphology is well illustrated in the most basal tetrapods, the amphibians (Figure 8.5).

Curiously Enough . . . In adult humans, the pituitary gland is about the size of a pea, varying in weight from 500 to 700 mg.

The **adenohypophysis**, often called the "anterior" lobe in humans, is made up of epithelial tissue-derived glandular endocrine cells and is thus a classical endocrine gland. The adenohypophysis controls body growth, lactation (in mammals), the gonads, adrenal, and thyroid glands, as well as skin pigmentation via the secretions of growth hormone (GH), prolactin (PRL), gonadotropins (follicle-stimulating hormone, FSH, and luteinizing

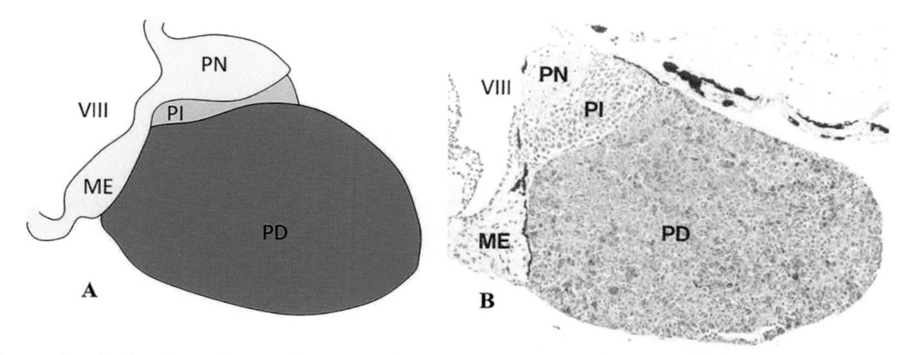

Figure 8.5 The pituitary gland of a frog, *Rana nigromaculata*, depicted in schematic (A) and histological section (B). Pars distalis (PD), pars intermedia (PI), pars nervosa (PN), median eminence (ME), and third ventricle (VIII).

Source of Images: (B) Matsumoto, A. and Ishii, S., 1992. Atlas of endocrine organs: vertebrates and invertebrates. Springer. Ch 3 figs 12-5. Used with permission.

hormone, LH), adrenocorticotropin (ACTH), thyroid-stimulating hormone (TSH), and melanocyte-stimulating hormone (MSH), respectively. In addition to their *trophic*, or growth-promoting effects, and their *tropic* (ability to stimulate the release of other hormones) effects, most of these hormones also exert pronounced effects on metabolism. The release of hormones from these cells into general circulation is controlled by release and release-inhibiting hormones transported from parvocellular hypothalamic nuclei via hypophyseal portal vessels.

In mammals, the adenohypophysis constitutes most of the pituitary's secretory cells and is divided into three regions. The **pars distalis** (distal part) produces at least six major hormones, the greatest diversity of hormones of any pituitary region. Cells of the so-called *intermediate lobe* (also known as the **pars intermedia**) form a distinct region in many tetrapod vertebrates. However, despite its misleading name, the intermediate lobe is not considered to be a distinct lobe of the pituitary. In many mammals these cells form a boundary layer between the adenohypophysis and neurohypophysis (Figure 8.6 and Figure 8.7). The pars intermedia develops posterior to the pituitary *cleft*, a vestige of a fetal developmental structure called **Rathke's pouch**. Although the pars intermedia is prominent in the adults of many vertebrates, in humans it is only transiently present during fetal life[16] and disappears after birth.[17] When present, the most important hormone it produces is the pigmentation hormone, melanocyte-stimulating hormone (MSH).

The **pars tuberalis** is a collar of tissue that forms the outer portion of the pituitary stalk connecting the pituitary to the hypothalamus. The endocrine function of the pars tuberalis remains uncertain. The primary endocrine cell type of the pars tuberalis are thyrotropes that express thyroid-stimulating hormone (TSH). However, in marked contrast with the thyrotropes located in the pars distalis, these cells lack receptors for the hypothalamic thyrotropin-releasing hormone (TRH)[18] and do not respond to conventional hypothalamic outputs. TSH of pars tuberalis origin may play an important role in modulating

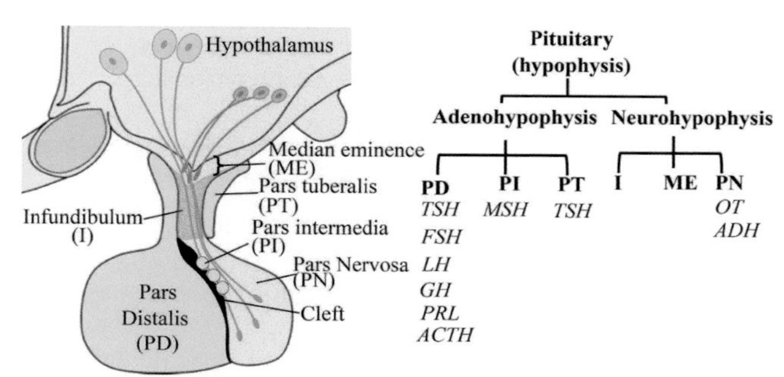

Figure 8.6 Gross morphology features and subdivisions of the human pituitary gland. The pars intermedia is a fetal structure that is not present in adult humans.

biological rhythms associated with seasonal reproduction in some birds and mammals.

The human *neurohypophysis* consists of the **median eminence**, *pars nervosa* and the *infundibulum* (Latin for "funnel", this is the neural component of the pituitary stalk that connects the pituitary gland to the hypothalamus) (Figure 8.6 and Figure 8.7). These regions consist primarily of unmyelinated axonal projections emanating from magnocellular bodies of hypothalamic supraoptic and paraventricular nuclei, whose terminals, which are housed in the pars nervosa, secrete the peptide hormones oxytocin and vasopressin into general circulation. In contrast to the adenohypophysis, which is made up of classical glandular endocrine cells, the neurohypophysis consists of neurosecretory cells.

Microanatomy

Individual pituitary cells can be characterized based on different features including

- *Morphology*: cell size and shape.
- *Histology*: affinity to acidic and basic dyes.
- *Ultrastructure*: shapes and sizes of intracellular structures, such as secretory granules.

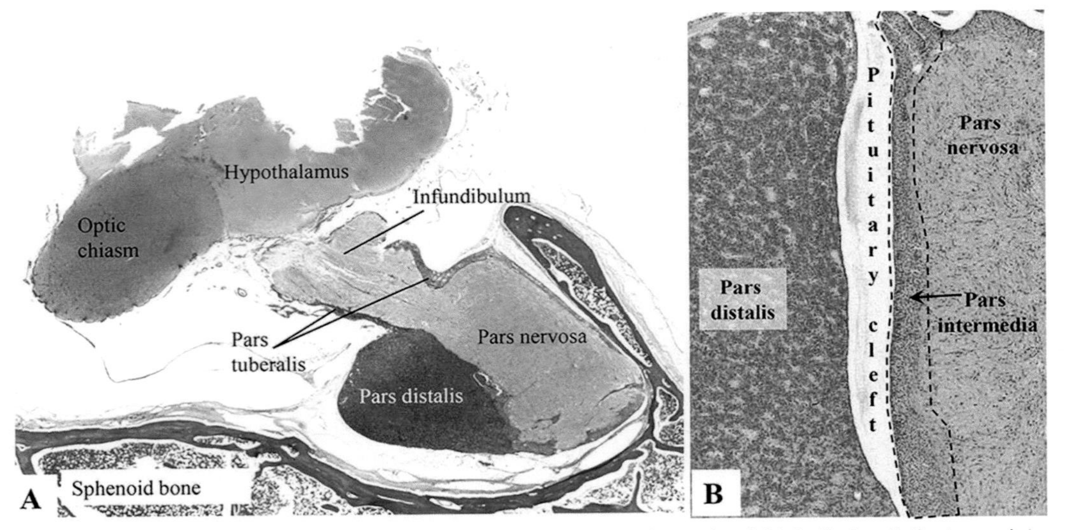

Figure 8.7 Histology of mammalian pituitary glands. (A) An adult human pituitary gland. Note that a distinct pars intermedia cannot be easily identified. **(B)** The pars intermedia of a cat develops posterior to the pituitary cleft, a vestige of a fetal developmental structure called Rathke's pouch. Rathke's pouch typically disappears in adults or becomes reduced to a small series of fluid-filled cysts.

Source of Images: (A) Miller, A., 2010. Functional Histology (Jeffrey B Kerr). The New Zealand Medical Journal, 123(1321); used with permission. (B) Courtesy of Prof. Dr. Hany E Marei, Mansoura University.

- *Immunohistochemistry*: recognition of specific peptides with labeled antibodies.
- In situ *hybridization*: recognition of specific mRNAs with labeled complementary oligonucleotides.

These methods for characterizing cells are useful not only for mapping the pituitary-wide distributions of the different cellular constituents, but also for diagnosing many pituitary dysfunctions. This section describes some of the most useful ways in which the cells of the pituitary are distinguished.

Cytology

Compared with the primarily neuronal composition of the neurohypophysis, the glandular nature of the adenohypophysis is easily appreciated in histological sections of the pituitary stained with hematoxylin, a dark blue nuclear stain, and eosin, which stains other processes pink/purple (Figure 8.8). The neurohypophysis is filled with non-myelinated neuronal axons and their nerve endings, as well as with non-endocrine, modified glial cells called *pituicytes*. Pituicytes are the most numerous and least understood cells of the neurohypophysis, are thought to have supporting and nutritive functions, and may serve a role in controlling rate of neuropeptide secretion.[19] They are morphologically elongated and form one or more processes that associate with adjacent blood vessels or connective tissues, similar to glial cells of other areas of the central nervous system. The nerve endings of the neurohypophysis also possess morphologically distinct features called *Herring bodies* (Figure 8.8), which are axonal swellings packed with secretory granules containing vasopressin and oxytocin. Like the

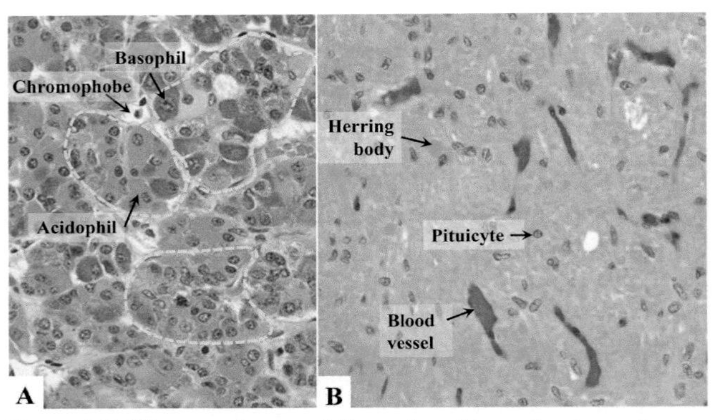

Figure 8.8 Microanatomy of the adenohypophysis and neurohypophysis. (A) Secretory cells of the adenohypophysis are arranged in clusters called "cords" (yellow lines) surrounded by capillaries (white spaces). Cell types stain differentially with hematoxylin and eosin based on the acidity of their cytoplasm. Pink-staining acidophils secrete GH and PRL, whereas purple-staining basophils secrete TSH, FSH, LH, or ACTH. Chromophobes are resistant to stain, and may represent stem cells. **(B)** The neurohypophysis consists of unmyelated axons of neurosecretory cells, modified glial cells called pituicytes, and blood vessels. Herring bodies are acidophilic accumulations of neurosecretory granules at the dilated terminal ends of axons.

Source of Images: (A) Courtesy of Dr. Peter Takizawa, Yale University; (B) Miller, A., 2010. Functional Histology (Jeffrey B Kerr). The New Zealand Medical Journal; Elsevier; used with permission.

adenohypophysis, the nerve endings of the neurohypophysis associate with an extensive vascular network. By contrast, the stained sections of the adenohypophysis reveal the presence of large classical endocrine cells

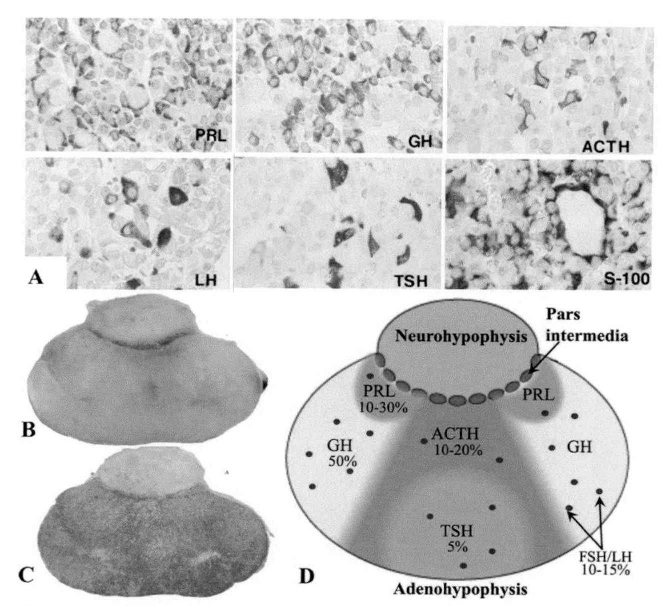

Figure 8.9 Regional distribution of adenohypophysis cells in rats and primates. (A) Rat adenohypophysis cell type and distribution identified by antibody labeling against peptide products unique to each cell type. Prolactin (PRL), growth hormone (GH), adrenocorticotropic hormone (ACTH), luteinizing hormone (LH), thyroid-stimulating hormone (TSH), folliculostellate cells (S-100). **(B)** Horizontal view of the dissected human pituitary gland. **(C)** Horizontal section of the human pituitary stained with hematoxylin and eosin. **(D)** Map of human pituitary cell distribution (the pars intermedia is a transient fetal structure in humans). Note how unlike other pituitary cells that cluster into distinct zones in humans, the gonadotropes are uniformly scattered throughout the adenohypophysis. The "mucoid wedge" is the triangular-shaped central region of the adenohypophysis occupied primarily by ACTH-secreting corticotropes.

Source of Images: (A) Courtesy of Dr. Jennifer Steel; (B-C) de León, A.B., 2022. Otolaryngol. Clin. North Am., 55(2), pp. 265-285; (D) Lowe, J.S., et al. 2018. Stevens & Lowe's Human Histology-E-Book. Elsevier Health Sciences. Used with permission.

grouped into clusters of cords and follicles surrounded by capillaries (Figure 8.8).

Immunocytology and Regional Distribution

Pituitary cell types can be characterized based on techniques that identify their synthesis of specific proteins (immunohistochemistry) (Figure 8.9) or mRNAs (*in situ* hybridization). Such methods are valuable not only for characterizing morphological features of specific cell types, but also for generating maps depicting the regional distributions of different pituitary cells (Figure 8.9). The cells of the adenohypophysis are named according to the tropic hormones that they produce. Next are descriptions of these cells, primarily as characterized for the human pituitary gland.

Somatotropes: The most abundant of the pituitary cells, these GH-producing cells occur in greatest density in the lateral wings of the adenohypophysis and comprise approximately 50% of all adenohypophyseal cells.

Lactotropes: These PRL-secreting cells are concentrated in the posterior portions of the lateral wings. Lactotropes comprise between 9% of hormone-secreting adenohypophysis cells in males and nulliparous (never pregnant) women and up to 25% in multiparous (given birth to more than one child) females.

Curiously Enough . . . Pregnancy results in a permanent doubling of the size of the pituitary, due primarily to a striking *hypertrophy* (growth of an organ caused by increases in individual cell sizes) of lactotropes.[20]

Thyrotropes: TSH-producing cells comprise only 5% of adenohypophyseal cells and are located primarily in the anterior part of the "mucoid wedge" (the midline region).

Gonadotropes: These FSH- and LH-secreting cells comprise 10% of the adenohypophysis. Unlike other pituitary cells that are clustered into distinct zones, gonadotropes are typically evenly distributed throughout the adenohypophysis.

Corticotropes: These ACTH-producing cells comprise 15% to 20% of adenohypophyseal cells and are most numerous in the mid- and posterior portions of the mucoid wedge. In addition to producing ACTH, corticotropes have been shown to produce other POMC derivatives, including MSH, lipotropin, endorphin, and enkephalin.

Melanotropes: Although virtually absent in adult humans, these MSH-secreting cells are located in the pars intermedia of many other adult vertebrates, as well as in fetal humans.

Folliculostellate (FS) cells: FS cells are star-shaped and follicle-forming cells in the adenohypophysis. In contrast with the classical endocrine pituitary cells described earlier, FS cells are non-endocrine cells that closely associate with hormone-producing cells of the adenohypophysis. Although these poorly described cells comprise less than 5% of the adenohypophysis, their long cytoplasmic processes and gap junctions form an extensive network of intercellular communication. They appear to participate in regulating the activity of adenohypophysis endocrine cells through the paracrine secretion of cytokines and growth factors[21,22] and also display scavenger activity by engulfing degenerated cells. They may also represent an adult stem cell progenitor population.[23]

Importantly, these staining techniques not only allow the mapping of specific adenohypophysis cell locations to aid in human pituitary surgery, for example in the removal of tumors, but also provide a tool to evolutionary endocrinologists for visualizing how the pituitary gland evolved from its most ancestral vertebrate forms in the jawless fishes to the more derived forms seen in most vertebrate species today (Figure 8.10).

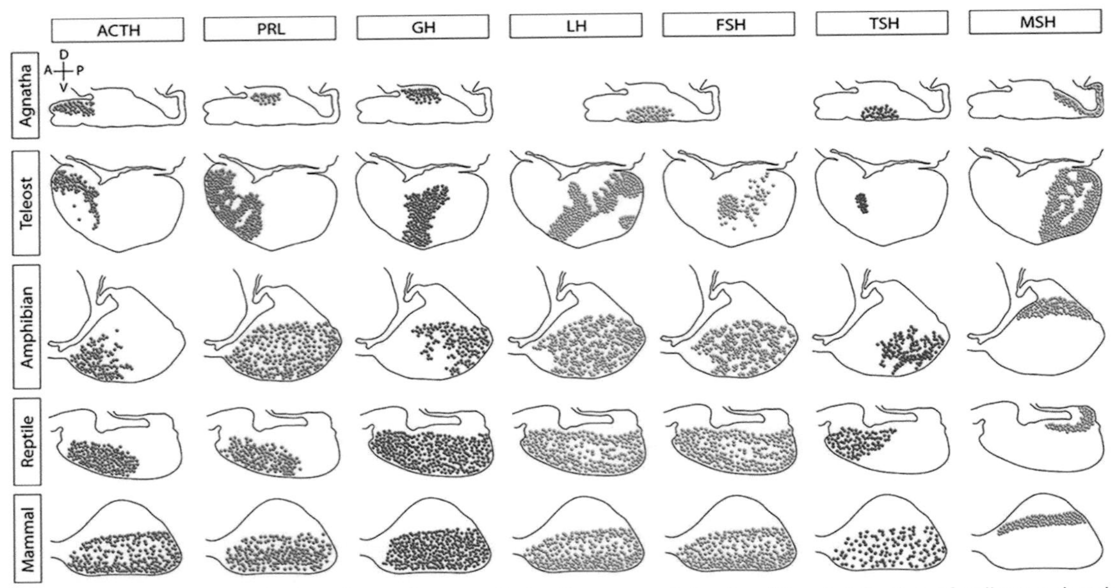

Figure 8.10 Regional distribution of adenohypophysis cells in diverse vertebrates. The distribution of cell types has been modified over the evolution of vertebrates. In agnathans and teleosts, here represented by the sea lamprey and Nile tilapia, respectively, cells are markedly regionalized. Amphibians and reptiles present a regionalized distribution of some cell types, but others are scattered through the pars distalis. In contrast, the mammalian secreting cells are widely scattered throughout the lobe. In lampreys there is only one cell type producing glycoprotein hormone (GPH, a gonadotropin homolog) and thyrostimulin (TSN, a TSH homolog). A, anterior or rostral; P, posterior or caudal; D, dorsal; V, ventral.

Source of Images: Santiago-Andres, Y., et al. 2021. Front. Endocrinol., 11, p. 619352.

SUMMARY AND SYNTHESIS QUESTIONS

1. List the major categories of endocrine/neuroendocrine cells or their processes that are present in the mammalian pars distalis, intermedia, and nervosa.
2. Whereas in humans and other primates the adenohypophysis and neurohypophysis are also referred to as the "anterior" and "posterior" pituitary lobes, this terminology is not appropriate when referring to these regions in non-primates. Why not?
3. Whereas adenohypophysis tissue is packed with the nuclei of secretory cells, the neurohypophysis has a far lower density of cell nuclei (all of which are non-secretory cells), but the lobe is still highly secretory. Explain.
4. How might the water balance of the body respond if there is damage specifically to the infundibulum during surgery?
5. An adenoma that releases ACTH could most likely also release which ONE of the following hormones: GH, PRL, TSH, FSH, LH, MSH. Explain your reasoning.

KEY CONCEPTS:

- The hypophyseal placode, which gives rise to the pituitary's adenohypophysis, is located in a region of the early embryo called the anterior neural ridge (ANR).
- Later in embryogenesis the neurohypophysis derives from neuroectodermal tissue, whereas the adenohypophysis derives from oral ectoderm.

Early Development and Tissue Fate Mapping

LEARNING OBJECTIVE Describe the developmental origins of the pituitary gland.

An understanding of pituitary development during embryogenesis is essential both for fully appreciating its normal functional anatomy and for diagnosing congenital disorders. The pituitary gland is an organ of dual origin in all vertebrates. While pituitary morphology and organization vary among taxa, the general principles of pituitary development and the molecular machinery involved in generating the different endocrine cell types are similar in all vertebrate species studied.

Cranial placodes are early embryonic structures of ectodermal origin that will give rise to diverse components of the sensory nervous system and endocrine/neuroendocrine systems of the head (Figure 8.11). For example, the lens placode gives rise to the lens of the eye, the otic placode gives rise to organs of hearing and equilibrium, and the olfactory placode gives rise to cells that mediate the sense of smell. Importantly, the olfactory placode also gives rise to future hypothalamic neurons that will produce GnRH. Additionally, studies in birds,[24,25] amphibians,[26] and mammals[27] have identified the *hypophyseal placode*. Located in a region called the *anterior neural ridge* (ANR), this placode gives rise to the pituitary's adenohypophysis, which is juxtaposed to cells that will form the neuroendocrine hypothalamus and pituitary neurohypophysis[28,29] (Figure 8.11). In contrast with the adenohypophysis, which originates in a neural fold placode, the hypothalamic primordium develops from neural plate tissue within a region called the *prosencephalon*.[30]

In the course of morphogenesis, the head undergoes extensive tissue rearrangements that move the hypophyseal placode and surrounding oral ectoderm from its initial position at the tip of the embryo's head to a location under the forebrain floor (Figure 8.12). This portion of the forebrain floor is the ventral portion of a structure called the *diencephalon*, a subdivision of the prosencephalon,

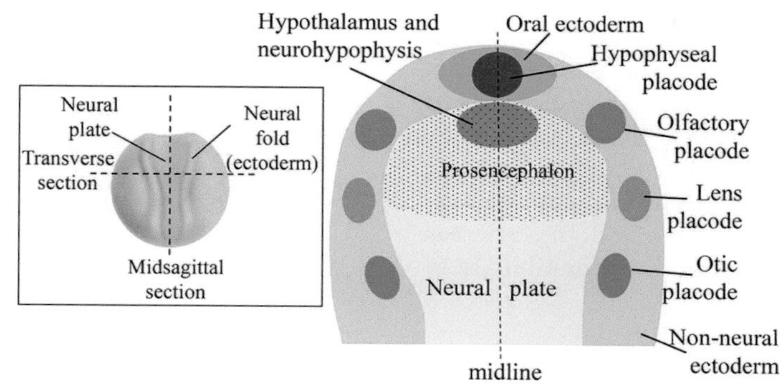

Figure 8.11 A dorsal view of an early neurula stage frog embryo depicting the relative positions of the neural plate and different placodes that will give rise to diverse components of the sensory nervous system and endocrine/neuroendocrine systems of the head. The hypophyseal placode, which gives rise to the pituitary's adenohypophysis, is located in a region called the anterior neural ridge (ANR) juxtaposed to cells that will form the neuroendocrine hypothalamus and pituitary neurohypophysis.

Source of Images: Suzuki, M., et al. 2012. Dev. Growth Diff., 54(3), pp. 266-276. Used with permission. Inset modified from Suga, H., 2016. Recapitulating Hypothalamus and Pituitary Development Using Embryonic Stem/Induced Pluripotent Stem Cells. In Stem Cells in Neuroendocrinology (pp. 35-50).

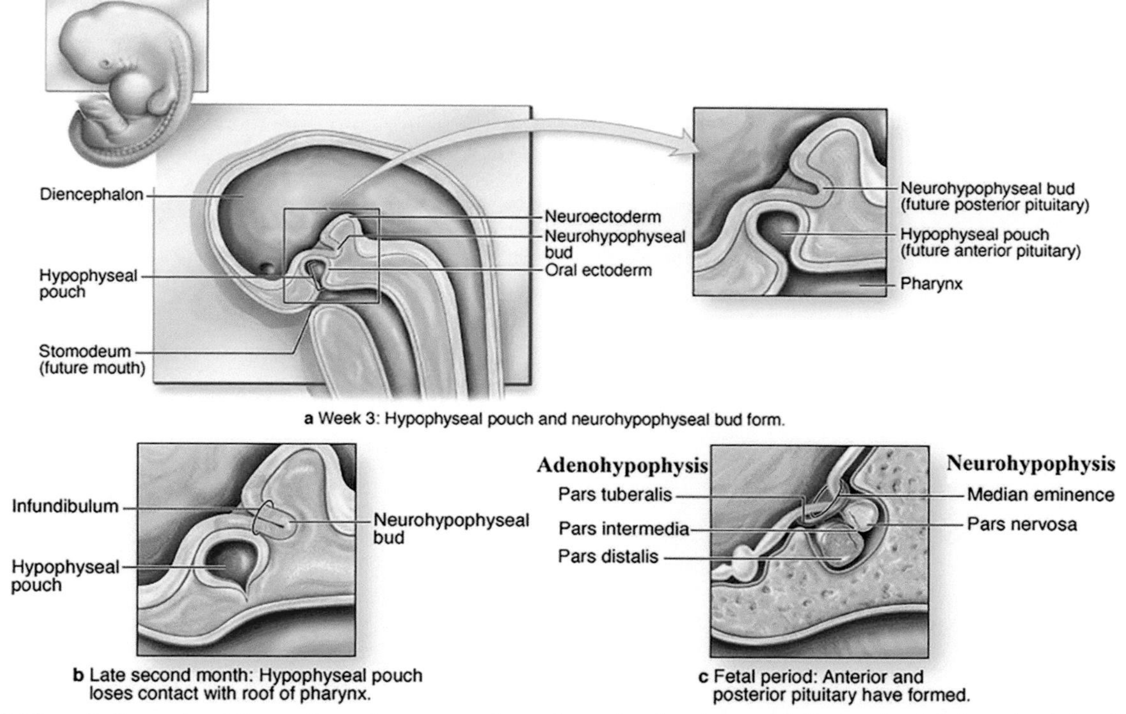

Figure 8.12 The pituitary gland forms from two separate embryonic structures. (a) During the third week of development, a hypophyseal pouch (or Rathke's pouch, the future anterior pituitary) grows from the roof of the pharynx, while a neurohypophyseal bud (future posterior pituitary) forms from the diencephalon. (b) By late in the second month, the hypophyseal pouch detaches from the roof of the pharynx and merges with the neurohypophyseal bud. (c) During the fetal period, the anterior and posterior parts of the pituitary complete development.

Source of Images: Junqueira, L.C. and Mescher, A.L., 2013. Junqueira's basic histology: text & atlas/ McGraw-Hill Education. Used with permission.

and will form most of the hypothalamus. The initial steps of pituitary organogenesis include a thickening and invagination of the oral ectoderm, forming the pituitary anlage called **Rathke's pouch.**[31–33] Simultaneously, Rathke's pouch associates with neurectoderm tissue budding off the ventral diencephalon. This budding neurectoderm tissue will develop into the neurophypophysis. Once Rathke's pouch is fully developed, it detaches from the underlying oral ectoderm, still juxtaposed to the neurohypophyseal anlage. The lumen (inner lining) of the pouch persists as the *pituitary cleft*, separating the adenohypophysis and intermediate lobes in the mature gland. Ultimately, the three main regions of the pituitary gland have a dual embryonic origin: the adenohypophysis and intermediate lobes are derived from the ANR/oral ectoderm, whereas the neurohypophysis originates from the infundibulum of the ventral diencephalon.

SUMMARY AND SYNTHESIS QUESTIONS

1. Compare and contrast the developmental origins of the adenohypophysis with that of the hypothalamus and neurohypophysis.

Hormones of the Adenohypophysis

LEARNING OBJECTIVE Describe the distinguishing biochemical features, functions, and evolutionary origins of the principal families of adenohypophyseal hormones.

KEY CONCEPTS:

- The mammalian pituitary gland produces nine major hormones, and homologs to these hormones are present in all vertebrates.
- Two rounds of genomic duplication are thought to have taken place at the ancestral base of vertebrate phylogeny, giving rise to pituitary hormone diversity among taxa.
- Pituitary hormones fall into four distinct families: the somatomammotropic family, the glycoprotein hormone family, the opiomelanocortin family, and the neurohypophyseal nonapeptides.
- The somatomammotropic hormone family constitutes four related protein hormones: growth hormone, prolactin, placental lactogen, and somatolactin.
- The members of the pituitary glycoprotein hormone family (thyroid-stimulating hormone, follicle-stimulating hormone, luteinizing hormone, chorionic gonadotropin) are heterodimers that each share an identical alpha subunit and possess a unique beta subunit.
- The opiomelanocortin hormone family consists of several different peptides (including adrenocorticotropic hormone, opioids, and melanocyte-stimulating hormone) that all derive from a single large prohormone called the proopiomelanocortin (POMC).

Pituitary Hormone Evolution: The Big Picture

Based upon their structural similarities and evolutionary origins, pituitary hormones fall into four distinct families: the somatomammotropic (growth hormone/prolactin) family, the glycoprotein hormone family (TSH, FSH, LH), the opiomelanocortin family (ACTH, α-MSH, ß-LPH, ß-endorphin), and the neurohypophyseal nonapeptides (OT, AVP) (summarized in Table 8.1).

Table 8.1 List of the Major Pituitary Mammalian Hormones and Their Receptor Characteristics

Pituitary Location	Hormone Family	Hormone Name	Receptor Type	Transducer/Effector
Adenohypophysis	Glycoprotein	Thyroid-stimulating hormone (thyrotropin) (TSH)	GPCR	$G\alpha_s$/AC $G\alpha_q$/PLC
		Follicle-stimulating hormone (FSH)	GPCR	$G\alpha_s$/AC
		Luteinizing hormone (LH)	GPCR	$G\alpha_s$/AC
	Somatomammotropic	Growth hormone (somatotropin) (GH)	Cytokine	Jak/Stat
		Prolactin (PRL)	Cytokine	Jak/Stat
	Opiomelanocortin	Adrenocorticotropic hormone (ACTH)	GPCR (MC2R)	$G\alpha_s$/AC
		*Beta-endorphin	GPCR (mu opiate receptor)	$G\alpha_i$
Pars intermedia	Opiomelanocortin	Melanocyte-stimulating hormone (MSH)	GPCR (MC4R)	$G\alpha_s$/AC
Neurohypophysis	Nonapeptides	Oxytocin (OT)	GPCR	$G\alpha_q$/PLC
		Vasopressin (VP)/antidiuretic hormone (ADH)	GPCR (V1A)	$G\alpha_q$/PLC $G\alpha_s$/AC
			GPCR (V1B)	$G\alpha_q$/PLC
			GPCR (V2)	$G\alpha_s$/AC

Note: *In contrast with endorphins produced by other regions of the brain, pituitary endorphins exert no known physiological roles.

Importantly, although birds, reptiles, amphibians, and bony fishes all share the complete repertoire of pituitary hormones that are present in mammals, the lamprey, an ancestral fish, has only four anterior pituitary hormones and possesses only a single member of the posterior pituitary's VP/OT superfamily[34] (Figure 8.13). Two rounds of genomic duplication are thought to have taken place at the base of the vertebrates, which may explain why in the cartilaginous fishes the GTHβ gene appears to have duplicated into LHβ and FSHβ genes.[35] Furthermore, duplication also appears to have taken place in an ancestral growth hormone-like gene that gave rise to prolactin (PRL) and somatolactin (SL). Interestingly, one ancestral peptide, thyrostimulin, is present in both protochordates and vertebrates. Following its duplication in early vertebrate lineages, it appears to have given rise to FSH, LH, and TSH.

Somatomammotropic Family

Growth hormone (GH), prolactin (PRL), placental lactogen (PL, also called chorionic somatomammotropin, CS), and somatolactin (SL) represent four *paralogous* lineages

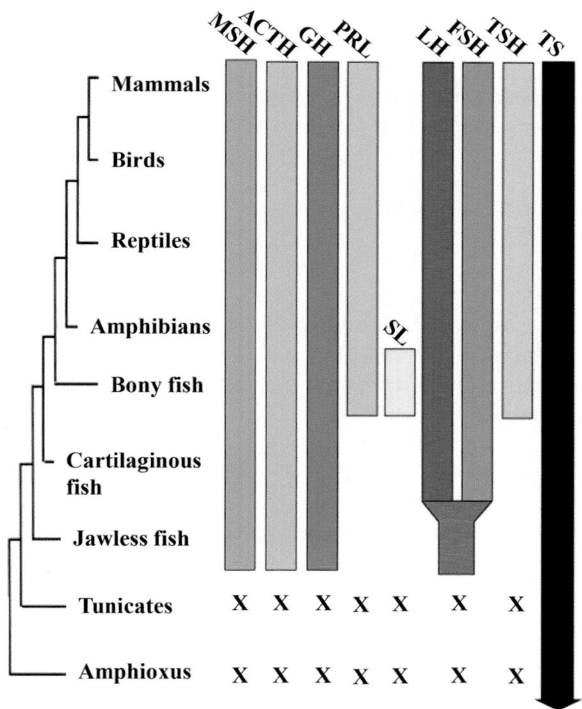

Figure 8.13 The hypothetical evolution of the eight pituitary hormones and thyrostimulin. The genes of the eight pituitary hormones are absent from the genomes of the protochordates, amphioxus and tunicates. ACTH, adrenocorticotropic hormone; FSH, follicle-stimulating hormone; GH, growth hormone; LH, luteinizing hormone; MSH, melanocyte-stimulating hormone; PRL, prolactin; SL, somatolactin; TSH, thyroid-stimulating hormone; TS, thyrostimulin.

Source of Images: Roch, G.J., et al. 2011. Gen. Comp. Endocrinol., 171(1), pp. 1-16. Used with permission.

of related hormones thought to have arisen by duplication and mutation of an ancestral gene (Figure 8.14). The members of this family are produced by distinct tissues: GH and PRL are each secreted by the somatotropes and lactotropes, respectively, of the pars distalis, whereas SL is secreted by the intermediate lobe of the fish pituitary and PL is produced by the mammalian placenta. All four protein hormones are structurally similar, share comparable sizes, and have two to three disulfide bonds in similar positions. The hormones are all classified as growth factors that activate cytokine receptors (Table 8.1). PRL and GH have similar molecular identities and are thought to have diverged from one another about 350 million years ago.[36] PL evolved most recently, less than 80 million years ago, with the advent of mammals. In mammals, GH, PRL, and PL each possess both lactogenic and growth-promoting activities. However, the principal role of GH is postnatal growth, and in mammals PRL is primarily involved in the initiation and maintenance of lactation. PL, which in humans has about 85% homology to GH, may provide GH-like activities to the developing fetus, stimulate mammary gland development, and may also alter maternal metabolism by counteracting maternal insulin to ensure adequate glucose, amino acids, and mineral availability for the fetus.

SL, a distant relative of this family, is present along with GH and PRL in ray-finned fishes and lungfishes, but has not yet been identified in any tetrapods. Structurally, SL has characteristics resembling both GH and PRL, hence its name. Although well-defined physiological roles have yet to be established for SL, the hormone has been linked to several physiological functions including reproduction and the stress response.[37]

Growth Hormone (GH)

GH is synthesized and secreted by the **somatotropes** of the adenohypophysis. Its secretion is under dual control by hypothalamic GH-releasing hormone (GHRH) and GH-inhibiting hormone (GHIH, somatostatin), as well as from GH secretagogues, such as ghrelin. GH exerts both direct and indirect effects on growth and metabolism of diverse tissues, including muscle, bone, cartilage, adipocytes, and the immune system. GH's indirect effects are mediated through stimulation of insulin-like growth factor-I from the liver. In non-mammalian vertebrates, GH generally exerts similar effects on growth and metabolism as in mammals.[38] In some teleost fish, such as salmonids, GH has also been implicated with adaptation to seawater.

Prolactin (PRL)

The prolactin protein is secreted by the **lactotropes** of the adenohypophysis. The release of PRL is thought to primarily be under negative control from hypothalamic dopamine, though TRH has also been shown to stimulate its release under pathologically high concentrations. Although the human pituitary typically contains five-fold less PRL than GH, the number and size of pituitary

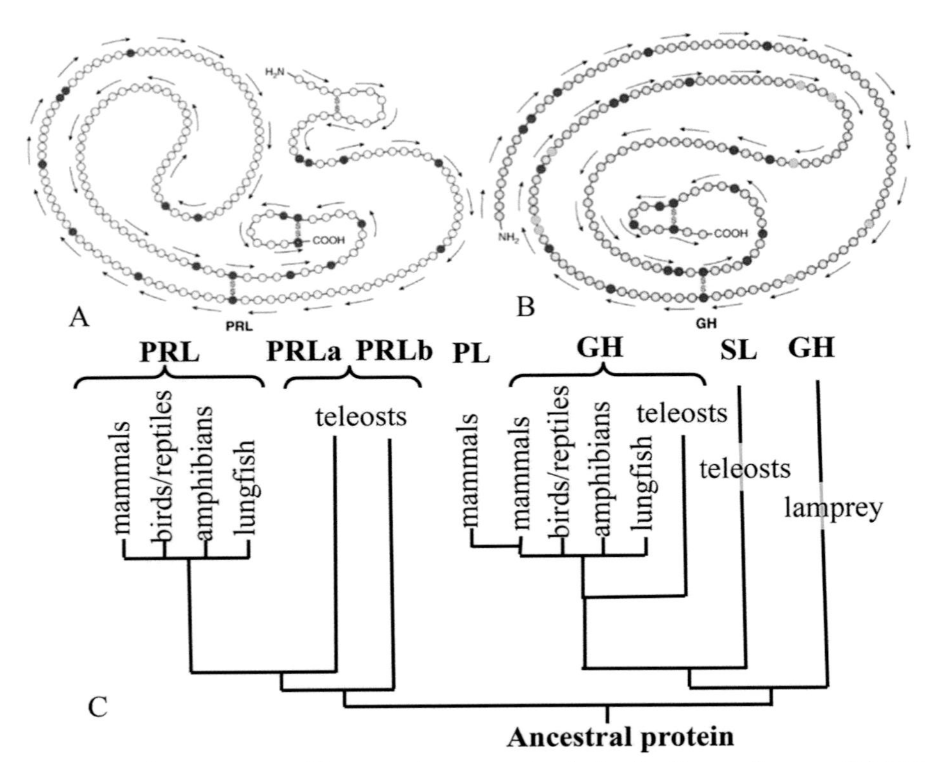

Figure 8.14 A phylogenetic tree for members of the somatomammotropic family. Mammalian growth hormone (GH) **(A)** and prolactin (PRL) **(B)** are of similar size, share considerable amino acid sequence homology (red amino acids), and have disulfide bonds in similar positions. **(C)** Members of this family are thought to have arisen via gene duplication and mutation from a common ancestral gene. Placental lactogen (PL), present only in mammals, has an 85% sequence similarity to GH. Somatolactin (SL) is found only in teleost fish.

Source of Images: (A-B) Norris, D.O., 2018. Integr Comp Biol, 58(6), pp. 1033-1042; used with permission. (C) modified from Daza, et al. (2009). Trends Comp. Endocrinol. Neurobio.: Ann. N.Y. Acad. Sci. 1163: 491–493.

lactotropes can increase significantly in response to elevated estrogen levels, particularly during pregnancy. The best-described action of PRL in mammals is the promotion of regulation of the milk secretory products during the process of lactation by the mammary glands. PRL stimulates both the proliferation and the synthesis of milk proteins, as well as the mobilization of milk lipids and carbohydrates by the glandular epithelium of the mammary gland.

The human PRL gene is expressed not only in the pituitary, but also by the placenta and immune system. In addition to its role in promoting lactation, PRL influences diverse aspects of physiology in both male and female mammals (Table 8.2). PRL also has remarkably diverse functions in non-mammalian vertebrates (Table 8.2). In contrast to GH, which is implicated with seawater adaptation in some fish, PRL is known to promote adaptation of fish to freshwater. Some species of newts (a type of amphibian) migrate from land to ponds to breed, a process accompanied by decreased cutaneous water permeability that facilitates adaptation to freshwater. Interestingly, both of these behavioral and physiological processes can be induced by injection of PRL into these animals, in a process called the "newt water-drive effect", and can be inhibited by administering antiserum to prolactin.[39]

Several families of birds, including pigeons and doves, flamingos, and penguins, secrete a milk-like substance called "crop milk" for their young, who put their bills inside the parent's mouth to nurse. Prolactin is known to not only induce the synthesis of this "milk", which has a higher protein and fat content than cow milk and also contains immune-enhancing factors,[40] but also induces parental nurturing behaviors in these birds. In pigeons and flamingoes, both the male and female adults produce crop milk and share in the feeding and care of the young. Interestingly, in penguins only the males lactate. PRL is also associated with nurturing and parental behavior in some fish. For example, the blue discus (*Symphysodon aequifasciata*) is a freshwater cichlid whose free-swimming larvae feed on the protein-rich mucosal secretions of parental fish skin for nutrition and presumably immunity.[41] Khong and colleagues[42] have shown that PRL receptor mRNA is upregulated in the skin of both male and female parental fish compared to non-parental fish, indicating the possibility of a role of PRL signaling in the regulation of mucus production in relation to parental care behavior.

Table 8.2 Some Differing Actions of Prolactin Among Vertebrate Taxa

Among mammals		**Among non-mammalian vertebrates**	
Female-specific	Mammary development and lactation	Fish	Freshwater osmoregulatory adaptation Skin pigmentation
	Vaginal mucification (rats)		Fin regeneration
	Anti-ovulatory actions (rats)		Parental behavior
Male-Specific	Increased androgen binding in prostate		Gonadal development Mucus synthesis
	Spermatogenic actions	Amphibians	Immune regulation
	Male sex accessory development		Salt and water balance
Both sexes	Upregulates immune function Osmoregulation		Gonadal development Limb regeneration
	Hair and sebaceous gland growth		Gill and tail fin development Immune regulation
	Melanocyte and keratinocyte proliferation	Reptiles	Tail regeneration Skin molting
	Intestinal mucosa growth		Gonadal development
	Increases body mass		Immune regulation
		Birds	Feather growth
			Proliferation of crop sac epithelium
			Crop sac milk production
			Intestinal mucosa growth
			Decreased gonad mass
			Immune regulation

Heterodimeric Glycoprotein Hormones

The adenohypophysis hormones, TSH, LH, and FSH, along with the mammalian placental hormone chorionic gonadotropin (CG), together constitute a closely related family of glycoprotein hormones. These hormones share a common alpha subunit, glycoprotein A (GPA1), and each also contains a unique beta subunit, glycoprotein B (GPB), which confers receptor specificity. Another broadly expressed but poorly characterized hormone closely related to the four aforementioned hormones, called thyrostimulin, has also been discovered.[43] Not only is thyrostimulin present in all chordates, but homologs occur in arthropods, nematodes, and cnidarians, implying that the heterodimeric glycoprotein hormone system existed prior to the emergence of bilateral metazoans.[44] Unlike the aforementioned GH-PRL hormone family, which is composed of simple proteins, this family of pituitary and placental glycoprotein hormones constitutes the most chemically complex family of polypeptide hormones. The nature of this complexity manifests in two ways (Figure 8.15):

1. All of these proteins have covalently bound carbohydrate moieties at one or more positions within their structure, containing up to 33% carbohydrate by weight.[45] The glycosylated moieties are thought to increase biological half-life, as well as promote dimer assembly and receptor binding.
2. With the exception of thyrostimulin, family members all share an identical alpha subunit, known as

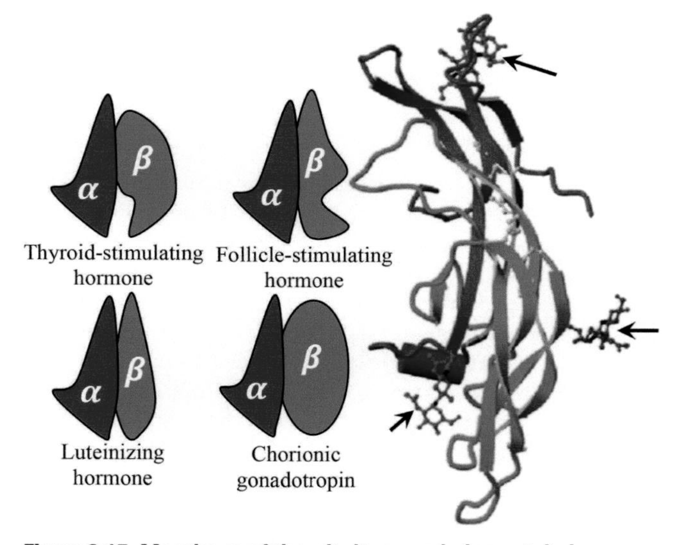

Figure 8.15 Members of the pituitary and placental glycoprotein hormone family. Each hormone in this family shares the same alpha subunit, and has a unique beta subunit. The image on the right shows the three-dimensional structure of FSH. The α-subunit is depicted with red and β-subunit with green. Note the carbohydrate side chains (arrows).

Source of Images: Three-dimensional structure of FSH: Alevizaki, M. and Huhtaniemi, I., 2002. Structure-function relationships of glycoprotein hormones: lessons from mutations and polymorphisms of the thyrotrophin and gonadotrophin subunit genes. HORMONES-ATHENS-, 1, pp. 224-232. Used by permission.

the alpha glycoprotein subunit (αGSU), which forms a non-covalent heterodimer with a hormone-specific beta subunit (i.e. βTSH, βFSH, βLH, and βCG). The beta subunits, which determine the receptor specificity and biological activity of the hormone, are structurally related to their alpha counterparts, but have different amino acid sequences and variable amino acid lengths.

The common alpha subunit and each of the hormone-specific beta subunits are encoded by different genes. These hormones are also characterized by the presence of several disulfide bridges. All members of the glycoprotein family transduce their intracellular effects via their respective GPCR receptors in association with the G protein/adenylyl cyclase and/or G protein/phospholipase C second-messenger system (Table 8.1).

Thyroid-Stimulating Hormone (TSH)

TSH, also known as *thyrotropin*, is made in the **thyrotropes** of the pituitary. In mammals TSH synthesis is under control of hypothalamic thyrotropin-releasing hormone (TRH). However, in the larvae of non-mammalian vertebrates TSH production can also be strongly influenced by corticotropin-releasing hormone (CRH).[46] As its name implies, the primary function of TSH is to promote thyroid gland function, specifically as it relates to the synthesis and release of thyroid hormones. The functions of thyroid hormones themselves are extremely diverse, ranging from the regulation of growth and development to vertebrate metamorphosis, as well as various aspects of metabolism and thermogenesis.

Thyrostimulin

As its name suggests, thyrostimulin activates the TSH receptor.[43] Indeed, thyrostimulin has been shown to be an even more potent ligand of the TSH receptor than TSH itself. Unlike other adenohypophysis hormones, thyrostimulin exhibits a wider distribution across diverse tissues, where it may act as a local but as of yet uncharacterized regulator.[47,48] Like the other glycoprotein hormones, thyrostimulin is a heterodimer consisting of an alpha and a beta subunit. Unlike the other pituitary glycoproteins, the thyrostimulin alpha subunit (called GPA2) is coded by a separate gene that is structurally similar to αGSU, but is not identical. As such, GPA2 does not form heterodimers with the beta subunits of TSH, FSH, LH, or CG. Rather, it dimerizes with its beta subunit called GPB5. Importantly, thyrostimulin is thought to be the ancestral pituitary glycoprotein from which the others are derived.

The Gonadotropins (FSH, LH, and CG)

FSH and LH are synthesized by the **gonadotropes** of the adenohypophysis, and secretion of both is controlled by the hypothalamic hormone GnRH. Chorionic gonadotropin (CG), a hormone that closely resembles LH, is made by the embryo and placenta in mammals. Though for historical reasons FSH and LH are each named after their functions in female reproduction, both hormones in fact play equally important roles in male reproduction as well. In females, FSH promotes the growth of ovarian follicles, whereas in males it induces spermatogenesis. LH is steroidogenic in the gonads of both females and males.

CG, a hormone that closely resembles LH and binds to LH receptors in the luteal cells of the ovary, is produced in mammals only during pregnancy. Initially, the developing embryo synthesizes and secretes CG, and following implantation of the embryo into the uterus a subset of placental cells continues to produce and secrete CG. The appearance of CG in the plasma and urine is one of the earliest signals of pregnancy and the basis of many pregnancy tests. The role of CG during pregnancy is to maintain the corpus luteum to sustain progesterone synthesis by this tissue.

Developments & Directions: The evolution of pituitary glycoprotein hormones

Kaoru Kubokawa's research group[49–51] has proposed that thyrostimulin is the ancestral molecule from which the other vertebrate pituitary glycoprotein hormones (GPH) diverged. Both thyrostimulin subunits are found in the protochordate amphioxus,[52] lamprey, and all vertebrates. The discovery of thyrostimulin GPA2 and GPB5 homologs in invertebrates including the fly, nematode, and sea urchin[53,54] suggests that an ancestral glycoprotein existed before the divergence of chordates and invertebrates, with later gene duplication events producing thyrostimulin in protochordates (Figure 8.16). Findings from Sower and colleagues[55] suggest that the ancestral GPB5 diverged into the single GTHβ (gonadotropin β) subunit found in lamprey, followed by duplication and divergence into the three GPHβ subunits (FSHβ, LHβ, and TSHβ) in found in the more derived vertebrates. The most recent glycoprotein hormone family member to appear was CGβ in placental mammals. The ancestral GPA2 is thought to have diverged to form the common GSUα subunit found in the more derived vertebrates, although the GSUα subunit has not yet been identified in lamprey.

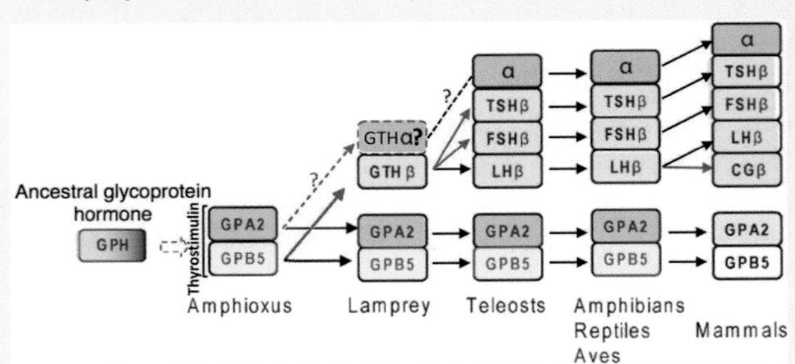

Figure 8.16 Refer to text for descriptions. Red arrows denote gene duplication events. Dashed arrows denote hypothesized gene duplication events.

Source of Images: Kubokawa, K., et al. 2010. Integr. Comp. Biol., 50(1), pp. 53-62.; used with permission.

Opiomelanocortins

The opiomelanocortin hormone family consists of several different peptides that all derive from a single large precursor glycoprotein prohormone called **proopiomelanocortin** (POMC) (Figure 8.17). POMC is expressed within the pars distalis and pars intermedia of the pituitary gland. POMC is also expressed in the arcuate nucleus of the hypothalamus, as well as in several peripheral tissues such as the skin and reproductive organs. Depending on the cell type it is expressed in, POMC undergoes variable posttranslational processing and proteolytic cleavage by tissue-specific enzymes called **prohormone convertases** (PCs) that reside in the intracellular secretory granules. PC1 and PC2 are differentially expressed in the pars distalis and pars intermedia. The exact peptides generated are dependent on the relative levels of the two enzymes (Figure 8.17).

The name "opio-melano-cortin" describes the three major hormone categories derived from POMC, namely endogenous *opiates* (endorphins and enkephalins), *melanotropins* (melanocyte-stimulating hormones), and *corticotropin* (i.e. adrenocorticotropic hormone). The primary hormone synthesized by the corticotropes of the pars distalis is adrenocorticotropic hormone (ACTH), a hormone that controls the synthesis of corticosteroids by the adrenal cortex. The pars intermedia secretes predominantly alpha-melanocyte-stimulating hormone (α-MSH), whose primary role in many vertebrates is to influence skin pigmentation via melanin production and intracellular distribution. Both the pars distalis and pars intermedia synthesize two opiates, or compounds with analgesic, or anti-pain, qualities. These are β-endorphin and its precursor peptide β-lipotropic peptide hormone (β-LPH). As its name implies, β-LPH also has lipid-mobilizing ability associated with lipolysis and steroidogenesis. Another POMC-derived peptide present in the pars intermedia, corticotropin-like intermediate lobe peptide (CLIP), has no known function. The non-opiate hormones derived from POMC exert their actions by binding to a family of paralogous GPCR receptors collectively called melanocortin receptors (MCRs) (Table 8.1). By contrast, the POMC-derived opiates bind to another family of GPCR receptors called the opiate receptors.

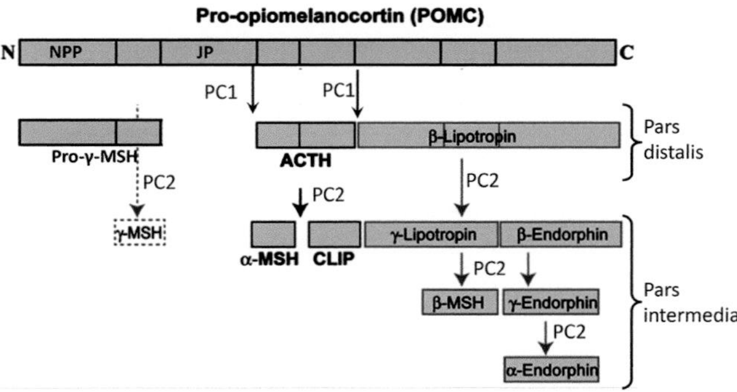

Figure 8.17 Processing of POMC in the pars distalis and pars intermedia of the rat pituitary. This 241 amino acid prohormone is cleaved into several mature hormones by enzymes that are called prohormone convertases (PC1 and PC2) that are differentially expressed in the pituitary. The types of peptides generated are dependent on the relative levels of these two enzymes, and different peptides can be generated in different tissues. In the corticotropic cells of the pars distalis, PC1 is predominately expressed and results in processing of POMC to ACTH, Pro-γ-MSH and β-LPH. In the pars intermedia, PC2 is also expressed and results in further processing to the shorter MSH and endorphin peptides. Corticotropin-like intermediate peptide (CLIP). N-terminal peptide/fragment of proopiomelanocortin (NPP). Joining peptide (JP).

ACTH

Release of ACTH from POMC by corticotropes of the pars distalis is regulated primarily by hypothalamic corticotropin-releasing hormone (CRH), although arginine vasopressin (AVP) of parvocellular hypothalamic origin also exerts important effects on its release. As its name implies, the primary effect of ACTH is to stimulate the production of adrenal cortex steroids, principally the glucocorticoids cortisol and corticosterone, which alter fuel metabolism and are involved in modulating the stress response by influencing blood glucose levels and the immune system. Since the first 13 amino acids in ACTH are common to those of α-MSH (Figure 8.17), it is not surprising that ACTH possesses some melanotropic capacity, and when produced at pathologically high levels such as that associated with Cushing's syndrome, hyperpigmentation can result. ACTH exerts its effects on the adrenal cortex by binding to a specific receptor, the melanocortin-2 receptor (MC2R), that is a member of the melanocortin receptor family (Figure 8.18).

MSH

There are several different POMC-derivatives that exhibit melanotropic activity: α-MSH, β-MSH, γ-MSH, and ACTH. These peptides, along with the antagonist peptides **agouti** and **agouti-related peptide** (AgRP), the melanocortin receptors, and the melanocortin receptor accessory proteins (MRAPs) constitute a functional group of peptides and proteins together known as the *melanocortin system*. AgRP is a neuropeptide produced specifically by cell bodies located in the ventromedial part of the arcuate nucleus in the hypothalamus in the brain and is one of the most potent and long lasting of appetite stimulators. Of the five vertebrate melanocortin receptors (Figure 8.18), one (MC2R) specifically binds ACTH, and the four others preferentially bind several different types of MSH with varying affinity, with α-MSH generally considered the most biologically important of the three. The melanocortin system, and α-MSH in particular, is known to influence at least three different physiological systems in vertebrates: (1) skin pigmentation, whose effects are modulated by MC1R; (2) appetite and metabolism regulation in humans and other vertebrates, mediated via hypothalamic MC3R and MC4R; and (3) the immune response, mediated by MC1R, MC3R, and MC5R on immunocytes.

The integument (skin) of many vertebrates can alter pigmentation and coloration to accommodate changing environment or behavior for the purposes of camouflage, social interaction, or protection of organs from solar radiation.

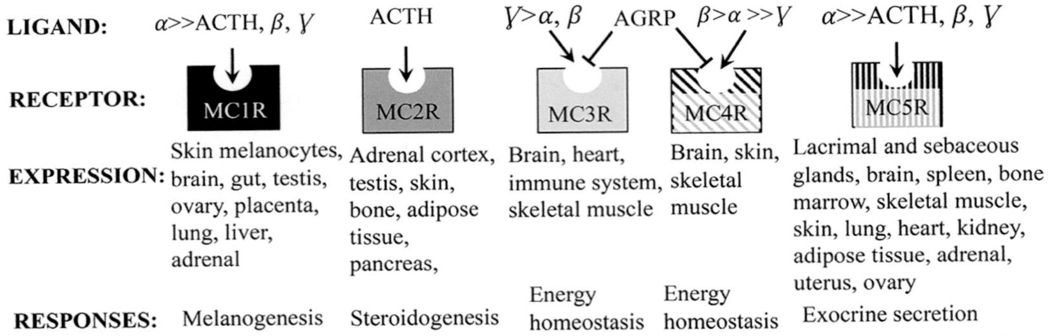

LIGAND:	α>>ACTH, β, γ	ACTH	γ>α, β / AGRP	β>α>>γ / AGRP	α>>ACTH, β, γ
RECEPTOR:	MC1R	MC2R	MC3R	MC4R	MC5R
EXPRESSION:	Skin melanocytes, brain, gut, testis, ovary, placenta, lung, liver, adrenal	Adrenal cortex, testis, skin, bone, adipose tissue, pancreas,	Brain, heart, immune system, skeletal muscle	Brain, skin, skeletal muscle	Lacrimal and sebaceous glands, brain, spleen, bone marrow, skeletal muscle, skin, lung, heart, kidney, adipose tissue, adrenal, uterus, ovary
RESPONSES:	Melanogenesis	Steroidogenesis	Energy homeostasis	Energy homeostasis	Exocrine secretion

Figure 8.18 Distribution and function of mammalian melanocortin receptors. α, β, γ denote α-, β-, γ-MSH ligands. AgRP, agouti-related peptide (a melanocortin receptor antagonist). Relative levels of ligand affinity to receptor is denoted by the greater than sign, ">".

Source of Images: Modified from Yeo, G.S. and Heisler, L.K., 2012. Nat. Neurosci., 15(10), pp. 1343-1349.

When these changes take place in a relatively slow manner, such as color changes associated with tanning in humans or seasonal changes in coat and plumage in some mammals and birds, this is known as *morphological color change*. In addition, many fishes, amphibians, and reptiles have the ability to alter pigmentation within minutes to hours in a response called *physiological color change*. A common black-brown pigment, called *melanin*, is present in a class of skin cells known as *melanophores* in fish and amphibians, and called *melanocytes* in mammals. Both categories of pigment cells contain hundreds of melanin-filled pigment granules, termed *melanosomes*. Morphologically, these cells resemble neurons and possess extensive processes. They also share a common neural crest developmental origin with neurons. These cells function to either promote pigment aggregation in the center of the cell, resulting in a light coloration, or dispersion of the melanosomes throughout the cytoplasm and processes, resulting in a dark coloration (Figure 8.19). In the cases of both morphological and physiological color change, melanosome dispersion is known to be mediated by the hormone α-MSH, although in fish the process is also thought to be under neural control. This is easily visualized in amphibians, which respond to α-MSH injections within seconds (Figure 8.19).

Similar to melanophores, melanosome dispersion in mammalian melanocytes is also mediated by the hormone α-MSH. Mammalian melanocytes can produce either *eumelanin* (black/brown pigment) or *pheomelanin* (yellow/red pigment). The activation of the MC1R in melanocytes by α-MSH leads to the production of eumelanin. Unlike skin, which is only transiently darkened by melanocytes, hair is permanently darkened. In contrast, the local overproduction of agouti, a melanocortin receptor antagonist, by specialized cells in the dermis of "yellow" agouti mouse mutants causes follicular melanocytes to produce pheomelanin, and their coats develop with very light pigmentation (Figure 8.20). Interestingly, in addition to a yellow coat, agouti mice that carry this dominant mutation also display a high incidence of obesity, insulin resistance, infertility, and high susceptibility to developing cancer, reflecting antagonism of the MC3R and MC4R categories of melanocortin receptor. Notably, red hair in humans, guinea pigs, horses, pigs, and dogs is caused by a loss-of-function mutation in MC1R that inhibits normal interaction with α-MSH (Figure 8.20). In humans, mutation of the POMC gene (from which α-MSH is derived) leads both to red hair coloration and to the development of severe early onset obesity, as well as ACTH insufficiency.[56]

> **Curiously Enough . . .** Human skin produces its own "pituitary" hormones autonomously of the pituitary itself. Although adult humans lack a pars intermedia and their pituitaries produce insignificant amounts of α-MSH, human skin itself not only synthesizes POMC-derived α-MSH, which forms the basis of sun tanning,[57] but also synthesizes POMC-derived β-endorphin, which alleviates pain associated with sunburns.[58]

Another pituitary hormone worth describing in this section, despite the fact that it is neither POMC-derived nor produced by the adenohypophysis, is *melanin-concentrating hormone* (MCH), whose effects on both pigmentation and appetite regulation are antagonistic to those of MSH. MCH is a highly conserved cyclic neuropeptide released by the pituitary neurohypophysis. MCH neurons occur in the hypothalamus of all vertebrates, and in teleost fish MCH functions as a neurohypophyseal hormone that is released by the neurohypophysis into the circulation when fish move onto a pale background (Figure 8.21). As the name implies, MCH causes a rapid aggregation of the pigment granules within melanophores of fish scales, resulting in skin lightening.[59] In mammals, the roles of MCH are numerous[60] and include the central regulation of feeding behavior. In contrast to MSH, which suppresses appetite, MCH stimulates food intake and weight gain when injected into the brains of rodents;

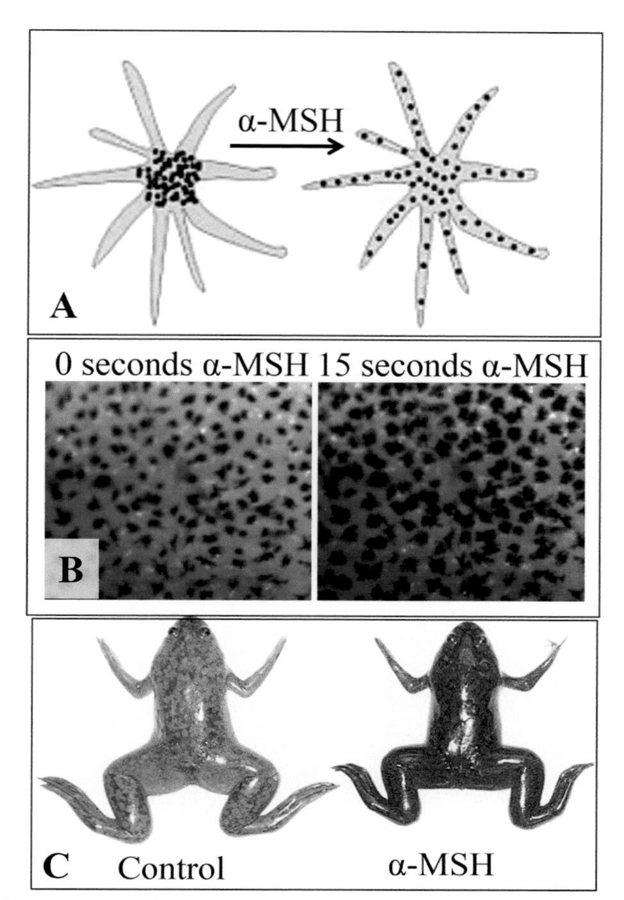

Figure 8.19 Melanosome dispersion within melanophores of amphibian skin is induced by alpha-MSH. (A) Melanophores are skin-localized cells with extensive processes, each of which contains hundreds of melanin-filled pigment granules, termed melanosomes. Following exposure to alpha-MSH, these granules disperse throughout the cell processes, darkening the appearance of the cells. **(B)** Dispersal of granules in response to alpha-MSH is rapid. This panel shows the same region of *Xenopus laevis* frog skin before and 15 seconds following injection with alpha-MSH. **(C)** An *X. laevis* frog injected with alpha-MSH (right) becomes much darker compared to a control (left).

Source of Images: (C) Heggland SJ, et al. Am Biol Teach 62: 597–601, 2000. Used with permission.

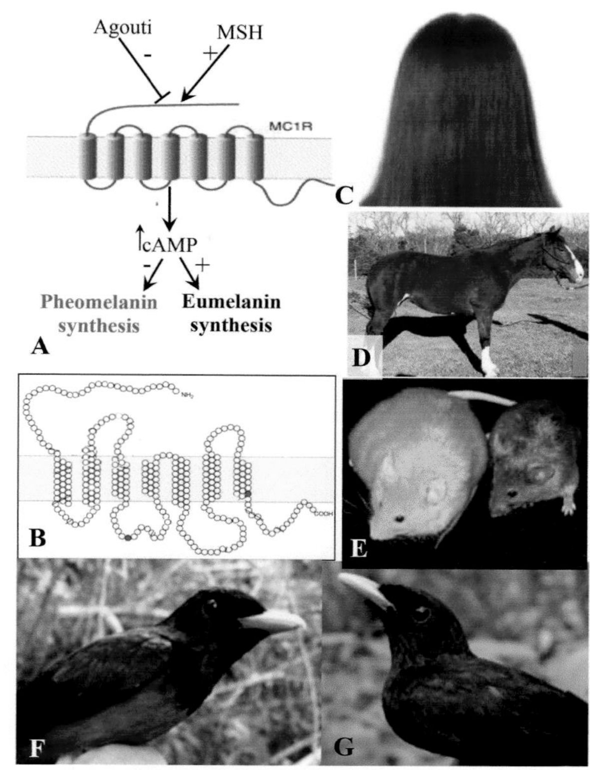

Figure 8.20 Melanocortin mutations in mice, humans, horses, and birds. (A) Control of melanin synthesis by melanocyte-stimulating hormone (MSH) and agouti signaling peptide. Whereas the binding of MSH to its receptor (MC1R) stimulates eumelanin (dark pigment) synthesis, agouti inhibits eumelanin synthesis, and instead promotes pheomelanin (red pigmentation) production. **(B)** Loss-of-function mutations in MC1R (red amino acid residues depict the mutation locations) that inhibit normal interaction with α-MSH cause the red hair phenotype in humans **(C)**, as well as in horses **(D)**, guinea pigs, pigs, and dogs. **(E)** "Yellow" mouse mutants that overproduce agouti (a melanocortin receptor antagonist) develop coats with very light pigmentation due to a blocked MCR1 receptor, compared with dark wild-type controls. **(F–G)** Two subspecies of Solomon Island flycatchers, *Monarcha castaneiventris*. The chestnut-bellied flycatcher **(F)** has an MC1R protein that is responsive to alpha-MSH, so black melanin is produced only when alpha-MSH is present. The black-bellied flycatcher **(G)** has an MC1R protein that is always active, regardless of whether alpha-MSH is present, so black melanin is always produced. This phenotypic difference is due to a single amino acid difference in the MC1R protein.

Source of Images: (A-B) Majerus, M.E. and Mundy, N.I., 2003. Trends Genet., 19(11), pp. 585-588; used with permission. (C) Shakeel, C.S., et al. 2021. Computational and mathematical methods in medicine; (D) Jacobs, L.N., et al. 2016. J. Heredity, 107(3), pp. 214-219; used with permission. (E) Murphy, G. Nature (2003); used with permission. (F-G) Uy, J.A.C., et al. 2016. Proc. Roy. Soc. B: Biol. Sci., 283(1834), p. 20160731. Used with permission.

MCH knockout mice are hypophagic (eat less) and are lean.[61] Very little is known about the physiological functions of MCH in humans, whose mRNA and peptide are synthesized in neurons of the dorso-lateral hypothalamus of adults.

Opioids

Opiates are natural or synthetic substances that bind to human GPCR opioid receptors and mimic the endogenously produced peptide hormones enkephalins and endorphins. These molecules are named for their first discovered receptor agonist, opium, a substance that is derived from the poppy plant (*Papaver somniferum*) and is the source of many narcotics, such as heroine, codeine, and morphine. Ever since the *analgesic*, or anti-pain, effects of opiates were discovered, the existence of endogenous opiate compounds (termed *opioids*) similar to

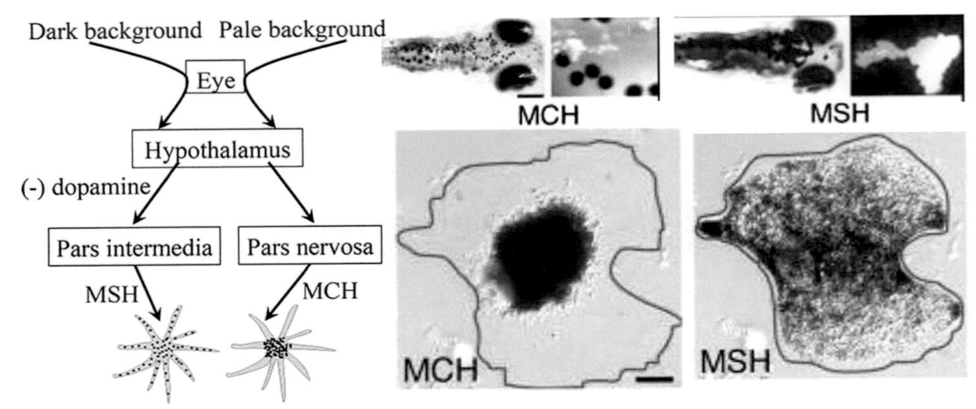

Figure 8.21 The fish model of effects of MSH and MCH on melanosome dispersion within melanophores. The diagram depicts how dark background sensory input via the eyes is transmitted to the hypothalamus, which responds with reduced dopamine synthesis, a signal that causes the pituitary pars intermedia to increase MSH synthesis, ultimately inducing dispersal of melanosomes within skin melanophores. Pale background sensory input via the eyes is transmitted to the hypothalamus, which responds by increasing MCH synthesis and release via the pituitary pars nervosa, ultimately inducing melanosome concentration within skin melanophores. The panel on the right shows aggregation and dispersion of melanosomes in 5-day-old wild-type zebrafish larvae. In response to MCH, pigment appears aggregated. In response to MSH, melanocytes disperse their pigment fully. The red line denotes the plasma membrane boundary of an individual melanophore cell. (The scale bar represents 200 μm in top images and 50 μm in bottom.)

Source of Images: Micrographs: Sheets, L., et al. 2007. Curr. Biol., 17(20), pp. 1721-1734. Used by permission.

morphine was postulated. In 1975 two endogenous opioids called enkephalins were discovered in pig brains,[62] and soon after a much larger peptide with potent opiate activity was isolated from the pituitary glands of camels and was named β-endorphin.[63] Although no evidence currently exists supporting a physiological role for pituitary-derived β-endorphin, POMC-derived β-endorphin has a broad distribution outside of the pituitary and has been identified in other regions of the brain and in the placenta, skin, and gastrointestinal tract.[64]

Developments & Directions: Ancient whole genome duplications and the evolution of POMC

Whole-genome duplications in prevertebrate and vertebrate lineages are known to have occurred three times: round 1 (called 1R) is thought to have occurred either immediately prior to or after the development of the agnathans (the jawless vertebrates, lamprey and hagfish); round 2 (2R) occurred after the divergence of gnathostomes (jawed vertebrates) from agnathans; these were followed by a third round (3R) early in teleost fish evolution[65–67] (Figure 8.22). Importantly, the two rounds of genome duplication in the vertebrate stem (1R and 2R) were followed by a period of rapid morphological and physiological innovation, which led to more complex endocrine, circulatory, nervous, sensory, immune, and skeletal systems.[68]

The complex neuroendocrine machinery of the opiomelanocortin system is thought to have diversified as a result of these genome duplication events. Based on genomic sequence analysis of different vertebrate lineages, Robert Dores and his colleagues[69,70] evaluated trends in the organization of the POMC gene during the radiation of the vertebrates to generate a hypothesis that accounts for the origin of the multiple melanocortins during the evolution of phylum Chordata (Figure 8.22). The common existence of α-, β-, and γ-MSH in the same order suggests that they appeared before the bifurcation of invertebrates and

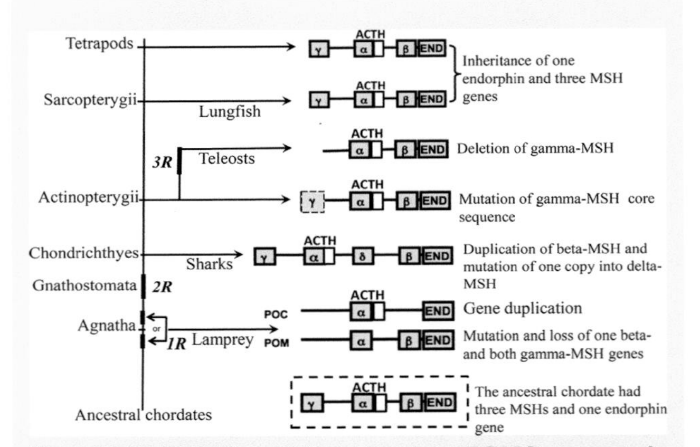

Figure 8.22 Evolution of the chordate POMC proprotein organizational plan. The ancestral POMC gene is hypothesized to have undergone several duplication, mutation, and deletion events that gave rise to the extant diversity in POMC organization among species. These genomic changes are thought to have been promoted by three whole genome duplication events (1R, 2R, 3R) in the chordate lineage whose possible positions are denoted by red lines. The dashed-lined rectangle represents the proposed ancestral form of the POMC proprotein. The gray boxes denote MSH (alpha, beta, gamma, delta) and endorphin (END) domains. The white box adjacent to most MSH-alpha boxes represents the domain for CLIP, which together with MSH-alpha constitutes ACTH. POC: proopiocortin gene (expressed in the pars distalis) encodes an ACTH sequence and a beta-endorphin sequence. POM: proopiomelanotropin gene (expressed in the pars intermedia) encodes the alpha-MSH and beta-MSH sequences and a beta-endorphin sequence.

Source of Images: Adapted from Kawauchi and Sower (2006). Gen. Comp. Endocrinol. Volume 148:3–14. Used by permission.

vertebrates. They propose that the ACTH/α-MSH sequence may have been the original melanocortin, and that POMC arose from an opioid-encoding gene following the 1R genome duplication event, appearing as a single sequence in an early chordate lineage ancestral to the Agnatha (likely a protochordate). Lampreys possess two POMC genes that are differentially expressed in the pituitary: (1) the proopiocortin (POC) gene expressed in the pars distalis encodes an ACTH sequence and a beta-endorphin sequence; (2) the proopiomelanotropin (POM) gene expressed in the pars intermedia encodes the α-MSH and β-MSH sequences and a β-endorphin sequence. The α- and β-MSH and β-endorphin emerged with the subsequent radiation of jawless vertebrates. The 2R genome duplication event gave rise to a γ-MSH peptide, but the accumulated mutations in the γ-MSH region over time may have led to its eventual deletion in teleosts in association with the 3R genome duplication.

SUMMARY AND SYNTHESIS QUESTIONS

1. What effect would ovariectomy or the inhibition of estrogen synthesis have on pituitary lactotrope development? Why?
2. Like adult humans, birds lack an intermediate lobe of the pituitary. However, the bird pituitary still synthesizes MSH. For the chicken, Hayashi et al.[71] have shown that MSH is not produced by dedicated melanocytes but is co-secreted with another hormone in the cephalic half of the pars distalis. What cell type has the highest probability of being able to secrete both MSH and a second hormone (what is the second hormone?)? Explain your reasoning.
3. List the four families of pituitary hormones and describe the biochemical features that make the categories distinctive.
4. Describe examples of prolactin's effects in two non-mammalian taxa that are analogous to its effects in mammals.
5. The POMC protein is expressed in both corticotropes of the pars distalis and melanotropes of the pars intermedia, but the two cell types secrete completely different proteins. Describe what makes this possible.
6. The deletion of the gene for which prohormone convertase enzyme would inhibit the synthesis of ACTH? Why?
7. Why would a mutation in the gene coding for the α-glycoprotein subunit (αGSU) affect the functions of multiple pituitary hormones?
8. ACTH is a pituitary tropic hormone that induces the production of cortisol from the adrenal cortex. You inject ACTH into a frog and see that its skin becomes darker—why does this happen?
9. You want to make a map of the ACTH-producing cells of the pituitary of a fish. Which approach would be most effective: using *in situ* hybridization to localize mRNA or immunocytochemistry against a protein. Explain.
10. Adult humans do not produce pituitary alpha-MSH, yet they still have the ability to tan. How?
11. The "yellow" mouse mutation and red hair in humans are both caused by mutations in the melanocortin system. However, whereas yellow mouse mutants are also prone to obesity, red-haired humans are not. Explain.
12. Would fish raised in tanks with white backgrounds be expected to have higher or lower levels of MSH expression compared with those raised in black backgrounds? Why?
13. An adenoma that releases high levels of ACTH would most likely also release which ONE of the following hormones: GH, PRL, TSH, FSH, LH, MSH. Explain your reasoning.
14. Although corticotropes of the pars distalis and melanotropes of the pars intermedia both express POMC, they produce different functional hormones. Explain why.
15. What POMC-derived hormone does not bind to melanocortin receptors?
16. A person has been diagnosed as completely lacking the gene that codes for the protein proopiomelanocortin (POMC). Of the eight hormones listed below (a–h),

    List the one(s) whose synthesis will be low or absent (explain your reasoning).
    List the one(s) whose synthesis would increase (explain your reasoning).
    List the one(s) whose synthesis would be least affected (explain your reasoning).

a. CRH b. FSH c. MSH
d. beta endorphins e. ACTH f. GnRH
g. TRH h. cortisol

Hormones of the Neurohypophysis

LEARNING OBJECTIVE Describe the distinguishing biochemical features, functions, and evolutionary origins of the neurohypophyseal family of hormones.

KEY CONCEPTS:

- The principal neurohypophyseal hormones consist of a single family of nonapeptides that are classified as either vasotocin (vasopressin)-like or oxytocin-like.
- Vasopressin and oxytocin not only have diverse physiological roles, but also play key roles in promoting social-bonding behaviors.

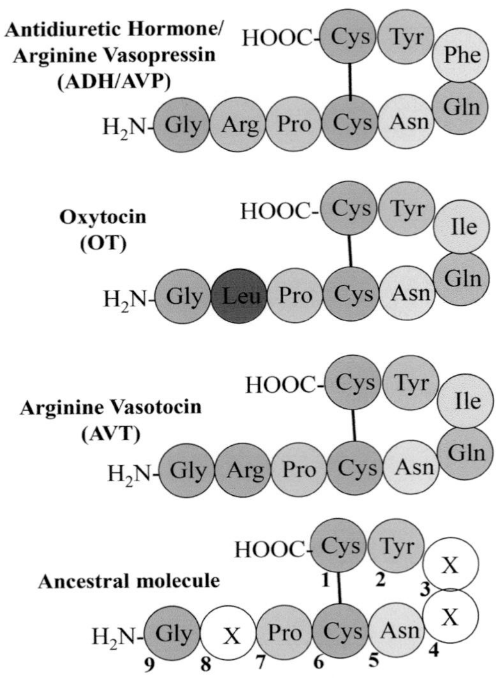

Figure 8.23 Amino acid sequences of some vertebrate neurohypophysis nonapeptide hormones. In human adults, only oxytocin and arginine vasopressin (AVP) are present. Arginine vasotocin (AVT) is present in the human fetus, as well as fish, amphibians, reptiles, and birds. The predicted amino acid sequence for the ancestral peptide that gave rise to the family is shown (X denotes a non-conserved amino acid). The numbers (1–9) adjacent to each amino acid (shown for the ancestral molecule) denote the amino acid position in the carboxyl to amino-terminus. Substitutions in this nonapeptide family have occurred only in positions 3, 4, and 8, which suggests that the amino acid residues the other positions (1, 2, 5, 6, 7, and 9) are essential for peptide functions and could constitute the original ancestral molecule that has evolved to become oxytocin, vasopressin, and vasotocin.

In contrast with the hormones of the adenohypophysis, the hormones of the neurohypophysis are all neurosecretory in origin—that is, they are synthesized by hypothalamic neurons and are stored and released by axonal termini housed in the neurohypophysis. One neurohypophyseal hormone, MCH, was already described in the context of its effects that oppose MSH. Here we will focus on the principal neurohypophyseal hormones, which consist of a single superfamily of cysteine-bridged *nonapeptides* (Figure 8.23).

Evolution of Neurohypophyseal Nonapeptides and Their Receptors

The evolution of nonapeptides and their receptors is one of the best-studied examples of hormone and receptor duplications and divergences. Nonapeptides are among the oldest families of neuropeptides, whose lineage traces through invertebrates and includes members in virtually all vertebrate taxa. Vertebrate nonapeptides are classified as either vasotocin (vasopressin)-like or oxytocin-like, with most chordates possessing both variants. The oxytocin and vasopressin peptides for humans and several other species all have similar structural organization and differ by only two amino acids (Figure 8.23). Vasotocin, which as the name implies possesses characteristics of both vasopressin and oxytocin, is the ancestral nonapeptide molecule.[72] In most vertebrate lineages the oxytocin and vasopressin genes are found on the same chromosome and often next to each other, reflecting a gene duplication event followed by a mutation in one of the genes. An important exception to this is the lamprey, one of the most ancestral jawless vertebrates, which has only one gene from this family (vasotocin). This suggests that this gene duplication event took place in a common ancestor of the more derived gnathostomes (jawed vertebrates).

The vertebrate oxytocin and vasopressin receptors form a family of G protein-coupled receptors (GPCRs) that mediate a large variety of functions, including social behavior and the regulation of blood pressure, water balance, and reproduction. Among mammals, there exists one receptor for OT and three receptors for AVP, each with distinct target tissue profiles and intracellular second messenger signaling pathways[73] (Table 8.1).

Arginine vasotocin (AVT) is the primary neurohypophyseal hormone associated with the regulation of urine production among vertebrates; the exceptions are adult mammals, which predominantly use the orthologous *arginine vasopressin* (AVP, also called antidiuretic hormone, ADH). Unlike the adults, fetal mammals also use AVT. AVT functions as an antidiuretic in marine fish, and also, along with AVP, in amphibians, birds, and mammals, increasing the ability of the kidneys to reabsorb water and produce a concentrated urine. AVT also plays an important role in inducing smooth muscle contractions of the oviduct[74,75] and can induce parturition and oviposition in a number of amphibian and reptile species.[76,77]

In mammals, the primary physiological functions of *oxytocin* (OT, which in Greek means "quick birth") include contraction of smooth muscle in the uterus, facilitating labor, and smooth muscle associated with the mammary glands, facilitating milk ejection. In males, oxytocin stimulates the contraction of the sperm duct (vas deferens). There are several oxytocin-like neurohypophysis hormones among vertebrates. Homologs of OT have even been found in invertebrates like annelid worms, snails, and insects. Remarkably, the snail homolog of OT, conopressin, has the analogous role of mammalian OT modulating ejaculation in males and egg-laying in females.

Curiously Enough . . . In addition to acting as hormones with peripheral targets and physiological roles, nonapeptides also act as neurotransmitters, with OT- and AVP-synthesizing neurons projecting to diverse targets throughout the brain. OT- and AVP-like peptides play important roles in modulating many sexually dimorphic social and reproductive behaviors in species as diverse as hydra, worms, insects, and vertebrates.[78]

SUMMARY AND SYNTHESIS QUESTIONS

1. A mutant strain of female mice has had the oxytocin gene deleted ("knocked-out") but still retains the ability to synthesize milk normally. How is this possible?

Evolution of the Pituitary Gland

LEARNING OBJECTIVE Describe distinctive features of pituitary anatomy among vertebrate taxa and their evolutionary origins from presumed homologous structures in ancestral protochordates.

KEY CONCEPTS:

- Pituitary-like structures that secrete some hormones present in vertebrates are present in extant non-vertebrate chordates known as "protochordates".
- The pituitary gland is a vertebrate innovation that emerged prior to or during the differentiation of the most ancestral living vertebrates, the jawless fishes.
- The evolution of the pituitary gland is characterized by variations in occurrence of a median eminence and hypothalamo-hypophyseal portal system, and the differentiation of morphologically distinct lobes.

In their classic book published in 1962, *A Textbook of Comparative Endocrinology*, Aubrey Gorbman and Howard Bern preface a section on pituitary evolution with a humorous cautionary disclaimer: "The reader who does not care for speculation should avoid this section, since it is based on only the most tenuous and indirect reasoning".[79] At the time, the study of evolution was based entirely on morphological and histological evidence, and progress in endocrine evolution was limited by a lack of appropriate fossils. The advent of molecular tools to study evolutionary and developmental biology, however, has advanced our modern understanding of pituitary origins. Indeed, Professors Gorbman and Bern would each go on to contribute enormously to this field later in the 20th and early 21st centuries. While considerable morphological variability exists in pituitary glands within any given vertebrate taxon, this section will describe some of the more distinctive features of pituitary anatomy among taxa and their evolutionary origins from presumed homologous structures in ancestral protochordates.

Protochordates

The pituitary gland is considered to be a vertebrate innovation that emerged prior to or during the differentiation of the most ancestral living vertebrates, the cyclostomes (jawless fishes). However, pituitary-like structures have been described for the most ancestral living non-vertebrate chordates, the cephalochordates (lancelets, such as amphioxus) and ascidians (sea squirts, such as *Ciona*). This group of animals are known as **protochordates.**

Amphioxus

In amphioxus, a structure termed "Hatschek's pit" (Figure 8.24) shares some important developmental similarities with Rathke's pouch, the adenohypophysis primordium[80,81]:

- Like Rathke's pouch, Hatschek's pit is an evagination of the pharynx that forms at the anterior terminus of the notochord. As such, this region of the notochord provides a landmark from where the pituitary develops in other chordates, including vertebrates.
- The amphioxus homolog of the vertebrate *Pit-1* gene (which, as its name implies, plays key roles in the development of the adenohypophysis) is also expressed in Hatschek's pit of amphioxus embryos and larvae.[82]
- A functional GnRH peptide has been identified,[83] and immunocytochemistry with antibodies raised against vertebrate LH and GnRH gave positive reactions in Hatschek's pit.[84]
- Analysis of the amphioxus genome has demonstrated the presence of homologs to regulators of the vertebrate hypothalamic-pituitary axis, including kisspeptin (which in vertebrates regulates release of GnRH), thyrostimulin

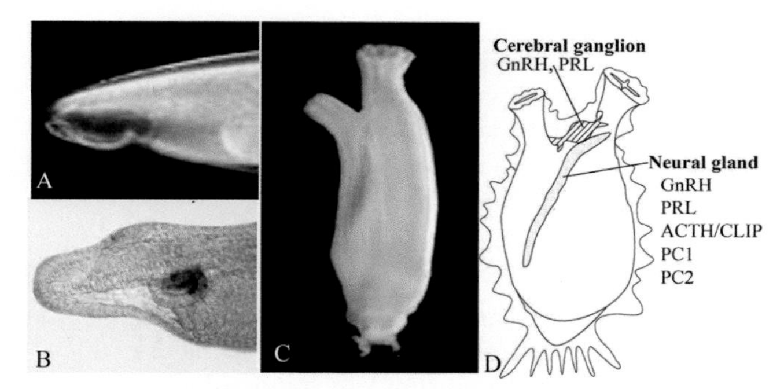

Figure 8.24 Homologous structures of the vertebrate hypothalamus and adenohypophysis in protochordates. In amphioxus **(A)**, "Hatschek's pit" **(B)** has developmental similarities to Rathke's pouch in vertebrates, including immunoreactivity against the vertebrate Pit-1 transcription factor (purple stain). In the adult tunicate **(C)**, the neural complex is composed of the cerebral ganglion juxtaposed to the neural gland **(D)**, a structure thought to represent a homolog of the vertebrate pituitary gland. These structures produce several peptides and proteins found in vertebrate endocrine systems, including gonadotropin-releasing hormone (GnRH), prolactin (PRL), adrenocorticotropic hormone (ACTH) and CLIP, as well as the prohormone convertases PC1 and PC2.

Source of Images: (A) Onai, T. et al. 2015. Zool. Let., 1(1), pp. 1-11. (B) Candiani, S., et al. 2008. Brain Res. Bull., 75(2-4), pp. 324-330; used by permission. (C) Holland, L.Z., 2016. Curr. Biol., 26(4), pp. R146-R152; used by permission. (D) Kawamura, K., et al. 2002. Gen. Comp. Endocrinol., 126(2), pp. 130-135. Used by permission.

(which may be ancestral to FSH, LH, and TSH), as well as receptors for GnRH, and diverse enzymes and other components of the vertebrate thyroid hormone biosynthesis pathway.[52]

Ascidians

Ascidians, or marine tunicates, are the closest living protochordate relatives of vertebrates. The central nervous system of larval ascidians shares basic organizational characteristics with vertebrates and has even been shown to produce some neurohormone peptides, including homologs of the vertebrate hypothalamic peptides GnRH and oxytocin/vasopressin[85] (Figure 8.24). At metamorphosis, the vertebrate-like nervous system remodels into an adult neural complex, composed of the cerebral ganglion juxtaposed to the *neural gland*, a structure thought to represent a homolog of the vertebrate pituitary gland[86] (Figure 8.24). As with amphioxus, genes for TSH, GH, PRL, and ACTH do not appear to be present in the ascidian genome, suggesting that these hormones may be vertebrate novelties.

Vertebrates

Although in all vertebrates the pituitary gland is comprised of two major anatomical regions, the neurohypophysis and the adenohypophysis, there exists considerable variability among the vertebrate classes in aspects of pituitary morphology, such as the extent of development of all lobes, the degree of interdigitation between the neurohypophysis and adenohypophysis, variations in occurrence of a hypothalamo-hypophyseal portal system, and distribution of cell types within the adenohypophysis.

The major anatomical differences among the vertebrate classes are discussed later, and are summarized in Figure 8.25 and Table 8.3.

The Agnathan (Jawless) Fishes: Hagfish and Lamprey

Hagfish, which appeared over 550 million years ago, represent the most ancestral vertebrate, either living or extinct.[87] They also possess the simplest hypothalamic-pituitary system, even compared with their closest living relatives, the lamprey. Whereas in hagfishes a neurohypophysis-like structure appears as a relatively undifferentiated part of the hypothalamus, in lampreys it is differentiated into a discrete lobular structure (Figure 8.25). Unlike hagfish, lamprey possess a well-developed pars intermedia with distinct regional distributions of adenohypophysis cell types.

Important similarities between hagfish and lamprey include the absence of both a median eminence and a hypothalamo-hypophyseal portal system, as well as a morphological separation between the adenohypophysis and neurohypophysis[88] (Figure 8.25). Both agnathans produce hypothalamic GnRH,[89–92] which likely diffuses from the hypothalamus to the pituitary across the thin connective tissue layer that separates the neural from the glandular tissues.[93–96] A single gonadotropin (GTH) is present in both hagfish and lamprey pituitaries[55,96,97] (see

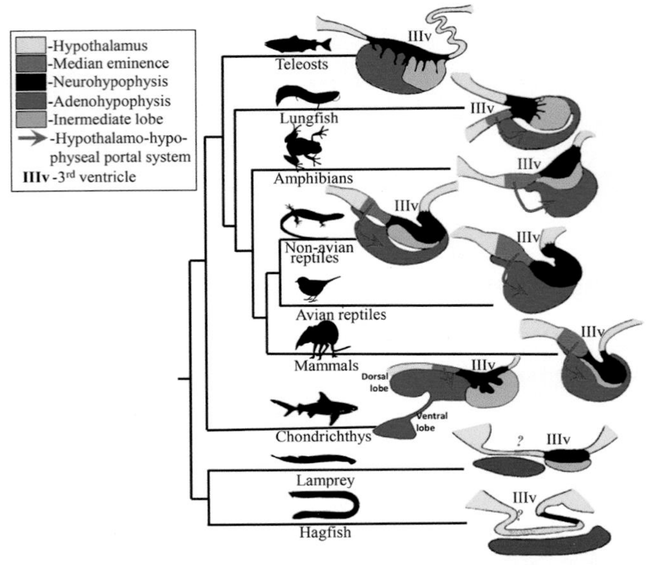

Figure 8.25 Comparative anatomy of the pituitary and evolutionary relationships of the component parts. Key elements are the (1) variations in occurrence of a hypothalamo-hypophyseal portal system (and associated median eminence, not shown), (2) differentiation of distinct lobes, (3) extent of the development of the intermediate lobe, and (4) degree of separation of the neurohypophysis.

Source of Images: Henderson, I.W., 2010. Endocrinology of the vertebrates. Comprehensive Physiology, pp. 623-749.

Figure 8.13), and both agnathans have functional hypothalamic-pituitary-gonadal axes similar to that of the more derived vertebrates.[98] No additional hormones have yet been isolated from the hagfish pituitary.[99] However, in addition to GTH, the lamprey pituitary has been shown to produce at least three more hormones: ACTH, GH, and MSH.[99] In general, these morphological and biochemical observations provide strong evidence that agnathans share common structural and functional neuroendocrine features with the more derived vertebrates, discussed later, suggesting that the ancestral vertebrate pituitary lacked a vascular connection between the neurohypophysis and the adenohypophysis, relying instead on diffusional control.

Chondrichthyes (Cartilaginous Fishes)

The cartilaginous fishes, which include sharks, rays, and ratfishes, evolved 50–100 million years after the lineage leading to the lampreys (400 million years before present) and are the phylogenetically oldest group of "gnathostomes", or jawed vertebrates. Compared with agnathan pituitaries, those of cartilaginous fishes have developed several evolutionary novelties. The elasmobranch adenohypophysis is juxtaposed to the neurohypophysis, forming a zone characterized by a considerable intermingling of the two tissues (the "pars neurointermedia") (Figure 8.25). The adenohypophysis consists of several additional subdivisions, including an elongated pars distalis (the dorsal lobe) and a ventral lobe, a structure unique

Table 8.3 Summary of Presence of Major Hypothalamic-Pituitary Features Among Vertebrate Taxa

				ADENOHYPOPHYSIS			
Taxa	**Neurohypo-physis**	**Median eminence**	**Portal vessels**	**Ventral lobe**	**Pars tuberalis**	**Pars intermedia**	**Pars distalis**
Hagfish	Indistinct	–	–	–	–	–	+
Lamprey	Distinct, but not lobular	–	–	–	–	+ Discrete lobe	+
Elasmobranchs	Distinct, but not lobular	+	+	+	–	+ Discrete lobe	+
Teleosts	Distinct, but not lobular	–	–	–	–	+ Discrete lobe	+
Lungfish	Discrete lobe	+	+	–	–	+ Cells present, but not a distinct lobe	+
Amphibians	Discrete lobe	+	+	–	+	+ Cells present, but not a distinct lobe	+
Non-avian reptiles	Discrete lobe	+	+	–	+	+ Cells present, but not a distinct lobe	+
Birds	Discrete lobe	+	+	–	+	–	+
Mammals	Discrete lobe	+	+	–	+	+ Absent in some species	+

to elasmobranchs attached by a stalk to the pars distalis. Also present are a true median eminence and a well-developed portal vasculature connecting the hypothalamus to the pars distalis. ACTH[100,101] and GH[102–104] are produced in at least some elasmobranchs, presumably by the pars distalis. The ventral lobe produces thyrostimulin and both gonadotropins. This suggests that portal communication is an ancient shared characteristic of gnathostomes. The intermediate lobe contains one cell type associated with α-MSH production. Arginine vasotocin (AVT) is produced by the shark and ray neurohypophysis, whereas both AVT and oxytocin are present in the ratfishes.

Teleosts (Bony Fishes)

Recall that teleosts are divided into the lobe-finned teleost fishes (Sarcopterygii, which ultimately gave rise to land-dwelling vertebrates), and the ray-finned teleost fishes (Actinopterygii). Importantly, the sarcopterygian pituitary resembles that of elasmobranchs as it retains a functional portal system and median eminence, structures that would later become even more developed in the terrestrial vertebrate lineage.

Strikingly, the portal system and median eminence were entirely lost in the ray-finned fishes (Figure 8.25). The loss of the portal system in the ray-finned teleosts was complemented by the evolution of a direct neural connection between the hypothalamus and pars distalis, and there is also extensive interdigitation of the pars nervosa with both the pars intermedia and the pars distalis (Figure 8.26). In this system, neurohormones are released directly to the adenohypophysis in a paracrine manner, circumventing the need for transport by a hypothalamo-hypophyseal blood portal system. Furthermore, within the pars distalis cells are not dispersed but are grouped into discrete clusters of similar cells, which presumably accommodates the direct innervation from hypothalamic nuclei[105] While the teleost pituitary morphology differs from that of tetrapods in these important ways, their strategy has clearly not impeded their success, as teleosts are by far the most numerous and diverse of all the vertebrate taxa! The teleost pars distalis possesses the same five categories of cells present in the pars distalis of most mammals. The teleost pars intermedia, in addition to containing MSH-producing cells, also produces another hormone, somatolactin (SL), a paralog of prolactin (see Figure 8.13). As with terrestrial vertebrates, the neurohypophysis produces vasotocin (vasopressin) and isotocin (oxytocin), as well as melanin concentrating hormone.

Terrestrial Tetrapods (Amphibians, Reptiles, Birds, Mammals)

The pituitary of amphibians, the most ancestral of the vertebrate tetrapods, closely resembles the general features of other tetrapod pituitary glands, with the pars intermedia forming a distinct band of tissue between the pars distalis and the pars nervosa (Figure 8.25; also refer to Figure 8.5). The median eminence and portal vessels are more elaborated compared with that of the lobe-finned lungfish, and the presence of a pars tuberalis is seen in amphibians for the first time in vertebrate phylogeny. Amphibians also possess a distinct lobular neurohypophysis with less intermingling with the pars intermedia compared with teleosts. In comparison with the highly regionalized pituitaries of teleosts, specific cell types of the amphibian pars distalis appear

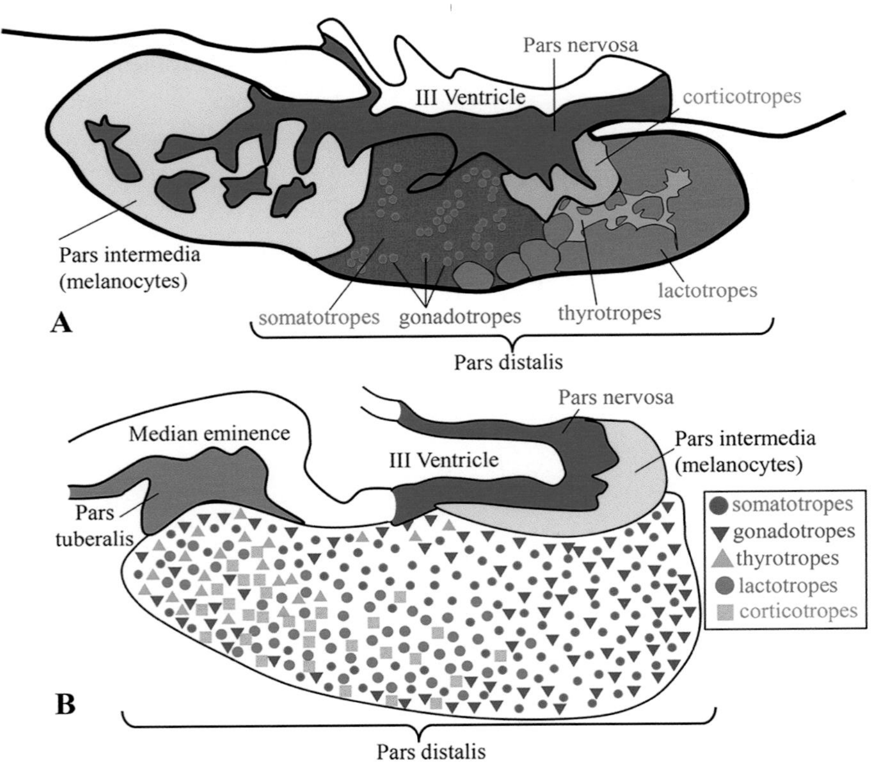

Figure 8.26 The pituitary gland of a teleost fish and of a reptile. (A) In the adenohypophysis of the teleost fish (an eel), the hormone-producing cells of the pars distalis are arranged in discreet zones. **(B)** In contrast to the fish, the hormone-producing cells of the pars distalis of a reptile (a Japanese soft-shelled turtle) are relatively scattered. In contrast to the fish, note the presence of a distinct pars tuberalis and median eminence in the reptile.

Source of Images: (A) Kubokawa, K., et al. 2010. Integr. Comp. Biol., 50(1), pp. 53-62; used with permission. (B) Kawauchi and Sower (2006). Gen. Comp. Endocrinol. 148:3–14. Used with permission.

relatively scattered, a feature that amphibians share in common with reptiles (Figure 8.26). The general arrangement of the reptilian pituitary is similar to that of amphibians, with the most variable components being the pars tuberalis, which is absent in some reptiles, and the neurohypophysis, which interdigitates with the pars intermedia in the phylogenetically older reptiles. Interestingly, the avian reptiles (birds) all appear to entirely lack a pars intermedia, although MSH is still produced in some species. As discussed previously, the pars intermedia is also absent in some mammals (e.g. adult humans, whales, dolphins, elephants, and armadillos). In comparison with amphibians and reptiles, the mammalian pars distalis tends to display a more regionalized distribution of cell types (refer to Figure 8.9). Finally, all reptiles, birds, and mammals have a more expanded portal system in comparison to amphibians.

The complex variability of pituitary morphology and function among the vertebrates is summarized eloquently by Trudeau and Somoza:[106]

> There appears to be three main hypothalamo-pituitary communication systems: 1. Diffusion, best exemplified by the agnathans; 2. Direct innervation of the adenohypophysis, which is most developed in teleost fish, and 3. The median eminence/portal blood vessel system, most conspicuously developed in tetrapods, showing also considerable variation between classes. Upon this basic classification, there exists various combinations possible, giving rise to taxon and species-specific, multimodal control over major physiological processes.

SUMMARY AND SYNTHESIS QUESTIONS

1. From the perspective of pituitary gland evolution, how does it make sense that tetrapods evolved from ancestors that were lobe-finned fishes, and not ray-finned fishes?
2. In the evolutionary sense, there have been three methods of brain regulation of the adenohypophysis in the vertebrates. Describe these.

LEARNING OBJECTIVE Summarize the effects of hypophysectomy on peripheral tissues in early experiments with tadpoles and rodents, and discuss their implications.

- Hypophysectomized tadpoles and rats display atrophy of the thyroid gland, reduced skin pigmentation, adrenal cortex atrophy, reduced growth rate, underdeveloped brains, and degenerated reproductive tissues.

- These phenotypes of hypophysectomy could all be reversed following injections of pituitary extracts or pituitary implants, demonstrating that the pituitary released mysterious substances into the bloodstream that stimulated diverse organs.
- It is now well established that the mammalian pituitary gland produces nine major hormones.

LEARNING OBJECTIVE Describe the gross anatomy, microanatomy and cytology of the various regions of the pituitary gland.

- The adenohypophysis of the pituitary consists of classical endocrine glandular cells, while the neurohypophysis is made of neurosecretory axonal termini that extend from hypothalamic nuclei.
- Adenohypophysis cell types can be functionally differentiated based upon the immunohistochemical labeling of specific neuropeptides.

LEARNING OBJECTIVE Describe the developmental origins of the pituitary gland.

- The hypophyseal placode, which gives rise to the pituitary's adenohypophysis, is located in a region of the early embryo called the anterior neural ridge (ANR).
- Later in embryogenesis the neurohypophysis derives from neuroectodermal tissue, whereas the adenohypophysis derives from oral ectoderm.

LEARNING OBJECTIVE Describe the distinguishing biochemical features, functions, and evolutionary origins of the principal families of adenohypophyseal hormones.

- The mammalian pituitary gland produces nine major hormones, and homologs to these hormones are present in all vertebrates.
- Two rounds of genomic duplication are thought to have taken place at the ancestral base of vertebrate phylogeny, giving rise to pituitary hormone diversity among taxa.
- Pituitary hormones fall into four distinct families: the somatomammotropic family, the glycoprotein hormone family, the opiomelanocortin family, and the neurohypophyseal nonapeptides.
- The somatomammotropic hormone family constitutes four related protein hormones: growth hormone, prolactin, placental lactogen, and somatolactin.
- The members of the pituitary glycoprotein hormone family (thyroid-stimulating hormone, follicle-stimulating hormone, luteinizing hormone, chorionic gonadotropin) are heterodimers that each share an identical alpha subunit and possess a unique beta subunit.
- The opiomelanocortin hormone family consists of several different peptides (including adrenocorticotropic hormone, opioids, and melanocyte-stimulating hormone) that all derive from a single large prohormone called proopiomelanocortin (POMC).

LEARNING OBJECTIVE Describe the distinguishing biochemical features, functions, and evolutionary origins of the neurohypophyseal family of hormones.

- The principal neurohypophyseal hormones consist of a single family of nonapeptides that are classified as either vasotocin (vasopressin)-like or oxytocin-like.
- Vasopressin and oxytocin not only have diverse physiological roles, but also play key roles in promoting social-bonding behaviors.

LEARNING OBJECTIVE Describe distinctive features of pituitary anatomy among vertebrate taxa and their evolutionary origins from presumed homologous structures in ancestral protochordates.

- Pituitary-like structures that secrete some hormones present in vertebrates are present in extant non-vertebrate chordates known as "protochordates".
- The pituitary gland is a vertebrate innovation that emerged prior to or during the differentiation of the most ancestral living vertebrates, the jawless fishes.
- The evolution of the pituitary gland is characterized by variations in occurrence of a median eminence and hypothalamo-hypophyseal portal system, and the differentiation of morphologically distinct lobes.

LITERATURE CITED

1. Harrower H. *Adventures in Endocrinology*. The Literary Department of the Harrower Laboratory; 1922.
2. Foster M. *A Textbook of Physiology* (6th American ed.). Lea Brothers; 1895.
3. Langdon-Brown W. Recent observations on the pituitary body. *Practitioner*. 1931;127:614–625.
4. Cushing H. *The Pituitary Body and Its Disorders. Clinical States Produced by Disorders of the Hypophysis Cerebri*. J.B. Lippincott Co.; 1912.
5. Cushing H. Lecture II. The pituitary gland as now known. *Lancet*. 1925;2:899–906.
6. Cushing H. *Papers Relating to the Pituitary Body, Hypothalamus, and Parasympathetic Nervous System*. C.C. Thomas; 1932:3–57.
7. Smith PE. *The Effect of Hypophysectomy in the Early Embryo Upon the Growth and Development of the Frog*. Waverly Press; 1916.
8. Smith PE. On the effects of ablation of the epithelial hypophysis on the other endocrine organs. *Proc Soc Exp Biol NY*. 1919;16:81–82.
9. Smith PE, Smith IP. The repair and activation of the thyroid in the hypophysectomized tadpole by the parenteral administration of fresh anterior lobe of the bovine hypophysis. *J Int Med Res*. 1922;43:267–283.
10. Allen BM. Extirpation experiments in Rana pipiens larvae. *Science*. 1916;44:755–757.
11. Allen BM. Effects of the extirpation of the anterior lobe of the hypophysis of Rana pipiens. *Biol Bull Wood's Hole*. 1917;32:117–130.
12. Allen BM. Brain development in anuran larvae after thyroid or pituitary gland removal. *Endocrinology*. 1924;8(5):639–651.
13. Smith PE, Walker AT, Graeser JB. The production of the adiposogenital syndrome in the rat, with preliminary notes upon the effects of a replacement therapy. *Exp Biol Med*. 1924;21:204–206.

14. Smith PE, Engle ET. Experimental evidence regarding the role of the anterior pituitary in the development and regulation of the genital system. *Amer J Anat.* 1927;40:159–217.
15. Smith PE. Hypophysectomy and replacement therapy in the rat. *Amer J Anat.* 1930;45:205–273.
16. Wingstrand KG. Comparative anatomy and evolution of the hypophysis. In: Harris GW, Donovan BT, eds. *The Pituitary Gland: Anterior Pituitary*, Vol 1. University of California Press; 1966:58–126.
17. Visser M, Swaab DF. Life span changes in the presence of alpha-melanocyte-stimulating-hormone-containing cells in the human pituitary. *J Develop Physiol.* 1979;1(2):161–178.
18. Bockmann J, Bockers TM, Winter C, et al. Thyrotropin expression in hypophyseal pars tuberalis-specific cells is 3,5,3'-triiodothyronine, thyrotropin-releasing hormone, and pit-1 independent. *Endocrinology.* 1997;138(3):1019–1028.
19. Lopes MBS, Pernicone PJ, Scheithauer BW, Horvath E, Kovacs K. Pituitary and sellar region. In: Mills SE, ed. *Histology for Pathologists* (3rd ed.). Lippincott Williams & Wilkins; 2007.
20. Castrique E, Fernandez-Fuente M, Le Tissier P, Herman A, Levy A. Use of a prolactin-Cre/ROSA-YFP transgenic mouse provides no evidence for lactotroph transdifferentiation after weaning, or increase in lactotroph/somatotroph proportion in lactation. *J Endocrinol.* 2010;205(1):49–60.
21. Horvath E, Kovacs K. Folliculo-stellate cells of the human pituitary: a type of adult stem cell? *Ultrastruct Pathol.* 2002;26(4):219–228.
22. Inoue K, Mogi C, Ogawa S, Tomida M, Miyai S. Are folliculo-stellate cells in the anterior pituitary gland supportive cells or organ-specific stem cells? *Arch Physiol Biochem.* 2002;110(1–2):50–53.
23. Devnath S, Inoue K. An insight to pituitary folliculo-stellate cells. *J Neuroendocrinol.* 2008;20(6):687–691.
24. Couly GF, Le Douarin NM. Mapping of the early neural primordium in quail-chick chimeras. II. The prosencephalic neural plate and neural folds: implications for the genesis of cephalic human congenital abnormalities. *Dev Biol.* 1987;120(1):198–214.
25. Couly GF, Le Douarin NM. Mapping of the early neural primordium in quail-chick chimeras. I. Developmental relationships between placodes, facial ectoderm, and prosencephalon. *Dev Biol.* 1985;110(2):422–439.
26. Kawamura K, Kikuyama S. Induction from posterior hypothalamus is essential for the development of the pituitary proopiomelacortin (POMC) cells of the toad (Bufo japonicus). *Cell Tissue Res.* 1995;279(2):233–239.
27. Kouki T, Imai H, Aoto K, et al. Developmental origin of the rat adenohypophysis prior to the formation of Rathke's pouch. *Development.* 2001;128(6):959–963.
28. Baker CV, Bronner-Fraser M. Vertebrate cranial placodes I. Embryonic induction. *Dev Biol.* 2001;232(1):1–61.
29. Rizzoti K, Lovell-Badge R. Early development of the pituitary gland: induction and shaping of Rathke's pouch. *Rev Endocr Metab Disord.* 2005;6(3): 161–172.
30. Saper CB, Lowell BB. The hypothalamus. *Curr Biol.* 2014;24(23):R1111–R1116.
31. Aujla PK, Bogdanovic V, Naratadam GT, Raetzman LT. Persistent expression of activated notch in the developing hypothalamus affects survival of pituitary progenitors and alters pituitary structure. *Dev Dyn.* 2015;244(8):921–934.
32. Dale JK, Vesque C, Lints TJ, et al. Cooperation of BMP7 and SHH in the induction of forebrain ventral midline cells by prechordal mesoderm. *Cell.* 1997;90(2):257–269.
33. Manning L, Ohyama K, Saeger B, et al. Regional morphogenesis in the hypothalamus: a BMP-Tbx2 pathway coordinates fate and proliferation through Shh downregulation. *Develop Cell.* 2006;11(6):873–885.
34. Roch GJ, Busby ER, Sherwood NM. Evolution of GnRH: diving deeper. *Gen Comp Endocrinol.* 2011;171(1):1–16.
35. Querat B, Tonnerre-Doncarli C, Genies F, Salmon C. Duality of gonadotropins in gnathostomes. *Gen Comp Endocrinol.* 2001;124(3):308–314.
36. Miller WL, Baxter JD, Eberhardt NL. Peptide hormone genes: Structure and evolution. In: Krieger DT, Brownstein MJ, Martin JB, eds. *Brain Peptides.* Wiley-Interscience; 1983:16–78.
37. Agustsson T, Sundell K, Sakamoto T, Ando M, Bjornsson BT. Pituitary gene expression of somatolactin, prolactin, and growth hormone during Atlantic salmon parr—smolt transformation. *Aquaculture.* 2003;222:229–238.
38. Norris DO, Carr JA. *Vertebrate Endocrinology* (5th ed.). Academic Press; 2013.
39. Brown PS, Brown SC. Osmoregulatory actions of prolactin and other adenohypophysial hormones. In: Pang PKT, Schreibman MP, eds. *Vertebrate Endocrinology: Fundamentals and Biomedical Implications, Vol. 2, Regulation of Water and Electrolytes.* Academic Press; 1987:45–84.
40. Gillespie MJ, Stanley D, Chen H, et al. Functional similarities between pigeon "milk" and mammalian milk: induction of immune gene expression and modification of the microbiota. *PLoS One.* 2012;7(10):e48363.
41. Buckley J, Maunder RJ, Foey A, Pearce J, Val AL, Sloman KA. Biparental mucus feeding: a unique example of parental care in an Amazonian cichlid. *J Exp Biol.* 2010;213(Pt 22):3787–3795.
42. Khong HK, Kuah MK, Jaya-Ram A, Shu-Chien AC. Prolactin receptor mRNA is upregulated in discus fish (Symphysodon aequifasciata) skin during parental phase. *Comp Biochem Physiol B Biochem Mol Biol.* 2009;153(1):18–28.
43. Nakabayashi K, Matsumi H, Bhalla A, et al. Thyrostimulin, a heterodimer of two new human glycoprotein hormone subunits, activates the thyroid-stimulating hormone receptor. *J Clin Invest.* 2002;109(11):1445–1452.
44. Paluzzi JP, Vanderveken M, O'Donnell MJ. The heterodimeric glycoprotein hormone, GPA2/GPB5, regulates ion transport across the hindgut of the adult mosquito, Aedes aegypti. *PLoS One.* 2014;9(1):e86386.
45. Stockell Hartree A, Renwick AG. Molecular structures of glycoprotein hormones and functions of their carbohydrate components. *Biochem J.* 1992;287(Pt 3):665–679.
46. Denver RJ. Neuroendocrinology of amphibian metamorphosis. *Curr Top Dev Biol.* 2013;103:195–227.
47. Okada SL, Ellsworth JL, Durnam DM, et al. A glycoprotein hormone expressed in corticotrophs exhibits unique binding properties on thyroid-stimulating hormone receptor. *Mol Endocrinol.* 2006;20(2):414–425.
48. Nagasaki H, Wang Z, Jackson VR, Lin S, Nothacker HP, Civelli O. Differential expression of the thyrostimulin subunits, glycoprotein alpha2 and beta5 in the rat pituitary. *J Mol Endocrinol.* 2006;37(1):39–50.
49. Kubokawa K, Tando Y, Roy S. Evolution of the reproductive endocrine system in chordates. *Integr Comp Biol.* 2010;50(1):53–62.
50. Tando Y, Kubokawa K. A homolog of the vertebrate thyrostimulin glycoprotein hormone alpha subunit (GPA2) is expressed in Amphioxus neurons. *Zoolog Sci.* 2009;26(6):409–414.
51. Tando Y, Kubokawa K. Expression of the gene for ancestral glycoprotein hormone beta subunit in the nerve cord of amphioxus. *Gen Comp Endocrinol.* 2009;162(3):329–339.
52. Holland, LZ, Albalat, R, Azumi, K. et al. The amphioxus genome illuminates vertebrate origins and cephalochordate biology. *Genome Res.* 2008;18(7): 1100–1111.
53. Sudo S, Kuwabara Y, Park JI, Hsu SY, Hsueh AJ. Heterodimeric fly glycoprotein hormone-alpha2 (GPA2) and glycoprotein hormone-beta5 (GPB5) activate fly leucine-rich repeat-containing G protein-coupled receptor-1 (DLGR1) and stimulation of human thyrotropin receptors by chimeric fly GPA2 and human GPB5. *Endocrinology.* 2005;146(8): 3596–3604.

54. Park JI, Semyonov J, Chang CL, Hsu SY. Conservation of the heterodimeric glycoprotein hormone subunit family proteins and the LGR signaling system from nematodes to humans. *Endocrine.* 2005;26(3):267–276.
55. Sower SA, Moriyama S, Kasahara M, et al. Identification of sea lamprey GTHbeta-like cDNA and its evolutionary implications. *Gen Comp Endocrinol.* 2006;148(1):22–32.
56. Krude H, Biebermann H, Luck W, Horn R, Brabant G, Gruters A. Severe early-onset obesity, adrenal insufficiency and red hair pigmentation caused by POMC mutations in humans. *Nat Genet.* 1998;19(2):155–157.
57. Tsatmali M, Ancans J, Thody AJ. Melanocyte function and its control by melanocortin peptides. *J Histochem Cytochem.* 2002;50(2):125–133.
58. Fell GL, Robinson KC, Mao J, Woolf CJ, Fisher DE. Skin β-endorphin mediates addiction to UV light. *Cell.* 2014;157(7):1527–1534.
59. Kawauchi H, Kawazoe I, Tsubokawa M, Kishida M, Baker BI. Characterization of melanin-concentrating hormone in chum salmon pituitaries. *Nature.* 1983;305(5932):321–323.
60. Griffond B, Baker BI. Cell and molecular cell biology of melanin-concentrating hormone. *Int Rev Cytol.* 2002;213: 233–277.
61. Ludwig DS, Mountjoy KG, Tatro JB, et al. Melanin-concentrating hormone: a functional melanocortin antagonist in the hypothalamus. *Am J Physiol.* 1998;274(4 Pt 1):E627–E633.
62. Hughes J, Smith TW, Kosterlitz HW, Fothergill LA, Morgan BA, Morris HR. Identification of two related pentapeptides from the brain with potent opiate agonist activity. *Nature.* 1975;258(5536):577–580.
63. Li CH, Chung D. Isolation and structure of an untriakontapeptide with opiate activity from camel pituitary glands. *Proc Natl Acad Sci U S A.* 1976;73(4): 1145–1148.
64. Liu J, Mershon J. Neurosecretory peptides. *Glob Libr Women's Med.*, 2008. doi:10.3843/GLOWM.10287/Update due. (ISSN: 1756-2228).
65. Steinke D, Hoegg S, Brinkmann H, Meyer A. Three rounds (1R/2R/3R) of genome duplications and the evolution of the glycolytic pathway in vertebrates. *BMC Biol.* 2006;4:16.
66. Dehal P, Boore JL. Two rounds of whole genome duplication in the ancestral vertebrate. *PLoS Biol.* 2005;3(10):e314.
67. Cortes R, Navarro S, Agulleiro MJ, et al. Evolution of the melanocortin system. *Gen Comp Endocrinol.* 2014;209:3–10.
68. Van de Peer Y, Maere S, Meyer A. The evolutionary significance of ancient genome duplications. *Nat Rev Genet.* 2009;10(10):725–732.
69. Dores RM, Baron AJ. Evolution of POMC: origin, phylogeny, posttranslational processing, and the melanocortins. *Ann N Y Acad Sci.* 2011;1220:34–48.
70. Dores RM, Lecaude S. Trends in the evolution of the proopiomelanocortin gene. *Gen Comp Endocrinol.* 2005;142 (1–2):81–93.
71. Hayashi H, Imai K, Imai K. Characterization of chicken ACTH and alpha-MSH: the primary sequence of chicken ACTH is more similar to Xenopus ACTH than to other avian ACTH. *Gen Comp Endocrinol.* 1991;82(3):434–443.
72. Mahlmann S, Meyerhof W, Hausmann H, et al. Structure, function, and phylogeny of [Arg8]vasotocin receptors from teleost fish and toad. *Proc Natl Acad Sci U S A.* 1994;91(4):1342–1345.
73. Ocampo Daza D, Lewicka M, Larhammar D. The oxytocin/vasopressin receptor family has at least five members in the gnathostome lineage, inclucing two distinct V2 subtypes. *Gen Comp Endocrinol.* 2012;175(1):135–143.
74. Guillette LJ, Norris DO, Norman MF. Response of amphibian (Ambystoma tigrinum) oviduct to arginine vasotocin and acetylcholine in vitro: influence of steroid hormone pretreatment in vivo. *Comp Biochem Physiol.* 1985;8:151–154.
75. Guillette LJ, Jones REJ. Arginine vasotocin-induced in vitro oviductal contractions in Anolis carolinensis: effects of steroid hormone pretreatment in vivo. *J Exp Zool.* 1980;212:47–152.
76. Moore FL, Wood RE, Boyd SK. Sex steriods and vasotocin interact in a female amphibian (Taricha granulosa) to elicit female-like egg-laying behaviour or male-like courtship. *Horm Behav.* 1992;26:156–166.
77. Guillette LJ, Jones RE. Further observations on arginine vasotocin-induced oviposition and parturition in lizards. *J Herpetol* 1982;16:140–144.
78. Donaldson ZR, Young LJ. Oxytocin, vasopressin, and the neurogenetics of sociality. *Science.* 2008;322(5903):900–904.
79. Gorbman A, Bern HA. *A Textbook of Comparative Endocrinology* (1st ed.). Wiley; 1962.
80. Gorbman A, Nozaki M, Kubokawa K. A brain-Hatschek's pit connection in amphioxus. *Gen Comp Endocrinol.* 1999;113(2):251–254.
81. Holland LZ, Holland ND. Evolution of neural crest and placodes: amphioxus as a model for the ancestral vertebrate? *J Anat.* 2001;199(Pt 1–2):85–98.
82. Candiani S, Holland ND, Oliveri D, Parodi M, Pestarino M. Expression of the amphioxus Pit-1 gene (AmphiPOU1F1/ Pit-1) exclusively in the developing preoral organ, a putative homolog of the vertebrate adenohypophysis. *Brain Res Bull.* 2008;75(2–4):324–330.
83. Chambery A, Parente A, Topo E, Garcia-Fernandez J, D'Aniello S. Characterization and putative role of a type I gonadotropin-releasing hormone in the cephalochordate amphioxus. *Endocrinology.* 2009;150(2):812–820.
84. Nozaki M, Gorbman A. The question of functional homology of Hatschek's pit of amphioxus (Branchiostoma helcheri) and the vertebrate adenohypophysis. *Zoolog Sci.* 1992;9:387–395.
85. Hamada M, Shimozono N, Ohta N, et al. Expression of neuropeptide- and hormone-encoding genes in the Ciona intestinalis larval brain. *Dev Biol.* 2011;352(2):202–214.
86. Manni L, Agnoletto A, Zaniolo G, Burighel P. Stomodeal and neurohypophysial placodes in Ciona intestinalis: insights into the origin of the pituitary gland. *J Exp Zool B Mol Dev Evol.* 2005;304(4):324–339.
87. Janvier P. *Early Vertebrates.* Clarendon Press; 1996.
88. Gorbman A, Kobayashi H, Uemura H. The vascularization of the hypophysial structures of the hagfish. *Gen Comp Endocrinol.* 1963;14:505–514.
89. Sherwood NM, Sower SA, Marshak DR, Fraser BA, Brownstein MJ. Primary structure of gonadotropin-releasing hormone from lamprey brain. *J Biol Chem.* 1986;261(11):4812–4819.
90. Sower SA, Chiang YC, Lovas S, Conlon JM. Primary structure and biological activity of a third gonadotropin-releasing hormone from lamprey brain. *Endocrinology.* 1993;132(3):1125–1131.
91. Suzuki K, Gamble RL, Sower SA. Multiple transcripts encoding lamprey gonadotropin-releasing hormone-I precursors. *J Mol Endocrinol.* 2000;24(3):365–376.
92. Sower SA, Nozaki M, Knox CJ, Gorbman A. The occurrence and distribution of GnRH in the brain of Atlantic hagfish, an agnatha, determined by chromatography and immunocytochemistry. *Gen Comp Endocrinol.* 1995;97(3):300–307.
93. Nozaki M, Fernholm B, Kobayashi H. Ependymal absorption of peroxidase in the third ventricle of the hagfish Eptatretus burgeri (Girard). *Acta zool (Stockh).* 1975;56:265–269.
94. Tsukahara T, Gorbman A, Kobayashi H. Median eminence equivalence of the neurohypophysis of the hagfish, Eptatretus burgeri. *Gen Comp Endocrinol.* 1986;61(3):348–354.
95. Kavanaugh SI, Nozaki M, Sower SA. Origins of gonadotropin-releasing hormone (GnRH) in vertebrates: identification of a novel GnRH in a basal vertebrate, the sea lamprey. *Endocrinology.* 2008;149(8):3860–3869.
96. Nozaki M. Hypothalamic-pituitary-gonadal endocrine system in the hagfish. *Front Endocrinol (Lausanne).* 2013;4:200.

97. Uchida K, Moriyama S, Chiba H, et al. Evolutionary origin of a functional gonadotropin in the pituitary of the most primitive vertebrate, hagfish. *Proc Natl Acad Sci U S A*. 2010;107(36):15832–15837.
98. Sower SA, Freamat M, Kavanaugh SI. The origins of the vertebrate hypothalamic-pituitary-gonadal (HPG) and hypothalamic-pituitary-thyroid (HPT) endocrine systems: new insights from lampreys. *Gen Comp Endocrinol*. 2009;161(1):20–29.
99. Nozaki M. The hagfish pituitary gland and its putative adenohypophysial hormones. *Zoolog Sci*. 2008;25(10):1028–1036.
100. Amemiya Y, Takahashi A, Suzuki N, Sasayama Y, Kawauchi H. Molecular cloning of proopiomelanocortin cDNA from an elasmobranch, the stingray, Dasyatis akajei. *Gen Comp Endocrinol*. 2000;118(1):105–112.
101. Takahashi A, Itoh T, Nakanishi A, et al. Molecular cloning of proopiomelanocortin cDNA in the ratfish, a holocephalan. *Gen Comp Endocrinol*. 2004;135(1):159–165.
102. Hayashida T. Biological and immunochemical studies with growth hormone in pituitary extracts of elasmobranchs. *Gen Comp Endocrinol*. 1973;20(2):377–385.
103. Hayashida T, Lewis UJ. Immunochemical and biological studies with antiserum to shark growth hormone. *Gen Comp Endocrinol*. 1978;36(4):530–542.
104. Moriyama S, Oda M, Yamazaki T, et al. Gene structure and functional characterization of growth hormone in dogfish, Squalus acanthias. *Zoolog Sci*. 2008;25(6):604–613.
105. Pogoda H-M, Hammerschmidt M. How to make a teleost adenohypophysis: molecular pathways of pituitary development in zebrafish. *Mol Cell Endocrinol*. 2009;312(1–2):2–13.
106. Trudeau VL, Somoza GM. Multimodal hypothalamo-hypophysial communication in the vertebrates. *Gen Comp Endocr*. 2020:113475.

CHAPTER

9

Central Control of Biological Rhythms

CHAPTER LEARNING OBJECTIVES:

- Define "biological clock" and describe how it is essential for the healthy life of an animal.
- Describe the mechanism by which the suprachiasmatic nucleus (SCN) receives circadian information, and how it disperses this information to peripheral organs.
- Describe the functions of melatonin and the evolution of the pineal gland as a photoreceptive and neuroendocrine organ.
- Outline the molecular basis of mammalian circadian clock function and its mechanisms for modulating endocrine activity, and describe how hormones influence clock function.
- Discuss the pathological effects of circadian disruption on wildlife and human physiology.

OPENING QUOTATIONS:

"You have heard of the pineal gland? I laugh at the shallow endocrinologist, fellow-dupe and fellow-parvenu of the Freudian. That gland is the great sense-organ of organs—I have found out. It is like sight in the end, and transmits visual pictures to the brain."

—H.P. Lovecraft, *From Beyond*,
short science fiction story (1934)
The Fantasy Fan (Vol. 1, No. 10).

"Man, I'll try just about anything, but I'd never in hell touch a pineal gland."

—Dr. Gonzo, in Hunter S. Thompson's *Fear and Loathing in Las Vegas: A Savage Journey to the Heart of the American Dream.*[1]

Introduction: Biological Clocks for a Rhythmic World

LEARNING OBJECTIVE Define "biological clock" and describe how it is essential for the healthy life of an animal.

KEY CONCEPTS:

- Biological clocks are endogenously generated mechanisms present in virtually all cells within an organism that allow the tracking of time.
- An ability to predict periodic geophysical events allows synchronization of time- and energy-consuming behavioral and physiological processes with the environment to optimize energy utilization, reproduction, and survival.
- "Circadian" clocks maintain a periodicity of about (but never exactly) 24 hours.
- In the absence of any external time entrainment signals, biological clock time will begin to drift, or "free run".
- Many hormones oscillate in a circadian pattern and may demonstrate rhythmicity across multiple time scales.

Characteristics of Biological Clocks

Our world is a rhythmic place—the rising and setting of the sun, the changing phases of the moon, the ebb and flow of ocean tides, the changing seasons, even the movements of the stars. As the earth rotates along its axis and orbits around the sun, periodic exposure to solar radiation causes predictable changes in light and ambient temperatures, as well as associated changes in the relative humidity of the air and in the oxygen levels of aquatic habitats. Virtually all organisms must adapt to such rhythms to thrive, and it is therefore advantageous for them to be able to anticipate, rather than simply react to, environmental changes. Effective prediction of these rhythms allows synchronization of time- and energy-consuming behavioral and physiological processes to the environment in order to optimize energy utilization, reproduction, and survival. For example, the genomic actions of steroid hormones can take several hours to manifest, and these hormones must therefore be secreted prior to the time during which they will exert their effects. As such, all organisms possess internal mechanisms to track time. These mechanisms, loosely termed *biological clocks*, operate on daily (Figure 9.1), as well as tidal, lunar[2], and even annual, cycles associated with seasonal reproduction, migration, and hibernation (Figure 9.3).

DOI: 10.1201/9781003359241-12

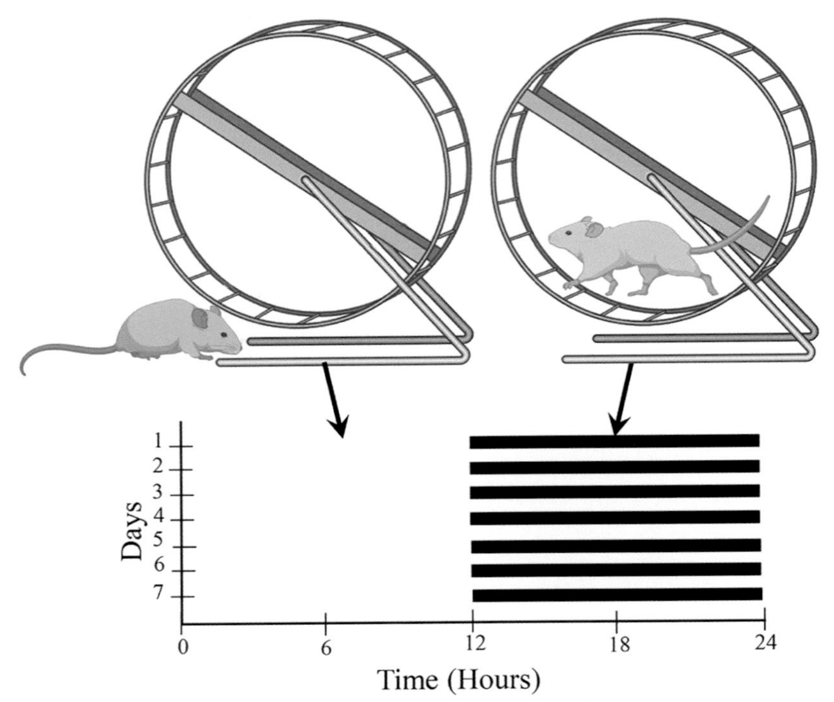

Figure 9.1 Collection of data on behavioral biological rhythms. A common method for measuring locomotor activity in mice is a computer-linked wheel-running device that activates data output with each revolution of the wheel. Sustained wheel running activity appears as a solid bar across the time line for each day. The period of the daily biological rhythm can be measured from the onset of activity on one day to the onset of activity on the next.

Biological clocks that operate on a daily cycle are called **circadian clocks** (from the Latin *circa*, meaning "approximately", and *diem*, "day"), which maintain a periodicity of about (but never exactly) 24 hours to follow the earth's rotation around its axis. Importantly, the rhythms of genuine biological clocks are generated *endogenously*, or within the organism, and these rhythms therefore persist even in the absence of external time cues. However, because the periodicity of the endogenous circadian clock is not exactly 24 hours, in the absence of external time cues the onset of an animal's circadian activity pattern will begin to drift, a phenomenon called a **free-running rhythm**[3] (Figure 9.4). Under laboratory conditions, free-running rhythms have also been observed for circalunar,[4] circatidal,[5] and even circannual cycles.[6] To remain synchronized with the environment, biological clocks are reset, or *entrained*, on a regular basis by a **zeitgeber** ("time giver", in German), an environmental cue such as light or temperature that provides information about the external time.

Hormones and Biological Clocks

Biological clocks have many uses and regulate daily sleep-wake cycles, locomotor activity, body temperature, and metabolic patterns, as well as trigger seasonal reproduction and migrations and even aid in navigation. Importantly, the generation, secretion, and abundance of diverse endocrine factors, as well as the sensitivity of target organs to these signals, are strongly influenced by these endogenous clocks. Some well-characterized hormones that oscillate in a circadian pattern are described in Figure 9.5 and Table 9.1, and some mechanisms controlling their fluctuations will be addressed in this chapter. Some hormones, particularly those associated with reproduction such as gonadotropin-releasing hormone (GnRH) and the gonadotropins, luteinizing hormone (LH) and follicle-stimulating hormone (FSH), may demonstrate rhythmicity across multiple time scales. For example, in hamsters, LH release ranges from pulsatile over the course of minutes to daily surges and seasonal peaks (Figure 9.6). Changes in gonadotropin pulse frequency and amplitude over different time scales play critical roles in the timing of puberty and reproduction, a topic that will be addressed in detail in Chapter 23: The Timing of Puberty and Seasonal Reproduction.

Curiously Enough . . . With respect to human reproduction, although the female menstrual cycle lasts an average of 27 days, or approximately the length of a lunar month (29.5 days), the menstrual period is not synchronized to any phase of the moon.

Foundations & Fundamentals: The Language of Chronobiology

Chronobiology, the field of biology that examines periodic phenomena in living organisms and their adaptation to environmental rhythms, has borrowed language and concepts from engineering disciplines that study physical oscillators (Figure 9.2). Some of these terms are summarized as follows:

- **Rhythm**: A recurrent event that is characterized by its period, frequency, amplitude, and phase.
- **Period (tau)**: the length of time for one cycle of an oscillation to repeat.
- **Frequency**: the number of complete cycles per unit of time.
- **Amplitude**: the size of an oscillation as measured by half the distance from the peak to the trough.
- **Phase**: the timing of a consistent point in the cycle, such as the peak or trough. If the periods of two cycles coincide, the two rhythms are said to be "in phase". If the periods of two cycles do not coincide, the two rhythms are said to be "out of phase".
- **Phase shift**: a change in phase such that it occurs earlier or later, owing to a displacement of the entire cycle.
- **Zeitgeber**: literally means "time giver" and refers to any resetting stimulus that serves as a time cue for the external environment.
- **Entrainment**: The synchronization of a biological rhythm to a periodic environmental time cue.
- **Free-running rhythm**: A biological rhythm that is not synchronized to its natural zeitgeber and expresses its own endogenous rhythm.
- **Circadian rhythm**: a rhythm with a period of about 24 hours.
- **Circatidal rhythm**: a rhythm with a period of about 12.5 hours that is closely tied to changes in tides.
- **Circalunar rhythm**: A rhythm with a period of about 29.5 days that is closely tied with phases of the moon.
- **Circannual rhythm**: A rhythm with a period of about 12 months.
- **Ultradian rhythm**: A biological rhythm with a frequency of greater than once per day (or a period less than 24 hours).
- **Infradian rhythm**: A biological rhythm with a frequency less than once per day (or a period of greater than 24 hours).
- **Diurnal**: Active during the day.
- **Nocturnal**: Active during the night.
- **Crepuscular**: Active at dawn and dusk.

Circadian biological rhythms typically meet several general criteria:[7]

1. The rhythm has an endogenous free-running period that lasts approximately 24 hours.
2. The rhythms are entrainable.
3. The rhythms exhibit temperature compensation.
4. The rhythms are inherited.

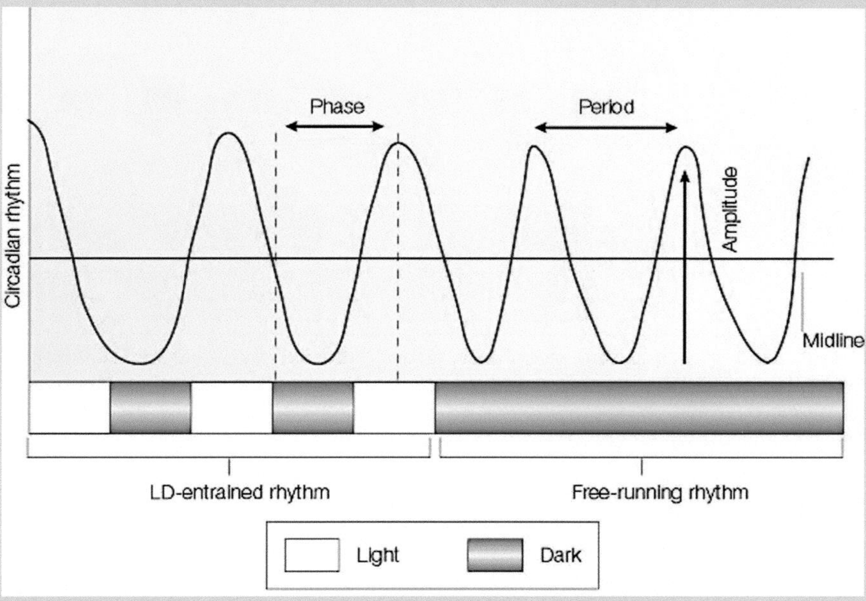

Figure 9.2 Some defining properties of oscillating biological rhythms.

Source of Images: Bell-Pedersen, D., et al. 2005. Nat. Rev. Genet., 6(7), pp. 544-556. Used with permission.

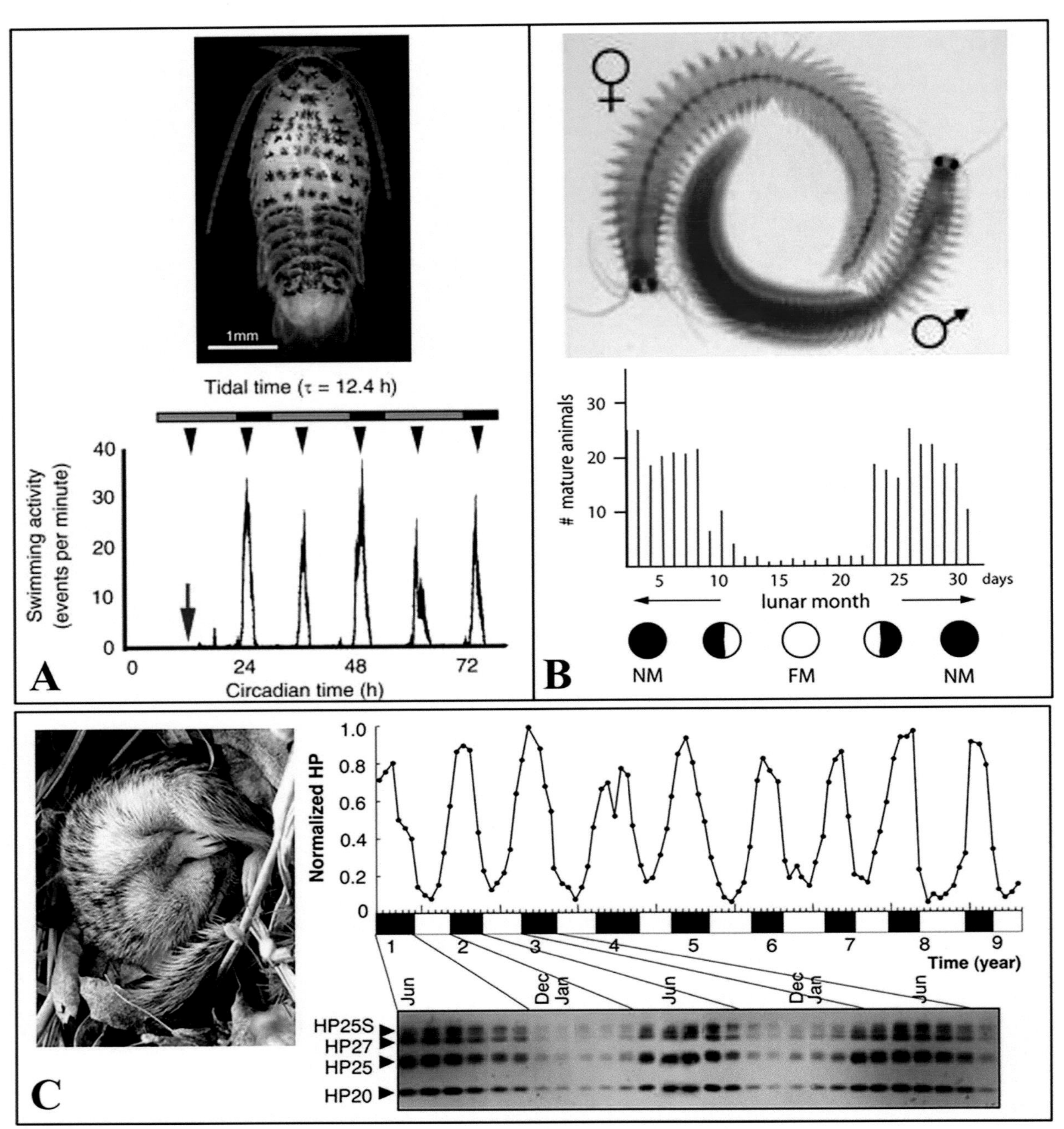

Figure 9.3 Examples of animals using circatidal, circalunar, and circannual rhythms. (A) The intertidal isopod, *Eurydice pulchra*, shows rhythmic swimming activity that coincides with high tide. The actogram illustrates the mean behavior of ten individuals taken from the same beach over five complete tidal cycles. Black arrowheads indicate the time of expected high water and the red arrow shows when the animals were captured. The black and gray bars indicate time of expected day and night. Note that the amplitude of alternate activity peaks modulates according to expected nighttime high water when the tidal range is greatest. **(B)** Circalunar reproductive periodicity of the bristle worm, *Platynereis dumerilii*. Numbers of mature animals displaying synchronized spawning behavior oscillate in accordance with the lunar cycle. **(C)** Circannual rhythm in hibernation-inducing protein (HP) in ground squirrels over the course of 9 years. HP, a protein produced by the brain, is thought to induce hibernation. HP content in plasma collected monthly was determined by Western blotting (bottom panel), normalized to its maximum content, and the normalized values were plotted as a function of time (top panel). A pair of alternating dark and light bars denotes one year, with the dark bar representing the time of hibernation. Arrowheads in Western blots denote the positions of varying sizes of different HP proteins.

Source of Images: (A) Wilcockson, D. and Zhang, L., 2008. Curr. Biol., 18(17), pp. R753-R755; used with permission. (B) Zantke, J., et al. 2013. Cell Rep., 5(1), pp. 99-113. (C) Kondo, N., et al. 2006. Cell, 125(1), pp. 161-172; used with permission. Ground squirrel: Feng, N.Y., et al. 2019. Curr. Biol., 29(18), pp. 3053-3058. Used with permission.

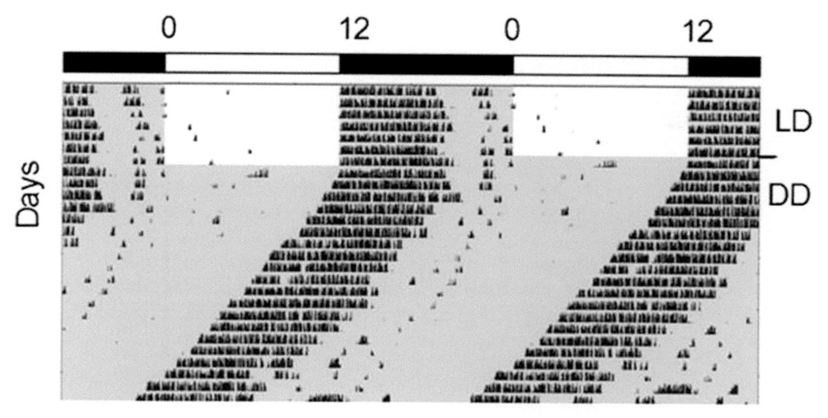

Figure 9.4 Entrained and free-running rhythms in mice. Actograms of a wild-type mouse entrained under a 12:12 light-dark (LD) cycle followed by continuous darkness (DD). The bar above each actogram represents the lighting regimen in LD (white boxes indicate presence of light, dark boxes indicate nighttime). Transition to DD is indicated by the line on the right of the actogram. Note the development of a free rhythm drifting pattern in the circadian rhythm following transition to DD cycle.

Source of Images: Yang, Y., et al. 2012. Biol. Open, 1(8), pp. 789-801.

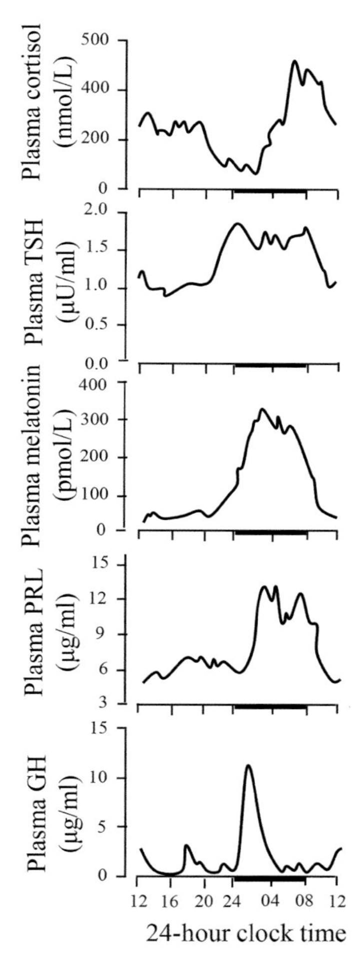

Figure 9.5 Mean 24-hour profiles of plasma cortisol, thyrotropin (TSH), melatonin, prolactin (PRL), and growth hormone (GH) levels in eight human subjects (20 to 27 years). Data were sampled at 15-minute intervals.

Source of Images: van Coevorden A, et al. Am J Physiol 260:E651–E661, 1991.

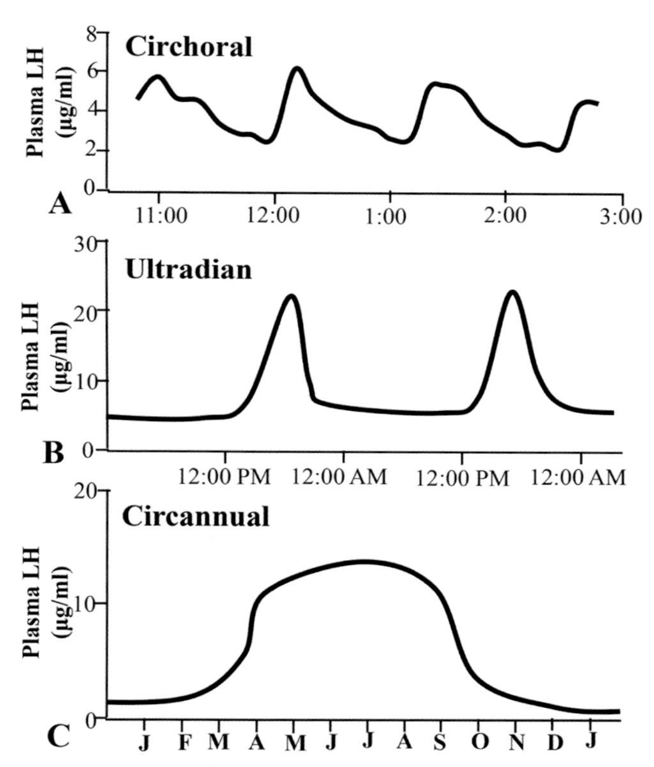

Figure 9.6 LH is released periodically over several different time scales in hamsters. (A) LH is released in an approximate circhoral (hourly) manner throughout the day. **(B)** LH is also released in an ultradian (more than once per day) fashion, in this case manifesting as programmed increases above the hourly concentrations approximately every 12 hours. **(C)** Hamsters housed in simulated winter conditions do not detect circhoral or ultradian pulses, but instead display a circannual (once per year) pattern of elevated LH secretion during the breeding season in spring and summer, compared with lower concentrations during autumn and winter.

Source of Images: Nelson, R.J., 2005. An introduction to behavioral endocrinology. Sinauer Associates.

Developments & Directions: Chronotherapy and the Timing of Medicine Delivery

Not only do diverse aspects of human physiology fluctuate over a 24-hour period, but morbidity and mortality events also peak at predictable times (Figure 9.7). For example, the occurrence of heart attacks occurs most frequently between 6:00 am and 12:00 noon.[8] Other diseases, including asthma, cancer, osteoarthritis, cardiovascular diseases, and peptic ulcers show circadian rhythmicity in the occurrence or intensity of symptoms.[9] *Chronotherapy*, an emerging concept in the field of therapeutics, involves the administration of medication in coordination with the body's circadian rhythms to maximize therapeutic effectiveness and minimize/avoid adverse effects.[9] For example, increased cholesterol biosynthesis occurs nocturnally, and for cholesterol-lowering medicines, such as statins, an evening administration is recommended. Technological advances now allow for complex drug delivery through programmable pumps and oral multiple-unit preparations.[10–12] Novel drug delivery systems could enable the personalization of chronotherapeutics with oral anticancer drugs through patient- and drug-specific preparations, thus contributing to improvement of the currently limited tolerability and efficacy of these agents.[13]

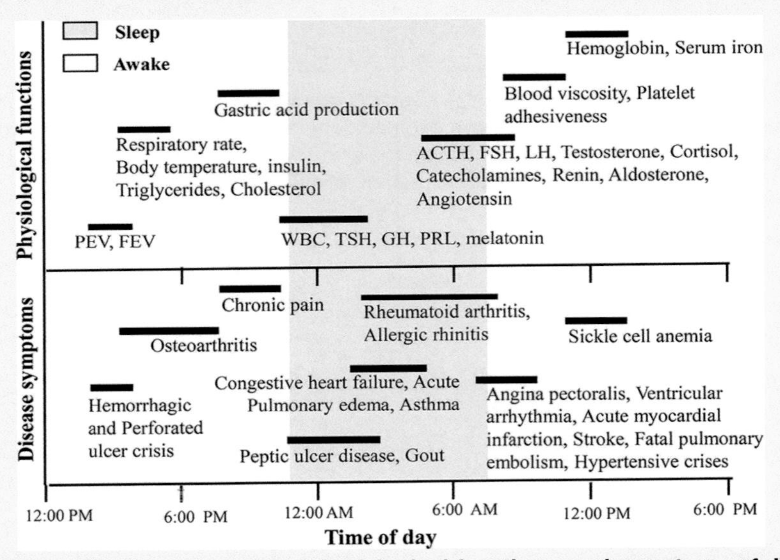

Figure 9.7 Time of day when endocrine parameters, physiological functions, and symptoms of diseases peak. PEV, peak expiratory flow rate; FEV, forced expiratory volume; TSH, thyroid-stimulating hormone; ACTH, adrenocortical tropic hormone; FSH, follicle-stimulating hormone; LH, luteinizing hormone.

Source of Images: Kaur, G., et al. 2016. Pharmaceutics, 8(2), p. 13.

Table 9.1 Endocrine Factors That Oscillate with a Periodicity of About 24 Hours in Humans

Hormone	Time of Peak (h)
Cortisol	0700–0800
GH	Pulsatile secretion with increased amplitude at night
Testosterone (males)	0700
Prolactin	0200 (amplitude greater in females)
TSH	0100–0200
T_3	0230–0330
RAAS	Early morning
FGF21	0500–0800
Ghrelin	0200–0430 in fed state
Adiponectin	1200–1400
Leptin	0100
Vasopressin	midnight
Insulin	1700
Melatonin	midnight

Source: After Gamble, K.L., Berry, R., Frank, S.J. and Young, M.E., 2014. Circadian clock control of endocrine factors. *Nature Reviews Endocrinology*, 10(8), pp. 466–475.

SUMMARY AND SYNTHESIS QUESTIONS

1. If all animals have endogenous circadian clocks, why will their biological rhythms "free run" if they don't receive environmental information about day length?

The Mammalian Master Clock

LEARNING OBJECTIVE Describe the mechanism by which the suprachiasmatic nucleus (SCN) receives circadian information, and how it disperses this information to peripheral organs.

KEY CONCEPTS:

- The mammalian circadian clock is organized as a hierarchy of multiple peripheral oscillators, all controlled by a central pacemaker, the SCN.

- Individual SCN neurons are inherently rhythmic.
- The SCN receives nonvisual light information from intrinsically photosensitive retinal ganglion cells (ipRGC).
- The SCN communicates with peripheral oscillators via both hormonal and neural signals.
- A desynchrony among peripheral oscillators and the SCN contributes to the symptoms of "jet lag".

A Hierarchy of Dispersed Oscillators

In mammals, light serves as the primary zeitgeber that entrains the circadian clock. The mammalian circadian clock can be described as having three components: inputs, oscillators, and outputs (Figure 9.8). Retinal photoreceptors in the eyes constitute the primary source of input. Light information is transmitted directly to the **suprachiasmatic nucleus** (SCN) of the hypothalamus. SCN temporal output in the form of both endocrine and neural signals in turn synchronizes the timing of diverse peripheral oscillators that ultimately regulate local circadian physiological outputs. Importantly, these peripheral clocks can function autonomously without central or systemic cues.[14,15] The circadian clock in mammals is currently conceptualized as a hierarchy of dispersed oscillators in which the SCN acts as a master pacemaker to synchronize or entrain peripheral clocks distributed throughout the body.[16] This section describes interactions among this multi-oscillator network and their roles in synchronizing endocrine and other physiological outputs.

The SCN Is the Master Oscillator in Mammals

The SCN Is Necessary for Mammalian Circadian Rhythmicity

In mammals, the suprachiasmatic nucleus (SCN) (Figure 9.9A) consists of two small, bilaterally paired nuclei in the anterior hypothalamus, each of which

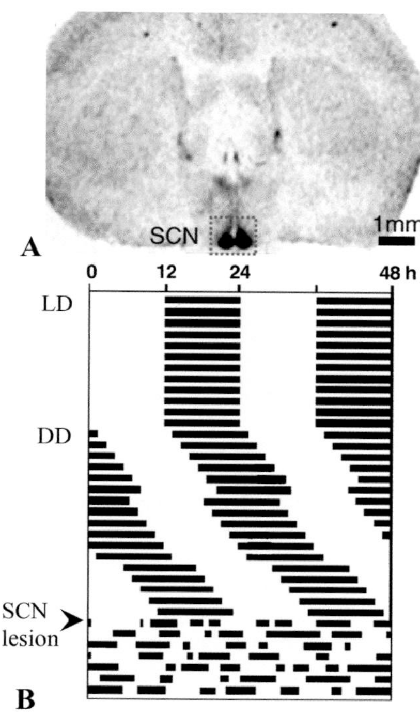

Figure 9.9 Location of the suprachiasmatic nuclei (SCN) in the rat brain, and the effects of constant darkness and SCN ablation on circadian activity. (A) A Nissl-stained coronal section of the rat brain. The SCN consists of two small, bilaterally paired nuclei in the anterior hypothalamus, situated at the base of the brain. **(B)** Schematic representation of a circadian rhythm as normally entrained to a 24-hour light-dark cycle (LD), as free-running pattern in constant darkness (DD), and SCN lesion condition. The dark bars indicate the active (movement, drinking, eating for rodents) periods on successive days, presented in an actogram. Note in particular the point where there is a lesion of the SCN results in a dramatic change in circadian rhythm. The fragmented activity pattern is consistent with the loss of master pacemaker function of the SCN.

Source of Images: (A) Patton, A.P. and Hastings, M.H., 2018. Curr. Biol., 28(15), pp. R816-R822; used with permission. (B) Stiller, J.W. and Postolache, T.T., 2005. Clin Sports Med., 24(2), pp. 205-235. Used with permission.

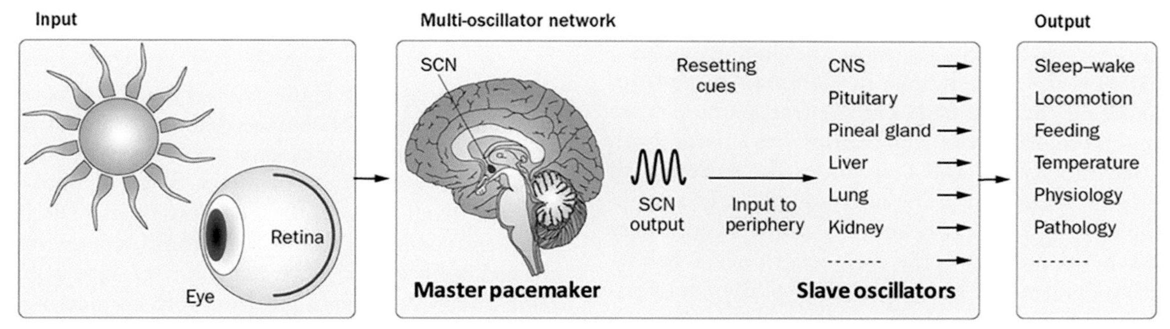

Figure 9.8 The components of the mammalian circadian clock. The mammalian circadian clock is organized in a hierarchy of multiple oscillators, in which the SCN is the central pacemaker at the top of the hierarchy. The SCN is synchronized by the external 24 h cycle and in turn coordinates the physiological outputs. The multi-oscillator network is synchronized through multiple lines of communication. For the SCN, light represents the primary input. Peripheral oscillators are reset by SCN outputs, which regulate local circadian physiology. Abbreviations: CNS, central nervous system; SCN, suprachiasmatic nucleus.

Source of Images: Koch, B.C., et al. 2009. Nat. Rev. Nephrol., 5(7), pp. 407-416. Used with permission.

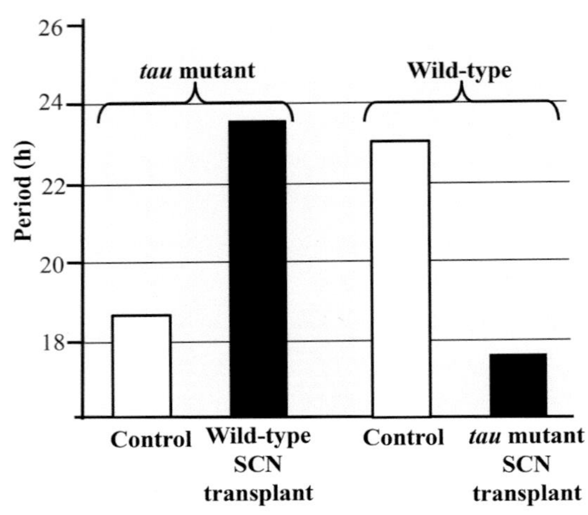

Figure 9.10 Restoration of circadian rhythms with the period of the donor by SCN transplants. Wild-type hamsters (with approximately 24-hour free-running periods) or *tau* mutant hamsters (with approximately 20-hour free-running periods) first received SCN lesions, then received SCN transplants from the donor genotype. In each case, the recipient hamsters gained the free-running periods of the SCN donors.

Source of Images: Ralph, M.R., et al. 1990. Science, 247(4945), pp. 975-978.

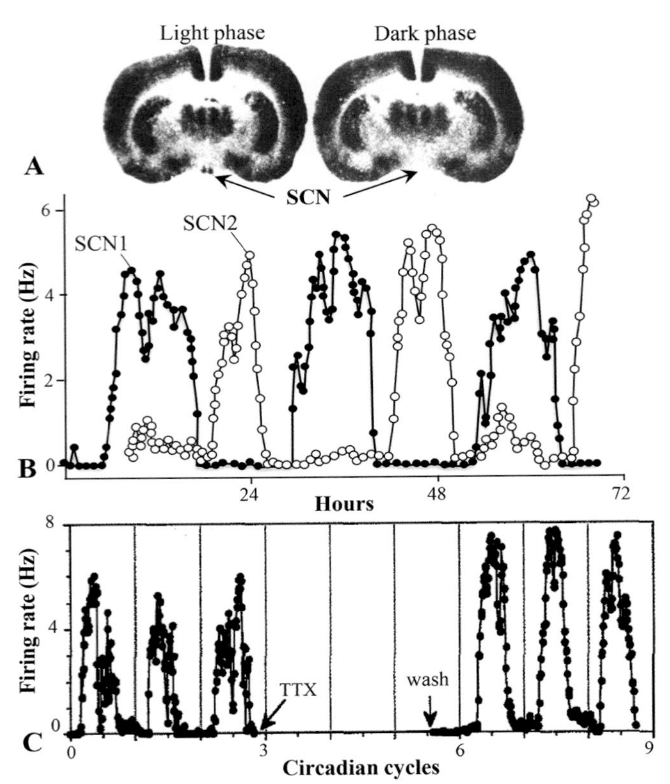

Figure 9.11 Activity of the SCN of mammals varies according to the time of day. (A) Autoradiographs of coronal sections through the SCN or rats injected with carbon 14-labeled 2-deoxyglucose during the light (left) or dark (right) part of the daily light-dark cycle show that energy use by the SCN in these nocturnal animals is higher during the day (dark staining) than at night. **(B)** Two individual SCN neurons show precise circadian firing rates in cell culture (one cell in closed circles and the other in open circles). The different periods and phases of these two cells provided the first evidence that individual SCN cells maintain a characteristic rhythmicity. **(C)** Circadian firing rhythm of a cell-cultured SCN neuron before and after treatment with tetrodotoxin (TTX, a sodium channel blocker) showing that the recovered rhythm is in phase with the pretreatment rhythm. The horizontal time axis is marked in multiples of the cell's circadian period length (23.25 hr).

Source of Images: (A) Schwartz, W.J. and Gainer, H., 1977. Science, 197(4308), pp. 1089-1091; used with permission. (B-C) Welsh, D.K., et al. 1995. Neuron, 14(4), pp. 697-706.

contains about 10,000 cells in rodents. The SCN is situated at the base of the brain directly above the optic chiasm and surrounding the third ventricle and functions as the master circadian clock. Following the identification of the SCN, a role of the SCN as the control center for the circadian system was further suggested by lesion studies that showed disruptions in circadian patterning of activity (Figure 9.9B), hormone secretion, and other physiological parameters like body temperature and heart rate.[17–20] Furthermore, when the SCN is isolated into slices and cultured *in vitro* it continues to display circadian rhythms of electrical activity.[21–23] Importantly, these *in vitro* experiments clearly demonstrate that the SCN neither requires association with a larger neural network in order to produce circadian rhythms nor functions merely as a site to which fibers from a central clock located elsewhere are projected through. Perhaps the most convincing evidence that the SCN is the master clock in mammals came from studies showing that the SCN can be transplanted from one animal to another and this transplanted tissue both restores rhythmicity[24–27] and determines the cycle length of the behavioral rhythm in the host organism. In a striking example, Ralph and colleagues restored rhythmicity in a wild-type hamster with the cycle length of the behavioral rhythm determined by the genotype of the donor tissue[28] (Figure 9.10).

Individual SCN Neurons Are Inherently Rhythmic

The SCN is rhythmic *in vivo*, as determined by both extracellular recordings of action potential frequency and metabolic measurements using carbon 14-labeled 2-deoxyglucose[29,30] (Figure 9.11). Furthermore, recordings from individual SCN neurons demonstrate that each cell is capable of generating its own specific circadian rhythm[31] (Figure 9.11), a unique property that distinguishes SCN neurons from many other neuron types. Distinct from peripheral neurons, the SCN neurons form interconnected circuits with other SCN neurons via intercellular couplings, such as through gap junctions connecting adjacent neurons, which comprise a special attribute of the SCN that promotes synchronization and precision in circadian electrical activity and rhythms.[32–34] As a result, the SCN clock is more robust against genetic and environmental perturbations than are peripheral oscillators whose cells are not as interconnected.

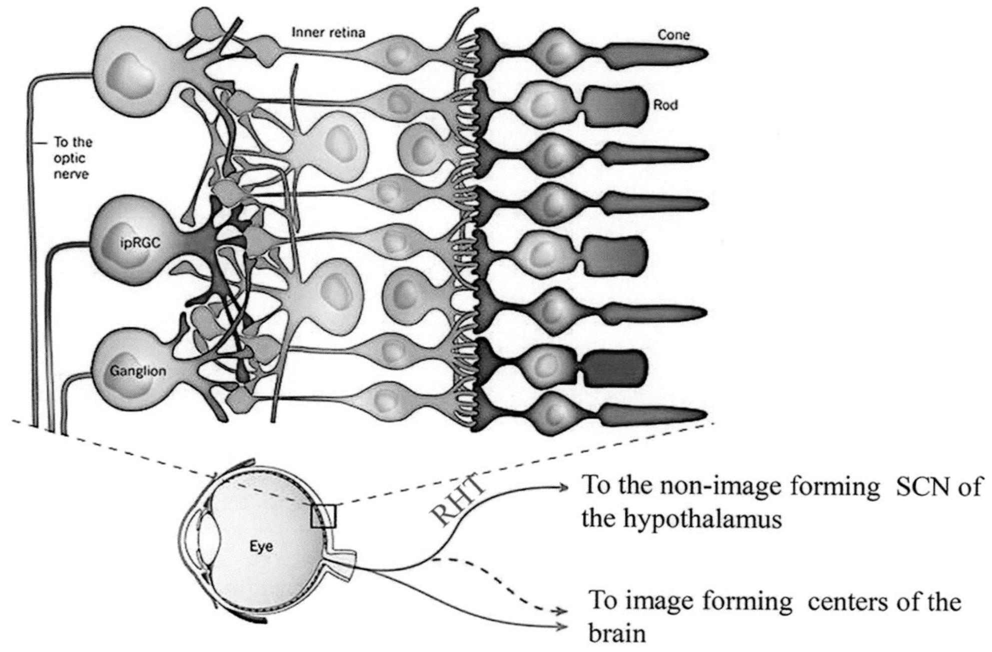

Figure 9.12 Intrinsically photosensitive retinal ganglion cells (ipRGC) of the retina send nonvisual information to the suprachiasmatic nuclei (SCN) of the hypothalamus via the retinohypothalamic tract (RHT) in mammals. Light passes through the ganglion layer (blue/green cells) and inner retina (gray) to rod and cone photoreceptors (purple and maroon). The rods and cones send information to ganglion cells, which transmit light information to the visual centers of the brain (green arrow pathway). A subset of these ganglion cells, called ipRCGs (blue cells), contain a photo pigment called melanopsin that can encode and transmit light information directly to the non-image-forming centers of the brain (blue arrow pathway). Projections of ipRGCs to the SCN form the bulk of the retinohypothalamic tract and contribute to photic entrainment of the circadian clock.

Source of Images: Lok, C., 2011. Nature, 469(284-285), p. 525. Used with permission.

Remarkably, circadian clocks in SCN cultures continue to operate in the absence of neuronal firing, since after action potentials are reversibly blocked for several days of application of the Na^+ channel blocker tetrodotoxin (TTX), circadian rhythms not only resume upon removal of the drug, but the recovered rhythm was in phase with the pretreatment rhythm (Figure 9.11). This result demonstrates that the neuronal firing is not an essential part of the clock mechanism itself.[31] Interestingly, regardless of whether the animal is diurnal or nocturnal in terms of behavior, cells in the SCN are electrically active in the day and show circadian rhythms in action potential firing rates in the middle of the day.[35,36] During the night, the cells are electrically inactive.

SCN Input: Retinal Photoreception

The SCN was first described by laboratories that used neural tract tracing techniques to follow anatomical projections along the retina from the eye to the hypothalamus, discovering a novel pathway called the **retinohypothalamic tract** (RHT)[37,38] (Figure 9.12). Importantly, this nonvisual pathway contains *intrinsically photosensitive retinal ganglion cells* (ipRGC) containing the photopigment *melanopsin*, which is a homolog of a photoreceptor in amphibian skin.[39] These ipRGCs are both necessary and sufficient for normal photic entrainment and are distinct from the classical rhodopsin-containing rod and cone cells of the visual pathway. Indeed, if the primary visual pathway is transected at the level of the optic tract beyond the optic chiasm and caudal to the SCN, then the mammal is visually blind, but the circadian system continues to respond to photic cues and the animal remains entrained.[20,40] An early puzzle where mice lacking both rod and cone photoreceptors (rd/rd mice) exhibited entrainment even though they were visually blind[41] was resolved with the identification of ipRGCs and the RHT. In mice, targeted deletion of the melanopsin gene impairs circadian entrainment,[42,43] and the loss of melanopsin, rods, and cones eliminates the ability of mice to synchronize to light.[44,45] Furthermore, the loss of melanopsin cells produces the same result, showing that all light signaling must go through the ipRGCs for circadian entrainment.[46] Together, these findings suggest that rods and cones project to light-sensitive, melanopsin-containing ganglion cells to entrain circadian rhythms.[47]

SCN Output

The SCN uses a variety of outputs to provide temporal patterning and partitioning of key behavioral and physiological processes throughout the body. Two forms of SCN output have been described: (1) neurosecretion, or chemical messengers are released from neurons specifically into the cerebrospinal fluid, and (2) innervation, or connection to various components of the autonomic nervous system.

SCN Neurons Are Neurosecretory Cells

The SCN is adjacent to the third ventricular space and some of its neurons secrete directly into the cerebrospinal fluid, thereby potentially gaining access to the central nervous system. The function of diffusible signals within the brain was first revealed in experiments where isolated SCN tissue was encapsulated in a semipermeable membrane to prevent neural outgrowth while allowing diffusion of permeable hormonal signals.[48] The SCN capsule, when implanted into the third ventricle of an SCN-ablated animal, was able to restore and sustain locomotor activity rhythms by virtue of diffusible signals, without synaptic connections. One study estimated that as many as 35 neuropeptides are rhythmically secreted into the cerebral spinal fluid.[49] However, SCN transplants do not restore melatonin and corticosterone rhythmicity,[26,50] suggesting that SCN efferent projections control these rhythms.

SCN Neurons Project to Other Regions of the Brain and to Peripheral Tissues

The connections of the SCN with other brain structures have been characterized by injecting antero- and retrograde tracers, or dyes that travel from the neural cell body towards axonal termini and vice versa, respectively. These findings show that SCN projections terminate in multiple brain sites.[51] Within the hypothalamus, SCN projections innervate various nuclei that generate release and release-inhibiting hormones that communicate with the adenohypophysis of the pituitary gland, as well as hormones released by the neurohypophysis, controlling the release of diverse pituitary hormones (Figure 9.13). These nuclei include the *paraventricular nucleus* (PVN), *dorso/ventromedial hypothalamus* (DNH/VMH), *arcuate nucleus* (Arc), *supraoptic nuclei* (SON), *medial preoptic area* (MPOA), and the *anteroventral periventricular nucleus* (AVPV).

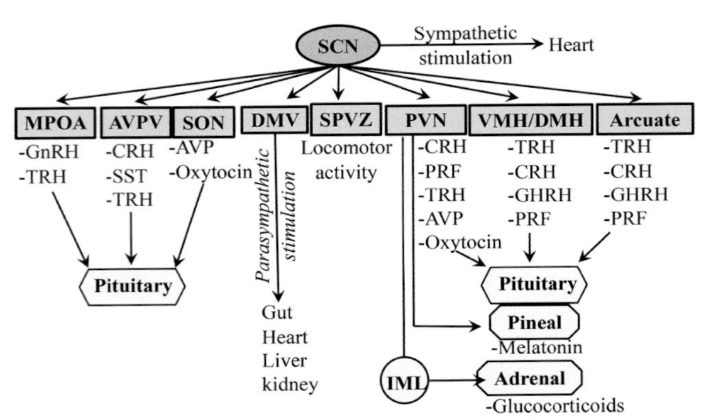

Figure 9.13 Efferent projections of the rodent SCN to its neuroendocrine targets in the hypothalamus and pineal. Below each targeted hypothalamic nuclei (denoted by squares) is a list of hormones whose release may be regulated by direct projections from the SCN. The pineal gland is indirectly innervated by the SCN via projections from the PVN. The adrenal gland is indirectly innervated by the SCN via projections from the autonomic system's intermediolateral column (IML). Abbreviations: SCN (suprachiasmatic nucleus), MPOA (medial preoptic area), AVPV (anteroventral periventricular nucleus), SON (supraoptic nucleus), DMV (dorsomedial hypothalamic nucleus), SPVZ (subparaventricular zone), PVN (paraventricular nucleus), VMH/DMH (ventro-/dorso-medial hypothalamic nucleus).

Outside of the hypothalamus, the SCN indirectly controls several tissues, such as the pineal gland and various digestive organs, through multisynaptic pathways. The dorsal motor nucleus of the vagus (DMV), which is considered to be the main source of the vagal innervation of various organs within the gastrointestinal tract, including the stomach and the pancreas (Figure 9.13), receives a range of projections from other areas within the CNS, including the SCN, paraventricular hypothalamic nucleus (PVN), the lateral hypothalamic area (LHA), the dorsomedial hypothalamic nucleus (DHN), and the posterior hypothalamus.[52] Given that numerous organs and endocrine glands (e.g. testes, adipose tissue, adrenal gland) investigated receive autonomic innervation from the SCN,[53–57] these multisynaptic connections may provide a rapid means of clock resetting in peripheral tissues.[58]

Developments & Directions: Jet Lag: A Desynchrony among Circadian Clocks

The term "jet lag", also known as *circadian desynchrony*, refers to a number of pathologies resulting from rapid changes in the day–night cycle incurred following jet travel across multiple time zones. These pathologies include daytime fatigue, insomnia, impaired cognitive performance, gastrointestinal disturbances, headaches, and irritability.[59] Physiologically, jet lag is thought to result from misalignments among internal circadian clocks with geophysical time, resulting in overall desynchrony and subsequent pathology.[60–62] Rodent models of jet lag have demonstrated that the condition is accompanied not only by a mismatch between external and internal time, but also by a marked internal desynchrony that stems from differential adaptation speeds of diverse cellular clocks dispersed throughout the body.[63–65] Whereas SCN clocks appear to be the first to be reset, most peripheral oscillators adapt at slower rates.

Within tissues, a multilevel internal desynchrony is observed both among cells and also among the clock genes that constitute the circadian molecular clock. Importantly, besides these time-restricted acute symptoms, significant long-term effects may also arise from a chronic misalignment of internal and external time. Studies exposing rodents to repeated experimental jet lag promoted tumor development, increased mortality rates, and impaired immune function.[61] In a particularly striking example of the long-term effects of repetitive jet lag in humans, chronic jet lag has been shown to cause cognitive deficits and even temporal lobe atrophy[66,67] (Figure 9.14).

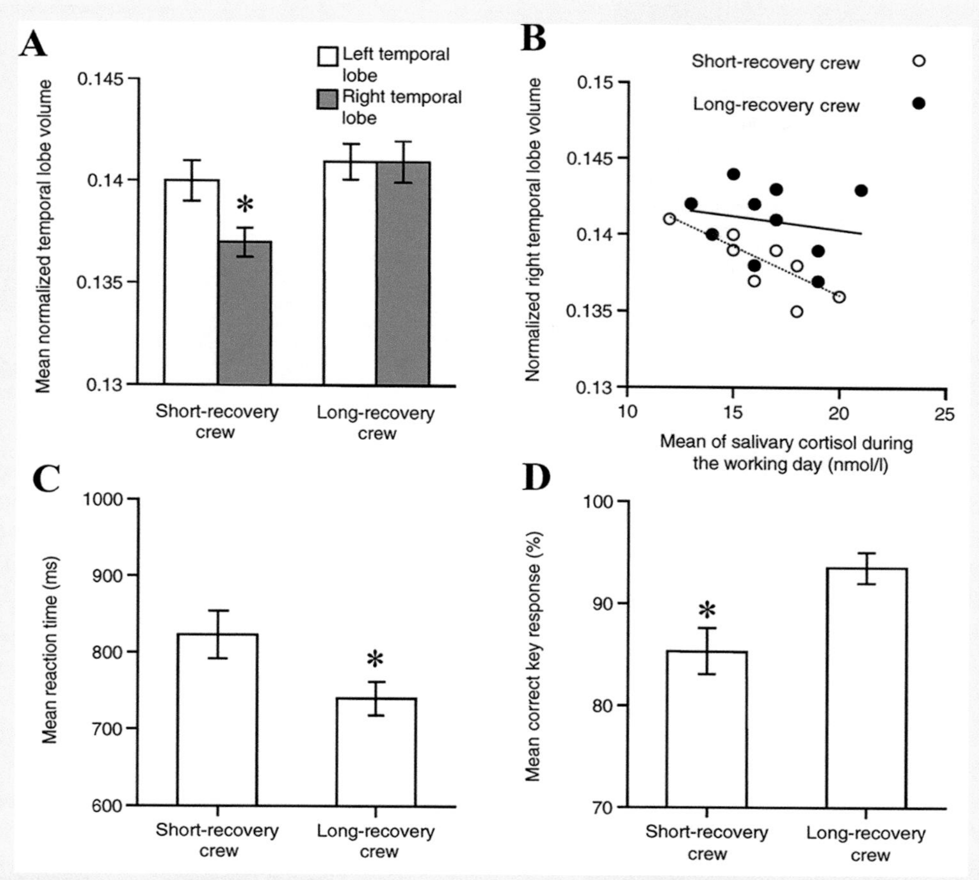

Figure 9.14 Chronic jet lag produces reduced temporal lobe volume (measured from coronal MRI slices) (A), increased salivary cortisol (B), and spatial-cognitive deficits (C, D) in short-recovery flight crews. Two flight attendant groups were compared: "short-recovery" (flight crews with less than 5 days to recover from an international flight over more than seven time zones), and "long-recovery" (flight crews with more than 15 days to recover from such flights). N = 10 flight attendants per group. *Denotes a significant group difference at $p < 0.05$.

Source of Images: Cho, K., 2001. Nat. Neurosci., 4(6), pp. 567-568. Used with permission.

SUMMARY AND SYNTHESIS QUESTIONS

1. If all cells have their own biological clocks, why is a master clock, like the SCN, needed?
2. What effect, if any, would deletion of the following retinal cell types in a mammal have on circadian entrainment? Cones only; rods only; ipRGCs only; rods and cones; rods, cones, and ipRGCs.
3. Why does selective lesion of the RHT fibers emanating from the eyes result in "circadian blindness" but not in loss of vision?
4. What effect, if any, would the addition of opaque contact lenses on both eyes have on melatonin circadian rhythmicity in a: a. frog, b. mammal?
5. This section describes an experiment whereby the placement of an encapsulated SCN graft inside the third ventricle that restored circadian locomotor activity in an SCN-lesioned hamster. Why did the researchers place this capsule specifically inside the third ventricle, instead of in a location more accessible to a blood supply?
6. Propose an explanation as to why, following a desynchronization of the body's biological clocks due to jet lag, the SCN clock is relatively fast to re-entrain compared with peripheral clocks that follow more slowly.

The Pineal Gland and Melatonin

LEARNING OBJECTIVE Describe the functions of melatonin and the evolution of the pineal gland as a photoreceptive and neuroendocrine organ.

KEY CONCEPTS:

- The pineal gland synthesizes the hormone melatonin at night.
- In addition to supporting sleep in humans, melatonin synchronizes peripheral clocks.
- Whereas the pineal gland is an autonomous photoreceptive organ in non-mammalian vertebrates, in mammals its activity is controlled entirely by the SCN.
- Melatonin biosynthesis is regulated by rate-limiting enzymes whose activities are inhibited by light.
- Melatonin relays light information regarding day length to trigger seasonal endocrine changes such as reproduction, migration, hibernation, and molting.

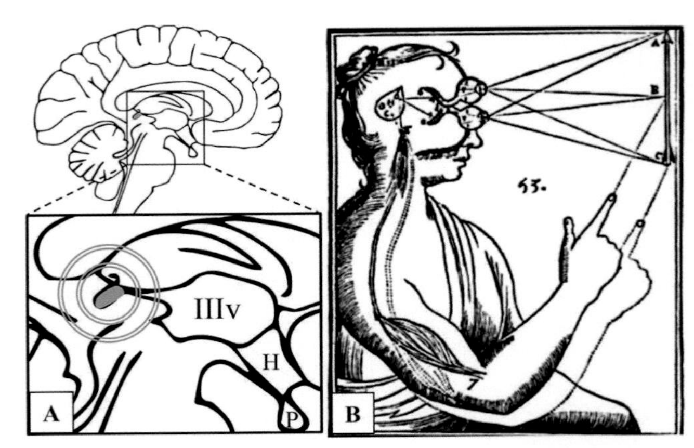

Figure 9.15 Anatomical relationships of the pineal gland. **(A)** The pineal gland is attached to the posterior end of the roof of the third ventricle (IIIv) of the brain, here shown in sagittal section. **(B)** Drawing from the *De Homine* of Descartes (1662), showing the pathway of the light through the ocular globe, retina, and collaterals of the optic nerves that project to the third ventricle to stimulate the pineal gland to release the "animal spirit" to the peripheral muscles. Descartes attached significance to the gland because he believed it to be the only section of the brain to exist as a single part rather than one-half of a pair. H, hypothalamus; P, pituitary.

Source of Images: (B) Lechan, R.M. and Toni, R., 2016. Functional anatomy of the hypothalamus and pituitary. In: Feingold KR, Anawalt B, Boyce A, et al, editors. Endotext. Used with permission.

The Pineal Gland and Melatonin

In humans, the pineal gland rests proximally on the posterior aspect of the brain's diencephalon (Figure 9.15). The pineal's centralized location and status as the only unpaired organ in the human brain led the influential French philosopher, mathematician, and scientist René Descartes (1596–1650) to conceptualize the pineal as the "seat of the soul" and the place in which all thoughts are formed. Descartes also asserted that in indirect response to light, the pineal gland released "animal spirit" into the nerves, which facilitated all sensation and movement (Figure 9.15). Setting aside the limitations in understanding of neural physiology by 17th-century scholars, Descartes' notion that the pineal responds to light information indirectly from the eyes in humans was indeed correct.

Curiously Enough . . . Galen of Pergamon (AD 130–200) provided the first recorded description of the pineal's location in the human brain, naming it "conarium" for its pine cone shape.[68] The organ was subsequently named "glandula pinealis", hence "pineal". Galen correctly identified the pineal as a "gland", though its function remained a mystery.

The shroud of mystery surrounding pineal physiology began to lift in the late 19th century when advances in microscopy and in histological techniques led to the observation of similarities between the pineal and the eyes of lower vertebrates and to speculation that the mammalian pineal was evolutionarily linked to such photosensory organs.[69] Bioassay advances resulted in the discovery of active pineal "extracts" capable of lightening the color of frog skin,[70] leading to the isolation of the pineal hormone *melatonin* in 1958.[71,72] Melatonin lightens frog skin by acting as an antagonist of α-melanocyte-stimulating hormone (α-MSH).[73] Subsequent measurement of levels of circulating melatonin and its precursor, serotonin, in rodents showed large day–night variations in their levels,[74,75] with light exposure reducing melatonin synthesis.[76] Indeed, whether the mammal is a nocturnal rat or a diurnal human, blood melatonin levels show similar circadian profiles, where levels are elevated at night and lowered during the day (Figure 9.16). Therefore, the conserved use of melatonin as

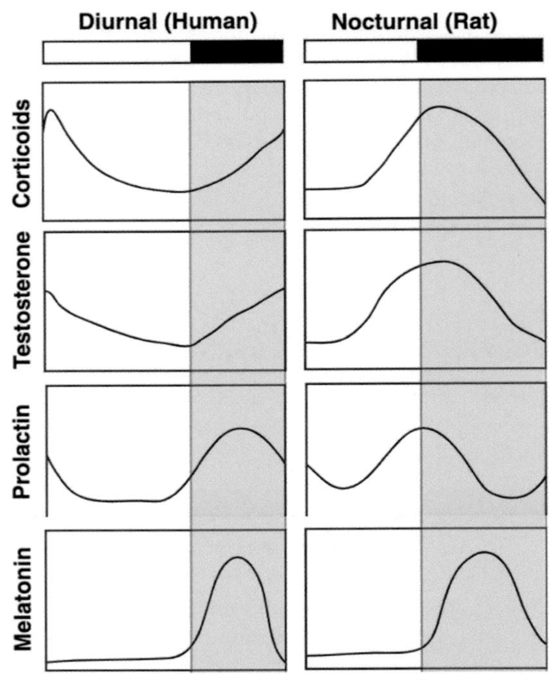

Figure 9.16 Daily patterns of glucocorticoids, testosterone, prolactin, and melatonin in diurnal humans and nocturnal rats. Note that although day/night patterns of the first three hormones are reversed in diurnal vs. nocturnal mammals, melatonin synthesis remains unchanged.

Source of Images: Pfaff, D.W., et al. 2018. Principles of hormone/behavior relations. Academic Press. Used with permission.

a nighttime signal must be used differently in different species, just as nighttime has different implications for activity in a nocturnal animal and a diurnal animal. Today we know that, in addition to inducing and supporting sleep in humans, other reported roles of melatonin include setting the timing of peripheral clocks, modulation of vascular reactivity, and anticonvulsant and antioxidant activity.[77]

Clinical Considerations: Melatonin as a sleep aid and for the treatment of jet lag

The circadian signal produced by the suprachiasmatic nucleus (SCN) promotes wakefulness during the day and consolidation of sleep at night.[78] In humans, sleep is initiated during a rise in the concentration of melatonin, and both the circadian and sleep-promoting effects of melatonin are attributed to two types of melatonin receptors, MT1 and MT2, present in the SCN. Melatonin acutely inhibits SCN neuronal firing, an effect presumably linked to the activation of GABAergic mechanisms in the SCN.[79,80] Suppression of SCN neuronal activity by melatonin represents a likely mechanism that contributes to the regulation of sleep in diurnal species.[81] Importantly, the effects of melatonin administration on sleep are known to depend on the timing of administration (Figure 9.17).

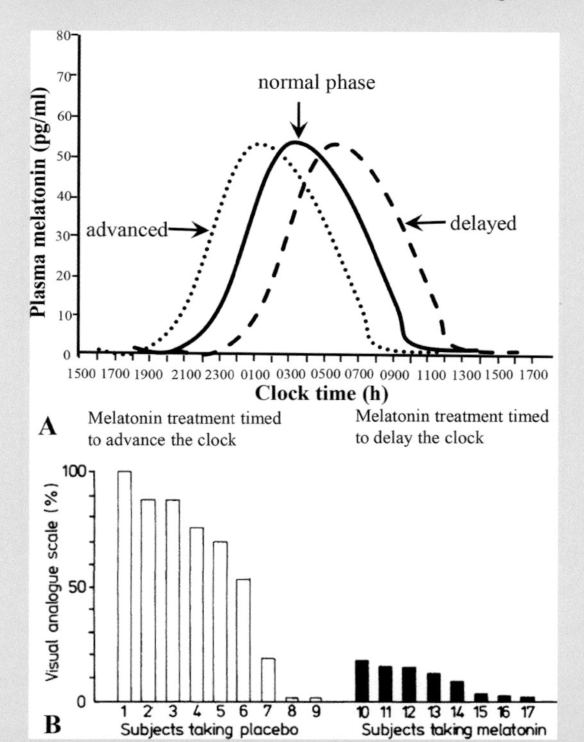

Figure 9.17 Resetting the biological clock with melatonin treatment. (A) A diagram illustrating the timing of oral melatonin treatment, relative to the endogenous melatonin rhythm, to induce phase advances or phase delays. The maximum advance shift obtainable with a single treatment of 3–5 mg is approximately 1–1.5 h. **(B)** Melatonin alleviates symptoms of jet lag in flights from San Francisco to London.

Source of Images: (A) Modified from Arendt, J., 2011. The pineal gland and pineal tumours.In: Feingold KR, Anawalt B, Boyce A, et al, editors. Endotext; (B) Arendt, J., Aldhous, M. and Marks, V., 1986. Brit. Med. J., 292(6529), p. 1170. Used with permission.

Whereas morning administration of melatonin delays the onset of evening sleepiness by delaying the phase of circadian rhythms, evening melatonin treatment promotes a phase shift that advances the circadian rhythms including the time of sleep onset.[82,83]

The promotion of a phase shift by administering melatonin can also be used to alleviate symptoms of jet lag (refer earlier to Developments & Directions: Jet Lag: A Desynchrony among Circadian Clocks for the symptoms of jet lag). The first double-blind, placebo-controlled trial of melatonin in jet lag using travelers taking an eight-hour eastbound trans-meridian flight was conducted by Arendt and colleagues.[84] They found that travelers taking 5 mg/day beginning three days prior to the flight in the early evening and for four days post-flight at the bedtime hour of the new local time zone experienced significantly fewer severe symptoms compared with controls (Figure 9.14). Combining melatonin treatment with light therapy at appropriate times can mitigate the symptoms of jet lag even further.[85]

The Pineal Gland Is a Photoreceptive Organ in Non-Mammalian Vertebrates

In the early 1900s, the young Karl von Frisch and Ernest Scharrer (co-founder with his wife, Berta, of the field of "neuroendocrinology") trained blinded fish to come to the surface of the water for food in response to presentation of a light signal.[86,87] This led to the discovery of photosensitivity in the pineal gland and associated structures in fish and later in amphibians and reptiles,[88] giving rise to one of its popular names, "the third eye" (Figure 9.18). Using a particularly creative experimental design, Groos[89] demonstrated extraretinal photoreception in birds, but only when the feathers on their head above the pineal gland were removed to allow light to cross into the brain. With the key exception of mammals, it was subsequently established that extraretinal or nonvisual photoreception was an invariant feature of all vertebrate classes.[90]

Pinealocytes, the primary cell type of the pineal gland, possess three categories of cells among the vertebrates[91]: (1) photoreceptive cells, which primarily release chemical signals into synaptic spaces juxtaposed to neurons; (2) rudimentary photoreceptive cells that signal via both classical neural synapses and by neuro-endocrine secretion; and (3) non-photoreceptive cells that signal entirely by neuro-endocrine secretion (Figure 9.19). Throughout vertebrate radiation, pinealocytes appear to have gradually lost their photoreceptive characteristics and gained neuroendocrine features. Whereas the pineal gland of fishes and amphibians consist of well-developed photoreceptive cells, those of reptiles and birds possess rudimentary photoreceptor cells, and the pinealocytes of adult mammals have entirely lost their photoreceptive capacity over the course of evolution.

The pineal glands of non-primate vertebrates are located near the surface of the skull, a position suitable for detecting light. This superficial location of the pineal, however, does not aid in photoreception in mammals since their pinealocytes have lost the ability to detect light directly. Instead, the mammalian pineal gland receives light information indirectly along a linear axis starting

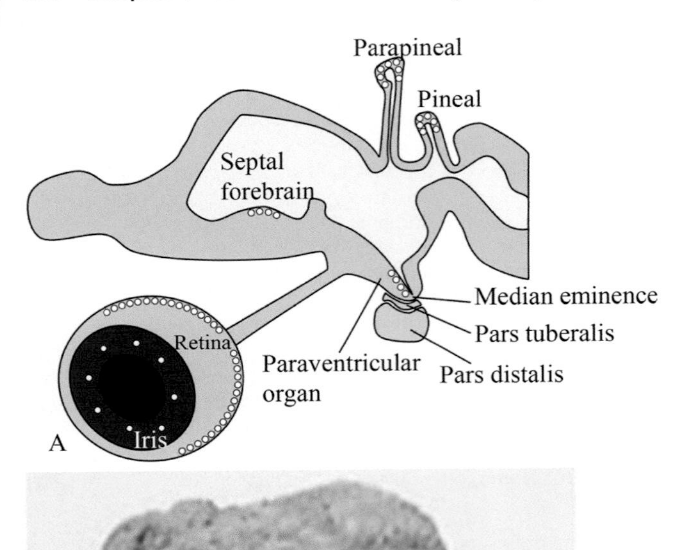

B

Figure 9.18 Nonvisual photoreceptors in the vertebrate brain. (A) The parapineal and similar pineal-associated structures are only found in some fish, amphibians, and reptiles, although the pineal itself is photoreceptive in all non-mammalian vertebrates. In some lizards, such as the tuatara (not shown), photoreceptive structures associated with the pineal have even developed that are analogous to the cornea, lens, and retina. The iris is intrinsically photoreceptive in these groups as well and perhaps in some mammals. The putative locations of nonvisual photoreceptors (shown in white) in the deep brain varies among the non-mammalian vertebrates. The adult mammalian pineal is not photoreceptive although it contains opsin. The only nonvisual photoreceptors in mammals are intrinsically photosensitive ganglion cells in the retina. **(B)** Parietal eye of the zebra-tailed lizard *Callisaurus draconoides* (arrow).

Source of Images: (A) Menaker, M., 2014. Curr. Biol., 24(13), pp. R613-R615; used with permission. (B) Falcón, J., et al. 2020. Front. Neurosci., p. 1183; used with permission.

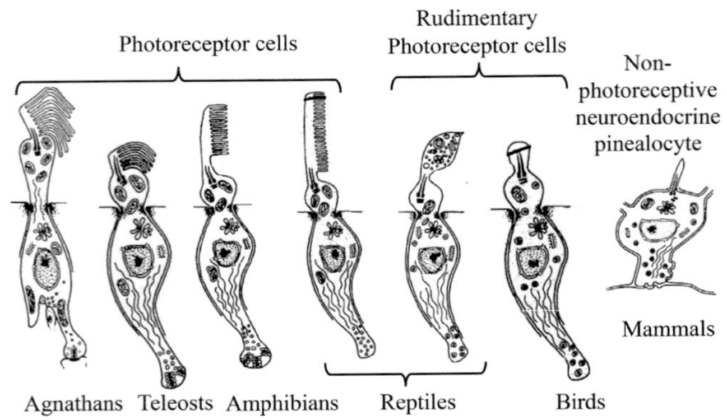

Figure 9.19 Differences in pinealocyte morphology among vertebrate classes. The anamniotes (fish and amphibians) possess photoreceptive pinealocytes. While some reptiles have photoreceptor pinealocytes, most possess rudimentary photoreceptor cells. Mammalian pinealocytes have lost all photoreceptive capacity.

Source of Images: Ekström, P. and Meissl, H., 2003. Philos. Trans. Royal Soc. B: Biol. Sci., 358(1438), pp. 1679-1700. Used with permission.

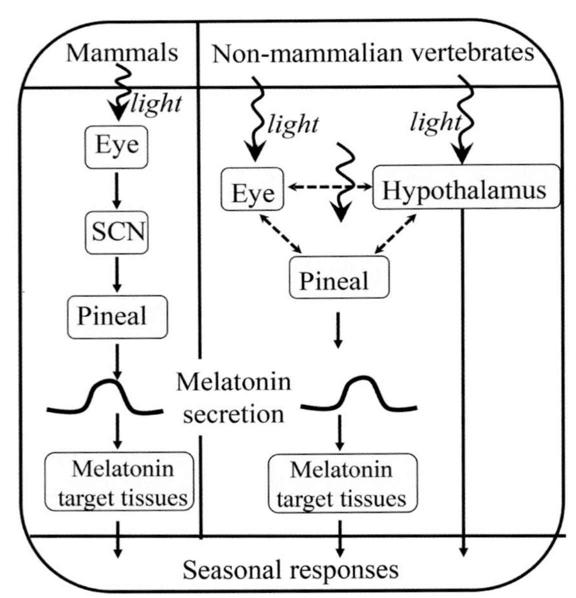

Figure 9.20 The photoperiodic axis in mammals and birds. In mammals, photoperiodic responses depend on a linear flow of information that starts with light perception in the retina, which synchronizes and modulates the waveform of circadian activity in the central hypothalamic pacemaker in the suprachiasmatic nuclei (SCN). The SCN, in turn, governs the nocturnal secretion of melatonin by the pineal gland. This signal, through its effects on neuroendocrine function, controls seasonal photoperiodic responses. In non-mammalian vertebrates, the hypothalamus and pineal gland respond independently to light, and jointly maintain circadian rhythmicity. The information pathways through which these components communicate with each other to coordinate circadian function are less well understood than in mammals (dashed arrows) and might vary among species.

Source of Images: Modified from Hazlerigg, D.G. and Wagner, G.C., 2006. Trends Endocrinol. Metab., 17(3), pp. 83-91.

from light reception in the retina, through a core circadian pacemaker in the SCN that affects pineal melatonin production, and culminates with cells that express the melatonin receptor (Figure 9.20). In birds and reptiles, the photoperiodic axis regulating melatonin release is not linear, since the eye, hypothalamus, and pineal gland each respond independently to light and jointly maintain circadian rhythmicity (Figure 9.20).[92] Not only is the pineal gland of non-mammalian vertebrates directly light-sensitive, but in birds and reptiles (and also likely in amphibians and fish) it also possesses an endogenous clock function[93,94] and, therefore, may not rely on the SCN for circadian information. By contrast, in mammals, pinealocytes neither are light-sensitive nor possess an endogenous clock.

Regulation of Melatonin Biosynthesis in Mammals

Whereas the pineal glands of fish, amphibians, and many reptiles are photoreceptive, have endogenous clock function, and synthesize melatonin, in mammals these attributes are located in separate but interconnected brain structures that form the *photoneuroendocrine system*[95] (Figure 9.21). This consists of the neuroretina, which perceives environmental light information, the retinohypothalamic tract, which transmits light signals to the SCN, the site of the endogenous master circadian oscillator, and the pineal gland. SCN-efferent circadian cues are first transmitted via the paraventricular nuclei (PVN) to the intermediolateral column of the spinal cord (IMC), then to the superior cervical ganglia (SCG). The release of norepinephrine by the SCG activates β-adrenoceptors of pinealocytes, which upregulates two rate-limiting enzymes involved in the synthesis of melatonin from its serotonin precursor: these are N-acetyltransferase (AANAT) and N-acetylserotonin O-methyltransferase (HIOMT) (Figure 9.22). Once released into circulation, melatonin feeds back on the SCN, inhibiting melatonin secretion in a classic negative feedback loop. In contrast to darkness, the SCN responds to light by releasing GABA, which inhibits PVN depolarization, ultimately downregulating AANAT and HIOMT, and reducing melatonin synthesis within 5 minutes of light exposure.[96] Importantly, the rhythmic release of melatonin by the pineal gland in mammals is not driven exclusively by light/dark cycling but is a true endogenous circadian rhythm generated by the SCN. That is, in rats the rhythm is retained during constant dark exposure but is abolished upon exposure to light.[97]

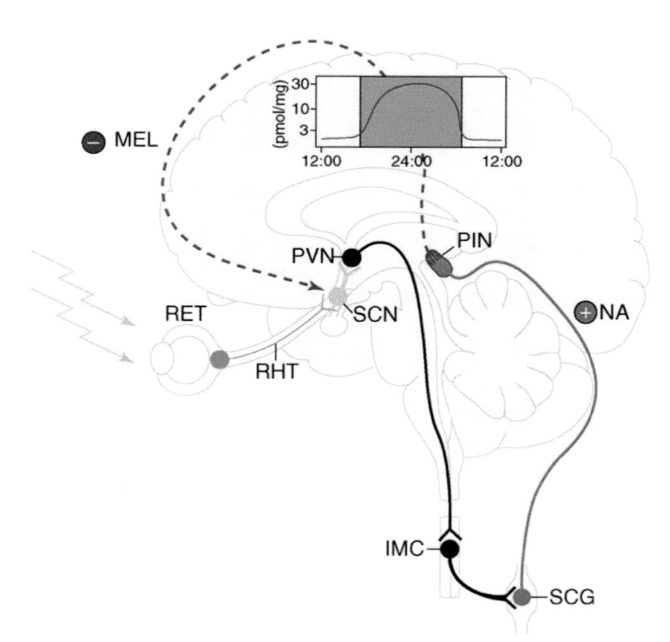

Figure 9.21 The mammalian photoneuroendocrine system. The graph in the inset shows the oscillation of melatonin levels in the human pineal gland with respect to clock time. Abbreviations: IMC, intermediolateral column of the spinal cord; MEL, melatonin; NA, noradrenaline; PIN, pineal; PVN, paraventricular nuclei; RET, neuroretina; RHT, retinohypothalamic tract; SCG, superior cervical ganglia; SCN, suprachiasmatic nucleus.

Source of Images: Maronde, E. and Stehle, J.H., 2007. Trends Endocrinol. Metab. 18(4), pp. 142-149. Used with permission.

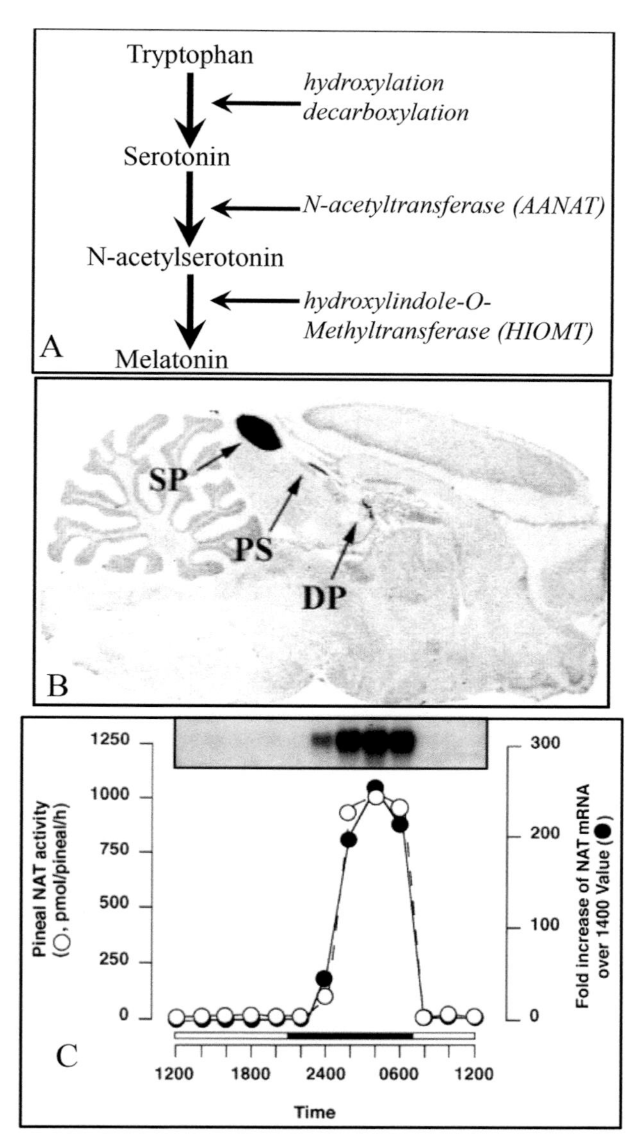

Figure 9.22 Melatonin synthesis is regulated by two rate-limiting enzymes, HIOMT and AANAT, that are expressed in a circadian manner. (A) Pathway showing the biosynthesis of melatonin from tryptophan and serotonin precursors by pinealocytes. **(B)** A parasagittal section of rat brain showing expression of HIOMT mRNA in the three parts of the pineal complex: SP, superficial pineal; PS, pineal stalk; DP, deep pineal. **(C)** AANAT mRNA rhythm parallels the AANAT catalytic activity rhythm in the rat pineal gland. Inset shows the northern blot analysis using rat AANAT cDNA as the probe at the times indicated below.

Source of Images: (B) Simonneaux, V. and Ribelayga, C., 2003. Pharmacol. Rev., 55(2), pp. 325-395; used with permission. (C) Foulkes, N.S., et al. 1997. Biol. Cell, 89(8), pp. 487-494. Used with permission.

Melatonin and Seasonal Breeding

Melatonin's primary biological functions include the synchronization of peripheral clocks and the relay of information regarding the environmental light–dark cycle, especially in response to changes in day length, to trigger relevant endocrine and other physiological changes. Annual changes in day length alter the duration of nocturnal secretion of melatonin (Figure 9.23),[98] and seasonal changes in melatonin exposure play an important role in preparing many animals physiologically and behaviorally for seasonal events such as reproduction, migration, hibernation, and molting. In particular, melatonin strongly affects changes along the hypothalamic-pituitary-gonad (HPG) axis of many seasonally breeding animals, influencing the timing of gonad maturation. Some seasonal species, such as sheep, goats, and elk, are *short day breeders*, mating in the winter when they are exposed to increasing durations of nights and of melatonin. Other species, such as quail, hamsters, and horses, are *long day breeders*, mating in the spring as nights and melatonin exposure decrease in length. The influence of melatonin on HPG axis is addressed further in Chapter 23: The Timing of Puberty and Seasonal Reproduction.

Curiously Enough . . . Melatonin deficiency is associated with an increased incidence of colorectal and breast cancer, and polymorphisms in melatonin receptors are associated with an increased risk of type 2 diabetes mellitus (T2DM).[77]

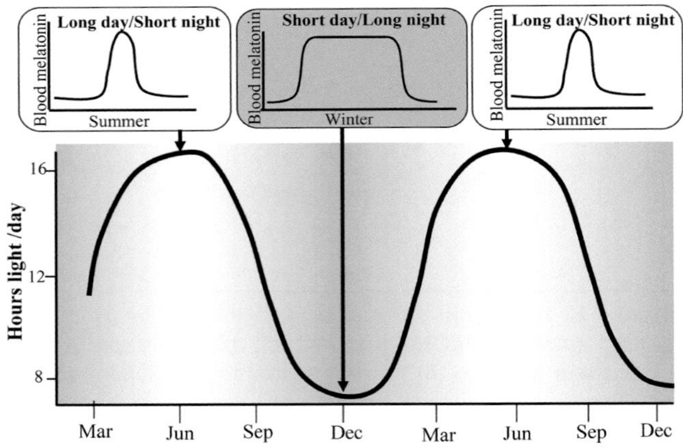

Figure 9.23 Melatonin secretion profiles change with seasonal alterations in day length (hours of light) in a generalized vertebrate. The shorter the day, the longer melatonin is present in the bloodstream, as indicated by the areas under the blood melatonin curves. Seasonal changes in melatonin are associated with initiation of reproduction in some species.

Source of Images: Nelson, R.J., 2005. An introduction to behavioral endocrinology. Sinauer Associates.

Developments & Directions: Melatonin puts breast cancer cells to sleep

Melatonin is an important regulator of not only the central circadian clock, but also of peripheral oscillators in tissues and organs, including mammary glands.[99] Oscillator genes and proteins are involved in diverse cellular processes, including cell cycle regulation, cell proliferation, cell survival, apoptosis, DNA damage/repair, and tumor suppression/promotion.[100,101] It is now clear that the central and peripheral oscillators play important roles in the regulation of intermediary metabolism and various types of cancers.[102–105] Specifically, melatonin appears to have an oncostatic effect in different types of cancers, including estrogen receptor (ER) positive breast cancers.[106–108]

Lopes and colleagues[109] investigated one potential pathway through which melatonin may exert its oncostatic effect, the regulation of the transcription factor OCT4* by estrogen receptor alpha (ERα). To study this, the researchers grew tiny tumors *in vitro*, known as "mammospheres", from breast cancer stem cells. Whereas treatments with ER agonists such as the natural hormone estradiol or the synthetic bisphenol A (BPA, a compound found in many types of plastic food packages) increased the number and size of mammospheres, these were significantly reduced in the presence of melatonin (Figure 9.24). Furthermore, the binding of the ERα to OCT4 was reduced, accompanied by a reduction of both OCT4 and ERα expression. Thus, melatonin treatment was effective against proliferation of breast cancer stem cells *in vitro* and impacts the ER pathway, demonstrating its potential therapeutic use in breast cancer.

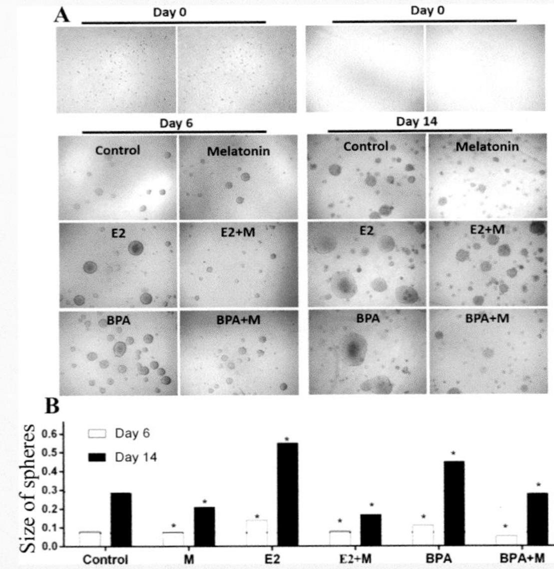

Figure 9.24 MCF-7 breast cancer cells grown as mammospheres, treated with estradiol or the estrogen mimic bisphenol A with or without melatonin. Melatonin (M) treatment significantly reduces mammosphere size when added to cultures treated with estradiol (E2) or the estrogen mimic bisphenol A (BPA). **(A)** Representative photographs of cultured mammospheres. **(B)** Quantitation of mammosphere size (asterisks denote statistical differences).

Source of Images: Lopes, J., et al. 2016. Genes Cancer, 7(5-6), p. 209.

* Octamer binding 4

Clinical Considerations: Light treatment for seasonal affective disorder (SAD)

Seasonal affective disorder (SAD) refers to major depression that regularly occurs during specific seasons of the year, most commonly during the short day lengths associated with winter.[110] First described by Rosenthal and colleagues in 1984,[111] SAD remits when the photoperiods are longer in the spring and summer and, as such, SAD is more common in subjects living in northern latitudes than in equatorial countries where the difference of time duration between seasonally short and long days is not as significant.[112] While the etiology and pathophysiology of SAD are not clearly understood, one leading theory is that in some people a circadian phase shift caused by the later arrival of dawn during the winter may trigger SAD depression via a misalignment of circadian rhythms with respect to sleep/wake cycle.[113,114] In accordance with this phase shift hypothesis, the use of bright light as a phase resetting agent can be used to successfully treat SAD.[115] The standard light therapy involves positioning patients about 12 to 18 inches from a white, fluorescent light source at a standard dosage of 10,000 lux for 30 minutes per day in the early morning. Two mechanisms by which light therapy may work are by (1) entraining the daily melatonin rhythm and (2) diminishing the levels of melatonin in the bloodstream by inhibiting melatonin synthesis (presumably by downregulating two enzymes of the melatonin biosynthetic pathway, AANAT and HIOMT) when natural light isn't sufficient to do so.[116–118]

SUMMARY AND SYNTHESIS QUESTIONS

1. Why might eating food high in tryptophan, such as a turkey dinner, make a person feel sleepy?
2. What effect would electrical stimulation of a mouse's superior cervical ganglia (SCG) during the daytime have on melatonin secretion? What effect would be seen following injection of a beta-adrenergic receptor antagonist prior to the onset of darkness?
3. The blind cavefish *Astyanax mexicanus* has undergone a million or more years of evolution in complete darkness. Furthermore, it undergoes bilateral eye degeneration during embryonic development and completely lacks eyes after hatching. Nonetheless, when raised in a lighted laboratory environment, cavefish larvae swim upward in a response to shading.[119] Propose a mechanism by which cavefish larvae can perceive shaded regions of an aquarium. How could you test this hypothesis?
4. In rats, the circadian rhythmicity of melatonin production is retained during constant dark exposure, though the rhythm begins to free run. By contrast, if rats are exposed to constant light exposure, melatonin rhythmicity is abolished completely. Explain these observations.
5. A Clinical Considerations box in this chapter described how light therapy can be used to treat SAD. Propose how one could combine light therapy with melatonin treatment to treat SAD. Specifically, describe what time of the day these treatments would be administered.

Molecular Cogs of the Circadian Clock

LEARNING OBJECTIVE Outline the molecular basis of mammalian circadian clock function and its mechanisms for modulating endocrine activity, and describe how hormones influence clock function.

KEY CONCEPTS:

- The molecular control of circadian rhythms is conserved in all animals studied, and virtually every circadian gene that has been found in flies or mice appears to have a counterpart in humans.
- The principles of circadian transcriptional circuits share the common network motif of "negative feedback with delay".
- In mammals, the positive elements (transcriptional activators), CLOCK and BMAL1, are generally active in the daytime, promoting the transcription of clock genes and clock-controlled genes.
- The negative elements (transcriptional repressors), CRY and PER, are generally active at night, suppressing the activity of the positive elements.
- Clock proteins bind to DNA response elements in genes coding for hormones and proteins that regulate hormone synthesis.
- Some hormones and their receptors influence the expression of clock genes themselves by binding to regulatory DNA response elements in their genes.

The first breakthroughs in unraveling the molecular basis of circadian rhythms came in a series of experiments with *Drosophila* fruit flies in the late 1960s and early 1970s when Seymour Benzer and one of his graduate students, Ronald Konopka, identified three mutant flies that exhibited abnormal patterns of circadian behavior.[120] One type of mutant fly appeared to have a circadian clock that kept a short day (a period of about 19 hours), the second had a clock with a long period (of about 29 hours), and the third mutant appeared to have no clock at all (its activity rhythms were random). These mutant flies were named "*period*" mutants, and all of their circadian timekeeping abnormalities were traced to different mutations in a single gene named "*per*". In 1984, using the new tools of molecular genetics, the laboratories of Michael Rosbash[121] and Michael Young[122] cloned and sequenced per, the first "clock gene", and in 2017 the *Nobel Prize in Physiology or Medicine* was been jointly awarded to Michael Rosbash, Michael Young, and Jeffrey Hall for their work on circadian rhythms. Since the identification of per, many new clock genes have since been discovered in *Drosophila*, such as timeless (tim), double-time (dbt), cycle (cyc), and others.[123] Importantly, some of the clock genes originally discovered in *Drosophila*, such as per, tim, and cryptochrome (cry), have been shown to have homologs in mammals and other vertebrates.[124] In 1997, Joseph Takahashi cloned the first mammalian circadian gene, which he called **Clock** (Clk), a humorous acronym for "circadian locomotor output cycles kaput".[125] Clk was later found to also be present in *Drosophila*. Indeed, the control of circadian rhythms at the molecular level is now known to be very similar in all animals studied, and virtually every circadian gene that has been found in flies or mice appears to have a counterpart in humans.[126] This section will focus on the mammalian model to discuss the molecular mechanisms of circadian clock function in general and describe specifically how endocrine parameters are controlled by these clocks.

The Mammalian Molecular Clock

Virtually all cells in the body express clock genes,[127] though in mammals their expression in the SCN is essential for integrated temporal rhythms. Although the specific components diverge across the kingdoms of life, the principles of circadian transcriptional circuits share the common network motif of "negative feedback with delay"[128] (Figure 9.25). The molecular mechanism of the circadian clock in mammals is generated by a cell-autonomous transcriptional/translational autoregulatory feedback loop that takes approximately 24 hours to complete.[16] In the mouse, the **positive elements** (transcriptional activators) CLOCK and BMAL1† are active in the daytime, and the **negative elements** (transcriptional repressors) CRY and PER are active at night. The molecular machinery of the clock consists of two components called "core" and "auxiliary" loops.

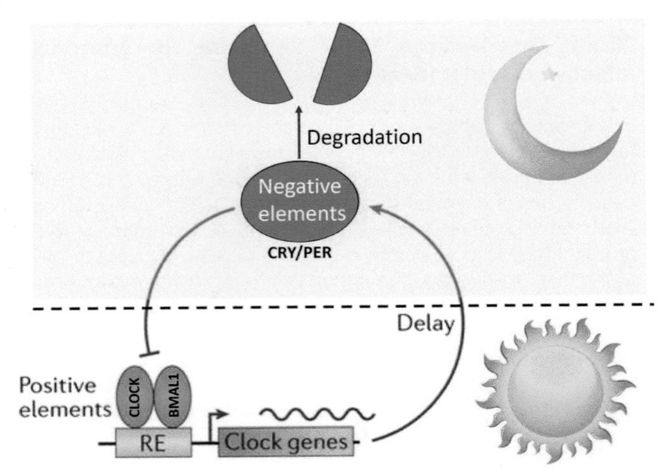

Figure 9.25 A conserved network motif of circadian clocks based on the mouse model involves a transcription–translation negative-feedback loop with delay. During the day, "positive elements" (transcription factors, such as CLOCK and BMAL1) bind to DNA response elements (RE) to promote the transcription of diverse genes, including genes that code for "negative elements" (transcriptional repressors, such as CRY and PER) of the molecular clock. These negative elements are translated in the evening and inhibit positive element-mediated transcription. Gradual degradation of the negative elements toward the end of the night releases the positive elements from inhibition, thus reinitiating the clock cycle by induction of clock gene transcription.

Source of Images: Takahashi, J.S., 2017. Nat. Rev. Genet., 18(3), pp. 164-179. Used with permission.

Core Loop of the Mammalian Circadian Clock

The activator protein CLOCK and its heterodimeric partner BMAL1 are transcription factors that bind to regulatory DNA elements called *E-boxes*, which contain the motif CACGTG located in the promoter or enhancer regions of two categories of genes. The first gene category is called *clock genes*, or genes whose proteins function as part of the machinery of the clock itself. The second category is called *clock-controlled genes*, or genes that perform endocrine or other physiological functions that are regulated by clock genes (Figure 9.26). In mice, CLOCK–BMAL1 activation occurs in the daytime, leading to the transcription of the PER and CRY genes in the afternoon and the accumulation of the PER and CRY proteins in the late afternoon or evening.[129] The PER and CRY proteins interact with each other and translocate into the nucleus at night, where they interact with CLOCK–BMAL1 to repress their own transcription, as well as that of other clock and CCG genes.[129–131] As repression progresses, PER and CRY transcription declines and the PER and CRY protein levels decrease. Importantly, the half-lives of PER and CRY proteins are relatively short. Once negative-feedback repression is relieved by turnover of the repressor complex, a new cycle of transcription by CLOCK–BMAL1 can begin anew the next morning.

† Brain and muscle aryl-hydrocarbon receptor nuclear translocator-like 1

Curiously Enough . . . Mutations of core clock genes, such as those that encode *Per2* and its regulatory kinase CK1δ, have been found to underlie a human sleep-timing disorder called "**familial advanced sleep phase syndrome**", a condition where sleep is advanced by 2–4 hours.[132,133] Furthermore, polymorphisms in the human gene for the PER3 protein may explain why some people have a "morningness" chronotype ("larks") while others have an "eveningness" chronotype ("owls").[134]

Auxiliary Loops

In addition to the PER and CRY target genes, the CLOCK–BMAL1 complex also activates auxiliary (additional) positive and negative regulatory loops involving nuclear receptors, such as REV-ERBα and retinoic acid-related orphan receptor (ROR), which compete for the DNA response element RORE (Figure 9.26). Whereas ROR is a transcriptional activator of *Bmal1*, the REV-ERBα protein suppresses *Bmal1* transcription. Since the *Clock* gene is constitutively expressed, the rhythmic transcription of relevant genes is driven by *REV-ERBa/ROR* rhythms of *Bmal1* expression.[135] The rhythmic

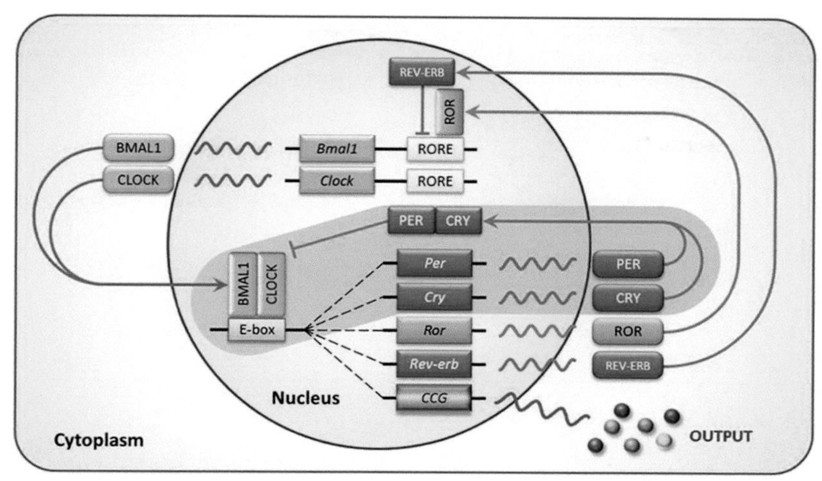

Figure 9.26 Transcriptional feedback loops of the mammalian circadian clock. In the core loop (purple background), BMAL1/CLOCK heterodimer activates transcription of the PER and CRY genes via binding to the E-box elements in their promoter regions. The resulting PER and CRY proteins heterodimerize, translocate to the nucleus, and interact with the BMAL1/CLOCK complex to inhibit their own transcription. In addition, ROR activates and REV-ERB‡ represses RORE-mediated transcription, forming the secondary autoregulatory feedback loops. This clock mechanism also controls rhythmic expression of numerous genes, called clock controlled genes (CCG), to perform biochemical or physiological roles in a circadian manner. The half-lives of PER and CRY proteins are short, and once negative-feedback repression is relieved by turnover of the repressor complex, a new cycle of transcription by CLOCK–BMAL1 can begin anew.

Source of Images: Chen, L. and Yang, G., 2015. Front. Pharmacol., 6, p. 71.

‡ REV-ERB is named because it is on the opposite strand of the ERB oncogene

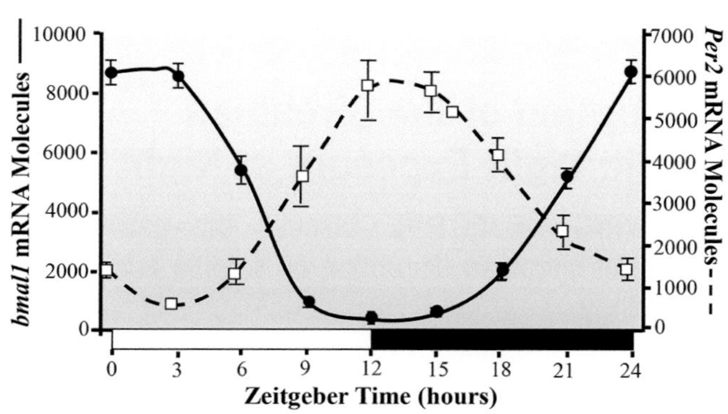

Figure 9.27 Diurnal expressions in *bmal1* and *per2* circadian clock genes are in anti-phase with one another in the intact rat heart. Gene expression of *bmal1* (solid line), and *per2* (dashed line) was measured in rat hearts isolated at 3 h intervals. Values are means ± SE for between 19 and 22 separate observations at each time point. Data are represented as number of mRNA molecules per 50 ng total RNA.

Source of Images: Durgan, D.J., et al. 2005. Am. J. Physiol.-Heart and Circulatory Physiology, 289(4), pp. H1530-H1541.

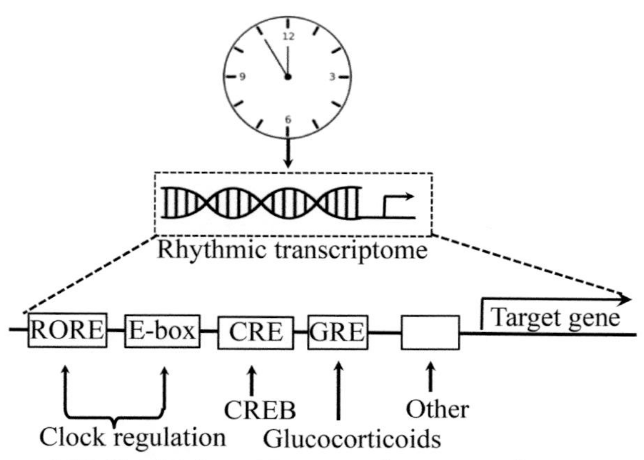

Figure 9.28 Rhythmic gene expression occurs in many central and peripheral cells. Within a cell, a given gene may be clock controlled or regulated by other cues. This is illustrated by a target gene with several possible upstream promoters, including an E-box, a retinoic acid–receptor-related orphan receptor response element (RORE), a cAMP response element (CRE), and a glucocorticoid response element (GRE).

Source of Images: Butler, M.P., et al. 2009. Circadian regulation of endocrine functions. Hormones, Brain and Behavior Online; Elsevier. pp. 473-507.

expression of REV-ERBa protein leads to the repression of *Bmal1* and *Clock* transcription, which induces a rhythm in these genes that is in *anti-phase* (peaks about 12 hours out of phase) with PER expression rhythms (Figure 9.27). Together, these interlocked core and auxiliary feedback loops of activators and repressors can generate cycles of transcription with various phases of expression.[16,136]

Control of Endocrine Function by Clock Proteins

Within a cell, a given gene may be clock-controlled and/or regulated by other variables, such as neural stimuli or hormones, depending on the combination of DNA response elements in the promoters and enhancers of specific target genes[42] (Figure 9.28). This complex interaction

Table 9.2 Examples of Clock Genes and Endocrine Regulators Containing Genes with E-Box Regulatory Elements

Hormones	Receptors	Hormonogenic Enzymes	Clock Genes
Insulin-like growth factor-1	GnRH receptor	Renin	Period (Per)
Vasopressin Growth hormone Prolactin Erythropoietin Luteinizing hormone β Ghrelin	FSH receptor Leptin receptor Estrogen β receptor REV-ERBa orphan nuclear receptor	Prostaglandin endoperoxidase synthase-2 Steroidogenic acute regulatory protein	Cryptochrome (Cry)
Insulin Adiponectin Glucagon			

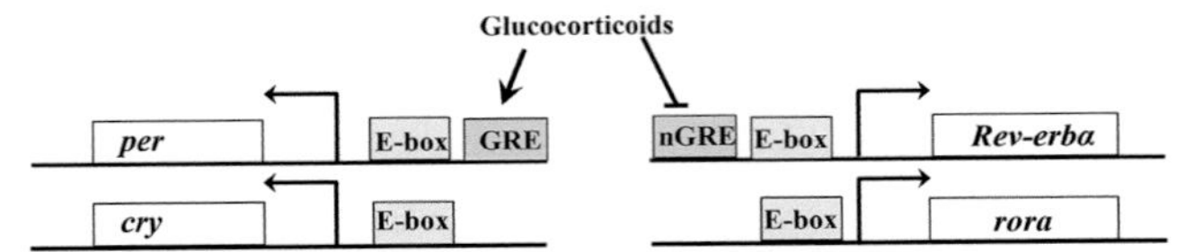

Figure 9.29 Glucocorticoids (GC) may modulate transcription of certain clock genes by binding to glucocorticoid response elements (GRE) in their promoters. A positive GRE element appears to mediate *per1* induction by glucocorticoids, whereas the promoter of *REV-ERBa* has appears to contain a negative GRE (nGRE) that mediates glucocorticoid induced repression. The CLOCK/BMAL1 dimer (not shown) also regulates the transcription of certain clock genes (e.g. per, cry, REV-ERBa, and RORa) by binding to E-boxes in the promoters of these genes.

Source of Images: Dickmeis, T., 2009. J. Endocrinol., 200(1), p. 3.

of response elements with clock genes and other factors likely plays a key role in determining the timing and tissue-specific expression of many genes. Some examples of hormones, receptors, and regulators of hormone synthesis whose genes are known to contain E-box regulatory elements are listed in Table 9.2.

Control of Clock Protein Synthesis by Hormones

Some hormones can modulate the expression of clock-related genes themselves. The most widely studied endocrine regulator of clock gene expression is by glucocorticoids. In mice, clock-related genes such as *Per*, *ROR*, and *REV-ERBa* have been shown to be phase-shifted by glucocorticoids in several peripheral organs.[137–141] This regulation is accomplished by the presence of two categories of **glucocorticoid response elements** (GREs) that associate with the hormone-bound glucocorticoid receptor in regulatory regions of these genes: 1) positive GREs that promote gene expression in the *Per1* and *Per2* genes, and 2) negative GREs that suppress gene expression in *REV-ERBa* and *RORa* genes (Figure 9.29). By changing the expression of these clock genes, glucocorticoids can transiently override and reset the peripheral clock system. Interestingly, while glucocorticoid treatment was found to synchronize peripheral rat cells and cause phase shifts in peripheral oscillators, it does not affect SCN rhythms.[137] This inability of the SCN to be phase shifted by its "downstream oscillator" (the adrenal gland) ensures that the SCN retains its "master clock" status.

SUMMARY AND QUESTIONS

1. Propose an explanation for how three different mutations on the same PER gene in flies produces three distinct phenotypes: short period, long period, random periodicity.
2. How could a dysfunction of ubiquitylation influence the molecular functioning of the circadian clock?
3. Which treatment would be the most effective for rapidly and specifically re-entraining the SCN following jet lag, melatonin or dexamethasone (a synthetic glucocorticoid)? Explain your reasoning.

Disruption of the Circadian Clock: Light Exposure at Night

LEARNING OBJECTIVE Discuss the pathological effects of circadian disruption on wildlife and human physiology.

KEY CONCEPTS:

- Disruptions of natural biological clock rhythms in animals can increase their susceptibility to disease and alter seasonal patterns of migration and reproduction.
- Because the circadian system is actively involved in maintaining nutrient metabolism and energy homeostasis, circadian disruption can lead to the development of metabolic syndrome, obesity, and other metabolic disorders.

Endocrine rhythms are an essential part of physiological timekeeping, and their disruption leads to a variety of social, ecological, behavioral, and health consequences. For example, mice with genetically disrupted circadian rhythms have reduced gonadotropin production, irregular estrous cycles, and high pregnancy failure rates.[142] **Light exposure at night** (LEN), or "light pollution" caused by city lights (Figure 9.30), has been shown to alter reproductive behavior in sea turtles near illuminated beaches,[143]

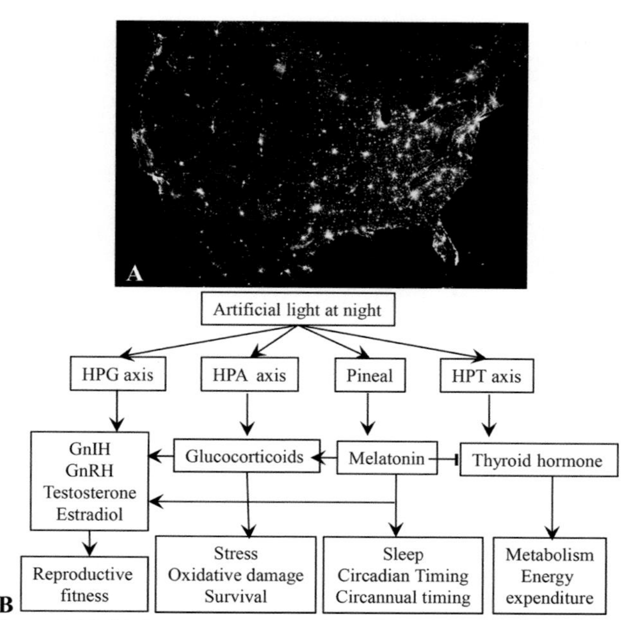

Figure 9.30 Light exposure at night (LEN) disrupts multiple endocrine axes simultaneously. (A) Light pollution's visual impact on earth's night sky in North America. **(B)** Different endocrine axes affected by artificial light at night that lead to changes in potential fitness components. GnIH, gonadotrophin-inhibitory hormone; GnRH, gonadotrophin-releasing hormone; HPA, hypothalamic-pituitary-adrenal; HPG, hypothalamic-pituitary-gonadal; HPT, hypothalamic-pituitary-thyroid.

Source of Images: (A) Courtesy of NASA's Earth Observatory; (B) Modified from Ouyang, J.Q., et al. 2018. J. Exp. Biol., 221(6), p. jeb156893.

promote misorientation/disorientation of migratory birds near urban sky glow,[144–146] cause daily desynchronization of biological rhythms in a nocturnal primate,[147] and may contribute to the extinction of some firefly species.[148] LEN has particularly promiscuous effects, as it simultaneously disturbs multiple interacting endocrine axes, including the hypothalamic-pituitary-gonadal, -adrenal, and -thyroid axes, as well as production of melatonin by the pineal gland (Figure 9.30).[149]

Today, 99% of the human population in the United States and Europe, and 62% of the world's remaining population, are exposed to LEN.[150] Importantly, disturbance of circadian hormone production in humans through night-shift work and LEN has been associated with increased incidence of obesity, diabetes, cardiovascular disease, rheumatoid arthritis, peptic ulcers, cancer, depression, and other disorders.[151–154] Indeed, in 2019 the World Health Organization's International Agency for Research on Cancer (IARC) officially classified night-shift work as Group 2A carcinogen that, based on strong mechanistic evidence in experimental animals, is "probably carcinogenic to humans".[155]

Circadian Disruption and Metabolic Disease

The circadian system is actively involved in maintaining nutrient metabolism and energy homeostasis.[154] For example, metabolic hormones such as insulin, glucagon, leptin, ghrelin, and corticosterone have been shown to fluctuate in a circadian manner.[156–158] Research with genetically

Table 9.3 Circadian Clock Disruptions and Metabolism in Genetically Altered Mice

Gene	Tissue	Main Findings
Bmal1	Whole body	Altered endocrine signaling, reduced islet size
Bmal1	Whole body	Lack insulin, rhythmicity, ectopic fat formation
Bmal1	Pancreas	Impaired glucose tolerance, defective insulin production
Bmal1	Liver	Altered glucose regulation
Bmal1	Adipocytes	Prone to obesity, altered feeding rhythm
Bmal1	CNS	Altered feeding rhythm
Clock	Whole body	Decreased body mass, fat absorption impairments
Clock $^{\Delta19}$	Whole body	Prone to obesity, increased daytime food intake, altered endocrine signaling
Clock $^{\Delta19}$	Whole body	Altered endocrine signaling, reduced pancreatic islet size
Clock $^{\Delta19}$	Liver & skeletal muscle	Altered glucose regulation
Cry1	Whole body	Diabetic-like phenotype
Cry1/2	Whole body	Prone to obesity, hyperinsulemic, altered lipid storage
Per1/2/3^{tm1Drw}	Whole body	Prone to obesity
Per1^{Brd}	Whole body	Decreased body mass, altered feeding, increased glucose efficiency
Per2	Whole body	Prone to obesity, altered feeding rhythm, altered endocrine signaling
Per3^{tm1Drw}	Whole body	Prone to obesity
REV-ERBa and β	Whole body	Altered endocrine signaling

Source: After Fonken, L.K. and Nelson, R.J., 2014. The effects of light at night on circadian clocks and metabolism. *Endocrine Reviews*, 35(4), pp. 648–670.

9

altered mice producing a disrupted clock gene network has shed new light on the influence of the circadian system in maintaining metabolic homeostasis (Table 9.3). Mice possessing mutations in several components of the circadian clock, for example, are susceptible to developing metabolic syndrome and obesity. Turek and colleagues[159] first reported that *Clock* mutant mice are prone to developing diet-induced obesity. Importantly, these dramatic changes in circadian rhythmicity in *Clock* mouse mutants are accompanied by disruptions in both diurnal food intake and increased body mass.

In humans, several epidemiological studies have linked exposure to light at night to metabolic impairments in night-shift workers.[160] For example, compared with healthcare day-shift workers, healthcare night-shift workers display a higher risk for developing metabolic syndrome.[161–166] Male night-shift workers have been shown to have increased levels of blood cholesterol and triglycerides, elevated blood pressure, and higher rates of obesity compared with day-shift workers.[167] Similarly, female night-shift workers display an increased risk for developing high blood pressure, hypertriglyceridemia, and diabetes.[168] Even a brief circadian disruption in a laboratory setting has been shown to produce altered metabolic hormone secretion and also postprandial glucose responses that are similar to patients in a prediabetic state.[169] Furthermore, considering that former shift workers have demonstrated an increased risk for developing obesity compared with non-shift workers,[170] alterations in metabolic signaling caused by shift work may be long lasting. Therefore, exposure to light at night, seemingly an innocuous environmental factor, appears to in fact have clear consequences for the circadian system and human health.

SUMMARY AND SYNTHESIS QUESTIONS

1. How does it make sense that night-shift workers are more likely to suffer from metabolic disorders than typical workers are?

Summary of Chapter Learning Objectives and Key Concepts

LEARNING OBJECTIVE Define "biological clock" and describe how it is essential for the healthy life of an animal.

- Biological clocks are endogenously generated mechanisms present in virtually all cells within an organism that allow the tracking of time.
- An ability to predict periodic geophysical events allows synchronization of time- and energy-consuming behavioral and physiological processes with the environment to optimize energy utilization, reproduction, and survival.
- "Circadian" clocks maintain a periodicity of about (but never exactly) 24 hours.
- In the absence of any external time entrainment signals, biological clock time will begin to drift, or "free run".
- Many hormones oscillate in a circadian pattern and may demonstrate rhythmicity across multiple time scales.

LEARNING OBJECTIVE Describe the mechanism by which the suprachiasmatic nucleus (SCN) receives circadian information, and how it disperses this information to peripheral organs.

- The mammalian circadian clock is organized as a hierarchy of multiple peripheral oscillators, all controlled by a central pacemaker, the SCN.
- Individual SCN neurons are inherently rhythmic.
- The SCN receives nonvisual light information from intrinsically photosensitive retinal ganglion cells (ipRGC).
- The SCN communicates with peripheral oscillators via both hormonal and neural signals.
- A desynchrony among peripheral oscillators and the SCN contributes to the symptoms of "jet lag".

LEARNING OBJECTIVE Describe the functions of melatonin and the evolution of the pineal gland as a photoreceptive and neuroendocrine organ.

- The pineal gland synthesizes the hormone melatonin at night.
- In addition to supporting sleep in humans, melatonin synchronizes peripheral clocks.
- Whereas the pineal gland is an autonomous photoreceptive organ in non-mammalian vertebrates, in mammals its activity is controlled entirely by the SCN.
- Melatonin biosynthesis is regulated by rate-limiting enzymes whose activities are inhibited by light.
- Melatonin relays light information regarding day length to trigger seasonal endocrine changes such as reproduction, migration, hibernation, and molting.

LEARNING OBJECTIVE Outline the molecular basis of mammalian circadian clock function and its mechanisms for modulating endocrine activity, and describe how hormones influence clock function.

- The molecular control of circadian rhythms is conserved in all animals studied, and virtually every circadian gene that has been found in flies or mice appears to have a counterpart in humans.
- The principles of circadian transcriptional circuits share the common network motif of "negative feedback with delay".
- In mammals, the positive elements (transcriptional activators), CLOCK and BMAL1, are generally active in the daytime, promoting the transcription of clock genes and clock-controlled genes.
- The negative elements (transcriptional repressors), CRY and PER, are generally active at night, suppressing the activity of the positive elements.
- Clock proteins bind to DNA response elements in genes coding for hormones and proteins that regulate hormone synthesis.

- Some hormones and their receptors influence the expression of clock genes themselves by binding to regulatory DNA response elements in their genes.

LEARNING OBJECTIVE Discuss the pathological effects of circadian disruption on wildlife and human physiology.

- Disruptions of natural biological clock rhythms in animals can increase their susceptibility to disease and alter seasonal patterns of migration and reproduction.
- Because the circadian system is actively involved in maintaining nutrient metabolism and energy homeostasis, circadian disruption can lead to the development of metabolic syndrome, obesity, and other metabolic disorders.

LITERATURE CITED

1. Thompson HS. *Fear and Loathing in Las Vegas: A Savage Journey to the Heart of the American Dream*. Vintage; 1998.
2. Grau EG, Dickhoff WW, Nishioka RS, Bern HA, Folmar LC. Lunar phasing of the thyroxine surge preparatory to seaward migration of salmonid fish. *Science*. 1981;211(4482):607–609.
3. Yang Y, Duguay D, Bedard N, et al. Regulation of behavioral circadian rhythms and clock protein PER1 by the deubiquitinating enzyme USP2. *Biol Open*. 2012;1(8):789–801.
4. Zantke J, Ishikawa-Fujiwara T, Arboleda E, et al. Circadian and circalunar clock interactions in a marine annelid. *Cell Rep*. 2013;5(1):99–113.
5. Palmer JD. Review of the dual-clock control of tidal rhythms and the hypothesis that the same clock governs both circatidal and circadian rhythms. *Chronobiol Int*. 1995;12(5):299–310.
6. Pengelley E, Asmundson SJ, Barnes B, Aloia RC. Relationship of light intensity and photoperiod to circannual rhythmicity in the hibernating ground squirrel, Citellus lateralis. *Comp Biochem Physiol A Physiolo*. 1976;53(3):273–277.
7. Johnson C. *Chronobiology: Biological Timekeeping*. Sinauer; 2004.
8. Willich SN, Linderer T, Wegscheider K, Leizorovicz A, Alamercery I, Schroder R. Increased morning incidence of myocardial infarction in the ISAM Study: absence with prior beta-adrenergic blockade. ISAM Study Group. *Circulation*. 1989;80(4): 853–858.
9. Kaur G, Phillips CL, Wong K, McLachlan AJ, Saini B. Timing of administration: for commonly-prescribed medicines in Australia. *Pharmaceutics*. 2016;8(2).
10. Roy P, Shahiwala A. Multiparticulate formulation approach to pulsatile drug delivery: current perspectives. *J Control Release*. 2009;134(2):74–80.
11. Alvarez-Lorenzo C, Concheiro A. Intelligent drug delivery systems: polymeric micelles and hydrogels. *Mini Rev Med Chem*. 2008;8(11):1065–1074.
12. Bikram M, West JL. Thermo-responsive systems for controlled drug delivery. *Expert Opin Drug Deliv*. 2008;5(10):1077–1091.
13. Levi F, Okyar A, Dulong S, Innominato PF, Clairambault J. Circadian timing in cancer treatments. *Annu Rev Pharmacol Toxicol*. 2010;50:377–421.
14. Kowalska E, Brown SA. Peripheral clocks: keeping up with the master clock. *Cold Spring Harb Symp Quant Biol*. 2007;72:301–305.
15. Takeda N, Maemura K. Circadian clock and cardiovascular disease. *J Cardiol*. 2011;57(3):249–256.
16. Takahashi JS. Transcriptional architecture of the mammalian circadian clock. *Nat Rev Genet*. 2016;1(8):789–801.
17. Moore RY, Eichler VB. Loss of a circadian adrenal corticosterone rhythm following suprachiasmatic lesions in the rat. *Brain Res*. 1972;42(1):201–206.
18. Stephan FK, Zucker I. Circadian rhythms in drinking behavior and locomotor activity of rats are eliminated by hypothalamic lesions. *Proc Natl Acad Sci U S A*. 1972;69(6):1583–1586.
19. Rusak B. The role of the suprachiasmatic nuclei in the generation of circadian rhythms in the golden hamster, Mesocricetus auratus. *J Comp Physiol*. 1977;118:145–164.
20. Klein DC, Moore RY. Pineal N-acetyltransferase and hydroxyindole-O-methyltransferase: control by the retinohypothalamic tract and the suprachiasmatic nucleus. *Brain Res*. 1979;174(2):245–262.
21. Green DJ, Gillette R. Circadian rhythm of firing rate recorded from single cells in the rat suprachiasmatic brain slice. *Brain Res*. 1982;245(1):198–200.
22. Groos G, Hendriks J. Circadian rhythms in electrical discharge of rat suprachiasmatic neurones recorded in vitro. *Neurosci Lett*. 1982;34(3):283–288.
23. Shibata S, Oomura Y, Kita H, Hattori K. Circadian rhythmic changes of neuronal activity in the suprachiasmatic nucleus of the rat hypothalamic slice. *Brain Res*. 1982;247(1):154–158.
24. Drucker-Colin R, Aguilar-Roblero R, Garcia-Hernandez F, Fernandez-Cancino F, Bermudez Rattoni F. Fetal suprachiasmatic nucleus transplants: diurnal rhythm recovery of lesioned rats. *Brain Res*. 1984;311(2):353–357.
25. Sawaki Y, Nihonmatsu I, Kawamura H. Transplantation of the neonatal suprachiasmatic nuclei into rats with complete bilateral suprachiasmatic lesions. *Neurosci Res*. 1984;1(1):67–72.
26. Lehman MN, Silver R, Gladstone WR, Kahn RM, Gibson M, Bittman EL. Circadian rhythmicity restored by neural transplant. Immunocytochemical characterization of the graft and its integration with the host brain. *J Neurosci*. 1987;7(6):1626–1638.
27. DeCoursey PJ, Buggy J. Circadian rhythmicity after neural transplant to hamster third ventricle: specificity of suprachiasmatic nuclei. *Brain Res*. 1989;500 (1–2):263–275.
28. Ralph MR, Foster RG, Davis FC, Menaker M. Transplanted suprachiasmatic nucleus determines circadian period. *Science*. 1990;247(4945):975–978.
29. Schwartz WJ, Gross RA, Morton MT. The suprachiasmatic nuclei contain a tetrodotoxin-resistant circadian pacemaker. *Proc Natl Acad Sci U S A*. 1987;84(6):1694–1698.
30. Newman GC, Hospod FE, Patlak CS, Moore RY. Analysis of in vitro glucose utilization in a circadian pacemaker model. *J Neurosci*. 1992;12(6):2015–2021.
31. Welsh DK, Logothetis DE, Meister M, Reppert SM. Individual neurons dissociated from rat suprachiasmatic nucleus express independently phased circadian firing rhythms. *Neuron*. 1995;14(4):697–706.
32. Herzog ED. Neurons and networks in daily rhythms. *Nat Rev Neurosci*. 2007;8(10):790–802.
33. Liu AC, Lewis WG, Kay SA. Mammalian circadian signaling networks and therapeutic targets. *Nat Chem Biol*. 2007;3(10):630–639.
34. Long MA, Jutras MJ, Connors BW, Burwell RD. Electrical synapses coordinate activity in the suprachiasmatic nucleus. *Nat Neurosci*. 2005;8(1):61–66.
35. Kuhlman SJ, McMahon DG. Encoding the ins and outs of circadian pacemaking. *J Biol Rhythms*. 2006;21(6):470–481.
36. Colwell CS. Linking neural activity and molecular oscillations in the SCN. *Nat Rev Neurosci*. 2011;12(10):553–569.
37. Hendrickson AE, Wagoner N, Cowan WM. An autoradiographic and electron microscopic study of retino-hypothalamic connections. *Zeitschrift fur Zellforschung und mikroskopische Anat*. 1972;135(1):1–26.
38. Moore RY, Lenn NJ. A retinohypothalamic projection in the rat. *J Comp Neurol*. 1972;146(1):1–14.
39. Berson DM. Strange vision: ganglion cells as circadian photoreceptors. *Trends Neurosci*. 2003;26(6):314–320.
40. Johnson RF, Moore RY, Morin LP. Loss of entrainment and anatomical

plasticity after lesions of the hamster retinohypothalamic tract. *Brain Res.* 1988;460(2):297–313.

41. Foster RG, Provencio I, Hudson D, Fiske S, De Grip W, Menaker M. Circadian photoreception in the retinally degenerate mouse (rd/rd). *J Comp Physiol A.* 1991;169(1):39–50.
42. Panda S, Antoch MP, Miller BH, et al. Coordinated transcription of key pathways in the mouse by the circadian clock. *Cell.* 2002;109(3):307–320.
43. Ruby NF, Brennan TJ, Xie X, et al. Role of melanopsin in circadian responses to light. *Science.* 2002;298(5601):2211–2213.
44. Hattar S, Liao HW, Takao M, Berson DM, Yau KW. Melanopsin-containing retinal ganglion cells: architecture, projections, and intrinsic photosensitivity. *Science.* 2002;295(5557):1065–1070.
45. Panda S, Provencio I, Tu DC, et al. Melanopsin is required for non-image-forming photic responses in blind mice. *Science.* 2003;301(5632):525–527.
46. Guler AD, Ecker JL, Lall GS, et al. Melanopsin cells are the principal conduits for rod-cone input to non-image-forming vision. *Nature.* 2008;453(7191):102–105.
47. Lok C. Vision science: Seeing without seeing. *Nature.* 2011;469(7330):284–285.
48. Silver R, LeSauter J, Tresco PA, Lehman MN. A diffusible coupling signal from the transplanted suprachiasmatic nucleus controlling circadian locomotor rhythms. *Nature.* 1996;382(6594):810–813.
49. Kramer A, Yang FC, Kraves S, Weitz CJ. A screen for secreted factors of the suprachiasmatic nucleus. *Methods Enzymol.* 2005;393:645–663.
50. Meyer-Bernstein EL, Morin LP. Electrical stimulation of the median or dorsal raphe nuclei reduces light-induced FOS protein in the suprachiasmatic nucleus and causes circadian activity rhythm phase shifts. *Neuroscience.* 1999;92(1):267–279.
51. Watts AG, Swanson LW, Sanchez-Watts G. Efferent projections of the suprachiasmatic nucleus: I. Studies using anterograde transport of Phaseolus vulgaris leucoagglutinin in the rat. *J Comp Neurol.* 1987;258(2):204–229.
52. Mussa BM, Verberne AJ. The dorsal motor nucleus of the vagus and regulation of pancreatic secretory function. *Exp Physiol.* 2013;98(1):25–37.
53. Bamshad M, Aoki VT, Adkison MG, Warren WS, Bartness TJ. Central nervous system origins of the sympathetic nervous system outflow to white adipose tissue. *Am J Physiol.* 1998;275(1 Pt 2):R291–299.
54. Bartness TJ, Song CK, Demas GE. SCN efferents to peripheral tissues: implications for biological rhythms. *J Biol Rhythms.* 2001;16(3):196–204.
55. Buijs RM, Wortel J, Van Heerikhuize JJ, et al. Anatomical and functional demonstration of a multisynaptic suprachiasmatic nucleus adrenal (cortex) pathway. *Eur J Neurosci.* 1999;11(5):1535–1544.
56. Kalsbeek A, Fliers E, Franke AN, Wortel J, Buijs RM. Functional connections between the suprachiasmatic nucleus and the thyroid gland as revealed by lesioning and viral tracing techniques in the rat. *Endocrinology.* 2000;141(10):3832–3841.
57. Olcese J, Domagalski R, Bednorz A, et al. Expression and regulation of mPer1 in immortalized GnRH neurons. *Neuroreport.* 2003;14(4):613–618.
58. Kriegsfeld LJ, Silver R. The regulation of neuroendocrine function: Timing is everything. *Horm Behav.* 2006;49(5):557–574.
59. American_Academy_of_Sleep_Medicine. Diagnostic and coding manual. In: *International Classification of Sleep Disorders* (2nd ed.). American Academy of Sleep Medicine; 2005.
60. Sack RL. The pathophysiology of jet lag. *Travel Med Infect Dis.* 2009;7(2): 102–110.
61. Barclay J, Husse J, Oster H. Adrenal Glucocorticoids as a Target for Jet Lag Therapies. *Expert Rev Endocrinol Metab.* 2011;6(5):673–679.
62. Choy M, Salbu RL. Jet lag: current and potential therapies. *P T.* 2011;36(4): 221–231.
63. Davidson AJ, Castanon-Cervantes O, Leise TL, Molyneux PC, Harrington ME. Visualizing jet lag in the mouse suprachiasmatic nucleus and peripheral circadian timing system. *Eur J Neurosci.* 2009;29(1):171–180.
64. Kiessling S, Eichele G, Oster H. Adrenal glucocorticoids have a key role in circadian resynchronization in a mouse model of jet lag. *J Clin Invest.* 2010;120(7):2600–2609.
65. Yamazaki S, Numano R, Abe M, et al. Resetting central and peripheral circadian oscillators in transgenic rats. *Science.* 2000;288(5466):682–685.
66. Cho K. Chronic "jet lag" produces temporal lobe atrophy and spatial cognitive deficits. *Nat Neurosci.* 2001;4(6):567–568.
67. Cho K, Ennaceur A, Cole JC, Suh CK. Chronic jet lag produces cognitive deficits. *J Neurosci.* 2000;20(6):RC66.
68. Zrenner C. Theories of pineal function from classical antiquity to 1900: a history. In: Reiter RJ, ed. *Pineal Research Reviews III.* Alan R. Liss; 1985:1–40.
69. Macchi MM, Bruce JN. Human pineal physiology and functional significance of melatonin. *Front Neuroendocrinol.* 2004;25(3–4):177–195.
70. McCord CP, Allen FP. Evidences associating pineal gland functions with alterations in pigmentation. *J Exp Zool.* 1917;23:207–224.
71. Lerner AB, Case JD, Heinzelmann RV. Structure of melatonin. *J Am Chem Soc.* 1959;81:6084–6085.
72. Lerner AB, Case JD, Takahashi Y, Lee TH, Mori N. Isolation of melatonin, pineal factor that lightens melanocytes. *J Am Chem Soc.* 1958;80:2587.
73. Lerner AB, Case JD, Takahashi Y. Isolation of melatonin and 5-methoxyindole-3-acetic acid from bovine pineal glands. *J Biol Chem.* 1960;235:1992–1997.
74. Quay WB. Circadian rhythm in rat pineal serotonin and its modifications by estrous cycle and photoperiod. *Gen Comp Endocrinol.* 1963;3:473–479.
75. Quay WB. Circadian and estrous rhythms in pineal melatonin and 5-hydroxy indole-3-acetic acid. *Proc Soc Exp Biol Med.* 1964;115:710–713.
76. Fiske V. Discovery of the relation between light and pineal function. In: Altschule M, ed. *Frontiers of Pineal Physiology.* MIT Press; 1975.
77. Gamble KL, Berry R, Frank SJ, Young ME. Circadian clock control of endocrine factors. *Nat Rev Endocrinol.* 2014;10(8):466–475.
78. Srinivasan V, Pandi-Perumal SR, Trahkt I, et al. Melatonin and melatonergic drugs on sleep: possible mechanisms of action. *Int J Neurosci.* 2009;119(6):821–846.
79. Liu C, Weaver DR, Jin X, et al. Molecular dissection of two distinct actions of melatonin on the suprachiasmatic circadian clock. *Neuron.* 1997;19(1):91–102.
80. van den Top M, Buijs RM, Ruijter JM, Delagrange P, Spanswick D, Hermes ML. Melatonin generates an outward potassium current in rat suprachiasmatic nucleus neurones in vitro independent of their circadian rhythm. *Neuroscience.* 2001;107(1):99–108.
81. von Gall C, Stehle JH, Weaver DR. Mammalian melatonin receptors: molecular biology and signal transduction. *Cell Tissue Res.* 2002;309(1):151–162.
82. Lewy AJ. Melatonin as a marker and phase-resetter of circadian rhythms in humans. *Adv Exp Med Biol.* 1999;460:425–434.
83. Rajaratnam SM, Dijk DJ, Middleton B, Stone BM, Arendt J. Melatonin phase-shifts human circadian rhythms with no evidence of changes in the duration of endogenous melatonin secretion or the 24-hour production of reproductive hormones. *J Clin Endocrinol Metab.* 2003;88(9):4303–4309.
84. Arendt J, Aldhous M, Marks V. Alleviation of jet lag by melatonin: preliminary results of controlled double blind trial. *Br Med J (Clin Res Ed).* 1986;292(6529):1170.
85. Parry BL. Jet lag: minimizing its effects with critically timed bright light and melatonin administration. *J Mol Microbiol Biotechnol.* 2002;4(5):463–466.
86. von Frisch K. Beitrage zur Physiologie der Pigmentzellen in der Fischhaut. *Pfluger's Archges Physiol.* 1911;138: 319–387.
87. Scharrer E. Die Lichtempfindlichkeit blinder Elritzen. I Untersuchungen über das Zwischenhirn der Fische. *Z Vergl Physiol.* 1928;7:1–38.
88. Dodt E. The parietal eye (pineal and parietal organs of lower vertebrates). In: Autrum H, Jung R, Lowenstein WR, Mckay DM, Tueber HL, eds. *Hand-*

book of Sensory Physiology, Vol VII/3B. Springer-Verlag; 1973:113–140.
89. Groos G. The comparative physiology of extraocular photoreception. *Experientia.* 1982;38(9):989–991.
90. Gerkema MP, Davies WI, Foster RG, Menaker M, Hut RA. The nocturnal bottleneck and the evolution of activity patterns in mammals. *Proc Biol Sci.* 2013;280(1765):20130508.
91. Ekström P, Meissl H. Evolution of photosensory pineal organs in new light: the fate of neuroendocrine photoreceptors. *Philos Trans Royal Soc Lond B Biol Sci.* 2003;358(1438):1679–1700.
92. Hazlerigg DG, Wagner GC. Seasonal photoperiodism in vertebrates: from coincidence to amplitude. *Trends Endocrinol Metab.* 2006;17(3):83–91.
93. Takahashi JS, Hamm H, Menaker M. Circadian rhythms of melatonin release from individual superfused chicken pineal glands in vitro. *Proc Natl Acad Sci U S A.* 1980;77(4):2319–2322.
94. Menaker M, Wisner S. Temperature-compensated circadian clock in the pineal of Anolis. *Proc Natl Acad Sci U S A.* 1983;80(19):6119–6121.
95. Maronde E, Stehle JH. The mammalian pineal gland: known facts, unknown facets. *Trends Endocrinol Metab.* 2007;18(4):142–149.
96. Korf HW, Von Gall C, Stehle J. The circadian system and melatonin: lessons from rats and mice. *Chronobiol Int.* 2003;20(4):697–710.
97. Snyder SH, Zweig M, Axelrod J, Fischer JE. Control of the circadian rhythm in serotonin content of the rat pineal gland. *Proc Natl Acad Sci U S A.* 1965;53:301–305.
98. Wood S, Loudon A. Clocks for all seasons: unwinding the roles and mechanisms of circadian and interval timers in the hypothalamus and pituitary. *J Endocrinol.* 2014;222(2):R39–59.
99. Stehle JH, von Gall C, Korf HW. Melatonin: a clock-output, a clock-input. *J Neuroendocrinol.* 2003;15(4):383–389.
100. Kohsaka A, Bass J. A sense of time: how molecular clocks organize metabolism. *Trends Endocrinol Metab.* 2007; 18(1):4–11.
101. Truong T, Liquet B, Menegaux F, et al. Breast cancer risk, nightwork, and circadian clock gene polymorphisms. *Endocr Relat Cancer.* 2014;21(4):629–638.
102. Blask DE, Brainard GC, Dauchy RT, et al. Melatonin-depleted blood from premenopausal women exposed to light at night stimulates growth of human breast cancer xenografts in nude rats. *Cancer Res.* 2005;65(23):11174–11184.
103. Slominski RM, Reiter RJ, Schlabritz-Loutsevitch N, Ostrom RS, Slominski AT. Melatonin membrane receptors in peripheral tissues: distribution and functions. *Mol Cell Endocrinol.* 2012;351(2):152–166.
104. Kelleher FC, Rao A, Maguire A. Circadian molecular clocks and cancer. *Cancer Lett.* 2014;342(1):9–18.
105. Hastings MH, Reddy AB, Maywood ES. A clockwork web: circadian timing in brain and periphery, in health and disease. *Nat Rev Neurosci.* 2003;4(8):649–661.
106. La Rosa P, Pellegrini M, Totta P, Acconcia F, Marino M. Xenoestrogens alter estrogen receptor (ER) alpha intracellular levels. *PLoS One.* 2014;9(2):e88961.
107. Vijayalaxmi, Thomas CR, Jr., Reiter RJ, Herman TS. Melatonin: from basic research to cancer treatment clinics. *J Clin Oncol.* 2002;20(10):2575–2601.
108. Cos S, Alvarez-Garcia V, Gonzalez A, Alonso-Gonzalez C, Martinez-Campa C. Melatonin modulation of crosstalk among malignant epithelial, endothelial and adipose cells in breast cancer (Review). *Oncol Lett.* 2014;8(2):487–492.
109. Lopes J, Arnosti D, Trosko JE, Tai MH, Zuccari D. Melatonin decreases estrogen receptor binding to estrogen response elements sites on the OCT4 gene in human breast cancer stem cells. *Genes Cancer.* 2016;7(5–6):209–217.
110. American_Psychiatric_Association. *Diagnostic and Statistical Manual of Mental Disorders* (5th ed., DSM-v). American Psychiatric Association; 2013.
111. Rosenthal NE, Sack DA, Gillin JC, et al. Seasonal affective disorder. A description of the syndrome and preliminary findings with light therapy. *Arch Gen Psychiatry.* 1984;41(1):72–80.
112. Magnusson A, Partonen T. The diagnosis, symptomatology, and epidemiology of seasonal affective disorder. *CNS Spectr.* 2005;10(8):625–634; quiz 621–614.
113. Lewy AJ, Lefler BJ, Emens JS, Bauer VK. The circadian basis of winter depression. *Proc Natl Acad Sci U S A.* 2006;103(19):7414–7419.
114. Lewy AJ, Sack RL, Singer CM, White DM. The phase shift hypothesis for bright light's therapeutic mechanism of action: theoretical considerations and experimental evidence. *Psychopharmacol Bull.* 1987;23(3):349–353.
115. Golden RN, Gaynes BN, Ekstrom RD, et al. The efficacy of light therapy in the treatment of mood disorders: a review and meta-analysis of the evidence. *Am J Psychiatry.* 2005;162(4):656–662.
116. Lewy AJ, Sack RL, Singer CM, White DM, Hoban TM. Winter depression and the phase-shift hypothesis for bright light's therapeutic effects: history, theory, and experimental evidence. *J Biol Rhythms.* 1988;3(2):121–134.
117. Lewy AJ, Ahmed S, Sack RL. Phase shifting the human circadian clock using melatonin. *Behav Brain Res.* 1996; 73(1–2):131–134.
118. Wurtman RJ, Wurtman JJ. Carbohydrates and depression. *Sci Am.* 1989;260(1): 68–75.
119. Yoshizawa M, Jeffery WR. Shadow response in the blind cavefish Astyanax reveals conservation of a functional pineal eye. *J Exp Biol.* 2008;211(Pt 3):292–299.
120. Konopka RJ, Benzer S. Clock mutants of Drosophila melanogaster. *Proc Natl Acad Sci U S A.* 1971;68(9):2112–2116.
121. Reddy P, Zehring WA, Wheeler DA, et al. Molecular analysis of the period locus in Drosophila melanogaster and identification of a transcript involved in biological rhythms. *Cell.* 1984;38(3):701–710.
122. Bargiello TA, Jackson FR, Young MW. Restoration of circadian behavioural rhythms by gene transfer in Drosophila. *Nature.* 1984;312(5996):752–754.
123. Peschel N, Helfrich-Forster C. Setting the clock--by nature: circadian rhythm in the fruitfly Drosophila melanogaster. *FEBS Lett.* 2011;585(10):1435–1442.
124. Lakin-Thomas PL. Circadian rhythms: new functions for old clock genes. *Trends Gen.* 2000;16(3):135–142.
125. King DP, Zhao Y, Sangoram AM, et al. Positional cloning of the mouse circadian clock gene. *Cell.* 1997;89(4):641–653.
126. Young MW, Kay SA. Time zones: a comparative genetics of circadian clocks. *Nat Rev Genet.* 2001;2(9):702–715.
127. Lowrey PL, Takahashi JS. Mammalian circadian biology: elucidating genome-wide levels of temporal organization. *Annu Rev Genomics Hum Genet.* 2004;5:407–441.
128. Alon U. Network motifs: theory and experimental approaches. *Nat Rev Genet.* 2007;8(6):450–461.
129. Lee C, Etchegaray JP, Cagampang FR, Loudon AS, Reppert SM. Post-translational mechanisms regulate the mammalian circadian clock. *Cell.* 2001;107(7):855–867.
130. Lowrey PL, Takahashi JS. Genetics of circadian rhythms in Mammalian model organisms. *Adv Genet.* 2011;74:175–230.
131. Gallego M, Virshup DM. Post-translational modifications regulate the ticking of the circadian clock. *Nat Rev Mol Cell Biol.* 2007;8(2):139–148.
132. Toh KL, Jones CR, He Y, et al. An hPer2 phosphorylation site mutation in familial advanced sleep phase syndrome. *Science.* 2001;291(5506):1040–1043.
133. Xu Y, Padiath QS, Shapiro RE, et al. Functional consequences of a CKIdelta mutation causing familial advanced sleep phase syndrome. *Nature.* 2005;434(7033):640–644.
134. Hida A, Kitamura S, Katayose Y, et al. Screening of clock gene polymorphisms demonstrates association of a PER3 polymorphism with morningness-eveningness preference and circadian rhythm sleep disorder. *Sci Rep.* 2014;4:6309.
135. Buhr ED, Takahashi JS. Molecular components of the Mammalian circadian clock. *Handb Exp Pharmacol.* 2013(217):3–27.
136. Novak B, Tyson JJ. Design principles of biochemical oscillators. *Nat Rev Mol Cell Biol.* 2008;9(12):981–991.
137. Balsalobre A, Brown SA, Marcacci L, et al. Resetting of circadian time in peripheral tissues by glucocorticoid

signaling. *Science*. 2000;289(5488): 2344–2347.
138. Yamamoto T, Nakahata Y, Tanaka M, et al. Acute physical stress elevates mouse period1 mRNA expression in mouse peripheral tissues via a glucocorticoid-responsive element. *J Biol Chem*. 2005;280(51):42036–42043.
139. So AY, Bernal TU, Pillsbury ML, Yamamoto KR, Feldman BJ. Glucocorticoid regulation of the circadian clock modulates glucose homeostasis. *Proc Natl Acad Sci U S A*. 2009;106(41): 17582–17587.
140. Surjit M, Ganti KP, Mukherji A, et al. Widespread negative response elements mediate direct repression by agonist-liganded glucocorticoid receptor. *Cell*. 2011;145(2):224–241.
141. Torra IS, Tsibulsky V, Delaunay F, et al. Circadian and Glucocorticoid Regulation of Rev-erbalpha Expression in Liver. *Endocrinology*. 2000;141(10): 3799–3806.
142. Miller BH, Olson SL, Turek FW, Levine JE, Horton TH, Takahashi JS. Circadian clock mutation disrupts estrous cyclicity and maintenance of pregnancy. *Curr Biol*. 2004;14(15):1367–1373.
143. Salmon M. Protecting sea turtles from artificial night lighting at Florida's oceanic beaches. In: Rich C, Longcore T, eds. *Ecological Consequences of Artificial Night Lighting*. Island Press; 2006:141–168.
144. Gauthreaux SA, Belser CG. Effects of artificial night lighting on migrating birds. In: Rich C, Longcore T, eds. *Ecological Consequences of Artificial Night Lighting*. Island Press; 2006:67–93.
145. Montevecchi WA. Influences of artificial light on marine birds. In: Rich C, Longcore T, eds. *Ecological Consequences of Artificial Night Lighting*. Island Press; 2006:94–113.
146. Becker DJ, Singh D, Pan Q, et al. Artificial light at night amplifies seasonal relapse of haemosporidian parasites in a widespread songbird. *Proc Royal Soc B Biol Sci*. 2020;287(1935):20201831.
147. Le Tallec T, Perret M, Thery M. Light pollution modifies the expression of daily rhythms and behavior patterns in a nocturnal primate. *PLoS One*. 2013;8(11):e79250.
148. Lewis SM, Wong CH, Owens A, et al. A global perspective on firefly extinction threats. *BioScience*. 2020;70(2):157–167.
149. Ouyang JQ, Davies S, Dominoni D. Hormonally mediated effects of artificial light at night on behavior and fitness: linking endocrine mechanisms with function. *J Exp Biol*. 2018;221(Pt 6).
150. Navara KJ, Nelson RJ. The dark side of light at night: physiological, epidemiological, and ecological consequences. *J Pineal Res*. 2007;43(3):215–224.
151. Marciano DP, Chang MR, Corzo CA, et al. The therapeutic potential of nuclear receptor modulators for treatment of metabolic disorders: PPARgamma, RORs, and Rev-erbs. *Cell Metab*. 2014;19(2):193–208.
152. Altman BJ. Cancer clocks out for lunch: disruption of circadian rhythm and metabolic oscillation in cancer. *Front Cell Dev Biol*. 2016;4:62.
153. Golombek DA, Casiraghi LP, Agostino PV, et al. The times they're a-changing: effects of circadian desynchronization on physiology and disease. *J Physiol Paris*. 2013;107(4):310–322.
154. Fonken LK, Nelson RJ. The effects of light at night on circadian clocks and metabolism. *Endocr Rev*. 2014;35(4):648–670.
155. Ward EM, Germolec D, Kogevinas M, et al. Carcinogenicity of night shift work. *Lancet Oncol*. 2019;20(8):1058–1059.
156. Ruiter M, La Fleur SE, van Heijningen C, van der Vliet J, Kalsbeek A, Buijs RM. The daily rhythm in plasma glucagon concentrations in the rat is modulated by the biological clock and by feeding behavior. *Diabetes*. 2003;52(7): 1709–1715.
157. Kalsbeek A, Fliers E, Romijn JA, et al. The suprachiasmatic nucleus generates the diurnal changes in plasma leptin levels. *Endocrinology*. 2001;142(6):2677–2685.
158. Sinha MK, Ohannesian JP, Heiman ML, et al. Nocturnal rise of leptin in lean, obese, and non-insulin-dependent diabetes mellitus subjects. *J Clin Invest*. 1996;97(5):1344–1347.
159. Turek FW, Joshu C, Kohsaka A, et al. Obesity and metabolic syndrome in circadian Clock mutant mice. *Science*. 2005;308(5724):1043–1045.
160. Wang XS, Armstrong ME, Cairns BJ, Key TJ, Travis RC. Shift work and chronic disease: the epidemiological evidence. *Occup Med (Lond)*. 2011;61(2):78–89.
161. Pietroiusti A, Neri A, Somma G, et al. Incidence of metabolic syndrome among night-shift healthcare workers. *Occup Environ Med*. 2010;67(1):54–57.
162. Sookoian S, Gemma C, Fernandez Gianotti T, et al. Effects of rotating shift work on biomarkers of metabolic syndrome and inflammation. *J Intern Med*. 2007;261(3):285–292.
163. Esquirol Y, Bongard V, Mabile L, Jonnier B, Soulat JM, Perret B. Shift work and metabolic syndrome: respective impacts of job strain, physical activity, and dietary rhythms. *Chronobiol Int*. 2009;26(3): 544–559.
164. De Bacquer D, Van Risseghem M, Clays E, Kittel F, De Backer G, Braeckman L. Rotating shift work and the metabolic syndrome: a prospective study. *Int J Epidemiol*. 2009;38(3):848–854.
165. Lin YC, Hsiao TJ, Chen PC. Persistent rotating shift-work exposure accelerates development of metabolic syndrome among middle-aged female employees: a five-year follow-up. *Chronobiol Int*. 2009;26(4):740–755.
166. Lin YC, Hsiao TJ, Chen PC. Shift work aggravates metabolic syndrome development among early-middle-aged males with elevated ALT. *World J Gastroenterol*. 2009;15(45):5654–5661.
167. Ha M, Park J. Shiftwork and metabolic risk factors of cardiovascular disease. *J Occup Health*. 2005;47(2):89–95.
168. Karlsson B, Knutsson A, Lindahl B. Is there an association between shift work and having a metabolic syndrome? Results from a population based study of 27,485 people. *Occup Environ Med*. 2001;58(11):747–752.
169. Scheer FA, Hilton MF, Mantzoros CS, Shea SA. Adverse metabolic and cardiovascular consequences of circadian misalignment. *Proc Natl Acad Sci U S A*. 2009;106(11):4453–4458.
170. Puttonen S, Viitasalo K, Harma M. Effect of shiftwork on systemic markers of inflammation. *Chronobiol Int*. 2011;28(6): 528–535.

Unit Overview
Unit IV

Developmental Endocrinology

Dr. Johann Jakob Guggenbühl (1842) at his "training school" for cretins, called the Abdenberg, in Switzerland. Most of the patients depicted in this image who have diminutive stature and rounded features are not children, but adults. The three tall women to the right are Guggenbühl's medical assistants.

DOI: 10.1201/9781003359241-13

OPENING QUOTATION:

"I asked the first person I met what the name of the village was, and when he did not reply I asked a second, and then a third; but a dismal silence, or a few inarticulate noises were the only response I received. The stupid amazement with which they looked at me, their enormous goiters, their fat, parted lips, their heavy, drooping eyelids, their hanging jaws, their doltish expressions, were quite terrifying. It was as if an evil spirit had transformed every inhabitant into a dumb animal, leaving only the human form to show that they had once been men. I left with an impression of fear and sadness which will never be erased from my memory."

—Horace Bénédict de Saussure, Swiss physicist, geologist, and early Alpine explorer describing his encounter with the inhabitants of a remote village in the Italian Alps who, unknown at the time, were suffering from cretinism caused by iodine deficiency and hypothyroidism (from *Voyages dans les Alpes, 1786*)

The image of Dr. Johann Jakob Guggenbühl at his "training school" for cretins and the opening quotation by Horace Bénédict de Saussure together forcefully describe the profound developmental influence of a single hormone, thyroid hormone, on diverse aspects of development. Thyroid hormone is critical for the normal growth and the development of multiple organs, including the long bones and the brain. When born in its absence, individuals may develop severe cognitive impairment, have greatly impeded linear growth, and display many other developmental hindrances. *Unit IV* addresses developmental endocrinology, a field whose specialists study the influences of hormones on periods of dramatic growth and developmental change that take place throughout the life of an organism. Examples of human life stages studied by developmental endocrinologists include fetal development, sexual differentiation, adolescent growth and puberty, pregnancy, and senescence. Dramatic hormone-mediated events among many animals also include metamorphosis, molting, migration, seasonal breeding, and environmental acclimatization. *Chapter 10* discusses the profound influences of thyroid hormones on the development and growth of diverse tissues and organs, vertebrate metamorphosis, and the pathophysiology of the major disorders of thyroid function.

Chapter 11 outlines the pioneering roles that studies on insects have played in developing our modern understanding of endocrinology and also describes the modes of action of the major hormones that mediate insect molting and metamorphosis, as well as key similarities and differences with vertebrate endocrine counterparts. *Chapter 12* describes the endocrine regulation of bone growth and remodeling, the maintenance of calcium and phosphate homeostasis, and associated pathologies caused by endocrine dysfunction.

***Chapter 10*: Thyroid Hormones: Development and Growth**
***Chapter 11*: Insect Molting and Metamorphosis**
***Chapter 12*: Calcium/Phosphate Homeostasis, Skeletal Remodeling, and Growth**

CHAPTER 10

Thyroid Hormones
Development and Growth

CHAPTER LEARNING OBJECTIVES:

- Explain the connections between environmental iodine distribution and its correlation with cretinism, goiter, and other types of hypothyroidism in humans worldwide.
- Describe the developmental and functional morphology of the mammalian thyroid gland and the different hormones that it produces.
- Trace the evolutionary origins of the vertebrate thyroid gland from protochordate ancestors.
- Explain the mechanisms of thyroid hormone biosynthesis and its regulation.
- Describe the roles of blood- and intracellular binding proteins, as well as peripheral intracellular deiodinases in modulating peripheral thyroid hormone activity.
- Describe the molecular mechanisms of action of thyroid hormone and its receptors in modulating gene expression.
- Describe the developmental and growth effects of thyroid hormones on diverse target tissues.
- Explain how vertebrate metamorphosis is a useful model for studying the mechanisms of thyroid hormone action.
- Discuss the etiology and pathophysiology of the major disorders of thyroid function.
- Defend the statement, and place it into the context of endocrine-disrupting compounds: "The single hormone system with the most influence on brain and cognitive development is, arguably, that of thyroid hormone".

OPENING QUOTATIONS:

"Well, I am satisfied, I have seen the principal features of Swiss scenery—Mont Blanc and the goiter—now for home!"

—Mark Twain (1880) from *A Tramp Abroad*

"It is possible to define the thyroid gland in a fairly simple way as the vertebrate tissue that can accumulate iodide in great excess and can combine it chemically in the organic compound thyroxine. Defined in this way, the thyroid gland appears refreshingly simple. . . . Unfortunately, much of the apparent simplicity of the thyroid gland disappears upon closer study, as we shall see, and some of the knottiest problems in comparative endocrinology involve this gland."

—Professor Aubrey Gorbman, a founder of the field of comparative endocrinology, from his classic text, *Comparative Endocrinology*[1]

Iodine and the Discovery of Thyroid Hormones

LEARNING OBJECTIVE Explain the connections between environmental iodine distribution and its correlation with cretinism, goiter, and other types of hypothyroidism in humans worldwide.

KEY CONCEPTS:

- Iodine is essential for thyroid hormone synthesis.
- Low dietary iodine is the most common cause of cretinism, goiter, and other types of hypothyroidism in humans worldwide.
- Over hundreds of millions of years of vertebrate evolution, the structure of thyroid hormone has remained entirely unchanged.

DOI: 10.1201/9781003359241-14

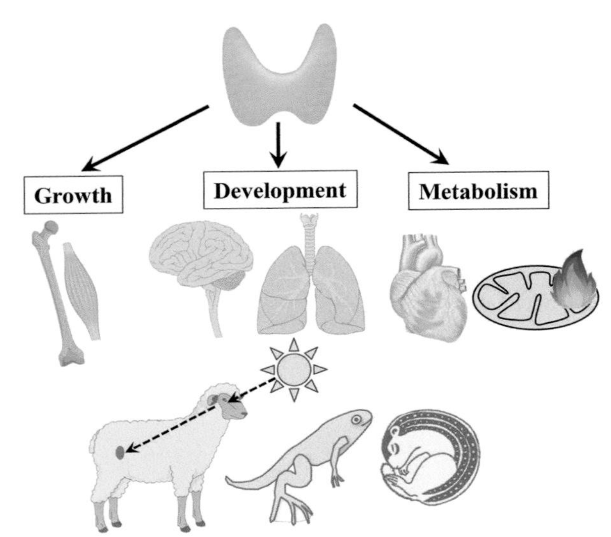

Figure 10.1 Some of the diverse actions of thyroid hormones in vertebrates include effects on growth, development, and metabolism. Thyroid hormone (TH) potentiates the effects of growth hormone, and in its absence animals will display stunted growth. TH facilitates many aspects of fetal and neonatal development, including maturation of the central nervous system, lungs, heart, gut, and bone. Post-embryonic developmental effects of TH include the promotion of reproduction in seasonally breeding animals, as well as metamorphosis in amphibians and fish, and hibernation in small mammals. In adult birds and mammals, TH strongly influences metabolic rate by affecting heart rate, energy mobilization, and thermogenesis.

Figure 10.2 Modern cretinism in the Democratic Republic of Congo, a region with severe iodine deficiency. Four inhabitants aged 18 to 20 years: a normal male and three cretinous females with severe hypothyroidism with dwarfism, retarded sexual development, puffy features, dry skin, and severe mental retardation (courtesy of Professor E. Delange, Brussels, Belgium).

Source of Images: Eastman CJ, Zimmermann MB. The Iodine Deficiency Disorders. 2018 Feb 6. In: Feingold KR, Anawalt B, Boyce A, et al, editors. Endotext. Used with permission.

Thyroid hormones are endocrine products of the thyroid gland and display a remarkable variety of physiological actions affecting virtually every tissue in the vertebrate body. Thyroid hormones have such a profound influence on diverse aspects of growth, development, and metabolism (Figure 10.1) that the functions of this organ defy any simple description. Indeed, thyroid hormones are so integral to vertebrate function that physiology and endocrinology textbooks typically devote far more pages to thyroid hormone than any other single hormone, and this textbook is no exception. This chapter focuses specifically on the influence of thyroid hormones on development and growth, as well as pathologies associated with thyroid dysfunction and disruption. The profound effects of thyroid hormone on metabolism and heat production will be addressed in Chapter 18: Energy Homeostasis.

Of all the organs influenced by the thyroid gland during mammalian development, perhaps none is as profoundly affected as the brain. Indeed, congenital hypothyroidism is the predominant cause of preventable intellectual dysfunction worldwide. So profound are the impacts of hypothyroidism on fetal brain development that since 1972 mandatory screening of newborns for hypothyroidism has been implemented in developed countries and is today considered one of the major achievements of preventive medicine.[2] **Cretinism** is the most severe manifestation of a spectrum of hypothyroid disorders that is, in part, characterized by a severely stunted physical stature and mental development caused by a deficiency of thyroid hormone during fetal and early postnatal development (Figure 10.2). Cretinism and **goiter**, an enlarged thyroid gland that may accompany hypothyroidism and other thyroid disorders, were widespread and well documented in the literature and artwork of 1st- through 18th-century Europe and elsewhere and particularly prevalent in remote mountainous valleys (see this unit's opening quotation and art). It would be a century before medical science discovered that the element iodine is an essential component of thyroid hormone, and that most instances of cretinism and goiter are, even today, caused by iodine deficiency.

The Iodine Connection

Iodine, the heaviest essential element, is highly soluble in water as iodate (IO_3) and iodide (I^-). Oceans are a rich iodine sink, and the Chinese were using burnt seaweed and sponge to treat goiter as early as 1600 BC.[3] However, the soils and drinking water of many regions of the world, particularly inland areas far from oceans, have a low iodine content, and inhabitants of these regions who consume only locally produced foods are at particular risk for iodine deficiency and accompanying thyroid disorders. Although it had been known for over a thousand years that goiter can be successfully treated with seaweed and other marine products now known to be rich in iodine, the hypothesis that iodine deficiency specifically induced goiter was not put forth until the 19th century.[4,5]

Curiously Enough . . . Inhabitants of mountainous regions are particularly prone to developing hypothyroidism because such areas were heavily glaciated during the Pleistocene. Water runoff from the melting glaciers leeched iodine out of the alpine soils, and the soils no longer provide adequate levels of iodine necessary for thyroid hormone synthesis.

The Discovery of Thyroid Hormones

Major steps forward in elucidating the function and composition of the thyroid gland were taken in 1895 when German physician and physiologist Adolf Magnus-Levy demonstrated that metabolic rate increased in healthy people who ate thyroid gland extract,[6] and the following year Eugen Baumann[7] discovered the presence of iodine in the thyroid gland, but not in other tissues. Friedrich Gudernatsch[8] made the striking observation that tadpoles fed horse thyroid gland were induced to precociously metamorphose into frogs, demonstrating that the thyroid gland contains a powerful developmental morphogen. Next, on Christmas morning at the Mayo Clinic in 1914, American biochemist Edward C. Kendall crystallized the first thyroid hormone from thyroid extracts and named it *thyroxine* (T_4) (Figure 10.3).[9] Charles R. Harington synthesized thyroxine in 1926,[10] and it was not until the mid-20th century that Gross and Pitt-Rivers[11] synthesized the hormone 3,5,3'-L triiodothyronine (T_3) (Figure 10.3) and discovered that this was in fact the most biologically active form of thyroid hormone in mammals. In France, Roche and colleagues[12,13] also synthesized T_3 and bolstered the hypothesis that T_3 could be derived from T_4. Since those early, heady days of rapid discovery, a multitude of thyroid hormone derivatives and metabolites have been identified (Figure 10.3). Remarkably, over hundreds of millions of years of vertebrate evolution, the structure of thyroid hormone has remained entirely unchanged, a curiously unique feat among hormones!

Curiously Enough . . . Kendall's purification of thyroxine required 3 tons of porcine thyroid glands to obtain just 33 grams of thyroxine.

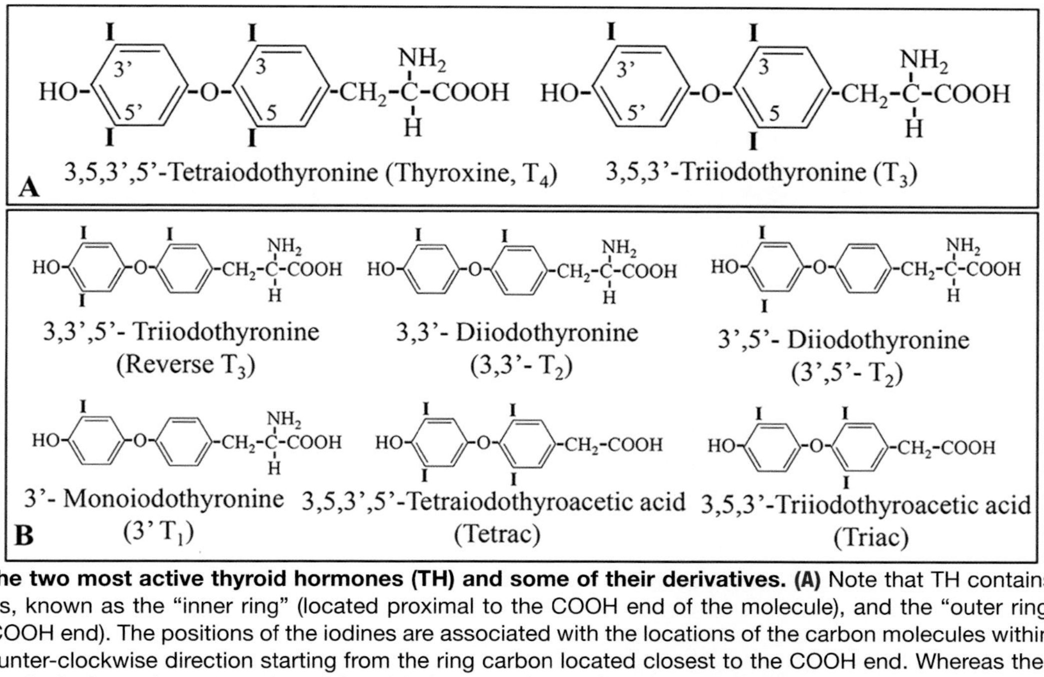

A

3,5,3',5'-Tetraiodothyronine (Thyroxine, T_4)

3,5,3'-Triiodothyronine (T_3)

B

3,3',5'- Triiodothyronine (Reverse T_3)

3,3'- Diiodothyronine (3,3'- T_2)

3',5'- Diiodothyronine (3',5'- T_2)

3'- Monoiodothyronine (3' T_1)

3,5,3',5'-Tetraiodothyroacetic acid (Tetrac)

3,5,3'-Triiodothyroacetic acid (Triac)

Figure 10.3 The two most active thyroid hormones (TH) and some of their derivatives. (A) Note that TH contains two distinct ring structures, known as the "inner ring" (located proximal to the COOH end of the molecule), and the "outer ring" (positioned distal to the COOH end). The positions of the iodines are associated with the locations of the carbon molecules within a ring, numbered in a counter-clockwise direction starting from the ring carbon located closest to the COOH end. Whereas the two possible iodine positions in the inner ring are numbered 3 and 5, the outer ring positions are designated 3' and 5'. 3,5,3'-triiodothyronine (T_3) is considered the most biologically active form of TH in vertebrates, with thyroxine (T_4) having approximately ten-fold less activity. **(B)** With some exceptions discussed in the text, the other metabolic derivatives of TH are generally considered to be physiologically inactive.

SUMMARY AND SYNTHESIS QUESTIONS

1. It has been known for over a thousand years that goiter can be successfully treated with seaweed and other marine products. What is the biological basis of this observation?

The Morphology and Development of the Mammalian Thyroid Gland

LEARNING OBJECTIVE Describe the developmental and functional morphology of the mammalian thyroid gland and the different hormones that it produces.

KEY CONCEPTS:

- The mammalian thyroid gland synthesizes two categories of hormones: (1) thyroid hormones that are produced by thyroid follicles and modulate development, growth, and metabolism, and (2) calcitonin, which is made by parafollicular "C cells" and is involved in calcium metabolism.
- Parathyroid glands that secrete another molecule involved in calcium metabolism, "parathyroid hormone", form distinct islands of tissue embedded within the mammalian thyroid gland, but these are not considered thyroid tissue.

The *thyroid gland* was so named in 1656 by physician Thomas Wharton because in humans the shape of the neighboring thyroid cartilage (the Adam's apple) resembles an ancient Greek peltast light infantry shield (*thyreos*) with a notch at the top for the soldier's chin (Figure 10.4). The gland itself wraps around the anterior surface of the trachea and contacts the larynx. The two lobes of the gland are connected anteriorly by the *isthmus*. Embedded in the posterior surface of the thyroid are two pairs of *parathyroid* glands whose secretion (*parathyroid hormone*) plays a crucial role in calcium homeostasis.

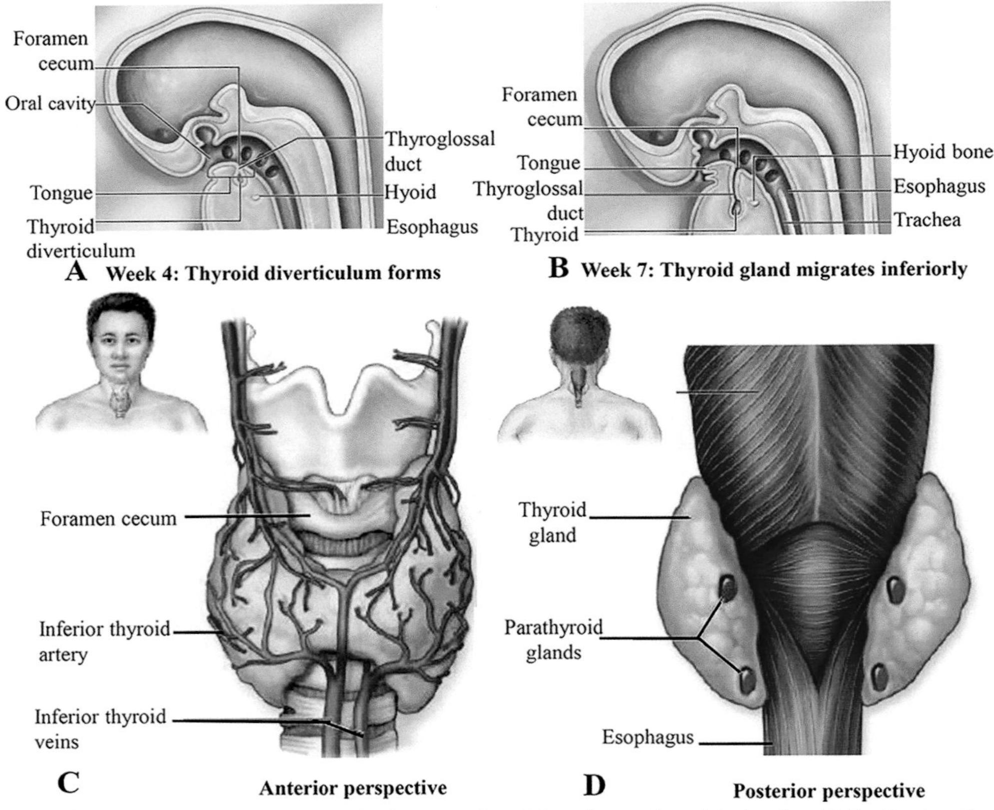

Figure 10.4 Development and gross anatomy of the human thyroid and parathyroid glands. (A) By the fourth week of development, the thyroid primordium derives from an endodermal thickening of the foregut that forms a diverticulum in the midline floor of the primitive pharynx between the first and second pharyngeal pouches. **(B)** This will develop into the thyroid, which will migrate caudally along the thyroglossal duct by the seventh week. **(C)** In adults, the gland itself wraps around the anterior surface of the trachea and contacts the larynx. **(D)** In contrast to thyroid gland tissue, which develops from tissue derived from the anterior pharyngeal floor and ultimobranchial cells of the neural crest, parathyroid tissue located on the dorsal side of the thyroid derives from brachial pouch tissue.

Source of Images: Junqueira, L.C. and Mescher, A.L., 2013. Junqueira's basic histology: text & atlas. McGraw Hill LLC; used with permission.

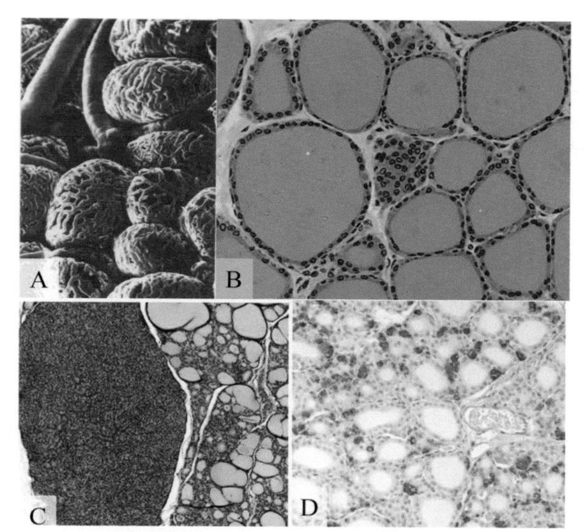

Figure 10.5 Major histological features of mammalian thyroid tissue. (A) Scanning electron micrograph depicting the three-dimensional structure of thyroid follicles and corresponding vasculature. **(B)** Cross section of thyroid follicles stained with hematoxylin and eosin. The thyroid follicles that synthesize thyroid hormones surround the protein-rich colloid where thyroglobulin is stored. **(C)** In contrast with thyroid follicles (right side of the image), parathyroid tissue (left side) is dense and lacks a colloid matrix. **(D)** The parafollicular C cells are labeled with an antibody against calcitonin (brown).

Source of Images: (A) Fujita, H., 1984. Fine structure of the thyroid follicle. In Ultrastructure of endocrine cells and tissues (pp. 265-275). Springer, Boston, MA; used with permission. (B) Courtesy of Dr. Peter Takizawa, Yale University. (C) Mense, M.G. and Rosol, T.J., 2018. Parathyroid Gland. In Boorman's Pathology of the Rat (pp. 687-693). Academic Press; used with permission. (D) Fernández-Santos, J.M., et al. 2012. Paracrine regulation of thyroid-hormone synthesis by C cells. Thyroid hormone. IntechOpen.

Curiously Enough . . . If the thyroid gland is completely removed from a mammal, the animal will die. However, its death is not due to the lack of thyroid hormone synthesis, but is rather caused by the removal of the parathyroid glands and a subsequent inability to maintain proper blood calcium levels. As such, when conducting a surgical thyroidectomy, precautions must be taken to ensure that the patient's parathyroid glands remain intact.

The thyroid is the first endocrine organ to begin to form during human organogenesis at 9 weeks of gestation, though complete maturation takes place throughout gestation.[14] In mammals, the thyroid primordium is derived from an endodermal thickening between the first and second pharyngeal arches, which migrates caudally along the thyroglossal duct (Figure 10.4).[15] In adults, a so-called *pyramidal lobe* near the isthmus may remain as a remnant of the thyroglossal stalk, the solidified duct connecting the thyroid to the pharynx, which usually atrophies during development.

In contrast to the thyroid gland, the juxtaposed parathyroid glands originate from the third and fourth brachial pouches.

The mammalian thyroid gland proper consists of two major types of cells that derive from different regions of the embryonic endoderm. Approximately 90% of the thyroid originates from the anterior pharyngeal floor (base of the tongue) and is made up of cuboidal epithelial cells that form large numbers of *thyroid follicles*, spherical structures surrounding a viscous, protein-rich fluid called the *colloid*[15](Figure 10.5). The thyroid follicles, which can reach a diameter of almost 1 cm,[16] capture iodide and synthesize thyroid hormones. The thyroid follicles are highly vascularized structures, facilitating the release and dispersal of thyroid hormones. The second cell type that constitutes less than 10% of the thyroid gland in humans are parafollicular "*C cells*", where "C" denotes "clear"-staining cells (Figure 10.5). These originate from ultimobranchial cells of the neural crest and produce the hormone *calcitonin*, which also plays a role in calcium homeostasis.

SUMMARY AND SYNTHESIS QUESTIONS

1. How do you explain the fact that if a human (or any mammal) undergoes a *total* thyroidectomy, it will die within days—even if supplemented with both T_3 and T_4?
2. Describe the developmental origins of the different cell types that constitute or are juxtaposed onto the mammalian thyroid gland.

Evolution of the Thyroid Gland

LEARNING OBJECTIVE Trace the evolutionary origins of the vertebrate thyroid gland from protochordate ancestors.

KEY CONCEPTS:

- In the most ancestral living protochordates, thyroid hormone is secreted into the gut via an *exo*crine gland (the endostyle), after which the hormone is subsequently absorbed into the bloodstream.
- During metamorphosis of the lamprey, which is among the most basal living vertebrates, the larval exocrine endostyle remodels into *endo*crine thyroid follicles that secrete hormone directly into the bloodstream.
- Whereas the thyroid follicles of all tetrapods are condensed into distinct organs by the end of embryogenesis, the thyroid follicles of many fishes are dispersed throughout various vascularized regions of the body.

A critical first step in the evolution of the thyroid gland was the development of the ability to collect iodide ions and iodinate protein. Among the most efficient accumulators of iodide are marine brown algal kelps, in which the iodine content can reach up to 5% of the dry weight.[17] Marine plants are a bountiful source of iodine and iodinated proteins that can be used efficiently by animals as substrates from which to synthesize TH.[18] Interestingly, there is evidence suggesting that the larvae of some echinoderms (non-chordate deuterostomes such as sea urchins and sand dollars) feed on iodinated protein-rich microalgae and potentially metabolize it to other organic forms such as thyroid hormones.[18–21] Therefore, in echinoderms, and possibly in the ancestors to deuterostomes, there may exist a cross-kingdom interaction whereby THs and/or their substrates are first synthesized exogenously in abundance by algae, and following consumption of the algae by the larvae these TH compounds are further metabolized or directly bind to receptors that will ultimately mediate the expression of genes necessary to facilitate developmental actions like metamorphosis.[22–25]

Protochordates and the Endostyle

All chordates synthesize TH, and the first evidence of an organ related to the vertebrate thyroid gland is found in the protochordates. The protochordate, amphioxus, is a cephalochordate and the most basal of the living chordates) (Figure 10.6), and ascidians are urochordates, the closest living relatives to vertebrates. Adult protochordates possess an **endostyle**, a grooved, mucus-secreting pharyngeal organ that traps food particles and facilitates filter feeding (Figure 10.6). The endostyle also has iodine-concentrating and TH biosynthesis activity and likely represents the ancestral form of the thyroid gland. These features of the endostyle reflect the evolutionary link between thyroid hormone and diet, and most of the molecular components of vertebrate thyroid hormone synthesis are present in the endostyle. Interestingly, the endostyle is not an endocrine gland, but instead appears to function as an *exocrine* gland, secreting TH and mucus into the alimentary canal where TH is subsequently absorbed into circulation and mediates developmental events, such as metamorphosis of the free-swimming larva into the adult form that is attached to the surface of the ocean floor.

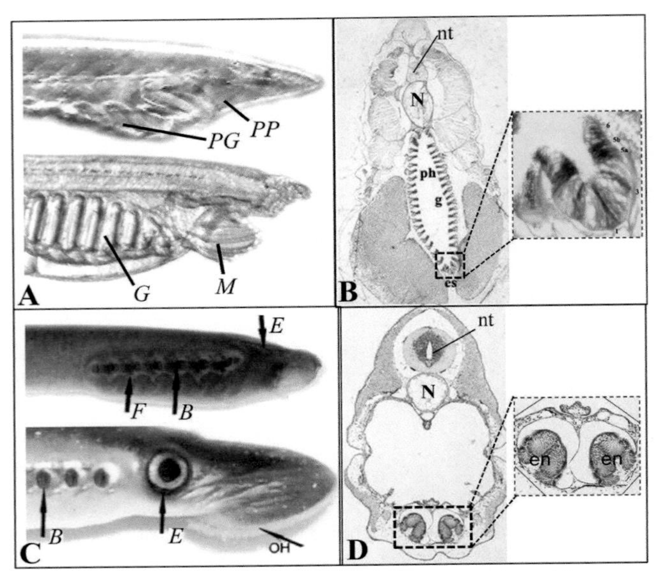

Figure 10.6 Morphology of the amphioxus and larval lamprey endostyle gland. **(A)** Metamorphosis of a larval amphioxus (top) into an adult (bottom) is accompanied by changes in the development of the mouth (*M*) from a preoral pit (*PP*), aa well as development of gills (*G*) from primary gill slit (*PG*) precursors. **(B)** The adult amphioxus endostyle, a grooved structure located along the ventral region of the alimentary canal, is shown in cross section at low magnification (left, boxed) and high magnification (right). The purple stain localizes thyroid hormone receptor (TR) mRNA transcripts by *in situ* hybridization. N, notochord; nt, neural tube; es, endostyle; g, gill; ph, pharynx. **(C)** Metamorphosis of the sea lamprey, *Petromyzon marinus*. The freshwater ammocoete larva (top) is a substrate-dwelling filter feeder with undeveloped eyes (*E*). Metamorphosis into a juvenile (bottom) is accompanied by the completion of eye development, morphological remodeling of the branchiopores (*B*), and silvering of the skin. The larval mouth adapted for filter feeding remodels into a suction cup shaped mouth (oral hood, *OH*) with tooth-like structures that is used to parasitize other fish. **(D)** The larval lamprey endostyle (en), also a grooved structure located along the ventral region of the alimentary canal, is shown in cross section at low magnification (left, boxed) and high magnification (right).

Source of Images: (A) Holland, L.Z. and Li, G., 2021. Laboratory culture and mutagenesis of amphioxus (Branchiostoma floridae). In Developmental Biology of the Sea Urchin and Other Marine Invertebrates (pp. 1-29). Humana, New York, NY; used with permission. (B) Wang, S. et al. 2009. Molec. Cell. Endocrinol., 313(1-2), pp. 57-63; used with permission. (C) Gelman, S., et al. 2008. J. Comp. Physiol. A, 194(11), pp. 945-956; used with permission. (D) Takagi, W., et al. 2022.. BMC Biol., 20(1), pp. 1-10.

Cyclostomes: Developmental Transition from Exocrine Endostyle to Endocrine Thyroid Follicles

The cyclostomes (lamprey and hagfish) constitute the most basal of all vertebrates, living or extinct.[26] Interestingly, the larval form of the lamprey, known as an ammocoete, resembles the *adult* protochordate form in two important ways: both are sessile filter feeders, and both possess endostyle organs that synthesize and secrete TH in an exocrine manner (Figure 10.6). However, during lamprey metamorphosis the exocrine endostyle loses its connection with the pharynx and transforms into internalized *endocrine* follicles, like those found in both the larvae and adults of teleost fishes. These follicles are not encapsulated into a distinct gland, as is the case for mammals, but rather are scattered throughout the pharyngeal area in a manner similar to that observed in most fishes (Figure 10.7).

From Scattered Follicles to Distinct Organs

In contrast to lamprey, whose larval form possesses an endostyle, the larval, juvenile, and adult forms of all

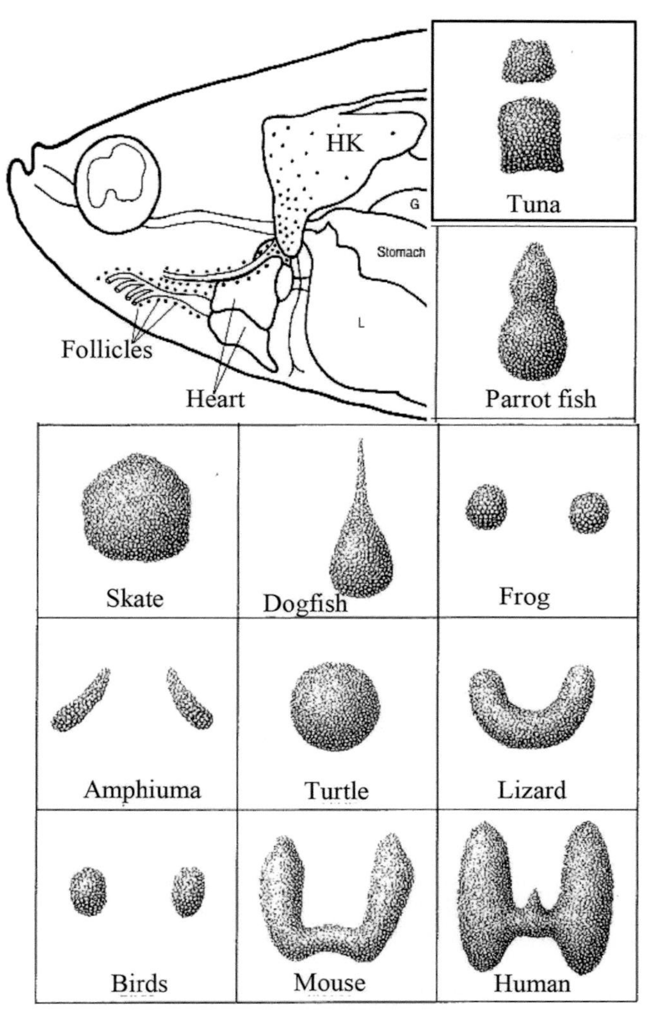

Figure 10.7 The distribution of thyroid follicles among vertebrates varies significantly. In most teleost fishes, thyroid follicles are dispersed throughout the body (top left) in regions such as the gills, head kidney (HK), and heart rather than condensed into a distinct organ. Exceptions to this are found in the tuna, parrot fish, skate, and dogfish, whose thyroid follicles are condensed into distinct organs. The thyroid gland shapes of tetrapods varies, but all are distinct, condensed organs.

Source of Images: John Gorbman, A., and H. A. Bern. A Textbook of Comparative Endocrinology. New York: Wiley, 1962. Wiley & Sons. Used with permission.

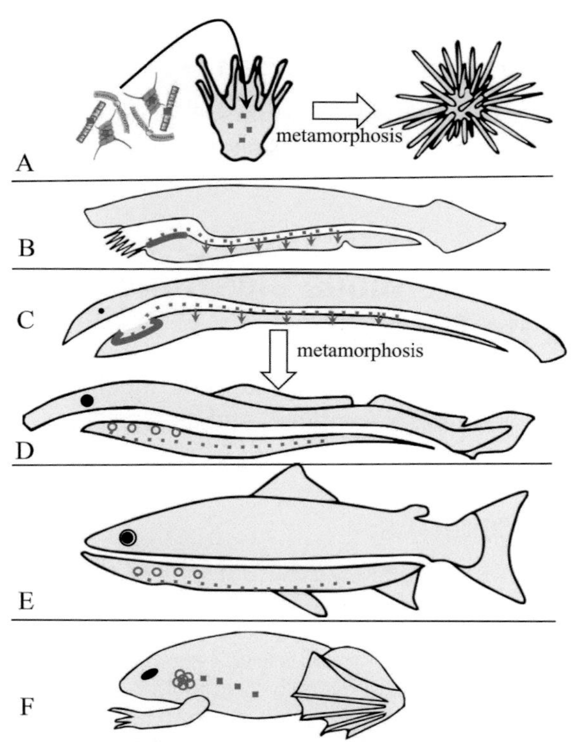

Figure 10.8 Hypothetical evolution of thyroid hormone acquisition/synthesis. (A) Echinoderm (sea urchins and sand dollars) larvae eat algae rich in iodinated substrates that may be metabolized into thyroactive compounds that facilitate metamorphosis into the adult. **(B)** In protochordates, like amphioxus, the *exo*crine endostyle gland secretes TH into the gut, where it is absorbed into the body. **(C)** In cyclostomes, like the lamprey, the larval exocrine endostyle secretes TH into the gut, and by the end of metamorphosis the endostyle has become internalized and remodeled into dispersed endocrine thyroid follicles **(D)** that secrete TH into the bloodstream. **(E)** In most teleost fish, dispersed endocrine thyroid follicles are present in the larval, juvenile, and adult stages. **(F)** In all tetrapods, as well as some of the more derived teleosts, the thyroid follicles are clustered into a distinct and compact gland. The endostyle is depicted as a thick red line in B and C. Internalized follicles are depicted as red circles in **D–F**. Thyroactive compounds are shown as red dots.

Source of Images: Norris, David O., and James A. Carr. Vertebrate endocrinology. Academic Press, 2020.

other vertebrates possess true endocrine thyroid follicles whose range of follicular organization varies considerably. With the exception of the cartilaginous fishes and the more derived teleosts (e.g. tuna, swordfish, and parrot fish) that do possess distinct encapsulated glands in lobes surrounding the branchial arteries, the thyroid tissue of most teleosts is arranged as scattered follicles or tubules associated with vascular tissue in the subpharyngeal or basibranchial regions (Figure 10.7).[27] In teleosts, ectopic thyroid follicles are also commonly located in the ventral aorta, kidneys, brain, intestine, and other regions.[28] Non-mammalian thyroid glands do not contain parafollicular "C" cells, and in these animals calcitonin production instead takes place in the separate "ultimobranchial" gland. Parathyroid cells in many non-mammalian species are also excluded from the thyroid gland. The extrathyroidal locations of the ultimobranchial and parathyroid glands for a bird can be seen in Figure 12.8 of Chapter 12: Calcium/Phosphate Homeostasis, Skeletal Remodeling, and Growth. A sequence of events summarizing the hypothetical evolution of the thyroid gland is portrayed in Figure 10.8.

SUMMARY AND SYNTHESIS QUESTIONS

1. Describe the major changes accompanying the evolution of the vertebrate thyroid gland from ancestral protochordates.

Thyroid Hormone Biosynthesis and Its Regulation

LEARNING OBJECTIVE Explain the mechanisms of thyroid hormone biosynthesis and its regulation.

KEY CONCEPTS:

- Thyroid hormone biosynthesis is a process by which two modified tyrosine molecules are joined and covalently bound to iodine.
- Thyroid hormone synthesis is under central control by hypothalamic thyrotropin-releasing hormone (TRH), which stimulates the production of thyroid-stimulating hormone (TSH) by the anterior pituitary gland. TSH, in turn, induces the thyroid gland to synthesize and release thyroid hormone.
- The hypothalamic-pituitary axis is under negative feedback influence from circulating thyroid hormones.

Thyroid hormone biosynthesis is a process by which the 6-carbon aromatic rings of two tyrosine molecules become joined and covalently bound to iodine (summarized in Figure 10.9). This complex process takes place as a series of interactions between follicular epithelial cells and the extracellular colloid component of thyroid follicles and can be divided into several generalized steps:

1. Sequestration of iodine by the epithelial cells and export to the extracellular colloid.
2. Incorporation of iodine into a large extracellular thyroid hormone precursor protein called **thyroglobulin.**
3. Oxidative coupling of mono- and di-iodinated tyrosine residues into thyroglobulin-linked tetra- and tri-iodothyronine residues.
4. Uptake and proteolytic processing of iodinated thyroglobulin by the epithelial cells.
5. Export of free thyroid hormones into circulation.

Knowledge of the steps of TH biosynthesis is critical for gaining an understanding of the diverse diseases of the thyroid system, as well as their treatments with pharmacological intervention. Elaborate interplay among diverse cellular actors, such as membrane transport proteins, hormone receptors, intracellular and extracellular enzymes, transcriptional and translational regulators, lysosomal processing, and endocytosis is sure to remind the student of endocrinology just how fundamental a comprehensive understanding of cell biology is to the discipline!

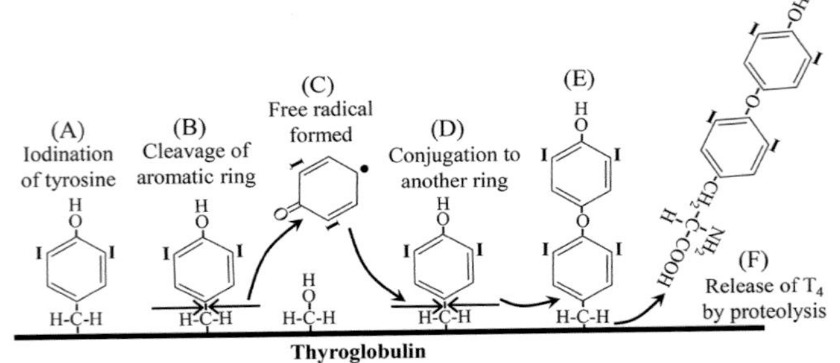

Figure 10.9 Formation of thyroid hormone via oxidative coupling of iodine to tyrosine residues on thyroglobulin. (A) Either 3,5-diiodotyrosine (DIT, shown) or 3-monoiodotyrosine (MIT, not shown) is synthesized via incorporation of iodide into the phenolic ring of a tyrosine residue. Following the cleavage of the DIT or MIT phenolic ring **(B)**, the ensuing free radical **(C)** becomes conjugated to another DIT ring **(D)**. **(E)** The coupling of two DIT rings forms thyroglobulin-bound 3,5,3',5'-tetraiodothyronine (T_4 shown), whereas the addition of MIT to a DIT residue yields 3,5,3'-triiodothyronine (T_3, not shown). **(F)** The proteolysis of thyroglobulin releases free T_4 or T_3.

Iodine Uptake

The details of thyroid hormone biosynthesis are depicted in Figure 10.10. Depending on the level of thyroid activity, iodine concentrations inside thyroid follicle cells range from 25 to 250 times that of blood plasma, and follicle cells therefore concentrate iodine against a very large gradient. Inorganic iodide from the extracellular fluid is taken up into follicular cells via a basolateral membrane-localized *sodium iodide symporter* (NIS). Iodide transport via NIS is competitively inhibited by anions such as thiocyanate and perchlorate (Table 10.1), which have similar ionic radii and solubility to iodide; as such these compounds are useful for the research and treatment of hyperthyroidism. Iodide uptake via NIS is a form of "secondary active transport", since the symporter relies on the presence of a Na^+ gradient maintained actively by the *Na^+/K^+-ATPase pump,* also present on the basolateral membrane of these epithelial cells. Once iodide has been taken up via the basolateral membrane symporter, it is then exported out of the cell into the colloid via an apical membrane-localized anion channel termed *pendrin.* Pendrin is also expressed in the kidney and inner ear, and the protein was first identified in association with *Pendred syndrome,* characterized by impaired thyroid hormone synthesis and hearing loss.

Thyroglobulin Iodination

The next step in thyroid hormone synthesis is the iodination of the colloid protein **thyroglobulin,** a large 660 kDa protein that functions as the primary substrate for thyroid hormone synthesis. This process is catalyzed by the apical membrane-bound enzyme, *thyroid peroxidase* (TPO), which uses hydrogen peroxide to convert iodide to an oxidized iodine species that becomes covalently bound to tyrosine residues on thyroglobulin forming the precursor amino acids, mono- and diiodotyrosine (MIT and DIT, respectively) (Figure 10.9 and Figure 10.10). Pharmacological inhibitors of thyroid peroxidase include thionomide drugs, such as methimazole, thiourea, and related compounds (Table 10.1), and are useful for treating **thyrotoxicosis**, a condition of excess thyroid hormone production. Iodinated thyroglobulin functions as an important storage source for iodine and contains enough protein and iodine substrate to make several weeks' worth of thyroid hormones. The iodinated tyrosines within thyroglobulin then undergo an *oxidative coupling* process (Figure 10.9 and Figure 10.10) resulting in the formation of mostly T_4, as well as smaller amounts of T_3, both of which remain bound to thyroglobulin. In humans, the ratio of T_4:T_3 synthesized by the thyroid is approximately 3:1.[29]

Thyroglobulin Proteolysis and Thyroid Hormone Export

The iodinated thyroglobulin is next engulfed by the follicular cell apical membrane and transported into the cells by phagocytosis, where it is sequestered in intracellular vesicles (Figure 10.10). These vesicles fuse with *lysosomes* to become *secondary lysosomes*, wherein diverse hydrolytic reactions generate free thyroid hormones. These free hormones are thought to be released into the cytoplasm through lysosomal membrane iodotyrosine transporters,[30] although the details of thyroid hormone export from lysosomes remain unclear. The enzyme *iodotyrosine dehalogenase 1* (DEHAL1) is involved in the deiodination of free MIT and DIT, which are released along with the thyroid hormones T_4 and T_3 through thyroglobulin proteolysis.[31] DEHAL1 action on DIT and MIT results in the liberation of iodide and tyrosine, which can be recycled for thyroid hormone synthesis. Thyroglobulin itself is recycled and not normally released into circulation in large amounts, although it can be measured in small quantities in normal subjects, probably reaching the blood via leakage into the lymphatic system.[14]

The details of thyroid hormone export across the basolateral plasma membrane and into circulation also remain mysterious. However, several important peripheral tissue thyroid hormone transporters have been identified, including MCT8,[32–35] and it is likely that thyroid hormones are secreted out of the thyroid gland via this or an as of yet undiscovered transporter.[36]

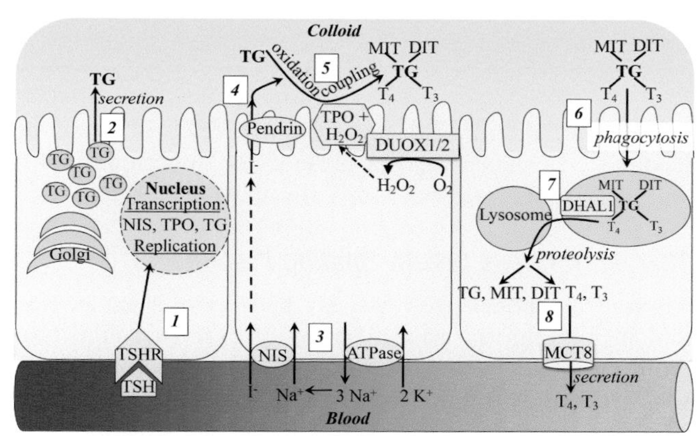

Figure 10.10 A stepwise overview of thyroid hormone biosynthesis. (1) The binding of TSH to its basal membrane-localized receptor (TSHR) initiates an intracellular signal cascade that induces replication of thyroid follicle cells, as well as the synthesis of enzymes and other proteins needed by the cell for TH synthesis (e.g. the sodium-iodide symporter, NIS; thyroid peroxidase, TPO; thyroglobulin, TG). (2) TG is translated, processed by the Golgi, and secreted across the apical membrane into the extracellular colloid. (3) Iodide is actively transported from the blood into the follicle cells across the NIS symporter using the Na+ gradient established by the Na/K-ATPase pump. (4) Iodide diffuses down its concentration gradient across transapical membrane pendrin channels into the colloid space. (5) Iodide becomes enzymatically coupled to TG (as T_3, T_4, MIT, and DIT) via the activity of the TPO and dual peroxidase 2 (DUOX2) enzymes. (6) Iodinated TG is taken up by phagocytosis. (7) Following fusion of the phagocytic vesicle with a lysosome, T_4 and T_3 are cleaved off through the actions of iodothyronine dehalogenase (DHAL1). (8) Whereas TG, MIT, and DIT are recycled back to the colloid, T_4 and T_3 are secreted into the bloodstream via the basolateral transmembrane protein, monocarboxylate transporter 8 (MCT8).

Central Control of Thyroid Hormone Synthesis

Release of thyroid hormone by the thyroid gland is controlled by the hypothalamus and pituitary gland. In mammals, the primary regulator of thyroid hormone synthesis is *TSH* (thyroid-stimulating hormone), also known as "thyrotropin", secreted from the pituitary adenohypophysis (Figure 10.11). TSH release is under hypothalamic control via the tri-peptide *TRH* (thyrotropin-releasing hormone). Interestingly, in the larvae of non-mammalian vertebrates, TSH production can also be strongly influenced by corticotropin-releasing hormone (CRH)[37] (discussed later in Section 10.8: Thyroid Hormones and Vertebrate Metamorphosis). TRH is the smallest known peptide-releasing hormone and the first hypothalamic-releasing factor to be purified. Both TRH synthesis and TSH synthesis are under negative feedback control from circulating thyroid hormones.

Table 10.1 Chemical Compounds That Inhibit Thyroid Hormone Synthesis

Inhibitors of Thyroid Peroxidase (Thionomide drugs)	Inhibitors of the Sodium-Iodide Transporter
Thiourea	Potassium perchlorate
Thiouracil	Thiocyanate
Propylthiouracil	
Methimazole	

Note: In addition to its central mechanism of inhibition, propylthiouracil also has peripheral inhibitory effects by inhibiting the enzyme type 1 deiodinase, which converts T4 to the more active T3.

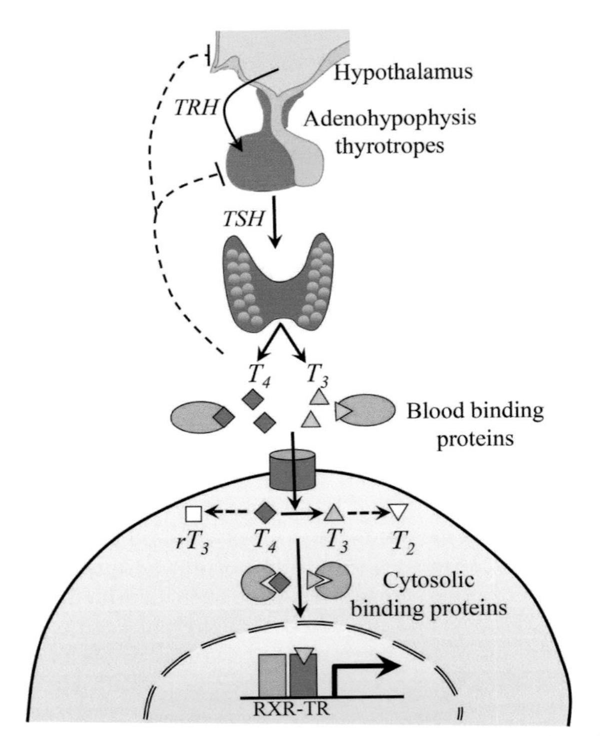

Figure 10.11 Central and peripheral regulation of thyroid hormone and the classical genomic signaling pathway in mammals. Central regulation of TH synthesis begins with secretion of thyrotropin-releasing hormone (TRH) by hypothalamic nuclei into local portal vessels. TRH binds to receptors on thyrotropes of the adenohypophysis that respond by releasing thyroid-stimulating hormone (TSH) into general circulation. TSH stimulates receptors on thyroid follicle cells that in turn secrete TH into general circulation. THs are peripherally regulated at multiple levels, including blood thyroid hormone-binding proteins, plasma membrane transporters on target cells, and intracellular cytosolic binding proteins. TH actions are ultimately regulated by nuclear-localized thyroid hormone receptor (TR)-RXR heterodimers that function as ligand-activated transcription factors, as well as the availability of transcriptional coactivators and corepressors (not shown). Genes transcribed and translated in response to TH go on to mediate diverse developmental and metabolic processes.

SUMMARY AND SYNTHESIS QUESTIONS

1. Outline the major events leading to the synthesis and release of thyroid hormones by thyroid follicles.
2. The pharmacological compounds methimazole and potassium perchlorate each inhibit the synthesis of thyroid hormones, but via different mechanisms. Describe their mechanisms of action.
3. A nuclear powerplant meltdown releases radioactive iodine into the air, which can be inhaled by people, resulting in thyroid cancer. Explain why in the event of such catastrophic occurrences, people are instructed to take potassium iodide tablets.

Peripheral Regulation of TH Activity

LEARNING OBJECTIVE Describe the roles of blood- and intracellular binding proteins, as well as peripheral intracellular deiodinases in modulating peripheral thyroid hormone activity.

KEY CONCEPTS:

- Blood serum thyroid hormone-binding proteins serve as a reservoir from which to maintain and resupply the unbound and biologically active free fraction of TH.
- Thyroid hormone-specific transmembrane transport proteins modulate thyroid hormone uptake by cells.
- Cytosolic T_3-binding proteins function as a cytoplasmic hormone reservoir and regulate transport between the cytoplasm and the nucleus.
- Intracellular T_3 concentrations are modulated by enzymes called deiodinases that activate or deactivate TH.

The Hypothalamic-Pituitary-Thyroid Axis

A summary of the central and peripheral regulation of TH activity is depicted in Figure 10.11. Hypothalamic TRH stimulates TSH secretion from the adenohypophysis. TSH, in turn, increases TH secretion, which suppresses both hypothalamic TRH and pituitary TSH via long-loop feedback. Evidence for the presence of functional short and ultra-short feedback loops also exists.[38]

Thyroid Hormone Serum-Binding Proteins

Thyroid hormones are relatively hydrophobic molecules, and in humans the vast majority of extrathyroidal hormone circulates bound to serum proteins with only a small free (unbound) fraction of T_4 (0.03%) and T_3 (0.30%).[39] The free hormone is thought to be the active fraction of the total hormone concentration, and a primary function of thyroid hormone-binding protein is to serve as a reservoir from which to maintain and resupply this free fraction. Serum-binding proteins ensure there is an even distribution of THs to all cells and tissues, and in their absence THs have been shown to be taken up disproportionately into the first cells that they encounter.[40] T_4 binds to serum-binding proteins with a higher affinity compared with T_3, resulting in a lower metabolic clearance rate and a longer serum half-life for T_4.

In humans there are three major blood plasma thyroid hormone-binding proteins, and each is present at a different concentration and has a different affinity for T_4 and T_3 (Table 10.2). The primary thyroid hormone-binding serum protein is *thyroxine-binding globulin* (TBG), which carries approximately 70% of serum T_4 and the majority of T_3. Of the three proteins, TBG is present in the lowest concentration but has the highest affinity for T_4 and T_3. *Transthyretin* (TTR; the name is derived from "*trans*ports *thy*roxine and *retin*ol") binds approximately 20%

Table 10.2 Distribution of Circulating Thyroid Hormone

	Bound to TBG	Bound to TTR	Bound to Alb	Unbound
T_4	70%	10%	20%	0.03%
T_3	80%	10%	10%	0.3%

Note: Alb, albumin; TBG, thyroid-binding globulin; TTR, transthyretin; T4, thyroxine; T3, triiodothyronine.

Source: After Kovacs, W.J. and Ojeda, S.R. eds., 2011. *Textbook of endocrine physiology*. OUP USA

of blood plasma T_4 and also transports retinol (vitamin A) by forming a complex with retinol-binding protein. The remaining serum T_4 and T_3 are bound by *albumin*. All three proteins are synthesized by the liver, though TTR is also made by the brain.

Plasma Membrane Transporters and Cytosolic Binding Proteins

Based on the hydrophobic structure of thyroid hormones, it had long been presumed that TH crosses the plasma membrane of cells via passive diffusion, similar to steroid hormone diffusion. However, it has become increasingly clear that passive diffusion not only plays little or no role in TH transmembrane transport, but that in fact there are specific thyroid hormone active transporters that mediate this process via expenditure of energy, and that the activities of these transporters play critical roles in determining intracellular thyroid hormone concentration.[41] Several families of transmembrane TH transport molecules have been identified, including *monocarboxylate transporters* (MCTs) and *organic anion transporter proteins* (OATPs),[35,42,43] and the MCT8 transporter has been shown to demonstrate selectivity for T_3 transport.[34] In fact, human mutations in the MCT8 gene cause a syndrome of severe psychomotor retardation and high serum T_3 levels in affected male patients, known as the *Allan–Herndon–Dudley syndrome*. The neurological deficits are likely explained by an impeded uptake of T_3 in MCT8-expressing cells that form the blood-brain barrier, as well as inhibited uptake by central neurons, resulting in impaired brain development.[44] Once thyroid hormone has been transported into the cytoplasm, it associates with specific proteins called *cytosolic T_3-binding proteins* (CTBP). CTBP plays an important role not only as a cytoplasmic TH reservoir, but also as a regulator of TH transport between the cytoplasm and nucleus, suggesting that CTBP levels directly influence gene transcription.[45]

Thyroid Hormone Metabolism by Deiodinases

In mammals, the majority of thyroid hormone released into circulation by the thyroid gland itself is in the form of T_4, with T_3 and reverse T_3 (rT_3) comprising less than 10% of the total fraction.[46] However, about 80% of the T_4 released by the thyroid is converted by peripheral tissues into one of two forms, either T_3, the more active hormone that has a ten-fold higher affinity to its receptors compared with T_4, or rT_3, a form generally considered to be biologically inactive. The majority of circulating T_3, therefore, is derived from peripheral organs that have converted it locally from its precursor, T_4. Although some tissues may depend entirely on blood circulation as their primary source for T_3, evidence derived from transgenic mouse models and findings from humans with mutations affecting thyroid hormone metabolism support a model in which regulated local adjustment of T_3 concentrations is much more important than circulating plasma hormone levels.[47] As is the case with most vertebrates, the thyroid follicles of teleost fish secrete predominately T_4. However, unlike mammals, where circulating T_4 is often two orders of magnitude higher than T_3, fish serum T_3 is typically comparable to and can even exceed T_4 concentrations,[28] suggesting that localized conversion of T_4 to T_3 within large organs, such as the liver, influences circulating levels of T_3 in these animals.

Intracellular T_3 concentrations are modulated locally by a class of enzymes called **deiodinases**. Three deiodinases have been identified in mammals, with distinct physiological functions, regulations, and tissue distributions (Figure 10.12). They are all transmembrane selenoproteins (selenium-containing enzymes with a selenocysteine in their catalytic site), and are designated deiodinase types 1, 2, and 3 (D1, D2, and D3, respectively). Both *D1* and *D2* convert T_4 to the more biologically active T_3 and are thus considered thyroid hormone activation enzymes. *D3* catalyzes the direct degradation of T_4 to the inactive reverse T_3 (rT_3), as well as T_3 to 3,3' T_2, and is therefore considered a TH deactivation enzyme. Both D1 and D3 are plasma membrane proteins, with their active sites located in the cytoplasm, whereas D2 is expressed in the endoplasmic reticulum. Therefore, the transport of thyroid hormones across the cell membrane is essential for its activation or inactivation, and peripheral tissues can modulate their sensitivity to thyroid hormone by adjusting the relative activities of these enzymes. *Iopanoic acid* pharmacologically inhibits the activity of all three deiodinases, whereas *propylthiouracil* specifically inhibits D1 in addition to its central inhibition of thyroid peroxidase activity (Figure 10.12).

Curiously Enough . . . Like iodine, selenium is a rare element that is an essential component of the diet for normal thyroid signaling and function. In its absence, deiodinases will not function properly.

During fetal development, thyroid hormone signaling is regulated both spatially and temporally via the expression pattern of deiodinases. For example, in mice there is a 3-day peak of D2 expression and activity in the cochlea of the ear from postnatal days 7–10, and if absent the mice are severely hearing impaired.[48–50] Furthermore, placental

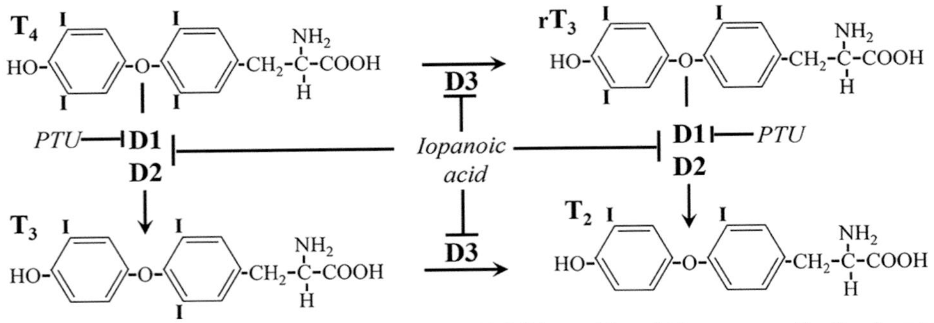

Figure 10.12 The deiodinases and some of their pharmacological inhibitors. Thyroid hormone activation is mediated by both deiodinase type I (D1) and type II (D2), which convert T_4 to T_3 through the removal of an iodine atom on the outer ring. Inactivation of TH occurs by removal of an iodine from the inner ring by type III deiodinase (D3), which catalyzes the direct degradation of T_4 to reverse T_3 (rT_3), as well as T_3 to T_2. Iopanoic acid inhibits the activity of all three deiodinases, whereas propylthiouracil (PTU) specifically inhibits D1. In addition to inhibiting D1, PTU also inhibits TH synthesis by inhibiting thyroid peroxidase activity.

D3 plays a crucial role in protecting fetal tissues from premature exposure to surplus thyroid hormones during critical periods of development.

SUMMARY AND SYNTHESIS QUESTIONS

1. Describe how TH synthesis is under central control by other hormones.
2. Name three rare elements from the periodic table that are essential for thyroid hormone functioning. Describe what roles they play.
3. Qi and colleagues[51] have shown that when the body's temperature rises, thyroxine-binding globulin's (TBG) affinity for T_4 decreases. How could this process be useful in maintaining a fever?
4. The compound iopanoic acid is not a goitrogen but can produce physiological effects that resemble hypothyroidism. How does the compound's mechanism of action yield these results?

Molecular Mechanisms of Thyroid Hormone Action

LEARNING OBJECTIVE Describe the molecular mechanisms of action of thyroid hormone and its receptors in modulating gene expression.

KEY CONCEPTS:

- Nuclear thyroid hormone receptors (TRs) are members of the steroid receptor superfamily and function as ligand-dependent transcription factors.
- TRs exert "dual control" over transcription of TH-responsive genes: in the absence of thyroid hormone they function as transcriptional repressors of positively regulated genes, and in the presence of TH they become transcriptional activators.

Since the initial description by Adolf Magnus-Levy over a century ago of TH's effects on metabolic rate,[6] different theories have been put forth to explain the molecular basis of TH action. Most investigators currently accept that the effects of TH are primarily genomic and initiated within the cell nucleus, and this will be the focus of this section. However, it is increasingly clear that TH also exerts some of its effects via relatively less characterized "nongenomic" receptors located in the mitochondria (described in Chapter 18: Energy Homeostasis), as well as in the plasma membrane and cytoplasm.[52] The physiological relevance of many of these nongenomic receptors still remains to be fully characterized.

To attempt to gain a working understanding of thyroid hormone action is no trivial task, and in writing a review article on the subject eminent thyroid hormone researcher Jack Oppenheimer belied his mixture of wonder and frustration on the topic by stating flatly, "The rapidly accumulating new information relating to the molecular basis of thyroid hormone action frequently appears bewildering, abstract, and disconnected from biological reality".[53] The student of endocrinology should take these words to heart when considering the sum of what follows.

Genomic Actions

TRs Are Nuclear Receptors

In 1966, Tata and Widnell[54] demonstrated that TH stimulates DNA-dependent RNA-polymerase activity, and high-affinity nuclear binding sites for TH were subsequently documented in 1974,[55,56] suggesting that TH nuclear receptors mediate transcriptional activity. Proof that thyroid hormone receptors (TRs) are members of the nuclear receptor superfamily was provided in 1986 after the receptors were cloned from chickens and mammals.[57,58] As with other members of the nuclear receptor superfamily, TRs share a conserved DNA-binding domain (DBD) and a ligand-binding domain (LBD) that interacts with diverse *coactivators* (CoAs) and *corepressors* (CoRs) (Figure 10.13). Notably, TRs typically form heterodimers with other members of the nuclear receptor superfamily, in particular with *retinoid X receptors* (RXRs).

TRs are highly conserved among vertebrates, and putative nuclear TRs with high degrees of homology to vertebrate thyroid hormone receptor genes have even been recognized in ascidian and amphioxus ancestral chordates.[59]

The Dual Function Model

TRs bind preferentially to their DNA response elements as heterodimers with retinoid X receptor (RXR). As members of the Class II family of nuclear receptors, TRs are capable of binding to their DNA response elements in the absence of thyroid hormone, and generally function to repress basal transcription in this unliganded form. Therefore, TRs are thought to exert "dual control" over transcription of positively regulated TH-responsive genes: in the absence of thyroid hormone (the "**aporeceptor**" form) they function as transcriptional repressors, and in the presence of TH (the "**holoreceptor**" form) they become transcriptional activators (Figure 10.13).

TR Splice Variants

In humans, TRs are derived from two genes (*THRA* and *THRB*; note that in rodents these genes are called *Thra* and *Thrb*) located on different chromosomes. Each TR gene can be alternatively spliced into different isoforms (Figure 10.14), reflecting the use of alternate promoters and exons. The main functional TRs are TRα1, TRβ1, and TRβ2, whereas the functions of other isoforms remain less certain. One isoform, TRα2, which lacks a functional ligand-binding domain but retains its DNA-binding domain, has been proposed to competitively inhibit transcriptional activation by both TRα and TRβ by binding to the same DNA response element (TRE) and functioning as a "dominant negative" repressor.[60,61] Additional and even more mysterious truncated forms of TRα and TRβ give rise to TRΔα1, TRΔα2, and TRΔβ3, which entirely lack the DNA-binding domain but still retain most of the ligand-binding domain. In general, differential expression of the TR isoforms among tissues and/or developmental stages likely helps mediate differential tissue responsivity to TH, and in some cases their mis-expression may lead to disorders of resistance to TH.[62]

Lessons from TR Knockout Mice

Targeted gene inactivation, or *knockout* (KO), of TR isoforms in mice has provided valuable information on

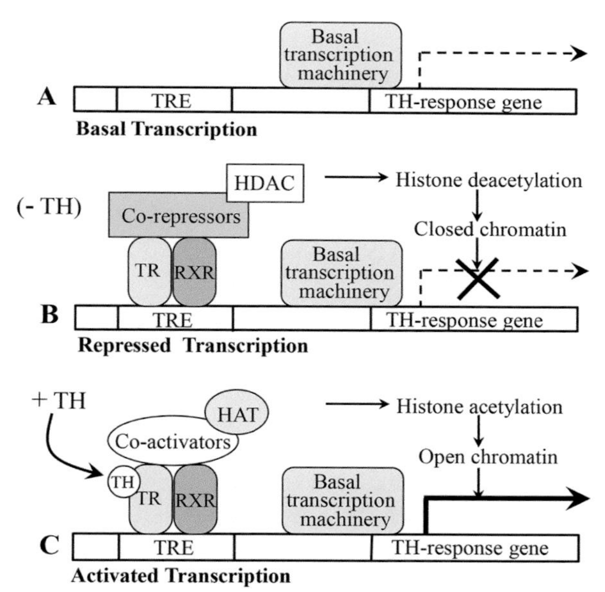

Figure 10.13 The thyroid hormone receptor dual function model of gene repression and activation (A) In the absence of thyroid hormone receptor (TR), there are basal levels of gene transcription. (B) When TR and its heterodimer RXR are present in the absence of thyroid hormone (TH), the TR-RXR complex interacts with a series of corepressor proteins and histone deacetlyases (HDAC), resulting in the tight packaging of chromatin and the repression of basal gene expression. **(C)** The binding of TH to TR causes the release of corepressors and the recruitment of coactivators and histone acetyltransferase (HAT) that relax the chromatin structure and activate gene transcription.

Source of Images: Manzon RG. Thyroidal regulation of life history transitions in fish. In: Flatt T, Heyland A, eds. Mechanisms of life history transitions: The genetics and physiology of life history traits and trade-offs. Oxford University Press; 2011.

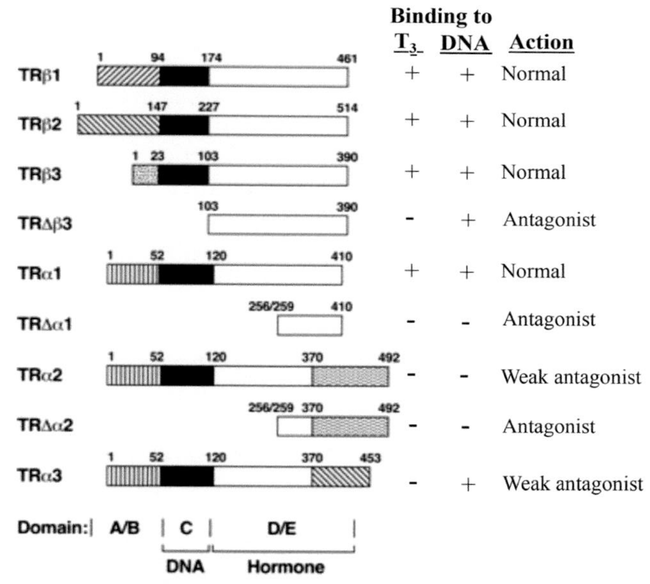

Figure 10.14 Schematic representation of thyroid hormone receptor (TR) isoforms and their actions in mammals. TRs are encoded by the TRα and TRβ genes located on different chromosomes. Alternative splicing of the primary transcripts gives rise to four receptor isoforms with normal actions, and other isoforms with varying degrees of negative antagonistic actions. TRs share high sequence homology in the DNA-binding domain C (solid bar) and the hormone-binding domain D/E (open bar). The amino-terminal A/B domains are variable in length and amino acid sequence. The amino acids of the truncated TRs (TRΔβ3, TRΔα1, and TRΔα2) at the amino and carboxy termini are indicated by numbers.

Source of Images: Cheng, S.Y., et al. 2010. Endo. Rev., 31(2), pp. 139-170.

Table 10.3 Phenotypes Caused by Knockouts of *Thra* and *Thrb* Genes Compared with Congenital Hypothyroidism in Mice

	Receptor Gene			
Phenotype	**Thra**	**Thrb**	**Combined *Thra/Thrb***	**Congenital Hypothyroidism**
Sensory deficits (auditory, visual)	–	++	+++	+++
Hyperactive HPT axis	–	+	+++	+++
Goiter	–	+	+++	+++
Liver, defective response to T_3	+/–	++	+++	+++
Dwarfism	+/–	–	++	+++
Infertility (female)	–	–	++	+++
Low body temperature	+	–	+	+++
Bradycardia	+	–	+	+++
Intestinal defects	+	–	++	+++
Osteosclerosis	+	–	++	+++
Post-weaning lethality	–	–	+	+++

Note: Phenotype severity: – no phenotype, + least severe, +++ most severe.

the mechanisms of TR function. When phenotypes from mice with congenital hypothyroidism are compared with either mice with individual TR isoform KOs (i.e. all transcripts from either *Thra* or *Thrb* gene loci are deleted) or mice with double TR isoform KO (both *Thra* and *Thrb* have been deleted), several important conclusions can be derived regarding the physiology of different TR isoforms (summarized in Table 10.3).

First, the different isoforms display some developmental stage and organ-specific functions.[63] For example, whereas *Thrb* plays a specific role in cochlea and retina development during the postnatal to pre-weaning period, in the adult this isoform is the main effector of hypothalamic and pituitary feedback regulation, as well as liver metabolism. *Thra* plays the major role in proper development of the brain, bone, and intestine during the weeks prior to weaning. In adults, *Thra* plays a more important role in regulation of heart rate and body temperature.

Second, compared with the severe phenotypes of combined *Thra* and *Thrb* KOs, the relatively mild phenotypes displayed by mice with individual *Thra* or *Thrb* KOs suggest that in general both isoforms have similar roles in the transcriptional regulation of tissues, and the absence of one isoform can be compensated to some extent by the second isoform.

Third, compared with the severely impaired congenitally hypothyroid mice that display high post-weaning mortality, infertility, and metabolic disorders, the double TR KO mice (*Thra*$^{-/-}$*Thrb*$^{-/-}$) are surprisingly viable, with relatively milder fertility and metabolic phenotypes. This observation that the complete absence of TRs produces several milder phenotypes compared with lack of the hormone suggests that basal gene transcription occurs in the absence of the receptor, whereas when the receptor is present gene repression takes place in the absence of the hormone. The fact that TR suppresses gene transcription in the absence of thyroid hormone has important implications to thyroid physiology. In particular, under conditions of hypothyroidism, rather than functioning as an inactive or passive receptor the TR actively represses transcription.

SUMMARY AND SYNTHESIS QUESTIONS

1. Describe various peripheral mechanisms that regulate the activity of TH on its target cells.
2. Compare and contrast the general mechanisms of action of classical steroid hormone and thyroid hormone receptors. In particular, explain the "dual function" model for TR action.
3. How do thyroid hormone receptor coactivators and corepressors modulate gene expression?
4. What are thyroid hormone response elements (TREs), where are they found, and what are their functions?
5. What is meant by "positively" versus "negatively" regulated TH-responsive genes?
6. Although there are only two different genes that code for thyroid hormone receptors (TRs) in mammals, there are many different kinds of TRs with varying functions. How is this diversity generated from two genes, and what are some known differential actions of TRs?
7. TRs are expressed in such organs as liver and lung before the onset of thyroid function and in the unliganded, aporeceptor form in developing humans and other vertebrates. What functions, if any, do these aporeceptors have?
8. Mutant mice lacking all thyroid hormone receptors produce a milder developmental phenotype compared with hypothyroid mice with TRs present. Explain this observation.
9. Suppose you create a transgenic line of mice that expresses very high levels of a modified TR gene

that specifically lacks the ligand-binding domain. Would this result in thyroid responsive gene overexpression or suppression, or would it have no effect? Explain your reasoning.

Development and Growth Effects of TH

LEARNING OBJECTIVE Describe the developmental and growth effects of thyroid hormones on diverse target tissues.

KEY CONCEPTS:

- TH is essential for both fetal and postnatal development of the mammalian brain, heart, skeletal muscle, lungs, and gut.
- TH exerts profound effects on vertebrate growth, and its presence is required to maintain normal levels of growth hormone.

The fetal and early postnatal periods of mammalian development are accompanied by significant longitudinal growth, as well as extensive remodeling and development of diverse organs, including the brain, heart, gut, lungs, and skeleton. The normal growth and development of these and other organs requires the presence of TH, which in mammals undergoes a dramatic surge around the time of birth (Figure 10.15). Importantly, TRs are expressed before the onset of thyroid function in developing humans, as well as in rodents, birds, and amphibians. More specifically, the TR protein starts accumulating in the fetal human brain from the tenth week of gestation, and increases up to the 16th–18th week.[64] TR protein also increases in other tissues, such as liver and lung, but in these organs the TR is mainly in the unliganded **aporeceptor** state, in contrast to the brain, where significant saturation with receptor-bound T_3 **holoreceptor** state occurs from the tenth week. Increased brain T_3 is due to selective expression of type 2 deiodinase, which produces T_3 from maternally derived T_4 in the cerebral cortex, whereas low activity of D2 and high activity of D3 maintains low concentrations of T_3 in other organs, as well as some regions as the cerebellum, compared to the cortex.[65,66] The fact that the developing brain before the fifth month is dependent on maternal T_4 as a D2 substrate explains the detrimental developmental effect of maternal hypothyroidism. This relationship between thyroid hormone action and development of the human brain is described in Figure 10.15.

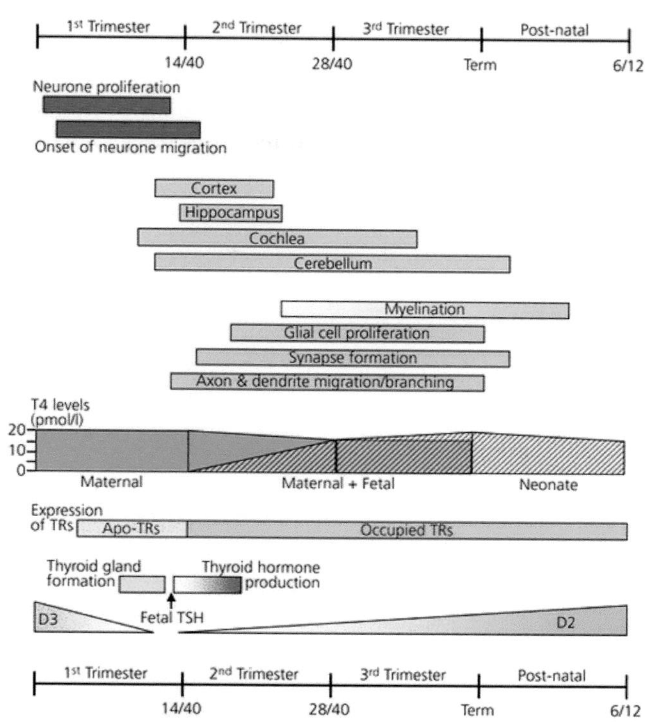

Figure 10.15 Relationship between thyroid hormone action and development of the human brain. In the first trimester of pregnancy, early neuronal proliferation and migration is dependent on maternal thyroxine (T_4). In fetal tissues, TH-inactivating type 3 deiodinase (D3) enzyme expression falls and development of the thyroid gland commences. By the end of the first trimester, development of the hypothalamic-pituitary axis has occurred and a surge in thyroid-stimulating hormone (TSH) secretion results in the onset of fetal thyroid hormone production, expression of the activating type 2 iodothyronine deiodinase enzyme (D2), and increasing occupation of thyroid hormone receptors (TRs) by 3,5,3'-l-triiodothyronine (T_3). Continuing development of the brain in the second and third trimesters relies increasingly on T_4 produced by both the fetus and the mother. Continued postnatal development is entirely dependent on neonatal thyroid hormone production. Apo-TR, unliganded thyroid hormone receptor, serves a repressive function.

Source of Images: Williams, G.R., 2008. J. Neuroendocrinol., 20(6), pp. 784-794. Used with permission.

Nervous System

TH is essential for both fetal and postnatal development of the mammalian brain and influences a vast range of processes, such as cell division, migration, neuronal and glial cell differentiation, synaptogenesis, and myelination, as well as synaptic pruning and cell death.[67] Gene regulation by TH occurs in a strict spatial-temporal fashion, such that different brain regions respond to TH with cell proliferation, neuronal migration, or synapse formation for only a limited period of time (Figure 10.15). Pathologically low maternal TH levels, particularly between the third and fifth month of gestation in humans when the fetal thyroid is not yet functional, results in severe and irreversible neurological impairment known as "neurological cretinism", which produces impaired cerebellar Purkinje cell (Figure 10.16) and cochlear development.[68]

Heart and Skeletal Muscle

Rising TH levels in rodents during the postnatal transition promotes important changes in cardiomyocyte morphology and is required for cardiac cells to obtain their characteristic elongated shape with parallel-oriented myofibrils[69] (Figure 10.16). Mai and colleagues[70] have shown that in the mouse fetus, unliganded TRα aporeceptors repress heart rate as well as the expression of several genes encoding ion channels involved in cardiac contractile activity. However, immediately following birth, when T_3 concentration sharply increases, liganded TRα holoreceptors turn on the expression of some of these same genes concomitantly with heart rate increase, demonstrating that the TRα receptor is a molecular switch controlling heart function near birth time. Furthermore, in the offspring of hypothyroid maternal mice, a postnatal switch in heart and skeletal muscle from embryonic and neonatal isoforms of myosin heavy chain to expression of the adult myosin isoforms is delayed or incomplete.[71,72]

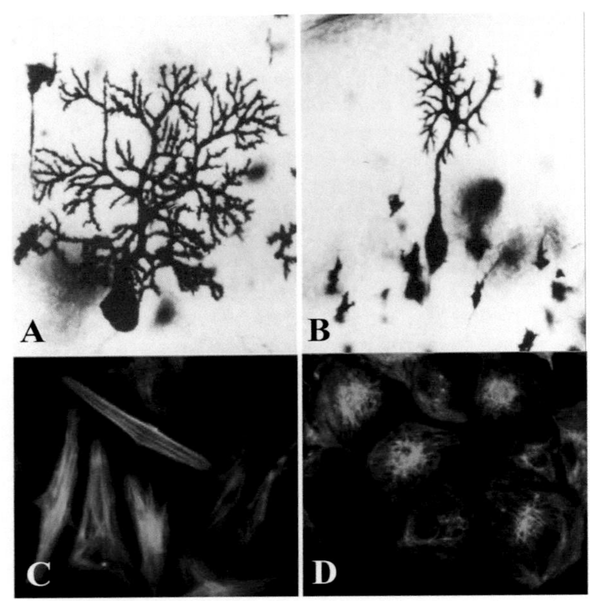

Figure 10.16 Morphology of developing rodent cerebellar Purkinge cells and cardiomyocytes under normal and hypothyroid conditions. Cerebellar Purkinje cell in a normal 14-day-old rat **(A)** and in a rat made hypothyroid with propylthiouracil treatment from day 18 of gestation **(B)**. Note the marked hypoplasia of the dendritic tree in the hypothyroid rat. Cardiomyocytes stained for actin filaments (fluorescent green) cultured in the presence of T_3 **(C)** have developed a typical elongated shape with well oriented myofibrils. By contrast, cardiomyocytes cultured in the absence of T_3 **(D)** have failed to reorganized their cytoskeleton into its fully differentiated state.

Source of Images: (A-B) Legrand J 1986 Thyroid hormone effects on growth and development. In: Hennemann G (ed) Thyroid Hormone Metabolism. Marcel Dekker, Inc, New York, pp 503–534.; InformaUK (Taylor & Francis Books); used with permission. (C-D) Pantos, C., et al. 2008. Pharmacol. Ther., 118(2), pp. 277-294. Used with permission.

Lungs and Gut

TH functions synergistically with adrenal glucocorticoids to influence perinatal lung development, with glucocorticoids stimulating the differentiation of type 2 pneumocytes that are involved with surfactant synthesis, and TH promoting lung septation and the differentiation of type 1 pneumocytes that are involved in gas exchange.[71] Indeed, lung septation and surfactant expression is impaired in hypothyroid mouse pups.[71]

The morphogenesis of the adult intestine is conserved among vertebrates, with TH playing a major regulatory role.[73,74] In mammals, a functional intestine forms during embryogenesis but undergoes extensive maturation around birth when TH levels are high. During this postnatal period the mature villus-crypt axis is established, where the adult stem cells are localized in the crypts. The influence of TH in intestinal development is evident in mutant mice lacking all TR isoforms, whose offspring completely fail to form villi (Figure 10.17). Similarly, following metamorphosis transgenic *Xenopus laevis* frogs expressing a mutant TR that lacks the TH-binding domain and functions as a transcriptional repressor also fail to develop villi-like intestinal folds (Figure 10.17).[75]

Growth

Physiological concentrations of circulating TH are critical for normal postnatal and pubertal growth and skeletal development in humans and other mammals.[76] Under conditions

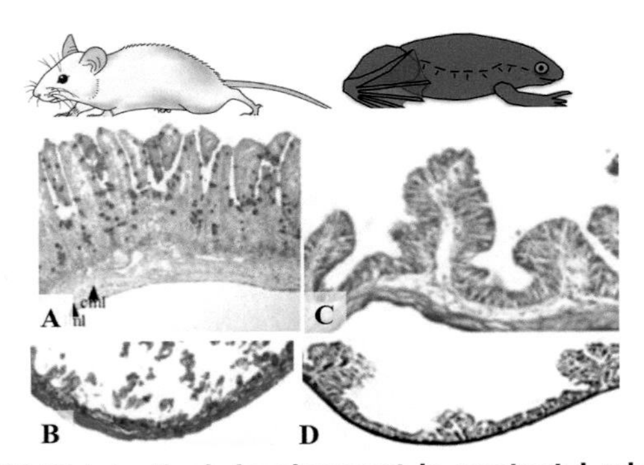

Figure 10.17 Intestinal development in postnatal mice and metamorphosing frogs is severely impaired in animals with mutated thyroid hormone receptors (TRs). Compared with the wild-type neonatal mouse intestine **(A)**, the intestine of the double TRα-/-TRβ-/- mutant **(B)** has dramatically impaired epithelial villi development and reduced thickening of mesenchyme and smooth muscle layers. After metamorphosis of the *Xenopus laevis* frog, a wild-type intestine is thick, possessing distinct epithelial, smooth muscle, and connective tissue layers **(C)**. However, a transgenic animal expressing a dominant negative TR (TRDN) that lacks the ligand-binding domain displays impaired epithelial villi development and reduced thickening of mesenchyme and smooth muscle layers, similar to the double TR mouse mutant phenotype.

Source of Images: (A-B) Karine Gauthier, et al. (1999). EMBO J. 18, 623 – 631; used with permission. (C-D) Alexander M. Schreiber, et al. (2009). Dev. Biol. 331: 89–98. Used with permission.

of subclinical hypothyroidism, growth rates in children are reduced. In severe cases of untreated childhood hypothyroidism, linear growth is almost completely halted but can resume with thyroid hormone treatment (Figure 10.18A). TH levels directly influence pituitary growth hormone (GH) content, with values of thyroidectomized adult rats dropping to less than 1% of normal rats.[77,78] TH has been shown to potentiate the pituitary response to hypothalamic growth hormone-releasing hormone (GHRH) in humans (Figure 10.18B). In addition to potentiating the effects of GHRH on the pituitary, TH also directly stimulates GH synthesis by somatotropes of the anterior pituitary. The rat GH gene is known to contain several regulatory sequences with both positively and negatively regulated TREs located in both the 5' and 3' flanking regions, as well as within introns throughout the gene.[79] Indeed, the rat GH TRE was the first TRE to be characterized in detail.[80]

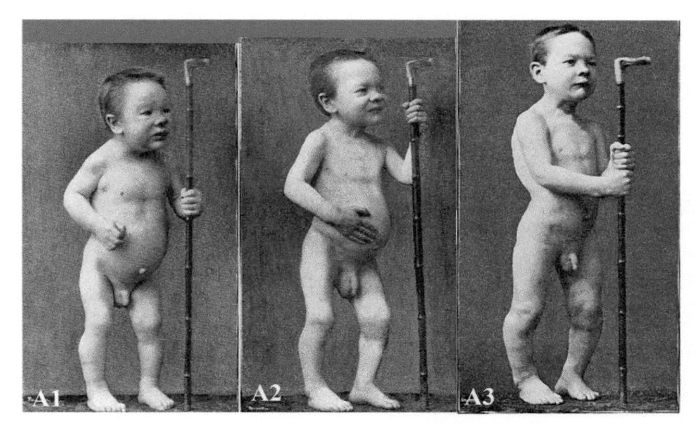

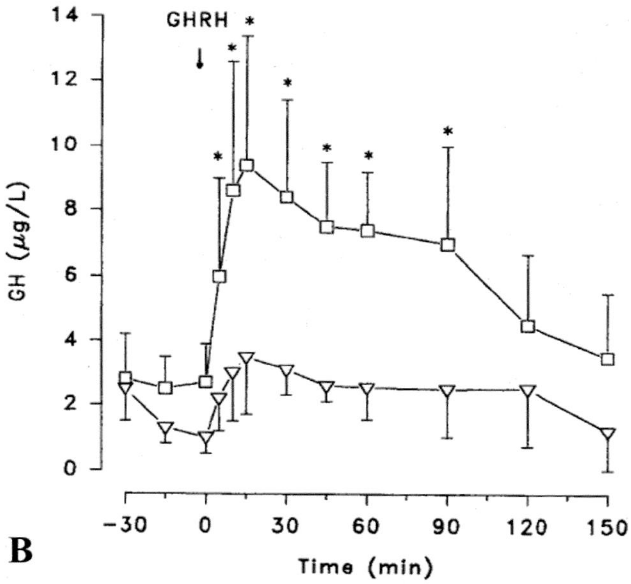

Figure 10.18 Thyroid hormone is required for normal growth. Panel A: In 1893, John Thomson, an English doctor, met an 18-year-old patient, known as "A.C.", who suffered from infantile cretinism **(A1)**. At the start of thyroid hormone treatment, A.C. was 33.5 inches tall. Following three months of treatment, A.C. grew in height by two inches **(A2)**. After a year of thyroid hormone treatments, Thomson reported a remarkable 4-3/8 inches of growth in his patient **(A3)**. Note the changes in height over time relative to the same walking cane in each photograph. Although bone growth was dramatic, the hormone treatment did not reverse A.C.'s severe cognitive impairments. Panel B: TH potentiates the pituitary response to hypothalamic growth hormone-releasing hormone (GHRH) in humans. Plasma GH levels in response to GHRH in five patients with hypothyroidism studied while untreated (triangles) and during thyroid hormone substitution (squares).

Source of Images: (A) Courtesy of Mark Rowley. (B) Williams T., et al. 1985. JCEM 61: 454-456. Used by permission.

SUMMARY AND SYNTHESIS QUESTIONS

1. Physiological concentrations of circulating TH are critical for normal postnatal growth. By what mechanisms does TH facilitate growth?

Thyroid Hormones and Vertebrate Metamorphosis

LEARNING OBJECTIVE Explain how vertebrate metamorphosis is a useful model for studying the mechanisms of thyroid hormone action.

KEY CONCEPTS:

- In all metamorphosing vertebrates, the larval to juvenile transformation is mediated by changing levels of thyroid hormone.
- As a naturally inducible system where exogenously administered TH initiates profound developmental changes in virtually every tadpole organ, the amphibian metamorphosis model lends itself to investigating the molecular mechanisms of TH actions, organogenesis, and tissue remodeling.
- TH exerts differential effects on tissues by targeting TH-responsive genes that code for transcription factors.

The term "**metamorphosis**" (derived from Greek *meta*, "beyond, changed"; *morphe*, "shape") describes a change in shape, form, or substance. In the most common biological usage of the term, metamorphosis refers to a period of spectacular development that defines the transition from a morphologically distinct larval form to a new juvenile (adult-like) form. The dramatic morphological changes that occur in many species are accompanied by concurrent changes in physiology and behavior, all of which accommodate the transition of the organism from a distinct larval ecological niche to one specific to the juvenile and adult. This is well illustrated in the flatfishes (e.g. flounder, sole, halibut) whose metamorphosis not only is accompanied by a dramatic remodeling of the skull and migration of one eye to the opposite side of the head, but the ensuing juvenile settles to the bottom of the ocean where it lies and swims on its "blind" side for the rest of its life[81,82] (Figure 10.19). As such, the post-metamorphic flatfish is the most asymmetrically shaped and behaviorally lateralized vertebrate to ever exist.[82] Among vertebrates, metamorphosis is confined to many species of fishes and amphibians but is absent in reptiles, birds, and mammals.

Curiously Enough . . . The larval forms of some vertebrates, such as eels, are so morphologically different compared to the adult that 19th-century taxonomists erroneously classified larval eels into a separate genus, *Leptocephalus* ("slim head"). Today, eel larvae are still referred to as leptocephali.[83]

A Naturally Inducible Developmental System

If metamorphosis is the trick of generating two different phenotypes from the same genome, then in vertebrates it is thyroid hormone that has the magic to pull it off. A remarkable feature of vertebrate metamorphosis is that the larval to juvenile transformation is accompanied by dramatic increases in the production of thyroid hormone and the expression of its receptors in almost every species examined.[84] This is particularly evident when comparing flounder and frog metamorphosis (Figure 10.19). An intriguing exception to the rule of rising TH levels accompanying metamorphosis is the lamprey (see Figure 10.6), whose metamorphosis is mysteriously accompanied by *decreasing* thyroid hormone levels.[85] Lamprey aside, rising TH concentrations not only typically correlate with natural vertebrate metamorphosis, but the treatment of

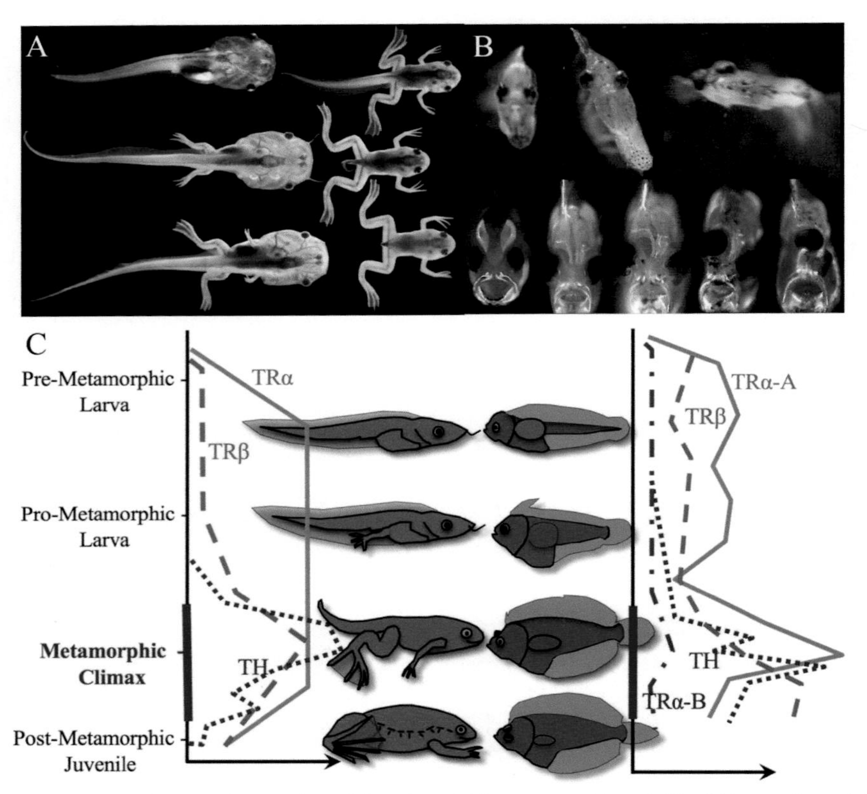

Figure 10.19 Metamorphosis in vertebrates is mediated by thyroid hormone. Panel **(A)**: African clawed frog (*Xenopus laevis)* metamorphosis. Top left: premetamorphosis (Nieuwkoop-Faber stage 53 stage); middle left: prometamorphosis (NF stage 57); bottom left: early metamorphic climax (NF stage 59); top right: mid-climax (NF stage 62); middle right: late climax (NF stage 65); bottom right: post-metamorphic juvenile (NF stage 66). Panel **(B)**: Southern flounder (*Paralichthys lethostigma*) metamorphosis. The top row depicts how larval flatfish initially possess bilateral symmetry and swim with an upright posture. However, they develop a lateralized swimming posture and morphological asymmetry as one eye translocates to the opposite side of the head during metamorphosis. The bottom row shows the asymmetric remodeling of the skull that accommodates eye translocation. Bone is visualized *in vivo* via the fluorescent calcium binding stain calcein. Panel **(C)**: A comparison of developmental profiles for thyroid hormone and its receptors in the Japanese flounder (*Paralichthys olivaceus*) and in the African clawed frog (*Xenopus laevis*). Thyroid hormone (TH, dotted lines), thyroid hormone receptor a (TRα/TRα-A, solid lines), TRα-B (alternating dash-dot lines), and TRβ (dashed lines). Data are derived from Shi (2000) (*Xenopus*) and Yamano and Miwa (1998) (flounder).

Source of Images: Schreiber (2013) Flatfish: an Asymmetric Perspective on Metamorphosis. In: "Animal Metamorphosis" (ed. Yun-Bo Shi), Current Topics in Developmental Biology, volume 103, pp. 167-194. Used by permission.

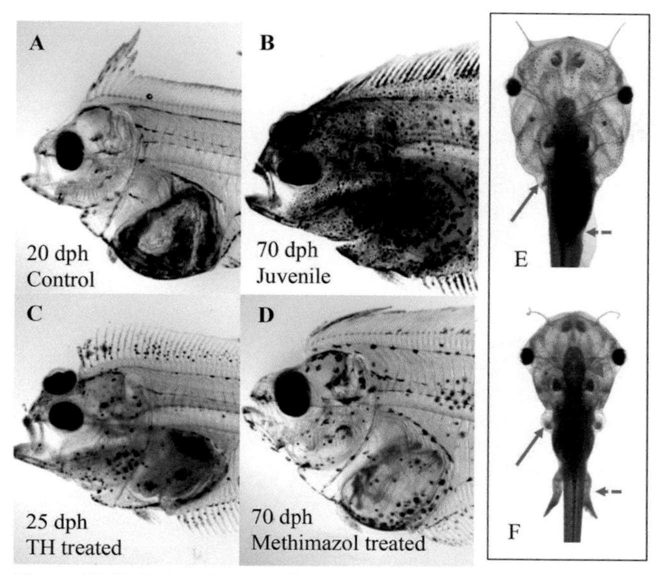

Figure 10.20 Induction and inhibition of metamorphosis with exogenous thyroid hormone treatment or chemical ablation of thyroid hormone synthesis in frogs and flounder. (A–D) Southern flounder, *Paralichthys lethodtigma* **(A)** A 20-day post-hatch (dph) fish is a bilaterally symmetrical pre-metamorphic larvae. **(B)** Metamorphosis, typified by the migration of the right eye to the left side of the head, has taken place by 50 dph, and a 70 dph fish has been growing as a post-metamorphic juvenile for several weeks. **(C)** Eye migration, as well as all other aspects of metamorphosis, can be induced precociously following treatment of a 20 dph larva with exogenous thyroid hormone for five days. **(D)** In contrast, treatment of a pre-metamorphic larva with methimazol (an inhibitor of thyroid hormone synthesis) inhibits eye migration and all other aspects of metamorphosis. **(E–F)** African clawed frog, *Xenopus laevis*. **(E)** Pro-metamorphic NF stage 54 tadpole (untreated). **(F)** NF 54 stage tadpole following treatment with exogenous thyroid hormone for 5 days. Morphological changes induced by TH visible in this figure include gill resorption (smaller head size), change in jaw shape, thickening of the brain due to cell proliferation (arrow head), and growth of the forelimbs (solid arrow) and hindlimbs (dashed arrow).

pre-metamorphic larvae with TH is sufficient to precociously induce virtually all developmental programs of metamorphosis (Figure 10.20). By contrast, the chemical ablation of TH synthesis inhibits metamorphosis altogether. As such, vertebrate metamorphosis is a naturally inducible system, one whose diverse and complex developmental programs can be easily initiated or halted for the purposes of studying TH's molecular mechanisms of action, its effects on vertebrate organogenesis, tissue remodeling, and other basic physiological questions. In this manner, the actions of TH on metamorphic development are strikingly similar to its actions in neonatal mammalian development, described previously.[86–88]

Although thyroid hormones are required to initiate metamorphosis, the transformation cannot be completed in the absence of glucocorticoids produced by the interrenal gland (in frogs and fish the adrenal gland is called the "interrenal").[89,90] Glucocorticoids are thought to increase tissue sensitivity to TH and to modulate the rate of developmental progress induced by TH.[91,92] Furthermore, glucocorticoids may also exert direct actions on tissues independent of their effects on TH signaling that are required for the completion of metamorphosis.[90,93] The interactions of TH with glucocorticoids in mediating metamorphosis will be addressed in Chapter 14: Adrenal Hormones and the Stress Response.

Amphibians as Models for Studying Thyroid Hormone Mechanisms of Action

Amphibians are the most widely studied models for the endocrine regulation of vertebrate metamorphosis, with virtually every tissue in the tadpole body responding to thyroid hormone by changing.[94–96] Whereas some tissues respond to TH by growth and differentiation (e.g. the limbs and lungs), others undergo programmed cell death and complete resorption (e.g. tail and gills), and yet others remodel from a tadpole-specific to an adult-specific form (e.g. the gastrointestinal tract, brain, and skin). Of the many mysteries surrounding metamorphosis, perhaps none is as profound as determining the mechanisms by which TH almost simultaneously exerts different, and sometimes opposite, effects on different tissues. For example, how does TH induce the same tissue, skeletal muscle, to grow in the limbs but die in the tail? Microarray analysis of thyroid hormone-responsive genes from multiple organs of tadpoles induced to metamorphose with exogenous TH treatment has shed some light on similarities and differences among developmental programs.[97,98] When comparing differentially regulated TH-responsive genes in the tail, brain, and limbs, several intriguing observations have been made (summarized in Figure 10.21):

1. For all programs there are far fewer downregulated genes compared with upregulated.
2. Even though the brain, limb, and tail are developmentally affected by TH in dramatically different ways, they all share a core set of genes regulated by TH.
3. Transcription factors represent the category containing the most upregulated genes in common among the organs.
4. Excluding the core sets of commonly regulated genes, the majority of genes regulated by TH are organ-specific.[99,100]

These findings support a model whereby the core set of commonly regulated genes contain thyroid hormone response elements (TREs) in their regulatory regions and are directly TH-responsive. Following their expression, these TH direct-response genes, most of which are transcription factors, then go on to regulate the tissue-specific expressions of other genes. A fascinating and still unresolved question asks, how did TH gain direct control over so many genes of amphibian metamorphosis? More precisely, by what mechanism did TREs spread and insert themselves into the

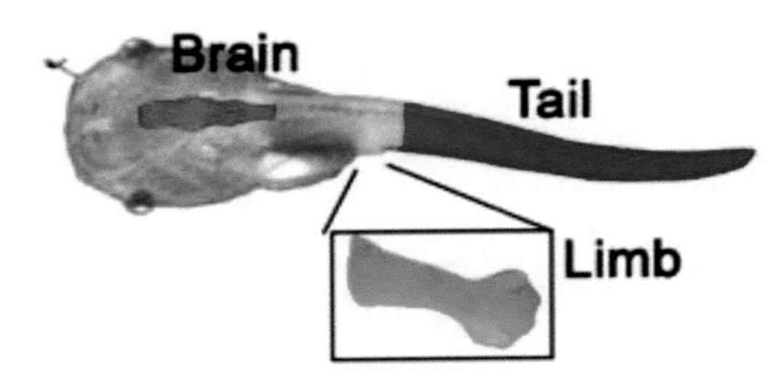

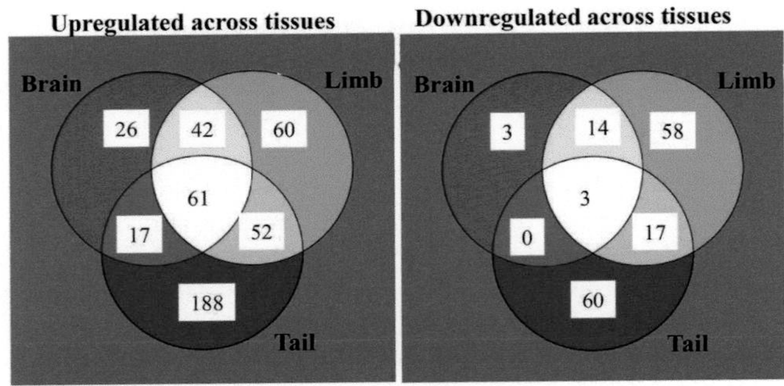

Figure 10.21 Venn diagram comparison of genes regulated by T_3 across brain, hind limb, and tail. The number of genes upregulated or downregulated in pre-metamorphic tadpoles after 2 days of 100 nM T_3 treatment in brain, hind limb, and tail are indicated in the white boxes.

Source of Images: Tadpole image: Das, B., et al. 2006. Dev. Biol., 291(2), pp. 342-355 used by permission. Venn diagrams: Buchholz, D.R., et al. 2007. Dev. Biol., 303(2), pp. 576-590. Used by permission.

amphibian genome to co-opt the expression of these genes, and why is such control confined only to some fishes and amphibians and not the more derived vertebrates like mammals and reptiles?

Developments & Directions: Thyroid hormones in the mediation of adaptive radiation and phenotypic plasticity in fish

Sometimes the degree of morphological, physiological, and/or behavioral phenotypic divergence among populations may become significant enough to produce varying degrees of *adaptive radiation*, or the rapid diversification of an ancestral population into several ecologically different species, associated with adaptive morphological or physiological divergence.[101] An example of this can be seen in the adaptive radiation of the three-spine stickleback species complex (*Gasterosteus aculeatus*), a group of fish whose marine ancestors invaded freshwater habitats, giving rise to a remarkable adaptive radiation (Figure 10.22[A–C]).[102,103] Indeed, a compelling study by Kitano and colleagues[104] has implicated evolutionary modifications of thyroid hormone signaling in the adaptive divergence of marine and freshwater stickleback populations. Considering that thyroid hormone is known to regulate metabolism, swimming behavior, salinity preference and tolerance, olfactory cellular proliferation, and sexual maturation in many fish, Kitano and colleagues hypothesized that thyroid hormone physiology may have diverged between marine and stream-resident sticklebacks with contrasting migratory behaviors. The researchers found that stream-resident fish exhibit a lower plasma concentration of thyroid hormone and a lower metabolic rate, which is likely adaptive for permanent residency in small streams (Figure 10.22[D–E]). Furthermore, the thyroid-stimulating hormone-β2 (TSHβ2) gene exhibited significantly lower mRNA expression in pituitary glands of genetically crossed stream-resident sticklebacks relative to crossed marine sticklebacks due to regulatory differences at the TSHβ2 gene locus caused by divergent natural selection (Figure 10.22[F]). The researchers therefore concluded that the marine and stream-resident sticklebacks have genetically based differences in TH concentration, and the low TH concentration in stream-resident fish does not simply result from environmental factors in freshwater, such as lower environmental levels of iodine, but from adaptive evolutionary changes in thyroid hormone signaling.

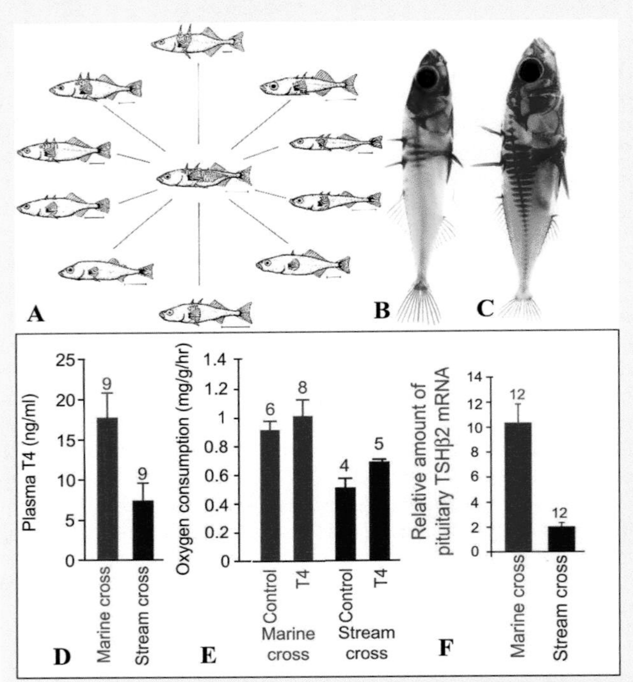

Figure 10.22 The adaptive radiation of the threespine stickleback, *Gasterosteus aculeatus*, and divergence in the thyroid hormone signaling pathway. (A) The center image represents the ancestral oceanic type that is the most widespread throughout its distribution. Peripheral images are freshwater derivatives that exhibit armor modification and loss as well as other changes in morphology. Stickleback fish from a lake **(B)** and from the ocean **(C)** stained red for bone to show differences in bony plates and spines between the ecotypes. Plasma T_4 concentration **(D)**, oxygen consumption **(E)**, and amount of pituitary TSHβ2 mRNA **(F)** are all significantly higher in marine crosses than stream crosses of laboratory-raised and genetically crossed sticklebacks raised under identical conditions.

Source of Images: (A) Foster, S.A., et al. 2015. Heredity, 115(4), pp. 335-348; used by permission. (B-C) Bolotovskiy, A.A., et al. 2018. PLoS One, 13(3), p. e0194040; (D-F) Kitano, J., et al. 2010. Curr. Biol., 20(23), pp. 2124-2130.

SUMMARY AND SYNTHESIS QUESTIONS

1. How does the endostyle reflect the evolutionary link between thyroid hormone and diet?
2. Suppose you raise frog tadpoles in the presence of the goitrogen thiourea for one year and generate giant tadpoles. If you then add thyroid hormone to the water of these thiourea-treated tadpoles, will they begin to metamorphose? Explain your reasoning.
3. Explain the significance of the observation that TRs are present in newly hatched amphibian tadpoles despite the fact that they do not yet have a functioning thyroid gland or circulating TH.
4. Thyroid hormone receptors are members of the steroid hormone receptor superfamily. However, whereas thyroid hormone receptor function accommodates the "dual function" model critical for modulating the larval state and inducing metamorphosis, steroid hormone receptors do not accommodate this model. Why not?
5. Tadpole development is characterized by the growth of the limbs and brain during prometamorphosis (when circulating TH levels are very low), followed by gill resorption and intestinal remodeling during early metamorphic climax (when TH levels begin to rise steeply), and lastly by tail resorption during late climax (when TH levels are highest). (A) What could explain the high sensitivity of the limbs to low levels of TH during prometamorphosis? (B) What could prevent the tail from resorbing during prometamorphosis and early climax?

Thyroid Disorders

LEARNING OBJECTIVE Discuss the etiology and pathophysiology of the major disorders of thyroid function.

KEY CONCEPTS:

- Hypothyroidism is characterized by a decreased production of or reduced responsivity to thyroid hormones, and is accompanied by high levels of serum TSH caused by negative feedback.
- If fetuses and neonates experience untreated hypothyroidism, this can lead to a severe retardation of growth and development known as cretinism.
- Congenital hypothyroidism due to iodine deficiency is the predominant cause of preventable intellectual disability.
- After dietary iodine deficiency, the leading cause of hypothyroidism in adults is Hashimoto's thyroiditis, an autoimmune condition that destroys the thyroid gland.
- Hyperthyroidism refers to increased thyroid hormone levels resulting from overactivation of the thyroid gland.
- Graves' disease, the most common form of hyperthyroidism, is an autoimmune disorder where antibodies activate the TSH receptor, stimulating increased TH synthesis.

Thyroid dysfunctions are among the most commonly observed human endocrine disorders, with hypothyroidism being the most common thyroid disorder, followed by hyperthyroidism, thyroid carcinoma, and thyroid hormone resistance. A summary of human thyroid pathophysiology is listed in Table 10.4.

Assessment of Thyroid Status

Radioactive Iodine Uptake

As described earlier, the thyroid gland concentrates iodine via the transmembrane sodium-iodine symporter (NIS). Therefore, the uptake of orally administered or injected radioactive iodine isotopes (typically ^{123}I or ^{131}I) or the radioactive compound ^{99m}Tc-pertechnetate (which is also transported by the NIS) by the thyroid can be measured and imaged using a scintillation counter. The ability to assess the thyroid's iodine uptake capacity and functional anatomy due to isotope uptake is a useful diagnostic tool for assessing thyroid status (Figure 10.23). A high uptake of iodine is seen in cases of hyperthyroidism due to Graves' disease or toxic nodular goiters, focal lesions present as "hot" nodules. A non-surgical method commonly used to ablate the thyroid to treat these conditions is to administer larger doses of ^{131}I, which induces localized tissue damaged caused by the emitted β-radiation. The presence of a "cold" nodule can denote a region of thyroidal tissue lacking NIS expression and could be indicative of a malignant tumor.

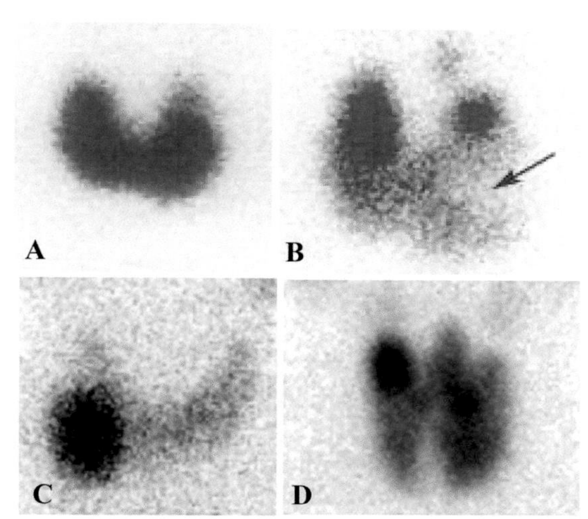

Figure 10.23 Thyroid imaging using a scintillation counter following administration of radioactive iodine isotopes. (A) 24 hours after ingestion of iodine-123 by a patient with a normal thyroid. **(B)** A scan of a "cold" thyroid nodule (arrow) failing to accumulate iodide isotope. **(C)** Scan of a "hot" thyroid nodule that accumulates excess iodide isotope. **(D)** Patient with a multinodular goiter shows multiple hot nodules involving both lobes of the gland.

Source of Images: (A-B) Nussey, S. and Whitehead, S., 2001. The thyroid gland in Endocrinology. London, GB: An Integrated Approach by Published by BIOS Scientific Publishers Ltd.; Informa UK (Taylor and Francis); used with permission. (C) Ahmed, S., et al. 2020. Cureus, 12(8); (D) Sitasuwan, T., et al. 2015. BMC Endocr. Disorders, 15(1), pp. 1-6.

Immunoassays for TSH and Free TH

Assuming there is no dysfunction of the hypothalamus or pituitary gland, the measurement of serum TSH via immunoassay is currently the best screening test for evaluating human thyroid status.[105] The usefulness of this method in assessing the thyroid lies in the log-linear relationship that exists between serum TSH and the levels of free TH (unbound to plasma TH binding-proteins) due to the negative feedback relationship between plasma TH and TSH synthesis (Figure 10.24). Although it is also possible to measure plasma T_4 and T_3 levels directly via immunoassay, the measurement of total circulating TH is strongly influenced by the concentration of TH binding proteins. Recall that over 99% of circulating TH is bound to serum-binding proteins, and the inherently high variability in TH binding protein levels can obscure the diagnosis of a thyroid dysfunction, minimizing the usefulness of direct serum TH measurements in a clinical setting. Furthermore, the actual bioavailable fraction of circulating TH, which consists in large part of the unbound TH, cannot be distinguished from the total serum TH content using immunoassays alone.

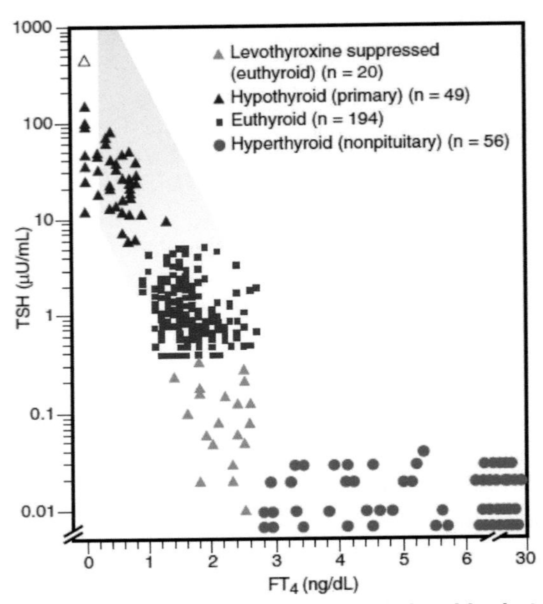

Figure 10.24 The log/linear inverse relationship between TSH (on the vertical axis) and the free T_4 concentrations (FT_4). Typical free T_4 concentrations in hypothyroid, euthyroid, and hyperthyroid patients are shown. A two-fold change in the serum FT_4 results in a 160-fold change in serum TSH.

Source of Images: Barrett, K.E., et al. 2010. Ganong's review of medical physiology twenty. McGraw Hill LLC. Used with permission.

Hypothyroidism

Hypothyroidism is one of several clinical states characterized by a decreased production of or reduced responsivity to thyroid hormones (Table 10.4) and is accompanied

Table 10.4 Human Thyroid Pathophysiology

Super-Category	Subcategory	Features

Hypothyroidism (hypothyroxinemia)	Primary	• Familial or congenital thyroid dysgenesis
		• Cretinism (fetal/infant onset)
		• Goitrous cretinism (lack of dietary iodide)
		• Athyreotic cretinism (congenital absence of thyroid gland)
		• Simple goiter (lack of dietary iodide)
		• Myxedema (adult onset)
		• Primary myxedema
		  • Hashimoto's thyroiditis (autoimmune disorder)
		  • Idiopathic (atrophic thyroiditis of unknown origin)
		• Iatrogenic (surgical removal or chemical inactivation)

	Secondary	Failure of pituitary to produce TSH
	Tertiary	Failure of hypothalamus to produce TRH
	Familial peripheral (end-organ) resistance	Lack of or defective thyroid hormone receptors
	Familial thyroid hormone-binding protein defects	
	Low T3 syndrome	Failure of peripheral conversion of T_4 to T_3 by deiodinases
	Allan–Herndon–Dudley syndrome	Severe psychomotor retardation due to mutations in transmembrane TH transporters and reduced TH in the developing brain
	Thyroid gland resistance to TSH	Defective TSH receptor
Hyperthyroidism (thyrotoxicosis)	Primary	Adenoma producing excessive thyroid hormone
	Secondary	Excessive production of TSH, pituitary thyrotrope, unresponsiveness to T3 feedback inhibition
	Tertiary	Excess production of TRH
	Graves' disease	Autoimmune agonistic antibodies of TSH receptors

by reduced uptake of iodine by the thyroid gland. Both human and animal studies indicate that moderate degrees of thyroid hormone insufficiency, particularly during late fetal and postnatal development, can impact brain structure and function and suggest that this may be continuous across the range of subclinical hypothyroidism.[106] The diverse mental and physiological symptoms of hypothyroidism (summarized in Table 10.5), which include reduced metabolic rate, weight gain, and depression, are generally easily treated with oral thyroxine supplements.

Primary hypothyroidism specifically refers to an inability of the thyroid gland to synthesize normal levels of TH due to a destruction or dysfunction of the organ, while the hypothalamus and pituitary gland are functioning normally. *Congenital hypothyroidism*, often caused by maternal and/or fetal iodine insufficiency, describes a condition of thyroid hormone deficiency present at birth. This condition afflicts approximately one in 4,000 newborn infants and is the predominant cause of preventable intellectual disability. Treatment, which consists of daily oral doses of thyroxine, is simple, effective, and inexpensive and underscores the importance of newborn screening to detect and treat this condition within the first few weeks following birth.

Iodine Deficiency

Dietary iodine deficiency is the most common cause of hypothyroidism and preventable mental impairment worldwide,[107] and since the 1920s an effective method of public iodine supplementation comes in the form of iodized salt. Nonetheless, over 2 billion individuals worldwide still have insufficient iodine intake, and cretinism and goiter are still prevalent conditions, particularly in south Asia and sub-Saharan Africa.[108] Importantly, even mild iodine deficiency during pregnancy can lead to fetal brain damage, decreased intelligence quotient (IQ), and impaired cognitive functions in school-age children.[109–111]

Curiously Enough . . . Unlike iodized salt, the popular culinary additive "sea salt", which consists of unprocessed salt produced through the evaporation of ocean water, does not contain high levels of iodine. Though this may seem counterintuitive given the high iodine concentrations in the ocean, iodine's high level of volatility causes it to be vaporized during the heating and drying process of "sea salt" synthesis. Likewise, when cooking with iodized table salt, heating food for too long at a high temperature will also cause the iodine to be volatilized.

Cretinism

If fetuses and neonates experience untreated hypothyroidism, this can lead to severe retardation of growth and development known as cretinism (refer to Figure 10.2 and Figure 10.18). This syndrome is still common in many nations due to dietary iodide deficiency and inadequate health care. If hypothyroid neonates are supplemented at

Table 10.5 Some Features Accompanying Human Hypothyroidism and Hyperthyroidism

	Hypothyroidism	Hyperthyroidism
Thyroid	Reduced T4 and T3 levels	Elevated T4 and T3 levels
	Goiter may or may not be present	Goiter present
Metabolism	Low basal metabolic rate (BMR)	High basal metabolic rate (BMR)
	Bradycardia, and hypotension	Tachycardia, and hypertension
	Low body temperature	High body temperature
	Cold intolerance	Heat intolerance
	Reduced perspiration	Increased perspiration
	Weight gain	Weight loss
Skin/hair/nails/face	Pale, dry, puffy skin (myxedema)	Pink, moist skin
	Brittle hair and nails, hair loss	Onycholysis (detachment of nails from bed)
	Swelling of face and eyelids	Bulging eyes
Bone development	Poor skeletal growth, reduced urinary excretion of Ca^{2+}	Bone demineralization, hypercalcemia, increased urinary excretion of Ca^{++} and PO_4^{3-}
Muscle	Stiffness, aching	Weakness, fatigability, wasting
Gastrointestinal	Constipation, decreased appetite	Hyperdefecation, increased appetite
Sleep and behavior	Depression	Anxiety
	Lethargy, sleepiness	Restlessness, insomnia
Reproduction	Delayed puberty, menorrhagia, infertility	Delayed puberty, amenorrhea, pregnancy loss
Central nervous system	*In children*: poor myelination and neuronal development (cretinism) *In adults*: slowed intellectual functions	

birth with thyroid hormone, normal growth and development can be reestablished from that point onwards. However, if the initiation of thyroid hormone treatment is delayed for too long, subsequent thyroid hormone treatment will not restore cognitive development, although growth rate may still be enhanced (Figure 10.18). In the case of neurological cretinism where the fetus has developed under hypothyroid conditions, supplementation of TH or iodine after birth will not reverse brain damage.

Goitrous Hypothyroidism

An inability to synthesize TH results in increased TSH production caused by reduced feedback inhibition. The thyroid gland can grow markedly in size (Figure 10.25), forming a goiter, because elevated TSH induces the thyroid gland to synthesize more thyroglobulin. However, the lack of iodine intake prevents the iodination of thyroglobulin and the synthesis of TH. Compared with a healthy thyroid gland, the continual synthesis and storage of thyroglobulin results in greatly enlarged follicles due to the larger colloid space; furthermore, the follicular epithelial cells appear unusually thin due to their inactive state and inability to synthesize TH.

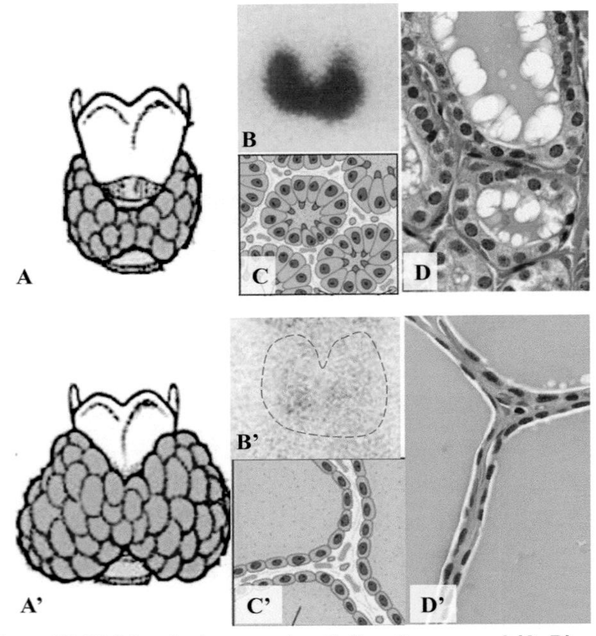

Figure 10.25 Morphology and activity of a normal (A–D) versus a hypothyroid (A'–D') thyroid gland. (A) The thyroid gland of a healthy person who is actively secreting thyroid hormone. **(B)** Thyroid autoradiography following administration of radioactive iodine isotopes. **(C)** Diagram and **(D)** histology from a normal active thyroid consists of active follicles lined by thick columnar epithelial cells surrounding a relatively small colloid. **(A')** The thyroid gland of a person with a hypothyroid goiter is considerably larger compared with normal. **(B')** Hypothyroid follicles have reduced radioactive iodine uptake and possess enlarged colloid with flattened epithelium (**C'** and **D'**), consistent with reduced activity.

Source of Images: (A, A') Stojsavljević, A., et al. 2020. Expos. Health, 12(2), pp. 255-264; used with permission. (B, B') Nussey, S. and Whitehead, S., 2001. The thyroid gland in Endocrinology. London, GB: An Integrated Approach by Published by BIOS Scientific Publishers Ltd.; Informa UK (Taylor & Francis). Used with permission. (C, C') Barrett, K.E., et al. 2010. Ganong's review of medical physiology twenty; McGraw Hill LLC, used with permission. (D, D') Courtesy of Andrey Bychkov MD, PhD.

Autoimmune Hypothyroidism: Hashimoto's Thyroiditis

After dietary iodine deficiency, the leading cause of hypothyroidism is a disease known as **Hashimoto's thyroiditis.**[112] Named after Japanese physician Hakaru Hashimoto, who first described the condition in 1912, the disease is characterized by the development of several cell- and antibody-mediated immune processes that destroy the thyroid gland (Figure 10.26), with antibodies most frequently targeting thyroid peroxidase and thyroglobulin proteins.[113–115] Thyroiditis ultimately leads to hypothyroidism and is

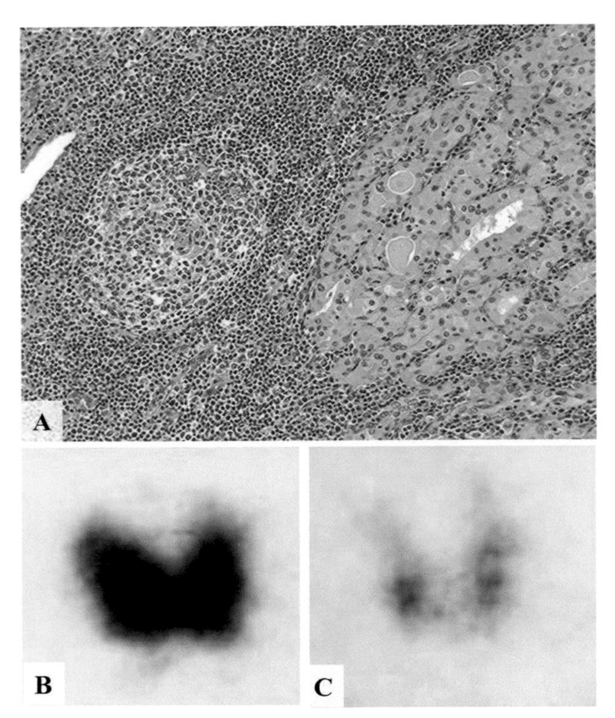

Figure 10.26 Histology and scintigraphy of patients with Hashimoto's thyroiditis. (A) A dense infiltrate of plasma cells and lymphocytes (most of the punctate purple nuclear staining) with germinal center formation is seen in this thyroid of a patient with Hashimoto's thyroiditis. **(B)** Fluorescent thyroid scan in a patient with a normal thyroid with stable I-127 uptake throughout both lobes. **(C)** In a patient with Hashimoto's thyroiditis, a marked reduction in I-127 uptake is apparent throughout the entire gland.

Source of Images: (A) Courtesy of Dr. Peter Takizawa, Yale University; (B) Zaletel, K. and Gaberscek, S., 2011. Curr. Genom., 12(8), pp. 576-588.

characterized by reduced uptake of radioactive iodine revealed by a thyroid scan (Figure 10.26). Considering that thyroid function is impaired, it may at first seem counterintuitive to note that the disease symptoms often include goiter formation. However, in contrast with goiter caused by low iodide uptake, the goiter of thyroiditis results in large part from lymphocytic infiltration, as well as TSH-stimulated hyperplasia of surviving thyroid tissue due to loss of feedback inhibition from TH.

Curiously Enough . . . Hashimoto's thyroiditis was the first recognized autoimmune disease.

Myxedema

Myxedema is a term often used synonymously with severe adult-onset hypothyroidism, typically associated with late stages of Hashimoto's thyroiditis. W.M. Ord first coined the term in 1878 in reference to the dermatological change accompanied by severe hypothyroidism caused by a "jelly-like swelling of the connective tissue". It is now known that hypothyroidism is accompanied by the accumulation of mucus-like material consisting of protein complexed with chondroitin sulfate and hyaluronic acid in extracellular spaces, particularly in the skin. Due to its osmotic effect, these proteins cause water to accumulate in these spaces, inducing the characteristic swelling. In 1891, British physician George Redmayne Murray presented a case of myxedema to his colleagues at the Northumberland and Durham Medical Society in England. His patient, "Mrs. H.", was a 42-year-old woman whose symptoms included swelling in the face and hands, hypersensitivity to cold, extreme lethargy, and slowness of speech (Figure 10.27).

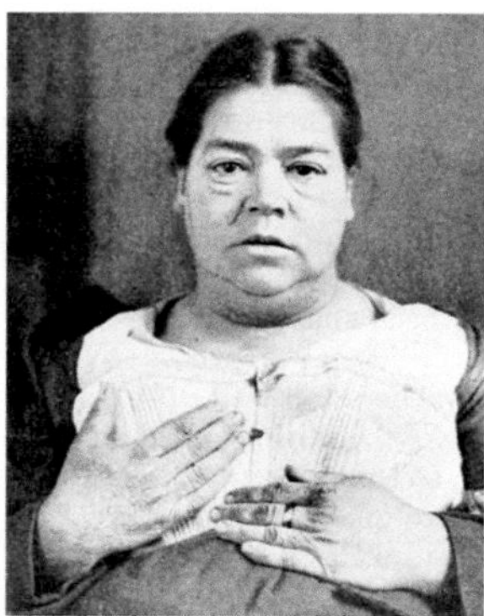

Myxedema before treatment   After 6 months treatment

Figure 10.27 Myxedema before and after treatment. In 1891 Dr. G.R. Murray presented at a meeting of a medical society in England a woman with myxedema, known as "Mrs. H." Murray treated his patient with subcutaneous injections of the extract of the thyroid gland of the sheep. When about six months later he showed the same patient, improved in every way, the success of the treatment was established.

Source of Images: Cattell, J.M., 1876. The Popular Science Monthly (Vol. 8). Wentworth Press.

Murray noticed that these general symptoms of myxedema corresponded exactly to a condition suffered by patients following surgical removal of the thyroid gland. Working against the advice of his colleagues, Murray treated his patient with injections of sheep thyroid gland extract, and Mrs. H.'s condition soon improved dramatically. Importantly, Murray's treatment was the first example of a human endocrine disease to be successfully treated with hormone replacement therapy.

Secondary and Tertiary Hypothyroidism

Less frequently, hypothyroidism may also result from a loss of tropic stimulation due to damage or disease of the hypothalamic-pituitary axis, reducing the production of TRH and/or TSH. A reduced ability of the pituitary to produce TSH, referred to as *secondary hypothyroidism*, may arise from trauma to or the accidental surgical ablation of thyrotrope cells. This condition may also arise due to the formation of a large adenoma that compresses and damages the thyrotropes. *Tertiary hypothyroidism* caused by similar damages to TRH-secreting hypothalamic nuclei will also inhibit the production of both TSH and TH. A particularly useful method for distinguishing primary versus secondary and tertiary thyroid disorders is a *TRH-TSH stimulation test*. Whereas the administration of TRH will normally increase plasma levels of TSH within 30 minutes after administration, no increase in plasma TSH is observed in hypothyroidism caused by a pituitary dysfunction (secondary hypothyroidism). Since there is no available reliable assay for serum TRH, hypothalamic malfunction (i.e. tertiary hypothyroidism) can be inferred from increasing TSH levels following TRH administration.

Paradigm Shifts in the Treatment of Hypothyroidism

Since ancient times when the Chinese and Romans used burnt seaweed and sponge to treat goiter, medical science has undergone two major paradigm shifts in the treatment of hypothyroidism. The first paradigm shift, which started in the 1890s, was to either inject mammalian thyroid gland extracts into hypothyroid patients or have them eat various forms of mammalian thyroid glands or extracts, typically desiccated thyroid glands, or extracts in tablet form. This treatment was so successful that it persisted until the 1970s, several decades after the discovery and purification of thyroxine. Beginning in the 1960s, synthetic *levothyroxine* (sodium L-thyroxine) became a popular and effective alternative to thyroid extracts, which were highly variable in potency. Indeed, levothyroxine monotherapy, the second paradigm shift, remains the current standard for management of primary, as well as central, hypothyroidism.[116] Today, synthetic thyroid hormones are among the most widely prescribed medications in the world.[117]

Hyperthyroidism

Whereas *hyperthyroidism* specifically refers to increased thyroid hormone levels resulting from overactivation of the thyroid gland, the broader term **thyrotoxicosis** describes a generalized presence of elevated serum thyroid hormones independent of the source—that is, high TH may be of endogenous, exogenous, or ectopic origin. Both conditions are diagnosed by low levels of serum TSH that are unresponsive to a TRH-TSH stimulation test, as well as elevated iodine uptake by the thyroid gland. The diverse physiological symptoms of hyperthyroidism are generally opposite to those that typify hypothyroidism (Table 10.5). Common treatments for hyperthyroidism include the use of thionomide drugs, such as methimazol and propylthiouracil, that inhibit TH synthesis (see Table 10.1). Other treatments include radioiodine therapy to ablate overactive thyroid tissue and surgical ablation of the thyroid gland. Interestingly, the fastest acting antithyroid compound is iodine itself, which reduces TH synthesis via an autoregulatory feedback mechanism known as the Wolff-Chaikoff effect.[118] Although no longer widely used, the treatment still remains effective as a rapid response to life threatening conditions such as "thyroid storm", described next.

Cardiotoxicosis and "Thyroid Storm"

The thyrotoxic state is accompanied by profound circulatory alterations, including increased total blood volume, decreased systemic vascular resistance, tachycardia (increased heart rate), and a generalized hyperdynamic circulatory state that imposes a hypervolemic stress on the heart resulting in increased cardiac work.[119] Thyroid hormones are known to control several genes coding for structural and regulatory cardiac proteins and hence strongly influence heart contractility.[120] Despite normal or low serum concentrations of catecholamines, patients with hyperthyroidism also display symptoms of increased adrenergic activity, which include increased heart rate, widened pulse pressure, and increased cardiac output. Studies of the various components of the adrenergic-receptor complex in plasma membranes have shown that β-adrenergic receptors, GPCRs, and adenylyl cyclases are altered by changes in thyroid status, the net effect of which promotes sensitivity of the heart to adrenergic stimulation.[120] Cardiovascular disease caused by thyrotoxicosis, referred to as **cardiotoxicosis**, may manifest itself as atrial fibrillation and congestive heart failure and is associated with a mortality of 20%–50%.[121] Severe thyrotoxicosis (or "*thyroid storm*") is a rare but life-threatening condition that develops in cases of untreated thyrotoxicosis and may develop following either the suspension of antithyroid drug treatment or an incomplete thyroidectomy. The condition is characterized by a rapid and dramatic increase in TH secretion accompanied by a hypermetabolic state and high fever.

Autoimmune Hyperthyroidism: Graves' Disease

Graves' disease is the underlying cause of 50%–80% of cases of hyperthyroidism and impacts 0.5% of the population, with women affected at rates five- to ten-fold higher than men.[122] The cause of Graves' disease is neither an inherently dysfunctional thyroid gland nor a direct pathology of the hypothalamic-pituitary axis. Rather, it is an autoimmune disorder that results in the production of antibodies that bind to and activate the TSH receptor, stimulating follicular hypertrophy and hyperplasia, thyroid gland enlargement, and increased TH synthesis (Figure 10.28). Unlike TSH, however, the autoantibodies are not regulated by negative feedback, and consequently they overstimulate the thyroid. For this reason these autoantibodies are called "thyroid-stimulating immunoglobulins" (TSI).

In addition to the typical symptoms of hyperthyroidism, such as weight loss, heat intolerance, and tachycardia, Graves' disease can also be accompanied by conditions known as Graves' ophthalmopathy and Graves' dermopathy. *Graves' ophthalmopathy* (Figure 10.29) is a potentially sight-threatening ocular disease characterized by the enlargement of extraocular muscle due to the accumulation of collagen fibrils and glycosaminoglycans. Current evidence suggests that intramuscular fibroblast

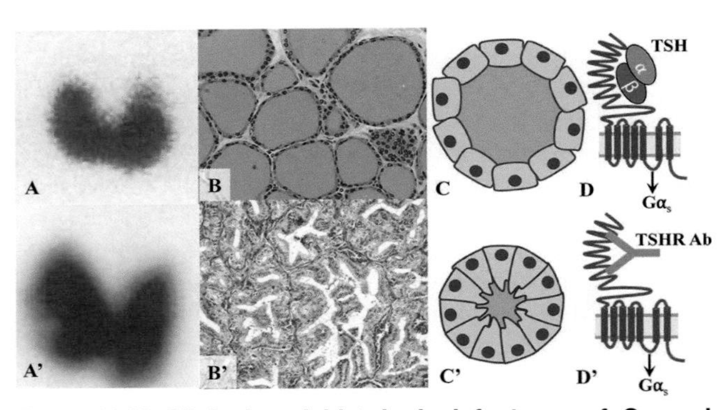

Figure 10.28 Clinical and histological features of Graves' disease. Radioiodine scans of the thyroid obtained 24 hours after ingestion of iodine-123 by a patient with a normal thyroid **(A)** and a patient with Graves' disease **(A')**. The thyroid of the patient with Graves' disease is larger and concentrates a higher fraction of radioiodine. (Images courtesy of Dr. Jerome Hershman, David Geffen School of Medicine at UCLA, Los Angeles.) Compared with the thyroid follicles from a healthy person **(B)**, the follicle epithelial cells in this patient **(B')** no longer surround round follicles of colloid. Instead, these follicles are disappearing as the colloid is becoming rapidly depleted and converted into thyroid hormone. Compared with euthyroid follicles **(C)**, whose large colloid region is surrounded by cuboidal epithelial cells, hyperthyroid follicles **(C')** are smaller with less colloid and columnar epithelia. Whereas TSH normally stimulates the TSH receptor (TSHR) to synthesize and release TH via a $G\alpha_s$ intracellular signaling pathway **(D)**, in Graves' disease the TSHR is stimulated by the presence of autoimmune antibodies that function as TSHR agonists **(D')**.

Source of Images: (A, A') Nussey, S. and Whitehead, S., 2001. The thyroid gland in Endocrinology. London, GB: An Integrated Approach by Published by BIOS Scientific Publishers Ltd.; Informa UK (Taylor & Francis), used with permission. (B) Courtesy of Dr. Peter Takizawa, Yale University; (B') Erickson, L.A., 2014. Graves Disease (Diffuse Hyperplasia). In Atlas of Endocrine Pathology (pp. 21-24). Springer, New York, NY; used with permission.

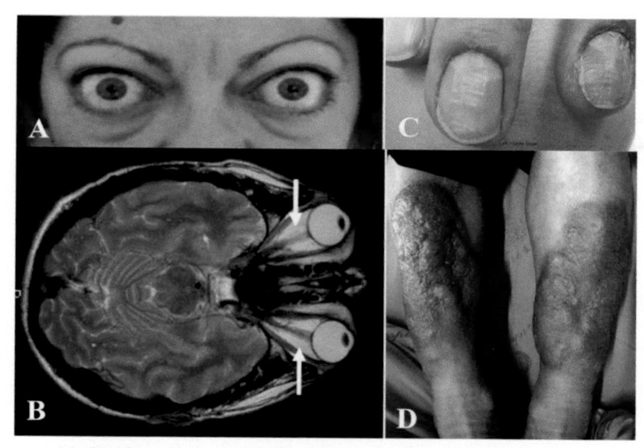

Figure 10.29 Some morphological features of severe hyperthyroidism: Graves' opthalmology and dermopathy. (A) A patient with severe Graves' ophthalmopathy. **(B)** Eye displacement results from the thickening of the extraocular retroorbital muscle and adipose tissue (arrows) caused by circulating TSH receptor autoantibodies. **(C)** "Onycholysis": fingernails become detached from the nail beds at the free margins. **(D)** Pretibial myxedema: red, swollen skin, usually on the shins and tops of the feet.

Source of Images: (A-B) Bruscolini, A., et al. 2018. J. Immunol. Res., 2018; (C) Reinecke, J.K. and Hinshaw, M.A., 2020. Int. J. Wom. Dermatol., 6(2), pp. 73-79; (D) Lan, C., et al. J. Thyroid Res., 2016.

proteins are autoantigens in Graves' ophthalmopathy.[123] The enlargement of ocular muscles is also accompanied by the inflammation and expansion of orbital fat, whose combined effect is to cause the eyes to bulge significantly. Interestingly, adipocytes are known to contain TSH receptors, which are thought to induce fat cell proliferation under hyperthyroid conditions, partially explaining this phenotype. *Graves' dermopathy* (Figure 10.29), a complication of Graves' disease, is a condition characterized by red, swollen skin, usually on the shins and tops of the feet (pretibial myxedema). Pretibial myxedema may be accompanied by swelling of the soft tissues that surround the bones and muscles in the fingers, hands, and feet, a condition called *acropachy*. Like Graves' ophthalmopathy, Graves' dermopathy is caused by increased deposition of connective tissue components (like glycosaminoglycans), but in subcutaneous tissue. The condition may also be accompanied by detachment of fingernails from the nail beds (onycholysis) and the growth of thin, brittle hair.

Primary, Secondary, and Tertiary Hyperthyroidism

The overproduction of TH by either TSH-dependent or autonomously hyperactive adenoma or multinodular goiter produces a condition called *toxic goiter.*

These goiters fall under the category of *primary hyperthyroidism*, since the thyroid tissue itself is abnormal but the hypothalamus and pituitary gland are functioning normally. *Secondary hyperthyroidism*, or the oversecretion of TSH by the anterior pituitary, can be caused by a rare TSH-secreting pituitary adenoma. *Tertiary hyperthyroidism*, a hypothetical condition induced by oversecretion of TRH, is theoretically possible, although actual cases of tertiary hyperthyroidism remain undocumented to date.

SUMMARY AND SYNTHESIS QUESTIONS

1. An inverse log/linear relationship exists between circulating TH and serum TSH concentrations, such that accurate estimates of TH levels can be derived from TSH measurements. What is the biological basis of the inverse (as opposed to directly proportional) nature of the relationship between these two hormones?
2. Explain how relative levels of circulating TRH and TSH would change under conditions of primary, secondary, and tertiary hypothyroidism.
3. Explain how relative levels of circulating TRH and TSH would change under conditions of primary, secondary, and tertiary hyperthyroidism.
4. In contrast with most hypo-endocrine conditions in which the gland becomes smaller in total size, hypothyroidism caused by an iodine deficiency is accompanied by an increased total thyroid gland size. Explain this observation.
5. Under conditions of Graves' disease (a form of hyperthyroidism), the colloid space within thyroid follicles is typically much smaller compared to euthyroid follicles. Why?
6. Compared with the euthyroid (healthy) state, do you expect the sizes of individual thyroid follicle cells to be smaller or larger in a patient with Graves' disease? Explain your reasoning.
7. Imagine that you treat a normal mouse with the thyroid hormone synthesis inhibitor methimazole. Compared with an untreated normal mouse, do you expect the sizes of individual thyroid follicle cells to be smaller or larger? Explain your reasoning.
8. Now, imagine that you treat a normal mouse with exogenously administered thyroid hormone. Compared with an untreated normal mouse, do you expect the sizes of individual thyroid follicle cells to be smaller or larger? Explain your reasoning.
9. Compare and contrast the etiology and pathophysiology of Hashimoto's thyroiditis versus Graves' disease.
10. Why are women more prone to developing a goiter during pregnancy than at other times?
11. Why are only small improvements in cognition evident when hypothyroid children suffering from cretinism are administered thyroid hormones?

12. You are a doctor and you have an infant patient exhibiting all of the physiological signs of hypothyroidism, yet the patient has normal levels of TRH, TSH, and thyroid hormones. Explain how this is possible.
13. In the late 19th and early 20th centuries, when the function of the thyroid gland was not well understood, goiters were often surgically removed. Indeed, Emil Theodor Kocher, a Swiss physician and medical researcher, received the 1909 Nobel Prize in Physiology or Medicine for his work in the physiology, pathology and surgery of the thyroid. After performing over 5,000 thyroidectomies and finding signs of physical and mental decay in many patients, Kocher ultimately came to the conclusion that a complete removal of the thyroid was not advisable. Explain why surgical removal of a goitrous thyroid is generally not a healthy option.

Thyroid-Disrupting Chemicals

LEARNING OBJECTIVE Defend the statement, and place it into the context of endocrine-disrupting compounds: "The single hormone system with the most influence on brain and cognitive development is, arguably, that of thyroid hormone".

KEY CONCEPTS:

- Low maternal thyroid hormone levels during the first trimester of pregnancy predict a higher risk of delays in a child's psychomotor development.
- Many EDCs with structural similarity to thyroid hormones disrupt gene transcription by functioning as thyroid receptor agonists or antagonists.
- Other thyroid-disrupting EDCs inhibit thyroid hormone synthesis or transport, or accelerate thyroid hormone clearance from the body.

The single hormone system with the most influence on brain and cognitive development is, arguably, that of thyroid hormone.[117] Recall from earlier that congenital hypothyroidism is the predominant cause of preventable intellectual dysfunction worldwide, and that even mild iodine deficiency during pregnancy can lead to fetal brain damage, decreased IQ, and impaired cognitive functions in school-age children.[109–111] Considering TH's profound impacts on the brain's development of normal neuronal myelination, branching, connectivity, and cellular architecture during the fetal and early childhood stages, it is not surprising that endocrine-disrupting chemicals (EDCs) that disrupt TH signaling can also have major impacts on cognitive development. Furthermore, because thyroid hormone synthesis and signaling is among the most complex of any hormone, it provides a particularly broad array of potential targets for different categories of EDC to disrupt (Figure 10.30). This section describes

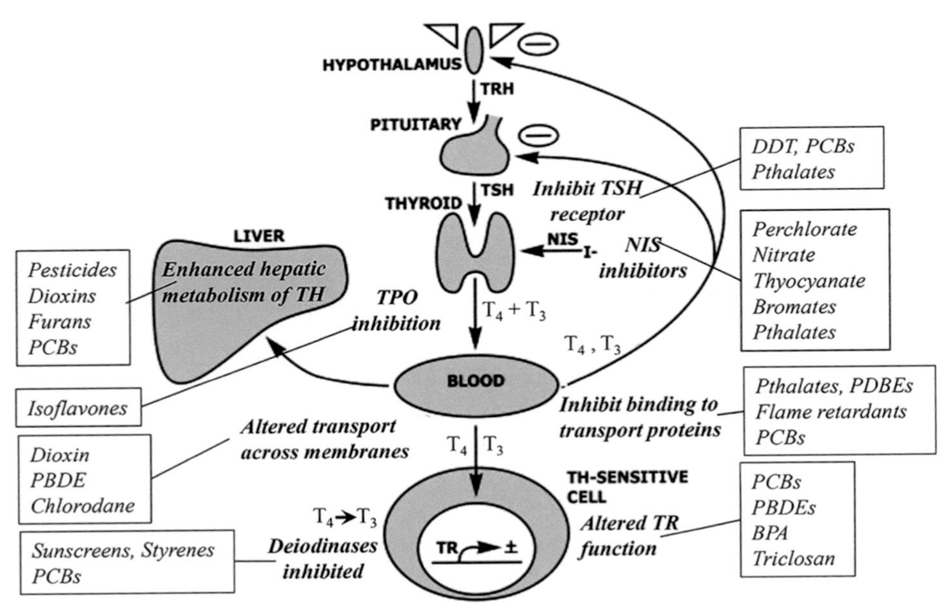

Figure 10.30 Endocrine-disrupting chemicals (EDCs) act at multiple levels of the hypothalamus-pituitary-thyroid (HPT) axis. Italicized names in boxes are known EDCs. Abbreviations: BPA, bisphenol A; NIS, sodium iodide symporters; PBDE, polybrominated diphenyl ethers; PCB, polychlorinated biphenyl; TRH, thyroid-releasing hormone; TSH, thyroid-stimulating hormone; TR, thyroid receptor; T_4, thyroxine; T_3, triiodothyronine.

Source of Images: Pearce, E.N. and Braverman, L.E., 2009. Best Pract. Res. Clin. Endocrinol. Metab., 23(6), pp. 801-813. Used with permission.

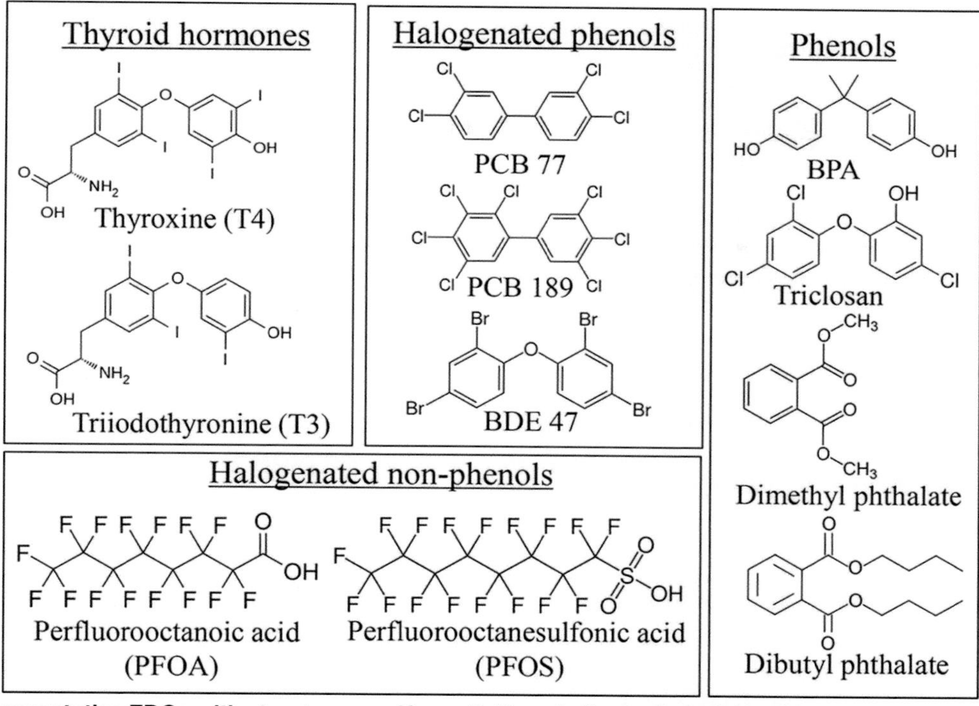

Figure 10.31 Representative EDCs with structures and/or activities similar to that of thyroid hormone.

molecular mechanisms by which some of the best-studied EDCs disrupt TH function.

Disrupted Gene Expression: Structural Similarities between THs and Key EDCs

Recall that THs are the only vertebrate hormones that possess halogenated (in this case complexed with the halogen iodine) biphenol groups that allow them to interact with TRs. Many EDCs bear striking structural similarities with THs, with some consisting of halogenated (chlorinated and brominated) biphenols that are eerily similar to TH, others with halogenated monophenolic structures, and some with non-halogenated biphenols (Figure 10.31). Importantly, these and other compounds, as well as their metabolites, can mimic TH and bind to TRs as agonists or antagonists, thereby disrupting diverse aspects of TH-mediated gene expression and feedback.[124]

Hypothyroidism: Disruption of TH Transportation and Metabolism

In addition to binding to TRs, in mammalian animal models PCBs, PBDEs, phenol compounds, phthalates, and their respective metabolites have been shown bind to *transthyretin* (TTR), a T_4 blood carrier and storage protein, displacing its binding of T_4 and interfering with its transport and storage.[125–130] In contrast to other TH transport proteins that are only present in the blood, TTR is also synthesized by the fetal central nervous system (CNS), and CNS-derived TTR subsequently resides in the cerebrospinal fluid, providing the developing CNS with TH.[131] Since maximal TTR synthesis by the developing mammalian fetal brain begins immediately before the period of rapid brain growth, this suggests that brain-derived TTR plays a role in transporting T_4 from the blood into the cerebrospinal fluid to facilitate brain development.[132] As such, EDCs that interfere with the binding of TH to TTR may inhibit the uptake of T_4 by the developing fetal brain, resulting in cognitive and other developmental impairment.[133]

Possibly the most common mechanism of TH disruption by PCBs and other EDCs is not through competitive binding to TRs and TTR, but through their promotion of hypothyroidism by increasing the rates of TH degradation by the liver. Specifically, dioxins and PCBs have been shown to bind to *aryl hydrocarbon receptors* (AhR) that function as "environmental xenobiotic toxin sensors". The activation of these receptors promotes increased transcription of several enzymes, including *uridine diphosphate glucuronyl transferases* (UDPGTs), that promote the *glucuronidation* (addition of a glucuronic acid group) of xenobiotic compounds and drugs, increasing their solubility and rates of excretion.[134,135] Importantly, UDPGTs also play a role in metabolizing and excreting excess T_4 from the body. In essence, the presence of EDCs stimulates the liver to go into detoxification "overdrive", which can cause excess T_4 to be excreted in bile, promoting hypothyroidism.[117] The combined effects of EDCs disrupting TH transportation at the level of TTR and increasing rates of TH metabolism by the liver appear to interact to produce a significant reduction in both total and free TH levels.[136–138]

SUMMARY AND SYNTHESIS QUESTIONS

1. In adult humans, albumin and thyroid-binding globulin are thought to play more important roles in transporting TH in the blood compared with transthyretin (TTR). Why, then, are EDCs that displace TH from TTR of significant concern?
2. Some EDCs do not interfere with TH's binding to the TR or TTR but nonetheless reduce circulating levels of TH. How?
3. The element selenium is essential for the function of deiodinase enzymes. Mercury is an EDC that, among other effects, interferes with the ability of selenium to associate with deiodinase enzymes. Considering this, what effect would mercury poisoning have on thyroid hormone signaling?

Summary of Chapter Learning Objectives and Key Concepts

LEARNING OBJECTIVE Explain the connections between environmental iodine distribution and its correlation with cretinism, goiter, and other types of hypothyroidism in humans worldwide.

- Iodine is essential for thyroid hormone synthesis.
- Low dietary iodine is the most common cause of cretinism, goiter, and other types of hypothyroidism in humans worldwide.
- Over hundreds of millions of years of vertebrate evolution, the structure of thyroid hormone has remained entirely unchanged.

LEARNING OBJECTIVE Describe the developmental and functional morphology of the mammalian thyroid gland and the different hormones that it produces.

- The mammalian thyroid gland synthesizes two categories of hormones: (1) thyroid hormones that are produced by thyroid follicles and modulate development, growth, and metabolism, and (2) calcitonin, which is made by parafollicular "C cells" and is involved in calcium metabolism.
- Parathyroid glands that secrete another molecule involved in calcium metabolism, "parathyroid hormone", form distinct islands of tissue embedded within the mammalian thyroid gland, but these are not considered thyroid tissue.

LEARNING OBJECTIVE Trace the evolutionary origins of the vertebrate thyroid gland from protochordate ancestors.

- In the most ancestral living protochordates, thyroid hormone is secreted into the gut via an *exo*crine gland (the endostyle), after which the hormone is subsequently absorbed into the bloodstream.
- During metamorphosis of the lamprey, which is among the most basal living vertebrates, the larval exocrine endostyle remodels into *endo*crine thyroid follicles that secrete hormone directly into the bloodstream.
- Whereas the thyroid follicles of all tetrapods are condensed into distinct organs by the end of embryogenesis, the thyroid follicles of many fishes are dispersed throughout various vascularized regions of the body.

LEARNING OBJECTIVE Explain the mechanisms of thyroid hormone biosynthesis and its regulation.

- Thyroid hormone biosynthesis is a process by which two modified tyrosine molecules are joined and covalently bound to iodine.
- Thyroid hormone synthesis is under central control by hypothalamic thyrotropin-releasing hormone (TRH), which stimulates the production of thyroid-stimulating hormone (TSH) by the anterior pituitary gland. TSH, in turn, induces the thyroid gland to synthesize and release thyroid hormone.
- The hypothalamic-pituitary axis is under negative feedback influence from circulating thyroid hormones.

LEARNING OBJECTIVE Describe the roles of blood- and intracellular binding proteins, as well as peripheral intracellular deiodinases in modulating peripheral thyroid hormone activity.

- Blood serum thyroid hormone-binding proteins serve as a reservoir from which to maintain and resupply the unbound and biologically active free fraction of TH.
- Thyroid hormone-specific transmembrane transport proteins modulate thyroid hormone uptake by cells.
- Cytosolic T_3-binding proteins function as a cytoplasmic hormone reservoir and regulate transport between the cytoplasm and the nucleus.
- Intracellular T_3 concentrations are modulated by enzymes called deiodinases that activate or deactivate TH.

LEARNING OBJECTIVE Describe the molecular mechanisms of action of thyroid hormone and its receptors in modulating gene expression.

- Nuclear thyroid hormone receptors (TRs) are members of the steroid receptor superfamily and function as ligand-dependent transcription factors.
- TRs exert "dual control" over transcription of TH-responsive genes: in the absence of thyroid hormone they function as transcriptional repressors of positively regulated genes, and in the presence of TH they become transcriptional activators.

LEARNING OBJECTIVE Describe the developmental and growth effects of thyroid hormones on diverse target tissues.

- TH is essential for both fetal and postnatal development of the mammalian brain, heart, skeletal muscle, lungs, and gut.
- TH exerts profound effects on vertebrate growth, and its presence is required to maintain normal levels of growth hormone.

LEARNING OBJECTIVE Explain how vertebrate metamorphosis is a useful model for studying the mechanisms of thyroid hormone action.

- In all metamorphosing vertebrates, the larval to juvenile transformation is mediated by changing levels of thyroid hormone.
- As a naturally inducible system where exogenously administered TH initiates profound developmental changes in virtually every tadpole organ, the amphibian metamorphosis model lends itself to investigating the molecular mechanisms of TH actions, organogenesis, and tissue remodeling.
- TH exerts differential effects on tissues by targeting TH-responsive genes that code for transcription factors.

LEARNING OBJECTIVE Discuss the etiology and pathophysiology of the major disorders of thyroid function.

- Hypothyroidism is characterized by a decreased production of or reduced responsivity to thyroid hormones, and is accompanied by high levels of serum TSH caused by negative feedback.
- If fetuses and neonates experience untreated hypothyroidism, this can lead to a severe retardation of growth and development known as cretinism.
- Congenital hypothyroidism due to iodine deficiency is the predominant cause of preventable intellectual disability.
- After dietary iodine deficiency, the leading cause of hypothyroidism is a disease known as Hashimoto's thyroiditis, an autoimmune condition that destroys the thyroid gland.
- Hyperthyroidism refers to increased thyroid hormone levels resulting from overactivation of the thyroid gland.
- Graves' disease, the most common form of hyperthyroidism, is an autoimmune disorder where antibodies activate the TSH receptor, stimulating increased TH synthesis.

LEARNING OBJECTIVE Defend the statement, and place it into the context of endocrine-disrupting compounds: "The single hormone system with the most influence on brain and cognitive development is, arguably, that of thyroid hormone".

- Low maternal thyroid hormone levels during the first trimester of pregnancy predict a higher risk of delays in a child's psychomotor development.
- Many EDCs with structural similarity to thyroid hormones disrupt gene transcription by functioning as thyroid receptor agonists or antagonists.
- Other thyroid-disrupting EDCs inhibit thyroid hormone synthesis or transport, or accelerate thyroid hormone clearance from the body.

LITERATURE CITED

1. Gorbman A, Dickhoff WW, Vigna SR, Clark NB, Ralph CL. *Comparative Endocrinology*. John Wiley and Sons; 1983.
2. Büyükgebiz A. Newborn screening for congenital hypothyroidism. *J Clin Res Pediatr Endocrinol*. 2013;5(Suppl 1):8.
3. Medvei VC. *The History of Clinical Endocrinology: A Comprehensive Account of Endocrinology from Earliest Times to the Present Day*. CRC Press; 1993.
4. Hoskins RG. *The Tides of Life*. W.W. Norton; 1946.
5. Lason AH. *The Thyroid Gland in Medical History*. Frogen Press; 1946.
6. Magnus-Levey A. Ueber den respiratorischen Gaswechsel unter Einfluss der Thyroidea sowie unter verschiedenen pathologische Zustand. *Berliner Klinische Wochenschrift*. 1895;32:650–652.
7. Baumann E. Uber den Jodgehalt der Schilddr ?usen von Menschen und tieren. *Hoppe-Seyler's Z Physiol Chem*. 1896;22:1–17.
8. Gudernatch. Feeding experiments on tadpoles. I. The influence of specific organs given as food on growth and differentiation. A contribution to the knowledge of organs with internal secretion. *Arch Entwicklungsmech Org*. 1912(35):457–483.
9. Kendall EC. Landmark article, June 19, 1915. The isolation in crystalline form of the compound containing iodin, which occurs in the thyroid. Its chemical nature and physiologic activity. By E.C. Kendall. *JAMA*. 1983 [1915];250(15):2045–2046.
10. Harington CR. Chemistry of thyroxine: isolation of thyroxine from the thyroid gland. *Biochem J*. 1926;20:293–299.
11. Gross J, Pitt-Rivers R. The identification of 3:5:3'-L-triiodothyronine in human plasma. *Lancet*. 1952;1(6705):439–441.
12. Roche J, Lissitzky S, R. M. Sur la triiothyronine, produit intermédiaire de la transformation de la diiodothyronine en thyroxine. *C R Acad Sci Paris*. 1952;234:997–998.
13. Roche J, Lissitzky SRM. Sur la présence de la triodothyronine dans la thyroglobuline. *C R Acad Sci Paris*. 1952;234:1228–1230.
14. Kovacs WJ, Ojeda SR. *Textbook of Endocrine Physiology* (6th ed.). Oxford University Press; 2011.
15. Policeni BA, Smoker WRK, Reede DL. Anatomy and embryology of the thyroid and parathyroid glands. *Semin Ultrasount CT MRI*. 2012;33:104–114.
16. Studer H, Peter HJ, Gerber H. Natural heterogeneity of thyroid cells: the basis for understanding thyroid function and nodular goiter growth. *Endocr Rev*. 1989;10(2):125–135.
17. Dumont JE, Opitz R, Christophe D, Vassart G, Roger PP, Maenhaut C. Ontogeny, anatomy, metabolism and physiology of the thyroid. https://www.thyroidmanager.org/. 2011.
18. Heyland A, Hodin J. Heterochronic developmental shift caused by thyroid hormone in larval sand dollars and its implications for phenotypic plasticity and the evolution of nonfeeding development. *Evolution Int J Org Evol*. 2004;58(3):524–538.
19. Chino Y, Saito M, Yamasu K, Suyemitsu T, Ishihara K. Formation of the adult rudiment of sea urchins is influenced by thyroid hormones. *Dev Biol*. 1994;161(1):1–11.
20. Eales JG. Iodine metabolism and thyroid-related functions in organisms lacking thyroid follicles: are thyroid hormones also vitamins? *Proc Soc Exp Biol Med*. 1997;214(4):302–317.
21. Miller AE, Heyland A. Iodine accumulation in sea urchin larvae is dependent on peroxide. *J Exp Biol*. 2012;216(Pt 5):915–926.
22. Heyland A, Reitzel AM, Degnan S. Emerging patterns in the regulation of marine invertebrate settlement and metmorphosis. In: Flatt TH, Heyland A, eds. *Mechanisms of Life History Evolution*. Oxford University Press; 2011:29–42.
23. Heyland A, Moroz LL. Signaling mechanisms underlying metamorphic transitions in animals. *Integr Comp Biol*. 2006;46(6):743–759.
24. Heyland A, Price DA, Bodnarova-Buganova M, Moroz LL. Thyroid hormone metabolism and peroxidase function in two

non-chordate animals. *J Exp Zool B Mol Dev Evol*. 2006;306(6):551–566.

25. Saito M, Yamasu K, Suyemitsu T. Binding properties of thyroxine to nuclear extract from sea urchin larvae. *Zoolog Sci*. 2012;29(2):79–82.
26. Heimberg AM, Cowper-Sal-lari R, Semon M, Donoghue PC, Peterson KJ. microRNAs reveal the interrelationships of hagfish, lampreys, and gnathostomes and the nature of the ancestral vertebrate. *Proc Natl Acad Sci U S A*. 2010;107(45):19379–19383.
27. Schmidt F, Braunbeck T. Alterations along the hypothalamic-pituitary-thyroid axis of the zebrafish (danio rerio) after exposure to propylthiouracil. *J Thyroid Res*. 2011;2011:376243.
28. Manzon RG. Thyroidal regulation of life history transitions in fish. In: Flatt T, Heyland A, eds. *Mechanisms of Life History Transitions: The Genetics and Physiology of Life History Traits and Trade-Offs*. Oxford University Press; 2011.
29. Salvatore D, Davies TF, Schlumberger M-J, Hay ID, Larsen PR. Thyroid physiology and diagnostic evaluation of patients with thyroid disorders. In: Melmed S, Polonsky KS, Larsen PR, Kronenberg HM, eds. *William's Textbook of Endocrinology* (12th ed.). Saunders-Elsevier; 2011:327–361.
30. Tietze F, Kohn LD, Kohn AD, et al. Carrier-mediated transport of monoiodotyrosine out of thyroid cell lysosomes. *J Biol Chem*. 1989;264(9):4762–4765.
31. Krause K, Karger S, Gimm O, et al. Characterisation of DEHAL1 expression in thyroid pathologies. *Eur J Endocrinol*. 2007;156(3):295–301.
32. Dumitrescu AM, Refetoff S. Novel biological and clinical aspects of thyroid hormone metabolism. *Endocr Dev*. 2007;10:127–139.
33. Bernal J. The significance of thyroid hormone transporters in the brain. *Endocrinology*. 2005;146(4):1698–1700.
34. van der Deure WM, Peeters RP, Visser TJ. Molecular aspects of thyroid hormone transporters, including MCT8, MCT10, and OATPs, and the effects of genetic variation in these transporters. *J Mol Endocrinol*. 2010;44(1):1–11.
35. Bernal J, Guadano-Ferraz A, Morte B. Thyroid hormone transporters-functions and clinical implications. *Nat Rev Endocrinol*. 2015;11(7):406–417.
36. Brix K, Fuhrer D, Biebermann H. Molecules important for thyroid hormone synthesis and action—known facts and future perspectives. *Thyroid Res*. 2011;4(Suppl 1):S9.
37. Denver RJ. Neuroendocrinology of amphibian metamorphosis. *Curr Top Dev Biol*. 2013;103:195–227.
38. Prummel MF, Brokken LJ, Wiersinga WM. Ultra short-loop feedback control of thyrotropin secretion. *Thyroid*. 2004;14(10):825–829.
39. Brent GA, Mestman JH. Physiology and tests of thyroid function. In: Sciarra JJ, ed. *Gynecology and Obstetrics*, Vol 5. Lippincott, Williams and Wilkins; 2004.
40. Mendel CM, Weisiger RA, Jones AL, Cavalieri RR. Thyroid hormone-binding proteins in plasma facilitate uniform distribution of thyroxine within tissues: a perfused rat liver study. *Endocrinology*. 1987;120(5):1742–1749.
41. Jansen J, Friesema EC, Milici C, Visser TJ. Thyroid hormone transporters in health and disease. *Thyroid*. 2005;15(8):757–768.
42. Visser WE, Friesema EC, Jansen J, Visser TJ. Thyroid hormone transport in and out of cells. *Trends Endocrinol Metab*. 2008;19(2):50–56.
43. Groeneweg S, van Geest FS, Peeters RP, Heuer H, Visser WE. Thyroid hormone transporters. *Endocr Rev*. 2020;41(2):bnz008.
44. Roberts LM, Woodford K, Zhou M, et al. Expression of the thyroid hormone transporters monocarboxylate transporter-8 (SLC16A2) and organic ion transporter-14 (SLCO1C1) at the blood-brain barrier. *Endocrinology*. 2008;149(12):6251–6261.
45. Mori J, Suzuki S, Kobayashi M, et al. Nicotinamide adenine dinucleotide phosphate-dependent cytosolic T(3) binding protein as a regulator for T(3)-mediated transactivation. *Endocrinology*. 2002;143(4):1538–1544.
46. Schimmel M, Utiger RD. Thyroidal and peripheral production of thyroid hormones. Review of recent findings and their clinical implications. *Ann Intern Med*. 1977;87(6):760–768.
47. Schweizer U, Weitzel JM, Schomburg L. Think globally: act locally. New insights into the local regulation of thyroid hormone availability challenge long accepted dogmas. *Mol Cell Endocrinol*. 2008;289(1–2):1–9.
48. Ng L, Goodyear RJ, Woods CA, et al. Hearing loss and retarded cochlear development in mice lacking type 2 iodothyronine deiodinase. *Proc Natl Acad Sci U S A*. 2004;101(10):3474–3479.
49. Campos-Barros A, Amma LL, Faris JS, Shailam R, Kelley MW, Forrest D. Type 2 iodothyronine deiodinase expression in the cochlea before the onset of hearing. *Proc Natl Acad Sci U S A*. 2000;97(3):1287–1292.
50. Ng L, Kelley MW, Forrest D. Making sense with thyroid hormone—the role of T(3) in auditory development. *Nat Rev Endocrinol*. 2013;9(5):296–307.
51. Qi X, Chan WL, Read RJ, Zhou A, Carrell RW. Temperature-responsive release of thyroxine and its environmental adaptation in Australians. *Proc Royal Soc B Biol Sci*. 2014;281(1779).
52. Davis PJ, Goglia F, Leonard JL. Nongenomic actions of thyroid hormone. *Nat Rev Endocrinol*. 2016;12(2):111–121.
53. Oppenheimer JH, Schwartz HL, Strait KA. An integrated view of thyroid hormone actions in vivo. In: Weintraub BD, ed. *Molecular Endocrinology: Basic Concepts and Clinical Correlations*. Raven Press; 1995:249–268.
54. Tata JR, Widnell CC. Ribonucleic acid synthesis during the early action of thyroid hormones. *Biochem J*. 1966;98(2):604–620.
55. Samuels HH, Tsai JS, Casanova J, Stanley F. Thyroid hormone action: in vitro characterization of solubilized nuclear receptors from rat liver and cultured GH1 cells. *J Clin Invest*. 1974;54(4):853–865.
56. Oppenheimer JH, Schwartz HL, Surks MI. Tissue differences in the concentration of triiodothyronine nuclear binding sites in the rat: liver, kidney, pituitary, heart, brain, spleen, and testis. *Endocrinology*. 1974;95(3):897–903.
57. Sap J, Munoz A, Damm K, et al. The c-erb-A protein is a high-affinity receptor for thyroid hormone. *Nature*. 1986;324(6098):635–640.
58. Weinberger C, Thompson CC, Ong ES, Lebo R, Gruol DJ, Evans RM. The c-erb-A gene encodes a thyroid hormone receptor. *Nature*. 1986;324(6098):641–646.
59. Laudet V. The origins and evolution of vertebrate metamorphosis. *Curr Biol*. 2011;21(18):R726–737.
60. Yang YZ, Burgos-Trinidad M, Wu Y, Koenig RJ. Thyroid hormone receptor variant alpha2. Role of the ninth heptad in DNA binding, heterodimerization with retinoid X receptors, and dominant negative activity. *J Biol Chem*. 1996;271(45):28235–28242.
61. Lazar MA, Hodin RA, Darling DS, Chin WW. Identification of a rat c-erbA alpha-related protein which binds deoxyribonucleic acid but does not bind thyroid hormone. *Mol Endocrinol*. 1988;2(10):893–901.
62. Ortiga-Carvalho TM, Sidhaye AR, Wondisford FE. Thyroid hormone receptors and resistance to thyroid hormone disorders. *Nat Rev Endocrinol*. 2014;10(10):582–591.
63. Flamant F, Samarut J. Thyroid hormone receptors: lessons from knockout and knock-in mutant mice. *Trends Endocrinol Metab*. 2003;14(2):85–90.
64. Bernal J, Pekonen F. Ontogenesis of the nuclear 3,5,3'-triiodothyronine receptor in the human fetal brain. *Endocrinology*. 1984;114(2):677–679.
65. Kester MH, Martinez de Mena R, Obregon MJ, et al. Iodothyronine levels in the human developing brain: major regulatory roles of iodothyronine deiodinases in different areas. *J Clin Endocrinol Metab*. 2004;89(7):3117–3128.
66. Bernal J, Morte B. Thyroid hormone receptor activity in the absence of ligand: physiological and developmental implications. *Biochim Biophys Acta*. 2013;1830(7):3893–3899.
67. Bernal J. Thyroid hormone receptors in brain development and function. *Nat Clin Pract Endocrinol Metab*. 2007;3(3):249–259.
68. Fisher DA, Nelson JC, Carlton EI, Wilcox RB. Maturation of human hypothalamic-

pituitary-thyroid function and control. *Thyroid*. 2000;10(3):229–234.

69. Pantos C, Mourouzis I, Xinaris C, Papadopoulou-Daifoti Z, Cokkinos D. Thyroid hormone and "cardiac metamorphosis": potential therapeutic implications. *Pharmacol Ther*. 2008;118(2):277–294.
70. Mai W, Janier MF, Allioli N, et al. Thyroid hormone receptor alpha is a molecular switch of cardiac function between fetal and postnatal life. *Proc Natl Acad Sci U S A*. 2004;101(28):10332–10337.
71. van Tuyl M, Blommaart PE, de Boer PA, et al. Prenatal exposure to thyroid hormone is necessary for normal postnatal development of murine heart and lungs. *Dev Biol*. 2004;272(1):104–117.
72. Schiaffino S, Reggiani C. Fiber types in mammalian skeletal muscles. *Physiol Rev*. 2011;91(4):1447–1531.
73. Sun G, Shi YB. Thyroid hormone regulation of adult intestinal stem cell development: mechanisms and evolutionary conservations. *Int J Biol Sci*. 2012;8(8):1217–1224.
74. Ishizuya-Oka A, Shi YB. Evolutionary insights into postembryonic development of adult intestinal stem cells. *Cell Biosci*. 2011;1:37.
75. Schreiber AM, Mukhi S, Brown DD. Cell-cell interactions during remodeling of the intestine at metamorphosis in Xenopus laevis. *Dev Biol*. 2009;331(1):89–98.
76. Giustina A, Wehrenberg WB. Influence of thyroid hormones on the regulation of growth hormone secretion. *Eur J Endocrinol*. 1995;133(6):646–653.
77. Martin D, Epelbaum J, Bluet-Pajot MT, Prelot M, Kordon C, Durand D. Thyroidectomy abolishes pulsatile growth hormone secretion without affecting hypothalamic somatostatin. *Neuroendocrinology*. 1985;41(6):476–481.
78. Katakami H, Downs TR, Frohman LA. Decreased hypothalamic growth hormone-releasing hormone content and pituitary responsiveness in hypothyroidism. *J Clin Invest*. 1986;77(5):1704–1711.
79. Harvey S, Scanes CG, Daughaday WH. *Growth Hormone* (1st ed.). CRC Press; 1994.
80. Koenig RJ, Brent GA, Warne RL, Larsen PR, Moore DD. Thyroid hormone receptor binds to a site in the rat growth hormone promoter required for induction by thyroid hormone. *Proc Natl Acad Sci U S A*. 1987;84(16):5670–5674.
81. Schreiber AM. Asymmetric craniofacial remodeling and lateralized behavior in larval flatfish. *J Exp Biol*. 2006;209(4):610–621.
82. Schreiber AM. Flatfish: an asymmetric perspective on metamorphosis. *Curr Top Dev Biol*. 2013;103:167–194.
83. Miller M. Ecology of anguilliform leptocephali: remarkable transparent fish larvae of the ocean surface layer. *AquaBioScience Mono*. 2009;2:1–94.
84. McMenamin SK, Parichy DM. Metamorphosis in teleosts. *Curr Top Devl Biol*. 2013;103:127–165.
85. Youson JH, Manzon RG. Lamprey metamorphosis. In: Doufour S, Rousseau K, Kapoor BG, eds. *Metamorphosis in Fish*. Science Publishers; 2012.
86. Shi YB, Ritchie JW, Taylor PM. Complex regulation of thyroid hormone action: multiple opportunities for pharmacological intervention. *Pharmacol Ther*. 2002;94(3):235–251.
87. Buchholz DR. More similar than you think: frog metamorphosis as a model of human perinatal endocrinology. *Dev Biol*. 2015;408(2):188–195.
88. Sachs LM, Buchholz DR. Frogs model man: in vivo thyroid hormone signaling during development. *Genesis*. 2017;55(1–2):e23000.
89. Kulkarni SS, Buchholz DR. Beyond synergy: corticosterone and thyroid hormone have numerous interaction effects on gene regulation in Xenopus tropicalis tadpoles. *Endocrinology*. 2012;153(11):5309–5324.
90. Sterner ZR, Shewade LH, Mertz KM, Sturgeon SM, Buchholz DR. Glucocorticoid receptor is required to survive through metamorphosis in the frog Xenopus tropicalis. *Gen Comp Endocr*. 2020:113419.
91. Bonett RM, Hoopfer ED, Denver RJ. Molecular mechanisms of corticosteroid synergy with thyroid hormone during tadpole metamorphosis. *Gen Comp Endocr*. 2010;168(2):209–219.
92. Kulkarni SS, Buchholz DR. Beyond synergy: corticosterone and thyroid hormone have numerous interaction effects on gene regulation in Xenopus tropicalis tadpoles. *Endocrinology*. 2012;153(11):5309–5324.
93. Sachs LM, Buchholz DR. Insufficiency of thyroid hormone in frog metamorphosis and the role of glucocorticoids. *Front Endocrinol*. 2019;10:287.
94. Berry DL, Rose CS, Remo BF, Brown DD. The expression pattern of thyroid hormone response genes in remodeling tadpole tissues defines distinct growth and resorption gene expression programs. *Dev Biol*. 1998;203(1):24–35.
95. Berry DL, Schwartzman RA, Brown DD. The expression pattern of thyroid hormone response genes in the tadpole tail identifies multiple resorption programs. *Dev Biol*. 1998;203(1):12–23.
96. Brown DD, Cai L. Amphibian metamorphosis. *Dev Biol*. 2007;306(1):20–33.
97. Das B, Cai L, Carter MG, et al. Gene expression changes at metamorphosis induced by thyroid hormone in Xenopus laevis tadpoles. *Dev Biol*. 2006;291(2):342–355.
98. Buchholz DR, Heimeier RA, Das B, Washington T, Shi Y-B. Pairing morphology with gene expression in thyroid hormone-induced intestinal remodeling and identification of a core set of TH-induced genes across tadpole tissues. *Dev Biol*. 2007;303(2):576–590.
99. Das B, Cai L, Carter MG, et al. Gene expression changes at metamorphosis induced by thyroid hormone in Xenopus laevis tadpoles. *Dev Biol*. 2006;291(2):342–355.
100. Buchholz DR, Heimeier RA, Das B, Washington T, Shi YB. Pairing morphology with gene expression in thyroid hormone-induced intestinal remodeling and identification of a core set of TH-induced genes across tadpole tissues. *Dev Biol*. 2007;303(2):576–590.
101. Seehausen O. Hybridization and adaptive radiation. *Trends Ecol Evol*. 2004;19(4):198–207.
102. Jones FC, Grabherr MG, Chan YF, et al. The genomic basis of adaptive evolution in threespine sticklebacks. *Nature*. 2012;484(7392):55.
103. Foster S, Wund M, Graham M, et al. Iterative development and the scope for plasticity: contrasts among trait categories in an adaptive radiation. *Heredity*. 2015;115(4):335.
104. Kitano J, Lema SC, Luckenbach JA, et al. Adaptive divergence in the thyroid hormone signaling pathway in the stickleback radiation. *Curr Biol*. 2010;20(23):2124–2130.
105. Brent G, Mestman J. Physiology and tests of thyroid function. In: Sciarra JJ, ed. *Gynecology and Obstetrics CD-ROM*, Vol 5: Lippincott Williams and Wilkins; 2004.
106. Gilbert ME, Rovet J, Chen Z, Koibuchi N. Developmental thyroid hormone disruption: prevalence, environmental contaminants and neurodevelopmental consequences. *Neurotoxicology*. 2012;33(4):842–852.
107. Zimmermann MB, Jooste PL, Pandav CS. Iodine-deficiency disorders. *Lancet*. 2008;372(9645):1251–1262.
108. Andersson M, Karumbunathan V, Zimmermann MB. Global iodine status in 2011 and trends over the past decade. *J Nutr*. 2012;142(4):744–750.
109. Bath SC, Steer CD, Golding J, Emmett P, Rayman MP. Effect of inadequate iodine status in UK pregnant women on cognitive outcomes in their children: results from the Avon Longitudinal Study of Parents and Children (ALSPAC). *Lancet*. 2013;382(9889):331–337.
110. Taylor PN, Okosieme OE, Dayan CM, Lazarus JH. Therapy of endocrine disease: impact of iodine supplementation in mild-to-moderate iodine deficiency: systematic review and meta-analysis. *Eur J Endocrinol*. 2014;170(1):R1–R15.
111. Trumpff C, De Schepper J, Tafforeau J, Van Oyen H, Vanderfaeillie J, Vandevijvere S. Mild iodine deficiency in pregnancy in Europe and its consequences for cognitive and psychomotor development of children: a review. *J Trace Elem Med Biol*. 2013;27(3):174–183.
112. Ross DS. Treatment of primary hypothyroidism in adults. *UpToDate*; 2018. https://www.uptodate.com/contents/7855#H3269250293 accessed April 12, 2023.
113. Pearce EN, Farwell AP, Braverman LE. Thyroiditis. *N Engl J Med*. 2003;348(26):2646–2655.

114. Quaratino S, Badami E, Pang YY, et al. Degenerate self-reactive human T-cell receptor causes spontaneous autoimmune disease in mice. *Nat Med.* 2004;10(9):920–926.
115. McLachlan SM, Rapoport B. Autoimmune hypothyroidism: T cells caught in the act. *Nat Med.* 2004;10(9): 895–896.
116. Wiersinga WM. Paradigm shifts in thyroid hormone replacement therapies for hypothyroidism. *Nat Rev Endocrinol.* 2014;10(3):164–174.
117. Demeneix B. *Toxic Cocktail: How Chemical Pollution Is Poisoning Our Brains.* Oxford University Press; 2017.
118. Nussey S, Whitehead S. *Endocrinology: An Integrated Approach.* BIOS Scientific Publishers Limited; 2001.
119. Woeber KA. Thyrotoxicosis and the heart. *N Engl J Med.* 1992;327(2):94–98.
120. Klein I, Ojamaa K. Thyroid hormone and the cardiovascular system. *N Engl J Med.* 2001;344(7):501–509.
121. Ross DS. Radioiodine therapy for hyperthyroidism. *N Engl J Med.* 2011;364(6):542–550.
122. Brent GA. Clinical practice. Graves' disease. *N Engl J Med.* 2008;358(24):2594–2605.
123. Bahn RS. Graves' ophthalmopathy. *N Engl J Med.* 2010;362(8):726–738.
124. Mughal BB, Fini J-B, Demeneix BA. Thyroid-disrupting chemicals and brain development: an update. *Endoc Connect.* 2018;7(4):R160–R186.
125. Yamauchi K, Ishihara A, Fukazawa H, Terao Y. Competitive interactions of chlorinated phenol compounds with 3, 3′, 5-triiodothyronine binding to transthyretin: detection of possible thyroid-disrupting chemicals in environmental waste water. *Toxicol Appl Pharmacol.* 2003;187(2):110–117.
126. Purkey HE, Palaninathan SK, Kent KC, et al. Hydroxylated polychlorinated biphenyls selectively bind transthyretin in blood and inhibit amyloidogenesis: rationalizing rodent PCB toxicity. *Chem Biol.* 2004;11(12):1719–1728.
127. Meerts IA, Van Zanden JJ, Luijks EA, et al. Potent competitive interactions of some brominated flame retardants and related compounds with human transthyretin in vitro. *Toxicol Sci.* 2000;56(1):95–104.
128. Kudo Y, Yamauchi K. In vitro and in vivo analysis of the thyroid disrupting activities of phenolic and phenol compounds in Xenopus laevis. *Toxicol Sci.* 2004;84(1):29–37.
129. Ishihara A, Nishiyama N, Sugiyama S-i, Yamauchi K. The effect of endocrine-disrupting chemicals on thyroid hormone binding to Japanese quail transthyretin and thyroid hormone receptor. *Gen Comp Endocr.* 2003;134(1):36–43.
130. Van den Berg K. Interaction of chlorinated phenols with thyroxine binding sites of human transthyretin, albumin and thyroid binding globulin. *Chem-Biol Interact.* 1990;76(1):63–75.
131. Rabah SA, Gowan IL, Pagnin M, Osman N, Richardson SJ. Thyroid hormone distributor proteins during development in vertebrates. *Front Endocrinol.* 2019;10:506.
132. Richardson SJ, Wijayagunaratne RC, D'Souza DG, Darras VM, Van Herck SL. Transport of thyroid hormones via the choroid plexus into the brain: the roles of transthyretin and thyroid hormone transmembrane transporters. *Front Neurosci.* 2015;9:66.
133. Ulbrich B, Stahlmann R. Developmental toxicity of polychlorinated biphenyls (PCBs): a systematic review of experimental data. *Arch Toxicol.* 2004;78(5):252–268.
134. Fisher JW, Campbell J, Muralidhara S, et al. Effect of PCB 126 on hepatic metabolism of thyroxine and perturbations in the hypothalamic-pituitary-thyroid axis in the rat. *Toxicol Sci.* 2005;90(1):87–95.
135. Kolaja KL, Klaassen CD. Dose—Response Examination of UDP-Glucuronosyltransferase Inducers and Their Ability to Increase both TGF-β Expression and Thyroid Follicular Cell Apoptosis. *Toxicol Sci.* 1998;46(1): 31–37.
136. Brouwer A, Morse DC, Lans MC, et al. Interactions of persistent environmental organohalogens with the thyroid hormone system: mechanisms and possible consequences for animal and human health. *Toxicol Ind Heal.* 1998;14(1–2):59–84.
137. Barter RA, Klaassen CD. UDP-glucuronosyltransferase inducers reduce thyroid hormone levels in rats by an extrathyroidal mechanism. *Toxicol Appl Pharmacol.* 1992;113(1):36–42.
138. Morse DC, Wehler EK, Wesseling W, Koeman JH, Brouwer A. Alterations in rat brain thyroid hormone status following pre-and postnatal exposure to polychlorinated biphenyls (Aroclor 1254). *Toxicol Appl Pharmacol.* 1996;136(2): 269–279.

CHAPTER

11

Insect Molting and Metamorphosis

CHAPTER LEARNING OBJECTIVES:

- Discuss the importance of insect metamorphosis to the study of endocrinology.
- Describe the three major developmental strategies insects use for reaching the adult stage.
- Delineate the roles played by major hormones that mediate insect molting and metamorphosis, and compare and contrast insect neuroendocrine systems with those of vertebrates.
- Discuss current theories regarding the evolutionary origins of insect metamorphosis.

OPENING QUOTATIONS:

"For the normal process of metamorphosis the presence of the of the brain, at least up to a certain moment, is indispensable."

—Stefan Kopeć (1922) Studies on the Necessity of the Brain for the Inception of Insect Metamorphosis. *Biological Bulletin*, Vol. 42, pp. 323–342

"The earth-bound early stages built enormous digestive tracts and hauled them around on caterpillar treads. Later in the life-history these assets could be liquidated and reinvested in the construction of an entirely new organism—a flying machine devoted to sex."

—Carrol M. Williams (1958). Hormonal regulation of insect metamorphosis. Pages 794–806 In W. D. McElroy and B. Glass, Eds. *A symposium on the chemical basis of development*. John Hopkins Press, Baltimore

Importance of Insect Molting and Metamorphosis to Endocrinology

LEARNING OBJECTIVE Discuss the importance of insect metamorphosis to the study of endocrinology.

KEY CONCEPTS:

- Insects provided the first hints of how steroid hormones exert transcriptional control over gene expression.
- The ability of insects to survive radical surgeries, like decapitations, neck tourniquets, and brain explants, provided valuable tools for founding the field of neuroendocrinology.
- There are more insect species (and therefore different insect endocrine systems) than all other animal, plant, and fungal species combined.
- Insect metamorphosis is triggered by changes in the timing of secretion of a small handful of hormones that are also involved in larval molting.

In "The Voyage of the H.M.S. Beagle", Charles Darwin recounts the story of a German naturalist named Renous whom he met in Chile.[1] In the 1830s Renous was arrested in San Fernando on the charge of heresy. His crime? Rumors suggested Renous had the power to turn caterpillars into butterflies. He was eventually released from jail, presumably after his collection of caterpillars metamorphosed into butterflies of their own accord. At the time, neither Renous nor any other scientist understood that hormones drive insect metamorphosis and virtually all other post-embryonic developmental, physiological, and behavioral processes. Insects are *arthropods*, a phylum of invertebrate animals that possess an exoskeleton, a segmented body plan, and jointed appendages. In addition to insects, phylum Arthropoda also includes arachnids (e.g. spiders, scorpions, and ticks), crustaceans (e.g. crabs, lobsters, shrimp), and myriapods (e.g. millipedes and centipedes).

In contrast to the dozens and perhaps hundreds of chemical signals that control early stages of embryonic specification and differentiation in insects, only a small handful of hormones control an extraordinarily diverse array of postembryonic developmental processes including growth, metamorphosis, and *molting*, or the replacement of an old exoskeleton with a new one.[2] When insects molt, they not only grow but can also change form, either

DOI: 10.1201/9781003359241-15

gradually with small morphological changes from molt to molt, or dramatically, with profound changes in morphology during metamorphosis. In the most dramatic examples of metamorphosis, the nervous system is completely rewired, the musculature is digested and rebuilt, the gut is replaced with a new epithelium, and the epidermis of the adult displaces the larval epidermal cells.[3] As we will see in this chapter, metamorphosis is triggered by changes in the timing of secretion of only a handful of hormones that are also involved in larval molting.

There are many reasons why insects are both fascinating and relevant to the field of endocrinology as a whole. From a developmental perspective, it is interesting to consider the mechanisms by which a single genome can generate multiple consecutive and often radically different morphological, physiological, and behavioral phenotypes. For example, maggots are very different than flies, but they are still the same organism. As you ponder the relevancy of insects to the field of endocrinology, here are some facts to consider:

1. Recall from Chapter 7: Neurosecretion and Hypothalamic Control of the Pituitary the pioneering roles that studies on insects have played in developing our understanding of neuroendocrinology. Studying the effects of decapitations, neck tourniquets, and brain explants on post-embryonic development could not have been accomplished with vertebrates.
2. Recall from Chapter 5: Receptors how studies showing that the "puffing" of polytenic chromosomes in larval insect salivary glands is regulated by the steroid molting hormone, ecdysone, providing the first indication that steroid hormone signals regulate transcription rather than translation.
3. There are more insect species, and therefore different insect endocrine systems, than all other animal, plant, and fungal species *combined*![4]
4. The evolutionary ancestry of insects and their endocrine systems is ancient, dating back to the Cambrian period, 540 million years ago.
5. Insect pests cost the global economy billions of dollars each year. Insect endocrinology is currently an active area of research because it offers the potential for disrupting the life cycle of a pest without harm to the environment.
6. Mosquitos alone transmit several major human diseases, including malaria, yellow fever, and dengue fever virus. Efforts to inhibit the reproduction of mosquitos and other pathogenic insect vectors by disrupting the hormones that mediate their life cycle is an emerging strategy for reducing disease in humans.

Therefore, insects are notable as basic endocrine research models, and an understanding of their hormonal functions is important for combating a large variety of agricultural pests and human disease vectors. As the largest and most varied assemblage of organisms on earth, insects also offer a wealth of knowledge on the evolution of endocrine systems.

An Overview of Insect Development

LEARNING OBJECTIVE Describe the three major developmental strategies insects use for reaching the adult stage.

KEY CONCEPTS:

- The exoskeleton plays an instrumental role in determining an insect's body form.
- In order to both grow and change form, the old exoskeleton must be removed and replaced by a new one in a process called "molting".
- Insect metamorphosis is a partition in the life of an individual that separates an early feeding and growth stage from a later reproductive stage.
- Insects use one of three major developmental strategies for reaching the adult stage: ametabolous development (no metamorphosis), hemimetabolous development (partial metamorphosis), and holometabolous development (complete metamorphosis).
- Ametabolous insect development represents the ancestral form, and hemimetaboly gave rise to holometaboly.
- In holometabolous insects, the precursor cells that will form many adult structures (e.g. legs, wings) exist in larvae as compact, internal structures called imaginal discs.

The endocrinology of insect molting and metamorphosis is both a fascinating and complex process. Prior to considering the intricacies of their hormonal control, it is useful to first become familiar with the general morphological and physiological changes that take place during their development, as well as some key terminology unique to insects.

Importance of the Exoskeleton

The *exoskeleton*, also called the integument, is a primary contributor to the success of insects. It serves not only as a protective covering over the body, but also as a surface for muscle attachment, a water-tight barrier against desiccation, and a sensory interface with the environment. The insect exoskeleton consists of two major parts, the epicuticle and procuticle (Figure 11.1). The *epicuticle* is a thin, dense layer of tough lipoprotein material called cutilin, and the surface is coated with wax. The *procuticle* is a thicker layer of a nitrogenous polysaccharide called chitin in which proteins are deposited for hardening. Beneath the exoskeleton is the *epidermis*, a secretory tissue formed by a single layer of epithelial cells. It is responsible for producing the overlying layers of cuticle, as well as the underlying basement membrane that supports the exoskeleton. Importantly, since the exoskeleton plays such an instrumental structural role in determining an insect's body form, the growth and development of an insect is largely a function of the growth and development of its exoskeleton, which also makes possible the dramatic changes in form that can accompany metamorphosis.

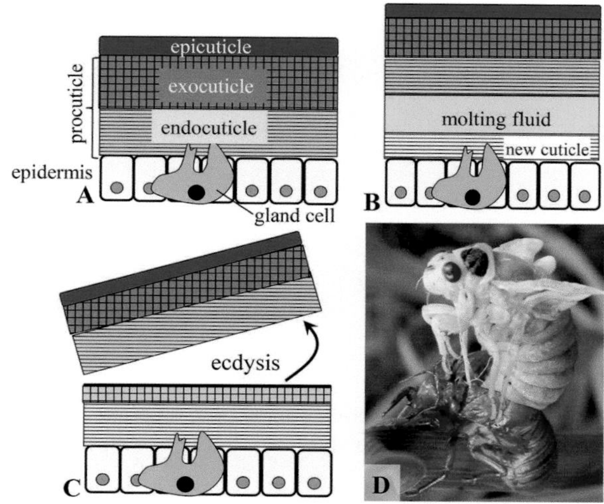

Figure 11.1 Molting of the insect exoskeleton. (A) The insect exoskeleton consists of a stratified cuticle layer on top of a epidermis. **(B)** Molting begins with the separation of the cuticle from the underlying epidermal cells via the production of molting fluid by molting glands in the epidermis. **(C)** Enzymes in the molting fluid digest the old cuticle, and it is shed (ecdysis). The epidermis begins to secrete the components of the new cuticle, which expands and hardens as it incorporates new proteins. **(D)** In a nymph-to-adult molt, the pink adult cicada emerges from its old brown cuticle.

Source of Images: (D) Drage, H.B. and Daley, A.C., 2016. BioEssays, 38(10), pp. 981-990. Used by permission.

Despite the many advantages of having an exoskeleton, the rigidity of its hardened portion constrains growth. To solve this problem, a new, larger exoskeleton must periodically be synthesized and the older one discarded, a process called *molting* or **ecdysis** (from the Greek "to strip off"). During molting, the epidermis separates from the procuticle and produces enzymes that digest it (Figure 11.1). Simultaneously, a new, temporarily stretchable cuticle is laid down. To shed the old exoskeleton, the insect gulps in air or water to build up internal body pressure. This eventually splits open the old exoskeleton along points of weakness and the soft-shelled insect wriggles free. Upon emerging from its old exoskeleton, the insect expands the new cuticle and hardens it through the incorporation of new proteins into the procuticle. Prior to adulthood, all insects undergo successive molts as they grow. However, with one exception discussed later, adult insects no longer molt or continue to grow.

Developmental Strategies

Insect metamorphosis is a partition in the life of an individual that separates an early feeding and growth stage from a later reproductive stage. Importantly, insect metamorphosis is functionally tied to molting, since that is the only time when an exoskeleton with a new morphology can be made.[5] Insects use three major developmental strategies for reaching the adult stage (Figure 11.2), with the degree of metamorphosis dependent on the degree of divergence between the immature stages and the adult.

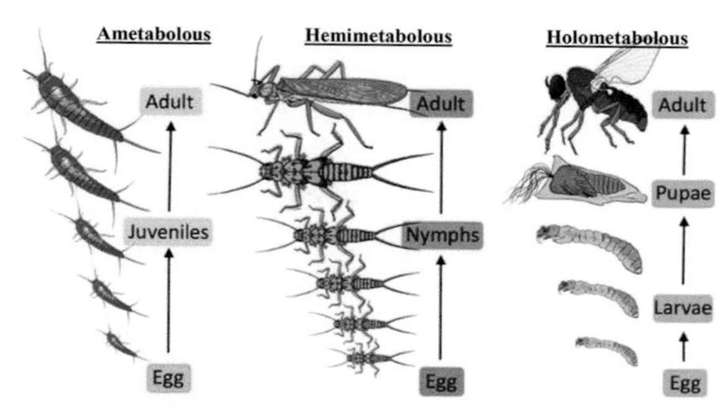

Figure 11.2 Major developmental strategies leading to the adult insect. In ametabolous development, there is no metamorphosis and the immature stages simply increase in size to produce the adult. In hemimetaboly (incomplete metamorphosis), the immature stages (called nymphs) resemble the ultimate adult, except that the completion of wing and genitalia development occurs at the final molt. With holometaboly (complete metamorphosis), the mobile larval stages, which often differ dramatically from that of the adult, pass through an immobile pupal stage immediately prior to metamorphosis.

Source of Images: Villar-Argaiz, M., et al. 2021. Sci. Rep., 11(1), pp. 1-8.

- **Ametabolous development** ("no metamorphosis"): A genitalia-lacking miniature version of the adult emerges from the egg and simply increases in size throughout the succeeding immature *nymphal* stages. Both the nymphs and adults occupy the same habitats, and, unlike other insects, these may continue to molt even after they reach sexual maturity. This form of development is called *direct development*, where the hatched animal is morphologically very similar to the adult. Direct development is confined to some wingless taxa and is considered the most ancestral form of insect development. An example of these insects is the silverfish (*Lepisma saccharina*).
- **Hemimetabolous development** ("incomplete metamorphosis"): Hemimetabolous development is also largely a form of direct development, except that in these winged orders the external wing primordia grow sequentially larger throughout the nymphal stages and are visible on the outside of the body in the later nymphal stages. Therefore, the extent of their metamorphosis is defined by the completion of wing and genitalia development. Similar to ametabolous insects, in hemimetabolous insects both the nymphs and the adults may occupy the same ecological habitats. Examples of hemimetabolous insects include cockroaches, dragonflies, grasshoppers and Hemiptera (true bugs).
- **Holometabolous development** ("complete metamorphosis"): The immature stages of these insects, called *larval* stages, are typically widely divergent from the adult form. Holometabolous larvae undergo a radical transformation by first passing from the larval to the *pupal* stage, an outwardly quiescent but developmentally active stage when larval structures are replaced with adult legs, wings, features of the head, and genitalia, as well as a distinct adult cuticle. Therefore, in

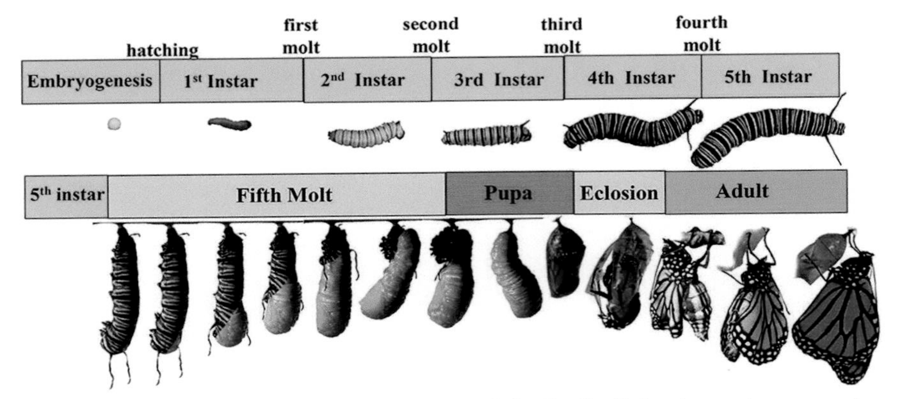

Figure 11.3 Larval development and metamorphosis of the monarch butterfly. Following embryogenesis and egg hatch, the larva develop through five instars, each separated by a molt. The fifth and final molt transforms the larva into an immobile pupa, where adult features develop. Finally, an eclosion event takes place, with the adult emerging.

Source of Images: Images courtesy of Ba Rea, barea@basrelief.org

contrast to the nymphs of hemimetabolous insects, larvae of holometabolous insects are often distinguished in part by a *vermiform* (wormlike) morphology with reduced or absent legs and the presence of internally developing wings. Importantly, these large differences between the adults and their larvae allow them to avoid intraspecific competition for food and habitat. The advantages of a holometabolous strategy are reflected in the relatively large number of insects that engage in holometabolous development, about 770,000 species, or 88% of all known insects.[5] This explosion in holometabolous diversity is attributable to one innovation in particular: by uncoupling larval and adult development, the same insect is able to produce two distinct forms, enabling larvae and adults to occupy separate ecological niches. In addition, the vermiform body permits the burrowing of larvae into ephemeral food sources, such as fruits and decomposing detritus. Examples of holometabolous insects include moths, butterflies, flies, beetles, bees, wasps, and ants.

The larval stages in between molts are known as *instars*. The number of larval instars is species specific and varies widely across insect taxa. Most commonly, however, insects tend to have three to eight larval instars. Importantly, in holometabolous insects the penultimate molt always yields the pupal stage. In most species, the exoskeleton of the pupa hardens into a protective structure, which in butterflies is called a *chrysalis* (Figure 11.3). The final molt yields the adult, which emerges (*eclosion*) after metamorphosis.

Curiously Enough . . . Immediately prior to pupation, some insect larvae, most notably the silkworm (*Bombyx mori*), spin an additional protective encasement made of silk called a *cocoon*.

Imaginal Discs

Recall from earlier that the larvae of holometabolous insects are distinguished from the nymphs of ametabolous and hemimetabolous insects in part by the presence of some internally developing adult structures that do not complete their development until pupation. These internally developing adult structures are formed from special epidermal cells whose growth is suppressed during larval life and have no functions in the survival of the larvae. These cells are often grouped into distinct structures called **imaginal discs**, which may be set aside as early as embryogenesis, or created *de novo* at metamorphosis. The term "imaginal" (derived from the Latin *imago*, meaning "image" or "likeness") was co-opted by entomologists to refer to the adult counterpart of the insect. These structures are "disc"-shaped thickenings of the inner surface of the larval skin that, upon receiving a specific endocrine signal, will give rise to adult organ systems.

Most of what we know about imaginal discs comes from research with the fruit fly, *Drosophila*. Each of the discs differentiates into a variety of cell types and organs that constitute the adult structures. There are 19 discs in total, each made up of columnar epithelial cells and attached to the larval epidermis by a stalk: the epidermis of the head, thorax, and limbs of the fly derive from nine bilateral pairs of discs and the genitalia derive from a medial disc (Figure 11.4). In *Drosophila*, imaginal discs arise as invaginations or thickenings of the ectoderm during late embryogenesis. During larval development, the cells of the discs undergo division and increase in size, attaining maturity during the last larval instar. At the end of the last larval instar, the cells switch from proliferation to differentiation. During metamorphosis, the discs evert (turn inside out), elongate, and continue the final differentiation of the external structures of the adult (Figure 11.4). The eversion of each disc causes the cells from each appendage to fuse with the cells from other discs and form the continuous epidermis of the body wall. Importantly, although the hormones of metamorphosis induce the final

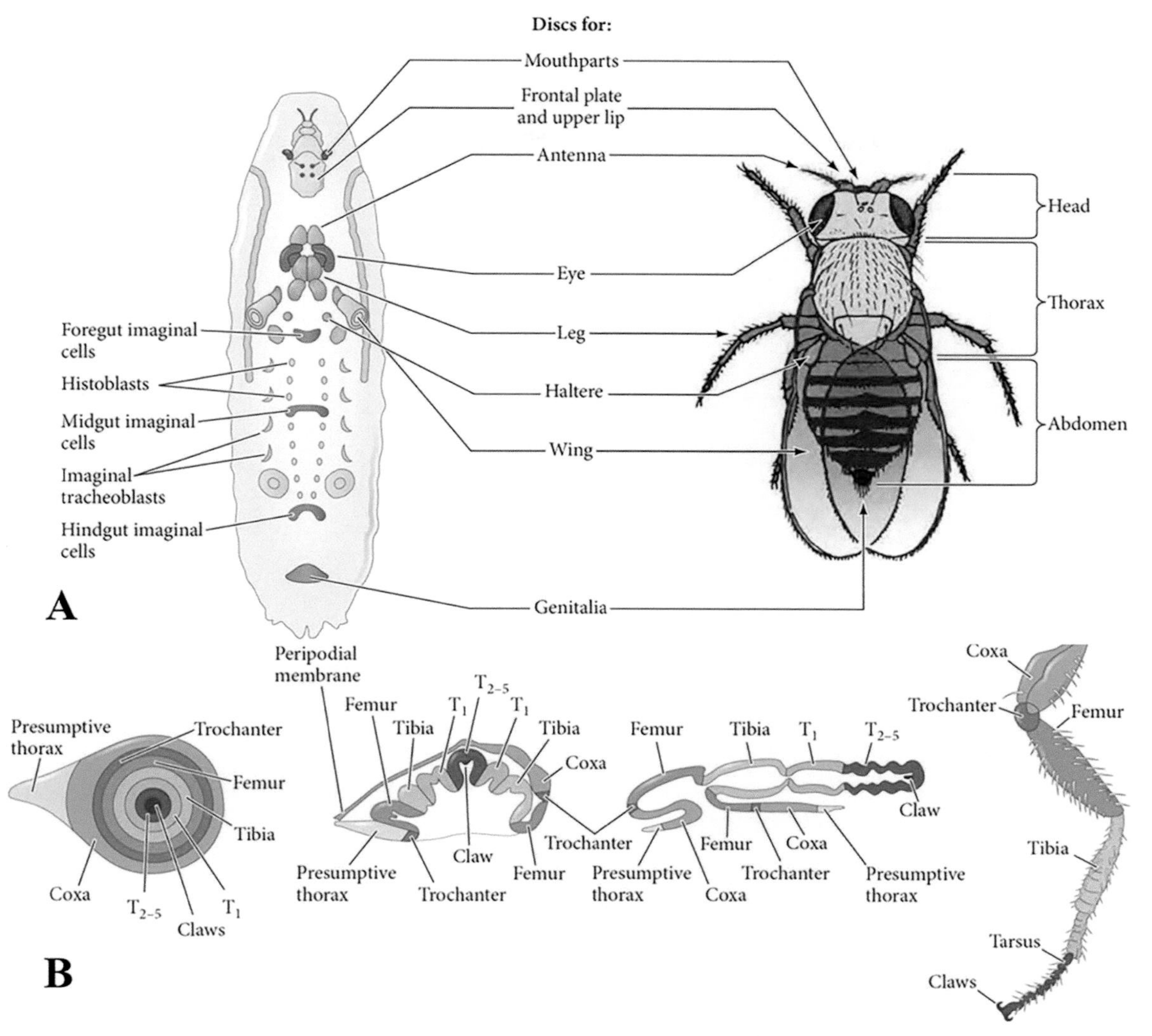

Figure 11.4 The imaginal discs and corresponding adult structures in *Drosophila*. (A) Diagram showing the localization of the imaginal discs inside the third (final) instar larva, left, and the adult structures they will form, right. **(B)** A third instar leg disc (left) subsequently undergoes metamorphic eversion and elongation (middle) to produce an adult leg (right). The disc and leg have been colored to make a simplified fate map, showing which regions in the disc will form which structures in the leg.

Source of Images: Gilbert, S.F., 2013. Developmental Biology. 10th edition. Sinauer Associates. Used by permission.

process of imaginal disc differentiation into their adult structures, the actual developmental fates of imaginal discs have already been determined by the end of embryogenesis.

SUMMARY AND SYNTHESIS QUESTIONS

1. Considering that the procuticle and epicuticle of the exoskeleton are entirely acellular structures, how can hormones mediate ecdysis?
2. What is the major advantage of holometaboly over hemi- or ametaboly in terms of increasing the likelihood of offspring survival?

Hormones of Insect Molting and Metamorphosis

LEARNING OBJECTIVE Delineate the roles played by major hormones that mediate insect molting and metamorphosis, and compare and contrast insect neuroendocrine systems with those of vertebrates.

KEY CONCEPTS:

- Ecdysone, a steroid hormone produced by the prothoracic gland, modulates the timing of a molt.

- Prothoracicotropic hormone, a neuropeptide produced by the brain, induces ecdysone synthesis.
- Juvenile hormone, a fatty acid derivative produced in the head, determines whether the molt will be larval or metamorphic.
- Allatotropin and allatostatin are neuropeptides produced by the brain that induce and inhibit, respectively, juvenile hormone synthesis.
- A nymphal/larval ecdysone peak occurring in the presence of JH maintains the current developmental state, whereas if JH is absent during an ecdysone peak the insect will progress to a more mature developmental state.
- Ecdysone receptors are nuclear transcription factors that modulate gene expression.
- Based upon functional, structural, and developmental similarities, homologies between the vertebrate adenohypophysis/neurohypophysis and insect corpora allata/corpora cardiaca, respectively, have been suggested.

Almost a dozen hormones have been shown to contribute to molting and metamorphosis in insects (Table 11.1), and here we will focus on several of the major hormones: ecdysone, juvenile hormone, and the neuroendocrine hormones prothoracicotropic hormone, allatostatin, and allatotropin.

Early History of Insect Endocrinology

In Chapter 7: Neurosecretion and Hypothalamic Control of the Pituitary, we learned that the beginnings of the study of insect endocrinology coincided with the birth of neuroendocrinology when Polish scientist Stefan Kopeć demonstrated that the head, and more specifically, the brain, was required for metamorphosis of the holometabolous gypsy moth to commence.[6,7] Kopeć showed that debraining a caterpillar completely inhibits pupation, and that placing a blood-tight ligature around the mid-body of caterpillars inhibited pupation in the posterior part. This finding provided early evidence that the brains of animals secreted factors that promoted physiological events and, hence, acted in an endocrine manner.

Eminent entomologist Sir Vincent B. Wigglesworth repeated Kopec's observations using decapitation experiments with nymphs of the hemimetabolous South American blood sucking bug, *Rhodnius prolixus*, also showing that the head was required for molting.[8] Wigglesworth chose *Rhodnius* because its development is closely timed to its blood meals, and he found that 3 days after a blood meal hormones are released into the *Rhodnius* system, stimulating the molt to the next stage. Thus, this is an inducible developmental system. Wigglesworth further demonstrated that by transplanting a specific portion of the brain of a blood-meal-fed nymph into the abdomens of unfed, decapitated nymphs that the recipient nymph molted, demonstrating that neurosecretory cells were indeed the source of the brain's endocrine effect. We now know the brain-derived hormone that induces metamorphosis is *prothoracicotropic hormone* (PTTH).

Table 11.1 Some Key Hormones and Neurohormones That Regulate Insect Molting and Metamorphosis

Signal	Chemical Class	Type of Secretion	Site of Secretion	Target Tissues	Actions
Prothoracicotropic hormone (PTTH)	Protein	Neuroendocrine	Soma in brain, axon terminals in corpora cardiaca	Prothoracic glands	Initiates molting by stimulating release of ecdysone from prothoracic glands
Ecdysone	Steroid	Endocrine	Prothoracic glands in larva/nymph; ovary in adult	Epidermis in larva/nymph; fat body in adult	Following conversion to 20-hydroxyecdysone, promotes destruction of old cuticle and synthesis of a new one; stimulates yolk protein synthesis on adult
Juvenile hormone (JH)	Terpene (fatty acid derivative)	Endocrine	Corpora allata	Epidermis in larva/nymph; ovary in adult	Opposes formation of adult structures and promotes formation of larval/nymph ones; acts like a gonadotropin in adults
Allotropin	Peptide	Neuroendocrine	Brain	Axon terminals extending to corpora allata	Induces JH release from corpora allata
Allostatin	Peptide	Neuroendocrine	Brain	Axon terminals extending to corpora allata	Inhibits JH release from corpora allata
Corazonin	Peptide	Neuroendocrine	Brain	Inka cells	Promotes PETH and ETH secretion by Inka cells
Eclosion hormone	Peptide	Neuroendocrine	Brain	Inka cells	Promotes PETH and ETH secretion by Inka cells
Pre-ecdysis triggering hormone (PETH)	Peptide	Endocrine	Inka cells of tracheae	Brain	Prepares motor programs for cuticle shedding
Ecdysis triggering hormone (ETH)	Peptide	Endocrine	Inka cells of tracheae	Brain	Prepares motor programs for escaping from old cuticle
Buriscon	Protein	Neuroendocrine	Brain and nerve cord	Cuticle and epidermis	Tans and hardens new cuticle

Source: After Hill, Wyse, Anderson (2012) *Animal Physiology*, Fourth edition, Sinauer Associates

Curiously Enough . . . *Rhodnius prolixus* is also known as the "kissing bug" because of its habit of sucking blood from the lips of sleeping humans. This bug is a vector for Chagas' disease, caused by the parasite *Trypanosoma cruzi.*

Soichi Fukuda[9,10] and Carroll Williams[11] further extended knowledge of insect endocrinology by providing evidence for the requirement of a second signal that was released in response to the PTTH. The second signal emanated from the *prothoracic gland* within the thorax of the insect (Figure 11.5). Williams demonstrated this with the silkmoth, *Hyalophora cecropia*, which has the relevant characteristic of *diapausing*, or entering a developmental stasis, as a pupa until it experiences a prolonged chill that terminates the diapause and stimulates it to molt to the adult stage. This was, essentially, a cold-inducible developmental system. Brains were removed from chilled pupae and implanted into isolated diapausing (unchilled) pupal abdomens, terminating their diapause only when a prothoracic gland was also added. This demonstrated that the neurosecretory function of the brain was to activate the prothoracic glands to produce another hormone, which we now know to be *ecdysone*. With these experiments, Kopeć, Wigglesworth, Fukuda, and Williams established a foundation for studying and understanding the endocrine activation of molting and metamorphosis, consisting of the brain, whose nervous tissue produces PTTH that acts on the prothoracic glands to produce the molting hormone ecdysone (Figure 11.5).

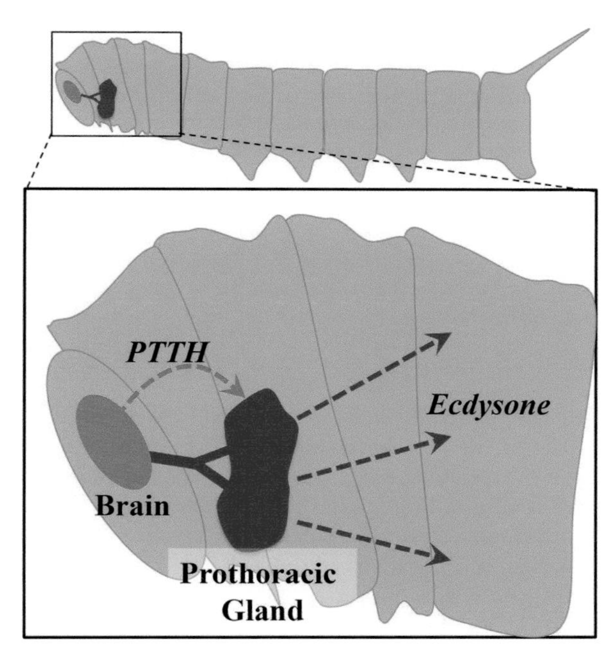

Figure 11.5 Model depicting the release of the "molting hormone", ecdysone, by the prothoracic gland following induction by the "brain hormone", prothoracicotropic hormone (PTTH).

Ecdysone: The Molting Hormone

Discovery

Ecdysone was the first insect hormone to be isolated, and its identification as the key molting hormone is a milestone in the history of endocrinology. Recall from Chapter 4: Hormone Classes and Biosynthesis German biochemist Adolf Friedrich Johann Butenandt, who had shared the *Nobel Prize for Chemistry* in 1939 for his work on vertebrate sex hormones. Clearly no stranger to hard work, Butenandt and his colleague, Peter Karlson, also isolated 25 mg of *ecdysone* (the molting hormone previously shown to be produced by the prothoracic glands) from 500 kg (a half ton!) of dried silkworm

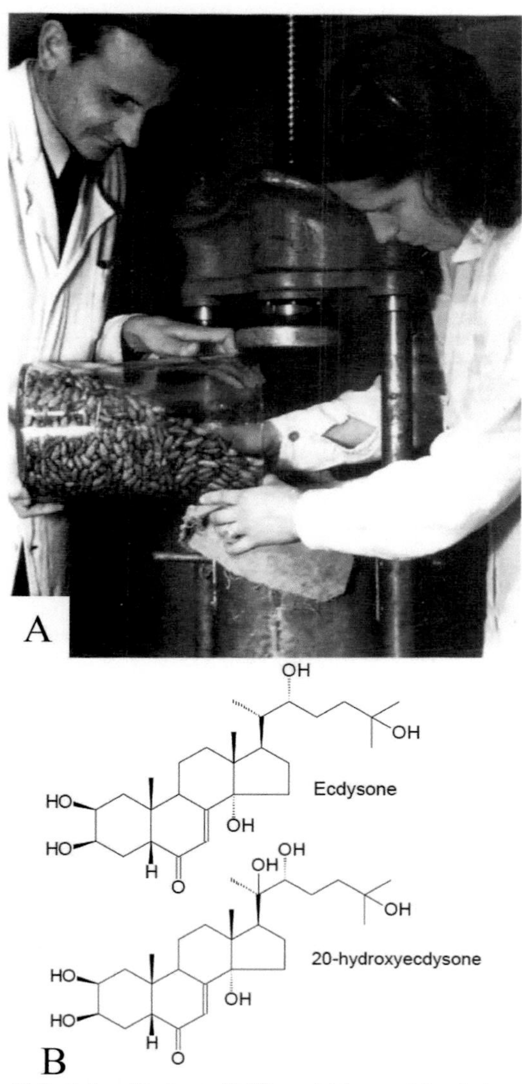

Figure 11.6 Peter Karlson (left) and his assistant, Ingeborg Brachmann, preparing a crude ecdysteroid extract from silkworm pupae using a press (A). Ecdysone, produced by the prothoracic gland, is generally considered to be an inactive prohormone that is converted into the active 20-hydroxyecdysone inside the target cells of peripheral tissues **(B)**.

Source of Images: Karlson, P., 2003. Int. J. of Dev. Biol., 40(1), pp. 93-96.

moth pupae[12] (Figure 11.6). Ecdysone was later shown by Karlson and others to be a steroid hormone[13,14] and to exert a direct role in regulating gene expression based on the puffing of salivary gland polytene chromosome.[15] In contrast with vertebrates that can synthesize the precursor substrate to steroid hormones, cholesterol, endogenously, insects lack a cholesterol-synthesis enzyme and thus have a dietary requirement for cholesterol in order to synthesize ecdysone. Ecdysone biosynthesis occurs in the *prothoracic gland* via a series of steps involving cytochrome P450 enzymes.

Several hundred other variants of ecdysone, collectively called *ecdysteroids*, have since been discovered in animals (more than 70 in insects alone), plants, and fungi (http://ecdybase.org/). Interestingly, many *phytoecdysteroids* produced by plants may function as feeding deterrents or toxic substances that affect the survival of herbivorous insects.[5] The first ecdysteroid to be identified, now referred to simply as "ecdysone", has historically been considered an inactive prohormone. The second ecdysteroid identified, *20-hydroxyecdysone*, is hydroxylated from ecdysone by target tissues and is generally considered the active hormone (Figure 11.6). 20-hydroxyecdysone binds to intracellular ecdysteroid receptors. Interestingly, more recent studies suggest that ecdysone itself may regulate a distinct set of genes from those controlled by 20-hydroxyecdysone.[16] In addition to modulating molting and metamorphosis, ecdysteroids have many other functions, particularly in insect embryogenesis and reproduction.

Molecular Mode of Action

Recall from Chapter 5: Receptors that the classical mechanism of steroid hormone action is to modulate gene transcription by binding to nuclear receptors that function as transcription factors. Indeed, the notion of steroid hormones as modulators of transcriptional activity was first demonstrated in insects by Clever and Karlson,[15] who showed that within two hours of injection into larvae of the midge *Chironomus tentans*, ecdysone prematurely induced chromosome "puffs" that would normally be seen only later in development. Clever[17] suggested that the "early reacting genes are involved in those processes leading to the sequential activation of the puffs which appear later" (Figure 11.7). Importantly, Clever and Michael Ashburner[18] later showed that blocking the protein translation of early puff RNAs with cycloheximide (a translational inhibitor) treatment prevents the regression of some early puffs and simultaneously prevents the appearance of many late puffs. From these findings, Ashburner postulated that an as of yet unidentified intracellular ecdysteroid receptor directs a transcriptional response at early puff sites in the presence of 20-hydroxyecdysone. Additionally, the early puff gene products regulate the appearance of later puffs and also feedback to repress their own expression. This original model for ecdysteroid action (Figure 11.7) has proven remarkably durable over the years.

Early in the 1990s, David Hogness and colleagues[19–22] discovered a class of nuclear receptors in fruit flies that they called "*ecdysteroid receptors*" (EcRs), which are distant relatives of the vertebrate farnesol X receptor. The EcR exists in three isoforms, each one having a different biological function. Similar to other members of the nuclear receptor family, EcR affects gene transcription upon binding its ligand, 20-hydroxyecdysone, and after forming a heterodimer with another nuclear receptor, *ultraspiracle protein* (USP), the insect ortholog of the vertebrate *retinoid X receptor* (RXR) (Figure 11.7). Recall that in vertebrates, RXR is a heterodimeric partner of receptors for thyroid hormone, vitamin D, retinoic acid, and other nuclear receptors. The ligated EcR-USP dimer makes

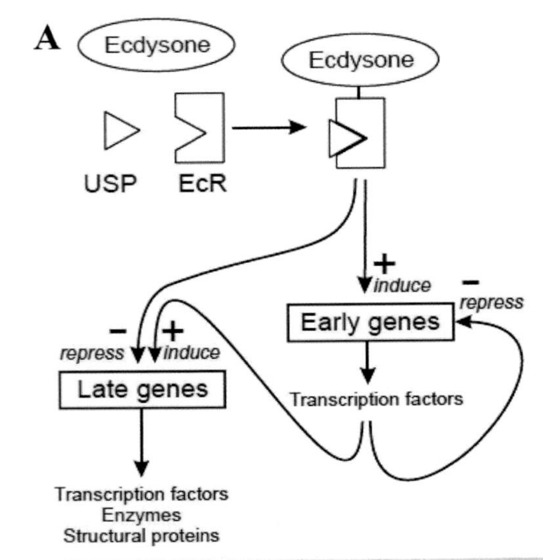

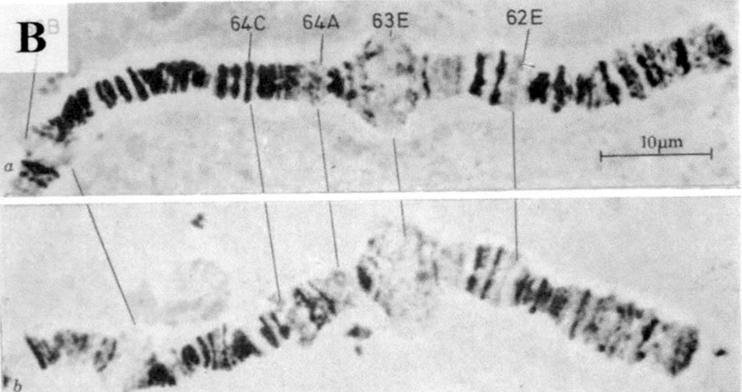

Figure 11.7 A model originally proposed by Ashburner (1974) for the action of ecdysteroids in the *Drosophila* salivary gland. (A) The ecdysteroid receptor (EcR) and the product of the ultraspiracle gene (USP) bind to the hormone. The ligated EcR-USP heterodimer complex binds to early genes, stimulating their transcription. The early gene products subsequently inhibit early gene expression but stimulate late genes' transcription, demonstrating the cascade of gene activity that is involved in salivary gland morphogenesis. **(B)** The transcription of early genes in response to ecdysone in *Drosophila* polytene chromosomes is accompanied by "puffs" (arrows) representing a localized chromatin decondensation.

Source of Images: (A) Klowden MJ. Physiological Systems in Insects, Third Edition. 3 ed: Academic Press; 2013; used by permission. (B) Ashburner, M., 1970. Proc. Roy. Soc. B. Biol. Sci., 176(1044), pp. 319-327. Used by permission.

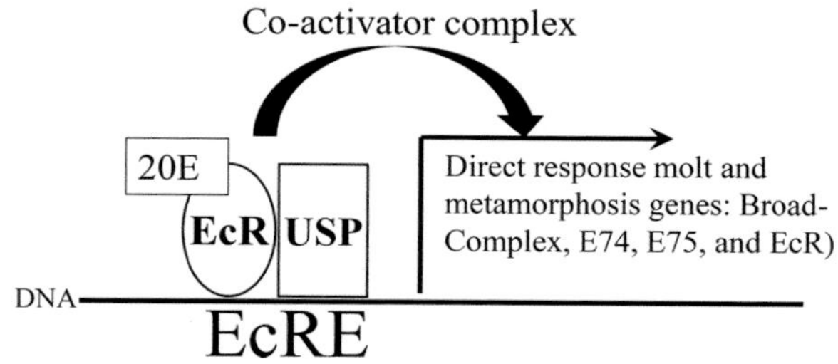

Figure 11.8 20-hydroxyecdysone (20E) signaling is mediated by a nuclear receptor heterodimer consisting of the ecdysone receptor (EcR) and the ultraspiracle (USP) protein. The DNA-binding domains of EcR and USP make direct contacts to nucleotide bases contained in the ecdysone response element (EcRE) in gene promoter regions. This promotes interaction of the EcR-USP complex with coactivator (shown) or corepressor (not shown) proteins that activate or repress the transcription of early response genes.

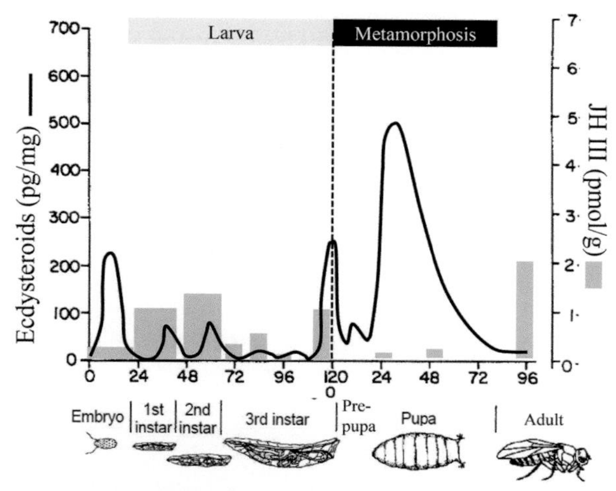

Figure 11.9 Juvenile hormone (JH; gray bars) and 20-hydroxyecdysone (20E; black line) titers during *Drosophila* development. The balance between 20E and JH determines the normal course of insect development. All developmental transitions, such as larva-to-larva, larva-to-pupa, and pupa-to-adult molts, are initiated by 20E. JH determines the type of transition. That is, larval molting and growth occur in the presence of a high JH titer. At the end of larval development, the JH titer falls and 20E signals pupal commitment. Metamorphosis occurs when JH levels are either very low or absent.

Source of Images: Riddiford, L.M. (1993).The development of drosophila melanogaster, vol 2, pp. 899-939. Cold Spring Harbor Laboratory Press.

direct contact with specific nucleotide base sequences of the *ecdysone response element* (EcRE) (Figure 11.8), located in the regulatory gene promoter region. Upon associating with transcriptional coactivator or corepressor proteins, ecdysone-responsive gene transcription is induced or repressed, respectively. It is now known that only a small number of genes are directly upregulated by the ecdysone-EcR-USP complex, and these react by puffing within minutes of exposure to ecdysone. By contrast, a much larger number (>100) of late response genes are indirectly upregulated within hours. Several direct response genes have been identified, and these include three regulatory genes (Broad-Complex, E74, and E75), each of which encodes a transcription factor that regulates the expression of late responsive genes. Another gene that is directly upregulated in response to ecdysone is the EcR itself. Therefore, in order to increase the output of ecdysone, EcR provides an autoregulatory loop to activate its own transcription and further increase receptor levels in response to the ecdysone ligand. This hierarchy of gene activation is required for modulating expression of the many cell death, cell cycle, and differentiation genes required for metamorphosis.[23,24] Although the genes in this ecdysone-activated hierarchy have been extensively described, their specific functions remain poorly understood.

Developmental Profile

Ecdysone pulses emanating from the prothoracic gland are required for all aspects of morphogenesis, starting with the formation of the body plan during late embryogenesis, hatching, molting at the end of each larval instar, and pupation (Figure 11.9). Importantly, the determination of whether a molt will be a larval-to-larval molt or a larval-to-pupal metamorphic molt is based not solely on ecdysone concentrations, but also on the presence and relative sensitivity of target tissues to a second hormone, *juvenile hormone* (JH) (Figure 11.9). As a general rule, if JH is present concurrently with ecdysone during a JH-sensitive period, the current developmental state is maintained, whereas if JH is absent during that period, the insect will progress to a more mature developmental state.[2] The importance of ecdysone and JH can be summarized succinctly: whereas peaks of ecdysone determine the timing of the molt, JH determines the nature of the molt.[3]

During pupation, ecdysone triggers the aforementioned eversion and elongation of the imaginal discs through cell shape changes (in *Drosophila*) or cell proliferation (in Lepidoptera). An additional level of complexity in the control of gene expression by ecdysone has been shown to occur in the tobacco hornworm (*Manduca sexta*), where the various EcR and USP isoforms are expressed in complex tissue-specific and temporal patterns. In this way, a complex profile of ecdysone pulses can target some tissues whilst leaving others unaffected.[25–27]

Neuroendocrine Control of Ecdysone Synthesis

Prior to the first two decades of the 20th century, it was believed that insects, when compared to vertebrates, have simple endocrine systems with no neurosecretory capacity. Today, however, more than 50 neuropeptides have been isolated and identified from the nervous system of locusts alone.[28] Several neuropeptides produced by the insect brain mediate molting and metamorphosis, and here we will discuss one: *prothoracicotropic hormone* (PTTH). Indeed, PTTH is the "brain hormone" first identified by Stefan Kopec and others, and over 70 years elapsed between its discovery and structural characterization. We now know that PTTH is a homodimer of two polypeptides of 109 amino acids. Like the vertebrate hypothalamus, populations of neurosecretory neurons in the insect brain produce peptide hormones that target accessory organs that either store the brain-derived neuropeptides for controlled release or stimulate the target organ to release another hormone (Figure 11.10). The *corpus cardiacum* is a major neurohemal organ in insects and releases a large number of neuropeptides, including PTTH. In some insects another accessory organ, the *corpus allatum*, releases PTTH. When released into circulation, PTTH binds to its receptors on the prothoracic glands, promoting the synthesis of ecdysone. Almost a century after this brain hormone was discovered, Rewitz and colleagues[29] identified the PTTH receptor as *Torso*, a receptor tyrosine kinase that signals through the Ras/Raf/Erk pathway (Figure 11.11). The activation of this pathway by PTTH ultimately induces the transcription of a group of genes known as the *Halloween genes*, with thematic names like *phantom*, *disembodied*, *shadow*, *shade*, *spook*, *spookier*, and *spookiest*. Halloween gene products are all P450 enzymes that are bound to the membranes of mitochondria and endoplasmic reticulum in cells of the prothoracic glands. These enzymes are involved in the synthesis of ecdysone from cholesterol. Ecdysone, generally considered a prohormone, is converted into the active 20-hydroxyecdsyone by a variety of peripheral tissues.

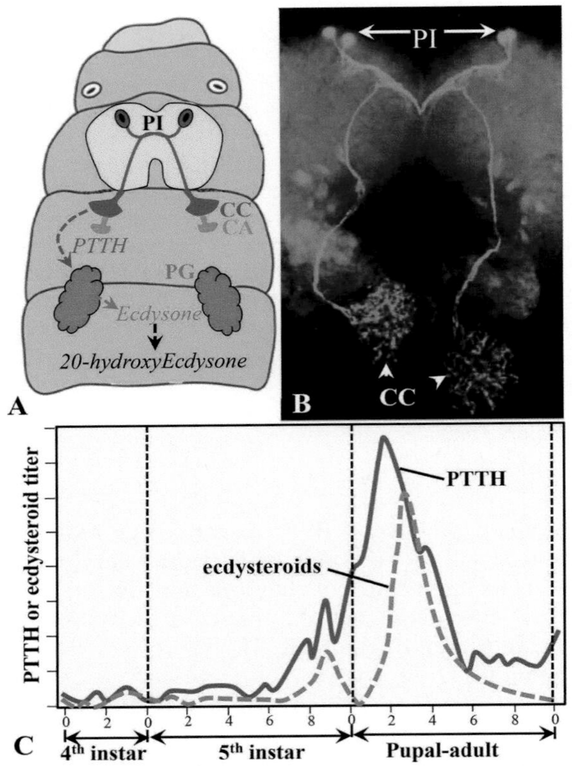

Figure 11.10 The PTTH-ecdysone neurosecretory axis of the larval insect. (A) PTTH-producing nuclei (white neurons) within the brain's pars intercereberalis region (PI) transport the peptide along axons to the corpus cardiacum (CC), where it is stored or released into circulation. When PTTH binds to its receptors on the prothoracic glands (PG), this promotes the synthesis and release of ecdysone, a steroid hormone, into the hemolymph. Ecdysone, a prohormone, is converted to the active 20-hydroxyecdysone inside target tissue cells. **(B)** The brain of an insect stained with an antibody that labels PTTH (white stain). Note the presence of PTTH in nuclei within the PI, along their axons that project into the CC, and within the CC. **(C)** Typical PTTH and ecdysteroid titers in *Bombyx mori* hemolymph. Note how the PTTH peaks always precede that of ecdysteroids. From Dai et al. (1995) and Mizoguchi et al. (2001, 2002).

Source of Images: (B) Hill, R.W., Wyse, G.A., Anderson, M. and Anderson, M., 2004. Animal physiology (Vol. 2, pp. 150-151). Used by permission. (C) Smith and Rybczynski (2012) Prothoracicotropic Hormone. In: Insect Endocrinology. Ed: L.I. Gilbert. Elsevier. Used by permission.

Juvenile Hormone

Discovery and Function

Vincent B. Wigglesworth, one of the early discoverers of the presence of the molt-inducing "brain hormone", PTTH, went on to identify another hormone that exerts profound control over molting and metamorphosis.

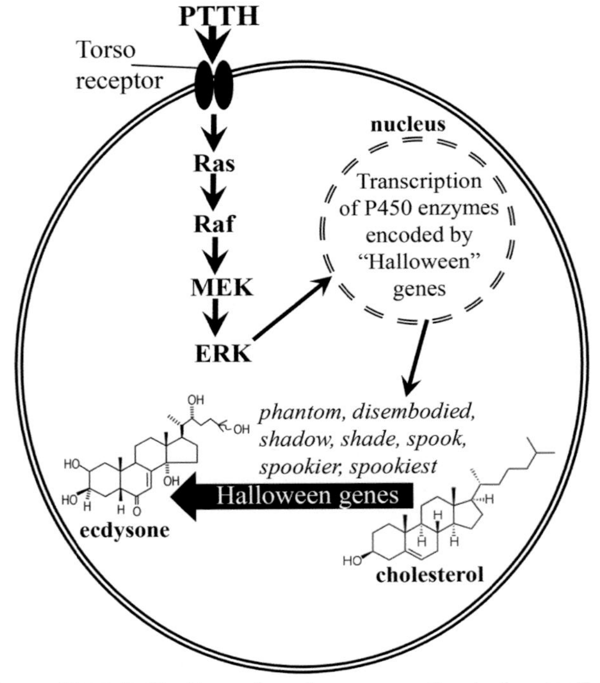

Figure 11.11 Activation of ecdysone synthesis by prothoracicotropic hormone in the prothoracic gland. The binding of PTTH to its receptor, Torso (a member of the receptor tyrosine kinase family), activates a MAP kinase phosphorylation cascade. Upon activation, ERK translocates to the nucleus and affects the expression of P450 enzymes encoded by the "Halloween genes", a group of genes whose enzyme products modulate the synthesis of ecdysone from cholesterol.

As before, he did this using creative parabiosis and explant experiments with his favorite research model, the hemimetabolous blood sucking *Rhodnius*.[30,31] First, Wigglesworth observed that when a fourth (penultimate)-instar nymph was joined with a fifth (final)-instar nymph in parabiosis, the fourth-instar nymph underwent a normal molt into a fifth-instar nymph (Figure 11.12, left panel). However, instead of molting into a winged adult, the original fifth-instar nymph molted into a never-before-seen supernumerary sixth-instar nymph lacking adult features like wings or mature genitalia. A mysterious factor in the blood of the fourth-instar nymph must have inhibited the metamorphosis of its parabiotic partner and caused it to molt into a sixth-instar nymph. This factor came to be known as *juvenile hormone* (JH) because it appeared to cause nymphs to retain their immature or "juvenile" morphology. Importantly, we now know that the general developmental effect of JH is not to "juvenilize" insects, as the name might imply, but rather to maintain the current developmental state. This interpretation comes from the finding that when JH is injected into a holometabolous pupa, it causes the synthesis of a pupal cuticle at the next molt, thus producing a second pupal stage, and not an adult.[32,33] As such, JH is often referred to as a "status quo" hormone.[34] In addition to its developmental effects during larval life and metamorphosis, JH is also used by embryos to modulate yolk uptake, and in adults to influence reproduction, migration, diapause, behavior, and caste determination.[5,35] Considering the versatility of this hormone throughout virtually every developmental stage, the name "juvenile hormone" is a rather unfortunate misnomer.

The source of JH was eventually traced to one of the brain's aforementioned accessory glands, the corpora allata. To show that the corpora allata was indeed the source of JH, Wigglesworth surgically removed the gland from fourth-instar nymphs and implanted them into the abdomens of final fifth-instar nymphs (Figure 11.12, right panel). Not only did the fifth-instar organ recipients

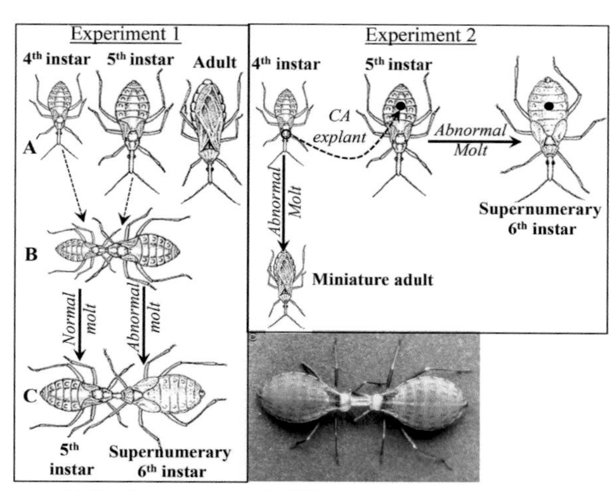

Figure 11.12 Summary of Wiggleworth's parabiosis and explant experiments that led to his discovery of juvenile hormone and its synthesis by the corpora allata. *Experiment 1*: **(A)** The kissing bug, *Rhodnius prolixus*, undergoes four nymph-nymph molts. Upon reaching the fifth nymph instar stage, the insect consumes a blood meal, which initiates its final molt into a winged adult. **(B)** Following the administration of a blood meal to the final fifth-instar nymph, the insect was parabiosed to a penultimate fourth-instar nymph. **(C)** The parabiotic sharing of hemolymph between a penultimate fourth-instar and a final fifth-instar nymph does not affect the progression of the fourth-instar nymph to its fifth instar. However, instead of metamorphosing into an adult, the nymph initially in its fifth instar undergoes a supernumerary sixth-instar molt (note the absence of wings). These findings demonstrate that a hormone present in the fourth-instar larva inhibits the metamorphic molt into an adult. *Experiment 2*: A supernumerary sixth-instar larva is produced when the corpora allata (CA) from a fourth-instar larva is implanted into the abdomen of a fifth-instar larva. The corpora allota-deficient fourth-instar larva goes on to metamorphose precociously into a miniature adult. These experimental findings demonstrate that the corpora allota are the source of JH. After Wigglesworth (1939). *Inset:* Photograph of parabiosed *Rhodnius* by Wiggleworth.

Source of Images: Bug drawings: Oswaldo Cruz Foundation, et al. 2009. Memórias do Instituto Oswaldo Cruz, 104, pp. 1072-1082; used by permission. Photograph: Riddiford, L.M., 2020. Front. Cell Dev. Biol., 8, p. 679. Used with permission.

Developments & Directions: Vincent B. Wigglesworth demonstrates the function of juvenile hormone

Although adult insects do not normally molt, the kissing bug, *Rhodnius*, can be induced into a supernumerary molt if administered PTTH after its last nymphal molt. Eminent English insect physiologist Vincent B. Wigglesworth (Figure 11.13(A)) demonstrated that if chemically purified juvenile hormone (JH) is first painted onto to a specific segment of the insect's exoskeleton during the penultimate molt, this segment reverts back to a nymphal morphology by developing into a cuticle with a dark dimpled band following this extra molt. Compare this with the normal lighter and smoother adult cuticle surrounding it (Figure 11.13(B)). True to his creative form, Dr. Wigglesworth replicated his experiment by painting his initials, "VBW", onto one of his bugs (Figure 11.13(C)).

Figure 11.13 Insect physiologist Vincent B. Wigglesworth (A) demonstrated that juvenile hormone (JH) inhibits the nymph-to-adult cuticle molt **(B–C)**.

Source of Images: (A) Riddiford, L.M., 2020. Front. Cell Dev. Biol., 8, p. 679. Used with permission. (B-C) Courtesy of J.W. Kimball.

undergo an abnormal molt to a supernumerary sixth instar, as expected, but the fourth-instar organ donors (which now lacked corpora allata as well as the ability to produce JH) initiated a precocious nymph-to-adult molt, producing a miniature adult with wings and genitalia.

Chemical Class and Neuroendocrine Regulation

Several JHs are now known to exist in insects, and these constitute a family of *sesquiterpenes* that are synthesized from acetyl CoA, a building block of fatty acids and cholesterol, and are structurally most similar to the vertebrate morphogenic molecule *retinoic acid* (Figure 11.14). Most species make more than one of these forms, and many insect taxa have characteristic mixes of JHs. *Methoprene*, a synthetic analog of JH developed in the 1970s (Figure 11.14), was the first successful insecticide designed to specifically inhibit reproduction by disrupting metamorphosis. It remains the most widely applied mosquito management compound, used to control the spread dengue fever, malaria, and West Nile virus.

JH is synthesized by a small pair of glands called the *corpora allata* (juxtaposed to the aforementioned corpora cardiaca) situated just below the insect brain (Figure 11.16). The stimulation of JH secretion by the corpora allata is known to be modulated by two neuroendocrine peptides that are delivered either indirectly through the hemolymph or directly by axons associated with neurosecretory neurons with cell bodies in the brain. The neuropeptides are *allatotropin*, which stimulates JH synthesis, and *allatostatin*, which inhibits JH synthesis (Figure 11.16). Once secreted by the corpora allata, JH is transported to target tissues through the hemolymph. However, because of its lipophilic nature, in order to move through the aqueous medium JH must first bind to one of a variety of transport proteins, collectively termed the *juvenile hormone-binding proteins* (JHBP). In a remarkable demonstration of the metamorphosis-suppressing ability of JH, Daimon and colleagues[36] have shown that a loss-of-function mutation in a gene whose expression is required for JH synthesis in the silkworm, *Bombyx mori*, is responsible for a precocious metamorphosis in the third instar, producing a tiny, but still sexually viable, adult (Figure 11.17).

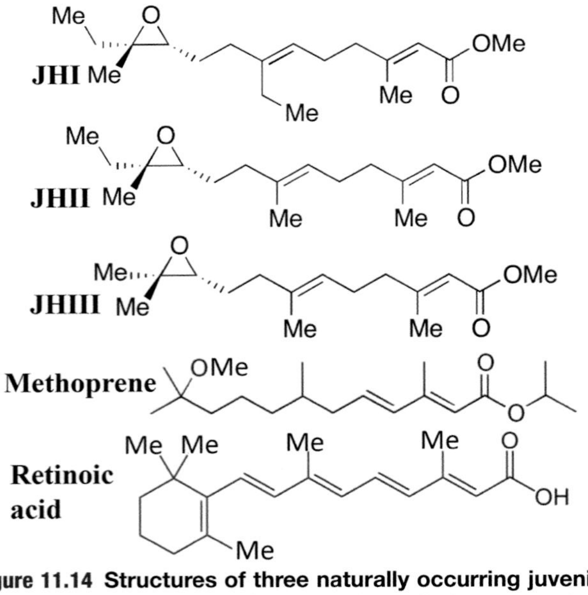

Figure 11.14 Structures of three naturally occurring juvenile hormones (JHI, JHII, JHIII), and the juvenile hormone pharmacological analog methoprene. JHIII is the most widespread form of juvenile hormone. Structurally, juvenile hormones are most similar to the vertebrate developmental signaling molecule retinoic acid.

Mode of JH Action

As a developmental hormone, JH controls tissue-level switches between alternative developmental pathways during metamorphosis. As such, there appear to be *critical sensitive periods* in a tissue when the presence or absence of JH during an ecdysone peak determines whether a tissue is retained at the same stage or changes to a more mature

Developments & Directions: Plants defend themselves from insects with juvenile hormone mimics and inhibitors

Karel Sláma, a researcher from Prague, moved to Boston to work in Carroll Williams' laboratory. Sláma brought his primary research animal, the European plant bug *Pyrrhocoris apterus*, with him. Interestingly, although these bugs developed normally in Czechoslovakia, for some mysterious reason they developed abnormally in North America:

> Instead of metamorphosing into normal adults, at the end of the 5th larval instar all molted into 6th instar larvae or into adultoid forms preserving many larval characters . . . Without exception, all individuals died without completing metamorphosis or attaining sexual maturity. . . . Evidently, when reared at Harvard University, the bugs had access to some unknown source of juvenile hormone.[37]

In the lab, larvae of these bugs are reared on paper, and the source of juvenile hormone activity was finally tracked down to a "paper factor" present only in American paper, such as from the journal *Science*, which the researchers tested, but not European paper such as from the journal *Nature*, which was also tested. The source of the American paper was the balsam fir, a tree indigenous to the northern United States and Canada, and it was eventually determined that this tree synthesizes a compound that closely resembles juvenile hormone,[38,39] which it appears to use as an insecticide.

Other plants, instead of producing JH mimics, produce chemicals that inhibit JH production by insects. A survey of plants for JH and anti-JH activity by Bowers[40] yielded the *precocenes* (Figure 11.15), compounds that destroy the JH-secreting cells of the corpora allata. When the larvae or nymphs of insects are dusted with precocenes, they undergo one more molt and then metamorphose into a sterile adult form. This has two major advantages for the plant: (1) it causes the herbivorous larva (which is damaging to the plant) to transform into a nonherbivorous adult, and (2) the resulting miniature adult is sterile, and therefore cannot generate new larvae that would otherwise further damage the plant.

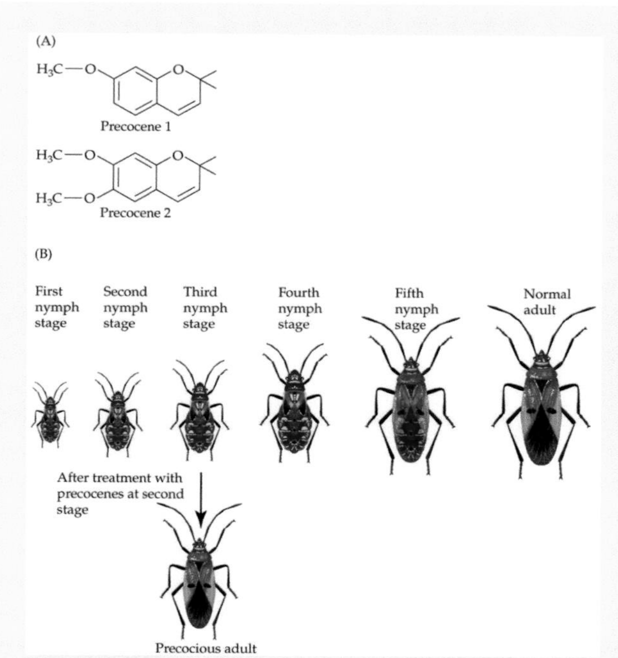

Figure 11.15 Precocious metamorphosis in the bug *Dysdercus* caused by precocenes. (A) Structures of two active precocenes found in plants. **(B)** When second nymph stages are treated with precocenes, they metamorphose into sterile precocious adults rather than continuing their normal developmental molting sequence. (After Bowers et al., 1976.)

Source of Images: Gilbert, S.F. and Epel, D., 2009. Ecological developmental biology: integrating epigenetics, medicine, and evolution. Oxford Publishing Limited. Used by permission.

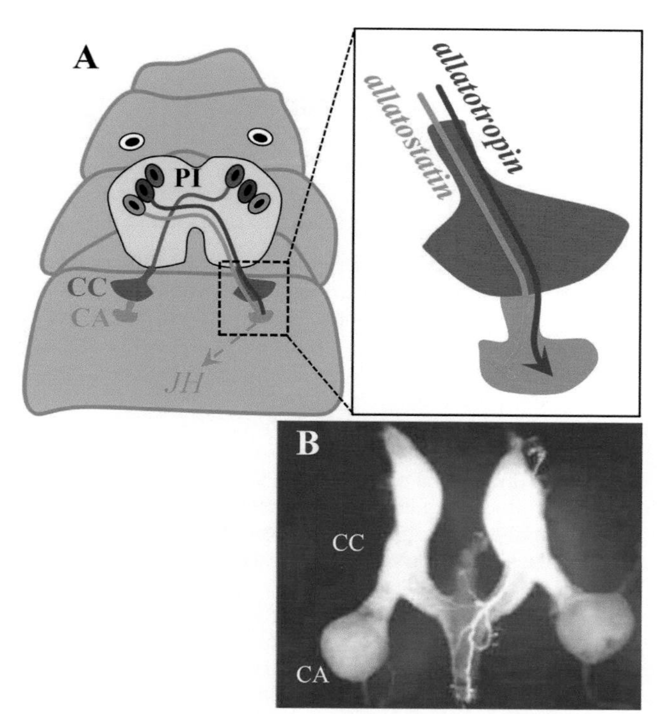

Figure 11.16 Neurosecretory control of JH synthesis in the larval insect. (A) Axons emanating from neurosecretory neurons in the pars intercereberalis (PI) of the brain innervate the corpora allata (CA, pink), a gland that releases JH into the hemolymph. **(B)** Whereas release of the neuropeptide allatotropin (blue neurons) stimulates JH release by the CA, release of allatostatin (green neurons) inhibits JH synthesis. **(C)** The CA and corpora cardiaca (CC, which secretes PTTH) are positioned posterior to the brain, where they release hormones into the blood. Structures shown were dissected from the cockroach *Periplaneta americana*.

Source of Images: (B) Resh, V.H. and Cardé, R.T. eds., 2009. Encyclopedia of insects. Academic press. Used by permission.

state (Figure 11.18). Different tissues have different sensitive periods. If JH is present during a particular sensitive period, then that tissue maintains its developmental state, and if JH is absent during that sensitive period, the pattern of gene expression changes and the tissue is set on a new developmental path. Importantly, JH alone has no overt developmental effect but typically requires the presence of ecdysone to express or reveal its effects. Using the cuticle as an example, a molt is always induced when ecdysteroids act on the epidermal cells to express a characteristic set of early genes whose products are transcription factors (Figure 11.20). For example, if JH is present during a critical sensitive period, the epidermal cells briefly express *E75A* and *Broad*, which promote the upregulation of JH-inducible late genes that regulate synthesis of the same type of cuticle when they molt. However, *Broad* expression is transient and ultimately suppressed by *E75A*. By contrast, if JH is absent during this critical sensitive period and the JH receptors remain in an unliganded state, then the epidermal cells alter their developmental commitment by expressing another set of transcription factors (e.g. *E74*, *E75*, and *Broad*) to generate the next developmental cuticle that they have been genetically programmed to synthesize. In the presence of ecdysone and the absence of JH, the

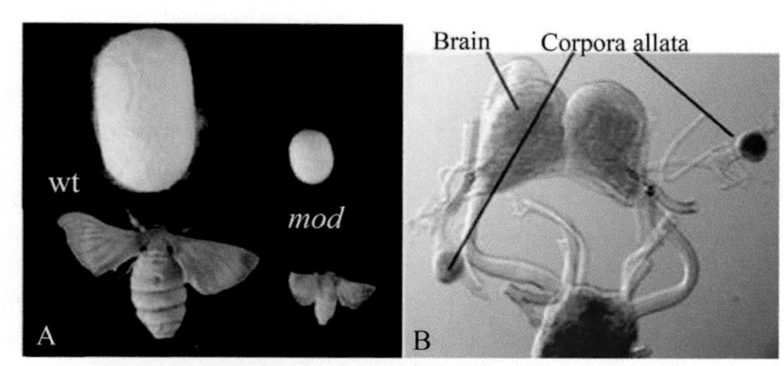

Figure 11.17 Precocious metamorphosis in *dimolting* (*mod*) mutants of the silkworm, *Bombyx mori*, is due to a loss of function of a gene encoding a cytochrome P450 enzyme in the JH synthesis pathway. (A) Whereas wild-type (wt) silkworms typically metamorphose after the fifth instar, *mod* mutants metamorphose precociously after the third instar, spinning tiny cocoons and emerging as small adults. **(B)** mRNA for CYP15C1 (purple stain), the gene encoding a cytochrome P450 enzyme, is specifically expressed in the corpus allatum where it is involved in the JH synthesis pathway. In *mod* mutants this gene produces a nonfunctional product.

Source of Images: Daimon, T., et al. 2012. PLoS genetics, 8(3), p. e1002486.

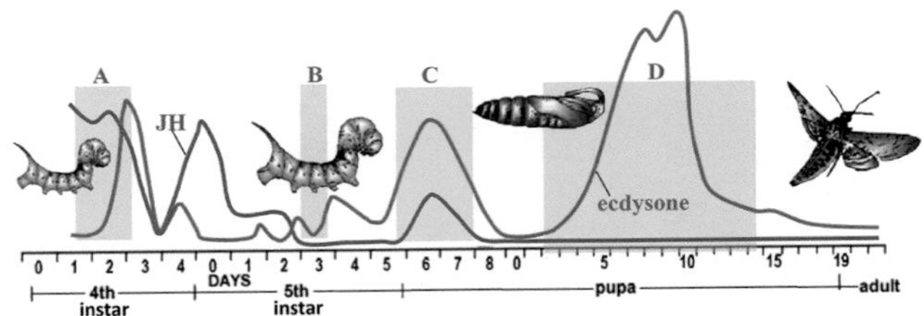

Figure 11.18 Some critical periods of sensitivity to juvenile hormone (green shading) during development of *Manduca sexta*. The presence or absence of JH during a critical period determines whether a tissue is retained at the same stage or changes to a more mature state. Critical periods during normal development: **(A)** Elevated JH is required to prevent a larval-to-pupal transition of the epidermis; **(B)** low or absent JH levels are necessary to allow a larval-to-pupal transition of the epidermis; **(C)** elevated JH is required to prevent the differentiation of imaginal discs into the adult form; **(D)** absence of JH is necessary to allow larval-to-adult transition of both the epidermis and the imaginal discs. Green line: GH levels. Red line: Ecdysone levels.

Source of Images: Lynn M Riddiford, et al. (2003). Insights into the molecular basis of the hormonal control of molting and metamorphosis from Manduca sexta and Drosophila melanogaster. Insect Biochemistry and Molecular Biology Volume 33, Pages 1327–1338

continued expression of *Broad*, in particular, is key to promoting a developmental commitment to a pupal molt.[33] In the epidermis *Broad* is constantly expressed only during the pupa stage, and not in either the larva or the adult; indeed, the presence of *Broad* suppresses the expression of both larval and adult genes in the pupa. Therefore, *Broad* expression appears to be the first step in the specification of pupal traits, and this gene may function as a master regulator of polymorphisms.

Developments & Directions: JH, polyphenism, and the rise of the supersoldiers

Many insect species have the ability to molt into one of two or more alternative morphologies, depending on nutritional and other environmental signals they receive. This is a kind of phenotypic plasticity called polyphenism (Figure 11.19(A) and Figure 11.19[B]). A *sequential polyphenism* occurs in holometabolous insects that undergo differing forms of larvae, pupae, and adults during their development. An *alternative polyphenism* occurs in the castes found in many social insects, where workers, soldiers, and reproductive castes (queens) can all arise from the same genotype (Figure 11.19[C]). In both cases, environmental factors affect hormone titers that in turn trigger developmental switches that alter the pattern of gene expression. In all cases that have been studied, the developmental switch that sets a larva onto the developmental pathway that leads to one or another caste is mediated by JH.[2] For example, in honeybees, *Apis mellifera*, there is a JH-sensitive period during the fourth and early fifth larval instars when exogenous JH can induce queen traits in any larva, and larvae that have been fed copious amounts of *royal jelly* (a secretion that is used in the nutrition of larvae) have a naturally elevated JH titer that induces them to develop into queens.[41,42]

In the ant, *Pheidole bicarinata*, there is a brief JH-sensitive period in the last larval instar during which exogenous JH can induce a larva to develop into a soldier instead of a worker.[43,44] JH is elevated at that time in larvae receiving high-nutrient food, and this causes a developmental switch that raises the critical weight of the larva and also reprograms the development of the imaginal disc of the head to produce a large-bodied soldier with a disproportionally large head (Figure 11.19[C]). Abouheif and his colleagues[45] found that in some species of ants there is an additional developmental switch. If JH levels exceed this additional threshold, the ants delay metamorphosis even further, grow even larger than regular soldiers and become *supersoldiers* (Figure 11.19[D]). Remarkably, by dabbing larvae with the JH mimic methoprene, the researchers could induce the supersoldier phenotype even in some species that normally lack them.

Figure 11.19 Polyphenism in insects. (A) Solitary and gregarious morphs of migratory locusts (*Locusta migratoria*). **(B)** Polyphenic expression of horns in minor (major) and minor (right) male dung beetles (*Onthophagus nigriventris*). **(C)** Caste determination occurs at JH-mediated switch points in response to environmental cues in the ant, *Pheidole obtusospinosa*, generating (from left to right) a queen, supersoldier, soldier, and minor worker. Queen determination occurs at an early developmental stage, whereas soldier determination occurs later. **(D)** In some species of ants there is an additional developmental switch whereby very high JH levels produce giant supersoldiers (left: normal soldier, right: supersoldier).

Source of Images: (A) Hartfelder and Emlen (2012) Endocrine control of insect polyphenism. In: Insect Endocrinology. Ed: L.I. Gilbert. Elsevier; used by permission. (B) Valena, S. and Moczek, A.P., 2012. Genet. Res. Int., 2012; (C) Lillico-Ouachour, A. and Abouheif, E., 2017. Curr. Opin. Insect. Sci., 19, pp. 43-51; used by permission. (D) Rajakumar, R., et al. 2018. Nature, 562(7728), pp. 574-577. Used by permission.

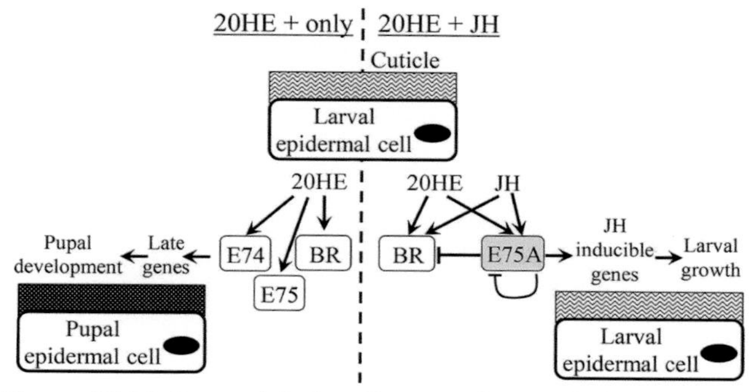

Figure 11.20 A model for the regulatory interaction of 20-hydroxyecdysone (20E) and juvenile hormone (JH) during insect cuticle development. A simple model only begins to describe the very complex system. For a larval-to-larval molt, both 20HE and JH are present and stimulate the production of the nuclear receptor E75A (right side). This receptor is responsible for the activation of several JH-inducible genes that are involved with larval growth, and it represses the 20HE-responsive gene, BR, as well as its own expression. Once larval development is complete, 20HE in the absence of JH activates a different group of early genes—BR, E74, and E75—that in turn activate a set of late genes responsible for pupal metamorphosis (left side).

Source of Images: Klowden MJ. Physiological Systems in Insects, Third Edition. 3 ed: Academic Press; 2013.

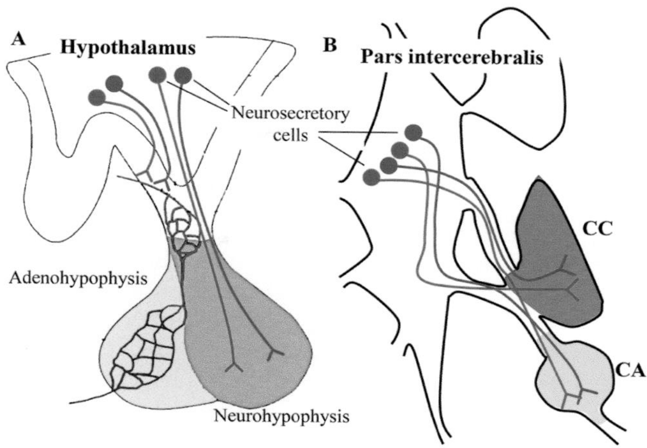

Figure 11.21 Comparison of the brain-pituitary complex of vertebrates and the brain-corpora cardiaca (CC)/allata (CA) complex of insects. (A) Parvocellular nuclei in the hypothalamus release peptides into portal vessels that transmit these hormones to the anterior pituitary gland, which responds by synthesizing and releasing its own hormones into general circulation. Magnocellular nuclei in the hypothalamus project axons directly into the posterior pituitary, which stores and releases these peptides into circulation. **(B)** Connections between the brain (pars intercereberalis) and corpora cardiaca/allata complex. Neurosecretory cells in the brain (pars intercereberalis) project their axons to the corpus cardiacum/allata complex, which store the neuropeptides and/or synthesize/release their own hormones in response. Similar to the juxtaposition of the vertebrate adenohypophysis and neurohypophysis, the CA is closely associated with the CC. Homologies between the adenohypophysis and corpora allata (yellow shading) and the neurohypophysis and corpora cardiaca (orange shading) have been suggested.

Source of Images: Wirmer, A., et al. 2012. Arthropod Struct Dev, 41(5), pp. 409-417.

Insect and Vertebrate Neuroendocrine Systems: More Alike than You Think

Ernst and Berta Scharrer, the founders of the field of neurosecretion, were the first researchers to point out the functional similarities between vertebrate and insect brain neurosecretion in their seminal 1944 publication[46]:

> A comparison of the hypothalamo-hypophyseal system of vertebrates with the intercerebralis-cardiacum-allatum system of insects reveals a parallelism which is the more striking because insect and vertebrate organs differ so greatly that no true organ homology can exist between these phyla. The hypothalamic nuclei of the vertebrates have their equivalent in the pars intercerebralis of the insects [refer to Figure 11.21] . . . . The parallelism in the organization of the two systems here compared could be merely a coincidence. However, it seems more likely that the comparison is significant in that it indicates a fundamentally similar relationship between the "master glands" and the central nervous system in invertebrates and vertebrates.

Despite the fact that vertebrates and insects are separated by more than 500 million years of evolution, in both taxa the highest command center of the neuroendocrine system is comprised of groups of neurosecretory cells located in the brain. Besides innervating brain centers and influencing neural circuits as "neuromodulators", these cells project their axons to peripheral neurohemal organs in which the hormones produced by the neurosecretory neurons are stored and released.[47] Functional neuroendocrine similarities between the two taxa include the regulation of energy metabolism, growth, water retention, and reproduction.[28,48]

However, do these functional similarities actually represent the existence of a shared ancestry (*homology*) between these structures? Based upon various functional, structural, and developmental similarities, researchers have proposed a homology between the brain endocrine systems of invertebrates and vertebrates. Specifically, homologies between the adenohypophysis and corpora allata,[47,49–52] and the neurohypophysis and corpora cardiaca,[50,51] have been suggested (Figure 11.21). During vertebrate embryogenesis, hypothalamic neurosecretory neurons and pituitary adenohypophysis cells derive from neurectoderm and pharynx (oral)-associated ectoderm, respectively. Interestingly, during *Drosophila* (fruit fly) embryogenesis, the neurosecretory pars intercerebralis cells of the brain and the corpus allata-cardiaca similarly derive from neurectoderm and pharynx-associated ectoderm.[53]

SUMMARY AND SYNTHESIS QUESTIONS

1. Decide whether the following statement is true or false, and explain your reasoning: 20-hydroxyecdysone plays a critical role in determining the type of organ (leg, wing, genitalia) that each imaginal disc will differentiate into.
2. List four different chemical categories of hormones involved in insect molting, and provide examples of each.
3. In this chapter it was mentioned that "phytoecdysteroids" produced by plants may function as feeding deterrents or toxic substances that affect the survival of herbivorous insects. By what mechanism might this occur?
4. Early studies with ecdysone showed that its application in larvae induces "puffing" of polytene chromosomes in *Drosophila* salivary glands. What family of receptor does the EcR belong to, and how does this explain chromosome puffing?
5. Explain the following observation: Clever and Ashburner showed that blocking the protein translation of ecdysone early puff RNAs with cycloheximide treatment prevents both the regression of early puffs and the appearance of late puffs.
6. What effect on molting would surgical ablation of the following organs have on a larva in its penultimate instar? A. Corpora allata (CA), B. Corpora cardiaca (CC).
7. Although at first glance the endocrinology of insect metamorphosis may seem quite different compared with vertebrate metamorphosis, there are some remarkable similarities between ecdysone receptors and thyroid hormone receptors (TR) and their signaling pathways. In fact, developmental biologist Jamshed Tata has pointed out, "Of all members of the nuclear receptor family, TR is closest to insect ecdysteroid receptor, to such an extent that the different receptor domains of mammalian and *Drosophila* receptors can be swapped without any loss of function in human or insect cells".[54] Describe these similarities.

The Origins of Holometabolous Development

LEARNING OBJECTIVE Discuss current theories regarding the evolutionary origins of insect metamorphosis.

KEY CONCEPTS:

- The evolution of holometabolous metamorphosis followed an advancement of JH production into earlier stages of embryonic development, resulting in an extended postembryonic phase of development (the larval phase).

The oldest fossilized insects appear to have developed much like modern ametabolous and hemimetabolous insects—they were direct developers devoid of a true larval period, possessing nymphs that looked similar to the adults. Fossils dating from 280 to 307 million years ago, however, record the emergence of a new developmental strategy. Some insects began to hatch from their eggs not as miniature adults, but with vermiform (wormlike) bodies possessing many short legs[55,56] (Figure 11.22). These insect larvae lived in a tropical environment, and their vermiform body shape would have permitted them to effectively rummage through moist, decaying vegetable matter. Holometabolous development in insects is thought to have arisen from hemimetabolous ancestors during the Permian,[57,58] and these intriguing insect larvae may represent the dawn of holometaboly.

Aristotle (322 BC)[59] and Harvey (1651)[60] described the holometabolous insect larval state as a "crawling egg", and the pupal stage as a "second egg". In 1913, Berlese[61] suggested that the holometabolous larva was a free-living embryo resulting from a premature egg hatch, and that the pupal stage represented a number of nymphal stages compressed into one. Truman and Riddiford[62,63] revitalized

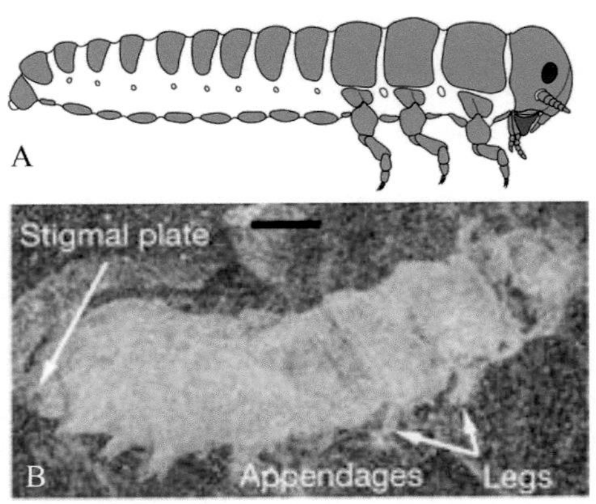

Figure 11.22 Fossils of the earliest known holometabolous insect larvae date from 280 to 307 million years. These insect larvae hatched from their eggs not as miniature adults, but with wormlike bodies possessing many short legs. This vermiform body shape would have permitted them to effectively rummage through moist, decaying vegetable matter and exploit an ecological niche different than their adult form. **(A)** Reconstructed groundplan of an ancient holometabolous larva. **(B)** Fossil of *Metabolarva bella* (Moscovian, Piesberg, Lower Saxony, Germany).

Source of Images: (A) Beutel, R.G., et al. 2022. Cladistics, 38(2), pp. 227-245; used by permission. (B) Nel, A., et al. 2013. Nature, 503(7475), pp. 257-261. Used by permission.

and extended Berlese's theory, suggesting that holometabolous larvae specifically correspond to the "pronymph" stage of hemimetabolous insects, and that the pronymph in an ancestral hemimetabola is the forerunner of the holometabolous larva. The *pronymph* is a developmental stage partitioning the end of embryogenesis and the first nymphal instar in hemimetabolus insects: in some hemimetabolous insects, the pronymph molts to the first nymphal instar prior to hatching from the egg, whereas in others the insect hatches while still in the pronymph stage. Pronymphs feed off of their yolk stores. Similar to holometabolous larvae, pronymphs have an underdeveloped nervous system and a soft cuticle with a similar ultrastructure to that of a larva. Truman and Riddiford hypothesize that this pronymph stage evolved into the larval stage of complete metamorphosis (Figure 11.23). Specifically, in holometabolous insects the pronymph stage developed into exogenously feeding larvae, specializing in feeding behaviors and diets that differed from those of their parents—namely, burrowing into and consuming fruits and detritus instead of nectar or other smaller insects. Over the generations, these insects may have postponed the transition to the more adult-like nymphal stage, extending the pronymphal stage for longer and longer periods of time, and becoming even more vermiform until they resembled modern larvae. Eventually, the nymphal phase was condensed into what is now a pupal stage (as was proposed by Berlese, 1913), transitioning from a gradual development of wings and other adult structures to an abrupt metamorphosis.

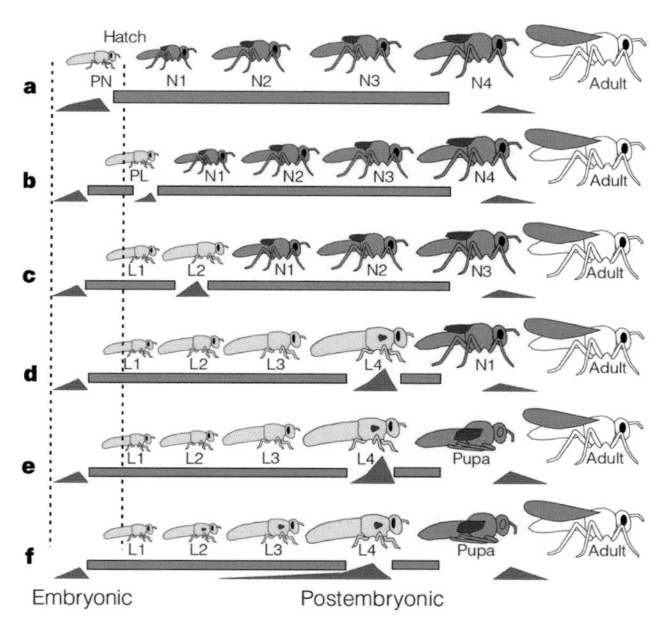

Figure 11.23 Hypothetical steps in the transition from a hemimetabolous to a holometabolous life history. The pronymphal (PN) and protolarval (PL) instars are in yellow; the nymphal (N) and pupal stages are in green; the wing buds, wing imaginal discs, and wings are in purple, and relative JH levels are in red. During hemimetabolous development there is a brief "pronymph" stage at the end of embryogenesis **(a)**. The pronymph developed into a "protolarva", which retains pronymph features but eats food exogenously that is different than the diet of nymphs or adults (e.g. leaves, instead of yolk, nectar or other insects) **(b)**. The resource-partitioning advantages of the protolarval form led to its extension and development into a specialized larval form (loss of legs and specialization in eating a lot) **(c–d)**. Eventually the multiple nymph stages were replaced by that of a pupa, a final larval molt that accommodates the abrupt transformation of one body plan into a now very different body plan **(e–f)**. The four most important trends depicted are (1) extension of the single pronymph stage into multiple larval stages, (2) condensation of the multiple nymphal stages into a single pupal stage, (3) replacement of nymphal wings with larval wing imaginal discs, and (4) advancement of JH production into earlier stages of embryonic development, followed by heterochronic shifts in JH expression among subsequent larval stages.

Source of Images: Truman, J.W. and Riddiford, L.M., 1999. Nature, 401(6752), pp. 447-452. Used by permission.

Truman and Riddiford[62,63] have put forth a functional endocrine explanation supporting this hypothesis, whereby the evolution of holometabolous metamorphosis followed an advancement of JH production into earlier stages of embryonic development, resulting in an extended postembryonic phase of pronymph development (the larval phase) (Figure 11.23). Indeed, when comparing the early developmental profiles of modern hemimetabolous and holometabolous insects, it is clear that whereas hemimetanolous insects produce a late peak of JH (at about 70% embryogenesis), holometabolous insects produce JH earlier at about 50% of embryogenesis (Truman and Riddiford 1999).

Developments & Directions: Crustacean Molting and Metamorphosis

Like insects, many crustaceans (e.g. crabs, lobsters, shrimp, crayfish) also undergo metamorphosis and molting (Figure 11.24). Unlike insects, which do not grow after metamorphosis, many crustaceans continue to periodically molt to accommodate growth long after metamorphosis is completed. Interestingly, molting and metamorphosis in crustaceans are also mediated by *ecdysone*, produced by the *Y organ* (a homolog of the insect prothoracic gland), as well as by *methyl farnesoate* (MF, a hormone similar to juvenile hormone in insects) produced by the *mandibular organ* (MO).[64,65] Importantly, these hormones are under regulatory control by diverse neurohormones secreted from the brain. The most studied crustacean neurosecretory system is the *X organ* (XO)-*sinus gland* complex located within the eyestalks[66] (Figure 11.24). Cells from the XO produce and transport a variety of neurohormones along axons to terminals in the sinus gland, which stores and releases the hormones into the hemolymph (blood). These neurohormones include *molt-inhibiting hormone* (MIH), which inhibits ecdysone synthesis by the Y organ, *mandibular organ-inhibiting hormone* (MOIH), which inhibits MF secretion by the MO, and *crustacean hyperglycemic hormone* (CHH), which regulates energy balance during molting. Considering that molting is under negative control by the sinus gland, it is well known that removal of the eyestalks in some crustaceans accelerates the next molt.

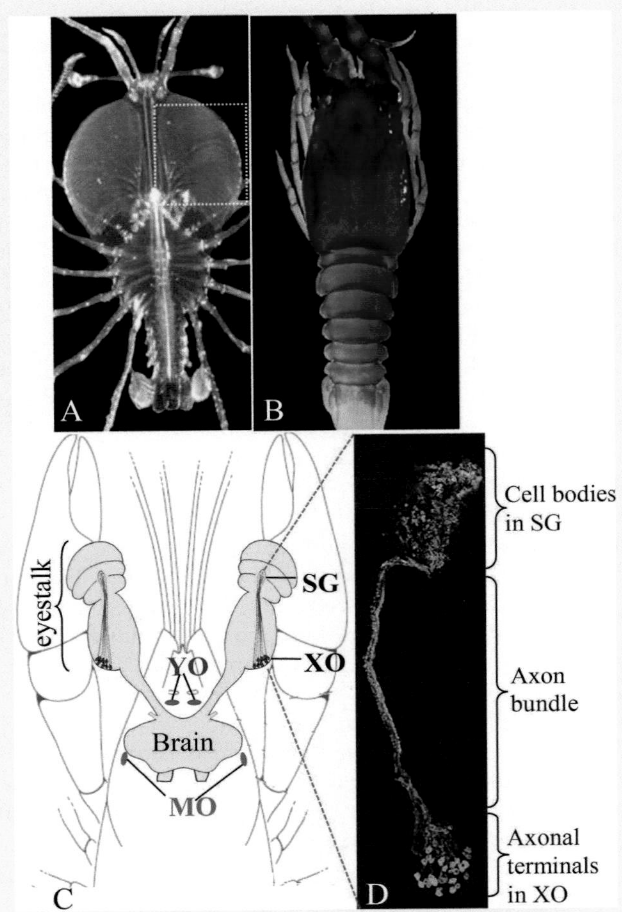

Figure 11.24 Metamorphosis and organization of the crustacean neuroendocrine X-organ (XO)/sinus gland (SG) complex in the eyestalk. The pre-metamorphic larval **(A)** and post-metamorphic juvenile **(B)** forms of the spiny lobster (*Sagmariasus verreauxi*). **(C)** The distribution of crustacean hyperglycemic hormone immunoreactivity (green) and immunoreactive molt inhibiting hormone (red) structures within the sinus gland of the shrimp, *Pagurus bernhardus*. Notice the segregated distribution of the two peptides, each restricted to one neuroendocrine neuron **(D)**. Abbreviations: SG (sinus gland), XO (X-organ), YO (Y-organ), MO (mandibular organ).

Source of Images: (A-B) Ventura, T., et al. 2015. Sci. Rep., 5(1), pp. 1-14. (D) N. Montagné, et al. Fed. Eur. Biochem. Soc. J., 275 (2008), pp. 1039–1052. Used by permission.

SUMMARY AND SYNTHESIS QUESTIONS

1. According to Truman and Riddiford's hypothesis for the evolutionary origins of holometaboly, what did the (1) single pronymph stage and (2) multiple nymph stages of the hemimetabolous ancestor each give rise to?

Summary of Chapter Learning Objectives and Key Concepts

LEARNING OBJECTIVE Discuss the importance of insect metamorphosis to the study of endocrinology.

- Insects provided the first hints of how steroid hormones exert transcriptional control over gene expression.
- The ability of insects to survive radical surgeries, like decapitations, neck tourniquets, and brain explants, provided valuable tools for founding the field of neuroendocrinology.
- There are more insect species (and therefore different insect endocrine systems) than all other animal, plant, and fungal species combined.
- Insect metamorphosis is triggered by changes in the timing of secretion of a small handful of hormones that are also involved in larval molting.

LEARNING OBJECTIVE Describe the three major developmental strategies insects use for reaching the adult stage.

- The exoskeleton plays an instrumental role in determining an insect's body form.
- In order to both grow and change form, the old exoskeleton must be removed and replaced by a new one in a process called "molting".
- Insect metamorphosis is a partition in the life of an individual that separates an early feeding and growth stage from a later reproductive stage.
- Insects use one of three major developmental strategies for reaching the adult stage: ametabolous development (no metamorphosis), hemimetabolous development (partial metamorphosis), and holometabolous development (complete metamorphosis).
- Ametabolous insect development represents the ancestral form, and hemimetaboly gave rise to holometaboly.
- In holometabolous insects, the precursor cells that will form many adult structures (e.g. legs, wings) exist in larvae as compact, internal structures called imaginal discs.

LEARNING OBJECTIVE Delineate the roles played by major hormones that mediate insect molting and metamorphosis, and compare and contrast insect neuroendocrine systems with those of vertebrates.

- Ecdysone, a steroid hormone produced by the prothoracic gland, modulates the timing of a molt.
- Prothoracicotropic hormone, a neuropeptide produced by the brain, induces ecdysone synthesis.
- Juvenile hormone, a fatty acid derivative produced in the head, determines whether the molt will be larval or metamorphic.
- Allatotropin and allatostatin are neuropeptides produced by the brain that induce and inhibit, respectively, juvenile hormone synthesis.
- In general, a nymphal/larval ecdysone peak occurring in the presence of JH maintains the current developmental state, whereas if JH is absent during an ecdysone peak the insect will progress to a more mature developmental state.

- Ecdysone receptors are nuclear transcription factors that modulate gene expression.
- Based upon functional, structural, and developmental similarities, homologies between the vertebrate adenohypophysis/neurohypophysis and insect corpora allata/corpora cardiaca, respectively, have been suggested.

LEARNING OBJECTIVE Discuss current theories regarding the evolutionary origins of insect metamorphosis.

- The evolution of holometabolous metamorphosis followed an advancement of JH production into earlier stages of embryonic development, resulting in an extended postembryonic phase of development (the larval phase).

LITERATURE CITED

1. Darwin C. *The Voyage of HMS Beagle*. Рипол Классик; 1910.
2. Nijhout HF. Arthropod developmental endocrinology. In: Minelli A, Boxshall G, Fusco G, eds. *Arthropod Biology and Evolution*. Springer-Verlag; 2013:123–148.
3. Erezyilmaz DF. The genetic and endocrine basis for the evolution of metamorphosis. In: Flatt T, Heyland A, ed. *Mechanisms of Life History Evolution: The Genetics and Physiology of Life History Traits and Trade-Offs*. Oxford University Press; 2011:504.
4. Wheeler WC. Insect diversity and cladistic constraints. *Ann Ent Soc Am*. 1990;83:91–97.
5. Klowden MJ. *Physiological Systems in Insects* (3rd ed.). Academic Press; 2013.
6. Kopeć S. Experiments on metamorphosis of insects. *Bull Int Acad Sci Cracovie*. 1917;:57–60.
7. Kopeć S. Studies on the necessity of the brain for the inception of insect metamorphosis. *Biol Bull*. 1922;42 323–342.
8. Wigglesworth V. The physiology of ecdysis in Rhodnius prolixus (Hemiptera). II. Factors controlling moulting and "metamorphosis". *Quart J Microsc Sci*. 1934;77:191–222.
9. Fukuda S. Induction of pupation in silkworms by transplanting the prothoracic glands. *Proc Imp Acad Tokyo*. 1940;16:411–416.
10. Fukata S. Hormonal control of molting and pupation in the silkworm. *Proc Imp Acad Tokyo* 1940;16:417–420.
11. Williams CM. Physiology of insect diapause. IV. The brain and prothoracic glands as an endocrine system in the cecropia silkworm. *Biol Bull*. 1952;103:120–138.
12. Butenandt A, Karlson P. Über die Isolierung eines Metamorphose-Hormons der Insekten in Kristallisierter Form. *z Naturforsch* 1954;9b:389–391.
13. Karlson P, Hoffineistcr H, Hummel H, Hocks P, Spiteller G. On the chemistry of ecdysone. VI. Reactions of ecdysone molecules. *Chem Beri*. 1965;98:2394–2402.
14. Huber R, Hoppe W. Die kristall- und moleku¨lstrukturanalyse des insektenverpuppungshormones ecdyson mit der automatisierten faltmoleku¨lmethode. *Chem Beri*. 1965;98:2403–2404.
15. Clever U, Karlson P. Induktion von puff-vera¨nderungen in den speicheldru¨senchromosomen von Chironomus tentans durch ecdyson. *Exp Cell Res*. 1960;20:623–626.
16. Beckstead RB, Lam G, Thummel CS. Specific transcriptional responses to juvenile hormone and ecdysone in Drosophila. *Insect Biochem Mol Biol*. 2007;37(6):570–578.
17. Clever U. Actinomycin and puromycin: effects on sequential gene activation by ecdysone. *Science*. 1964;146(3645):794–795.
18. Ashburner M, Chihara C, Meltzer P, Richards G. Temporal control of puffing activity in polytene chromosomes. *Cold Spring Harb Symp Quant Biol*. 1974;38:655–662.
19. Koelle MR, Talbot WS, Segraves WA, Bender MT, Cherbas P, Hogness DS. The Drosophila EcR gene encodes an ecdysone receptor, a new member of the steroid receptor superfamily. *Cell*. 1991;67(1):59–77.
20. Fletcher JC, Burtis KC, Hogness DS, Thummel CS. The Drosophila E74 gene is required for metamorphosis and plays a role in the polytene chromosome puffing response to ecdysone. *Development*. 1995;121(5):1455–1465.
21. Burtis KC, Thummel CS, Jones CW, Karim FD, Hogness DS. The Drosophila 74EF early puff contains E74, a complex ecdysone-inducible gene that encodes two ets-related proteins. *Cell*. 1990;61(1):85–99.
22. White KP, Hurban P, Watanabe T, Hogness DS. Coordination of Drosophila metamorphosis by two ecdysone-induced nuclear receptors. *Science*. 1997;276(5309):114–117.
23. Thummel CS. Flies on steroids—Drosophila metamorphosis and the mechanisms of steroid hormone action. *Trends Gen: TIG*. 1996;12(8):306–310.
24. Thummel CS. Molecular mechanisms of developmental timing in C. elegans and Drosophila. *Dev Cell*. 2001;1(4):453–465.
25. Jindra M, Malone F, Hiruma K, Riddiford LM. Developmental profiles and ecdysteroid regulation of the mRNAs for two ecdysone receptor isoforms in the epidermis and wings of the tobacco hornworm, Manduca sexta. *Dev Biol*. 1996;180(1):258–272.
26. Jindra M, Huang JY, Malone F, Asahina M, Riddiford LM. Identification and mRNA developmental profiles of two ultraspiracle isoforms in the epidermis and wings of Manduca sexta. *Insect Mol Biol*. 1997;6(1):41–53.
27. Riddiford L, Cherbas P, Truman JW. Ecdysone receptors and their biological actions. In: Litwack G, ed. *Vitamins and Hormones*, Vol 60. Academic Press; 2000:1–73.
28. Veelaert D, Schoofs L, De Loof A. Peptidergic control of the corpus cardiacum-corpora allata complex of locusts. *Int Rev Cytol*. 1998;182:249–302.
29. Rewitz KF, Yamanaka N, Gilbert LI, O'Connor MB. The insect neuropeptide PTTH activates receptor tyrosine kinase torso to initiate metamorphosis. *Science*. 2009;326(5958):1403–1405.
30. Wigglesworth V. The determination of characters at metamorphosis in Rhodnius prolixus (Hemiptera). *J Exp Biol*. 1940;17:201–223.
31. Wigglesworth VB. The function of the corpus allatum in the growth and reproduction of Rhodnius prolixus (Hemiptera). *Q J Microsc Sci*. 1936;79:91–123.
32. Williams CM, Moorhead LV, Pulis JF. Juvenile hormone in thymus, human placenta and other mammalian organs. *Nature*. 1959;183(4658):405.
33. Zhou X, Riddiford LM. Broad specifies pupal development and mediates the "status quo" action of juvenile hormone on the pupal-adult transformation in Drosophila and Manduca. *Development*. 2002;129(9):2259–2269.
34. Riddiford LM. Juvenile hormone: the status of its "status quo" action. *Arch Insect Biochem Physiol*. 1996;32(3–4):271–286.
35. Nijhout HF. *Insect Hormones*. Princeton University Press; 1994.
36. Daimon T, Kozaki T, Niwa R, et al. Precocious metamorphosis in the juvenile hormone-deficient mutant of the silkworm, Bombyx mori. *PLoS Genet*. 2012;8(3):e1002486.
37. Slama K, Williams CM. Juvenile hormone activity for the bug Pyrrhocoris apterus. *Proc Natl Acad Sci U S A*. 1965;54(2):411–414.
38. Bowers WS, Fales HM, Thompson MJ, Uebel EC. Juvenile hormone: identification of an active compound from balsam fir. *Science*. 1966;154(3752):1020–1021.
39. Sláma K, Williams CM. The juvenile hormone. V. The sensitivity of the bug, Pyrrhocoris apterus, to a hormonally active factor in American paper-pulp. *Biol Bull*. 1966;130:235–246.

40. Bowers WS, Ohta T, Cleere JS, Marsella PA. Discovery of insect anti-juvenile hormones in plants. *Science*. 1976;193(4253):542–547.
41. Wirtz P, Beetsma J. Induction of caste differentiation in the honeybee (Apis mellifera) by juvenile hormone. *Ent Exp Appl*. 1972;15:517–520.
42. Rachinsky A, Hartfelder K. Corpora allata activity, a prime regulating element for caste-specific juvenile hormone titre in honey bee larvae (Apis mellifera carnica). *J Insect Physiol*. 1990;36:189–194.
43. Wheeler DE, Nijhout HF. Soldier determination in ants: new role for juvenile hormone. *Science*. 1981;213(4505):361–363.
44. Wheeler D. The developmental basis of worker caste polymorphism in ants. *Am Nat*. 1991;138:1218–1238.
45. Rajakumar R, San Mauro D, Dijkstra MB, et al. Ancestral developmental potential facilitates parallel evolution in ants. *Science*. 2012;335(6064):79–82.
46. Scharrer E, Scharrer B. Neurosecretion VI. A comparison between the intercerebralis-cardiacum-allatum system of the insects and the hypothalamo-hypophyseal system of the vertebrates. *Biol Bull*. 1944;87:242–251.
47. Hartenstein V. The neuroendocrine system of invertebrates: a developmental and evolutionary perspective. *J Endocrinol*. 2006;190(3):555–570.
48. Nassel DR. Neuropeptides in the nervous system of Drosophila and other insects: multiple roles as neuromodulators and neurohormones. *Prog Neurobiol*. 2002;68(1):1–84.
49. Wirmer A, Bradler S, Heinrich R. Homology of insect corpora allata and vertebrate adenohypophysis? *Arthropod Struct Dev*. 2012;41(5):409–417.
50. Tessmar-Raible K. The evolution of neurosecretory centers in bilaterian forebrains: insights from protostomes. *Semin Cell Dev Biol*. 2007;18(4):492–501.
51. Tessmar-Raible K, Raible F, Christodoulou F, et al. Conserved sensory-neurosecretory cell types in annelid and fish forebrain: insights into hypothalamus evolution. *Cell*. 2007;129(7):1389–1400.
52. De Loof A, Lindemans M, Liu F, De Groef B, Schoofs L. Endocrine archeology: do insects retain ancestrally inherited counterparts of the vertebrate releasing hormones GnRH, GHRH, TRH, and CRF? *Gen Comp Endocrinol*. 2012;177(1):18–27.
53. Wang S, Tulina N, Carlin DL, Rulifson EJ. The origin of islet-like cells in Drosophila identifies parallels to the vertebrate endocrine axis. *Proc Natl Acad Sci U S A*. 2007;104(50):19873–19878.
54. Tata JR. A hormone for all seasons. *Perspect Biol Med*. 2007;50(1):89–103.
55. Shear WA, Kukalova-Peck J. The ecology of Paleozoic terrestrial arthropods: the fossil evidence. *Can J Zool*. 1990;68:1807–1834.
56. Nel A, Roques P, Nel P, et al. The earliest known holometabolous insects. *Nature*. 2013;503(7475):257–261.
57. Kukalova-Peck J. Fossil history and the evolution of hexapod structures. In: *The Insects of Australia: A Textbook for Students and Research Workers* (2nd ed.). Cornell University Press; 1991:141–179.
58. Labandeira CC, Phillips TL. A Carboniferous insect gall: insight into early ecologic history of the Holometabola. *Proc Natl Acad Sci U S A*. 1996;93(16):8470–8474.
59. Aristotle. *De Partibus Animalium and De Generatione Animalium I (with passages from II. 1–3)*. Clarendon Press; 332 BC.
60. Harvey W. *Disputations Touching the Generation of Animals*. Blackwell Scientific Publications; 1651.
61. Berlese A. Intorno alle metamorfosi degli insetti. *Redia*. 1913;9:121–136.
62. Truman JW, Riddiford LM. The origins of insect metamorphosis. *Nature*. 1999;401(6752):447–452.
63. Truman JW, Riddiford LM. Endocrine insights into the evolution of metamorphosis in insects. *Annu Rev Entomol*. 2002;47:467–500.
64. Chang ES. Comparative endocrinology of molting and reproduction: insects and crustaceans. *Ann Rev Entomol*. 1993;38(1):161–180.
65. Chang ES, Mykles DL. Regulation of crustacean molting: a review and our perspectives. *Gen Comp Endocr*. 2011;172(3):323–330.
66. Hartenstein V. The neuroendocrine system of invertebrates: a developmental and evolutionary perspective. *J Endocrinol*. 2006;190(3):555–570.

CHAPTER

12

Calcium/Phosphate Homeostasis, Skeletal Remodeling, and Growth

CHAPTER LEARNING OBJECTIVES:

- Describe the biosynthetic pathway of vitamin D, and use historical context to describe how vitamin D can be considered both a vitamin and a hormone.
- Discuss some key roles that calcium and phosphate play in normal physiology and the physiological mechanisms by which calcium and phosphate homeostasis are maintained.
- Review the endocrine mechanisms by which various hormones maintain calcium and phosphate homeostasis.
- Describe the endocrine bases of various pathologies of calcium/phosphate homeostasis.
- Describe the roles of various hormones in regulating longitudinal bone growth.
- Describe the endocrine bases for disorders of bone growth.

OPENING QUOTATIONS:

"If the parents' financial status permits, it is best to take the children out into the country and keep them as much as possible in the dry, open and pure air. If not, at least they should be carried about in the open air especially in the sun, the direct action of which on our bodies must be regarded as one of the most efficient methods for the prevention and the cure of this disease [referring to Rickets]."

—Jędrzej Śniadecki was a Polish physician and physiologist and professor of chemistry at the University of Wilno. Quoted from his book *On the Physical Education of Children*, written in 1822, in the chapter titled "English Disease" (Śniadecki, Jędrzej, "Dziela", vol. 1, 273–274, Warszawa [1840])

"Bone growth is necessary for the health, not just of the present child, but of the future adult."

—Michael Parfitt, from Parfitt, A.M., Travers, R., Rauch, F. & Glorieux, F.H. Structural and cellular changes during bone growth in healthy children. *Bone* 27, 487–494 (2000)

Discovery and Importance

LEARNING OBJECTIVE Describe the biosynthetic pathway of vitamin D, and use historical context to describe how vitamin D can be considered both a vitamin and a hormone.

KEY CONCEPTS:

- Rickets (in children) and osteomalacia (in adults) are among the most prevalent endocrine disorders in human history.
- Vitamin D is a calcitropic hormone that can be both generated by the skin in response to sunlight and obtained via diet. An inability to produce vitamin D in children causes rickets (weak bones), and in adults osteomalacia.
- Calcium and phosphorous are the most abundant of inorganic elements in vertebrates and are the primary structural elements of bones, teeth, scales, and the outer shells of reptile and bird eggs.

Rickets and the Discovery of Vitamin D

One of the most prevalent endocrine-related childhood disorders in human history, which remains common in many developing countries today, is **rickets**, a condition characterized by the defective mineralization of bones and the development of weak and toneless muscles due to impaired metabolism of vitamin D, phosphorus, or calcium. This malady may lead to the bending and potential fracture of bones due to weight bearing and muscular tension (Figure 12.1). When this condition develops in adults, it is called *osteomalacia*. Although the condition was first documented in humans as early as the 2nd century AD, the disease was not considered a significant health problem until the industrialization of northern Europe in the 17th century, when it was recognized that many of the children who lived in the crowded, polluted, and dark (even during daylight) cities developed the disease.

DOI: 10.1201/9781003359241-16

The incidence of rickets in northern Europe and North America increased dramatically during the industrial revolution as populations moved from farms to smoke-filled cities, and by the latter part of the 19th century about 60% of the children living in London had rickets.[1] This disease was particularly devastating for young women of childbearing age who often had a deformed pelvis, resulting in a high incidence of infant and maternal morbidity and mortality.

The Sunlight Cure

The first hints to both the cause and cure of rickets emerged in 1822, when Polish physician and scientist Jędrzej Śniadecki observed that children living in urban Warsaw had a high incidence of rickets, whereas children living in rural areas outside Warsaw did not[2] (see this chapter's opening quotation). Also, Theobald Palm, a medical missionary, collected observations from a number of physicians throughout the British Empire and the Orient revealing that whereas the children of the middle class and the poor who lived in industrialized cities in the British Isles had a high incidence of rickets, the disease was rare in children living in impoverished cities in non-industrialized China, Japan, and India where people received poor nutrition and lived in relative squalor.[1] Based on his and Palm's observations, Śniadecki urged systematic sunbathing as a preventive and therapeutic measure for rickets. The specific component of sunlight responsible for the curative effects of sunlight on rickets, ultraviolet B radiation (UVB), was discovered by Kurt Huldschinsky, a Berlin pediatrician, who effectively treated children by irradiating them with quartz mercury-vapor lamps[3,4] (Figure 12.1). He correctly theorized that an unidentified chemical in the skin activated by UVB light diffused throughout the body, a chemical that would ultimately turn out to be vitamin D.

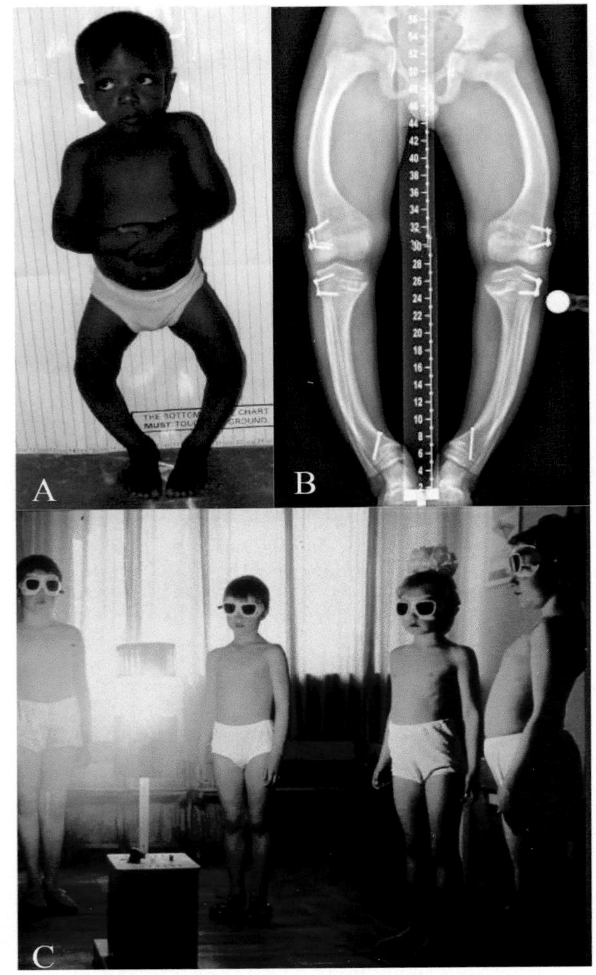

Figure 12.1 Rickets and the sunlight cure. (A) A young child suffering from rickets, a childhood disorder involving softening and weakening of the bones, primarily caused by lack of vitamin D, calcium, and/or phosphate. It is often associated with lack of exposure to sunlight. **(B)** Radiograph of a 2-year-old rickets sufferer, with a marked bowing of the femurs and decreased bone opacity, suggesting poor bone mineralization. **(C)** By the early 1920s it was recognized that irradiation with either sunlight or ultraviolet B light from lamps helps prevent rickets. This photograph shows irradiation of German children with quartz-mercury vapor lamps. It was recognized that the eyes should be protected from UVB rays.

Source of images: (A) Thacher, T.D., 2003. Endocr. Dev., 6, pp. 105-125; used by permission. (B) Adams, J.E., 2018. Radiology of rickets and osteomalacia. In Vitamin D (pp. 975-1006). Academic Press; used by permission. (C) Russell W. Chesney. 2012. Nutrients, 4, 42-51. Used by permission.

The Diet Cure

At the same time that the "sunlight cure" was taking effect, a completely different dietary approach to treating rickets was also being developed. Mellanby[5] in Great Britain and McCollum[6] in the United States developed animal models for rickets and showed that the condition could be cured by ingesting cod liver oil. McCollum named the antirachitic factor *vitamin D*, the fourth discovered after vitamins A, B, and C. Subsequently, the laboratories of Alfred Hess[7] and Harry Steenbock[8] simultaneously demonstrated that UV irradiation of various foods, such as milk, oils, and cereals, could render them with antirachitic properties through the activation of an unidentified molecule. Windaus and Hess[9] discovered the exact molecular identity to be *ergosterol* (Figure 12.2A), a member of a subgroup of steroid molecules found in plants and fungi, called *sterols*. The *Nobel Prize for Chemistry* for 1928 was awarded to Adolf Windaus "for his studies on the constitution of the sterols and their connection with vitamins", and the ultraviolet-irradiated secosteroid (broken ring structure) product of ergosterol was later purified and named *vitamin D2* (*ergocalciferol*) (Figure 12.2B).

These findings, together with those of sunbathing therapy, strongly suggested that UV rays in sunlight must convert a sterol-like compound present in vertebrate skin into the active vitamin D. Ultimately, Windaus et al.[10] determined the structure of this vertebrate precursor molecule

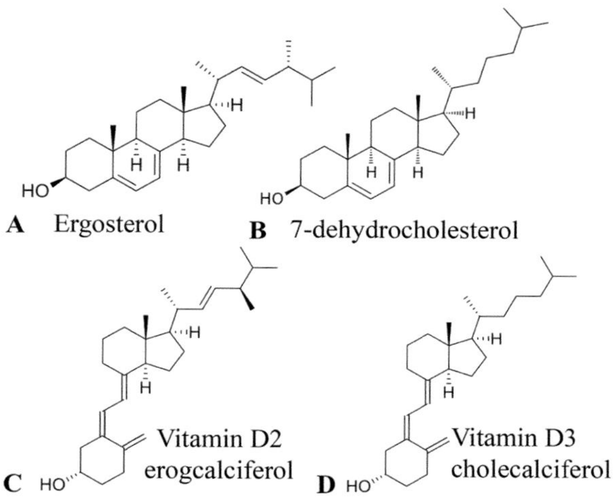

Figure 12.2 Some structures of vitamin D and its chemical precursors.

Table 12.1 Some Biological Functions of Calcium and Phosphate

Calcium	Phosphate
Structural component of bones, teeth, scales, and bird eggshells Blood coagulation Muscle contraction Maintains neuronal membrane potential Intracellular cell signaling second messenger Exocytosis	Structural component of bones, teeth, and scales Blood buffer Component of membrane phospholipids Component of some intracellular cell signaling second messengers (e.g. cAMP, inositol triphosphate) Component of nucleic acids (DNA, RNA) Activator of intracellular enzymes (phosphorylation) Necessary for glucose and glycogen metabolism Intracellular anion

Curiously Enough . . . The public health initiative of the fortification of cow's milk–based infant formulas with vitamin D2 and also vitamin D2 supplementation of breastfed infants became common practice by the 1930s, resulting in the virtual disappearance of rickets in the United States by the 1960s.[11]

(*7-dehydrocholesterol*) (Figure 12.2C) from porcine skin and the pathway by which it is converted to *vitamin D3* (*cholecalciferol*) (Figure 12.2D). Irradiated ergosterol soon became readily available as a potent vitamin D2 source for food fortification and the treatment of rickets. Note how the hydroxyl-containing ring structure of vitamin D2 (Figure 12.2B) is identical to that of D3 (Figure 12.2D), explaining its biological potency in vertebrates.

Vitamin D Is a Hormone

Despite its name, vitamin D is actually not a vitamin but a hormone. It is largely through historical accident that vitamin D became classified as a vitamin. The formal definition of a vitamin is that it is a trace dietary constituent required to produce the normal function of a physiological process or processes. The emphasis here is on the fact that the vitamin must be supplied regularly in the diet, and that the body is unable to independently synthesize the vitamin in question. However, the ultraviolet exposure of 7-dehydrocholesterol, which functions as the vitamin D prohormone, present in the skin results in the photochemical production of the active vitamin D hormone. Thus, vitamin D only becomes a true vitamin when the individual does not have regular access to sunlight. Under normal physiological circumstances, all mammals can generate, via ultraviolet exposure of 7-dehydrocholesterol present in the skin, adequate quantities of vitamin D to meet their nutritionally defined requirements. As will be explored in this chapter, several hormones, including vitamin D, have profound roles in maintaining the body's calcium and phosphate homeostasis.

Distribution and Importance of Calcium and Phosphorous to Life

Calcium and phosphorous are the most abundant inorganic elements in vertebrates and are the primary structural elements of bones, teeth, scales, and the outer shells of reptile and bird eggs. In addition to their roles in structural support, both ions also play critical physiological roles in other diverse cellular processes (Table 12.1). In fact, it would be difficult to name a physiological process that does not depend in some way on calcium or phosphate. As the endocrinology of calcium and phosphate regulation cannot be fully appreciated without a sufficient understanding of their basic physiology, an overview of their essential physiological dynamics is provided in the next section.

SUMMARY AND SYNTHESIS QUESTIONS

1. At the molecular level, what does UV irradiation do to the precursors of vitamin D2 and D3 that initiates their activation?
2. How can vitamin D be considered both a vitamin and a hormone?

Calcium and Phosphate Dynamics

LEARNING OBJECTIVE Discuss some key roles that calcium and phosphate play in normal physiology and the physiological mechanisms by which calcium and phosphate homeostasis are maintained.

KEY CONCEPTS:

- Calcium is involved in diverse and critical physiological functions, such as cellular protein secretion, nerve impulse conduction, intracellular signaling, blood coagulation, and muscle contraction.
- Phosphate is integral to life and an essential component of nucleotides, phospholipids in cell membranes, intracellular second messengers, phosphorylated intermediates of intracellular signal transduction pathways, and high-energy compounds (e.g. ATP).
- Blood serum concentrations of free calcium and phosphate ions are inversely proportional to each other.
- In mammals, the skeleton, intestine, and kidney are the most important organs involved in maintaining calcium and phosphate homeostasis.
- The intestine takes up calcium and phosphate via active transcellular and passive paracellular pathways.
- Whereas most calcium in the kidney filtrate is reabsorbed, 25% of the phosphate load is typically excreted in the urine.
- Bone is the primary storage site for calcium and phosphate.
- The continual process of bone turnover, called remodeling, is facilitated by four cell types: osteogenic cells, osteoblasts, osteocytes, and osteoclasts.

Physiological Functions and Interactions

Calcium

Some key physiological functions that calcium is involved in include hormone and neurotransmitter secretion, nerve impulse conduction, intracellular signaling, blood coagulation, and muscle contraction, with many of these effects brought about by the binding of calcium to proteins that alter their structures and functions. In humans, the vast amount of total body calcium is found in the skeleton (99%), with about 0.9% of the fraction located in the intracellular compartment, and the remaining 0.1% found in the extracellular fluid and blood. Although at any given time only a small fraction of total body calcium is found in the extracellular fluid and blood plasma, it is this fraction that ultimately transports calcium to the skeleton and cells over time. Despite broad fluctuations of dietary calcium intake, cytosolic and extracellular fluid concentrations are tightly maintained within narrow limits for optimal physiology. For example, in a healthy human, levels of blood plasma Ca^{++} are typically maintained at 8.5–10.5 mg/100 ml, and deviation from these levels may result in broad impairments in physiological functioning, with hypocalcemia resulting in seizures and tetany and hypercalcemia producing muscle weakness, muscle twitching, calcium precipitation in soft tissues (e.g. kidney stones), and psychiatric disturbances.

In the blood plasma, calcium is normally found in three forms: free ionized Ca^{++}, albumin-bound Ca^{++}, and Ca^{++} complexed to organic ions, such as lactate and bicarbonate (Table 12.2). Free ionized Ca^{++} is considered the most diffusible, biologically active and regulated form. However, due to its free nature, it is prone to ultrafiltration by the kidney and subsequent excretion via the urine. Therefore, as will be discussed in detail later, to prevent hypocalcemia the kidney tubules have the critical ability to reabsorb Ca^{++} back into the blood.

Table 12.2 Distribution of Calcium and Phosphate in Normal Human Plasma

State	Percentage of Total
Calcium	
Free Ca^{2+}	49
Protein-bound	47
$CaHPO_4$	3
Phosphorous	
Free HPO_4^{2-}	44
Free $H_2PO_4^-$	10
Protein-bound	12
$NaHPO_4^-$	28
$CaHPO_4$	3
$MgHPO_4$	3

Source: Adapted from Walser, M. (1961). Ion association. VI. Interaction between calcium, magnesium, inorganic phosphate, citrate, and protein in normal human plasma. *J. Clin. Invest*. 40:723

Phosphorous

About 85% of the body's phosphorous exists in the structure of bone and teeth combined with calcium in the form of *hydroxyapatite* ($Ca_{10}(PO_4)_6(OH)_2$), with the remaining 15% dissolved in the intra- and extracellular fluids as phosphate ($H_2PO_4^-$ and HPO_4^{2-}). Phosphate is an essential component of nucleotides like DNA and RNA, phospholipids in cell membranes, intracellular second messengers such as cAMP and inositol phosphates, phosphorylated intermediates of intracellular signal transduction pathways (e.g. kinases), and high-energy compounds such as ATP and creatinine phosphate. The blood plasma concentrations of the various forms of phosphate (free, protein-bound, and cation-bound) (Table 12.2) normally range from about 2.5 to 4.5 mg/100 ml. However, in contrast to Ca^{++}, phosphate is rarely a limiting variable to vertebrates, and, thus, phosphate levels are not as stringently regulated.

Interrelationship between Free Blood Plasma Calcium and Phosphate

The small but physiologically most active fraction of calcium and phosphate is found in the free (aqueous) blood plasma. Free calcium and phosphate bind to form a calcium phosphate precipitate:

$$3Ca^{+}_{2\,(aq)} + 2PO_4^{3-}{}_{(aq)} \rightleftarrows Ca_3\left(PO_4\right)_{2(s)}$$

where aq denotes an aqueous state, s denotes a solid precipitate state, and the double arrows denote a state of equilibrium.

Importantly, the relative concentrations of free serum calcium and phosphate generally have an inverse relationship such that as free calcium levels increase, free phosphate concentrations decrease, and vice versa.

The Principal Organs of Calcium and Phosphate Homeostasis

The most important organs involved in maintaining blood calcium and phosphate homeostasis in mammals and reptiles are the skeleton, kidneys, and intestine (Figure 12.3). Whereas bone is the primary source of calcium and phosphate storage in the body, the kidneys regulate the retention and excretion of excess ions via the urine and the intestine is involved in calcium and phosphate uptake. Additionally the intestine also excretes any unabsorbed calcium and phosphate in the feces.

> **Curiously Enough . . .** In fish and amphibians, calcium is taken up via gill and skin epithelial tissues by molecular mechanisms similar to those of the intestine and the kidney.

Intestine

In addition to absorbing nutrients for metabolic fuel and cellular maintenance, the morphology and high surface area-to-volume ratio of the intestine is equally well suited for the uptake of essential ions and elements, including calcium and phosphate. In mammals and reptiles virtually all body calcium and phosphate are ultimately derived from the diet, and adequate intestinal absorption is critical for regulating calcium and phosphate balance.

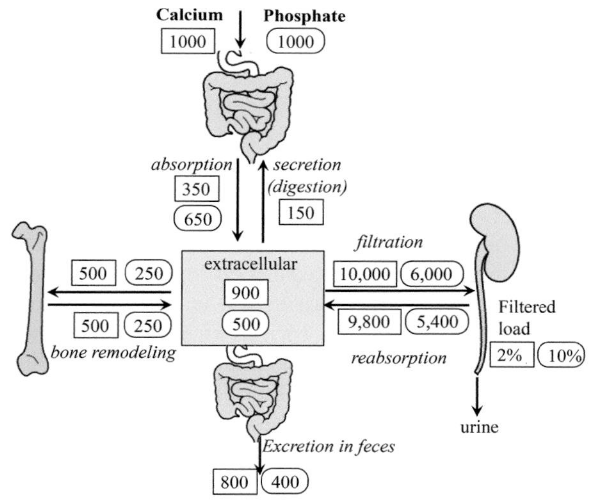

Figure 12.3 Daily turnover of calcium and phosphate in a human with a dietary intake of 1,000 mg/day. Other forms of calcium and phosphate loss not depicted in the figure include sweat, lactation, and across the placenta.

Source of images: Nussey, S. and Whitehead, S., 2001. The thyroid gland in Endocrinology. London, GB: An Integrated Approach by Published by BIOS Scientific Publishers Ltd.

Calcium

Although the stomach itself does not absorb calcium, its acidic environment solubilizes the mineral, thus facilitating subsequent uptake by the intestine. Intestinal calcium absorption occurs via two mechanisms (Figure 12.4): (1) the active *transcellular* pathway occurring across the plasma membrane, and (2) the passive *paracellular* transport pathway taking place through intercellular tight junctions. The active transcellular pathway predominates when relatively low concentrations of calcium are present in the intestinal lumen, whereas in the presence of high dietary calcium the paracellular pathway predominates. At the molecular level,

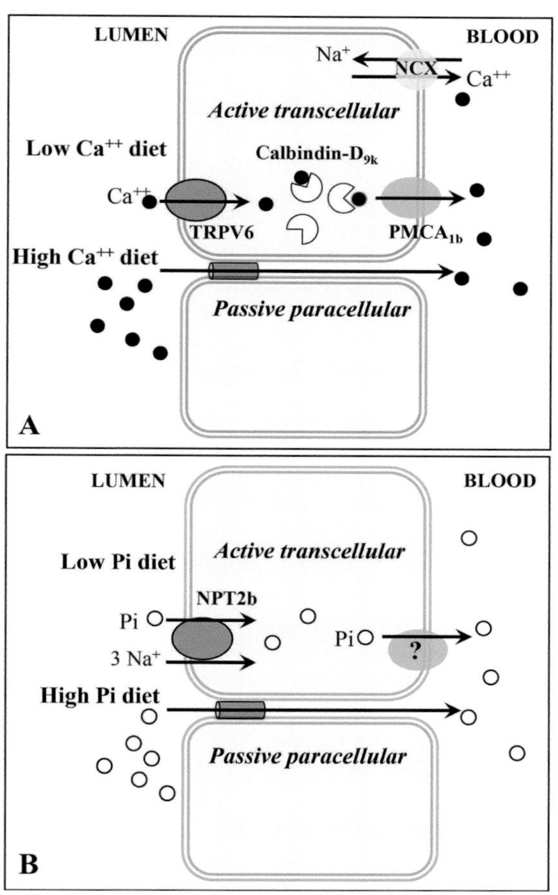

Figure 12.4 Intestinal calcium and phosphate absorption. **(A)** Model of intestinal calcium transport comprising a transcellular, active mechanism that transports calcium when dietary calcium intake is normal/low and a paracellular, passive pathway that functions under high calcium intake. Calcium transporters include apical membrane localized TRPV6, cytosolic calbindin-D9k, and basal membrane localized NCX and PMCA1b. **(B)** A large fraction of dietary phosphate intake is considered to be transported by a passive, paracellular pathway. The transcellular pathway consists of NPT2b localized at the intestinal brush border membrane. The expression of this transporter is increased when dietary phosphate intake is low. The mechanism of transmembrane phosphate into circulation is poorly understood.

Source of images: International Bone and Mineral Society (2012). BoneKEy Reports.

the rate-limiting step for the active transcellular pathway is calcium crossing the luminal membrane via the epithelial calcium channels TRPV5 and TRPV6, where *TRPV* signifies "transient receptor potential cation channel subfamily V". Once in the cell cytoplasm, *calbindin-*D_{9k} binds free calcium and prevents the levels of free intracellular calcium from accumulating in intestinal cells, and it also facilitates diffusion within the cytosol. Plasma membrane Ca^{++} ATPase (PMCA) pumps are localized to the basolateral membrane and pumps calcium out of the cell and into the extracellular fluid and blood. Additionally, calcium may be pumped out of the cell via Na^{+}/Ca^{++} exchangers (NCX).

Phosphate

Similar to calcium uptake, intestinal phosphate uptake also occurs via transcellular and paracellular pathways (Figure 12.4). The transcellular mechanism of intestinal phosphate transport depends on the sodium-dependent phosphate cotransporter IIb (NaPiIIb) present at the luminal surface (apical membranes) of intestinal epithelial cells. Energy for this transport process is provided by an inward downhill sodium gradient, maintained by export of Na^{+} from the cell via a Na^{+}/K^{+}-ATPase cotransporter at the basolateral membrane. The phosphate incorporated into the enterocytes by this mechanism is transferred to the circulation by poorly understood mechanisms

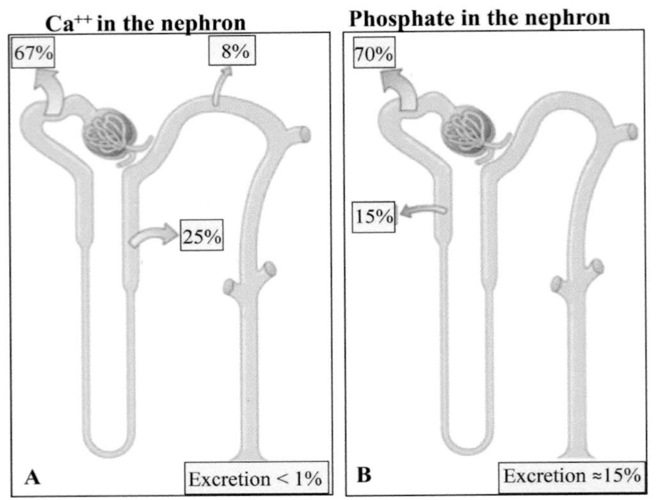

Figure 12.5 Calcium and phosphate reabsorption by kidney nephrons. Percentages refer to the amount of filtered Ca^{++} and PO_4– reabsorbed by each nephron segment. **(A)** Calcium is reabsorbed along much of the nephron, and very little is excreted. **(B)** Under normal conditions, ~75% of the filtered load of phosphate is reabsorbed, with all of the reabsorption occurring in the proximal tubule. However, low-phosphate diets significantly increase inorganic phosphate (P_i) reabsorption, recruiting transporters in sites distal to the proximal convoluted tubule, which can reduce phosphate excretion to 5% to 10%. Ca^{++} and phosphate transporter distribution varies with nephron region, but the transporters are believed to be similar to those described for the intestine.

Source of images: Costanzo, L.S., 2017. Physiology, E-Book. Elsevier Health Sciences. Used by permission.

Kidney

Calcium is reabsorbed throughout the tubules of the kidneys' *nephrons*, with very little excreted (normally less than 2%) (Figure 12.5). In contrast, under typical conditions, about 25% of dietary phosphate is excreted from the nephron, with 75% of the filtered load reabsorbed specifically by the proximal tubule. Low-phosphate diets significantly increase phosphate reabsorption, activating additional nephron regions (loop of Henle, distal tubule) that can reduce phosphate excretion to as low as 5%. At the cellular level, the mechanisms of calcium and phosphate transport by the nephron tubule epithelial cells are very similar to those described for transport by the intestinal epithelium (see Figure 12.4).

Bone

Composition and Maintenance

Bone is a complex tissue composed of both cellular and extracellular, as well as organic and inorganic, matrix components. *Bone matrix* is a composite of inorganic calcium and phosphate ions in an arrangement called *calcium hydroxyapatite* that gives bones rigidity, in combination with an organic component called *osteoid* that is rich in proteins like collagen, an elastic protein that imparts fracture resistance. About 65%–70% of bone dry weight consists of calcium hydroxyapatite crystals, and 30%–35% is comprised of osteoid. In addition to playing a structural role that is essential for maintaining normal bone mass and bone quality, calcium and phosphate are taken up or released by bone, thus buffering blood serum concentrations for these ions. Importantly, the preservation of serum calcium and phosphate levels may occur at the expense of skeletal integrity. Although the cellular component of bone represents only a tiny fraction of bone mass, these cells are ultimately entirely responsible for the biosynthesis, secretion, and mineralization of the bone matrix, as well as its resorption.

Bone Modeling and Remodeling

The term **bone modeling** refers to bone lengthening and shape changes taking place from birth to puberty. However, even in adults who have ceased their longitudinal growth, bone is far from a static tissue and is in a constant state of both formation and resorption. This continual process of bone turnover, as well as the strengthening or repairing of damaged bone, is termed **bone remodeling,** a process facilitated by four cell types (Figure 12.6). Estimates of the rate of bone remodeling in adult humans suggest that the equivalent of the entire skeleton is destroyed and rebuilt by this process every 10 years. Importantly, dysfunctions among these cells underlie many clinical diseases affecting calcium and phosphate homeostasis.

The steps of typical bone remodeling can be divided into five phases:

1. **Initial synthesis of the organic bone matrix:** The protein components of the extracellular matrix (osteoid)

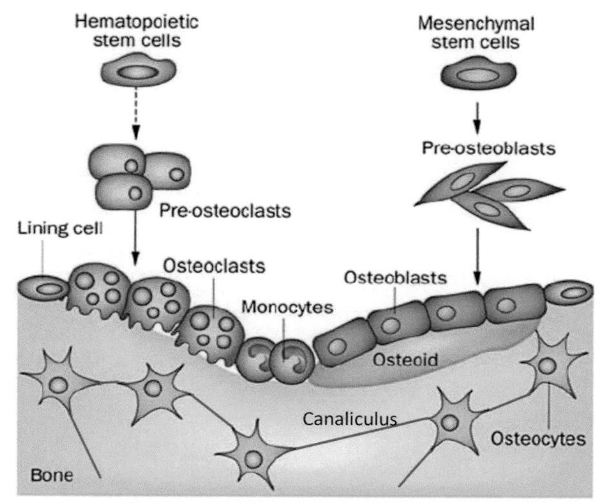

Figure 12.6 Four cell types that mediate bone remodeling. Osteogenic cells are undifferentiated stem cells. These develop into osteoblasts, which subsequently secrete extracellular matrix proteins (collectively termed "osteoid") and calcium and phosphate salts. When osteoblasts become trapped within the mineralized matrix, their structure and function changes, and they develop into osteocytes. Osteocytes maintain the mineral concentration of the matrix and communicate with each other and receive nutrients via long cytoplasmic processes that extend through canaliculi (tiny channels), channels within the bone matrix. Osteoclasts are large, multinucleated cells derived from a non-osteogenic cell lineage. These cells facilitate bone demineralization and degradation of osteoid (uncalcified bone matrix) and mediate bone resorption.

Source of images: Norman, A.W. and Henry, H.L., 2022. Hormones. Academic Press. Used by permission.

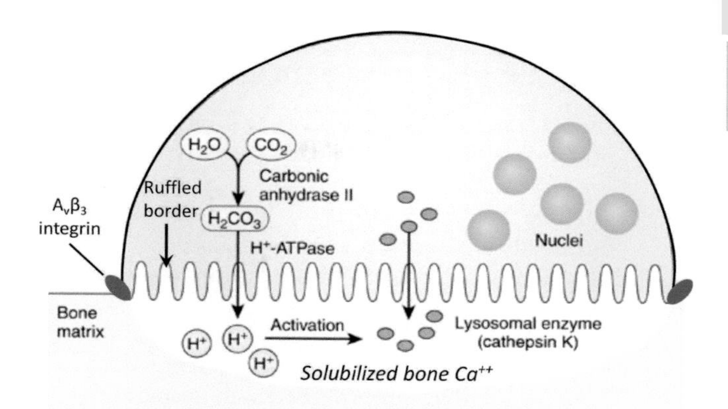

Figure 12.7 Classification of bone cells based on source, resorption, and formation function. A remodeling cycle, regulated by parathyroid hormone and 1,25-dihydroxyvitamin D3, is initiated with a resorption phase by activated osteoclasts that solubilize bone mineral and degrade the matrix. Osteoclasts originate from hematopoietic stem cells that first differentiate (dashed arrow) to a committed mononuclear preosteoclast cell that fuses to form the multinucleated osteoclasts. Attachment of osteoclasts to the bone surface creates a local acidic environment forming a resorption pit. The activities of monocytes or macrophages remove debris, followed by a bone formation phase by osteoblasts that produce osteoid matrix that will mineralize. When osteoblasts become trapped within the mineralized matrix, their structure and function changes, and they develop into osteocytes. Osteocytes maintain the mineral concentration of the matrix and communicate with each other and receive nutrients via long cytoplasmic processes that extend through canaliculi, "tiny channels" within the bone matrix.

Source of images: Lian, J.B., et al. 2012. Nat. Rev. Endocrinol. 8(4), pp. 212-227. Used by permission.

are synthesized and secreted by **osteoblasts** that develop from stem cells called *osteogenic cells*.

2. **Mineralization and hardening of the bone matrix:** The matrix becomes mineralized via secretion of calcium and phosphate ions by the osteoblasts.
3. **Maturation and maintenance of the bone matrix:** As the secreted matrix surrounding the osteoblasts calcifies, the osteoblasts become trapped within it. As a result, osteoblasts change in structure and differentiate into *osteocytes*, the primary cell of mature bone and the most common type of cell in bone. Each osteocyte is located in a space called a *lacuna* and is surrounded by bone tissue. Osteocytes maintain the mineral concentration of the matrix and communicate with each other and receive nutrients via long cytoplasmic processes that extend through *canaliculi*, channels within the bone matrix.
4. **Resorption of the bone matrix:** Chemical factors released by damaged or old bone are thought to recruit **osteoclasts** (literally, bone-destroying cells) to the bone. Osteoclasts mediate bone resorption by attaching to bone, forming a sealed space between the osteoclast and the bone (Figure 12.7). Proton pumps located below the ruffled bottom cell membrane acidify the space to dissolve the hydroxyapatite crystals, and proteolytic enzymes cleave the structural proteins of bone, creating pits. Osteoclasts are large multinucleated cells that originate from smaller mononuclear *preosteoclast* cells that fuse to form large multinucleated cells. Osteoclast maturation and activation requires close interaction with osteoblasts, a process coordinated by several hormones.
5. **Reformation of bone matrix:** Osteoblasts follow the path of the burrowing osteoclasts, depositing new osteoid and mineralizing it.

Curiously Enough . . . Virtually all bone found in mammals, birds, reptiles, and amphibians is comprised of *cellular bone*, or bone that contains osteocytes embedded within the bone matrix. However, the skeletons of the majority of teleost fish species (and thus a large proportion of vertebrates) are comprised of *acellular* bone, which contains no osteocytes. Despite the absence of osteocytes, acellular bone is still known to remodel in response to physical stimuli.[12,13]

SUMMARY AND SYNTHESIS QUESTIONS

1. Describe the function(s) of Ca^{++} and phosphate in hormone-mediated intracellular signaling pathways.
2. Blood serum calcium levels are stringently regulated, with even small deviations from the set point resulting in severe physiological distress or death. How can unusually high or low serum calcium levels cause death?
3. Osteopetrosis (literally "stone bone") is a rare disorder whereby the bones harden excessively, becoming denser, due to increased mineralization. Ironically, like osteoporosis (reduced bone mass and density), osteopetrosis makes bones more prone to fracture. Why?
4. Osteoclasts secrete protons, as well as lysosomal enzymes (e.g. cathepsin K). What specific roles do each of these osteoclast secretions have in bone resorption?
5. How can having elevated levels of free phosphate in the blood produce symptoms of hypocalcemia?

KEY CONCEPTS:

- Parathyroid hormone (PTH) is the primary hormone involved in increasing blood calcium.
- The active form of vitamin D, called calcitriol, is produced via a complex pathway involving the skin, liver, and kidneys.
- The vitamin D receptor (VDR) is a member of the superfamily of ligand-activated nuclear transcription factors, which mediate gene transcription.
- Vitamin D increases intestinal calcium and phosphate uptake and plays important roles in mediating both bone deposition and bone resorption.
- Calcitonin generally opposes the effects of PTH.
- Fibroblast growth factor 23 (FGF23) reduces blood phosphate by stimulating urinary phosphate secretion.
- Whereas the sex steroids, estrogens and androgens, maintain bone mass, excess levels of glucocorticoids and thyroid hormones promote bone resorption.

Endocrinology of Ca^{++} and Phosphate Homeostasis

LEARNING OBJECTIVE Review the endocrine mechanisms by which various hormones maintain calcium and phosphate homeostasis.

Disorders of calcium and phosphate metabolism are among the most common groups of diseases in humans. While the intestine, kidney, and bone, as well as gills and skin in amphibians and fish, are the primary effector organs responsible for maintaining blood serum calcium and phosphate homeostasis, most of the cellular and physiological processes described earlier are ultimately controlled by a variety of hormones (Table 12.3). As such, disorders of calcium and phosphate metabolism frequently result from an underlying endocrine dysfunction.

Table 12.3 A Summary of the Major Hormones that Influence Calcium and Phosphate Metabolism

Hormone	Action on target organ
Parathyroid hormone (PTH)	*Kidney*: increase calcium/decrease phosphate reabsorption; induces calcitriol synthesis *Bone*: induces osteoclast differentiation and bone resorption
Parathyroid hormone-related protein (PTHrP)	Effects are similar to those of PTH, though PTHrP signals in an autocrine/paracrine, and not a classical endocrine, manner
Calcitonin (CT)	*Kidney*: stimulates urinary excretion of both calcium and phosphate *Bone*: inhibits bone resorption by osteoclasts
Vitamin D3, calcitriol	*Intestine*: increases both calcium and phosphate absorption *Bone*: at low levels promote bone synthesis and mineralization; at high levels promotes osteoclast differentiation and bone resorption
Fibroblast growth factor 23 (FGF23)	*Kidney*: increases urinary phosphate excretion; inhibits calcitriol synthesis *Intestine*: indirectly inhibits dietary phosphate uptake by inhibiting kidney calcitriol synthesis *Parathyroid glands*: inhibits PTH synthesis
Estrogens and androgens	*Bone*: promote bone growth and maintain bone mass by stimulating osteoblast function and inhibiting osteoclast activity
Glucocorticoids	*Bone*: stimulate bone resorption
Thyroid hormone	*Bone*: promotes bone development and growth in children, but under pathologically high levels in adults (thyrotoxicosis) it stimulates bone resorption
Osteocalcin	*Bone*: promotes mineralization
Growth hormone (GH) and insulin-like growth factor-I (IGF-I)	*Bone*: stimulate of longitudinal bone growth in children and also the maintenance of bone mass in adults

Parathyroid Hormone and Parathyroid Hormone-Related Peptide

The *parathyroid glands* are so named because in many mammals, such as humans, dogs, and mice, they are juxtaposed onto the thyroid gland (Figure 12.8). In other mammals, such as goats, pigs, and rabbits, as well as in non-mammalian tetrapods, the parathyroid glands are anatomically separate from the thyroid gland (Figure 12.8) and are altogether absent in fish. The most abundant cell type in these glands, the chief cells, synthesize *parathyroid hormone* (PTH). In humans PTH is an 84-amino acid protein hormone whose main action is to increase serum calcium levels. So important are the parathyroid glands to maintaining blood serum calcium homeostasis that in mammals complete thyroidectomy accompanied by removal of all parathyroid glands can result in death due to lowered blood calcium levels.

Curiously Enough . . . Although the parathyroid glands are absent in fish, the hormone *stanniocalcin* appears to serve an analogous function. Stanniocalcin is produced by the *corpuscles of Stannius*, islands of cells found on the surface of the teleost kidney. Indeed, stanniocalcin is now recognized as a major hypocalcemic factor produced by diverse tissues (including the brain, heart, kidney, gut, gonads, liver, and pancreas) in all vertebrates, including humans.

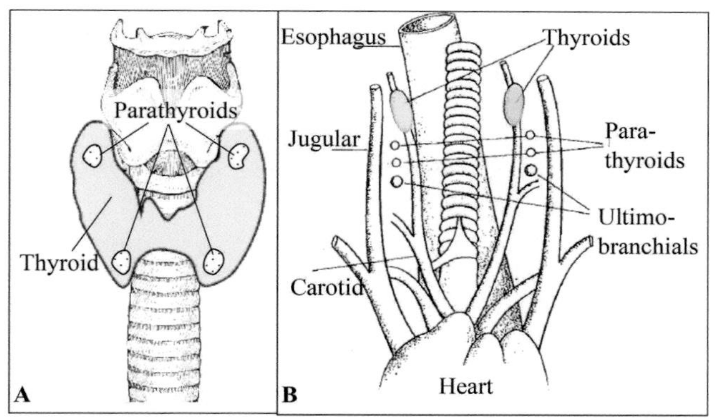

Figure 12.8 Parathyroid glands. (A) In humans and many other mammals, the parathyroid glands, which produce parathyroid hormone (PTH), are embedded in the posterior surface of the thyroid gland. There are normally four parathyroids in a human, each the size of a pea. Although the wings of the thyroid gland normally wrap around the trachea such that the parathyroids are on the posterior side, note that for simplicity this diagram portrays the thyroid gland with its wings expanded and the parathyroids thus appear to be localized to the anterior side. **(B)** In other mammals (e.g. goats, pigs, and rabbits), as well as in all non-mammalian tetrapods, the parathyroid glands are anatomically separate from the thyroid gland. The drawing depicts the neck anatomy of a chicken, where the parathyroid glands are clearly separate from the thyroid. The ultimobranchial organs produce calcitonin, a hormone whose effects oppose those of PTH. In mammals, calcitonin cells are dispersed amongst the thyroid follicle cells and are not condensed into distinct glands.

Source of images: (B) Copp, et al. (1967). Can. J. Physiol. Pharmacol. 45: 1095-99. Used by permission.

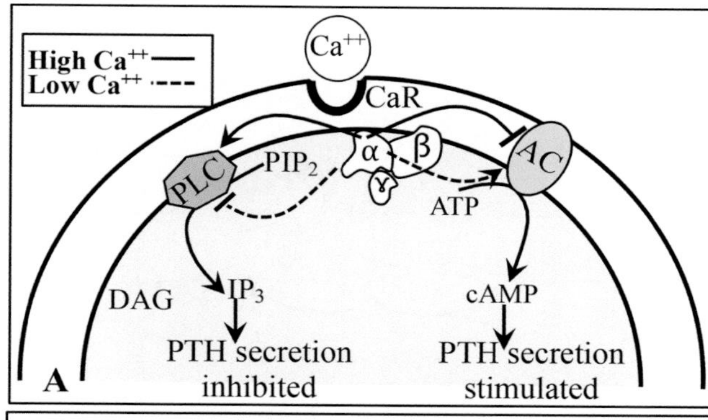

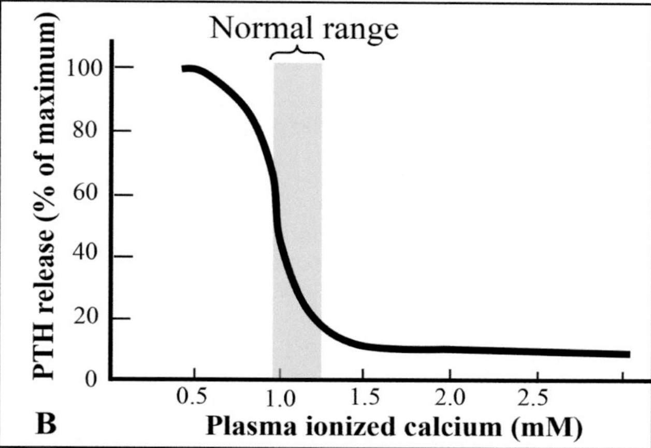

Figure 12.9 Regulation of PTH release by circulating blood calcium levels. (A) calcium-sensing receptors are GPCRs localized to the plasma membranes of PTH-secreting cells, and other calcium-sensing tissues. When blood Ca^{++} concentration is high, the α-subunit of the G protein preferentially stimulates phospholipase C (PLC) to hydrolyze phosphatidylinositol-4,5-bisphosphate (PIP2) to inositol 1,4,5-trisphosphate (IP_3) and diacylglycerol (DAG), which inhibits PTH secretion by the cells. Simultaneously, in the presence of high Ca^{++} concentrations, adenylate cyclase (AC)-induced cAMP generation is inhibited, also attenuating PTH secretion. The reverse occurs when serum Ca^{++}concentration is low, with elevated cAMP stimulating PTH secretion. When calcium concentrations are within their normal limits, the two second messenger pathways are balanced and basal secretions of PTH are maintained. **(B)** Relation between plasma ionized calcium concentration and PTH secretion in humans.

Source of images: (A) modified from Nussey, S. and Whitehead, S., 2001. The thyroid gland in Endocrinology. London, GB: An Integrated Approach by Published by BIOS Scientific Publishers Ltd.; (B) modified from Brown, E.B. (1983) J. Clin. Endocrinol. Metab. 56: 572–581.

The most important regulator of PTH secretion is serum ionized calcium concentration as measured by its binding to *calcium-sensing receptors* (CaSR), plasma membrane-localized GPCR receptors found in all calcium-sensing tissues. Under low serum calcium concentrations, the CaSR in the parathyroid gland chief cells act via an intracellular cAMP signaling cascade to promote PTH gene expression and PTH export (Figure 12.9). In contrast, high serum calcium levels suppress PTH release, primarily through

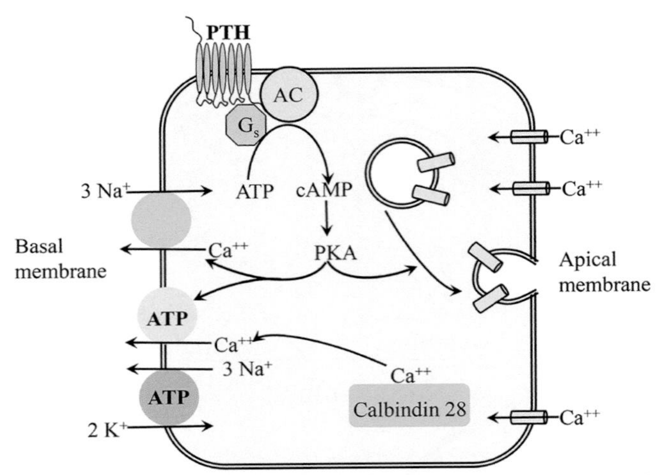

Figure 12.10 Effects of PTH on the principal cells in the distal tubule of kidney nephrons. PTH increases calcium reabsorption in the distal tubule by (1) stimulating the insertion of calcium channels (stored in recycling endosomes) into the apical membrane of epithelial cells, and (2) upregulating basolateral membrane-localized Na/Ca antiporter and Ca-ATPase pumps. G_s, G protein; AC, adenylate cyclase; PKA, protein kinase A.

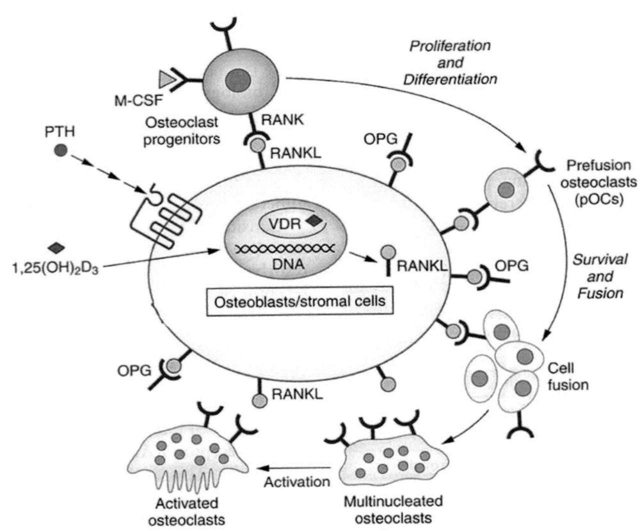

Figure 12.11 PTH-stimulated osteoblasts transform osteoclast precursors cells into mature osteoclasts through direct contact. PTH (in conjunction with vitamin D3 calcitriol) acts on PTH/PTHrP receptors on precursors of osteoblasts to increase the production of macrophage colony-stimulating factor (M-CSF) and RANKL, and decrease the production of osteoprotegerin (OPG is a decoy soluble receptor for RANKL produced by osteoblasts that acts as a decoy receptor for RANKL and thereby inhibits osteoclastogenesis and osteoclast activation). M-CSF and RANKL bind to receptors on osteoclast progenitor cells, promoting their differentiation into multinucleated active osteoclasts. The vitamin D receptor (VDR) is a nuclear transcription factor that, in the presence of vitamin D3 calcitriol, promotes the transcription of RANKL mRNA in osteoblasts.

Source of images: Norman, A.W. and Henry, H.L., 2022. Hormones. Academic Press. Used by permission.

the increased degradation of PTH prior to its secretion. Thus, plasma PTH and calcium concentrations normally display an inverse relationship.

PTH Actions on Effector Organs

The general effects of PTH on effector organs are summarized in Table 12.3. PTH exerts several effects on kidney function. These include:

1. Increased calcium reabsorption from the distal convoluted tubule of the nephron facilitated by the insertion of calcium channels into the apical membrane, and upregulation of basolateral membrane-localized Na^+/Ca^{++} antiporter and Ca^{++}-ATPase pumps (Figure 12.10).
2. Inhibition of phosphate uptake by removal of sodium-phosphate cotransporters from the apical membrane of the nephron's proximal tubule.
3. Activation of the kidney enzyme 1-hydroxylase, which converts circulating 25(OH) D3 into the most active form of vitamin D3, calcitriol (1,25(OH)D), which increases calcium uptake by the intestine and increases calcium release by bones.

Although PTH does not directly affect intestinal function, it does indirectly enhance intestinal calcium uptake by increasing the synthesis of calcitriol by the kidneys. Calcitriol directly increases calcium uptake by the gut (see later). Although no receptors for PTH are present in bone osteoclasts, PTH-stimulated osteoblasts transform osteoclast precursors cells into mature osteoclasts through direct contact (Figure 12.11). Specifically, PTH binding to osteoblasts increases their expression of macrophage colony-stimulating factor (M-CSF) and RANKL,* which when bound to its receptor (RANK) stimulates these osteoclast precursors to fuse, forming new osteoclasts, which ultimately enhances bone resorption. Concurrently, PTH inhibits the expression of *osteoprotegerin* (OPG), which when present binds to RANKL and blocks it from interacting with RANK and promoting osteoclast differentiation.

Calcitonin

Calcitonin (CT) is a polypeptide hormone that generally acts to reduce blood calcium, opposing the effects of PTH (Figure 12.12). In mammals it is produced by the

Curiously Enough . . . Compared with mammalian calcitonin, non-mammalian calcitonins are more stable and have 10–50 times more potent an effect in mammals. In fact, salmon calcitonin has been used to treat a human disorder called *Paget's disease*, a malady caused by increased osteoclast activity resulting in accelerated bone resorption and abnormal remodeling in the face and limbs.

* receptor activator of nuclear factor kappa-B ligand

parafollicular cells (also known as C cells) of the thyroid gland (Figure 12.12). However, in many other animals the hormone is of extrathyroidal origin, produced by the ultimobranchial body (refer to Figure 12.8). Although calcitonin's importance in regulating blood serum calcium homeostasis is well established in most vertebrates, its influence in regulating human calcium balance under normal physiological levels remains unclear, as no known calcium imbalances accompany either excess calcitonin (e.g. from malignancy) or C cell ablation (e.g. thyroidectomy). However, at pharmacological doses calcitonin is used to decrease bone resorption in osteoporosis, Paget's bone disease, and hypercalcemia of malignancy. Additionally, it is hypothesized that calcitonin may function to protect from excess maternal bone loss caused by breastfeeding-induced *parathyroid hormone-related protein* (PTHrP) production.[14]

Among vertebrates in general, CT is thought to mediate reductions in blood serum calcium by interacting with its plasma membrane-localized receptor (a GPCR) located in bone, kidney, and other tissues:

1. The binding of CT to receptors located on osteoclasts inhibits bone resorption by the osteoclasts.
2. The binding of CT to receptors located in the kidney stimulates the urinary excretion of both calcium and phosphate.
3. Protects newborn animals from hypercalcemia.

Interestingly, CT secretion is controlled by serum calcium through the same CaSR that regulates PTH secretion, but in an inverse manner, with high concentrations of calcium promoting the release of CT.

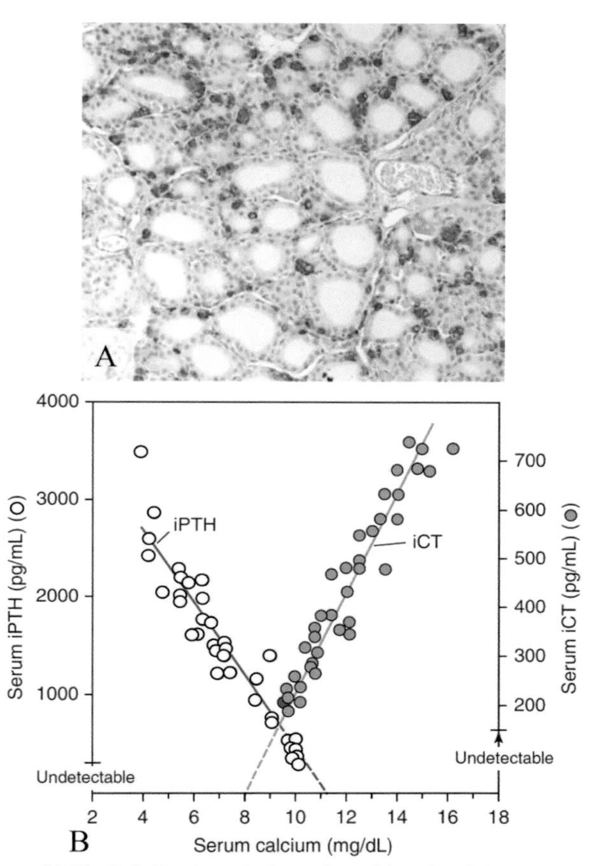

Figure 12.12 Calcitonin acts to reduce blood calcium, opposing the effects of PTH. (A) In mammals, calcitonin is produced by the parafollicular cells of the thyroid. Note how calcitonin immunoreactivity (brown stain) in humans is confined to C cells located in between the thyroid follicles. **(B)** Changes in plasma levels of immunoreactive parathyroid hormone (iPTH) and calcitonin (iCT) as a function of plasma total calcium. The data were obtained from pigs given EDTA (a calcium chelator) to decrease plasma calcium or given calcium infusions to increase plasma calcium. Note that, as serum calcium increases, iPTH falls and serum iCT increases; as serum calcium decreases the reverse occurs.

Source of images: (A) Fernández-Santos, J.M et al. 2012. Paracrine regulation of thyroid-hormone synthesis by C cells. Thyroid hormone; IntechOpen(B) Norman, A.W. and Henry, H.L., 2022. Hormones. Academic Press. Used by permission.

Vitamin D

In vertebrates the active form of vitamin D3, called *calcitriol* (1,25-$(OH)_2$ D), is produced via a complex pathway involving the skin, liver, and kidneys (Figure 12.13). First, 7-dehydrocholesterol in the skin is converted to the secosteroids previtamin D3 and its isomer vitamin D3 (also

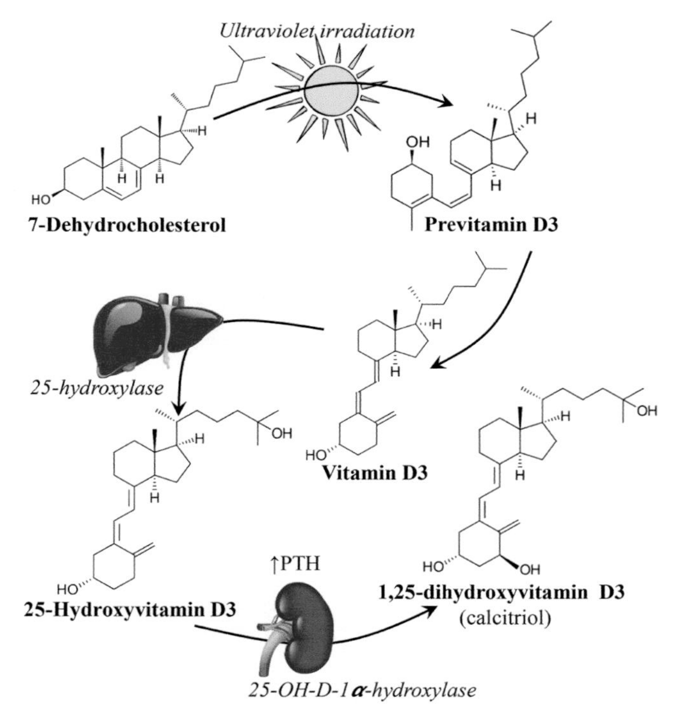

Figure 12.13 The vitamin D3 synthesis and activation pathway. The major source of vitamin D3 is through ultraviolet irradiation of 7-dehydrocholesterol in skin. The liver 25-hydroxylase enzyme then converts vitamin D3 to 25-hydroxyvitamin D3 (25-OH-D3), the major circulating form of the vitamin. Generation of 1α,25-dihydroxyvitamin D3 (calcitriol) occurs primarily in the kidney by the 25-OH-D-1α-hydroxylase enzyme. Note that the transcription of 25-OH-D-1α-hydroxylase is induced by parathyroid hormone (PTH).

called *cholecalciferol*) through ultraviolet light irradiation. Vitamin D3 is subsequently transported in the blood bound to vitamin D-binding protein (VDBP) to the liver, where it becomes hydroxylated into 25-hydroxyvitamin D3 (25-OH D). 25-hydroxyvitamin D3 is bound to blood VDBP, where it is transported to the kidneys. Calcitriol is then generated by the kidney enzyme, 25-OH-D-1α-hydroxylase, which is itself upregulated by the previously described action of parathyroid hormone on the kidney.

Natural nutritional sources of vitamin D3 are limited primarily to fatty ocean fish, like sardines, and dairy milk fortified with vitamin D; human and unfortified dairy milk contains very little vitamin D. Recall that another exogenous form of vitamin D, vitamin D2 (ergocalciferol), is produced by ultraviolet irradiation of the plant sterol ergosterol and is made available through the diet. Both forms of vitamin D require the aforementioned further metabolism to be activated, and their respective metabolism is indistinguishable.

Vitamin D Receptor

The vitamin D receptor (VDR) is a member of the superfamily of ligand-activated nuclear transcription factors. As such, secreted calcitriol functions by crossing the plasma membrane of target cells, where it binds to intracellular VDR that forms a heterodimer with its partner, retinoid X receptor (RXR) and accessory coactivator proteins (Figure 12.14). This heterodimeric complex modulates gene transcription by associating with DNA response elements located in the gene promoters of VDR-regulated genes. Interestingly, although vitamin D itself is found throughout the animal and plant kingdoms, the VDR is only present in vertebrates.

Vitamin D Actions on Effector Organs

The general effects of vitamin D on effector organs are summarized in Table 12.3. Of the three primary calcium and phosphate balance effector organs, the effects of vitamin D have been most studied in the intestine and bone.

Intestine

Vitamin D increases intestinal calcium and phosphate absorption, primarily in the jejunum and ileum, by increasing calcium uptake through the brush border membrane of the enterocyte (Figure 12.15). The vitamin D-bound VDR transcriptional regulation complex upregulates the transcription of several key genes:

1. The apical membrane-localized $3Na/PO_4$ symporter, which facilitates phosphate uptake from the gut lumen.
2. Two types of apical epithelial calcium channels (TRPV5 and TRPV6), which mediate diffusion of calcium from the lumen into the enterocytes.
3. The mammalian cytosolic calbindin protein (CaBP 9K), which binds to incoming calcium and moves it to the basolateral membrane where calcium is off-loaded from the CaBP and pumped out by calcium ATPase into the extracellular fluid and blood.

Bone

Vitamin D plays important roles in mediating both bone deposition and bone resorption. Under relatively low concentrations, vitamin D promotes the expression of several bone proteins, notably osteocalcin and osteopontin, and also suppresses the synthesis of type I collagen. Low levels

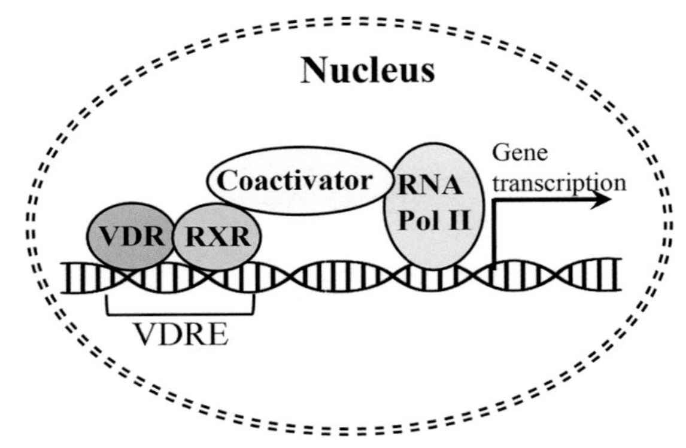

Figure 12.14 The vitamin D receptor (VDR) modulates calcium and phosphate metabolism at the genomic level. Upon binding of calcitriol to the VDR, the membrane-localized receptor binds to its heterodimeric partner retinoid X receptor (RXR), which both bind to vitamin D receptor DNA response elements (VDRE), regulating the transcription of genes that directly or indirectly modulate calcium metabolism.

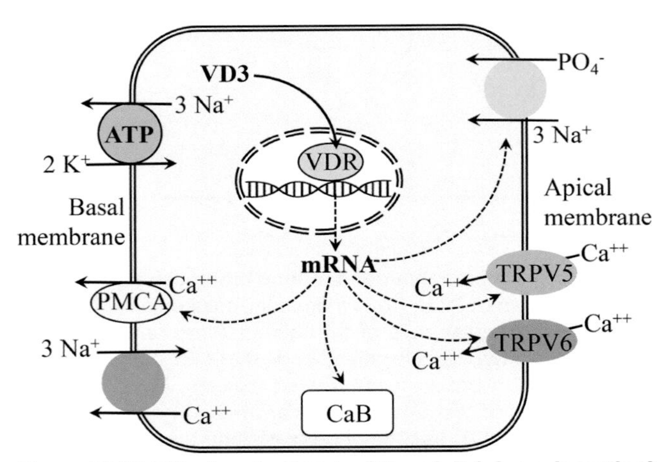

Figure 12.15 Effects of vitamin D3 calcitriol on intestinal transport of calcium and phosphate. Vitamin D3 increases intestinal calcium and phosphate absorption by increasing calcium and phosphate uptake through the brush border membrane of enterocytes. The calcitriol-bound VDR transcriptional regulation complex upregulates the transcription of several key genes, including the apical membrane-localized calcium channels (TRPV5 and TRPV6) and Na/PO4 cotransporter, the basal membrane-localized calcium-ATPase pump (PMCA), and the cytosolic calcium binding proteins (calbindin, CaB).

of vitamin D also promote the mineralization of osteoid via an unknown mechanism.

Under high concentrations, vitamin D, in conjunction with PTH, is known to facilitate bone resorption. This is in part caused by the upregulation of the gene for RANK ligand by osteoblasts, which induces osteoclast differentiation and activation. In the absence of vitamin D, the effect of PTH in causing bone resorption is greatly reduced or even prevented.

Non-Calciotropic Effects of Vitamin D

In addition to its central role as a mediator of calcium metabolism, vitamin D plays critical and diverse non-calciotropic developmental roles among vertebrates. For example, vitamin D can modulate the innate and adaptive immune responses.[15,16] Deficiency in vitamin D is associated with increased autoimmunity, like rheumatoid arthritis, multiple sclerosis, and inflammatory bowel disease, as well as an increased susceptibility to infection. The stimulation of vitamin D receptors can also inhibit cancer cell proliferation and reduce the metastatic potential of breast, colon, and prostate cancers. Further, vitamin D is critical for normal nervous system development, and its deficiency has been linked to numerous developmental disorders including attention deficit hyperactive disorder (ADHD),[17] schizophrenia,[18] and autism.[19]

Integrated Responses of the Major Hormones and Effector Organs to Calcium and Phosphate Imbalance

Serum calcium and phosphate homeostasis has evolved to maintain extracellular ion levels in the physiologic range while simultaneously allowing the flow of calcium and phosphate to and from essential stores. A decrease in serum calcium, for instance, inactivates the CaSR in the parathyroid glands to increase PTH secretion, which acts on the PTHR in the kidney to increase calcium reabsorption and in bone to increase net bone resorption (Figure 12.16). The increased PTH also stimulates the kidney to increase secretion of calcitriol, which activates the VDR in the gut to increase calcium uptake, in the parathyroid glands to decrease PTH secretion, and in bone to increase resorption. The decrease in serum calcium also inactivates the CaSR in the kidneys to increase calcium reabsorption, potentiating the effect of PTH. This integrated hormonal response restores serum calcium and closes the negative feedback loop. With a rise in serum calcium, these actions (with the help of calcitonin) are reversed and the integrated hormonal response reduces serum calcium.

Other Hormones Involved in Calcium and Phosphate Balance

In addition to the primary calciotropic hormones described earlier, other hormones play important roles in calcium and phosphate metabolism (summarized in Table 12.3). In particular, the *sex steroids*, estrogens and androgens, help to maintain skeletal mass in both females and males. Significant decreases in these circulating hormones, such as during menopause in females, are associated with bone loss and osteoporosis. Estrogens promote osteoblast function and inhibit osteoclast activity[20] (Figure 12.17), and androgens function similarly in this regard. In addition to promoting bone mass, both sex steroids are also known to promote longitudinal bone growth both directly and through the induction of growth hormone and insulin-like growth factor-I (IGF-I). Importantly, estrogens, but not androgens, also play a key role in terminating long bone growth in both males and females.

Osteocalcin is the most abundant non-collagenous protein in bone, comprising almost 2% of total protein in the human body. Secreted by osteoblasts, the protein is implicated in promoting bone mineralization and calcium ion homeostasis. Interestingly, osteocalcin also acts as a hormone, causing beta cells in the pancreas and adipocytes to release insulin and adiponectin, respectively.

In contrast to sex steroids, *glucocorticoids* are deleterious to bone mass. In fact, high-dose glucocorticoid therapy for the treatment of inflammation is almost

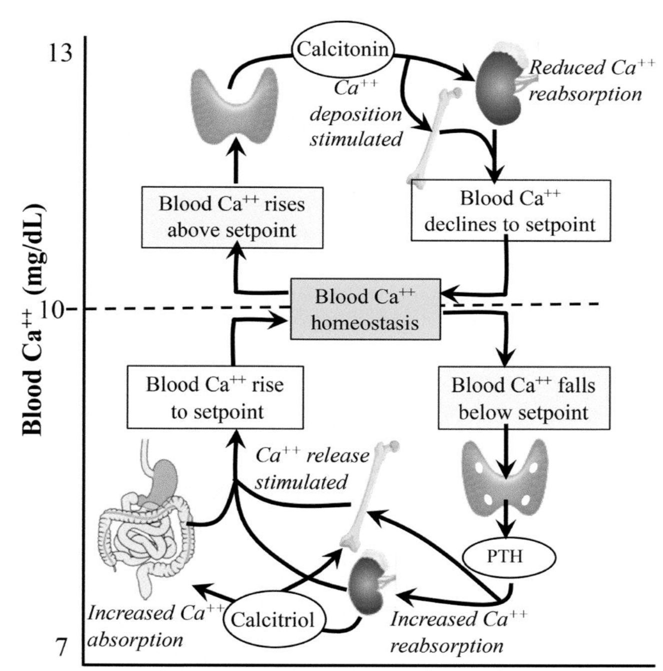

Figure 12.16 Integrated responses of the major hormones and effector organs to calcium imbalance in mammals. A reduction in plasma calcium stimulates parathyroid hormone (PTH) secretion. PTH in turn induces a rise in blood calcium by promoting calcium reabsorption via the kidneys and calcium release by bone. PTH also stimulates the synthesis of the most active form of vitamin D, calcitriol (1,25-dihydroxycholecalciferol) by the kidneys, which increases calcium uptake in the intestine and further stimulates bone to release calcium into the blood. A rise in blood calcium concentrations simultaneously inhibits the release of PTH and induces the secretion of calcitonin by interfollicular cells of the thyroid gland. Calcitonin promotes the deposition of calcium into bone and inhibits calcium reabsorption by the kidneys.

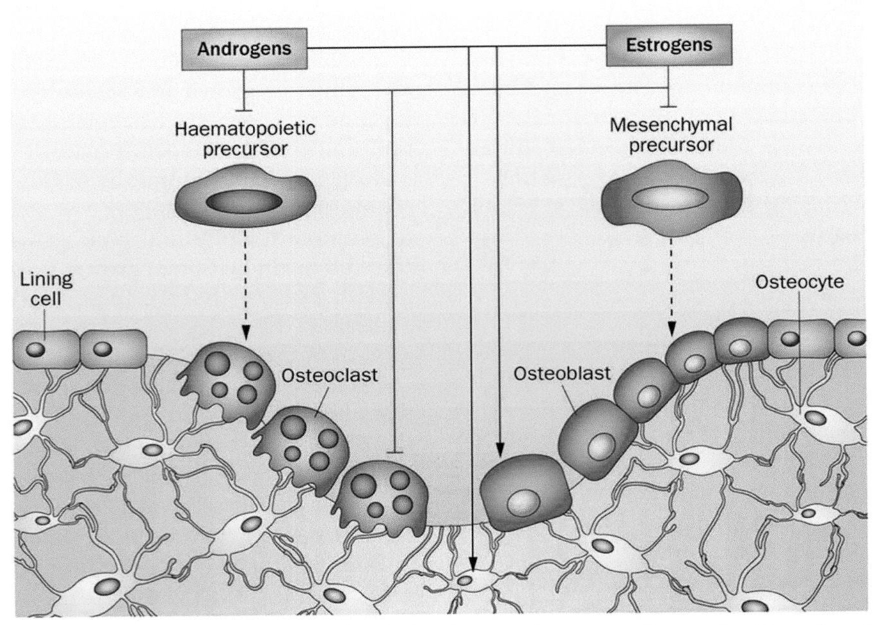

Figure 12.17 Estrogens and androgens both maintain skeletal mass by promoting osteoblast function and by inhibiting osteoclast activity. Osteoclasts and osteoblasts are derived from hematopoietic and mesenchymal precursors, respectively. During the process of bone remodeling, bone matrix excavated by osteoclasts is replaced with new matrix produced by osteoblasts. Both estrogens and androgens influence the differentiation of osteoclast and osteoblast precursors and the lifespan of mature osteoclasts and osteoblasts, as well as the lifespan of osteocytes. Positive (black arrows) and negative (red bars) effects on the cells are depicted as well as differentiation of cells (dashed arrows).

Source of images: Manolagas, S.C., et al. 2013. Nat. Rev. Endocrinol., 9(12), p. 699. Used by permission.

universally associated with bone loss, causing osteoporosis.[21] Osteoporosis is also a feature of Cushing's syndrome, which is characterized by high levels of circulating glucocorticoids. The mechanisms of glucocorticoid-mediated bone loss are not well understood but appear to be mediated by the inhibition of osteoblastogenesis and the promotion of apoptosis of osteoblasts and osteocytes.[22]

Fibroblast growth factor 23 (FGF23) is produced by osteoclasts and osteoblasts and plays a major role in phosphate and, indirectly, calcium homeostasis.[23] FGF23 synthesis is stimulated by active vitamin D (calcitriol) and also in response to an increase in serum phosphate concentration. The hormone functions by binding to kidney tubule-localized receptors where it suppress the expression of NaPi-2a and NaPi-2c cotransporters (Figure 12.4), inducing increased urinary phosphate excretion.

Thyroid hormones exert profound anabolic actions during skeletal development and growth in children. Hypothyroidism in children causes reduced bone turnover and cessation of linear growth resulting in a marked delay of bone age and epiphyseal dysgenesis (for example, refer to Figure 10.2 in Chapter 10: Thyroid Hormones: Development and Growth). Childhood thyrotoxicosis (excess TH production), by contrast, accelerates linear growth and advances bone maturation resulting in premature fusion of the growth plates and persistent short stature. In adults, TH regulates bone turnover and mineralization. In contrast to its effects on children, thyrotoxicosis elicits a catabolic responses in adult bone, producing increased bone turnover and bone loss, which may result in osteoporosis and an increased risk of bone fracture.

Developments & Directions: Osteocalcin: endocrine regulation of male fertility by the skeleton

It has been known for some time that through synthesis of sex steroids the reproductive system influences bone growth and development. Gerard Karsenty of Columbia University and his colleagues wondered whether the influence might occur in the reverse direction as well. Given the strong links between estrogen, menopause, and osteoporosis, his team anticipated that if such a connection exists, it would most likely turn up in females. Interestingly, Karsenty and colleagues[24] discovered that instead *osteocalcin*, a hormone produced by bone, promotes the synthesis of testosterone in male mice but does not influence estrogen production by ovaries. The researchers showed that osteocalcin works through a receptor found on testosterone-producing Leydig cells in the testes, but that this receptor is absent in ovaries (Figure 12.18). These findings provide the first evidence that the skeleton is an endocrine regulator of reproduction and expand possibilities of treating male infertility.

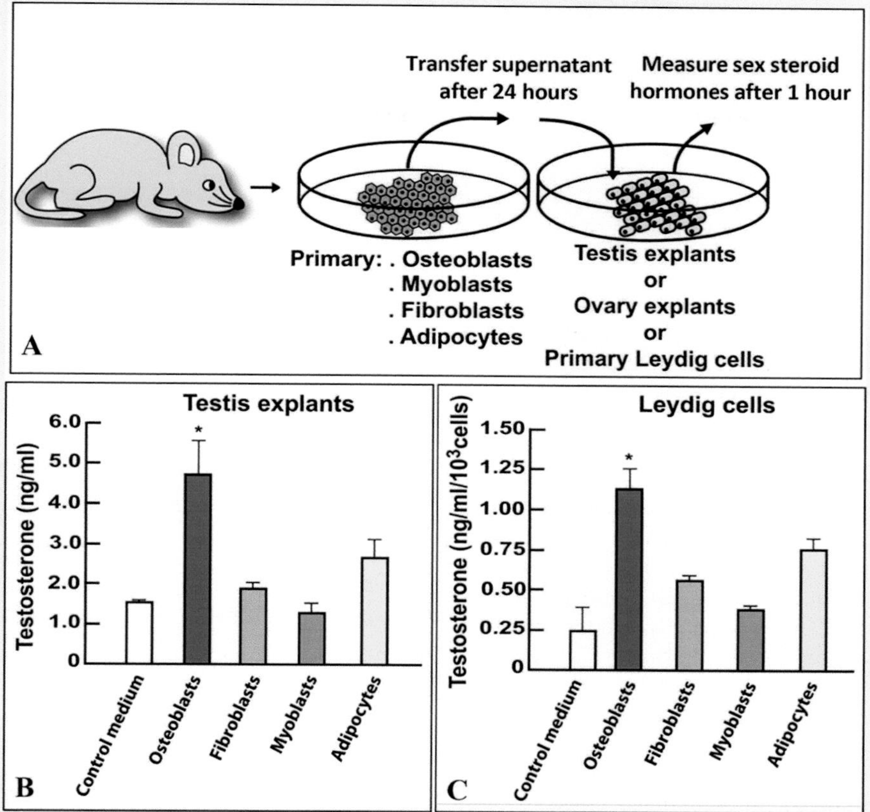

Figure 12.18 The osteoblast hormone osteocalcin enhances testosterone biosynthesis by Leydig cells of the testis. (A) Various primary mesenchymal cells, including osteoblasts, from mice were cultured for 24 hours, and supernatants were collected. Then, testis or ovary explants or primary Leydig cells were cultured for 1 hour with these supernatants, and radioimmunoassays were performed to measure levels of testosterone, estradiol, or progesterone. Testis **(B)** and Leydig cell **(C)** explants produced significantly elevated testosterone levels only when cultured with osteoblast supernatant. In contrast, testosterone, estradiol, and progesterone levels did not change when ovary explants were cultured with osteoblast supernatant (not shown). The active osteoblast factor in the supernatant was identified as osteocalcin.

Source of images: Oury, F., et al. 2011. Cell, 144(5), pp. 796-809. Used by permission.

SUMMARY AND SYNTHESIS QUESTIONS

1. In humans, a complete thyroidectomy is lethal without hormone supplementation. In contrast, complete thyroidectomy in birds is non-lethal. Why can birds, but not humans, survive a complete thyroidectomy?
2. PTHrP is known to be released by the human breast into maternal circulation during breastfeeding. How is this nutritionally useful to the breastfeeding infant?
3. Drugs that mimic the effect of calcium *at the calcium receptor* are referred to as calcimimetics. The drug cinacalcet is a calcimimetic that has been approved to treat which of the following conditions: hyperparathyroidism or hypoparathyroidism. Explain your reasoning.

Endocrine Pathologies of Calcium and Phosphate Homeostasis

LEARNING OBJECTIVE Describe the endocrine bases of various pathologies of calcium/phosphate homeostasis.

KEY CONCEPTS:

- The symptoms of overt hypercalcemia and hypocalcemia are severe and can lead to seizures, coma, or death.
- The most common causes of hypocalcemia are a deficiency of PTH secretion, as well as vitamin D deficiency.
- The most common cause of hypercalcemia is hyperparathyroidism.
- The most common form of osteoporosis (decrease in bone mass and density) is due to a deficiency of sex steroid production, particularly reduced estrogen synthesis after menopause.

Table 12.4 Some Clinical Manifestations of Hypercalcemia, Hypocalcemia, and Hypophosphatemia

Organ System	Hypercalcemia	Hypocalcemia	Hypophosphatemia
Neuropsychiatric	Fatigue, lethargy, coma, depression	Paresthesia (sensation of tingling, numbness, or burning with no obvious cause), seizures, bronchospasm, laryngospasm, Trousseau sign	Progressive encephalopathy, seizures, coma
Musculoskeletal	Weakness, pain in joints or bone, bone easily fractured	Muscle twitching, tetany, weakness	Muscle weakness, bone softening
Cardiovascular	Short QT interval on ECG, bradycardia, hypertension	Prolonged QT interval on ECG, tachycardia, hypotension	-
Renal	Polyuria, kidney stones	-	May be accompanied/caused by renal wasting

Source: Adapted from Kovacs, W.J. and Ojeda, S.R. eds., 2011. *Textbook of endocrine physiology*. OUPUSA.

Due to the inverse relationship between blood phosphate and calcium concentrations, abnormal blood phosphate levels can result in altered calcium metabolism. Indeed, *hypercalcemia* (elevated blood plasma calcium) is typically accompanied by reduced serum phosphate, and *hypocalcemia* (reduced blood calcium) is typically accompanied by elevated serum phosphate. The symptoms of hyper- and hypocalcemia are severe[25,26] (Table 12.4), and because vertebrates cannot tolerate great deviations in plasma calcium before death occurs, this makes diagnosing hypercalcemia and hypocalcemia *in vivo* difficult based exclusively on blood calcium measurements. A common test for neuromuscular excitability caused by hypocalcemia is accomplished by inflating a blood pressure cuff on the arm 20 mm above systolic blood pressure for 3 minutes. Under hypocalcemic conditions, a characteristic carpal spasm in the hand is produced called the *Trousseau sign*, as well as the *Chvostek sign*, which is evidenced by the twitching of muscles innervated by the facial nerve. In contrast to calcium, broad variability in blood phosphate levels is generally tolerated.

Hypocalcemia

The typical conditions leading to hypocalcemia are summarized in Table 12.5. The most common manifestation of hypocalcemia is due to a deficiency in PTH secretion, or *hypoparathyroidism*.[27] This typically results from damage to the parathyroid glands via surgical excision or autoimmune destruction. Reduced PTH secretion can also be caused by an activating mutation of the CaSR, which inactivates PTH synthesis even under low calcium levels. In the case of *pseudohypoparathyroidism*, PTH levels are normal, but there is either end organ resistance to PTH (e.g. mutated PTH receptor) or an abnormal PTH is synthesized that does not activate the receptors. *Hypomagnesmia* (reduced blood serum magnesium) is also known to inhibit PTH release, indirectly inducing hypocalcemia.

Another common cause of hypocalcemia is inadequate or impaired vitamin D synthesis or intake. Such a condition may manifest from inadequate sunlight or dietary consumption, or mutations of vitamin D converting enzymes in the liver and/or kidney. Resistance to vitamin D at the level of the VDR also produces hypocalcemia. Because ionized serum calcium readily binds with phosphate to form the precipitate calcium-phosphate, elevated serum phosphate (caused, for example, by renal failure) may promote calcium deposition into soft tissues, also resulting in hypocalcemia. **Rickets** (discussed earlier), a disease of

Table 12.5 Mechanisms of Hypercalcemia, Hypocalcemia, and Hypophosphatemia

General Cause	Mechanism
Hypocalcemia	
Reduced PTH action	Decreased PTH synthesis
	Resistance to PTH
Reduced vitamin D action	Decreased vitamin D
	Reduced conversion to cholecalciferol or calcitriol
	Resistance to vitamin D
Reduced free Ca^{++}	Alkalosis
	Hyperphosphatemia
Hypercalcemia	
Increased Ca^{++} input	Increased bone resorption
	Increased intestinal calcium absorption
Decreased Ca^{++} output	Reduced renal Ca^{++} excretion
	Reduced bone formation
Hypophosphatemia	
Renal dysfunction	Reduced renal reabsorption due to excess FGF23 synthesis, chronic kidney disease or alcohol use disorder
Decreased intestinal absorption	Inadequate intake (starvation), antacids containing aluminum or magnesium, diarrhea, vitamin D deficiency
Increased urinary excretion	Hyperparathyroidism, vitamin D deficiency, diuretics
Shift from extracellular phosphate into the intracellular space	Severe respiratory alkalosis, alcohol use disorder

Source: Adapted from Kovacs, W.J. and Ojeda, S.R. eds., 2011. *Textbook of endocrine physiology*. OUP USA growing bone that is unique to children and adolescents, is caused by failure of osteoid to calcify due to hypocalcemia. Failure of osteoid to calcify in adults is called *osteomalacia*.

Hypercalcemia

The typical conditions leading to hypercalcemia are summarized in Table 12.5. The most common cause of hypercalcemia is due to *hyperparathyroidism*, most often caused by excess PTH production from a parathyroid adenoma.[28] Because excess PTH increases the rate of calcium removal from bone, PTH-induced hypercalcemia can lead to osteoporosis. Although parathyroid carcinomas are very rare, non-parathyroid malignancies that secrete PTHrP, such as of the breast and kidney, are known to cause hypercalcemia since PTHrP binds to the same receptor as PTH. Excess vitamin D synthesis or uptake can also produce hypercalcemia, primarily due to elevated intestinal absorption of calcium.

Osteoporosis

Osteoporosis, one of the most common endocrine disorders, is a progressive bone disease that is characterized by a decrease in bone mass and density, which can lead to an increased risk of fracture.[29,30] It occurs when bone resorption exceeds formation, and it differs from osteomalacia in that collagen, as well as mineral, is also lost from bone. In the United States osteoporosis accounts for 70% of bone fractures in adults over 45 years old.

The most common form of osteoporosis is age related and characterized by a gradual loss of bone from the age of 30–40 years onwards. As such, the maximization of bone mass from childhood through the late twenties with exercise, vitamin D, and a calcium-rich diet helps to offset bone loss in the later years. Regular exercise and proper diet during these later years has also been shown to reduce the rate of bone loss. In women, bone loss is accelerated in the postmenopausal years due to the loss of estrogen synthesis (Figure 12.19). Importantly, estrogens also regulate skeletal homeostasis in men, with strong correlations between bioavailable estrogens and bone remodeling. Although bone loss also occurs as men age, the rate of bone loss in males is typically not as high as in females, presumably because in males elevated androgen levels reduce bone loss. Sex steroid deficiency at any age promotes bone resorption. As such, diseases associated with **hypogonadism** (reduced sex steroid synthesis), such as Turner syndrome, Klinefelter syndrome, and other hypogonadal states, are accompanied by increased risk of osteoporosis. Until recently, estrogen replacement was the primary therapeutic agent for both preventing and treating osteoporosis. However, complications associated with estrogen therapy, such as an increased risk of breast cancer, have led to the development of new approaches to the treatment of osteoporosis. These include endocrine modulators, such as *selective estrogen response modulators* (SERMS), or designer estrogens that bind to the same receptor in separate tissues to elicit different outputs by recruiting distinct coregulators. Another class of drugs used to treat osteoporosis are "biphosphonates", compounds that inhibit the digestion of bone by encouraging osteoclasts to undergo cell death, thereby slowing bone loss. In addition to reduced steroid hormone levels, other endocrine factors that may promote osteoporosis include elevated glucocorticoid levels (of exogenous or endogenous origin), elevated thyroid hormones (thyrotoxicosis), and elevated PTH.

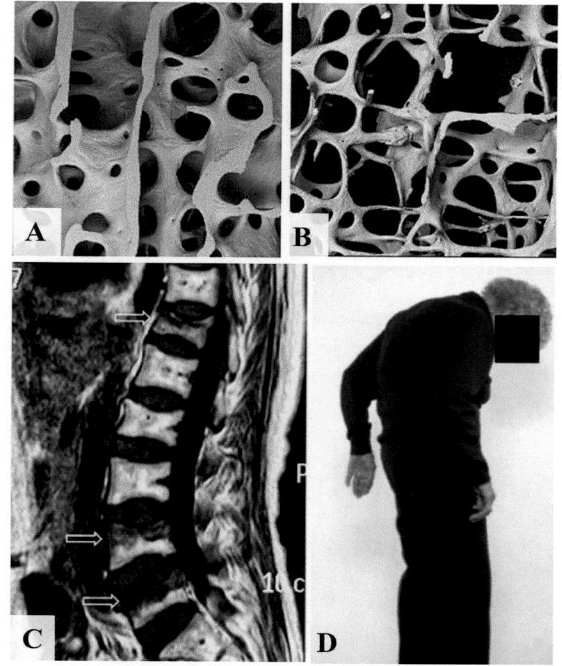

Figure 12.19 Osteoporosis reduces bone density, leading to compression and curvature of the spine and increased susceptibility to bone fracture. Scanning electron micrographs of a normal vertebrae from a 31-year-old man **(A)** and an osteoporotic vertebrae from an 89-year-old woman **(B)** showing extensive loss of trabecular bone architecture with conversion of plates to rods and a microfracture. **(C–D)** Osteoporosis makes vertebrae susceptible to compression and curvature with age, and all bones in the body become more susceptible to fracture. A 73-year-old woman with severe osteoporosis, curved vertebrae, and three compression fractures (arrows).

Source of images: (A-B) Boyd A. Endocrine 2002;17:5-14; used by permission. (C) Yang, W., et al. 2019. Clin. Spine Surg, 32(2), p. E99; used by permission. (D) Cifu, D.X., 2020. Braddom's physical medicine and rehabilitation E-book. Elsevier Health Sciences. Used by permission.

SUMMARY AND SYNTHESIS QUESTIONS

1. One drug used to treat osteoporosis, called denosumab, is a human monoclonal antibody against the protein RANKL. How can the inhibition of RANKL function be used to treat osteoporosis?
2. "Humoral hypercalcemia of malignancy" is not caused by parathyroid gland tumor, but rather is most often caused by a tumor in the breast. How can this lead to hypercalcemia?

Endocrinology of Linear Bone Growth

LEARNING OBJECTIVE Describe the roles of various hormones in regulating longitudinal bone growth.

KEY CONCEPTS:

- The long bones grow via endochondral ossification.
- During puberty osteoblasts produce bone faster than chondrocytes produce cartilage, stopping bone growth.
- Linear bone growth is ultimately regulated by the hypothalamus and pituitary, whose hormones affect bone growth by both direct and indirect mechanisms.
- GH promotes growth by stimulating the release of IGF-I by the liver and bone.
- TH promotes bone growth directly and indirectly by potentiating GH release.
- Sex steroids stimulate bone growth by promoting the release of GH.

Endochondral and Intramembranous Ossification

Osteogenesis, the process of bone formation, occurs by two distinct mechanisms: intramembranous and endochondral ossification. Flat bones, such as those found in the skull and scapula, as well as the exoskeleton of tortoises and turtles, and in the gills, fins, and scales of fish form by *intramembranous ossification*, a process in which condensations of mesenchyme differentiate into osteoblasts, which secrete and mineralize osteoid to form bone directly without an intermediate template. Bone formed in this manner, called *dermal bone* because it forms within the dermis, grows by accretion with the outer portion of the bone deposited by osteoblasts. In contrast, long bones (arms and legs), as well as virtually all other human bones not associated with the skull, form via *endochondral ossification*, a process by which bone gradually replaces a preexisting cartilaginous template (Figure 12.20). In the embryo, mesenchymal stem cells differentiate into chondrocytes that proliferate and secrete cartilage matrix to form a scaffold. Chondrocytes undergo hypertrophic differentiation, and this process is followed by cartilage mineralization, chondrocyte apoptosis, and vascular invasion. This calcified cartilage forms a template for bone formation by invading osteoblasts that lay down and mineralize bone matrix (osteoid). Importantly, chondrocytes at both ends of the bone organize to form **epiphyseal growth plates** and secondary ossification centers. The ordered process of growth plate chondrocyte proliferation, hypertrophic differentiation, apoptosis, and subsequent new bone formation at these plates mediates linear growth until adulthood. The normal age-dependent decrease in growth rate is related primarily to the senescence of growth plate chondrocytes, which are programmed to undergo a finite number of cell divisions. This causes a decrease in the rate of proliferation of these cells, a process referred to as *growth plate senescence*. During puberty, osteoblasts begin to produce bone faster than chondrocytes produce cartilage. This, in combination with growth plate senescence, causes

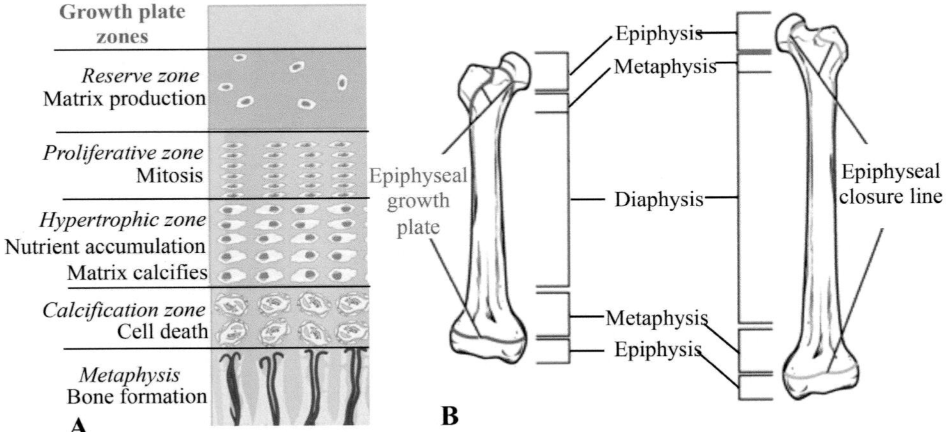

Figure 12.20 Endochondral ossification in the epiphyseal growth plate from childhood through the end of puberty. (A) The epiphyseal growth plate consists of several principal zones: reserve (resting), proliferative, hypertrophic, and calcification. The resting zone lies adjacent to the epiphyseal bone and contains infrequently dividing chondrocytes. The proliferative zone contains replicating chondrocytes arranged in columns parallel to the long axis of the bone. The proliferative chondrocytes located farthest from the resting zone stop replicating and enlarge to become hypertrophic chondrocytes. The calcification zone is closest to the metaphysis, or ossified bone border. Simultaneously, the metaphyseal border of the growth plate remodels and mineralizes into bone. **(B)** The synchronized processes of chondrogenesis and cartilage ossification lead to longitudinal bone growth and elongation. During puberty osteoblasts produce bone faster than chondrocytes produce cartilage. This, combined with growth plate chondrocyte senescence, causes the epiphyseal cartilage to disappear (epiphyseal closure), forming an epiphyseal line, and bone growth stops.

Source of images: (A) Biga, L.M., et al. 2020. Bone formation and development. Anatomy & Physiology. OpenStax/Oregon State, 6.

epiphyseal cartilage to eventually disappear (*epiphyseal closure*) and long bone growth stops (Figure 12.20). The progression of endochondral ossification and the rate of linear growth are tightly regulated by multiple hormones, several of which will be discussed in this section.

Endocrine Control of Linear Bone Growth by the Brain

In humans, linear growth (height) velocity decreases from birth onwards, punctuated by a short period of growth acceleration, known as the adolescent "growth spurt", just before completion of growth (Figure 12.21). During puberty, which begins earlier in girls than in boys, linear growth velocity approximately doubles for one year or more. The brain directly influences linear bone growth through hormones released from the adenohypophysis, an organ under control of hypothalamic release factors. These hormones are growth hormone (GH), thyroid-stimulating hormone (TSH), and follicle-stimulating hormone (FSH). These hormones stimulate bone growth indirectly by promoting the release of downstream hormones that modulate bone development (IGF-I, thyroid hormone, and sex steroids, respectively). Both GH and TSH also exert direct effects on bone growth, binding to receptors on bone cells.

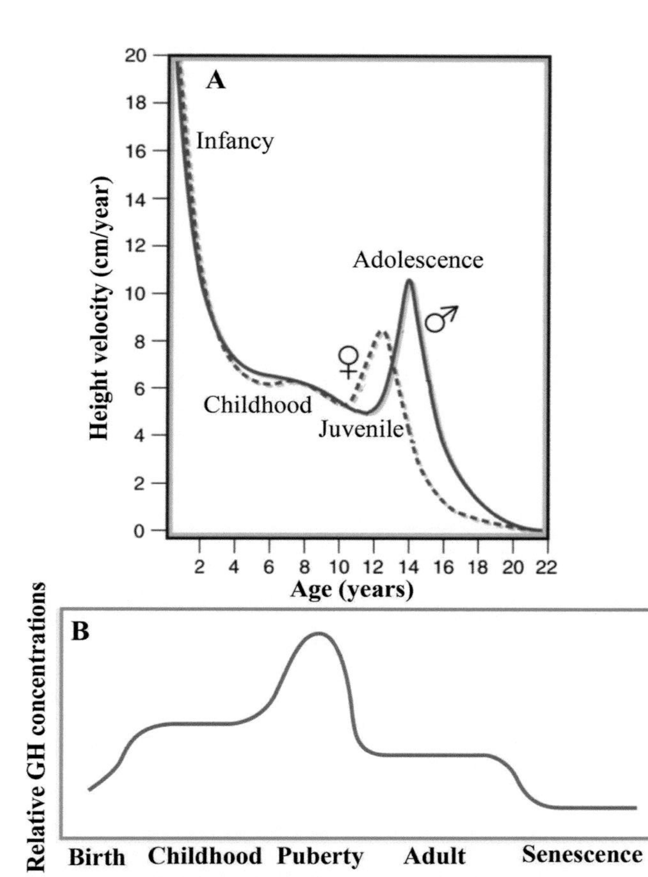

Figure 12.21 The adolescent growth spurt in girls and boys (growth velocity curves). (A) Note the later onset of the pubertal growth spurt in boys and the approximately 2-year difference in peak height velocity and the greater magnitude of peak height velocity compared with girls. Progressive epiphyseal fusion terminates the growth spurt and leads to final adult height. **(B)** The adolescent growth spurt is accompanied by increased levels of growth hormone (GH) secretion. Subsequently, GH secretion declines with age.

Source of images: (A) Melmed, S., Polonsky, K.S., Larsen, P.R. and Kronenberg, H.M., 2015. Williams textbook of endocrinology E-Book. Elsevier Health Sciences. Used by permission. (B) Koeppen, B.M. and Stanton, B.A., 2009. Berne & Levy Physiology, Updated Edition E-Book. Elsevier Health Sciences. Used by permission.

The Hypothalamic-Pituitary-GH-IGF Axis

As its name clearly implies, growth hormone is a potent stimulator of both general organ growth and longitudinal bone growth. An incremental rise in GH secretion by the adenohypophysis from childhood through puberty corresponds with a doubling of linear growth velocity during this time, and the post-pubertal decline in GH corresponds with dramatically reduced growth rate (Figure 12.21). The major end organ of the hypothalamic-pituitary-growth hormone-IGF (HP-GH-IGF) axis is the liver, which responds to pituitary GH by releasing *insulin-like growth factor I* (IGF-I, also called "somatomedin"), a potent stimulator of longitudinal bone growth, as well as general organ growth (Figure 12.22). The binding of IGF-I to receptors on the bone epiphyseal growth plates stimulates chondrocyte proliferation and growth, as well as synthesis of chondrocyte matrix proteins such as chondroitin sulfate and collagen. In addition to upregulating

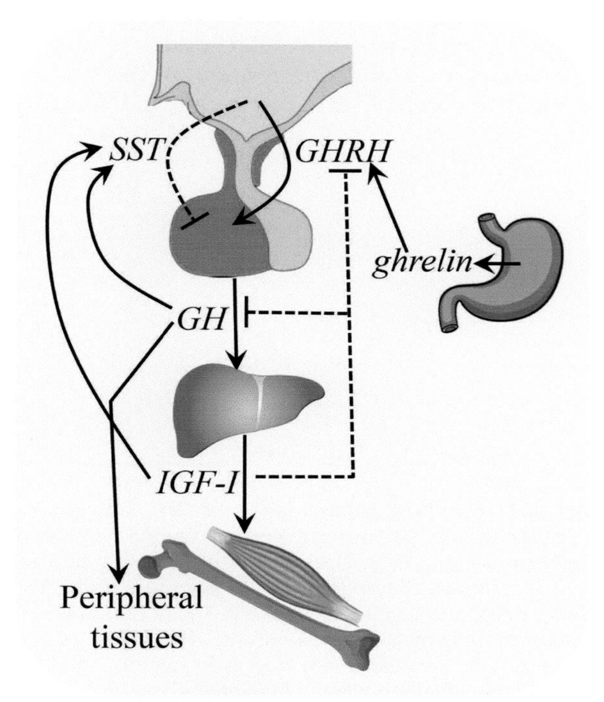

Figure 12.22 The hypothalamus-pituitary-growth hormone-IGF (HP-GH-IGF) axis. Stimulatory and inhibitory actions are denoted by solid arrows and dashed lines, respectively.

IGF-I production, GH also contributes directly to linear growth by inducing the differentiation of the precursor cells within the growth plate towards chondrocytes.

The HP-GH-IGF axis is under complex control by virtue of the fact that the release of GH is strongly regulated by two antagonistic hypothalamic factors: growth hormone-releasing hormone (GHRH), which induces GH release, and somatostatin (SST), an inhibitor of GH release (Figure 12.22). Importantly, the reciprocal relationship between the activity of GHRH neurons and somatostatin neurons leads to a pulsatile secretion of growth hormone[31,32] (Figure 12.23), whereby the peaks of secretion are mediated by GHRH release and the troughs by somatostatin output. In addition to its characteristic pulsatile release, another feature of GH release in humans is its distinct circadian profile, whereby GH production surges at night during deep sleep.

An additional level of complexity for this axis is that it is strongly influenced by molecules that reflect nutrient status, such as the growth hormone secretagogue, ghrelin (Figure 12.22). Although ghrelin is produced by the arcuate nucleus of the hypothalamus where it regulates feeding behavior, the most important source of this hormone is the stomach, which releases ghrelin during fasting in part to stimulate hunger. Indeed, ghrelin is a much more potent stimulant of GH secretion than is GHRH itself.

The Hypothalamic-Pituitary-Gonad Axis and the Adolescent Growth Spurt

Although growth rates in humans and other mammals generally decrease from infancy onwards, the adolescent/pubertal growth spurt (refer to Figure 12.21) is a transient exception to this trend. The adolescent growth spurt yields two contrasting results: on the one hand, it doubles the prepubertal growth rate and contributes more than 15% to the total adult height; on the other hand, it initiates epiphyseal fusion, which terminates linear growth.[33]

In both males and females, the adolescent growth spurt is primarily induced by low levels of estrogen, which act to increase the activity of the GH–IGF-I axis.[33,34] However, at higher levels, estrogen is also the key hormone that directly promotes growth plate fusion, thus terminating growth.[35] Specifically, at high levels estrogen has been found to accelerate the growth plate senescence program, thereby exhausting the proliferative potential of growth plate chondrocytes sooner, consequently triggering earlier epiphyseal fusion.[35] Therefore, estrogen appears to have a biphasic dose-response relationship for epiphyseal growth, with stimulation at low levels and inhibition at high levels. Indeed, premature estrogen exposure is well known to lead to precocious puberty and premature epiphyseal fusion.[36] Interestingly, androgens appear to stimulate growth by a direct effect on growth plate chondrocytes,[37–39] but in contrast with estrogens does not fuse the epiphyses.[40,41]

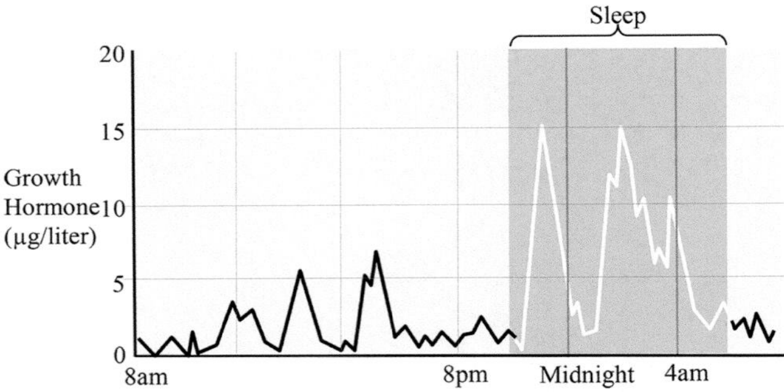

Figure 12.23 Bursts in plasma levels of GH, sampled in the blood plasma of a 23-year-old woman every 5 minutes over a 24-hour period. Each peak in the plasma GH concentration reflects bursts of hundreds of GH secretory pulses by the somatotrophs of the anterior pituitary. These bursts are most common during the first few hours of sleep. The integrated amount of GH secreted each day is higher during pubertal growth than in younger children or in adults.

Source of images: Hartman ML, et al. J Clin Endocrinol Metab 1990; 70:1375.

SUMMARY AND SYNTHESIS QUESTIONS

1. Estrogen exerts two seemingly opposing effects on long bones during puberty, functioning to both promote linear growth and to permanently stop it. How does estrogen manage to do this?
2. GH has a profound effect on bone and cartilage growth, yet these tissues possess very few GH receptors. How can GH influence the growth of these tissues with so few receptors?

Disorders of Longitudinal Growth

LEARNING OBJECTIVE Describe the endocrine bases for disorders of bone growth.

KEY CONCEPTS:

- Extremely rapid or slow growth rates may sometimes be indicative of an underlying pathology, such as a pituitary tumor or genetic disorder.
- Gigantism is a condition characterized by elevated GH secretion and excessive growth and height during childhood before the epiphyses have fused.
- Acromegaly occurs when elevated GH secretion begins in adulthood, after the epiphyses have fused.
- Growth hormone deficiency (GHD) is a condition of GH hyposecretion that generally results in proportionate dwarfism.
- Achondroplasia is a form of short-limbed disproportionate dwarfism caused by a mutation in the FGFR3 receptor that inhibits growth of the long bones.

"Short" or "tall" stature is medically defined as height that is less than the third percentile or greater than the 97th percentile, respectively, for persons of the same sex, age, and population. Deviations from typical height are usually caused by variants of a normal growth pattern reflecting a family history of height. Having short or tall stature in itself is not, of course, a disorder. Typical basketball players do not have abnormal endocrine physiology—they are tall due to normal interactions among genetics, nutrition, and environment. Similarly, Pygmy populations in Africa and Asia, whose members have average adult male heights of less than 150 cm (4 feet 11 inches), also have normal endocrine physiology within their group. However, some patients with extremely tall or short stature may have serious underlying pathologies, such as a pituitary tumor or genetic disorder, and a child whose growth velocity is outside the 25th to 75th percentile range for her/his comparison group may be considered abnormal in terms of height.[42] Next we will consider the endocrine bases for some of the most common longitudinal growth disorders.

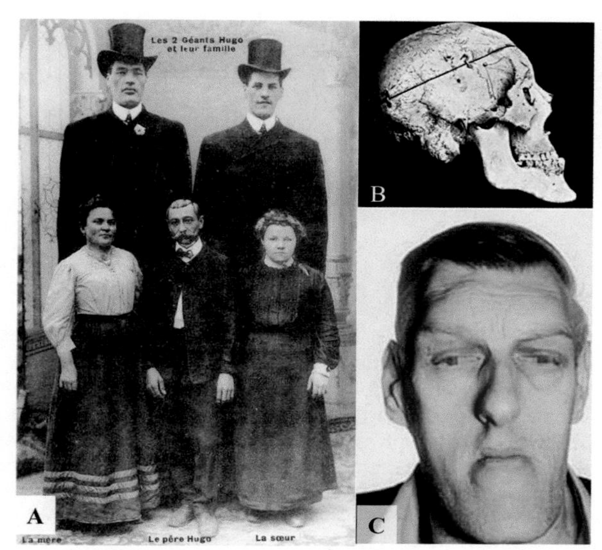

Figure 12.24 Some features of gigantism and acromegaly. (A) Familial gigantism in the two Hugo brothers. Top row: Battista Ugo (Baptiste Hugo), 1876–1916, reached a height of 2.30 m (7 ft, 7 in) and Paolo Antonio Ugo (Antoine Hugo), 1887–1914, reached a height of 2.25 m (7 ft, 5 in). Bottom row: the parents of the Ugo brothers, Teresa Chiardola (1849–1905) and Antonio Ugo (1840–1917), and their sister, Maddalena Ugo (1885–1960). Picture from the collection of Dr. W. W. de Herder. **(B)** A skull from the Irish giant, Cornelius Magrath (2.26 m, 7 ft 5 in). **(C)** A patient with acromegaly exhibiting the classic facial appearance, including an elongated lower jaw.

Source of images: (A) Herder, W.W.D., 2012. Familial gigantism. Clinics, 67, pp. 29-32; (B) S. Karger AG, De Herder, W.W., 2016. Neuroendocrinology, 103(1), pp. 7-17; used by permission. (C) Greenwood, M. and Meechan, J.G., 2003. Brit. Dent. J., 195(3), pp. 129-133. Used by permission.

Gigantism and Acromegaly

Gigantism

Gigantism is a condition characterized by excessive growth and height that is not just in the upper 1% of the population in question, but several standard deviations above the mean.[43] The condition is most typically caused by hypersecretion of GH before puberty due to a tumor on the adenohypophysis (as such, the condition is sometimes called *pituitary gigantism*), which produces very high rates of bone growth before the epiphyses have fused, resulting in adult heights ranging between 2.13 m (7 feet) and 2.74 m (9 feet) in height (Figure 12.24). These elevated GH levels are accompanied by increased IGF-I levels. Using the amphibian model, *Xenopus laevis*, Huang and Brown[44] have shown how overexpression of GH in transgenic tadpoles produces giant frogs that still possess open long bone epiphyses (Figure 12.25).

Acromegaly

Excessive GH secretion that begins after puberty, and hence after epiphyseal fusion of the long bones has taken place, results in a condition known as **acromegaly** (Greek for "extremities" and "enlargement"). Although long bone growth has ceased, appositional, cartilage, and membranous bone growth continues. Therefore, acromegaly is typically accompanied by growth of membranous bones of the skull, forehead, and jaw, enlargement of the hands and feet, as well as growth of cartilaginous structures, such as the nose and ears (Figure 12.24), potentially causing severe disfigurement. The aforementioned transgenic frogs overexpressing GH also possess many features of acromegaly, such as a disproportionally large head and feet, protruding lower jaw, and bent spine (Figure 12.25).[44]

Importantly, because GH levels cycle significantly due to diurnal variation (see Figure 12.23), diagnosing GH hypersecretion for either gigantism or acromegaly with a random blood sample will not always indicate abnormal secretion. In contrast, circulating IGF-I levels do not undergo diurnal variation and are consistently elevated under conditions of GH hypersecretion, and IGF-I secretion is therefore an effective diagnostic test.[45]

Both gigantism and acromegaly can produce multiple and severe complicating conditions, such as hypertension, diabetes, heart failure, kidney failure, colorectal cancer, and vision loss, and may cause premature death if unchecked. The most common treatment for GH hypersecretion is surgical removal of the pituitary adenoma. For situations in which surgery cannot completely remove the tumor, treatments with pharmacological analogs of somatostatin (SST, the hypothalamic inhibitor of GH release) may be effective. The development of new GH receptor antagonist drugs is also showing great promise for treatment of GH hypersecretion.

Pituitary Dwarfism and Achondroplasia

Dwarfism, or short stature resulting from a genetic or medical condition, is generally defined as an adult height of 4 feet 10 inches (147 centimeters) or less.[46] Dwarfism

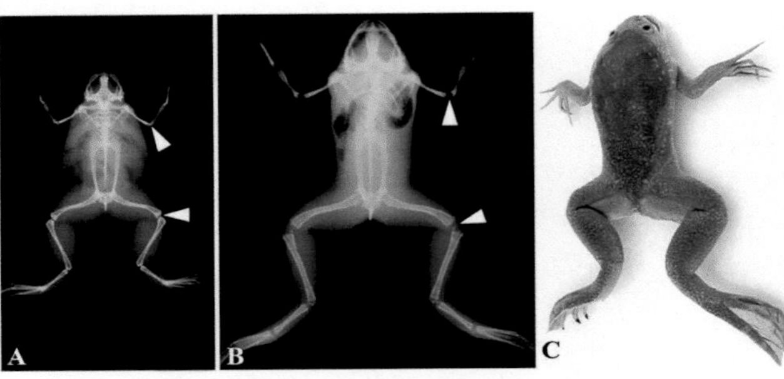

Figure 12.25 Transgenic frogs overexpressing GH possess features of both gigantism and acromegaly. X-ray radiography of a large normal adult male frog **(A)** and a sexually immature male transgenic frog overexpressing GH at 1 year of age **(B)**. Note how the epiphyses (arrows) in the transgenic frog are open, in contrast to the adult control. **(C)** An 8-month-old transgenic frog overexpressing GH. Note the disproportionally large head and feet, protruding lower jaw, and bent spine, all features that accompany of acromegaly.

Source of images: Huang, H. and Brown, D.D., 2000. Overexpression of Xenopus laevis growth hormone stimulates growth of tadpoles and frogs. Proc Nat Acad Sci USA, 97(1), pp. 190-194. Used by permission.

can result from over 100 different medical conditions, each with its own separate symptoms and causes. Dwarfism can be conceptualized as a linear growth dysfunction of the skeletal growth plate chondrocytes, and a vast array of genetic defects outside the GH–IGF-I axis can produce this effect.[47] Here the most common endocrine and paracrine bases of dwarfism will be addressed, and they are divided into two broad categories.

Proportionate dwarfism: **all parts of the body are small to the same degree and appear to be proportioned like a body of average stature. Sexual development is often delayed or impaired into adulthood, and in some cases intellectual disability may accompany this condition.**

Disproportionate dwarfism: **some parts of the body are small, and others are of average size or above-average size. Orthopedic problems, such as progressive development of bowed legs, swayed lower back, and hip dysplasia can develop. Virtually all people with disproportionate dwarfism have normal intellectual capacity.**

Two disorders, growth hormone deficiency and achondroplasia, are responsible for the majority of human dwarfism cases.

Growth Hormone Deficiency

Also known as "pituitary dwarfism", growth hormone deficiency (GHD) is a condition of GH hyposecretion that generally results in proportionate dwarfism. GHD may result from congenital hypopituitarism, physical damage to the pituitary somatotropes, resistance to GH in the form of mutated or deleted GH receptors (Laron-type dwarfism), IGF-I defects, hypothyroidism, hypogonadism (e.g. Turner syndrome), poor nutrition, and even stress-induced psychogenic dwarfism caused by a state of irreversible GH hyposecretion that is generally linked to abuse, neglect, and sensory deprivation resulting in stunted physical, intellectual, and social growth. In the case of GHD due to congenital GH hyposecretion, a damaged pituitary gland, or Turner syndrome, growth can be restored with GH injection treatments, typically over a period of years terminating after puberty. GHD caused by hypothyroidism can be reversed with thyroid hormone therapy, which restores GH secretion.

Clinical Considerations: Adam Rainer: the dwarf who became a giant

Adam Rainer (1899–1950) is the only person in medical history to have been classified as both a giant and a dwarf.[48] Born in Austria, when WWI began Rainer tried enlisting in the army at the age of 18, but at 4 feet 6 inches tall he was deemed too short for service as he was below the cut-off of 4 feet 10 inches. However, by the age of 21, things began to change rapidly as Rainer began growing at a remarkable pace. Over the next decade he experienced a growth spurt, growing to a stature of 7 feet 2 inches. Rainer was diagnosed with an adult onset pituitary gland tumor, causing overproduction of growth hormone. In addition to displaying the rapid linear growth characteristic of gigantism, Rainer also had many symptoms of acromegaly, including abnormally large hands and feet, protruding forehead and jaw, thicker lips and teeth that are more widely spaced than normal. When he died in 1950 he had reached a height of 7 feet 8 inches.

Achondroplasia

Achondroplasia, a form of short-limbed disproportionate dwarfism, is the most common and recognizable form of human dwarfism and occurs in 1 in 15,000 to 40,000 newborns.[49] In addition to limbs that are proportionately shorter than the trunk, the condition is accompanied by increased spinal curvature, distortion of skull growth, and a larger than average head (Figure 12.26). The word achondroplasia literally means "without cartilage formation", and the condition is caused by a glycine mutation in the transmembrane domain of the fibroblast growth factor receptor 3 (FGFR3). In normal development FGFR3 has a negative regulatory effect on epiphyseal cartilage and bone growth. In achondroplasia, the mutated form of the receptor is constitutively active, even in the absence of its ligand, leading to severely shortened bones. Specifically, a mutation in the gene for FGFR3 causes the premature constitutive activation of the STAT pathway and the production of phosphorylated Stat1 protein (Figure 12.26). This transcription factor activates genes that cause the premature termination of chondrocyte cell division. At present, there is no known endocrine treatment for achondroplasia. Although GH treatment is effective in pituitary dwarfism, it remains controversial in achondroplasia, where the response to treatment is only moderate, at best.[50]

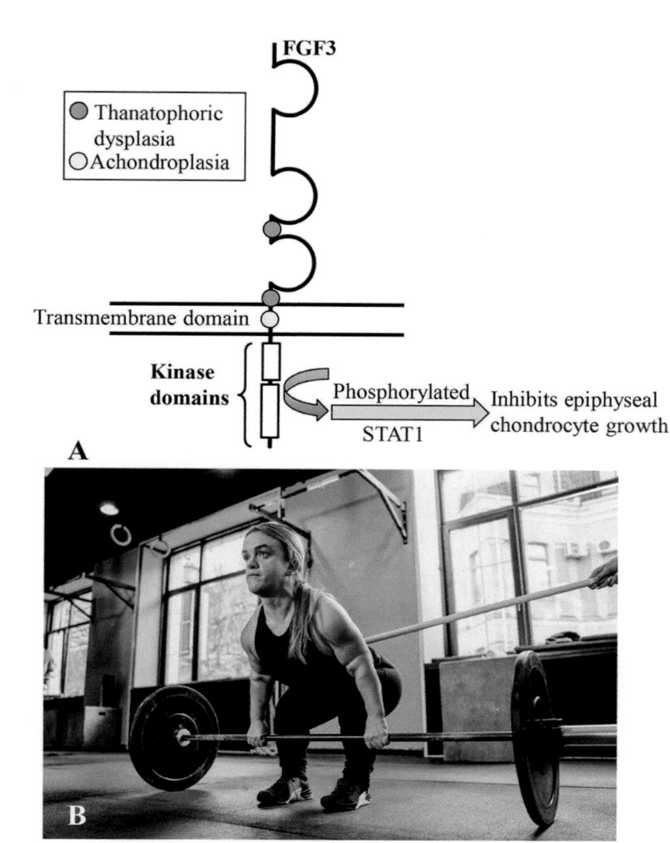

Figure 12.26 Achondroplasia is caused by a mutation in the transmembrane domain of the fibroblast growth factor receptor 3 (FGFR3). (A) In achondroplasia, the mutated form of the FGFR3 receptor (light gray circle) is constitutively active, even in the absence of its ligand, leading to activation of the STAT signaling pathway and development of shortened bones. Thanatophoric dysplasia, caused by different mutations of the FGFR3 receptor (dark gray circles) is a much more severe skeletal disorder characterized by extremely short limbs and folds of extra skin on the arms and legs. **(B)** A woman with achondroplasia.

Source of images: (A) Modified from Yamaguchi, T. P. and Rossant, J. 1995. Curr. Opin. Genet. Devel. 5: 485-491. (B) Shutterstock.

SUMMARY AND SYNTHESIS QUESTIONS

1. Whereas linear growth can be successfully treated with growth hormone supplements in people with pituitary dwarfism, growth hormone treatment is not very effective in promoting linear growth in achondroplasia dwarfism. Why not?
2. In Chapter 1: The Scope and Growth of Endocrinology, the effects of castration on humans to produce eunuchs was described. One type of eunuch, known as Castrati, possessed the singing range of a soprano, but with larger and more powerful lungs. Compared with eunuchs castrated as young men (e.g. harem guards and political advisors), Castrati were castrated as prepubescent boys and therefore developed distinct morphological characteristics, tending to be taller than average, with long slender arms, unusually wide hips, and a lack of facial hair. Explain the endocrine basis for why, compared with eunuchs castrated as young men, Castrati are taller than average, with long slender arms, unusually wide hips, and a lack of facial hair.

Summary of Chapter Learning Objectives and Key Concepts

LEARNING OBJECTIVE Describe the biosynthetic pathway of vitamin D, and use historical context to describe how vitamin D can be considered both a vitamin and a hormone.

- Rickets (in children) and osteomalacia (in adults) are among the most prevalent endocrine disorders in human history.
- Vitamin D is a calcitropic hormone that can be both generated by the skin in response to sunlight and obtained via diet. An inability to produce vitamin D in children causes rickets (weak bones), and in adults osteomalacia.
- Calcium and phosphorous are the most abundant of inorganic elements in vertebrates and are the primary structural elements of bones, teeth, scales, and the outer shells of reptile and bird eggs.

LEARNING OBJECTIVE Discuss some key roles that calcium and phosphate play in normal physiology and the physiological mechanisms by which calcium and phosphate homeostasis are maintained.

- Calcium is involved in diverse and critical physiological functions, such as cellular protein secretion, nerve impulse conduction, intracellular signaling, blood coagulation, and muscle contraction.
- Phosphate is integral to life and an essential component of nucleotides, phospholipids in cell membranes, intracellular second messengers, phosphorylated intermediates of intracellular signal transduction pathways, and high-energy compounds (e.g. ATP).
- Blood serum concentrations of free calcium and phosphate ions are inversely proportional to each other.
- In mammals, the skeleton, intestine, and kidney are the most important organs involved in maintaining calcium and phosphate homeostasis.
- The intestine takes up calcium and phosphate via active transcellular and passive paracellular pathways.
- Whereas most calcium in the kidney filtrate is reabsorbed, 25% of the phosphate load is typically excreted in the urine.
- Bone is the primary storage site for calcium and phosphate.
- The continual process of bone turnover, called remodeling, is facilitated by four cell types: osteogenic cells, osteoblasts, osteocytes, and osteoclasts.

LEARNING OBJECTIVE Review the endocrine mechanisms by which various hormones maintain calcium and phosphate homeostasis.

- Parathyroid hormone (PTH) is the primary hormone involved in increasing blood calcium.
- The active form of vitamin D, called calcitriol, is produced via a complex pathway involving the skin, liver, and kidneys.
- The vitamin D receptor (VDR) is a member of the superfamily of ligand-activated nuclear transcription factors, which mediate gene transcription.
- Vitamin D increases intestinal calcium and phosphate uptake and plays important roles in mediating both bone deposition and bone resorption.
- Calcitonin generally opposes the effects of PTH.
- Fibroblast growth factor 23 (FGF23) reduces blood phosphate by stimulating urinary phosphate secretion.
- Whereas the sex steroids, estrogens and androgens, maintain bone mass, excess levels of glucocorticoids and thyroid hormones promote bone resorption.

LEARNING OBJECTIVE Describe the endocrine bases of various pathologies of calcium/phosphate homeostasis.

- The symptoms of overt hypercalcemia and hypocalcemia are severe and can lead to seizures, coma, or death.
- The most common causes of hypocalcemia are a deficiency of PTH secretion, as well as vitamin D deficiency.
- The most common cause of hypercalcemia is due to hyperparathyroidism.
- The most common form of osteoporosis (decrease in bone mass and density) is due to a deficiency of sex steroid production, particularly reduced estrogen synthesis after menopause.

LEARNING OBJECTIVE Describe the roles of various hormones in regulating longitudinal bone growth.

- The long bones grow via endochondral ossification.
- During puberty osteoblasts produce bone faster than chondrocytes produce cartilage, stopping bone growth.
- Linear bone growth is ultimately regulated by the hypothalamus and pituitary, whose hormones affect bone growth by both direct and indirect mechanisms.
- GH promotes growth by stimulating the release of IGF-I by the liver and bone.
- TH promotes bone growth directly and indirectly by potentiating GH release.
- Sex steroids stimulate bone growth by promoting the release of GH.

LEARNING OBJECTIVE Describe the endocrine bases for disorders of bone growth.

- Extremely rapid or slow growth rates may sometimes be indicative of an underlying pathology, such as a pituitary tumor or genetic disorder.
- Gigantism is a condition characterized by elevated GH secretion and excessive growth and height during childhood before the epiphyses have fused.
- Acromegaly occurs when elevated GH secretion begins in adulthood, after the epiphyses have fused.
- Growth hormone deficiency (GHD) is a condition of GH hyposecretion that generally results in proportionate dwarfism.
- Achondroplasia is a form of short-limbed disproportionate dwarfism caused by a mutation in the FGFR3 receptor that inhibits growth of the long bones.

LITERATURE CITED

1. Chesney RW. Environmental factors in Tiny Tim's near-fatal illness. *Arch Pediatr Adolesc Med*. 2012;166(3): 271–275.
2. Mozoåowski W. Jäccaron; drzej Sniadecki (1768–1838) on the cure of rickets. *Nature* 1939;143(3612):121–121.
3. Huldschinsky K. Heilung von rachitis durch kunstliche hohensonne. *Dtsch Med Wochenschr* 1919;45:712–713.
4. Huldschinsky K. Die behandlung der rachitis durch ultraviolettbestrahlung. *Ztschr f Orthop Chir*. 1920;39:426.
5. Mellanby E. An experimental investigation on rickets. *Lancet*. 1919;1:407–412.
6. McCollum EV, Simmonds N, Becker JE, Shipley PG. An experimental demonstration of the existence of a vitamin which promotes calcium deposition. *J Biol Chem*. 1922;53:293–298.
7. Hess AF, Weinstock M. Antirachitic properties imparted to inert fluids by ultraviolet irradiation. *Proc Soc Exp Biol Med*. 1924;22:6–7.
8. Steenbock H, Black A. Fat-soluble vitamins. XVII. The induction of growth-promoting and calcifying properties in a ration by exposure to ultraviolet light. *J Biol Chem*. 1924;61:405–422.
9. Windaus A, Hess A. Sterine und antirachitisches Vitamin. *Nachr Ges Wiss Göttingen*. 1927:175–184.
10. Windaus A, Schenck F, von Werder F. Uber das antirachitisch wirksame bestrahlungs-produkt aus 7-dehydro-cholesterin. *Hoppe Seylers Z Physiol Chem*. 1936;241:100–103.
11. Rajakumar K, Thomas SB. Reemerging nutritional rickets: a historical perspective. *Arch Pediatr Adolesc Med*. 2005;159(4):335–341.
12. Parenti LR. The phylogenetic significance of bone types in euteleost fishes. *Zool J Linn Soc*. 1986;87:37–51.
13. Kranenbarg S, van Cleynenbreugel T, Schipper H, van Leeuwen J. Adaptive bone formation in acellular vertebrae of sea bass (Dicentrarchus labrax L.). *J Exp Biol*. 2005;208 (Pt 18):3493–3502.
14. Kovacs CS. Vitamin D in pregnancy and lactation: maternal, fetal, and neonatal outcomes from human and animal studies. *Am J Clin Nutr*. 2008;88(2):520S–528S.
15. Aranow C. Vitamin D and the immune system. *J Inv Med*. 2011;59(6):881–886.
16. Maalouf NM. The noncalciotropic actions of vitamin D: recent clinical developments. *Curr Opin Nephrol Hypertens*. 2008;17(4):408–415.
17. Johnson SR, Zelig R, Parker A. Vitamin D status of children with attention-deficit hyperactivity disorder. *Top Clin Nut*. 2020;35(3):222–239.
18. Cui X, McGrath JJ, Burne TH, Eyles DW. Vitamin D and schizophrenia: 20 years on. *Mol Psych*. 2021:1–13.
19. Lee BK, Eyles DW, Magnusson C, et al. Developmental vitamin D and autism spectrum disorders: findings from the

Stockholm Youth Cohort. *Mol Psych*. 2021;26(5):1578–1588.
20. Manolagas SC, O'Brien CA, Almeida M. The role of estrogen and androgen receptors in bone health and disease. *Nat Rev Endocrinol*. 2013;9(12):699–712.
21. Kim HJ, Zhao H, Kitaura H, et al. Glucocorticoids suppress bone formation via the osteoclast. *J Clin Invest*. 2006;116(8):2152–2160.
22. Weinstein RS, Jilka RL, Parfitt AM, Manolagas SC. Inhibition of osteoblastogenesis and promotion of apoptosis of osteoblasts and osteocytes by glucocorticoids. Potential mechanisms of their deleterious effects on bone. *J Clin Invest*. 1998;102(2):274–282.
23. Razzaque MS. The FGF23-Klotho axis: endocrine regulation of phosphate homeostasis. *Nat Rev Endocrinol*. 2009;5(11):611–619.
24. Oury F, Sumara G, Sumara O, et al. Endocrine regulation of male fertility by the skeleton. *Cell*. 2011;144(5):796–809.
25. Auron A, Alon US. Hypercalcemia: a consultant's approach. *Pediatr Nephrol*. 2018;33(9):1475–1488.
26. Pepe J, Colangelo L, Biamonte F, et al. Diagnosis and management of hypocalcemia. *Endocrine*. 2020:1–11.
27. Bilezikian JP. Hypoparathyroidism. *J Clin Endocrinol Metab*. 2020;105(6):1722–1736.
28. Bilezikian JP, Bandeira L, Khan A, Cusano NE. Hyperparathyroidism. *Lancet*. 2018;391(10116):168–178.
29. Johnston CB, Dagar M. Osteoporosis in older adults. *Med Clin Nor Am*. 2020;104(5):873–884.
30. Aspray TJ, Hill TR. Osteoporosis and the ageing skeleton. *Biochem Cell Biol Age II Clin Sci*. 2019:453–476.
31. Hartman ML, Faria AC, Vance ML, Johnson ML, Thorner MO, Veldhuis JD. Temporal structure of in vivo growth hormone secretory events in humans. *Am J Physiol*. 1991;260(1 Pt 1):E101–110.
32. Farhy LS, Veldhuis JD. Joint pituitary-hypothalamic and intrahypothalamic autofeedback construct of pulsatile growth hormone secretion. *Am J Physiol Regul Integr Comp Physiol*. 2003;285(5):R1240–1249.
33. Cutler GB, Jr. The role of estrogen in bone growth and maturation during childhood and adolescence. *J Steroid Biochem Mol Biol*. 1997;61(3–6):141–144.
34. Veldhuis JD, Bowers CY. Three-peptide control of pulsatile and entropic feedback-sensitive modes of growth hormone secretion: modulation by estrogen and aromatizable androgen. *J Pediatr Endocrinol Metab*. 2003;16 Suppl 3:587–605.
35. Weise M, De-Levi S, Barnes KM, Gafni RI, Abad V, Baron J. Effects of estrogen on growth plate senescence and epiphyseal fusion. *Proc Natl Acad Sci U S A*. 2001;98(12):6871–6876.
36. Sigurjonsdottir TJ, Hayles AB. Precocious puberty. A report of 96 cases. *Am J Dis Child*. 1968;115(3):309–321.
37. Keenan BS, Richards GE, Ponder SW, Dallas JS, Nagamani M, Smith ER. Androgen-stimulated pubertal growth: the effects of testosterone and dihydrotestosterone on growth hormone and insulin-like growth factor-I in the treatment of short stature and delayed puberty. *J Clin Endocrinol Metab*. 1993;76(4):996–1001.
38. Stanhope R, Buchanan CR, Fenn GC, Preece MA. Double blind placebo controlled trial of low dose oxandrolone in the treatment of boys with constitutional delay of growth and puberty. *Arch Dis Child*. 1988;63(5):501–505.
39. Nilsson KO, Albertsson-Wikland K, Alm J, et al. Improved final height in girls with Turner's syndrome treated with growth hormone and oxandrolone. *J Clin Endocrinol Metab*. 1996;81(2):635–640.
40. Morishima A, Grumbach MM, Simpson ER, Fisher C, Qin K. Aromatase deficiency in male and female siblings caused by a novel mutation and the physiological role of estrogens. *J Clin Endocrinol Metab*. 1995;80(12):3689–3698.
41. Smith EP, Boyd J, Frank GR, et al. Estrogen resistance caused by a mutation in the estrogen-receptor gene in a man. *N Engl J Med*. 1994;331(16):1056–1061.
42. Nwosu BU, Lee MM. Evaluation of short and tall stature in children. *Am Fam physician*. 2008;78(5):597–604.
43. Bello MO, Garla VV. *Gigantism and Acromegaly*. StatPearls [Internet]; 2020.
44. Huang H, Brown DD. Overexpression of Xenopus laevis growth hormone stimulates growth of tadpoles and frogs. *Proc Natl Acad Sci U S A*. 2000;97(1):190–194.
45. Mais DD. *Quick Compendium of Clinical Pathology* (2nd ed.). American Society for Clinical Pathology; 2007.
46. Grunauer M, Jorge AA. Genetic short stature. *Growth Horm IGF Res*. 2018;38:29–33.
47. Baron J, Savendahl L, De Luca F, et al. Short and tall stature: a new paradigm emerges. *Nat Rev Endocrinol*. 2015;11(12):735–746.
48. McDermott MT, Interesting endocrine facts and figures. In McDermott MT. (ed.), *Endocrine Secrets*. WB Saunders. 2013:521–524.
49. Pauli RM. Achondroplasia: a comprehensive clinical review. *Orphanet J Rare Dis*. 2019;14(1):1–49.
50. Mehta A, Hindmarsh PC. The use of somatropin (recombinant growth hormone) in children of short stature. *Paediatr Drugs*. 2002;4(1):37–47.

Unit Overview
Unit V

Stress, Blood Pressure, and Ion Balance

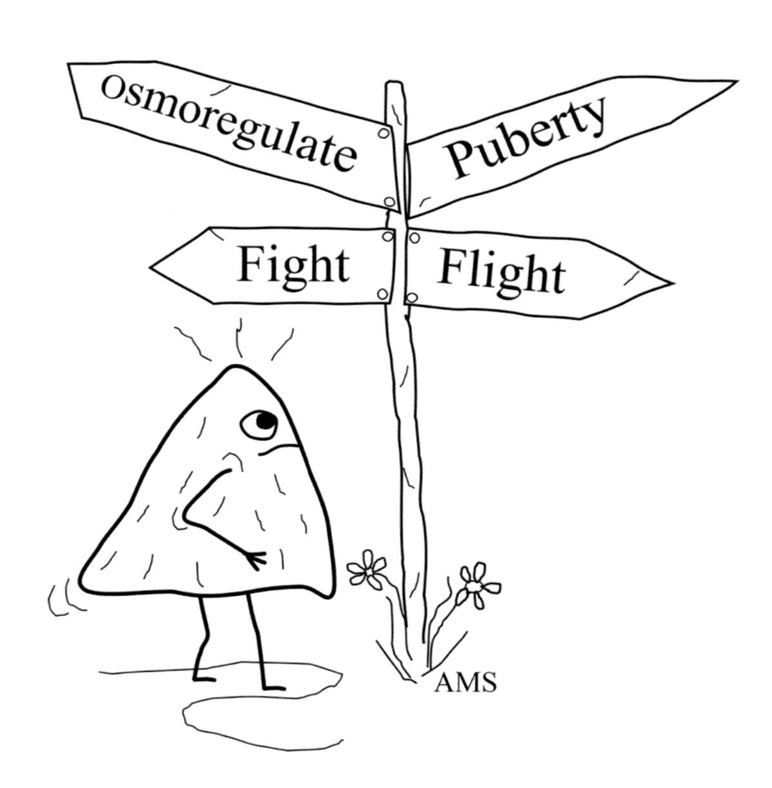

"Adrenal Decision", by A.M. Schreiber.

DOI: 10.1201/9781003359241-17

OPENING QUOTATION:

> "These changes—the more rapid pulse, the deeper breathing, the increase of sugar in the blood, the secretion from the adrenal glands—were very diverse and seemed unrelated. Then, one wakeful night, after a considerable collection of these changes had been disclosed, the idea flashed through my mind that they could be nicely integrated if conceived as bodily preparations for supreme effort in flight or in fighting."
>
> —Walter Bradford Cannon (Cannon WB. *The Way of an Investigator: A Scientist's Experiences in Medical Research*. W.W. Norton & Company, Inc.; 1945.)

Unit V is devoted to telling the story of the broad physiological reach of the adrenal gland (called the "interrenal" gland in non-mammalian taxa) and its many hormones. In addition to modulating the vertebrate stress response, adrenal hormones affect virtually every cell in the body, profoundly influencing osmoregulation, cardiovascular function, and puberty, as well as many aspects of perinatal development and immune function. In many animals adrenal hormones interact with thyroid hormones to modulate metamorphosis and sex hormones to regulate mating. Indeed, I am hard-pressed to think of a physiological system that isn't, in some fashion, significantly influenced by adrenal hormones. Chapter 13 provides an overview of the functional anatomy of the gland's medullary and cortical tissues, the regulation of adrenal/interrenal hormone biosynthesis and action and the physiological roles of the various hormones it produces. Chapter 14 describes the prominent roles that adrenal/interrenal hormones play in the vertebrate stress response, the mechanisms of developmental programming and transgenerational transmission of stress, and the major disorders associated with the hyper- and hyposecretion of adrenal hormones. Chapter 15 discusses the roles of adrenal and other hormones in modulating blood pressure and ion balance among diverse taxa, the molecular and cellular mechanisms of action by which various hormones promote diuresis and antidiuresis, and the endocrine bases of hypertension and other cardiovascular disorders.

CHAPTER 13

The Multifaceted Adrenal Gland

CHAPTER LEARNING OBJECTIVES:

- Describe the earliest experiments and observations suggesting that the adrenal glands are essential to vertebrate life.
- Discuss the developmental origins and functional anatomy of the adrenal gland among vertebrates.
- Outline the regulation of biosynthesis and action of hormones from the adrenal medulla.
- Outline the regulation of biosynthesis and action of hormones from the adrenal cortex.

OPENING QUOTATIONS:

"The suprarenal capsules appear to be organs essential to life."

—Brown-Séquard, E. (1856). Recherches experimentales sur la physiologie et la pathologie des capsules surrenales. *C R Acad Sci*, 43:422–425

"No other class of steroid is used so routinely and equally by men and women—or children. No other steroids are injected, swallowed, rubbed on the skin, dropped in the eyes, or inhaled as often as the various forms of cortisone."

—Thom Rooke *The Quest for Cortisone* (2012)

Importance and Discovery of Adrenal Hormones

LEARNING OBJECTIVE Describe the earliest experiments and observations suggesting that the adrenal glands are essential to vertebrate life.

KEY CONCEPTS:

- Mammals cannot survive adrenalectomy without hormone supplementation.
- Under clinical conditions, glucocorticoids, such as cortisone, are potent therapeutic agents for suppressing inflammation and for treating autoimmune disorders such as rheumatoid arthritis.

The Adrenals Are Essential for Life

The modern understanding of adrenal physiology began in 1849 when doctor Thomas Addison of London's Guy's Hospital described several patients with adrenal destruction caused by tuberculosis. In 1855 he published a monograph, *The Constitutional and Local Effects of Disease of the Supra-renal Capsules*,[1] the first suggestion that these seemingly obscure bits of tissue had any clinical relevancy. In this work, Addison described with remarkable detail and accuracy the symptoms of adrenal failure in the disease that now bears his name, "Addison's disease". The importance of the adrenal glands was confirmed a year later by Charles Brown-Séquard, who demonstrated that a complete surgical removal of these glands (adrenalectomy) in dogs, cats, rabbits, and guinea pigs produced conditions similar to those of Addison's disease in humans and was invariably lethal.[2] Hans Selye would also go on to show that adrenal hormones are necessary for the survival of mammals under stressful conditions, such as starvation.[3] These pioneering works suggested that the adrenals synthesize chemicals essential to life. But what were these mysterious compounds?

The Search for Adrenal Hormones

Crystallized in 1900 by Jokichi Takamine,[4] the first adrenal hormone to be chemically identified was the adrenal medullary catecholamine *epinephrine* ("epi" denotes above, and "nephros" signifies kidney), also known as *adrenaline*. At the time that epinephrine was discovered, it was considered the sole active factor of the adrenal gland. However, definitive evidence for the existence of an adrenocortical hormone appeared in 1930 when Swingle and Pfiffner[5] and Hartman and Brownell[6] reported the use of lipid-soluble extraction procedures to prepare adrenal extracts that could sustain the life of adrenalectomized animals. As long as the extract was supplied, the glandless animals survived;

DOI: 10.1201/9781003359241-18

when it ran out, they died. These adrenocortical extracts, termed "cortin" by Hartman, would also relieve the symptoms of Addison's disease. Indeed, the Swingle–Pfiffner preparation was the first such extract used to save the life of an adrenally insufficient, moribund human patient.[7]

In the early 1930s, Edward Kendall at the Mayo Clinic, who had already gained fame for having isolated thyroxine, turned his efforts to purifying the hormones of the adrenal cortex.[8] Kendall succeeded in isolating six adrenocortical hormones and identified each by a letter, A through F. Several other groups joined the quest for corticoids, including one led by Tadius Reichstein in Zurich. Working in intense competition with one another they quickly isolated, purified, and determined chemical structures for 29 different steroids, including what would eventually be characterized as the most biologically active glucocorticoid in humans, *hydrocortisone* (Kendall's "compound F", now called cortisol) and its inactive precursor, *cortisone* ("compound E"), as well as various androgens.[9–13] The journey towards the isolation and identification of the third major hormone class produced by the adrenal cortex, the mineralocorticoids *aldosterone* and *11-deoxycorticosterone* (DOC), would not begin until 1953, led by the wife and husband team of Sylvia Simpson and James Tait.[14,15] Importantly, these findings indicated that, unlike the catecholamines of the adrenal medulla, the adrenocortical hormones were all steroids. The researchers of the 1930s knew that adrenal steroids were physiologically important, but it took clinical investigations to show that steroids had incredible potential as therapeutic agents.

Curiously Enough . . . For 5 years, the Kendall laboratory processed a remarkable 900 pounds of bovine adrenals *each week*, preparing "cortin" extract used to treat the patients with Addison's disease.[16]

The "Miracle Cure" for Inflammation and Autoimmune Disorders

The first evidence to suggest that an adrenocortical hormone can exert immunosuppressive effects was put forth by Hans Selye, the founder of research concerning the stress response. Selye, who coined the terms "glucocorticoid" and "mineralocorticoid" and emphasized that both were needed for survival,[3] noted that the exposure of rats to a variety of stress-inducing substances and environments elicited both adrenal cortex hyperplasia (increased cell number) and the regression of the *thymus gland*, a key component of the immune system.[17]

At about the same time that Selye began correlating an enlarged adrenal cortex with immunosuppression in rats, Philip Hench, a rheumatologist at the Mayo Clinic, obtained several grams of cortisone and began treating a 29-year-old, wheelchair-bound arthritic woman. Incredibly, after receiving 100 mg daily for 4 days, she was able to walk out of the hospital to enjoy a 3-hour shopping spree.[18] Hench treated another 15 patients with similar spectacular results, this time filming them before and after treatment. Fifty years after watching the movie, one audience member still emotionally recalled the event, "It was like God had touched them".[18] The results were announced in 1949,[19] the world lauded the "miracle cure", and Kendall, Hench, and Reichstein received the *Nobel Prize in Physiology or Medicine* for their work with adrenal steroids in 1950. Today, synthetic glucocorticoids remain among the most widely prescribed medications and are the basis of a multibillion dollar industry whose products are widely used for treating various diseases with an inflammatory or autoimmune basis, such as allergies, asthma, fibromyalgia, Addison's disease, and rheumatoid arthritis.

Cortisone Treatment: A Two-Edged Sword

Though initially hailed as a panacea, it soon became clear that with high dosage and prolonged treatment, some of the remarkable therapeutic effects of glucocorticoids were accompanied by undesirable side effects, such as increased blood sugar, excessive salt and water retention, high blood pressure, potassium depletion, reduced immune function, fatigue, muscle weakness, increased gastric acidity, osteoporosis, and psychosis[20]—all symptoms of *Cushing's syndrome*, a condition caused by overproduction of glucocorticoids. Indeed, the researchers were well aware of these side effects, with Hench using the analogy that "Cortisone is the fireman who puts out the fire, it is not the carpenter who rebuilds the damaged house".[21]

SUMMARY AND SYNTHESIS QUESTIONS

1. If not supplemented with saline solution, the complete removal of the adrenal glands from mammals causes them to die within days. Why, specifically, do they die so quickly?
2. What was the significance of the use of *lipid-soluble* extraction methods by Swingle and Pfiffner[5] and Hartman and Brownell[6]?
3. Philip Hench was the first to note the symptomatic remission of chronic rheumatoid arthritis in patients who had liver failure. By what mechanism can liver failure partially reverse the symptoms of rheumatoid arthritis?

Adrenal Gland Functional Morphology and Development

LEARNING OBJECTIVE Discuss the developmental origins and functional anatomy of the adrenal gland among vertebrates.

Located on the anterior poles of the kidneys, in mammals the *adrenal glands* are so named for their positions adjacent to each kidney (Figure 13.1). In humans and other primates

KEY CONCEPTS:

- Adrenal hormones influence diverse physiological systems, including fuel metabolism, immune function, cardiac function, blood pressure, the stress response, and the timing of puberty.
- Mammalian adrenal glands consist of two functionally distinct tissues, the adrenal cortex, which produces steroid hormones, and the adrenal medulla, which synthesizes catecholamine hormones.
- The catecholamines, epinephrine and norepinephrine, mediate key aspects of the stress response.
- The three categories of adrenal steroid hormones, glucocorticoids, mineralocorticoids, and androgens, modulate fuel metabolism, mineral balance, and the peripheral synthesis of sex steroids, respectively.
- The adrenal cortex is subdivided into functionally unique steroidogenic cell zones under the control of different endocrine signals.
- The adrenal medulla is composed of modified neural tissue.

Table 13.1 A Summary of the Hormones Produced by the Two Major Zones of the Mammalian Adrenal Gland

Zone	Hormones	Hormone Class	General Hormone Effects
Medulla	Epinephrine and norepinephrine	Catecholamine	Mediate aspects of the stress response, including increased cardiac activity, blood pressure, and mobilization of fuels for energy
Cortex	Mineralcorticoids	Steroid	Increased renal reabsorption of Na^+ and water
	Glucocorticoids	Steroid	Mediate aspects of the stress response, including mobilization of fuels for energy, and anti-inflammatory effects
	Androgen precursors	Steroid	Modulate the timing of adrenarche and puberty

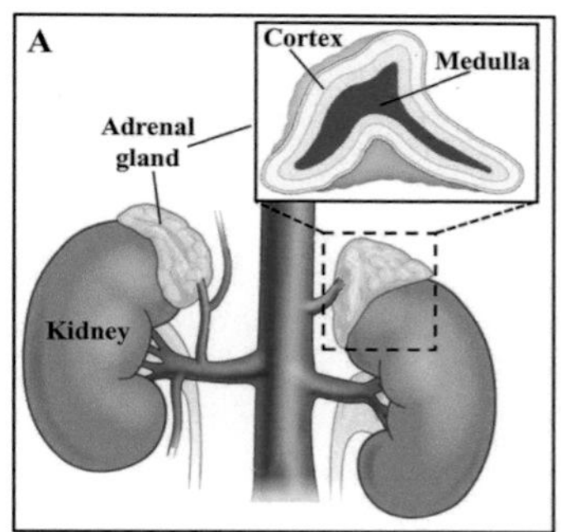

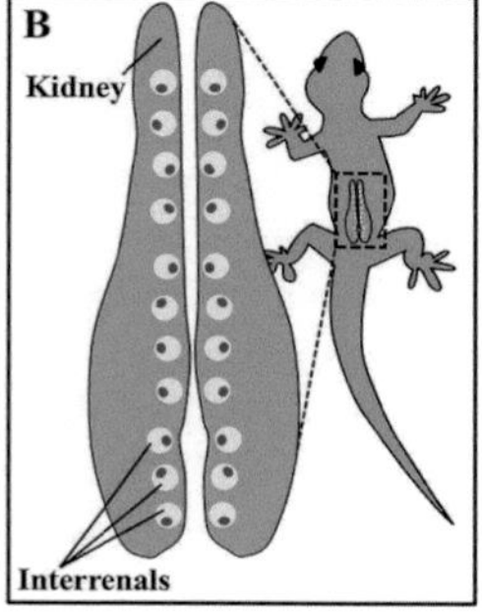

Figure 13.1 Gross anatomy of the adrenal glands in a human and the interrenal glands of a salamander. (A) The mammalian adrenal glands are juxtaposed anterior to the kidneys, and consist of two functionally distinct tissues, an outer cortex that secretes corticosteroid hormones and an inner medulla that produces catecholamines. **(B)** In non-mammalian vertebrates, adrenal glands are known as "interrenal" glands, as these tissues are often interspersed throughout and between the kidneys instead of being clustered towards one end.

Source of images: (A) Sherlock, M., et al. 2020. Endocr. Rev., 41(6), pp. 775-820.

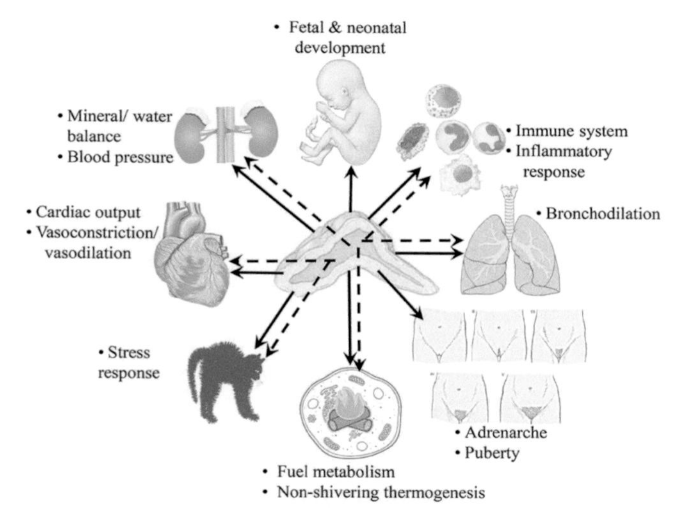

Figure 13.2 Some of the diverse physiological effects modulated by adrenal hormones. Solid arrows denote the effects of corticoid hormones from the adrenal cortex origin, and dashed arrows show effects of catecholamine hormones from the adrenal medulla.

they are sometimes called *suprarenal glands* ("above the kidney"). In non-mammalian vertebrates the clusters of adrenocortical producing cells are referred to as **interrenal cells**, as they are often found between or within the kidneys. Adrenal glands consist of two functionally distinct tissues that secrete different classes of hormones (summarized in Table 13.1). The *adrenal medulla* (inner zone) secretes the catecholamines *epinephrine* and *norepinephrine*, which are primarily involved in mediating aspects of the stress response. By contrast, the *adrenal cortex* (outer zone) is steroidogenic and secretes three major categories of steroid hormones: (1) *glucocorticoids*, which modulate the metabolism of glucose, as well as that of protein and lipid, and also mediate aspects of the stress response; (2) *mineralocorticoids*, which influence blood Na^+ and K^+ regulation, as well as water balance; (3) *adrenal androgen precursors*, which are converted by peripheral tissues like the brain, gonads, breast, prostate gland, skin, and others into active androgen and estrogen sex steroids and neurosteroids. Though the position and distribution of interrenal tissues in relation to the kidneys varies broadly among non-mammalian vertebrates, the hormones produced by these tissues are homologous to those of mammals.

Although in humans each gland is no bigger than a walnut, the myriad of hormones produced by adrenal tissues have diverse and critical functions, such as modulating mineral balance and blood pressure, mobilizing energy resources for metabolism, facilitating the stress response, and contributing to puberty (Figure 13.2). In fact, the adrenal glands exert such profound influences on different physiological systems that a meaningful description of their functions simply cannot be contained in a single chapter. This chapter will serve as a general overview to adrenal gland functional anatomy, development, and control of hormone biosynthesis and secretion. The specific effects of adrenal hormones on different physiological systems and some associated pathologies will be addressed in Chapter 14: Adrenal Hormones and the Stress Response and Chapter 15: Blood Pressure and Osmoregulation.

Mammalian Adrenal Anatomy and Zonation

The mammalian adrenal glands receive a rich arterial blood supply from the inferior, middle, and superior suprarenal arteries (Figure 13.3). In fact, on a per gram basis the adrenal glands receive one of the highest rates of blood flow in the body.[22] The adrenal gland consists of several morphologically and biochemically distinct regions, summarized in Figure 13.4:

Adrenal Capsule

The adrenal gland is covered by a fibrous capsule that serves as both a support structure and, critically, as a reservoir of stem/progenitor cells for the underlying steroidogenic cortex.

Adrenal Cortex

In adults the adrenal cortex, which comprises most (90%) of the gland, is subdivided into discrete histological and functional steroidogenic cell zones under the control of distinct endocrine signals:

- **Zona glomerulosa:** the most superficial layer of the adrenal cortex, lying directly beneath the renal capsule. Its cells are organized into spherical clusters (*glomus* is Latin for "ball") that secrete primarily the mineralocorticoid aldosterone under control by the renin-angiotensin-aldosterone system (RAAS).
- **Zona fasciculata:** the middle zone, where constituent cells are organized into bundles or "fascicles". The largest zone of the adrenal cortex, it primarily produces glucocorticoids, stimulated by the adenohypophysis hormone, adrenocorticotropic hormone (ACTH).
- **Zona reticularis:** the innermost zone of the cortex, the cells are arranged in cords that project in different directions giving a net-like (*reticulum* is Latin for "net") appearance. These cells synthesize the androgen precursors, androstenedione, dehydroepiandrosterone (DHEA), and its sulfated derivative DHEA-S. The release of these androgens is also regulated primarily by ACTH.

In addition to the three primary zones of the adult adrenal cortex, the fetal and neonatal adrenal cortex of humans (and primates, in general) contains a transient fourth zone (the **fetal zone**) that produces large quantities of DHEA and DHEA-S.

Arterial blood enters the cortex, flows through capillaries between adrenocortical cells, and drains inwardly into veins exiting the adrenal medullary region. This *centripetal* (i.e. from the outside toward the center) arrangement of blood flow within the gland is important in developing and maintaining the morphological and functional zonation of the gland. As a result each zone is exposed to increasing levels of adrenal steroids, with zona glomerulosa cells differentiating on the arterial side and zona reticularis cells on the venous side. This layout is particularly important to the adrenal medulla, which requires a high glucocorticoid concentration for chromaffin cell postnatal maintenance and survival,[23,24] and to induce a key enzyme necessary for epinephrine biosynthesis, *phenylethanolamine N-methyltransferase* (PNMT).[25]

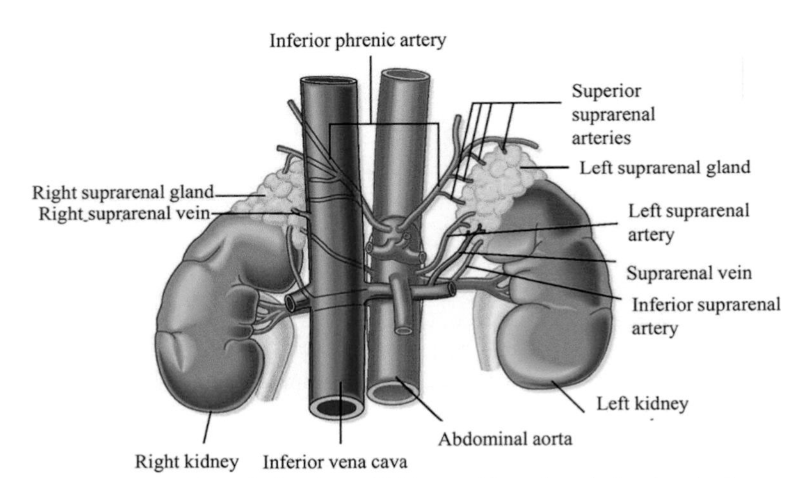

Figure 13.3 Gross anatomy of the human adrenal glands and kidneys. The adrenal glands, also known as the suprarenal glands in humans, are positioned on the superior poles of the kidneys and receive a rich arterial supply from the inferior, middle, and superior suprarenal arteries. The adrenals are drained by a single suprarenal vein.

Source of images: Koeppen, B.M. and Stanton, B.A., 2009. Berne & Levy Physiology, Updated Edition E-Book. Elsevier Health Sciences. Used by permission.

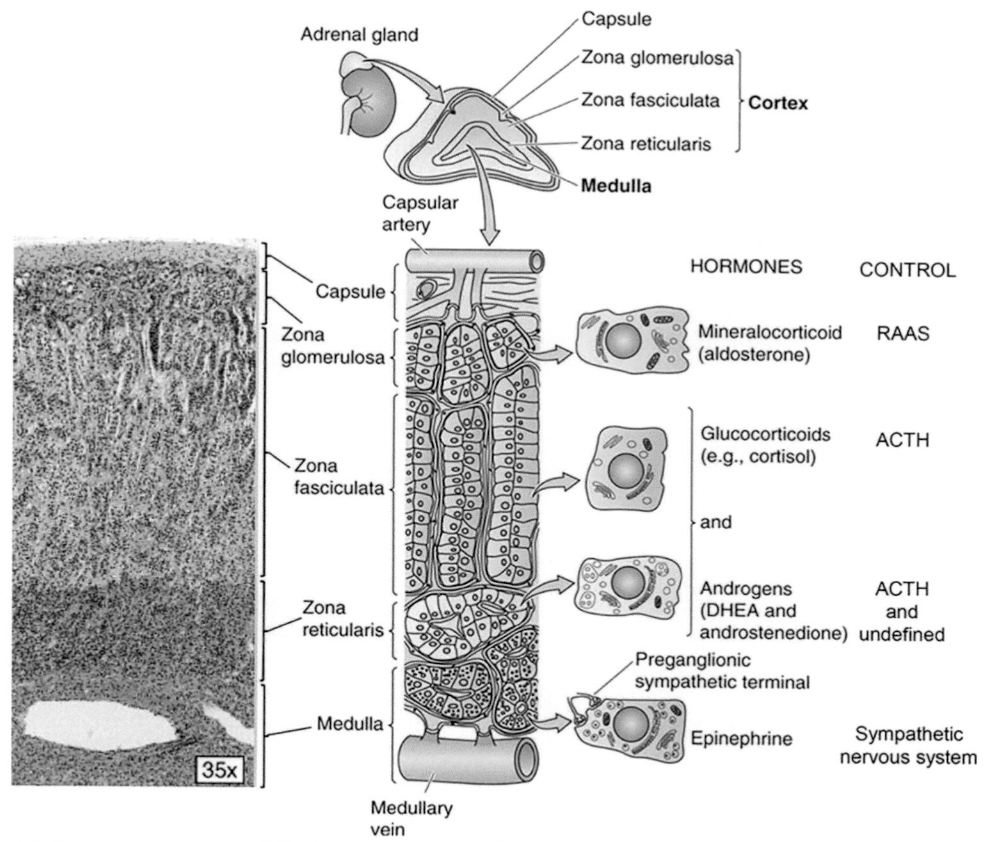

Figure 13.4 Fine histology and anatomy of the human adrenal gland, and major subdivisions, hormones secreted, and control mechanisms. The left panel portrays a hematoxylin and eosin-stained cross section of the adrenal cortex. The capsule forms the outermost layer, the adrenal cortex constitutes the middle layers, and the adrenal medulla is situated in the center. Cortical cells are arranged as three layers: the zona glomerulosa near the capsule, the zona fasciculata (the thickest layer), and the zona reticularis. Arterial blood enters the cortex, flows through capillaries between adrenocortical cells, and drains inwardly into veins exiting the adrenal medullary region. RAAS: renin-angiotensin-aldosterone system; ACTH: adrenocorticotropic hormone (pituitary origin).

Source of images: Photographs: Junqueira, L.C. and Mescher, A.L., 2013. Junqueira's basic histology: text & atlas/ McGraw Hill LLC; used by permission. Drawing: Boron, W.F. and Boulpaep, E.L., 2016. Medical physiology E-book. Elsevier Health Sciences. Used by permission.

Adrenal Medulla

The most interior region of the adrenal gland, the medulla (Figure 13.4), is composed of modified neural tissue called **chromaffin cells**, so named because they have a high staining affinity to the element chromium. These cells secrete catecholamines, primarily epinephrine (80%), as well as norepinephrine (20%), directly into the blood. Importantly, unlike the adrenocortical tissues whose steroid hormone synthesis is regulated hormonally either by ACTH from the pituitary (in the case of glucocorticoids and androgens) or by the renin-angiotensin-aldosterone system (mineralocorticoids), the synthesis of catecholamines by chromaffin tissue is under sympathetic neural control.

Mammalian Adrenal Organogenesis and Development

The adrenal cortex and medulla may be regarded as two discrete endocrine organs with distinct embryological origins. Whereas the adrenal medulla forms from cells derived from the embryonic *neural crest*, the adrenal cortex is derived from a specialized region of intermediate mesoderm called the *urogenital ridge* (Figure 13.5), which also gives rise to the kidney and reproductive organs. Following their embryonic development, the morphological and functional features of adrenal glands change dramatically during different stages of prenatal and postnatal development.

Adrenal Cortex Development

Embryogenesis and Fetal Development

In the 3rd-fourth week of human gestation, cells from the urogenital ridge destined to become adrenocortical cells condense to form a shared embryological structure, the *adrenogonadal primordium* (AGP). The AGP is the precursor of two different steroidogenic organs: the adrenal cortex and the gonads (Figure 13.5). As development proceeds, progenitors of the adrenal cortex and the gonad separate and activate different transcription programs. The neural crest-derived precursors of chromaffin cells in the medulla begin migrating into the fetal adrenal gland in the ninth week of human gestation, the adrenal primordium becomes encapsulated by mesenchymal cells, and by the 20th week of gestation, the fetal adrenal cortex contains three morphologically distinct layers[26] (Figure 13.5):

1. *Fetal zone*: a thick zone that accounts for 80%–90% of the fetal adrenal gland.[27] Functionally, the fetal zone resembles the adult zona reticularis, producing large amounts of DHEA and DHEA-S, which are converted by the sequential actions of the liver and placenta into estrogens.
2. *Definitive zone*: expresses the enzyme 3β-hydroxysteroid dehydrogenase (3β-HSD) that produces aldosterone. This zone is thought to give rise to the postnatal zona glomerulosa.[28]
3. *Transitional zone*: develops between the fetal and definitive zone and gives rise to the zona fasciculata that produces glucocorticoids.[29]

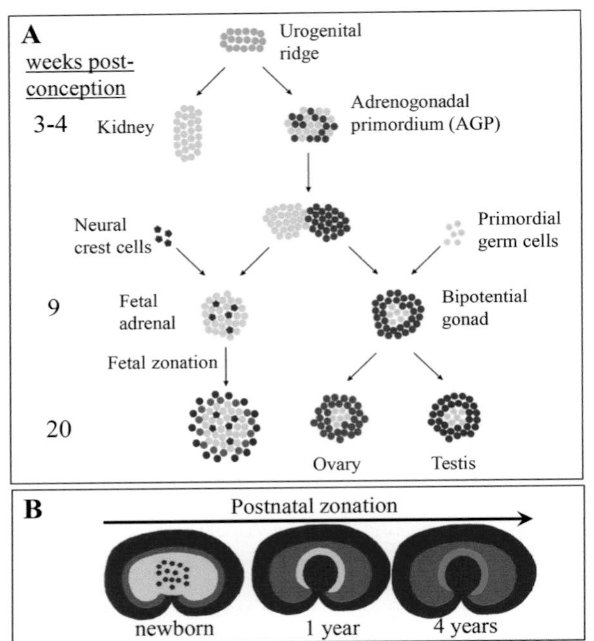

Figure 13.5 Development of the human adrenal gland and gonads. (A) In the third to fourth week of human gestation, cells from the urogenital ridge destined to become adrenocortical cells condense to form a shared embryological structure, the adrenogonadal primordium (AGP). The AGP is the precursor of both the adrenal cortex and the gonads. As development proceeds, the fetal adrenal cortex (yellow tissue) and gonad progenitors separate and follow different developmental paths. By the ninth week of human gestation, neural crest-derived precursors of chromaffin cells (dark gray cells) begin migrating into the fetal adrenal gland. By the 20th week of gestation, the fetal adrenal cortex contains three morphologically distinct layers: the fetal zone (yellow dots), the transitional zone (red dots), and the definitive zone (green dots). **(B)** By birth, the zona glomerulosa (solid green) and zona fasciculata (solid red) have differentiated, and the adrenal gland is almost as large as the kidney due to the large fetal zone (solid yellow). The size of the fetal zone decreases dramatically and typically disappears by one year of age, and the formerly dispersed chromaffin cells have become clustered into a distinct medulla. A zona reticularis (solid blue) appears in humans by four years of age, and the cortex and medulla continue to grow until puberty.

Source of images: (A) Pihlajoki, M., et al. 2015. Front. Endocrinol., 6, p. 27.

By birth, the zona glomerulosa and zona fasciculata have differentiated, and the adrenal gland is almost as large as the kidney due to the large fetal zone (Figure 13.6). Subsequently, the size of the fetal zone decreases dramatically via apoptosis over the first two weeks of neonatal life, and typically disappears by one year of age. A distinct zona reticularis appears in humans by four years of age, and the cortex and medulla continue to grow until puberty.

Unlike the adrenal cortex, the adrenal medulla does not exist as a discrete structure throughout gestation. Instead it is composed of small islands of chromaffin cells that aggregate within the fetal zone (see Figure 13.5). The disappearance of the fetal zone at the beginning of postnatal life stimulates the further condensation of the chromaffin elements around the central vein to form an elementary medulla, and by the fourth week of postnatal life almost all of the chromaffin cells have clustered in the central region of the adrenal gland. However, it is not until one to two years postpartum that the morphological development of the adrenal medulla resembles its adult counterpart.[30]

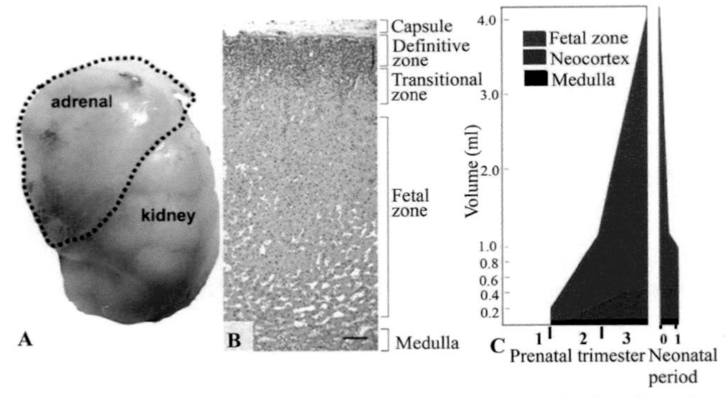

Figure 13.6 The size of the human fetal adrenal glands relative to the kidney is large due in part to the large fetal zone of the adrenal cortex. (A) Fetal adrenal gland at 18 weeks' gestation. **(B)** Fetal adrenal gland histology and zonation. Compare with the adult histology and zonation in Figure 13.4. (C) The fetal zone size increases throughout gestation, then regresses rapidly after birth.

Source of images: (A and C): Rainey, W.E., et al. 2004, November. The human fetal adrenal: making adrenal androgens for placental estrogens. In Seminars in reproductive medicine (Vol. 22, No. 04, pp. 327-336). Thieme Medical Publishers, Inc. Used by permission. (B) Monticone, S., et al. 2012. Nat. Rev. Endocrinol., 8(11), pp. 668-678. Used by permission.

Ectopic Adrenal Tissue in Humans

The presence of ectopic chromaffin and steroidogenic tissues is a common feature in newborn humans. Extra-adrenal chromaffin tissue can be found adjacent to blood vessels, such as the aorta, and are generally referred to as *para-aortic bodies*, with larger bodies known as the *organs of Zuckerkandl*. These groups of ectopic chromaffin cells may represent neural crest cell derivatives that did not migrate properly.[31] Although these ectopic tissues normally disappear before childhood, the retention of ectopic adrenal tissues in the adult can lead to overproduction of steroid and catecholamine hormones, causing hypertension and other disorders.

Clinical Considerations: Pheochromocytomas and hypertension

Pheochromocytomas are rare but potentially very dangerous neuroendocrine tumors that typically arise from adrenomedullary chromaffin tissue within the adrenal gland or ectopically in various extra-adrenal tissues[32,33] (Figure 13.7). Such tumors are generally characterized by the excessive production and release into circulation of one or more catecholamines (epinephrine, norepinephrine, and dopamine). "Pheochromocytomas" (from the Greek *phaios*, meaning dark, and *chroma*, meaning color), are so named due to the black-colored staining caused by the oxidation of catecholamines within chromaffin cells. Considering that elevated levels of blood-borne catecholamines can increase blood pressure by elevating heart rate and promoting vasoconstriction, if unrecognized pheochromocytoma can lead to a fatal hypertensive crisis associated with stroke or heart attack. When identified, a surgical removal of the tumor may correct the cardiovascular pathologies associated with hypertension.

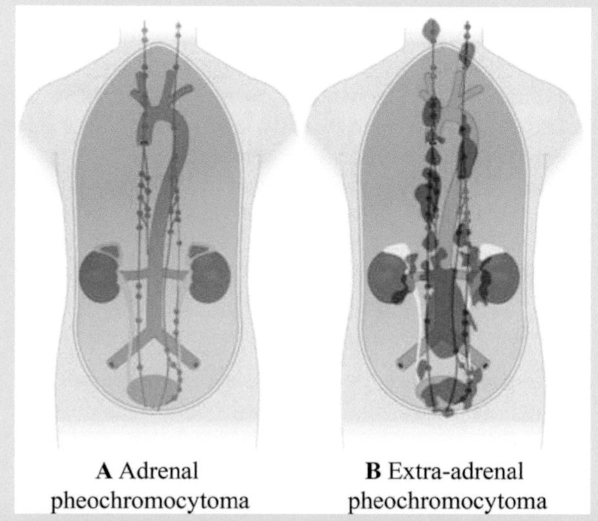

Figure 13.7 Locations of adrenal (A) and extra-adrenal (B) pheochromocytomas. Red denotes location of the tumor.

Source of images: Kasper, D., et al. 2015. Harrison's principles of internal medicine, 19e (Vol. 1, No. 2). New York, NY, USA:: Mcgraw-hill. Used by permission.

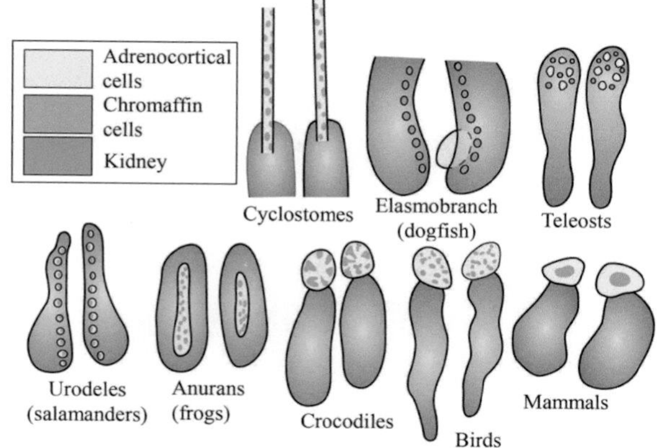

Figure 13.8 Arrangements of adrenal steroidogenic and chromaffin tissues among vertebrates range from scattered chromaffin separate from steroidogenic tissue (elasmobranchs), to scattered chromaffin integrated within adrenocortical tissue (cyclostomes, teleosts, amphibians), to clustered chromaffin integrated within adrenocortical tissue (reptiles, birds, mammals).

Source of images: Norris, D.O. and Carr, J.A., 2020. Vertebrate endocrinology. Academic Press. Used by permission.

Distribution of Adrenal Tissues among Vertebrates

Non-Mammalian Vertebrates

In most vertebrate groups there is an obvious intermingling of chromaffin and steroidogenic tissues, and discrete adrenal glands are present in the non-mammalian tetrapods (Figure 13.8). However, unlike in mammals, there is no clear division of a steroidogenic cortex and a chromaffin cell-enriched medulla.[34,35] In non-mammalian vertebrates, the clusters of adrenocortical producing cells are referred to as *interrenal cells* (they are often found between or within the kidneys) and are the homologs of the mammalian adrenal cortex. An interesting evolutionary trend in the arrangement of adrenal tissues among the vertebrates is the gradual transition from scattered adrenocortical and chromaffin cells in the more ancestral taxa into more distinct and condensed clusters in the more derived taxa[34,36] (Figure 13.8). Such a transition is reminiscent of a similar trend observed in phylogenetic changes in thyroid follicle distributions. In amphibians, reptiles, birds, and mammals, the close association and condensation of chromaffin and steroidogenic tissues may be important, considering that the epinephrine-synthesizing enzyme, PNMT, is controlled by glucocorticoids.

SUMMARY AND SYNTHESIS QUESTIONS

1. Describe the significance of the blood vessel organization and direction of blood flow through the adrenal gland to adrenal function.
2. The mammalian adrenal gland can be considered as two distinct tissues organized into a single organ. How so?

3. The urogenital system comprises the urinary and reproductive organs. Why are these two different systems often studied and described in combination?

Adrenal Medulla: Regulation of Hormone Biosynthesis and Action

LEARNING OBJECTIVE Outline the regulation of biosynthesis and action of hormones from the adrenal medulla.

KEY CONCEPTS:

- The adrenal medulla is a modified sympathetic ganglion regulated by the autonomic nervous system.
- Three major categories of adrenergic receptors are differentially expressed among diverse tissues and mediate various aspects of the stress response.
- The variety of adrenergic responses to catecholamines revolve around a central theme: the mobilization of energy resources to cope with a real or perceived stressful situation that threatens to disrupt physiological homeostasis and survival.

Earlier we have seen how the adrenal cortex and medulla are histologically and functionally different tissues with distinct developmental origins. In the next two sections we will learn how the release of hormones by these two tissues is also regulated by very different mechanisms. Whereas the release of steroid hormones by all zones of the adrenal cortex is under hormonal control, the release of catecholamines by the adrenal medulla is controlled by the nervous system (Figure 13.8). Despite the developmental, functional, and regulatory differences between the adrenal cortex and medulla, we will learn how these tissues often behave in concert as a functional unit that creates a remarkable capacity to cope with diverse internal or external threats to physiological homeostasis.

Adrenal Medulla

The term *sympathoadrenal* means "involving the adrenal medulla and sympathetic nervous system". It normally relates to increased activity of the sympathetic nervous system, which acts in part on the adrenal medulla to regulate the release of epinephrine and norepinephrine into the blood. As sympathetic neurons and chromaffin cells are both integral components of the autonomic nervous system (ANS), in order to understand adrenal medullary functioning it is useful to begin by describing the ANS in the context of general nervous system architecture.

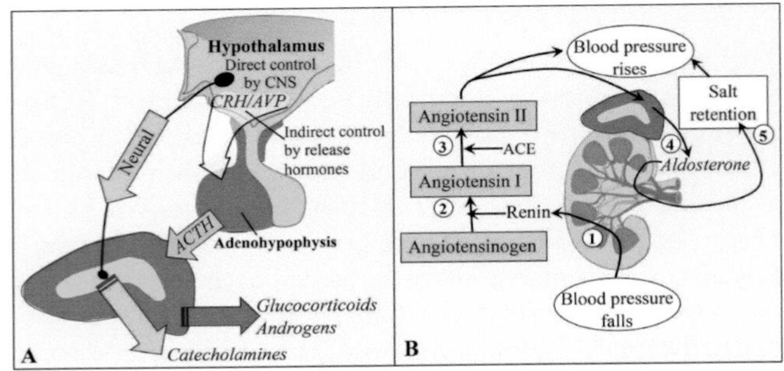

Figure 13.9 The secretion of adrenal hormones is under both neural and endocrine control. (A) The release of catecholamines (epinephrine and norepinephrine) by the adrenal medulla is controlled by the sympathetic nervous system. The release of glucocorticoids and androgenic precursors from the adrenal cortex is regulated by ACTH from the adenohypophysis, which itself is controlled by the hypothalamic release factors CRH and AVP. **(B)** The release of the mineralcorticoid, aldosterone, from the adrenal cortex is regulated by the renin-angiotensin-aldosterone system (RAAS). (1) Low blood pressure is sensed by the kidneys juxtaglomerular cells, which respond by releasing the protease, renin, into circulation. (2–3) Renin and angiotensin-converting enzyme (ACE) convert circulating angiotensinogen into its active form, angiotensin II. (4) Angiotensin II signals the adrenal cortex to produce aldosterone, which functions to retain salt and water (5) and increase blood pressure.

Adrenal Medulla: A Modified Sympathetic Ganglion

The adrenal medulla is a modified sympathetic ganglion whose post-ganglionic cells (chromaffin) are distinguished from typical post-ganglionic sympathetic neurons in several important ways (see **Figure 13.3b**):

1. Whereas post-ganglionic sympathetic neurons primarily secrete the catecholamine norepinephrine, chromaffin cells secrete primarily epinephrine (80%), as well as some norepinephrine (20%).
2. Unlike typical sympathetic neurons that release catecholamines as neurotransmitters, chromaffin cells release their products directly into systemic circulation where they function as hormones.
3. Chromaffin cells are distinguished by their very large neuroendocrine storage vesicles, the "chromaffin granules".
4. Chromaffin cells lack neural processes, such as axons and dendrites.

Curiously Enough . . . Because one preganglionic splanchnic nerve fiber innervates many chromaffin cells, just a few nerve impulses can result in a massive catecholamine discharge that mediates the rapid "fight or flight" component of the stress response.

Catecholamine Synthesis and Degradation

The biosynthetic pathway of catecholamines, summarized in Figure 13.13, is highly conserved throughout the animal kingdom. Human chromaffin consists of two primary

Foundations & Fundamentals: Overview of Vertebrate Nervous System Divisions

The vertebrate nervous system can be broadly divided into a *central nervous system* composed of neurons located in the brain and spinal cord and a *peripheral nervous system* with nerves that are located outside the brainstem and spinal cord (Figure 13.10). The peripheral nervous system is further subdivided into *sensory* and *motor pathways*, the latter of which is subdivided into two divisions: the *somatic nervous system*, with cranial and spinal nerves that directly innervate skeletal muscle and are under voluntary control, and the *autonomic nervous system* (ANS), with nerves under subconscious control that innervate all other structures, including endocrine organs, smooth and cardiac muscle, blood vessels, and the gastrointestinal tract.

The ANS is subdivided into nerves that comprise the *sympathetic* and *parasympathetic* divisions. The sympathetic division generally mediates processes related to arousal or the active state (including the "fight or flight" stress response), resulting in increased heart rate and blood flow to muscles, increased mobilization of metabolic fuels, and decreased gastrointestinal activity. By contrast, the parasympathetic division mediates resting state processes, such as digestion, nutrient storage, and energy conservation. As such, the sympathetic and parasympathetic divisions often exert antagonistic (opposing) effects when they both innervate the same tissue. For example, whereas sympathetic stimulation causes pupil dilation and increased heart rate, parasympathetic stimulation exerts the opposite effects on these organs (Figure 13.11).

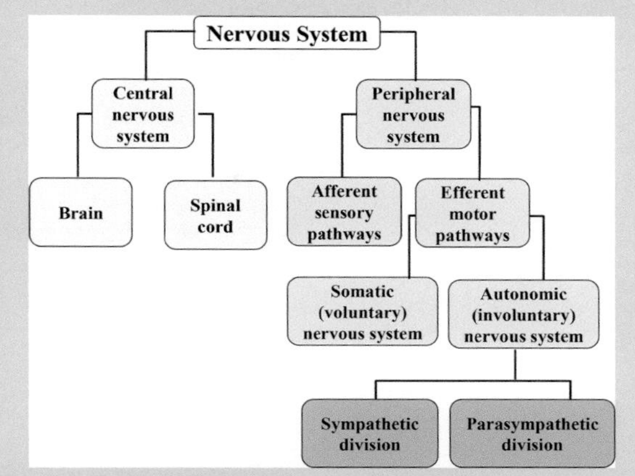

Figure 13.10 The divisions of the vertebrate nervous system.

The divisions of the nervous system not only are functionally different, but also have important structural and biochemical differences as well. Whereas somatic neurons innervate their skeletal muscle targets directly, autonomic neurons emanating from the CNS always form

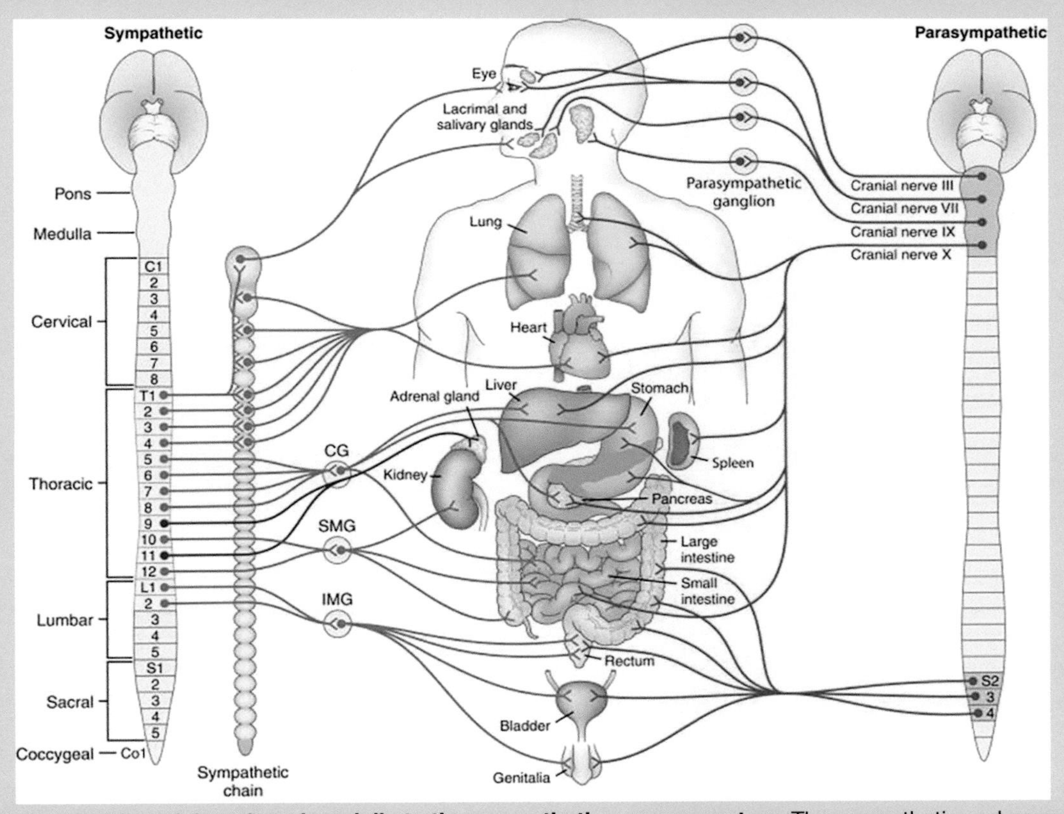

Figure 13.11 Relationship of the adrenal medulla to the sympathetic nervous system. The sympathetic and parasympathetic nervous systems generally exert antagonistic effects on their target organs. The adrenal medulla is part of the sympathetic nervous system, and in contrast to other organs it is not under antagonistic parasympathetic control. Sympathetic thoracic spinal nerves #9 and #11 (purple) transmit information to the adrenal medulla passing through the celiac ganglion (CG) and directly form an acetylcholine-releasing synapse with the adrenal medulla.

Source of images: Norman, A.W. and Henry, H.L., 2022. Hormones. Academic Press. Used by permission.

synapses with clusters of other neurons outside the CNS in structures called *ganglia* (singular, ganglion) (Figure 13.12). ANS neuronal cell bodies within a ganglion project postganglionic axons to the target tissues. Importantly, though preganglionic sympathetic and parasympathetic neurons are both *cholinergic* (use acetylcholine as a neurotransmitter), their postganglionic neurotransmitters differ: sympathetic postganglionic neurons are *noradrenergic* (use norepinephrine) and parasympathetic postganglionic neurons are primarily cholinergic.

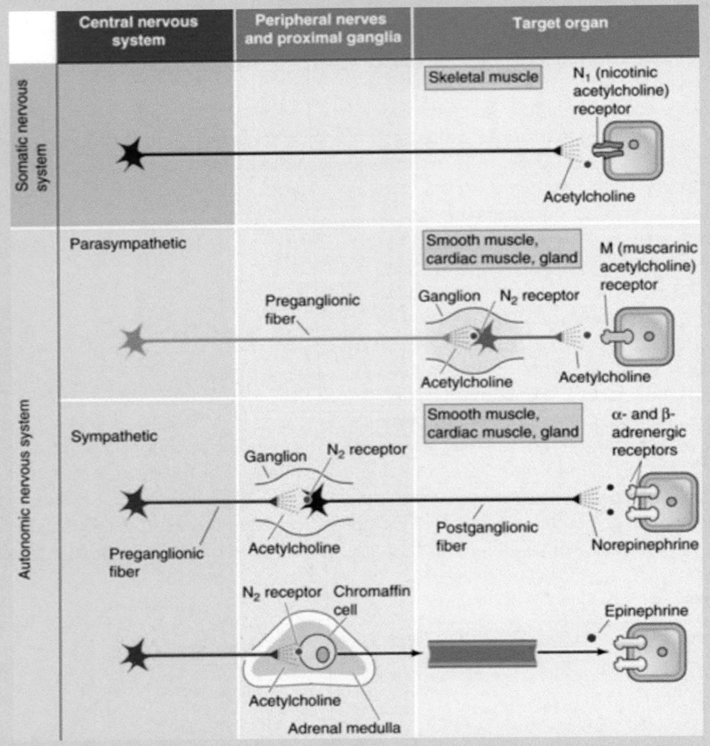

Figure 13.12 Some structural and biochemical features of the sympathetic adrenal pathway compared with other nervous system signaling pathways. A somatic neuronal pathway between the CNS and effector cell (i.e. skeletal muscle cell) is monosynaptic, with the neuron releasing acetylcholine (Ach), which binds to N_1-type nicotinic receptors on the postsynaptic membrane. Parasympathetic and sympathetic divisions utilize multisynaptic pathways that form ganglia, or clusters of neuronal synapses upstream of the final effector synapse. In the parasympathetic nervous system, the preganglionic neuron releases Ach, which acts at N_2-type nicotinic receptors on the postsynaptic membrane of the postganglionic neuron. In the case of the postganglionic parasympathetic neuron, the neurotransmitter is also Ach. In the case of most postganglionic sympathetic neurons, the neurotransmitter is norepinephrine. In the sympathetic adrenal pathway, the adrenal medulla forms a modified sympathetic ganglion. The postsynaptic cells are chromaffin cells that release primarily epinephrine directly into the bloodstream.

Source of images: Boron, W.F. and Boulpaep, E.L., 2005. Medical Physiology: A Cellular and Molecular Approach, updated 2nd ed. Elsevier. Used by permission.

cell types: the majority are those that secrete epinephrine, with a minority secreting norepinephrine. Critically, the epinephrine cells possess the cytoplasmic enzyme phenylethanolamine-O-methyltransferase (PNMT, an enzyme upregulated by glucocorticoid signaling) that converts norepinephrine to epinephrine where it is returned to the storage granule. The encasement of the mammalian adrenal medulla by a glucocorticoid-secreting adrenal cortex ensures that PNMT is highly expressed.

Catecholamine Receptors and Their Actions

Adrenergic Receptors

Adrenergic receptors are those that bind the catecholamines epinephrine and norepinephrine. Three major categories of adrenergic receptors originally distinguished by their ligand binding characteristics have been identified in mammals, and these are called α1, α2, and β. Furthermore, each of these categories consists of three subtypes (Table 13.2), each encoded by a separate gene. All adrenergic receptors are plasma membrane-localized G protein-coupled receptors (GPCRs), which each interact with effector proteins that modulate excitatory or inhibitory intracellular signaling cascades. Although receptors α1, α2, and β1 bind nonselectively to both epinephrine and norepinephrine, epinephrine has about a ten-fold higher affinity to β2 compared with norepinephrine, and norepinephrine has a higher affinity to β3 than epinephrine (summarized in Table 13.2). At first glance the varied effects of different tissues to adrenergic stimulation may appear chaotic, but as we will see in the next chapter (Chapter 14: Adrenal Hormones and the Stress Response), their responses all revolve around one central theme: the

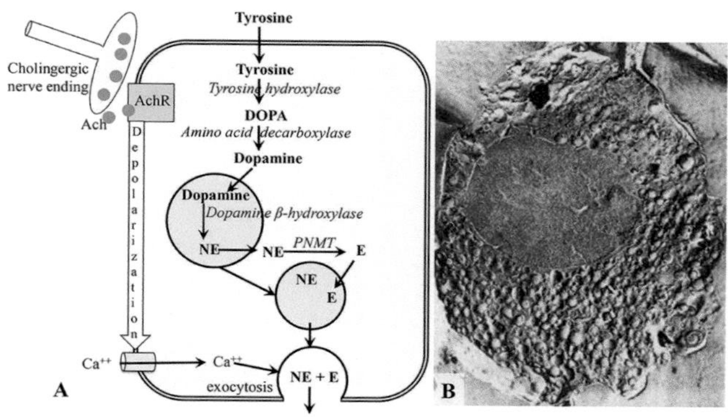

Figure 13.13 The biosynthetic pathway of catecholamine synthesis by chromaffin cells. (A) The synthesis of epinephrine and norepinephrine by chromaffin tissue begins with the amino acid tyrosine. Following the uptake of tyrosine by chromaffin cells, it is converted to dihydroxyphenylalanine (DOPA), the rate-limiting step. DOPA is subsequently decarboxylated to dopamine, which is taken up into chromaffin granules where the enzyme dopamine β-hydroxylase catalyzes the conversion of dopamine to norepinephrine (NE). Epinephrine (E) cells possess the cytoplasmic enzyme phenylethanolamine-*O*-methyltransferase (PNMT) that converts norepinephrine to epinephrine where it is returned to the storage granule. The secretion of epinephrine and norepinephrine from chromaffin cells is stimulated directly by acetylcholine released from preganglionic sympathetic fibers innervating the adrenal medulla. Association of acetyl choline with its receptor (AchR) depolarizes the cell membrane, which increases calcium influx and induces exocytosis of the secretory granules. **(B)** The subcellular features of a chromaffin cell, highlighted by a freeze-fracture technique, include hundreds of chromaffin vesicles, or intracellular sacs containing epinephrine, norepinephrine, and a variety of other proteins and peptides.

Source of images: ((B) Oheim, M., 2001. Lasers Med Sci. 16(3), pp. 149-158. Used by permission.

mobilization of energy resources to cope with stressful situations that threaten to disrupt physiological homeostasis and survival, such as a change in body temperature, reduced blood pressure, decreased blood oxygen or glucose, and the presence of a predator, injury, disease, and so on.

Table 13.2 Relative Ligand Specificity and Functions of Adrenergic Receptors

Receptor Type	Gpcr Transducer/ Effector	Ligand Preference	Physiological Effects
α1	G_q/PLC	E = NE	Hepatic gluconeogenesis Hepatic glycogenolysis Increased cardiac contractility Increased artiolar vasoconstriction and blood pressure Pupil dilation Intestinal and bladder sphincter contraction
α2	G_i/AC	E = NE	Decreased insulin secretion Increased platelet aggregation
β1	G_s/AC	E = NE	Cardioacceleration Increased myocardial strength Calorigenesis
β2	G_s/AC	E >>NE	Arteriolar dilation and reduced blood pressure Smooth muscle relaxation (gut, urinary, uterus) Bronchodilation Hepatic glycogenolysis
β3	G_s/AC	NE>>E	Lipolysis, non-shivering thermogenesis by brown adipose tissue (BAT)

SUMMARY AND SYNTHESIS QUESTIONS

1. Compare and contrast chromaffin cells with postganglionic sympathetic neurons.
2. Provide a biochemical explanation for why, in virtually all vertebrates, chromaffin tissues are juxtaposed to adrenocortical tissue.

Adrenal Cortex: Regulation of Hormone Biosynthesis and Action

LEARNING OBJECTIVE Outline the regulation of biosynthesis and action of hormones from the adrenal cortex.

KEY CONCEPTS:

- Synthesis of steroid hormones by different zones of the adrenal cortex is under endocrine control from the hypothalamic-pituitary axis and by the renin-angiotensin-aldosterone system.
- Pituitary ACTH induces the synthesis of both glucocorticoids and adrenal androgen precursors.
- Although ACTH synthesis is under negative feedback control by glucocorticoids, it is not affected by circulating levels of adrenal androgens.
- Adrenarche is characterized by increased rates of adrenal androgen precursor synthesis that promote pubic hair development and other components of puberty that occur independently of the gonads.

In contrast to the catecholamines of the adrenal medulla whose synthesis and release are under sympathetic neural control, the production of steroid hormones by the adrenal cortex is under endocrine control (see Figure 13.9). Specifically, the release of glucocorticoids and androgen precursors by the zona fasciculata and zona reticularis, respectively, of the adrenal cortex occurs in response to ACTH and is therefore regulated by the hypothalamic-pituitary axis. The release of mineralocorticoids by the zona glomerulosa is regulated by the renin-angiotensin-aldosterone system (RAAS). Later we will specifically address the actions of ACTH and angiotensin II on the adrenal cortex. A review of other aspects of RAAS physiology is found in Chapter 15: Blood Pressure and Osmoregulation.

Though the biosynthetic pathways regulating the three zones of the adult mammalian adrenal cortex share many common features, a key component of this functional zonation is the regional expression of specific *cytochrome P450* (CYP) and **hydroxysteroid dehydrogenase** (HSD) enzymes that are located in the mitochondrial and smooth endoplasmic reticulum subcellular compartments of steroidogenic cells (summarized in Appendix 8). Their sequential actions convert the common steroid substrate, *cholesterol*, into mineralocorticoids, glucocorticoids, and androgenic precursors. Although some cholesterol is synthesized by cells in the adrenal cortex from acetyl-coenzyme A, it is obtained primarily from the blood complexed to cholesterol-rich low-density lipoproteins (LDLs) and high-density lipoproteins (HDLs). In this section, we will address the regulation of steroidogenesis in different zones by two hormones, ACTH and angiotensin II, as well as the actions of these hormones on their target tissues.

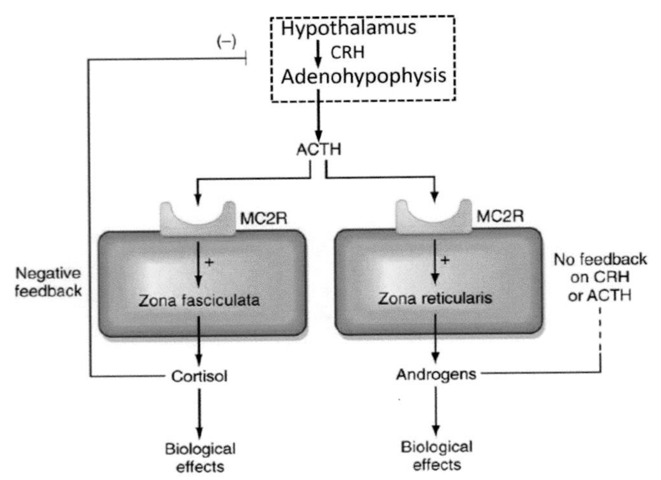

Figure 13.14 The hypothalamus-pituitary-adrenal axis "loophole". ACTH stimulates the production of both cortisol and adrenal androgens, but only cortisol feeds back negatively on ACTH and CRH. Thus, if cortisol production is blocked by an adrenal enzymatic deficiency, both ACTH and adrenal androgen levels rise.

Source of images: Koeppen, B.M. and Stanton, B.A., 2009. Berne & Levy Physiology, Updated Edition E-Book. Elsevier Health Sciences. Used by permission.

Control of Adrenal Steroidogenesis

ACTH Regulates Adrenal Glucocorticoid and Androgenic Precursor Synthesis

Adrenocorticotropic hormone (ACTH), a hormone produced by the adenohypophysis of the pituitary gland in response to hypothalamic stimulation, binds to the *melanocortin 2 receptor* (MC2R) located on the plasma membrane of cells in both the zona fasciculata and the zona reticularis to stimulate the synthesis of glucocorticoids and androgenic precursors, respectively (Figure 13.14). Glucocorticoids exert negative feedback on the hypothalamus. Importantly, although ACTH stimulates the synthesis of adrenal androgen precursors, neither these precursors nor their more active metabolites, like testosterone and estradiol, feedback negatively to inhibit pituitary ACTH or hypothalamic CRH synthesis (Figure 13.14). Therefore, if cortisol synthesis is inhibited by an enzymatic defect in the adrenal cortex, this "loophole" in the hypothalamus-pituitary-adrenal axis can give rise to an important clinical condition called **congenital adrenal hyperplasia** that is associated with both a dramatic increase in ACTH production and adrenal androgen precursors (see the following **Clinical Considerations** box). The binding of ACTH to MC2R activates the adenylyl cyclase effector enzyme via G_s, resulting in increased cAMP synthesis and protein kinase A (PKA) activity (Figure 13.15). PKA, in turn, initiates key components of the steroidogenic pathway:

1. Increased cholesteryl ester import by increasing the availability of low-density lipoprotein (LDL) receptors on the plasma membrane.
2. Increased activity of *hormone-sensitive lipase*, which cleaves cholesteryl esters stored in lipid droplets into free cholesterol.
3. Increased expression of **steroidogenic acute regulatory** (StAR) protein, a mitochondria membrane protein that transfers cholesterol from the outer mitochondrial membrane to the inner mitochondria membrane where the cholesterol side-chain cleavage enzyme (CYP11A) (which converts cholesterol to pregnenalone) is located.
4. Phosphorylation of the transcription factor, CREB,* that promotes the expression of CYP11A1 and other enzymes that facilitate the conversion of cholesterol into precursor steroids such as pregnenalone and progesterone.

Enzymes expressed preferentially in glucocorticoid producing cells of the zona fasciculata (3β HSD, CYP21A2 and CYP11B1) subsequently convert pregnenalone into primarily cortisol via progesterone, 17-hydroxyprogesterone, and 11-deoxycortisol intermediaries. In contrast, steroidogenic cells of the zona reticularis have much lower 3β HSD activity, and as a result primarily synthesize the androgen precursor DHEA, as well as DHEAS through the preferential expression of SULT2A1. Although small amounts

* cAMP response element binding

of active androgens like testosterone are also normally produced by the zona reticularis, most active sex steroids derived from the adrenal cortex are synthesized primarily by the peripheral conversion of DHEA and DHEAS. In the long term, ACTH promotes increased size and number of adrenocortical cells, as well as increased size and functional complexity of cellular organelles such as mitochondria and smooth endoplasmic reticulum.

Angiotensin II Regulates Adrenal Mineralocorticoid Synthesis

The zona glomerulosa, the thin, outermost layer of the adrenal cortex, synthesizes the mineralocorticoid *aldosterone*, which regulates salt and blood volume homeostasis. Although ACTH receptors are present in the zona glomerulosa, ACTH is not a key regulator of aldosterone synthesis. Rather, aldosterone production is regulated primarily by renin-angiotensin signaling (see Figure 13.9). In response to angiotensin II, zona glomerulosa cells secrete aldosterone, which induces the retention of Na^+ and water and promotes the excretion of K^+ by the kidney. In addition to angiotensin II, elevated plasma K^+ concentrations can also stimulate aldosterone synthesis by depolarizing the zona glomerulosa cells. In contrast to cells of the zona fasciculata and zona reticularis, the zona glomerulosa cannot synthesize cortisol or androgen precursor. Instead, progesterone is converted into 11-deoxycorticosterone (DOC). DOC is catalyzed to aldosterone by the presence of CYP11B2 (*aldosterone synthase*), an enzyme unique to the zona glomerulosa. The endocrine physiology of the renin-angiotensin-aldosterone system (RAAS) will be discussed in Chapter 15: Blood Pressure and Osmoregulation.

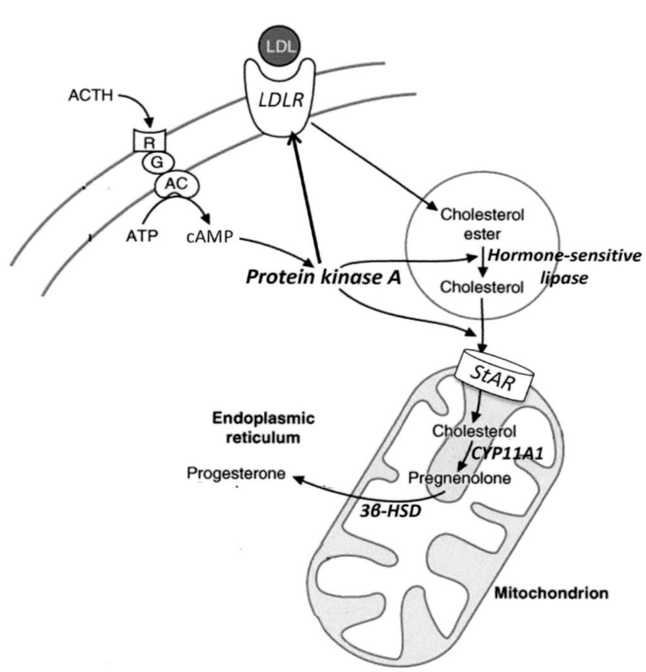

Figure 13.15 The early steps of steroid hormone synthesis in zona fasciculata and zona reticularis adrenal cortex cells are promoted by ACTH. The binding of ACTH to its MC2R receptor activates the adenylyl cyclase effector enzyme via G_s resulting in increased cAMP synthesis. The subsequent induction of protein kinase A activity promotes increases in various receptors and enzymes (in italics) that facilitate the production of pregnenolone and progesterone. These substrates can then be converted into cortisol or androgenic precursors via additional cell-type specific enzymes in the endoplasmic reticulum and mitochondria.

Source of images: Antoniou-Tsigkos, A., et al. 2019. Adrenal androgens. In: Feingold KR, Anawalt B, Boyce A, et al, editors. Endotext. Used by permission.

Fates and Functions of Adrenal Androgen Precursors

Adrenarche

In humans the zona reticularis begins to form as early as 4 years of age[37] and begins to synthesize DHEA and DHEA-S at approximately age 5 to 6. An increased rate of DHEA and DHEA-S synthesis by the zona reticularis that in humans typically begins at around 10 to 11 years of age denotes a developmental period called **adrenarche**, with increasing rates of synthesis continuing through puberty (Figure 13.16). Adrenarche occurs only in some primates, with chimpanzees and gorillas showing a pattern of adrenarche development similar to humans.[38] Pubertal increases in DHEA also occur in rabbits and dogs but are of a magnitude lower than what is seen in humans, and this DHEA is predominantly derived from the gonads.[39] In humans DHEA and DHEA-S synthesis peaks at about age 20 and declines slowly thereafter in a process called *adrenopause* (Figure 13.16). The stimulus

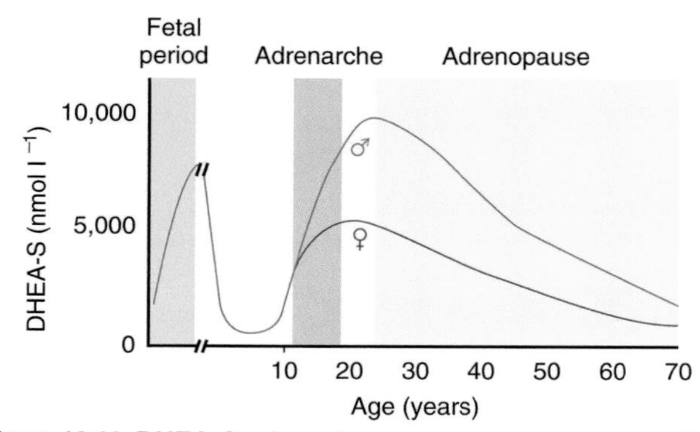

Figure 13.16 DHEA-S adrenal androgen precursor secretion throughout the human lifespan. There is a notable regression of DHEA-S secretion that accompanies the regression of the fetal zone after birth. An abrupt increase in secretion occurs during adrenarche and puberty, followed by a gradual decline beginning after 20 years of age (adrenopause). The secretion of other adrenal androgen precursors follows a similar pattern to that shown for DHEA-S.

Source of images: Norris, D.O. and Carr, J.A., 2020. Vertebrate endocrinology. Academic Press. Used by permission.

that triggers the onset of adrenarche remains unknown, but the crossing of a critical body weight threshold may be an important factor.

The primary effects of adrenarche in humans are the promotion of pubic hair development, the development of apocrine glands in the skin and an accompanying change of sweat composition that produces adult body odor, and increased oiliness of the hair and skin, which can lead to acne. In girls, the adrenal androgens of adrenarche are responsible for producing most of these changes, but the estrogenic effects of puberty such as breast development and growth acceleration are mediated independently by estrogens from the ovaries. In boys, the changes mediated by adrenarche are indistinguishable from the early testicular testosterone effects occurring at the beginning of gonadal puberty. Importantly, although adrenarche is a developmental period related to puberty and normally precedes **gonadarche**, or increased sex hormone synthesis by the gonads during puberty, the initiation and progression of adrenarche is independent of the maturation of the hypothalamic-pituitary-gonadal axis and gonadal steroidogenesis.[40,41]

Curiously Enough . . . Patients with Addison's disease, a form of adrenal insufficiency, still undergo gonadarche and puberty in the absence of adrenarche, though with minimal pubic hair growth.[42] Additionally, girls with Turner syndrome, characterized, in part, by defective ovary development, still have normal adrenarche and normal pubic hair development, although gonadarche and true puberty do not occur.

Adult

DHEAS is the main product of the adult zona reticularis and is also the most abundant steroid hormone in adult human circulation, with serum concentrations 250–500 times higher than DHEA, 100–500 times higher than testosterone, and 1,000–10,000 times higher than estradiol.[43] About half of the body's DHEA is produced in the adrenal cortex, with the rest derived from gonads, adipose tissue, and the brain.[44] Both DHEAS and its close relative, androstenedione, are androgenic precursors, but

Clinical Considerations: Congenital adrenal hyperplasia

Glucocorticoids feedback negatively on the hypothalamus and pituitary gland to regulate ACTH synthesis. Therefore, if glucocorticoid synthesis is blocked for any reason, such as a deficiency in **21-hydroxylase** (CYP21A2) function or expression (Figure 13.17), then the resulting increase in ACTH synthesis by the pituitary produces abnormally high levels of progesterone and 17-hydroxyprogesterone, precursors common to both glucocorticoid and androgen synthesis, causing adrenocortical tissue to expand greatly through increased cell division (hyperplasia) (Figure 13.18(A)). Since a 21-hydroxylase deficiency prevents the conversion of this substrate into glucocorticoids, it is instead diverted into the synthesis of abnormally high levels of active adrenal androgens, like testosterone, and their precursors DHEA and DHEAS, giving rise to a condition called **congenital adrenal hyperplasia** (CAH). In females, CAH results in varying degrees of virilization, or the acquisition of male sexual features. These range from hirsutism (excess facial hair) to the development of ambiguous genitalia (Figure 13.18(B-D)).

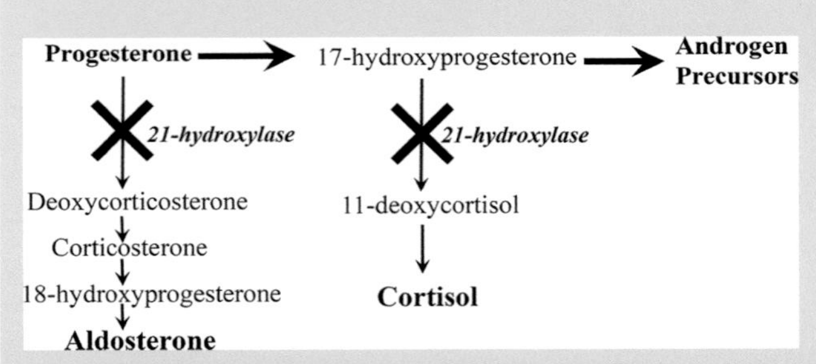

Figure 13.17 21-hydroxylase deficiency results in reduced aldosterone and cortisol synthesis, and elevated androgen synthesis.

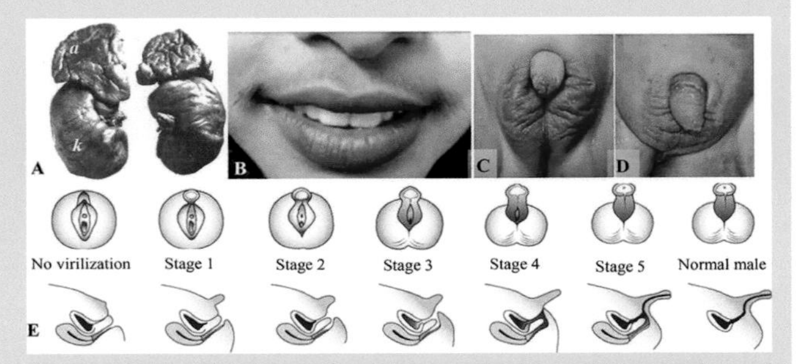

Figure 13.18 Adrenocortical hyperplasia and examples of female virilization (adrenal gland, *a*, kidney, *k*). (A) Bilateral diffuse adrenocortical hyperplasia of the adrenal glands in a 5-year-old girl with congenital adrenal hyperplasia. **(B)** Mild hirsutism on the upper lip of a female patient. External genitalia of 46,XX patients with congenital adrenal hyperplasia (CAH) due to 21-hydroxylase deficiency: Prader stage 3 **(C)** and stage 5 **(D)**. **(E)** Schematic of Prader staging for patients with CAH.

Source of images: (A) Norman, A.W. and Henry, H.L., 2022. Hormones. Academic Press; Used by permission. (B) Faghihi, G., et al. 2015. Evid Based Complement Alternat Med.; (C-D) Finkielstain, G.P., et al. 2021. Front. Endocrinol., 12; (E) Allen, L., 2009. Obstet. Gynecol. Clin. North Am., 36(1), pp. 25-45. Used by permission.

neither binds to the androgen receptor with high affinity. However, these steroids are converted to active male and female sex hormones within peripheral target cells such as the gonads, breast, prostate gland, and skin. In those locations DHEAS is desulfated enzymatically to produce DHEA and oxidized to make androstenedione, which is in turn converted into various active estrogenic and androgenic compounds (Figure 13.19) that are capable of binding to the androgen or estrogen receptors. Interestingly, these peripherally formed androgens and estrogens are not released into general circulation but exert their action locally only in the same cells where their synthesis takes place, and they are released from these target cells only after being inactivated.

Mineralocorticoid and Glucocorticoid Receptors

Receptor Distributions and Functions

The most important actions of all steroid hormones are mediated primarily by intracellular receptors that function as ligand-activated nuclear transcription factors that activate or repress gene transcription, though plasma membrane-localized steroid hormone receptors have also been described. The distribution of mineralocorticoid receptors (MR) is confined to specific tissues that play important roles in regulating salt and water balance, such as the kidneys, brain (thirst-promoting centers), colon, sweat glands, and salivary glands. Aldosterone promotes salt and water retention by orchestrating synergistic physiological and behavioral responses among these organs to loss of blood volume or elevated blood osmotic pressure. In contrast to the relatively limited distribution and function of MRs, glucocorticoid receptors (GRs) are expressed ubiquitously. While originally named for their ability to stimulate glucose formation in the liver, glucocorticoids exert a variety of effects on target cells and play important roles in development, physiology, and metabolism (see Figure 13.2 and Table 13.3), especially as pertains to the stress response.

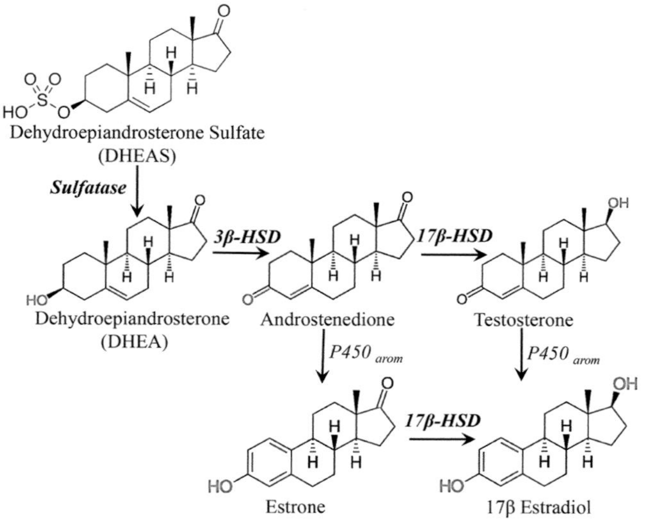

Figure 13.19 Pathways of extra-adrenal synthesis of testosterone and estrogens from DHEA-S (dehydroepiandrosterone sulfate). Enzyme-catalyzed changes are shown in italics.

The Paradox of Mineralocorticoid and Glucocorticoid Receptor-Ligand Interactions

Importantly, the mineralocorticoid receptor (MR) is unique among the steroid nuclear receptors in being a physiologically important receptor for *two* classes of hormones: (1) the mineralocorticoids, aldosterone and, to a lesser extent, DOC, and (2) the glucocorticoids, cortisol (in humans), and corticosterone in rodents. In fact, the glucocorticoid cortisol induces equal amounts of glucocorticoid and mineralocorticoid activity in biological assays (Table 13.4). *In vitro*, MR has a high and very similar binding affinity (K_d ~0.5–1 nM) for corticosterone, cortisol, and aldosterone, but, notably, low affinity to cortisone, the precursor to cortisol.[45] Moreover, the same DNA sequence serves as a hormone response element (HRE) for the activated forms of both mineralocorticoid and glucocorticoid receptors. In contrast, the glucocorticoid receptor (GR) has high affinity for only glucocorticoids, but not

Table 13.3 Some Physiological Effects Mediated by Glucocorticoids

Carbohydrates
Increases blood glucose levels
Promotes hepatic gluconeogenesis
Decreased glycolysis
Protein
Decreased protein synthesis
Lipids
Increased lipolysis and fatty acid mobilization
Hematopoietic
Stimulates red blood cell production
Immune
Suppress antibody synthesis
Inhibits inflammatory response

Table 13.4 Relative Glucocorticoid and Mineralcorticoid Potency of Natural Adrenal Steroids and Some Pharmacological Derivatives in Bioassays

Steroid	Glucocorticoid Activity	Mineralcorticoid Activity
Cortisol	1.0	1.0
Corticosterone	0.3	15
Aldosterone	0.3	3000
Deoxycorticosterone	0.2	100
Cortisone	0.7	0.8
Prednisolone	4	0.8
9α-Fluorocortisol	10	125
Dexamethasone	25	<0.01

Source: Barrett KE, Barman SM, Boitano S, Brooks HL, eds. *Ganong's Review of Medical Physiology* (25th ed.). McGraw Hill; 2018. https://accessmedicine.mhmedical.com/content.aspx?bookid=1587§ionid=96462506

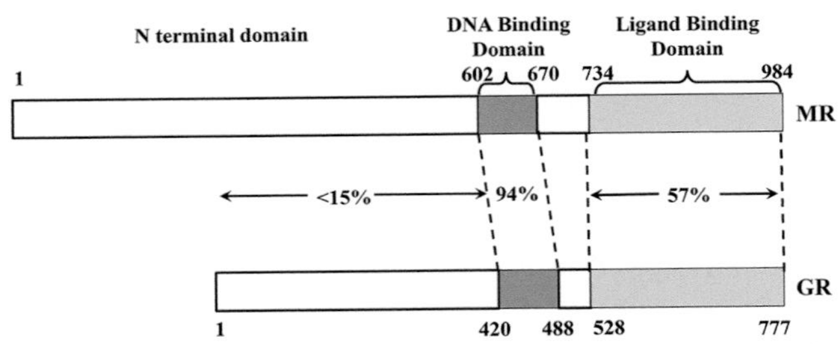

Figure 13.20 Domains of the human mineralocorticoid receptor (MR) and glucocorticoid receptor (GR). The position of amino acids is shown by the numbers on the top and bottom, and the percentage amino acid identity between each of the domains is shown in the middle.

Source of images: Arriza, J.L., et al. 1987. Science, 237(4812), pp. 268-275.

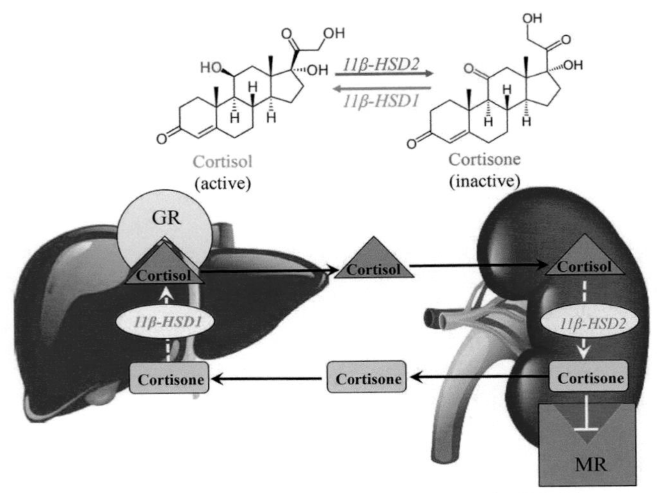

Figure 13.21 The cortisol-cortisone shunt. 11β-HSD2 is present primarily in classical aldosterone target tissues (kidney, colon, sweat glands, salivary glands, placenta), where it oxidizes cortisol to an inactive form, cortisone, preventing it from activating the mineralcorticoid receptor. In contrast, 11β-HSD1 is present in many other tissues (e.g. liver, skin, adipose tissue, central nervous system, placenta) where it acts to convert cortisone to its active form, cortisol, maintaining levels of active glucocorticoid in these tissues. GR, glucocorticoid receptor; MR, mineralcorticoid receptor.

aldosterone. These observations make sense considering that the cloning of the MR revealed that its sequence is highly homologous to that of the GR, with the steroid-binding domains of the human GR and MR being 56% identical[46] (Figure 13.20). However, this generates an intriguing paradox: considering that circulating levels of glucocorticoids are typically over 1,000-fold higher than aldosterone, how is it possible for the MR to preferentially bind to aldosterone in the primary target tissues of mineralocorticoids, such as the kidney tubules? Another way of phrasing this paradox is as follows: under normal conditions why don't the much higher levels of circulating glucocorticoids constantly stimulate the MR and induce excess water retention by the kidneys?

Paradox Resolved: The Cortisol-Cortisone Shunt

A large part of the answer to this paradox is that, in mineralocorticoid-responsive cells, cortisol is effectively inactivated, allowing aldosterone to bind its receptor without competition. Specifically, target cells for aldosterone express the enzyme *11-beta-hydroxysteroid dehydrogenase II* (11β-HSD II) that inactivates cortisol by oxidizing it to cortisone, which has a very low affinity for the mineralocorticoid receptor (Figure 13.21). Therefore, organs that are targets of aldosterone are major sources of circulating cortisone. In contrast to aldosterone target tissues, glucocorticoid target tissues, such as the liver, express high levels of a related HSD called *11-beta-hydroxysteroid dehydrogenase I* (11β-HSD I), which reduces cortisone back to its active form, cortisol (Figure 13.21). As such, 11β-HSD isoforms function as intracellular gatekeepers of tissue glucocorticoid and mineralocorticoid action.[45] This system of extrarenal glucocorticoid interconversion is known as the *cortisol-cortisone shunt*. In pathological excess such as Cushing's syndrome, cortisol exerts aldosterone-like effects in the kidney causing salt and water retention because the capacity of 11β-HSD I, which inactivates cortisol, has become overwhelmed. In such a scenario, cortisol is then available to interact with the aldosterone receptor, for which it has equal affinity, and may cause hypertension. As the prescription of high doses of glucocorticoids by a physician can also potentially cause unwanted side effects through stimulation of the mineralocorticoid receptor, various synthetic steroid analogs are available with modifications that render higher affinity to the glucocorticoid receptor and reduced affinity to the mineralocorticoid receptor (see Table 13.4). It is important to emphasize here that glucocorticoids are not inactivated in all tissues containing mineralocorticoid receptors. For example, the hippocampus of the brain, a key region involved in the stress response pathway, expresses predominantly 11β-HSD I, and both mineralocorticoid and glucocorticoid receptors bind glucocorticoids at physiological levels to modulate the stress response.

SUMMARY AND SYNTHESIS QUESTIONS

1. Metyrapone is a potent inhibitor of the enzyme, 11-hydroxylase, coded for by the CYP11B1 gene. This compound inhibits the synthesis of which hormone, and can be used to treat what category of endocrine diseases?
2. ACTH regulates both the synthesis of glucocorticoids from cells of the zona fasciculata, and also androgenic precursors from the zona reticularis. Why do these two zones respond differently to ACTH?

3. Although both ACTH and angiotensin II receptors are GPCRs, they initiate different intracellular signaling cascades following binding of their ligands. Why?
4. A number of reports have documented a syndrome of excessive water and sodium retention coupled with low plasma concentrations of potassium in individuals that ingested excessive amounts of licorice. The basis of this effect is that licorice contains glycyrrhizinic acid, a molecule that inhibits a particular enzyme responsible for adrenal hormone synthesis. The inhibition of which specific enzyme by glycyrrhizinic acid would produce the aforementioned effects? Explain your reasoning.
5. Parents and many physicians often infer (incorrectly) the onset of puberty from the first appearance of pubic hair. Why is this incorrect?
6. In diseases of pathological excess (such as Cushing's syndrome), cortisol exerts aldosterone-like effects in the kidney, causing salt and water retention that can lead to hypertension. Under such conditions, how does cortisol mimic aldosterone?
7. In humans, 21-hydroxylase deficiency can cause hypoglycemia, hyponatremia, and in females virilization of external genitalia. Explain how this enzymatic deficiency results in such a diversity of phenotypes.

Summary of Chapter Learning Objectives and Key Concepts

LEARNING OBJECTIVE Describe the earliest experiments and observations suggesting that the adrenal glands are essential to vertebrate life.

- Mammals cannot survive adrenalectomy without hormone supplementation.
- Under clinical conditions, glucocorticoids, such as cortisone, are potent therapeutic agents for suppressing inflammation and for treating autoimmune disorders such as rheumatoid arthritis.

LEARNING OBJECTIVE Discuss the developmental origins and functional anatomy of the adrenal gland among vertebrates.

- Adrenal hormones influence diverse physiological systems, including fuel metabolism, immune function, cardiac function, blood pressure, the stress response, and the timing of puberty.
- Mammalian adrenal glands consist of two functionally distinct tissues, the adrenal cortex, which produces steroid hormones, and the adrenal medulla, which synthesizes catecholamine hormones.
- The catecholamines, epinephrine and norepinephrine, mediate key aspects of the stress response.
- The three categories of adrenal steroid hormones, glucocorticoids, mineralocorticoids, and androgens, modulate fuel metabolism, mineral balance, and the peripheral synthesis of sex steroids, respectively.
- The adrenal cortex is subdivided into functionally unique steroidogenic cell zones under the control of different endocrine signals.
- The adrenal medulla is composed of modified neural tissue.

LEARNING OBJECTIVE Outline the regulation of biosynthesis and action of hormones from the adrenal medulla.

- The adrenal medulla is a modified sympathetic ganglion regulated by the autonomic nervous system.
- Three major categories of adrenergic receptors are differentially expressed among diverse tissues and mediate various aspects of the stress response.
- The variety of adrenergic responses to catecholamines revolve around a central theme: the mobilization of energy resources to cope with a real or perceived stressful situation that threatens to disrupt physiological homeostasis and survival.

LEARNING OBJECTIVE Outline the regulation of biosynthesis and action of hormones from the adrenal cortex.

- Synthesis of steroid hormones by different zones of the adrenal cortex is under endocrine control from the hypothalamic-pituitary axis and by the renin-angiotensin-aldosterone system.
- Pituitary ACTH induces the synthesis of both glucocorticoids and adrenal androgen precursors.
- Although ACTH synthesis is under negative feedback control by glucocorticoids, it is not affected by circulating levels of adrenal androgens.
- Adrenarche is characterized by increased rates of adrenal androgen precursor synthesis that promote pubic hair development and other components of puberty that occur independently of the gonads.

LITERATURE CITED

1. Addison T. *The Constitutional and Local Effects of Disease of the Supra-renal Capsules*. Highley; 1855.
2. Brown-Sequard E. Recherches experimentales sur la Physiologie et la Pathologic des Capsules surrenales. *C R Acad Sci.* 1856;43:542–546.
3. Selye H. The general adaptation syndrome and the diseases of adaptation. *J Clin Endocrinol Metab.* 1946;6:117–230.
4. Takamine J. *The Isolation of the Active Principle of the Suprarenal Gland*. Cambridge University Press; 1901.
5. Swingle WW, Pfiffner JJ. The survival of comatose adrenalectomized cats with an

extract of the suprarenal cortex. *Science*. 1930;72:75–76.
6. Hartman FA, Brownell KA. The hormone of the adrenal cortex. *Science*. 1930;72:76.
7. Rowntree LG, Greene CH, Swingle WW, Pfiffner JJ. The treatment of patients with Addison's Disease with the "cortical hormone" of Swingle and Pfiffner. *Science*. 1930;72:482–483.
8. Kendall EC, Mason HL, McKenzie BF, Myers CS, Koelsche GA. Isolation in crystalline form of the hormone essential to life from the supranetal cortex: its chemical nature and physiologic properties. *Trans Assoc Am Physicians*. 1934;48:147–152.
9. Wintersteiner O, Pfiffner JJ. Chemical studies of the adrenal cortex. III. Isolation of two new physiologically inactive compounds. *J Biol Chem*. 1936;116:291–305.
10. Mason HL, Meyers CS, Kendall EC. Chemical studies of the suprarenal cortex. II. The identification of a substance which possesses the qualitative action of cortin; its conversion into a diketone closely related to androstenedione. *J Biol Chem*. 1936;116:267–276.
11. Reichstein T. "Adrenosteron". Über die Bestandteile der Nebennierenrinde II (vorläufige Mitteilung). *Helv Chim Acta*. 1936;19:223–225.
12. Reichstein T. Über die Bestandteile der Nebennierenrinde IV. *Helv Chim Acta*. 1936;19:402–412.
13. Steiger M, Reichstein T. Desoxy-corticosteron (21-oxyprogesterone) aus Δ5–3 -xoy-atio-cholensaure. *Helv Chim Acta*. 1937;20:1164–1179.
14. Simpson SA, Tait JF, Wettstein A, et al. Konstitution des Aldosterons, des neuen Mineralocorticoids. *Experientia*. 1954;10:132–133.
15. Tan LB, Schlosshan D, Barker D. Fiftieth anniversary of aldosterone: from discovery to cardiovascular therapy. *Int J Cardiol*. 2004;96(3):321–333.
16. Ingle DJ. Biographical memoir of Edward C. Kendall. *Nat Acad Sci*. 1975;47.
17. Selye H. A syndrome produced by diverse nocuous agents. *Nature*. 1936:32.
18. Rooke T. *The Quest for Cortisone*. Michigan State University Press; 2012.
19. Hench PS, Kendall EC, Slocumb CH, Polley HF. The effect of a hormone of the adrenal cortex (17-hydroxy-11-dehydrocorticosterone [compound E]) and of the pituitary adrenocorticotropic hormone on rheumatoid arthritis. *Proc Staff Meet Mayo Clin*. 1949;24:181–197.
20. Schacke H, Docke WD, Asadullah K. Mechanisms involved in the side effects of glucocorticoids. *Pharmacol Ther*. 2002;96(1):23–43.
21. Hench PS. *Interview with Philip Showalter Hench by a Cuban Newspaper: Answers to Newspaper Questions, Dr. Philip S. Hench, the Mayo Clinic, Rochester, Minnesotta*. Rector and Visitors of the University of Virginia; 1998–2001:1952.
22. Kovacs WJ, Ojeda SR. *Textbook of Endocrine Physiology* (6th ed.). Oxford University Press; 2011.
23. Parlato R, Otto C, Tuckermann J, et al. Conditional inactivation of glucocorticoid receptor gene in dopamine-beta-hydroxylase cells impairs chromaffin cell survival. *Endocrinology*. 2009;150(4):1775–1781.
24. Schober A, Parlato R, Huber K, et al. Cell loss and autophagy in the extra-adrenal chromaffin organ of Zuckerkandl are regulated by glucocorticoid signalling. *J Neuroendocrinol*. 2013;25(1):34–47.
25. Evinger MJ, Towle AC, Park DH, Lee P, Joh TH. Glucocorticoids stimulate transcription of the rat phenylethanolamine N-methyltransferase (PNMT) gene in vivo and in vitro. *Cell Mol Neurobiol*. 1992;12(3):193–215.
26. Ishimoto H, Jaffe RB. Development and function of the human fetal adrenal cortex: a key component in the feto-placental unit. *Endocr Rev*. 2011;32(3):317–355.
27. Liggins GC. Adrenocortical-related maturational events in the fetus. *Am J Obstetr Gynecol*. 1976;126(7):931–941.
28. Kaludjerovic J, Ward WE. The Interplay between Estrogen and Fetal Adrenal Cortex. *J Nutr Metab*. 2012;2012:837901.
29. Sucheston ME, Cannon MS. Development of zonular patterns in the human adrenal gland. *J Morphol*. 1968;126(4):477–491.
30. Mesiano S, Jaffe RB. Developmental and functional biology of the primate fetal adrenal cortex. *Endocr Rev*. 1997;18(3):378–403.
31. Hadley ME, Levine JE. *Endocrinology* (6th ed.). Pearson Prentice Hall; 2007.
32. Manger WM, Eisenhofer G. Pheochromocytoma: diagnosis and management update. *Curr Hypertens Rep*. 2004;6(6):477–484.
33. Sbardella E, Grossman AB. Pheochromocytoma: an approach to diagnosis. *Best Pract Res Clin Endocrinol Metab*. 2019:101346.
34. Perry SF, Capaldo A. The autonomic nervous system and chromaffin tissue: neuroendocrine regulation of catecholamine secretion in non-mammalian vertebrates. *Auton Neurosci Basic Clin*. 2011;165(1):54–66.
35. Perry SF, Capaldo A. The autonomic nervous system and chromaffin tissue: neuroendocrine regulation of catecholamine secretion in non-mammalian vertebrates. *Auton Neurosci*. 2011;165(1):54–66.
36. Bentley PJ. *Comparative Vertebrate Endocrinology* (3rd ed.). Cambridge University Press; 1998.
37. Hui XG, Akahira J, Suzuki T, et al. Development of the human adrenal zona reticularis: morphometric and immunohistochemical studies from birth to adolescence. *J Endocrinol*. 2009;203(2):241–252.
38. Cutler GB, Jr., Glenn M, Bush M, Hodgen GD, Graham CE, Loriaux DL. Adrenarche: a survey of rodents, domestic animals, and primates. *Endocrinology*. 1978;103(6):2112–2118.
39. Schiebinger RJ, Albertson BD, Barnes KM, Cutler GB, Jr., Loriaux DL. Developmental changes in rabbit and dog adrenal function: a possible homologue of adrenarche in the dog. *Am J Physiol*. 1981;240(6):E694–699.
40. Ducharme JR, Forest MG, De Peretti E, Sempe M, Collu R, Bertrand J. Plasma adrenal and gonadal sex steroids in human pubertal development. *J Clin Endocrinol Metab*. 1976;42(3):468–476.
41. Sizonenko PC, Paunier L. Hormonal changes in puberty III: Correlation of plasma dehydroepiandrosterone, testosterone, FSH, and LH with stages of puberty and bone age in normal boys and girls and in patients with Addison's disease or hypogonadism or with premature or late adrenarche. *J Clin Endocrinol Metab*. 1975;41(5):894–904.
42. Sklar CA, Kaplan SL, Grumbach MM. Evidence for dissociation between adrenarche and gonadarche: studies in patients with idiopathic precocious puberty, gonadal dysgenesis, isolated gonadotropin deficiency, and constitutionally delayed growth and adolescence. *J Clin Endocrinol Metab*. 1980;51(3):548–556.
43. Kroboth PD, Salek FS, Pittenger AL, Fabian TJ, Frye RF. DHEA and DHEA-S: a review. *J Clin Pharmacol*. 1999;39(4):327–348.
44. Shah SB. *Allergy-Hormone Links* (1st ed.). Jaypee Brothers Medical Pub; 2011.
45. Chapman K, Holmes M, Seckl J. 11beta-hydroxysteroid dehydrogenases: intracellular gate-keepers of tissue glucocorticoid action. *Physiol Rev*. 2013;93(3):1139–1206.
46. Arriza JL, Weinberger C, Cerelli G, et al. Cloning of human mineralocorticoid receptor complementary DNA: structural and functional kinship with the glucocorticoid receptor. *Science*. 1987;237(4812):268–275.

CHAPTER

14

Adrenal Hormones and the Stress Response

CHAPTER LEARNING OBJECTIVES:

- Define stress in the context of homeostasis and allostasis.
- Describe the three main components of the integrative physiological response to stress.
- Explain the endocrine mechanisms behind fetal developmental programming and the transgenerational transmission of stress.
- Describe the synergistic interactions between stress hormones of the hypothalamic-pituitary-adrenal axis and thyroid hormones in regulating the timing of amphibian metamorphosis.
- Explain the major disorders associated with the hyper- and hyposecretion of adrenal hormones.

OPENING QUOTATIONS:

"Everybody knows what stress is and nobody knows what it is."

—Hans Selye[1] The evolution of the stress concept. *American Scientist*. 1973;61(6):692–699

"Stress itself is one of those terms that we use to shield us from our ignorances. We would be better off without it. It survives because it is a convenient term to indicate the general topic under discussion. Attempts to provide such a vague concept with a precise physiological definition engender confusion and misunderstanding."

—Jeffrey Rushen[2] Some problems with the physiological concept of "stress". *Australian Veterinary Journal*. 1986;63(11):359–361

What Is Stress?

LEARNING OBJECTIVE Define stress in the context of homeostasis and allostasis.

KEY CONCEPTS:

- Hans Selye coined the term "stress" in the 1930s and defined it as the nonspecific response of the body to any demand for change.
- Modern definitions of stress often emphasize any increased energy consumption by an organism that is required to restore a disruption to homeostasis.
- The term "allostasis" describes the process of achieving homeostasis through endocrine and other physiological changes that actively promote adaptation.
- The pathological effects of stress result from the prolonged exposure to stress hormones associated with the stress response.

Art, pornography, good, and evil are all examples of concepts that are famously difficult to define with precision; nonetheless, most people recognize them when they see them. Add to this list the notion of "stress". The word stress has a variety of meanings and is used in many contexts. For instance, stress is commonly used to describe the resulting physiological and psychological changes that occur in response to an actual or perceived threat to an organism, a concept known as the *stress response*.[3] Additionally, it is often used to describe various forms of stressful stimuli called *stressors*, whose categories include environmental (e.g. predators, weather, food shortage), physiological (e.g. exertion, trauma, inflammation, starvation), and psychological (e.g. fear, anxiety, social defeat). Moreover, there is a large degree of subjectivity associated with stress, since the level of the stressor determines the magnitude of the consequential stress response, and stress is experienced differently by different individuals. For example, some people consider bungee jumping to be a positively thrilling experience, whereas others experience this event with fear and anxiety. In the 1930s–1950s, when definitions for stress were first being developed, many scientists complained about the confusion surrounding the term's usage, and in a 1951 issue of the *British Medical Journal* one physician concluded, rather tongue-in-cheek, "Stress in addition to being itself, was also the cause of itself, and the result of itself".[4] This section examines the historical underpinnings of the notion of stress and also considers some modern approaches to understanding stress. The endocrine mechanisms that mediate the stress response will then be described, as well as the physiological effects that stress hormones exert on various organs and tissues in normal and pathological states.

Stress as a Syndrome

The term "stress" was coined in the 1930s by Hungarian medical doctor and researcher Hans Selye (Figure 14.1),

DOI: 10.1201/9781003359241-19

Figure 14.1 Early pioneers of the stress response. (A) Endocrinologist Hans Selye popularized the idea of stress, and his experiments with rats showed that persistent stress could cause these animals to develop various diseases similar to those seen in humans. **(B)** Walter Bradford Cannon coined the terms "homeostasis" and "fight or flight". He posited that epinephrine plays a central role in maintaining physiological equilibrium in the long term when "emergency situations" threaten to disrupt it in the short term.

Source of images: (A) Ribatti, D., 2019. Inflamm. Res., 68(2), pp. 177-180; Used by permission. (B) Davies, K.J., 2016. Mol. Aspects Med. 49, pp. 1-7. Used by permission.

Table 14.1 Pathological Effects of Long-Term Stress Responses

Acute Stress Response	Pathological State
Shift from energy storage to use	Fatigue; myopathy; steroid diabetes
Increased cardiovascular tone	Hypertension
Inhibited digestion	Peptic ulcers
Inhibited growth	Psychosocial dwarfism
Inhibited reproduction	Impotence; anovulation; loss of libido
Altered immune function and inflammatory responses	Immune suppression; cancer
Enhanced cognition	Accelerated neural degeneration during aging
Enhanced analgesia	Weight loss

who borrowed the word from physics. Selye defined biological stress as "the non-specific response of the body to any demand for change".[5] Selye would go on to add, "whether it is caused by, or results in, pleasant or unpleasant conditions", and termed "positive" stress *eustress* and "negative" stress *distress*. As a medical student, Selye observed that patients suffering from different diseases often exhibited many identical signs and symptoms. At the time, it was believed that all disease symptoms were unique to specific pathogens. For example, all symptoms of tuberculosis were due to the bacterium *Mycobacterium tuberculosis*, all symptoms of syphilis by the spirochete *Treponema pallidum*, and so on. However, in seminal experiments with rats, Selye demonstrated,

> if the organism is severely damaged by acute non-specific nocuous agents such as exposure to cold, surgical injury . . . excessive muscular exercise, or intoxications with sublethal doses of diverse drugs . . . a typical syndrome appears, the symptoms of which are independent of the nature of the damaging agent or the pharmacological type of the drug employed, and represent rather a response to damage as such.[6]

Therefore, regardless of the type of insult that Selye subjected his rats to, they all exhibited many of the same nonspecific pathological changes (Table 14.1), such as enlargement of the adrenal glands and immune suppression. He also demonstrated that persistent stress could cause these animals to develop various diseases similar to those seen in humans, such as heart attacks, stroke, kidney disease, and rheumatoid arthritis. Selye proposed the existence of a generalized physiological syndrome that occurs in response to a great diversity of threats to the integrity of the organism, describing the process of coping with stressors as a multistaged *general adaptation syndrome* (GAS).[7]

Integrating Stress with Homeostasis

The concept of stress is inextricably linked with that of **homeostasis**, the actively maintained constancy of the internal environment. To maintain steady states, the homeostatic process is self-limiting and incorporates negative feedback to return physiological parameters to a resting state. Both stress and homeostasis can ultimately be viewed as biological states, the latter representing the optimal physiological state or "comfort zone", and the former the state that arises when homeostasis is disrupted.[8] Most modern definitions of stress take homeostasis into account. For example, Chrousos and Gold define stress as "a state of disharmony, or threatened homeostasis", and the stress response as an "attempt to counteract the effects of the stressors in order to reestablish homeostasis".[9] Organisms can be thought of as occupying a multidimensional physiological space in which they have optimal zones or set points for key body parameters, such as blood pH, blood glucose, blood oxygen tension, temperature, metabolic activity, ion balance, and so on. When stressors cause (or threaten to cause) a significant deflection from homeostasis in one or more parameters, the organism must mount a stress response that restores homeostasis, repairs any resulting damage, and, if appropriate, recalibrates the homeostatic set points and tolerance zones in the light of the new environmental circumstances.[8] Therefore, **stress** can be broadly defined as any state of threatened or actual disruption to homeostasis and the resulting increased energy consumption by an organism that is required to restore or recalibrate the homeostatic state. Importantly, whereas in the short term the stress response is adaptive and enhances an organism's ability to cope with emergency situations, a prolonged stress response tends to be maladaptive.

Curiously Enough . . . The hypothesis of homeostasis was first described by French physiologist Claude Bernard in 1865, but the word itself was coined by Walter Bradford Cannon in 1926 (Figure 14.1). Cannon's impetus for coining the term was to demonstrate the critical role that the adrenal medullary hormone, epinephrine, plays in maintaining physiological equilibrium in the long term when "emergency situations" threaten to disrupt it in the short term.[10,11] According to Cannon, the brain responds to all emergencies in the same way, by evoking increased secretion of epinephrine via the "sympatho-adreno-medullary" system. Cannon also coined the phrase, "*fight or flight*",[12] asserting that not only physical emergencies, but also psychological crises evoke release of epinephrine into the bloodstream.

Allostasis and Allostatic Load

You were introduced to the notion of allostasis in Chapter 2: Fundamental Features of Endocrine Signaling (Figures 2.14–2.15), and a review of that content is recommended before proceeding further. Although homeostasis refers to a physiological state of equilibrium, the term does not take into account the actual processes or energetic costs that are used to achieve that state. For example, consider the maintenance of one homeostatic variable, body temperature, by experimental rats raised under two conditions: room temperature and exposure to cold temperature for several days. In both circumstances the rats' body temperatures are similar and maintained at the optimal homeostatic state. The key difference between the two scenarios is that the maintenance of homeostasis in the cold-stressed rats comes at a much larger energetic expense compared with that of the non-cold-stressed controls; that is, they must increase heat production through shivering and non-shivering thermogenesis, as well as minimize heat loss via piloerection and behavioral changes. Furthermore, the classical concept of homeostasis assumes a return to a constant physiological set point. However, there are many instances when set points clearly do not remain constant but operate at an elevated or reduced level throughout the day (e.g. circadian cycling of hormones), season (e.g. physiological changes associated with hibernation), or life history stage (e.g. changes in food intake, osmoregulatory processes, and metabolism that lactating females undergo). Indeed, the stress response itself is accompanied by transient changes in homeostatic set points that promote elevated heart rate, blood pressure, and generation of stress hormones.

Some useful terminology highlights the physiological mechanisms and energetic costs of reestablishing or recalibrating homeostatic set points and also the notion that homeostatic set points can be variable. These are:

Allostasis (**"stability through change"): The *process* of achieving homeostasis through physiological change that actively promotes adaptation.**[13,14]

Allostatic mediators: **The modulators of the stress response whose actions regulate homeostasis. These include neural and endocrine regulators and their target tissues.**

Allostatic load: **The physiological *effort* (energetic cost) expended when a regulatory system is in a state of allostasis.**[3] **This can also be considered the cumulative wear and tear on the brain and body caused by the allostatic state.**

Allostatic overload: **A state when the energetic expenditure of the allostatic load exceeds energetic availability, resulting in pathology such as mental or physiological dysfunction.**

From this perspective, homeostasis refers specifically to the maintenance of the physiological parameter in question (e.g. blood pressure), whereas allostasis considers both the physiological mechanism(s) that mediate that parameter (e.g. vasoconstriction and increased heart rate mediated by epinephrine release) and the associated energetic costs required to implement these physiological changes. An allostatic state can help to overcome challenges and ensure survival either by temporarily forcing systems to function outside their normal ranges and set points or by establishing new set points to accommodate changes in life history (Figure 14.2).

Animals gradually prepare physiologically in advance of anticipated predictable environmental or social

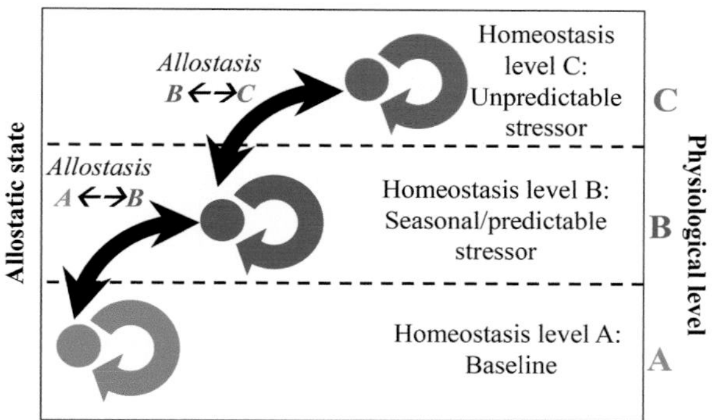

Figure 14.2 Environmental or life history demands: allostasis and homeostasis. Physiological level A represents basic physiological and behavioral processes necessary for baseline existence. Homeostatic mechanisms (level A, green arrow) operate around specific set points to maintain physiological balance (green circle). Predictable or manageable demands of the environment or life history events, such as reproduction, activate allostatic mechanisms (two-headed arrow A←→B) that raise the physiological level to B involving a new physiological balance (level B, blue circle) maintained by new homeostatic set points (blue arrow). In the face of unpredictable and potentially life-threatening events, additional allostatic mechanisms (two-headed arrow B←→C) drive the physiological level to C, and hence new homeostatic mechanisms (level C, red arrow) are required to maintain the new physiological balance (level C, red circle) to ensure survival.

Source of images: Landys, MM et al., Gen. Comp. Endocrinol., 148, 132-149, 2006.

changes, such as seasonal changes in temperature and photoperiod and preparations for migration and seasonal breeding. By contrast, responses to *unpredictable* events, such as injury, disease, and predation, generally require rapid changes in physiology and behavior both during and after the unexpected stress event. The classic "stress response" generally occurs when an unpredictable challenge is encountered and serves to increase the likelihood of survival in the short term, but may have longer-term trade-offs. As such, predictable demands of the annual cycle and demands associated with unpredictable perturbations represent different ends of an allostatic continuum to which animals respond very differently. Some parameters commonly used to measure allostatic load in humans include blood cortisol, blood pressure, body mass index (BMI), blood cholesterol, and glycosylated hemoglobin. When these parameters fall outside of the normal range for a prolonged period, this may result in chronic illnesses, such as atherosclerosis, hypertension, allergies, asthma, diabetes, myocardial infarction, obesity, autoimmune disorders, memory loss, and depression.

SUMMARY AND SYNTHESIS QUESTIONS

1. Hans Selye's proposal of a multistage "general adaptation syndrome" in response to stress is a testament to careful scientific observation and synthesis. However, Selye considered the final GAS stage, "exhaustion", to be caused by the depletion of stress hormones and the termination of the stress response. We now know that this is incorrect. Explain why.
2. How does the concept of allostasis differ from that of homeostasis?

The Integrated Stress Response

LEARNING OBJECTIVE Describe the three main components of the integrative physiological response to stress.

As a major center for the processing and assimilation of sensory information, the hypothalamus plays a central role in integrating the stress response. The integrated stress response has been classically described as having two main components that operate in parallel in response to any stressor: a **sympathetic component** (the "flight or fight response") that is mediated by the adrenal medulla and the sympathetic nervous system (Figure 14.3) and an **adrenocortical component** (the "general adaptation" response) mediated by the hypothalamus-pituitary-adrenal cortex/interrenal (HPAc/I) axis (Figure 14.4). A third component, the **neurohypophyseal component**, is activated specifically in response to hypovolemic and/or osmoregulatory stress (Figure 14.5). The latter two components are not independent, and the neurohypophyseal component can potentiate

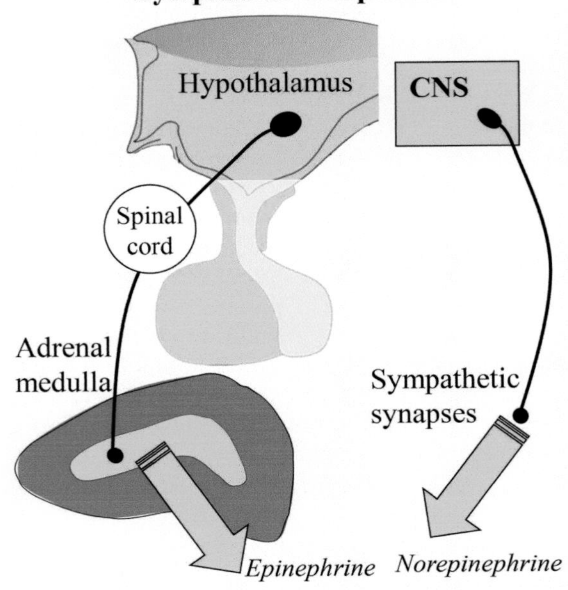

Figure 14.3 The sympathetic component of the integrated stress response. Innervation of the adrenal medulla from the hypothalamus via spinal cord ganglia stimulates chromaffin cells of the medulla to secrete primarily epinephrine into circulation. Sympathetic synapses throughout the body emanating from the central nervous system (CNS) release primarily norepinephrine, some of which diffuses out of the synaptic spaces and into circulation.

KEY CONCEPTS:

- The sympathetic component of the stress response is mediated by catecholamines released from the adrenal medulla and the sympathetic nervous system.
- The adrenocortical component of the stress response is mediated by the hypothalamus-pituitary-adrenal cortex/interrenal (HPAc/I) axis and circulating glucocorticoids.
- The neurohypophyseal component of the stress response is activated specifically in response to hypovolemic and/or osmoregulatory stress.
- The short-term biological responses to sympathetic stimulation, which include increased nutrient mobilization, cardiac, renal, and pulmonary activities, are mediated by plasma membrane-localized adrenergic receptors.
- Longer-term metabolic, developmental, and immunological aspects of adrenocortical stimulation are mediated by nuclear-localized glucocorticoid receptors.

the adrenocortical component. Upon perception of stress, the sympathetic and adrenocortical responses are activated simultaneously, though their response time courses differ, eliciting a multivariate response to stress by hormones with different temporal profiles. Catecholamines released by the sympathetic component act immediately: the adrenal medulla releases primarily epinephrine, which enters general circulation to affect distant target cells, whereas

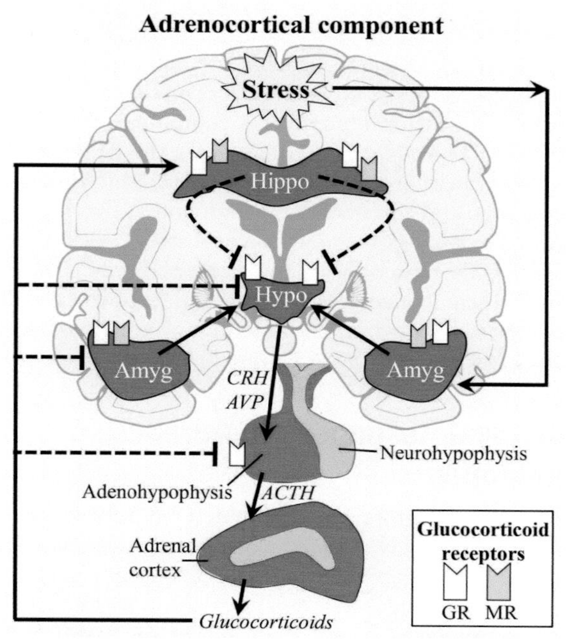

Figure 14.4 The adrenocortical component of the integrated stress response. Stress perceived by the brain's cortex stimulates the amygdala (Amyg), which in turn induces the hypothalamus (Hypo) to release CRH and AVP (of parvocellular origin), which promotes ACTH synthesis by the adenohypophysis, and ultimately induces the adrenal cortex to secrete glucocorticoids. In response to rising levels of circulating glucocorticoids, the hippocampus (Hippo) suppresses the HPA axis, inhibiting the synthesis if glucocorticoids. The negative feedback response to glucocorticoids is mediated by the glucocorticoid receptor (GR) expressed in the hypothalamus and pituitary, and by both GR and mineralcorticoid receptors (MR) in the hippocampus and amygdala.

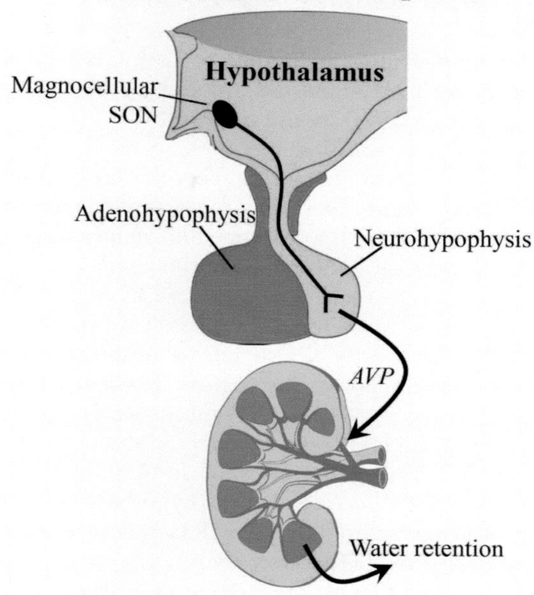

Figure 14.5 The neurohypophyseal component of the integrated stress response. This neuroendocrine pathway is only activated in response to hypovolemic and/or hyperosmotic stress. Note that in contrast to the adrenocorticoid component of the stress response where AVP is released by parvocellular hypothalamic nuclei to stimulate the adenohypophysis via local pituitary portal circulation, in the neurohypophyseal component AVP is synthesized by hypothalamic magnocellular nuclei that release the hormone at the neurohypophysis into general circulation. AVP targets the kidneys to promote water retention, which increases blood volume and blood pressure.

sympathetic ganglia release mostly norepinephrine, which acts on target cells at the point of release. In contrast, the actions of glucocorticoids released by the adrenal cortex begin to take effect about 5–30 minutes later in many vertebrates, and also have longer-lasting (hours–days) effects. Both components act to promote physiological and behavioral actions that remove the animal from the stressful situation and/or facilitate coping.

The Sympathetic Component

Although 80% of adrenal medullary catecholamine secretion is epinephrine, epinephrine constitutes only 10% of all circulating catecholamines, suggesting that the majority of circulating catecholamines are of extra-adrenal origin,[15] most likely from postganglionic sympathetic nerve terminals (Figure 14.3). Thus, at any given time, target cells are exposed to circulating catecholamines derived from both neural and glandular sources.

The Adrenocortical Component

The adrenocortical component (Figure 14.4) is a particularly prominent and complex part of the stress response. Following a stressful stimulus, the hypothalamic peptide, corticotropin-releasing hormone (CRH), is secreted from parvocellular paraventricular nuclei (PVN) into the portal system surrounding the median eminence. CRH subsequently induces the corticotropes of the adenohypophysis to release adrenocorticotropic hormone (ACTH) into the systemic blood, which stimulates the synthesis and release of glucocorticoids from the adrenal cortex/interrenal gland. Glucocorticoids act via negative feedback to inhibit the stress response. In short, the forward loop prepares the organism to anticipate and respond optimally to a threat, while the feedback loop ensures returning efficiently to a homeostatic balance when it is no longer challenged. Importantly, note that HPA-axis activity is also regulated by other regions of the brain, including the hippocampus, amygdala, and the prefrontal cortex (Figure 14.4). In particular, the hippocampus contains high concentrations of both the glucocorticoid receptor (GR) and the mineralocorticoid receptor (MR) and has an important inhibitory role on both basal HPA-axis activity and termination of the stress response.[16] Recall from the previous chapter that glucocorticoids have high affinity to both the GR and the MR. In contrast to osmoregulatory tissues, such as the kidney, whose MRs are protected from activation by glucocorticoids due to the expression of the cortisol-inactivating enzyme *11-beta-hydroxysteroid dehydrogenase*

II (11β-HSD II), the hippocampus expresses *11-beta-hydroxysteroid dehydrogenase I* (11β-HSD I), which promotes the conversion of cortisone to the more active cortisol and promotes its binding to both GRs and MRs. Therefore, both MRs and GRs modulate the glucocorticoid-mediated stress response pathway.

Importantly, in addition to CRH, in most vertebrates vasopressin- and oxytocin-like peptides also stimulate ACTH release (Figure 14.4). This important nuance is often a source of confusion for students, and rightly so as vasopressin was previously described in this textbook as being a *neurohypophyseal* hormone produced by *magnocellular* nuclei. However, in mammals a distinct subset of *parvocellular* cells of the PVN also release arginine vasopressin (AVP) into the portal vasculature associated with the median eminence, and this AVP subsequently acts synergistically with CRH on the adenohypophysis corticotropes to control ACTH secretion[17–19] (Figure 14.6). Indeed, early studies proposed that AVP was the hypothalamic corticotropin-releasing factor,[20] though CRH later proved to be a more powerful ACTH inducer than AVP alone[21] and has since been shown to play a major role in the regulation of the HPA axis.[22] Similar to mammals, arginine vasotocin (AVT) promotes ACTH release in birds, and in the rainbow trout, *Oncorhynchus mykiss*, CRH and arginine vasotocin also synergize to release ACTH.[23]

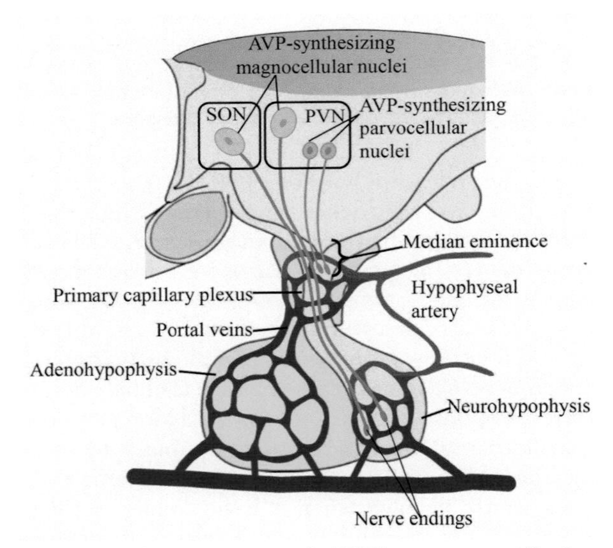

Figure 14.6 Arginine vasopressin (AVP) neurons and the hypothalamic-pituitary system. Axons from magnocellular neurons in the paraventricular nucleus (PVN) and the supraoptic nucleus (SON) of the hypothalamus terminate in the neurohypophysis, which receives blood supply from the inferior hypophyseal artery forming "the hypothalamo-neurohypophysial system". AVP is synthesized in the body of the neuron, is transported through the nerve axons, and accumulates in the nerve terminals in the neurohypophysis. Parvocellular AVP neurons in the PVN, on the other hand, project their neuronal axes to the portal capillary plexus juxtaposed to the median eminence. The hypothalamo-neurohypophysial portal vessels transfer released AVP from the median eminence to the adenohypophysis, where AVP stimulates ACTH release from corticotropes.

The Neurohypophyseal Component

The release of AVP into general circulation by the neurohypophysis can function as a third component of the stress response, the **neurohypophyseal component** (Figure 14.5 and Figure 14.6). Note that in this case the AVP is of *magnocellular* nuclei origin. However, in contrast to the classical sympathetic and adrenocortical components that are always activated in response to any perceived stress, AVP of neurohypophyseal origin appears to be released only in response to two specific categories of stress: (1) hypovolemic stress resulting from dehydration or blood loss, and (2) osmoregulatory stress manifesting as elevated blood ion content.

Biological Responses to Adrenergic Stimulation

Adrenergic receptors are distributed broadly throughout the body, and a given tissue's response to circulating epinephrine and norepinephrine will depend not only on the catecholamine concentrations, but also on the types and relative numbers of receptors present in the tissue. Some of the distributions and physiological effects mediated by different adrenergic receptors are summarized in Figure 14.7. Note that similar types of tissues may respond in opposite ways to the same catecholamine because they express different receptors. For example, smooth muscle surrounding the bronchioles of the lungs, rich in β2 receptors, responds to epinephrine by dilation, whereas smooth muscle associated with veins and lymphatic vessels, rich in α2 receptors, undergo vasoconstriction. Responses to adrenergic stimulation generally revolve around one central theme: the mobilization of energy resources to cope with a stressful situation that threatens to disrupt physiological homeostasis and survival. Two broad categories of responses to adrenergic receptor stimulation will be considered.

Cardiovascular, Renal, and Pulmonary Responses to Adrenergic Stimulation

In the heart, the predominant adrenoreceptor type in number and function is β1. These receptors primarily bind norepinephrine that is released from sympathetic adrenergic nerves, as well as epinephrine and norepinephrine of adrenal origin that circulate in the blood. The binding of these catecholamines to β1 receptors in the heart results in elevated heart rate (*chronotropic effect*) and strength of contraction (*ionotropic effect*), together resulting in increased cardiac output and blood pressure (Figure 14.7). Epinephrine also binds to β1 receptors in the kidney's renal juxtaglomerular cells, increasing renin synthesis and further increasing blood pressure. Simultaneously, α2 receptors in the veins promote vasoconstriction, reducing blood perfusion to the lungs, increasing the rate of venous return to the heart, and further increasing cardiac output. In addition, β2 receptors in smooth muscle surrounding the pulmonary bronchioles promote bronchodilation that facilitates gas exchange. β2 receptors in the arterioles of skeletal and cardiac muscle, as well as those in the liver,

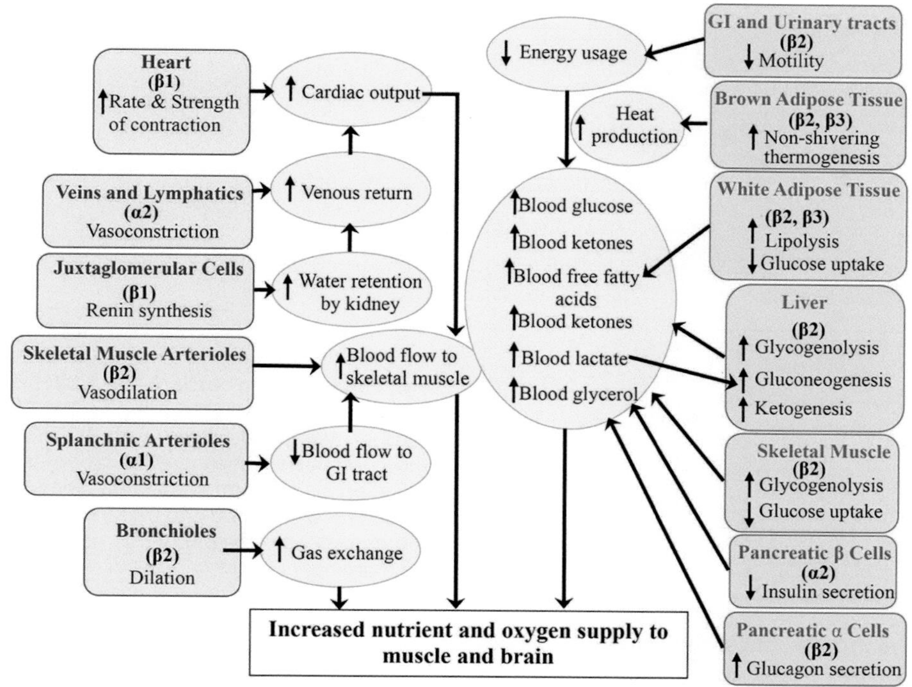

Figure 14.7 Summary of physiological responses of mammals to adrenergic stimulation. Two categories of responses are shown: on the left (pink boxes) are cardiovascular actions of catecholamines, and on the right (green boxes) are metabolic actions. In each box, the primary adrenergic receptor (α1, α2, β1, β2, β3) responsible for the actions is shown. The center ovals (yellow) denote the specific final physiological effects whose combined actions serve to increase nutrient and oxygen supply to brain and muscle.

Source of images: Porterfield SP, White BA: Endocrine Physiology, 3rd ed. Philadelphia, Mosby, 2007.

vasodilate and divert blood to some of the organs that need oxygen and nutrients the most. In contrast, blood flow to organs that are not of immediate importance (gastrointestinal tract, skin, and kidneys) is reduced via vasoconstriction of the splanchnic arterioles rich in α1 receptors.

Metabolic Responses to Adrenergic Stimulation

The diverse metabolic effects of adrenergic receptor stimulation are summarized on the right side of Figure 14.7. In general, epinephrine and norepinephrine promote the mobilization of stored carbohydrates and fats to provide readily available energy to fuel muscular work. Specifically, epinephrine increases blood glucose levels by promoting increased hepatic **glycogenolysis**, or the breakdown of stored glycogen polymers into glucose monomers, and increased **gluconeogenesis**, or the synthesis of glucose from non-carbohydrate substrates, both mediated by β2 receptors. Epinephrine also stimulates glycogenolysis in skeletal muscle, but in contrast to the liver the glycogen is not released as glucose into the blood but rather broken down locally via anaerobic glycolysis, which provides a rapid source of ATP for the muscles. An additional source of hepatic energy promoted by the binding of epinephrine to β2 receptors is the synthesis of **ketone bodies** derived from increased fatty acid catabolism. Ketones are additional fuels that can be used by the brain and other organs.

In addition to blood glucose, catecholamines also raise blood fatty acid levels by promoting lipolysis within adipocytes (fat storage cells), a process mediated by both β3 and β2 receptors. Since β3 receptors have a higher affinity to norepinephrine than epinephrine, norepinephrine plays an important role in stimulating lipolysis in β3-rich adipocytes. Thus, whereas epinephrine tends to mediate glycogenolysis and increased plasma glucose levels, norepinephrine is primarily responsible for the plasma-free fatty acid rise due to the lipolysis in the adipocytes. Although blood fatty acids can be metabolized by muscle, the brain cannot metabolize this fuel, and it must instead rely on blood glucose and ketones for energy.

Curiously Enough . . . A neurotransmitter closely related to norepinephrine, called *octopamine* (it was first discovered in octopi), is present in arthropods and mollusks. In response to stressful situations, octopamine mobilizes energy stores from fat bodies for release into the hemolymph (invertebrate blood).

Responses to Glucocorticoids

Glucocorticoids influence diverse aspects of development and physiology (Figure 14.8). The primary glucocorticoid

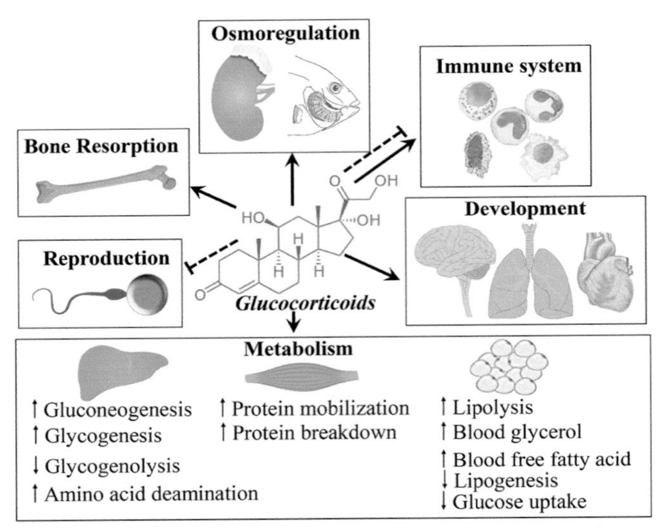

Figure 14.8 A summary of physiological responses to glucocorticoid stimulation in vertebrates.

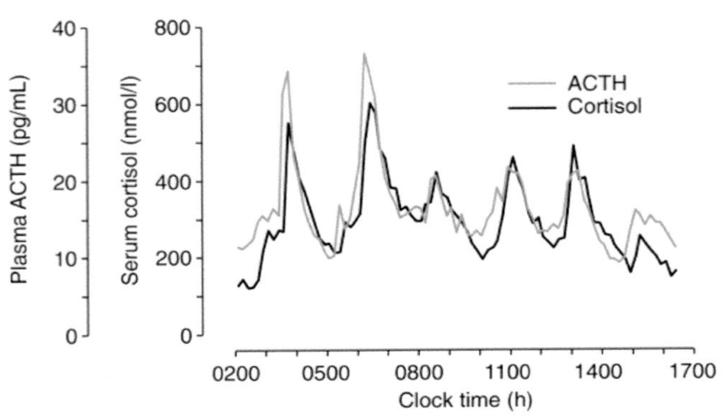

Figure 14.9 Human cortisol and ACTH dynamics throughout a day in a healthy male. Ultradian ACTH and cortisol oscillations are tightly correlated, with ACTH leading cortisol. ACTH and cortisol were measured in blood samples collected at 10 min intervals using an automated blood sampling system. Light was off between 2300 and 0700. Note the oscillations in secretion as well as the major surge near the time of awakening in the morning.

Source of images: Spiga, F., et al. 2011. Compr. Physiol., 4(3), pp. 1273-1298. Used by permission.

in fishes and large mammals (including humans) is cortisol, while corticosterone is the more biologically active glucocorticoid in amphibians, birds, reptiles, and small mammals. Glucocorticoids are essential for life, but, remarkably, we do not fully understand why. The *pharmacological actions* of glucocorticoids, or the responses following administration of concentrations that exceed normal physiological levels, are far better characterized than are their *biological actions* under physiological levels. Compared with catecholamine actions, glucocorticoids act over a longer time frame to prepare the organism to face adversity. Glucocorticoid receptors (GRs) are expressed ubiquitously. While originally named for their ability to stimulate glucose formation in the liver, glucocorticoids exert a variety of effects on target cells and play important roles in development, physiology, and metabolism, especially as pertains to the stress response. In humans, levels of circulating cortisol vary dramatically with time of day, with a major burst of activity at about 8:00 am (Figure 14.9). Glucocorticoids are also potently induced by stress. Most of the effects of glucocorticoids are *permissive*, meaning that they do not solely initiate these processes, but their presence is required to permit the full response by inducing or increasing the activity of enzymes or modulating the action of other hormones. The direct effects of corticosteroids are often difficult to separate from effects of other hormones, in part due to the permissive action of low levels of corticosteroid on the effectiveness of other hormones, such as catecholamines, glucagon, and thyroid hormones. Nonetheless, glucocorticoids have been shown to directly affect diverse physiological systems, two of which will be considered below.

Anti-Inflammatory and Immunosuppressive Actions

Inflammatory and immune responses are often components of the stress response, with their combined effect being to increase blood flow to a damaged region in order to more effectively transport nutrients and immune cells to eliminate the initial cause of cell injury, clear out necrotic cells and tissues damaged both from the original insult and also by the inflammatory process itself, and initiate tissue repair. However, if not held in homeostatic check, these responses have the potential to cause significant harm in several ways:

1. *Low blood pressure*: The inflammatory response to injury consists of an increased local permeability of capillaries that promotes localized *edema*, or swelling due to the build-up of extracellular fluid and white blood cells. Excess swelling caused by the loss of too much fluid from the blood to extracellular sites can lead to a lethal drop in systemic blood pressure. *Anaphylaxis*, also called anaphylactic shock, is a serious allergic reaction commonly caused by insect bites and stings, foods, and medications whose rapid onset may cause blood pressure to plummet, potentially leading to coma or death.
2. *Airway constriction*: Excess swelling of tissues surrounding the upper airway (trachea) can restrict oxygen uptake, causing suffocation and respiratory distress.
3. *Autoimmunity*: An autoimmune disorder is an aberrant immune response whereby the immune system attacks healthy body tissues. Although the ultimate cause of autoimmune disorders remains unknown, an unchecked immune response may promote or exacerbate the development of autoimmune diseases such as rheumatoid arthritis.

A key role of glucocorticoids is to maintain immune system homeostasis, in part by repressing a full inflammatory

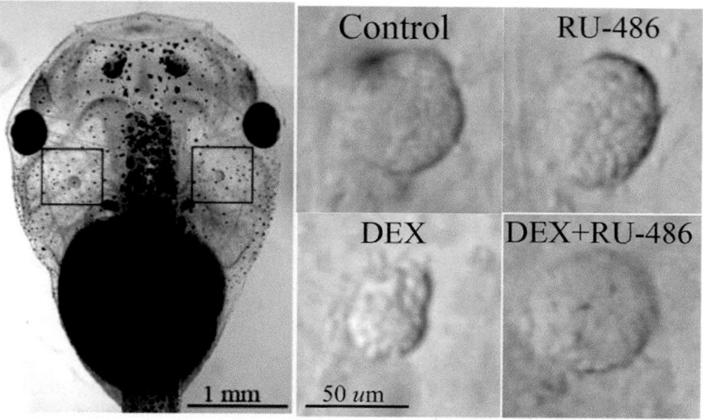

Figure 14.10 Regression of tadpole thymus glands following treatment with the glucocorticoid analog, dexamethasone, is rescued by concurrent treatment with the glucocorticoid receptor antagonist, RU-486. Seven-day-old *Xenopus laevis* tadpoles (left panel) were treated with dexamethasone (2μM) or RU-486 (150 nM), either separately or in combination, for six days. Boxes (left panel) on tadpole head denote location of each thymus gland. Right panel depicts representative high magnification images of thymus glands following treatments.

Source of images: Schreiber (2011) Adv Physiol Educ 35: 445–453. Used by permission.

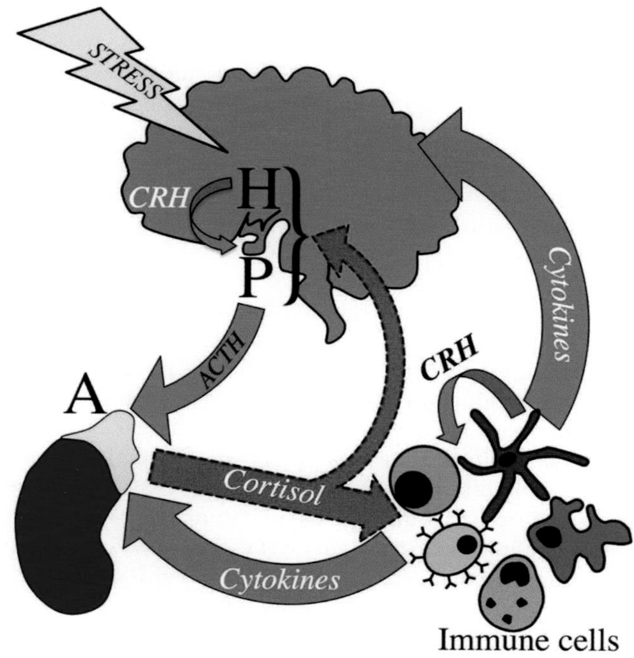

Figure 14.11 The neuroendocrine glucocorticoid stress response pathway exhibits complex interactions with the immune system. The green arrows denote stimulatory signals, and the red arrows denote inhibitory signals. H, hypothalamus; P, pituitary gland; A, adrenal gland; CRH, corticotropin-releasing hormone; ACTH, adrenocorticotropic hormone.

Source of images: Schreiber (2011) Adv Physiol Educ 35: 445–453. Used by permission.

response via the suppression of pro-inflammatory cytokine synthesis and the promotion of anti-inflammatory cytokine synthesis. As such, glucocorticoids and their pharmacological analogs are routinely used as *immunosuppressants* following organ transplants and to treat various autoimmune disorders and inflammatory conditions. Glucocorticoids induce rapid apoptosis in lymphatic tissue,[24] and the regressive effects of glucocorticoids on the thymus gland, a primary lymphoid organ of the immune system that specializes in the maturation of T lymphocytes, can be easily visualized in experiments with tadpoles where treatment with dexamethasone (a synthetic glucocorticoid) shrinks the thymus gland (Figure 14.10). This effect is rescued by concurrent treatment with mifepristone (also called RU-486), an antagonist of both glucocorticoid and progesterone nuclear receptors.

The immune response is modulated by a complex network of bidirectional signals among the nervous, endocrine, and immune systems (Figure 14.11). For example, **cytokines**, which are small cell-signaling proteins, such as interferons, interleukins, and growth factors secreted by the glial cells of the nervous system and by numerous cells of the immune system, stimulate CRH production by both the hypothalamus and immune cells themselves. CRH has peripheral pro-inflammatory effects mediated by immune cell-localized CRH receptors that stimulate immune activity.[25] However, CRH also exerts central anti-inflammatory effects by promoting glucocorticoid synthesis by the adrenal cortex,[26] which inhibits cytokine production in a classic negative feedback loop. Cytokines from immune cells also stimulate the direct production of cortisol by the adrenal cortex, further inhibiting the immune system. Virtually all immune cells have receptors for one or more of the hormones that are associated with the HPA axis, as well as for catecholamines produced by the sympathetic stress response.[27] At the molecular level, the best characterized mechanisms of action by glucocorticoids on the immune system are for the suppression of the inflammatory response, which manifests in at least three ways: (1) promoting the transcription of anti-inflammatory genes that inhibit the functioning of inflammatory proteins; (2) activating dormant anti-inflammatory proteins via nongenomic mechanisms; and (3) inhibiting the transcription of inflammatory proteins.[28]

Curiously Enough . . . Although glucocorticoid's anti-inflammatory actions have been the best studied and most widely reported, glucocorticoids can also exert pro-inflammatory effects in response to acute stress.[29] Examples of this include an exacerbation of inflammatory reactions initiated by mononuclear leukocytes,[30–32] increases in pro-inflammatory cytokines in the central nervous system,[30,33,34] and increased systemic trafficking of lymphocytes and monocytes.[35]

Central Nervous System

Corticosteroid levels influence human mood, behavior, electroencephalograph (EEG) patterns, memory consolidation, and brain excitability.[36] Chronic glucocorticoid treatment causes cell death in hippocampal neurons in rats (Figure 14.12), and elevated glucocorticoid in the hippocampus is thought to play a role in altered cognition, dementia, and depression in aging humans.[37,38] Patients with Addison's disease (adrenal insufficiency) are subject to apathy, depression, irritability, and psychosis, symptoms that are alleviated by glucocorticoid therapy.[39] In contrast, Cushing's disease patients who suffer from excess glucocorticoid production sometimes develop neuroses and psychoses that are reversible with the removal of excess hormone.[40] Glucocorticoids affect the central nervous system indirectly by maintaining normal plasma glucose levels and optimal cardiovascular parameters. The direct effects of glucocorticoids on the central nervous system are just beginning to be elucidated, and some of these effects, such as epigenetic imprinting, are discussed later.

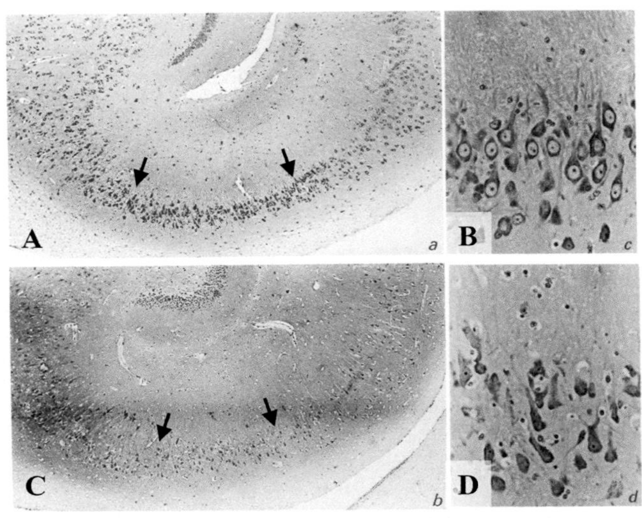

Figure 14.12 Neural degeneration in the hippocampus associated with stress. The hippocampal regions between the arrows in a normal Vervet monkey **(A)**. **(B)** A representative magnified region depicting normal pyramidal neuron morphology. **(C)** The hippocampal regions between the arrows in a Vervet monkey subjected to stress show large deficits in the number of pyramidal cells (black spots) present in the stressed monkey compared with the normal monkey. **(D)** Representative magnified region showing shrunken and sparse pyramidal neurons in the stressed animal compared with the normal monkey **(B)**.

Source of images: Uno, H., et al. 1989. J. Neurosci., 9(5), pp. 1705-1711.

Developments & Directions: Non-stressful methods for measuring stress hormones

The most direct way to measure circulating glucocorticoid levels is, of course, to analyze samples of blood. Glucocorticoids are typically extracted from blood plasma and quantified using either a radiolabeled competitive binding assay or an alternative called an *enzyme-linked immunosorbent assay* (ELISA) (described in Figures 6.13–6.14 of Chapter 6: Receptor Binding Kinetics). However, blood collection itself is a stressful event that can be a confounding variable causing an overestimation of glucocorticoid secreting activity. Furthermore, the prominent circadian cycling aspect of glucocorticoid secretion may necessitate the collection of multiple samples throughout the day, a challenging prospect when studying stress in wild animal populations. Noninvasive methods for measuring glucocorticoids from saliva (a blood product), which contains steroid hormones in proportion to their concentrations found in the blood, allow greater flexibility with reduced stress. These approaches are commonly used for collecting samples from humans (Figure 14.13[A]) and large domestic animals, such as horses and cattle. Similar methods are also even used to collect and analyze whale and dolphin "blow" (fluid sprayed out of their blowholes) samples (Figure 14.13[B]).[41] Glucocorticoids and their metabolites are relatively stable and are now routinely measured from urine and fecal samples of wildlife, such as elephant (Figure 14.13[C]),[42] for the noninvasive assessment of stress and adrenocortical function. Lifetime glucocorticoid exposure profiles have even been measured from earwax plugs extracted from blue whales.[43] Glucocorticoids have been measured from integument structures such as bird feathers[44] and mammalian fur.[45] Van Uum and colleagues explored the measurement of cortisol extracted from human hair as a biomarker for chronic stress and determined that systemic cortisol levels are consistently higher in chronically stressed individuals than in a healthy control group.[46] Remarkably, this method has even been used to assess cortisol levels in mummified human archaeological hair samples from AD 550–1532[47] (Figure 14.13[D]), demonstrating that hair cortisol analysis has the potential to be a valuable stress indicator that will permit the reconstruction of increasingly detailed life histories.

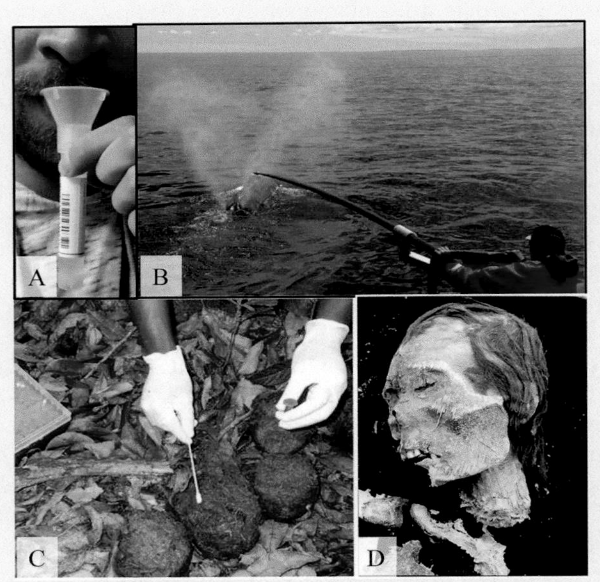

Figure 14.13 Glucorticoids can be collected and measured non-invasively from body fluids and excretions, and even from hair.

Source of images: (A) Wyoming Dept health; (B) Burgess, E.A., et al. 2018. Sci. Rep., 8(1), pp. 1-14; (C) Bourgeois, S., et al. 2019. PloS one, 14(1), p. e0210811; (D) Boutellis, A., et al. 2013. PLoS One, 8(10), p. e76818.

SUMMARY AND SYNTHESIS QUESTIONS

1. While mountain biking in Kenya's Masai Mara, you were attacked by a pack of wild dogs. You managed to escape, but are now bleeding profusely from a bite wound on one leg. Describe the signaling pathways of the sympathetic, adrenocortical, and neurohypophyseal components of the integrated stress response that have been activated in your body. Name the specific regions of the organs that secrete each hormone, the location(s) of their target(s) organs, and the relevant physiological responses that ensue.
2. Describe the role of mineralocorticoid receptors in the stress response.
3. Whereas the metabolic effects of catecholamines begin rapidly, the metabolic effects of glucocorticoids generally take longer to initiate. Why?
4. Why do you think that when physicians prescribe *prednisone* (an artificial glucocorticoid) for the treatment of inflammation, they specifically prescribe that it be ingested in the morning (as opposed to later in the day)?
5. You are a cardiologist treating a patient who suffers from both heart arrhythmia and asthma. Your primary concern is to manage the heart arrhythmia; however, you don't want to aggravate the asthmatic condition in the process. The following are some drugs available to use for treatment: epinephrine, propranolol (nonselective beta blocker), atenolol (selective beta-1 receptor blocker), LF16–0687M (selective beta-2 receptor blocker). Describe the effect of each drug would have on both heart arrhythmia and asthma (improves, worsens, or has no effect) and explain your reasoning. Which drug should be most effective for this patient, and why?
6. Adrenalectomized animals are able to function normally as long as they are administered a saline solution and food is readily available. However, adrenalectomized animals do not handle starvation well and die much sooner than adrenal gland intact animals. Why?
7. You are an ecologist studying wild porcupines and you hypothesize that porcupine populations that live close to roads experience greater levels of stress compared with populations that live far away from roads. You decide to measure "baseline stress levels" in two suitably located populations of porcupines by capturing porcupines in cage traps, collecting blood samples in the field, and measuring glucocorticoid levels later in the lab. What major challenges will you face in obtaining meaningful data?
8. Regarding Figure 14.10, how does RU-486 rescue the effects of dexamethasone treatment on the tadpole thymus gland?

Fetal Programming and the Transgenerational Transmission of Stress

LEARNING OBJECTIVE Explain the endocrine mechanisms behind fetal developmental programming and the transgenerational transmission of stress.

KEY CONCEPTS:

- During pregnancy, glucocorticoids developmentally program and adapt the fetus to the predicted postnatal environment.
- The developmental effects of glucocorticoids on fetal and postnatal developmental programming are mediated by epigenetic modifications that can manifest in offspring for multiple generations.
- Some of the most important epigenetic changes carried out by glucocorticoids include histone modifications and DNA methylation.

Fetal Programming

Glucocorticoids exert a powerful influence on growth, maturation, and tissue remodeling during mammalian fetal development. Among diverse species of mammals, a prepartum rise in glucocorticoids (Figure 14.14) stimulates a wide variety of changes in preparation for birth, including fetal lung maturation and surfactant production, brain cell differentiation and synaptogenesis, hepatic gluconeogenic enzyme expression, renal development, and maturation of the immune system.[48,49]

Curiously Enough . . . Synthetic glucocorticoids are now routinely administered to women who are at risk of preterm delivery to accelerate fetal development and improve neonatal viability.

Glucocorticoids are also key mediators between the maternal environment and the fetus, and as such are involved in adapting the fetus to the predicted postnatal environment. This phenomenon, known as *developmental programming*, reflects the concentration-dependent action of a chemical signal during a sensitive developmental period or "window" to affect the development of specific tissues, producing permanent effects that persist throughout life.[50–52] If uterine conditions are suboptimal, a premature rise in glucocorticoids produced by the fetus (e.g. due to restricted space for movement or maternal undernutrition) and/or the mother (e.g. due to chronic psychosocial stress) may reduce fetal growth and promote maturational processes that result in an altered

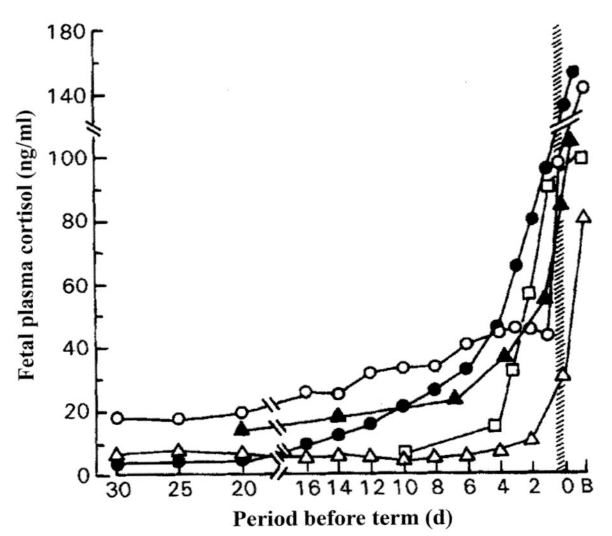

Figure 14.14 Mean fetal concentrations of plasma cortisol with respect to time before delivery in sheep (closed circles), pig (open circles), human (solid triangles), guinea pig (open squares), and horse (open triangles). Hatched vertical line represents birth.

Source of images: Fowden AL, et al. Proc Nutr Soc. 1998;57:113–122. Used by permission.

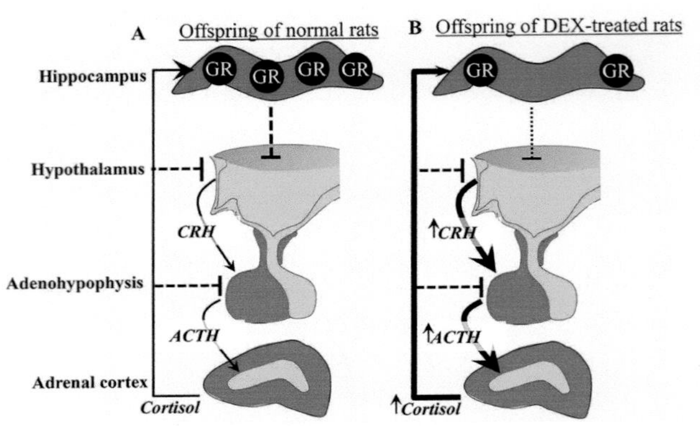

Figure 14.15 Programming of the hypothalamic-pituitary-adrenal (HPA) axis by early life stress in fetal rats. (A) In normal rats, glucocorticoid receptors (GR) in the hippocampus respond to rising levels of glucocorticoids by suppressing hypothalamic CRH secretion, bringing circulating glucocorticoid concentrations down to baseline levels. **(B)** However, in rats exposed to high levels of the synthetic glucocorticoid dexamethasone (DEX) during pregnancy, hippocampal GR concentrations are reduced, leading to a loss of feedback inhibition and an overactive HPA axis, both in the basal state and under conditions of stress.

developmental trajectory in order to adapt the fetus to an adverse postnatal environment and ensure the maximum chances of survival at birth.[53] Augmented maternal glucocorticoid deposition in bird eggs under periods of environmental stress also appears to enhance aspects of the offspring's fitness.[54]

However, if there is a mismatch between the predicted environment and the actual environment experienced postnatally, the developmental modifications can instead result in the dysregulation of metabolic function and the development of diseases in the adult, such as hypertension, ischemic heart disease, glucose intolerance, insulin resistance, type 2 diabetes, altered sensitivity of the HPA axis, and anxiety-related behaviors.[55] An example of the effects of elevated glucocorticoids during the fetal period on subsequent adult physiology is illustrated by experiments with pregnant rats administered the synthetic glucocorticoid dexamethasone. In these experiments, maternal dexamethasone treatment not only programmed the development of hypertension in the offspring, but also the negative feedback that normally turns off the stress response is blunted because of a reduction in hippocampal glucocorticoid receptor (GR) expression[56] (Figure 14.15). Similar studies have shown that glucocorticoid exposure in gestating rats is associated with decreased hippopcampal mineralcorticoid receptor (MR) concentrations and a slower return to baseline glucocorticoid levels following recovery from stress compared with controls.[57–59] Altogether, such studies suggest that late-gestational exposure to elevated glucocorticoid concentrations in rats can produce an altered "set point" for negative feedback sensitivity, permanently altering HPA axis function in response to stress, resulting in hyper-anxiety and impaired coping in stressful situations.

Another illustration of the importance of the HPA axis in mediating the effects of fetal stress is provided by studies using adrenalectomized pregnant rats.[60] Whereas the offspring of stressed adrenal intact rats exhibit an elevated stress response as adults, the stress responses of offspring from stressed adrenalectomized rats were no different than those from non-stressed controls (Figure 14.16). As such, the hypothalamic secretion of corticotropin-releasing hormone (CRH) at the top of the HPA axis has been elegantly described as a "homeostatic rheostat of feto-maternal symbiosis and developmental programming in utero and neonatal life".[61] The importance of stress hormones to the timing of mammalian birth are discussed in Chapter 24: Pregnancy, Birth, and Lactation.

Epigenetic Modifications as Mediators of Developmental Programming

Remarkably, the effects of glucocorticoid exposure and maternal stress on the fetus can manifest in offspring for multiple generations without further exposure of the F1 generation (Figure 14.17). For example, offspring of female rats administered dexamethasone during the last week of pregnancy exhibit lower birth weight and dysregulated glucose homeostasis in adulthood.[62] When these F1 animals are mated, their offspring (F2) also had lower birth weight and disrupted glucose homeostasis, even in the absence of any gestational manipulation (the F3 generation, however, is unaffected).[63] In order for such effects to be inherited, glucocorticoid exposure likely involves **epigenetic** modifications, or heritable changes in gene expression that do not affect underlying DNA sequence, that are maintained

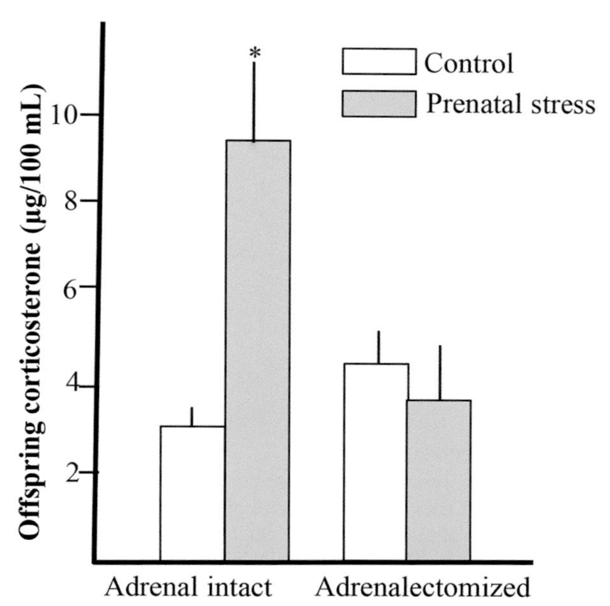

Figure 14.16 Adrenalectomy in the mother prevents the elevated stress response in offspring stressed *in utero*. In adult rats, corticosterone concentrations are elevated in response to restraint stress in rats whose mothers were stressed while pregnant. However, adrenalectomy of the mothers prevented this elevated stress response in the offspring. An asterisk (*) denotes a significant difference from the control.

Source of images: Barbazanges A, et al. J Neurosci. 1996; 16(12):3943-3949.

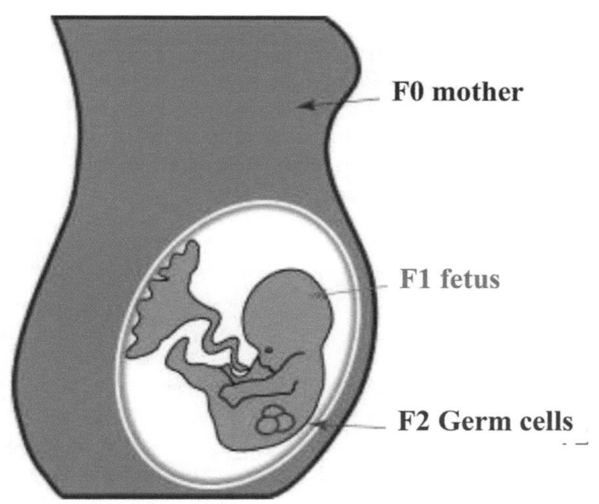

Figure 14.17 Multigenerational exposure to an environmental effect. An environmental insult during pregnancy to a mother (F0 generation) might affect not only the developing fetus (F1 generation) but also the germ cells that will go on to form the F2 generation.

Source of images: Drake and Liu (2010) Trends Endocrinol. Metab. 21:206-213. Used by permission.

through germ cell maturation. Epigenetic regulation of gene expression therefore allows the integration of intrinsic and environmental signals in the genome, thus facilitating

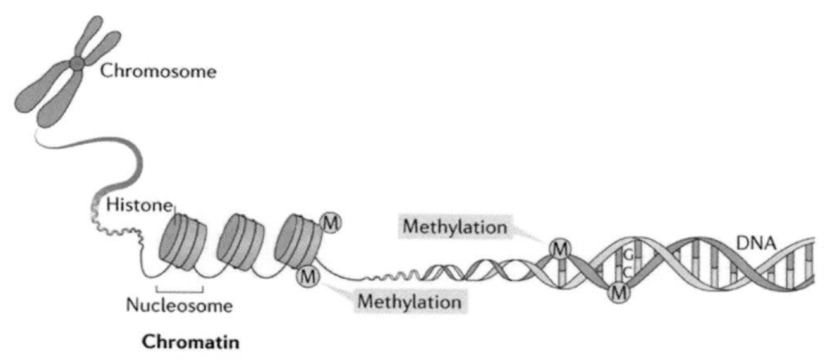

Figure 14.18 Categories of epigenetic modifications modulated by glucocorticoids include the covalent modifications of histone protein tails ("histone modifications") and DNA methylation.

Source of images: Niederberger, E., et al. 2017. Nat. Rev. Neurol., 13(7), pp. 434-447. Used by permission.

the adaptation of an organism to a changing environment through alterations in gene activity. In this way, epigenetic changes provide mechanisms by which early experiences can be integrated into the genome, conferring additional plasticity to the hard-coded genome.[64]

In general, epigenetic modifications govern the accessibility of DNA to the machinery driving gene expression such that accessible genes are actively transcribed, whereas inaccessible genes become silenced. Two of the best-understood epigenetic marks are the covalent modification of core histones that package the DNA into chromatin (discussed in detail in Chapter 5: Receptors), and the methylation of DNA at the cytosine side-chain in cytosine–guanine (CpG) dinucleotides (Figure 14.18). A basic understanding of epigenetics is essential to grasping many fundamental aspects of modern endocrinology, and this topic will be revisited in different forms throughout this textbook.

Foundations & Fundamentals: Epigenetic Mechanisms: DNA Methylation

DNA methylation by enzymes known as **DNA methyltransferases** (DNMTs) refers to the chemical transfer of a methyl group to the 5-position of cytosine rings, usually in the context of GC-rich regions (Figure 14.19). In general, patterns of DNA methylation tend to correlate with chromatin structure, with active regions of the chromatin associated with hypomethylated DNA, whereas hypermethylated DNA is associated with inactive chromatin. Most (about 85%) GC-rich DNA in mammalian genomes is located outside of promoter regions and is associated with highly repetitive non-coding DNA. However, approximately 15% of CG dinucleotides cluster within GC-rich regions known as "*CpG islands*" that locate within or around the promoters of approximately 40% of genes in the genome.[65] When the "CpG islands" associated with the promoters of such genes are methylated, this tends to inhibit gene transcription. About 70% of the promoters in the human genome are characterized by a high frequency of CpG islands and are susceptible to epigenetic modifications.[66]

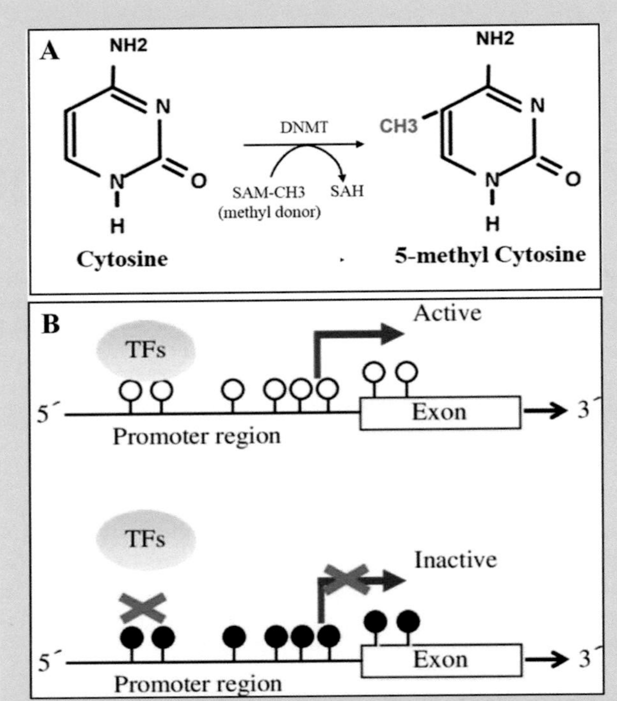

Figure 14.19 Transcriptional silencing by DNA methylation. (A) DNA methylation occurs at the cytosine bases of DNA, which are converted to 5-methylcytosine by DNA methyltransferase (DNMT) enzymes. **(B)** Methylated regions, referred to as "CpG islands", often localize within and adjacent to gene promoters, inhibiting the binding of transcription factors (TF) to DNA, and reducing or silencing transcription. Open circles denote non-methylated cytosine, and closed circles denote methylated cytosine residues.

Source of images: Mahmoud, A.M. and Ali, M.M., 2019. Nutrients, 11(3), p. 608.

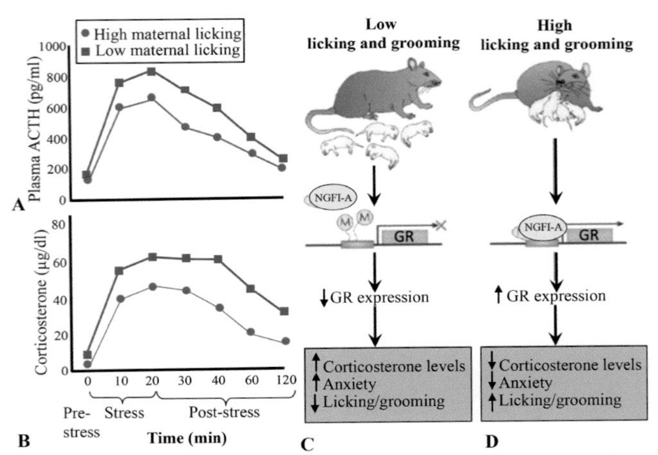

Figure 14.20 Epigenetic transmission of stress responsiveness by mother rats to their offspring. (A–B) Stress responses are reduced in offspring of rats that received maternal attention as pups. Both ACTH **(A)** and corticosterone **(B)** concentrations were lower before, during, and after a 20-minute restraint stress in rats that had received a large amount of maternal licking as pups (red circles) than in rats that had received a small amount (green circles). (C–D) Epigenetic mechanisms of stress responsiveness. **(C)** Receiving low levels of grooming results in increased methylation of nerve growth factor-inducible protein (NGFI-A) transcription factor binding sites in the promoter region of the glucocorticoid receptor (GR) gene, resulting in low expression of glucocorticoid receptor (GR) in the hippocampus. Lower levels of GR expression in the hippocampus contribute to several traits in adulthood: higher levels of baseline and post-stress glucocorticoid (corticosterone) secretion, higher levels of anxiety-like behavior and, in females, lower levels of grooming behavior towards their own offspring. **(D)** The offspring of high-grooming mothers have decreased methylation of NGFI-A transcription factor binding sites in the promoter region of the glucocorticoid receptor (GR) gene, allowing increased expression of GR in the hippocampus. In adulthood this is associated with lower levels of baseline and post-stress corticosterone secretion, low anxiety-like behavior and, in females, high levels of grooming of offspring.

Source of images: (A-B) Modified from Liu, D., et al. 1997. Science, 277(5332), pp. 1659-1662; (C-D) Federer, et al. (2009) Psychobiology and molecular genetics of resilience. Nat Rev Neurosci. 10(6): 446–457. Used by permission.

Postnatal Programming

As with the fetal environment, stress experienced during neonatal and early childhood development can also influence adult phenotypes. Animal models that use maternal separation from infants demonstrate that early life experiences have an impact on the way the adult responds to stress later in life. For example, young rats that are separated from their mothers as infants daily for extended periods of time without any human handling produces adults that are highly fearful in novel environments and show increased startle responses to sudden noise.[67] Remarkably, exactly opposite effects are found in adult animals that are removed from their mothers as infants and handled by humans for brief periods of time on a daily basis; these rats displayed reduced fearful behavior in novel environments and reduced HPA axis activation compared with their unhandled siblings.

Another well-characterized early life stress programming model examines variations in the quality of early postnatal maternal care in rats, where some rat mothers naturally display high levels of nurturing behaviors (e.g. licking, grooming, and arched-back nursing) and others display low levels of such behaviors. Interestingly, the offspring of high-nurturing mothers are less anxious and display more nurturing maternal behavior towards their own pups, and they also have attenuated corticosterone responses to stress[68] (Figure 14.20). Furthermore, these more nurtured offspring also express higher levels of glucocorticoid receptor (GR) in the hippocampus, which may promote a greater ability to suppress the HPA axis following stress. Importantly, this enhanced hippocampal GR expression is mediated in part by the transcription factor nerve growth factor-inducible protein A (NGFI-A), and pups that received little nurturing show increased methylation and decreased acetylation of the GR gene promoter at the NGFI-A binding site in the hippocampus[69] (Figure 14.20). Recall from earlier that increased

DNA methylation and reduced DNA acetylation are epigenetic changes that are associated with reduced gene transcription. These differences in epigenetic markings emerge in the first week following birth and persist into adulthood. As a result, adult offspring of low-nurturing mothers have reduced hippocampal GR expression, which contributes to the behavioral deficits that these animals exhibit and pass on to their offspring.[70]

Importantly, the transmission of stress to offspring occurs not just through direct exposures to maternal glucocorticoids and other hormones of the stress axis, but also paternally via the sperm. Specifically, a stressful paternal preconception environment has been documented to promote diverse epigenetic marks in germ cells that can be transmitted to the offspring at fertilization.[71–73] Such epigenetic alterations can be maintained in the male germ cells up to the F3 generation, even though the subsequent generations are not exposed to the initial (F0) environmental stress. Epigenetic transmission of preconception stress inflicted on the oocyte also likely occurs, but few studies have addressed this.

SUMMARY AND SYNTHESIS QUESTIONS

1. Of the two categories of stress hormones (glucocorticoids and catecholamines), only glucocorticoids are thought to induce epigenetic changes in brain development. Why?
2. You inject a drug that specifically inhibits the enzyme "DNA methyltransferase" into the hippocampus regions of the brains of half a litter of rat pups. Compared with the control rat pups, what effect would you expect the drug treatment to have on expression of glucocorticoid receptors in the hippocampus, as well as circulating levels of glucocorticoids? Explain your reasoning.

Stress and the Timing of Amphibian Metamorphosis

LEARNING OBJECTIVE Describe the synergistic interactions between stress hormones of the hypothalamic-pituitary-adrenal axis and thyroid hormones in regulating the timing of amphibian metamorphosis.

KEY CONCEPTS:

- In contrast with the mammalian model for regulation of the hypothalamic-pituitary-thyroid axis, in tadpoles it is CRH, and not TRH, that regulates thyroid hormone synthesis.
- Glucocorticoids modulate the rate of developmental progress induced by TH by increasing tissue sensitivity to TH.
- Stress and the productions of stress hormones biases the developmental trajectory towards metamorphosis.

Hypothalamic Control of the HPT Axis in Tadpoles

The primary hormone that drives the diverse developmental changes that occur during metamorphosis in amphibians and in fish is thyroid hormone, a topic addressed in Chapter 10: Thyroid Hormones: Development and Growth. The timing of the initiation of metamorphosis, however, is ultimately under control by the hypothalamus. So, what hypothalamic neuropeptide ultimately induces a tadpole to transform into a frog? Like most endocrinologists, you probably just guessed *thyrotropin-stimulating hormone* (TRH), and if so, let me be the first to forgive you for this incorrect choice. The correct choice, in fact, appears to be *corticotropin-releasing hormone* (CRH).

In juvenile and adult mammals the hypothalamic-pituitary thyroid axis and the hypothalamic-pituitary-adrenal axis each operates independently, with thyroid hormone synthesis prompted through thyroid-stimulating hormone (TSH) produced by the adenohypophysis, and TSH synthesis is promoted by the release of TRH from the hypothalamus (Figure 14.21). Notably, evidence also suggests that this model is applicable to non-mammalian adult vertebrates, including amphibians.[74–77] However, it is becoming increasingly clear that this conventional model of H-P-T axis regulation is not applicable to *larval* amphibians. Specifically, Robert Denver and his

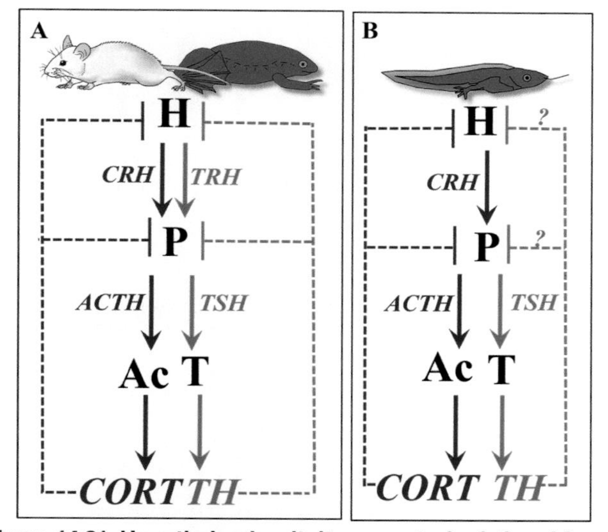

Figure 14.21 Hypothalamic-pituitary control of thyroid and adrenal function in adult and larval vertebrates. In adult amphibians, the H-P-T and H-P-A/I axes are thought to be independently regulated, similar to the model that has been established for mammals. In larval amphibians, however, CRH (but not TRH) controls both interrenal (adrenal) glucocorticoid and thyroid hormone synthesis. Additionally, both TRH and CRH can induce TSH synthesis in adult amphibians and birds. It remains unclear if thyroid hormones exert negative feedback along the H-P-T axis in tadpoles. H: hypothalamus; P: pituitary, T: thyroid gland, Ac: adrenal cortex (mammals) or interrenals (fish, amphibians, reptiles).

colleagues[77–80] have shown that injections of TRH into tadpoles neither stimulates thyroid hormone synthesis nor accelerates the start of metamorphosis, but actually delays metamorphosis (Figure 14.22). However, injection of CRH, a hypothalamic hormone previously thought to stimulate only ACTH and the stress response pathway, was found to stimulate both TSH and, indirectly, thyroid hormone synthesis, as well as induce an earlier metamorphosis. Therefore, the initiation of amphibian metamorphosis is indeed under hypothalamic control, but in contrast with the mammalian model of H-P-T regulation, in tadpoles CRH stimulates both ACTH and TSH (Figure 14.21). CRH has since been shown to stimulate TSH synthesis in fish, salamanders, and chickens.[80] Importantly, the endocrine notion that stress can accelerate the timing of metamorphosis integrates well with ecological models postulating that sudden stressful changes in environmental parameters, such as reduced food availability, evaporating water levels, increased conspecific density, and increased predation, may also promote tadpole metamorphosis.[77,81–88] Furthermore, hormones of the stress axis are now also implicated in modulating the timing of birth in humans and other mammals, a fascinating topic explored further in Chapter 24: Pregnancy, Birth and Lactation.

Glucocorticoids and Thyroid Hormones: A Synergistic Relationship in Amphibian Metamorphosis

Although thyroid hormones are required to initiate metamorphosis, the transformation cannot be completed in the absence of glucocorticoids produced by the interrenal gland (in frogs and fish the adrenal gland is called the "interrenal").[89,90] Glucocorticoids are thought to increase tissue sensitivity to TH and to modulate the rate of developmental progress induced by TH.[91,92] Furthermore, glucocorticoids may also exert direct actions on tissues independent of their effects on TH signaling that are required for the completion of metamorphosis.[90,93] Indeed, rising levels of corticosterone, the primary glucocorticoid stress hormone in frogs, peak concurrently with TH during metamorphic climax and actually begin to rise prior to the increase in TH[94] (Figure 14.23). Interestingly, rising levels of glucocorticoids during metamorphosis have been shown to synergize with thyroid hormone to promote and accelerate morphogenesis.[95,96] Using cultured tadpole tail explants, Bonnet and colleagues[97] have shown that although treatment with very low levels of T_3 alone fails to induce tail resorption, concurrent treatment with corticosterone promotes tail resorption (Figure 14.24). Kulkarni and

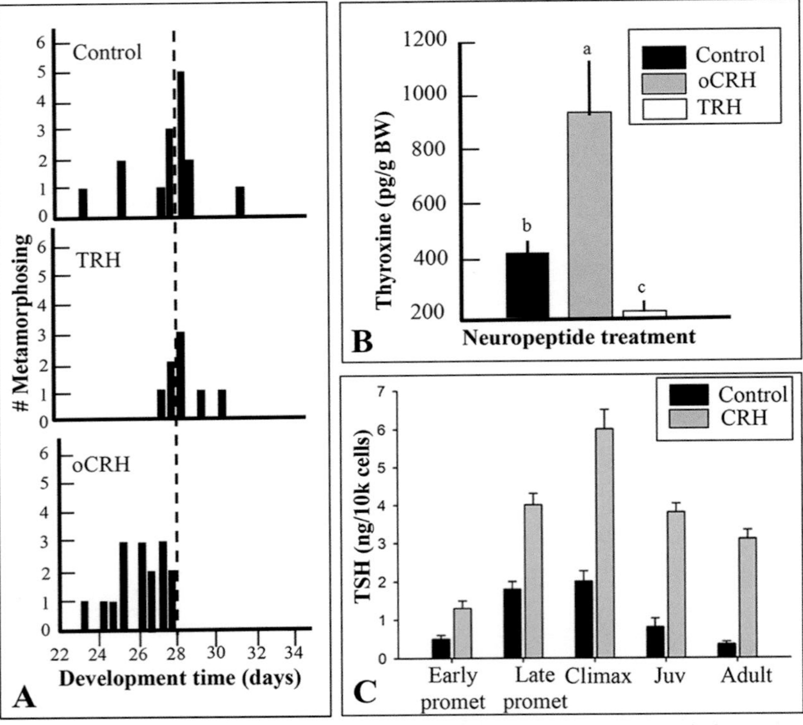

Figure 14.22 CRH, but not TRH, stimulates thyroid hormone production and metamorphosis in anuran tadpoles. (A) Frequency distribution of development time of *Scaphiopus hammondi* tadpoles injected with neuropeptides for 3 weeks. The start of metamorphosis was defined as front limb emergence. **(B)** 6-hours post-injection of prometamorphic tadpoles with neuropeptides, oCRH increased thyroxine synthesis compared with TRH. **(C)** Thyroid-stimulating hormone (TSH) secretion of dispersed tadpole or frog pituitary cells is induced with CRH treatment.

Source of images: Denver (1993). Acceleration of amphibian metamorphosis by cortocotropin-releasing hormone-like peptides. Gen. Comp. Endocrinol., 91: 38-51.

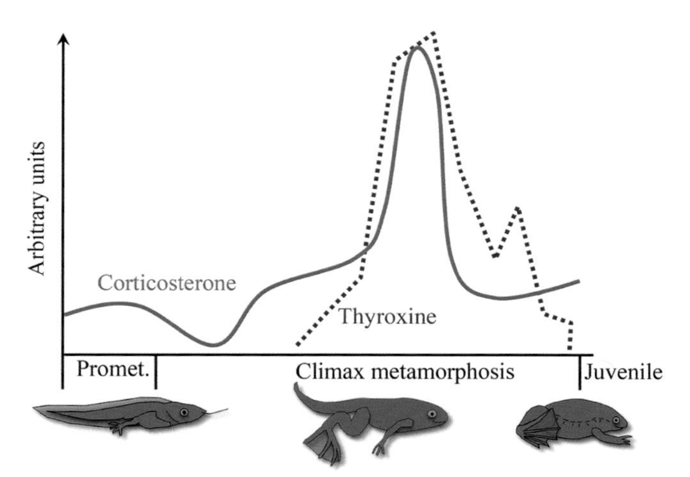

Figure 14.23 Changes in blood plasma corticosterone and thyroid hormone during *Xenopus laevis* metamorphosis. The rise in corticosterone levels during metamorphic climax precedes that of thyuroxine, and the two hormones peak concurrently. Thyroxine parameters adapted from Leloup and Buscaglia (1977); corticosterone profile from Jolivet Jaudet and Leloup Hatey, 1984.

Source of images: Leloup, J., 1977. La triiodothyronine, hormone de la métamorphose des amphibiens. Gen. Comp. Endocrinol., 56(1), pp. 59-65.

Buchholz[89] performed microarray experiments on tadpole tails treated with thyroid hormone and/or corticosterone and showed that of the 5,432 genes whose expressions were altered by either or both hormones, the majority of the differentially expressed genes were in response to co-treatment with thyroid hormone + corticosterone (Figure 14.24). At the level of the whole organism, Kuhn and colleagues[98] used Mexican axolotl salamanders (*Ambystoma mexicanum*), a species that does not metamorphose in nature, to demonstrate that low, individually submetamorphic doses of T_4 and dexamethasone were sufficient to induce complete metamorphosis when both hormones are administered concurrently. These findings emphasize the importance of considering the combined effects of TH and glucocorticoids when studying the mechanistic basis of amphibian metamorphosis. Additional mechanisms of interaction between TH and stress hormones of the HPA axis will be explored further in Chapter 24: Pregnancy, Birth and Lactation.

A Salamander's Choice: To Metamorphose or Not?

Salamanders are a diverse group of amphibians that have incorporated several developmental strategies into their life history. The most common strategy among salamanders is

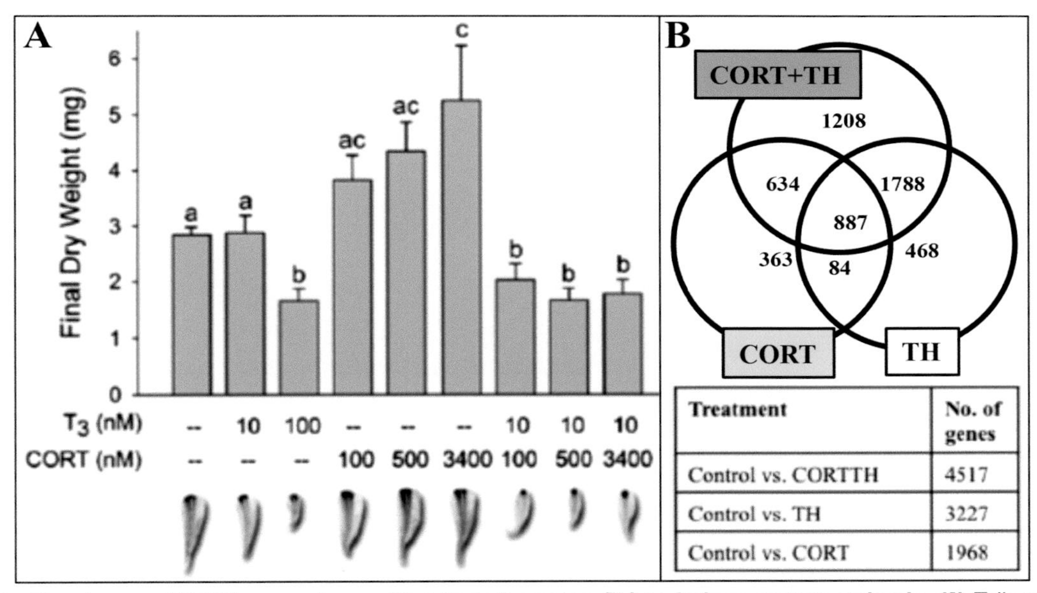

Treatment	No. of genes
Control vs. CORTTH	4517
Control vs. TH	3227
Control vs. CORT	1968

Figure 14.24 Corticosterone (CORT) synergizes with triiodothyronine (T_3) to induce metamorphosis. (A) Tails were harvested from premetamorphic *X. laevis* tadpoles and cultured for 1 week. Hormones were added at the indicated doses every 12 h. Data were analyzed by one-way ANOVA and letters indicate significant differences among the group means. Representative images of tails from each treatment group at the 7-day time point are shown below the graphs. In contrast to the synergistic effect of combined treatment with T_3 plus CORT, treatment of tadpole tail explants with CORT alone tended to increase final tail dry weight. The mechanism for this effect is unknown, but it may have functional importance with regard to predation. Tadpoles of many amphibian species exposed to predation develop larger tails (Relyea, 2007), and chronic predator presence elevates CORT in premetamorphic tadpoles (J. Middlemis-Maher, E.E. Werner, and R.J. Denver, unpublished data). **(B)** Venn diagram showing results for microarray analysis from *Xenopus tropicalis* tadpole tails treated with CORT, TH, or CORT+TH for 18 hours. The expression of 5,432 genes was significantly altered in response to either or both hormones. 1,968 genes were differentially expressed in response to CORT, 3,227 genes were differentially expressed in response to TH, and 4,517 genes were differentially expressed in response to CORT-TH.

Source of images: (A) Bonett (2010) Gen Comp. Endocrinol. 168: 209–219; Used by permission. (B) Kulkarni and Buchholz (2012). Endocrinology 153(11). Used by permission.

Developments & Directions: Corticosterones and thyroid hormones in the mediation of adaptive radiation and phenotypic plasticity

Among frogs, spadefoot toads (pelobatoid frogs) have evolved the largest differences in rates of tadpole development. Old World spadefoot toads (species in family Pelobatidae) tend to breed in long-lasting ponds and have relatively long larval periods lasting 93–186 days. By contrast, the closely related New World spadefoot toads (species in family Scaphiopodidae) are desert-dwelling amphibians that breed opportunistically in short-lived pools filled by periodic rainfall and have evolved much faster developmental rates to avoid desiccation. *Scaphiopus couchii*, in particular, has the fastest developmental rate known for any frog, with a larval period lasting 7–30 days (Figure 14.25[A]). A study by Kulkarni and colleagues[99] compared hormonal variation underlying differences in the timing of metamorphosis among three species of spadefoot toads (one Old World species, *Pelobates cultripes*, and two New World species, *Spea multiplicata* and *Scaphiopus couchii*) with different larval period durations and responsiveness to pond drying (Figure 14.25[B]). The authors found that among the three species, *S. couchii* not only had the shortest larval period, but that their rapid development corresponded with the highest whole-body contents of both thyroid hormone and corticosterone. Furthermore, in *S. couchii* these trait values were least affected by pond drying among the three species, suggesting that the rapid development of *S. couchii* evolved by genetic accommodation of endocrine pathways controlling metamorphosis. This is a remarkable demonstration of how within species *phenotypic plasticity*, or the capacity of the same genotype to produce variable phenotypic outcomes depending upon inputs received from the internal or external environment during earlier development,[100,101] may evolve into trait variation among species.

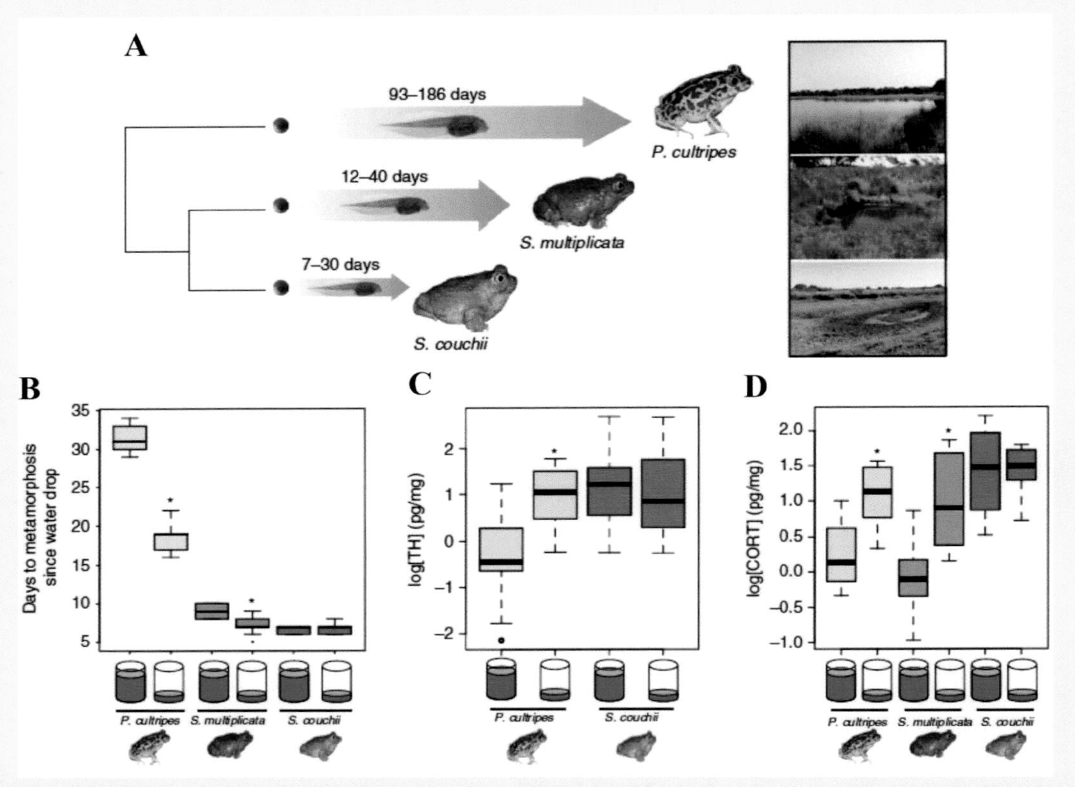

Figure 14.25 An endocrine basis for species differences in spadefoot toad larval period and plasticity. (A) Old World spadefoot toads (*Pelobates* spp.) tend to breed in long-lasting ponds and have a relatively long larval period. The closely related New World species have evolved faster developmental rates, and *S. couchii* in particular shows the fastest developmental rate known in any frog. Phylogenetic relationships among spadefoot genera with corresponding larval period durations and a typical pond for each genus are shown. **(B)** Simulated pond drying initiated at early prometamorphosis (Gosner stage 35) and maintained until measurement at metamorphic climax (Gosner stage 42) induced marked developmental acceleration in *P. cultripes* (in yellow), an intermediate response in *S. multiplicata* (in blue), and no significant response in *S. couchii* (in red). Boxes in the boxplots indicate the median and the upper and lower quartiles, whereas the whiskers indicate the minimum and maximum values excluding outliers. Developmental acceleration in *P. cultripes* in response to simulated pond drying is regulated by **(C)** increased thyroid hormone (TH; $p = 0.00019$) and **(D)** increased corticosterone (CORT; $p = 0.0027$). A similar degree of CORT increase regulating adaptive plastic responses in *S. multiplicata* was observed (d; $p = 0.00083$). However, *S. couchii* exhibited a higher and invariant tissue content of both hormones c and d underlying its minimal adaptive developmental plasticity. Simulated pond drying was initiated at early prometamorphosis (Gosner stage 35), and hormone tissue contents were assessed using tadpoles at late prometamorphosis (Gosner stage 38). Asterisks indicate statistically significant differences between tadpoles within a species raised in high vs. low water. Photo credits: I. Gomez-Mestre.

Source of images: Kulkarni, S.S., et al. 2017. Nat. Comm., 8(1), p. 993.

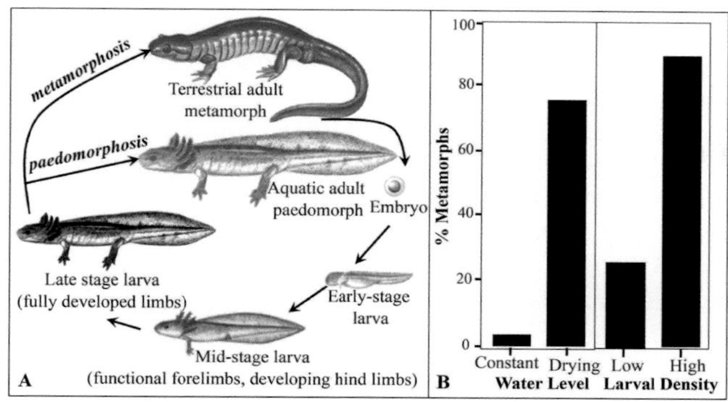

Figure 14.26 Metamorphic and paedomorphic life history strategies of salamanders. (A) Whereas some species consist entirely of either obligate metamorphs or obligate paedomorphs, others exhibit the strategy of facultative metamorph/paedomorph development, whereby an individual tadpole either undergoes metamorphosis or paedomorphosis, depending on environmental conditions. **(B)** Percentages of metamorphs observed in the facultatively paedomorphic salamander *Ambystoma talpoideum* exposed to different environmental conditions. Larvae were reared under conditions of constant or decreasing (drying) water level or high or low population density. Data are adapted from Semlitsch (1987a) and Harris (1987).

Source of images: (B) Modified from Semlitsch (1987a) and Harris (1987).

direct development, with the complete absence of a tadpole form.[102] In a second strategy, **metamorphosis**, all members of the species undergo a complete tadpole-to-juvenile transformation that is characterized in part by degeneration of larval features (gills and tailfins), *de novo* development of adult-specific structures (eyelids, skin glands), and remodeling of various tissues (skeleton, intestine, and alterations in nitrogen excretion, urogenital function, and oxygen transport), all features that accommodate the transition from an aquatic to terrestrial existence (Figure 14.26). As with all other metamorphosing vertebrates, salamander metamorphosis requires the actions of thyroid hormones. In a radically different third developmental strategy, many salamander species incorporate "paedomorphosis" into their life cycle. In the specific developmental context of salamander ontogeny, the term **paedomorphosis** describes the retention of larval morphological features (e.g. presence of gills and tail fins, and the absence of lungs) in the adult[103] (Figure 14.26). Paedomorphic salamanders neither undergo metamorphosis nor transition to a terrestrial existence but mature into breeding adults and maintain a completely aquatic life.[86] From an endocrine perspective, the loss of metamorphosis in "obligate" paedomorphic salamander species (i.e. all species members are paedomorphic) may ultimately result from dramatic alterations in thyroid hormone signaling pathways reflecting either a reduced ability to produce thyroid hormones and/or a resistance to the hormone at the receptor level.[104]

Interestingly, some salamander species are "facultative" paedomorphs, with a variable percentage of tadpoles within a species that undergo metamorphosis, and the remainder commit to paedomorphosis. In this striking example of phenotypic plasticity, each member of a facultative paedomorphic species must eventually make the fateful and irreversible "choice" to either metamorphose or not. What influences the tadpole's choice? In a word, "stress", which can be induced by environmental changes such as reduced food, lowering water levels, the presence of predators, and other variables. Experiments by Semlitsch[105] and Harris[106] in which water levels and larval densities for two species of facultative paedomorphic salamander were manipulated clearly illustrate the nature of this relationship, whereby stress and the production of stress hormones bias the developmental trajectory towards metamorphosis (Figure 14.26), similar to what was described earlier for anurans.

Curiously Enough . . . The popular pet salamander, the Mexican axolotl (*Ambystoma mexicanum*), is entirely paedomorphic in its natural environment. However, its tadpoles can be artificially induced to metamorphose if treated with exogenous thyroid hormone, producing an adult phenotype not seen in nature for thousands of years!

SUMMARY AND SYNTHESIS QUESTIONS

1. From the perspective of a tadpole, name some environmental stressors and postulate how it makes sense that glucocorticoid stress hormones would accelerate metamorphosis.
2. Compare and contrast the mammalian H-P-T model with that of larval amphibians.
3. Once metamorphosis has already initiated, the steroid hormone cortisol is known to accelerate the rate of metamorphosis in peripheral tissues. Interestingly, if cortisol is administered to tadpoles *before* their metamorphosis has initiated, this will delay or completely inhibit the start of metamorphosis. What is the physiological mechanism by which early cortisol treatment would inhibit the start of metamorphosis?

Adrenal Gland Disorders

LEARNING OBJECTIVE Explain the major disorders associated with the hyper- and hyposecretion of adrenal hormones.

KEY CONCEPTS:

- Whereas primary adrenal disorders of glucocorticoid hypo- or hypersecretion are caused by damage to the adrenal gland itself, secondary disorders are characterized by damage to the corticotropes of the adenohypophysis, and tertiary disorders by damage to the CRH-secreting cells of the hypothalamus.

- The symptoms of glucocorticoid hyposecretion disorders, such as Addison's disease, include fatigue, weight loss, and hypoglycemia.
- The symptoms of glucocorticoid hypersecretion disorders, such as Cushing's syndrome, include hyperglycemia, weight gain, fat redistribution, and skeletal muscle wasting.
- Catecholamine hypersecretion can be a serious condition that promotes hypertension.
- Post-traumatic stress disorder (PTSD) manifests as a failure to engage the biological mechanisms associated with recovery and return to physiologic homeostasis in response to stress.

Glucocorticoid and Catecholamine Hyposecretion

The modern understanding of adrenal physiology began in 1849 when Thomas Addison described patients with adrenal cortex destruction caused by tuberculosis. Worldwide, the most common cause of **Addison's disease** (also called *primary adrenal insufficiency* due to the destruction of all adrenocortical tissue) is still tuberculosis, though in Western countries autoimmune disease is now a more common cause of adrenal failure. *Secondary adrenal insufficiency* arises as a dysfunction of the HPA axis stemming from pituitary tumors or their surgical removal, and *tertiary adrenal insufficiency* may be caused by hypothalamic lesions. Since pituitary ACTH can also induce the synthesis of adrenocortical androgen precursors, patients with secondary adrenal insufficiency may display symptoms associated with reductions in both glucocorticoids and androgen precursors. However, because the adrenal gland is intact, it is still amenable to regulation by the renin-angiotensin-aldosterone system, and mineralocorticoid synthesis is not affected. Primary adrenal insufficiency, by contrast, manifests as reductions in all three major adrenal steroids, including mineralocorticoids. The importance of glucocorticoids in modulating diverse aspects of metabolism (e.g. gluconeogenesis, lipolysis, protein catabolism) is reflected by many of the physiological symptoms of glucocorticoid insufficiency, such as fatigue, weight loss, and hypoglycemia (Table 14.2).

Table 14.2 Manifestations of Abnormal Cortisol Synthesis

Hypocortisolism	Hypercortisolism
Hypoglycemia	Hyperglycemia
Increased insulin sensitivity	Insulin resistance
Hypotension	Hypertension
Weight loss Hyponatremia Impaired gluconeogenesis Fatigue Hyperpigmentation (with primary, but not secondary adrenal insufficiency)	Weight gain (fat redistribution, increased appetite) Hypokalemia Impaired wound healing Immunosuppression Protein wasting (thin skin, muscle weakness) Osteoporosis Capillary fragility

A striking feature of primary (but not secondary or tertiary) adrenal insufficiency is *hyperpigmentation* caused by the presence of elevated ACTH levels that compete for melanocortin 1 receptor (MC1R) binding in melanocytes (Figure 14.27). Recall from Chapter 8: The Pituitary Gland and Its Hormones that at normal circulating levels ACTH preferentially binds to the MC2R receptor in adrenocortical tissues. However, as a POMC-derivative ACTH contains nearly the complete α-MSH (melanocyte-stimulating hormone) sequence within it and, at high levels, ACTH can also bind to MC1R and stimulate melanin production. Another condition leading to primary adrenal glucocorticoid insufficiency, called **congenital adrenal hyperplasia** (CAH), was previously described in Chapter 13: The Multifaceted Adrenal Gland. Caused by a deficiency

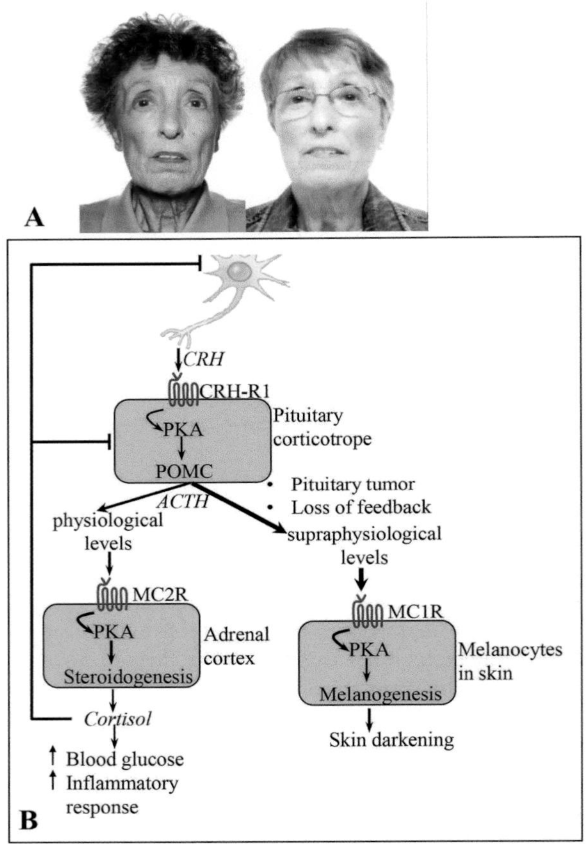

Figure 14.27 Hyperpigmentation associated with Addison's disease. (A) Hyperpigmentation representing an accentuation of normal pigmentation. A patient's facial appearance at presentation (left) versus after treatment with glucocorticoids. **(B)** Reduced cortisol secretion associated with Addison's disease promotes increased ACTH production due to negative feedback on corticotropes of the adenohypophysis. Whereas normal levels of ACTH act on the melanocortin-2 receptor (MC2R) to increase cortisol, supraphysiological levels of ACTH act on both the MC2R and the melanocortin-1 receptor (MC1R) on melanocytes and cause skin darkening.

Source of images: (A) Perros, P., 2005. PLoS Medicine, 2(8), p. e229; (B) Koeppen, B.M. and Stanton, B.A., 2009. Berne & Levy Physiology, Updated Edition E-Book. Elsevier Health Sciences. Used by permission.

in an enzyme required for the synthesis of both glucocorticoids and mineralocorticoids (e.g. *21-hydroxylase*), this disease is accompanied by reduced production of both steroid hormone classes. However, unlike classical primary adrenal insufficiency, the enzyme deficiency results in an excess of 17-hydroxyprogesterone synthesis, which promotes the hypersecretion of adrenal androgen precursors.

In contrast with adrenal cortical hyposecretion, adrenal medullary syndromes characterized by catecholamine hypofunction are not clinically significant.[15] Because the adrenal medulla accounts for only 10% of total sympathetic nervous activation, humans appear to be able to survive quite well without it.

Clinical Considerations: JFK: A strong spirit, in spite of Addison's disease

John F. Kennedy (Figure 14.28) was probably the most famous person to suffer from Addison's disease. While visiting England in 1947, Kennedy was hospitalized and diagnosed with Addison's disease in London, the same city where Thomas Addison had practiced medicine a century earlier and discovered the disease. Kennedy's adrenal glands were most likely permanently shrunken due to years of cortisone treatment for his colitis (inflammation of the inner lining of the colon).[107,108] As long as he took large daily doses of cortisone, his Addison's disease was controlled. However, if he stopped taking the steroids, even for brief periods, he would quickly develop the symptoms of Addison's disease, including fatigue, nausea, and a bronzing of the skin due to hyperpigmentation. Ironically, the prolonged treatment of Kennedy's condition with cortisone likely produced some symptoms of Cushing's disease (overproduction of glucocorticoids), such as osteoporosis, which caused a severe weakening of his backbone, sometimes making it difficult for him to walk. Despite his disease and crumbling backbone, JFK, the 35th president of the United States, outwardly maintained a vigorous spirit and a metaphorical "spine of steel" to the end of his presidency in 1963.

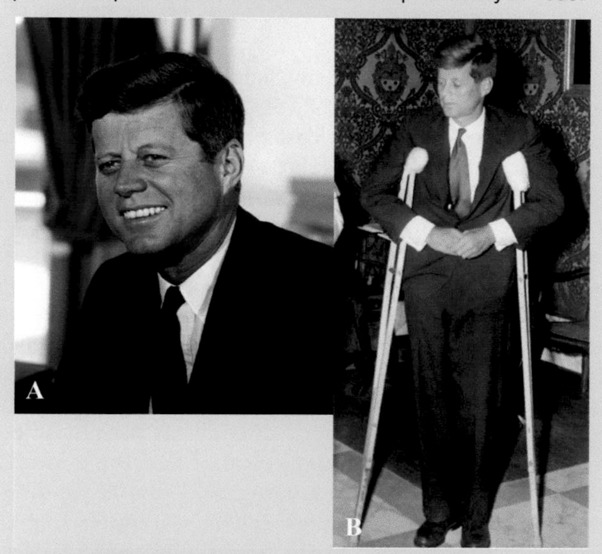

Figure 14.28 John F. Kennedy, the 35th president of the United States. **(A)** JFK's tanned appearance may have resulted from hyperpigmentation caused by Addison's disease. **(B)** JFK walks on crutches as he leaves his limousine to board the presidential yacht, "Honey Fitz", for a cruise down the Potomac River with Japanese prime minister Ikeda, in Washington on June 21, 1961 (AP). Prolonged treatment of Kennedy's condition with cortisone likely produced osteoporosis, which caused a severe weakening of his backbone, making it difficult for him to walk.

Source of images: (A) Courtesy of JFK archives; (B) Shutterstock.

Glucocorticoid and Catecholamine Hypersecretion

It is clear that in the short term the stress response is adaptive and enhances an organism's ability to cope with emergency situations and survive another day to reproduce or care for offspring. However, a prolonged chronic stress response tends to have longer-term trade-offs, as the same hormones recruited to protect the organism can instead damage it, resulting in a protection-versus-damage paradox.

Cushing's Syndrome

First described in 1932 by Harvey Cushing, *Cushing's syndrome* refers to diverse symptoms of chronic exposure to glucocorticoids. **Cushing's disease** specifically denotes Cushing's syndrome caused by a pituitary corticotrope adenoma (secondary disorder). The most prominent symptoms relate to the metabolic effects of glucocorticoids and include higher levels of gluconeogenesis, lipolysis, and protein catabolism (Table 14.2). In particular, weight gain and fat redistribution promoted by hypercortisolism may manifest as *central (truncal) obesity*, the formation of *intrascapular fat distribution* ("buffalo hump"), and a rounding of the face (*moon face*) (Figure 14.29). In contrast, *skeletal muscle wasting* (due to increased protein catabolism) causes the limbs to appear thin, with muscle weakness evident. Damage to connective tissues due to proteolysis results in *reddened cheeks* (skin thinning), the formation of purple abdominal *striae* (skin damage), and easy bruising caused by *capillary fragility*. **Osteoporosis** of the vertebrae results from glucocorticoid-induced calcium leaching. Furthermore, **hyperglycemia** and increased blood free fatty acids contribute to the development of metabolic disturbances, such as insulin resistance and other *type 2 diabetes*-like symptoms. In the case of Cushing's disease, elevated ACTH production may also be associated with excess adrenal androgen precursor secretion, and in women this may cause *hirsutism* (excess facial hair), *male pattern baldness*, and *androgenital syndrome* (clitoral enlargement).

Pheochromocytoma

In contrast with adrenal medullary hyposecretion, which does not appear to have any clinically significant manifestations, the hypersecretion of adrenal medullary catecholamines are potentially life-threatening causes of hypertension. The term **pheochromocytoma**, from the Greek words *phaios* (dark), *chromo* (color), and *cytoma* (tumor),

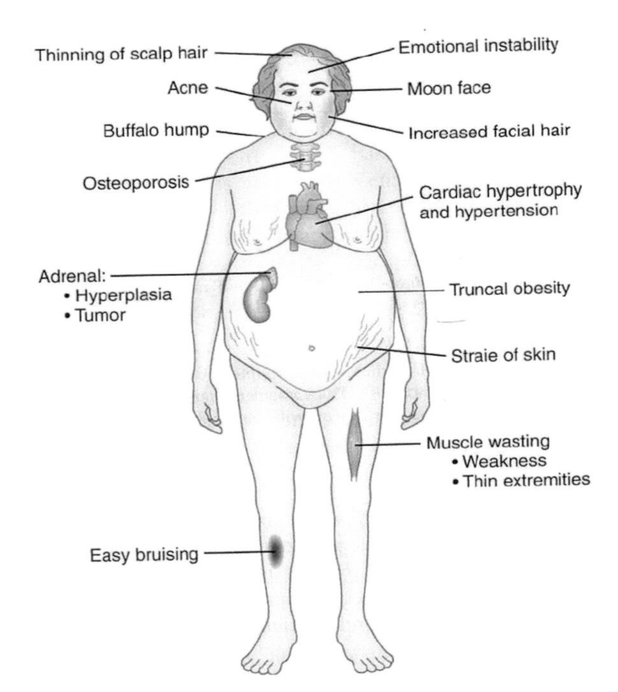

Figure 14.29 Effects of long-term hypercortisolism (Cushing's syndrome).

Source of images: Norman, A.W. and Henry, H.L., 2022. Hormones. Academic Press. Used by permission.

Table 14.3 Signs and Symptoms of Pheochromocytoma

Episodic	Chronic
Headache	Hypertension
Palpitations	Tachycardia
Sweating	Postural hypotension
Anxiety	Pallor
Nausea	Tremor
Chest/abdominal pain	Weight loss
Hypertension	Fasting hyperglycemia
	Increased respiratory rate
	Decreased gut motility

together describe the coloration that catecholamine-secreting tumors render tissues when stained with dichromium salts. Pheochromocytomas are typically located in the adrenal medulla, but may also be derived from ectopic chromaffin tissue (see Figure 13.1 from Chapter 13: The Multifaceted Adrenal Gland). The symptoms of pheochromocytoma parallel the known actions of catecholamines (see Figure 14.4), and are summarized in Table 14.3. Surgical removal of the tumor is the treatment of choice for pheochromocytoma and usually results in cure of the hypertension. Preoperative treatment to control blood pressure may include the use of alpha and beta blockers.

Post-Traumatic Stress Disorder

Post-traumatic stress disorder (PTSD) is a psychological and physiological condition resulting from exposure to a significantly traumatic event or stressor to which the

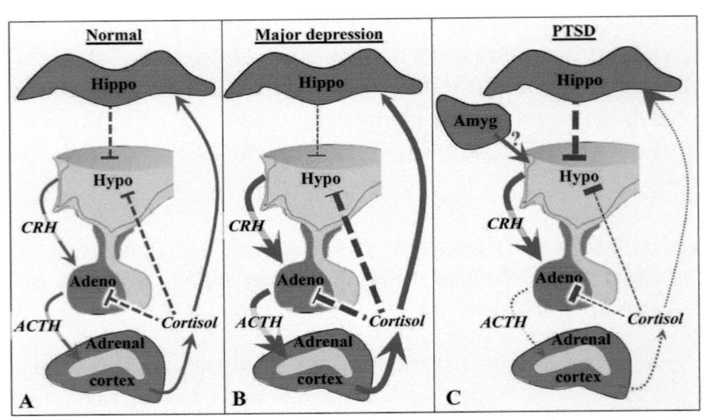

Figure 14.30 The hypothetical endocrine profiles describing the "cortisol deficit model" in PTSD patients compared with that of patients with major depression. With PTSD classically associated with increased GR responsivity along the HP axis and reduced levels of circulating cortisol, its endocrine profile is distinct from that observed in major depression. The thickness of the arrow shafts is proportional to the amount of hormone in the blood. The thickness of the arrow heads is proportional to the hormone receptor concentration in the target tissue. Pointed arrow heads (associated with solid green lines) denote a positive stimulus. Blunted arrow heads (associated with dashed red lines) denote an inhibitory stimulus. The question mark (?) denotes the hypothetical influence of increased hypothalamic stimulation by the amygdala in patients with PTSD, despite augmented inhibitory signaling by the hippocampus.

Source of images: Yehuda R. N Engl J Med. 2002;346(2):108-114.

person responded with fear, helplessness, or horror.[109] Patients diagnosed with PTSD display three distinct types of symptoms for at least one month: (1) unwanted re-experiencing of the traumatic event in the form of distressing images, nightmares, or flashbacks; (2) avoidance of reminders of the event, including persons, places, and thoughts associated with the incident; and (3) physiological manifestations of hyperarousal, such as insomnia, irritability, impaired concentration, hypervigilance, and increased startle reactions. Historically, PTSD has been regarded as primarily a psychological phenomenon, with emphases on parameters such as learning, conditioning, memory, and fear. However, research on the physiological and neuroendocrine manifestations of the condition constitute an important emerging field that is yielding a wealth of important findings and is revolutionizing the way PTSD is diagnosed, treated, and prevented.

Altered Hypothalamic-Pituitary-Adrenal Axis: The Cortisol Deficit Model

Alterations of the HPA axis are a predominant feature of PTSD pathophysiology. While there are exceptions in the literature, it appears that PTSD is associated with a uniquely paradoxical profile whereby corticotropin-releasing hormone (CRH) levels are increased, while urinary and plasma levels of cortisol are lower or at least not elevated, compared to non-exposed persons without PTSD[110,111] (Figure 14.30). This pattern differs from the

Table 14.4 Summary of PTSD Neuroendocrine Parameters from Most Studies

- Hyperactive amygdala in response to emotional stimuli
- Lower hippocampus volume
- Lower cortisol levels
- Increased CRH levels in cerebrospinal fluid
- Increased glucocorticoid receptor number in hippocampus
- Increased cortisol suppression (negative feedback inhibition) in response to low dose of dexamethasone
- Increased catecholamine/cortisol ratio

patterns associated with brief and sustained periods of stress and with major depression, which are typically associated with increased levels of both cortisol and corticotropin-releasing factor. Importantly, lowered cortisol in PTSD is not thought to be caused by adrenal or pituitary insufficiency but rather by enhanced responsiveness of the hippocampus, hypothalamus, and pituitary to glucocorticoids or increased glucocorticoid receptor sensitization. As such, increased cortisol suppression in response to dexamethasone (DEX, the synthetic glucocorticoid) treatment is also typically observed, reflecting an enhanced sensitivity to negative feedback inhibition. At the same time, sympathetic catecholamine secretion levels responsible for the fight or flight response are high, with an increased urinary catecholamine/cortisol ratio.[112] Therefore, the HPA profile of PTSD patients is distinct from that observed in acute and chronic stress and depression, which has been classically associated with increased CRH and cortisol levels, reduced cortisol suppression to DEX, and reduced GR responsiveness (Figure 14.30). Thus, rather than representing the normative response to trauma of increased cortisol levels as might be expected in a state of chronic stress, PTSD appears to manifest as a failure to engage the biological mechanisms associated with recovery and physiologic homeostasis.[110] The mechanism by which the paradoxically high CRH levels are sustained in the presence of increased feedback inhibition in PTSD patients remains unclear but may result, in part, from increased hypothalamic stimulation by the amygdala. A summary of PTSD neuroendocrine parameters from most studies is shown in Table 14.4.

The proposition that a cortisol deficit may have a role in the pathogenesis of PTSD has been supported by findings that administration of supplemental doses of cortisol to acutely ill medical patients reduces the PTSD outcome,[113] and a study found that high-dose cortisol administration within hours of a traumatic event reduced the subsequent development of PTSD.[114] These therapeutic results may represent a new direction in the treatment and prevention of PTSD.

Glucocorticoid Receptor Sensitization

In humans it has been hypothesized that the sensitization of glucocorticoid receptors at the level of the pituitary gland and/or the hippocampus via epigenetic mechanisms promotes glucocorticoid hypersensitivity.[115,116] Consequently, the pituitary output of ACTH, and subsequently adrenal cortisol, is diminished, while the hypothalamus attempts to rectify this by increasing the output of CRH. Since CRH also acts as a neurotransmitter, this affects other neural regions that in turn promote increased catecholamine production by the sympatho-adreno-medullary axis. Consistent with glucocorticoid receptor hypersensitivity postulated for patients with PTSD, increased hippocampal glucocorticoid receptor levels have been confirmed in animal PTSD models.[111]

Clinical Considerations: Computational modeling of cortisol dynamics in the neuro-endocrine axis of PTSD patients

The activity of the HPA axis normally exhibits a prominent 24-hour circadian rhythm with the circadian peak and trough occurring in the morning (waking) and evening respectively (refer to Figure 14.9). This biological rhythm has many important functions in the timing and distribution of energy reserves in addition to regulating the availability of other hormones important for growth and reproduction. Rachel Yehuda and colleagues at the Mount Sinai School of Medicine[117,118] have described patients suffering from PTSD who display a "flattened" circadian secretion of cortisol, with low waking and elevated evening concentrations that overall results in a decreased cortisol output (Figure 14.31). One major symptom resulting from this loss of endocrine rhythmicity is the disruption of the sleep-wake cycle, subsequently leading to insomnia, and increased incidence of emotional disturbance.

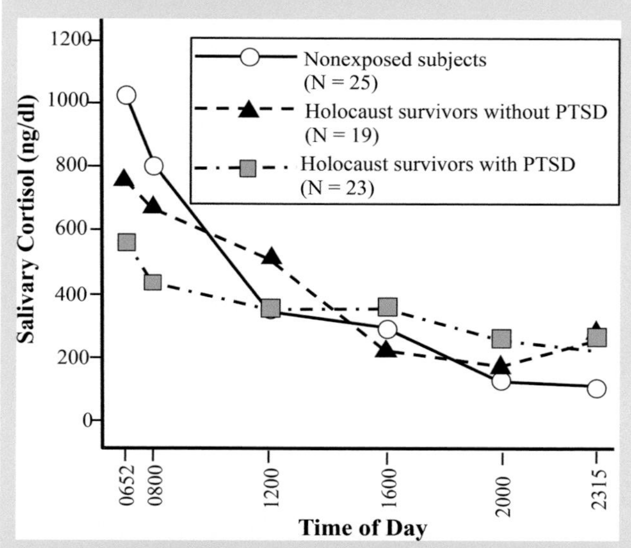

Figure 14.31 Salivary cortisol levels from awakening to bedtime in Holocaust survivors with and without PTSD.

Source of images: Yehuda, R., et al., 2005. Am. J. Psychi., 162(5), pp. 998-1000.

Sriram and colleagues[116] complemented Yehuda's hypothesis by constructing a mathematical model for cortisol dynamics in the HPA axis and estimated the kinetic parameters that fit the cortisol time series obtained from the clinical data of normal, depressed, and PTSD patients (Figure 14.32). The parameters obtained from the simulated phenotypes strongly support the hypothesis that hypocortisolemia in PTSD is due to a strong negative feedback loop as hypothesized by Yehuda, and that the weak negative feedback loop is responsible for depression. Importantly, this computational modeling approach predicted transitions from normal to various diseased states, and these transitions were shown to occur due to changes in the strength of the negative feedback loop and the stress intensity in the neuro-endocrine axis.

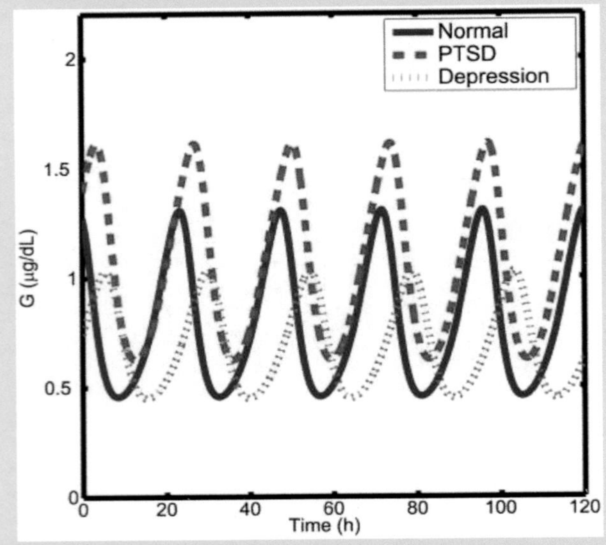

Figure 14.32 Simulated free glucocorticoid receptor time series of normal, depressed, and PTSD subjects.

The free glucocorticoid receptor (G) in PTSD is much higher than in normal and depressed subjects, indicating a stronger negative feedback loop.

Source of images: Sriram K, et al. PLoS Comput Biol. 2012;8(2):e1002379.

SUMMARY AND SYNTHESIS QUESTIONS

1. Whereas secondary adrenal insufficiency is accompanied specifically by reduced glucocorticoid synthesis, primary adrenal insufficiency manifests as both reduced glucocorticoid and mineralocorticoid synthesis. Why?
2. Why is primary, but not secondary or tertiary, adrenal insufficiency accompanied by hyperpigmentation?
3. Why is a condition characterized by the hyposecretion of epinephrine by adrenal medulla not considered a significant health concern?
4. How would the endocrine profiles along the hypothalamus-pituitary-adrenal axis differ between a patient suffering from Cushing's syndrome caused by a tumor on the adrenal cortex and one with Cushing's disease caused by a pituitary adenoma?
5. Provide physiological explanations for the following symptoms of Cushing's disease (secondary hypercortisolism):
   a. Central (trunkal) obesity, "buffalo hump", and a rounding of the face (moon face)
   b. Thin limbs and weak muscles
   c. Reddened cheeks and purple abdominal striae
   d. Osteoporosis of the vertebrae
   e. Hyperglycemia
   f. Hirsutism (excess facial hair), male pattern baldness, and androgenital syndrome (clitoral enlargement) in women
6. Compared with the endocrine profile for major depression, the endocrine profile for PTSD has been described as "paradoxical". Distinguish between the endocrine profiles of PTSD vs. major depression. What is the nature of the PTSD paradox, and how is it explained?

Summary of Chapter Learning Objectives and Key Concepts

LEARNING OBJECTIVE Define stress in the context of homeostasis and allostasis.

- Hans Selye coined the term "stress" in the 1930s and defined it as the nonspecific response of the body to any demand for change.
- Modern definitions of stress often emphasize any increased energy consumption by an organism that is required to restore a disruption to homeostasis.
- The term "allostasis" describes the process of achieving homeostasis through endocrine and other physiological changes that actively promote adaptation.
- The pathological effects of stress result from the prolonged exposure to stress hormones associated with the stress response.

LEARNING OBJECTIVE Describe the three main components of the integrative physiological response to stress.

- The sympathetic component of the stress response is mediated by catecholamines released from the adrenal medulla and the sympathetic nervous system.
- The adrenocortical component of the stress response is mediated by the hypothalamus-pituitary-adrenal cortex/interrenal (HPAc/I) axis and circulating glucocorticoids.
- The neurohypophyseal component of the stress response is activated specifically in response to hypovolemic and/or osmoregulatory stress.
- The short-term biological responses to sympathetic stimulation (which include increased nutrient mobilization, cardiac, renal, and pulmonary activities) are

mediated by plasma membrane-localized adrenergic receptors.
- Longer-term metabolic, developmental, and immunological aspects of adrenocortical stimulation are mediated by nuclear-localized glucocorticoid receptors.

LEARNING OBJECTIVE Explain the endocrine mechanisms behind fetal developmental programming and the transgenerational transmission of stress.

- During pregnancy, glucocorticoids developmentally program and adapt the fetus to the predicted postnatal environment.
- The developmental effects of glucocorticoids on fetal and postnatal developmental programming are mediated by epigenetic modifications that can manifest in offspring for multiple generations.
- Some of the most important epigenetic changes carried out by glucocorticoids include histone modifications and DNA methylation.

LEARNING OBJECTIVE Describe the synergistic interactions between stress hormones of the hypothalamic-pituitary-adrenal axis and thyroid hormones in regulating the timing of amphibian metamorphosis.

- In contrast with the mammalian model for regulation of the hypothalamic-pituitary-thyroid axis, in tadpoles it is CRH, and not TRH, that regulates thyroid hormone synthesis.
- Glucocorticoids modulate the rate of developmental progress induced by TH by increasing tissue sensitivity to TH.
- Stress and the productions of stress hormones biases the developmental trajectory towards metamorphosis.

LEARNING OBJECTIVE Explain the major disorders associated with the hyper- and hyposecretion of adrenal hormones.

- Whereas primary adrenal disorders of glucocorticoid hypo- or hypersecretion are caused by damage to the adrenal gland itself, secondary disorders are characterized by damage to the corticotropes of the adenohypophysis, and tertiary disorders by damage to the CRH-secreting cells of the hypothalamus.
- The symptoms of glucocorticoid hyposecretion disorders, such as Addison's disease, include fatigue, weight loss, and hypoglycemia.
- The symptoms of glucocorticoid hypersecretion disorders, such as Cushing's syndrome, include hyperglycemia, weight gain, fat redistribution, and skeletal muscle wasting.
- Catecholamine hypersecretion can be a serious condition that promotes hypertension.
- Post-traumatic stress disorder (PTSD) manifests as a failure to engage the biological mechanisms associated with recovery and return to physiologic homeostasis in response to stress.

LITERATURE CITED

1. Selye H. The evolution of the stress concept. *Am Scient*. 1973;61(6):692–699.
2. Rushen J. Some problems with the physiological concept of "stress". *Aust Vet J*. 1986;63(11):359–361.
3. McEwen BS, Stellar E. Stress and the individual. Mechanisms leading to disease. *Arch Int Med*. 1993;153(18):2093–2101.
4. Humphrey JH. *Anthology of Stress Revisited: Selected Works of James H. Humphrey*. Nova Science Publishers; 2005.
5. Selye H. *The Stress of Life*. McGraw-Hill; 1956.
6. Selye H. A syndrome produced by diverse nocuous agents. *Nature*. 1936:32.
7. Selye H. *Stress*. Acta, Inc; 1950.
8. Monaghan P, Spencer KA. Stress and life history. *Curr Biol*. 2014;24(10):R408–412.
9. Chrousos GP, Gold PW. The concepts of stress and stress system disorders. Overview of physical and behavioral homeostasis. *JAMA*. 1992;267(9):1244–1252.
10. Cannon WB. *The Wisdom of the Body*. W.W. Norton; 1932.
11. Cannon WB. Stresses and strains of homeostasis. *Am J Med Sci*. 1935;189:1–14.
12. Cannon WB. *Bodily Changes in Pain, Hunger, Fear and Rage*. D. Appleton and Co; 1915.
13. Sterling P, Eyer J. Allostasis: a new paradigm to explain arousal pathology. In: Fisher S, Reason J, eds. *Handbook of Life Stress, Cognition and Health*. Wiley and Sons; 1988.
14. McEwen BS, Wingfield JC. The concept of allostasis in biology and biomedicine. *Horm Behav*. 2003;43(1):2–15.
15. Kovacs WJ, Ojeda SR. *Textbook of Endocrine Physiology* (6th ed.). Oxford University Press; 2011.
16. Jacobson L, Sapolsky R. The role of the hippocampus in feedback regulation of the hypothalamic-pituitary-adrenocortical axis. *Endocr Rev*. 1991;12(2):118–134.
17. Aguilera G, Subburaju S, Young S, Chen J. The parvocellular vasopressinergic system and responsiveness of the hypothalamic pituitary adrenal axis during chronic stress. *Prog Brain Res*. 2008;170:29–39.
18. Gillies GE, Linton EA, Lowry PJ. Corticotropin releasing activity of the new CRF is potentiated several times by vasopressin. *Nature*. 1982;299(5881):355–357.
19. Tilbrook AJ, Clarke IJ. Neuroendocrine mechanisms of innate states of attenuated responsiveness of the hypothalamo-pituitary adrenal axis to stress. *Front Neuroendocrinol*. 2006;27(3):285–307.
20. McCann S, Brobeck JR. Evidence for a role of the supraopticohypophyseal system in regulation of adrenocorticotrophin secretion. *Proc Soc Exp Biol Med*. 1954;87(2):318–324.
21. Vale W, Spiess J, Rivier C, Rivier J. Characterization of a 41-residue ovine hypothalamic peptide that stimulates secretion of corticotropin and beta-endorphin. *Science*. 1981;213(4514):1394–1397.
22. Aguilera G, Harwood JP, Wilson JX, Morell J, Brown JH, Catt KJ. Mechanisms of action of corticotropin-releasing factor and other regulators of corticotropin release in rat pituitary cells. *J Biol Chem*. 1983;258(13):8039–8045.
23. Baker BI, Bird DJ, Buckingham JC. In the trout, CRH and AVT synergize to stimulate ACTH release. *Regul Pept*. 1996;67(3):207–210.
24. Schwartzman RA, Cidlowski JA. Glucocorticoid-induced apoptosis of lymphoid cells. *Int Arch Allergy Immunol*. 1994;105(4):347–354.
25. Webster EL, Torpy DJ, Elenkov IJ, Chrousos GP. Corticotropin-releasing hormone and inflammation. *Ann N Y Acad Sci*. 1998;840:21–32.
26. Gaillard RC. *Neuroendocrine-Immune Interactions*, Vol 29: Karger; 2002.
27. Glaser R, Kiecolt-Glaser JK. Stress-induced immune dysfunction: implications for health. *Nat Rev Immunol*. 2005;5(3):243–251.
28. Rhen T, Cidlowski JA. Antiinflammatory action of glucocorticoids—new

mechanisms for old drugs. *N Engl J Med.* 2005;353(16):1711–1723.

29. Cruz-Topete D, Cidlowski JA. One hormone, two actions: anti-and pro-inflammatory effects of glucocorticoids. *Neuroimmunomodulation.* 2015;22(1–2):20–32.
30. Sorrells SF, Sapolsky RM. An inflammatory review of glucocorticoid actions in the CNS. *Brain Behav Immun.* 2007;21(3):259–272.
31. Dhabhar FS. Stress-induced augmentation of immune function—the role of stress hormones, leukocyte trafficking, and cytokines. *Brain behav Immun.* 2002;16(6):785–798.
32. Dhabhar FS, McEwen BS. Enhancing versus suppressive effects of stress hormones on skin immune function. *Proc Nat Acad Sci.* 1999;96(3):1059–1064.
33. Deinzer R, Granrath N, Stuhl H, et al. Acute stress effects on local Il-1β responses to pathogens in a human in vivo model. *Brain Behav Immun.* 2004;18(5):458–467.
34. O'Connor KA, Johnson JD, Hansen MK, et al. Peripheral and central proinflammatory cytokine response to a severe acute stressor. *Brain Res.* 2003;991(1–2): 123–132.
35. Bowers SL, Bilbo SD, Dhabhar FS, Nelson RJ. Stressor-specific alterations in corticosterone and immune responses in mice. *Brain Behav Immun.* 2008;22(1):105–113.
36. McKay LI, Cidlowski JA. Corticosteroids. In: Kufe D, Pollock R, Weichselbaum R, et al., eds. *Cancer Medicine* (6th ed.). BC DEcker; 2003.
37. Lupien SJ, McEwen BS, Gunnar MR, Heim C. Effects of stress throughout the lifespan on the brain, behaviour and cognition. *Nat Rev Neurosci.* 2009;10(6):434–445.
38. Lupien SJ, Nair NP, Briere S, et al. Increased cortisol levels and impaired cognition in human aging: implication for depression and dementia in later life. *Rev Neurosci.* 1999;10(2):117–139.
39. Carpenter WT, Jr., Gruen PH. Cortisol's effects on human mental functioning. *J Clin Psychopharmacol.* 1982;2(2):91–101.
40. Hall RC, Popkin MK, Stickney SK, Gardner ER. Presentation of the steroid psychoses. *J Nerv Ment Dis.* 1979;167(4):229–236.
41. Thompson LA, Spoon TR, Goertz CE, Hobbs RC, Romano TA. Blow collection as a non-invasive method for measuring cortisol in the beluga (Delphinapterus leucas). *PLoS One.* 2014;9(12):e114062.
42. Ganswindt A, Palme R, Heistermann M, Borragan S, Hodges JK. Non-invasive assessment of adrenocortical function in the male African elephant (Loxodonta africana) and its relation to musth. *Gen Comp Endocrinol.* 2003;134(2):156–166.
43. Trumble SJ, Robinson EM, Berman-Kowalewski M, Potter CW, Usenko S. Blue whale earplug reveals lifetime contaminant exposure and hormone profiles. *Proc Natl Acad Sci U S A.* 2013;110(42):16922–16926.
44. Bortolotti GR, Marchant T, Blas J, Cabezas S. Tracking stress: localisation, deposition and stability of corticosterone in feathers. *J Exp Biol.* 2009;212(Pt 10):1477–1482.
45. Koren L, Mokady O, Geffen E. Social status and cortisol levels in singing rock hyraxes. *Horm Behav.* 2008;54(1):212–216.
46. Van Uum SH, Sauve B, Fraser LA, Morley-Forster P, Paul TL, Koren G. Elevated content of cortisol in hair of patients with severe chronic pain: a novel biomarker for stress. *Stress.* 2008;11(6):483–488.
47. Webb E, Thomson S, Nelson A, et al. Assessing individual systemic stress through cortisol analysis of archaeological hair. *J Archaeol Sci.* 2010;37(4):807–812.
48. Fowden AL, Forhead AJ. Hormones as epigenetic signals in developmental programming. *Exp Physiol.* 2009;94(6):607–625.
49. Fowden AL, Li J, Forhead AJ. Glucocorticoids and the preparation for life after birth: are there long-term consequences of the life insurance? *Proc Nutr Soc.* 1998;57(1):113–122.
50. Barker DJ, Gluckman PD, Godfrey KM, Harding JE, Owens JA, Robinson JS. Fetal nutrition and cardiovascular disease in adult life. *Lancet.* 1993;341(8850):938–941.
51. Seckl JR. Physiologic programming of the fetus. *Clin Perinatol.* 1998;25(4):939–962, vii.
52. McGowan PO, Matthews SG. Prenatal stress, glucocorticoids, and developmental programming of the stress response. *Endocrinology.* 2018;159(1):69–82.
53. Barker DJ, Eriksson JG, Forsen T, Osmond C. Fetal origins of adult disease: strength of effects and biological basis. *Int J Epidemiol.* 2002;31(6):1235–1239.
54. Noguera JC, da Silva A, Velando A. Egg corticosterone can stimulate telomerase activity and promote longer telomeres during embryo development. *Mol Ecol.* 2020;31(23):6252–6260.
55. Harris A, Seckl J. Glucocorticoids, prenatal stress and the programming of disease. *Horm Behav.* 2011;59(3):279–289.
56. Levitt NS, Lindsay RS, Holmes MC, Seckl JR. Dexamethasone in the last week of pregnancy attenuates hippocampal glucocorticoid receptor gene expression and elevates blood pressure in the adult offspring in the rat. *Neuroendocrinology.* 1996;64(6):412–418.
57. Shoener JA, Baig R, Page KC. Prenatal exposure to dexamethasone alters hippocampal drive on hypothalamic-pituitary-adrenal axis activity in adult male rats. *Am J Physiol Regul Integr Comp Physiol.* 2006;290(5):R1366-R1373.
58. Noorlander C, De Graan P, Middeldorp J, Van Beers J, Visser G. Ontogeny of hippocampal corticosteroid receptors: effects of antenatal glucocorticoids in human and mouse. *J Comp Neurol.* 2006;499(6):924–932.
59. Henry C, Kabbaj M, Simon H, Le Moal M, Maccari S. Prenatal stress increases the hypothalamo-pituitary-adrenal axis response in young and adult rats. *J Neuroendocrinol.* 1994;6(3):341–345.
60. Barbazanges A, Piazza PV, Le Moal M, Maccari S. Maternal glucocorticoid secretion mediates long-term effects of prenatal stress. *J Neurosci.* 1996;16(12):3943–3949.
61. Alcantara-Alonso V, Panetta P, de Gortari P, Grammatopoulos DK. Corticotropin-releasing hormone as the homeostatic rheostat of feto-maternal symbiosis and developmental programming in utero and neonatal life. *Front Endocrinol (Lausanne).* 2017;8:161.
62. Nyirenda MJ, Lindsay RS, Kenyon CJ, Burchell A, Seckl JR. Glucocorticoid exposure in late gestation permanently programs rat hepatic phosphoenolpyruvate carboxykinase and glucocorticoid receptor expression and causes glucose intolerance in adult offspring. *J Clin Invest.* 1998;101(10):2174–2181.
63. Drake AJ, Walker BR, Seckl JR. Intergenerational consequences of fetal programming by in utero exposure to glucocorticoids in rats. *Am J Physiol Regul Integr Comp Physiol.* 2005;288(1):R34–38.
64. Murgatroyd C, Spengler D. Epigenetics of early child development. *Front Psychiatry.* 2011;2:16.
65. Illingworth RS, Bird AP. CpG islands—"a rough guide". *FEBS Lett.* 2009;583(11):1713–1720.
66. Saxonov S, Berg P, Brutlag DL. A genome-wide analysis of CpG dinucleotides in the human genome distinguishes two distinct classes of promoters. *Proc Nat Acad Sci.* 2006;103(5):1412–1417.
67. Francis DD, Caldji C, Champagne F, Plotsky PM, Meaney MJ. The role of corticotropin-releasing factor--norepinephrine systems in mediating the effects of early experience on the development of behavioral and endocrine responses to stress. *Biol Psych.* 1999;46(9):1153–1166.
68. Liu D, Diorio J, Tannenbaum B, et al. Maternal care, hippocampal glucocorticoid receptors, and hypothalamic-pituitary-adrenal responses to stress. *Science.* 1997;277(5332):1659–1662.
69. Weaver IC, Cervoni N, Champagne FA, et al. Epigenetic programming by maternal behavior. *Nat Neurosci.* 2004;7(8):847–854.
70. Feder A, Nestler EJ, Charney DS. Psychobiology and molecular genetics of resilience. *Nat Rev Neurosci.* 2009;10(6):446–457.
71. Chan JC. *Intergenerational Mechanisms of Paternal Stress Transmission.* University of Pennsylvania; 2018.
72. Chan JC, Nugent BM, Bale TL. Parental advisory: maternal and paternal stress can impact offspring neurodevelopment. *Biol Psych.* 2018;83(10):886–894.

73. González CR, Gonzalez B. Exploring the stress impact in the paternal germ cells epigenome: can catecholamines induce epigenetic reprogramming? *Front Endocrinol.* 2020;11.
74. Darras VM, Kuhn ER. Increased plasma levels of thyroid hormones in a frog Rana ridibunda following intravenous administration of TRH. *Gen Comp Endocrinol.* 1982;48(4):469–475.
75. Darras VM, Kuhn ER. Difference of the in vivo responsiveness to thyrotropin stimulation between the neotenic and metamorphosed axolotl, Ambystoma mexicanum: failure of prolactin to block the thyrotropin-induced thyroxine release. *Gen Comp Endocrinol.* 1984;56(2):321–325.
76. Jacobs GF, Kuhn ER. Thyroid hormone feedback regulation of the secretion of bioactive thyrotropin in the frog. *Gen Comp Endocrinol.* 1992;88(3):415–423.
77. Denver RJ. Stress hormones mediate developmental plasticity in vertebrates with complex life cycles. *Neurobiol Stress.* 2021;14:100301–100301.
78. Denver RJ. Acceleration of anuran amphibian metamorphosis by corticotropin-releasing hormone-like peptides. *Gen Comp Endocrinol.* 1993;91(1):38–51.
79. Denver RJ. Neuroendocrinology of amphibian metamorphosis. *Curr Top Dev Biol.* 2013;103:195–227.
80. Denver RJ. Environmental stress as a developmental cue: corticotropin-releasing hormone is a proximate mediator of adaptive phenotypic plasticity in amphibian metamorphosis. *Horm Behav.* 1997;31(2):169–179.
81. Manzon RG, Denver RJ. Regulation of pituitary thyrotropin gene expression during Xenopus metamorphosis: negative feedback is functional throughout metamorphosis. *J Endocrinol.* 2004;182(2):273–285.
82. Hensley FR. Ontogenetic loss of phenotypic plasticity of age of metamorphosis in tadpoles. *Ecology.* 1993;74:2405–2412.
83. Hentschel BT. Complex life cycles in a variable environment: predicting when the timing of metamorphosis shifts from resource dependent to developmentally fixed. *Am Nat.* 1999;154:549–558.
84. Travis J. Anuran size at metamorphosis: experimental test of a model based on intraspecific competition. *Ecology.* 1984;65:1155–1160.
85. O'Laughlin BE, Harris RN. Models of metamorphic timing: an experimental evaluation with the pond-dwelling salamander Hemidactylium scutatum (Caudata: Plethodontidae). *Oecologia.* 2000;124:343–350.
86. Johnson CK, Voss SR. Salamander paedomorphosis: linking thyroid hormone to life history and life cycle evolution. *Curr Top Dev Biol.* 2013;103:229–258.
87. Whiteman HH. Evolution of facultative paedomorphosis in salamanders. *Q Rev Biol.* 1994;69:205–221.
88. Wilbur HM, Collins JP. Ecological aspects of amphibian metamorphosis. *Science.* 1973;182:1305–1314.
89. Kulkarni SS, Buchholz DR. Beyond synergy: corticosterone and thyroid hormone have numerous interaction effects on gene regulation in Xenopus tropicalis tadpoles. *Endocrinology.* 2012;153(11):5309–5324.
90. Sterner ZR, Shewade LH, Mertz KM, Sturgeon SM, Buchholz DR. Glucocorticoid receptor is required to survive through metamorphosis in the frog Xenopus tropicalis. *Gen Comp Endocr.* 2020:113419.
91. Bonett RM, Hoopfer ED, Denver RJ. Molecular mechanisms of corticosteroid synergy with thyroid hormone during tadpole metamorphosis. *Gen Comp Endocr.* 2010;168(2):209–219.
92. Kulkarni SS, Buchholz DR. Beyond synergy: corticosterone and thyroid hormone have numerous interaction effects on gene regulation in Xenopus tropicalis tadpoles. *Endocrinology.* 2012;153(11):5309–5324.
93. Sachs LM, Buchholz DR. Insufficiency of thyroid hormone in frog metamorphosis and the role of glucocorticoids. *Front Endocrinol.* 2019;10:287.
94. Jolivet Jaudet G, Leloup Hatey J. Variations in aldosterone and corticosterone plasma levels during metamorphosis in Xenopus laevis tadpoles. *Gen Comp Endocrinol.* 1984;56(1):59–65.
95. Denver RJ. Stress hormones mediate environment-genotype interactions during amphibian development. *Gen Comp Endocrinol.* 2009;164(1):20–31.
96. Kikuyama S, Kawamura K, Tanaka S, Yamamoto K. Aspects of amphibian metamorphosis: hormonal control. *Int Rev Cytol.* 1993;145:105–148.
97. Bonett RM, Hoopfer ED, Denver RJ. Molecular mechanisms of corticosteroid synergy with thyroid hormone during tadpole metamorphosis. *Gen Comp Endocrinol.* 2010;168(2):209–219.
98. Kuhn ER, De Groef B, Grommen SV, Van der Geyten S, Darras VM. Low submetamorphic doses of dexamethasone and thyroxine induce complete metamorphosis in the axolotl (Ambystoma mexicanum) when injected together. *Gen Comp Endocrinol.* 2004;137(2):141–147.
99. Kulkarni SS, Denver RJ, Gomez-Mestre I, Buchholz DR. Genetic accommodation via modified endocrine signalling explains phenotypic divergence among spadefoot toad species. *Nat Commun.* 2017;8(1):993.
100. West-Eberhard MJ. Phenotypic plasticity and the origins of diversity. *Ann Rev Ecol System.* 1989;20(1):249–278.
101. Whitman DW, Agrawal AA. What is phenotypic plasticity and why is it important. In: *Phenotypic Plasticity of Insects: Mechanisms and Consequences.* 2009:1–63.
102. Wake DB, Hanken J. Direct development in the lungless salamanders: what are the consequences for developmental biology, evolution and phylogenesis? *Int J Dev Biol.* 1996;40(4):859–869.
103. Gould SJ. *Ontogeny and Phylogeny.* Harvard University Press; 1977.
104. Laudet V. The origins and evolution of vertebrate metamorphosis. *Curr Biol.* 2011;21(18):R726-R737.
105. Semlitsch RD. Paedomorphosis in Ambystoma talpoideum: effects of density, food, and pond drying. *Ecology.* 1987;68:994–1002.
106. Harris RN. Density-dependent paedomorphosis in the salamander Notophthalmus viridescens dorsalis. *Ecology.* 1987;68:705–712.
107. Rooke T. *The Quest for Cortisone.* Michigan State University Press; 2012.
108. Dallek R. The medical ordeals of JFK. *Atlantic Monthly*, 2002:49–61.
109. Yehuda R. Post-traumatic stress disorder. *N Engl J Med.* 2002;346(2):108–114.
110. Yehuda R. Status of glucocorticoid alterations in post-traumatic stress disorder. *Ann N Y Acad Sci.* 2009;1179:56–69.
111. Pitman RK, Rasmusson AM, Koenen KC, et al. Biological studies of post-traumatic stress disorder. *Nat Rev Neurosci.* 2012;13(11):769–787.
112. Pervanidou P. Biology of post-traumatic stress disorder in childhood and adolescence. *J Neuroendocrinol.* 2008;20(5):632–638.
113. Schelling G, Kilger E, Roozendaal B, et al. Stress doses of hydrocortisone, traumatic memories, and symptoms of posttraumatic stress disorder in patients after cardiac surgery: a randomized study. *Biol Psych.* 2004;55(6):627–633.
114. Zohar J, Yahalom H, Kozlovsky N, et al. High dose hydrocortisone immediately after trauma may alter the trajectory of PTSD: interplay between clinical and animal studies. *Eur Neuropsychopharmacol.* 2011;21(11):796–809.
115. Yehuda R, Golier JA, Halligan SL, Meaney M, Bierer LM. The ACTH response to dexamethasone in PTSD. *Am J Psychiatry.* 2004;161(8):1397–1403.
116. Sriram K, Rodriguez-Fernandez M, Doyle FJ, 3rd. Modeling cortisol dynamics in the neuro-endocrine axis distinguishes normal, depression, and post-traumatic stress disorder (PTSD) in humans. *PLoS Comput Biol.* 2012;8(2):e1002379.
117. Yehuda R, Kahana B, Binder-Brynes K, Southwick SM, Mason JW, Giller EL. Low urinary cortisol excretion in Holocaust survivors with posttraumatic stress disorder. *Am J Psychiat.* 1995;152(7):982–986.
118. Yehuda R, Southwick SM, Nussbaum G, Wahby V, Giller EL, Jr., Mason JW. Low urinary cortisol excretion in patients with posttraumatic stress disorder. *J Nerv Ment Dis.* 1990;178(6):366–369.

CHAPTER
15

Blood Pressure and Osmoregulation

CHAPTER LEARNING OBJECTIVES:

- Define "osmoregulation" and explain why the hormones involved in osmoregulation are so fundamental to life.
- Provide an overview of vertebrate renal function and evolution, and describe how the kidney both produces and is the target of diverse osmoregulatory hormones.
- Describe how mammals sense both osmotic and blood volume disequilibrium.
- List the primary hormones responsible for regulating blood pressure, diuresis and antidiuresis in mammals, and describe their general effects on various target tissues, including the brain, cardiovascular system, and the kidney.
- Describe evolutionary trends in key osmoregulatory hormone functions.

OPENING QUOTATIONS:

"Crawling out on dry land some millions of years later, terrestrial forms were faced with diametrically opposite problems, at least with respect to water. Fluid conservation, rather than fluid elimination, was the major concern. Instead of discarding their now unnecessary pressure filters and redesigning their kidneys as efficient secretory organs, the terrestrial vertebrates modified and amplified their existing systems to salvage the precious water of the filtrate."

—Robert F. Pitts, *Physiology of the Kidney and Body Fluids*, Chicago, Year Book Medical Publishers, 1963

"The marriage of water and solutes is essential for the perpetuation of life, but can sometimes lead to disharmony. Osmoregulation is the process that attempts to forestall and correct such problems."

—P.J. Bentley (1971), *Endocrines and Osmoregulation: A Comparative Account in Vertebrates* (2nd ed.). Springer-Verlag; 2002

Throughout history, both salt and water have been regarded as vital elements of life. Indeed, in ancient times salt was a valuable currency, and the word *salary* derives from the Latin word *salarium* signifying "salt money", or the sum paid to Roman soldiers for purchasing salt. Water was incorporated into ancient Greek medical thinking as the solvent for each of the four Hippocratic humors: black bile, yellow bile, phlegm, and blood. Vertebrates consist of between one-half and two-thirds water by weight, in which is dissolved a myriad of nutrients, minerals, and other chemicals that are essential for life. A deviation from optimal salt and water balance can be catastrophic. In humans even a mild disequilibrium of ion-water balance is associated with feelings of headache, reduced levels of alertness, and difficulty in concentrating.[1] Larger perturbations can lead to lethargy, weakness, irritability, spasticity, confusion, and coma, and acute changes can cause seizures and death.

This chapter introduces the reader to the mechanisms by which diverse hormones regulate salt and water homeostasis, as well as systemic blood pressure. The chapter provides an overview of the major osmoregulatory organs among vertebrates and the hormones that regulate them, with emphasis on the endocrine regulation of mammalian kidney function.

Osmoregulation: A Fundamental Physiological Function

LEARNING OBJECTIVE Define "osmoregulation" and explain why the hormones involved in osmoregulation are so fundamental to life.

KEY CONCEPTS:

- A significant amount of energy is spent maintaining ion gradients between the intracellular and extracellular body compartments, as well with the external environment.
- Osmoregulation, or the maintenance of a constant internal osmotic condition, creates the necessary environment for diverse metabolic processes essential for survival.
- Osmoregulation plays a critical role in maintaining optimal blood pressure, which is essential for nutrient mobilization and the removal of cellular waste.
- The hormones that regulate osmoregulation are highly conserved among vertebrates.

DOI: 10.1201/9781003359241-20

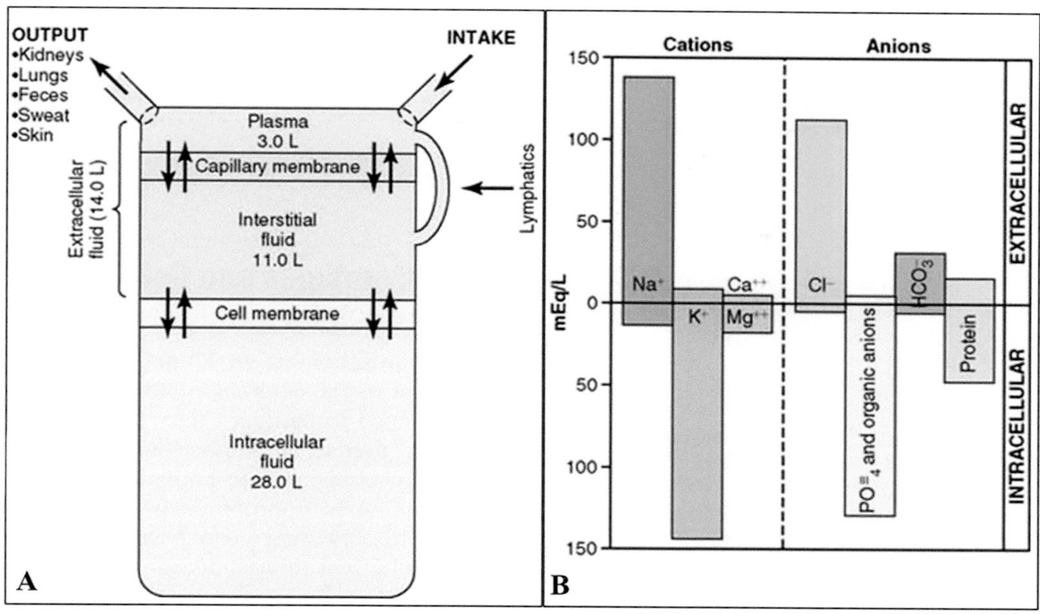

Figure 15.1 Distribution of water and ions among the various compartments of the human body. (A) Summary of body fluid regulation among fluid compartments and the membranes separating these compartments. The values shown are for an average 70-kilogram person. **(B)** Major cations and anions of the intracellular and extracellular fluids. The concentrations of Ca^{++} and Mg^{++} represent the sum of these two ions. The concentrations shown represent the total of free ions and complexed ions. The variable ion concentrations among the compartments are actively maintained by the actions of various transmembrane ion pumps, channels, and transport proteins.

Source of images: John, E., 2011. Guyton and Hall Textbook of Medical Physiology 12th Edition. Elsevier. Used by permission.

Approximately two-thirds of total body water is contained in the *intracellular* fluid compartment, or within cells (Figure 15.1). The remaining one-third is located in the *extracellular* compartment, or outside of the cells, consisting of three separate but interacting subregions: the *interstitial* fluid located between cells, the *blood plasma* contained within the circulatory system vasculature, other fluids such as *lymphatic fluid* contained within lymphatic vessels, and *cerebrospinal fluid* within the ventricles of the brain and around the spinal cord. Importantly, the composition of the fluids in the intracellular and extracellular compartments differ, with most of the body's sodium and chloride ions located in the extracellular compartment, and most potassium in the intracellular compartment (Figure 15.1). The maintenance of these intercompartmental ion gradients involves a significant expenditure of energy, one that is critical for generating cellular transmembrane potentials that are essential for such basic processes as cellular nutrient uptake, neuronal action potentials, and muscle contraction. **Osmoregulation**, or the maintenance of a constant intracellular and extracellular ionic and osmotic condition, creates the necessary environment for diverse metabolic processes and is essential for the normal functioning of all cells and the organisms they constitute. As such, osmoregulation is one of the most basic of the cell's regulatory needs, and only after this requirement is fulfilled can other aspects of cellular metabolism proceed. Because water movement across cell membranes can change the volume of the body's fluid compartments, including blood plasma volume, osmoregulation plays a critical role in modulating *blood pressure*, the force exerted by circulating blood upon the walls of blood vessels. The maintenance of optimal blood pressure is essential for the mobilization of nutrients to cells and the removal of their wastes.

Foundations & Fundamentals: Definitions Necessary to Understanding Osmoregulation

- *Diffusion*: The movement of molecules or ions from an area of high to low concentration.
- *Facilitated diffusion*: The passive diffusion of molecules or ions down an electrochemical gradient and across a cell's membrane via specific transmembrane proteins that function as regulatory channels.
- *Active transport*: The use of energy to move molecules or ions across a cell's membrane against an electrochemical gradient via specific transmembrane proteins.
- *Osmosis*: A type of diffusion specifically concerned with the movement of water molecules across a cell's membrane. Although most water molecules cross the cell membrane by facilitated diffusion through transmembrane proteins called "aquaporins", some water also moves across membranes independent of specific channels via poorly understood mechanisms. As a form of diffusion, water molecules move down their own concentration gradient from an area of high to low water molecule concentration. However, it can also be useful to think of osmosis as the movement of water from an area of low to high ion (or other soluble molecule) concentration. Importantly, unlike ions that can be transported across

a cell membrane via either facilitated diffusion or active transport, water molecules are not actively transported and always travel down their concentration gradient.

- *Osmotic pressure*: The hydrostatic pressure that would be required to prevent the inward movement of water across a membrane by osmosis. From a practical perspective, osmotic pressure of a solution is directly proportional to the concentration of dissolved particles in the aqueous solution being measured.
- *Hyperosmotic*: A condition in which the osmotic pressure of a solution is greater than that of a comparison solution.
- *Hypoosmotic*: A condition in which the osmotic pressure of a solution is lower than that of a comparison solution.
- *Isoosmotic*: A condition in which the osmotic pressure of a solution is the same as that of a comparison solution.
- *Osmolality*: The concentration of a solution expressed as the total number of solute particles per kilogram of water. Osmolality is measured in units called "osmoles" (Osm) and milliosmoles (mOsm) per kilogram of water, or mOsm/kg. Note that a similar term, "osmolarity", describes the concentration of a solution expressed as the total number of solute particles per *liter* of water. Because the volume of water varies with temperature, osmolality is the preferred term for biologic systems.
- *Osmoregulator*: A category of animal that actively maintains a constant osmotic pressure within its extracellular fluid compartment irrespective of that of the external environment. Most vertebrates, as well as many invertebrates, are osmoregulators.
- *Osmoconformer*: A category of animal whose extracellular fluids remain isosmotic with their marine environment. Many invertebrates and even some vertebrates, like sharks and hagfish, are osmoconformers. There are no known osmoconformers that live in freshwater.

The Advantages and Costs of Being an Osmoregulator

The composition of an animal's internal fluids differs from that of the external environment, and energy must be expended to maintain that composition. The amount of energy used to maintain cellular ion and water balance in humans and rats, for example, is considerable, and is estimated to be approximately 20% of basal metabolic rate (BMR).[2] However, energy used for this purpose varies among different tissues, accounting for up to 50%–70% of metabolism in the brain and kidney.[3]

Irrespective of whether they live in freshwater, in saltwater, or on land, most vertebrates (with some notable exceptions) maintain a remarkably similar salt content of their extracellular fluid, approximately one-third that of full-strength seawater[1] (Figure 15.2). These similarities

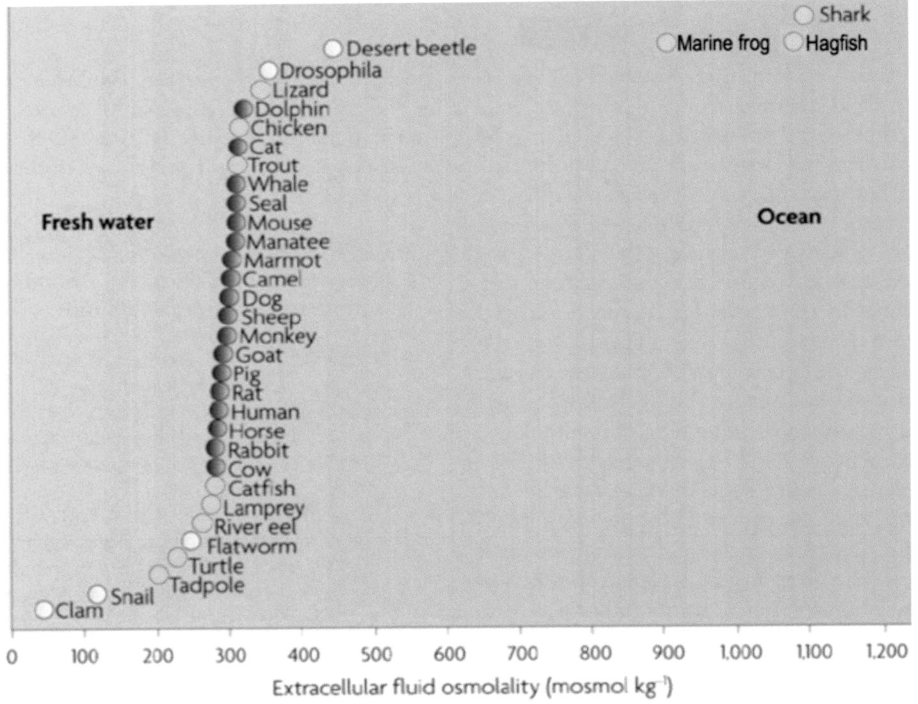

Figure 15.2 Extracellular fluid osmotic pressures of various animals. Although different types of organisms can display values that span almost the full range of environmental osmolalities, all mammals (blue circles) and most other vertebrates (yellow circles), irrespective of where they live, display osmotic set points that cluster around 300 mosmol kg^{-1}. These blood plasma osmotic pressures are actively maintained by regulating the concentrations of plasma salts and water. Interesting examples of vertebrates that deviate dramatically from the typical blood osmotic pressure are marine frogs, sharks, and hagfish. These animals are osmoconformers, which maintain blood plasma osmotic pressures similar to the environment that they live in.

Source of images: Bourque CW. Nat Rev Neurosci. 2008;9(7):519-531. Used by permission.

reflect a shared evolutionary heritage, a diluted remnant of the seas in which life evolved. Indeed, the great French physiologist Claude Bernard originally developed his theory of the *milieu intérieur*, or constancy of the internal environment,[4] as a way to explain the ability of ancestral organisms to move from the seas and radiate into diverse niches on land and in freshwater. He proposed that they did so by developing the ability to physiologically "carry the ocean with them" in their blood and interstitial fluids, which chemically resemble the oceans from which they evolved.[5] This general vertebrate strategy of maintaining constant extracellular fluid osmotic pressures among species produces divergent transport demands for animals depending on the specific environment they live in.

Osmoregulatory Hormones and Target Tissues

Although recognized as being essential to all life, the physiological mechanisms of salt and water balance in vertebrates remained a mystery well into the early 20th century. It is now known that hormones play critical roles in controlling the homeostatic and acclimation demands of salt and water balance, and despite the divergences in transport needs and capabilities among vertebrates, many of the hormones involved are remarkably similar, though their precise functions and target tissues have changed during evolution (Table 15.1).[6] Therefore, an effective understanding of osmoregulatory pathology, as well as knowledge of major evolutionary trends such as the evolution of terrestriality and acclimatization of species to new or severe environments, requires knowledge of the underlying endocrine control mechanisms for salt and water regulation.

SUMMARY AND SYNTHESIS QUESTIONS

1. Why are osmoregulatory hormones so fundamental to life?
2. The average human has a blood volume of about 5 liters. If you exercise heavily and sweat off 1 liter of water, does this mean that your blood volume has been reduced to 4 liters? Explain your reasoning.
3. What is the physiological basis for why a state of either severe dehydration or overhydration can be accompanied by neurological and muscular disorders?

Table 15.1 Major Vertebrate Osmoregulatory Organs and the Hormones That Stimulate Them

Organ	Phyletic Distribution of Organ	Stimulatory Hormone	Known Phyletic Distribution of Stimulatory Hormone
Kidney	All vertebrates	Aldosterone Angiotensin Natriuretic peptides Vasotocin Vasopressin	Mammals, birds, reptiles Most vertebrates Most vertebrates Teleosts, amphibians, reptiles, birds Mammals
Urinary bladder	• Present in mammals and amphibians • Absent in birds, some lizards, and most fish	Vasotocin Aldosterone	Amphibians Amphibians, reptiles, mammals
Gills	Fish Amphibian tadpoles Some amphibian adults	Glucocorticoids Growth Hormone Prolactin Natriuretic peptides Epinephrine Urotensins Vasotocin	FW and SW teleosts SW teleosts FW teleosts SW teleosts Teleosts Teleosts FW and SW teleosts
Skin	All vertebrates	Vasotocin Aldosterone Prolactin Epinephrine	Some amphibians Some amphibians Urodele amphibians Teleosts
Sweat glands	Only in mammals	Aldosterone Epinephrine	Mammals Mammals
Salt glands	• Nasal and orbital glands in some birds and reptiles • Rectal glands in sharks and other chondrichthyeans	Aldosterone Glucocorticoids Vasoactive intestinal peptide Natriuretic peptides	Lizards Chondrichthyeans Chondrichthyeans Chondrichthyeans and birds
Salivary glands	Only in mammals	Aldosterone Epinephrine	Mammals Mammals
Intestine	All vertebrates	Glucocorticoid Aldosterone	Teleosts Mammals, birds, amphibians

Source: Extrapolated from Bentley, P.J. 1998. *Comparative Vertebrate Endocrinology*. Cambridge University Press.

An Overview of Vertebrate Renal (Kidney) Osmoregulatory Function

LEARNING OBJECTIVE Provide an overview of vertebrate renal function and evolution, and describe how the kidney both produces and is the target of diverse osmoregulatory hormones.

KEY CONCEPTS:

- The nephron is the basic functional unit of the vertebrate kidney, which selectively secretes wastes and excess salts into the urine, reabsorbs nutrients, and maintains blood plasma osmotic pressure.
- The kidney both produces and is the target of diverse osmoregulatory hormones.
- The first kidneys evolved to maintain osmotic equilibrium in an aquatic environment and were modified to adapt to terrestrial life.

General Structure and Function of the Nephron

The *nephron* is the most basic structural and functional unit of the vertebrate kidney, whose primary function is to regulate the concentration of the blood's water and soluble substances by (1) *filtering* the blood, (2) selectively *reabsorbing* nutrients like glucose, ions, and water, (3) *secreting* wastes like urea, as well as excess ions and water, and (4) *excreting* the wastes as urine (Figure 15.3). The typical pair of human kidneys filters approximately 180 liters of blood plasma per day. Considering that the average human has a total blood volume of between 5 and 6 liters, this means that the entire blood plasma volume is filtered at least 30 times per day! The urinary solute content and rate of urine production affects ion and water balance throughout the whole organism. The term **diuresis** denotes a relatively high volume of urine production, whereas **antidiuresis** refers to a relatively low volume of urine production.

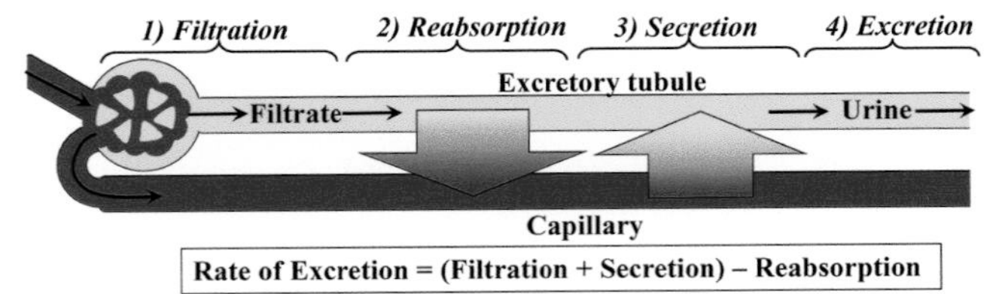

Figure 15.3 General strategy of the vertebrate nephron. (1) Filtration: "filtrate" from the blood is collected into excretory tubules of nephrons. "Tubules" are the functional units of most animal excretory systems. (2) Reabsorption: the transport epithelium of the excretory tubules reabsorbs most nutrients from the filtrate, as well as necessary salts and water, and returns it to the blood. (3) Secretion: toxins (e.g. urea) and excess ions are added to the filtrate from the blood. (4) Excretion: a filtrate of variable concentration leaves the body. The total rate of excretion for any urine component (e.g. water, ions, urea) is equal to the amount of filtrated and secreted component minus that reabsorbed over time.

Curiously Enough . . . The number of nephrons in a single kidney may range from as few as one in tadpoles and larval fishes to 12,400 in a mouse, 1.4 million in a human, and 192 million in a fin whale kidney[7] (Figure 15.4).

Fish, Amphibian, and Reptilian Nephrons

The nephrons of fishes, amphibians and non-avian reptiles (Figure 15.5) are able to reabsorb nutrients and excrete wastes, but the linear arrangements of their

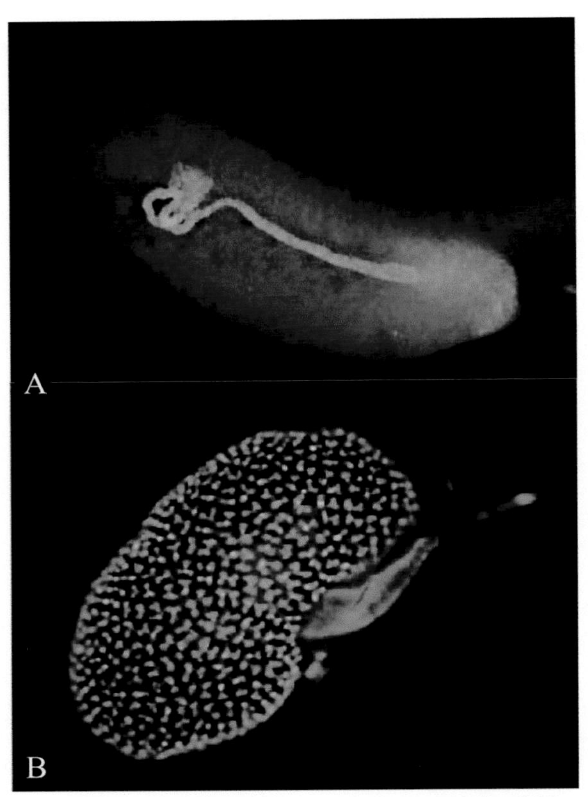

Figure 15.4 The pronephric and metanephric kidneys. (A) A single embryonic pronephros of a frog, *Xenopus laevis*, stained green for the expression of the sodium-potassium-ATPase pump. Frog embryos and tadpoles each have two pronephric nephrons. **(B)** A metanephric mouse kidney expressing green fluorescent protein in dozens of individual nephrons. There are over 12,000 nephrons in an entire mouse kidney.

Source of images: (A) Zhou, X. and Vize, P.D., 2004. Dev. Biol., 271(2), pp. 322-338; Used by permission. (B) Cullen-McEwen, L.A., et al. 2014. Imaging tools for analysis of the ureteric tree in the developing mouse kidney. In Confocal Microscopy (pp. 305-320). Humana Press, New York, NY. Used by permission.

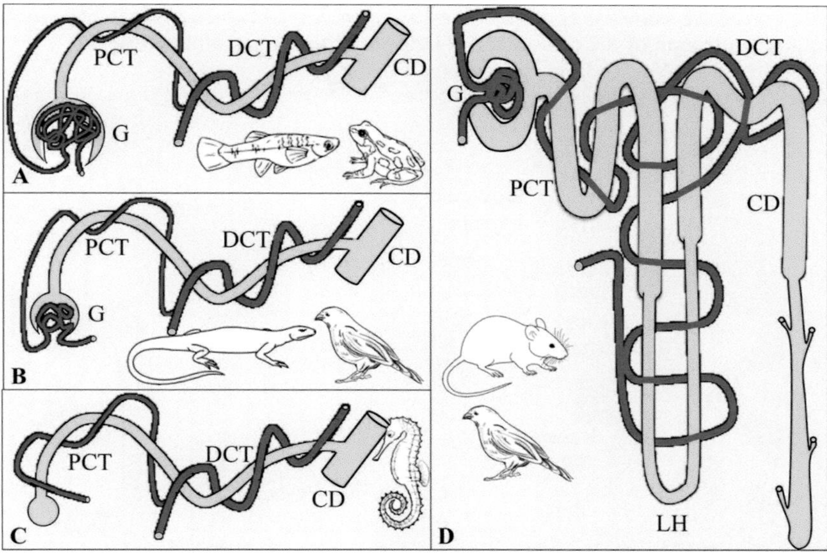

Figure 15.5 Conceptual diagram for nephron structure in different vertebrate taxa. Compared with nephron structure found in freshwater fishes and amphibians **(A)**, in reptiles **(B)** the glomerulus tends to be reduced in size to conserve water. In addition to possessing nephrons of typical reptilian morphology, birds (also known as avian reptiles) also possess a subset of nephrons that more closely resembles the mammalian nephron. Many fishes that live exclusively in seawater, like seahorses **(C)**, have lost the glomerulus entirely in an effort to avoid dehydration in the saltwater environment. An innovation of the avian and mammalian nephron **(D)** is the presence of a distinctive loop of Henle, which enables greater water reabsorption capacity than glomerular reduction alone can provide. G: glomerulus; PCT: proximal convoluted tubule; DCT: distal convoluted tubule; CD: collecting duct; LH: loop of Henle. Yellow: nephron tubules; Red lines: peritubular blood vessels.

nephrons are unable to generate a hyperosmotic urine compared with their blood. Therefore, these animals must rely on other osmoregulatory structures such as gills in teleost fish, rectal glands in sharks, and salt glands in reptiles and birds (Figure 15.6) to eliminate excess salts from the body.

The Avian and Mammalian Nephron

Birds and mammals evolved an important modification to the basic structure of the nephron, a novel hairpin loop localized between the proximal and the distal tubules, called the *loop of Henle* (see Figure 15.5 and Figure 15.1). The loop of Henle permits the generation of an osmotic gradient for water reabsorption in the collecting duct that enables the formation of a concentrated urine, which contributed to the successful establishment of life on land where water is often scarce. The long loops of Henle and collecting ducts form a deep medullary area that contributes to the pyramidal shape of the kidney structure. In mammals, the number of nephrons is considerably increased in comparison to other animal classes. The kidneys of birds also have distinct cortical and medullary areas formed by nephrons containing the long loops of Henle. However, most peripheral nephrons in birds are very short and similar to those of most reptiles, lacking loops of Henle. Therefore, the medullary zones are smaller compared with those of mammals, and birds (particularly marine birds) must compensate for their reduced ability to concentrate urine by excreting salts through salt glands in a manner identical to their reptilian relatives (Figure 15.6).

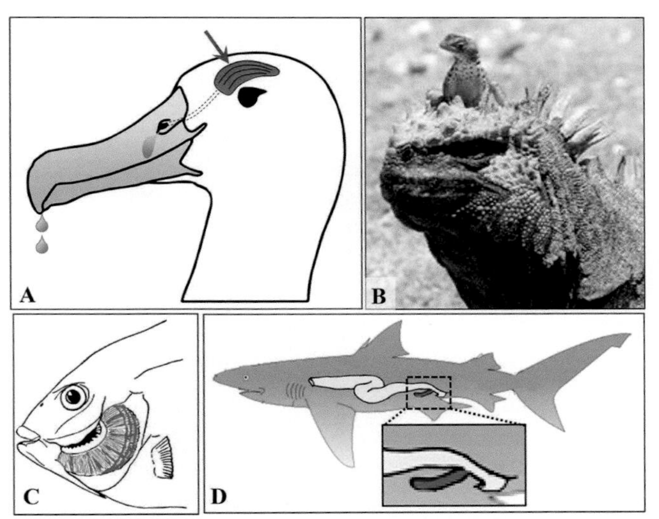

Figure 15.6 Some extrarenal accessory organs used by marine animals to expel excess salts from the blood. (A) Avian supraorbital salt glands (red arrow) produce a briny liquid (blue drops) that leaks onto the beaks of marine birds. **(B)** The Galapagos marine iguana, *Amblyrhynchus cristatus*, possesses a salt gland similar to birds, and the white coloration on top of the head is dried salt excreted by the glands. A smaller, insectivorous lava lizard, *Microlophus albemarlensis*, is perched atop the head of this iguana. **(C)** The marine teleost fish gill (red) expels ions from blood into the seawater. **(D)** The shark rectal gland (red), attached to the rectum, expels excess salt into the feces.

Source of images: (B) Rash, R. and Lillywhite, H.B., 2019. Mar. Biol., 166(10), pp. 1-21; Used by permission.

Foundations & Fundamentals: Synthesis of a Concentrated Urine by the Mammalian Nephron

A nephron, the smallest functional unit of a kidney, is comprised of several sequential structural components that collectively facilitate its function (Figure 15.7).

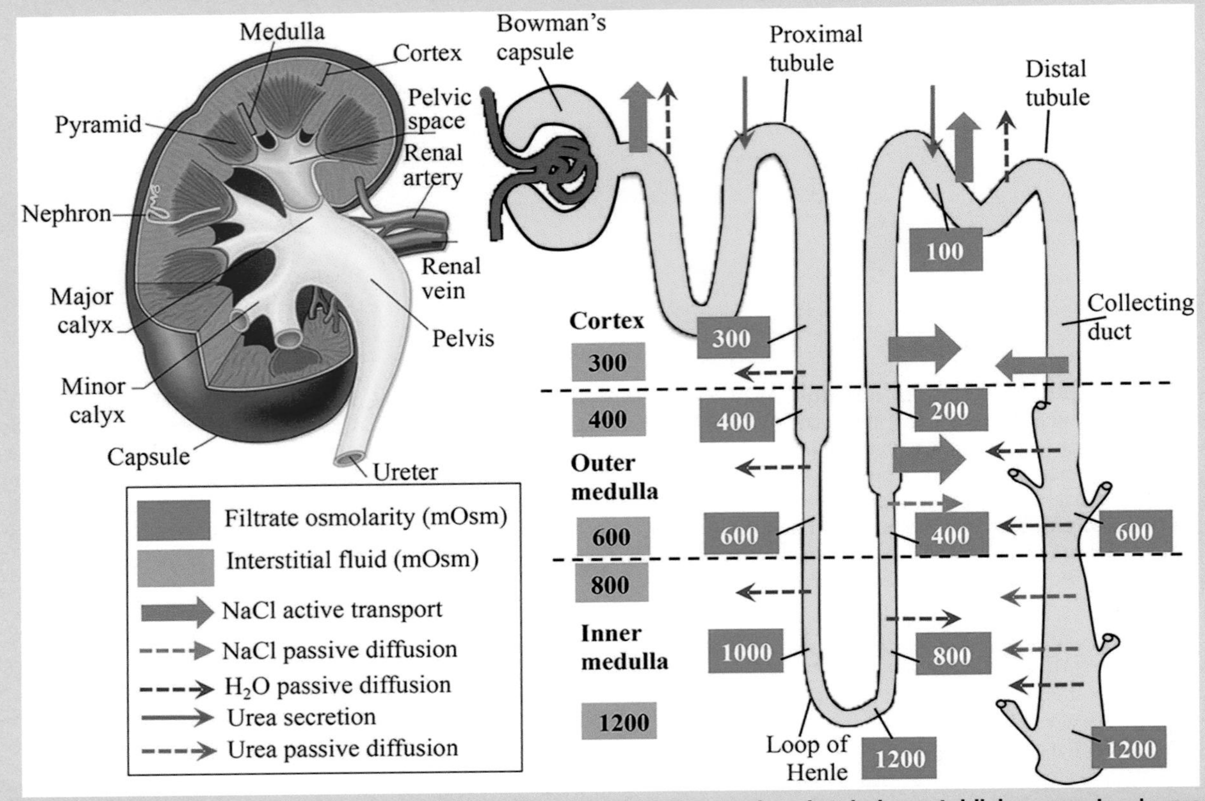

Figure 15.7 The differential movement of salts and urea throughout the nephron's tubules establishes a cortex-to-medulla osmotic gradient necessary for the reabsorption of water and formation of a concentrated urine. Note that water always moves passively across the walls of the tubules from an area of low to high osmotic pressure. An exception to this is the ascending loop of Henle, which is mostly impermeable to water movement.

1. *Glomerulus*: Hydrostatic blood pressure drives the *ultrafiltration* (movement between extracellular spaces of the glomerulus) of blood fluid and low molecular weight solutes such as salts, wastes like urea and excess protons, and nutrients like glucose and amino acids into the tubules.
2. *Proximal convoluted tubule (PCT)*: Some wastes are *secreted* from surrounding blood vessels into this tubule across transmembrane transport proteins embedded within the plasma membranes of tubule epithelial cells. Some water, salts, and nutrients are also *reabsorbed* back into the blood in a constitutive manner via specific transmembrane transporters.
3. *Loop of Henle*:
   a. *Descending loop of Henle*: Some water is *reabsorbed* from the filtrate and returned to circulation. This structure has only moderate permeability to salts and low numbers of mitochondria with little or no active reabsorption.
   b. *Ascending loop of Henle*: Some salts are *reabsorbed*, but this structure is less permeable to water.

   The hairpin morphology of the loop of Henle, in combination with the differential salt and water uptake properties of the descending and ascending regions, creates a powerful *counter current multiplier* that helps to generate the cortex-medulla osmotic pressure gradient necessary to form a concentrated urine. Note that whereas the cortical regions of the kidney tubules are isosmotic with blood (~300 mOsm), in a human the inner medullary regions corresponding with the loop of Henle and collecting duct have an osmotic pressure approximately equal to that of full-strength seawater (1,200 mOsm).
4. *Distal convoluted tubule (DCT)*: Similar to the proximal tubule, the distal tubule secretes wastes and reabsorbs some nutrients. In contrast to the proximal tubule, the rate at which salts and water are reabsorbed by the late distal tubule is not constitutive, but variable and under hormonal control.
5. *Collecting duct*: Similar to the late distal tubule, the collecting duct variably reabsorbs salts and water under hormonal control. Collecting ducts ultimately empty into the renal pelvis, which drains into the urinary bladder. The medullary regions of the collecting duct are permeable to urea, an osmolyte that along with Na^+ contributes to the formation of the aforementioned cortex-medulla osmotic pressure gradient.

Developments & Directions: Ion transport: from toad epithelia to the kidney

Amphibians represent the first vertebrates that emerged from aquatic habitats to terrestrial environments. To adapt to dryer environments, many adult anurans (frogs) have evolved two specialized osmoregulatory organs: the *ventral pelvic patch* to absorb water from the external environments, and the *urinary bladder* that stores water and reabsorbs it in times of need[8–10] (Figure 15.8). These relatively large and accessible organs have historically served as important model systems for studying hydromineral transport across epithelial surfaces.[11,12] Pioneered by Hans Ussing in several years following World War II, frog skin served as one of the preferred experimental preparations for the study of active ion transport by epithelia.[13] In fact, the first model of a NaCl-absorbing epithelium was developed using frog skin,[14] which greatly influenced future studies of epithelial ion transport in other organs and organisms.[15] To accomplish this, Ussing developed an apparatus (now called an "Ussing chamber") with paired compartments between which frog skin was mounted.[16] By putting the same solution on both sides and short-circuiting the transepithelial electrical potential, Ussing was able to simultaneously measure current and Na^+ flux across frog skin. This allowed him to calculate the electrical resistance to Na^+ diffusion and the electromotive force of the active Na^+ transport system. Later, electrophysiological and pharmacological studies revealed the presence of numerous key molecules in the toad skin and bladder epithelia such as the epithelial sodium channel (ENaC),[17,18] H^+-ATPase,[19] cystic fibrosis transmembrane conductance regulator (CFTR) chloride channel,[20] Na^+/K^+-ATPase,[21] aquaporins,[22] and other transporters. These transporters and channels are now known to also be present in the vertebrate kidney, though the simplicity of these toad epithelial organs make them popular models of study even today.

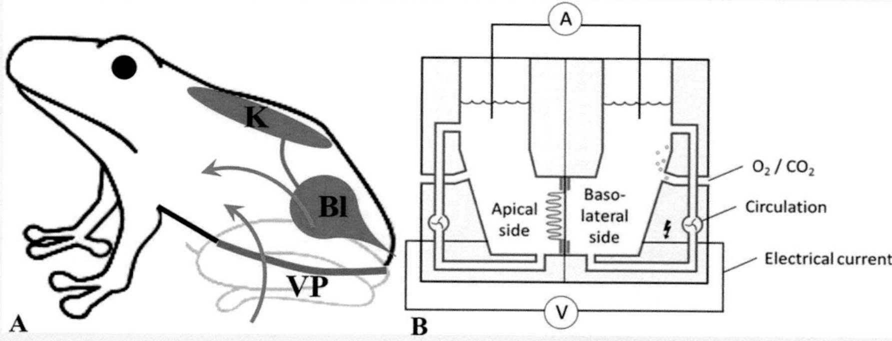

Figure 15.8 (A) In many adult frogs, ions and water are actively absorbed into the blood from the outside environment via epithelia associated with the ventral pelvic patch (VP) of skin, and also from urine in the urinary bladder (Bl). Blue arrows denote reabsorption of ions and water. In amphibians the kidney (K) also plays a limited role in ion and water reabsorption. **(B)** An Ussing's chamber. By short-circuiting the transepithelial electrical potential, Ussing could simultaneously measure radioactive Na^{24} flux and current across a given sample tissue. This allowed him to accurately assess active transport across epithelia. A, microammeter; V, voltmeter.

Source of images: (B) Westerhout, J., et al. 2015. Ussing chamber. The impact of food bioactives on health. Springer.

The kidney has an elaborate array of sodium transporters throughout the nephron[23,24] (Figure 15.9). In the proximal tubule, sodium reabsorption is linked to bicarbonate uptake by excreting H^+ ions with *Na^+/H^+ antiporters*. Antiporters are proteins that transport two ions or molecules in opposite directions. Sodium exchange also plays an important role in the reabsorption of glucose, sulfate, phosphate, and several amino acids. The remaining fraction of filtered sodium is reabsorbed by distinct transporters in each of the subsequent nephron segments. These transporters include the *Na/K/2Cl symporter* (a symporter cotransports multiple ions in the same direction) in the loop of Henle, the *sodium chloride cotransporter* (NCC) that is primarily in the DCT, and the *epithelial sodium channel* (ENaC) transporter that is located primarily in the collecting tubules. In addition to sodium transporters, channels for the transport of potassium, calcium, and water are also present. Importantly, note that present in all cells of the nephron tubules is the *sodium-potassium ATPase* pump (Na^+/K^+-ATPase), which generates the transmembrane electrochemical gradient necessary to drive ion and water reabsorption. The *cystic fibrosis transmembrane regulator* (CFTR) is a specific chloride ion channel that is expressed in all nephron segments and particularly in the distal tubule and collecting duct.[25] The morphology of the mammalian nephron singly and collectively imparts a distinct structure to the mammalian kidney. In a typical mammalian nephron, the cortex (outer layer) of the kidney consists primarily of glomeruli, proximal, and distal tubules, whereas the inner medulla is made up entirely of loops of Henle and collecting duct.

Curiously Enough . . . The cystic fibrosis transmembrane regulator (CFTR) protein is expressed on the lumen surfaces of diverse tissues, including the sweat glands, lungs, pancreas, intestines, and kidney where it helps to maintain ion and water balance. The absence of a functional CFTR protein promotes the formation of a thick mucus layer in these organs, resulting in the disease *cystic fibrosis*.

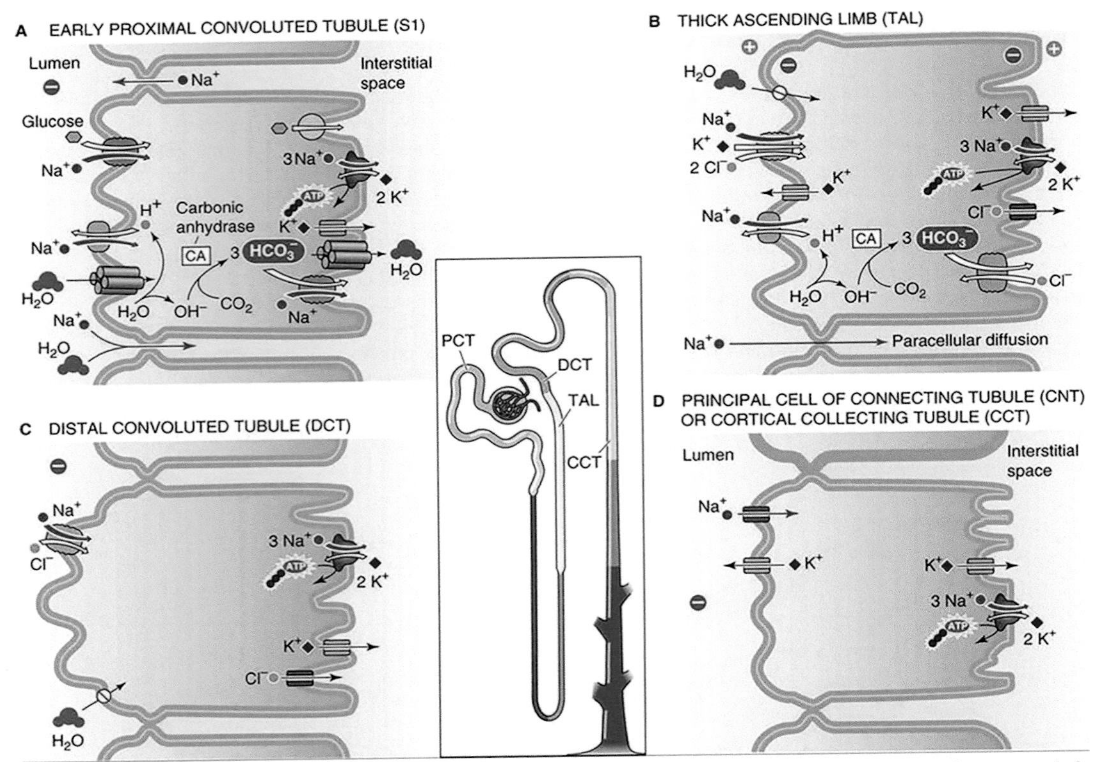

Figure 15.9 The unique ion and water transporters and cell structure of each segment of the nephron work in concert to generate the cortex-to-medulla osmotic gradient and maintain osmoregulatory homeostasis.

Source of images: Boron, W.F. and Boulpaep, E.L., 2005. Medical Physiology: A Cellular and Molecular Approach, updated 2nd ed. Elsevier. Used by permission.

Foundations & Fundamentals: The fish gill and the mammalian kidney share similar ion channels and osmoregulatory functions

Gills are aquatic respiratory organs that are present in fishes and larval amphibians. However, the gill is a multipurpose organ that also plays important roles in osmoregulation, acid-base regulation, and the excretion of nitrogenous wastes. While morphologically different from the kidney, at the cellular and tissue levels both organs possess similar types of ion transporters that are under endocrine control. Gills comprise an amazing 90% to 95% of the total surface area of the integument.[26] This, in combination with its unidirectional water flow and countercurrent vasculature arrangement, provide ideal mechanisms for gas and ion exchange (Figure 15.10(A)). Despite the fact that all fish groups have functional kidneys, the gill actually performs most of the functions that are controlled by pulmonary and renal processes in mammals. Indeed, many of the pathways that mediate these processes in mammalian renal tissue are present in the gill, as are receptors for their endocrine modulators.[27]

Freshwater physiology

The freshwater environment is strongly hypoosmotic compared with a fish's blood plasma, and fish living in freshwater therefore continually gain water from the environment and also lose ions to it. Freshwater fish address the problem of water gain by excreting a large volume of dilute urine that is hypotonic to their body fluids and by not drinking water (water enters the mouth for respiration, but it is not purposefully swallowed) (Figure 15.11). The problem of ion loss is mitigated by eating and by reabsorbing some ions across the nephron tubules from the glomerular filtrate back into the blood. However, food ingestion and renal salt reabsorption alone are insufficient to maintain blood ion homeostasis, and fish in freshwater must also obtain ions directly from the environment by actively transporting salts from the water into their blood against a high concentration gradient. This occurs across special cells in the gill epithelia called *ionocytes*, which are rich in ion transporters, including the Na^+/K^+-ATPase pump. Other gill ionocyte transporters that are also present in the mammalian kidney include chloride channels, Na^+/Cl^- cotransporters, Cl^-/bicarbonate transporters, and Na^+/H^+ antiporters, all of which facilitate ion uptake into the cells (Figure 15.12).

Seawater physiology

The majority of fish species that live in the marine environment are hypoosmotic to seawater and face the reverse osmotic and ionic gradients across the branchial epithelium compared to those found in the freshwater environment. Since teleost nephrons lack avian/mammalian-type loops of Henle, this precludes their production of a concentrated urine, and their urine is therefore isosmotic to their

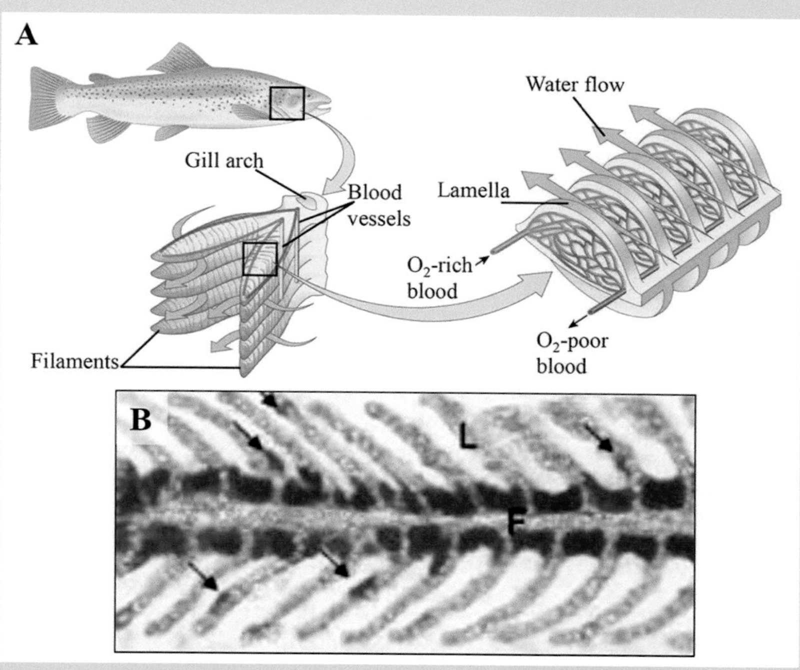

Figure 15.10 Functional morphology of the teleost gill and visualization of osmoregulatory cells. (A) Water flows across the gill lamella, thin structures supported by the gill filaments. **(B)** Sagittal sections of the gills stained with anti-Na^+/K^+-ATPase (dark red stain) in a freshwater-acclimated chum salmon denote the locations of ionocytes (osmoregulatory cells, arrows) localized to the gill lamellae (L) and filament (F).

Source of images: (A) Courtesy of openstax.org; (B) Kang, Chao-Kai, et al. Fish physiol. Biochem. 38, no. 3 (2012): 665-678. Used by permission.

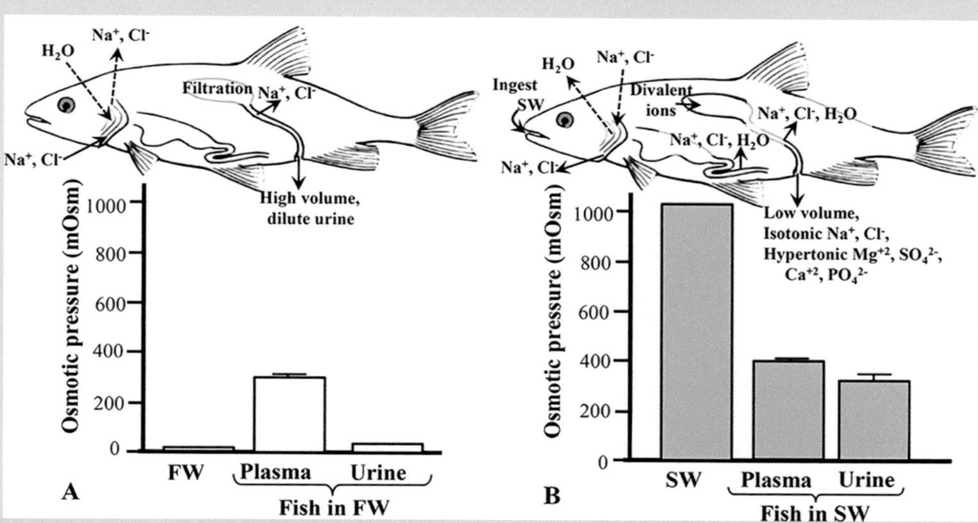

Figure 15.11 Mechanisms of osmoregulation by teleost fishes. (A) Freshwater teleosts are hyperosmotic to the surrounding solution, so they face osmotic gain of water and diffusional loss of NaCl across the permeable gill epithelium. These potentially disruptive osmotic and ionic movements are compensated for by excretion of relatively large volumes of a dilute urine and active uptake of NaCl across the gill epithelium. **(B)** Marine teleosts are hypoosmotic to seawater, so they face osmotic loss of water and diffusional gain of NaCl across the gill. Compensatory mechanisms include ingestion of seawater, intestinal absorption of NaCl and water, excretion of small volumes of blood-isotonic urine (after tubular reabsorption of Na, Cl, and water), and active secretion of NaCl across the gill epithelium. Whereas in saltwater the gills are the primary excretory organ for monovalent ions (Na^+ and Cl^-), the kidney is the primary excretory organ for divalent ions (Mg^{++}, Ca^{++}, SO_4^{2-}, PO_4^{2-}). Passive ion movements are denoted by dashed arrows; active by solid arrows.

Source of images: Evans, D.H. Front Physiol. 2010; 1: 13; and also Beyenbach, K.W., 2000. Front. Biosci., 5(3), pp. 712-719.

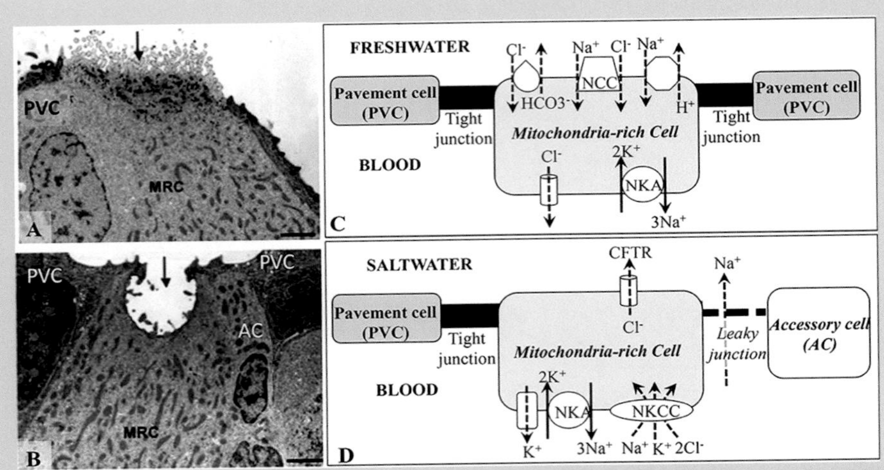

Figure 15.12 Model of ion transporters and the ultrastructure of teleost gill ionocytes, also called mitochondria-rich cells (MRCs), in freshwater and saltwater-acclimated fish. Compared with those of a freshwater-acclimated killifish **(A)**, the MRCs of saltwater-acclimated killifish **(B)** tend to be larger, with a greater number of mitochondria. The apical region of a freshwater MRC forms a convex surface studded with microvilli that extend above the surrounding pavement cells (PVCs). By contrast, a saltwater MRC typically forms a smooth, concave crypt that is recessed below the PVCs. Whereas both freshwater **(C)** and saltwater **(D)** MRCs adjoin with PVCs via tight junctions, the saltwater MRC also forms leaky junctions with accessory cells (AC). Compared with freshwater MRCs, saltwater MRCs have elevated numbers of the NKA transporter. PVC, pavement cell; NCC, Na^+/Cl^- cotransporter; CFTR, cystic fibrosis transmembrane regulator; NKA, Na^+/K^+-ATPase; NKCC, Na/K/2Cl channel. Bar = 1 µm.

Source of images: (A-B) Farrell, A.P., 2011. Encyclopedia of fish physiology: from genome to environment. Academic press. Used by permission.

blood plasma. Their osmotic loss of water is balanced by the ingestion of seawater and the subsequent uptake of ions and water by the intestine into the blood (Figure 15.11). The excess monovalent ions (Na^+ and Cl^-) that have been gained are then actively expelled via the gill ionocytes, relying on the Na^+/K^+-ATPase pump to drive this process. The basal-lateral membrane of saltwater ionocytes is involuted and rich in Na/K/2Cl uniporters that facilitate the uptake of ions from the blood and into the ionocyte. Chloride ions are subsequently expelled through CFTR channels localized on a well-developed apical pit (Figure 15.12). Na^+, which is pumped out of the basolateral ionocyte membrane via the Na^+/K^+-ATPase pump, diffuses between branchial cells through leaky junctions and is expelled to the environment. In contrast to monovalent ions that are excreted by the gills, divalent ions (Mg^{2+} and SO_4^{2-}) are excreted primarily by the kidney.

SUMMARY AND SYNTHESIS QUESTIONS

1. A drug specifically blocks the transport of chloride and sodium ions in the kidney's *ascending* loop of Henle. Would this drug be considered a diuretic or an antidiuretic? Why?
2. You suffer from a disease where your body is producing too much urea. To treat the disease, your doctor prescribes a drug that specifically makes the *collecting ducts* of nephrons *impermeable* to urea, allowing more urea to be excreted in the urine. Does taking this drug increase or decrease the ability of the kidney to reabsorb water and make a highly concentrated urine? Explain your reasoning.
3. The nephrons of some exclusively marine fish (like seahorses) are aglomerular—they completely lack a glomerulus. (A) How is this advantageous to an exclusively marine fish but would be lethal to a freshwater fish? (B) Despite the lack of a glomerulus, these fish still use their kidneys to excrete divalent ions and nitrogenous wastes in their urine. How is this possible?
4. Euryhaline fish can live in either freshwater or saltwater. In which environment is *elevated* hydrostatic pressure (higher blood pressure) more important for kidney function? Why?
5. In contrast to freshwater-acclimated fish, which don't drink water, saltwater-acclimated fish do actively drink water. Why must saltwater-acclimated fish drink saltwater?
6. Why are the osmoregulatory cells of the fish gill so rich in mitochondria?
7. Unlike mammals, whose nephrons possess a loop of Henle, the nephrons of marine fish and reptiles lack this structure. Although some bird nephrons possess loops of Henle, bird kidneys still are not as effective as the mammalian kidney in forming a concentrated urine. How, then, do marine fish, reptiles and birds handle the excess salt loads that accompany life in a salty environment?

15

Sensing Osmotic and Blood Volume Disequilibrium in Mammals

LEARNING OBJECTIVE Describe how mammals sense both osmotic and blood volume disequilibrium.

KEY CONCEPTS:

- An elevated blood osmotic pressure, sensed by hypothalamic osmoreceptors, promotes the sensation of "osmotic thirst".
- Low blood pressure, sensed by peripheral baroreceptors, promotes the sensation of "hypovolemic thirst".
- The activation of either osmoreceptors or baroreceptors stimulates the release of AVP/ADH from the pituitary neurohypophysis, increasing water retention and elevating blood pressure.

Both blood plasma osmotic pressure and blood volume are critical variables in the regulation of blood pressure. One key strategy to reversing low blood pressure caused by dehydration or blood volume loss in mammals is the release of *antidiuretic hormone (ADH)/arginine vasopressin* (AVP) by the neurohypophysis into general circulation. This potent hormone promotes the retention of water by the kidney through the production of a concentrated urine, which assists in the restoration of blood volume and tonicity.

Clues to the physiological mechanisms responsible for sensing changes in osmotic pressure and blood volume can be found by studying *thirst*, or the motivation to drink water. Specifically, there are two categories of thirst. **Osmotic thirst** (Figure 15.13) is a potent thirst that occurs in response to drops in *intracellular* fluid volume caused by water being drawn out of cells due to an increased blood plasma osmotic pressure. This could be caused, for example, by an increased blood salt concentration following the ingestion of salty food or by a disproportionately high water loss relative to electrolyte content through the production of a copious dilute urine, diarrhea, or severe sweating. In humans the release of ADH/AVP from the neurohypophysis is extremely sensitive to small changes in osmolality, and a change in plasma osmolality of only 1% is needed to increase this hormone (Figure 15.14). Therefore, osmotic thirst appears to be the day-to-day mechanism for regulating dehydration.

In contrast, **hypovolemic thirst** (Figure 15.13) is experienced following a reduction of blood volume due to bleeding, diarrhea, or excessive perspiration.[28–30] Thirst is, therefore, a powerful behavioral response that promotes the restoration of body fluids to their normal tonicity and volume following a hyperosmotic and/or hypovolemic insult to the blood that has caused a drop in blood pressure. Whereas osmotic thirst may be quenched by drinking water alone, hypovolemic thirst requires the replacement of not just the lost water, but also of lost ions and other solutes. Hypovolemic thirst requires a blood volume reduction of about 10% to increase ADH/AVP, making

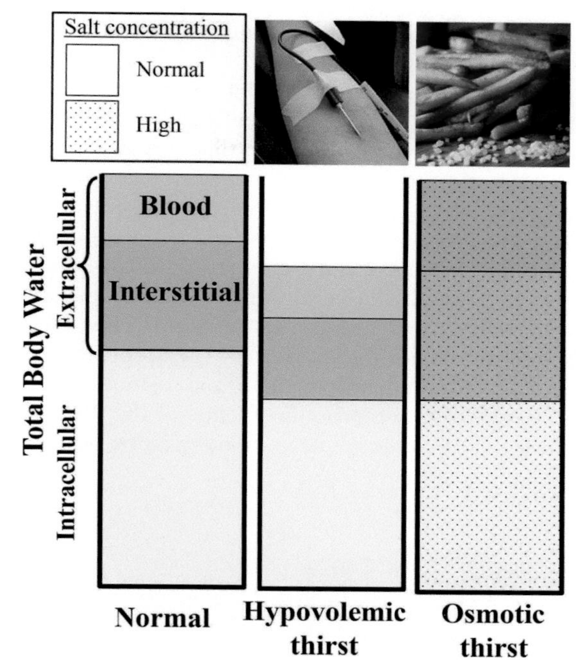

Figure 15.13 Two kinds of thirst. Hypovolemic thirst is caused by the loss of fluids and solute, such as following blood loss, diarrhea, or excessive sweating. Replacement of both water and salts is required to quench hypovolemic thirst. Osmotic thirst is caused by cellular dehydration, which can occur after ingesting a salty snack or drink, causing increased extracellular osmolarity, which draws water out of the intracellular compartment, increasing intracellular osmotic pressure as well. Ingestion of water alone can satisfy osmotic thirst.

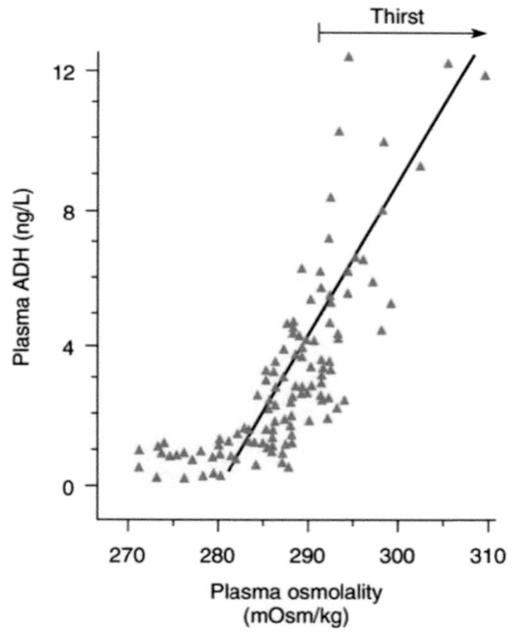

Figure 15.14 Relation between plasma osmolality and plasma antidiuretic hormone (ADH).

Source of images: White, B. and Porterfield, S., 2012. Endocrine and reproductive physiology: Mosby physiology monograph series. Elsevier Health Sciences. Used by permission.

it less sensitive compared with similar percent changes in blood osmotic pressure.

The induction of osmotic and hypovolemic thirst are mediated by two types of specialized receptors: brain-localized *central osmoreceptors* that sense changes in blood osmotic pressure, and *peripheral baroreceptors*, which sense changes in blood hydrostatic pressure. Each of these receptors innervate hypothalamic neurons that in turn release ADH/AVP into circulation via the pituitary neurohypophysis under conditions of high blood osmotic pressure and/or low hydrostatic pressure (Figure 15.15). Although the blood-brain barrier (BBB) protects most brain cells from alterations in blood osmotic pressure, these central osmoreceptors lie outside of the BBB, allowing them to sense alterations in blood osmotic pressure.

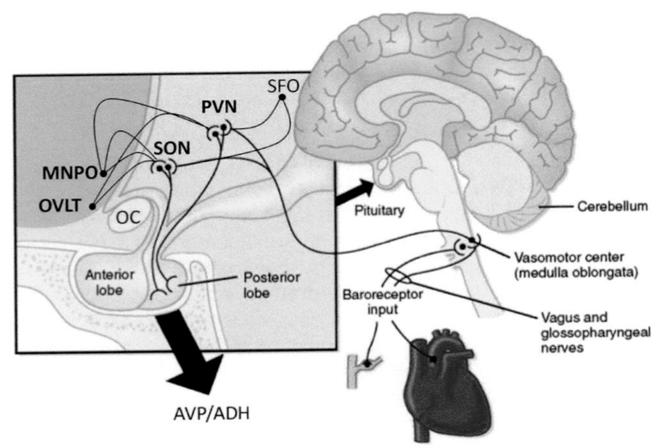

Figure 15.15 Osmoreceptor and baroreceptor pathways involved in regulating secretion of AVP/ADH. Afferent fibers from the baroreceptors, located in the walls of the carotid sinuses and aortic arch, are carried to the PVN and SON hypothalamic nuclei by the vagus and glossopharyngeal nerves. Central osmoreceptors located in the OVLT, MNPO, SFO, and other regions also innervate the SON and PVN. Peripheral baroreceptors are located in the walls of the carotid sinuses and aortic arch. SON, supraoptic nuclei; PVN, periventricular nuclei; OC, optic chiasm; OVLT, organum vasculosum of the lamina terminalis; SFO, subfornical organ; MNPO, median preoptic nucleus.

Source of images: Koeppen, B.M. and Stanton, B.A., 2009. Berne & Levy Physiology, Updated Edition E-Book. Elsevier Health Sciences. Used by permission.

SUMMARY AND SYNTHESIS QUESTIONS

1. Although the hypothalamic PVN and SON secrete ADH/AVP, they are themselves not osmoreceptors. What feature prevents them from being effective osmoreceptors, and where are the actual osmoreceptors located?
2. Hyperosmotic thirst caused by ingestion of salty food does not result in changes in total body water volume, but it does result in increased blood and interstitial fluid volume. Why?

Hormones of Mammalian Osmoregulation

LEARNING OBJECTIVE List the primary hormones responsible for regulating blood pressure, diuresis and antidiuresis in mammals, and describe their general effects on various target tissues, including the brain, cardiovascular system, and the kidney.

KEY CONCEPTS:

- The hormone AVP/ADH, released under conditions of dehydration, promotes synergistic vasoconstrictive and antidiuretic responses.
- AVP/ADH promotes antidiuresis by upregulating aquaporin-2 water channels in the kidney tubules.
- The renin-angiotensin-aldosterone system (RAAS), activated under conditions of dehydration, promotes sodium uptake and retention.
- Aldosterone is an adrenal mineralocorticoid that promotes renal salt reabsorption by upregulating several types of sodium transporters.
- Hormones that increase glomerular filtration rate (GFR) inhibit salt and water retention, thereby lowering blood pressure.
- Hormones that reduce GFR promote salt and water retention, thereby increasing blood pressure.

AVP/ADH

You may have noticed that throughout this chapter the dual-name status of the mammalian neurohypophyseal nonapeptide, AVP/ADH, has been emphasized. There is a good reason for this: each of its names describes a distinct physiological function of the hormone. The hypothalamic/neurohypophyseal hormone was originally called "vasopressin" for the ability of bovine pituitary extracts to increase blood pressure by inducing *vasoconstriction*, or its "vasopressor" effects.[31] Arginine vasopressin (AVP) is the most common mammalian form of the hormone, though lysine vasopressin (LVP) is found in pigs and marsupials. The *antidiuretic* property of pituitary extract, or its ability to reduce urine volume, was discovered soon thereafter,[32,33] suggesting the presence of a separate antidiuretic hormone. It was not until the mid-20th century that definitive proof that both vasopressin and antidiuretic hormone were in fact the same molecule was provided by Vincent du Vigneaud, who won the *1955 Nobel Prize in Chemistry* for the isolation, structural identification, and synthesis of the neurohypophysis nonapeptides AVP/ADH and oxytocin.

Lessons from Rodents

The antidiuretic properties of AVP/ADH and its storage in the neurohypophysis of the pituitary gland are elegantly demonstrated by studying various species and strains of rodents, both under laboratory conditions and in their natural habitats. For example, destruction of the neurohypophysis of rats causes them to produce large volumes of dilute urine that they are unable to concentrate, a condition called **diabetes insipidus**. Additionally, a natural mutant model for the AVP/ADH gene has been known for many years: the *Brattleboro rat*, which has a single base deletion in the AVP/ADH gene.[34] Though these rats display the expected diabetes insipidus (Figure 15.16), the injection of AVP/ADH into the rats results in a reduction of their urine volume and increases its concentration.[35] Interestingly, these rats also display atypical social development, like social play behavior, huddling, social investigation, and grooming of others.[36] This reinforces observations that AVP/ADH also targets behavioral regions of the brain.

AVP/ADH Receptors

AVP/ADH not only exerts vasopressor and antidiuretic effects in mammals, but also plays roles in modulating the stress response, elevating blood glucose levels, and increasing monogamous/polygamous social behavior. The multiple actions of AVP/ADH can be explained by its interaction with at least three different types of G protein-coupled receptors (GPCR):

1. *V1a receptor*: V1a receptors interact with the phospholipase C (PLC) effector enzyme, which hydrolyzes phosphatidyl inositol to produce the intracellular second messengers, inositol trisphosphate and diacylglycerol, that mobilize intracellular calcium (Figure 15.17). This receptor is expressed in vascular smooth muscle and accounts for the "vasopressor" effects of AVP/ADH.
2. *V1b receptor*: V1b receptors also utilize the PLC effector pathway to mobilize calcium (Figure 15.17). However, these receptors localize to various tissues (heart, lungs, thymus, mammary glands, and central nervous system), including the corticotropes (ACTH-producing cells) of the adenohypophysis. Recall from Chapter 14: Adrenal Hormones and the Stress Response that AVP/ADH of hypothalamic parvocellular origin is known to target

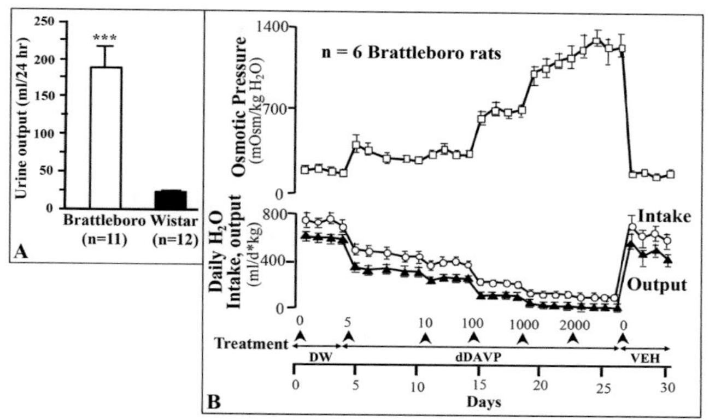

Figure 15.16 Brattleboro strain rats (also called diabetes insipidus rats) have a mutated gene for AVP/ADH and are always thirsty, producing large volumes of dilute urine. (A) Whereas standard Wistar rats drink about 20 ml of water per day, Brattleboro rats drink almost ten times as much water and excrete about ten times the urine volume as standard rats. **(B)** Treatment of Brattleboro rats with AVP (dDAVP) reverses their phenotype, reducing drinking and urination rates and increasing urinary osmotic pressure. Six rats were first given distilled water (DW, four days) then increasing concentrations of AVP (5–2,000 ug/L) at intervals of at least four days, followed an additional four days of vehicle (VEH).

Source of images: (A) Modified from Promeneur, D., et al. 2000. Am. J. of Physiol.-Renal Physiology, 279(2), pp. F370-F382; (B) modified from Kinter, L.B., 1982. Annal. NY Acad. Sci., 394(1), pp. 448-463.

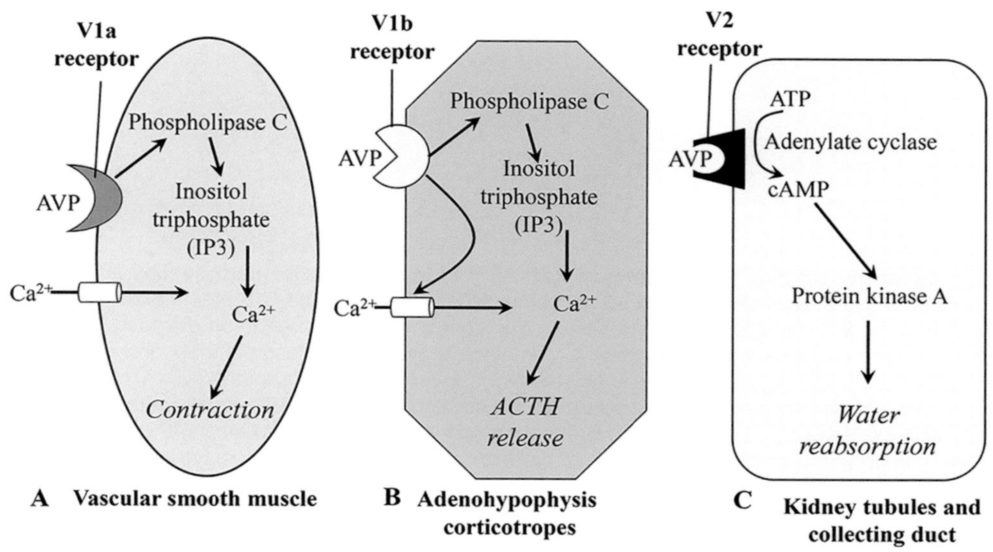

Figure 15.17 The different actions of AVP/ADH are mediated by different GPCR receptors and intracellular signaling pathways. (A) Whereas the V1a receptor mediates the pressor actions of AVP/ADH, **(B)** the V1b receptor facilitates ACTH release, and **(C)** the V2 receptor mediates the antidiuretic effects. The differential effects of the receptors signal through distinct G proteins that activate different effector enzymes.

the anterior pituitary gland to modulate the release of ACTH as part of the stress response.

3. *V2 receptor*: In contrast to the V1 receptor class, V2 receptors are coupled to the adenylate cyclase effector, which synthesizes cAMP as an intracellular second messenger (Figure 15.17). These receptors play a key role in modulating the antidiuretic response and are found in kidney cells that constitute portions of the distal tubules and collecting ducts of nephrons.

The Cell Biology of Renal Function

AVP/ADH exerts several primary actions on the kidney. These function collectively to promote the uptake of water from the nephron tubules into the blood, reduce the volume of urine formed, and allow the production of a concentrated urine. The actions of AVP/ADH on different regions of the nephron are described as follows.

Aquaporin Upregulation

While all cell plasma membranes are weakly permeable to water that diffuses slowly across the phospholipid bilayer, the rates of water transport across cell membranes are dramatically accelerated by the presence of *aquaporins*, integral membrane proteins that selectively transport water molecules in and out of the cell, while preventing the passage of ions and other solutes (Figure 15.18). The discovery of aquaporins, which are present in virtually all bacterial and eukaryotic cells, by Peter Agre earned him the *2003 Nobel Prize in Chemistry*, which he shared with Roderick MacKinnon for his work on the structure and mechanism of potassium channels.

The permeability of the late distal tubule and collecting duct to water is increased by AVP/ADH, which binds to V2 receptors localized to these regions of the nephron. The binding of the hormone to this receptor, a GPCR coupled to a stimulatory G protein (G_s), activates the adenylate cyclase effector that results in increased synthesis of the second messenger, cAMP. Elevated cAMP activates protein kinase A (PKA), which results in two synergistic actions: (1) the insertion of aquaporin-2 (AQP2)-containing vesicles into the apical cell membrane, as well as (2) the synthesis of more AQP2 (Figure 15.18). The upregulation of AQP on the apical membrane facilitates the shuttling of water from the tubule lumen into the cell. This water exits the cell on the basolateral membrane via AQP3 and AQP4 channels, which are constitutively

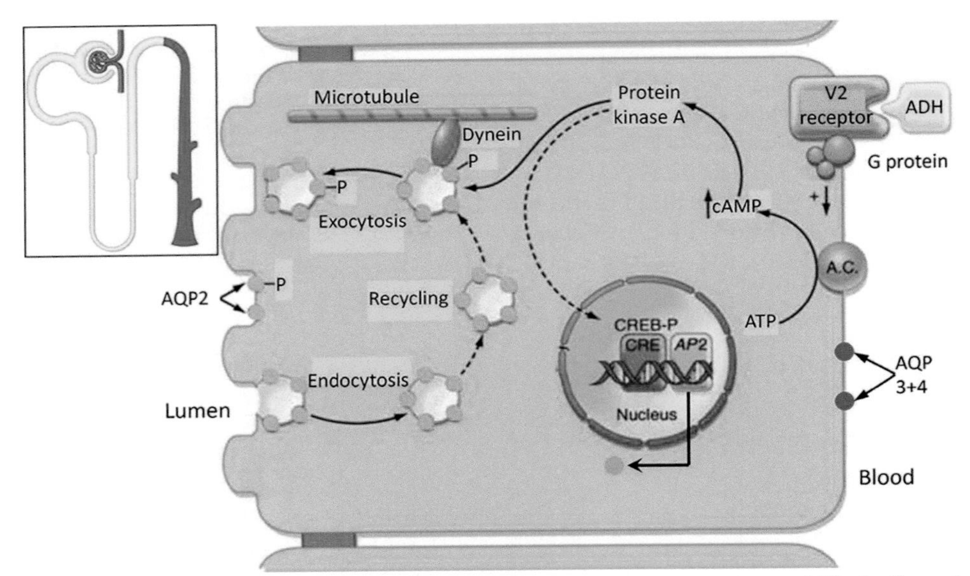

Figure 15.18 Schematic representation of the effect of AVP/ADH to increase water permeability in the principal cells of the collecting duct via the upregulation of aquaporin-2 proteins. AVP is bound to the G protein-linked V2 receptor located on the basolateral membrane of the late distal tubule and collecting duct (green shading on nephron diagram). This activates the effector, adenylate cyclase (A.C.) that interacts with dissociated G protein subunits to synthesize the second messenger, cAMP, which in turn activates protein kinase A (PKA). Activated PKA initiates two concurrent signaling pathways by phosphorylating two types of proteins. First, it phosphorylates cyclic AMP response element binding protein (CREBP), which translocates to the nucleus, binds to the cyclic AMP response element (a DNA sequence) and interacts with activator protein-2 (AP2) to induce the transcription of aquaporin-2 (AQP2). Subsequently, the AQP2 protein (light green circles) is sequestered in recycling endosome vesicles. Secondly, PKA phosphorylates AQP2, which facilitates its association with the cytoskeletal network, inducing the upregulation of AQP2 on the apical membrane. The upregulation of AQP2 promotes the uptake of water from the filtrate. In contrast with AQP2, aquaporin-3 (AQP3) and aquaporin-4 (AQP4) water channels are expressed constitutively at the basolateral membrane. "–P" denotes a phosphorylated protein.

Source of images: Koeppen, B.M. and Stanton, B.A., 2009. Berne & Levy Physiology, Updated Edition E-Book. Elsevier Health Sciences. Used by permission.

present on this membrane and not regulated by AVP/ADH. Therefore, the regulation of just one aquaporin (AQP2) by AVP/ADH on the apical membrane is sufficient to promote water absorption from the tubule into the blood (the process of antidiuresis). In the absence of AVP/ADH, the AQP2 channels are internalized from the apical membrane and sequestered in vesicles. In the absence of apical AQP2, less water is reabsorbed, and a higher volume, more dilute urine forms (the process of diuresis).

Urea and Sodium Transporter Upregulation

In order to create a concentrated urine, water must be reabsorbed from the kidney tubules, and for that to occur a high osmotic gradient in the medullary interstitial fluid must be established to draw the water out of the tubules. Two key osmolytes that contribute to the formation of a high osmotic pressure in medullary interstitial fluid that facilitate antidiuresis are sodium and urea. Although urea is excreted in the urine as a waste product, it is also an osmolyte and some urea is transported into the medullary interstitial fluid where it makes an important contribution to the formation of an osmotic gradient. As with the upregulation of AQP2, AVP/ADH acts through V2 receptors to upregulate the urea transport protein, *urea transporter A1* (UT-A1), in the apical plasma membrane of the inner medullary collecting duct, as well as apical membrane-localized *epithelial sodium channels* (ENaC) in the thick ascending tubule of the loop of Henle, the distal tubule, and the collecting duct.[7,37,38]

Clinical Considerations: Two flavors of urine: mellitus and insipidus

The word "diabetes" refers to one of several diseases characterized by *polyuria*, or a marked excessive urination, and *polydipsia*, a persistent thirst. As such, the diseases of diabetes manifest in whole or in part as osmoregulatory dysfunctions. The general condition has been known since the first century BC, when Greek physician Demetrius of Apameia named it *diabainein*, meaning "to pass through" (as in a siphon), referring to the excessive urination associated with the disease.[39] In the most common form of the disease, **diabetes mellitus** (where the Latin word for honey, *mellitus*, refers to the fact that the urine produced tastes sweet) is a complex metabolic disorder characterized by an inability of cells to absorb and process excess sugar from the blood due to a deficiency in or resistance to the hormone insulin. Diabetes mellitus is addressed in detail in Chapter 19: Metabolic Dysregulation and Disruption. In a diabetic with **hyperglycemia**, or elevated blood sugar, the load of glucose filtered by the kidneys can exceed the capacity of the kidney tubules to reabsorb glucose, since the glucose transport proteins have become saturated. Glucose is an osmolyte, and the resulting glucose in the filtrate draws water into the urine by osmosis. Thus, hyperglycemia causes a diabetic to produce a high volume of glucose-containing urine.

Another less common form of diabetes, **diabetes insipidus** (*insipidus* means "without taste" in Latin), manifests as the production of a copious urine due to a deficiency in pituitary ADH/AVP synthesis or function. The discovery of diabetes insipidus as a deficiency of pituitary ADH/AVP during the past century paved the way to our understanding of internal water balance when it was shown that the osmoreceptors that elicit responses from the hypothalamic-pituitary system are essential for the regulation of osmolality and volume of body fluids fundamental to survival.[40] Several pathophysiologic mechanisms can result in diabetes insipidus, including *central diabetes insipidus* caused by the inability to secrete ADH/AVP and *nephrogenic diabetes insipidus* caused by the inability of an otherwise normal kidney to respond to ADH/AVP.

The Renin-Angiotensin-Aldosterone System

The loss of extracellular fluid is a potent generator of thirst in mammals. However, under conditions of dehydration and/or hypovolemia the appetite for sodium also increases. The craving for sodium is present throughout a number of terrestrial vertebrates,[41] though the phenomenon has been most thoroughly studied in mammals and birds. In fact, large herds of some species, such as bison in North America and zebra and wildebeest in Africa, will travel great distances to ingest salt at mineral lick sites when the sodium content of their diet is reduced. As sodium is an osmolyte and an essential element required for the maintenance of electrochemical transmembrane potentials, sodium homeostasis is intimately tied to blood pressure regulation. The *renin-angiotensin-aldosterone system* (RAAS) (overview in Figure 15.19) is involved in regulating sodium uptake. Like AVP/ADH regulation, RAAS can be activated in response to low blood pressure caused by blood loss or by water and sodium deprivation. A product of this complex pathway is *aldosterone*, a mineralocorticoid produced by the adrenal cortex's zona glomerulosa that, among other functions, promotes salt reabsorption by

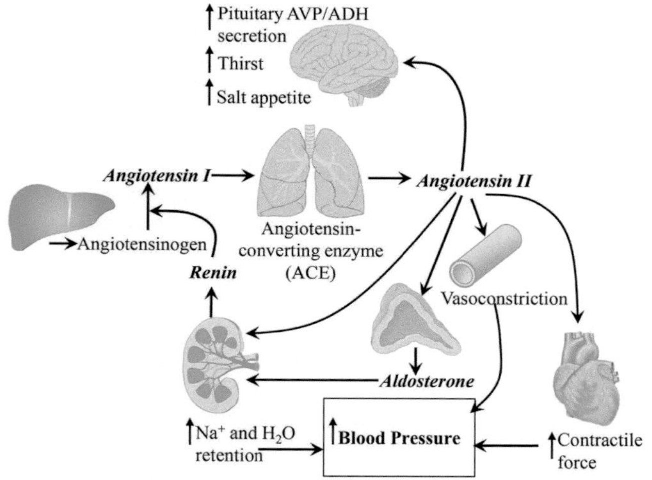

Figure 15.19 Schematic representation of the essential components of the renin-angiotensin-aldosterone system (RAAS). Activation of this system is initiated by the kidney under conditions of low blood pressure. *Note:* Angiotensin I is converted to angiotensin II by angiotensin-converting enzyme (ACE), which is present on all vascular endothelial cells. The endothelial cells within the lungs, due to the organ's high surface area, play a significant role in this conversion process.

the kidney. Recall that the adrenal cortex produces a diversity of steroids, including glucocorticoids and sex steroids. However, adrenalectomy is lethal to mammals after several days due specifically to *hyponatremia* (low blood sodium) and *hyperkalemia* (elevated blood potassium) caused by aldosterone deficiency.[42] Adrenalectomized animals can be kept alive with injections of adrenal cortical extracts[43] or high sodium chloride with cortisone supplementation.[44]

Curiously Enough . . . The importance of aldosterone in blood pressure homeostasis is particularly apparent considering that every known form of inherited hypertension in humans results from aberrant aldosterone signaling or hyperactivity of its final effectors.[45]

Natural inverse correlations between ecological sodium availability and aldosterone synthesis have also been observed. For example, elegant studies by Ken Meyers and colleagues[46,47] showed that wild Australian rabbits living in the extremely sodium-poor Snowy Mountains synthesize three to eight times more aldosterone and have greater sizes of adrenal zona glomerulosa regions (the zone that synthesizes aldosterone) compared with conspecific rabbits living in regions containing moderate (coastal habitat) or very high (desert habitat) sodium in plants and soil (Figure 15.20, Table 15.2). In a creative and illuminating touch to the study, Meyers also documented that when offered wooden stakes soaked with salt, the mountain rabbits would chew these to destruction, demonstrating a strong sodium appetite.

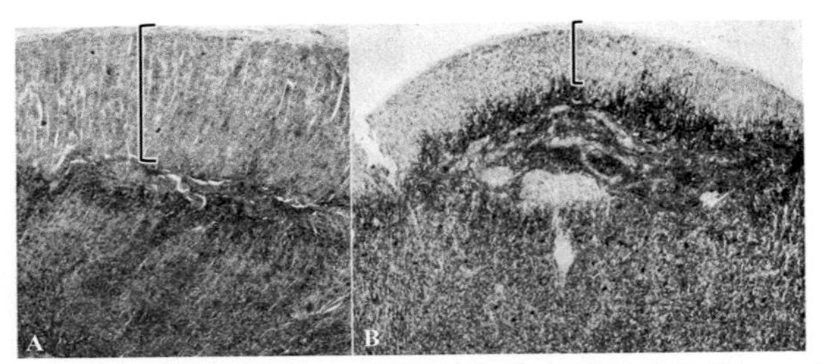

Figure 15.20 Histological section of adrenal glands from wild rabbits (*Oryctolagus cuniculus*) from sodium-poor alpine (A) and sodium-rich coastal (B) regions in Australia. Note how hypertrophy of the zona glomerulosa (brackets) of the glands from alpine-living rabbits is strikingly evident.

Source of images: Myers, K. Nature 213, 147–150 (1967). Used by permission.

Renin and the Angiotensins

Considering that a major function of aldosterone is to modulate body fluid volume by regulating sodium reabsorption through the kidneys, it is appropriate that the major stimulus for initiating aldosterone synthesis also takes place in the kidney. *Juxtaglomerular* cells, modified smooth muscle cells of the afferent arteriole that supplies blood to the glomerulus (Figure 15.21), produce this stimulus in the form of *renin*, a proteolytic enzyme secreted into the blood. Renin is secreted in response to three different, but related, conditions:

1. *Reduced kidney blood perfusion pressure*. The afferent arteriole behaves as a high-pressure baroreceptor, and reduced perfusion pressure stimulates the release of renin into the blood. In contrast, increased perfusion pressure inhibits the release of renin.
2. *Increased sympathetic nerve activity*. Juxtaglomerular cells are richly innervated by sympathetic nerve fibers that are activated reflexively by a decreased arterial blood pressure sensed by baroreceptors of the carotid sinuses and the aortic arch. The release of norepinephrine by the axonal termini and subsequent binding to beta-adrenergic receptors on the juxtaglomerular plasma membrane promotes the release of renin granules.
3. *Reduced NaCl delivery to the macula densa*. Cells in the *macula densa*, a region of the distal tubule in contact with the juxtaglomerular cells (Figure 15.21), take up NaCl in proportion to their availability via the Na/K/2Cl symporter in the luminal membrane. Reduced NaCl delivery to these cells stimulates the release of renin via a pathway that is not fully understood. Conversely, increased NaCl uptake by these cell inhibits renin release. Thus, the macula densa serves as the detector, while the juxtaglomerular apparatus acts as the effector.

The substrate for renin is *angiotensinogen*, which in humans is a 453 amino acid-long blood protein that is constitutively secreted by the liver. Following release into the blood, renin cleaves this prohormone, releasing a 10 amino acid-long peptide called *angiotensin I* (Figure 15.22). Angiotensin I is subsequently converted into the potent octapeptide hormone *angiotensin II* by the action of *angiotensin-converting enzyme* (ACE), which is a

Table 15.2 Comparison of Sodium Regulatory Indices of Rabbits Living in Three Different Habitats in Australia

Habitat	Urinary Na+ (mmol/L)	Plasma Aldosterone (ng/100 mL)	Adrenal Mass (g)	Zona Glomerulosa Size (% of total adrenal cortex)
Snowy mountains	0.53	69.0	0.12	22.6
Coast	18.00	21.0	0.11	15.6
Desert	139.00	9.0	0.13	15.3

Source: From Nelson, R.J., 2005. *An introduction to behavioral endocrinology*. Sinauer Associates.
Data extrapolated from Blair-West, J.R., Coghlan, J.P., Denton, D.A., Nelson, J.F., Orchard, E., Scoggins, B.A., Wright, R.D., Myers, K. and Junqueira, C.L., 1968. Physiological, morphological and behavioural adaptation to a sodium deficient environment by wild native Australian and introduced species of animals. *Nature*, *217*(5132), pp. 922–928.

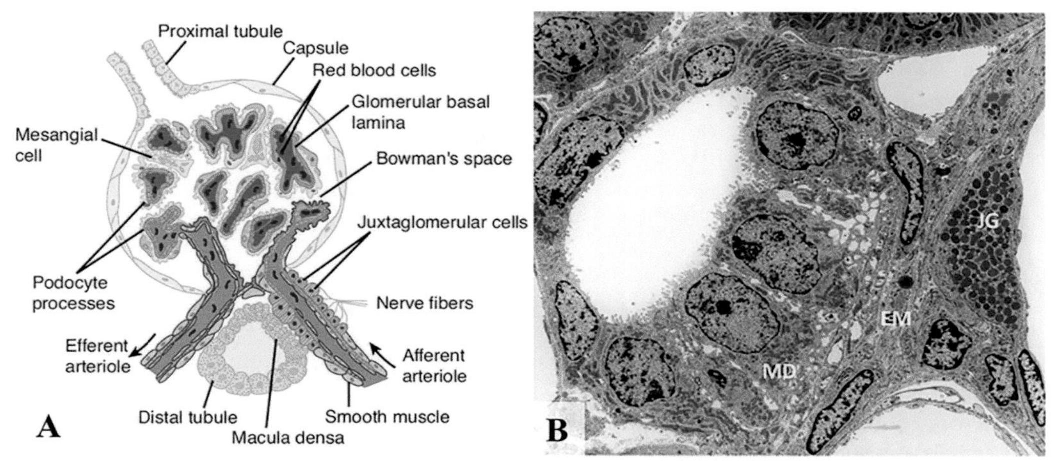

Figure 15.21 The juxtaglomerular apparatus consists of juxtaglomerular cells, macula densa cells, and extraglomerular mesangial cells all juxtaposed to the glomerulus of the nephron. (A) The macula densa, a specialized portion of the distal tubule, senses the amount of Na^+ flowing through the distal tubule. Decreased Na^+ sensed by the macula densa cells elicits a signal that stimulates renin secretion by the juxtaglomerular cells, as well as mesangial cells. **(B)** Transmission electron micrograph of juxtaglomerular apparatus from rabbit kidney, illustrating macula densa (MD), extraglomerular mesangium (EM), and a portion of a juxtaglomerular cell (JG) containing numerous electron-dense renin granules.

Source of images: (A) Barrett, K.E., et al. 2010. Ganong's review of medical physiology twenty; McGraw Hill LLC. Used by permission. (B) Taal, M.W., et al. 2011. Brenner and Rector's The Kidney E-Book. Elsevier Health Sciences. Used by permission.

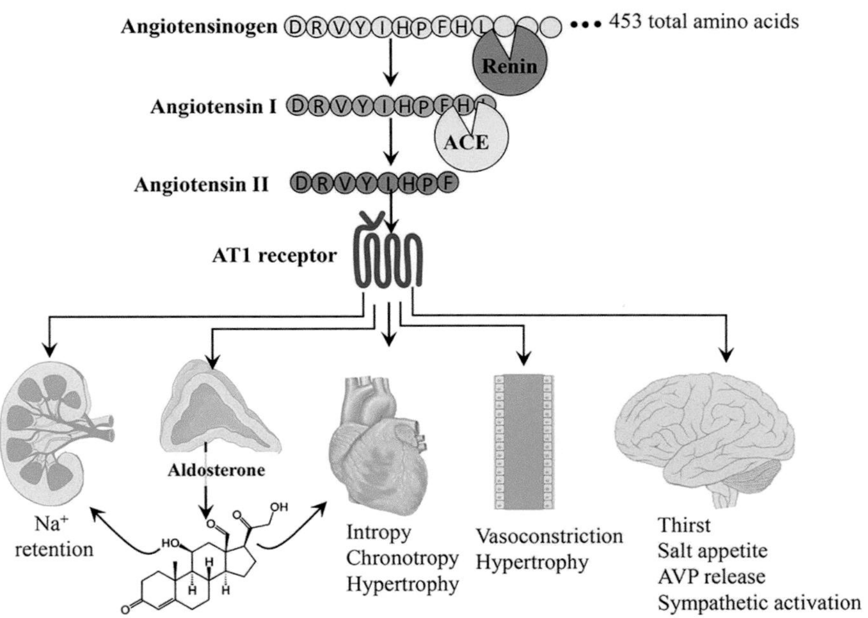

Figure 15.22 Biological actions of angiotensin II. Angiotensin II is formed by the cleavage of two precursor proteins (angiotensinogen and angiotensin I) by two proteases (renin and angiotensin-converting enzyme, respectively). In addition to stimulating aldosterone synthesis by the adrenal cortex, angiotensin II is also a powerful vasoconstrictor that raises blood pressure and influences renal flow rates both directly and indirectly. Its receptors are found in many regions of the body, including the adrenal gland, kidney, brain, and cardiovascular system.

transmembrane proteolytic enzyme localized to the plasma membranes of vascular epithelial cells throughout the body. ACE's enzymatic domain faces the lumen of blood vessels. Though initially thought to be present primarily in the vasculature of the lungs, ACE has since been found throughout the body's vasculature. However, the exceptionally high vascularity of the lungs makes it an important location for the synthesis of angiotensin II. The conversion of angiotensinogen into angiotensin I by renin is the rate-limiting step in the synthesis of angiotensin II. Another substrate of ACE is the potent vasodilator peptide hormone *bradykinin*. In this case, however, ACE degrades and inactivates bradykinin, which has the effect of potentiating an increase in blood pressure by suppressing vasodilation.

Biological Actions of Angiotensin II

Angiotensin II is perhaps the most powerful sodium-retention hormone produced by the body, and it exerts this effect by targeting diverse physiological and behavioral pathways (summarized in Figure 15.19).

Actions on the Adrenal Cortex

Angiotensin II is the primary hormone that stimulates aldosterone secretion by the glomerulosa cells of the adrenal cortex. The angiotensin II receptor is a GPCR whose intracellular signaling pathway ultimately increases the activity and synthesis of steroidogenic acute regulatory protein (StAR), the rate-limiting step in the synthesis of steroid hormones that facilitates cholesterol entry into the mitochondria.

Actions on the Kidney

Angiotensin II exerts direct effects on the kidney by stimulating renal sodium reabsorption. Angiotensin II receptors are located in the proximal tubule and likely other regions of the nephron tubules where they stimulate sodium proton exchange in the apical membrane of tubular cells, while in the basolateral membrane sodium bicarbonate cotransporters and the Na^+/K^+-ATPase pump are upregulated (Figure 15.23).

Cardiovascular Actions

Angiotensin II exerts direct effects on vascular smooth muscle, promoting vasoconstriction (Figure 15.24). Furthermore, angiotensin II acts directly on heart muscle to increase calcium influx, resulting in increased force of cardiac contraction. The combination of direct and indirect effects dramatically increases blood pressure, making angiotensin II one of the most potent pressor agents known.

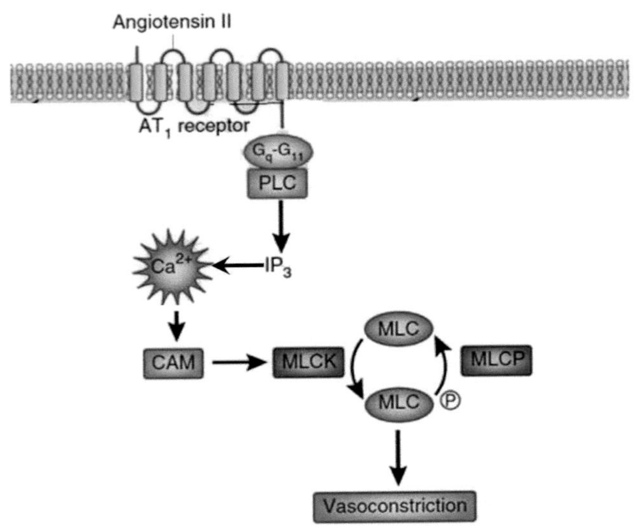

Figure 15.24 Angiotensin II exerts direct effects on vascular smooth muscle, promoting vasoconstriction. The binding of Ang II to its receptor (a GPCR) on vascular smooth muscle activates the phospholipase C (PLC) second messenger pathway, increasing the synthesis of inositol trisphosphate (IP_3). In turn, IP_3 promotes an increase in cytosolic Ca^{++} levels, activating calmodulin (CAM). CAM activates myosin light chain kinase (MLCK), which phosphorylates the myosin light chain (MLC) of smooth muscle, promoting contraction and vasoconstriction. Dephosphorylation of MLC by MLC-phosphatase (MLCP) promotes vasodilation.

Source of images: Coffman, T.M., 2011. Nat. Med., 17(11), pp. 1402-1409. Used by permission.

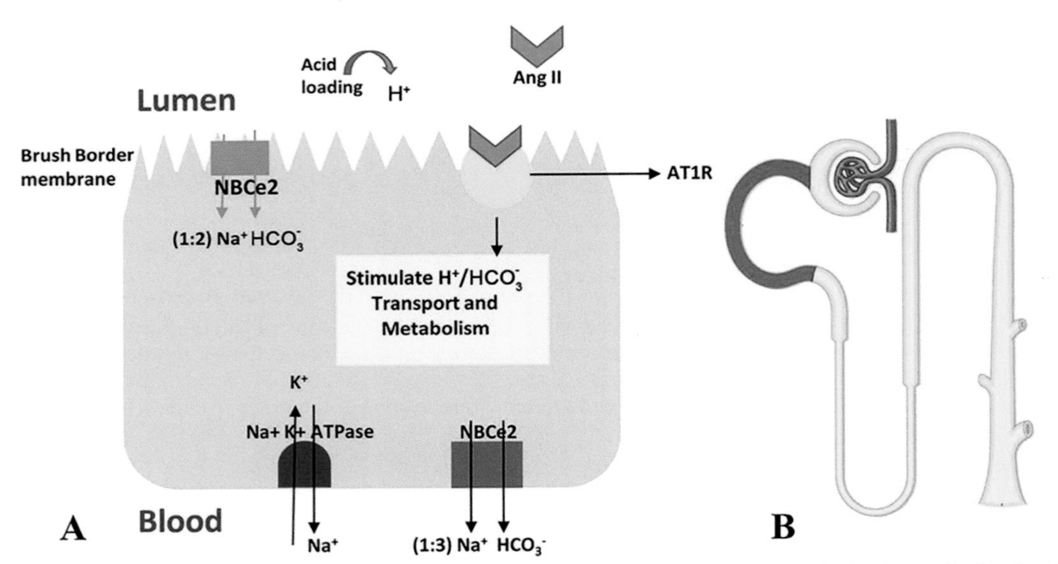

Figure 15.23 Angiotensin II increases sodium reabsorption in the proximal tubule. (A) Angiotensin II stimulates sodium-hydrogen exchange on the luminal membrane and the sodium-potassium ATPase transporter as well as sodium-bicarbonate cotransport on the basolateral membrane of the proximal tubule (green shading in **B**). These same effects of Ang II likely occur in several other parts of the renal tubule, including the loop of Henle, distal tubule, and collecting tubule (yellow regions in **B**).

Source of images: (A) Aryal, D. and Jackson, K.E., 2020. J. Biosci. Med., 8(04), p. 26.

Actions on Brain and Behavior

Angiotensin II acts as both a hormone and a neurotransmitter in the brain. Angiotensin II receptors are located in the paraventricular and supraoptic nuclei of the hypothalamus, where they stimulate the secretion of AVP/ADH in response to circulating angiotensin II. Angiotensin II receptors and neuronal cell bodies that produce the molecule as a neurotransmitter are localized in the brain's circumventricular organs where thirst and appetite for sodium are modulated. These include the subfornical organ (SFO) and organum vasculosum of the lateral terminalis (OVLT) (also see Figure 15.15 for organ locations). In fact, when angiotensin II is injected into the brain, mammals immediately begin to drink.[41]

Clinical Considerations: Pit vipers and the treatment of hypertension

The venom from the Brazilian pit viper, *Bothrops jararaca*, causes an instant drop in blood pressure in its prey, making thwarting escape impossible. The venom contains a nonapeptide shown to be a potent inhibitor of the *angiotensin-converting enzyme* (ACE), the enzyme catalyzing the conversion of the peptide angiotensin I into the active peptide hormone angiotensin II, which increases blood pressure. While the very rapid action of the venom peptide may be advantageous for a hungry snake, it could not possibly be beneficial to humans—or could it?

In 1968, studies carried out in the Royal College of Surgeons by John Vane showed that peptides from the Brazilian viper's venom inhibited the activity of ACE from dog lung.[48–50] *Captopril*, the first ACE-inhibitor drug for the treatment of hypertension, congestive heart failure, and chronic kidney disease, entered the market in 1981. Ironically, a snake venom designed through evolution to kill its prey has now found a new function: saving human lives. Vane was awarded the *1982 Nobel Prize in Physiology or Medicine* along with Sune Bergström and Bengt Samuelsson for another endocrine discovery, "their discoveries concerning prostaglandins and related biologically active substances".

Aldosterone

Produced by the glomerulosa cells of the adrenal cortex in response to angiotensin II, aldosterone is the principal biologically active mineralocorticoid in mammals, exerting its effects on ion balance primarily in the kidney, but also in sweat glands, salivary glands, and the gut. In addition to upregulation by angiotensin II, elevated blood potassium levels also stimulate the synthesis of aldosterone. This response to potassium is appropriate, considering that aldosterone promotes the uptake of sodium in exchange for potassium by Na^+/K^+-ATPase. Aldosterone exerts diverse effects on responsive cells of the kidney tubules, and some of the most important effects are listed below and depicted in Figure 15.25.

1. Increased transcription, translation, and insertion of *sodium* (ENaC) and potassium *channels* in the apical membranes of the late portion of the distal tubule and collecting duct.
2. Increased transcription, translation, and insertion of *Na^+/K^+-ATPase* in the basolateral membranes of the late portion of the distal tubule and collecting duct.

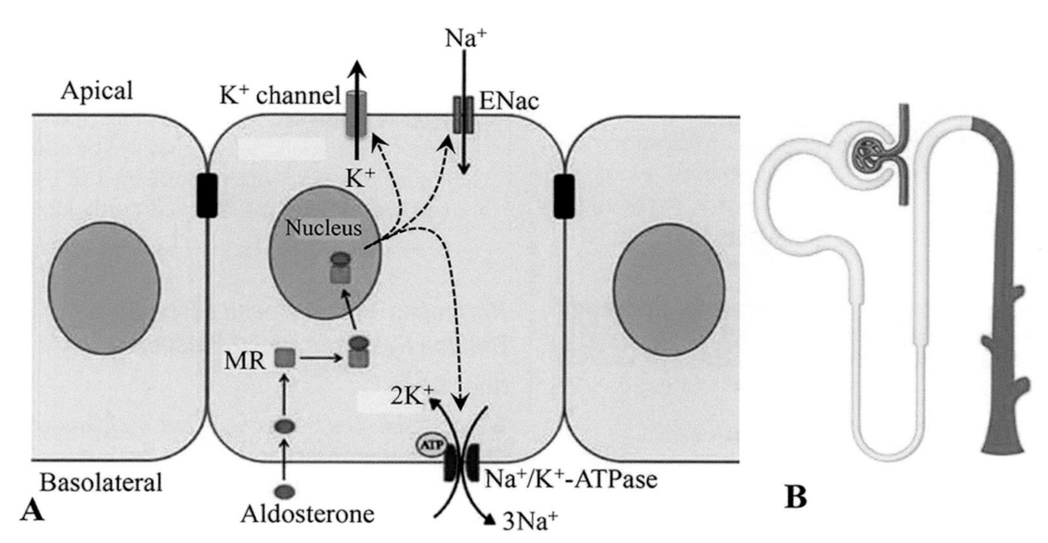

Figure 15.25 Mechanism of sodium retention by the action of aldosterone on the nephron. (A) After binding to the cytoplasmic mineralocorticoid receptor (MR), the MR translocates to the nucleus where it functions as a transcription factor. The nuclear-localized liganded MR promotes the transcription of various ion transporters, including the epithelial sodium channel (ENac), the K^+ channel, and the Na^+/K^+-ATPase pump. Following translation, both ENac and the K^+ channel are inserted into the apical membrane, whereas the Na^+/K^+-ATPase pump inserts into the basolateral membrane. **(B)** These effects of aldosterone target the late distal tubule and collecting duct (green shading).

Source of images: (A) Kostakis, I.D., et al. 2012. Hormones, 11(1), pp. 31-53; Used by permission.

3. Increased sodium reabsorption by the cells of the thick ascending tubule of the loop of Henle, likely by increasing the activity of the apical membrane symporter, Na/K/2Cl, as well as the basolateral Na^+/K^+-ATPase.

These actions of aldosterone increase entry of sodium into cells across the apical membrane and extrusion along the basolateral membrane of kidney tubule cells.

Hormones That Modulate Glomerular Filtration Rate (GFR) and Systemic Blood Pressure

Glomerular filtration is the process by which the kidneys filter blood plasma through the glomeruli of nephrons, transferring excess wastes and fluids to the nephron tubules where they will ultimately be excreted in the form of urine. *Glomerular filtration rate* (GFR) is a calculation that describes the speed at which fluid moves from the blood plasma into the kidney tubules. Whereas elevated GFR can result in reduced systemic blood pressure as more fluid from the blood is lost to the urine than is reabsorbed, reduced GFR increases systemic blood pressure as more fluid is being reabsorbed into the blood than is being filtered into the urine. A summary of the effects of various vasoactive (promote vasoconstriction or vasodilation) hormones on GFR is shown in Table 15.3, and the mechanisms of action for some are described next.

Hormones That Reduce GFR and Increase Systemic Blood Pressure

Angiotensin II

As described previously, angiotensin II is known to increase peripheral vasoconstriction and systemic blood pressure. However, it also induces renal vasoconstriction reducing both renal blood flow (RBF) and glomerular filtration rate (GFR).[51,52] By simultaneously promoting both vasoconstriction and antidiuresis, angiotensin II is a particularly powerful regulator of systemic blood pressure.

Table 15.3 Vasoactive Hormones and Their Influences on Glomerular Filtration Rate (GFR) and Blood Pressure (BP)

Category	Hormone	Effect on GFR	Effect on BP
Vasoconstrictor	Norepinephrine	↓	↑
	Epinephrine	↓	↑
	Endothelin	↓	↑
	Angiotensin II	↓	↑
Vasodilator	Endothelial-derived nitric oxide	↑	↓
	Atrial natriuretic peptide	↑	↓
	Bradykinin	↑	↓

Catecholamines

Recall that catecholamines promote peripheral vasoconstriction. This, in conjunction with stimulating increased cardiac output, has the effect of increasing mean arterial blood pressure. Catecholamines also promote vasoconstriction of the renal afferent arterioles, which serve to reduce both RBF and GFR. Note in Figure 15.21 that the afferent arteriole and juxtaglomerular cells are richly innervated by the sympathetic nervous system, and under pathological conditions such as severe hemorrhaging or dehydration, the sympathetic stress response can reduce GFR. As such, a severe drop in blood pressure due to excessive fluid loss will induce several concurrent events involving sympathetic activity that reduces GFR, thereby reducing the rate of urine production, increasing salt and water reabsorption, and promoting increased blood volume and blood pressure:

1. Decreased mean arterial pressure is detected by arterial baroreceptors of the carotid sinuses and the aortic arch, which leads to sympathetic nervous system (SNS) stimulation. The release of norepinephrine by the axonal termini and subsequent binding to beta-adrenergic receptors on the smooth muscle surrounding the afferent arteriole promotes its constriction, resulting in reduced renal blood flow and decreased GFR.
2. The juxtaglomerular cells are also richly innervated by SNS axonal termini, and SNS stimulation also promotes renin secretion by these cells, activating the renin-angiotensin-aldosterone system (RAAS). Subsequently, RAAS increases blood volume by increasing renal sodium and water reabsorption.
3. The rise in sympathetic activity also increases the release of epinephrine by chromaffin cells of the adrenal medulla, which causes further afferent arteriole vasoconstriction and decrease in GFR.

Curiously Enough . . . *Endothelins* (ETs) are a family of three related peptides synthesized by endothelial cells lining the blood vessels. Endothelin-1 (ET-1) is the most potent vasoconstrictor agent currently identified.[53]

Hormones That Increase GFR and Reduce Systemic Blood Pressure

Nitric Oxide

Nitric oxide (NO) is a gas that can function as a chemical messenger. Since it is uncharged, it can diffuse across plasma membranes and act as an intracellular or paracrine signal. NO functions as a messenger that induces a broad spectrum of biological responses and features prominently as an inducer of peripheral and renal vasodilation. As such, NO activity generally functions to both increase GFR and lower blood pressure. You were previously introduced to the role of NO as a modulator of vasodilation in Chapter 5: Receptors, where it functions as a second messenger in a pathway downstream of M3 muscarinic acetylcholine receptors. NO of endothelial origin diffuses into adjacent smooth muscle cells and promotes muscle relaxation.

Natriuretic Peptide Family

Natriuretic peptides are hormones that induce *natriuresis*, or the excretion of sodium through urine, and play key roles in both increasing GFR and lowering systemic blood pressure. The heart and brain are a source of a family of natriuretic peptides. The first of these peptides to be discovered, *atrial natriuretic peptide* (ANP) (Figure 15.26), is a 28-amino acid peptide produced primarily by the heart atria upon stretch. ANP is synthesized, stored, and released primarily by cardiac myocytes of the heart atria in response to atrial distension, angiotensin II stimulation, endothelin, and sympathetic stimulation. Therefore, elevated levels of ANP are found during hypervolemic (elevated blood volume) states, which occurs in congestive heart failure. As such, both ANP and BNP (of ventricular origin) serve as diagnostic markers for heart failure in patients. ANP receptors are present in several tissues that operate in synergy to lower blood pressure upon activation by ANP (Figure 15.26):

1. Brain: ANP inhibits AVP/ADH secretion by the pituitary neurohypophysis, decreasing water reabsorption by the distal tubule and collecting duct, and promoting diuresis. ANP also reduces thirst and sodium appetite.
2. Vascular system: ANP promotes vasodilation and reduced systemic vascular resistance by desensitizing smooth muscle contractility.
3. Adrenal gland: ANP directly inhibits aldosterone secretion by the glomerulosa cells of the adrenal cortex by blocking the stimulatory effects of angiotensin II. ANP also indirectly inhibits aldosterone production by decreasing ACTH secretion, leading to reduced steroidogenesis.
4. Kidney: ANP affects the kidneys by increasing glomerular filtration rate (GFR), which produces both natriuresis and diuresis. This is accomplished via multiple mechanisms.

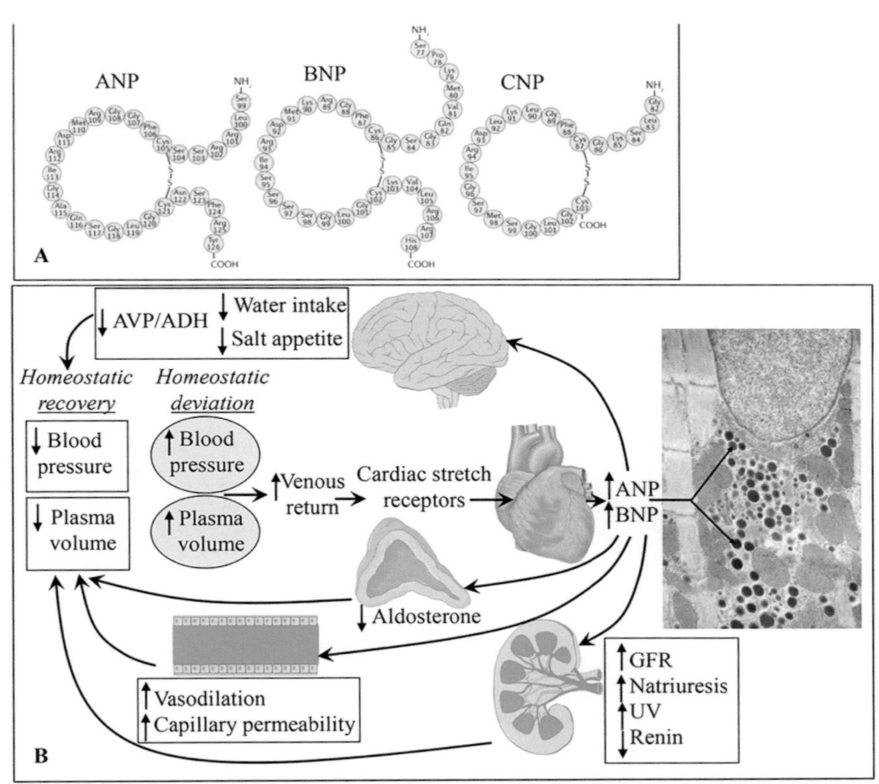

Figure 15.26 Natriuretic peptides and their biological responses following an increase in blood pressure. (A) There are three natriuretic peptides: atrial (ANP), brain (BNP) and C-type (CNP). All three hormones have a 17-member amino acid ring formed by a disulfide bridge between cysteine residues, though they are products of different genes. Despite its historical name, BNP is also released by heart tissue. **(B)** Following an increase in blood pressure, ANP and BNP are released from the heart's atria and ventricles into the circulatory system. The variety of responses include inhibition of renin secretion by the kidney, and also inhibition of the adrenal *zona glomerulosa* from producing aldosterone. ANP increases the glomerular filtration rate as well as the amount of natriuresis (Na^+ excretion in the urine) and urinary volume (UV). The capillaries experience an increase in permeability when exposed to ANP, thereby increasing the flow of blood. ANP also inhibits AVP/ADH release by the pituitary neurohypophysis, increasing water reabsorption by the kidneys. Inset: Electron micrograph of an atrial myocyte showing the location of ANP storage granules (arrows) in a mouse model.

Source of images: (A) Goetze, J.P., et al. 2020. Nat. Rev. Cardiology, 17(11), pp. 698-717; Used by permission.

1. In addition to inhibiting AVP/ADH release by the brain, ANP antagonizes AVP/ADH actions on collecting duct, further promoting antidiuresis.
2. ANP inhibits renin synthesis by the juxtaglomerular cells, further suppressing aldosterone production and blood pressure rise.
3. ANP promotes vasodilation of the afferent arteriole as well vasoconstriction of the efferent arteriole of the glomerulus, increasing GFR.
4. ANP directly inhibits sodium reabsorption by collecting duct cells.
5. ANP decreases sodium reabsorption by the proximal tubule by inhibiting the effects of angiotensin II on sodium bicarbonate reabsorption.

Therefore, the physiological effects of ANP essentially oppose those of both the AVP/ADH and the RAAS systems and function to protect the body from volume overload.

Developments & Directions: Hormones target the gill to prepare migrating salmon for changing salinity

Many species of salmon exhibit a complex life cycle where the early life stages take place in freshwater streams. They then migrate to seawater where they spend most of their adult lives, and return to freshwater to breed (Figure 15.27). The transformation of freshwater-acclimatized "parr" into seagoing "smolts" is accompanied by profound physiological changes that prepare the fish to move into the new

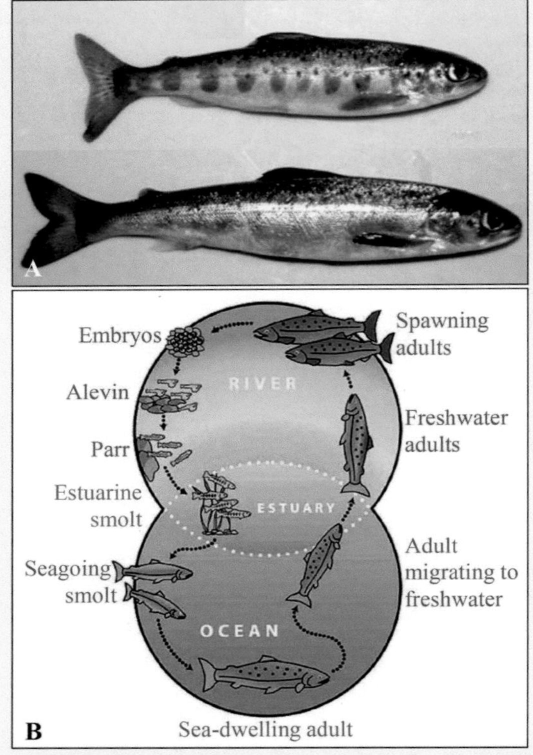

Figure 15.27 The complex life cycle of anadromous salmon is characterized by a parr-to-smolt transformation that physiologically prepares the fish for life in saltwater. (A) The freshwater parr is characterized by the presence of distinct dark "parr" marks (top fish). The beginning of the parr-smolt transformation, which initiates while the fish still lives in freshwater, is characterized morphologically by a "silvering" of the skin (right fish). Physiologically, the fish is preparing for life in saltwater. **(B)** After the transformation is complete, the adults will spend most of their lives in the ocean but will return to freshwater to spawn. After spawning, the adults of many salmon species soon die.

Source of images: (A) Courtesy of Steve McCormick; (B) Courtesy of Kings County Department of Natural Resources, WA.

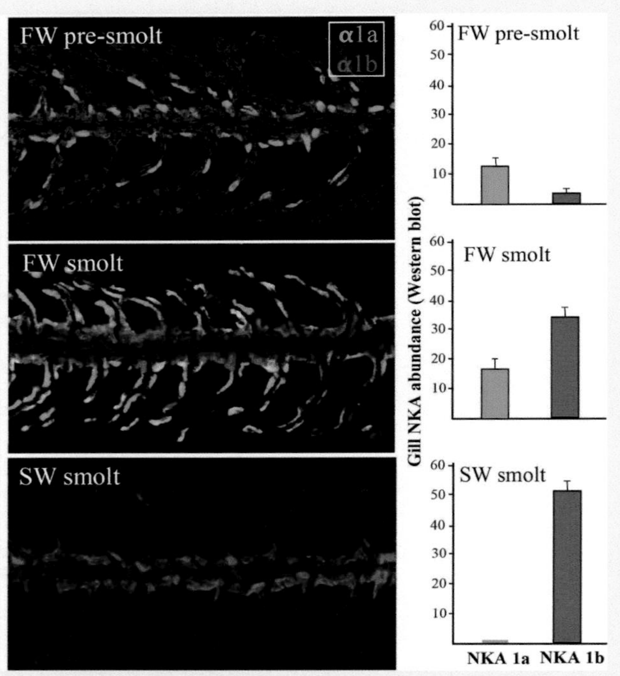

Figure 15.28 Localization and abundance of Na^+/K^+-ATPase α subunit isoforms in gill tissue of Atlantic salmon during the parr-smolt transformation. The histological images on the left depict NKAα1a (green), NKAα1b (red), and co-labeled NKAα1a and NKAα1b (yellow–orange) immunoreactive ionocytes in gill tissue of Atlantic salmon pre-smolts (19 Feb), freshwater smolts (6 May), and smolts adapted to seawater for two weeks (SW, 19 May). The histogram on the right shows the concentrations of NKA protein in gills measured via Western blot.

Source of images: Gill micrographs: Courtesy of Steve McCormick. Histograms: Modified from McCormick, S.D., et al. 2013. J. Exp. Biol. 216(7), pp. 1142-1151.

saltwater environment. Interestingly, the localization and abundance of two Na^+/K^+-ATPase α subunit isoforms changes in gill tissue of Atlantic salmon during the parr-smolt transformation (Figure 15.28), with NKAα1a and NKAα1b being the major isoforms expressed in freshwater and seawater, respectively. These observations suggest the isoforms are distinct in their regulation and physiological function, with NKAα1a presumably involved in ion uptake and NKAα1b in salt secretion.[54]

The parr-smolt transformation is accompanied by elevated levels of several hormones that mediate the acclimatization response to seawater (Figure 15.29). Specifically, elevated growth hormone (GH)/IGF-I axis activity and cortisol levels are thought to individually promote the differentiation of the seawater ionocytes from stem cell precursors and also interact to control epithelial transport capacity[55] (Figure 15.29). Interestingly, cortisol has a dual osmoregulatory function by maintaining transport proteins, like Na^+/K^+-ATPase, that are important for osmoregulation in both freshwater and seawater.[56] In contrast to GH, PRL inhibits the formation of seawater ionocytes and promotes the development of freshwater ionocytes in several other fish models, including the zebra fish (*Danio rerio*) and tilapia (*Oreochromis mossambicus*).[57–59] The role of thyroid hormones in salmon seawater acclimation is less clear than other hormones. While TH is elevated during the parr-smolt transformation and is necessary for normal seawater acclimation, it appears to play a supportive/permissive role in seawater acclimatization and may interact with both the GH/IGF-I and cortisol axes.

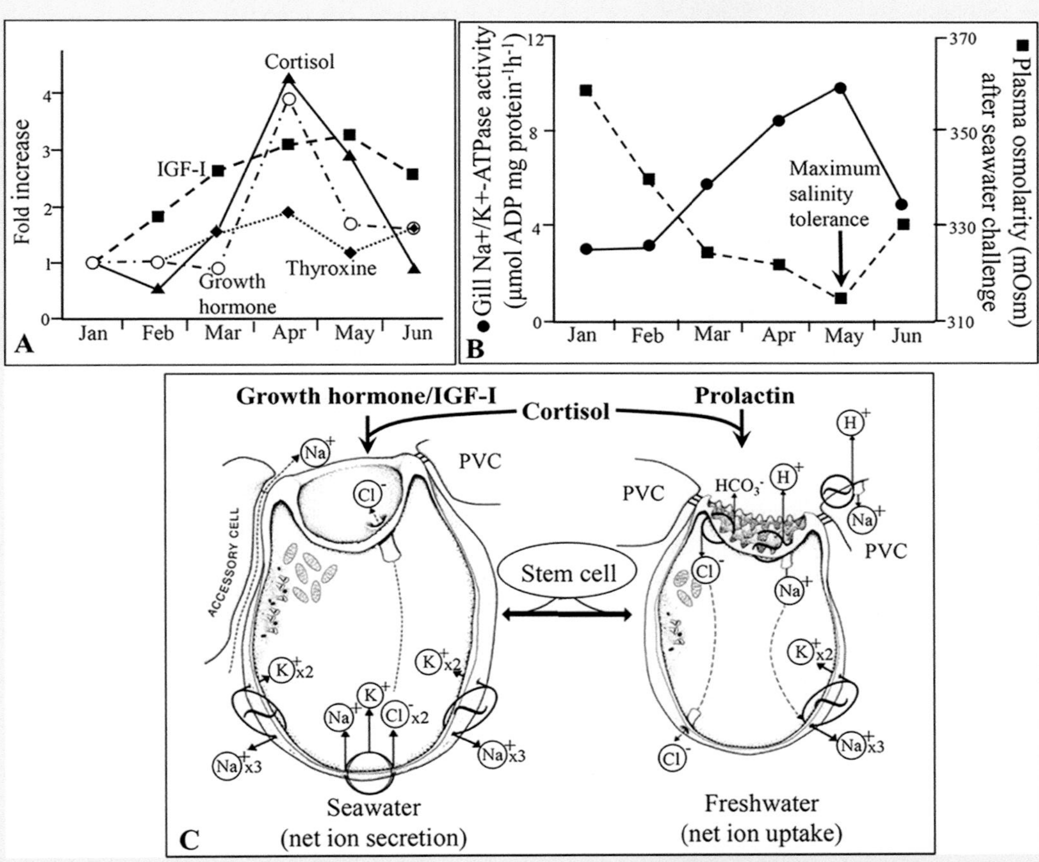

Figure 15.29 Endocrine regulation of osmoregulatory acclimatization to seawater and freshwater in Atlantic salmon. **(A)** Changes in plasma hormone levels during the parr–smolt transformation. Interactions among cortisol, growth hormone, IGF-I, and thyroid hormones results in physiological changes that are preparatory for seawater entry. **(B)** Changes in gill Na+/K+-ATPase (NKA) activity and plasma osmolarity during the parr–smolt transformation reflect the ability of the fish to tolerate seawater. **(C)** Working model for the effects of various hormones on the morphology and transport mechanisms of gill ionocytes in the gill epithelium of teleost fish. In seawater, ionocytes are generally larger and contain a deep apical crypt; in fresh water, the apical surface is broad and contains numerous microvilli. Growth hormone and cortisol can individually promote the differentiation of the seawater ionocyte, and also interact to control epithelial transport capacity. Prolactin inhibits the formation of seawater ionocytes and promotes the development of freshwater ionocytes. Cortisol also promotes acclimation to fresh water by maintaining ion transporters and ionocytes, and by interacting with prolactin.

Source of images: (A-B) modified from McCormick SD, et al. (2009) Taking it with you when you go: How perturbations to the freshwater environment, including temperature, dams and contaminants, affects seawater performance of anadromous fish. In: Haro AJ, Smith KL, Rulifson RA, et al. (eds.) Challenges For Diadromous Fishes in a Dynamic Global Environment. American Fisheries Society Symposium, vol. 69, pp. 195–214. Bethesda, MD, USA; (C) McCormick SD (2001) Am. Zool. 41: 781–794. Used by permission.

SUMMARY AND SYNTHESIS QUESTIONS

1. Describe the etymology of the two different names for the same hormone: AVP/ADH.
2. Would the Brattleboro rat be considered a xeric (dry environment-adapted) or humidic (requires the presence of water) species? Why?
3. Elegant studies by Ken Meyers and colleagues[46,47] showed that wild Australian rabbits living in the extremely sodium-poor Snowy Mountains have greater sizes of adrenal zona glomerulosa regions compared with conspecific rabbits living in regions containing moderate (coastal habitat) or very high (desert habitat) sodium in plants and soil. Explain the significance of these observations.
4. AVP/ADH binds to three receptor categories: V1a, V1b, and V2. If three mutant mouse strains, each lacking one of the receptor types, was produced, which would you predict would have the least effect on the hormones vasopressinergic and antidiuretic actions?
5. A patient producing an excessive volume of dilute urine has normal AVP/ADH synthesis and normal AVP/ADH receptors, but is found to have a dysfunctional aquaporin-2 gene. What category of disease would this patient be classified with?
6. During a surgery to remove a tumor from the adenohypophysis, the pituitary stalk was accidentally severed by the surgeon. The patient now subsequently suffers from polyuria (excess urine synthesis). Why?
7. When aldosterone is produced, what happens to urine K^+ levels? Why?
8. How do the AVP/ADH and RAAS systems work together to increase blood pressure?
9. How does it make sense that AVP/ADH upregulates both aquaporin-2 and urea transporter channels in the collecting duct of mammalian nephrons?
10. What would happen to GFR if the renal nerves were damaged or destroyed? Why?
11. One of your patient's kidneys has a constricted renal artery (renal artery stenosis). What effect would this produce on the body's mean arterial blood pressure and why?
12. How would a "fight or flight" response to acute stress affect glomerular filtration rate (GFR)? How is this change in GFR beneficial to a stressful condition?
13. Examine Figure 15.24 (Angiotensin II exerts direct effects on vascular smooth muscle, promoting vasoconstriction). Inhibiting the function of what single protein along this pathway might promote hypertension? Why?
14. Explain the osmoregulatory bases through which the following diseases alter the volume of urine production: chronic kidney disease, diabetes insipidus, diabetes mellitus.
15. Tsukada and Takei Y (2006) found that injection of atrial natriuretic peptide (ANP) into saltwater eels inhibited their drinking behavior (they drank less water). Given ANP's well-established functions in mammals, how do these findings with eels make sense?

Broad Evolutionary Trends in Osmoregulatory Hormone Function

LEARNING OBJECTIVE Describe evolutionary trends in key osmoregulatory hormone functions.

KEY CONCEPTS:

- Despite differences in osmoregulatory needs, capabilities, and ion transport organs among vertebrates, many of the hormones involved function in a similar manner.
- AVT's role in water conservation arose early in vertebrates, though the specific AVT–aquaporin response evolved with terrestriality.
- The angiotensin II receptor has been localized to the teleost gill and plays an important role in steroidogenesis in vertebrates as basal as elasmobranchs.
- Natriuretic peptides are active in the heart, brain, kidneys, and gills of every fish species studied, as well as in the rectal glands of sharks.
- In fish, glucocorticoids carry out both glucocorticoid and mineralocorticoid functions, whereas in tetrapods aldosterone has taken on the primary osmoregulatory role.
- Whereas prolactin plays a dominant role in osmoregulation in freshwater fishes and amphibians, in reptiles, birds, and mammals many of its fluid transport functions are associated with reproduction.
- Whereas thyroid hormones generally play supportive/permissive roles in fish osmoregulation, they exert profound effects on gill and kidney development during fish and amphibian metamorphosis.

The coevolution of hormones and their receptors, along with functional changes in physiological responses to those hormones, is a fascinating topic that comparative endocrinologists are uniquely positioned to investigate. Since hormones constitute the major signaling pathway for environmental osmotic stress, it is likely that the endocrine system will be a strong target of natural selection when animals are challenged by osmotically extreme environments.[60] This may result in differential endocrine responses and control even among closely related species. An understanding of broad evolutionary trends helps establish how evolution has shaped the endocrine system and its control of osmoregulatory physiology. In outlining the broad evolutionary trends in osmoregulatory hormone function (Table 15.4), it is important to note that these represent a functional presence in a particular phylum, and that there are likely to be exceptions within any given phylum. Remarkably, despite the differences in transport needs and capabilities among vertebrates, as well as the organs responsible for ion transport, many of the hormones involved function in a similar manner.

AVP/AVT

Like AVP in mammals, AVT functions as an antidiuretic in marine fish, amphibians, birds, and in the few species of reptiles that have been studied to date, increasing the ability of the kidneys to reabsorb water and produce a concentrated urine[6,60] (Table 15.4). Interestingly, in addition to its function in the kidney, in some fish species studied, AVT receptors are also upregulated in the gill in seawater where it promotes the excretion of Na^+ and Cl^-, suggesting a direct action of this peptide on the gills.[61–63] Therefore, AVT's role in water conservation appears to have arisen early in vertebrates. However, the specific AVT–aquaporin response detailed for mammals earlier may have evolved with terrestriality, since it has only been found to date in amphibians,[64] birds,[65] and mammals. Indeed, amphibians' successful colonization of terrestrial habitats approximately 200–300 million years ago may be attributed, in part, to their regulation of total body water balance by AVT.[41]

Angiotensin II

Angiotensin II has widespread effects on drinking behavior among vertebrates, promoting water uptake[66] (Table 15.4). An interesting exception is in adult amphibians where angiontensin II does not promote drinking (these animals apparently do not drink) but does promote behavioral water uptake by increasing the water absorption response, wherein the animals press a highly vascularized ventral skin patch into water or moist soil[64] (see Figure 15.1). The angiotensin II receptor has been localized to the teleost gill,[67] and exogenously administered angiotensin II also has a vasoconstrictor action and increased blood pressure in teleosts, lungfish, and elasmobranchs. Angiotensin II plays an important role in steroidogenesis in vertebrates as basal as elasmobranchs,[68,69] and the injection of angiotensin or renin can promote the release of glucocorticoids in teleosts and elasmobranchs, and aldosterone and corticosterone in the Australian lungfish.[41]

Natriuretic Peptides

Studies have identified natriuretic peptide (NP) and NP receptor immunoreactivity in the heart, brain, kidneys, and gills of every fish species studied,[70–74] as well as in the rectal glands of sharks,[75,76] demonstrating vasodilatory and osmoregulatory effects (Table 15.4). Interestingly, Tsukuda and Takei [77] demonstrated that in eels, atrial natriuretic peptide (ANP) inhibits both drinking behavior and intestinal absorption of Na^+ in saltwater, thereby limiting salt uptake. In amphibians, natriuretic peptides may function primarily to protect the animal from hypervolemia following periods of rapid rehydration.[78] NPs have also been identified in reptiles and birds,[79] though their specific functions have not yet been described.

Table 15.4 Overview of Major Physiological Functions and Target Tissues of Hormones Critical to Ion and Water Balance among Vertebrates

	Elasmobranch	Teleost	Amphibian	Reptile	Bird	Mammal
AVT/AVP	↓GFR	↑Cl secretion: G	↑Water absorption: K,S,UB ↓GFR	↑Tubular reabsorption ↓GFR	↑Tubular reabsorption ↓GFR	↑Tubular reabsorption
Angiotensin II	↑Drinking ↑1α-hydroxycort	↑Drinking ↑Cortisol	↑Water absorption: K ↑Aldosterone	↑Drinking ↑Aldosterone and corticoids	↑Drinking	↑Drinking ↑Aldosterone
Natriuretic peptide	↑Na secretion: RG	↓Drinking ↓Na uptake: I	↑GFR ↓Aldosterone	?	↑GFR ↓Aldosterone ↑Na secretion: SG	↑GFR ↓Aldosterone
Corticosteroid	↑Na secretion: RG	↑Na secretion: G ↑Na and water uptake: I	↑Na absorption: S,I,UB	↑Na reabsorption: K,I,UB	↑Na reabsorption: K,I	↑Na reabsorption: K,I,UB, SG,MG
Prolactin	?	↓Na and water permeability: G,I	↓Na and water permeability: G,I	?	Crop milk synthesis	Milk production: MG
GH/IGF-I	?	↑Na secretion: G ↑ Gill ionocytes	?	?	?	↑Kidney growth ↑Tubular Na reabsorption ↓GFR
VIP	↑Na secretion: RG	?	?	↑Na secretion: SG	↑Na secretion: SG	?

Note: K = kidney, I = intestine, UB = urinary bladder, S = skin, SG = sweat gland, MG = mammary gland, RG = rectal gland, GFR = glomerular filtration rate.
Source: After McCormick, S.D. and Bradshaw, D., 2006. Hormonal control of salt and water balance in vertebrates. *General and comparative endocrinology*, *147*(1), pp. 3–8.

Adrenocorticosteroids

In general, adrenocorticosteroids help regulate sodium metabolism by increasing sodium transport in the renal tubule of mammals, the skin and urinary bladder of amphibians, and the gills of many fish[6,80] (Table 15.4). The two primary adrenocorticosteroid classes among vertebrates are glucocorticoids and the mineralocorticoid aldosterone. Glucocorticoids are present in teleost fish, amphibians, reptiles, birds, and mammals and activate both the glucocorticoid receptor (GR) and the mineralocorticoid receptor (MR). However, aldosterone is an evolutionarily recent novelty specific to tetrapods. In fish, the glucocorticoid cortisol carries out both glucocorticoid and mineralocorticoid functions, whereas in tetrapods aldosterone has taken on primarily an osmoregulatory role.

It has been shown for many species of euryhaline fish that treatment with cortisol in freshwater improves their subsequent survival and capacity to maintain low levels of plasma ions after exposure to saltwater.[60] An increased drinking response after transfer to saltwater has been observed in salmonids treated with cortisol, and in the intestine, exogenous cortisol stimulates NKA activity, together with ion and water absorption, thus improving acclimatization to high environmental salinity. Cortisol appears to function synergistically with the GH/IGF-I axis (see later) to promote seawater acclimatization in many teleosts. In addition to promoting seawater acclimatization, there is an increasing body of evidence demonstrating that cortisol is also involved in ion uptake, thus giving it a dual osmoregulatory function. Cortisol treatment in freshwater fish increases the surface area of gill mitochondria-rich cells and the influx of sodium and chloride, affecting both renal and branchial functions.

In most terrestrial vertebrates, aldosterone has a critical role in regulating the long-term capacity for Na^+ retention, primarily through increased synthesis of renal, urinary bladder, and skin transport proteins, such as the epithelial Na^+ channel (ENaC), via classic genomic steroid action. Interestingly, though absent from most fish, aldosterone is present in ancestral sarcopterygii fish (coelacanths and lungfish),[80] and aldosterone may have evolved a mineralocorticoid function in conjunction with the evolutionary movement of these vertebrates to land. Aldosterone acts as a classical mineralocorticoid in the kidneys of all tetrapods (amphibians, reptiles, birds, and mammals) in remarkably similar ways[81,82] The fascinating story of the evolution of aldosterone-mineralocorticoid receptor complex was described in Chapter 3: Evolution of Endocrine Signaling and deserves to be looked at again through your now much more enlightened eyes.

Somatomammotropic Hormones

Recall from Chapter 8: The Pituitary Gland and Its Hormones that growth hormone (GH) and prolactin (PRL), protein hormones produced by the somatotropes and lactotropes/mammotropes (respectively) of the adenohypophysis, are closely related members of the somatomammotropic hormone family. Interestingly, in euryhaline teleosts these hormones have opposing acclimation effects, with GH promoting saltwater acclimation and PRL promoting acclimation to freshwater (Table 15.4). The relatively rapid saltwater acclimation effects of GH are independent of the hormone's growth-promoting role, though in salmon larger size also confers greater salinity tolerance. In the gills, GH promotes salinity tolerance by increasing the number and size of mitochondria-rich cells as well as the abundance of NKA and NKCC ion transporters, enhancing the ability of the gill to expel excess ions from the blood. The osmoregulatory effects of GH are mediated primarily through its capacity to increase circulating levels and local tissue production of insulin-like growth factor-I (IGF-I), which targets the gill directly. GH also affects renal function and kidney growth in mammals. For example, GH-supplementation, acting via IGF-I, increases glomerular filtration rate (GFR) and renal plasma flow (RPF) in GH-deficient as well as in normal human adults.[83,84] Furthermore, GFR and RPF are low in hypopituitarism and elevated in acromegaly, a disease caused by excess GH production. However, it is not yet clear if these effects on GFR and RPF are due indirectly to the hormone's growth-promoting effects on renal tissue, or due to direct effects on nephron physiology, or both.

In a landmark study, Grace Pickford[85] demonstrated that the pituitary hormone, PRL, was critical for the maintenance of ion balance of teleost fish in freshwater. She did this by showing that the removal of the pituitary (hypophysectomy) in freshwater-acclimatized killifish (*Fundulus heteroclitus*) resulted in loss of ions and death, which could be alleviated by replacing the lost PRL with injections. Prolactin was later shown to promote long-term effects on ion conservation and water secretion processes in a series of euryhaline teleosts by acting on the gill, kidney, and gut.[86] It has been hypothesized that prolactin stimulates ion uptake by the teleost gill in several ways: (1) direct stimulation of freshwater-type ionocyte differentiation from a progenitor/stem cell pool (see Figure 15.5); (2) promotion of the transdifferentiation of saltwater-type ionocytes into freshwater-type ionocytes; (3) direct transcriptional regulation of specific ionoregulatory genes within differentiated ionocytes.[55,58,87] In support of the latter hypothesis, a study by Breves and colleagues[59] using zebra fish (*Danio rerio*) has shown that PRL acts on gill ionocytes to upregulate the expression of the Na/Cl^--cotransporter (NCC). In a separate study using euryhaline tilapia (*Oreochromis mossambicus*), it was shown that PRL acts to stimulate the expression of gill aquaporin-3 (Aqp3),[57] though it remains unclear what function Aqp3 has in the gill of a freshwater-acclimated fish.

In urodele amphibians (newts and salamanders), prolactin also appears to promote freshwater acclimation by reducing salt and water permeability in the skin and inducing the migration of adults from a terrestrial habitat to a freshwater aquatic habitat (known as the "water drive" effect) for the purpose of reproduction.[6,80,88] Although there is no apparent role of PRL in the overall regulation of salt and water balance in reptiles and birds, this hormone does promote the movement of fluid and secretion of "crop milk" (which is used for feeding young chicks) in the crop sac of some birds. In mammals, PRL influences solute and water transport across renal, intestinal, mammary, and amniotic epithelial membranes.[89,90] It is tempting to speculate that this "transfer of function" from whole animal osmoregulation in fish to reproduction in tetrapods occurred in conjunction with the abandonment of freshwater during tetrapod evolution.[60] With no selection pressure to maintain its freshwater osmoregulatory function, prolactin in terrestrial vertebrates may have been "free" to adopt new functions. Since prolactin was already associated with the "water drive" and freshwater breeding in amphibians, it may have been predisposed to adopt a reproductive function as tetrapods became wholly terrestrial.

Thyroid Hormone

Although there is conflicting evidence regarding the role of thyroid hormones in fish osmoregulation, most studies have found that thyroid hormones alone cannot increase ion uptake or secretory capacity.[56] It is generally thought that thyroid hormones play a supportive/permissive role in osmoregulation, and in some salmon species thyroid hormones appear to support the action of growth hormone and cortisol in promoting seawater acclimation during the parr-smolt transformation. It is worth noting, however, that in some vertebrates that undergo a "spectacular" metamorphosis, such as amphibians and flatfishes (see Chapter 10: Thyroid Hormones: Development and Growth), thyroid hormone does dramatically influence osmoregulatory physiology during this larval to juvenile transformation. In flatfishes, thyroid hormones have been shown to promote the acquisition of salinity tolerance concurrent with changes in gill ionocytes.[91] In amphibians, thyroid hormones are well known to simultaneously promote the regression of the larval kidney (pronephros) and development of the adult kidney (mesonephros) during metamorphosis. Considering that thyroid hormones are thought to have co-opted the regulation of virtually every developmental program of metamorphosis in these spectacularly metamorphosing species,[92] it is not surprising that these hormones should have a particularly pronounced effect on osmoregulatory development as well. Though an influence of thyroid hormones on reptilian and avian osmoregulation has not been reported, in mammals thyroid hormones are known to influence renal development, glomerular filtration rate, and sodium and water homeostasis.[93] These effects of thyroid hormone are due to both direct renal actions and indirect cardiovascular and systemic hemodynamic effects that influence kidney function.

SUMMARY AND SYNTHESIS QUESTIONS

1. List three examples of how the functions of osmoregulatory hormones changed with the evolution of terrestriality.

Summary of Chapter Learning Objectives and Key Concepts

LEARNING OBJECTIVE Define "osmoregulation" and explain why the hormones involved in osmoregulation are so fundamental to life.

- A significant amount of energy is spent maintaining ion gradients between the intracellular and extracellular body compartments, as well with the external environment.
- Osmoregulation, or the maintenance of a constant internal osmotic condition, creates the necessary environment for diverse metabolic processes essential for survival.
- Osmoregulation plays a critical role in maintaining optimal blood pressure, which is essential for nutrient mobilization and the removal of cellular waste.
- The hormones that regulate osmoregulation are highly conserved among vertebrates.

LEARNING OBJECTIVE Provide an overview of vertebrate renal function and evolution, and describe how the kidney both produces and is the target of diverse osmoregulatory hormones.

- The nephron is the basic functional unit of the vertebrate kidney, which selectively secretes wastes and excess salts into the urine, reabsorbs nutrients, and maintains blood plasma osmotic pressure.
- The kidney both produces and is the target of diverse osmoregulatory hormones.
- The first kidneys evolved to maintain osmotic equilibrium in an aquatic environment and were modified to adapt to terrestrial life.

LEARNING OBJECTIVE Describe how mammals sense both osmotic and blood volume disequilibrium.

- An elevated blood osmotic pressure, sensed by hypothalamic osmoreceptors, promotes the sensation of "osmotic thirst".
- Low blood pressure, sensed by peripheral baroreceptors, promotes the sensation of "hypovolemic thirst".
- The activation of either osmoreceptors or baroreceptors stimulates the release of AVP/ADH from the pituitary neurohypophysis, increasing water retention and elevating blood pressure.

LEARNING OBJECTIVE List the primary hormones responsible for regulating blood pressure, diuresis and antidiuresis in mammals, and describe their general effects on various target tissues, including the brain, cardiovascular system, and the kidney.

- The hormone AVP/ADH, released under conditions of dehydration, promotes synergistic vasoconstrictive and antidiuretic responses.
- AVP/ADH promotes antidiuresis by upregulating aquaporin-2 water channels in the kidney tubules.
- The renin-angiotensin-aldosterone system (RAAS), activated under conditions of dehydration, promotes sodium uptake and retention.

- Aldosterone is an adrenal mineralocorticoid that promotes renal salt reabsorption by upregulating several types of sodium transporters.
- Hormones that increase glomerular filtration rate (GFR) inhibit salt and water retention, thereby lowering blood pressure.
- Hormones that reduce GFR promote salt and water retention, thereby increasing blood pressure.

LEARNING OBJECTIVE Describe evolutionary trends in key osmoregulatory hormone functions.

- Despite differences in osmoregulatory needs, capabilities, and ion transport organs among vertebrates, many of the hormones involved function in a similar manner.
- AVT's role in water conservation arose early in vertebrates, though the specific AVT–aquaporin response evolved with terrestriality.
- The angiotensin II receptor has been localized to the teleost gill, and plays an important role in steroidogenesis in vertebrates as basal as elasmobranchs.
- Natriuretic peptides are active in the heart, brain, kidneys, and gills of every fish species studied, as well as in the rectal glands of sharks.
- In fish, glucocorticoids carry out both glucocorticoid and mineralocorticoid functions, whereas in tetrapods aldosterone has taken on the primary osmoregulatory role.
- Whereas prolactin plays a dominant role in osmoregulation in freshwater fishes and amphibians, in reptiles, birds, and mammals many of its fluid transport functions are associated with reproduction.
- Whereas thyroid hormones generally play supportive/permissive roles in fish osmoregulation, they exert profound effects on gill and kidney development during fish and amphibian metamorphosis.

LITERATURE CITED

1. Bourque CW. Central mechanisms of osmosensation and systemic osmoregulation. *Nat Rev Neurosci.* 2008;9(7):519–531.
2. Rolfe DF, Brown GC. Cellular energy utilization and molecular origin of standard metabolic rate in mammals. *Physiol Rev.* 1997;77(3):731–758.
3. Clausen T, Van Hardeveld C, Everts ME. Significance of cation transport in control of energy metabolism and thermogenesis. *Physiol Rev.* 1991;71(3):733–774.
4. Bernard C. Leç ons sur les phénomenes de la vie communs aux animaux et aux végétaux. Paris: J. *B Baillière*. 1878.
5. Hoenig MP, Zeidel ML. Homeostasis, the milieu interieur, and the wisdom of the nephron. *Clin J Am Soc Nephrol.* 2014;9(7):1272–1281.
6. Bentley PJ. *Endocrines and Osmoregulation: A Comparative Account in Vertebrates* (2nd ed.). Springer-Verlag; 2002.
7. Sands JM, Layton HE. The urine concentrating mechanism and urea transporters. In: Alpern RJ, Caplan MJ, Ow M, eds. *Seldin and Giebisch's The Kidney: Physiology & Pathophysiology* (5th ed.). Academic Press; 2013:1463–1510.
8. Bentley PJ, Yorio T. Do frogs drink?. *J Exp Biol* 1979;79(41–46).
9. Bentley PJ. The amphibia. In: Bentley PJ, ed. *Endocrines and Osmoregulation. A Comparative Account in Vertebrates*, Vol 39. Springer; 2002:155–186.
10. Hillyard SD. Behavioral, molecular and integrative mechanisms of amphibian osmoregulation. *J Exp Zool.* 1999;283(7):662–674.
11. Macknight AD, DiBona DR, Leaf A. Sodium transport across toad urinary bladder: a model "tight" epithelium. *Physiol Rev.* 1980;60(3):615–715.
12. Jorgensen CB. 200 years of amphibian water economy: from Robert Townson to the present. *Biol Rev Camb Philos Soc.* 1997;72(2):153–237.
13. Larsen EH. Hans H. Ussing—scientific work: contemporary significance and perspectives. *Biochim Biophys Acta.* 2002;1566(1–2):2–15.
14. Koefoed-Johnsen V, Ussing HH, Zerahn K. The origin of the short-circuit current in the adrenaline stimulated frog skin. *Acta Physiol Scand.* 1952;27(1):38–48.
15. Larsen EH, Deaton LE, Onken H, et al. Osmoregulation and excretion. *Compr Physiol.* 2014;4(2):405–573.
16. Ussing HH, Zerahn K. Active transport of sodium as the source of electric current in the short-circuited isolated frog skin. *Acta Physiol Scand.* 1951;23:110–127.
17. Bentley PJ. Amiloride: a potent inhibitor of sodium transport across the toad bladder. *J Physiol.* 1968;195(2):317–330.
18. Garty H, Palmer LG. Epithelial sodium channels: function, structure, and regulation. *Physiol Rev.* 1997;77(2):359–396.
19. Harvey BJ. Energization of sodium absorption by the H(+)-ATPase pump in mitochondria-rich cells of frog skin. *J Exp Biol.* 1992;172:289–309.
20. Jensen LJ, Willumsen NJ, Amstrup J, Larsen EH. Proton pump-driven cutaneous chloride uptake in anuran amphibia. *Biochim Biophys Acta.* 2003;1618(2):120–132.
21. Koefoed-Johnsen V, Ussing HH. The nature of the frog skin potential. *Acta Physiol Scand.* 1958;42(3–4):298–308.
22. Ogushi Y, Tsuzuki A, Sato M, et al. The water-absorption region of ventral skin of several semiterrestrial and aquatic anuran amphibians identified by aquaporins. *Am J Physiol Regul Integr Comp Physiol.* 2010;299(5):R1150–1162.
23. Feraille E, Doucet A. Sodium-potassium-adenosinetriphosphatase-dependent sodium transport in the kidney: hormonal control. *Physiol Rev.* 2001;81(1):345–418.
24. Hoenig MP, Zeidel ML. Homeostasis, the milieu interieur, and the wisdom of the nephron. *Clin J Am Soc Nephrol.* 2014;9(7):1272–1281.
25. Morales MM, Falkenstein D, Lopes AG. The cystic fibrosis transmembrane regulator (CFTR) in the kidney. *An Acad Bras Cienc.* 2000;72(3):399–406.
26. Parry G. Osmotic adaptation in fishes. *Biol Rev.* 1966;41:392–444.
27. Evans DH, Piermarini PM, Choe KP. The multifunctional fish gill: dominant site of gas exchange, osmoregulation, acid-base regulation, and excretion of nitrogenous waste. *Physiol Rev.* 2005;85(1):97–177.
28. Stricker EM, Hosutt JA, Verbalis JG. Neurohypophyseal secretion in hypovolemic rats: inverse relation to sodium appetite. *Am J Physiol.* 1987;252(5 Pt 2):R889–896.
29. Fitzsimons JT. Angiotensin, thirst, and sodium appetite. *Physiol Rev.* 1998;78(3):583–686.
30. Stachenfeld NS. Acute effects of sodium ingestion on thirst and cardiovascular function. *Curr Sports Med Rep.* 2008;7(4 Suppl):S7–13.
31. Oliver G, Schäfer EA. On the physiological action of extracts of pituitary body and certain other glandular organs. *J Physiol (Lond).* 1895;18:277–279.
32. Farini F. Diabete insipido ed opoterapia. *Gazz Osped Clin.* 1913;34:1135–1139.
33. Vongraven D. Die nierenwirkung von hypophysenextrakten meschen. *Berl Klin Wochenscgr* 1913;50:2083–2086.

34. Valtin H, Schroeder HA, Benirschke K, Sokol HW. Familial hypothalamic diabetes insipidus in rats. *Nature*. 1962;196:1109–1110.
35. Möhring J, Kohrs G, Möhring B, Petri M, Homsy E, Haack D. Effects of prolonged vasopressin treatment in Brattleboro rats with diabetes insipidus. *Am J Physiol*. 1978;234(2):F106–111.
36. Paul MJ, Peters NV, Holder MK, et al. Atypical social development in vasopressin-deficient brattleboro rats. *eNeuro*. 2016;3(2).
37. Sands JM, Blount MA, Klein JD. Regulation of renal urea transport by vasopressin. *Trans Am Clin Climatol Assoc*. 2011;122:82–92.
38. Teoh CW, Robinson LA, Noone D. Perspectives on edema in childhood nephrotic syndrome. *Am J Physiol Renal Physiol*. 2015;309(7):F575–F582.
39. Eknoyan G. A history of diabetes insipidus: paving the road to internal water balance. *Am J Kidney Dis*. 2010;56(6):1175–1183.
40. Schrier RW. Body water homeostasis: clinical disorders of urinary dilution and concentration. *J Am Soc Nephrol*. 2006;17(7):1820–1832.
41. Lovejoy DA. *Neuroendocrinology: An Integrated Approach*. John Wiley & Sons; 2005.
42. Miller WL. A brief history of adrenal research: Steroidogenesis—the soul of the adrenal. *Mol Cell Endocrinol*. 2013;371(1–2):5–14.
43. Rogoff JM, Stewart GN. Further studies on adrenal insufficiency in dogs. *Science*. 1926;64(1649):141–142.
44. Burns TW. Endocrine factors in the water metabolism of the desert mammal, G. gerbillus. *Endocrinology*. 1956;58(2):243–254.
45. Lifton RP, Gharavi AG, Geller DS. Molecular mechanisms of human hypertension. *Cell*. 2001;104(4):545–556.
46. Blair-West JR, Coghlan JP, Denton DA, et al. Physiological, morphological and behavioural adaptation to a sodium deficient environment by wild native Australian and introduced species of animals. *Nature*. 1968;217(5132):922–928.
47. Myers K. Morphological changes in the adrenal glands of wild rabbits. *Nature*. 1967;213(5072):147–150.
48. Bakhle YS. Conversion of angiotensin I to angiotensin II by cell-free extracts of dog lung. *Nature*. 1968;220(5170):919–921.
49. Smith CG, Vane JR. The discovery of captopril. *FASEB J*. 2003;17(8):788–789.
50. Gavras H, Brunner HR, Turini GA, et al. Antihypertensive effect of the oral angiotensin converting-enzyme inhibitor SQ 14225 in man. *N Engl J Med*. 1978;298(18):991–995.
51. Zhuo JL, Ferrao FM, Zheng Y, Li XC. New frontiers in the intrarenal Renin-Angiotensin system: a critical review of classical and new paradigms. *Front Endocrinol (Lausanne)*. 2013;4:166.
52. Zhuo JL, Li XC. Proximal nephron. *Compr Physiol*. 2013;3(3):1079–1123.
53. Kawanabe Y, Nauli SM. Endothelin. *Cell Mol Life Sci*. 2011;68(2):195–203.
54. Blanco G, Mercer RW. Isozymes of the Na-K-ATPase: heterogeneity in structure, diversity in function. *Am J Physiol*. 1998;275(5 Pt 2):F633–650.
55. Sakamoto T, McCormick SD. Prolactin and growth hormone in fish osmoregulation. *Gen Comp Endocrinol*. 2006;147(1):24–30.
56. McCormick SD. Endocrine control of osmoregulation in teleost fish. *Am Zool*. 2001;41:781–794.
57. Breves JP, Inokuchi M, Yamaguchi Y, et al. Hormonal regulation of aquaporin 3: opposing actions of prolactin and cortisol in tilapia gill. *J Endocrinol*. 2016;230(3):325–337.
58. Breves JP, McCormick SD, Karlstrom RO. Prolactin and teleost ionocytes: new insights into cellular and molecular targets of prolactin in vertebrate epithelia. *Gen Comp Endocrinol*. 2014;203:21–28.
59. Breves JP, Serizier SB, Goffin V, McCormick SD, Karlstrom RO. Prolactin regulates transcription of the ion uptake Na+/Cl⁻ cotransporter (ncc) gene in zebrafish gill. *Mol Cell Endocrinol*. 2013;369(1–2):98–106.
60. McCormick SD, Bradshaw D. Hormonal control of salt and water balance in vertebrates. *Gen Comp Endocrinol*. 2006;147(1):3–8.
61. Avella M, Part P, Ehrenfeld J. Regulation of Cl⁻ secrsetion in seawater fish (Dicentrarchus labrax) gill respiratory cells in primary culture. *J Physiol*. 1999;516 (Pt 2):353–363.
62. Guibbolini ME, Henderson IW, Mosley W, Lahlou B. Arginine vasotocin binding to isolated branchial cells of the eel: effect of salinity. *J Mol Endocrinol*. 1988;1(2):125–130.
63. Balment RJ, Lu W, Weybourne E, Warne JM. Arginine vasotocin a key hormone in fish physiology and behaviour: a review with insights from mammalian models. *Gen Comp Endocrinol*. 2006;147(1):9–16.
64. Uchiyama M, Konno N. Hormonal regulation of ion and water transport in anuran amphibians. *Gen Comp Endocrinol*. 2006;147(1):54–61.
65. Goldstein DL. Regulation of the avian kidney by arginine vasotocin. *Gen Comp Endocrinol*. 2006;147(1):78–84.
66. Nishimura H. Role of the renin—angiotensin system in osmoregulation. In: Pang PKT, Schreibman MP, Sawyer WH, eds. *Endocrinology: Fundamentals and Biomedical Implications*, Vol. 2. Academic Press; 1987:157–187.
67. Marsigliante S, Muscella A, Vinson GP, Storelli C. Angiotensin II receptors in the gill of sea water- and freshwater-adapted eel. *J Mol Endocrinol*. 1997;18(1):67–76.
68. Anderson WG, Pillans RD, Hyodo S, et al. The effects of freshwater to seawater transfer on circulating levels of angiotensin II, C-type natriuretic peptide and arginine vasotocin in the euryhaline elasmobranch, Carcharhinus leucas. *Gen Comp Endocrinol*. 2006;147(1):39–46.
69. Anderson WG, Taylor JR, Good JP, Hazon N, Grosell M. Body fluid volume regulation in elasmobranch fish. *Comp Biochem Physiol A Mol Integr Physiol*. 2007;148(1):3–13.
70. Evans DH. An emerging role for a cardiac peptide hormone in fish osmoregulation. *Annu Rev Physiol*. 1990;52:43–60.
71. Evans DH, Takei Y. A putative role for natriuretic peptides in fish osmoregulation. *News physiol Sci*. 1992;7:15–19.
72. Hagiwara H, Hirose S, Takei Y. Natriuretic peptides and their receptors. *Zool Sci*. 1995;12:141–149.
73. Donald JA, Toop T, Evans DH. Localization and analysis of natriuretic peptide receptors in the gills of the toadfish, Opsanus beta (teleostei). *Am J Physiol*. 1994;267(6 Pt 2):R1437–1444.
74. Donald JA, Toop T, Evans DH. Distribution and characterization of natriuretic peptide receptors in the gills of the spiny dogfish, Squalus acanthias. *Gen Comp Endocrinol*. 1997;106(3):338–347.
75. Silva P, Stoff JS, Solomon RJ, et al. Atrial natriuretic peptide stimulates salt secretion by shark rectal gland by releasing VIP. *Am J Physiol*. 1987;252(1 Pt 2):F99–F103.
76. Silva P, Solomon RJ, Epstein FH. Mode of activation of salt secretion by C-type natriuretic peptide in the shark rectal gland. *Am J Physiol*. 1999;277(6 Pt 2):R1725–1732.
77. Tsukada T, Takei Y. Integrative approach to osmoregulatory action of atrial natriuretic peptide in seawater eels. *Gen Comp Endocrinol*. 2006;147(1):31–38.
78. Donald JA, Trajanovska S. A perspective on the role of natriuretic peptides in amphibian osmoregulation. *Gen Comp Endocrinol*. 2006;147(1):47–53.
79. Trajanovska S, Donald JA. Molecular cloning of natriuretic peptides from the heart of reptiles: loss of ANP in diapsid reptiles and birds. *Gen Comp Endocrinol*. 2008;156(2):339–346.
80. Bentley PJ. *Comparative Vertebrate Endocrinology* (3rd ed.). Cambridge University Press; 1998.
81. Bradshaw SD, Rice GE. The effects of pituitary and adrenal hormones on renal and postrenal reabsorption of water and electrolytes in the lizard, Varanus gouldii (gray). *Gen Comp Endocrinol*. 1981;44(1):82–93.
82. Shane MA, Nofziger C, Blazer-Yost BL. Hormonal regulation of the epithelial Na+ channel: from amphibians to mammals. *Gen Comp Endocr*. 2006;147(1):85–92.

83. Ogle GD, Rosenberg AR, Kainer G. Renal effects of growth hormone. I. Renal function and kidney growth. *Pediatr Nephrol.* 1992;6(4):394–398.
84. Hisano S, Latta K, Krieg RJ, Chan JCM. Growth hormone and renal function. *Nephrology*. 1997;3:309–314.
85. Pickford GE, Phillips JG. Prolactin, a factor in promoting survival of hypophysectomized killifish in fresh water. *Science*. 1959;130(3373):454–455.
86. Hirano T. The spectrum of prolactin action in teleosts. *Prog Clin Biol Res*. 1986;205:53–74.
87. Hiroi J, McCormick SD. New insights into gill ionocyte and ion transporter function in euryhaline and diadromous fish. *Respir Physiol Neurobiol.* 2012;184:257–268.
88. Brown PS, Brown SC. Osmoregulatory actions of prolactin and other adenohypophysial hormones. In: Pang PKT, Schreibman MP, eds. *Vertebrate Endocrinology: Fundamentals and Biomedical Implications, Vol. 2, Regulation of Water and Electrolytes*. Academic Press; 1987:45–84.
89. Bole-Feysot C, Goffin V, Edery M, Binart N, Kelly PA. Prolactin (PRL) and its receptor: actions, signal transduction pathways and phenotypes observed in PRL receptor knockout mice. *Endocr Rev*. 1998;19(3):225–268.
90. Freeman ME, Kanyicska B, Lerant A, Nagy G. Prolactin: structure, function, and regulation of secretion. *Physiol Rev*. 2000;80(4):1523–1631.
91. Schreiber AM. Metamorphosis and early larval development of the flatfishes (Pleuronectiformes): an osmoregulatory perspective. *Comp Biochem Physiol B Biochem Mol Biol.* 2001;129(2–3):587–595.
92. Schreiber AM, Das B, Huang H, Marsh-Armstrong N, Brown DD. Diverse developmental programs of Xenopus laevis metamorphosis are inhibited by a dominant negative thyroid hormone receptor. *Proc Natl Acad Sci U S A*. 2001;98(19):10739–10744.
93. Mariani LH, Berns JS. The renal manifestations of thyroid disease. *J Am Soc Nephrol*. 2012;23(1):22–26.

Unit Overview
Unit VI

Appetite, Digestion, and Metabolism

Caveman with Junk Food, by Banksy. Los Angeles, 2008, Stencil and colored spray on concrete wall.

DOI: 10.1201/9781003359241-21

OPENING QUOTATION:

> "Throughout evolution, humans and animals have evolved redundant mechanisms promoting the accumulation of fat during periods of feast to survive during periods of famine. However, what was an asset during evolution has become a liability in the current 'pathoenvironment' or 'obesogenic' environment."
>
> —Galgani J, Ravussin E. Energy metabolism, fuel selection and body weight regulation. *International Journal of Obesity*. 2009;32(S7):S109.

Perhaps no work of contemporary art most effectively captures the mismatch of the modern human lifestyle with our bodies than does Banksy's "caveman with junk food". While human cultures have changed drastically since Paleolithic times, particularly in the types and quantities of the foods that we eat, our physiology and metabolism have not changed commensurately from that of our Pleistocene ancestors.

Our ancestral physiology imparted an ability to efficiently balance short-term intermittent swings in food abundance and deficit with long-term storage of energy reserves for growth and activity. Prior to the advent of more secure and predictable sources of food, such as agriculture and animal domestication, an ability to rapidly and efficiently accumulate fat reserves gave early humans a significant survival advantage. However, today we live in an industrialized world where efficient food production and distribution methods provide much of the global population access to abundant and nutritionally rich foods, and sustaining an efficient metabolism under these conditions may come with long-term costs. For the first time in human history, such mismatches to modernity are now producing skyrocketing rates of obesity, type 2 diabetes, cardiovascular disorders, and other dangerous chronic metabolic diseases. Unit VI explores the endocrine regulation of appetite, digestion, and energy homeostasis in four chapters. Chapter 16 describes the integrated neural and endocrine regulation of digestion, and Chapter 17 addresses the hormonal regulation of short-term appetite signaling and long-term energy balance. Chapter 18 discusses the endocrine control of energy storage, mobilization, metabolism, and thermogenesis in the normal physiological state. Chapter 19 explores dysfunctions of metabolism in humans, focusing on the examples of obesity, type 1 diabetes, type 2 diabetes, and the roles that "obesogens" and "diabetogens" may play in these diseases.

Chapter 16: **Digestion**
Chapter 17: **Appetite**
Chapter 18: **Energy Homeostasis and Thermogenesis**
Chapter 19: **Metabolic Dysregulation and Disruption**

CHAPTER

16

Digestion

CHAPTER LEARNING OBJECTIVES:

- Describe the first experiments demonstrating that gut digestive function is under both neural and endocrine control and that gut hormones influence brain function.
- Provide an overview of the functional anatomy of the digestive tract and accessory organs from mouth to colon, and describe the major events of the cephalic, gastric, and intestinal phases of gastrointestinal secretion.
- Describe the integrated neural and endocrine regulation of digestion, listing the major hormones and their specific functions.

OPENING QUOTATIONS:

"The stomach has the liver below it like a fire underneath a cauldron; and thus the stomach is like a kettle of food, the gall-bladder its cook, and the liver is the fire."

—Master Nicolaus, ca. 1150–1200

"[I]t is clear that the mere act of eating, the food even not reaching the stomach, determines the stimulation of the gastric glands."

—Pavlov, 1904 Nobel Lecture

Introduction

LEARNING OBJECTIVE Describe the first experiments demonstrating that gut digestive function is under both neural and endocrine control and that gut hormones influence brain function.

KEY CONCEPTS:

- Ivan Pavlov discovered that the nervous system plays a key role in regulating digestion.
- With their discovery of secretin, the first hormone ever identified, William Bayliss and Ernest H. Starling showed that the gut was not solely under neural control but also functions as an endocrine organ and a target for endocrine regulation.
- Digestion reflects a highly integrative physiological system, a complex interaction among different digestive organs, peripheral tissues, and the brain.
- A bidirectional "brain-gut axis" regulates both appetite and digestion.

In Chapter 15: Blood Pressure and Osmoregulation, you learned how salt and water balance constitute primary physiological needs for all organisms. An equally fundamental physiological requirement is the procurement and distribution of nutrients for energy and cellular building blocks. In fact, osmoregulation and feeding are closely interrelated, as food ingestion introduces water, salt, and other essential ions into the organism. For example, in humans the average consumption of a meal is accompanied by the exchange of about 9 liters of water, and many transporters of the intestine rely on ion gradients for the uptake of nutrients, such as amino acids and carbohydrates (Figure 16.1). Considering this, it should not be surprising that many of the same hormones regulate both processes.

Most foods eaten by animals are chemically far more complex than can be used immediately and must first be transformed into simpler compounds that can be absorbed and metabolized by cells. The processing of complex nutrients into forms that can be readily assimilated by the body is called *digestion*. Digestion in vertebrates is a complex operation of mechanical and chemical breakdown of foodstuffs by the gastrointestinal tract and accessory organs (salivary glands, liver, gall bladder, pancreas), followed by the absorption of nutrients into the blood and the elimination of wastes in the feces. Ultimately, digestion reflects a highly integrative physiological system, a complex interaction among different digestive organs, peripheral tissues, and the brain. This chapter describes neural, endocrine, and neuroendocrine mechanisms that regulate digestion, focusing primarily on well-studied mammalian models. Considering that the first hormone to ever be identified, secretin, originates in the gut, the history of gastrointestinal physiology plays a particularly prominent role in the foundation of the field of endocrinology.

DOI: 10.1201/9781003359241-22

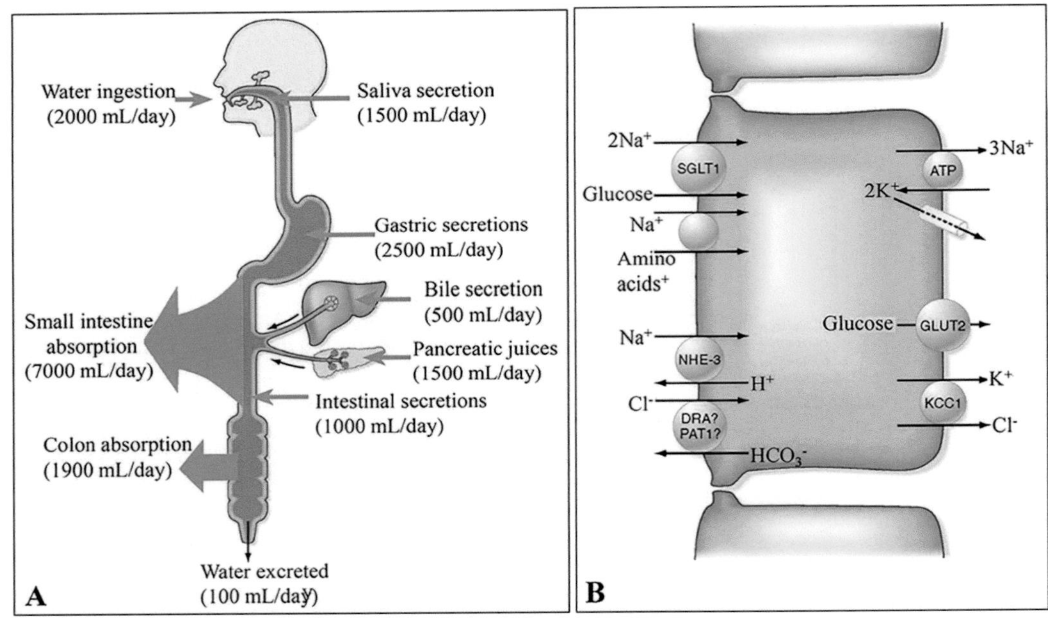

Figure 16.1 Fluid balance in the human gastrointestinal tract, and the absorption of glucose and amino acids by the gut, are linked to ion transport across the intestinal epithelium. (A) Overall fluid balance in the human gastrointestinal tract. On average, about 2 L of water is ingested per day, and 7 L of various secretions enters the gastrointestinal tract. Of this total, most is absorbed in the small intestine and the colon, leaving about 100 mL in the stool. **(B)** The general mechanism for ion-driven nutrient and water absorption in the small intestine. The uptake of water, nutrients, and ions is actively driven by the Na^+/K^+-ATPase pump. Sodium-glucose transport protein 1 (SGLT1), sodium-hydrogen antiporter 3 (NHE-3), putative anion transporter 1 (PAT1), and "downregulated in adenoma (DRA) HCO_3^-/Cl^- exchanger, glucose transporter 2 (GLUT2), K^+-Cl^- cotransporter type 1 (KCC1), Na^+/K^+-ATPase pump (ATP).

Source of images: Koeppen, B.M. and Stanton, B.A., 2009. Berne & Levy Physiology, Updated Edition E-Book. Elsevier Health Sciences. Used by permission.

Neural Control of Digestion

An event that catalyzed the modern understanding of gastrointestinal physiology that took place in the 1820s was a description by a then obscure American army surgeon, Captain William Beaumont, of a man with a fistula (a hole in between organs) in his stomach caused by a gunshot wound. In this case, the rare fistula manifested as an opening that allowed Beaumont access to the inside of the patient's stomach from outside the abdominal cavity. By collecting gastric contents from his patient over the course of many years, Beaumont discovered that gastric digestion is a chemical process that was controlled by nervous influences that affect gastric secretion. Importantly, Beaumont's experiments inspired Russian physiologist Ivan Petrovich Pavlov to conduct his famous experiments demonstrating that classical conditioning could spur dogs to salivate on cue. Even more important, Beaumont spurred Pavlov to conduct fistula operations of his own in dogs in 1890, producing discoveries that the brain directly stimulates the secretion of gastric and pancreatic juices via the vagus nerve. Pavlov went on to demonstrate that severing the vagus nerve just above the diaphragm abolished the gastric acid secretory response to sham feeding in dogs. These discoveries garnered Pavlov the *1904 Nobel Prize in Physiology or Medicine*, and the physiological climate at the turn of the century became dominated by Pavlov's concept that nerves were a controlling principle for bodily functions.

The Birth of Endocrinology and the "Brain-Gut Axis"

The seminal finding that the gut was not solely under neural control but also functions as an endocrine organ was provided by William Bayliss and Ernest H. Starling. These researchers reported their discovery of *secretin*, the first hormone ever identified, in 1902. Noting that acid in the gut stimulated secretion of the pancreas in dogs even when both organs were denervated, they hypothesized that the action of acid on the gut caused the release of a new category of blood-borne chemical messenger that targeted the pancreas. Their hypothesis was validated when they showed that injection of small intestinal epithelial lining extract into the bloodstream induced the pancreatic response. Notably, injection of acid into the stomach did not yield the pancreatic response. As a result, the notion of nerves as the only regulatory mechanism of the body waned, the term "hormone" was coined, and the field of modern endocrinology was born.

Importantly, Bayliss and Starling went on to show that intestinal extracts derived from a broad taxonomic distribution of animals, including many mammals, birds, reptiles, frogs, teleost fish, and cartilaginous fishes, elicited the same pancreatic response as when injected into dogs. This interesting paper published in 1903, titled "On the Uniformity of the Pancreatic Mechanism of Vertabrata",[1] is one of the earliest contributions to comparative endocrinology. The endocrine revolution continued in rapid succession, the gut still leading the charge, with the discovery in 1905 by John Edkins of *gastrin*, a hormone produced by the stomach that stimulates secretion of the gastric acid, HCl, by the parietal cells, and in 1928 with the discovery of *cholecystokinin* (CCK),[2] an intestinal hormone that stimulates gall bladder contraction. Though the existence of these hormones was originally based on the effects of injected tissue extracts, the isolation and chemical characterization of these compounds would not take place for several decades.

The three classical gut hormones, secretin, gastrin, and cholecystokinin, were not only discovered first but structurally identified first, and in 1970 endocrinologists thought that the entire hormonal regulation of digestion could be explained by this hormonal triumvirate alone.[3] That was quite an underestimation, and the late 20th and early 21st centuries have been characterized by an explosion in understanding of gastrointestinal endocrinology, with the discoveries of over 30 peptide hormone genes expressed in the stomach and intestines, and more than 100 different hormonally active peptides, including hormones whose genes are expressed as multiple peptides due to alternative splicing or differential posttranslational processing. Indeed, the gut constitutes the largest system of endocrine cells in the body, in terms of both number of cells and variety of hormones produced.[4,5] Hunger- and satiety-promoting hormones produced from gut, pancreas, and adipose tissue are now known to have receptors in the brain that directly regulate feeding behavior. Furthermore, many "gut hormones" are also expressed as neurotransmitters in the central nervous system, acting to translate metabolic information between the gut and the brain. The development of the notion of a bidirectional "*brain-gut axis*" that regulates appetite and digestion has profoundly influenced our understanding of not only gastrointestinal physiology and associated maladies, but also of eating disorders (particularly those associated with obesity) and their treatments.[6] An even more recent and influential concept that will be described at the end of this chapter is that of a bidirectional "**brain-gut-microbiome axis**", where not only is the gut's microbial makeup influenced by neural and endocrine signaling from the brain, but the microbes themselves also influence diverse neurodevelopmental processes, energy homeostasis, metabolic health, and associated diseases.

SUMMARY AND SYNTHESIS QUESTIONS

1. The ancient Greek physician Galen viewed the stomach as an animate organ that could both sense its own emptiness and generate the sensation of hunger. Provide neural and endocrine evidence that Galen was quite correct.
2. In 1902 William Bayliss and Ernest H. Starling demonstrated that the injection of small intestinal mucosal (epithelial lining) extract into the bloodstream induced a digestive response by the pancreas. How was this finding so important at the time?

Overview of Human Digestion

LEARNING OBJECTIVE Provide an overview of the functional anatomy of the digestive tract and accessory organs from mouth to colon, and describe the major events of the cephalic, gastric, and intestinal phases of gastrointestinal secretion.

KEY CONCEPTS:

- The lumen of the stomach and intestine consist of cells with both exocrine and endocrine functions.
- Whereas the lining of the stomach consists of gastric pits that secrete HCl and proteolytic enzymes, the small intestine lining is composed primarily of villi that promote the absorption of nutrients, electrolytes, and water.
- The small intestine has its own array of brush border-associated digestive enzymes, but it relies on accessory organs for the production of diverse enzymes and other compounds for the complete digestion of nutrients.
- The large intestine digests some compounds that cannot be processed by the stomach and small intestine through the fermenting actions of symbiotic bacteria.

As an integrative physiological system, the components of the digestive tract are in constant communication not only among themselves, but also with accessory digestive organs and the central nervous system. Before we can reach the ultimate goal of this chapter, to achieve an understanding of the complex neural and endocrine mechanisms that regulate these interactions, it is important to first review key aspects of the functional anatomy of the digestive system (summarized in Figure 16.2 and Appendix 9). The gross anatomy of the human digestive tract includes the oral cavity, esophagus, stomach, small intestine, large intestine, rectum, and anus. Each component, separated from each other at key locations by sphincters, makes an essential contribution to the breakdown of food into small molecules like glucose, amino acids, fatty acids, and glycerol that are absorbed into the blood and distributed to every cell in the body. Together, these constitute an approximately 8-meter-long muscular tube known as the *alimentary canal* that propels food distally via peristaltic contractions, and whose digestive actions are supplemented by several accessory organs, including the salivary glands, liver, gall bladder, and pancreas.

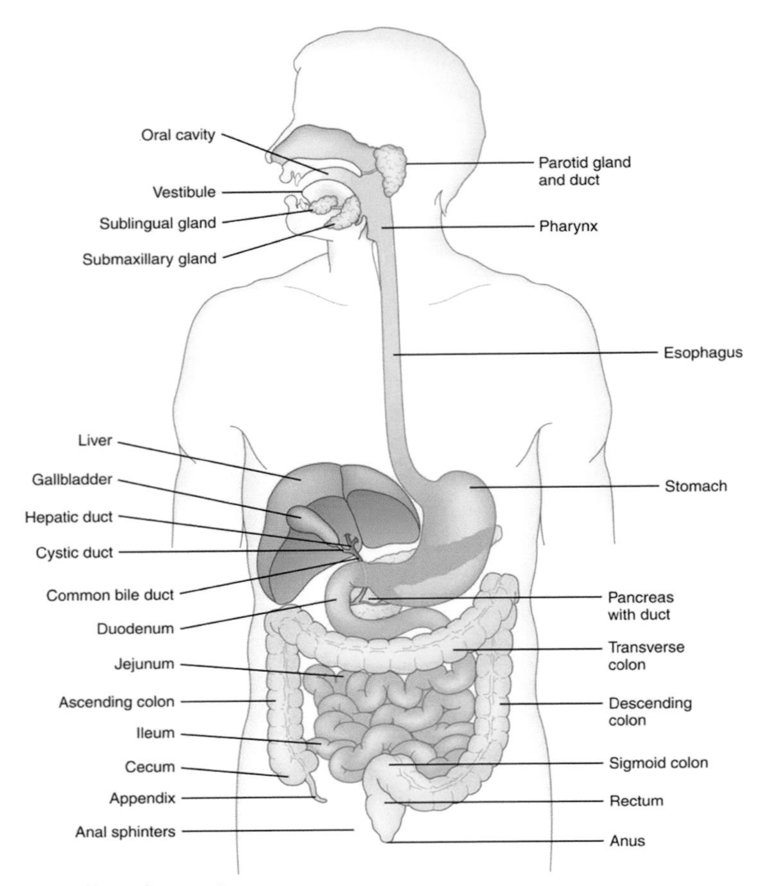

Figure 16.2 Organs of the human digestive system.

Source of images: Norman, A.W. and Henry, H.L., 2022. Hormones. Academic Press. Used by permission.

Mouth

In addition to *mastication* (chewing, or the physical breakdown of large food chunks into smaller pieces with higher surface area), salivary gland secretions begin the process of chemical digestion with *salivary amylases* breaking down starches into smaller carbohydrate chains. Also present in the saliva of human infants (but not in significant amounts in adults) is *salivary lipase*, which initiates the breakdown of lipids into fatty acids and glycerol. In addition, saliva plays an important role in hydrating and lubricating food, allowing it to travel effectively from the esophagus through the *cardiac sphincter* and into the stomach.

Stomach

The stomach is subdivided into several distinct regions based upon their physiological roles, which include both mechanical (muscular motility) and secretory functions (Figure 16.3). Much of the stomach lumen consists of *gastric pits* and *gastric glands* that provide HCl and other secretions to facilitate the processing of ingested nutrients into *chyme*, or partially digested acidic food. The gastric glands and pits have no absorptive capability, but rather are composed of several classes of secretory epithelial cells with differential distributions among the stomach regions. These cells include *parietal cells*, which secrete both HCl and intrinsic factor, a glycoprotein required for the absorption of vitamin B12 by the intestine; *chief cells*, which secrete pepsinogen and gastric lipase; *mucus neck cells*, which secrete protective mucus and bicarbonate ion; as well as **enteroendocrine cells**, which secrete various hormones that regulate digestion. The arrival of food in the stomach triggers the secretion of acid and the protease precursor pepsinogen. HCl creates an extremely low pH environment (pH of 1–2) that has two primary digestive functions: first, it denatures the tertiary structures of most food proteins, allowing greater access by proteolytic enzymes for cleavage; second, the low pH promotes the auto-conversion of pepsinogen into the active pepsin, which cleaves peptide bonds between hydrophobic and aromatic amino acids. The acidic environment also facilitates the absorption of iron, calcium, and vitamin B12, as well as prevents bacterial overgrowth and enteric infection. The acid chyme produced by stomach digestion is transferred via the *pyloric sphincter* into the small intestine.

Small Intestine and Accessory Organs

The human *small intestine* consists of three segments with motility, secretory, and absorptive functions. The most

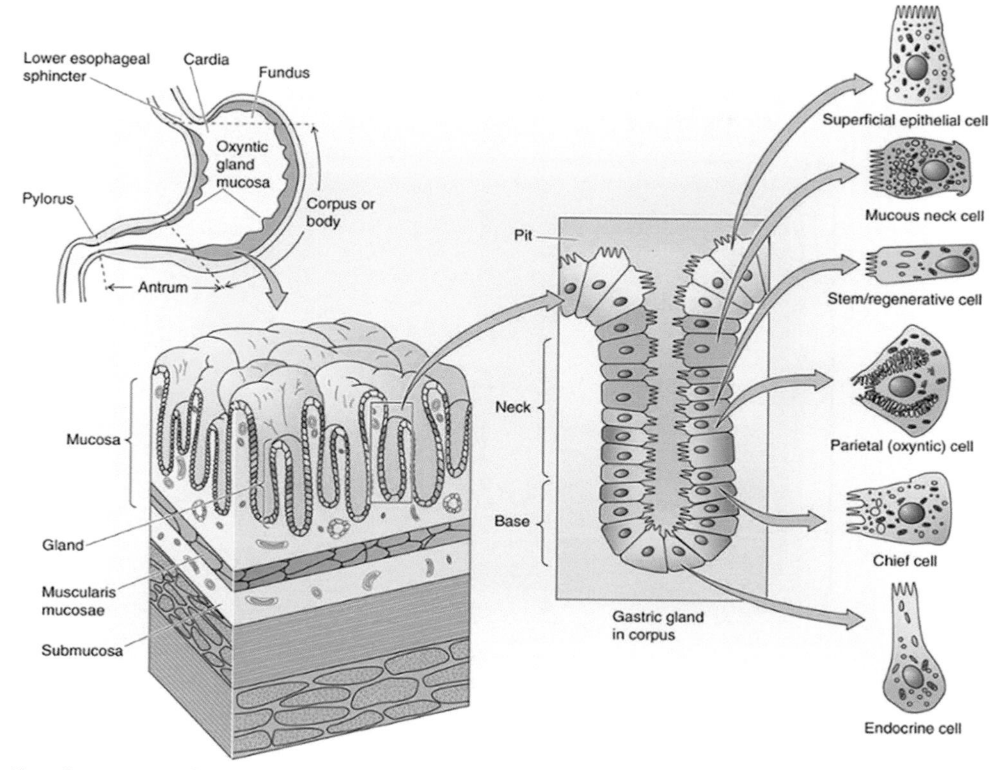

Figure 16.3 Functional anatomy of the stomach. The wall of the stomach consists of both mucosal and muscle layers. The surface area of the gastric mucosa is substantially increased by the presence of gastric glands, which consist of a pit, a neck, and a base. These glands contain several cell types, including mucous, parietal (secretes HCl), chief (secretes pepsinogen), superficial (secretes HCO_3^- and mucus), and endocrine cells. The stomach consists of three anatomic areas: the fundus and corpus (characterized by the presence of parietal cells) and the antrum (which is rich in G or gastrin endocrine cells).

Source of images: Boron, W.F. and Boulpaep, E.L., 2005. Medical Physiology: A Cellular and Molecular Approach, updated 2nd ed. Elsevier. Used by permission.

proximal segment is the *duodenum* (~25 cm length), followed by the *jejunum* (~275 cm length), and the *ileum* (~400 cm length) that connects to the large intestine via the *ileocecal sphincter*. In contrast to the stomach lumen, which consists of millions of gastric pits and glands with virtually no ability to absorb nutrients, the small intestine lumen is composed of abundant finger-like *villi* (Figure 16.4) that greatly increase the mucosal surface area and are superbly adapted for nutrient absorption. The lining of the intestine is renewed at an extraordinary rate, with stem cells replacing the entire population of differentiated cells that cover the intestinal villi every few days. In the adult intestine, stem cells and their transit-amplifying progeny reside in *crypts of Lieberkühn* and give rise to four classes of terminally differentiated cells: *villous absorptive cells*, which are enterocytes involved in nutrient absorption, *goblet cells*, which are mucus-secreting exocrine cells, *Paneth cells*, which are exocrine cells that secrete anti-microbial compounds, and *enteric endocrine cells* that secrete various hormones into the blood. These terminally differentiated cells migrate upwards from the depths of the crypts onto the surfaces of the villi where the cells are exposed to the gut contents and finally undergo apoptosis and are sloughed from the villus tips (Figure 16.4).

The small intestine possesses *brush border enzymes*, which are transmembrane proteins localized to the apical membranes of epithelial cells. These digest oligosaccharide carbohydrates into monosaccharides (e.g. lactase, glucoamylase, sucrose isomaltase) and oligopeptides to amino acids (e.g. aminopeptidases, carboxypeptidases, and dipeptidases). However, the small intestine relies heavily upon digestive enzymes and other chemicals supplied to it by accessory organs for the complete digestion of food. In particular, the pancreas not only produces sodium bicarbonate to neutralize the acid chyme arriving from the stomach, but it also supplies critical lipases responsible for 90% of lipid digestion, proteases that account for 50% of protein digestion, and amylases responsible for 50% of carbohydrate digestion (Table 16.1). The proteases are secreted in inactive proenzyme forms and must be converted to their active form in the intestinal lumen. These exocrine secretions are produced by specialized pancreatic *acini*, which are clusters of cells around a cul-de-sac area that marks the end of a duct system (Figure 16.5). These cells constitute most of the mass of the pancreas. The pancreas also has a critical endocrine function, with hormone-secreting cells located in *islets*, small clusters of cells scattered diffusely among the acini that play key

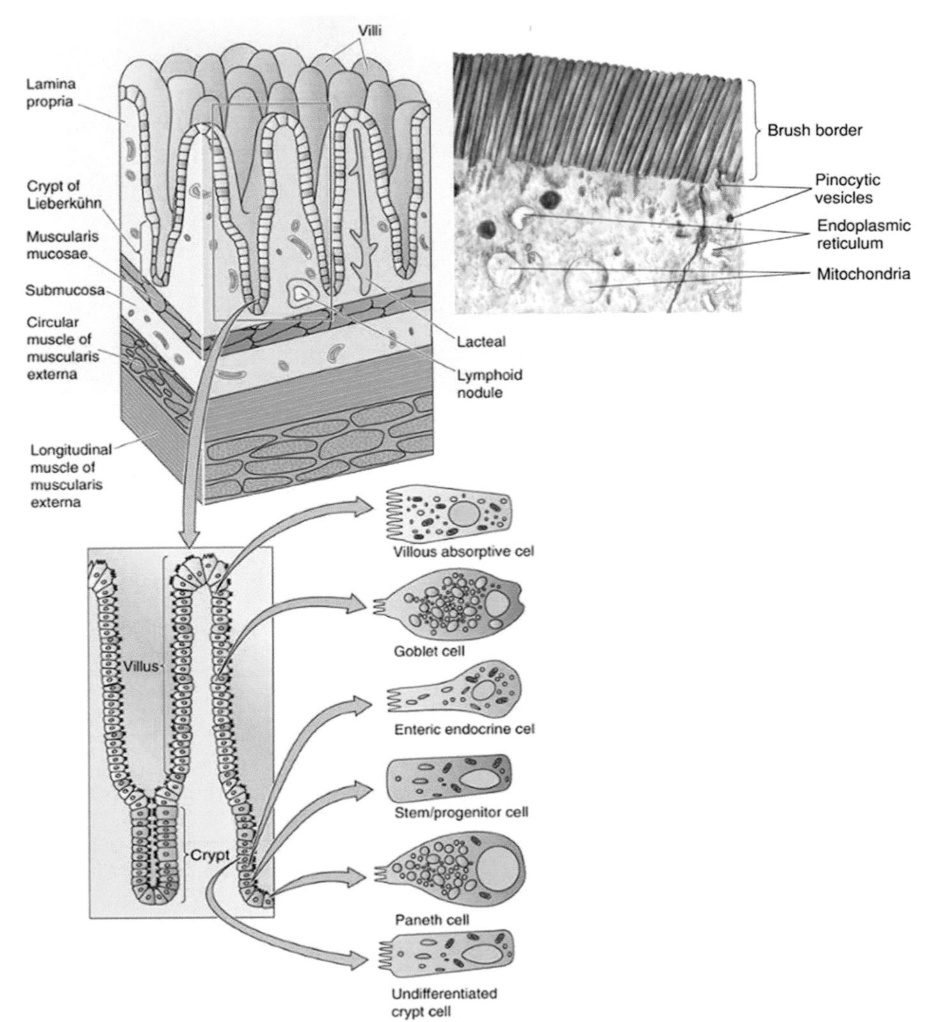

Figure 16.4 Functional anatomy of the small intestine. The small intestine has a specialized epithelial structure that correlates well with epithelial transport function. It consists of finger-like projections—villi—surrounded by the openings of glandular structures called crypts of Lieberkühn, both covered by columnar epithelial cells. The cells lining the villi are considered to be the primary cells responsible for both nutrient and electrolyte absorption, whereas the crypt cells primarily participate in endocrine and exocrine secretion, as well as cell proliferation. The electron micrograph depicts the brush border of a gastrointestinal epithelial cell, showing also absorbed pinocytic vesicles, mitochondria, and endoplasmic reticulum lying immediately beneath the brush border. (Courtesy Dr. William Lockwood.)

Source of images: Boron, W.F. and Boulpaep, E.L., 2005. Medical Physiology: A Cellular and Molecular Approach, updated 2nd ed. Elsevier. Used by permission.

Table 16.1 Enteroendocrine Families of the Mammalian Gastrointestinal Tract and Their Actions

Family/Class	Hormone	Enteric Secretory Cell Type and Location	Principal Effects
Secretin family	Secretin	S cells (duodenum and jejunum)	Reduction of acidity in upper small intestine by promoting release of pancreatic bicarbonate
	VIP	Enteric nerves	Relaxes smooth muscle of gut, blood vessels; increases water and electrolyte secretion from pancreas and gut
Glucagon family	Proglucagon	L cells (distal small intestine and colon)	Precursor to GLP-1, -2
	GLP-1, -2		Stimulation of carbohydrate uptake, slowing of intestinal transit, appetite regulation, insulin release
	GIP	K cells (duodenum and jejunum)	Stimulates insulin secretion by pancreas

Table 16.1 Enteroendocrine Families of the Mammalian Gastrointestinal Tract and Their Actions

Family/Class	Hormone	Enteric Secretory Cell Type and Location	Principal Effects
Gastrin family	Gastrin	G cells (pyloric antrum of the stomach, duodenum, and pancreas)	Stimulation of gastric acid secretion
	Cholecystokinin	I cells (duodenum and jejunum) and enteric neurons	Activation of gall bladder contraction and stimulation of pancreatic enzyme secretion
Ghrelin–motilin family	Ghrelin	X/A cells (stomach and pancreas)	Appetite control, growth hormone release
	Motilin	M cells (duodenum and jejunum)	Intestinal motility
Bombesin-like peptide family	Gastrin-releasing peptide	Enteric nerves	Promotes gastrin release
	Neurotensin	N cells (distal small intestine)	Inhibits intestinal motility
Neuropeptide Y family	Neuropeptide Y	Enteric neurons	Gastrointestinal motility and secretion.
	Peptide YY	L cells (distal small intestine and colon)	Stimulation of carbohydrate uptake, slowing of intestinal transit, appetite regulation, insulin release
Somatostatin	Somatostatin	D cells (pyloric antrum, duodenum and pancreatic islets)	Inhibits gastrin release
Non-peptide molecules	Histamine	ECL cells (stomach fundus)	Stimulates gastric acid secretion
	Serotonin	EC cells (entire gut length)	Regulates intestinal motility and secretion by modulating enteric nervous system signaling

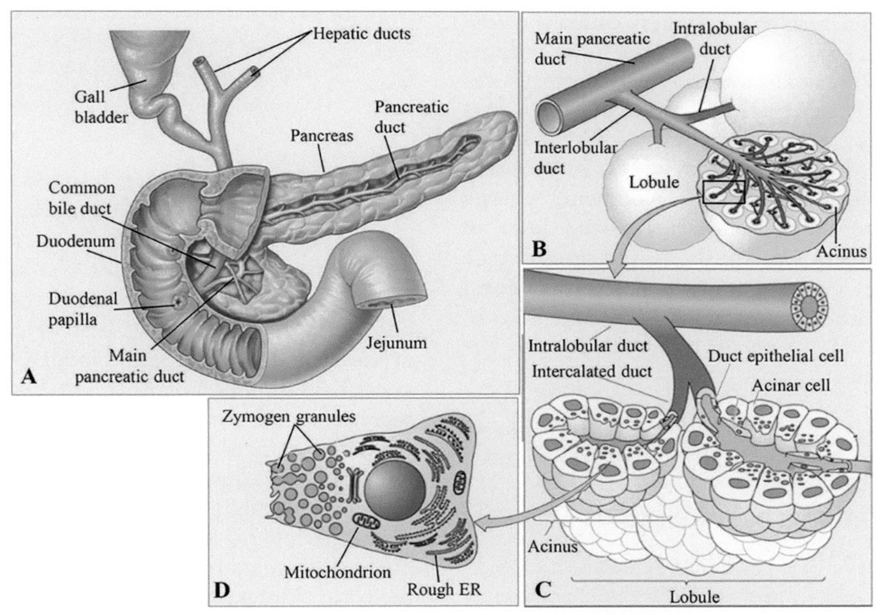

Figure 16.5 Functional anatomy of the biliary and exocrine pancreatic systems. (A) The biliary ductules of the liver drain into large bile ducts that coalesce into the right and left hepatic ducts to permit exit of bile from the liver. These, in turn, form the common hepatic duct, from which bile can flow into either the gall bladder, via the cystic duct, or the duodenum, via the common bile duct. **(B–C)** The fundamental secretory exocrine unit of the pancreas is composed of an acinus and an intercalated duct. Intercalated ducts merge to form intralobular ducts, which, in turn, merge to form interlobular ducts, and then the main pancreatic duct, which merges with the common bile duct. **(D)** The acinar cell is specialized for protein secretion. Large condensing vacuoles are gradually reduced in size and form mature zymogen granules that store digestive enzymes in the apical region of the acinar cell.

Source of images: (A) C. Shi, E. Liu, Anatomy, Histology, and Function of the Pancreas, Editor(s): Linda M. McManus, Richard N. Mitchell, Pathobiology of Human Disease, Academic Press, 2014, Pages 2229-2242; Used by permission (B-D) Boron, W.F. and Boulpaep, E.L., 2005. Medical Physiology: A Cellular and Molecular Approach, updated 2nd ed. Elsevier. Used by permission.

roles in post-digestive metabolism. Hormones produced by the islets include insulin, glucagon, somatostatin, pancreatic polypeptide, and ghrelin. The liver produces *bile*, a cholesterol-based compound that emulsifies (breaks apart into tiny droplets) fats in the small intestine and also helps neutralize acidic chyme. Most bile is stored in the gall bladder and secreted into the intestine, where it is ultimately absorbed into the blood and recycled to the liver. Both hepatic and pancreatic secretions enter the duodenum through the common bile and pancreatic ducts (Figure 16.5).

Large Intestine

The most distal region of the gastrointestinal tract is the *large intestine*, which is composed of the *cecum*, which connects to the *ileum* by the *ileocecal sphincter*. This is followed by the ascending, transverse, and descending portions of the *colon*, and then the *rectum* and the *anus*. The large intestine reabsorbs much of the fluid and electrolytes that were used to move food through the stomach and small intestine and stores and compacts undigested materials until they can be conveniently eliminated as feces. From a digestive perspective, the function of the large intestine is to digest and absorb components that cannot be processed by the stomach and small intestine. However, this digestion is accomplished not through the actions of endogenously produced enzymes, but by the actions of symbiotic bacteria via a process called *fermentation*, in which bacterial enzymes act on both endogenous substrates (e.g. bile acids) and exogenous substances (e.g. fiber) (Table 16.1). Interestingly, these microbes also produce diverse neuroactive substances, like GABA, noradrenaline, and dopamine, and other metabolites that influence digestion, neural development, energy homeostasis, and metabolic health (discussed later in Developments & Directions: The brain-gut-microbiome axis). Lastly, the large intestine sends and receives neural signals to and from other gastrointestinal segments, timing the evacuation of colon contents to create space for the residues of the next meal.

SUMMARY AND SYNTHESIS QUESTIONS

1. If the vagus nerve of a mammal is severed, how will this affect gastrointestinal secretion when a meal is consumed?
2. The gall bladder stores bile, a lipid emulsifier important for normal digestion. However, humans who gave had their gall bladder surgically removed can still digest lipids normally. How is this possible?
3. Why is an acidic pH optimal for digestion of protein in the stomach, but a more neutral pH is required for digestion of protein in the intestine?

Integrated Neural and Endocrine Control of Digestion

LEARNING OBJECTIVE Describe the integrated neural and endocrine regulation of digestion, listing the major hormones and their specific functions.

KEY CONCEPTS:

- The gastrointestinal tract differs from all other peripheral organs in that the majority of enteric neurons can function independently of the central nervous system.
- Unlike typical glandular endocrine organs, enteroendocrine cells are scattered diffusely throughout the gastrointestinal tract.
- Like neuroendocrine cells, enteroendocrine cells generate peptide hormones and neurotransmitters, many of which are also produced by neurons of the central nervous system.
- The cephalic phase of gastrointestinal secretion prepares the gastrointestinal tract to receive and digest incoming food by activating exocrine and endocrine outputs in response to visual, olfactory, and cognitive input.
- The gastric phase of gastrointestinal secretion is initiated by the arrival of food in the stomach and promotes additional exocrine, endocrine, and neural responses by the stomach, intestine, and accessory organs.
- The intestinal phase of gastrointestinal secretion is initiated by the arrival of acid chyme in the duodenum and inhibits gastric acid secretion, as well as promotes the release of digestive enzymes and fluids from accessory organs.

The coordinated digestive response to an ingested meal by each of the aforementioned gut segments and accessory organs is regulated by both neural and endocrine interactions. The *enteric nervous system* is a large neural network whose cell bodies lie in the submucosal and myenteric plexuses of gut wall (Figure 16.6), as well as in the pancreas and gall bladder. The enteric nervous system modulates not only gut exocrine and endocrine secretions, but also intestinal *motility* (or smooth muscle-mediated peristalsis and sphincter function), blood flow, and immune function. The enteric nervous system was originally thought to be part of the autonomic component of the peripheral nervous system, and the neurons in the gut wall were considered postganglionic parasympathetic neurons.[7] However, the idea that the gut has a "brain of its own" arose when it was demonstrated over a century ago that the intestine has the ability to function even when isolated from the central nervous system (CNS),[8] with peristaltic contractions emanating from coordinated local enteric neural reflexes. Thus, the gastrointestinal tract differs from all other peripheral organs in that the majority of enteric neurons do not contact the parasympathetic axons of the CNS directly. Enteric neurons use a wide variety of substances as neurotransmitters, although the majority use peptides once thought to function only as gut

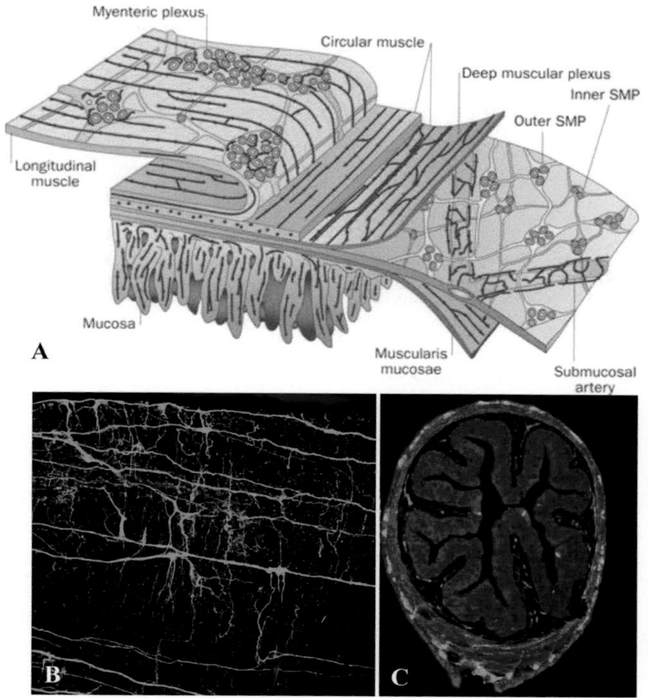

Figure 16.6 Organization of human and amphibian enteric nervous systems (ENS). **(A)** The human ENS has ganglionated plexuses, the myenteric plexus between the longitudinal and circular layers of the external musculature and the submucosal plexus that has outer and inner components. Nerve fiber bundles connect the ganglia and also form plexuses that innervate the longitudinal muscle, circular muscle, and other cells such as enteroendocrine cells (not shown). Sagittal view **(B)** and cross-sectional view **(C)** of the amphibian (*Xenopus laevis* frog) ENS visualized by fluorescent confocal microscopy (green: neurons; red: smooth muscle; blue: epithelial cells).

Source of images: (A) Furness, J.B., 2012. Nat. Rev. Gastroenterol. Hepatol., 9(5), pp. 286-294. Used by permission.

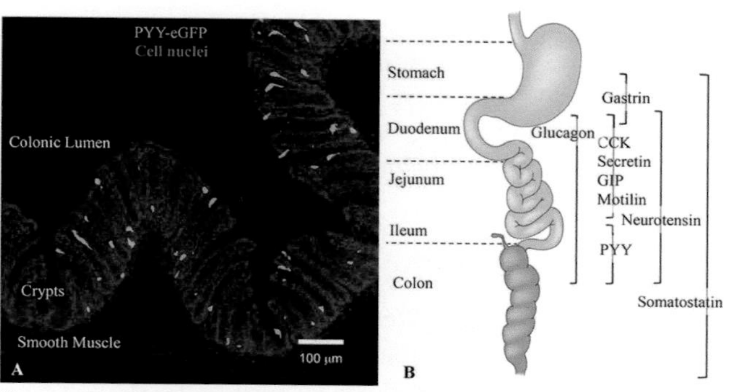

Figure 16.7 Diffuse and overlapping distribution of enteroendocrine cells in the mouse and human gastrointestinal tract. **(A)** Immunofluorescence photomicrograph of colonic tissue from a transgenic PYY-eGFP mouse. Cells expressing PYY-eGFP cells are green and cell nuclei stained with DAPI appear blue. **(B)** Locations of gastrointestinal hormone secretions in the human gut.

Source of images: (A) Bohórquez, D.V. and Liddle, R.A., 2011. Clin Transl Sci., 4(5), pp. 387-391. Used by permission. (B) Mulholland MW, Lillemoe KD, Doherty GM, Maier RV, Upchurch GR, eds. Greenfield's Surgery. 4th ed. Philadelphia, PA: Lippincott Williams & Wilkins; 2005. Used by permission.

hormones (Table 16.1). Interestingly, many of the peptides found in enteric neurons are also present in the brain, and for some peptides, such as cholecystokinin, brain concentration exceeds that of the gut.[9]

Curiously Enough . . . Because the enteric nervous system is so large (about 100 million nerves, approximately the number found in the spinal cord), shares many of the same neurotransmitters that are found in the brain, and can function independently of the central nervous system, it is often referred to as the body's "second brain".[10]

The gut epithelial lining is a rich source of secreted hormones that are grouped into families based on structural homology, modes of release, and mechanism of action via cell-surface receptors (summarized in Table 16.1). These hormones are synthesized by a diverse array of specialized **enteroendocrine cells**. Unlike typical glandular endocrine organs, like the thyroid, adrenal gland, and testes, enteroendocrine cells comprise a diffuse endocrine system (Figure 16.7) that is scattered along the gut lining from the stomach to the colon. Indeed, enteroendocrine cells only represent about 1% of the gut epithelial cell population.[5] Because they are diffuse and far outnumbered by non-endocrine epithelial cells, this has impaired the identification of the source of any single hormone as has been done for other endocrine organs, and enteroendocrine cells have therefore historically been difficult to study. Modern methods, in particular the use of transgenic enteroendocrine reporter mice[11] (Figure 16.7), are now lifting the veil of mystery surrounding the functions of these cells and their hormones. Interestingly, most of the hormones secreted by enteroendocrine cells are also produced by neurons of both the enteric nervous system and the CNS, where they serve as neurotransmitters. For example, serotonin, a molecule with well-established neurotransmitter functions in the CNS, is also broadly expressed throughout the gut epithelial lining, produced by a special category of enteroendocrine cells called *enterochromaffin cells*.

A striking biochemical similarity between enteroendocrine cells and neurons is their ability to generate peptide hormones or neurotransmitters from amine precursors, known as the *APUD* concept (this acronym stands for amine precursor uptake and decarboxylation).[12] That is, the molecular machinery that picks out the scattered subset of endodermal cells destined for an enteroendocrine fate is similar to that which picks out the subset of cells in the developing nervous system that are destined to be neurons. However, despite this biochemical similarity, enteroendocrine cells are not derived from embryonic neural crest cells, as are enteric nervous system cells. Instead, enteroendocrine cells share the same embryonic endoderm developmental ancestry as the more common absorptive and digestive epithelial cells among which they are dispersed.[13,14]

Gastrointestinal Cells

The enteroendocrine system consists of at least 15 different cell types traditionally classified according to the principal hormone they produce (Table 16.1). Whereas some enteroendocrine cell products, such as serotonin, are produced along the whole length of the gut by enterochromaffin cells, most others are produced primarily in a particular gut segment. Enteroendocrine cells can be divided into several categories based on morphology, and different subcategories based on postulated pathways by which their peptides are secreted[14] (summarized in Figure 16.8):

1. *Open-type cells*: The unifying feature of these cells is that they possess microvilli-covered apical membranes and processes in contact with (i.e. "open" to) the gut lumen and its contents. As they make direct contact with ingested and digested nutrients, these cells are ideally positioned to act as "sensors" that directly detect fluctuations in luminal nutrient concentrations and pH change. This nutrient-dependent enteroendocrine cell stimulation on the apical side promotes hormone secretion by the basal side.
2. *Closed-type endocrine cells*: These cells are located deep in the mucosal surface and are thus "closed" to (not in contact with) the intestinal lumen. As such, these cells are thought to be functionally modulated by hormones, paracrine factors, neuronal stimulation, or mechanical distension. Closed-type cells release hormone directly to the blood compartment and therefore cannot engage in paracrine activity.
3. *Neurocrines*: These cells are characterized by possession of a distinct basolaterally located neural-like process termed a "neuropod", which extends toward the enteric nervous system and glia as well as toward other underlying structures such as myofibroblasts. Although its function is currently unknown, the neuropod may enable unidirectional or bidirectional communication between enteroendocrine cells and the enteric nervous system.[15]

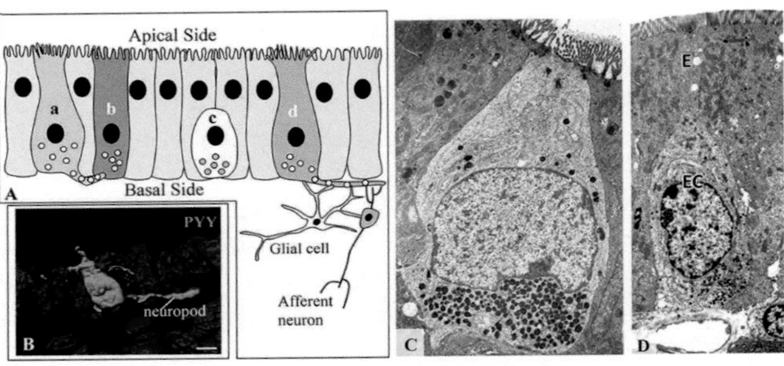

Figure 16.8 Gut epithelium showing different representative enteroendocrine cell (EEC) types. (A) (a) A gastric somatostatin-producing D-cell with basolateral process that communicates via a *paracrine process* with (b) a neighboring *open-type* gastrin-producing G-cell. (c) A *closed-type* EEC. (d) A *neurocrine* small intestinal/colonic-type open EEC with neuropod basolateral extension in contact with a neuronal terminus. **(B)** A confocal microphotograph of a PYY-expressing enteroendocrine cell (green) within the mouse small intestine. The cell possesses a neuropod that extends below the basal surface of surrounding mucosal cells. Nuclei are stained blue. **(C)** Transmission electron micrographs (TEM) of an open-type enteroendocrine cell in the epithelium of the duodenum showing microvilli at its apical end in contact with the lumen. (With permission, from A.G.E. Pearse, Department of Histochemistry, Royal Postgraduate Medical School, London.) **(D)** TEM of a closed-type enteroendocrine cell (EC) with secretory granules that can be distinguished along the basal lamina. Above it is a typical absorptive enterocyte **(E)**.

Source of images: (A) modified from Gribble, F.M. and Reimann, F., 2016. Ann. Rev. Physiol., 78, pp. 277-299; (B) Liddle, R.A., 2019. Cell. Mol. Gastroenterol. Hepatol, 7(4), pp. 739-747; Used by permission. (C-D) Junqueira, L.C. and Mescher, A.L., 2013. Junqueira's basic histology: text & atlas/Anthony L. Mescher. McGraw Hill LLC. Used by permission.

Phases of Gastrointestinal Secretion

In this section the integrated neural and endocrine processes that occur in the gastrointestinal tract in response to a meal will be described. Events that take place during digestion can be divided into different phases according to the part of the body that is stimulating gastrointestinal and accessory organ secretions in response to food: the cephalic, gastric, and intestinal phases.

Cephalic Phase

The **cephalic** ("head") **phase** of gastrointestinal secretion refers to a set of food intake-associated autonomic and endocrine responses to the stimulation of sight, smell, and taste sensory systems located in the head's eyes and oropharyngeal cavity. Recall from earlier the discoveries by Beaumont and Pavlov showing that the presence of food, or even just thoughts of food, elicits a nervous response conducted from the head to the gut via the vagus nerve that affects the secretion of digestive juices and hormones. The vagus nerve contains preganglionic neurons that synapse with postganglionic neurons within the wall of the stomach and that are part of the enteric nervous system. Higher brain sites like the limbic system, hypothalamus, and cerebral cortex are also involved in the cognitive components of this response. As such, certain psychological conditions, such as anxiety or fear, can alter the cognitive response to a meal and inhibit the ability to prepare the gut adequately for a meal.

Overall, the main function of the cephalic phase is to enhance the ability of the gastrointestinal tract to receive and digest incoming food by activating specific exocrine and endocrine responses through vagal stimulation. Importantly, in addition to its previously well-established effect of promoting salivary and gastric secretions, the cephalic phase is now also known to have important effects on stimulating

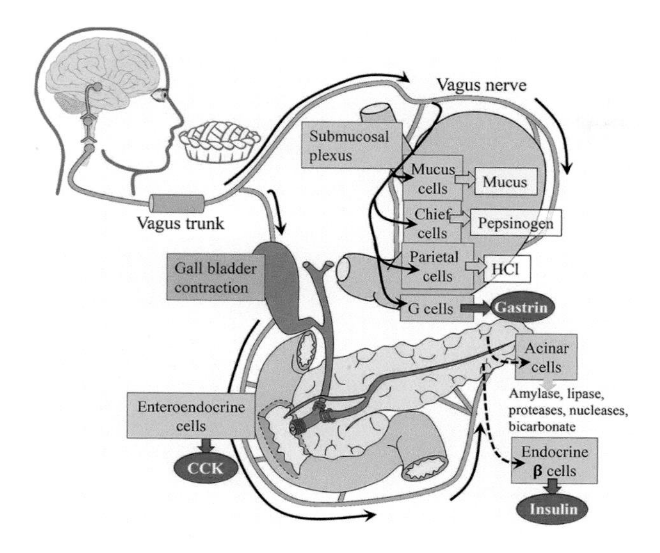

Figure 16.9 Cephalic phase of digestion. This phase occurs before food enters the stomach and especially while food is being eaten. Neurogenic signals that initiate this phase originate from the cerebral cortex and in the appetite centers of the amygdala and hypothalamus. They are transmitted through the vagus nerve to the stomach, intestine, pancreas, and gall bladder, which respond by releasing various exocrine (light boxes) and endocrine (red ovals) secretions that initiate digestion.

Source of images: Schreiber

pancreatic, gall bladder, and duodenum secretions.[16] These responses, summarized in Figure 16.9, are as follows:

1. *Salivary glands*: Parasympathetic nerves to salivary glands release vasoactive intestinal peptide (VIP), which stimulates the production of saliva and digestive enzymes such as salivary amylase and lipase.
2. *Gastric exocrine stimulation*: Excitatory parasympathetic vagal nerve outflow from the brainstem to the stomach is relayed via the gastric submucosal plexus to stimulate parietal, chief, and mucus cells. These secrete HCl and intrinsic factor, pepsinogen, and mucus and bicarbonate, respectively, in the fundal part of the stomach. HCl facilitates the cleavage of the proenzyme pepsinogen into the active pepsin. This phase of secretion normally accounts for about 30% of the gastric secretion associated with eating a meal.
3. *Gastric endocrine stimulation*: Vagal outflow also modulates the secretion of gastric hormones, such as gastrin from G cells, somatostatin from D cells, and histamine from enterochromaffin-like cells. These hormones will interact to further regulate the release of HCl from parietal cells in the gastric phase.
4. *Accessory organ exocrine stimulation*: Exocrine pancreatic secretions in response to oropharyngeal stimulation include the pancreatic digestive enzymes amylase, lipase, trypsin, and chymotrypsin, as well as bicarbonate ion. These are secreted through acinar cells in response to vagus nerve stimulation. About 20% of the total secretion of pancreatic enzymes after a meal occurs during the cephalic phase. Vagal stimulation during the cephalic phase also initiates gall bladder contraction and relaxation of the sphincter of Oddi, allowing bile and pancreatic juices to flow into the duodenum via the common bile duct.
5. *Accessory organ endocrine stimulation*: Endocrine pancreatic secretions produced by the cephalic phase include insulin, which is mediated by the vagus nerve. Insulin secretion results in uptake of glucose and other metabolites from the blood into target organs such as muscle and fat. Its anticipatory release suggests a preparation of the organism for the metabolites which will be absorbed from the gut into the bloodstream to maintain homeostatic circulating levels.
6. *Duodenum endocrine stimulation*: Cholecystokinin (CCK) is released by enteroendocrine cells of the duodenum during the cephalic phase of digestion, also mediated by vagus nerve. CCK will go on to influence the intestinal phase of secretion.

Clinical Considerations: Diagnosing (and misdiagnosing) Zollinger–Ellison syndrome

Zollinger–Ellison syndrome (ZES) is characterized by the hypersecretion of gastric acid, severe peptic ulcerations in the upper small intestine, and often diarrhea. It is usually caused by excess gastrin secretion via a tumor in the pancreas or small intestine (a gastrinoma) and diagnosed by measuring increased levels of gastrin in blood plasma. Interestingly, up to 3% of patients with ZES have been reported to have normal or low gastrin concentrations.[17–19] The consequences of missing a diagnosis of ZES can be serious, potentially leading to dangerous conditions such as esophagitis, peptic stenosis, and intestinal ulcerations. In these situations, clinicians may have erroneously disregarded the diagnosis of ZES because of low gastrin measurements and therefore failed to prescribe the antisecretory regimen that would have been appropriate.

Considering that gastrin is not a single molecule but circulates as several peptides of various lengths and amino acid modifications (Figure 16.10[A]), Rehfeld and colleagues[20] hypothesized that one explanation for initially missing a diagnosis of ZES is an inability of some commercial gastrin detection kits to correctly recognize and quantify all types of gastrin molecules. These kits are based on immunoassays, or the use of antibodies against gastrin molecules to measure their presence. An accurate identification of gastrin molecules is especially important as the peptide profile of circulating gastrins and their concentration vary considerably not only between healthy subjects and patients with ZES, but also among individual patients with ZES (Figure 16.10[B]). ZES appears to significantly change the processing of the 101 amino acid precursor form of progastrin into the family of gastrin peptides. The researchers characterized the specificity of various commercial gastrin immunoassay kits to all bioactive gastrin peptides, irrespective of size and modification such as tyrosyl sulfation.

Strikingly, they found that some kits giving false-low results are due to antibodies that bind only gastrin-17, but neither shorter nor longer forms. Furthermore, the antibody binding is also heavily influenced by modifications such as tyrosine sulfation, which is highly variable in patients with ZES. These results emphasize the necessity of accurately evaluating not only peptide length but also amino acid modifications when diagnosing ZES.

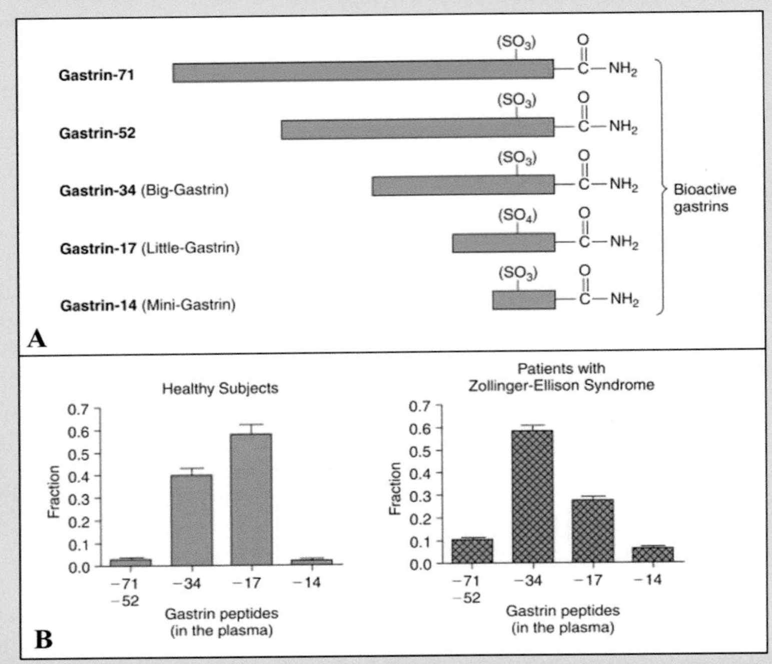

Figure 16.10 The relative amounts (fraction) of five blood circulating gastrin peptides are shown for healthy subjects and for comparison to patients with the Zollinger–Ellison syndrome (ZES). (A) Progastrin is enzymatically processed into five major peptides: the three principal biologically active circulating forms are G-34 (big gastrin), G-17 (little gastrin), and G-14 (mini-gastrin); other low concentration gastrins include gastrin-6, gastrin-52 and gastrin-71. **(B)** The disease process associated with Zollinger–Ellison syndrome is suggested to significantly alter the processing of progastrin, changing the ratio of gastrin peptides. Note that gastrin-71 and gastrin-52 are not distinguishable in this experiment.

Source of images: Norman, A.W. and Henry, H.L., 2022. Hormones. Academic Press. Used by permission.

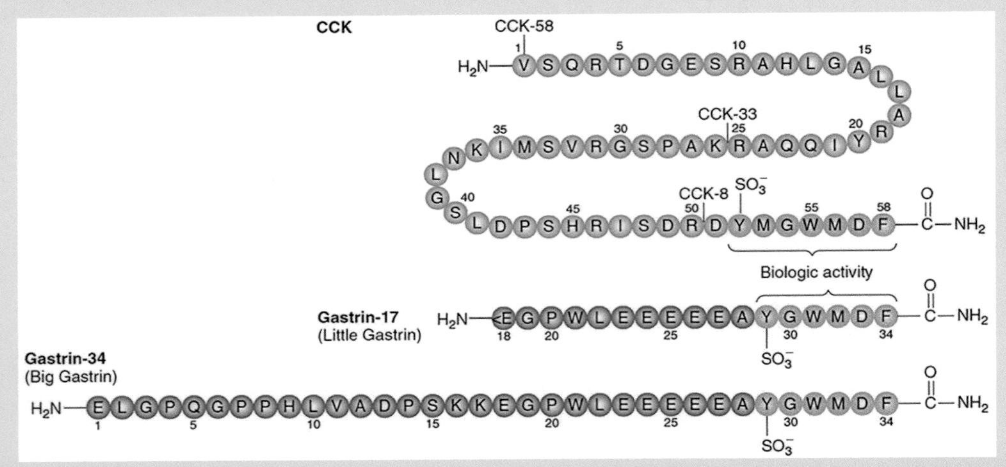

Figure 16.11 Identification of the minimal amino acid sequences of cholecystokinin (CCK) and gastrin that are required for biological activity. The CCK peptide and the gastrin-34 peptide are aligned such that the carboxy terminus amides of the phenylalanines are aligned together. The orange-colored amino acids on the N-terminus are identical and impart biological activity.

Source of images: Norman, A.W. and Henry, H.L., 2022. Hormones. Academic. Used by permission.

Curiously Enough . . . The genes of two gastrin family members, gastrin and cholecystokinin (CCK), are thought to have diverged from a common ancestral gene over 500 million years ago in the vertebrate lineage.[21] In vertebrates, gastrin and CCK possess a common C-terminal four amino-acid sequence, required for receptor activation (**Curiously Enough Figure 16.14**). In addition to sharing a common receptor with gastrin, CCK also binds to a second, distinct receptor.

Gastric Phase

The entry of food into the stomach from the esophagus initiates the **gastric phase of digestion**, which provides both mechanical and chemical stimulation to the gastric wall. Food in the stomach produces distention and stretch, which is detected by sensory nerve endings in the gastric wall, that in turn transmit information to the brainstem and drive activity in vagal efferent fibers, a process called *vagovagal reflex*. An example of such a reflex is the "gastric receptive relaxation reflex", in which distention of the stomach results in relaxation of the smooth muscle in the stomach, which allows filling of the stomach to occur without an increase in intraluminal pressure. In contrast to volume-induced stretch receptors, chemical stimulation of the gastric wall is predominantly from the digestion of proteins into oligopeptides and amino acids. These products are detected by chemosensors in the gastric mucosa that go on to stimulate vagal afferent activity. Diverse amino acid- and oligopeptide-sensing GPCRs have been identified in enteroendocrine cells of the stomach and intestine.[14] In addition to vagovagal reflexes, local enteric nervous system pathways release acetylcholine (Ach), stimulating parietal cells to secrete acid. Gastric phase mechanical and chemical stimulations build upon events already put into motion during the cephalic phase, promoting additional exocrine, endocrine, and neural responses not only by the stomach itself, but also by accessory organs and the intestine. These responses, summarized in Figure 16.12, are as follows:

1. *Appetite regulation*: Upon eating, the distention of stomach mechanoreceptors transmits stretch information to the hypothalamus and other forebrain structures via vagal afferents to hindbrain nuclei, which functions as a satiety signal regulating food intake.[22] In contrast, ghrelin secretion rises in the preprandial (before a meal) state, when it plays a role in promoting the sensation of hunger, and falls rapidly after a meal. The neural and endocrine regulation of appetite will be discussed in greater detail in Chapter 17: Appetite.
2. *Regulating gastric pH*: The gastric phase of secretion accounts for about 60% of the total gastric secretion associated with eating a meal. Mechanical and chemically induced vagovagal stimulations directly promote parietal cell acid secretion. The stimuli also cause antral G cells to release gastrin, the main hormonal stimulant of acid secretion. Gastrin travels in circulation to reach the acid-secreting parietal cells in the fundus, stimulating H^+ secretion directly by binding to cholecystokinin-2 (CCK-2) receptors coupled to the opening of calcium channels on cell membrane and release of intracellular calcium from the endoplasmic reticulum (Figure 16.13). Importantly, gastrin also acts indirectly by stimulating histamine secretion from enterochromaffin-like cells in the fundus. Histamine exerts paracrine actions on adjacent parietal cells, where it binds to histamine H2 receptors coupled to stimulation of cAMP and acid secretion. In fact, histamine is the strongest agonist of H^+ secretion, whereas gastrin and acetylcholine are much weaker agonists.

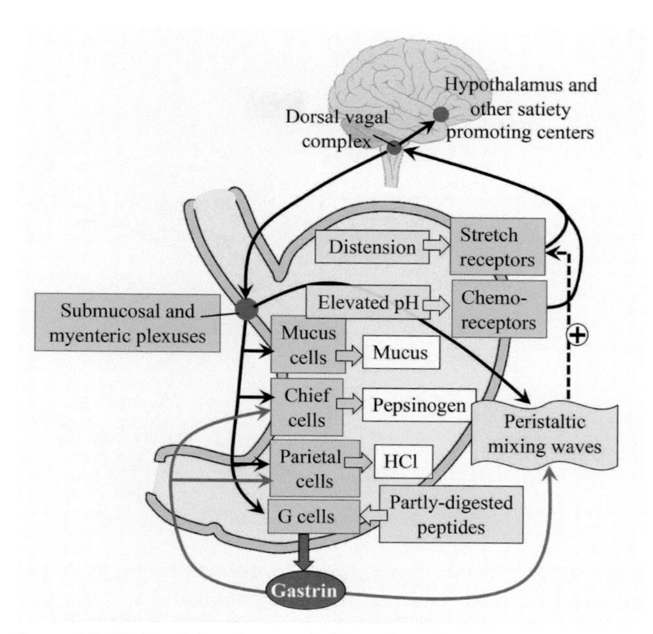

Figure 16.12 Gastric phase of digestion. The entrance of food into the stomach provides both mechanical and chemical stimulation to the gastric wall. Distention and the presence of food substrates, detected by mechanoreceptors and chemoreceptors, respectively, transmit information via a vagal-vagal reflex to exocrine, endocrine, and smooth muscle target cells in the stomach. This promotes the secretion of digestive juices and the generation of peristaltic mixing waves that churn the food in a positive feedback loop with the stretch receptors (dashed arrow).

In a negative feedback response to low luminal pH, somatostatin (SST) is released from D cells in both the antrum and fundus. SST in the antrum inhibits gastrin secretion from G cells via paracrine action, indirectly lowering HCl secretion by parietal cells. SST in the fundus binds to SSTR2 receptors coupled to inhibitory G proteins (G_i), directly inhibiting acid secretion by parietal cells (Figure 16.13). Note that you have seen SST before in a very different context and will encounter it again in yet another circumstance. Recall from Chapter 7: Neurosecretion and Hypothalamic Control of the Pituitary that in the brain SST is produced by hypothalamic neuroendocrine neurons that project to the median eminence, where it is carried to the anterior pituitary gland and inhibits the secretion of growth hormone from somatotrope cells.

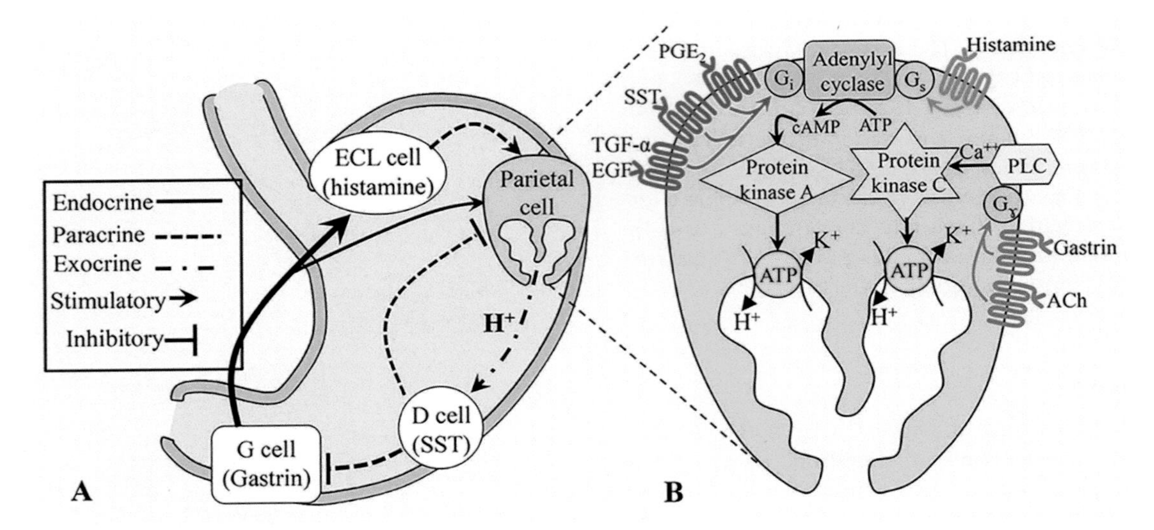

Figure 16.13 Models illustrating the regulation of gastric acid secretion by various hormones at the tissue and cellular levels. **(A)** Overall scheme showing regulation of gastric acid secretion by endocrine and paracrine pathways acting on cells of the gastric fundus and antrum. **(B)** Regulation of gastric acid secretion by diverse GPCR receptors of parietal cells. The adenylyl cyclase effector is positively regulated by the histamine receptor via interaction with the G_s alpha G protein subunit, ultimately activating protein kinase A (PKA), followed by activation of the H^+/K^+-ATPase pump that secretes protons into the stomach lumen to facilitate digestion. The adenylyl cyclase effector is negatively regulated by prostaglandin E2 (PGE2), somatostatin (SST), transforming growth factor α (TGF-α), and epidermal growth factor (EGF) receptors via interaction with the G_i alpha G protein subunit. The phospholipase C (PLC) effector is positively regulated by the gastrin and acetylcholine (Ach) receptors that interact with the G_γ alpha G protein subunit, ultimately activating protein kinase C (PKC) via a Ca^{++}-dependent pathway, followed by activation of the H^+/K^+-ATPase pump that secretes protons into the stomach lumen to facilitate digestion. Acetylcholine binds to muscarinic M3 receptors. Histamine acts via the H2 receptor. Gastrin binds to the cholecystokinin type 2 (CCK2) receptor. PGE2, SST, TGF-α, and EGF each bind to receptors named after them, respectively.

3. *Protecting the stomach lining*: Many prostaglandins (PGs), especially PGE2, have strong cytoprotective effects on the gastric epithelial lining. Prostaglandins protect the stomach lining from being damaged by proteases and low pH via at least two mechanisms.[23] First, PGs promote exocytosis of protective mucin proteins by mucus cells. Second, PGs inhibit HCl secretion by parietal cells by upregulating inhibitory G proteins (G_i), reducing cAMP synthesis and acid secretion (Figure 16.13). The importance of prostaglandins in mucosal defense is evident by the increased susceptibility of the stomach to the development of peptic ulcers (sores in the lining of the stomach or duodenum caused by damage to the epithelial layer) following chronic ingestion of aspirin and other non-steroidal anti-inflammatory drugs (NSAIDs). Importantly, prostaglandin synthesis is blocked by NSAIDs, and ulcer disease is becoming more common as the human population ages and has more need of NSAIDs for non-gastrointestinal complaints such as arthritis.
4. *Gastric and gastrointestinal motor responses*: Gastric motor responses refer to changes in activity by stomach smooth muscle. An important vagovagal-mediated smooth muscular reflex is the stimulation of motility of the distal part of the stomach, which causes antral peristalsis. This motility plays an important role in storage and mixing of the meal with secretions and is also involved in regulating the flow of contents out of the stomach. Stomach motility is also affected by gastrin, which inhibits its smooth muscle contractions in the proximal stomach but stimulates contractions in the distal stomach, causing antral peristalsis. Thus, hormone and vagovagal mechanisms of stomach motility function in a complementary manner to mix food and to regulate the flow of contents out of the stomach. Additionally, the presence of the meal in the stomach activates two gastrointestinal reflex arcs: (1) the *gastroileal reflex*, which induces peristalsis in the ileum and the opening of the ileocecal valve, allowing the emptying of the ileal contents into the large intestine, and (2) the *gastrocolic reflex*, that results in increased motility of the colon, promoting evacuation of the colonic contents to make way for the residues of the next meal.

Developments & Directions: The curious case of the Australian stomach-brooding frog

The first sentence of a 1983 research article published in the journal *Science* by Michael Tyler and colleagues[24] opens like a mystery novel: "In southeast Queensland there is a rare aquatic frog, *Rheobatrachus silus*. The female frog swallows fertilized eggs or early-stage larvae and broods the young in her stomach. The young frogs eventually emerge by way of her mouth" (Figure 16.14). The mystery lies in how the embryos and

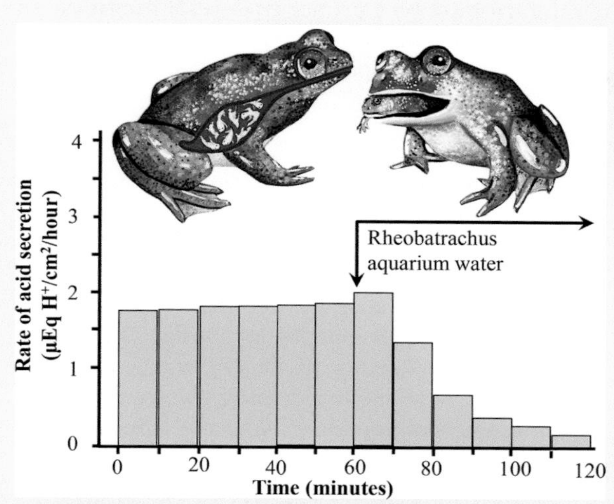

Figure 16.14 The Australian stomach-brooding frog, *Rheobatrachus silus***.** Tadpoles of this species, now extinct, developed through metamorphosis inside the mother's stomach, emerging as fully formed froglets. The secretion of prostaglandin E2 (PGE2) by embryos and tadpoles of *R. silus* may inhibit digestive activity by the mother's stomach. Inhibition of gastric acid secretion is by exposure of the *R. silus* gastric serosal surface to aquarium water containing PGE2 (histogram).

Source of images: Tyler, M.J., et al. 1983. Science, 220(4597), pp. 609-610.

tadpoles survive without being digested. The authors go on to propose, "A substance that inhibits gastric acid secretion in a toad stomach preparation *in vitro* appears to be secreted by the developing young". In order to demonstrate this, the researchers devised a creative way to monitor gastric acid secretion and its inhibition:

> We used the spontaneously secreting gastric mucosa of *Bufo marinus* as the principal bioassay for the potential inhibitor of gastric secretion. The toads were pithed, the stomach was removed, and the gastric mucosa was dissected from the muscle coats. Fundic mucosa was mounted as a diaphragm between two halves of a Lucite chamber to provide exposed mucosal and serosal surfaces measuring 2.0 cm^2.[24]

By adding water from the aquaria of the stomach brooding frogs into the serosal side (i.e. blood-side) of the chamber, they could monitor changes in the rate of acid secretion by the mucosal side (stomach lumen side). Their results showed that the addition of aquarium water from brooding *Rheobatrachus silus* dramatically reduced acid secretion by the stomach (Figure 16.18). Importantly, the inhibitory effect of *R. silus* aquarium water was reversed by replacement with a bathing buffer solution (not shown). Furthermore, water holding *Limnodynastes tasmaniensis* tadpoles (a non-stomach-brooding frog species) had no effect on the rate of acid secretion. Finally, the researchers used analytical chemical methods (radioimmunoassay and gas chromatography-mass spectrometry) to confirm that the chemical culprit present in the water of stomach-brooding *R. silus* was prostaglandin E2 (PGE2). Prostaglandins are known to exert cytoprotective effects on the stomach by inhibiting acid synthesis and promoting mucus secretion. This species of frog appears to have usurped the protective machinery of digestion to evolve a unique mechanism to control the environment in which its young develop. Sadly, many fascinating questions pertaining to the evolution of this lifestyle will remain unanswered, as this species became extinct in the 1980s.

Intestinal Phase

The delivery of chyme from the stomach to the duodenum of the small intestine initiates the **intestinal phase of digestion**. This triggers the inhibition of gastric acid secretion, acceleration of gastric emptying, release of digestive fluids from accessory organs, and deceleration of the rate of food transit in the intestine to enhance the absorption of food substrates. The stimuli that regulate these processes are both mechanical (distention of the intestinal wall) and chemical (the presence of protons and nutrients in the intestinal lumen). Key characteristics of the intestinal phase are as follows (summarized in Figure 16.15):

1. *Gastrointestinal motility*: When food enters the duodenum, several nervous reflexes and endocrine actions are initiated from the duodenal wall in response to stretch and chemical receptors. Examples of neural reflexes include the *enterogastric reflex*, in which distention and acidity in the duodenum suppresses secretion and motor activity in the stomach, and the *duodenocolic reflex*, in which colonic motility increases in response to duodenal stimulation to make way for incoming food. An intestinal hormone that influences gut motility in many mammals (though it is absent in rodents) is motilin, released cyclically by M cells of the small intestine in response to the ingestion of a meal.
2. *Nutrient detection and absorption pathways*: Enteroendocrine cells of the gut can directly detect and respond to fluctuations in concentrations of luminal nutrients and other compounds, such as sugars, lipids, oligopeptides, amino acids, and bile acids that are differentially distributed along the gut depending on the nature of the ingested food and the rates of digestion and absorption.[14] To achieve this, open-type cells often have a distinct cone-shaped morphology with one extremity adjacent to the basal lamina and the other possessing microvilli on apical processes (see Figure 16.8). Microvilli are thus in immediate contact with the luminal contents, the sensing of which can lead to the release of hormones from secretory granules directly into the nearby blood vessels. When the products of food breakdown move through the GI tract, specific macronutrients stimulate the chemosensors of a variety of enteroendocrine cells that respond by releasing compounds such as motilin, CCK, GIP, GLP-1 and GLP-2, and PYY, which ultimately influence intestinal motility and the efficiency of nutrient absorption by the intestine's absorptive epithelium.[25–28] Many of these compounds can also influence neuronal signaling in appetite centers of the brain to stimulate the appropriate feeding behaviors, such as the termination of hunger and the induction of satiety.

Stimulation of the Stomach

The distention of the duodenal wall upon entry of the chyme from the stomach, as well as the change in pH and presence of *peptones* (digested protein fragments), stimulates neural and endocrine signaling from the small intestine to the stomach that elicits a two-phase response by the stomach. First, the *excitatory phase* is characterized by increased gastric secretion, which accounts for about 10% of the stomach's secretory response to a meal, and increased stomach motility that promotes the further emptying of stomach contents into the small intestine. Stomach secretions during this phase are stimulated, in part, by the presence of peptones that stimulate duodenal G cells to secrete *gastrin*. Just as peptones stimulate antral G cells in the gastric phase, gastrin of duodenal origin also stimulates stomach secretion and motility.

The *inhibitory phase*, which follows the excitatory phase, is characterized by reduced gastric secretion, reduced gastric motility, and closing of the pyloric sphincter (Figure 16.15). These processes are mediated predominantly by the actions of three hormones released from the duodenum. *Cholecystokinin* (CCK) is released from I cells of the duodenum and upper jejunum in response to the presence of long-chain fatty acids and peptones in chyme. *Secretin* is released from S cells in response to reduced pH (less than 4.5) and, to a lesser extent, bile acids and lipids. Note that secretin not only targets the stomach to inhibit gastric secretions and motility, but also targets the pancreas, where it stimulates the release of sodium bicarbonate into the duodenum to neutralize incoming stomach acid. *Glucose-dependent insulinotropic polypeptide* (GIP) is released from K cells primarily in response to the presence of lipids and carbohydrates, and as much as 30% of the insulin secreted by the pancreas is stimulated by GIP.

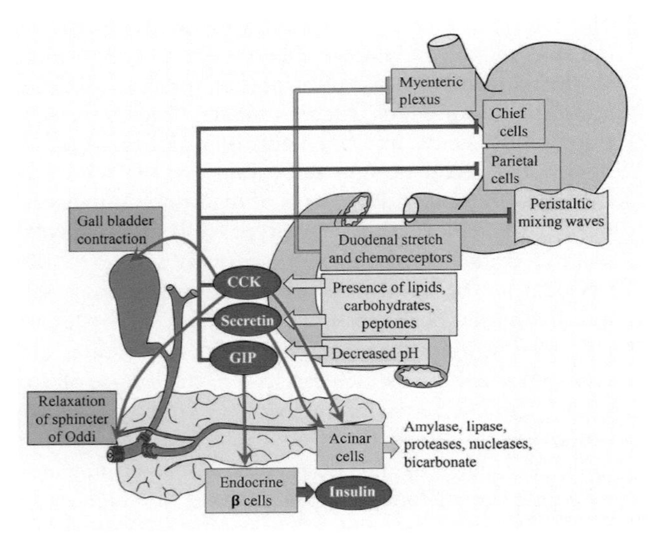

Figure 16.15 Intestinal phase of digestion. The arrival of chyme in the duodenum, the first segment of the small intestine, stimulates duodenal enteroendocrine cells to release secretin, cholecystokinin, and GIP. In many mammals, these hormones both stimulate the pancreas and gall bladder and suppress gastric secretion and motility. In humans, GIP stimulates insulin secretion by the pancreas, but has no known inhibitory effects on gastric activity.

Curiously Enough . . . GIP was originally called "gastric inhibitory peptide" do to its potency in inhibiting gastric function in dogs and rodents. Despite this, GIP is virtually devoid of gastric inhibitory activity in humans. When subsequent studies showed that GIP produced a glucose-dependent stimulation of insulin secretion in humans and other mammals, the GIP acronym's meaning was changed to its current form: glucose-dependent insulinotropic polypeptide.[29] Note the stimulation of pancreatic insulin release by GIP in Figure 16.15.

Pancreatic Stimulation

Like other organs associated with the digestive system, the pancreas has both exocrine and endocrine functions. In this section we will focus on the regulation of two categories of pancreatic exocrine secretions, digestive enzymes and bicarbonate ion, during the intestinal phase. Chapter 18: Energy Homeostasis will address the regulation of the endocrine secretion of insulin, glucagon, and other metabolic hormones. The exocrine secretory components of the pancreas, which constitute the overwhelming majority of pancreatic cells, are comprised of *acinar cells* that are specialized for the secretion of enzymes and *duct cells* that secrete a bicarbonate ion-rich fluid (refer to Figure 16.5).

Acinar cells have receptors for diverse ligands, though the most important in regulating protein secretion are the CCK and muscarinic Ach receptors. The CCK and muscarinic Ach receptors share important similarities: both are linked to the Gαq heterotrimeric G protein, both use the phospholipase C (PLC)/Ca^{++} signal transduction pathway, and both lead to increased enzyme secretion from the acinar cell (Figure 16.16). CCK stimulates acinar cells to release digestive enzymes, accounting for 70%–80% of the total secretion of the pancreatic digestive enzymes after a meal. Simultaneous to the release of CCK, a vagovagal *enteropancreatic reflex* releases Ach, which also stimulates acinar cells. In response to a meal, plasma CCK levels increase five- to ten-fold within 10 to 30 minutes. In contrast with CCK and Ach receptors, VIP and secretin receptors promote acinar cell secretion by activating a G*s* heterotrimeric G protein pathway (Figure 16.16). Therefore, activation of receptors that stimulate different signal transduction pathways may lead to an additive effect on secretion response.

Biliary Stimulation

Bile is synthesized by the liver, stored in the gall bladder, and transported to the duodenum by the bile duct (the close anatomical arrangements of these three structures is reviewed in Figure 16.5). The primary digestive role

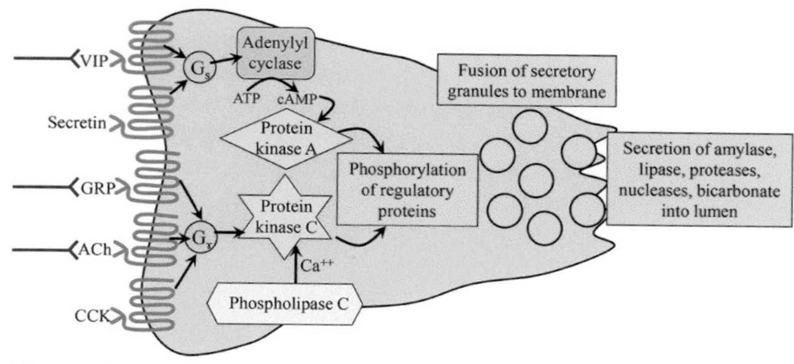

Figure 16.16 Stimulation of digestive enzyme secretions from the pancreatic acinar cell. The pancreatic acinar cell has at least two complementary pathways for stimulating the release of digestive enzyme-containing granules by multiple hormones. Acetylcholine (Ach), cholecystokinin (CCK), and gastrin-releasing peptide (GRP) each activate Gαq G protein signaling pathways that stimulates phospholipase C, which ultimately leads to an increase in intracellular Ca^{++} levels, activation of protein kinase C (PKC), and the phosphorylation of proteins that promote the fusion of granules to the plasma membrane. Vasoactive intestinal peptide (VIP) and secretin both promote acinar cell secretion by activating the Gs heterotrimeric G protein pathway. Whereas secretin and CCK are of endocrine origin, VIP, GRP, and Ach are released as neurotransmitters from axonal termini juxtaposed to the acinar cell.

Source of images: Koeppen, B.M. and Stanton, B.A., 2009. Berne & Levy Physiology, Updated Edition E-Book. Elsevier Health Sciences.

of bile is to emulsify fats into micelles to facilitate lipid uptake. Thus, when food (particularly fatty food) arrives in the upper gastrointestinal tract, the gall bladder begins to empty into the duodenum. In fact, when fat is not in the food, the gall bladder empties poorly, but when significant quantities of fat are present, the gall bladder normally empties completely in about 1 hour. The emptying of the gall bladder requires the simultaneous relaxation of the sphincter of Oddi, located at the exit of the common bile duct into the duodenum, and rhythmical contractions of smooth muscle in the wall of the gall bladder. The most important mediator of these responses is *CCK*, which acts directly on the gall bladder and also by binding to vagal afferents that stimulate gall bladder contraction and sphincter of Oddi relaxation (Figure 16.16). The net result of CCK action is ejection of a concentrated bolus of bile into the duodenal lumen, in addition to pancreatic juices described previously.

Curiously Enough . . . The aptly named word "cholecystokinin" is a Greek derivative, meaning "that which excites or moves the gall bladder".

Clearly, the endocrine regulation of digestion is quite complex. To assist you in digesting the content, an overview of the influences of enteroendocrine cell products on digestion and absorption of nutrients is provided in Figure 16.17.

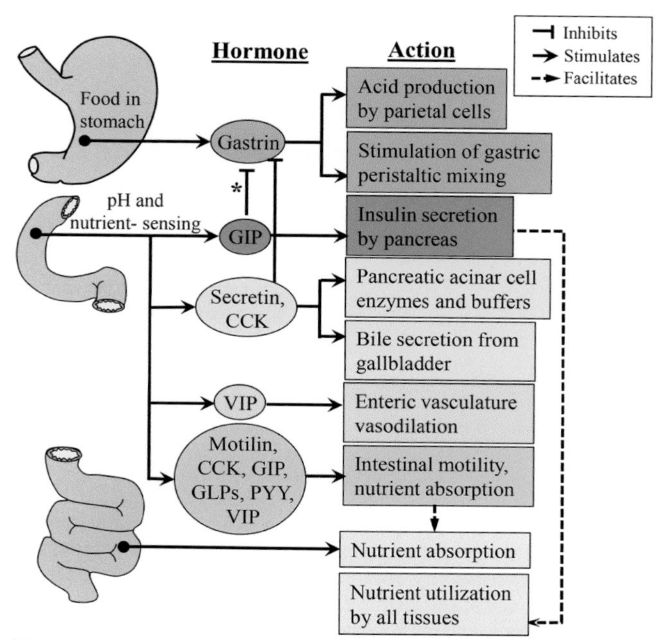

Figure 16.17 A summary of the influences of some enteroendocrine cell products on digestion and absorption of nutrients. *Note that GIP inhibits gastrin in some mammals, but not in humans.

Developments & Directions: The brain-gut-microbiome axis

"All disease begins in the gut" is a quote attributed to the ancient Greek physician Hippocrates nearly 2,500 years ago. As we will see, there is more than just a grain of truth to Hippocrates' insight. We now know that the gut microbiome has a far greater influence on physiology than just augmenting digestion. The *microbiome-gut-brain* axis refers to a network of connectivity that facilitates the bidirectional communication between gut microbes, the gastrointestinal system, and the brain, as well as other biological systems[30–32] (Figure 16.17). Microbial metabolites not only affect gut motility, nutrient absorption, and enteroendocrine cell activity, but they also profoundly influence immune system function, energy homeostasis, appetite regulation, and brain development. Indeed, this new outlook has generated a tremendous amount of research correlating the compositions of gut microbial communities with neuropsychiatric disorders associated with development (e.g. autism spectrum disorder and schizophrenia), mood (e.g. depression and anxiety) and neurodegeneration (e.g. Parkinson's disease, Alzheimer's disease, and multiple sclerosis).[30] For example, compared with neurotypical counterparts, in individuals with autism spectrum disorder gastrointestinal dysfunction is more prevalent, including altered gut permeability[33] and increased susceptibility to intestinal inflammation,[34] suggesting a neurodevelopmental link between the gut and the brain.

Not only does the microbiome influence the sympathetic and hypothalamic-pituitary-adrenal axis (HPA) stress response,[35,36] but the stress response can alter the nature of gut microbes themselves,[37–39] making the host more vulnerable to infectious disease. Specifically, the brain and gut communicate through the vagus nerve, which is a parasympathetic, cholinergic pathway that exerts anti-inflammatory effects. Under stress, the sympathetic nervous system predominates and vagal function is reduced. The increased sympathetic response elevates the concentration of norepinephrine in the gut lumen, which is known to alter gene expression in some bacteria, resulting in preferential growth of certain communities.[32,40] Elevated HPA activity, particularly adrenal glucocorticoids, also alters the microbial population, in part by altering the enteric immune cell inflammatory response and the integrity of the intestinal epithelial barrier and permeability to microbes and signaling molecules.[41–44]

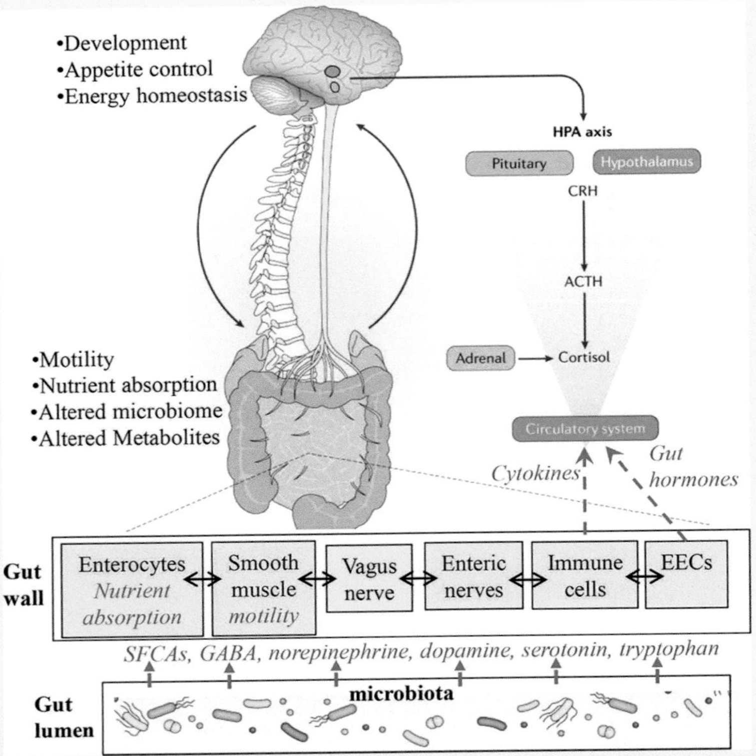

Figure 16.17 The bidirectional brain-gut-microbiome axis is mediated by several direct and indirect pathways. Crosstalk is mainly comprised of four components: (1) metabolic, facilitated by short-chain fatty acids (SFCAs, such as propionate, butyrate, and acetate) generated by bacteria; (2) neural, mediated by neuroactive substances produced by microbes (e.g. GABA, noradrenaline, dopamine, serotonin, tryptophan) and their actions on the vagus nerve and enteric nervous system; (3) immune, promoted by the effects of bacterial metabolites and stress hormones on immune cell and cytokine activity; and (4) endocrine, regulated by the actions of enteroendocrine hormones and stress hormones from the hypothalamus-pituitary-adrenal (HPA) axis and sympathetic pathways. Whereas the brain can influence microbial composition and function via endocrine and neural mechanisms, gut microbiota can influence various neurodevelopmental processes, including microglial maturation and function, blood-brain barrier formation and integrity, myelination, and neurogenesis.

Source of images: Morais, L.H., et al. 2021. Nat. Rev. Microbiol., 19(4), pp. 241-255. Used by permission.

SUMMARY AND SYNTHESIS QUESTIONS

1. How would you expect gastrin and somatostatin production by the stomach to change during starvation, and why?
2. You are taking antihistamine drugs for allergies. What side effect on digestion could this have?
3. Pernicious anemia is an autoimmune disorder that leads, in part, to the destruction of parietal cells. Another symptom of this disease is enterochromaffin-like cell hyperplasia (proliferation). How are the two symptoms related, and why do the enterochromaffin-like cells proliferate?
4. Gastric ulcers are a particularly painful malady. Though aspirin and ibuprofen are effective treatments for pain and inflammation, it would be inadvisable to take these medications to treat gastric ulcers. Why?

5. Surgical treatments for obesity include various types of bariatric surgeries that physically limit food consumption and may also promotes satiety. Adjustable gastric banding (AGB, **Summary and Synthesis Questions Figure 16.1**) involves the placement of an inflatable silicone band around the upper stomach to partition it into a small ~30 ml proximal pouch and a large distal remnant, connected through a narrow adjustable constriction. Propose a *neural* mechanism by which this method promotes satiety.
6. While AGB (**Summary and Synthesis Questions Figure 16.1**) effectively promotes satiety for some patients, on average it is the least effective form of satiety-promoting bariatric surgery. Interestingly, AGB increases levels of circulating ghrelin, which may actually induce hunger in these patients.[45] Propose a mechanism by which AGB increases ghrelin production and hunger.
7. A vertical sleeve gastrectomy (*VSG*, **Summary and Synthesis Questions Figure 16.18**) is a more radical bariatric surgery compared with AGB, and is also more successful at promoting satiety.[45] It involves the removal of 80% or more of the stomach, leaving a small diameter tube that connects to the duodenum (González-Muniesa, P., et al., 2017). Propose both neural and endocrine mechanisms by which VSG promotes satiety.

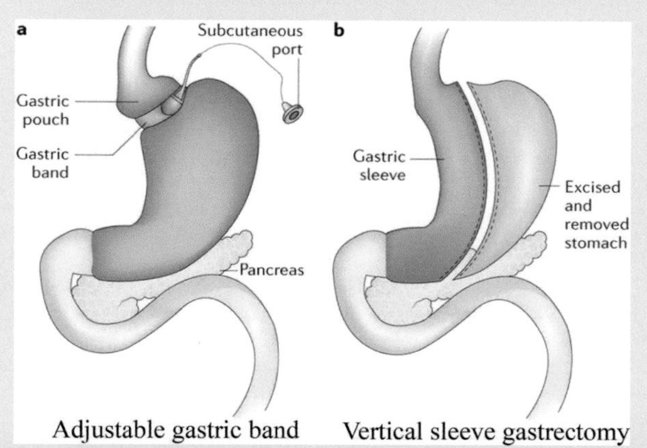

Figure 16.18 Adjustable gastric band and vertical sleeve gastrectomy.

Source of images: González-Muniesa, P., et al. Nat Rev Dis Primers 3, 17034 (2017). Used by permission.

Summary of Chapter Learning Objectives and Key Concepts

LEARNING OBJECTIVE Describe the first experiments demonstrating that gut digestive function is under both neural and endocrine control and that gut hormones influence brain function.

- Ivan Pavlov discovered that the nervous system plays a key role in regulating digestion.
- With their discovery of secretin, the first hormone ever identified, William Bayliss and Ernest H. Starling showed that the gut was not solely under neural control but also functions as an endocrine organ and a target for endocrine regulation.
- Digestion reflects a highly integrative physiological system, a complex interaction among different digestive organs, peripheral tissues, and the brain.
- A bidirectional "brain-gut axis" regulates both appetite and digestion.

LEARNING OBJECTIVE Provide an overview of the functional anatomy of the digestive tract and accessory organs from mouth to colon, and describe the major events of the cephalic, gastric, and intestinal phases of gastrointestinal secretion.

- The lumen of the stomach and intestine consist of cells with both exocrine and endocrine functions.
- Whereas the lining of the stomach consists of gastric pits that secrete HCl and proteolytic enzymes, the small intestine lining is composed primarily of villi that promote the absorption of nutrients, electrolytes, and water.
- The small intestine has its own array of brush border-associated digestive enzymes, but it relies on accessory organs for the production of diverse enzymes and other compounds for the complete digestion of nutrients.
- The large intestine digests some compounds that cannot be processed by the stomach and small intestine through the fermenting actions of symbiotic bacteria.

LEARNING OBJECTIVE Describe the integrated neural and endocrine regulation of digestion, listing the major hormones and their specific functions.

- The gastrointestinal tract differs from all other peripheral organs in that the majority of enteric neurons can function independently of the central nervous system.
- Unlike typical glandular endocrine organs, enteroendocrine cells are scattered diffusely throughout the gastrointestinal tract.
- Like neuroendocrine cells, enteroendocrine cells generate peptide hormones and neurotransmitters, many of which are also produced by neurons of the central nervous system.
- The cephalic phase of gastrointestinal secretion prepares the gastrointestinal tract to receive and digest incoming food by activating exocrine and endocrine outputs in response to visual, olfactory, and cognitive input.
- The gastric phase of gastrointestinal secretion is initiated by the arrival of food in the stomach and promotes additional exocrine, endocrine, and neural responses by the stomach, intestine, and accessory organs.
- The intestinal phase of gastrointestinal secretion is initiated by the arrival of acid chyme in the duodenum and inhibits gastric acid secretion, as well as promotes the release of digestive enzymes and fluids from accessory organs.

LITERATURE CITED

1. Bayliss WM, Starling EH. On the uniformity of the pancreatic mechanism in vertebrata. *J Physiol.* 1903;29(2):174–180.
2. Ivy AC, Oldberg E. A hormone mechanism for gallbladder contraction and evacuation. *Am J Physiol.* 1928;65:599–613.
3. Holst JJ, Fahrenkrug J, Stadil F, Rehfeld JF. Gastrointestinal endocrinology. *Scand J Gastroenterol Suppl.* 1996;216:27–38.
4. Rehfeld JF. Beginnings: a reflection on the history of gastrointestinal endocrinology. *Regul Pept.* 2012;177(Suppl):S1–S5.
5. Skipper M, Lewis J. Getting to the guts of enteroendocrine differentiation. *Nat Genet.* 2000;24(1):3–4.
6. Khamsi R. Metabolism in mind: new insights into the "gut-brain axis" spur commercial efforts to target it. *Nat Med.* 2016;22(7):697–700.
7. Goyal RK, Hirano I. The enteric nervous system. *N Engl J Med.* 1996;334(17):1106–1115.
8. Bayliss WM, Starling EH. The movements and innervation of the small intestine. *J Physiol.* 1899;24:99–143.
9. Yee LF, Mulvihill SJ. Neuroendocrine disorders of the gut. *West J Med.* 1995;163(5):454–462.
10. Johnson A. Gut feelings: the world of the second brain. *Lancet Gastroenterol Hepatol.* 2018;3(8):536.
11. Engelstoft MS, Egerod KL, Lund ML, Schwartz TW. Enteroendocrine cell types revisited. *Curr Opin Pharmacol.* 2013;13(6):912–921.
12. Pearse AG. 5-hydroxytryptophan uptake by dog thyroid "C" cells, and its possible significance in polypeptide hormone production. *Nature.* 1966;211(5049):598–600.
13. May CL, Kaestner KH. Gut endocrine cell development. *Mol Cell Endocrinol.* 2010;323(1):70–75.
14. Gribble FM, Reimann F. Enteroendocrine cells: Chemosensors in the intestinal epithelium. *Annu Rev Physiol.* 2016;78:277–299.
15. Bohorquez DV, Shahid RA, Erdmann A, et al. Neuroepithelial circuit formed by innervation of sensory enteroendocrine cells. *J Clin Invest.* 2015;125(2):782–786.
16. Zafra MA, Molina F, Puerto A. The neural/cephalic phase reflexes in the physiology of nutrition. *Neurosci Biobehav Rev.* 2006;30(7):1032–1044.
17. Wolfe MM, Jain DK, Edgerton JR. Zollinger–Ellison syndrome associated with persistently normal fasting serum gastrin concentrations. *Ann Intern Med.* 1985;103(2):215–217.
18. Zimmer T, Stolzel U, Bader M, et al. Brief report: a duodenal gastrinoma in a patient with diarrhea and normal serum gastrin concentrations. *N Engl J Med.* 1995;333(10):634–636.
19. Jais P, Mignon M. Normal serum gastrin concentration in gastrinoma. *Lancet.* 1995;346(8987):1421–1422.
20. Rehfeld JF, Gingras MH, Bardram L, Hilsted L, Goetze JP, Poitras P. The Zollinger–Ellison syndrome and mismeasurement of gastrin. *Gastroenterology.* 2011;140(5):1444–1453.
21. Dupre D, Tostivint H. Evolution of the gastrin-cholecystokinin gene family revealed by synteny analysis. *Gen Comp Endocrinol.* 2014;195:164–173.
22. Rolls BJ, Castellanos VH, Halford JC, et al. Volume of food consumed affects satiety in men. *Am J Clin Nutr.* 1998;67(6):1170–1177.
23. Hoshino T, Tsutsumi S, Tomisato W, Hwang HJ, Tsuchiya T, Mizushima T. Prostaglandin E2 protects gastric mucosal cells from apoptosis via EP2 and EP4 receptor activation. *J Biol Chem.* 2003;278(15):12752–12758.
24. Tyler MJ, Shearman DJ, Franco R, O'Brien P, Seamark RF, Kelly R. Inhibition of gastric acid secretion in the gastric brooding frog, Rheobatrachus silus. *Science.* 1983;220(4597):609–610.
25. Iwasaki M, Akiba Y, Kaunitz JD. Recent advances in vasoactive intestinal peptide physiology and pathophysiology: focus on the gastrointestinal system. *F1000Res.* 2019;8.
26. McCauley HA, Matthis AL, Enriquez JR, et al. Enteroendocrine cells couple nutrient sensing to nutrient absorption by regulating ion transport. *Nat Commun.* 2020;11(1):1–10.
27. McCauley HA. Enteroendocrine regulation of nutrient absorption. *J Nutr.* 2020;150(1):10–21.
28. Wu T, Rayner CK, Young RL, Horowitz M. Gut motility and enteroendocrine secretion. *Curr Opin Pharmacol.* 2013;13(6):928–934.
29. Goodman HM. *Basic Medical Endocrinology* (4th ed.). Academic Press; 2009.
30. Morais LH, Schreiber HL, Mazmanian SK. The gut microbiota—brain axis in behaviour and brain disorders. *Nat Rev Microbiol.* 2021;19(4):241–255.
31. Gonzalez-Santana A, Heijtz RD. Bacterial peptidoglycans from microbiota in neurodevelopment and behavior. *Trends Mol Med.* 2020;26(8):729–743.
32. Collins SM, Surette M, Bercik P. The interplay between the intestinal microbiota and the brain. *Nat Rev Microbiol.* 2012;10(11):735–742.
33. Coury DL, Ashwood P, Fasano A, et al. Gastrointestinal conditions in children with autism spectrum disorder: developing a research agenda. *Pediatrics.* 2012;130(Supplement 2):S160–S168.
34. McElhanon BO, McCracken C, Karpen S, Sharp WG. Gastrointestinal symptoms in autism spectrum disorder: a meta-analysis. *Pediatrics.* 2014;133(5):872–883.
35. Farzi A, Fröhlich EE, Holzer P. Gut microbiota and the neuroendocrine system. *Neurotherapeutics.* 2018;15(1):5–22.
36. Misiak B, Łoniewski I, Marlicz W, et al. The HPA axis dysregulation in severe mental illness: Can we shift the blame to gut microbiota? *Prog Neuro-Psychopharmacol Biol Psychiatry.* 2020;102:109951.
37. Maltz RM, Keirsey J, Kim SC, et al. Social stress affects colonic inflammation, the gut microbiome, and short chain fatty acid levels and receptors. *J Pediatr Gastroenterol Nutr.* 2019;68(4):533.
38. Molina-Torres G, Rodriguez-Arrastia M, Roman P, Sanchez-Labraca N, Cardona D. Stress and the gut microbiota-brain axis. *Behav Pharmacol.* 2019;30(2):187–200.
39. Du Y, Gao X-R, Peng L, Ge J-F. Crosstalk between the microbiota-gut-brain axis and depression. *Heliyon.* 2020;6(6):e04097.
40. Sandrini S, Aldriwesh M, Alruways M, Freestone P. Microbial endocrinology: host—bacteria communication within the gut microbiome. *J Endocrinol.* 2015;225(2):R21–R34.
41. Zhao L, Xiong Q, Stary CM, et al. Bidirectional gut-brain-microbiota axis as a potential link between inflammatory bowel disease and ischemic stroke. *J Neuroinflammation.* 2018;15(1):1–11.
42. Huang EY, Inoue T, Leone VA, et al. Using corticosteroids to reshape the gut microbiome: implications for inflammatory bowel diseases. *Inflamm Bowel Dis.* 2015;21(5):963–972.
43. Ünsal H, Balkaya M. Glucocorticoids and the intestinal environment. *Glucocorticoids.* 2012:107–150.
44. Ahmed A, Schmidt C, Brunner T. Extra-adrenal glucocorticoid synthesis in the intestinal mucosa: Between immune homeostasis and immune escape. *Front Immunol.* 2019;10:1438.
45. Meek CL, Lewis HB, Reimann F, Gribble FM, Park AJ. The effect of bariatric surgery on gastrointestinal and pancreatic peptide hormones. *Peptides.* 2016;77:28–37.
46. Wu V, Sumii K, Tari A, Sumii M, Walsh JH. Regulation of rat antral gastrin and somatostatin gene expression during starvation and after refeeding. *Gastroenterology.* 1991;101(6):1552–1558.

CHAPTER

17

Appetite

CHAPTER LEARNING OBJECTIVES:

- List the names, functions, and tissues of origin for peripheral hormones involved in short-term hunger/satiety signaling.
- List the names, functions, and tissues of origin for peripheral hormones involved in long-term energy balance and body weight.
- Describe the influence of the melanocortin system on the central (hypothalamus) regulation of appetite control and energy balance.

OPENING QUOTATION:

> "These discoveries not only changed the prevailing theory about obesity, from being caused by a lack of willpower to being caused by an imbalance of hormone signaling, but also showed that adipose tissue is not just a useless and unwanted fat storage site but rather an important and essential endocrine organ."
>
> —Douglas L Coleman (2010),[1] reminiscing on his contributions to the discovery of leptin

Introduction

Despite significant day-to-day fluctuations in food intake and physical activity, most healthy adult humans maintain a steady body weight and body fat content over many years.[2,3] This is due to a biological process termed *energy homeostasis* that matches cumulative energy intake and energy storage to energy expenditure over long periods of time. Energy homeostasis in an organism can be described with the following equation:

$$\text{Energy intake (food)} = \text{Energy stored} + \text{Energy expended (work + heat)}$$

Importantly, the left side of the equation, energy intake, is regulated in large part by *appetite*, or the desire to eat, and the regulation of appetite is the subject of this chapter. The right side of the equation, energy storage and expenditure, will be considered in Chapter 18: Energy Homeostasis. Appetite regulation is currently understood to operate on the balance of positive and negative circulating signals emanating from the gastrointestinal tract, pancreas, and adipose tissue that target neurons in the hypothalamus, brainstem, and reward centers of the brain (Figure 17.1). This system consists of multiple pathways that incorporate significant redundancy in order to regulate the drive to eat. Two distinct classes of peripheral signals modulate neuronal pathways in the brain that control meal initiation, meal termination, and the fate of ingested energy[4,5]: (1) **short-term hunger and satiety signals** are situational- and meal-related signals that are proportional to the size of the meal being consumed and are crucial to regulating the size and timing of individual meals, and (2) **long-term**

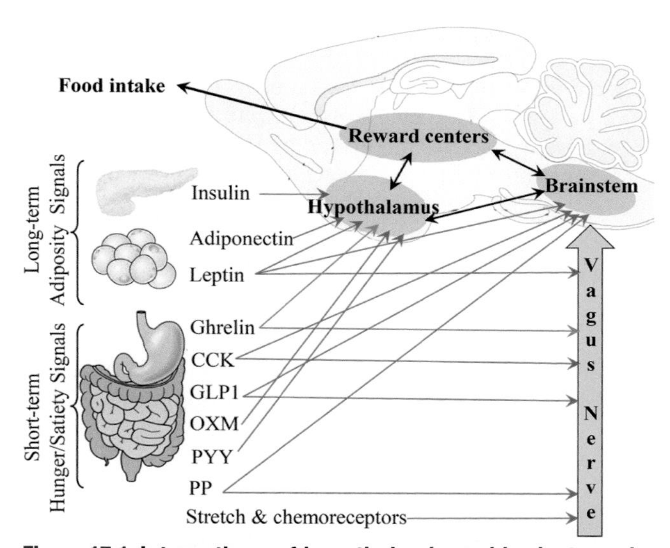

Figure 17.1 Interactions of hypothalamic and brainstem signaling with reward centers in response to peripheral endocrine and neural signals play important roles in modulating food intake in mammals. Specific hypothalamic nuclei have high densities of neurons that directly respond to long-term adiposity signals and to ghrelin. Brainstem neurons appear primarily involved in the short-term control of feeding control in response to satiety signals and stretch and chemoreceptor signals originating from the gastrointestinal tract. The central regulation of feeding also involves the neurons of the reward centers that modulate the need and desire to eat. Insulin and leptin can also act directly on their receptors present on dopamine neurons within the brain's reward centers (not shown). Red and green arrows denote anorexigenic and orexigenic signals, respectively.

DOI: 10.1201/9781003359241-23

adiposity signals that circulate in proportion to body *adiposity*—or adipose mass—and participate in the negative feedback control of fat stores for the maintenance of optimal energy balance and body weight (Figure 17.1). The two signal classes interact, with body-weight regulation occurring as long-term adiposity signals alter the efficacy of short-term meal-generated satiety signals. The concept and model of interaction between short- and long-term signals was introduced by Figlewicz and colleagues.[6] Using a baboon model, the researchers showed that a dose of the long-term satiety signal, insulin, that by itself had no effect on food intake, significantly enhanced the ability of short-term satiety signal, CCK, to reduce meal size. This was the first suggestion that a signal related to body energy stores could exert its effects on food intake by modulating within meal feedback satiety signaling.

In most cases, energy homeostasis regulates body weight tightly. However, the robust efficiency with which the energy homeostasis system works in normal-weight humans and in animal models seems to be at odds with the high prevalence of human obesity in modern society, particularly in Westernized populations. It has been argued that evolutionary pressure has resulted in a drive to eat without limit when food is readily available,[7] and the disparity between the environment in which these systems evolved and the current abundance of energy-rich foods in some human societies may contribute to overeating and the increasing prevalence of obesity. Relying on information derived from well-studied mammalian models, this chapter describes some of the various types of blood-borne and neural satiety and hunger signals emanating from peripheral organs and tissues, and it also discusses how these signals are integrated by the brain to modulate both short-term and long-term energy homeostasis.

Short-Term Hunger and Satiety Signals

LEARNING OBJECTIVE List the names, functions, and tissues of origin for peripheral hormones involved in short-term hunger/satiety signaling.

KEY CONCEPTS:

- Short-term hunger and satiety signals produced by the gut target the brain to regulate the size and timing of individual meals.
- Whereas the gut produces diverse satiety hormones, it only produces one hunger-inducing hormone, ghrelin.

Numerous hormones and neurotransmitters of central (brain) and peripheral origin have been discovered that specifically regulate appetite and feeding behavior (Table 17.1). Such signals that stimulate appetite and feeding are termed **orexigenic** (from *orexin*, meaning appetite in Greek), and those that suppress appetite and feeding are called **anorexigenic signals**. The majority of signals emanating from the gastrointestinal tract regulate "short-term" feeding, or the size of individual meals. The high redundancy of anorexogenic signals promoting the consumption of smaller meals during each feeding session allows food to pass through the digestive tract at an optimal pace for the efficient absorption of nutrients. Gastrointestinal filling inhibits feeding, with mechanoreceptors that quantify the stretch of the stomach and chemoreceptors that are activated by nutrients in the gut, each transmitting information via vagal and sympathetic afferents to the nucleus of the solitary tract (NTS) in the hindbrain nuclei (Figure 17.1). This information is then relayed to the hypothalamus and other forebrain structures for integration with additional endocrine signals regulating food intake. Some gastrointestinal hormones bind receptors in the gut and accessory digestive organs to activate vagal afferents to regulate appetite, but they also target the brain directly via the circulatory system. Next, some of the best-studied gut and pancreatic hormones with important short-term appetite-regulating roles are described.

Table 17.1 Representative Neuropeptides, Neurotransmitters, and Hormones That Influence Eating Behavior

Stimulate Eating	Inhibit Eating
Neuropeptide Y	Leptin[b]
Melanin-concentrating hormone	α-MSH
Agouti-related peptide	CRH/urocortin
Orexins	TRH
Ghrelin[a]	Cocaine- and amphetamine-regulated transcript peptide (CART)
Galanin	Peptide YY[c]
Opioids	Calcitonin gene-related peptide (CGRP)
Alpha2-noradrenergic	Insulin
GABA	Prolactin-releasing peptide
GHRH	Somatostatin
Opioid peptides	Cholecystokinin
	Glucagon-like peptide-1 and -2
	Neurotensin

Notes: [a]Secreted mainly from stomach and acts on the hypothalamus.
[b]Secreted mainly from adipocytes and acts on the hypothalamus.
[c]Secreted mainly from intestines and acts on the hypothalamus.
Source: After Takahashi, K., Murakami, O. and Mouri, T., 2010. Hypothalamus and neurohypophysis. In *Endocrine Pathology:* (pp. 45–72). Springer, New York, NY.

Cholecystokinin (CCK)

CCK's roles in regulating gastrointestinal digestive functions such as gall bladder contraction, gastric emptying, gut motility, and gastric acid secretion in response to food entering the duodenum were described in Chapter 16: Digestion. CCK also targets the central nervous system to promote satiety and is thus an important anorexigenic hormone. Expressed primarily by intestinal enteroendocrine cells, particularly in the duodenum, CCK is also present in the brain where it functions as a neurotransmitter. Release of CCK in the gut is stimulated by protein and fat. Although administration of CCK to hungry humans and other mammals decreases their food intake,[8] these effects can be blocked by severing the abdominal section of the vagus nerve[9] or via treatment with CCK antagonists.[10,11] This indicates that gastrointestinal CCK regulates food intake primarily through vagal afferent signals to the brain rather than through endocrine mechanisms. Interestingly, long-term administration of CCK does not result in weight loss since the reduction in food intake at each meal is offset by the consumption of more meals.[12] This emphasizes the fact that CCK is a short-term inhibitor of food intake, and that signals of long-term energy balance, such as leptin, can override the CCK signal. An important consideration is that these studies in which injected CCK was shown to inhibit food intake were achieved by injecting concentrations at supraphysiologic levels, and experiments mimicking the biological concentrations of CCK that occur with meal intake do not inhibit feeding (Robert V. Considine, personal communication). Therefore, the extent to which CCK functions as an appetite-regulating signal currently remains unresolved.

> **Curiously Enough . . .** Highlighting CCK's dual functions in regulating digestion and appetite, two receptors for CCK have been characterized: the CCKA receptor is located primarily in the gastrointestinal tract, and the CCKB receptor is found in the brain.

Proglucagon-Derived Peptides

Glucagon-like peptide 1 (GLP-1) and oxyntomodulin are products of the preproglucagon gene, produced by alternative posttranslational processing of proglucagon in the L-cells of the intestinal mucosa, as well as in the brainstem in response to the appearance of glucose or free fatty acids in the small intestine (refer to Figure 4.12 of Chapter 4: Hormone Classes and Biosynthesis for a review of proglucagon processing). Both GLP-1 and oxyntomodulin act as satiety signals through the GLP-1 receptor. GLP-1 inhibits gastric emptying in humans at physiological concentrations achieved after meal ingestion, and GLP-1 also suppresses appetite and food consumption with central and systemic administration in rats as well as humans, causing hypophagia (reduced feeding) and weight loss.[13–15] The interaction of GLP-1 with receptors in the hypothalamus and brainstem mediates these satiety effects. Furthermore, GLP-1 functions as an incretin, increasing insulin secretion during meals.

Intestinal glucose-dependent insulinotropic polypeptide (GIP) is secreted from k-cells in the duodenum and proximal jejunum as well as in the hippocampus in response to food intake.[16,17] Similar to GLP-1, intestinal GIP acts as an incretin by increasing glucose-dependent insulin release from pancreatic β cells and, therefore, it contributes to postprandial plasma glucose normalization.

Peptide Tyrosine Tyrosine (PYY)

In response to the intake of fat, PYY is secreted from the L cells of the intestine (particularly from the ileum and colon) and targets hypothalamic nuclei to reduce food intake.[18] Human studies have shown that PYY treatment reduces food intake in both normal and obese people, and that obese individuals show a dysregulation of PYY,[19–22] suggesting that PYY may be an important anorexigenic regulator of appetite.

Ghrelin

Ghrelin (from Greek *ghre*, grow), originally identified as a growth hormone-inducing substance, is a peptide hormone released from the stomach in anticipation of a meal. In marked contrast to the previously described gut hormones that have anorexigenic effects, ghrelin is one of the only known gut-derived signals that stimulates food intake rather than causing satiety. Ghrelin stimulates feeding by binding to specific receptors in the hypothalamus. As such, circulating ghrelin levels are normally elevated prior to meals, but drop soon after eating. Chronic administration of ghrelin results in obesity in rodents,[23] and if acutely administered it will cause animals and humans to consume larger meals than normal.[24,25] Under conditions of starvation or anorexia nervosa (a neurological disorder associated with a fear of eating), blood ghrelin levels are high, as would be predicted. However, considering ghrelin's hunger-promoting effects, researchers were surprised to find that most obese humans actually have low levels of circulating ghrelin.[26]

Ghrelin is produced predominately by the X/A cells of the stomach, but also in lesser amounts by the intestine and kidneys and in the hypothalamus. The processing of the 117 amino acid human preproghrelin peptide (Figure 17.2) produces, through differential proteolytic

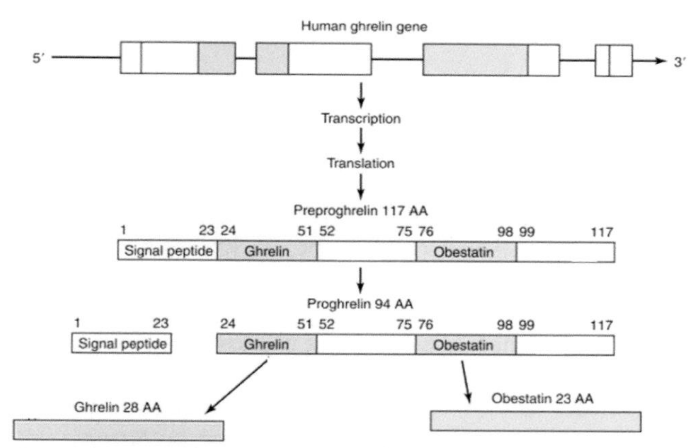

Figure 17.2 Translation and processing of preproghrelin. The 117 amino acid prepropeptide and the subsequent posttranslational steps that lead to the 94 amino acid proghrelin and the final two products, the 24 amino acid ghrelin (pink) and the 23 amino acid obestatin. Amino acid #3 (a serine) is acetylated by an 8- to 10-carbon fatty acid. This serine must be acetylated for ghrelin to stimulate both growth hormone (GH) and appetite.

Source of images: Norman, A.W. and Henry, H.L., 2022. Hormones. Academic Press. Used by permission.

processing, two functional peptides: ghrelin (28 amino acids) and obestatin (23 amino acids). Obestatin was initially reported to exert effects antagonistic to those of ghrelin in rats, reducing food intake, body weight gain, and gastric emptying and suppressing intestinal motility (hence, its name).[27] Although currently a precise biological function for obestatin remains unclear, accumulating evidence supports positive actions on both metabolism and cardiovascular function,[28] with promise for therapeutic potential of conditions associated with obesity, such as type 2 diabetes.

SUMMARY AND SYNTHESIS QUESTIONS

1. Compared with wild-type mice, what effects on appetite and weight would occur in rats lacking functional receptors for (1) CCK and (2) ghrelin?

Long-Term Adiposity Signals

LEARNING OBJECTIVE List the names, functions, and tissues of origin for peripheral hormones involved in long-term energy balance and body weight.

KEY CONCEPTS:

- Long-term adiposity signals produced by adipocytes (leptin) and the pancreas (insulin) circulate in proportion to adipose mass and target the brain to maintain optimal energy balance and body weight.

The biological system that regulates body adiposity, the so-called adipostat mechanism, was hypothesized by Gordon Kennedy in the mid-20th century[29] to involve then unknown humoral signals generated in proportion to body fat stores that act in the hypothalamus to alter food intake and energy expenditure. We now know these signals function in part by changing brain responsiveness to short-term satiety signals.[6,30,31] The CNS response to short-term signals is adjusted in proportion to changes in body adiposity, resulting in compensatory changes in food intake and energy expenditure that collectively favor the long-term (weeks to months) stability of fat stores.[5] The two best-described adiposity signals are leptin and insulin, produced by adipose tissue and the pancreas, respectively, which circulate in blood in amounts proportional to body fat and blood glucose. These long-term signals target a hunger center of the brain, the *hypothalamic arcuate nucleus*. Since leptin and insulin proteins are too large to cross the blood-brain barrier (BBB), an as of yet unidentified active transport mechanism is most likely responsible for their uptake into the brain.

Insulin and leptin inhibit hypothalamic orexigenic neurons and activate anorexigenic neurons, resulting in decreased food intake and increased energy expenditure. However, long-term increases in leptin or insulin may lead to receptor desensitization and insulin or leptin "resistance", increasing plasma glucose levels and fat accumulation, and can contribute to metabolic disorders such as obesity and diabetes. Although there are several good candidates for adiposity signals, including amylin (which is co-secreted with insulin by the pancreas) and adiponectin (secreted by adipose tissue),[32] the following overview will confine itself to the two best-studied adiposity signals, leptin and insulin.

Leptin

In 1949, a strain of mice that ate voraciously and were massively obese was identified,[33,34] and the putative gene mutation was termed "*ob*" for obese. In 1994, the location of this gene in rodents was identified,[35] and subsequent studies soon confirmed that the *ob* gene encoded a novel hormone that circulated in blood and could suppress food intake and body weight in both wild-type mice and mice heterozygous for the gene mutation (ob^-/ob^-)[36–38] (Figure 17.3). The hormone was named "leptin", from Greek *leptos*, meaning thin. Leptin is now known to be a 146 amino acid protein product of the Lep^{ob} gene (originally named the *ob* gene) belonging to the cytokine family of signals and is expressed and secreted primarily by white adipose tissue cells.[39] Subsequent studies in mice identified the gene coding for the leptin receptor, which is expressed in the hypothalamus.[40–43] The leptin receptor was originally called the *db*, or "diabetes", receptor, and is now called $Lepr^{db}$. The leptin receptor belongs to the cytokine family of receptors that are known to stimulate gene transcription via activation of the JAK/STAT signaling pathway (refer to Figure 5.14 in Chapter 5: Receptors).

The first discovered adiposity signal, leptin represents one of the core components of the physiological system

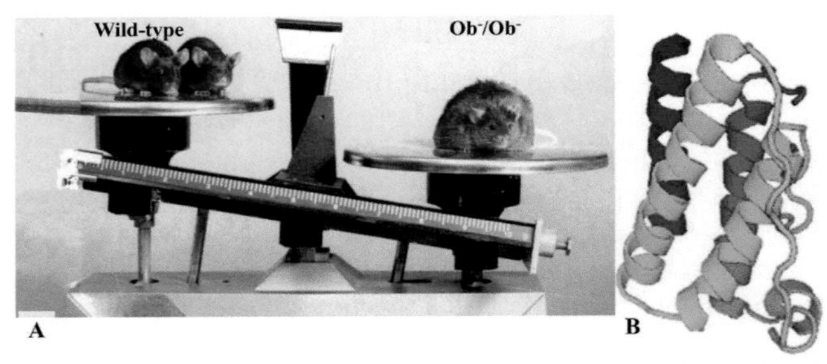

Figure 17.3 A leptin gene knockout mouse (ob^-/ob^-; right) stacking up against its lean counterparts (wild-type; left) (A). Leptin is a protein hormone comprising 167 amino acids in humans. **(B)** Its tertiary structure resembles that of other cytokines in its class, such as growth hormone and prolactin. Depicted is human leptin.

Source of images: (A) Used by permission of Jeffrey Friedman. (B) Han, D., et al. 2016. Fish Physiol. Biochem., 42(6), pp. 1665-1679. Used by permission.

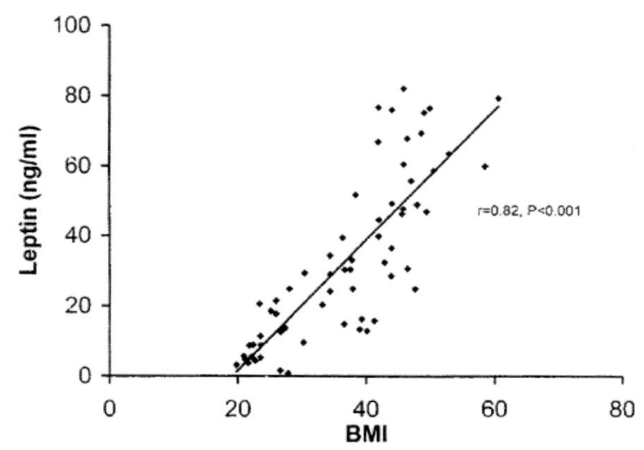

Figure 17.4 Leptin concentrations in blood plasma correlate with body mass index (BMI) in humans.

Source of images: Van Dielen, F.M.H., et al. 2001. Int. J. Obesity, 25(12), pp. 1759-1766. Used by permission.

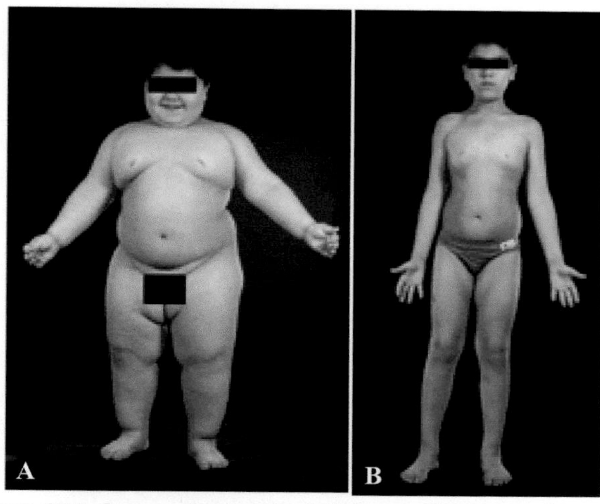

Figure 17.5 Congenital leptin deficiency. Because of a gene defect, the boy doesn't make leptin. Treatment with leptin, begun when he was 3.5 years old **(A)**, brought his weight down to normal levels, as shown at age 8 **(B)**.

Source of images: Farooqi, S.I., 2015. Clin. Endocrinol., 82(1), pp. 23-28.

that controls body weight in mammals, and in humans circulating levels of leptin are proportional to fat mass (Figure 17.4). Like the *ob⁻/ob⁻* mice, human children with rare congenital leptin deficiency also have a voracious appetite and become obese, and the administration of exogenous leptin ameliorates these abnormalities in both mice and humans[38,44–46] (Figure 17.5). Importantly, although circulating leptin levels are generally reflective of current energy stores, serum leptin levels can change independently of adipose status in response to extremes in caloric intake. For example, serum leptin falls dramatically during fasts of 24 hours or longer and will increase again within 4–5 hours of refeeding despite the fact that adipose tissue mass does not change over this time period.[47] Interestingly, the absence of leptin appears to convey a more powerful signal than does its excess. For example, the vast majority of obese humans have normally functioning leptin genes and often display elevated plasma leptin levels reflecting their high adipose mass.[48,49] However, despite high levels of circulating leptin, satiety is not induced in these obese humans. These types of observations have led to the suggestion that leptin, or more precisely, low levels or the absence of leptin, functions more as a "starvation" signal than a satiety signal.[50] Consistent with the notion that leptin is not a potent satiety signal in humans, years of research have shown that leptin treatments are unsuccessful in reversing obesity, with the exception of rare cases of congenital leptin deficiency.[51] A hypothesis for why leptin treatment does not reverse human obesity is that a form of leptin "resistance" develops in obese individuals, a poorly understood phenomenon that may involve reduced leptin receptor expression and/or suppressed intracellular signaling, mutations in the genes encoding leptin and its receptors, as well as reduced blood-brain barrier permeability to leptin.[52–54]

Beyond leptin's function as a strong starvation or weak satiety signal, there is mounting evidence that the hormone might play an important role in the timing of puberty, a topic further discussed in Chapter 23: The Timing of Puberty and Seasonal Reproduction. Because the onset of puberty occurs only if sufficient fuel stores are available, and leptin functions as an adiposity signal, threshold levels of leptin are thought to be necessary for puberty to proceed, particularly in females. For example, various animal models and human pathologies associated with low or absent leptin levels are often linked to a delay or absence of puberty onset and perturbed fertility.[55] However, while leptin is indispensable for puberty to proceed, leptin alone cannot trigger early puberty in humans, suggesting a major permissive role of this hormone in the metabolic gating of pubertal maturation.[56]

Insulin

In addition to playing a critical role in modulating blood glucose homeostasis (discussed in Chapter 18: Energy Homeostasis), the pancreatic β cell-derived hormone, insulin, is another important adiposity signal that conveys additional information about long-term changes of peripheral metabolism to the brain. Insulin is thought to have a similar lipostatic role to that of leptin, although its central effects on food intake and energy homeostasis are less pronounced.[57] By decreasing food intake, insulin contributes to decreased blood glucose concentrations. Insulin receptors are widely distributed in the brain, particularly in hypothalamic nuclei involved in food intake,[58–60] and central administration of insulin in rodents has been shown to reduce food intake and body weight.[61] In contrast, neuron-specific deletion of its receptor causes obesity and hyperinsulinemia,[62,63] and antibodies that inactivate insulin injected into the VMH of rats stimulate food intake[64] and weight gain.[65] As with leptin, increased adiposity can lead to a decrease in insulin sensitivity and a state of insulin resistance.[66] Furthermore, it is also believed that adiposity might in fact be a consequence of insulin resistance itself[67] and a potential underlying cause of various metabolic abnormalities, including obesity.[68] As such, approaches aimed at improving insulin sensitivity by way of pharmacological intervention remain a driving force in drug development.[69]

Developments & Directions: Gut microbial metabolites influence appetite

You were introduced to the notion of the bidirectional **brain-gut-microbiome axis** in Chapter 16: Digestion (refer to Developments & Directions Figure 16.17 of that chapter). Regarding their regulation of appetite, these microbes are able to manufacture various metabolites, including short-chain fatty acids (SCFAs) and neuroactive compounds that are able to exert their effects through direct interaction with receptors in gut enteroendocrine L cells and the vagus nerve, or by translocating through the intestinal epithelium into the peripheral circulation. Stimulation of L cells by SCFAs results in the release of anorexigenic hormones PYY and GLP-1, which can exert effects directly on the brain's appetite centers, and can increase concentrations of peripheral appetite regulatory hormones, such as insulin, leptin, and ghrelin. These observations suggest that imbalances in the gut microbiota could contribute to overeating and obesity and might be linked to poor glucose homoeostasis and potentially diabetes.[70]

SUMMARY AND SYNTHESIS QUESTIONS

1. Although leptin normally functions as an appetite suppressant, treatment of obese individuals with high levels of leptin does not inhibit the desire to eat in the long term. Why not?

Integrating Appetite and Energy Homeostasis Signals in the Brain

LEARNING OBJECTIVE Describe the influence of the melanocortin system on the central (hypothalamus) regulation of appetite control and energy balance.

KEY CONCEPTS:

- A key region of the brain that integrates endocrine and neural satiety and adiposity signals is the arcuate nucleus of the hypothalamus.
- Components of the central melanocortin system, such as POMC, MSH, melanocortin receptors, and receptor antagonists, play prominent roles in regulating appetite and energy homeostasis.

Insights from human and mouse genetics have shown that the brain is the primary integrator of peripheral appetite and energy homeostasis signals. The key regions that integrate these humoral and neural inputs include the hypothalamus, brainstem, and reward centers of the brain (see Figure 17.1). These regions signal via diverse neuropeptides (Table 17.1) that modulate the activities of higher-order neurons involved in regulating energy homeostasis.

Control of Feeding Behavior by the Hypothalamus

Near the time that Kennedy was forming his "adipostat" hypothesis, Eliot Stellar introduced the notion that dual centers in the hypothalamus function as regulators for feeding behaviors: an anorexigenic "satiety center" situated in the ventral medial hypothalamus (VMH) and an orexigenic "hunger center" in the lateral hypothalamus area (LHA).[71] Stellar's hypothesis was based on electrical stimulation and lesion studies of these regions in rats.[72] Whereas electrical stimulation of the VMH decreased food intake, lesions in this area produced a marked increase in food intake (hyperphagia). In contrast, electrical stimulation of the opposing LHA increased food intake, and lesions in this area resulted in decreased food intake. Furthermore, in *parabiosis* experiments consisting of the surgical union of two animals allowing exchange of blood, it was not only demonstrated that circulating substances were capable of regulating food intake, but rats with lesions to the VMH parabiosed to normal animals remained hyperphagic and obese, whereas the parabiosed partner ate less and lost weight (Figure 17.6).[73] These experiments were the first to provide substantive evidence that a blood-borne satiety factor that was produced in obese animals required an intact hypothalamus for its activity. After leptin was identified as a key satiety hormone, parabiosis experiments using mutant mice lacking the genes for either leptin (ob/ob) or the leptin receptor (db/db) generated similar findings, suggesting that leptin targeted hypothalamic satiety centers.

Although this hypothesis for the control of feeding behavior by the hypothalamus was important in shaping the framework for interpreting research findings for decades to come, additional research has replaced the dual-center hypothesis with multiple hypothalamic integrated neural circuits that regulate energy balance using specific neuropeptides. Importantly, these multiple circuits are thought to each control diverse behavioral and physiological aspects of energy intake, storage, and expenditure with little separation of function.[74–76]

The Arcuate Nucleus of the Hypothalamus

One particularly well-studied and important region of the hypothalamus that integrates peripheral signals

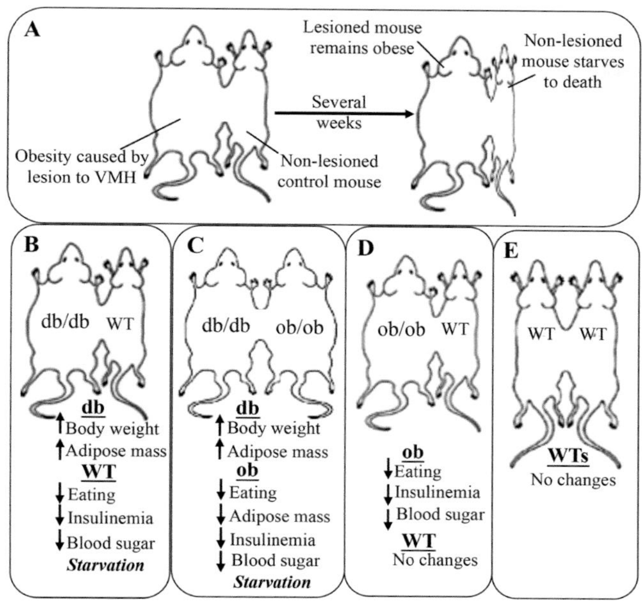

Figure 17.6 A summary of parabiosis experiment supporting the hypothesis that a circulating factor in the blood (later discovered to be leptin) suppresses food intake by targeting the hypothalamus in mice. (A) A parabiotic pairing an obese (caused by lesions to the VMH) with a normal lean mouse caused starvation in the lean mouse, whereas the lesioned mouse continued to overeat. This shows that the lesioned mouse overproduces a satiety factor (i.e. leptin) but does not respond to it. In contrast, the lean mouse responds to the overproduced satiety factor and starves. **(B–E)** Summary of parabiosis experiments using wild-type lean (WT), leptin receptor knockout (db/db) (the leptin receptor was originally named "db" for "diabetes"), and leptin gene knockout (ob/ob) (the leptin gene was originally named "ob" for "obese"). Images B–E depict initial mouse morphology at pairing and don't represent their final appearance at the end of the experiments.

Source of images: Coleman, D.L., 2010. Nature medicine, 16(10), pp. 1097-1099.

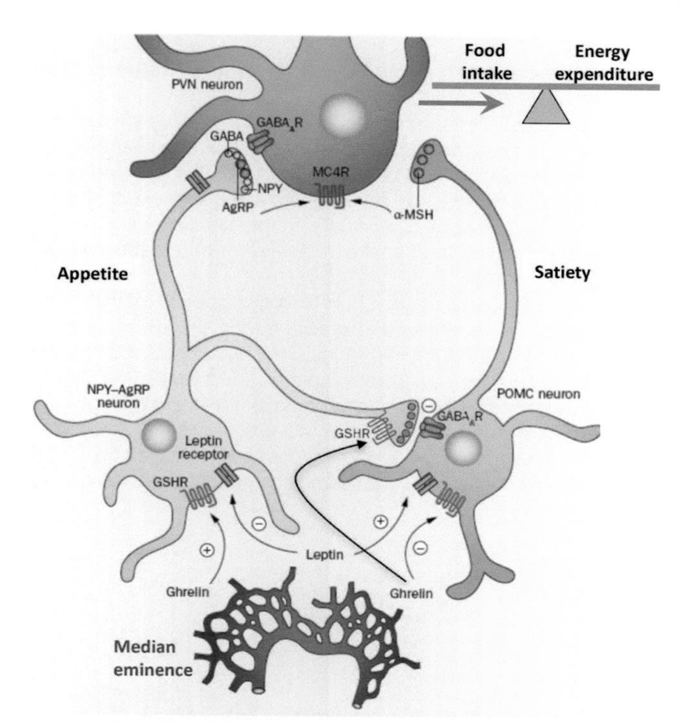

Figure 17.7 The melanocortin system in the arcuate nucleus of the hypothalamus. The primary neurons of the melanocortin system are the orexigenic NPY–AgRP neurons and the anorexigenic POMC neurons, which both send projections to the paraventricular nucleus. NPY–AgRP neurons also send inhibitory projections to neighboring POMC cells. Activation of POMC cells by leptin triggers the release of α-MSH, which binds to MC4R and promotes satiety. AgRP neurons are inhibited by leptin but stimulated by ghrelin, which promotes feeding and silences firing of POMC neurons. The effect of the NPY–AgRP cells is mediated by release of GABA, NPY, and AgRP. Abbreviations: α-MSH, α-melanocyte-stimulating hormone; AgRP, agouti-related protein; GABA, γ-aminobutyric acid; GABAAR, GABA receptor; GHSR, growth hormone secretagogue receptor (also known as ghrelin receptor); MC4R, melanocortin receptor 4; NPY, neuropeptide Y; POMC, proopiomelanocortin; PVN, paraventricular nucleus.

Source of images: Nasrallah, C. M. & Horvath, T. L. Nat. Rev. Endocrinol. 10, 650–658 (2014). Used by permission.

related to appetite and energy balance is the *arcuate nucleus* (ARC) (Figure 17.7). Situated between the third ventricle and the median eminence, the arcuate nucleus is uniquely positioned to sample factors in both blood and cerebrospinal fluid. The ARC functions as a relay hub that contains so-called first-order neurons that mediate and integrate the first contact of peripheral signals with the CNS. There are two distinct sets of hypothalamic neurons situated in the ARC that have opposing effects on food intake.[77–79] One population of neurons co-expresses the anorexigenic precursor peptide **proopiomelanocortin** (POMC) as well as *cocaine- and amphetamine-related transcript* (CART); the other co-expresses the orexigenic peptides *neuropeptide Y* (NPY) and *agouti-related protein* (AgRP) (Figure 17.7 and Figure 17.9). A majority of both POMC/CART and NPY–AgRP neurons have been found to co-express leptin and insulin receptors,[77,79–81] and both types of neurons are regulated by these hormones in an opposing manner.[82,83] As such, administration of exogenous leptin or insulin activates POMC/CART neurons but inhibits NPY–AgRP neurons. The net result is that conditions of starvation are commensurate with low levels of leptin and insulin, as well as with high ghrelin levels, resulting in promotion of the activity of NPY–AgRP neurons and the inhibition of POMC/CART neurons. Notably, the opposite occurs under conditions of elevated insulin and leptin and reduced ghrelin. The first-order neurons of the NPY–AgRP and POMC/CART nuclei not only interact with each other, but also communicate with orexigenic and anorexigenic "second-order neurons" located in the VMH, LHA, and PVN of the hypothalamus and other brain nuclei (Figure 17.7).

Developments & Directions: Thyroid hormone induces food intake via direct actions on the hypothalamus

It is well established that alterations in thyroid status are associated with changes in feeding in both humans and rodents. In fact, *hyperphagia* (increased feeding) is observed in 85% of hyperthyroid patients, despite the weight loss typically associated with hyperthyroidism. Until recently, it had been assumed that it is reduced body weight that induces hyperphagia. However, current evidence suggests that the hypothalamus-pituitary-thyroid (HPT) axis exerts a direct role in the hypothalamic regulation of appetite, independent of effects on energy expenditure. Specifically, in rodents peripheral or central administration of the thyroid hormone triiodothyronine (T_3) has been shown to stimulate food intake independent of metabolic effects, whereas both central and peripheral administration of TRH and TSH inhibit food intake.[84]

To investigate the role for T_3 in the physiological regulation of food intake and energy expenditure, Kong and colleagues[85] studied the effects of both peripheral and central administration of T_3 in rats, using low doses of T_3 that did not elevate plasma free T_3 (fT_3) levels outside the normal euthyroid range. The researchers found that not only did chronic peripheral administration of T_3 at concentrations as low as 4.5 nmol/kg significantly increase food intake (Figure 17.8[A]), but injections of extremely low concentrations of T_3 (as low as 0.5 *pico*moles) directly into the ventromedial nucleus (VMN) of the hypothalamus stimulates feeding (Figure 17.8[B]). Importantly, although T_3 injected at a high concentration (75 nmol/kg) stimulated both increased feeding and elevated metabolic rate (Figure 17.8[C]), injection with T_3 concentrations 4.5 nmol/kg or lower increased food intake without altering metabolic rate (Figure 17.8[D]). Together, these findings strongly implicate T_3 with stimulating appetite via central actions on the hypothalamus.

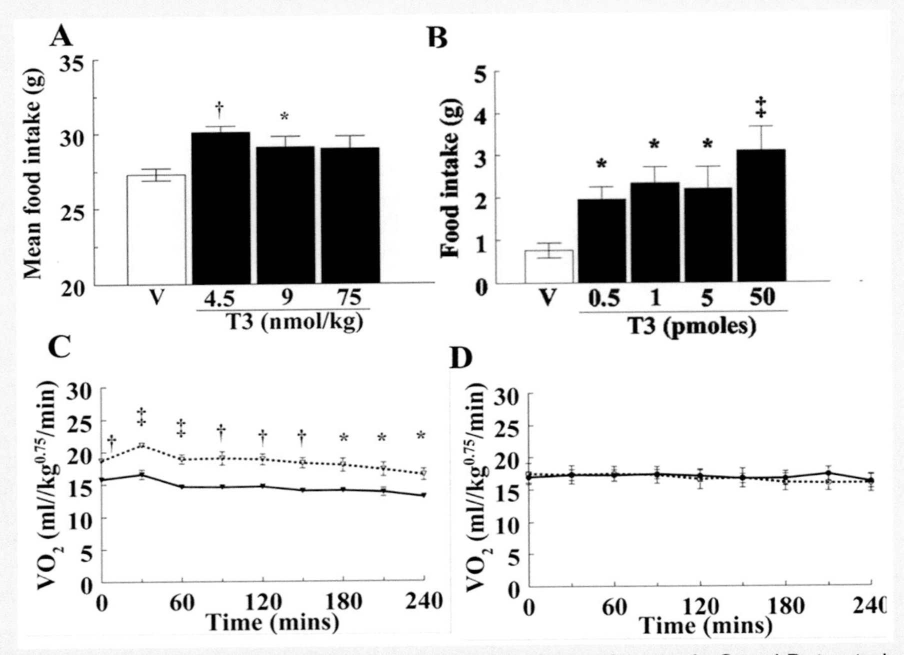

Figure 17.8 T_3 induces food intake via direct actions on the hypothalamus in rats. In C and D, treated groups are depicted by a dashed line, and controls by a solid line. *, $P < 0.05$; †, $P < 0.005$; ‡, $P < 0.0001$ vs. control.

Source of images: Kong, W.M., et al. 2004. Endocrinology, 145(11), pp. 5252-5258. Used by permission.

A Central Role for the Melanocortin System

As you can see from the prior description, the arcuate nucleus is composed of fibers expressing both agonists (e.g. POMC-derived melanocortins, such as α-MSH) and antagonists (e.g. AgRP) of the melanocortin receptors (MCRs), indicating that the melanocortin system plays a particularly prominent role in appetite regulation (refer to Chapter 8: The Pituitary Gland and Its Hormones for a review of the melanocortin system). Melanocortin system components are expressed not just in the ARC, but also in other hypothalamic regions implicated with appetite control, including the paraventricular nucleus (PVN), lateral hypothalamic area (LHA), and dorsomedial hypothalamus (DMH),[86,87] as well as in the *nucleus of the solitary tract* (NTS) in the caudal brainstem.[87] Stimulation of MC3/4 receptors induces a negative energy balance by decreasing food intake and increasing energy expenditure in animals and humans.[88–90] In contrast, by acting as a melanocortin antagonist, AgRP decreases at MC3/4R activity and induces positive energy balance by increasing food intake and decreasing energy expenditure. The NPY–AgRP system not only antagonizes anorexigenic melanocortin cells at their target sites via MC4 receptors, but it also projects axonal termini to POMC/CART

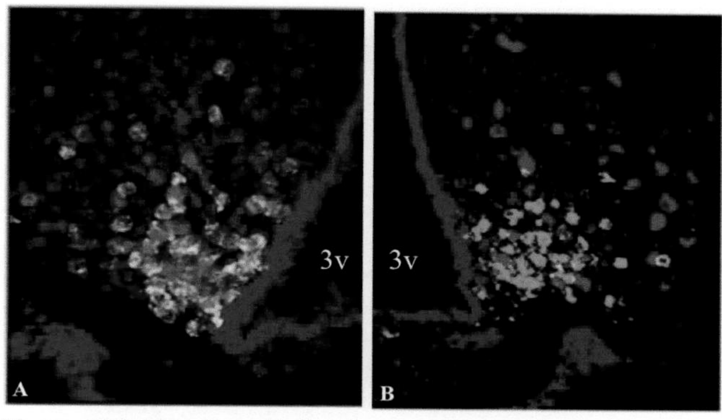

Figure 17.9 The hypothalamic arcuate nucleus after histochemical detection of messenger RNA using a method known as fluorescent *in situ* hybridization. (A) Neurons that express the gene encoding NPY (green) or AgRP (red) are shown. Cells in which both NPY and AgRP mRNA is present both show up as yellow-orange, due to the merging of the green and red staining of the two mRNA species. The blue DAPI counterstain detects cell nuclei. **(B)** Neurons that express the gene encoding NPY (green) or POMC (red) are shown. Note that unlike co-expression of NPY and AgRP genes (left), NPY and POMC genes are not co-expressed by the same cells (right). 3v: third ventricle. Images courtesy of the laboratory of Michael Schwartz, MD.

Source of images: Drs. Tina Hahn, John Breininger, Denis Baskin and Michael Schwartz, Division of Metabolism, Endocrinology and Nutrition, Dept. of Medicine, University of Washington.

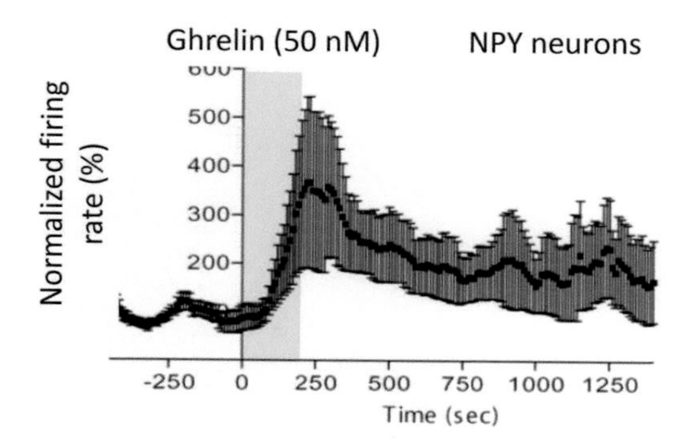

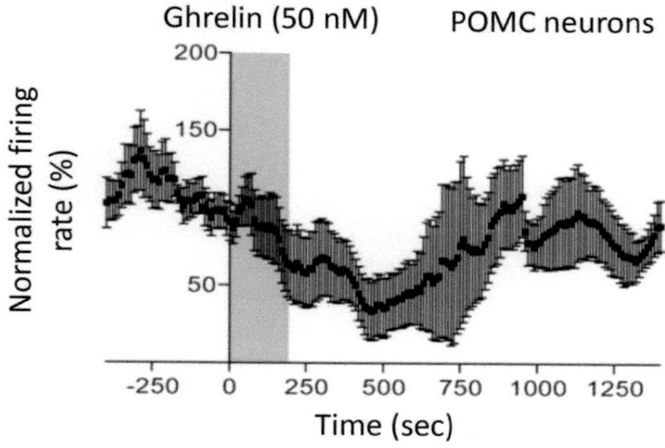

Figure 17.10 Ghrelin increases the spontaneous activity of NPY neurons and inhibits POMC neurons by increasing the inhibitory tone onto them.

Source of images: Cowley, M.A., et al. 2003. Neuron, 37(4), pp. 649-661. Used by permission.

neurons, inhibiting activity of these cells directly via synaptic release of the inhibitory transmitter γ-aminobutyric acid (GABA) (Figure 17.7).[91] This interaction provides a chronic inhibition of the POMC expressing neurons whenever NPY–AgRP neurons are active. In support of this model, the lack of the MC4R leads to hyperphagia and obesity in rodents.[88,92] Furthermore, these receptors are implicated in 1%–6% of severe early-onset human obesity.[93–95] A mutation in the AgRP gene in humans is associated with lower body weight and fat mass, whereas transgenic mice with ubiquitous overexpression of AgRP become obese.[96]

Some of the best evidence for a role for the hypothalamic melanocortin system in a direct response to acute signals of satiety and hunger comes from data on ghrelin, the growth and hunger-stimulating hormone whose levels are reduced with meal ingestion in both rodents and humans but increase after an overnight fast. In particular, ghrelin receptors (GHSRs) have been found on arcuate NPY neurons,[97] and pharmacological doses of ghrelin stimulate food intake and obesity, in part by stimulating NPY and AgRP expression,[98–100] and antagonizing leptin's anorexigenic effect[100] (Figure 17.7). On the cellular level, electrophysiological analyses suggest that ghrelin acts on the arcuate NPY–AgRP neurons to coordinately activate these orexigenic cells, while inhibiting the anorexigenic POMC cells by increasing GABA release onto them[101] (see Figure 17.10).

Developments & Directions: Melanin-concentrating hormone: another dual-regulator of pigmentation and appetite

We have discussed several peptides that regulate both pigmentation and appetite, namely melanocyte-stimulating hormone (MSH), agouti, and agouti-related peptide (AgRP) (see Chapter 8: The Pituitary Gland and Its Hormones for a review of the melanocortin system). Add to this list yet another dual functioning peptide: melanin-concentrating hormone (MCH). MCH is a cyclic peptide (Figure 17.11[A]) originally identified in teleost fish,[102] where it is synthesized by neurons in the hypothalamus that project primarily to the neurohypophysis of the pituitary, as well as other regions of the brain.[103] In teleost fish MCH is secreted in response to stress and environmental stimuli, where it has antagonistic actions to MSH and lightens skin color by stimulating the aggregation of melanosomes, pigment-containing granules in the melanophores of fish scales (Figure 17.11[B]).

Although the peptide structure is highly conserved among vertebrates, in mammals MCH has no demonstrable effects on pigmentation. Instead, MCH has emerged as a critical hypothalamic regulator of mammalian feeding behavior and energy expenditure.[104] In the rodents, MCH is expressed primarily in the lateral hypothalamic area (LHA) and the rostral zona incerta[105,106] (Figure 17.11[C]), two areas known to be critically involved in the regulation of feeding behaviors. Specifically, MCH is thought to function as an appetite stimulant. For example, injection of MCH into either the ventricular system of the brain or directly into the paraventricular nucleus (PVN) stimulates food intake in

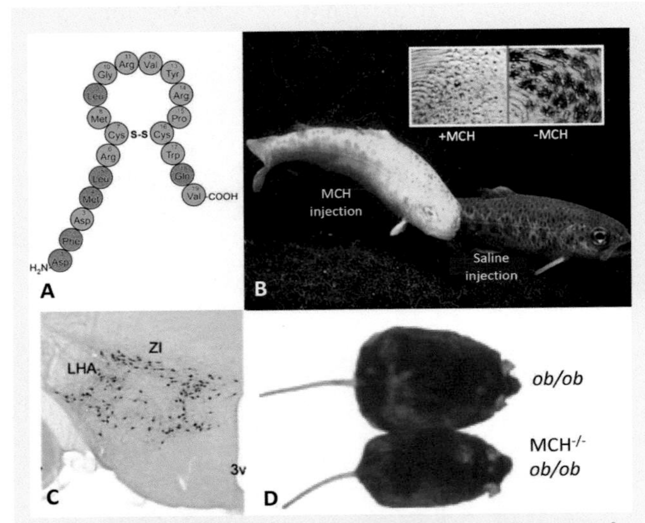

Figure 17.11 Diverse roles of MCH in fish and mammals. **(A)** The amino acid structure of mammalian MCH. Amino acid residues in green are conserved between mammals and salmon; acid residues in red are different in salmon. **(B)** Effects of MCH on body color of rainbow trout and on scales (inset). The fish were intraperitoneally injected with MCH at a dose of 1 pmole/g body weight or with saline. The scales were incubated in a buffer with or without 50nM MCH. **(C)** Immunohistochemical detection of MCH-expressing neurons in the lateral hypothalamus and zona incerta (ZI) of the mouse. **(D)** Photograph showing body size of $MCH^{-/-}$, *ob/ob*, and *ob/ob* mice at 16 weeks of age.

Source of images: (A) MacNeil, D.J., 2013. Front. Endocrinol., 4, p. 49; (B) Kawauchi, H., 2006. J. Exp. Zool. A: Comparative Experimental Biology, 305(9), pp. 751-760. Used by permission. (C) Pissios, P., et al. 2006. Endocr. Rev., 27(6), pp. 606-620; Used by permission. (D) Segal-Lieberman, G., et al. 2003. Proc. Nat. Acad. Sci. USA, 100(17), pp. 10085-10090. Used by permission.

rats,[107–109] and overexpression of MCH in the LHA of mice leads to increased food intake and obesity.[110] In addition, MCH mRNA is increased in the hypothalamus of fasting rats,[111] as well as in leptin-deficient ob^-/ob^- mice,[108] both models of hyperphagia. In contrast, knockout of the MCH gene produces mice that are lean, hypophagic and have an increased metabolic rate.[112] As was expected, mice lacking both leptin (ob^-/ob^-) and MCH genes (double-knockouts) had a dramatic reduction in body fat compared with mice lacking only the leptin gene (Figure 17.11[D]),[113] suggesting that MCH overexpression is a critical mediator of the leptin-deficient phenotype. Collectively, these observations suggest that endogenous MCH in the brain functions to both enhance appetite and to reduce energy expenditure, and is thus a potent mediator of positive energy balance.

SUMMARY AND SYNTHESIS QUESTIONS

1. An experiment was conducted whereby an obese mouse lacking leptin receptors ($lepr^-/lepr^-$) was parabiosed (circulatory systems were linked) to an obese mutant mouse lacking the leptin (lep^-/lep^-) gene. What effects on eating behavior and body weight, if any, would each mouse experience after several weeks of parabiosis? Why?
2. These two pharmaceutical compounds are being developed to inhibit overeating and obesity: (1) TTP435 (an AgRP inhibitor), (2) RM493 (an MC3/4R agonist). By what mechanism of action does each drug prevent overeating?

Summary of Chapter Learning Objectives and Key Concepts

LEARNING OBJECTIVE List the names, functions, and tissues of origin for peripheral hormones involved in short-term hunger/satiety signaling.

- Short-term hunger and satiety signals produced by the gut target the brain to regulate the size and timing of individual meals.
- Whereas the gut produces diverse satiety hormones, it only produces one hunger-inducing hormone, ghrelin.

LEARNING OBJECTIVE List the names, functions, and tissues of origin for peripheral hormones involved in long-term energy balance and body weight.

- Long-term adiposity signals produced by adipocytes (leptin) and the pancreas (insulin) circulate in proportion to adipose mass and target the brain to maintain optimal energy balance and body weight.

LEARNING OBJECTIVE Describe the influence of the melanocortin system on the central (hypothalamus) regulation of appetite control and energy balance.

- A key region of the brain that integrates endocrine and neural satiety and adiposity signals is the arcuate nucleus of the hypothalamus.
- Components of the central melanocortin system, such as POMC, MSH, melanocortin receptors, and receptor antagonists, play prominent roles in regulating appetite and energy homeostasis.

LITERATURE CITED

1. Coleman DL. A historical perspective on leptin. *Nat Med*. 2010;16(10):1097–1099.
2. Edholm OG, Fletcher JG, Widdowson EM, McCance RA. The energy expenditure and food intake of individual men. *Br J Nutr*. 1955;9(3):286–300.
3. Bray GA, Flatt JP, Volaufova J, Delany JP, Champagne CM. Corrective responses in human food intake identified from an analysis of 7-d food-intake records. *Am J Clin Nutr*. 2008;88(6):1504–1510.
4. Schwartz MW, Baskin DG, Kaiyala KJ, Woods SC. Model for the regulation of energy balance and adiposity by the central nervous system. *Am J Clin Nutr*. 1999;69(4): 584–596.

5. Kaiyala KJ, Woods SC, Schwartz MW. New model for the regulation of energy balance and adiposity by the central nervous system. *Am J Clin Nutr*. 1995;62(5 Suppl):1123S–1134S.
6. Figlewicz DP, Stein LJ, West D, Porte D, Jr., Woods SC. Intracisternal insulin alters sensitivity to CCK-induced meal suppression in baboons. *Am J Physiol*. 1986;250(5 Pt 2):R856–860.
7. Wynne K, Stanley S, McGowan B, Bloom S. Appetite control. *J Endocrinol*. 2005;184(2):291–318.
8. Morley J, Levine A, Bartness T, Nizielski S, MJ S, JJ H. Species differences in the response to cholecystokinin. *Ann N Y Acad Sci*. 1985;448:413–416.
9. Bloom SR, Polak JM, eds. *Gut Hormones* (2nd ed.). Churchill-Livingstone; 1981.
10. Beglinger C, Degen L, Matzinger D, D'Amato M, Drewe J. Loxiglumide, a CCK-A receptor antagonist, stimulates calorie intake and hunger feelings in humans. *Am J Physiol Regul Integr Comp Physiol*. 2001;280(4):R1149–1154.
11. Hewson G, Leighton GE, Hill RG, Hughes J. The cholecystokinin receptor antagonist L364,718 increases food intake in the rat by attenuation of the action of endogenous cholecystokinin. *Br J Pharmacol*. 1988;93(1):79–84.
12. Considine RV. Regulation of energy intake. *EndoText*; 2002. www.endotext.org/.
13. Tang-Christensen M, Larsen PJ, Goke R, et al. Central administration of GLP-1-(7–36) amide inhibits food and water intake in rats. *Am J Physiol*. 1996;271(4 Pt 2):R848–R856.
14. Naslund E, Barkeling B, King N, et al. Energy intake and appetite are suppressed by glucagon-like peptide-1 (GLP-1) in obese men. *Int J Obes Relat Metab Disord*. 1999;23(3):304–311.
15. Larsen PJ, Fledelius C, Knudsen LB, Tang-Christensen M. Systemic administration of the long-acting GLP-1 derivative NN2211 induces lasting and reversible weight loss in both normal and obese rats. *Diabetes*. 2001;50(11): 2530–2539.
16. Parker HE, Reimann F, Gribble FM. Molecular mechanisms underlying nutrient-stimulated incretin secretion. *Expert Rev Mol Med*. 2010;12:e1.
17. Nyberg J, Anderson MF, Meister B, et al. Glucose-dependent insulinotropic polypeptide is expressed in adult hippocampus and induces progenitor cell proliferation. *J Neurosci*. 2005;25(7):1816–1825.
18. Gale SM, Castracane VD, Mantzoros CS. Energy homeostasis, obesity and eating disorders: recent advances in endocrinology. *J Nutr*. 2004;134(2):295–298.
19. Batterham RL, Cowley MA, Small CJ, et al. Gut hormone PYY(3–36) physiologically inhibits food intake. *Nature*. 2002;418(6898):650–654.
20. Batterham RL, Cohen MA, Ellis SM, et al. Inhibition of food intake in obese subjects by peptide YY3–36. *N Engl J Med*. 2003;349(10):941–948.
21. Degen L, Oesch S, Casanova M, et al. Effect of peptide YY3–36 on food intake in humans. *Gastroenterology*. 2005;129(5):1430–1436.
22. le Roux CW, Batterham RL, Aylwin SJ, et al. Attenuated peptide YY release in obese subjects is associated with reduced satiety. *Endocrinology*. 2006; 147(1):3–8.
23. Tschop M, Smiley DL, Heiman ML. Ghrelin induces adiposity in rodents. *Nature*. 2000;407(6806):908–913.
24. Wren AM, Small CJ, Ward HL, et al. The novel hypothalamic peptide ghrelin stimulates food intake and growth hormone secretion. *Endocrinology*. 2000;141(11):4325–4328.
25. Wren AM, Seal LJ, Cohen MA, et al. Ghrelin enhances appetite and increases food intake in humans. *J Clin Endocrinol Metab*. 2001;86(12):5992.
26. Gottero C, Broglio F, Prodam F, et al. Ghrelin: a link between eating disorders, obesity and reproduction. *Nutr Neurosci*. 2004;7(5–6):255–270.
27. Zhang JV, Ren PG, Avsian-Kretchmer O, et al. Obestatin, a peptide encoded by the ghrelin gene, opposes ghrelin's effects on food intake. *Science*. 2005;310(5750):996–999.
28. Cowan E, Burch KJ, Green BD, Grieve DJ. Obestatin as a key regulator of metabolism and cardiovascular function with emerging therapeutic potential for diabetes. *Br J Pharmacol*. 2016;173(14):2165–2181.
29. Kennedy GC. The role of depot fat in the hypothalamic control of food intake in the rat. *Proc Royal Soc Lond B*. 1953;140(901):578–596.
30. McMinn JE, Sindelar DK, Havel PJ, Schwartz MW. Leptin deficiency induced by fasting impairs the satiety response to cholecystokinin. *Endocrinology*. 2000;141(12):4442–4448.
31. Riedy CA, Chavez M, Figlewicz DP, Woods SC. Central insulin enhances sensitivity to cholecystokinin. *Physiol Behav*. 1995;58(4):755–760.
32. Cancello R, Tounian A, Poitou C, Clement K. Adiposity signals, genetic and body weight regulation in humans. *Diabetes Metab*. 2004;30(3):215–227.
33. Dickie MM, Lane PW. Plus letter to Roy Robinson 7/7/70. *Mouse News Lett*. 1957;17.
34. Ingalls AM, Dickie MM, Snell GD. Obese, a new mutation in the house mouse. *J Hered*. 1950;41(12):317–318.
35. Zhang Y, Proenca R, Maffei M, Barone M, Leopold L, Friedman JM. Positional cloning of the mouse obese gene and its human homologue. *Nature*. 1994;372(6505):425–432.
36. Halaas JL, Gajiwala KS, Maffei M, et al. Weight-reducing effects of the plasma protein encoded by the obese gene. *Science*. 1995;269(5223):543–546.
37. Campfield LA, Smith FJ, Guisez Y, Devos R, Burn P. Recombinant mouse OB protein: evidence for a peripheral signal linking adiposity and central neural networks. *Science*. 1995;269(5223):546–549.
38. Pelleymounter MA, Cullen MJ, Baker MB, et al. Effects of the obese gene product on body weight regulation in ob/ob mice. *Science*. 1995;269(5223): 540–543.
39. Friedman JM. The function of leptin in nutrition, weight, and physiology. *Nutr Rev*. 2002;60(10 Pt 2):S1–14; discussion S68–84, 85–17.
40. Tartaglia LA, Dembski M, Weng X, et al. Identification and expression cloning of a leptin receptor, OB-R. *Cell*. 1995;83(7):1263–1271.
41. Chen H, Charlat O, Tartaglia LA, et al. Evidence that the diabetes gene encodes the leptin receptor: identification of a mutation in the leptin receptor gene in db/db mice. *Cell*. 1996;84(3):491–495.
42. Lee GH, Proenca R, Montez JM, et al. Abnormal splicing of the leptin receptor in diabetic mice. *Nature*. 1996;379(6566):632–635.
43. Chua SC, Jr., Chung WK, Wu-Peng XS, et al. Phenotypes of mouse diabetes and rat fatty due to mutations in the OB (leptin) receptor. *Science*. 1996;271(5251):994–996.
44. Farooqi IS, Matarese G, Lord GM, et al. Beneficial effects of leptin on obesity, T cell hyporesponsiveness, and neuroendocrine/metabolic dysfunction of human congenital leptin deficiency. *J Clin Invest*. 2002;110(8):1093–1103.
45. Montague CT, Farooqi IS, Whitehead JP, et al. Congenital leptin deficiency is associated with severe early-onset obesity in humans. *Nature*. 1997;387(6636): 903–908.
46. Paz-Filho GJ, Volaco A, Suplicy HL, Radominski RB, Boguszewski CL. Decrease in leptin production by the adipose tissue in obesity associated with severe metabolic syndrome. *Arq Bras Endocrinol Metabol*. 2009;53(9): 1088–1095.
47. Kolaczynski JW, Considine RV, Ohannesian J, et al. Responses of leptin to short-term fasting and refeeding in humans: a link with ketogenesis but not ketones themselves. *Diabetes*. 1996;45(11):1511–1515.
48. Maffei M, Halaas J, Ravussin E, et al. Leptin levels in human and rodent: measurement of plasma leptin and ob RNA in obese and weight-reduced subjects. *Nat Med*. 1995;1(11):1155–1161.
49. Considine RV, Sinha MK, Heiman ML, et al. Serum immunoreactive-leptin concentrations in normal-weight and obese humans. *N Engl J Med*. 1996;334(5):292–295.
50. Flier JS. Clinical review 94: What's in a name? In search of leptin's physi-

ologic role. *J Clin Endocrinol Metab.* 1998;83(5):1407–1413.
51. Dardeno TA, Chou SH, Moon HS, Chamberland JP, Fiorenza CG, Mantzoros CS. Leptin in human physiology and therapeutics. *Front Neuroendocrinol.* 2010;31(3):377–393.
52. Murphy KG, Bloom SR. Gut hormones in the control of appetite. *Exp Physiol.* 2004;89(5):507–516.
53. Rabe K, Lehrke M, Parhofer KG, Broedl UC. Adipokines and insulin resistance. *Mol Med.* 2008;14(11–12):741–751.
54. Gruzdeva O, Borodkina D, Uchasova E, Dyleva Y, Barbarash O. Leptin resistance: underlying mechanisms and diagnosis. *Diabetes Metab Syndr Obes.* 2019;12:191.
55. Roa J, Garcia-Galiano D, Castellano JM, Gaytan F, Pinilla L, Tena-Sempere M. Metabolic control of puberty onset: new players, new mechanisms. *Mol Cell Endocrinol.* 2010;324(1–2):87–94.
56. Sanchez-Garrido MA, Tena-Sempere M. Metabolic control of puberty: roles of leptin and kisspeptins. *Horm Behav.* 2013;64(2):187–194.
57. Garcia-San Frutos M, Fernandez-Agullo T, De Solis AJ, et al. Impaired central insulin response in aged Wistar rats: role of adiposity. *Endocrinology.* 2007;148(11):5238–5247.
58. Corp ES, Woods SC, Porte D, Jr., Dorsa DM, Figlewicz DP, Baskin DG. Localization of 125I-insulin binding sites in the rat hypothalamus by quantitative autoradiography. *Neurosci Lett.* 1986;70(1):17–22.
59. Marks JL, Porte D, Jr., Stahl WL, Baskin DG. Localization of insulin receptor mRNA in rat brain by in situ hybridization. *Endocrinology.* 1990;127(6):3234–3236.
60. Figlewicz DP. Adiposity signals and food reward: expanding the CNS roles of insulin and leptin. *Am J Physiol Regul Integr Comp Physiol.* 2003;284(4):R882–892.
61. Air EL, Benoit SC, Blake Smith KA, Clegg DJ, Woods SC. Acute third ventricular administration of insulin decreases food intake in two paradigms. *Pharmacol Biochem Behav.* 2002;72(1–2):423–429.
62. Niswender KD, Schwartz MW. Insulin and leptin revisited: adiposity signals with overlapping physiological and intracellular signaling capabilities. *Front Neuroendocrinol.* 2003;24(1):1–10.
63. Bruning JC, Gautam D, Burks DJ, et al. Role of brain insulin receptor in control of body weight and reproduction. *Science.* 2000;289(5487):2122–2125.
64. Strubbe JH, Mein CG. Increased feeding in response to bilateral injection of insulin antibodies in the VMH. *Physiol Behav.* 1977;19(2):309–313.
65. McGowan MK, Andrews KM, Grossman SP. Chronic intrahypothalamic infusions of insulin or insulin antibodies alter body weight and food intake in the rat. *Physiol Behav.* 1992;51(4):753–766.
66. Adam TC, Toledo-Corral C, Lane CJ, et al. Insulin sensitivity as an independent predictor of fat mass gain in Hispanic adolescents. *Diabetes Care.* 2009;32(11):2114–2115.
67. Morrison JA, Glueck CJ, Horn PS, Schreiber GB, Wang P. Pre-teen insulin resistance predicts weight gain, impaired fasting glucose, and type 2 diabetes at age 18–19 y: a 10-y prospective study of black and white girls. *Am J Clin Nutr.* 2008;88(3):778–788.
68. Eu CH, Lim WY, Ton SH, bin Abdul Kadir K. Glycyrrhizic acid improved lipoprotein lipase expression, insulin sensitivity, serum lipid and lipid deposition in high-fat diet-induced obese rats. *Lipids Health Dis.* 2010;9:81.
69. Wang ZQ, Ribnicky D, Zhang XH, Raskin I, Yu Y, Cefalu WT. Bioactives of Artemisia dracunculus L enhance cellular insulin signaling in primary human skeletal muscle culture. *Metabolism.* 2008;57(7 Suppl 1):S58–S64.
70. Van de Wouw M, Schellekens H, Dinan TG, Cryan JF. Microbiota-gut-brain axis: modulator of host metabolism and appetite. *J Nutr.* 2017;147(5):727–745.
71. Stellar E. The physiology of motivation. *Psychol Rev.* 1954;61(1):5–22.
72. King BM. The rise, fall, and resurrection of the ventromedial hypothalamus in the regulation of feeding behavior and body weight. *Physiol Behav.* 2006;87(2):221–244.
73. Hervey GR. The effects of lesions in the hypothalamus in parabiotic rats. *J Physiol.* 1959;145(2):336–352.
74. Grill HJ. Distributed neural control of energy balance: contributions from hindbrain and hypothalamus. *Obesity (Silver Spring).* 2006;14(Suppl 5):216S–221S.
75. Grill HJ. Leptin and the systems neuroscience of meal size control. *Front Neuroendocrinol.* 2010;31(1):61–78.
76. Grill HJ, Hayes MR. The nucleus tractus solitarius: a portal for visceral afferent signal processing, energy status assessment and integration of their combined effects on food intake. *Int J Obes (Lond).* 2009;33(Suppl 1):S11–S15.
77. Schwartz MW, Woods SC, Porte D, Jr., Seeley RJ, Baskin DG. Central nervous system control of food intake. *Nature.* 2000;404(6778):661–671.
78. Elias CF, Aschkenasi C, Lee C, et al. Leptin differentially regulates NPY and POMC neurons projecting to the lateral hypothalamic area. *Neuron.* 1999;23(4):775–786.
79. Elmquist JK, Elias CF, Saper CB. From lesions to leptin: hypothalamic control of food intake and body weight. *Neuron.* 1999;22(2):221–232.
80. Sipols AJ, Baskin DG, Schwartz MW. Effect of intracerebroventricular insulin infusion on diabetic hyperphagia and hypothalamic neuropeptide gene expression. *Diabetes.* 1995;44(2):147–151.
81. Baskin DG, Wilcox BJ, Figlewicz DP, Dorsa DM. Insulin and insulin-like growth factors in the CNS. *Trends Neurosci.* 1988;11(3):107–111.
82. Cheung CC, Clifton DK, Steiner RA. Proopiomelanocortin neurons are direct targets for leptin in the hypothalamus. *Endocrinology.* 1997;138(10):4489–4492.
83. Baskin DG, Breininger JF, Schwartz MW. Leptin receptor mRNA identifies a subpopulation of neuropeptide Y neurons activated by fasting in rat hypothalamus. *Diabetes.* 1999;48(4):828–833.
84. Amin A, Dhillo WS, Murphy KG. The central effects of thyroid hormones on appetite. *J Thyroid Res.* 2011;2011:306510.
85. Kong WM, Martin NM, Smith KL, et al. Triiodothyronine stimulates food intake via the hypothalamic ventromedial nucleus independent of changes in energy expenditure. *Endocrinology.* 2004;145(11):5252–5258.
86. Mountjoy KG, Mortrud MT, Low MJ, Simerly RB, Cone RD. Localization of the melanocortin-4 receptor (MC4-R) in neuroendocrine and autonomic control circuits in the brain. *Mol Endocrinol.* 1994;8(10):1298–1308.
87. Wan S, Browning KN, Coleman FH, et al. Presynaptic melanocortin-4 receptors on vagal afferent fibers modulate the excitability of rat nucleus tractus solitarius neurons. *J Neurosci.* 2008;28(19):4957–4966.
88. Fan W, Boston BA, Kesterson RA, Hruby VJ, Cone RD. Role of melanocortinergic neurons in feeding and the agouti obesity syndrome. *Nature.* 1997;385(6612):165–168.
89. Biebermann H, Castaneda TR, van Landeghem F, et al. A role for beta-melanocyte-stimulating hormone in human body-weight regulation. *Cell Metab.* 2006;3(2):141–146.
90. Lee YS, Challis BG, Thompson DA, et al. A POMC variant implicates beta-melanocyte-stimulating hormone in the control of human energy balance. *Cell Metab.* 2006;3(2):135–140.
91. Cowley MA, Smart JL, Rubinstein M, et al. Leptin activates anorexigenic POMC neurons through a neural network in the arcuate nucleus. *Nature.* 2001;411(6836):480–484.
92. Huszar D, Lynch CA, Fairchild-Huntress V, et al. Targeted disruption of the melanocortin-4 receptor results in obesity in mice. *Cell.* 1997;88(1):131–141.
93. Farooqi IS, Yeo GS, Keogh JM, et al. Dominant and recessive inheritance of morbid obesity associated with melanocortin 4 receptor deficiency. *J Clin Invest.* 2000;106(2):271–279.
94. Lubrano-Berthelier C, Cavazos M, Dubern B, et al. Molecular genetics of human obesity-associated MC4R mutations. *Ann N Y Acad Sci.* 2003;994:49–57.

95. Lubrano-Berthelier C, Durand E, Dubern B, et al. Intracellular retention is a common characteristic of childhood obesity-associated MC4R mutations. *Hum Mol Genet*. 2003;12(2):145–153.
96. Marks DL, Boucher N, Lanouette CM, et al. Ala67Thr polymorphism in the Agouti-related peptide gene is associated with inherited leanness in humans. *Am J Med Genet A*. 2004;126A(3):267–271.
97. Willesen MG, Kristensen P, Romer J. Co-localization of growth hormone secretagogue receptor and NPY mRNA in the arcuate nucleus of the rat. *Neuroendocrinology*. 1999;70(5):306–316.
98. Kamegai J, Tamura H, Shimizu T, Ishii S, Sugihara H, Wakabayashi I. Central effect of ghrelin, an endogenous growth hormone secretagogue, on hypothalamic peptide gene expression. *Endocrinology*. 2000;141(12):4797–4800.
99. Nakazato M, Murakami N, Date Y, et al. A role for ghrelin in the central regulation of feeding. *Nature*. 2001;409(6817):194–198.
100. Shintani M, Ogawa Y, Ebihara K, et al. Ghrelin, an endogenous growth hormone secretagogue, is a novel orexigenic peptide that antagonizes leptin action through the activation of hypothalamic neuropeptide Y/Y1 receptor pathway. *Diabetes*. 2001;50(2):227–232.
101. Cowley MA, Smith RG, Diano S, et al. The distribution and mechanism of action of ghrelin in the CNS demonstrates a novel hypothalamic circuit regulating energy homeostasis. *Neuron*. 2003;37(4):649–661.
102. Kawauchi H, Kawazoe I, Tsubokawa M, Kishida M, Baker BI. Characterization of melanin-concentrating hormone in chum salmon pituitaries. *Nature*. 1983;305(5932):321–323.
103. Kawauchi H, Baker BI. Melanin-concentrating hormone signaling systems in fish. *Peptides*. 2004;25(10):1577–1584.
104. Pissios P, Bradley RL, Maratos-Flier E. Expanding the scales: The multiple roles of MCH in regulating energy balance and other biological functions. *Endocr Rev*. 2006;27(6):606–620.
105. Bittencourt JC, Presse F, Arias C, et al. The melanin-concentrating hormone system of the rat brain: an immuno- and hybridization histochemical characterization. *J Comp Neurol*. 1992;319(2):218–245.
106. Bittencourt JC. Anatomical organization of the melanin-concentrating hormone peptide family in the mammalian brain. *Gen Comp Endocrinol*. 2011;172(2):185–197.
107. Della-Zuana O, Presse F, Ortola C, Duhault J, Nahon JL, Levens N. Acute and chronic administration of melanin-concentrating hormone enhances food intake and body weight in Wistar and Sprague-Dawley rats. *Int J Obes Relat Metab Disord*. 2002;26(10):1289–1295.
108. Qu D, Ludwig DS, Gammeltoft S, et al. A role for melanin-concentrating hormone in the central regulation of feeding behaviour. *Nature*. 1996;380(6571):243–247.
109. Rossi M, Beak SA, Choi SJ, et al. Investigation of the feeding effects of melanin concentrating hormone on food intake--action independent of galanin and the melanocortin receptors. *Brain Res*. 1999;846(2):164–170.
110. Ludwig DS, Tritos NA, Mastaitis JW, et al. Melanin-concentrating hormone overexpression in transgenic mice leads to obesity and insulin resistance. *J Clin Invest*. 2001;107(3):379–386.
111. Presse F, Sorokovsky I, Max JP, Nicolaidis S, Nahon JL. Melanin-concentrating hormone is a potent anorectic peptide regulated by food-deprivation and glucopenia in the rat. *Neuroscience*. 1996;71(3):735–745.
112. Shimada M, Tritos NA, Lowell BB, Flier JS, Maratos-Flier E. Mice lacking melanin-concentrating hormone are hypophagic and lean. *Nature*. 1998;396(6712):670–674.
113. Segal-Lieberman G, Bradley RL, Kokkotou E, et al. Melanin-concentrating hormone is a critical mediator of the leptin-deficient phenotype. *Proc Natl Acad Sci U S A*. 2003;100(17):10085–10090.

CHAPTER

18

Energy Homeostasis

CHAPTER LEARNING OBJECTIVES:

- List the primary hormones that influence fuel homeostasis and describe their metabolic actions on skeletal muscle, liver, and adipose tissue.
- Describe how glucose is the most tightly regulated fuel substrate.
- List the pancreatic hormones and their roles in maintaining energy homeostasis.
- Describe the complementary actions of glucagon with adrenal glycemic hormones in maintaining energy homeostasis.
- Describe the three main categories of adipose tissue, their functions and their influence on energy homeostasis.
- Explain the roles TH plays in the metabolism of glucose, lipid, and protein, as well as modulating appetite.
- Describe the hypermetabolic effects of thyroid hormones on mammals and birds, and their roles in facilitating non-shivering thermogenesis.

OPENING QUOTATIONS:

"I do not know where or when the various species of animals were given their basal metabolism. Perhaps Noah did it when they left the Ark. I suppose that the reason he could not do a uniform job with the animals was because he did not have scales small enough for the dwarf mice and large enough for the bull, the horse, and the elephant. The subject deserves more study."

—-Eugene F. DuBois,[1] Professor of Physiology and Biophysics, Cornell University

"The presence of the brain in the animals is necessary to the conservation of the entire organism . . . the brain regulates the heat and boiling occurring in the heart."

—Aristotle (3rd century BC) Book 2, chapter 7 of his *De Partibus Animalium*

Introduction

Physiology adheres to the first law of thermodynamics, which states that energy can be neither created nor destroyed but can be converted from one form to another. When a calorie of food is eaten, digested, and absorbed into the bloodstream, it has several energetic fates: (1) transformation into chemical energy for imminent cellular work, (2) immediate dissipation as heat, or (3) long-term storage as fat, glycogen, or protein. The fate of the ingested food is ultimately determined by the energetic needs of homeostasis. *Energy homeostasis*, or the homeostatic control of energy balance, can be described in an organism with the following equation:

$$\text{Energy intake (food)} = \text{Energy stored} + \text{Energy expended (work} + \text{heat)}$$

In order for animals to maintain a relatively constant body weight, the energy ingested in the form of food must be matched by an equal energy expenditure (Figure 18.1). Energy is expended primarily by the metabolic processes that are required for the obligatory performance of basic cellular and organ functions, known as *basal metabolic rate* (BMR), as well as through physical activity, *adaptive thermogenesis* (heat production in response to cold), and the *thermal effects of food* (energy expended to process and store food). Total body energy expenditure represents the conversion of oxygen and food in the forms of lipids, glucose, and protein into carbon dioxide, water, heat, and work on the environment. If energy intake and storage is lower than its expenditure, weight loss ensues

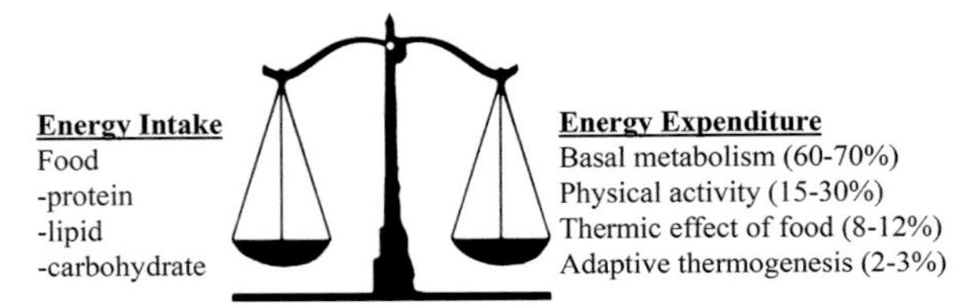

Figure 18.1 Energy homeostasis depends upon the balance between caloric intake and energy expenditure. Energy enters an organism as food and exits as heat and work. If energy intake exceeds energy expenditure by even a small margin, obesity will result from the cumulative net gain of stored fuel.

DOI: 10.1201/9781003359241-24

and a continued depletion of stored energy can ultimately lead to starvation. If long-term energy intake and storage is greater than its expenditure, even by a small margin, obesity develops, a condition that can lead to severe metabolic and physiological dysfunction.

Whereas the focus of Chapter 16: Digestion and Chapter 17: Appetite was on the regulation of energy intake, or the left side of the energy balance equation, this chapter will address the right side of the energy equation, specifically the endocrine control of normal energy storage, mobilization, and metabolism as pertains to mammals. Dysfunctions of energy regulation that can lead to severe metabolic disorders, including obesity and diabetes, are addressed in Chapter 19: Metabolic Dysregulation and Disruption. The endocrine basis and treatment of metabolic disorders constitutes a particularly large and active discipline within endocrinology.

The Principal Hormones of Energy Homeostasis

LEARNING OBJECTIVE List the primary hormones that influence fuel homeostasis and describe their metabolic actions on skeletal muscle, liver, and adipose tissue.

KEY CONCEPTS:

- Metabolic hormones exert anabolic (build up) and/or catabolic (breakdown) effects on fuel substrates.
- On a short-term basis, the most important endocrine regulators of fuel metabolism are insulin, glucagon, and catecholamines.
- On a more prolonged time basis, glucocorticoids and growth hormone exert effects that require hours to days to manifest.
- Activation of glucagon and adrenergic receptors increases blood glucose by promoting glycogenolysis.
- Hepatic gluconeogenesis is promoted by the synergistic actions of glucagon and glucocorticoids.

Whether a fuel is oxidized or stored is ultimately determined by the concentrations of various hormones in the blood. The general roles of six principal hormones are summarized in Table 18.1.

Insulin: Produced by beta cells of the pancreas, insulin is the most potent *anabolic* (builds new compounds) hormone known, promoting the storage and synthesis of lipids, protein, and carbohydrates and inhibiting their breakdown and release into circulation[2] (Figure 18.2). Though this hormone promotes the absorption of nutrients by the gut during digestion, it ultimately lowers the levels of circulating blood metabolic fuels, especially glucose, through several actions: (1) promoting the uptake of glucose into most cells, where it can either be oxidized or stored as glycogen by muscle and liver through a process called **glycogenesis**; (2) stimulating uptake of free fatty acids (FFAs) and triacylglycerol (TAG) from the blood into adipose tissue and muscle and promoting lipid storage by adipocytes; (3) increasing the rate of protein synthesis in muscle, liver, and other tissues, as well as enhancing the transport of amino acids into tissues; (4) inhibiting **glycogenolysis** (breakdown of glycogen into glucose) in liver and muscle, *lipolysis* (breakdown of triacycglycerols into FFAs + glycerol) in adipocytes, and protein degradation in muscle.

Glucagon: Produced by alpha cells of the pancreas, glucagon counters the effects of insulin by elevating the concentrations of glucose and FFA in the blood. As such, glucagon is a major *counterregulatory hormone*, or a hormone whose actions generally oppose that of insulin. Glucagon exerts *catabolic* effects that promote substrate breakdown by stimulating glycogenolysis, as well as by stimulating the release of amino acids and FFAs into circulation[3–5] (Figure 18.3). As glycogen stores become depleted, glucagon also exerts anabolic effects by encouraging the liver to synthesize additional glucose by **gluconeogenesis**, the assembly of glucose from non-carbohydrate precursors, such as lipids and amino acids.

Glucocorticoids: Produced by the adrenal cortex, particularly in times of stress, glucocorticoids exert both anabolic and catabolic actions that are synergistic with those of glucagon. They stimulate the catabolism of

Table 18.1 Major Hormones That Regulate Fuel Metabolism

	Muscle			Liver					Adipose tissue	
Hormones	Glucose uptake	Glucose utilization	Protein catabolism	Glucose output	Ketogenesis	Gluconeogenesis	Glycogenolysis	Glycogenesis	Fat synthesis	Lipolysis
Insulin	↑↑	↑↑	↓↓	↓↓	↓↓	↓↓	↓↓	↑	↑↑	↓↓
Glucagon	-	-	-	↑↑	↑	↑	↑↑	↓	-	↑
Epinephrine & Norepinephrine	-	↑	-	↑↑	-	↑	↑↑	↑↓	-	↑
Glucocorticoid	↓	↓	↑	↑	↑	↑	-	↑	-	↑
Growth hormone	↓	↓	↓	↑	↑	↑	-	-	-	↑
Thyroid hormone	↑	↑	↑	↑	↑	↑	↑	-	-	↑

Source: After Nelson, D.L., Lehninger, A.L. and Cox, M.M., 2008. *Lehninger Principles of Biochemistry*. Macmillan.

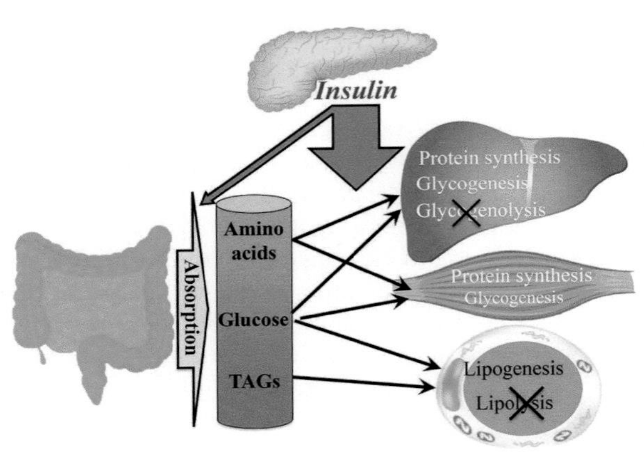

Figure 18.2 The most potent anabolic hormone known, insulin promotes the uptake from the blood and storage or assimilation of the various metabolic fuel substrates by target tissues. Following the absorption into the blood from digested food in the gut, insulin promotes(1) the uptake of amino acids by liver and muscle and their incorporation into proteins, (2) the uptake of glucose by the liver and muscle and its polymerization into glycogen, and (3) the uptake of both glucose and triacylglycerides (TAGs) by adipocytes where they undergo lipogenesis and are stored as fat. While only three primary metabolic tissues are shown in this diagram, insulin also promotes the uptake (but not storage) of glucose, amino acids, and fatty acids (a TAG derivative) by most other tissues.

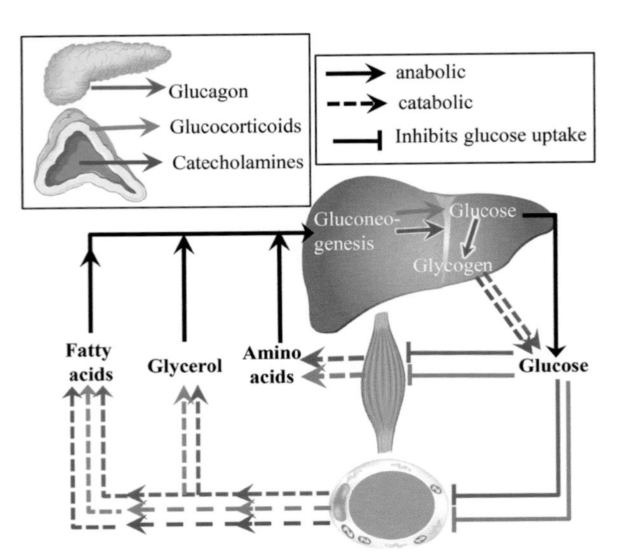

Figure 18.3 Glucocorticoids, glucagon, and catecholamines exert synergistic anabolic and catabolic effects on peripheral tissues. Both glucocorticoids (of adrenal cortex origin) and glucagon (made by the pancreas) affect muscle and adipose cells by inhibiting glucose uptake, as well as by inhibiting protein synthesis and lipogenesis. Simultaneously, both hormones promote muscle proteolysis, releasing amino acids into the blood. Glucocorticoids, glucagon, and catecholamines (norepinephrine and epinephrine of adrenal medulla origin) stimulate adipocyte lipolysis, releasing free fatty acids (FFA) and glycerol. Blood amino acids, FFA, and glycerol are transported via the circulatory system to liver cells and converted to glucose via gluconeogenesis under the influence of both glucocorticoids and glucagon. Newly synthesized glucose will either be released into the circulatory system to tissues that require it or stored as glycogen. Both glucagon and catecholamines promote glycogenolysis.

triacyglycerol into FFAs + glycerol, and the breakdown of proteins into amino acids (Figure 18.3), and thus exert counterregulatory activity. Simultaneously, glucocorticoids stimulate anabolic effects by upregulating liver enzymes that convert circulating amino acids and lipid substrates into glucose via gluconeogenesis. Therefore, glucocorticoids are another example of a counterregulatory hormone that opposes insulin by increasing blood glucose levels.

Catecholamines: Produced by sympathetic nerve endings and by the adrenal medulla in times of stress, epinephrine and norepinephrine (Figure 18.3) exert counterregulatory, glycogenolytic effects similar to that of glucagon on the liver and muscle. Norepinephrine, and to a lesser extent, epinephrine, is a key regulator of lipolysis in adipose tissue. Like glucagon, it also stimulates increased blood FFA levels.

Growth hormone (GH): Produced by the pituitary adenohypophysis, this hormone exerts both anabolic and catabolic actions on fuel metabolism, targeting tissues both directly and indirectly by stimulating insulin-like growth factor I (IGF-I). When subjects are well nourished, GH promotes protein retention in muscle and other tissues through inhibition of protein breakdown and stimulation of protein synthesis, and in the liver through inhibition of amino acid degradation and ureagenesis.[6] In the fed state GH also promotes anabolic storage of lipids in adipose tissue and glucose in liver and muscle glycogen reserves.

Thyroid hormone (TH): TH is a counterregulatory hormone that in mammals and birds exerts marked effects in the regulation of virtually all aspects of glucose homeostasis and metabolism. Indeed, thyroid hormones have such a profound influence on overall metabolic rate and fuel consumption in mammals and birds that much of this chapter is devoted to this topic.

On a short-term, moment-to-moment basis, the most important endocrine regulators of fuel metabolism are insulin and glucagon, as well as catecholamines released by the sympathetic nervous system and adrenal medulla. Note the generally opposing responses of insulin and glucagon following ingestion of a meal in Figure 18.4. However, note that immediately after eating a meal, glucagon levels temporarily rise with insulin. The purpose of the transient rise in glucagon is to prevent hypoglycemia. The actions of these hormones manifest within minutes to hours of the hormone–receptor interaction and usually results from changes in the activity of key preexisting enzymes caused by their phosphorylation or dephosphorylation. On a more prolonged time basis, glucocorticoids and growth hormone exert their effects genomically by modulating the transcription and translation of key metabolic enzymes, processes that require hours to days to manifest. Importantly, these differences in counterregulatory hormone mechanisms of action and time frames promote synergy among the hormones, whereby the combined effects of individual hormones may exert greater effects than would be predicted by their additive effects alone (Figure 18.5).

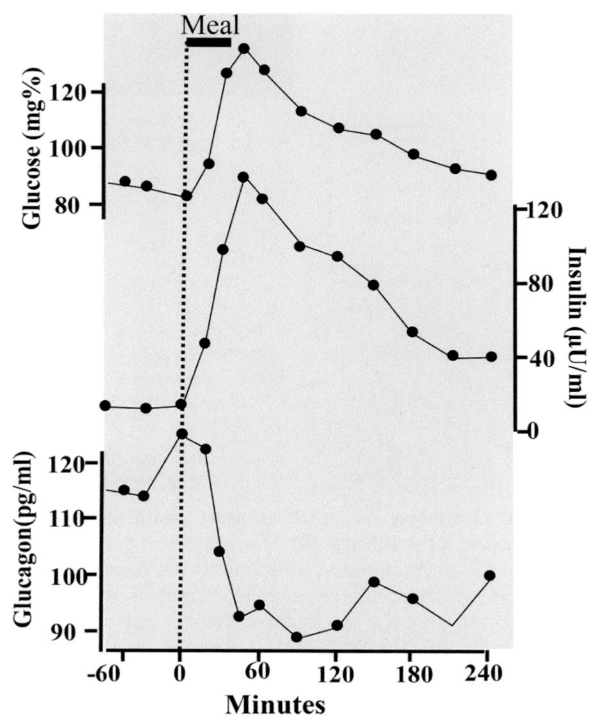

Figure 18.4 Effect of a large carbohydrate meal on plasma insulin, glucagon, and glucose levels in 11 normal humans.

Source of images: Unger, R.H. (1971). N Engl J Med. 285:443-449.

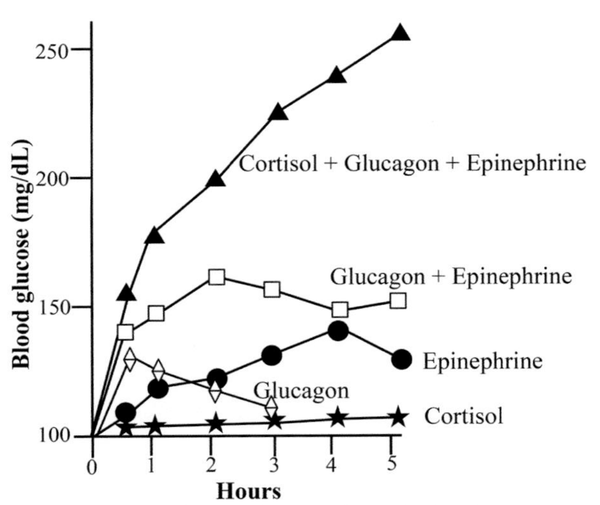

Figure 18.5 Synergistic effects of cortisol, glucagon, and epinephrine on increasing plasma glucose concentration. Note that the hyperglycemic response to the triple hormone infusion is far greater than the additive response of all three hormones given singly.

Source of images: Eigler, N., et al. (1979) J. Clin. Invest. 63: 114–123.

SUMMARY AND SYNTHESIS QUESTIONS

1. Describe the secretory responses of insulin and glucagon to eating either a pasta-rich meal or a meatball-rich meal.
2. Notice in Figure 18.4 that immediately after consuming a meal, glucagon levels rise briefly, along with insulin. What is the purpose of the transient secretion of glucagon?

Glucose as the Central Fuel Substrate

LEARNING OBJECTIVE Describe how glucose is the most tightly regulated fuel substrate.

KEY CONCEPTS:

- Whereas acute hypoglycemia can rapidly inhibit cerebral function and muscle coordination and cause death, many of the pathological effects of hyperglycemia require chronic, long-term exposure to manifest.
- A major role of the counterregulatory hormones in energy homeostasis is to maintain blood glucose levels above the hypoglycemic threshold of 55–60 mg/dl in humans.
- A key role for insulin is to ensure that fasting blood glucose levels remain below the hyperglycemic threshold of 110 mg/dl in humans.

Of the major circulating metabolic fuels, glucose is the most tightly regulated.[7] Although free fatty acids are the main fuel for most organs, glucose is the obligate metabolic fuel for the brain under normal conditions and is one important reason why its circulating levels are under such tight regulation. Indeed, a deviation from a normal 90 mg/dl value in either direction can prove disastrous. Glucose plasma concentrations below 55 mg/dl, a condition called *hypoglycemia*, can impair cerebral function and muscle coordination,[8] whereas more severe and prolonged hypoglycemia causes convulsions, loss of consciousness, permanent brain damage, coma, and even death. Therefore, a major role of the counterregulatory hormones in energy homeostasis is to maintain blood glucose levels above 60 mg/dl. Conversely, even mild chronically elevated plasma glucose concentrations, a condition called **hyperglycemia**, that occur in patients with impaired glucose tolerance, such as diabetes, increase risk for cardiac and peripheral vascular disease, renal failure, blindness, neuropathy, and other disorders.[9–11] A key role of insulin is to ensure that fasting blood glucose levels remain below the hyperglycemic threshold of 110 mg/dl.

SUMMARY AND SYNTHESIS QUESTIONS

1. Why do you think that there are several counter-regulatory hormones and redundant mechanisms that oppose the actions of insulin?

Pancreatic Hormones and Their Actions

LEARNING OBJECTIVE List the pancreatic hormones and their roles in maintaining energy homeostasis.

KEY CONCEPTS:

- The endocrine portion of the pancreas, called islets of Langerhans, are composed of five cell types. The most numerous of these, the α and β cells, secrete glucagon and insulin, respectively.
- While both rising glucose and glucagon levels stimulate insulin secretion, rising insulin reduces blood glucose and suppresses glucagon secretion.
- Activation of insulin receptors facilitates glucose uptake by promoting intracellular signaling pathways that translocate GLUT4 glucose transporters to the plasma membrane of skeletal muscle cells and adipose cells.
- Inhibition of insulin receptor-mediated intracellular signaling leads to a phenomenon termed "insulin resistance".

Pancreas Functional Anatomy and Development

The pancreas has both exocrine and endocrine functions. The exocrine pancreas, which constitutes most of the organ, is comprised of **acinar cells** that secrete enzymes and fluids associated with digestion and is connected to the intestine via a highly branched ductal network (Figure 18.6). The **islets of Langerhans** make up the endocrine portion of the organ with approximately one to two million islets scattered throughout the central regions of a human pancreas, comprising 1%–2% of the total pancreas mass.[12] The islets contain at least five types of secretory cells: α *cells*, the second most numerous cells, comprise as much as 30% of islets and secrete glucagon; β *cells*, the most numerous cells, comprise 50%–80% of the total islet cells, depending on species, and co-secrete insulin, proinsulin, C peptide, and amylin; δ *cells* secrete somatostatin; ε *cells* secrete ghrelin; and *F cells* secrete pancreatic polypeptide.[13–15] The other islet cells respond to various peripheral signals and also interact with each other in a paracrine manner to modulate the release of insulin and glucagon, the two primary pancreatic hormones involved in regulating blood glucose (Figure 18.6). The mechanisms by which the two most important pancreatic hormones,

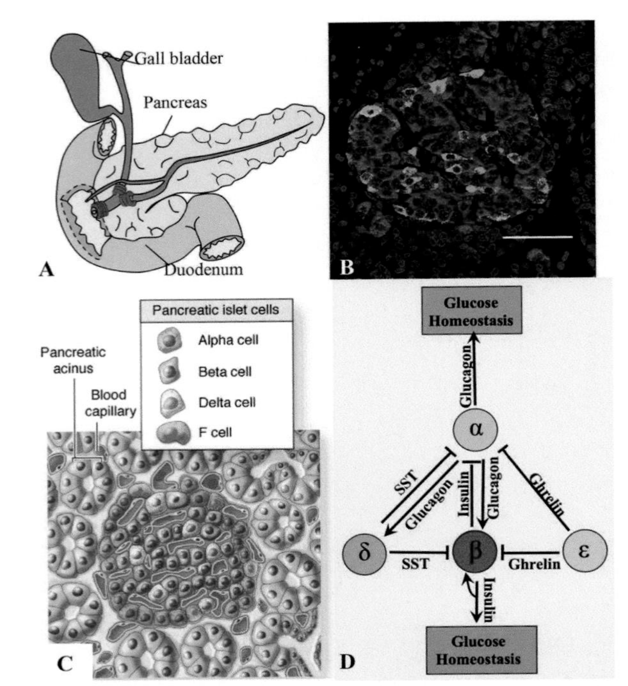

Figure 18.6 Overview of adult human pancreas anatomy and endocrine physiology. (A) The pancreas lies behind the stomach in the abdomen and attaches to the duodenum. Exocrine acinar cells secrete pancreatic digestive enzymes that are transported through a ductal network and released into the duodenum. **(B)** The endocrine pancreas consists of small cell clusters, called islets of Langerhans. These islets are mini endocrine organs scattered throughout the pancreas among the acinar cells. At least four endocrine cell types surround an extensive capillary system in a well-organized pattern. In this human islet, the insulin-producing β cells (in red) are clustered together, with the other endocrine cells (the glucagon-producing k-cells and the somatostatin-producing δ cells; in green) being more peripheral. **(C)** Islets contain five endocrine cell types, namely glucagon-producing α cells, insulin-producing β cells, somatostatin (SST)-producing δ cells, pancreatic polypeptide-producing PP cells, and ghrelin-producing ϵ cells. In human islets, the three primary endocrine cell types (insulin, glucagon, and somatostatin) are evenly distributed throughout the islet, and most β cells are in direct contact with the other types of endocrine cells. **(D)** Paracrine interactions among islet cells. Insulin secreted by β cells acts as a prime hormone of glucose homeostasis and inhibits glucagon secretion by α cells. Whereas glucagon activates insulin and somatostatin secretion, somatostatin secreted by δ cells and ghrelin by ε cells inhibit insulin secretion.

Source of images: (B) Bonner-Weir, S., 2020. Nat. Rev. Endocrinol., 16(2), pp. 73-74; Used by permission. (C) Junqueira, L.C. and Mescher, A.L., 2013. Junqueira's basic histology: text & atlas; McGraw Hill LLC. Used by permission.

insulin and glucagon, regulate glucose homeostasis are described next.

Insulin

Isolated in 1921 by Banting, Best, Collip, and Macleod,[16] the discovery of insulin was one of the most remarkable and consequential feats in the history of endocrinology. The purification of insulin largely brought an end to the suffering and death of countless people afflicted with one of the most heart-wrenching diseases, type 1 diabetes mellitus, a disease that typically begins in childhood. This devastating disease (discussed in detail in Chapter 19: Metabolic Dysregulation and Disruption) is characterized by hyperglycemia and other metabolic dysfunctions resulting from reduced insulin secretion and subsequent glucagon hypersecretion.

Curiously Enough . . . Because insulin is secreted into the hepatic portal vein, about half of the insulin is degraded by *insulinase* in the liver before the remainder enters peripheral circulation. Therefore, insulin has a short half-life of 5–8 minutes. In contrast, C peptide, which is co-secreted with insulin, is not metabolized by the liver, making it a useful blood or urinary diagnostic marker for insulin secretion.

Synthesis

Insulin is a protein hormone secreted exclusively by pancreatic β cells. Insulin is derived from a 110 amino acid-long preproinsulin molecule. The mature insulin protein is a dimer consisting of a 21 amino acid-long A chain and a 30 amino acid-long B chain linked by a pair of disulfide bonds. Proteases cleave the connecting C peptide to yield the mature hormone that consists of the A and B chains held together by disulfide linkages. The mature hormone is secreted into the blood along with the C peptide. The cellular processing and secretory pathway for insulin synthesis were previously described in Figure 4.9 of Chapter 4: Hormone Classes and Biosynthesis.

Secretion

The release of insulin by pancreatic beta cells is integral to the maintenance of glucose homeostasis and is under complex control by a variety of fuel substrates, neural, and endocrine parameters. A deficiency in any of these pathways can desensitize beta cells to elevated blood glucose levels, exacerbating hyperglycemia,[17] and an understanding of these pathways is essential to comprehend the mechanisms of action of diverse categories of pharmacological interventions for types 1 and 2 diabetes mellitus.

Glucose- and Nutrient-Stimulated Insulin Secretion

Blood glucose is the primary stimulus for insulin release. In response to eating a meal, insulin levels typically begin to rise concomitantly with blood glucose within 10 minutes after ingestion, peaking in 30 minutes to 1 hour (see Figure 18.4). The entry of glucose into β cells is facilitated by the *glucose transporter* 2 (GLUT2) and is rapidly phosphorylated by the glycolytic enzyme *glucokinase* into glucose-6-phosphate (G6P)[18] (Figure 18.7). Glucokinase is considered to function as the β cell glucose sensor. Subsequent metabolism of G6P increases the ATP/ADP ratio, which promotes the closure of *ATP-sensitive K^+ channels*. The resulting depolarization of the β cell plasma membrane opens *voltage-gated Ca^{++} channels*, and elevated intracellular Ca^{++} levels activate the exocytosis of insulin-containing secretory vesicles. The ATP/ADP switch can also be activated by the oxidation of amino acids and FFAs, though to a lesser extent compared with glucose. When glucose levels are high, cationic amino acids, like arginine and leucine, depolarize the β cell membrane when transported into the cells, accelerating Ca^{++} influx and potentiating insulin secretion.[19] Arginine is, in fact, the strongest insulin secretagogue and is often used to initiate insulin secretion in clinical testing of beta cell capacity.

Cholinergic Receptor and Incretin-Stimulated Insulin Secretion

Parasympathetic (vagal) cholinergic innervation also stimulates insulin release by increasing intracellular Ca^{++}, but in contrast to the glucose- and nutrient-stimulated pathway it does so by activating the phospholipase C pathway

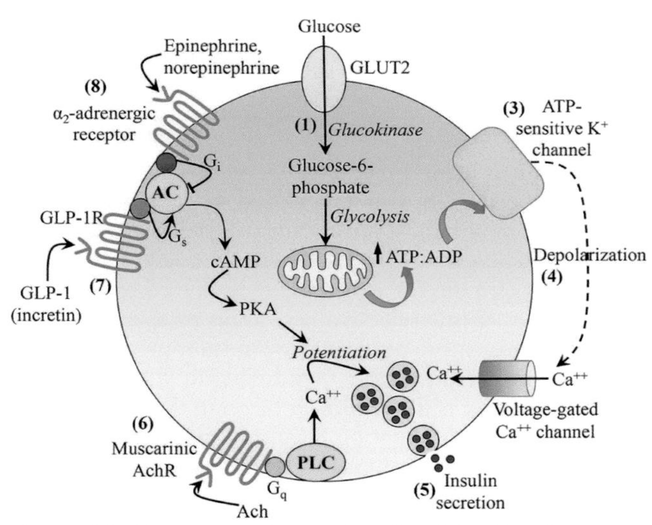

Figure 18.7 Glucose- and receptor-stimulated insulin secretion by pancreatic β cells. (1) Following the uptake of glucose and its metabolism by glucokinase, (2) an increase in the intracellular ATP:ADP ratio results in (3) closure of ATP-sensitive potassium channels, (4) depolarization of the cell membrane and subsequent opening of voltage-dependent Ca^{++} channels. (5) The resulting increase in cytosolic Ca^{2+} concentration triggers insulin release. (6) Cholinergic innervation also stimulates insulin release by increasing intracellular Ca^{++}, but by activating the phospholipase C/IP_3 pathway associated with the muscarinic acetylcholine receptor, a GPCR. (7) The binding of incretins, such as glucose-dependent insulinotropic polypeptides (GIP), to GPCR receptors raises intracellular cAMP, which augments the intracellular effects of Ca^{++} in the presence of glucose. (8) Stimulation of the α_2 adrenergic receptor inhibits cAMP synthesis, reducing insulin secretion.

Source of images: De León, D.D. and Stanley, C.A., 2007. Nature Clin. Pract. Endocrinol. Metab., 3(1), pp. 57-68.

associated with the muscarinic acetylcholine receptor (a GPCR) (Figure 18.7). Such neural stimulation occurs in response to ingestion of a meal during the cephalic phase of eating. Insulin release can also be enhanced by the binding of *incretins*, hormones that promote insulin release. Incretins, such as *glucose-dependent insulinotropic polypeptides* (GIP) and *glucagon-like peptides* (GLP), bind to GPCR receptors that raise intracellular cAMP, which augments the intracellular effects of Ca^{++} in the presence of glucose. As such, orally administered glucose is a greater stimulus to insulin secretion than intravenously injected glucose, as glucose in the gastrointestinal tract stimulates GIP and GLP-1 secretion.

Curiously Enough . . . Although GLP-1 is a potent incretin theoretically useful for the treatment of diabetes, GLP-1 has a very short half due to its rapid breakdown by peptidases. However, a much more stable analog of GLP-1, called exendin-4, was discovered in the saliva of the Gila monster and has been approved for use in treating diabetic patients,[20,21] a significant contribution of comparative endocrinology to medicine.

Adrenergic Receptor-Modulated Insulin Secretion

In addition to being innervated by parasympathetic neurons, islet cells are also innervated by norepinephrine-releasing sympathetic neurons and are rich in both α- and β-type adrenergic receptors. These adrenergic receptors also bind to circulating epinephrine produced by the adrenal medulla. While selective activation of α_1 and β adrenoceptors may stimulate insulin secretion, physiologically the most important effect of adrenergic stimulation on insulin secretion is its *inhibition* through the activation of α_2 adrenoceptors present on β cell plasma membranes[22,23] (Figure 18.7). This inhibition is mediated, in part, by interactions of the α_2 adrenoceptor with inhibitory G protein subunits (G_i) that prevent the increased intracellular cAMP required for the augmentation of stimulated insulin release. This α_2 adrenergic inhibition of insulin appears to protect against hypoglycemia during exercise, ensuring that glucose is readily available to active tissues, such as muscle and brain. Interestingly, pheochromocytoma, a condition characterized by chronic excess epinephrine synthesis by the adrenal medulla, has been attributed as a cause of elevated blood glucose levels due to blunted insulin secretion,[24–26] producing symptoms similar to those of diabetes mellitus.

Glucagon-Stimulated Insulin Secretion

Glucagon is a well-known insulin secretagogue[27–29] (see Figure 18.6 D). However, in contrast to incretins (GIP and GLP-1) that are released from the intestine during feeding to prepare pancreatic β cells for the imminent hyperglycemia, glucagon is released in between meals in response to hypoglycemia. Although the insulin-stimulating effects of glucagon may at first seem counterintuitive, some insulin presence is required for cells to absorb and utilize the glucose mobilized by glucagon in the first place. Eventually, elevated insulin and glucose levels inhibit glucagon release by the α cells.

Insulin Receptor Signaling Pathways

An appreciation for some of the complexity of insulin receptor signaling pathways is critical, as defects in one or more post-receptor insulin-signaling components may promote or exacerbate the development of **insulin resistance**, a defect in the ability of tissues to respond to insulin and an important medical condition that predisposes individuals to developing type 2 diabetes mellitus and other metabolic disorders. Consequently, elucidating the molecular components that mediate insulin action is a particularly active area of research, one that is essential to fully understand the nature of insulin resistance and its treatment.

The insulin receptor is a member of the receptor tyrosine kinase (RTK) family. It is expressed on the plasma membranes of target cells as a homodimer composed of two extracellular α subunits and two transmembrane β subunits linked together by disulfide bonds (Figure 18.8). The binding of insulin to the α subunit induces a conformational change resulting in the autophosphorylation of a number of tyrosine residues present in the β subunit. The resulting phosphotyrosine residues are recognized by phosphotyrosine-binding domains of adapter proteins such as members of the *insulin receptor substrate* family (IRS) and *Shc** protein. The phosphorylation of these adapter proteins initiates complex intracellular signaling cascades that generate both cytosolic and genomic responses. For example, the phosphorylation of the Shc protein by the activated insulin receptor is linked to the mitogen-activated protein kinase (MAPK) pathway, which promotes gene transcription, cell growth, and mitogenic actions of insulin. Receptor activation also leads to the phosphorylation of key tyrosine residues on IRS proteins, some of which are recognized by the SH2 domain of *phosphoinositide-3-kinase* (PI3K, a lipid kinase). Several key events mediated by PI3K include:

1. Stimulation of glucose uptake into cells (primarily skeletal muscle and adipose tissue) by inducing the translocation of the glucose transporter, GLUT4, from intracellular storage in vesicles to the plasma membrane.
2. The promotion of glucose storage as glycogen. This is accomplished by the phosphorylation and inactivation of *glycogen synthase kinase 3* (GSK3, an enzyme that inhibits the enzyme glycogen synthase).
3. Inhibition of the expression of genes encoding hepatic enzymes involved in gluconeogenesis and glycogenolysis.

* Src homology 2 domain containing

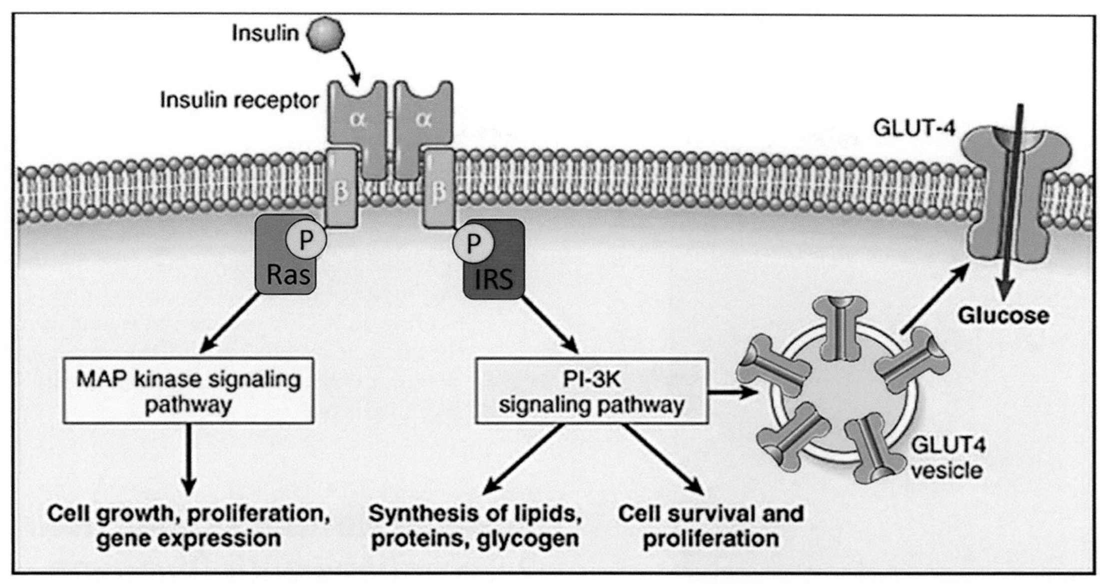

Figure 18.8 Insulin receptor signaling pathways. Upon binding by insulin, the insulin receptor undergoes autophosphorylation, leading to the creation of multiple docking sites for the adapter proteins, like IRS and Ras, that mediate activation of the PI3K and the MAP kinase signaling pathways, respectively. The PI3K pathway facilitates metabolic actions of insulin, including lipogenesis, glycogenesis, and protein synthesis. It also stimulates translocation of glucose transporters (e.g. GLUT4) from storage vesicles to the cell surface, which promotes glucose uptake by insulin-sensitive tissues such as skeletal muscle and fat. The MAP kinase pathway mediates cell growth, proliferation, and regulation of expression of various genes in insulin-responsive cells.

Source of images: Woyesa, S.B. and Taylor-Robinson, A.W., 2019. J. Molec. Pathophys., 8(1), pp. 1-13.

Glucagon

The existence of a second pancreatic hormone was first postulated in 1923 when Kimball and Murlin were studying methods of insulin extraction from the pancreas and found that injection of a largely insulin-free extract into dogs and rabbits caused rapid hyperglycemia. They inferred from this that the preparation contained a second pancreatic hormone and named this the "*gluc*ose *agon*ist", hence "glucagon".[30] Importantly, although it was initially thought that the hyperglycemia associated with diabetes mellitus was due entirely to reduced insulin production, it is now known that a concomitant elevation of glucagon expression is required to establish hyperglycemia in this disease. Recall from earlier that insulin from β cells inhibits the secretion of glucagon from α cells. As such, increased glucagon production by α cells reflects, in part, the loss of glucagon inhibition normally exerted by high local concentrations of insulin on α cells.[31,32] A remarkable finding supporting a glucagon signaling dysfunction contribution to diabetes mellitus symptoms was that glucagon receptor–null mice do not develop diabetes following complete beta cell destruction.[33] Furthermore, in mice with type 1 diabetes, the blocking of glucagon receptors with antibodies was found to normalize blood glucose levels, even without insulin treatment.[34] Pharmacological antagonists of glucagon signaling are currently being tested in human clinical trials for diabetes[35]

Synthesis

The proglucagon protein undergoes alternative posttranslational processing and is differentially expressed in pancreatic α cells (as glucagon) and brain/gut (as GLP-1 and -2). Both GIP and glucagon-like peptide-1 (GLP-1) stimulate insulin secretion and are thus classified as incretins. Glucagon is itself a potent insulin secretagogue; however, glucagon secretion is inhibited by rising levels of insulin (see Figure 18.6) and glucose in a classic negative feedback loop.

Secretion

As a hormone that generally opposes the actions of insulin, several factors that inhibit insulin promote glucagon secretion, and vice versa. For example, a key stimulus for glucagon secretion is reduced blood glucose levels, which, as mentioned earlier, is primarily an indirect effect of the removal of α cell inhibition by insulin. Circulating catecholamines, which inhibit insulin secretion by α_2 adrenergic receptors on β cells, stimulate glucagon release by binding to β receptors on α cells and promoting a cAMP signaling pathway.[36] Notably, whereas both glucose and amino acids stimulate insulin secretion by β cells, only amino acids stimulate glucagon secretion by α cells.[37] Therefore, a protein meal will increase both insulin and glucagon levels (Figure 18.9), whereas a carbohydrate meal stimulates insulin but inhibits glucagon (compare with Figure 18.4). The signaling mechanisms by which α cells recognize either glucose or amino acids are currently not well understood.

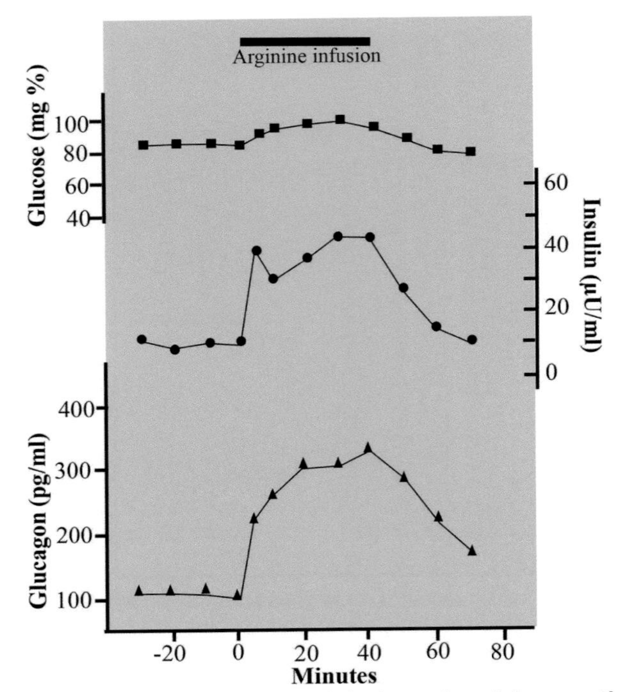

Figure 18.9 The effect of an infusion of arginine on the peripheral venous plasma levels of pancreatic glucagon, insulin, and glucose in 11 normal humans.

Source of images: Unger, R.H. (1971). N Engl J Med. 285:443-449.

SUMMARY AND SYNTHESIS QUESTIONS

1. The measurement of pancreatic beta cell-derived C peptide is a more accurate diagnostic measure for the rate of insulin secretion than is insulin itself. Why?
2. In theory, what effect would the deletion of pancreatic islet delta and epsilon cells have on blood glucose levels? Why?
3. Why is orally administered glucose a greater stimulus to insulin secretion than intravenously injected glucose?
4. Although we typically think of dysfunctional insulin synthesis or signaling with regards to diabetes mellitus, glucagon also plays a critical role in elevating blood glucose in diabetic patients. Describe how.
5. Pharmacological compounds that act as "incretin receptor agonists" are being studied as potential ways to treat metabolic dysfunctions such as obesity and type 2 diabetes. What are the cellular targets of these compounds, and how could they be useful in combating these dysfunctions?
6. Sulfonylurea drugs stimulate insulin secretion by binding to ATP-sensitive potassium channels on the surface of pancreatic beta cells, resulting in closure of the channel. How does this stimulate insulin release?

Adrenal Hormones and Their Interactions with Glucagon

LEARNING OBJECTIVE Describe the complementary actions of glucagon with adrenal glycemic hormones in maintaining energy homeostasis.

KEY CONCEPTS:

- Due to receptor and intracellular signaling pathway similarities, catecholamines and glucagon exert additive effects on glycogenolysis and lipolysis by hepatic, muscle, and adipose tissues.
- Glucocorticoids interact synergistically with glucagon signaling pathways to stimulate the transcription of key hepatic enzymes that mediate gluconeogenesis.

Catecholamines and Glucagon: Partners in Glycogenolysis

The metabolic actions of catecholamines are mediated primarily by α1 receptors in the liver, β2 receptors in skeletal muscle, and β3 receptors in adipose tissue. Note in Table 18.2 that both adrenergic and glucagon receptors

Table 18.2 Synergistic Adrenergic and Glucagon Receptor-Mediated Responses to Hypoglycemia

Receptor types	Tissue expressed	G-protein	GPCR Effector	2nd messenger	Target enzyme	Metabolic action
Adrenergic α1 and Glucagon	Liver	G_q	Phospholipase C	DAG/IP_3, Ca^{++}	↑Phosphorylase kinase ↓Glycogen synthase	↑Glycogenolysis, ↓Glycogenesis
Adrenergic β2 and Glucagon	Muscle	G_s	Adenylyl cyclase	cAMP	↑Phosphorylase kinase ↓Glycogen synthase	↑Glycogenolysis, ↓Glycogenesis
Adrenergic β3 and Glucagon	Adipose	G_s	Adenylyl cyclase	cAMP	↑Hormone sensitive lipase	↑Lipolysis

are GPCRs that interact with the same Gα proteins in target cells ($G\alpha_q$ in liver, and $G\alpha_s$ in muscle and adipose tissue), activating the same GPCR effector proteins, second messenger signals, and target enzymes. Therefore, the actions of both hormone classes by these mechanisms augment blood glucose levels via the liver, increase intracellular muscle glucose levels, and elevate blood lipid levels via adipocytes in an *additive* manner; that is, the combined effect of both hormones is equal to the sum of the effect of each given alone (also see Figure 18.5).[38]

Glucocorticoids and Glucagon: Partners in Gluconeogenesis

The enzymes phosphoenolpyruvate carboxykinase (PEPCK), glucose-6-phosphatase (G6Pase), and pyruvate

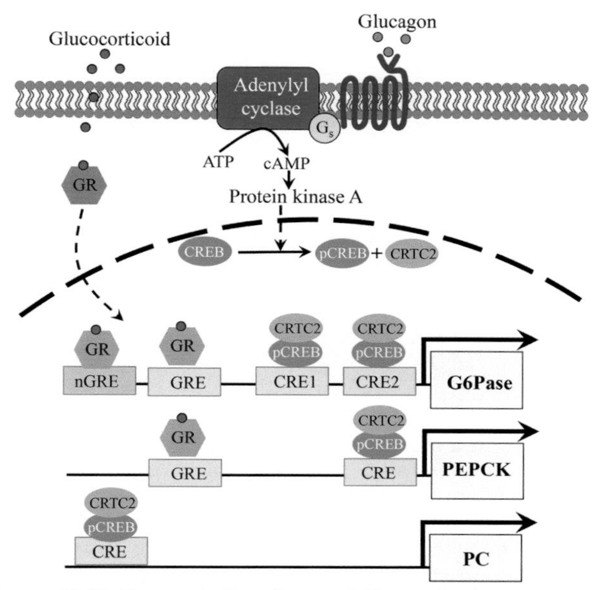

Figure 18.10 Transcriptional regulation of gluconeogenic enzymes in liver by glucocorticoid and glucagon. Glucocorticoids and glucagon are released from the adrenal cortex and pancreas, respectively, in response to hypoglycemia. Glucocorticoids bind the glucocorticoid receptor (GR) and the binary complex is translocated from the cytoplasm to the nucleus where it binds to glucocorticoid responsive elements (GREs) in the promoters of mouse G6Pase and rat PEPCK genes. nGREs denote negative elements whose association with GRs inhibits gene transcription. Binding of glucagon to its cognate G protein-coupled receptor stimulates activity of adenylyl cyclase which converts ATP to cAMP. The cAMP in turn stimulates protein kinase A (PKA) to phosphorylate the cAMP-responsive element binding protein (CREB). Phosphorylated CREB (pCREB) then interacts with its coactivator, CRTC2/TORC2, before binding to the cAMP-responsive element (CRE) in the promoters of the G6Pase, PEPCK, and PC genes and stimulating their transcription. These gene promoters also contain several binding sites for other hepatocyte nuclear factors (not shown) that synergize with the glucocorticoid and glucagon receptor signaling pathways to regulate their expression.

Source of images: Jitrapakdee, S., 2012. Int. J. Biochem. Cell Biol., 44(1), pp. 33-45.

carboxylase (PC) catalyze key steps in hepatic gluconeogenesis. During starvation, glucocorticoids released from the adrenal cortex bind the intracellular *glucocorticoid receptor* (GR) of target cells, after which the receptor-ligand complex is translocated from the cytoplasm to the nucleus where it binds to *glucocorticoid responsive elements* (GREs) in the promoters of mouse G6Pase and rat PEPCK genes (Figure 18.10) and helps promote their transcription.[39] The binding of glucagon to its G protein-coupled receptor stimulates activity of adenylyl cyclase (AC) which converts ATP to cAMP. The cAMP in turn stimulates protein kinase A (PKA) to phosphorylate the cAMP-responsive element binding protein (CREB). Phosphorylated CREB (pCREB) then interacts with its coactivator, CRTC2/TORC2, before binding to the cAMP-responsive element (CRE) in the promoters of the G6Pase, PEPCK, and PC genes and stimulating their transcription. Thus, the recruitment of GR by glucocorticoids and CREB by glucagon to the promoter/enhancer regions synergize to promote expression of the genes of gluconeogenic enzymes in a synergistic manner.

SUMMARY AND SYNTHESIS QUESTIONS

1. Explain the molecular (intracellular signaling) basis for the synergistic actions of glucagon, catecholamines, and cortisol on hepatic glucose synthesis and release.

Fat: A Spectrum of Color, Endocrine, and Metabolic Functions

LEARNING OBJECTIVE Describe the three main categories of adipose tissue, their functions and their influence on energy homeostasis.

KEY CONCEPTS:

- Adipose tissue falls into three categories: (1) energy-storing white adipose tissue (WAT), (2) thermogenic brown adipose tissue (BAT), and (3) thermogenic "beige" adipose tissue (BeAT).
- In addition to functioning as a site for lipid storage and release, WAT is also an active endocrine organ that influences whole-body metabolism.
- Thermogenic adipose tissue generates heat by the presence of uncoupling proteins (UCPs) in the mitochondria inner membrane.

Adipose tissue is a central metabolic tissue in the regulation of whole-body energy homeostasis. Far from merely being a storage site for lipids, this complex tissue exerts other key functions that include endocrine and thermogenic activities. Adipose tissue can be broadly characterized into three types: (1) energy-storing and -releasing **white adipose tissue** (WAT), (2) thermogenic **brown adipose tissue** (BAT), and (3) thermogenic **BAT-like adipocytes** that are found interspersed within WAT, which are often referred to as **beige adipose tissue** (BeAT). Some background on the regulation of adipose tissue development, provided next, will be important for understanding the endocrine basis of dysfunctions of energy homeostasis, as well as their treatments, discussed in Chapter 19: Metabolic Dysregulation and Disruption.

Adipogenesis

Adipogenesis is the process whereby fibroblast-like progenitor cells commit their fate to the adipogenic lineage, forming preadipocytes. Prior to commitment to the adipogenic line, these same progenitor cells also have the potential to develop into myoblasts (muscle progenitors), osteoblasts (bone progenitors), and chondroblasts (cartilage progenitors). The differentiation of the progenitor into the various types of adipose and other tissues is mediated by a variety transcription factors (Figure 18.11). In particular, expression of the nuclear hormone receptor *peroxisome proliferator-activated receptor* γ (PPARγ), has been described as the "master regulator" of WAT adipogenesis, as it is essential for adipocyte differentiation both in culture and *in vivo*[40] (Figure 18.12). Strikingly, PPARγ knockout mice fail to develop adipose tissue altogether.[41–43] The physiological ligands of PPARγ remain unclear but are thought to include endogenous nutrients, such as FFAs and their derivatives, like eicosanoids. Members of the synthetic thiazolidinedione class of drugs (TZDs, used to treat type 2 diabetes) are potent PPARγ ligands that stimulate whole-body lipid metabolism and insulin sensitivity.[44] PPARγ also plays a role in BAT differentiation, as well.

Adipocyte Morphology and Histology

Individual WAT adipocytes are characterized by the presence of a large, single lipid droplet and relatively few mitochondria and are devoid of thermogenic uncoupling protein 1 (UCP1) (Figure 18.11). By contrast, BAT adipocytes contain multiple small lipid droplets, a large number of mitochondria, and express high levels of thermogenic UCP1. BeAT adipocytes are phenotypic intermediates of BAT and WAT and express thermogenic UCP1. Histologically, BAT stains strongly with the cytosolic stain eosin due to its high mitochondria and small lipid droplets (Figure 18.11). WAT stains very weakly for eosin due to the presence of a large lipid droplet that displaces the cytoplasm. BeAT adipocytes are interspersed among WAT, displaying an intermediate eosin staining profile.

WAT as an Endocrine Organ

An innervated connective tissue, WAT is composed of a number of different cell types including adipocytes,

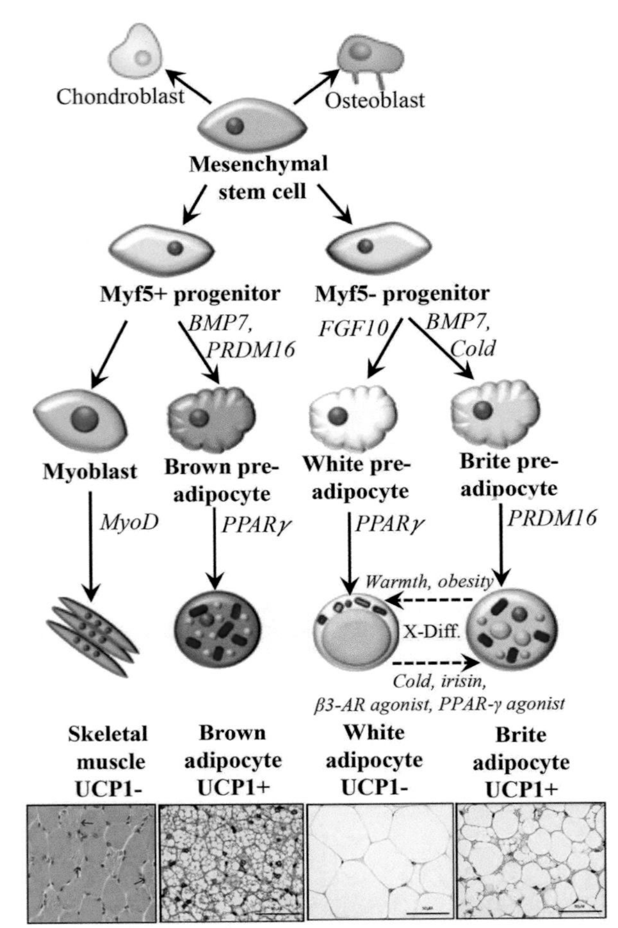

Figure 18.11 Differentiation of brown, beige, and white adipocytes from a mesenchymal precursor. Multipotent, fibroblast-like mesenchymal precursors function as precursors not only to adipocytes, but also to myoblasts, osteoblasts, and chondroblasts. Specific transcription factors present in progenitor and preadipocyte cells mediate their differentiation into one of the three adipocyte categories. In addition, the differentiation of Myf5-negative progenitors into beige/brite adipocytes is influenced by cold temperature. Whereas BAT and WAT develop from different progenitors and are not thought to be capable of transdifferentiation, WAT and beige/brite adipocytes share a common precursor and are capable of transdifferentiation. Cold temperature, as well as the skeletal muscle hormone, irisin, and β3-adrenergic receptor and PPAR-γ receptor agonists can induce the transdifferentiation of WAT into beige/brite adipocytes. By contrast, warm temperature and obesity can promote the transdifferentiation of beige/brite adipocytes into WAT. Colors on cartoon: blue circle (cell nucleus), yellow circle (lipid droplet), red (mitochondria). Abbreviations: β3-adrenergic receptor (β3-AR), bone morphogenetic protein 7 (BMP7), myogenin D (MyoD), myogenic factor 5 (Myf5), peroxisome proliferator-activated receptor γ (PPARγ), PR domain containing 16 (PRDM16), platelet-derived growth factor receptor-α (Pdgfr-α), fibroblast growth factor 10 (FGF10).

Source of images: Drawings adapted by permission of Jiménez, G., et al. 2016. Brown adipose tissue and obesity. In Obesity (pp. 13-28). Springer, Cham. Photographs used by permission of Keipert, S. and Jastroch, M., 2014. Biochim. Biophys. Acta (BBA)-Bioenergetics, 1837(7), pp. 1075-1082.

fibroblasts, vascular cells, and immune cells (Figure 18.13). Besides serving as the primary site for energy storage, WAT also acts as an energy source and releases lipids in the form of free fatty acids (FFAs) when the demand for energy increases. The processes of energy storage and release by WAT are carefully regulated by different energy states and hormonal cues. Far from being a passive reservoir for energy storage, WAT is a highly active metabolic and endocrine organ, secreting numerous steroid hormones (including glucocorticoids and estrogens), some components of the renin angiotensin system that regulate blood pressure (e.g. angiotensinogen, angiotensin-converting enzyme and angiotensin II receptor), various pro- and anti-inflammatory cytokines, metabolism-regulating microRNAs, leptin, and other peptides, as well as a diverse array of hormone receptors[45–48] (Table 18.3). Indeed, white adipose tissue may represent the largest endocrine tissue by mass, particularly in obese humans.

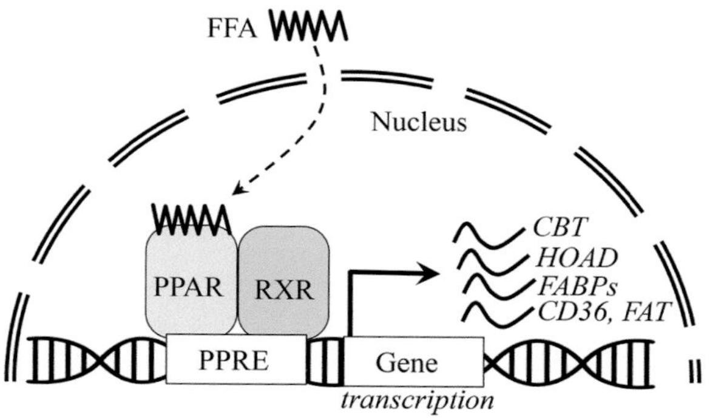

FFA
Nucleus
PPAR
RXR
CBT
HOAD
FABPs
CD36, FAT
PPRE
Gene
transcription

Figure 18.12 Free fatty acids regulate fat metabolism by binding directly to PPAR receptors. Ligand binding causes the dimerization of PPAR with the retinoid X receptor (RXR). The PPAR–RXR complex activates target genes by recognizing promoter regions called peroxisome proliferator response elements (PPREs). Target genes include the β-oxidation enzymes carnitine palmitoyl transferase (CPT) and hydroxyacyl dehydrogenase (HOAD), fatty acid binding proteins (FABPs), and the transmembrane fatty acid transporters CD36 and FAT.

BAT and BeAT Generate Heat by Reducing the Efficiency of ATP Synthesis

The process through which mitochondria synthesize ATP, known as *oxidative phosphorylation*, relies on the generation of a proton gradient, or "proton-motive force", via the *electron transport chain* located on the inner mitochondrial membrane (Figure 18.14). The subsequent facilitated diffusion of protons from the inner membrane space to the matrix across the transmembrane *ATP synthase* enzyme couples the proton-motive force with ATP synthesis. The efficiency of ATP synthesis, however, can be reduced by the presence of *uncoupling proteins* (UCPs) in the inner membrane that allow protons to flow very rapidly, or "leak", back into the mitochondrial matrix, bypassing ATP synthase and dissipating all of the proton-motive force as heat instead of capturing some in ATP synthesis (Figure 18.14). *UCP-1*, also known as *thermogenin*, appears to be essential to mammalian thermoregulation, as UCP-1 knockout mice are intolerant to cold.[49]

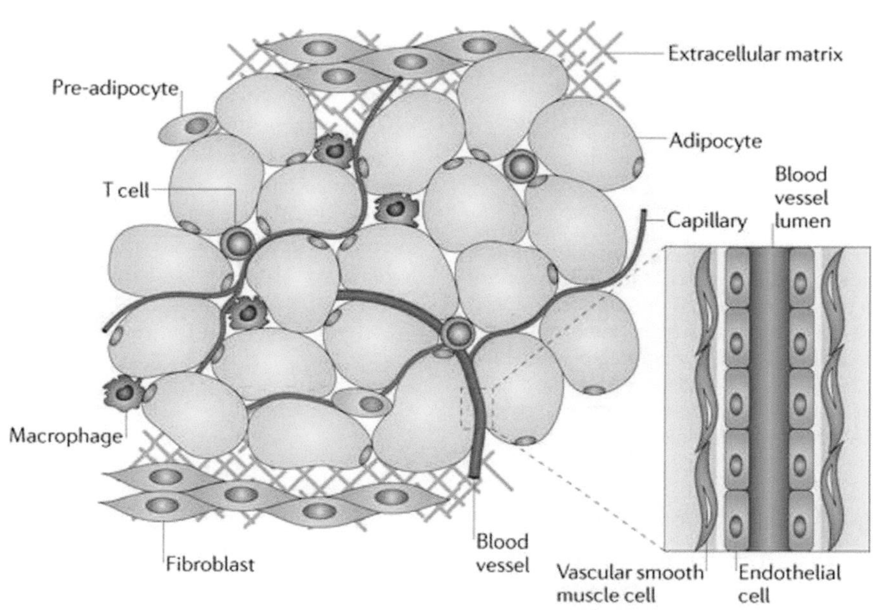

Figure 18.13 The components of adipose *tissue*—more than just fat. Adipose tissue consists of a variety of cell populations: blood vessels, stem cells, pericytes (cells that stabilize blood vessel walls), preadipocytes (precursors to fat cells), macrophages (immune cells), endothelial cells (form the inner lining of blood vessels), and fibroblasts (help form the extracellular matrix).

Source of images: Ouchi, N., et al. 2011. Nat Rev Immunol, 11(2), pp. 85-97. Used by permission.

Table 18.3 Examples of Endocrine-Related Compounds Secreted or Synthesized by White Adipose Tissue

Secreted compounds	Receptors	Enzymes & Intracellular signaling
Cytokines and immune-related proteins	***Peptide hormones***	***Glucose metabolism***
Leptin	Insulin	Insulin receptor substrates 1 and 2
Tumor necrosis factor α (TNF α)	Glucagon	GLUT4
Interleukin-6 (IL-6)	TSH	Glycogen synthase kinase 3-α
Monocyte chemoattractant protein 1 (MCP1)	GH	***Steroid hormone metabolism***
Adipsin	Angiotensin II	Cytochrome p450 Aromatase
Adiponectin	Gastrin	11β-hydroxysteroid dehydrogenase 1
Acylation stimulating protein	Adiponectin	17β-hydroxysteroid dehydrogenase
Fibrinolytic proteins	***Cytokines***	***Lipid metabolism***
PAI-1	Leptin	Lipoprotein lipase
Tissue Factor	TNF α	Hormone-sensitive lipase (HSL)
Other Proteins	Interleukin-6 (IL-6)	Cholesteryl ester transfer protein
Resistin	***Nuclear***	Apolipoprotein E
Angiotensinogen	Glucocorticoid	Adipocyte fatty acid binding protein
microRNAs	Estrogen	
	Androgen	
	Progesterone	
	TH	
	PPARγ	
	Vitamin D	
	Catecholamine	
	β1, β2, β3 receptors	
	α1, α2 receptors	

Source: After Kershaw, E.E. and Flier, J.S., 2004. Adipose tissue as an endocrine organ. *The Journal of Clinical Endocrinology & Metabolism*, *89*(6), pp. 2548–2556.

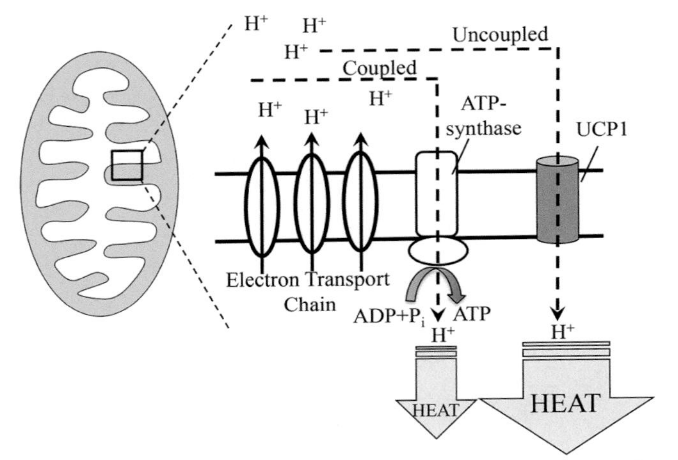

Figure 18.14 Uncoupling ATP synthesis to generate heat in brown and beige adipose tissue. In the oxidative phosphorylation system, proton-motive force can be coupled to ATP synthesis via the enzyme ATP synthase, which generates some heat, and ATP is then exported to the cytoplasm where it is used for cellular work. However, the expression of uncoupling protein 1 (UCP1) uncouples ATP synthesis from the proton-motive force and generates heat exclusively.

BAT/BeAT Physiology and Distribution

In small mammals, as well as in the newborns of larger mammals, the primary tissue involved in non-shivering thermogenesis is brown adipose tissue (BAT). Indeed, human newborns and infants have not yet developed the neurological ability to shiver and, especially considering their high surface area-to-volume ratios and rates of heat loss, must rely on BAT to provide heat. In contrast with WAT, BAT has a very high mitochondrial concentration, giving the tissue its distinctive brown color. BAT is also highly vascularized and contains multiple lipid droplets per cell (Figure 18.11). The newborns of virtually all mammals contain BAT, and in human infants BAT may constitute approximately 5% of the body mass. Analysis using positron emission tomography (PET) scans has also revealed the presence of cold-activated BAT in adults[50,51] (Figure 18.15). In hibernating adult mammals, such as ground squirrels and bats, BAT is known to play an important role in thermogenesis, particularly during times of periodic arousal from hibernation.

Similar to BAT, BeAT also possesses multiple lipid droplets per cell and abundant cristae-dense mitochondria that express UCP1,[52] and can be considered a phenotypic intermediate between WAT and BAT (Table 18.4). However, whereas in rodents and human infants BAT localizes to dedicated BAT depots, such as the interscapular and perirenal regions, BeAT adipocytes are interspersed within WAT depots. As such, BeAT adipocytes are also known as "*brite*" ("brown-in-white") adipocytes.

Developments & Directions: The "brite" side of fat: potential for combating obesity with inducible thermogenic beige fat

Until recently, adipose tissue had been broadly categorized into two distinct types, lipid-storing white adipose tissue (WAT) and thermogenic brown adipose tissue (BAT). Given BAT's thermogenic role and that BAT tissue mass is negatively correlated with body mass index (BMI) and age in humans,[53] the potential of using BAT as a therapeutic target to combat obesity and metabolic disorders has become increasingly important. Although both WAT and BAT each differentiate from mesenchymal/mesodermal stem cells, BAT shares a closer developmental lineage with skeletal muscle than with WAT. Currently there is no evidence that the two categories of fat can transdifferentiate, which limits BAT's therapeutic potential.

More recently, another category of thermogenic adipose tissue known as beige adipose tissue (BeAT) has drawn considerable interest. Similar to BAT, BeAT also possesses thermogenic properties,[52] but unlike BAT it has potential to transdifferentiate from white adipose tissue (WAT). Over 100 different pharmacological and natural compounds have been shown to induce beigeing,[54] and the activation of BeAT and BAT is known to reduce metabolic disease in rodent models.[55] However, many of these pharmacological compounds act by stimulating β3-adrenergic receptors, which often promotes dangerous cardiovascular side effects. The search for alternative therapies for enhancing BeAT to treat obesity and other metabolic disorders in humans is beginning to produce clinically promising findings. For example, stimuli that have been shown to increase BeAT activity and glucose metabolism, as well as decrease WAT mass in humans, include cold treatment[56–59] and exercise.[54,60] Two naturally occurring compounds of note shown to have stimulatory effects on BeAT and BAT in humans include *irisin* (a hormone secreted from muscles in response to exercise)[61] and the plant-derivative *berberine*.[62]

Links between Obesity and Reduced BAT/BeAT

BAT/BeAT is now known to be present in significant quantities in adult humans and appears to play an important role in non-shivering thermogenesis. Considering that non-shivering thermogenesis can constitute up to 30% of human adult basal metabolic rate,[63] BAT/BeAT very likely plays a key role in regulating metabolic homeostasis in adults. Yoneshiro and colleagues[64] have shown that BAT/BeAT activity decreases with age, suggesting that this decrease may accelerate the accumulation of body fat as we age. Interestingly, in obese humans, the prevalence of BAT/BeAT tissue may also be significantly reduced[50,63,65,66] (Figure 18.16). Using the mouse model, Sugatani and colleagues[67] have shown that decreased BAT activity actually promotes the development of obesity in mutant mice. Together, these observations suggest that BAT/BeAT play important metabolic roles, and the possibility of increasing resting metabolic rate by activating BAT/BeAT thermogenesis is being explored as a target for potential anti-obesity therapies.

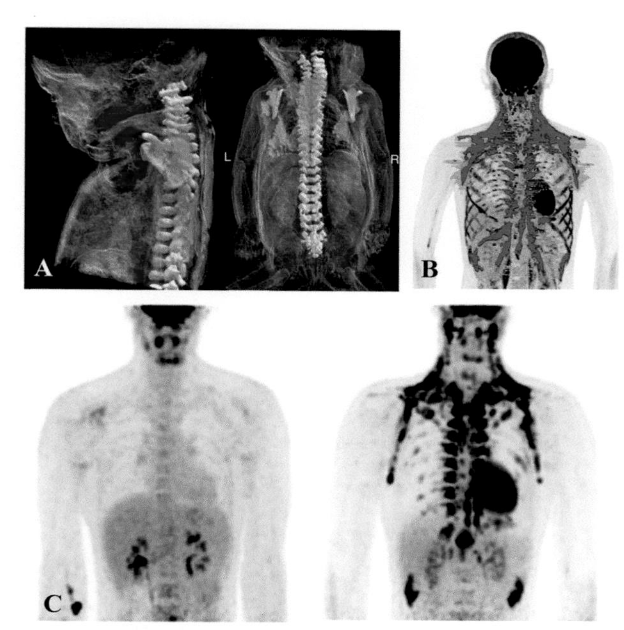

Figure 18.15 Distribution and activity of brown adipose tissue (BAT) in infants and adults. (A) Three-dimensional reconstruction of interscapular BAT (green) in a human infant. **(B)** Activated BAT (red) in adult adipose depots. Blue regions indicate remaining adipose tissue within defined BAT depots. **(C)** Comparative PET-CT scans reveal the pattern of 18F-fluorodeoxyglucose (18F-FDG) uptake (a technique that relies on the uptake of radiolabeled glucose by metabolically active tissues) in the same person under relatively warm conditions (left) and after cold exposure (right). Under warm conditions, the only uptake visible is that into the brain, heart, kidneys, and bladder. Under comparatively cold conditions, note that the symmetrical pattern of uptake into the supraclavicular, neck, and paravertebral areas of brown adipose tissue is now visible.

Source of images: (A) Lidell, M.E., et al. 2013. Nat. Med., 19(5), pp. 631-634; Used by permission. (B) Biyan Qin, et al. (2022) Endocr. J., 69, 1, (55-65),; (C) Courtesy of MJW Hanssen.

Table 18.4 Some Characteristics of Different Types of Adipocytes

	White adipocytes	Brown adipocytes	Beige adipocytes
Number of lipid droplets	**Uniocular**	**Multiocular**	**Uniocular → multiocular**
Lipid storage	+++	+	+++ → ++
Mitochondria	+	+++	+ → ++
Fatty acid oxidation	+	+++	+ → +++
Respiratory chain	+	+++	+ → +++
UCP1	-	+++	- → +++

-/+ indicates relative strength of phenotype

→ denotes that inducible phenotype ranges with conditions

Source: After Lizcano, F. and Vargas, D., 2016. Biology of beige adipocyte and possible therapy for type 2 diabetes and obesity. *International Journal of Endocrinology* , 2016, pp. 1-10.

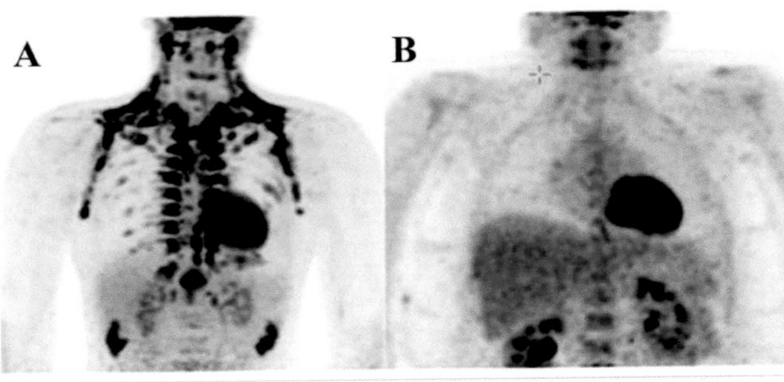

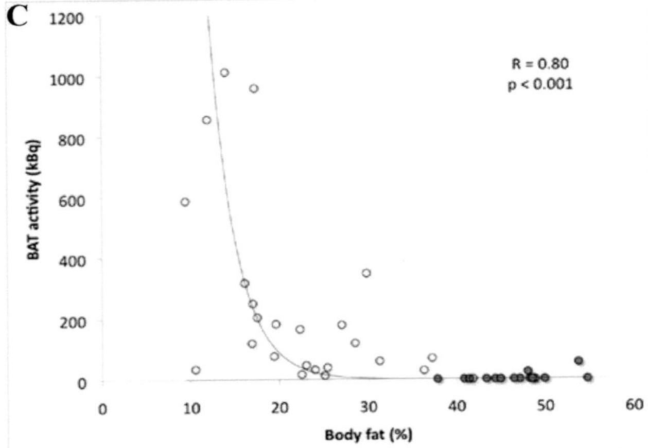

Figure 18.16 A negative relationship exists between BAT activity and body mass index. Comparative PET-CT scans conducted during cold exposure a lean person **(A)** and in a morbidly obese individual **(B)**. Whereas there is elevated BAT activity in the lean individual, very little BAT activity is present in the obese person following cold exposure. **(C)** Brown adipose tissue activity in relation to body fat percentage. The open dots indicate the study group ranging from lean to morbidly obese (age range: 18–32 years; van Marken Lichtenbelt et al., 2009) and the closed dots indicate a second group of morbidly obese subjects (age range: 25–51 years; Vijgen et al., 2011).

Source of images: van Marken Lichtenbelt, W., 2011. Front. Endocrinol., 2, p. 52.

SUMMARY AND SYNTHESIS QUESTIONS

1. Describe the mechanism by which UCP-1 proteins facilitate heat production in brown and beige adipose tissue.
2. Describe different lines of evidence that suggest there is a connection between impaired non-shivering thermogenesis capacity and the development of obesity in humans and mice.

Thyroid Hormones and Fuel Metabolism

LEARNING OBJECTIVE Explain the roles TH plays in the metabolism of glucose, lipid, and protein, as well as modulating appetite.

KEY CONCEPTS:

- TH's metabolic effects are mediated by nuclear and mitochondrial receptors.
- TH generally enhances appetite and the metabolic rates of turnover of fuel substrates.
- Hyperthyroidism is associated with higher metabolic turnover rates that favor fuel consumption and reduced muscle and fat mass.
- Hypothyroidism promotes reduced metabolic turnover rates that favor increased fat storage.
- TH's effects on appetite and fuel metabolism are mediated via both central (brain) and peripheral mechanisms.

Perhaps no single hormone exerts as profound an effect on metabolism in birds and mammals than does thyroid hormone (TH). From his quotation in the chapter opening, it is clear that Aristotle envisioned that a function of the brain is to maintain body temperature. Although Aristotle had no way of knowing the brain's mechanism of action in this regard, the production of thyrotropin-releasing hormone (TRH) by hypothalamus profoundly influences both metabolism and body temperature, particularly in mammals and birds, through its modulation of thyroid hormone production. Thyroid hormone affects key metabolic pathways that control energy storage and expenditure primarily through actions in the brain, white and brown fat, skeletal muscle, liver, and pancreas. This section will discuss the role of TH in enhancing the rate of turnover of fuel substrates, sometimes simultaneously in opposing metabolic pathways in a process called *metabolic cycling*. Considering that TH excess is associated with higher turnover rates that favor fuel consumption and weight loss, whereas hypothyroidism promotes reduced turnover rates that favor fuel storage and weight gain, it is clear that TH plays a central role in multiple aspects of fuel metabolism.[68]

TH's Metabolic Effects Are Mediated by Nuclear and Mitochondrial Receptors

A key finding by Jamshed Tata[69] indicated that the thermogenic activity of TH was reduced by the transcriptional inhibitor actinomycin D (Figure 18.17), suggesting that TH-induced heat production is mediated by modulating nuclear gene expression of mitochondrial and/or extra-mitochondrial proteins. Indeed, this was the first time that an inhibitor of transcription was demonstrated to block a metabolic response *in vivo*. Most physiological effects of TH are now thought to be mediated by receptors that function as nuclear transcription factors, and hypothyroid mice as well as mice lacking all functional thyroid hormone receptors (TR) show decreased metabolism.[70]

Control of Mitochondrial Activity by TRs

In mammals, thyroid hormones exert a profound influence on energy usage, metabolic rate, and heat production in large part by modulating mitochondrial biogenesis and activity. Mitochondria possess their own genome and mechanisms for replication, transcription, and protein synthesis but also rely extensively on genomic DNA for their functioning. TH modulates

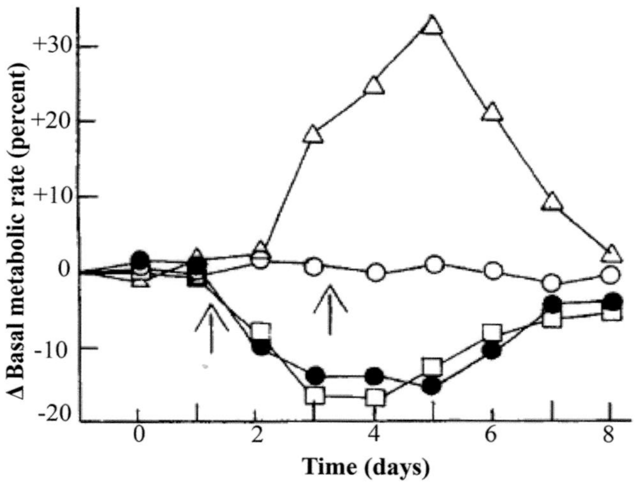

Figure 18.17 Changes in basal metabolic rate of thyroidectomized rats following administration of thyroid hormone and the transcriptional inhibitor actinomycin D, separately or in combination. Vertical arrows show the time of injection for each substance. Treatments: untreated (open circle); 26 ug T3/100 g (open triangle); 8 ug actinomycin D/100 g (open square); T_3 + actinomycin D (closed circle).

Source of images: Tata, J.R., 1963. Nature, 197(4873), pp. 1167-1168. Used by permission.

mitochondrial activity indirectly by regulating the nuclear expression of mitochondrion-specific transcription factors and nuclear-encoded components of the mitochondrial respiratory apparatus. TH also directly regulates mitochondrial gene expression and oxidative phosphorylation by binding to regulatory proteins in that organelle.

Indirect Control

Basal levels of mitochondrial gene transcription are regulated by the actions of the nuclear-derived *mitochondrial transcription factor A* (Tfam) and a *mitochondrial DNA-directed RNA polymerase* (Polrmt). Importantly, the expressions of both Tfam and Polrmt are themselves now known to be regulated by the classical genomic T_3 target genes, *nuclear respiratory factors 1* and 2 (NRF-1 and NRF-2) and the cofactor *peroxisome proliferator-activated receptor gamma coactivator 1 alpha* (PGC-1α) (Figure 18.18).

Direct Control

Although most of the thermogenic effects of TH on mitochondria are thought to be mediated indirectly by the binding of TH (primarily T_3) to nuclear-localized TRs, there is mounting evidence that both T_3 and T_2 also accumulate in the mitochondria itself and directly regulate mitochondrial biogenesis.[71,72] T_3 has a direct effect on the mitochondrion via the binding of a mitochondria-localized thyroid hormone receptor (p28, a truncated isoform of TRα that lacks a DNA-binding domain) to stimulate the mitochondrial uncoupling proteins (UCP)

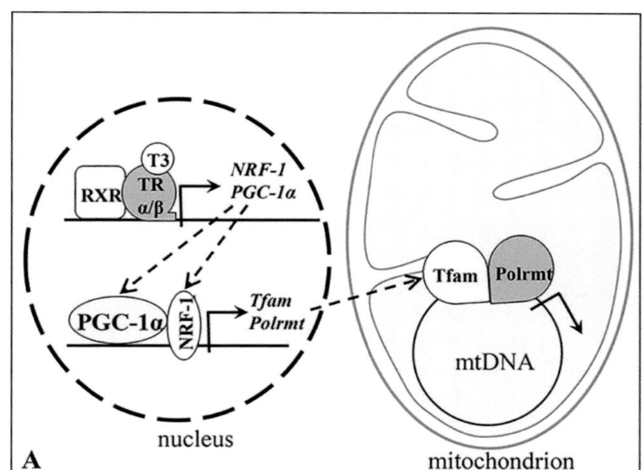

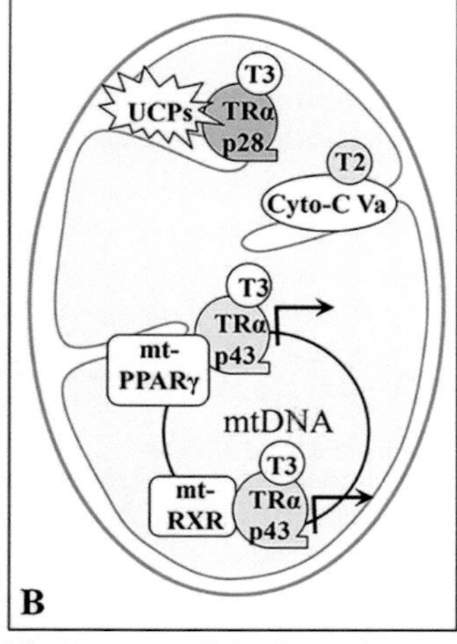

Figure 18.18 Direct and indirect control of mitochondrial activity by thyroid hormone. (A) In the cell nucleus, T_3 binds with classical TR/RXR heterodimers and mediates transcription of nuclear respiratory factor 1 (NRF-1) and the cofactor peroxisome proliferator-activated receptor gamma coactivator 1 alpha (PGC-1α). These T_3 target genes code for transcription factors that, in turn, modulate the transcription of mitochondrial transcription factors A (Tfam) and mitochondrial DNA-directed RNA polymerase (Polrmt), which regulate basal levels of mitochondrial gene transcription. **(B)** In the mitochondria, diiodothyronine (T_2) stimulates oxidative phosphorylation by binding to and activating the cytochrome-c oxidase Va subunit (Cyto-C Va). T_3 also binds with TRα p28 to stimulate the mitochondrial uncoupling proteins (UCP) on the inner mitochondrial membrane, generating a thermogenic response. T_3 also binds TRα p43 which heterodimerizes with either the mitochondrial peroxisome proliferator activator receptor γ2 (mtPPARγ) or the mitochondrial retinoid X receptor (mtRXR) and stimulates transcription of mitochondrial transcription factors and rRNAs.

on the inner mitochondrial membrane, generating a rapid thermogenic response[73] (Figure 18.18). *p43* is another TRα isoform that functions as a T_3-dependent mitochondrial transcription factor and heterodimerizes with either a mitochondrion-specific isoform of the peroxisome proliferator activator receptor γ2 (mtPPAR) or a mitochondrion-specific isoform of the retinoid X receptor (mtRXR). p43 functions as a TH-dependent transcription factor that regulates global mitochondrial transcription. Lastly, 3,5-diiodothyronine (T_2) has been shown to specifically bind and activate the cytochrome-c oxidase Va subunit[74,75] (Figure 18.18), and is thus an active effector of oxidative phosphorylation.

Glucose Metabolism

Thyroid hormones exert profound effects in the regulation of virtually all aspects of glucose homeostasis, including (1) promoting intestinal glucose absorption and uptake of glucose by white adipose tissue (WAT) and skeletal muscle; (2) influencing hepatic glycogen synthesis and glycogenolysis; and (3) modulating the responsiveness of liver, WAT, and muscle to other hormones such as insulin and catecholamines (Figure 18.19). Considering the close relationship of TH with carbohydrate metabolism, it is not surprising that there exists an association between hyperthyroidism and glucose intolerance in humans.[76] In contrast with hyperthyroidism, many physiological parameters regulating glucose metabolism are reversed in hypothyroidism, with some important exceptions (see Table 18.5 for a comparison of hyper- vs. hypothyroidism on fuel metabolism).

Table 18.5 Major Effects of Thyroid Hormone on Fuel Substrate Physiology Under Hyper- and Hypothyroid Conditions

Hyperthyroidism	Hypothyroidism
Hyperglycemia	Normal blood glucose
Insulin resistance	Insulin resistance
Increased pro-insulin (inactive) secretion	Reduced insulin secretion
Increased insulin clearance by kidney	?
Increased glucose absorption by intestine	Reduced glucose absorption by intestine
Increased liver gluconeogenesis/glycogenolysis	Reduced liver gluconeogenesis/glycogenolysis
Increased peripheral Glut4	Reduced peripheral Glut4
Increased glycolysis	Reduced glycolysis
Hypocholesterolemia (increased LDL receptor synthesis)	Hypercholesterolemia (reduced LDL receptor synthesis)
Increased free fatty acids	Increased free fatty acids
Increased muscle wasting	Muscle weakness and fatigue
Reduced bone density	Reduced bone density

Lipid Metabolism

Effects of TH on WAT

In general, TH has been shown to increase the activities of enzymes that promote both lipogenesis[77] and lipolysis[78,79] (Figure 18.19). In agreement with the notion of metabolic cycling, the promotion of lipogenesis by TH ensures that lipids will be available for lipolysis, as long as an adequate food supply is present. Additionally, thyroid hormones have been found to inhibit the proliferation of adipocyte progenitor cells but stimulate their differentiation to functional adipocytes,[80,81] thus promoting capacity for the storage and availability of lipids for metabolism.

Another well-known mechanism through which TH facilitates lipolysis in adipocytes is by sensitizing the tissue to catecholamines. Specifically, *in vitro* studies using

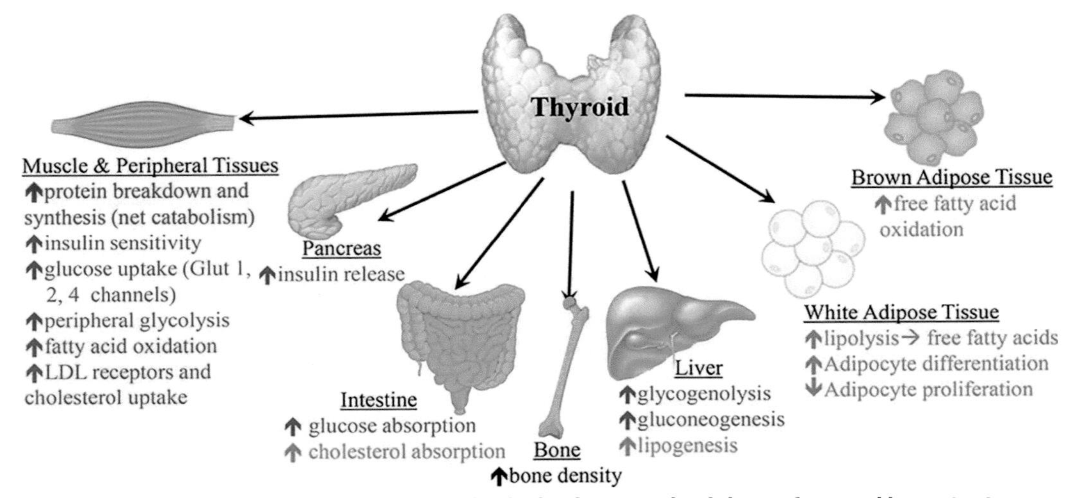

Figure 18.19 Some major effects of thyroid hormone on the fuel substrate physiology of several important organs and tissues under euthyroid conditions. Red highlighting denotes protein metabolism, blue denotes carbohydrate metabolism, green denotes lipid metabolism.

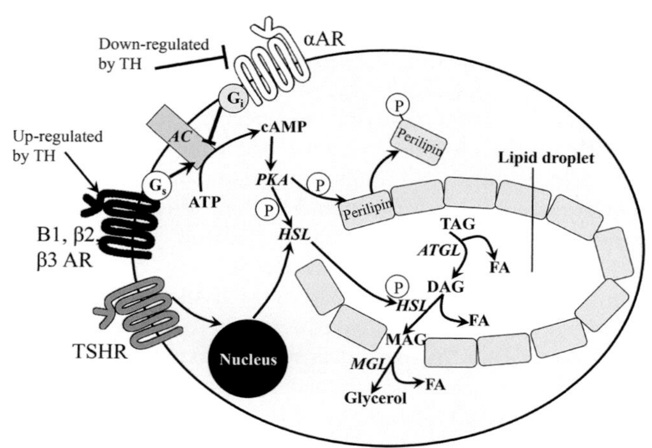

Figure 18.20 Regulation of lipolysis by TH, TSH, and catecholamines in adipocytes. Under basal conditions, perilipin (a protein associated with lipid droplets) prevents lipolysis. Under stimulation by catecholamines (via increased cAMP and protein kinase A), perilipin allows lipolysis. Thyroid hormones work synergistically with catecholamines by upregulating stimulatory beta-adrenergic receptors, and downregulating inhibitory alpha adrenergic receptors. Lipolysis is catalyzed by several lipases. Triacylglycerol (TAG) is catabolized by the consecutive action of adipose triglyceride lipase (ATGL), hormone-sensitive lipase (HSL), and monoglyceride lipase (MGL) to generate fatty acids (FA) and glycerol. The transcription of HSL is regulated by TSH via its cell membrane localized receptor, and activation of HSL is mediated through phosphorylation by PKA. Enzyme names are written in italics.

human adipocytes have shown that TH upregulates stimulatory β-adrenergic receptors and downregulates inhibitory α-adrenergic receptors in adipocytes.[82] This ultimately amplifies the effects of catecholamines, which stimulate a messaging cascade that results in the activation of hormone-sensitive lipase (HSL) and the release of free fatty acids (Figure 18.20). Consistent with these *in vitro* observations are findings that although basal lipolysis is unchanged in hyperthyroid patients, β-adrenergic-stimulated lipolysis is markedly enhanced in adipocytes from hyperthyroid subjects.[83]

Direct Effects of TSH on Adipocytes

Interestingly, thyroid-stimulating hormone (TSH) has also been shown to function directly as a lipolytic factor in WAT of mice and in cultured human adipocytes by increasing *hormone-sensitive lipase* (HSL) mRNA expression after binding to its receptors located on the WAT cell membrane[84–86] (Figure 18.20). Consistent with these observations are reports that human patients with subclinical primary hypothyroidism (i.e. elevated levels of blood TSH) have high levels of serum free fatty acids[87] and that TSH triggers lipolysis during the neonatal period,[88] a time when serum TSH levels surge dramatically.

Protein Metabolism

Whereas under euthyroid (healthy thyroid) conditions thyroid hormones stimulate both synthesis and degradation of proteins in a balanced manner (Figure 18.19), under hyperthyroid conditions the rates of degradation exceed synthesis,[89] and consequently urinary nitrogen (a byproduct of amino acid breakdown) excretion is enhanced.[90,91] Thyroid hormone has been shown to enhance rates of protein degradation in skeletal muscle and liver by increasing lysosomal enzyme activities.[92] Rates of amino acid release by skeletal muscle, a tissue that singly constitutes the majority of the body's protein, are known to increase in hyperthyroidism, providing a substrate for gluconeogenesis by the liver and resulting in reduced skeletal muscle mass.[93]

SUMMARY AND SYNTHESIS QUESTIONS

1. In 1963, Jamshed Tata discovered that the thermogenic activity of TH was reduced by the transcriptional inhibitor actinomycin D. What are the implications of this finding?
2. Describe effects of TH on carbohydrate metabolism that oppose the actions of insulin and effects that are synergistic with insulin's actions under euthyroid conditions.
3. What are three ways in which hyperthyroidism promotes glucose intolerance?
4. Describe the mechanisms by which catecholamines, TH, and TSH work together to promote lipolysis in white adipose tissue (WAT).
5. Human patients with subclinical primary hypothyroidism have high levels of serum free fatty acids. How does this make sense?
6. How do WAT and BAT interact to modulate metabolic homeostasis? What roles do TH and other hormones play in this interaction?
7. Why is hypothyroidism typically accompanied by elevated blood cholesterol levels, and hyperthyroidism by low cholesterol?
8. How does hyperthyroidism promote fat loss? Muscle loss?
9. Describe some indirect and direct molecular mechanisms by which TH is thought to induce mitochondria's thermogenic effects.
10. A patient suffering from primary hypothyroidism also has hyperlipidemia (high levels of serum free fatty acids). How are the two pathologies connected?

Hypermetabolic Effects of Thyroid Hormone in Mammals and Birds

LEARNING OBJECTIVE Describe the hypermetabolic effects of thyroid hormones on mammals and birds, and their roles in facilitating non-shivering thermogenesis.

KEY CONCEPTS:

- TH is the primary hormone involved in regulating metabolic rate and thermogenesis in mammals and birds.
- The metabolic effects of TH manifest at the cellular level, with TH treatment producing increased oxygen consumption by a variety of tissues.
- Thyroid hormones affect basal metabolic rate, with hyperthyroidism increasing metabolic heat production and hypothyroidism decreasing it.
- Brown and beige adipose tissue are targets of TH for facilitating non-shivering thermogenesis.
- The activation of BAT thermogenesis is induced primarily by the sympathetic nervous system (SNS), but requires the presence of TH for the full response to manifest.

The general effects of THs on development, growth, and fuel metabolism are properties shared among virtually all vertebrates. However, one major property of thyroid hormone, its thermogenic effect, is not shared equally among vertebrate taxa but instead appears to be restricted to birds and mammals that rely on this for maintaining body temperature homeostasis. Although many metabolic reactions produce heat as a byproduct, in this chapter the term "*thermogenesis*" will be referred to as the metabolic production of heat specifically for the purpose of regulating body temperature. A direct thermogenic role for TH in taxa other than birds and mammals has not yet been demonstrated.

Hypermetabolic Effects of TH in Humans

Direct observations that the thyroid gland modulates metabolism in humans have been documented for over 100 years, ever since German physician and physiologist Adolf Magnus-Levy asked in 1895 if the changes in weight observed in hyper- and hypothyroid patients were related to metabolism.[94] Using respiration calorimetry, he reported that not only were both oxygen consumption and carbon dioxide production increased in hyperthyroid patients, but that this imbalance disappeared when the condition was cured. Importantly, Magnus-Levy also demonstrated that hyperthyroid metabolism could be reproduced in healthy subjects when administered thyroid extract.

The metabolic effects of TH are not only visible in the whole animal, but also manifest at the tissue level, with TH treatment producing increased oxygen consumption by a variety of tissues in thyroidectomized rats (Figure 18.21). Thyroidectomy in humans is accompanied by up to a 40% decrease in metabolic rate,[95] and hypothyroid rats die after several hours when exposed to cold due to a thermogenesis deficiency that is restored by replacement with thyroid hormones.[96] Interestingly, Astrup and colleagues[97] have shown that human patients who are predisposed to obesity also have low resting metabolic rate and thyroid hormone levels. Thus, TH is generally considered the primary hormone involved in regulating energy expenditure in mammals.

Curiously Enough . . . THs and metabolism are so intertwined that prior to the development of the radioimmunoassay for quantifying TH concentrations, the measurement of resting metabolic rate was the most accepted method for assessing thyroidal status.[98]

Influence of Body Mass on Rate of Thyroid Hormone Secretion

Importantly, as the primary hormone influencing mammalian and avian metabolism, the rates of thyroid hormone secretion have been shown to correlate positively with body

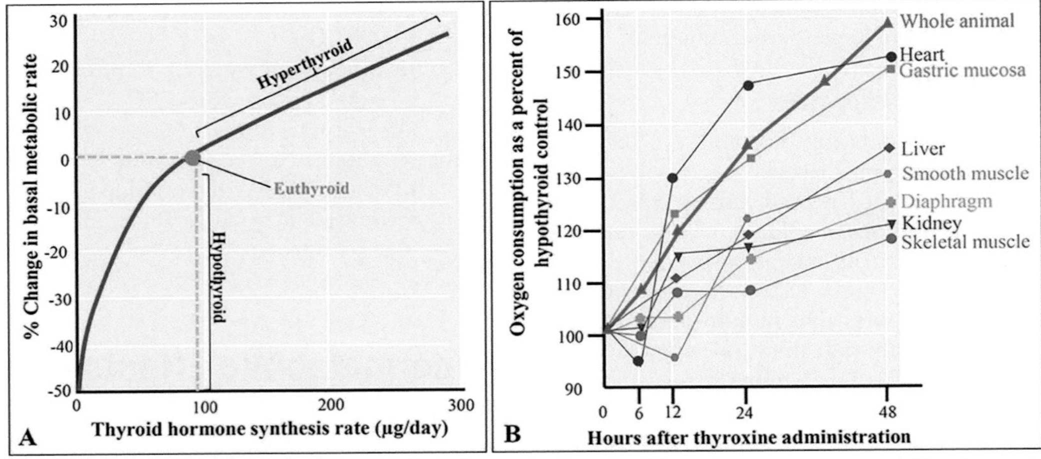

Figure 18.21 Effects of thyroid hormone on human basal metabolic rate (A), and oxygen consumption by various tissues of thyroidectomized rats (B).

Source of images: (A) Modified from Boron, W.F. and Boulpaep, E.L., 2005. Medical Physiology: A Cellular and Molecular Approach, updated 2nd ed.; (B) Modified from Barker, S.B. and Klitgaard, H.M. (1952) Am. J. Physiol. 170: 81.

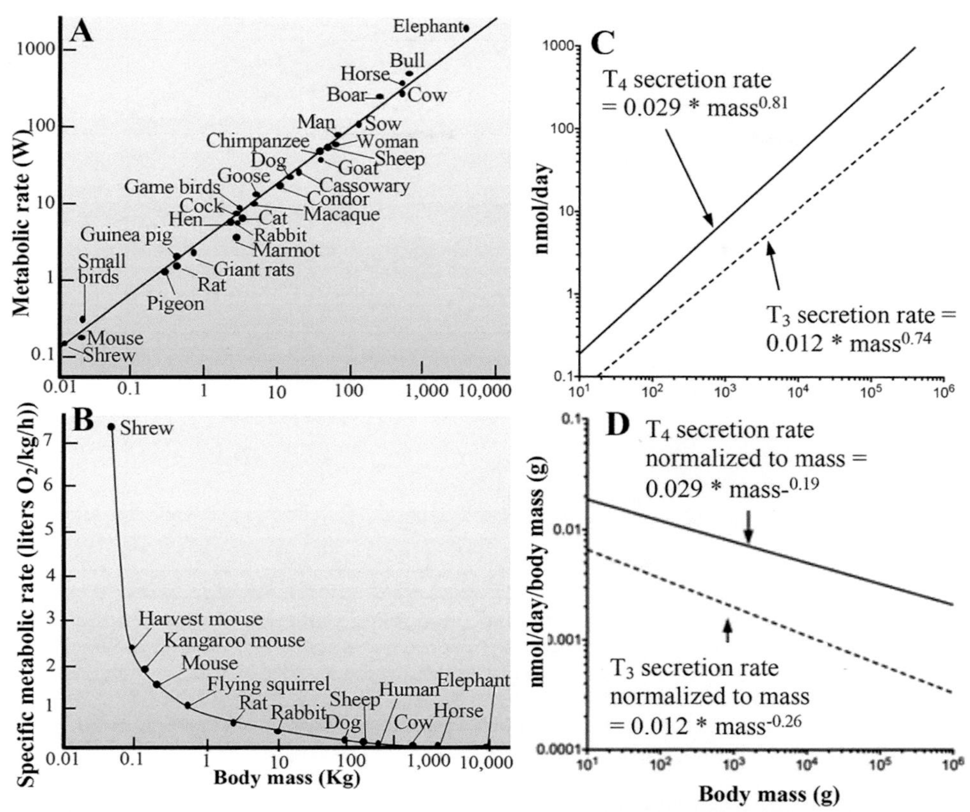

Figure 18.22 Relationship of metabolic rate and thyroid hormone secretion with body size among homeothermic vertebrates. Whereas total metabolic rate increases in proportion with body mass **(A)**, weight-specific metabolic rates are inversely proportional to body mass **(B)**. Similarly, although rates of thyroid hormone secretion correlate positively with body mass **(C)**, mass-specific rates of TH secretion correlate negatively with body mass **(D)**. Thyroxine (T_4), triiodothyronine (T_3). Analysis in D was extrapolated from data from Tomasi (1991) by A.M. Schreiber.

Source of images: (A-B) Modified from Schmidt-Nielsen (1984), Cambridge University Press; (C-D) Modified from Tomasi TE. Comp Biochem Physiol A Comp Physiol. 1991;100(3):503-516.

mass among species[99] (Figure 18.22), with an elephant having a larger total metabolism and rate of TH production than a shrew. Interestingly, the *mass-specific* rates of metabolism and thyroid hormone secretion correlate inversely with body mass, so compared with the elephant the higher rates of TH secretion by a shrew facilitate its higher mass-specific metabolism. Because the surface area-to-volume ratios of small animals are much higher compared with those of larger animals, the rate of heat loss for any animal will be inversely proportional to its body mass, with smaller animals losing heat more rapidly than large animals. To compensate for this disproportionate loss of heat, smaller animals must maintain higher mass-specific metabolic rates than larger animals, and hence increased rates of mass-specific TH secretion.

Synergistic Control of BAT Activity by TH and the Sympathetic Nervous System

The activation of BAT thermogenesis is induced primarily by the sympathetic nervous system (SNS) but requires the presence of TH for the full response to manifest.[100–102] Specifically, stimulation of *GPCR β $_3$-adrenergic receptors* on the plasma membranes of BAT cells by norepinephrine initiates an intracellular cAMP signaling response that facilitates two key processes. First, the activation of cAMP response element binding proteins (CREBS), which are nuclear transcription factors, induce the transcription of *type 2 deiodinase* (D2), the enzyme that converts T_4 to the more biologically active T_3 (Figure 18.23). Subsequently, the locally produced T_3 binds to receptors associated with TREs upstream of UCP-1, initiating its transcription. The second process initiated by cAMP is the liberation of fatty acids from triglycerides stored in fat vesicles, which will function as substrates combusted during UCP-1-catalyzed thermogenesis. An increase in mitochondria-localized UCP-1 in conjunction with elevated free fatty acid substrate facilitates the dissipation of energy in the form of heat. In contrast with TH's ability to upregulate β-adrenergic receptors in WAT,[82] it remains uncertain if TH exerts the same effect in BAT.[103]

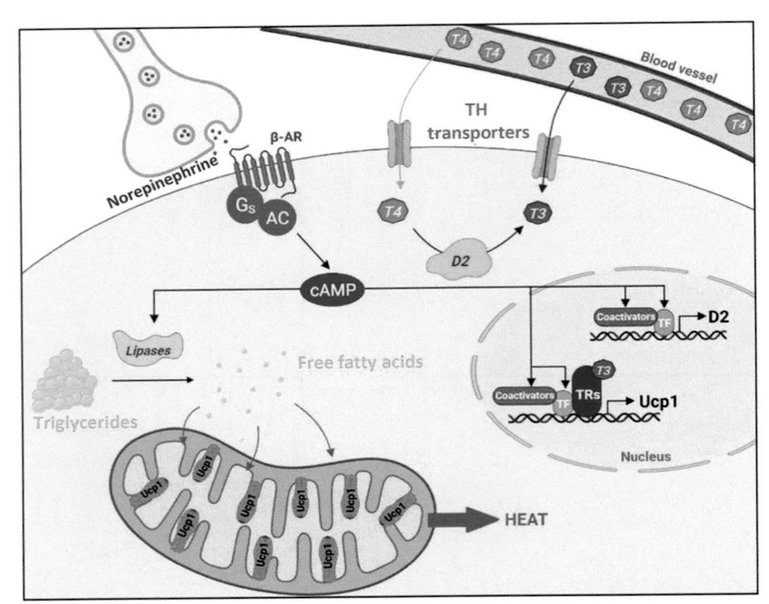

Figure 18.23 Classical view of thyroid hormones in modulating uncoupling oxidative phosphorylation and heat production in brown adipose tissue (BAT). Norepinephrine binding to the β3 adrenoreceptor (β-AR) elicits increased cAMP levels. Production of the second messenger, cAMP, promotes lipolysis and ultimately activates nuclear transcription factors called CREBS (cAMP response element binding proteins) that bind to their DNA elements (CRE) and induce the transcription of the enzyme "type 2 deiodinase" (D2). D2 converts the thyroid hormone thyroxine, T_4, into its more active form, T_3. T_3 binds to nuclear receptors attached to thyroid hormone response elements (TREs), inducing the transcription of uncoupling protein 1 (UCP1). The presence of free fatty acids promotes the activation of UCP1, causing the leakage of protons from the inner membrane of the mitochondria, hence dissipating energy in the form of heat. Thyroid hormones act on the brown adipocyte increasing the stimulatory action of norepinephrine on thermogenesis, as well as enhancing the cAMP-mediated acute rise in UCP1 gene expression by stimulating both gene transcription and mRNA half-life.

Source of images: Zekri, Y., et al. 2021. Central vs. Cells, 10(6), p. 1327.

Developments & Directions: 3,5-diiodothyronine (T_2) stimulates basal metabolic rate and BAT activity in hypothyroid rats

Long considered an inactive metabolite of thyroid hormones in mammals, *di*iodothyronine (T_2) has been shown to rapidly stimulate basal metabolic rate in hypothyroid rats without the thyrotoxic cardiac effects that accompany T_3 treatment and also prevent obesity

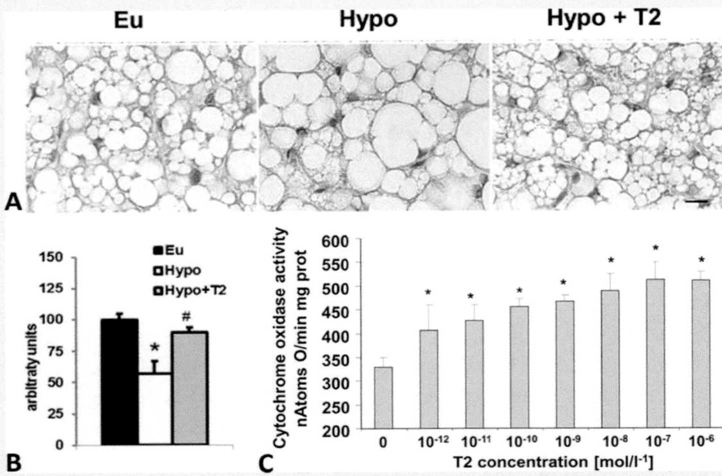

Figure 18.24 Effects of hypothyroidism and T_2 administration to Hypo rats. (A) BAT morphology. **(B)** BAT UCP1 content. **(C)** BAT cytochrome oxidase activity.

Source of images: Lombardi, A., et al. 2015. PLoS One, 10(2), p. e0116498.

when administered to rats receiving a high-fat diet.[104,105] Using euthyroid, hypothyroid, and T_2-treated hypothyroid rats, Lombardi and colleagues[106] found that hypothyroidism reduced features associated with BAT activity, such as enlarged cells and uniocular lipid droplets, reduced UCP1 content, and reduced mitochondrial oxidative capacity (Figure 18.24). Furthermore, *in vivo* administration of T_2 to hypothyroid rats activated BAT thermogenic features, suggesting that T_2, like T_3, is a thyroid hormone derivative able to activate BAT thermogenesis.[107]

SUMMARY AND SYNTHESIS QUESTIONS

1. What general effects does TH have on obligatory and facultative thermogenesis in birds and mammals?
2. Describe the molecular mechanisms by which TH interacts synergistically with catecholamines to modulate thermogenesis in BAT.

Summary of Chapter Learning Objectives and Key Concepts

LEARNING OBJECTIVE List the primary hormones that influence fuel homeostasis and describe their metabolic actions on skeletal muscle, liver, and adipose tissue.

- Metabolic hormones exert anabolic (build up) and/or catabolic (breakdown) effects on fuel substrates.
- On a short-term basis, the most important endocrine regulators of fuel metabolism are insulin, glucagon, and catecholamines.
- On a more prolonged time basis, glucocorticoids and growth hormone exert effects that require hours to days to manifest.
- Activation of glucagon and adrenergic receptors increases blood glucose by promoting glycogenolysis.
- Hepatic gluconeogenesis is promoted by the synergistic actions of glucagon and glucocorticoids.

LEARNING OBJECTIVE Describe how glucose is the most tightly regulated fuel substrate.

- Whereas acute hypoglycemia can rapidly inhibit cerebral function and muscle coordination and cause death, many of the pathological effects of hyperglycemia require chronic, long-term exposure to manifest.
- A major role of the counterregulatory hormones in energy homeostasis is to maintain blood glucose levels above the hypoglycemic threshold of 55–60 mg/dl in humans.
- A key role for insulin is to ensure that fasting blood glucose levels remain below the hyperglycemic threshold of 110 mg/dl in humans.

LEARNING OBJECTIVE List the pancreatic hormones and their roles in maintaining energy homeostasis.

- The endocrine portion of the pancreas, called islets of Langerhans, are composed of five cell types. The most numerous of these, the α and β cells, secrete glucagon and insulin, respectively.
- While both rising glucose and glucagon levels stimulate insulin secretion, rising insulin reduces blood glucose and suppresses glucagon secretion.
- Activation of insulin receptors facilitates glucose uptake by promoting intracellular signaling pathways that translocate GLUT4 glucose transporters to the plasma membrane of skeletal muscle cells and adipose cells.
- Inhibition of insulin receptor-mediated intracellular signaling leads to a phenomenon termed "insulin resistance".

LEARNING OBJECTIVE Describe the complementary actions of glucagon with adrenal glycemic hormones in maintaining energy homeostasis.

- Due to receptor and intracellular signaling pathway similarities, catecholamines and glucagon exert additive effects on glycogenolysis and lipolysis by hepatic, muscle, and adipose tissues.
- Glucocorticoids interact synergistically with glucagon signaling pathways to stimulate the transcription of key hepatic enzymes that mediate gluconeogenesis.

LEARNING OBJECTIVE Describe the three main categories of adipose tissue, their functions and their influence on energy homeostasis.

- Adipose tissue falls into three categories: (1) energy-storing white adipose tissue (WAT), (2) thermogenic brown adipose tissue (BAT), and (3) thermogenic "beige" adipose tissue (BeAT).
- In addition to functioning as a site for lipid storage and release, WAT is also an active endocrine organ that influences whole-body metabolism.
- Thermogenic adipose tissue generates heat by the presence of uncoupling proteins (UCPs) in the mitochondria inner membrane.

LEARNING OBJECTIVE Explain the roles TH plays in the metabolism of glucose, lipid, and protein, as well as modulating appetite.

- TH's metabolic effects are mediated by nuclear and mitochondrial receptors.
- TH generally enhances appetite and the metabolic rates of turnover of fuel substrates.
- Hyperthyroidism is associated with higher metabolic turnover rates that favor fuel consumption and reduced muscle and fat mass.
- Hypothyroidism promotes reduced metabolic turnover rates that favor increased fat storage.
- TH's effects on appetite and fuel metabolism are mediated via both central (brain) and peripheral mechanisms.

LEARNING OBJECTIVE Describe the hypermetabolic effects of thyroid hormones on mammals and birds, and their roles in facilitating non-shivering thermogenesis.

- TH is the primary hormone involved in regulating metabolic rate and thermogenesis in mammals and birds.
- The metabolic effects of TH manifest at the cellular level, with TH treatment producing increased oxygen consumption by a variety of tissues.
- Thyroid hormones affect basal metabolic rate, with hyperthyroidism increasing metabolic heat production and hypothyroidism decreasing it.
- Brown and beige adipose tissue are targets of TH for facilitating non-shivering thermogenesis.
- The activation of BAT thermogenesis is induced primarily by the sympathetic nervous system (SNS), but requires the presence of TH for the full response to manifest.

LITERATURE CITED

1. DuBois EF. Heat loss from the human body: harvey lecture, December 15, 1938. *Bull N Y Acad Med.* 1939;15(3):143–173.
2. Saltiel AR, Kahn CR. Insulin signalling and the regulation of glucose and lipid metabolism. *Nature.* 2001;414(6865):799–806.
3. Slavin BG, Ong JM, Kern PA. Hormonal regulation of hormone-sensitive lipase activity and mRNA levels in isolated rat adipocytes. *J Lipid Res.* 1994;35(9):1535–1541.
4. Duncan RE, Ahmadian M, Jaworski K, Sarkadi-Nagy E, Sul HS. Regulation of lipolysis in adipocytes. *Annu Rev Nutr.* 2007;27:79–101.
5. Adeva-Andany MM, Funcasta-Calderón R, Fernández-Fernández C, Castro-Quintela E, Carneiro-Freire N. Metabolic effects of glucagon in humans. *J Clin Transl Endocrinol.* 2019;15:45–53.
6. Moller N, Jorgensen JO. Effects of growth hormone on glucose, lipid, and protein metabolism in human subjects. *Endocr Rev.* 2009;30(2):152–177.
7. Gerich JE. Control of glycaemia. *Baillieres Clin Endocrinol Metab.* 1993;7(3):551–586.
8. Mitrakou A, Ryan C, Veneman T, et al. Hierarchy of glycemic thresholds for counterregulatory hormone secretion, symptoms, and cerebral dysfunction. *Am J Physiol.* 1991;260(1 Pt 1):E67–74.
9. Jarrett RJ, Keen H. Hyperglycaemia and diabetes mellitus. *Lancet.* 1976;2(7993):1009–1012.
10. Tominaga M, Eguchi H, Manaka H, Igarashi K, Kato T, Sekikawa A. Impaired glucose tolerance is a risk factor for cardiovascular disease, but not impaired fasting glucose. The Funagata Diabetes Study. *Diabetes Care.* 1999;22(6):920–924.
11. Tchobroutsky G. Relation of diabetic control to development of microvascular complications. *Diabetologia.* 1978;15(3):143–152.
12. Motta PM, Macchiarelli G, Nottola SA, Correr S. Histology of the exocrine pancreas. *Microsc Res Tech.* 1997;37(5–6):384–398.
13. Jorgensen MC, Ahnfelt-Ronne J, Hald J, Madsen OD, Serup P, Hecksher-Sorensen J. An illustrated review of early pancreas development in the mouse. *Endocr Rev.* 2007;28(6):685–705.
14. Brissova M, Fowler MJ, Nicholson WE, et al. Assessment of human pancreatic islet architecture and composition by laser scanning confocal microscopy. *J Histochem Cytochem.* 2005;53(9):1087–1097.
15. Cabrera O, Berman DM, Kenyon NS, Ricordi C, Berggren PO, Caicedo A. The unique cytoarchitecture of human pancreatic islets has implications for islet cell function. *Proc Natl Acad Sci U S A.* 2006;103(7):2334–2339.
16. Bliss M. *The Discovery of Insulin: Twenty-fifth Anniversary Edition.* University of Chicago Press; 2007.
17. Rustenbeck I, Wienbergen A, Bleck C, Jörns A. Desensitization of insulin secretion by depolarizing insulin secretagogues. *Diabetes.* 2004;53(Suppl 3):S140–150.
18. Iynedjian PB. Mammalian glucokinase and its gene. *Biochem J.* 1993;293(Pt 1):1–13.
19. Henquin JC, Meissner HP. Effects of amino acids on membrane potential and 86Rb+ fluxes in pancreatic beta-cells. *Am J Physiol.* 1981;240(3):E245–E252.
20. Furman BL. The development of Byetta (exenatide) from the venom of the Gila monster as an anti-diabetic agent. *Toxicon.* 2012;59(4):464–471.
21. Eng J, Andrews PC, Kleinman WA, Singh L, Raufman JP. Purification and structure of exendin-3, a new pancreatic secretagogue isolated from Heloderma horridum venom. *J Biol Chem.* 1990;265(33):20259–20262.
22. Yamazaki S, Katada T, Ui M. Alpha 2-adrenergic inhibition of insulin secretion via interference with cyclic AMP generation in rat pancreatic islets. *Mol Pharmacol.* 1982;21(3):648–653.
23. Peterhoff M, Sieg A, Brede M, Chao CM, Hein L, Ullrich S. Inhibition of insulin secretion via distinct signaling pathways in alpha2-adrenoceptor knockout mice. *Eur J Endocrinol.* 2003;149(4):343–350.
24. Nestler JE, McClanahan MA. Diabetes and adrenal disease. *Baillieres Clin Endocrinol Metab.* 1992;6(4):829–847.
25. Sherwin RS, Shamoon H, Hendler R, Sacca L, Eigler N, Walesky M. Epinephrine and the regulation of glucose metabolism: effect of diabetes and hormonal interactions. *Metabolism.* 1980;29(11 Suppl 1):1146–1154.
26. Colwell JA. Inhibition of insulin secretion by catecholamines in pheochromocytoma. *Ann Intern Med.* 1969;71(2):251–256.
27. Kawai K, Yokota C, Ohashi S, Watanabe Y, Yamashita K. Evidence that glucagon stimulates insulin secretion through its own receptor in rats. *Diabetologia.* 1995;38(3):274–276.
28. Dalle S, Fontes G, Lajoix AD, et al. Miniglucagon (glucagon 19–29): a novel regulator of the pancreatic islet physiology. *Diabetes.* 2002;51(2):406–412.
29. Dalle S, Smith P, Blache P, et al. Miniglucagon (glucagon 19–29), a potent and efficient inhibitor of secretagogue-induced insulin release through a Ca2+ pathway. *J Biol Chem.* 1999;274(16):10869–10876.
30. Kimball C, Murlin J. Aqueous extracts of pancreas III. Some precipitatoion reaction of insulin. *J Biol Chem.* 1923;58:337–346.
31. Kawamori D, Kurpad AJ, Hu J, et al. Insulin signaling in alpha cells modulates glucagon secretion in vivo. *Cell Metab.* 2009;9(4):350–361.
32. Lee YH, Wang MY, Yu XX, Unger RH. Glucagon is the key factor in the development of diabetes. *Diabetologia.* 2016;59(7):1372–1375.
33. Lee Y, Wang MY, Du XQ, Charron MJ, Unger RH. Glucagon receptor knockout prevents insulin-deficient type 1 diabetes in mice. *Diabetes.* 2011;60(2):391–397.
34. Wang MY, Yan H, Shi Z, et al. Glucagon receptor antibody completely suppresses type 1 diabetes phenotype without insulin by disrupting a novel diabetogenic pathway. *Proc Natl Acad Sci U S A.* 2015;112(8):2503–2508.
35. Campbell JE, Drucker DJ. Islet alpha cells and glucagon—critical regulators of energy homeostasis. *Nat Rev Endocrinol.* 2015;11(6):329–338.
36. Lacey RJ, Berrow NS, Scarpello JH, Morgan NG. Selective stimulation of glucagon secretion by beta 2-adrenoceptors

in isolated islets of Langerhans of the rat. *Br J Pharmacol.* 1991;103(3): 1824–1828.
37. Gannon MC, Nuttall FQ. Amino acid ingestion and glucose metabolism—a review. *IUBMB Life.* 2010;62(9):660–668.
38. Gustavson SM, Chu CA, Nishizawa M, et al. Interaction of glucagon and epinephrine in the control of hepatic glucose production in the conscious dog. *Am J Physiol-Endocrinol Metab.* 2003;284(4):E695–E707.
39. Jitrapakdee S. Transcription factors and coactivators controlling nutrient and hormonal regulation of hepatic gluconeogenesis. *Int J Biochem Cell Biol.* 2012;44(1):33–45.
40. Ghaben AL, Scherer PE. Adipogenesis and metabolic health. *Nat Rev Mol Cell Biol.* 2019:1.
41. Barak Y, Nelson MC, Ong ES, et al. PPARγ is required for placental, cardiac, and adipose tissue development. *Mol Cell.* 1999;4(4):585–595.
42. Rosen ED, Sarraf P, Troy AE, et al. PPARγ is required for the differentiation of adipose tissue in vivo and in vitro. *Mol Cell.* 1999;4(4):611–617.
43. Kubota N, Terauchi Y, Miki H, et al. PPARγ mediates high-fat diet—induced adipocyte hypertrophy and insulin resistance. *Mol Cell.* 1999;4(4):597–609.
44. Ahmadian M, Suh JM, Hah N, et al. PPARγ signaling and metabolism: the good, the bad and the future. *Nat Med.* 2013;19(5):557–566.
45. Kershaw EE, Flier JS. Adipose tissue as an endocrine organ. *J Clin Endocrinol Metab.* 2004;89(6):2548–2556.
46. Fischer-Posovszky P, Wabitsch M, Hochberg Z. Endocrinology of adipose tissue—an update. *Horm Metab Res.* 2007;39(5):314–321.
47. Coelho M, Oliveira T, Fernandes R. Biochemistry of adipose tissue: an endocrine organ. *Arch Med Sci.* 2013;9(2):191–200.
48. Thomou T, Mori MA, Dreyfuss JM, et al. Adipose-derived circulating miRNAs regulate gene expression in other tissues. *Nature.* 2017;542(7642):450.
49. Enerback S, Jacobsson A, Simpson EM, et al. Mice lacking mitochondrial uncoupling protein are cold-sensitive but not obese. *Nature.* 1997;387(6628):90–94.
50. van Marken Lichtenbelt WD, Vanhommerig JW, Smulders NM, et al. Cold-activated brown adipose tissue in healthy men. *N Engl J Med.* 2009;360(15):1500–1508.
51. Nedergaard J, Bengtsson T, Cannon B. Unexpected evidence for active brown adipose tissue in adult humans. *Am J Physiol Endocrinol Metab.* 2007;293(2):E444–452.
52. Ikeda K, Maretich P, Kajimura S. The common and distinct features of brown and beige adipocytes. *Trends Endocrinol Metab.* 2018;29(3):191–200.
53. Cypess AM, Lehman S, Williams G, et al. Identification and importance of brown adipose tissue in adult humans. *N Engl J Med.* 2009;360(15):1509–1517.
54. Vidal P, Stanford KI. Exercise-induced adaptations to adipose tissue thermogenesis. *Front Endocrinol.* 2020;11:270.
55. Chen Y, Pan R, Pfeifer A. Fat tissues, the brite and the dark sides. *Pflügers Archi Eur J Physiol.* 2016;468(11–12):1803–1807.
56. Yoneshiro T, Aita S, Matsushita M, et al. Recruited brown adipose tissue as an antiobesity agent in humans. *J Clin Investig.* 2013;123(8):3404–3408.
57. Chondronikola M, Volpi E, Børsheim E, et al. Brown adipose tissue improves whole-body glucose homeostasis and insulin sensitivity in humans. *Diabetes.* 2014;63(12):4089–4099.
58. Lee P, Smith S, Linderman J, et al. Temperature-acclimated brown adipose tissue modulates insulin sensitivity in humans. *Diabetes.* 2014;63(11):3686–3698.
59. Hanssen MJ, van der Lans AA, Brans B, et al. Short-term cold acclimation recruits brown adipose tissue in obese humans. *Diabetes.* 2016;65(5):1179–1189.
60. Dewal RS, Stanford KI. Effects of exercise on brown and beige adipocytes. *Biochim Biophys Acta Mol Cell Biol Lipids.* 2019;1864(1):71–78.
61. Zhang Y, Xie C, Wang H, et al. Irisin exerts dual effects on browning and adipogenesis of human white adipocytes. *Am J Physiol-Endocrinol Metab.* 2016;311(2):E530–E541.
62. Wu L, Xia M, Duan Y, et al. Berberine promotes the recruitment and activation of brown adipose tissue in mice and humans. *Cell Death Dis.* 2019;10(6):1–18.
63. van Marken Lichtenbelt W. Human brown fat and obesity: methodological aspects. *Front Endocrinol (Lausanne).* 2011;2:52.
64. Yoneshiro T, Aita S, Matsushita M, et al. Age-related decrease in cold-activated brown adipose tissue and accumulation of body fat in healthy humans. *Obesity (Silver Spring).* 2011;19(9):1755–1760.
65. van Marken Lichtenbelt WD. Human brown adipose tissue—a decade later. *Obesity (Silver Spring, Md).* 2021;29(7):1099.
66. Kulterer OC, Herz CT, Prager M, et al. Brown adipose tissue prevalence is lower in obesity but its metabolic activity is intact. *Front Endocrinol.* 2022;13.
67. Sugatani J, Sadamitsu S, Yamaguchi M, et al. Antiobese function of platelet-activating factor: increased adiposity in platelet-activating factor receptor-deficient mice with age. *FASEB J.* 2014;28(1):440–452.
68. Mullur R, Liu YY, Brent GA. Thyroid hormone regulation of metabolism. *Physiol Rev.* 2014;94(2):355–382.
69. Tata JR. Inhibition of the biological action of thyroid hormones by actinomycin D and puromycin. *Nature.* 1963;197:1167–1168.
70. Golozoubova V, Gullberg H, Matthias A, Cannon B, Vennstrom B, Nedergaard J. Depressed thermogenesis but competent brown adipose tissue recruitment in mice devoid of all hormone-binding thyroid hormone receptors. *Mol Endocrinol.* 2004;18(2):384–401.
71. Morel G, Ricard-Blum S, Ardail D. Kinetics of internalization and subcellular binding sites for T3 in mouse liver. *Biol Cell.* 1996;86(2–3):167–174.
72. Almeida A, Orfao A, Lopez-Mediavilla C, Medina JM. Hypothyroidism prevents postnatal changes in rat liver mitochondrial populations defined by rhodamine-123 staining. *Endocrinology.* 1995;136(10):4448–4453.
73. Wrutniak-Cabello C, Casas F, Cabello G. Thyroid hormone action in mitochondria. *J Mol Endocrinol.* 2001;26(1):67–77.
74. Arnold S, Goglia F, Kadenbach B. 3,5-Diiodothyronine binds to subunit Va of cytochrome-c oxidase and abolishes the allosteric inhibition of respiration by ATP. *Eur J Biochem.* 1998;252(2):325–330.
75. Goglia F, Lanni A, Barth J, Kadenbach B. Interaction of diiodothyronines with isolated cytochrome c oxidase. *FEBS Lett.* 1994;346(2–3):295–298.
76. Potenza M, Via MA, Yanagisawa RT. Excess thyroid hormone and carbohydrate metabolism. *Endocr Pract.* 2009;15(3):254–262.
77. Oppenheimer JH, Schwartz HL, Lane JT, Thompson MP. Functional relationship of thyroid hormone-induced lipogenesis, lipolysis, and thermogenesis in the rat. *J Clin Invest.* 1991;87(1):125–132.
78. Ito M, Takamatsu J, Matsuo T, et al. Serum concentrations of remnant-like particles in hypothyroid patients before and after thyroxine replacement. *Clin Endocrinol (Oxf).* 2003;58(5):621–626.
79. Fugier C, Tousaint JJ, Prieur X, Plateroti M, Samarut J, Delerive P. The lipoprotein lipase inhibitor ANGPTL3 is negatively regulated by thyroid hormone. *J Biol Chem.* 2006;281(17):11553–11559.
80. Darimont C, Gaillard D, Ailhaud G, Negrel R. Terminal differentiation of mouse preadipocyte cells: adipogenic and antimitogenic role of triiodothyronine. *Mol Cell Endocrinol.* 1993;98(1):67–73.
81. Hauner H, Entenmann G, Wabitsch M, et al. Promoting effect of glucocorticoids on the differentiation of human adipocyte precursor cells cultured in a chemically defined medium. *J Clin Invest.* 1989;84(5):1663–1670.
82. Viguerie N, Millet L, Avizou S, Vidal H, Larrouy D, Langin D. Regulation of human adipocyte gene expression by thyroid hormone. *J Clin Endocrinol Metab.* 2002;87(2):630–634.
83. Wahrenberg H, Wennlund A, Arner P. Adrenergic regulation of lipolysis in fat cells from hyperthyroid and hypothyroid patients. *J Clin Endocrinol Metab.* 1994;78(4):898–903.
84. Endo T, Kobayashi T. Expression of functional TSH receptor in white adipose

tissues of hyt/hyt mice induces lipolysis in vivo. *Am J Physiol Endocrinol Metab.* 2012;302(12):E1569–1575.

85. Gagnon A, Antunes TT, Ly T, et al. Thyroid-stimulating hormone stimulates lipolysis in adipocytes in culture and raises serum free fatty acid levels in vivo. *Metabolism.* 2010;59(4):547–553.
86. Rodbell M. Metabolism of isolated fat cells. I. Effects of hormones on glucose metabolism and lipolysis. *J Biol Chem.* 1964;239:375–380.
87. Caraccio N, Natali A, Sironi A, et al. Muscle metabolism and exercise tolerance in subclinical hypothyroidism: a controlled trial of levothyroxine. *J Clin Endocrinol Metab.* 2005;90(7): 4057–4062.
88. Marcus C, Ehren H, Bolme P, Arner P. Regulation of lipolysis during the neonatal period. Importance of thyrotropin. *J Clin Invest.* 1988;82(5): 1793–1797.
89. Goldberg AL, Tischler M, DeMartino G, Griffin G. Hormonal regulation of protein degradation and synthesis in skeletal muscle. *Fed Proc.* 1980;39(1):31–36.
90. Garrel DR, Todd KS, Pugeat MM, Calloway DH. Hormonal changes in normal men under marginally negative energy balance. *Am J Clin Nutr.* 1984;39(6):930–936.
91. Reed HL, Ferreiro JA, Mohamed Shakir KM, Burman KD, O'Brian JT. Pituitary and peripheral hormone responses to T3 administration during Antarctic residence. *Am J Physiol.* 1988;254(6 Pt 1):E733–739.
92. DeMartino GN, Goldberg AL. Thyroid hormones control lysosomal enzyme activities in liver and skeletal muscle. *Proc Natl Acad Sci U S A.* 1978;75(3):1369–1373.
93. Muller MJ, Seitz HJ. Thyroid hormone action on intermediary metabolism. Part III. Protein metabolism in hyper- and hypothyroidism. *Klin Wochenschr.* 1984;62(3):97–102.
94. Magnus-Levey A. Ueber den respiratorischen Gaswechsel unter Einfluss der Thyroidea sowie unter verschiedenen pathologische Zustand. *Berl Klin Wochenschr.* 1895;32:650–652.
95. DuBois E. *Basal Metabolism in Health and Disease.* Lea and Febiger; 1936.
96. Bianco AC, Silva JE. Intracellular conversion of thyroxine to triiodothyronine is required for the optimal thermogenic function of brown adipose tissue. *J Clin Invest.* 1987;79(1):295–300.
97. Astrup A, Buemann B, Toubro S, Ranneries C, Raben A. Low resting metabolic rate in subjects predisposed to obesity: a role for thyroid status. *Am J Clin Nutr.* 1996;63(6):879–883.
98. Harper ME, Seifert EL. Thyroid hormone effects on mitochondrial energetics. *Thyroid.* 2008;18(2):145–156.
99. Tomasi TE. Utilization rates of thyroid hormones in mammals. *Comp Biochem Physiol A Comp Physiol.* 1991;100(3):503–516.
100. Cannon B, Nedergaard J. Brown adipose tissue: function and physiological significance. *Physiol Rev.* 2004;84(1):277–359.
101. Silva JE. Thermogenic mechanisms and their hormonal regulation. *Physiol Rev.* 2006;86(2):435–464.
102. Silva JE. Physiological importance and control of non-shivering facultative thermogenesis. *Front Biosci (Schol Ed).* 2011;3:352–371.
103. Sentis SC, Oelkrug R, Mittag J. Thyroid hormones in the regulation of brown adipose tissue thermogenesis. *Endocr Connect.* 2021;10(2): R106-R115.
104. Moreno M, de Lange P, Lombardi A, Silvestri E, Lanni A, Goglia F. Metabolic effects of thyroid hormone derivatives. *Thyroid.* 2008;18(2):239–253.
105. Lombardi A, de Lange P, Silvestri E, et al. 3,5-Diiodo-L-thyronine rapidly enhances mitochondrial fatty acid oxidation rate and thermogenesis in rat skeletal muscle: AMP-activated protein kinase involvement. *Am J Physiol Endocrinol Metab.* 2009;296(3):E497–502.
106. Lombardi A, Senese R, De Matteis R, et al. 3, 5-Diiodo-L-thyronine activates brown adipose tissue thermogenesis in hypothyroid rats. *PLoS One.* 2015;10(2):e0116498.
107. Cioffi F, Gentile A, Silvestri E, Goglia F, Lombardi A. Effect of iodothyronines on thermogenesis: focus on brown adipose tissue. *Front Endocrinol.* 2018;9:254.

CHAPTER

19

Metabolic Dysregulation and Disruption

CHAPTER LEARNING OBJECTIVES:

- List the types of diseases that are associated with metabolic dysfunction, and define "metabolic syndrome."
- Describe the "developmental origins of health and disease", the "thrifty phenotype" hypothesis, and how these concepts are useful for explaining the global rise in metabolic disease.
- Define "obesity", explain how it is a risk factor for developing other metabolic disorders, and contrast the endocrine function of adipose tissue in the lean versus obese states.
- Define "diabetes mellitus" and describe the physiological implications of chronic hyperglycemia.
- Describe the etiology (cause) of type 1 diabetes mellitus (T1DM) and its symptoms and complications.
- Describe the etiology of type 2 diabetes mellitus (T2DM) and its symptoms, complications, and link to obesity.
- Define the terms "obesogen", "diabetogen", and "diabesogen", and provide some examples of these compounds and their mechanisms of action based on studies with humans and animal models.

OPENING QUOTATIONS:

"Corpulence is not only a disease itself, but the harbinger of others."

—Hippocrates (460–375 BCE),[1] recognizing that obesity is a medical disorder that also leads to many comorbidities

"Over the past 10,000 years, humans have led changes in life conditions. Our genotype has remained anchored in the Stone Age, while our phenotype advances through the 21st century at breakneck speed. The conflict has resulted in the so-called 'diseases of civilization', most of which belong to the realm of endocrinology."

—Carles Zafón[2]

Introduction

LEARNING OBJECTIVE List the types of diseases that are associated with metabolic dysfunction, and define "metabolic syndrome."

KEY CONCEPTS:

- The dysregulation of energy homeostasis may lead to the development of chronic metabolic disorders that include obesity, type 2 diabetes, and cardiovascular disease.
- Genetic susceptibility and "life-style" risk factors associated with overeating and sedentary behavior are contributors to metabolic disorders. However, these factors alone do not explain the rapidly rising rates of the global metabolic disease epidemic over just the last few decades.
- The "metabolic syndrome" describes the presence of three or more metabolic risk factors in a patient that put them at increased risk for developing type 2 diabetes mellitus, cardiovascular disease, and other metabolic disorders.

Metabolic disorders, such as obesity, type 2 diabetes, dyslipidemia, hypertension, atherosclerosis, and hepatic steatosis are examples of chronic conditions that manifest largely from the dysregulations of energy homeostasis. Importantly, these metabolic disorders often manifest as comorbidities, or the simultaneous presence of two or more diseases, with one disorder exacerbating or increasing the probability of developing others. The term *metabolic syndrome* describes the presence of three or more risk factors in a patient, which include abdominal obesity, high triglycerides, low- and high-density lipoprotein cholesterol, high blood pressure, and elevated fasting blood glucose (Table 19.1).[3] Individuals with metabolic syndrome are at increased risk for developing type 2 diabetes mellitus and cardiovascular disease. For example, when obesity occurs in conjunction with insulin resistance, dyslipidemia, and hypertension, these increase the risk of developing type 2 diabetes by about five-fold, and cardiovascular disease by three-fold.[4] Although genetic susceptibility and "life-style" risk factors associated with overeating and sedentary behavior are contributors to the aforementioned metabolic disorders, these factors alone do not explain the magnitude or rapidly rising rates of the

DOI: 10.1201/9781003359241-25

Table 19.1 Clinical Diagnosis of the Metabolic Syndrome

Risk Factors	Defining Level
Abdominal obesity as waist circumference	
Men	>102 cm (>40 in)
Women	>88 cm (>35 in)
Triglycerides	≥150 mg/dL
HDL cholesterol	
Men	<40 mg/dL
Women	<50 mg/dL
Blood pressure	≥130/≥85 mm Hg
Fasting glucose	≥110 mg/dL

Note: BMI, body mass index; HDL, high-density lipoprotein
Source: After Sherling DH, Perumareddi P, Hennekens CH. Metabolic Syndrome: Clinical and Policy Implications of the New Silent Killer. *Journal of Cardiovascular Pharmacology and Therapeutics*. 2017;22(4):365–367.

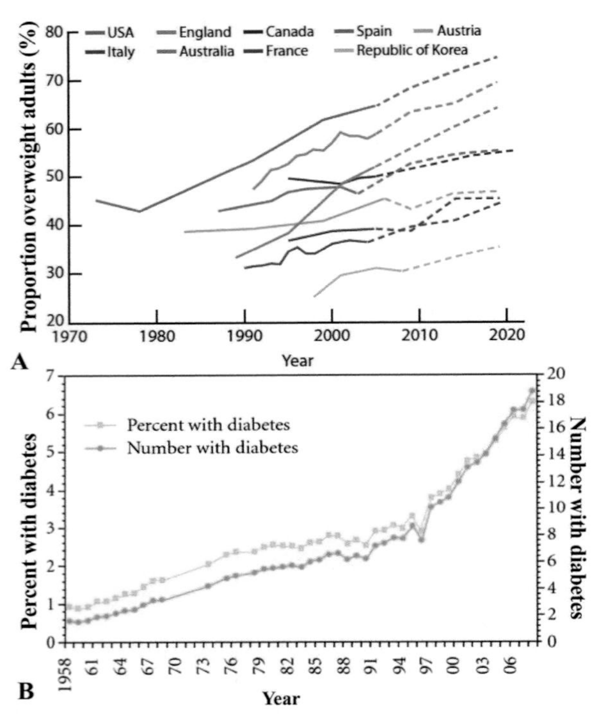

Figure 19.1 The steady increase in incidence of two metabolic disorders: overweight and type 2 diabetes. (A) Past (solid lines) and projected (dashed lines) overweight rates in selected countries. (B) Number and percentage of U.S. population diagnosed with type 2 diabetes from 1958 to 2013.

Source of images: (A) UNEP, 2013. State of the science of endocrine disrupting chemicals-2012. WHO-UNEP, Geneva; Used by permission. (B) Diabetes statistics. CDC Web site (https://www.cdc.gov/diabetes/statistics/slides/long_term_trends.pdf).

global metabolic disease epidemic over just the last few decades (Figure 19.1).

This chapter begins by describing some important hypotheses on how metabolic phenotypes may predispose humans to be maladapted to the modern world of caloric abundance. The chapter then considers the endocrine bases of three key metabolic disorders: obesity, type 1 diabetes mellitus, and type 2 diabetes mellitus. The chapter ends by considering the possible contributions of endocrine-disrupting compounds to the development of some of these conditions.

SUMMARY AND SYNTHESIS QUESTIONS

1. How do we know that genetic susceptibility alone cannot be the primary contributor to increasing rates of global obesity?

Early Life Programming of Metabolic Disorders

LEARNING OBJECTIVE Describe the "developmental origins of health and disease", the "thrifty phenotype" hypothesis, and how these concepts are useful for explaining the global rise in metabolic disease.

KEY CONCEPTS:

- "Adaptive developmental plasticity" describes environment-dependent developmental strategies that an organism can take to maximize its probability of survival and reproduction in a new environment.
- When a resulting developmental phenotype produces a mismatch with the actual environment, this can lead to a loss of fitness.
- The "thrifty phenotype" hypothesis proposes that in adapting to an adverse intrauterine environment, such as under- or overnutrition, the fetus permanently alters its physiology and metabolism, potentially increasing its risk of developing obesity, type 2 diabetes, cardiovascular disease, and other metabolic disorders later in life.

Adaptive Developmental Plasticity and Environmental Mismatch

Adaptive developmental plasticity is the capacity of the same genotype to produce variable phenotypic outcomes depending upon inputs received from the internal or external environment during earlier development. Such an environmental dependency of phenotype, whether expressed as variation in physiology, morphology, behavior, life history, or susceptibility to disease, is collectively termed *phenotypic plasticity*.[5,6] In mammals, the theory most often hinges upon the tenet that the quality of the future postnatal environment is forecasted on the basis of cues received during the prenatal period, with the goal of fine-tuning the fetus's phenotype to optimize its ability to survive the challenges of the predicted environment. For example, if a mouse fetus develops in a malnourished intrauterine environment, its resulting phenotype may be programmed for survival in an anticipated stressful postnatal world where nutrition is difficult to obtain. Examples of desirable phenotypes for surviving to adulthood and reproducing in such a challenging environment might include an enhanced propensity for storing fuel as fat in combination with a low metabolic rate for conserving energy, as well as elevated blood glucose, blood lipids, and blood pressure for rapidly mobilizing energy in a stressful world. However, if instead there exists a mismatch between the fetal and postnatal environments such that offspring developmentally programmed to exhibit "thrifty phenotypes" are instead born into a world characterized by excess nutrition and a sedentary lifestyle, then the aforementioned phenotypes become liabilities to long-term health. For example, under conditions of excess nutrition, an efficient fat storage and low metabolism increase the probability of developing obesity. Furthermore, elevated blood glucose levels may promote diabetes, and high blood lipids and blood pressure may lead to cardiovascular disease.

The notion that subtle functional changes in specific tissues during sensitive developmental windows of early life can result in increased susceptibility to disease or dysfunction later in life laid the foundation for an influential paradigm for understanding non-communicable disease known as the **developmental origins of health and disease** (DOHaD).[7] For example, over- or undernutrition of a pregnant woman is known to influence the fetus's propensity to develop obesity and diabetes, cardiovascular disease, and reproductive dysregulation later in life.[8–11] Similarly, the notion that perinatal exposure to endocrine-disrupting chemicals (EDCs) can also lead to the manifestation of disease later in the adult also features prominently in DOHaD theory,[12–14] a topic that is explored further in Unit VIII: Endocrine-Disrupting Compounds.

Fetal Programming and the "Thrifty Phenotype" Hypothesis

The idea that in adapting to an adverse intrauterine environment the fetus permanently alters its physiology and metabolism, potentially increasing its risk of developing disease later in life, was first proposed by David Barker and Charles Nicholas Hales as a way to explain the results of epidemiological studies showing associations between low fetal birth weight and elevated risk of developing type 2 diabetes mellitus, hypertension, elevated triacylglycerols, insulin resistance and metabolic syndrome later in adult life[15–17] (Figure 19.2). Hales and Barker originally proposed a **"thrifty phenotype" hypothesis** for the development of type 2 diabetes mellitus, whereby fetal malnutrition programs develop, at least in part, by promoting hypoplasia (reduced cell division) of the endocrine pancreas, resulting in insulin hyposecretion by the pancreatic beta cells (Figure 19.3). Reduced insulin secretion facilitates fetal adaptation to malnourishment by reducing its rates of somatic nutrient uptake, general growth, and metabolism, thereby conserving and prioritizing the limited supply of nutrients for brain development and future survival. Importantly, if the malnourished fetus is born into a world characterized by an abundance of food, then the child will exhibit a period of rapid "catch-up growth" and its thrifty phenotype programming may become a liability that contributes to the development of obesity and other metabolic dysfunctions later in life. The thrifty phenotype hypothesis has been expanded to also highlight the potential effects and consequences of fetal malnutrition to changes in growth, metabolism, and vascularization of other organs, including the kidney and the hypothalamic-pituitary-adrenal axis[18,19] (Figure 19.3). Indeed, research using non-human primate and rodent models has shown that adverse maternal and perinatal environments impair

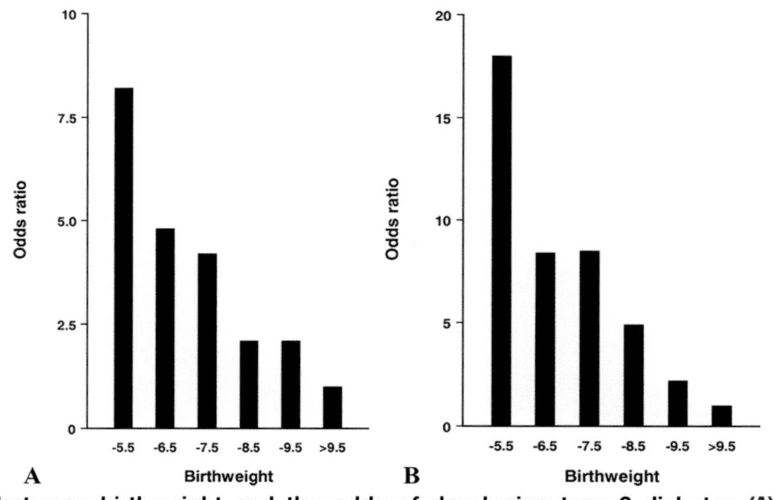

Figure 19.2 Relationship between birthweight and the odds of developing type 2 diabetes (A) or metabolic syndrome (B) among men aged 64 years born in Hertfordshire, United Kingdom (adjusted for adult body mass index). N = 370 (left) and 407 (right).

Source of images: Hales, C.N. and Barker, D.J., 2001. Brit. Med. Bull., 60(1), pp. 5-20. Used by permission.

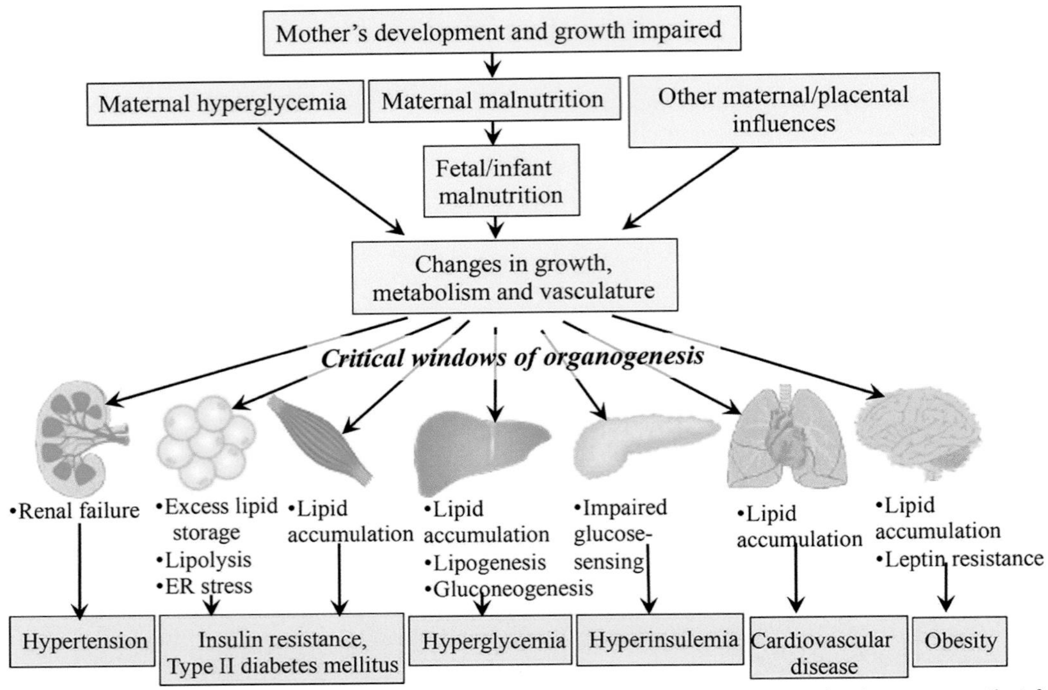

Figure 19.3 Diagrammatic representation of the "thrifty phenotype hypothesis". The hypothesis proposes that fetal and infant malnutrition promotes altered growth and development of key organs involved in maintaining energy homeostasis. These changes may manifest as diverse metabolic dysfunctions in the adult that increase the probability of developing metabolic syndrome, ultimately compounding the risk of developing type 2 diabetes, cardiovascular disease, and other metabolic disorders.

Source of images: Hales, C.N. and Barker, D.J., 2001. Brit. Med. Bull., 60(1), pp. 5-20.

the development of the hypothalamus, a region critical for energy balance regulation, potentially influencing altered metabolic phenotypes in later life.[20]

To study the thrifty phenotype hypothesis, scientists rely primarily on animal models and small sample sizes of correlational data gathered from human pregnancies where fetal growth has been abnormal for various reasons. However, retrospective analysis of records from horrific events associated with large-scale human starvations suggest that exposure to famine exerts both gestational stage dependent and independent effects on future adult health, supporting the hypothesis.[21–26] Although the thrifty phenotype hypothesis was originally developed to study the link between fetal undernutrition and the development of metabolic disease later in life, it is important to note that fetal *over*nutrition can also trigger metabolic disease in later life.[27–29]

SUMMARY AND SYNTHESIS QUESTIONS

1. Describe the basis of a field of medicine known as the developmental origins of health and disease (DOHaD).
2. How is the "thrifty phenotype" hypothesis an example of developmental plasticity?

Obesity and the Endocrinology of Fat

LEARNING OBJECTIVE Define "obesity", explain how it is a risk factor for developing other metabolic disorders, and contrast the endocrine function of adipose tissue in the lean versus obese states.

KEY CONCEPTS:

- Obesity stems from complex interactions among behavioral, environmental, physiological, genetic, epigenetic, social, and economic variables.
- Rather than simply arising from the passive accumulation of excess weight, obesity appears to be a disorder of the energy homeostasis system.
- White adipose tissue is an active endocrine organ, and in the obese state potentially the largest in the body.
- Under conditions of obesity, the white adipose tissue endocrine profile shifts, with increased production of hormones that promote insulin resistance, hyperglycemia, and inflammation.
- Due to its distinct morphological and biochemical features, visceral obesity is more dangerous to health than subcutaneous obesity.

What Is Obesity?

The most widespread and preventable dysregulation of energy homeostasis in humans today is likely **obesity**, broadly defined as excessive fat accumulation that presents a risk to health.[30] An accurate quantification of body-fat mass requires the use of sophisticated and expensive imaging tools such as magnetic resonance imaging, bio-electrical impedance, isotope dilution, and dual energy X-ray absorptiometry. However, a broadly used, albeit increasingly controversial, surrogate measure of adiposity is *body mass index* (BMI), which expresses total body weight as a function of body height:

Importantly, although BMI is a cheap, convenient, and widely used measurement, it does not take into account the distribution of fat mass in different body sites, muscle mass, bone density, overall body composition, and ethnic and sex differences, and is increasingly viewed as an inaccurate measure of body fat content that is of questionable clinical value.[31–37] Clearly, the need for accurate and affordable tools for assessing body composition and phenotyping obesity and related metabolic disorders is urgent.[38] In the absence of such tools, although the value of applying a standardized numerical BMI score for diagnosing any *individual* patient in an unqualified fashion with obesity may be questionable, at the broad *population* level the vast amount of epidemiological data quantifying the effects of increasing BMI nonetheless has some value for identifying the relative physiological progression of pathologies that become more likely as adiposity increases.[34]

According to the World Health Organization, the terms "overweight" and "obese" constitute a range of conditions of excessive body fat based on BMI[39] (Table 19.2). In 2016, an unprecedented 39% of the global population was obese or overweight, and it is predicted that by the year 2030 this number will rise to 57.8% of the world's adult population.[40,41] Indeed, globally there are now more people who are obese than underweight, and for the first time in human history overnutrition is killing more people worldwide than starvation.[42]

The obesity pandemic is a major health concern as the condition not only significantly reduces an individual's quality of life, but also greatly increases the risk of developing chronic diseases. These include type 2 diabetes mellitus, fatty liver disease, cardiovascular diseases, osteoarthritis, dementia, obstructive sleep apnea, depression, and infertility, as well as several types of cancers, such as breast, ovarian, prostate, colon, kidney, and liver.[43] The links between obesity and these chronic diseases are strong, though their mechanisms are complex and in many cases not well understood. In general, these diseases are thought to stem from several key characteristics of obesity that include (1) a constant state of low-grade inflammation, (2) dyslipidemia (elevated blood lipid concentration), and (3) an altered endocrine profile. For example, adipose tissue contains CYP19A1 aromatase, the enzyme that converts testosterone to estradiol, which simultaneously imparts lower levels of circulating testosterone and higher estradiol in obese individuals, contributing to infertility in males and to various cancers in both males and females. A state of low grade inflammation contributes to osteoarthritis and many other immunological conditions, and dyslipidemia contributes to aspects of cardiovascular disease and fatty liver disease. Much of this chapter will focus on how obesity predisposes individuals to developing type 2 diabetes.

Far from simply arising from a passive accumulation of excess calories, current evidence suggests that obesity stems from complex interactions among behavioral, environmental, physiological, genetic, epigenetic, social, and economic variables.[44] From a strictly biological perspective, obesity arises from two distinct but related processes: (1) a sustained positive energy balance, where energy intake exceeds energy expenditure (e.g. overeating in conjunction with a sedentary lifestyle), and (2) a genetic and/or epigenetic predisposition to recalibrating the body weight and energy homeostatic "set point" to a higher value.[45] Indeed, the latter is a significant impediment to the effective long-term treatment of obesity and accounts for why weight lost via dieting or exercise tends to be regained over time in many people. The question of whether obesity primarily arises from an excess caloric intake or from a reduced energy expenditure remains unclear, though the importance of each parameter will likely vary significantly from person to person. Although a genetic component to obesity susceptibility exists, within a population only a small percentage of BMI variation can be explained by known genetic variants, which points to a complex interaction between genes and the environment.[46] Furthermore, the rapid rise of global obesity over just a few decades suggests the increased incidence does not result entirely from genetic predisposition.[47]

Table 19.2 World Health Organization Body Mass Index (BMI) Classifications

Classification	BMI (kilogram/m^2)
Underweight	<18.5
Normal weight	18.5–24.9
Overweight	25–29.9
Obesity Class 1	30–34.9
Obesity Class 2	35–39.9
Extreme Obesity Class 3	>40

Importance of White Adipocyte Size and Growth

In Chapter 18: Energy Homeostasis, you learned about some distinct developmental and morphological features of normal white adipose tissue (WAT), including its status as a tissue that synthesizes and responds to a diverse set of hormones. In this section you will see how in the obese state WAT adipocytes radically alter their endocrine and metabolic profiles, in turn contributing to the genesis of various metabolic diseases.

In states of overnutrition, adipose tissue can store excess calories and expand in two very different ways: *hyperplasia*, where numbers of adipocytes increase by the differentiation of precursors to form new adipocytes, and *hypertrophy*, the enlargement of existing individual adipocytes (Figure 19.4). The potential of adipocytes to grow by hypertrophy is remarkable, with the capacity to increase in size from small (<50 micrometers diameter) to over 100 micrometers in diameter.[48,49] Importantly, adipose tissue growth by hyperplasia is typically considered healthy and adaptive, as the tissue is able to maintain proper vascularization and levels of insulin-sensitizing and anti-inflammatory hormones.[50] Newly differentiated, smaller adipocytes appear to act as a sink for the efficient absorption of excess FFAs and triglycerides following digestion of a meal. By contrast, when adipocytes individually grow too large by hypertrophy, this promotes increased mechanical and a hypoxic response by these cells due to their massively expanded size, which approaches the limits of oxygen diffusion, and they become dysfunctional. As such, WAT is a heterogeneous tissue, and large, hypertrophied adipocytes are biochemically different from smaller adipocytes. Indeed, small adipocytes have been shown to correlate with reduced susceptibility to developing diabetes[51,52] and may be critical for minimizing metabolic declines associated with obesity.[53,54]

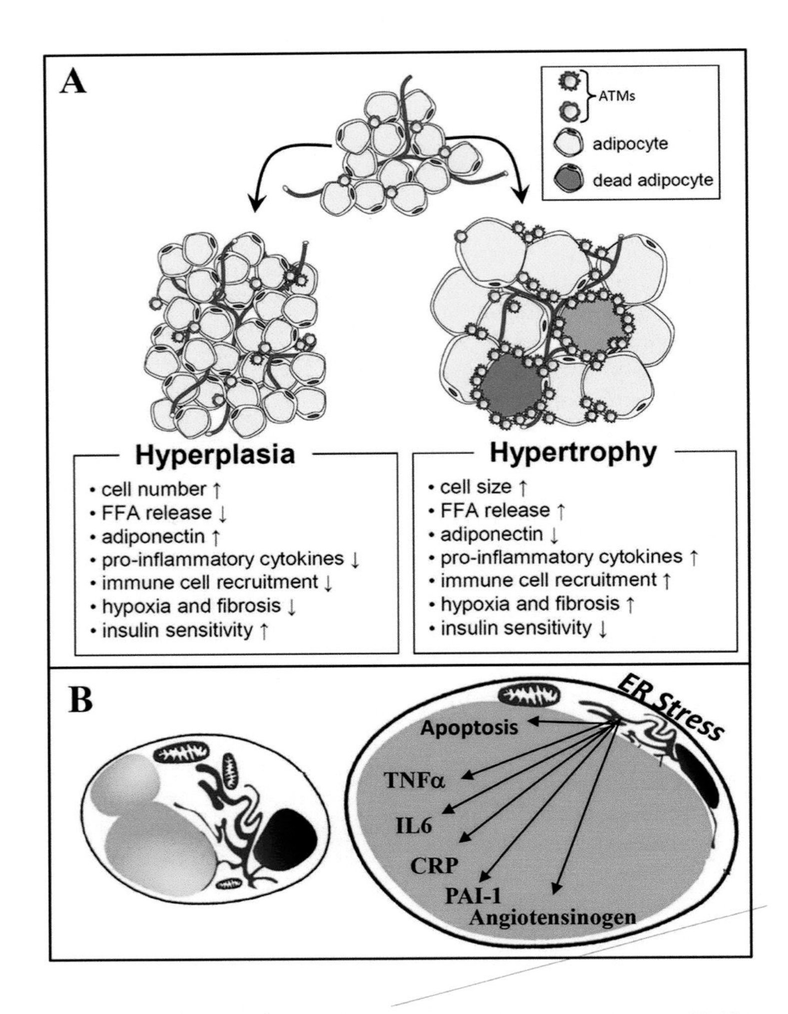

Figure 19.4 Features of hyperplastic and hypertrophic adipocytes and adipose tissue. (A) Hyperplastic adipocyte expansion though increased adipocyte number is linked to beneficial phenomena, such as decreases in fatty acid release, pro-inflammatory cytokine release, immune cell recruitment, hypoxia, and fibrosis and increased adiponectin synthesis and improved insulin sensitivity. By contrast, hypertrophic adipose expansion through increased adipocyte size is associated with harmful phenomena, such as increases in fatty acid release, pro-inflammatory cytokine release, inflammatory adipose tissue macrophage (ATM) cell recruitment, hypoxia, and fibrosis, as well as decreased adiponectin and impaired insulin sensitivity. **(B)** Adipocyte hypertrophy coincides with an expanding triglyceride droplet, which promotes a reduced mitochondrial function, as well as an endoplasmic reticulum (ER) stress response. The ER stress response manifests as multiple signaling cascades that may culminate in the release of inflammatory cytokines and apoptosis. Abbreviations: tumor necrosis factor α (TNFα), interleukin-6 (IL6), C-reactive protein (CRP), plasminogen activator inhibitor-1 (PAI-1).

Source of images: (A) Choe, S. S., et al. (2016). Front. Endocrinol. 7(30). (B) Gregor, M.F., 2007. J. Lipid Res. 48, 1905–1914.

Table 19.3 Characteristics of Lean vs. Obese Adipose Tissue

	Lean Adipose Tissue	Obese Adipose Tissue
Adipocyte size	small	large
Inflammation (macrophage count)	low	high
Vascular density	high	low
Apoptotic adipocytes	few	many
FFAs released	low	high

As adipocytes become individually larger in the obese state, the growing intracellular triacylglycerol droplet is thought to displace and damage adipocyte organelles, such as the smooth endoplasmic reticulum (Figure 19.4 and Table 19.3), inducing an *endoplasmic reticulum stress response* that results in a shift in hormone secretion by adipocytes.[55] In particular, *pro-inflammatory cytokines* that include tumor necrosis factor α (TNFα), interleukin-6 (IL6), C-reactive protein (CRP), angiotensinogen, and plasminogen activator inhibitor-1 (PAI-1) are produced in increasing amounts by expanding adipose depots, resulting in the recruitment of more macrophages that promote adipose tissue inflammation (Figure 19.4 and Table 19.3). Inflamed adipose tissue, in part through the actions of TNFα, secretes greater amounts of free fatty acids (FFAs) into the blood. Elevated FFAs promote insulin resistance in muscle and liver tissue, a process described later in Section 19.6: Type 2 Diabetes Mellitus. The hypersecretion of the hormone resistin by adipocytes in the obese state also contributes to insulin resistance. Furthermore, in the obese state adipocytes secrete lower amounts of potentially beneficial anti-inflammatory adipocytokines that also promote insulin sensitivity, such as leptin and adiponectin. Thus, adipose tissue inflammation promotes insulin resistance, and insulin resistance perpetuates adipose inflammation. Ultimately, this unbalanced production of pro- and anti-inflammatory adipocytokines by obese adipose tissue may contribute not only to the development of type 2 diabetes, but also to hyperlipidemia, hypertension, and atherosclerosis. When these disorders occur together they comprise the aforementioned metabolic syndrome, simultaneously increasing the risk of heart disease, stroke, and diabetes.[56]

WAT as a Contributor to Inflammation and Metabolic Disease

The increased stress experienced by hypertrophied and hypoxic adipocytes contributes to adipocyte *necrosis*, or death due to damage. Necrosis leads to infiltration of adipose tissue by immune cells, a process termed *inflammation*.[52,57] However, unlike a normal inflammatory response designed to fight off an infection, the inflammation marked by obesity does not resolve and can become chronic. A perpetual state of adipose tissue inflammation promotes the release of pro-inflammatory cytokines by adipose tissue and macrophages. Importantly, tissue inflammation in obesity can extend beyond adipose tissue and include the accumulation of immune cells in skeletal muscle, liver, gut, pancreatic islets, and brain, and may contribute to obesity-linked metabolic dysfunctions leading to insulin resistance and type 2 diabetes mellitus.[58]

Curiously Enough . . . Because of its accompanying chronic state of low-grade inflammation, obesity can impair the immune response making obese people much more susceptible to the debilitating and potentially lethal effects of influenza, SARS, COVID-19, and other contagious diseases.[59–61]

In the obese state, adipocyte function may promote insulin desensitization, yielding persistently elevated levels of blood sugars and FFAs, causing hyperglycemia and *lipotoxicity*, or excess lipid deposition and buildup of toxic compounds caused by incomplete oxidation in other tissues, such as heart, muscle, and liver. These factors contribute to the onset of metabolic disease.[50] For example, *non-alcoholic fatty liver disease* (NAFLD), which is becoming the most common chronic liver disease globally, is induced by an excess accumulation of fat caused by abnormal lipid metabolism and excessive *reactive oxygen species* (ROS) production in hepatocytes.[62]

Importance of WAT Location

A confounding variable in the study of obesity is the fact that the different regions of the body where WAT expands are remarkably heterogeneous among individuals, and WAT mechanisms of growth (hyperplasia vs. hypertrophy), adipocyte size, and endocrine function appear to vary significantly with their specific locations in the body. This is particularly evident between the sexes, where males tend to store fat viscerally (around abdominal organs, such as the stomach, small intestine, pancreas, spleen, omentum, and other organs) and females have a propensity to store fat subcutaneously (around the hips, thighs, and arms). Sexual dimorphisms in body fat distribution are influenced by androgen and estrogen sex steroid hormones.[63] For example, intra-abdominal fat varies inversely with estrogen levels,[64,65] and after the decline of estrogens at menopause, women develop increased intra-abdominal adiposity, but those who receive estrogen replacement therapy do not,[66,67] which suggests a specific role of estrogen in limiting intra-abdominal fat mass. By contrast, androgens favor abdominal fat deposition. For example, most women with polycystic ovary syndrome (PCOS), a hyperandrogenic disease, possess increased abdominal fat.[68] Reflecting the heterogeneous nature of fat distribution in a human population, there also exists a continuum between two broad obesity phenotypes: *subcutaneous obesity*, sometimes referred to as "pear-shaped" or "gynoid" obesity, since it is more common in women, and *visceral obesity*, sometimes referred to as "apple-shaped" or "android" obesity, since it is more common in men (Figure 19.5).

Importantly, visceral obesity appears to carry a higher risk of developing cardiovascular disease, type 2 diabetes, and other metabolic disorders compared with subcutaneous obesity. One large-scale study showed that whereas the amount of visceral fat is correlated with obesity-associated cardiovascular risks, these cardiovascular risks did not increase with similar increases of subcutaneous fat.[69] In another example, hyperandrogenic women with PCOS not only have increased abdominal fat, but also have increased risk for developing the metabolic syndrome.[70] Considering that obesity is not a homogenous condition among individuals, it has been suggested that clinicians should instead refer to "obesities".[44]

Reasons why visceral obesity is of greater health concern than subcutaneous obesity can be better appreciated after considering how adipocytes within *visceral adipose tissue* (VAT) are morphologically and biochemically distinct from those found within *subcutaneous adipose tissue* (SCAT). In particular, whereas VAT contains greater numbers of large, hypertrophied adipocytes, SCAT contains mostly small adipocytes. As discussed earlier, small adipocytes are more insulin sensitive and display greater uptake of sugars and FFAs, preventing their deposition in other tissues such as the heart, muscle, and liver. Indeed, VAT has been found to be more insulin-resistant than SCAT,[71,72] and VAT is infiltrated with more inflammatory cells capable of releasing TNF-α, IL-6, angiotensinogen, C-reactive protein, and other pro-inflammatory cytokines than SCAT.[73–77] SCAT appears to function as the zone for accumulation of normal amounts of excess calories, and its expansion via hyperplasia functions as a metabolic sink that protects the lean organs (heart, liver, kidneys, and skeletal muscles) from accumulating an excess of dangerous lipids. However, if the lipid storage capacity of SCAT is either exceeded or its ability to produce new adipocytes is impaired due to genetic or epigenetic predisposition, then lipids begin to accumulate in areas outside of SCAT, namely in VAT, but potentially also in the aforementioned lean organs.[78,79] As such, the amount of VAT, in particular, is a critical factor associated with variations in insulin resistance and risk of metabolic disease.[80–83]

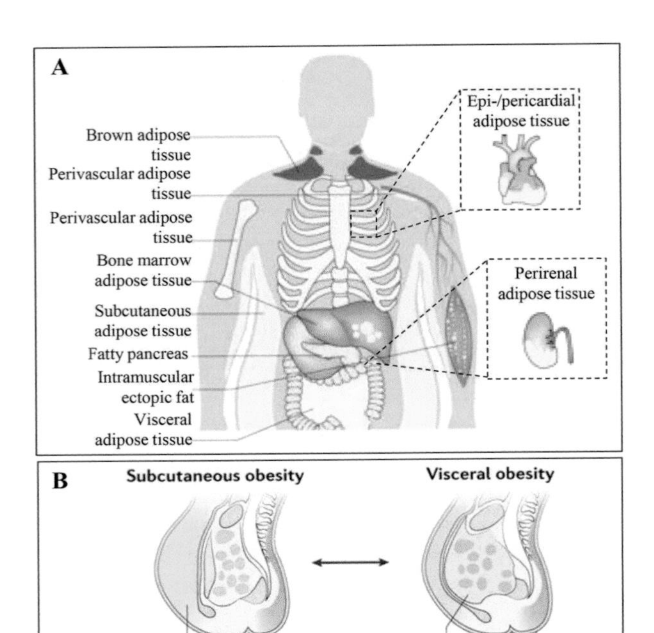

Figure 19.5 Distributions of human adipose tissues and different body shapes of obesity. (A) The diagram depicts the generalized locations of the two major categories of adipose tissue in adults, the metabolically more active brown adipose tissue (BAT) and white adipose tissue (WAT). WAT is subdivided into morphologically and biochemically distinct adipocytes located in two different regions of the body: subcutaneous adipocytes and visceral adipocytes. Ectopic adipocytes include perivascular, epi- and pericardial, perirenal, intramuscular, hepatic, and bone marrow localized adipose tissue. Part A depicts the general distribution of fat in humans from a forward view and part B depicts types of visceral fat depots from a side view. **(B)** The diagram shows a continuum between two obesity phenotypes: subcutaneous obesity, where excess fat is stored subcutaneously, and visceral obesity, where excess lipids are stored around abdominal organs, such as the stomach, small intestine, pancreas, spleen, omentum and other organs. Visceral obesity carries a higher risk of developing cardiovascular disease, type 2 diabetes, and other metabolic disorders compared with subcutaneous obesity.

Source of images: (A) Quail, D.F. and Dannenberg, A.J., 2019. Nat. Rev. Endocrinol., 15(3), pp. 139-154. Used by permission. (B) González-Muniesa, P., et al. 2015. Oxid Med Cell Longev., 2016. Used by permission.

Developments & Directions: Transplant studies in rodents demonstrate the protective metabolic effects of subcutaneous fat over visceral fat

The assessment of metabolic properties of subcutaneous adipose tissue (SCAT) versus visceral adipose tissue (VAT) depots has been studied in mice through whole-adipose-tissue transplantation. In a study by Foster and colleagues,[84] mice were raised on a control diet of chow ("chow control") or an obesity-inducing high fat diet (HFD) for 4 weeks following either the surgical removal of subcutaneous fat depots ("HFD lipx sub") or the transplantation of subcutaneous fat to into the visceral cavity of a recipient ("HFD trans sub") (Figure 19.6). Glucose tolerance tests were performed 4 weeks post-surgery, and circulating levels of insulin and leptin were measured at 5 weeks post-surgery. Obese recipients of subcutaneous fat transplants into their visceral cavities had significantly improved glucose tolerance, as well as decreased systemic insulin and leptin concentration. By contrast, a majority of other studies have shown that transplantation of visceral adipose tissue into donor mice has no significant effect on these parameters.[85–92] The benefits of SCAT versus VAT transplantation on glucose metabolism of recipients likely reflect the morphological and biochemical differences between these two tissues. Based on this and other studies with mice, it has been suggested that in humans the surgical removal of VAT (liposuction) followed by repopulation with SCAT could, in theory, be useful in the treatment of metabolic diseases such as type 2 diabetes, atherosclerosis, and fatty liver disease.[93]

19

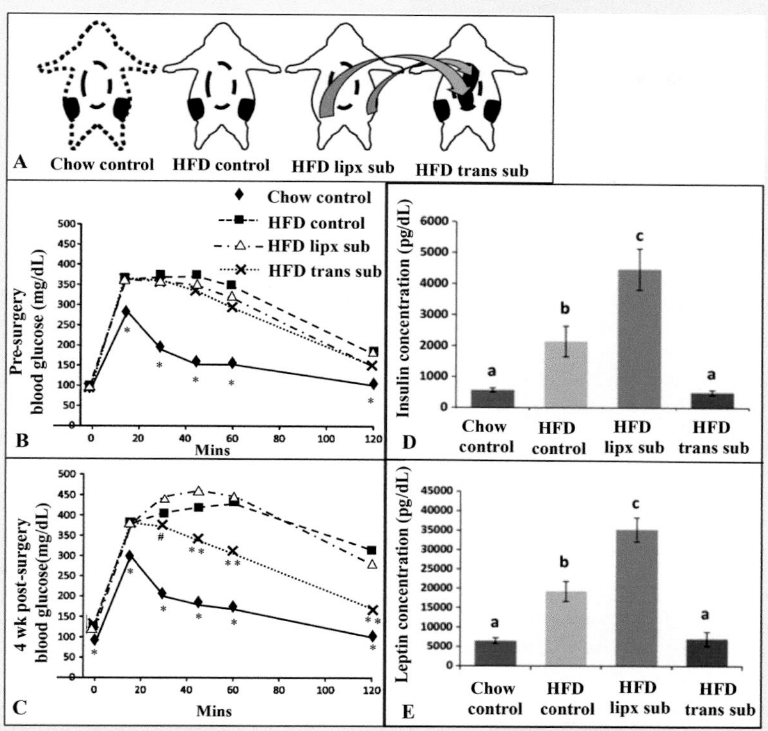

Figure 19.6 **(A) Experimental groups: solid animal outline represents high fat diet (HFD) fed, while square dot is chow fed.** Oval dash in the center is the visceral cavity. Black bilateral structures are the inguinal adipose depots. Arrows indicate where adipose tissue was removed and relocated to. **(B)** Glucose tolerance tests before surgery, and **(C)** 4 weeks after surgery. Concentrations of blood insulin **(D)** and leptin **(E)** were measured five weeks after surgery. Abbreviations: recipient of subcutaneous adipose tissue transplantation (trans sub), subcutaneous fat removed (lipx sub). Statistics: Panels B and C, *(P ≤ 0.05 chow control vs. with HFD control), **(P ≤ 0.05 HFD trans sub vs. HFD control), #(P ≤ 0.05 HFD trans sub vs. HFD control). For panels D and E, different letters indicate significance P ≤ 0.05.

Source of images: Foster, M.T., et al. 2013. Physiol. Rep., 1(2).

The accumulation of excess fat in organs that are normally lean is a condition referred to as *ectopic fat accumulation* (Figure 19.5). Ectopic fat distribution around the heart (epipericardial fat), the liver, and kidneys can lead to cardiovascular disorders, liver steatosis, and hypertension, respectively. Interestingly, studies with mice have shown that the transplant of SCAT into animals fed obesity-inducing high fat diets exerts protective metabolic effects, suggesting that such SCAT transplant approaches may be used in the future to treat metabolic diseases in humans.

SUMMARY AND SYNTHESIS QUESTIONS

1. How can obesity be considered a low-grade state of constant inflammation?
2. Two obese humans with the same BMI can have very different susceptibilities to developing metabolic diseases like type 2 diabetes and hypertension. Explain why.

Overview of Diabetes Mellitus

LEARNING OBJECTIVE Define "diabetes mellitus" and describe the physiological implications of chronic hyperglycemia.

KEY CONCEPTS:

- Diabetes mellitus is a group of metabolic diseases characterized by hyperglycemia resulting from defects in insulin secretion, insulin action, or both.
- Hyperglycemia causes a diabetic to produce a high volume of glucose-containing urine, resulting in constant thirst and drinking to replace the lost water.
- The inability of diabetics to effectively take up and metabolize glucose forces their tissues to rely on other fuels for energy, such as fatty acids, amino acids, and ketones.
- The chronic hyperglycemia of diabetes is associated with long-term damage, dysfunction, and failure of different organs, especially the eyes, kidneys, nerves, heart, and blood vessels.

Table 19.4 Categories of Diabetes Mellitus

Categories of Diabetes Mellitus
Type 1 (pancreatic beta cell destruction) A. Immune mediated B. Idiopathic
Type 2 (ranges from insulin resistance to insulin deficiency or both)
Gestational diabetes (pregnancy onset)
Other categories A. Genetic defects of beta cell function characterized by mutations in one of a variety of genes (e.g. insulin promoter factor 1, proinsulin, or insulin) B. Genetic defects of insulin action (e.g. insulin receptor mutation) C. Diseases of the exocrine pancreas (e.g. pancreatitis, cystic fibrosis) D. Drug or chemical induced (e.g. glucocorticoids, thiazides) E. Genetic syndromes associated with diabetes (e.g. Down's, Klinefelter, and Turner syndromes)

Source: Adapted from Harrison's Endocrinology and American Diabetes Association: Diabetes Care 34:S11, 2011

Diabetes mellitus is a group of metabolic diseases (Table 19.4) characterized by hyperglycemia resulting from defects in insulin secretion, insulin action, or both.[94] Type 1 diabetes mellitus (T1DM) results from a total or near-complete loss of insulin production, whereas type 2 diabetes mellitus (T2DM) constitutes a diversified group of disorders distinguished by varying degrees of insulin resistance, impaired insulin secretion, and elevated blood glucose.[95] Note that gestational diabetes, which may occur during pregnancy, is described in Chapter 24: Pregnancy, Birth, and Lactation.

The word *diabetes*, derived from the Greek for "siphon", refers to the formation of a copious urine, one clinical component of the disease. The word *mellitus*, the Latin word for honey, describes the high sugar content of the urine of diabetics. The chronic hyperglycemia of diabetes mellitus is associated with long-term damage, dysfunction, and failure of different organs, especially the eyes, kidneys, nerves, heart, and blood vessels. Several important physiological consequences of chronic hyperglycemia are described as follows.

1. *Dehydration*: Glucose is an osmolyte, and if the load of glucose filtered by the kidneys exceeds the renal threshold to reabsorb it, the resulting excess glucose in the filtrate draws water into the urine by osmosis. Thus, hyperglycemia causes a diabetic to produce a high volume of glucose-rich urine, resulting in constant thirst and drinking to replace the lost water. Uncorrected, systemic hyperosmolality can cause dehydration of both extracellular and intracellular fluid, resulting in hypotension, coma, and death.
2. *Metabolic change*: In spite of elevated blood glucose, the reduced ability of cells to take in glucose forces them to rely on other fuels for energy, such as fatty acids, amino acids, and ketones. The mobilization of large amounts of fatty acids may play a role in the development of insulin resistance in one form of type 2 diabetes. High levels of ketogenesis may result in **ketoacidosis**, or the severe reduction in blood pH, a condition that can lead to cardiac arrhythmia, coma, and death. Also, if lipid reserves are low, amino acids from skeletal muscle protein and other tissues are consumed, resulting in a "wasting of the flesh" at the cost of physiologically important organs.
3. *Protein glycation*: *Glycation* is the non-enzymatic, covalent bonding of a protein molecule with a sugar molecule, such as glucose, which interferes with the protein's normal functions by disrupting its molecular conformation. Glycation reactions with proteins associated with nerves, hemoglobin, blood plasma proteins, immune cells, the lens and retina of the eye, vasculature, and kidney tubules are thought to be the major causes of different diabetic complications such as retinopathy (impaired vision), nephropathy (impaired kidney function), neuropathy (impaired neural function), and cardiomyopathy (impaired heart function).[96]

Clinically, diabetes mellitus is characterized by hyperglycemia during both the fasted and fed states (Figure 19.7). The most common clinical diagnoses for diabetes mellitus are to assess glucose tolerance in three ways:

1. *Fasting plasma glucose*: Although with normal fasting (i.e. 8 hours with no food intake) blood glucose levels will vary among populations, plasma glucose levels should normally be below 100 mg/dL, and the diagnosis of diabetes mellitus is made if fasting plasma glucose exceeds 126 mg/dL on two successive days.
2. *2-hour plasma glucose (2-h PG) tolerance test*: Following overnight fasting, the patient is given a bolus of glucose (typically 75 g) orally and blood glucose levels are measured at 2-hours post-ingestion. The diagnosis of diabetes mellitus is made if a 2-hour plasma glucose greater than 200 nm/dL is measured on two consecutive days.
3. *Glycated hemoglobin count (hemoglobin A1c)*: A glycated hemoglobin count reflects the 3-month average plasma glucose concentration (the test is limited to a 3-month average because the lifespan of a red blood cell is four months). A glycated hemoglobin count greater than 6.5% is sufficient for the diagnosis of diabetes mellitus.

Figure 19.8 depicts values for the clinical diagnostic spectrum of diabetes mellitus. Globally, diabetes mellitus is a

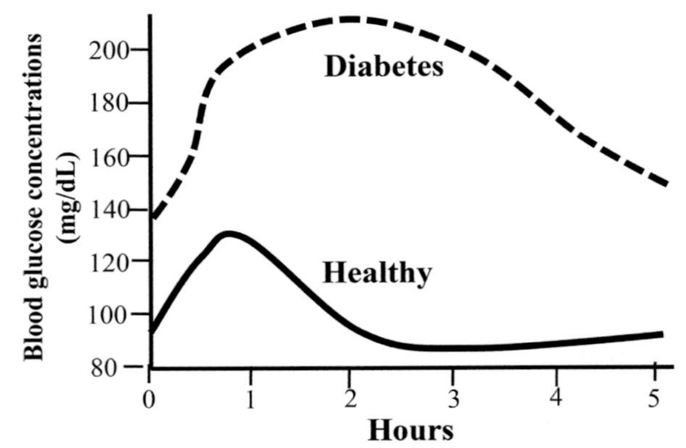

Figure 19.7 Glucose tolerance curves in a healthy person and in a diabetic person.

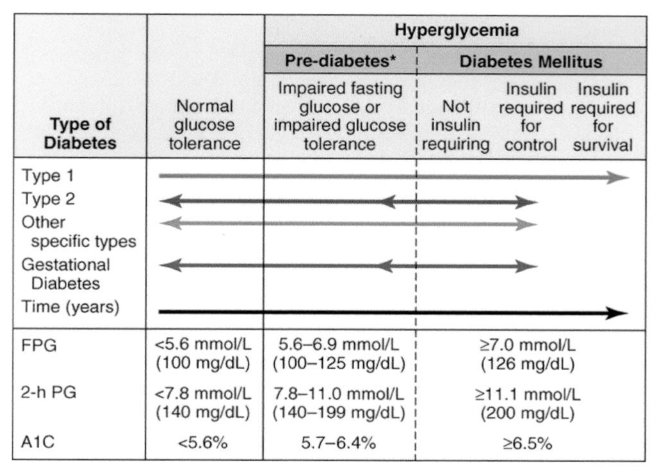

Figure 19.8 The clinical diagnostic spectrum of diabetes mellitus (DM). The spectrum from normal glucose tolerance to diabetes in type 1 DM, type 2 DM, other specific types of diabetes, and gestational DM is shown from left to right. In most types of DM, the individual traverses from normal glucose tolerance to impaired glucose tolerance to overt diabetes (these should be viewed not as abrupt categories but as a spectrum). Arrows indicate that changes in glucose tolerance may be bidirectional in some types of diabetes. For example, individuals with type 2 DM may return to the impaired glucose tolerance category with weight loss; in gestational DM, diabetes may revert to impaired glucose tolerance or even normal glucose tolerance after delivery. The fasting plasma glucose (FPG), the 2-h plasma glucose (PG) after a glucose challenge, and the A1C for the different categories of glucose tolerance are shown at the lower part of the figure. These values do not apply to the diagnosis of gestational DM. The World Health Organization uses an FPG of 110–125 mg/dL for the prediabetes category. Some types of DM may or may not require insulin for survival. *Some use the term "increased risk for diabetes" (ADA) or "intermediate hyperglycemia" (WHO) rather than "prediabetes".

Source of images: Harrison's Principles of Internal Medicine, 18th edition. McGraw-Hill [Adapted from Medical Management of Type 1 Diabetes, 3rd ed, JS Skyler (ed). American Diabetes Association, Alexandria, VA, 1998.] Used by permission.

common metabolic disease, and the occurrence of type 2 diabetes mellitus, in particular, has risen steeply to epidemic proportions from an estimated 30 million cases in 1985 to 285 million in 2010. Based on current trends, the International Diabetes Federation projects that 642 million people will be diagnosed with diabetes mellitus by the year 2040. In the United States, the Center for Disease Control and Prevention estimated in 2010 that 25.8 million Americans, or 8.3% of the population, had diabetes.

SUMMARY AND SYNTHESIS QUESTIONS

1. Why is diabetes mellitus associated with large volumes of urine production?
2. Many of the complications associated with high glucose levels in diabetes mellitus, such as retinopathy and neuropathy, stem from what, specifically?

Type 1 Diabetes Mellitus

LEARNING OBJECTIVE Describe the etiology (cause) of type 1 diabetes mellitus (T1DM) and its symptoms and complications.

KEY CONCEPTS:

- Type 1 diabetes mellitus (T1DM) typically results from the autoimmune destruction of the insulin-secreting pancreatic β cells in children, but it can occur at any age.
- In the absence insulin, target cells fail to express GLUT4 glucose transporters on their plasma membranes, inhibiting glucose uptake and metabolism.
- The loss of glucagon inhibition by insulin in T1DM promotes elevated blood glucagon levels, which results in increased glycogenolysis by the liver.
- In untreated T1DM, elevated ketone synthesis can lead to the potentially fatal condition of metabolic acidosis.
- The most common treatment for T1DM is the daily injection of insulin.

Type 1 diabetes mellitus (T1DM), which accounts for 5%–10% of those with diabetes mellitus, results from the destruction of the pancreatic β cells, often by autoimmune mechanisms. By contrast, islet cells secreting glucagon (α cells), somatostatin (δ cells), or pancreatic polypeptide (PP cells) are preserved. The gradual but persistent destruction of the β cells manifests as a failure to produce insulin; consequently, patients with T1DM remain insulin-dependent for their lifespan. It should be noted that the current classification of T1DM now excludes two terms previously used in association with the disease but still commonly encountered in published literature.[94] The first is "juvenile-onset diabetes". Although T1DM typically develops in children and in adults before the age of 30, an autoimmune destruction of the β cells, the salient feature of current T1DM classification, can occur at any age.[97] The second obsolete term is "insulin-dependent diabetes mellitus", associated with type 2 diabetes mellitus (T2DM). While T1DM patients indeed currently must be treated with insulin for their entire lives, many patients with T2DM must also be treated with insulin, so the terms "insulin-dependent" and "insulin-independent" are no longer used.

Clinical Considerations: The discovery of insulin

Untreated, type 1 diabetes is a terrifying disease. The first description of diabetes was in 100 CE from Aretaeus of Cappadocia, who described it as a "melting down of the flesh and limbs into urine".[98] In 1920, diabetes was a terminal disease, with a life expectancy of just 6 to 12 months from diagnosis.[99] Its treatment consisted of restricting caloric intake to less than 500 calories per day. Individuals who crossed that nutrition threshold in an attempt to appease their hunger would develop diabetic ketoacidosis (reduced blood pH due to excess ketones) and die within hours to days. Life for diabetic patients in the early 1900s was, in a word, horrific. Patients, most of them children, suffered from malnutrition, cataracts, blindness, gangrene, and immune-resistant infections. All too often, death from diabetes would be considered a blessing by those who long suffered the consequences of chronic hyperglycemia.

The discovery of insulin was among the most important early contributions to endocrinology and medicine, giving life and hope to countless people afflicted with diabetes. The road to its discovery involved multiple researchers from different countries, spanning two continents over the course of three decades. Arguably the biggest breakthrough came when Frederick Banting and Charles Best (Figure 19.9[A]) conducted a series of experiments over the summer of 1921 at the University of Toronto in the laboratory of J.J.R. Macleod. Like previous researchers, they showed that the removal of the pancreas

Figure 19.9 (A) Dr. Frederick Banting (right) and Charles H. Best (left) on the roof of the Medical Building, summer 1921, with their experimental dog, Marjorie, successfully treated with what would later be called insulin. (By courtesy of the Thomas Fisher Rare Book Library, University of Toronto) **(B)** In 1922, 10,000 pounds of pig pancreases (background pile) made one pound of insulin crystals (small white bottle in foreground).

Source of images: (A) Rydén, L. and Lindsten, J., 2021. Diabetes Res. Clin. Pract., 175, p. 108819. (B) Image courtesy of Eli Lilly Company Corporate Archives.

from dogs made them diabetic. However, they took the additional painstaking steps of preparing extracts from the islets of Langerhans of healthy dogs, injecting it into the diabetic dogs, and showed that the dogs returned to normalcy for as long as they had the extract. With the help of biochemist J.B. Collip, Banting and Best extracted a relatively pure form of insulin from dog pancreases and, ultimately, from the pancreases of cattle obtained from slaughterhouses. In January 1922, Leonard Thompson, a diabetic teenager in a Toronto hospital, became the first person to receive an injection of purified canine insulin extract. After his seemingly miraculous recovery, the news about insulin spread rapidly around the world. For their work, Banting and Macleod received the *Nobel Prize in Medicine* the very next year, in 1923. The University of Toronto immediately gave pharmaceutical companies license to produce insulin free of royalties, and the medical firm, Eli Lilly, began large-scale production of the extract (Figure 19.9[B]). In 1923, the firm was producing enough insulin to supply the entire North American continent, saving countless lives. The history of the discovery of insulin defined a modern paradigm for the integration of experimental physiology, medicine, and the pharmaceutical industry for the treatment of a disease on a large scale.

Etiology

Our current understanding is that the *etiology*, or physiological cause, of T1DM likely depends on environmental factors that interact with predisposing genes to induce an autoimmune assault against the pancreatic β cells[100,101] (Figure 19.10). Among the potential environmental factors

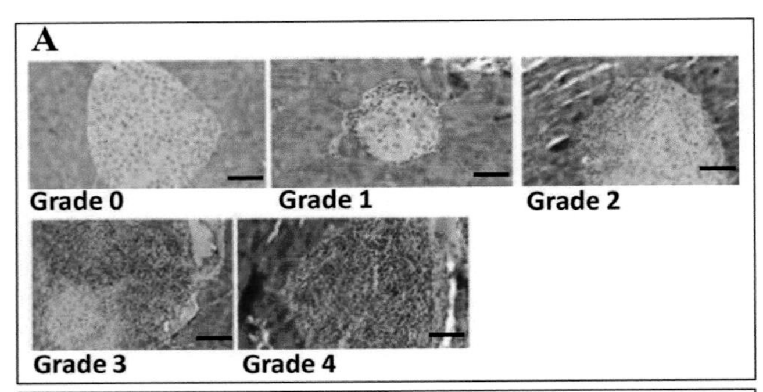

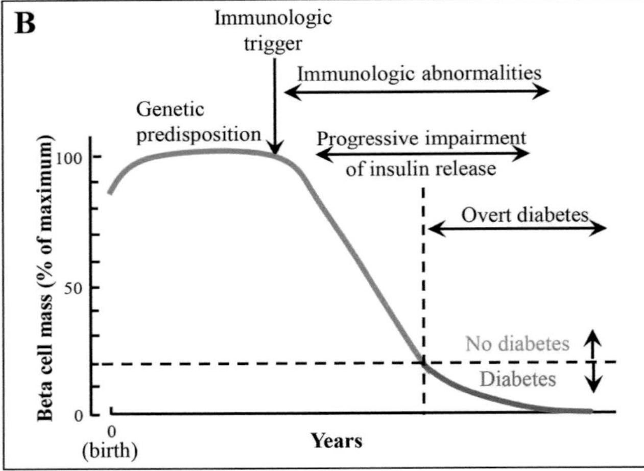

Figure 19.10 Model for the development of type 1 diabetes. (A) Images of a pancreatic islet showing the stages of infiltration by white blood cells (darker-staining cells) from grade 0 (no infiltration) to grade 4 in which most of the insulin producing cells have been destroyed. **(B)** Individuals with a genetic predisposition are exposed to an immunologic trigger that initiates an autoimmune process, resulting in a gradual decline in beta cell mass. The downward slope of the beta cell mass varies among individuals and may not be continuous. This progressive impairment in insulin release results in diabetes when ~80% of the beta cell mass is destroyed. A "honeymoon" phase may be seen in the first 1 or 2 years after the onset of diabetes and is associated with reduced insulin requirements.

Source of images: (A) Bodin, J., et al. 2016. Toxicol. Rep. 3, pp. 664-672. Used by permission. (B) Modified from Harrison's Principles of Internal Medicine, 18th edition. McGraw-Hill.

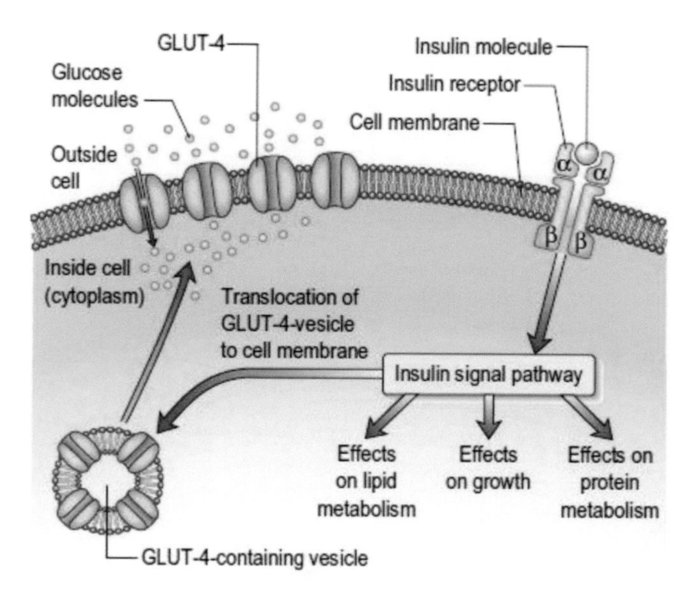

Figure 19.11 Activation of the glucose transporter GLUT4 by insulin. In the absence of insulin, GLUT4 remains sequestered in storage vesicles, inhibiting glucose uptake by cells.

Source of images: Woyesa, S.B. and Taylor-Robinson, A.W., 2019. J. Molec. Pathophysiol., 8(1), pp. 1-13.

as triggers for the development of T1DM, studies in humans support a role for viral infections, particularly by enteroviruses, or viruses that first enter the body through the gastrointestinal tract and then may attack other organs. In most cases, T1DM is characterized by *insulitis*, or pancreatic islet inflammation, and progressive β cell loss by apoptosis. The symptoms of diabetes mellitus are not evident until the majority of β cells (70%–80%) are destroyed, with the remaining cells insufficient in number to maintain glucose tolerance. The ensuing hyperglycemia that results from the destruction of the β cells is caused primarily by two factors:

1. *Reduced uptake of glucose by target tissues.* In the absence of circulating insulin, the insulin receptor remains unbound, resulting in a failure of target cells to activate the intracellular signaling pathways that transfer GLUT4 glucose transporters to their plasma membranes. Since the GLUT4 proteins remain sequestered in their cytosolic storage vesicles, glucose cannot be transferred from the blood into the cells for energy consumption or storage as glycogen (Figure 19.11).
2. *Increased glucagon production by α cells.* The loss of glucagon inhibition by insulin in T1DM promotes elevated blood glucagon levels, which results in increased glycogenolysis by the liver.

Symptoms and Complications of Untreated T1DM

Untreated T1DM is accompanied by diverse symptoms reflecting severe metabolic disruption, some of the more harrowing of which were described in Clinical Considerations: The discovery of insulin. Several primary pathophysiological symptoms of acute untreated T1DM (summarized in Figure 19.12 and Figure 19.13) are described next:

Fatigue: **Caused in part by hypoxia due to glycation of hemoglobin, which disrupts oxygen-carrying capacity. Other contributing variables include reductions in muscular glycogen stores and reduced glucose uptake by cells.**

Weight loss: **Increased breakdown of fat stores into fatty acids for fuel and ketone production. Breakdown of protein (primarily muscle) into amino acids for fuel and gluconeogenesis. An extremely emaciated appearance was common for untreated type 1 diabetics (Figure 19.14).**

Polyphagia **(increased hunger and rate of food consumption): The brain interprets reduced glucose uptake and utilization as starvation, greatly stimulating appetite.**

Osmotic diuresis: **Caused by** *glucosuria* **(elevated urinary glucose), osmotic diuresis is reflected by** *polyuria* **(increased urine output) in an attempt to excrete excess glucose.** *Polydipsia* **(increased thirst and drinking) results in an attempt to replace lost fluids.**

Hypotension: **Reduced blood pressure caused by dehydration may result in cardiac arrest and circulatory failure.**

Metabolic acidosis: **This is caused primarily by elevated ketogenesis but also by increased lactic acid production by muscles respiring anaerobically. The body responds to metabolic acidosis by hyperventilating. Specifically, the respiratory centers of the brain are stimulated by low blood pH, and hyperventilation ensues in an attempt to blow off perceived excess CO_2 to elevate blood pH.**

Coma and death: **Caused by reduced oxygen and nutrient transport to the brain due to circulatory failure, shock, and exhaustion.**

Curiously Enough . . . The breath of patients in the late stages of T1DM has been described as "a sickish sweet smell, like rotten apples, that sometimes pervaded whole rooms or hospital wards".[99] This smell results from the exhalation of acetone, a volatile ketone.

Screening and Treatments for T1DM

An early pre-symptomatic phase of T1DM can be identified by the presence of circulating autoantibodies against several different pancreatic islet β cell proteins. While historically such screens have been applied to relatives of patients with T1DM who have a higher

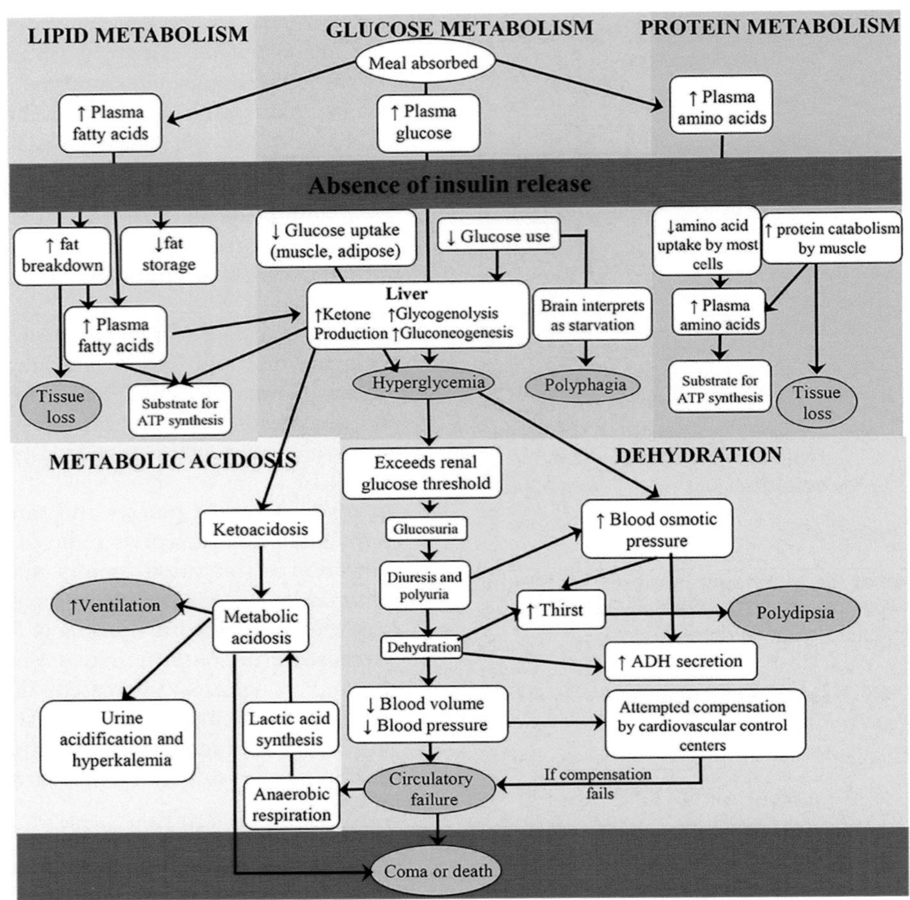

Figure 19.12 Summary of the symptoms and complications of untreated type 1 diabetes mellitus.

Source of images: Silverthorn, D.U., 2015. Human physiology. Jones & Bartlett Publishers.

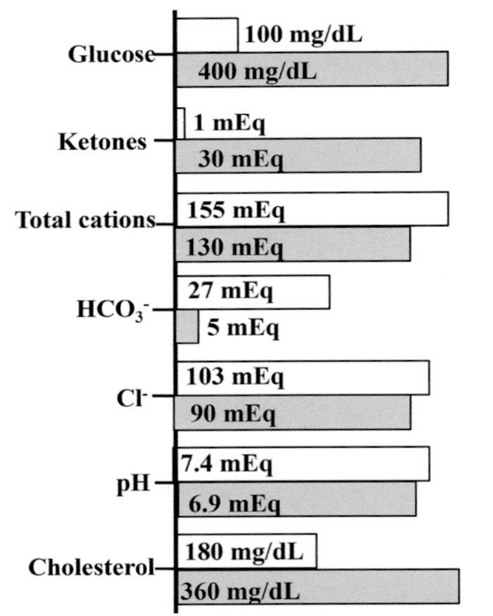

Figure 19.13 Changes in blood constituents in diabetic coma, showing normal values (white bars) and diabetic coma values (gray bars).

Source of images: Hall, J.E. and Hall, M.E., 2020. Guyton and Hall textbook of medical physiology e-Book. Elsevier Health Sciences.

risk of developing the disease, successful public health screens of a general population of children have also been successful at identifying pre-symptomatic phase T1DM individuals.[102]

The most common treatment for T1DM, one that has changed little since the 1920s, is the daily injection of insulin, though the administration of insulin via portable pumps is now becoming more widespread[103]. The source of insulin for the treatment of diabetics has progressed from livestock-derived pancreatic extracts in the 1920s to the larger-scale production of insulin using recombinant DNA grown in bacteria beginning in the early 1980s (Figure 19.15). In 1982, Eli Lilly's "Humulin" (human insulin) became the first genetically engineered drug approved by the U.S. Food and Drug Administration. It was made by inserting human genes responsible for insulin production into *Escherichia coli* bacteria, stimulating the bacteria to synthesize insulin in large quantities. Insulin therapy, given by injection or insulin pump, is life saving, but it is not perfect. Other examples of therapy for T1DM can include the administration of amylin agonists to reduce gastric emptying and postprandial glycemia, consumption of a low-calorie diet, as well as regular exercise, which is known

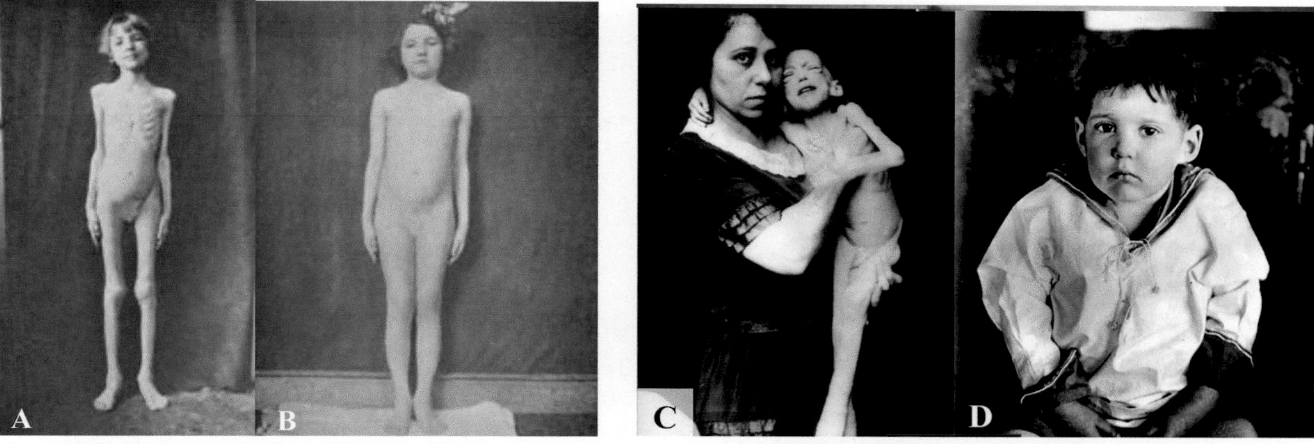

Figure 19.14 Photographed in 1922, this 13-year-old diabetic girl weighed just 45 lb (A) before treatment with insulin. A few months later **(B)** she had made a dramatic recovery. Patient J.L, age 3, before **(C)** and after **(D)** insulin in 1923.

Source of images: (A-B) Courtesy of Wellcome Library London; (C-D) Flier, J.S. and Kahn, C.R., 2021. Molec. Metab., 52, p. 101194. Used by permission.

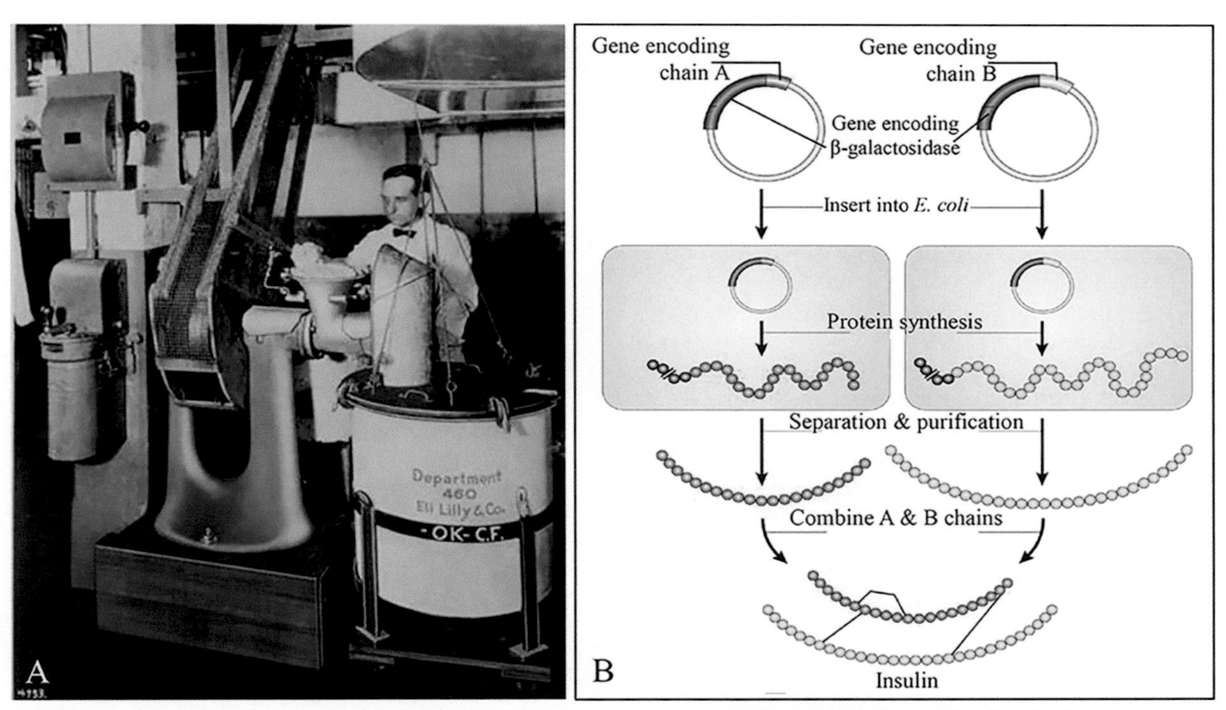

Figure 19.15 Insulin production, from ground livestock pancreases to genetic engineering. (A) In the 1920s, pancreases from livestock were run through grinders prior to extraction. **(B)** Modern insulin is synthesized using recombinant DNA technology. In the initial process for insulin manufacture, synthetic DNA coding for the A and B chains was inserted into two plasmids, which were then separately introduced into *Escherichia coli* bacteria. Following incubation, the translated A and B chain proteins were purified from the two bacterial cultures and then chemically recombined (disulfide bridges added) to generate the mature insulin molecule.

Source of images: (A) Gift of Eli Lilly, Division of Medicine and Science, National Museum of American History, Smithsonian Institution. (B) Johnson, I.S. (2003) Nat. Rev. Drug Discov. 2, 747-751. Used by permission.

to augment insulin synthesis (Table 19.5). Importantly, most people with T1DM still have blood glucose levels that are, on average, above normal. Because of this, long-term T1DM survivors are susceptible to developing vascular complications leading to conditions like diabetic retinopathy and nephropathy.

Table 19.5 Agents Used in the Treatment of Type 1 Diabetes Mellitus

Category	Examples	A1C Reduction (%)	Mechanism of Action
Insulin	-	Not limited	Insulin receptor agonist, increases glucose utilization, reduces hepatic glucose synthesis
Amylin agonists	Pramlintide	0.25–0.5	Slows gastric emptying, reduces glucagon synthesis, reduces postprandial glycemia
Nutrition therapy and physical activity	Low calorie diet, exercise	1–3	Augments insulin secretion

Note: A1C reduction (absolute) depends partially on starting A1C levels.
Source: Adapted from Harrison's *Endocrinology*, 3rd edition table 19–11. McGraw-Hill

SUMMARY AND SYNTHESIS QUESTIONS

1. The hyperglycemia associated with T1DM was originally thought to be due exclusively to a dysfunction of insulin secretion by the pancreatic beta cells. However, it is now known that pathologically elevated glucagon levels also play an important role in promoting T1DM hyperglycemia. Explain why glucagon levels are abnormally high in untreated T1DM.
2. The breath of dying type 1 diabetics has been described as "The sickish, sweet smell of rotten apples". What explains this particular smell?
3. Individuals with T1DM use daily insulin injections to maintain blood glucose within normal physiological range. Although they are treating themselves to prevent hyperglycemia, these diabetics must also constantly be on guard for symptoms of *hypo*glycemia, which can be lethal. What makes these patients susceptible to experiencing hypoglycemia?
4. T1DM is a disease characterized by the loss of pancreatic beta cells from the islets of Langerhans. Imagine a disease that instead targets the pancreatic alpha cells for destruction. Would the loss of glucagon production be as important as the loss of insulin? Why or why not?

Type 2 Diabetes Mellitus

LEARNING OBJECTIVE Describe the etiology of type 2 diabetes mellitus (T2DM) and its symptoms, complications, and link to obesity.

KEY CONCEPTS:

- Type 2 diabetes mellitus (T2DM) symptoms may include hyperglycemia, insulin resistance, excessive hepatic glucose synthesis, abnormal fat metabolism, and impaired insulin secretion.
- The most common cause of T2DM is obesity.
- Globally, the occurrence of T2DM, together with obesity, has risen steeply to epidemic proportions.

Definition of Type 2 Diabetes Mellitus

Type 2 diabetes mellitus (T2DM), which accounts for 90%–95% of those with diabetes, is a heterogeneous collection of progressive metabolic disorders characterized by hyperglycemia, insulin resistance in the main target organs (liver, skeletal muscle, adipose tissue), excessive hepatic glucose synthesis, abnormal fat metabolism, and impaired insulin secretion.[97] T2DM patients can range from those with predominantly an insulin resistance phenotype but with sufficient beta cell reserve to remain insulin independent to those who may require early insulin treatment during the course of their disease.[104] Because insulin resistance may initially be compensated for by the increased production of insulin (*hyperinsulinemia*) (Figure 19.16), overt diabetes does not occur until there is β cell failure, and T2DM may remain undiagnosed for years.[105] Most patients at the

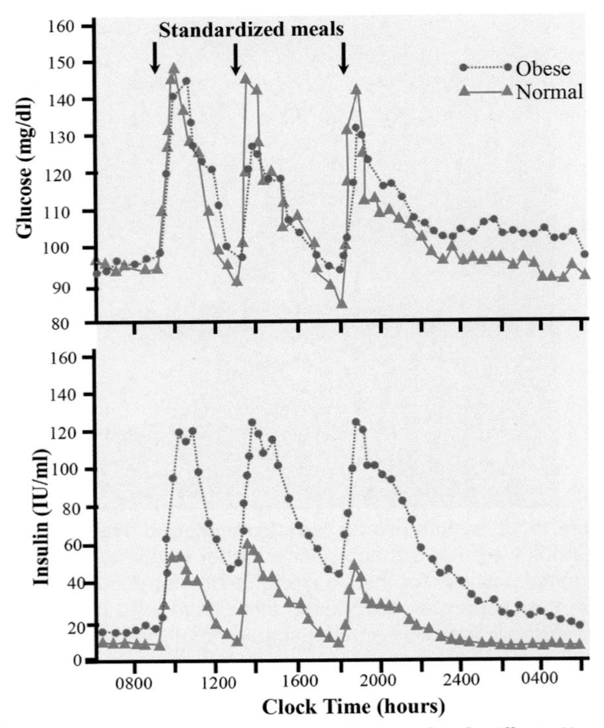

Figure 19.16 Twenty-four-hour profiles and pulsatile patterns of insulin secretion in normal and obese subjects. Note that although obese subjects maintain normal blood glucose levels, they produce higher than normal amounts of insulin, indicating the onset of insulin resistance.

Source of images: Polonsky KS, et al. J Clin Invest 1988;81:442-448.

time of diagnosis of T2DM have some degree of impaired insulin secretion.[106] Unlike T1DM, T2DM patients rarely develop lethal ketoacidosis in the absence of insulin therapy. As such, this form of diabetes has previously been termed "insulin-independent diabetes", reflecting the observation that in the initial phases of the disease insulin treatment is usually not essential for immediate survival in these patients. However, insulin administration has become an important treatment option in T2DM.

"Diabesity": Obesity as a Trigger for T2DM Development

T2DM and obesity are disorders of glucose homeostasis and energy homeostasis, respectively. The majority of T2DM patients are obese, and the two conditions are so tightly linked that the term "diabesity" was coined to reflect this close relationship.[107,108] Indeed, the risk of T2DM increases by two- to eight-fold at BMI 25, ten- to 40-fold at BMI > 30, and over 40-fold at BMI > 35 depending on age, gender, duration and distribution of adiposity, and ethnicity.[108,109] Although excess fat in any region of the body is associated with increased risk of T2DM and cardiovascular disease,[110] it is generally held that an accumulation of abdominal fat ("central" obesity), as indicated by an increased waist-to-hip ratio, is a risk for T2DM irrespective of the extent of obesity.[111] This is mainly attributed to increased intra-abdominal (visceral) adiposity. As discussed earlier, visceral adipose tissue (VAT) displays a different endocrine and immunological profile compared with subcutaneous adipose tissue (SCAT). Obesity can promote insulin resistance through mechanisms of *intracellular lipotoxicity*, or detrimental cellular effects of chronically elevated concentrations of fatty acids and excess lipid accumulation in tissues other than adipose tissue. Since the secreted products of VAT enter the hepatic portal system, they are conveyed directly to the liver, making hepatocytes particularly prone to the effects of lipotoxicity.

Reports of T2DM were rare in the literature prior to the 19th century, with the disease primarily affecting people who were wealthy enough to be able to overfeed themselves. The recent global epidemic of T2DM indicates the importance of behavioral and environmental triggers such as sedentary lifestyle and dietary changes over last several decades in contributing to this disease. While T2DM was virtually nonexistent in children in the early 1980s, it now accounts for about 45% of new cases of diabetes in the pediatric population.[112,113] As with obesity, other important contributors to T2DM include genetics, epigenetics, and possibly exposure to chemical "obesogens", "diabetogens", and "diabesogens"[114] (discussed later in Section 19.7: Metabolism-Disrupting Compounds).

Etiology of T2DM

Obesity-Induced Insulin Resistance

Glucose homeostasis and lipid metabolism are inherently related processes, as insulin normally both promotes glucose uptake and reduces blood triglyceride levels through its inhibition of hormone-sensitive lipase.[115] The development of impaired glucose and lipid metabolism is typically a gradual process, beginning with excess weight gain and the development of obesity. An important consequence of obesity-induced increases in free fatty acid (FFA) availability is their uptake and accumulation in skeletal muscle.[116] Within muscle, FFAs impair insulin signaling by activating isoforms of *protein kinase C* (PKC) that promote serine phosphorylation of insulin signaling proteins that normally undergo tyrosine phosphorylation (Figure 19.17) such as insulin receptor substrates (e.g. IRS-1, IRS-2), thereby reducing their signaling activity.[117–119] This results

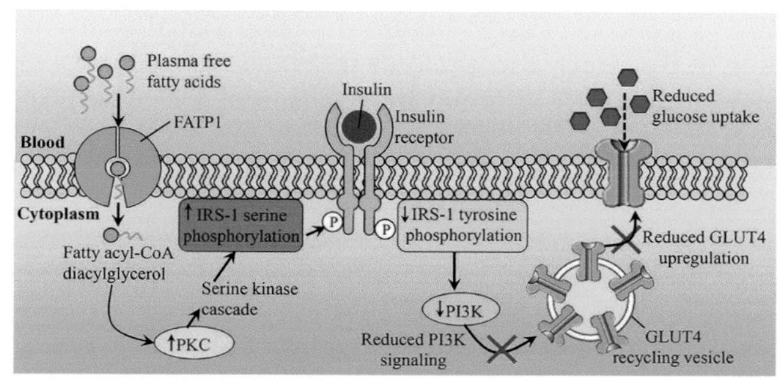

Figure 19.17 Mechanism of fatty acid-induced insulin resistance in human skeletal muscle. Increases in FFAs due to increased delivery and/or decreased mitochondrial oxidation of fatty acids triggers the activation of PKC, a serine/threonine kinase. The serine kinase cascade results in increased serine phosphorylation of IRS-1 on sites that are normally tyrosine phosphorylated. This reduces the ability of IRS-1 to bind and activate phosphoinositol 3-kinase (PI3K), resulting in reduced GLUT4 upregulation and reduced glucose uptake by muscle.

Source of images: Shulman, G.I., 2004. Physiology, 19(4), pp. 183-190.

Clinical Considerations: Thiazolidinediones drugs in the treatment of metabolic disease

Peroxisome proliferator-activated receptors (PPARs) are ligand-activated transcription factors belonging to a nuclear hormone receptor superfamily, containing three isoforms (α, β/δ, and γ). PPARs are activated by naturally occurring fatty acids or fatty acid derivatives (e.g. eicosanoids) and play critical physiological roles in the regulation of numerous biological processes, including lipid and glucose metabolism and overall energy homeostasis (for a review of PPAR function and roles in normal metabolism refer to Figure 18.12 of Chapter 18: Energy Homeostasis). Heterodimers of PPARγ and retinoid X receptor (RXR) form transcription activators that upon binding PPAR response elements can modulate many important cell functions, such as adipocyte differentiation, lipid metabolism, and glucose homeostasis, as well as the expression of many adipose secreted factors. Clinically, PPAR ligands like *thiazolidinediones* (TZDs) are used for treatment of type 2 diabetes by decreasing insulin resistance in the hormone's three primary target tissues: adipose tissue, liver, and skeletal muscle. In addition, PPARs also have been implicated as modulators of obesity-induced inflammation, making them interesting therapeutic targets to mitigate obesity-induced inflammation and its consequences.

in impaired translocation of the glucose transporter GLUT-4 to the cell surface with a consequent decrease in the ability to clear a given glucose load. Since skeletal muscle accounts for > 70% of glucose uptake by the body, the development of insulin resistance in muscle is a major feature of obesity-induced insulin resistance.[120,121]

Excessive Hepatic Glucose and Lipid Synthesis

In the liver, impaired insulin signal transduction due to increased circulating FFAs, probably via a mechanism similar to that described for skeletal muscle, can decrease glycolysis (glucose utilization) and glycogenesis (glucose storage) and instead promote glucose synthesis via glycogenolysis and gluconeogenesis.[122] Inhibited insulin signal transduction in the pancreatic islet cells results in reduced suppression of glucagon secretion by α cells, and the subsequent overproduction of glucagon promotes glycogenolysis and hyperglycemia. Simultaneously, elevated lipolysis leads to increased synthesis of very low-density lipoprotein (VLDL) and triacylglycerol storage by the liver. This increased liver lipid storage (*steatosis*) may lead to *nonalcoholic fatty liver disease*, which can progress to liver inflammation, cirrhosis (liver damage and scar tissue), and liver failure.

Gradually Impaired Insulin Secretion

The pancreas responds to increased hepatic glucose output by increasing insulin output (hyperinsulinemia), which is a hallmark of the early insulin-resistant state. However, over time, the pancreatic β cells can no longer keep pace with the need for insulin, and a progressive decline of β cell function leads to a phenomenon termed "β *cell exhaustion*", which precedes β cell death.[123,124] As such, loss of β cell mass and function are central to the progression of both T1DM and T2DM. Some likely contributors to β cell decline include:

1. *Insulitis*: A perpetual state of adipose tissue inflammation under obese conditions promotes the release of pro-inflammatory cytokines by adipose tissue and macrophages. The cytokine, interleukin-1 (IL-1), produced by macrophages, is thought to participate in β cell death by promoting islet inflammation (*insulitis*) and β cell necrosis.[125]
2. *Oxidative stress*: The accumulation of elevated cytoplasmic FFA in pancreatic islets may increase the formation of *ceramide*, a toxic phospholipid derivative. Ceramide promotes oxidative stress by increasing the cell's radical oxygen species (ROS) content, which induces β cell apoptosis.[126,127]
3. *Rough endoplasmic reticulum stress*: Under conditions of hyperglycemia, β cells may experience an increase of up to 20-fold in insulin and total protein synthesis. It has been proposed that this increase in proinsulin biosynthesis generates a heavy load of accumulated unfolded/misfolded proteins in the RER lumen,[128] resulting in the disruption of RER homeostasis and induction of RER stress.[129] A failure to correct the production of unfolded proteins can promote β cell death.[130,131] An inability to fold another secreted protein, amylin (which is co-secreted with insulin), has also been implicated with the onset of β cell dysfunction and death.

Clinical Considerations: Islet amyloidosis

The pancreatic islets in T2DM patients share an intriguing feature in common with the pathology of neurodegenerative diseases such as Alzheimer's disease, Parkinson's disease, amyotrophic lateral sclerosis, and Huntington's disease—namely, cell loss associated with an abnormal aggregation of locally expressed proteins.[132] The proteins that constitute these aggregates form insoluble "amyloid" (starch-like) fibrils arranged in a β-pleated sheet structure. Interestingly, insulin is co-secreted by β cells with another protein called *amylin*, a hormone that reduces blood glucose by promoting satiety, reduction of the rate of gastric emptying, and reducing glucagon secretion (Figure 19.2). Under conditions of hyperglycemia, the excessive synthesis of amylin is associated with endoplasmic reticulum stress, and a failure to properly fold amylin can result in the formation of amylin plaques within islets. The amyloid formed may cause β cell apoptosis and dysfunction of remaining cells, and amylin secretion could be an important link between RER stress and β cell death in T2DM.[133]

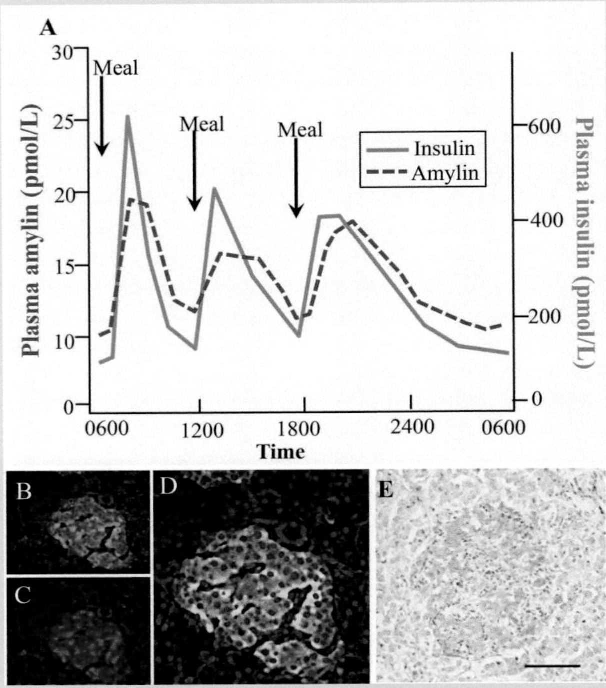

Figure 19.18 Amylin secretion and pancreatic islet cell amyloid plaques. (A) Amylin and insulin are co-secreted by pancreatic beta cells in response to food intake. Staining of paraffin embedded human pancreas using guinea pig anti-pig insulin antibody (**B**, green stain) and mouse anti-human amylin antibody (**C**, red stain). Merged image **(D)** shows co-localization of the two proteins. DAPI was used as nuclear counterstain (blue stain). **(E)** Human islet from a diabetic subject. Most of the islet has been converted into amyloid (Congo red stain), but there are still cords of living cells, a majority being β cells, most probably dysfunctional. Bar: 50 μm.

Source of images: (A) Modified from Koda, J. E., et al. (1995). Diabetes 44(suppl. 1), 238A. (B-D) Courtesy of Novus Biologicals. (E) Westermark, P., 2011. Upsala J. Med. Sci, 116(2), pp. 81-89.

Chronic Complications, Treatments, and Management of T2DM

Because in T2DM hyperglycemia typically develops gradually and patients retain some level of insulin synthesis and sensitivity prior to its diagnosis, it is often not severe enough for the patient to notice any of the classic symptoms of diabetes described earlier for type 1 diabetes mellitus (T1DM). In particular, it is unusual for a T2DM patient to suffer life-threatening ketoacidosis. Nevertheless, after many years of T2DM, patients are at increased risk of developing diverse circulatory, neural, and kidney pathologies caused by micro- and macrovascular damage associated with hyperglycemia (Table 19.6). As such, T2DM is a leading cause of retinopathy and blindness (Figure 19.19), kidney failure, amputations, strokes, and heart attacks, complications that make it such a devastating disease. Therefore, the goals of therapy for T2DM include not only the management of hyperglycemia, but

Table 19.6 Chronic Complications of Diabetes Mellitus

Microvasculature	Cerebrovascular disease
Eye disease	***Other***
Retinopathy	Gastrointestinal (gastroparesis, diarrhea)
Macular edema	Genitourinary (uropathy/sexual dysfunction)
Neuropathy	Dermatologic
Sensory and motor	Infectious
Autonomic	Cataracts
Nephropathy	Glaucoma
Macrovascular	Periodontal disease
Coronary artery disease	
Peripheral arterial disease	

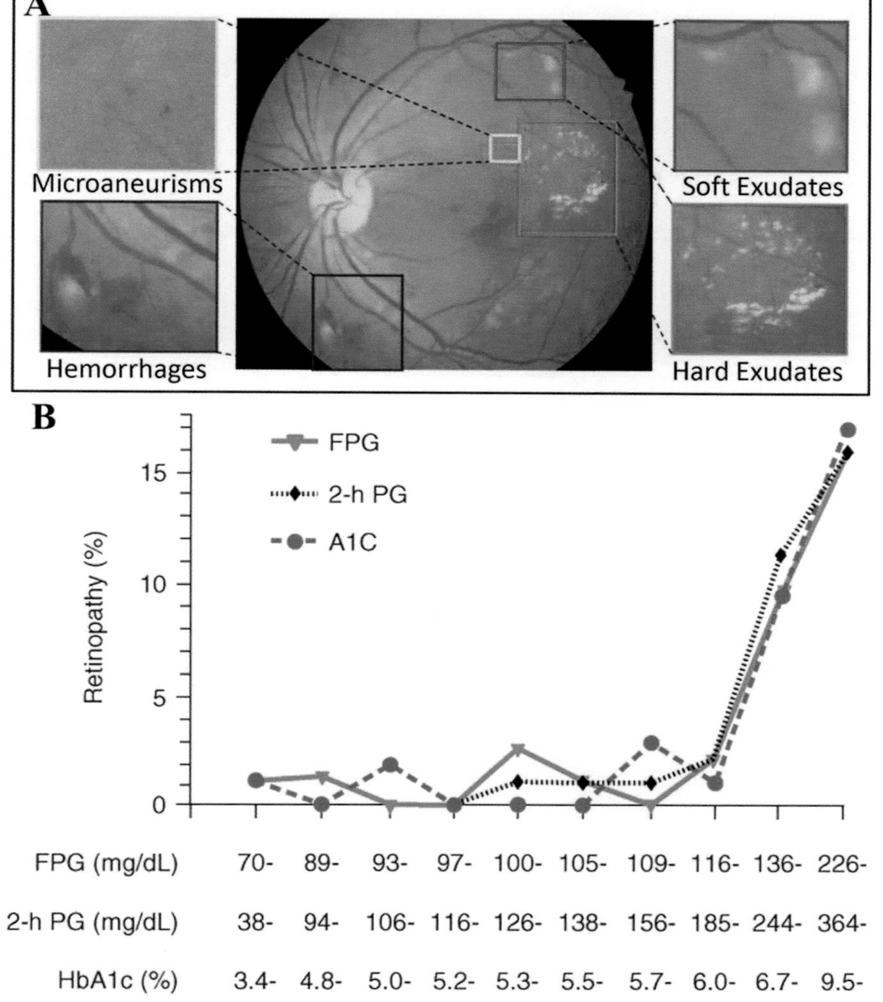

Figure 19.19 Relationship of diabetes-specific retinopathy and glucose tolerance. (A) Different retinal lesions associated with diabetic retinopathy. (B) This figure shows the incidence of retinopathy in Pima Indians as a function of the fasting plasma glucose (FPG), the 2-h plasma glucose after a 75-g oral glucose challenge (2-h PG), or the A1C. Note that the incidence of retinopathy greatly increases at a fasting plasma glucose >116 mg/dL, or a 2-h plasma glucose of 185 mg/dL, or an A1C > 6.5%.

Source of images: (A) Porwal, P., et al. 2018. Data, 3(3), p. 25. (B) Harrison's Principles of Internal Medicine, 18th edition. McGraw-Hill. Used by permission.

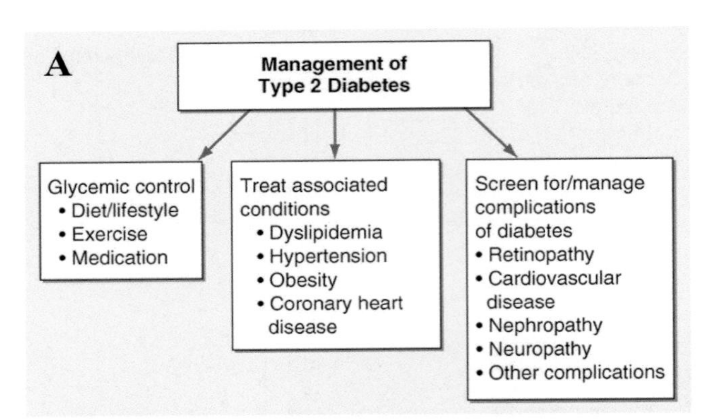

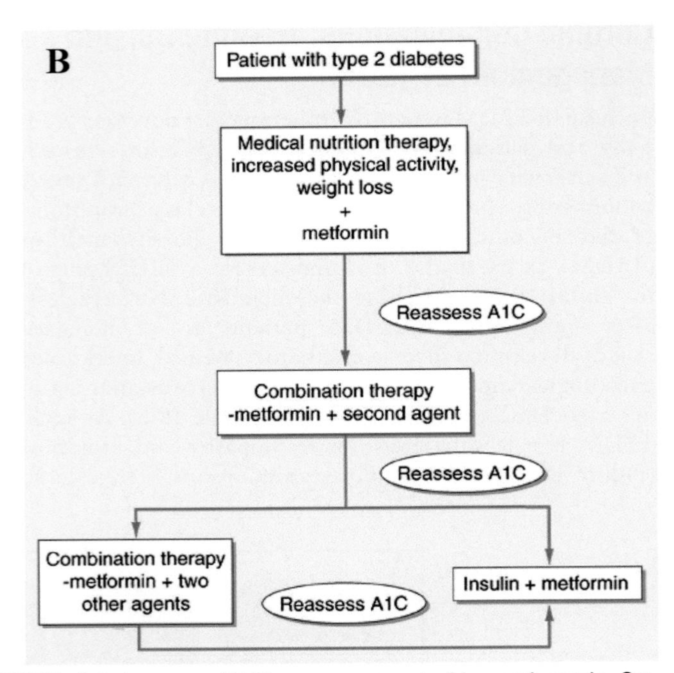

Figure 19.20 The management of T2DM. (A) Critical elements of T2DM diabetes care. **(B)** The management of hyperglycemia. See text for discussion of treatment of severe hyperglycemia or symptomatic hyperglycemia. Agents that can be combined with metformin include insulin secretagogues, thiazolidinediones, α-glucosidase inhibitors, and GLP-1 receptor agonists. A1C, hemoglobin A1C.

Source of images: Harrison's Principles of Internal Medicine, 18th edition. McGraw-Hill. Used by permission.

"Sir, so for Type 1 diabetes in which insulin levels are low, you give insulin?"
"Yes"
"And for Type 2, in which insulin levels are high, you also give insulin?
"Yes"
"That doesn't make sense."

Figure 19.21 Some humor depicting the complexity associated with the treatment of diabetes.

Source of images: Cartoon courtesy of Walter J. Pories, MD

also treatment of accompanying conditions such as obesity, hypertension, dyslipidemia, and cardiovascular and renal diseases (Figure 19.20). In addition to prescribing exercise and diet control, pharmacological approaches to T2DM management include multiple categories of glucose-lowering treatments, including insulin and agents that increase insulin secretion (insulin secretagogues and incretins), agents that reduce glucose synthesis (inhibit gluconeogenesis and glycogenolysis), and agents that increase insulin sensitivity (see Figure 19.20, Figure 19.21, and Table 19.7).

Developments & Directions: BAT transplantation improves glucose metabolism and insulin sensitivity in rodent models

In Chapter 18: Energy Homeostasis, you learned about another category of adipose tissue called brown adipose tissue (BAT) that is involved in thermogenesis. In mouse models, there is growing evidence for BAT as a potential therapeutic target in the prevention or treatment of obesity and diabetes.[134] Specifically, many studies have shown that transplantation of BAT into recipient mice results in dramatic improvements in glucose tolerance and reductions in adiposity, even reversing obesity and diabetes.[135–139] These effects appear to be mediated, at least in part, through the release of various BAT-derived hormones that improve metabolic health, such as interleukin-6 (IL6), insulin-like growth factor-1 (IGF-1), vascular endothelial growth factor A (VEGFA), and bone morphogenetic proteins (BMPs). An interesting alternative to BAT transplants is the *ex vivo* culturing of BAT adipocytes from precursor cells that are reimplanted into the donor to increase thermogenic capacity. Such experiments have been performed successfully in mice[140,141] and may hold therapeutic promise for humans.

Table 19.7 Some Agents Used in the Treatment of Type 2 Diabetes Mellitus

Category	A1C Reduction (%)	Mechanism of Action
Oral administration		
Sulfonylureas	1–2	Stimulate insulin secretion by binding to and closing ATP-sensitive potassium channels on the surface of pancreatic beta cells resulting in closure of the channel and depolarization of the beta cell.
Metformin	1–2	Reduces hepatic glucose production by activating AMPK by uncoupling mitochondria oxidative phosphorylation and increasing cellular AMP levels. Increased hepatic AMPK activity reduces gluconeogenesis and lipogenesis.
Thiazolidinediones	0.5–1.4	PPAR-gamma agonists that reduce insulin resistance in part by enhancing GLUT expression, reducing free fatty acid levels, decreasing hepatic glucose output, increasing adiponectin and reducing resistin release from adipocytes, and increasing the differentiation of preadipocytes into adipocytes.
Alpha-glucosidase inhibitors	0.5–0.8	Reduce rates of intestinal glucose absorption by inhibiting intestinal brush border alpha-glucosidase activity.
Non-oral administration		
Insulin	Not limited	Insulin receptor agonist, increases glucose utilization, reduces hepatic glucose synthesis.
GLP-1 receptor agonists	0.5–1.0	GLP-1 is an incretin, which increases insulin secretion by binding to beta cell GLP-1 receptors.
Amylin agonists	0.25–0.5	Slows gastric emptying, reduces glucagon synthesis, reduces post-prandial glycemia.
Nutrition therapy and physical activity	1–3	Low-calorie diet and exercise reduces insulin resistance, augments insulin secretion.

Note: A1C reduction (absolute) depends partially on starting A1C levels.
Source: Adapted from Harrison's *Endocrinology*, 3rd edition table 19–11. McGraw-Hill

SUMMARY AND SYNTHESIS QUESTIONS

1. The tissues of individuals with untreated T1DM and T2DM each have impaired abilities of tissues to take up blood glucose. Compare and contrast the cellular basis for this impairment in both diseases.
2. In T2DM, the early stages of the disease may be characterized by hyperinsulemia, but the later stages characterized by reduced insulin secretion. What explains the transition from one state to the next?
3. Whereas late-stage untreated T1DM is often accompanied by a lethal drop in blood pressure, late stage T2DM patients are generally at an elevated risk for developing high blood pressure. Explain why these symptoms differ so dramatically in the two diseases.
4. Thiazolidinedione drugs are PPAR-gamma agonists. How can these be effective treatments for type 2 diabetes?
5. Refer to Figure 19.21, which is a cartoon. Explain why it does indeed make sense to treat both type 1 and type 2 diabetes mellitus with insulin.
6. The drug, Ozempic, is a GLP-1 analog that is used to treat type 2 diabetes. First, refer back to **Chapter 17: Appetite** to remind yourself how GLP-1 functions. Then, describe the diverse ways in which this type of drug could be useful in treating type 2 diabetes.

Metabolism-Disrupting Compounds

LEARNING OBJECTIVE Define the terms "obesogen", "diabetogen", and "diabesogen", and provide some examples of these compounds and their mechanisms of action based on studies with humans and animal models.

KEY CONCEPTS:

- "Obesogens" are molecules that inappropriately alter lipid metabolism and adipogenesis and promote obesity.
- "Diabetogens" induce type 2 diabetes by promoting obesity, altering pancreatic β cell insulin synthesis, and/or inducing insulin resistance in target tissues such as adipocytes, skeletal muscle, and liver.

Although genetic susceptibility and "life-style" risk factors associated with overeating and sedentary behavior are major contributors to obesity, type 2 diabetes, heart disease, and other metabolic disorders, these factors alone do not explain the magnitude or rapidly rising rates of the global metabolic disease epidemic. Endocrine-disrupting compounds (EDCs) have been proposed to play a role in their etiologies.[45,142–144] The term **metabolism-disrupting chemical** (MDC) refers to any EDC that alters susceptibility to developing metabolic disorders, and encompasses several more specific terms: (1) **obesogen**, an EDC that alters the regulation of energy balance to favor weight gain and obesity; (2) **diabetogen**, an EDC that promotes the development of diabetes, and (3) **diabesogen**, an EDC that can induce both diabetes and obesity.[145] This chapter section describes the links of some metabolism-disrupting compounds with obesity and type 2 diabetes in humans and model organisms.

EDC Links to Obesity

Mechanisms of Obesogen Action

EDCs have been proposed to be an important risk factor that contributes to the global obesity epidemic.[146–148] In 2006, Bruce Blumberg and Felix Grün at the University of California, Irvine, brought attention to the role of EDCs in the global obesity epidemic and coined the term "obesogen", defining the term as "molecules that inappropriately regulate lipid metabolism and adipogenesis to promote obesity".[149] Some broad mechanisms by which obesogens promote obesity have been summarized,[150] and include:

- Increasing the number of adipocytes
- Increasing the size of adipocytes, storage of fat per cell, or both
- Altering endocrine pathways responsible for control of adipose tissue commitment and differentiation from mesenchymal stem cell precursors
- Altering hormones that regulate appetite, satiety, and food preferences
- Altering basal metabolic rate
- Altering energy balance to favor storage of calories
- Altering insulin synthesis by pancreatic beta cells
- Altering insulin sensitivity of target tissues such as adipose tissue, liver, gastrointestinal tract, brain, and muscle
- Disrupting intestinal epithelial transport, gut microbiota composition, and the release of gut peptides that may alter serum levels of nutrients

Regarding adipogenesis, *multipotent mesenchymal stem cells* have the potential to develop into chondroblasts, myoblasts, osteoblasts, fibroblasts, stromal cells, or adipocytes (Figure 19.22). After committing to the adipogenic lineage, the progeny of these stem cells become preadipocytes and ultimately differentiate into mature adipocytes. This complex process is mediated by diverse transcription factors, described in Chapter 18: Energy Homeostasis (refer to Figure 18.11 of that chapter). Importantly, in both *in vitro* and *in vivo* experiments, numerous

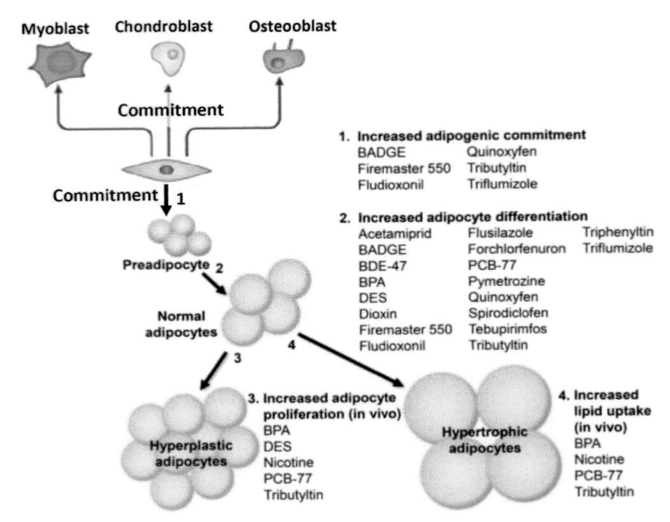

Figure 19.22 Mechanisms of adipogenesis and actions of some known metabolism disrupting chemicals (MDCs) in rodent models. 1. Mesenchymal stem cells first commit to the adipogenic lineage by becoming preadipocytes. 2. Preadipocytes then differentiate into adipocytes. 3. Some MDCs can induce adipose tissue to grow via hyperplasia. 4. MDCs can also promote adipose tissue growth via hypertrophy.

Source of images: Preadipocyte lineage: Ghaben, A.L. and Scherer, P.E., 2019. Nat. Rev. Molec. Cell Biol., 20(4), pp. 242-258. Used by permission. Adipose cell lineage: Heindel, J.J., et al. 2017. Reproduct. Toxicol., 68, pp. 3-33. Used by permission.

obesogenic synthetic compounds are not only capable of biasing mesenchymal stem cell developmental commitment towards preadipocytes and differentiation into mature adipocytes, but also can promote their expansion via hyperplastic or hypertrophic growth, in many cases by

Developments & Directions: The "vicious upwards spiral" of obesogens and obesity

Because most obesogens and EDCs are lipophilic and can bioaccumulate in body fat over time, it is likely that obese individuals retain greater amounts of lipophilic pollutants in their higher fat volume compared to a person of normal weight. This notion is important and suggests that the link of obesity to chronic disease relates not only to the deposition of fat, but also to an elevated obesogen body burden. Figure 19.23 illustrates the potential for such a "vicious upwards spiral" to be generated whereby obesogens first act to increase the amount of fat stored, which is followed by greater retention of lipophilic obesogens in the fat, which subsequently leads to an increasing spiral of greater body fat and even more lipophilic pollutant retention.[150] As such, obesogens may be able to increase capacity for their own accumulation and retention. Importantly, the obesogenic nature of such EDCs may also enable increased storage of a wide range of other *non*-obesogenic lipophilic EDCs that have other adverse effects on the human body, such as the promotion of breast cancer, prostate cancer, and infertility. Therefore, the actions of obesogens may have an even broader impact on the development of chronic diseases than might be expected from merely the accumulation of excess fat.

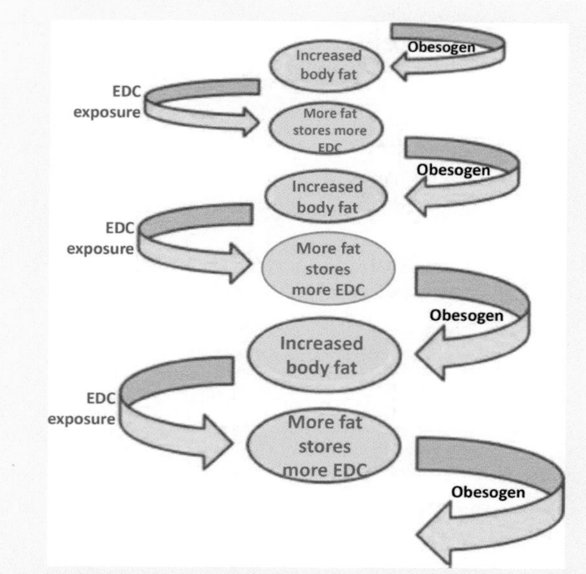

Figure 19.23 The "vicious spiral" of obesogenic activity and lipophilic properties of endocrine disruptors. See text for details.

Source of images: Darbre, P.D., 2017. Curr Obesity Rep, 6(1), pp. 18-27.

activating or mimicking the respective transcription factors[151] (Figure 19.22).

One of the more nefarious effects of certain obesogens is their ability to predispose offspring to developing type 2 diabetes in adulthood. In both animal and human studies, exposure to the estrogen mimic, DES, during fetal or neonatal development leads to both adipocyte hyperplasia and increased weight gain, predisposing the offspring to developing obesity during adulthood.[152–154] In a particularly dramatic experiment, Newbold and colleagues[4,153] showed that female mice treated during the perinatal period with as little as 1 part per billion of DES did not affect body weight during treatment, but it induced grotesque obesity in the mice upon reaching adulthood (Figure 19.24).

EDC Links to Type 2 Diabetes Mellitus (T2DM)

Mounting epidemiological data suggest a strong dose-response relation between blood or urinary concentrations of EDCs and the prevalence of T2DM in humans[144,151,155,156] (Figure 19.25). Therefore, in addition to obesity, the body burden of particular EDCs

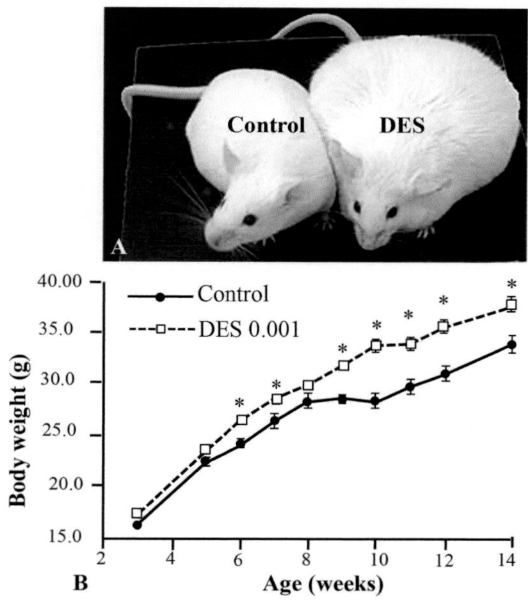

Figure 19.24 Neonatal treatments of mice with DES induce obesity in the adult. Mice treated with a low dose (1 μg/kg/day) of DES on days 1–5 of neonatal life did not affect body size during treatment, but the mice became obese starting at 6 weeks of age near adulthood. **(A)** A representative image of control and DES-treated mice as 4- to 6-month-old adults is shown. **(B)** Each point represents a minimum of 20 mice per dose per age. *Statistically significant difference from untreated.

Source of images: (A) Newbold, R.R., et al. 2009. Molec. Cell. Endocrinol., 304(1-2), pp. 84-89. Used by permission. (B) Modified from Newbold, R.R et al. 2005. Birth Defects Res A: Clin. Mol. Teratol., 73(7), pp. 478-480.

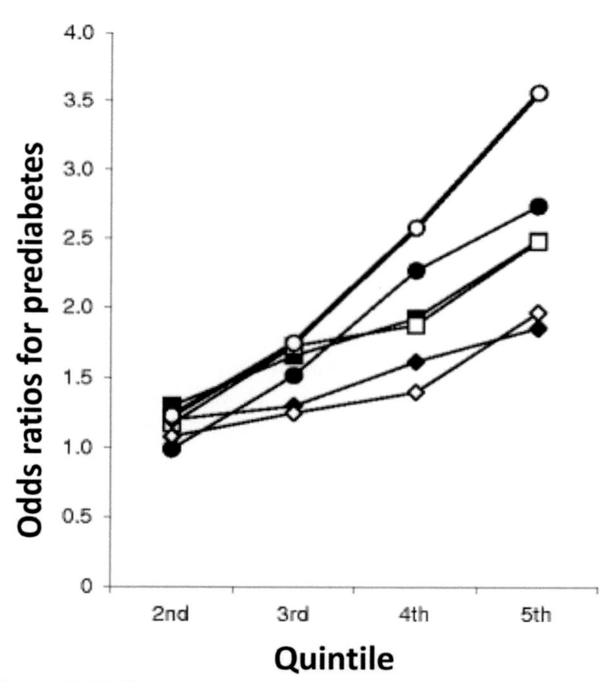

Figure 19.25 The prevalence of prediabetes increases with increased circulating levels of endocrine-disrupting chemicals. Black circles, polychlorinated biphenyls (PCBs, 15 congeners); black squares, dichlorodiphenyldichloroethylene (p,p′-DDE); white squares, dichlorodiphenyltrichloroethane (p,p′-DDT); black diamonds, hexachlorobenzene (HCB); white diamonds, hexachlorocyclohexane (HCH); white circles, POLL5 (variable expressing the cumulative effect of all five EDCs).

Source of images: Ukropec J, et al. 2010; Diabetologia. 53(5):899-906. Used by permission.

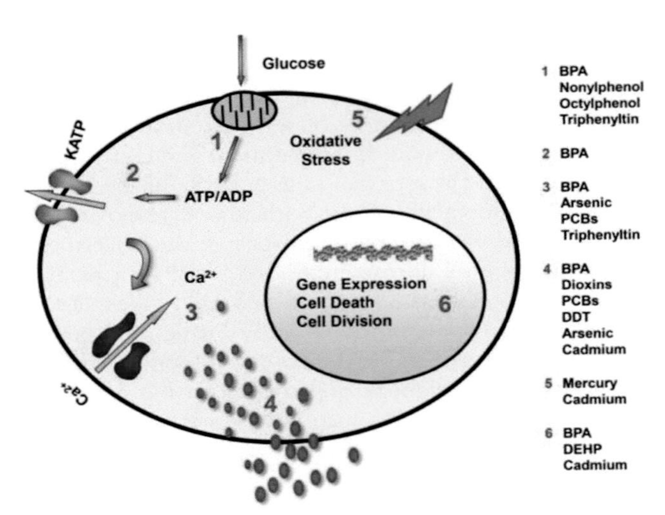

Figure 19.26 Actions of metabolism disrupting chemicals (MDCs) on pancreatic beta cells. (1) Blood glucose enters pancreatic beta cells through glucose GLUT1 transporters in humans, where it is metabolized in mitochondria resulting in an increase in the ATP/ADP ratio. (2) This promotes closure of membrane ATP-sensitive K^+ channels (K_{ATP}) that are responsible for maintaining the resting membrane potential. (3) K_{ATP} channel closure results in cellular depolarization that opens voltage-gated calcium channels, triggering Ca^{2+} signals that (4) induce insulin granule exocytosis and a subsequent rise in circulating insulin levels. This secretory pathway is disrupted by different EDCs at different points (1–4). In addition, some EDCs can damage or kill beta cells by producing oxidative stress (5), and others promote cell death and reduced beta cell mass by altering nuclear function (6).

Source of images: Heindel, J.J., et al. 2017. Reprod. Toxicol., 68, pp. 3-33. Used by permission.

can also be used to predict the likelihood of developing T2DM. In particular, diabetogens may induce T2DM by promoting obesity, altering pancreatic β cell insulin synthesis (Figure 19.26), and/or inducing insulin resistance in target tissues such as adipocytes, skeletal muscle, and liver.

Some of the first studies linking human chemical exposure to the onset of diabetes focused on patients exposed to diabetogens occupationally. Such exposures include (1) exposure of military personnel to dioxins during the Vietnam War; (2) pesticide exposures of agricultural workers; (3) exposure via industrial accidents, such as a 1977 explosion of a chemical plant in Seveso, Italy, that released the highest known exposures of residential populations to dioxin; (4) a 1979 contamination of rice with PCBs and furans in Yucheng, Taiwan; and (5) recreational exposure, such as consumption of the DDT byproduct, DDE, from sport fish in the Great Lakes.[151,157] Such events led to the study of these and other chemicals in model organisms, particularly rodents, to determine their diabetogenic mechanisms of actions, and their findings indicate that the insulin-secreting pancreatic β cells are a main target for disruption.

SUMMARY AND SYNTHESIS QUESTIONS

1. List three broad mechanisms by which diabetogens act.
2. Beyond just promoting the accumulation of body fat, obesogens may also promote the storage of other EDCs that contribute to other chronic diseases. How?
3. Whereas some diabetogens, like cadmium, arsenic, and Hg, cause type 1 diabetes (i.e. reduced insulin production), others with estrogenic effects, such as BPA and other estrogen mimics, can promote type 2 diabetes (e.g. insulin resistance). How/why?
4. The focus of this chapter was on energy dysregulation caused by obesity and diabetes. Provide some examples of other notable metabolic disorders from other chapters of this book.

Summary of Chapter Learning Objectives and Key Concepts

LEARNING OBJECTIVE List the types of diseases that are associated with metabolic dysfunction, and define "metabolic syndrome."

- The dysregulation of energy homeostasis may lead to the development of chronic metabolic disorders that include obesity, type 2 diabetes, and cardiovascular disease.
- Genetic susceptibility and "life-style" risk factors associated with overeating and sedentary behavior are contributors to metabolic disorders. However, these factors alone do not explain the rapidly rising rates of the global metabolic disease epidemic over just the last few decades.
- The "metabolic syndrome" describes the presence of three or more metabolic risk factors in a patient that put them at increased risk for developing type 2 diabetes mellitus, cardiovascular disease, and other metabolic disorders.

LEARNING OBJECTIVE Describe the "developmental origins of health and disease", the "thrifty phenotype" hypothesis, and how these concepts are useful for explaining the global rise in metabolic disease.

- "Adaptive developmental plasticity" describes environment-dependent developmental strategies that an organism can take to maximize its probability of survival and reproduction in a new environment.
- When a resulting developmental phenotype produces a mismatch with the actual environment, this can lead to a loss of fitness.
- The "thrifty phenotype" hypothesis proposes that in adapting to an adverse intrauterine environment, such as under- or overnutrition, the fetus permanently alters

its physiology and metabolism, potentially increasing its risk of developing obesity, type 2 diabetes, cardiovascular disease, and other metabolic disorders later in life.

LEARNING OBJECTIVE Define "obesity", explain how it is a risk factor for developing other metabolic disorders, and contrast the endocrine function of adipose tissue in the lean versus obese states.

- Obesity stems from complex interactions among behavioral, environmental, physiological, genetic, epigenetic, social, and economic variables.
- Rather than simply arising from the passive accumulation of excess weight, obesity appears to be a disorder of the energy homeostasis system.
- White adipose tissue is an active endocrine organ, and in the obese state potentially the largest in the body.
- Under conditions of obesity, the white adipose tissue endocrine profile shifts, with increased production of hormones that promote insulin resistance, hyperglycemia, and inflammation.
- Due to its distinct morphological and biochemical features, visceral obesity is more dangerous to health than subcutaneous obesity.

LEARNING OBJECTIVE Define "diabetes mellitus" and describe the physiological implications of chronic hyperglycemia.

- Diabetes mellitus is a group of metabolic diseases characterized by hyperglycemia resulting from defects in insulin secretion, insulin action, or both.
- Hyperglycemia causes a diabetic to produce a high volume of glucose-containing urine, resulting in constant thirst and drinking to replace the lost water.
- The inability of diabetics to effectively take up and metabolize glucose forces their tissues to rely on other fuels for energy, such as fatty acids, amino acids, and ketones.
- Chronic hyperglycemia of diabetes is associated with long-term damage, dysfunction, and failure of different organs, especially the eyes, kidneys, nerves, heart, and blood vessels.

LEARNING OBJECTIVE Describe the etiology (cause) of type 1 diabetes mellitus (T1DM) and its symptoms and complications.

- Type 1 diabetes mellitus (T1DM) typically results from the autoimmune destruction of the insulin-secreting pancreatic β cells in children, but it can occur at any age.
- In the absence insulin, target cells fail to express GLUT4 glucose transporters on their plasma membranes, inhibiting glucose uptake and metabolism.
- The loss of glucagon inhibition by insulin in T1DM promotes elevated blood glucagon levels, which results in increased glycogenolysis by the liver.
- In untreated T1DM, elevated ketone synthesis can lead to the potentially fatal condition of metabolic acidosis.
- The most common treatment for T1DM is the daily injection of insulin.

LEARNING OBJECTIVE Describe the etiology of type 2 diabetes mellitus (T2DM) and its symptoms, complications, and link to obesity.

- Type 2 diabetes mellitus (T2DM) symptoms may include hyperglycemia, insulin resistance, excessive hepatic glucose synthesis, abnormal fat metabolism, and impaired insulin secretion.
- The most common cause of T2DM is obesity.
- Globally, the occurrence of T2DM, together with obesity, has risen steeply to epidemic proportions.

LEARNING OBJECTIVE Define the terms "obesogen", "diabetogen", and "diabesogen", and provide some examples of these compounds and their mechanisms of action based on studies with humans and animal models.

- "Obesogens" are molecules that inappropriately alter lipid metabolism and adipogenesis and promote obesity.
- "Diabetogens" induce type 2 diabetes by promoting obesity, altering pancreatic β cell insulin synthesis, and/or inducing insulin resistance in target tissues such as adipocytes, skeletal muscle, and liver.

LITERATURE CITED

1. Complètes d'Hippocrate Œ, complètes d'Hippocrate Œ. Traduction nouvelle, avec le texte grec en regard, collationné sur les manuscrits et toutes les éditions; accompagnée d'une introduction de commentaires médicaux, de variantes et de notes philologiques: suivie d'une table générale des matières. *Par É Littré*. 1839;4.
2. Zafón C. Evolutionary endocrinology: a pending matter. *Endocrinol Nutr (English Edition)*. 2012;59(1):62–68.
3. Grundy SM, Cleeman JI, Daniels SR, et al. Diagnosis and management of the metabolic syndrome. An American Heart Association/National Heart, Lung, and Blood Institute Scientific Statement. Executive summary. *Cardiol Rev*. 2005;13(6):322–327.
4. Newbold RR, Padilla-Banks E, Snyder RJ, Jefferson WN. Developmental exposure to estrogenic compounds and obesity. *Birth Defects Res A Clin Mol Teratol*. 2005;73(7):478–480.
5. West-Eberhard MJ. Phenotypic plasticity and the origins of diversity. *Ann Rev Ecol Syst*. 1989;20(1):249–278.
6. Whitman DW, Agrawal AA. What is phenotypic plasticity and why is it important. In: Whitman DW, Ananthakrishnan TN, eds. *Phenotypic Plasticity of Insects: Mechanisms and Consequences*. Science Publishers; 2009:1–63.
7. Barker DJ. The origins of the developmental origins theory. *J Int Med*. 2007;261(5):412–417.
8. Barker DJ, Osmond C, Winter P, Margetts B, Simmonds SJ. Weight in infancy and death from ischaemic heart disease. *Lancet*. 1989;334(8663):577–580.
9. Barker DJ, Osmond C, Golding J, Kuh D, Wadsworth M. Growth in utero, blood pressure in childhood and adult life, and mortality from cardiovascular disease. *BMJ*. 1989;298(6673):564–567.

10. Barker DJ. Maternal nutrition, fetal nutrition, and disease in later life. *Nutrition*. 1997;13(9):807–813.
11. Saad M, Abdelkhalek T, Haiba M, et al. Maternal obesity and malnourishment exacerbate perinatal oxidative stress resulting in diabetogenic programming in F1 offspring. *J Endocrinol Investig*. 2016;39(6):643–655.
12. Newbold RR, Padilla-Banks E, Snyder RJ, Phillips TM, Jefferson WN. Developmental exposure to endocrine disruptors and the obesity epidemic. *Reprod Toxicol*. 2007;23(3):290–296.
13. Gluckman PD, Hanson MA. Developmental origins of disease paradigm: a mechanistic and evolutionary perspective. *Pediatr Res*. 2004;56(3):311.
14. Heindel JJ, Balbus J, Birnbaum L, et al. Developmental origins of health and disease: integrating environmental influences. *Endocrinology*. 2016;2016(1):17–22.
15. Barker DJ. Fetal origins of coronary heart disease. *BMJ*. 1995;311(6998):171–174.
16. Hales CN, Barker DJ. Type 2 (non-insulin-dependent) diabetes mellitus: the thrifty phenotype hypothesis. *Diabetologia*. 1992;35(7):595–601.
17. Barker DJ, Hales CN, Fall C, Osmond C, Phipps K, Clark P. Type 2 (non-insulin-dependent) diabetes mellitus, hypertension and hyperlipidaemia (syndrome X): relation to reduced fetal growth. *Diabetologia*. 1993;36(1):62–67.
18. Hales CN, Barker DJP. The thrifty phenotype hypothesis: Type 2 diabetes. *Brit Med Bull*. 2001;60(1):5–20.
19. Gruber T, Pan C, Contreras RE, et al. Obesity-associated hyperleptinemia alters the gliovascular interface of the hypothalamus to promote hypertension. *Cell Metab*. 2021;33(6):1155–1170.e1110.
20. Dearden L, Ozanne SE. Early life origins of metabolic disease: Developmental programming of hypothalamic pathways controlling energy homeostasis. *Front Neuroendocrinol*. 2015;39:3–16.
21. Schulz LC. The Dutch Hunger Winter and the developmental origins of health and disease. *Proc Nat Acad Sci*. 2010;107(39):16757–16758.
22. Boekelheide K, Blumberg B, Chapin RE, et al. Predicting later-life outcomes of early-life exposures. *Environ Heal Perspect*. 2012;120(10):1353–1361.
23. Roseboom T, de Rooij S, Painter R. The Dutch famine and its long-term consequences for adult health. *Early Hum Dev*. 2006;82(8):485–491.
24. Roseboom TJ, Van Der Meulen JH, Ravelli AC, Osmond C, Barker DJ, Bleker OP. Effects of prenatal exposure to the Dutch famine on adult disease in later life: an overview. *Twin Res Hum Genet*. 2001;4(5):293–298.
25. Roseboom TJ, van der Meulen JH, Osmond C, Barker DJ, Ravelli AC, Bleker OP. Plasma lipid profiles in adults after prenatal exposure to the Dutch famine. *Am J Clin Nutr*. 2000;72(5):1101–1106.
26. Painter R, Osmond C, Gluckman P, Hanson M, Phillips D, Roseboom TJ. Transgenerational effects of prenatal exposure to the Dutch famine on neonatal adiposity and health in later life. *BJOG*. 2008;115(10):1243–1249.
27. Lin X, Lim IY, Wu Y, et al. Developmental pathways to adiposity begin before birth and are influenced by genotype, prenatal environment and epigenome. *BMC Med*. 2017;15(1):50.
28. Yu Z, Han S, Zhu G, et al. Birth weight and subsequent risk of obesity: a systematic review and meta-analysis. *Obes Rev*. 2011;12(7):525–542.
29. Larqué E, Labayen I, Flodmark C-E, et al. From conception to infancy—early risk factors for childhood obesity. *Nat Rev Endocrinol*. 2019:1.
30. Purnell JQ. Definitions, classification, and epidemiology of obesity. *Endotext [Internet]*. South Dartmouth, MA; 2018.
31. Nordqvist C. Why BMI is inaccurate and misleading. *STAT*. 2013;408:608.
32. Rothman KJ. BMI-related errors in the measurement of obesity. *Int J Obes*. 2008;32(3):S56–S59.
33. Adab P, Pallan M, Whincup PH. *Is BMI the Best Measure of Obesity?* British Medical Journal Publishing Group; 2018.
34. Nuttall FQ. Body mass index: obesity, BMI, and health: a critical review. *Nutr Today*. 2015;50(3):117.
35. Batsis JA, Mackenzie TA, Bartels SJ, Sahakyan KR, Somers VK, Lopez-Jimenez F. Diagnostic accuracy of body mass index to identify obesity in older adults: NHANES 1999–2004. *Int J Obes*. 2016;40(5):761–767.
36. Vanderwall C, Randall Clark R, Eickhoff J, Carrel AL. BMI is a poor predictor of adiposity in young overweight and obese children. *BMC Pediatr*. 2017;17(1):1–6.
37. Tomiyama AJ, Hunger JM, Nguyen-Cuu J, Wells C. Misclassification of cardiometabolic health when using body mass index categories in NHANES 2005–2012. *Int J Obes*. 2016;40(5):883–886.
38. Ahima RS, Lazar MA. The health risk of obesity—better metrics imperative. *Science*. 2013;341(6148):856–858.
39. Nuttall FQ. Body mass index: obesity, BMI, and health: a critical review. *Nutr Today*. 2015;50(3):117–128.
40. Finkelstein EA, Khavjou OA, Thompson H, et al. Obesity and severe obesity forecasts through 2030. *Am J Prev Med*. 2012;42(6):563–570.
41. Kelly T, Yang W, Chen C-S, Reynolds K, He J. Global burden of obesity in 2005 and projections to 2030. *Int J Obes*. 2008;32(9):1431.
42. Organization WH. Obesity and overweight. *Fact Sheets* 2018.
43. Blüher M. Obesity: global epidemiology and pathogenesis. *Nat Rev Endocrinol*. 2019;15(5):288–298.
44. González-Muniesa P, Mártinez-González M-A, Hu FB, et al. Obesity. *Nat Rev Dis Prim*. 2017;3:17034.
45. Schwartz MW, Seeley RJ, Zeltser LM, et al. Obesity pathogenesis: an Endocrine Society scientific statement. *Endocr Rev*. 2017;38(4):267–296.
46. Speliotes EK, Willer CJ, Berndt SI, et al. Association analyses of 249,796 individuals reveal 18 new loci associated with body mass index. *Nat Gen*. 2010;42(11):937–948.
47. Maher B. Personal genomes: The case of the missing heritability. *Nature*. 2008;456(7218):18–21.
48. Tandon P, Wafer R, Minchin JE. Adipose morphology and metabolic disease. *J Exp Biol*. 2018;221(Suppl 1):jeb164970.
49. Verboven K, Wouters K, Gaens K, et al. Abdominal subcutaneous and visceral adipocyte size, lipolysis and inflammation relate to insulin resistance in male obese humans. *Sci Rep*. 2018;8(1):4677.
50. Ghaben AL, Scherer PE. Adipogenesis and metabolic health. *Nat Rev Mol Cell Biol*. 2019:1.
51. Yang J, Eliasson B, Smith U, Cushman SW, Sherman AS. The size of large adipose cells is a predictor of insulin resistance in first-degree relatives of type 2 diabetic patients. *Obesity*. 2012;20(5):932–938.
52. Lonn M, Mehlig K, Bengtsson C, Lissner L. Adipocyte size predicts incidence of type 2 diabetes in women. *The FASEB J*. 2010;24(1):326–331.
53. McLaughlin T, Sherman A, Tsao P, et al. Enhanced proportion of small adipose cells in insulin-resistant vs insulin-sensitive obese individuals implicates impaired adipogenesis. *Diabetologia*. 2007;50(8):1707–1715.
54. Lundgren M, Svensson M, Lindmark S, Renström F, Ruge T, Eriksson JW. Fat cell enlargement is an independent marker of insulin resistance and "hyperleptinaemia". *Diabetologia*. 2007;50(3):625–633.
55. Gregor MF, Hotamisligil GS. Thematic review series: Adipocyte Biology. Adipocyte stress: the endoplasmic reticulum and metabolic disease. *J Lipid Res*. 2007;48(9):1905–1914.
56. Coelho M, Oliveira T, Fernandes R. Biochemistry of adipose tissue: an endocrine organ. *Arch Med Sci*. 2013;9(2):191–200.
57. Halberg N, Khan T, Trujillo ME, et al. Hypoxia-inducible factor 1α induces fibrosis and insulin resistance in white adipose tissue. *Mol Cell Biol*. 2009;29(16):4467–4483.
58. Wu H, Ballantyne CM. Metabolic Inflammation and Insulin Resistance in Obesity. *Circ Res*. 2020;126(11):1549–1564.
59. Green WD, Beck MA. Obesity impairs the adaptive immune response to influenza virus. *Ann Am Thorac Soc*. 2017;14(Supplement 5):S406–S409.
60. Luzi L, Radaelli MG. Influenza and obesity: its odd relationship and the lessons for COVID-19 pandemic. *Acta Diabetologica*. 2020:1–6.
61. Honce RR, Schultz-Cherry S. Impact of obesity on influenza A virus pathogenesis,

immune response, and evolution. *Front Immunol.* 2019;10:1071.
62. Kim MH, Seong JB, Huh J-W, Bae YC, Lee H-S, Lee D-S. Peroxiredoxin 5 ameliorates obesity-induced non-alcoholic fatty liver disease through the regulation of oxidative stress and AMP-activated protein kinase signaling. *Redox Biol.* 2020;28:101315.
63. Shi H, Seeley RJ, Clegg DJ. Sexual differences in the control of energy homeostasis. *Front Neuroendocrinol.* 2009;30(3):396–404.
64. Bouchard C, Després J-P, Mauriège P. Genetic and nongenetic determinants of regional fat distribution. *Endocr Rev.* 1993;14(1):72–93.
65. Bjørntorp P. Hormonal effects on fat distribution and its relationship to health risk factors. *Acta Paediatr Suppl.* 1992;383:59–60; discussion 61.
66. Gambacciani M, Ciaponi M, Cappagli B, et al. Body weight, body fat distribution, and hormonal replacement therapy in early postmenopausal women. *J Clin Endocrinol Metab.* 1997;82(2):414–417.
67. Haarbo J, Marslew U, Gotfredsen A, Christiansen C. Postmenopausal hormone replacement therapy prevents central distribution of body fat after menopause. *Metabolism.* 1991;40(12):1323–1326.
68. Diamanti-Kandarakis E. Role of obesity and adiposity in polycystic ovary syndrome. *Int J Obes (Lond).* 2007;31 Suppl 2:S8–13; discussion S31–12.
69. Hiuge-Shimizu A, Kishida K, Funahashi T, et al. Absolute value of visceral fat area measured on computed tomography scans and obesity-related cardiovascular risk factors in large-scale Japanese general population (the VACATION-J study). *Ann Med.* 2012;44(1):82–92.
70. Sanchez-Garrido MA, Tena-Sempere M. Metabolic dysfunction in polycystic ovary syndrome: Pathogenic role of androgen excess and potential therapeutic strategies. *Mol Metab.* 2020;35:100937.
71. Abate N, Garg A, Peshock RM, Stray-Gundersen J, Adams-Huet B, Grundy SM. Relationship of generalized and regional adiposity to insulin sensitivity in men with NIDDM. *Diabetes.* 1996;45(12):1684–1693.
72. Frayn KN. Visceral fat and insulin resistance—causative or correlative? *Brit J Nutr.* 2000;83(S1):S71–S77.
73. Weisberg SP, McCann D, Desai M, Rosenbaum M, Leibel RL, Ferrante AW. Obesity is associated with macrophage accumulation in adipose tissue. *J Clin Investig.* 2003;112(12):1796–1808.
74. Lemieux I, Pascot A, Prud'homme D, et al. Elevated C-reactive protein: another component of the atherothrombotic profile of abdominal obesity. *Arterioscler Thromb Vasc Biol.* 2001;21(6):961–967.
75. Pepys MB, Hirschfield GM. C-reactive protein: a critical update. *J Clin Investig.* 2003;111(12):1805–1812.
76. Forouhi N, Sattar N, McKeigue P. Relation of C-reactive protein to body fat distribution and features of the metabolic syndrome in Europeans and South Asians. *Int J Obes.* 2001;25(9):1327.
77. Pou KM, Massaro JM, Hoffmann U, et al. Visceral and subcutaneous adipose tissue volumes are cross-sectionally related to markers of inflammation and oxidative stress. *Circulation.* 2007;116(11):1234–1241.
78. Danforth Jr E. Failure of adipocyte differentiation causes type II diabetes mellitus? *Nat Gen.* 2000;26(1):13.
79. Bray GA, Redman LM, de Jonge L, Rood J, Smith SR. Effect of three levels of dietary protein on metabolic phenotype of healthy individuals with 8 weeks of overfeeding. *J Clin Endocrinol Metab.* 2016;101(7):2836–2843.
80. Mårin P, Andersson B, Ottosson M, et al. The morphology and metabolism of intraabdominal adipose tissue in men. *Metabolism.* 1992;41(11):1242–1248.
81. Misra A, Vikram NK. Clinical and pathophysiological consequences of abdominal adiposity and abdominal adipose tissue depots. *Nutrition.* 2003;19(5):457–466.
82. Kadowaki T, Hara K, Yamauchi T, Terauchi Y, Tobe K, Nagai R. Molecular mechanism of insulin resistance and obesity. *Exp Biol Med.* 2003;228(10):1111–1117.
83. Zhang P, Sun X, Jin H, Zhang F-L, Guo Z-N, Yang Y. Association between obesity type and common vascular and metabolic diseases: a cross-sectional study. *Front Endocrinol.* 2019;10:900.
84. Foster MT, Softic S, Caldwell J, Kohli R, de Kloet AD, Seeley RJ. Subcutaneous Adipose Tissue Transplantation in Diet-Induced Obese Mice Attenuates Metabolic Dysregulation While Removal Exacerbates It. *Physiol Rep.* 2013;1(2).
85. Stanford KI, Middelbeek RJ, Townsend KL, et al. A novel role for subcutaneous adipose tissue in exercise-induced improvements in glucose homeostasis. *Diabetes.* 2015;64(6):2002–2014.
86. Tran TT, Yamamoto Y, Gesta S, Kahn CR. Beneficial effects of subcutaneous fat transplantation on metabolism. *Cell Metab.* 2008;7(5):410–420.
87. Hocking S, Chisholm D, James D. Studies of regional adipose transplantation reveal a unique and beneficial interaction between subcutaneous adipose tissue and the intra-abdominal compartment. *Diabetologia.* 2008;51(5):900–902.
88. Foster M, Shi H, Softic S, Kohli R, Seeley R, Woods S. Transplantation of non-visceral fat to the visceral cavity improves glucose tolerance in mice: investigation of hepatic lipids and insulin sensitivity. *Diabetologia.* 2011;54(11):2890.
89. Foster MT, Softic S, Caldwell J, Kohli R, deKloet AD, Seeley RJ. Subcutaneous adipose tissue transplantation in diet-induced obese mice attenuates metabolic dysregulation while removal exacerbates it. *Physiol Rep.* 2013;1(2).
90. Foster MT, Shi H, Seeley RJ, Woods SC. Transplantation or removal of intra-abdominal adipose tissue prevents age-induced glucose insensitivity. *Physiol Behav.* 2010;101(2):282–288.
91. Chen L, Wang L, Li Y, et al. Transplantation of normal adipose tissue improves blood flow and reduces inflammation in high fat fed mice with hindlimb ischemia. *Front Physiol.* 2018;9:197.
92. Torres-Villalobos G, Hamdan-Pérez N, Díaz-Villaseñor A, et al. Autologous subcutaneous adipose tissue transplants improve adipose tissue metabolism and reduce insulin resistance and fatty liver in diet-induced obesity rats. *Physiol Rep.* 2016;4(17).
93. Sanal MG. Adipose tissue transplantation may be a potential treatment for diabetes, atherosclerosis and nonalcoholic steatohepatitis. *Med Hypotheses.* 2009;72(3):247–249.
94. American Diabetes Association. (2) Classification and diagnosis of diabetes. *Diabetes Care.* 2015;38(Suppl):S8–S16.
95. American Diabetes Association. Standards of medical care in diabetes—2007. *Diabetes Care.* 2007;30(Suppl 1):S4–S41.
96. Singh VP, Bali A, Singh N, Jaggi AS. Advanced glycation end products and diabetic complications. *Korean J Physiol Pharmacol.* 2014;18(1):1–14.
97. Powers AC. *Harrison's Endocrinology* (3rd ed.). McGraw Hill; 2013.
98. Loriaux L. *A Biographical History of Endocrinology.* John Wiley & Sons; 2016.
99. Bliss M. *The Discovery of Insulin: Twenty-fifth Anniversary Edition.* University of Chicago Press; 2007.
100. de Beeck AO, Eizirik DL. Viral infections in type 1 diabetes mellitus—why the beta cells? *Nat Rev Endocrinol.* 2016;12(5):263–273.
101. Mathis D, Vence L, Benoist C. Beta-cell death during progression to diabetes. *Nature.* 2001;414(6865):792–798.
102. Ziegler A-G, Kick K, Bonifacio E, et al. Yield of a public health screening of children for islet autoantibodies in Bavaria, Germany. *JAMA.* 2020;323(4):339–351.
103. Abramson A, Caffarel-Salvador E, Khang M, et al. An ingestible self-orienting system for oral delivery of macromolecules. *Science.* 2019;363(6427):611–615.
104. American Diabetes Association. Diagnosis and classification of diabetes mellitus. *Diabetes Care.* 2014;37(Suppl 1):S81–90.
105. Kahn BB. Type 2 diabetes: when insulin secretion fails to compensate for insulin resistance. *Cell.* 1998;92(5):593–596.
106. Defronzo RA. Banting Lecture. From the triumvirate to the ominous octet: a new paradigm for the treatment of type 2 diabetes mellitus. *Diabetes.* 2009;58(4):773–795.
107. Eckel RH, Kahn SE, Ferrannini E, et al. Obesity and type 2 diabetes: what can be unified and what needs to be individualized? *J Clin Endocrinol Metab.* 2011;96(6):1654–1663.

108. Chan JM, Rimm EB, Colditz GA, Stampfer MJ, Willett WC. Obesity, fat distribution, and weight gain as risk factors for clinical diabetes in men. *Diabetes Care.* 1994;17(9):961–969.
109. Colditz GA, Willett WC, Rotnitzky A, Manson JE. Weight gain as a risk factor for clinical diabetes mellitus in women. *Ann Intern Med.* 1995;122(7):481–486.
110. Emerging Risk Factors C, Wormser D, Kaptoge S, et al. Separate and combined associations of body-mass index and abdominal adiposity with cardiovascular disease: collaborative analysis of 58 prospective studies. *Lancet.* 2011;377(9771):1085–1095.
111. Montague CT, O'Rahilly S. The perils of portliness: causes and consequences of visceral adiposity. *Diabetes.* 2000;49(6):883–888.
112. Pinhas-Hamiel O, Zeitler P. Clinical presentation and treatment of type 2 diabetes in children. *Pediatr Diabetes.* 2007;8(Suppl 9):16–27.
113. Mohamadi A, Cooke DW. Type 2 diabetes mellitus in children and adolescents. *Adolesc Med State Art Rev.* 2010;21(1):103–119.
114. Das SK, Elbein SC. The genetic basis of Type 2 diabetes. *Cellscience.* 2006;2(4):100–131.
115. Evans RM, Barish GD, Wang Y-X. PPARs and the complex journey to obesity. *Nat Med.* 2004;10(4):355.
116. Hallsten K, Yki-Jarvinen H, Peltoniemi P, et al. Insulin- and exercise-stimulated skeletal muscle blood flow and glucose uptake in obese men. *Obes Res.* 2003;11(2):257–265.
117. Shulman GI. Cellular mechanisms of insulin resistance. *J Clin Invest.* 2000;106(2):171–176.
118. Saltiel AR, Kahn CR. Insulin signalling and the regulation of glucose and lipid metabolism. *Nature.* 2001;414(6865):799–806.
119. Kahn BB, Flier JS. Obesity and insulin resistance. *J Clin Invest.* 2000;106(4):473–481.
120. DeFronzo RA. Lilly lecture 1987. The triumvirate: beta-cell, muscle, liver. A collusion responsible for NIDDM. *Diabetes.* 1988;37(6):667–687.
121. Kahn SE, Hull RL, Utzschneider KM. Mechanisms linking obesity to insulin resistance and type 2 diabetes. *Nature.* 2006;444(7121):840–846.
122. Marchesini G, Bugianesi E, Forlani G, et al. Nonalcoholic fatty liver, steatohepatitis, and the metabolic syndrome. *Hepatology.* 2003;37(4):917–923.
123. Ferrannini E. The stunned beta cell: a brief history. *Cell Metab.* 2010;11(5):349–352.
124. Talchai C, Xuan S, Lin HV, Sussel L, Accili D. Pancreatic beta cell dedifferentiation as a mechanism of diabetic beta cell failure. *Cell.* 2012;150(6):1223–1234.
125. Steer SA, Scarim AL, Chambers KT, Corbett JA. Interleukin-1 stimulates beta-cell necrosis and release of the immunological adjuvant HMGB1. *PLoS Med.* 2006;3(2):e17.
126. Cnop M, Hannaert JC, Hoorens A, Eizirik DL, Pipeleers DG. Inverse relationship between cytotoxicity of free fatty acids in pancreatic islet cells and cellular triglyceride accumulation. *Diabetes.* 2001;50(8):1771–1777.
127. Boslem E, Meikle PJ, Biden TJ. Roles of ceramide and sphingolipids in pancreatic beta-cell function and dysfunction. *Islets.* 2012;4(3):177–187.
128. Harding HP, Zeng H, Zhang Y, et al. Diabetes mellitus and exocrine pancreatic dysfunction in perk-/- mice reveals a role for translational control in secretory cell survival. *Mol Cell.* 2001;7(6): 1153–1163.
129. Goodge KA, Hutton JC. Translational regulation of proinsulin biosynthesis and proinsulin conversion in the pancreatic beta-cell. *Semin Cell Dev Biol.* 2000;11(4):235–242.
130. Oslowski CM, Urano F. The binary switch that controls the life and death decisions of ER stressed beta cells. *Curr Opin Cell Biol.* 2011;23(2):207–215.
131. Szegezdi E, Logue SE, Gorman AM, Samali A. Mediators of endoplasmic reticulum stress-induced apoptosis. *EMBO Rep.* 2006;7(9):880–885.
132. Haataja L, Gurlo T, Huang CJ, Butler PC. Islet amyloid in type 2 diabetes, and the toxic oligomer hypothesis. *Endocr Rev.* 2008;29(3):303–316.
133. Hoppener JW, Ahren B, Lips CJ. Islet amyloid and type 2 diabetes mellitus. *N Engl J Med.* 2000;343(6):411–419.
134. White JD, Dewal RS, Stanford KI. The beneficial effects of brown adipose tissue transplantation. *Mol Aspect Med.* 2019;68:74–81.
135. Gunawardana SC, Piston DW. Reversal of type 1 diabetes in mice by brown adipose tissue transplant. *Diabetes.* 2012;61(3):674–682.
136. Gunawardana SC, Piston DW. Insulin-independent reversal of type 1 diabetes in nonobese diabetic mice with brown adipose tissue transplant. *Am J Physiol-Endocrinol Metab.* 2015;308(12):E1043–E1055.
137. Liu X, Wang S, You Y, et al. Brown adipose tissue transplantation reverses obesity in Ob/Ob mice. *Endocrinology.* 2015;156(7):2461–2469.
138. Stanford KI, Middelbeek RJ, Townsend KL, et al. Brown adipose tissue regulates glucose homeostasis and insulin sensitivity. *J Clin Investig.* 2012;123(1).
139. Zhu Z, Spicer EG, Gavini CK, Goudjo-Ako AJ, Novak CM, Shi H. Enhanced sympathetic activity in mice with brown adipose tissue transplantation (transBATation). *Physiol Behav.* 2014;125:21–29.
140. Min SY, Kady J, Nam M, et al. Human "brite/beige" adipocytes develop from capillary networks, and their implantation improves metabolic homeostasis in mice. *Nat Med.* 2016;22(3):312–318.
141. Silva FJ, Holt DJ, Vargas V, et al. Metabolically active human brown adipose tissue derived stem cells. *Stem Cells.* 2014;32(2):572–581.
142. Magueresse-Battistoni L, Vidal H, Naville D. Environmental pollutants and metabolic disorders: the multi-exposure scenario of life. *Front Endocrinol.* 2018;9:582.
143. Papalou O, Kandaraki EA, Papadakis G, Diamanti-Kandarakis E. Endocrine disrupting chemicals: an occult mediator of metabolic disease. *Front Endocrinol.* 2019;10.
144. Nadal A, Quesada I, Tudurí E, Nogueiras R, Alonso-Magdalena P. Endocrine-disrupting chemicals and the regulation of energy balance. *Nat Rev Endocrinol.* 2017;13(9):536.
145. Heindel JJ, Blumberg B, Cave M, et al. Metabolism disrupting chemicals and metabolic disorders. *Reprod Toxicol (Elmsford, NY).* 2017;68:3–33.
146. Baillie-Hamilton PF. Chemical toxins: a hypothesis to explain the global obesity epidemic. *J Altern Complement Med.* 2002;8(2):185–192.
147. Holtcamp W. *Obesogens: An Environmental Link to Obesity.* National Institute of Environmental Health Sciences; 2012.
148. Kelishadi R, Poursafa P, Jamshidi F. Role of environmental chemicals in obesity: a systematic review on the current evidence. *J Environ Public Health.* 2013;2013.
149. Grün F, Blumberg B. Environmental obesogens: organotins and endocrine disruption via nuclear receptor signaling. *Endocrinology.* 2006;147(6):s50-s55.
150. Darbre PD. Endocrine disruptors and obesity. *Curr Obes Rep.* 2017;6(1): 18–27.
151. Heindel JJ, Blumberg B, Cave M, et al. Metabolism disrupting chemicals and metabolic disorders. *Reprod Toxicol.* 2017;68:3–33.
152. Newbold RR. Lessons learned from perinatal exposure to diethylstilbestrol. *Toxicol Appl Pharmacol.* 2004;199(2):142–150.
153. Newbold RR, Padilla-Banks E, Jefferson WN. Environmental estrogens and obesity. *Mol Cell Endocrinol.* 2009;304(1–2):84–89.
154. Hatch E, Troisi R, Palmer J, et al. Prenatal diethylstilbestrol exposure and risk of obesity in adult women. *J Dev Orig Health Dis.* 2015;6(3):201–207.
155. Gore AC, Chappell V, Fenton S, et al. EDC-2: the Endocrine Society's second scientific statement on endocrine-disrupting chemicals. *Endocr Rev.* 2015;36(6):E1-E150.
156. Pizzorno J. Is the diabetes epidemic primarily due to toxins? *Integr Med Clin J.* 2016;15(4):8.
157. Neel BA, Sargis RM. The paradox of progress: environmental disruption of metabolism and the diabetes epidemic. *Diabetes.* 2011;60(7):1838–1848.

Unit Overview
Unit VII

Reproduction

DOI: 10.1201/9781003359241-26

OPENING QUOTATION:

"For as long as men and women have been making babies, they've been trying not to."

—Vondráčková, L., 2017. Jonathan Eig: The Birth of the Pill: How Four Crusaders Reinvented Sex and Launched a Revolution. *Sociologický časopis/Czech Sociological Review*, 53(01), pp. 138–140.

"The management of the process of sexual reproduction in higher vertebrates is undoubtedly the most complicated phenomenon in integrative physiology. Not only are the roles of each of the hormones involved in this integration themselves complex, but there also are systems for coordinating or gearing the actions of the separate hormones with each other as well as with environmental cues. Moreover, the mechanisms in the two sexes are different."

—A. Gorbman et al., 1983. Comparative Endocrinology.

There are good reasons why chapters on reproduction, the topic of Unit VII, are typically among the last in endocrinology textbooks. Not only is the topic among the most complex in physiology, but reproductive endocrinology content is highly integrative, requiring a substantive understanding of not just the fundamental mechanisms of hormone action and neuroendocrine signaling, but also of other hormone systems that interact extensively with sex hormones to modulate reproductive development and behaviors. Other hormone systems that significantly influence reproduction include thyroid, adrenal, and metabolic hormones, as well as hormones that regulate bone growth and development, biological clocks, and many others. Chapter 20 begins the unit by discussing the influences of genotype, environment, and hormones on sex determination among vertebrates. The chapter also describes the different endocrine factors that contribute to sexual differentiation in mammals and outlines various categories of human disorders/differences of sexual development. Chapter 21 addresses the role of the testis in facilitating both steroidogenesis and spermatogenesis, and the effects of various androgens on fetal and adult target tissues are assessed. The implications of globally decreasing sperm counts and the influence of obesity on male reproductive health are also evaluated. Chapter 22 focuses on the endocrine regulation of the ovarian-uterine cycles and contrasts the human menstrual cycle with the estrous cycle in other mammals. The endocrine bases of endometriosis and polycystic ovary syndrome, as well as changes in health parameters associated with menopause and their treatments, are also discussed. Chapter 23 discusses rhythmic environmental and endogenous factors that influence the timing of puberty, fertility, and seasonal reproduction among animals. Emphasis is placed on the cyclical and pulsatile natures of hypothalamic-pituitary-gonadal signaling that form the basis of negative and positive feedback signaling throughout the mammalian ovarian cycle. Chapter 24 begins by describing the evolution of pregnancy from an endocrine perspective. The chapter then outlines the major fetal-placental-maternal endocrine interactions during mammalian pregnancy, as well as the neuroendocrine mechanisms responsible for the timing of parturition and lactation.

CHAPTER

20

Sexual Determination and Differentiation

CHAPTER LEARNING OBJECTIVES:

- Provide an overview of different strategies of sex determination among animals.
- Compare and contrast the molecular mechanisms of genotypic and environmental sex determination in vertebrates.
- Explain the developmental fates of the mammalian primordial internal and external genitalia in response to endocrine signaling.
- Provide an overview of the different factors that contribute to the sexual differentiation of the mammalian brain and behavior.
- Outline the various categories of human disorders and differences of sexual development (DSDs), and provide detailed examples of each.

OPENING QUOTATIONS:

> "If, then, the male element [fire] prevails, it draws the female element [water] into itself; but if it is prevailed over, it changes into the opposite or is destroyed."[1]
>
> —Aristotle (384–322 BC) asserted that heat gave rise to males and cold to females

> "I was born twice: first, as a baby girl, on a remarkably smogless Detroit day in January of 1960; and then again, as a teenage boy, in an emergency room near Petoskey, Michigan, in August of 1974."
>
> —Jeffrey Eugenides, *Middlesex*

Sex determination is the process that establishes whether the gonads of a sexually reproducing organism will develop into either an ovary or a testis, a course of development that begins early during embryogenesis. **Sexual differentiation**, by contrast, is a more prolonged process that follows gonadal determination whereby internal structures, external genitalia, the brain, and secondary sex characteristics that arise at puberty acquire their female or male phenotypes. In vertebrates, the process of sexual differentiation begins during fetal development and is completed at puberty. Sex determination and sexual differentiation are central to the propagation of a species, affecting not only a population's male-to-female sex ratio, but also the success of both sexes and their species.[2] Since reproduction directly affects fitness, the mechanisms of sex determination are under strong selective pressure. However, despite its critical importance in the survival of a species, the mechanisms of sex determination are remarkably variable among organisms, with some determined by internal factors, such as sex chromosomes, others by environmental parameters, like temperature, and others via a combination of the two. In contrast to sex determination, the hormones and pathways of gene expression that mediate sexual differentiation and development of male and female sexual phenotypes are highly conserved among vertebrates. This chapter addresses the strategies and mechanisms of sex determination and sexual differentiation among vertebrates, the key roles of endocrine signaling, as well as certain disorders/differences of human sexual development.

Strategies of Sex Determination among Animals

LEARNING OBJECTIVE Provide an overview of different strategies of sex determination among animals.

KEY CONCEPTS:

- In organisms with genotypic sex determination (GSD) the sex of an individual is determined entirely by its genotype, usually by genes located on sex chromosomes.
- In the X/Y chromosomal sex determination system of mammals, males are the heterogametic sex (XY) and females the homogametic sex (XX).
- In the Z/W chromosomal sex determination system of birds, males are the homogametic sex (ZZ) and females are the heterogametic sex (ZW).
- In organisms with environment-dependent sex determination, the sex of an individual is determined during egg incubation by external variables, such as temperature.

DOI: 10.1201/9781003359241-27

How is sex determined? The ancient Greek philosopher Aristotle thought that males had an abundance of a particular element, fire, whereas females had an abundance of the cooler element, water. In 335 BCE, Aristotle proposed that the sex of the offspring is determined by the heat of the male during intercourse: if the male's heat was sufficient, then a male child would form, whereas if the male's heat was insufficient, a female child would form.[3] Though at first glance Aristotle's hypothesis may seem quaint or outlandish, Aristotle puts forth two concepts that are remarkably relevant to modern sex determination theory. The first is that a "male principle" drives sex determination, and if that principle is absent or insufficiently expressed then the fetus develops as a female. We now know that in many animals, including humans, that principle is not "fire", but, rather, the Y chromosome. A second concept that Aristotle touches upon is the importance of temperature in determining the sex of the offspring. Indeed, environmental temperature is a key determinant of sex in many reptiles, amphibians, and fish.

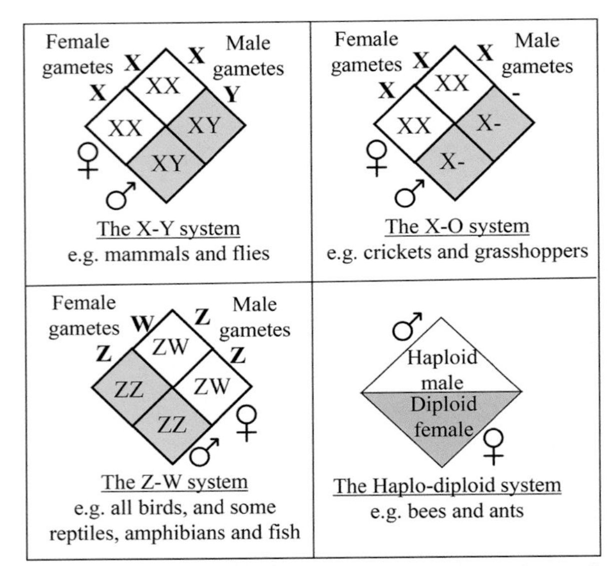

Figure 20.1 Some examples of chromosomal sex determination strategies in animals.

Genotypic Sex Determination

Prior to 1900, it was commonly assumed that sex was environmentally determined. In the first edition of Edmund Beecher Wilson's book *The Cell in Development and Inheritance* (1896),[4] he wrote, "the determination of sex is not by inheritance, but by the combined effect of external conditions". However, only a few years later, in 1905, he and Nettie Stevens would independently make the first major breakthrough in centuries in understanding the mechanisms of sex determination that would replace the dogma of environmental sex determinism with internal sex determinism. Studying the chromosomes of insects, Stevens and Wilson found that although most chromosomes were present in equal numbers in both males and females, some chromosomes were represented differentially in the two sexes. Specifically, males of these species possess both an X and a smaller Y chromosome that induces the development of the testes, whereas females always have two X chromosomes and no Y. Together, Wilson and Stevens established the "XY sex-determination system" that is present in some insects (e.g. flies) and in all mammals (Figure 20.1). In such animals, the males are the *heterogametic sex* that possess two different sex chromosomes and are, hence, capable of producing two types of gametes, and females the *homogametic sex*, which produce only one type of gamete. By contrast, the system is reversed in all birds, as well as in some reptiles, amphibians, and fishes, and also in insects of the order Lepidoptera (butterflies and moths). In these animals the males are the homogametic sex (ZZ) and females heterogametic (ZW) (Figure 20.1). Not all animals use two sex chromosomes for sex determination. For example, cockroaches and some other insects use an X-O system whereby individuals with two copies of the X chromosome develop into females, and those with one copy develop into males, and the sex of the offspring is determined by whether an X chromosome is present in the sperm that fertilizes the egg. Finally, in bees and ants, sex is not determined at all by the presence or absence of a specific chromosome, but by the total number of chromosomes present. Specifically, whereas males develop from unfertilized eggs and are haploid, females develop from fertilized eggs and are diploid. The term **genotypic sex determination** (GSD) denotes that the sex of an individual is determined entirely by its genotype, usually by genes located on sex chromosomes, and is fixed at fertilization.

Curiously Enough . . . Not all vertebrates reproduce sexually. For example, some species of reptiles, like the Sonoran spotted whiptail lizard (*Aspidoscelis sonorae*), are all female and generate offspring via a strategy termed *parthenogenesis*, whereby embryogenesis initiates without the activation or fertilization of the egg by sperm.[5] Ovulation is stimulated by "pseudocourtship" between female pairs, with one female changing color and behavior to resemble the ancestral male. Unfertilized eggs produce hatchlings genetically identical to the mothers.

An accurate description of human chromosome number had to wait until 1956 when the diploid number of chromosomes was found to be 46,[6] and in males this included the presence of an X and a Y chromosome.[7] Soon thereafter, two categories of human sex-chromosome abnormalities and their effects on development were described: (1) a patient with **Klinefelter syndrome**, an infertile male phenotype with an XXY chromosome set,[8] and (2) two patients with **Turner syndrome**, an infertile female phenotype with just a single X chromosome and no Y.[9,10] These observations suggested that in humans and other mammals the presence of a Y chromosome determines the male sex, and that female development requires the absence of the Y chromosome, regardless of the number of X chromosomes present.

Environment-dependent Sex Determination

Reptiles

The rapid development of genetic research during the 20th century introduced a powerful new explanatory tool for many aspects of animal development, and a genotypic basis for sex determination became the new dogma. It came as quite a surprise, then, when it was discovered that in most turtles and crocodilians and in some lizards sex is determined not by chromosome differences, but by the environmental temperature during egg incubation.[11] It soon became clear that sex determination was not always influenced by internal factors alone, but *environment-dependent sex-determination* (ESD) factors could also play an important role in some animals. **Temperature-dependent sex determination** (TSD) is a prime example, whereby the temperature of incubation that exists during the time of embryonic gonad formation biases the percentage of male or female offspring.[12] Whereas in some reptile species, such as sea turtles, higher temperatures generate primarily females, in others, like tuatara, higher temperatures generate mostly males (Figure 20.2). Still other species, like alligators, exhibit another strategy, with increasing temperatures initially promoting the development of females, then transitioning back towards a male preference. Importantly, there are many reported cases in which both GSD and TSD are simultaneously present in the same species.[13–15] Furthermore, in the case of at least one reptile that uses GSD (the Australian central bearded dragon, *Pogona vitticeps*), chromosomal male sex determination (ZZ) is overridden at high temperatures to produce sex-reversed female offspring.[16] Considering that the range of temperature between all-male and all-female clutches can be less than 1°C,[17] the ecological and evolutionary impact of climate change becomes quite apparent.[18]

Fish

In fish, environmental factors, such as temperature and water pH, can influence sex determination during embryogenesis. However, in some species, environmentally induced post-embryonic sex change in juveniles and adults also constitutes an importanvw t component of their life history strategies. Intriguingly, such species display **sequential hermaphroditism**, whereby one individual changes its sex during its lifetime.[19] In the case of **protogyny**, a female transforms into a reproductively viable male, and with **protandry** a male transforms into a female. These transformations are typically influenced by variables such as relative juvenile size and social interactions.[20] A well-studied example of socially controlled sex change is the protogynous Caribbean bluehead wrasse (*Thalassoma bifasciatum*). These fish are sexually dimorphic, with dominant, reproductively active males typically having a blue coloration, and the females and subordinate (non-breeding) males both possessing yellow and black stripes (Figure 20.3).[21] The natural social groups typically consist of one dominant male and a female harem, as well as some subordinate males. When the dominant male is removed from the group, one of the largest females ascends to the top of the social hierarchy and transforms into the dominant male, exhibiting rapid changes in coloration and external morphology, as well as dramatic alterations in reproductive behavior and gonadal anatomy. In the newly protogynous dominant fish, ovarian tissues regress within the first 3 days, and the new testes become fully functional and filled with sperm by days 8–10.[22] Based on studies with the congeneric species, the saddle wrasse (*Thalassoma duperrey*), it is thought that these morphological changes are mediated by increasing levels of 11-ketotestosterone, the primary androgen in fish, and declining concentrations of estradiol[23] (Figure 20.3).

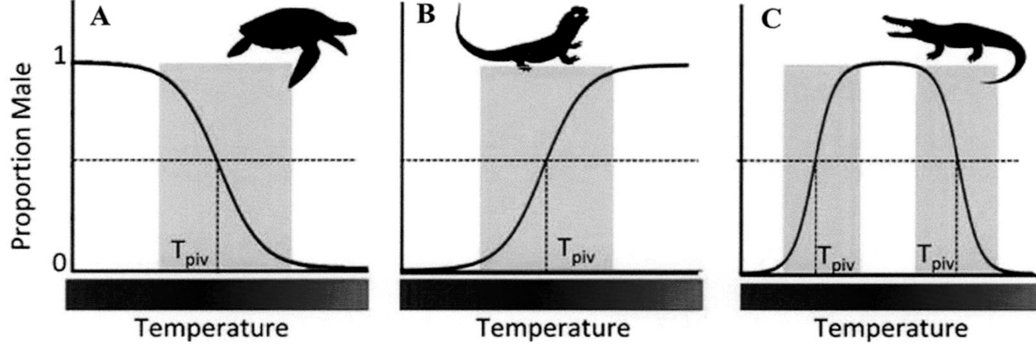

Figure 20.2 Three patterns of temperature-dependent sex determination among reptiles: (A) as seen in sea turtles; (B) as known in tuatara; and (C) as present in crocodilians. The pivotal temperature (T piv) is the temperature at which an even proportion of males and females is produced. Red denotes the transitional range of temperatures, where both sexes can be produced, generally defined as between 5% and 95% of one sex.

Source of images: Lockley, E.C. and Eizaguirre, C., 2021. Evol. Appl., 14(10), pp. 2361-2377.

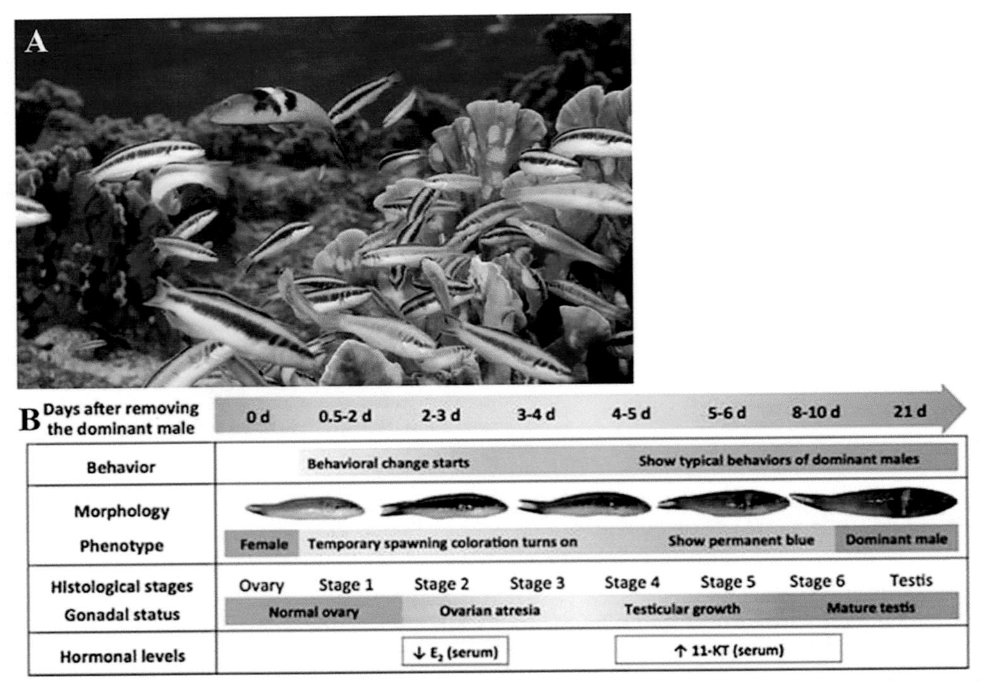

Figure 20.3 Timeline of socially controlled protogynous sex change of bluehead wrasses. (A) A dominant male (the largest fish with blue coloration) and his harem of females and subordinate males (smaller fish with yellow and black stripes). **(B)** A timeline of behavioral, morphological, and hormone changes during the transformation of a female to a dominant male following the removal of the original dominant male from the group. Hormonal changes are predicted based on patterns in the congener *Thalassoma duperrey* (Nakamura et al., 1989). Abbreviations: E_2 (estradiol), 11-KT (11-ketotestosterone).

Source of images: Liu H, et al. Mol Reprod Dev. 2017 Feb;84(2):171-194. doi: 10.1002/mrd.22691. Used by permission.

Curiously Enough . . . The existence of *simultaneous hermaphroditism*, whereby one individual concurrently possesses both functional male and female gonads and is capable of self-fertilization as part of the species' natural life cycle, is exceptionally rare in vertebrates and has only been documented in two species of *Kryptolebias* mangrove rivulus fish in Florida.[24]

SUMMARY AND SYNTHESIS QUESTIONS

1. Compare and contrast genetic mechanisms of male sex determination in placental mammals and birds.

Molecular Mechanisms of Sex Determination in Vertebrates

LEARNING OBJECTIVE Compare and contrast the molecular mechanisms of genotypic and environmental sex determination in vertebrates.

KEY CONCEPTS:

- The embryonic gonadal primordium is sexually indifferent and has the potential to differentiate into either a testis or an ovary.
- In mammals, the SRY gene on the Y chromosome is the testis-determining factor that is responsible for masculinizing the embryo.
- Ovary-determining factors on the mammalian X chromosome suppress testis development.
- In birds, a double dose of DMRT1 gene expression from the ZZ karyotype promotes male development, whereas a half dose of the gene's expression from the ZW karyotype promotes female development.
- Following the divergence of birds and mammals from a common ancestor, one mammalian X chromosome experienced dramatic gene loss and shortening, retaining only a fraction of the genes from the ancestral X, and is now the Y chromosome.
- In order to compensate for imbalances in gene dosage between the sexes, mammals evolved the process of X chromosome inactivation, the epigenetic silencing of one of a female's two X chromosomes in each somatic cell.
- In vertebrates that use temperature-dependent sex determination, the development of testes or ovaries is determined by the androgen-to-estrogen ratio.
- The enzyme CYP19 aromatase converts androgens into estrogens, and its activity helps establish sex in vertebrates that use temperature-dependent sex determination.

An important tenet of vertebrate sex development that will be explored throughout this chapter is that the embryonic gonadal primordium is *sexually indifferent*, or uncommitted and morphologically indistinguishable between both sexes. Another adjective frequently used to describe the gonadal primordium is that it is *bipotential*, or has the ability to differentiate into either a testis or an ovary. As such, the process of sex determination normally activates one of the two developmental pathways and simultaneously shuts down the other. What determines which developmental trajectory the sexually indifferent gonadal primordium will ultimately take?

Genotypic Sex Determination

Sex Determination in Mammals

The X and Y Chromosomes

The mammalian sex chromosomes are thought to have evolved from a pair of *autosomes* (non-sex chromosomes) about 180 million years ago, prior to the divergence of the marsupial and placental mammalian lineages.[25] Compared to the X chromosome, which more closely resembles the shape and size of the autosomes, the Y chromosome is only one-third the size (Figure 20.4). It is hypothesized that over the course of millions of years, an ancestral X chromosome experienced dramatic gene loss and shortening, retaining only 3% of the genes from its ancestral form, and is now represented by the Y chromosome.[26,27] Today the human Y chromosome bears only 33 genes, the majority of which are concerned with development of the testis and spermatogenesis[28] but also exert non-gonadal effects, including differential brain development[29] and differences in susceptibility to autoimmune disease between the sexes.[30] Despite its different size and appearance, the Y chromosome does share two small regions of homology with the X chromosome. These are located on the chromosome tips and are called *pseudo-autosomal regions* (PAR), which play crucial roles in mediating X and Y chromosomal segregation during meiosis. Importantly, the rest of the Y chromosome contains primarily male-specific genes and neither participates in chromosomal segregation or homologous recombination with the X.

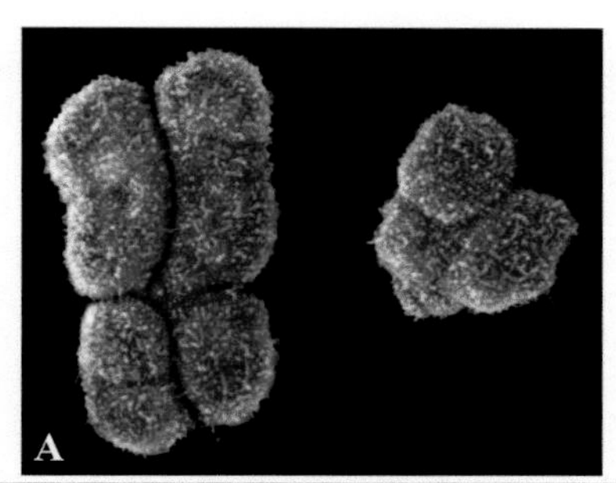

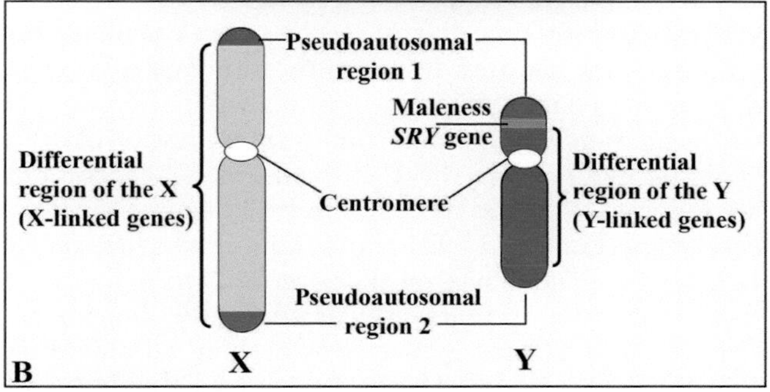

Figure 20.4 The human X and Y chromosomes. (A) The human Y chromosome (right) is much smaller than the X chromosome (left), as a result of significant degeneration early in Y chromosome evolution. **(B)** Both chromosomes retain homology only at pseudoautosomal regions 1 and 2, which are essential for chromosome pairing in the male during meiosis.

Source of images: (A) Clark, A.G., 2014. The vital Y chromosome. Nature, 508(7497), pp. 463-464. (B) Modified from Griffiths, et al. (2010). Introduction to Genetic Analysis 10th Edition. W. H. Freeman. Used by permission.

With the deletion of most genes from the Y chromosome over millions of years came imbalances in *gene dosage*, or the number of copies of a gene and the amount of gene product in a cell, between the sexes. Specifically, in order to compensate for the absence of now unpartnered X genes in males (XY), females (XX) evolved mechanisms to restore balanced expression between the X chromosome and the autosomes. One important mechanism involves the transcriptional silencing of most of one X chromosome from each pair in every female somatic cell, a process called *X chromosome inactivation* (Figure 20.5). The inactivated chromosome, called a **Barr body**, is only reactivated when cells replicate their DNA and divide. Since in humans the selection of the maternally derived versus paternally derived X chromosome for inactivation during embryogenesis occurs at random, this means that female bodies are "X chromosome mosaics", consisting of patches of inactivated maternal or paternal X chromosomes. Compared with males, whose somatic cells always possess the same functional X chromosome, the random nature of X chromosome inactivation in females protects them from developing severe phenotypes associated with X-linked diseases, such as androgen insensitivity syndrome, hemophilia, red-green color blindness, and Duchenne muscular dystrophy. Interestingly, as many as 15% of X genes in humans escape inactivation,[31] suggesting that female cells have a "double dose" of some X genes.

SRY, the Testis-Determining Factor

In animals that use genotypic sex determination (GSD) as a developmental strategy, such as mammals, genetic inequalities in the sex chromosomes must play a key role as they are presumably the only factors that differ. Since the Y chromosome is necessary for the development of the testis and male sex determination, it was postulated that it possessed an as of yet uncharacterized "testis-determining factor", a hypothetical gene or genes whose expression causes males to develop differently than females.[32] In humans, this region is now known to be a 35 kb piece of the Y chromosome called **SRY** (for "sex-determining region on the Y chromosome")[33] (Figure 20.4), which encodes a transcription factor that induces a testis-forming pathway beginning at about 7 weeks post-fertilization. Specifically, the SRY protein is a transcription factor that binds to DNA

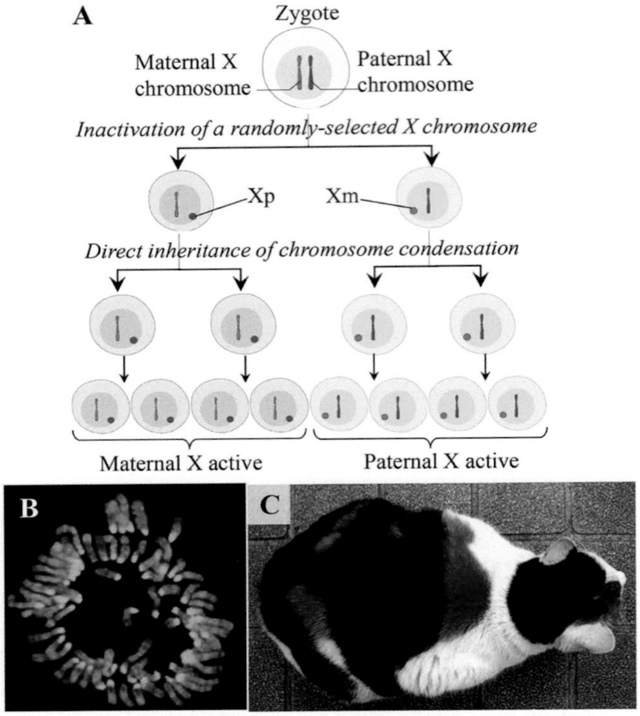

A
Zygote
Maternal X chromosome
Paternal X chromosome
Inactivation of a randomly-selected X chromosome
Xp
Xm
Direct inheritance of chromosome condensation
Maternal X active
Paternal X active

B

C

Figure 20.5 X chromosome inactivation in female mammals. **(A)** Each somatic cell of the female embryo contains two X chromosomes, one inherited paternally (Xp) and the other inherited maternally (Xm). Beginning at approximately the 20-cell stage, one X chromosome is inactivated at random by condensation of the chromatin into transcriptionally inactive heterochromatin. The inactivated X chromosome is called a Barr body. **(B)** Of all the chromosomes (blue stain) visible in this female mouse nucleus, only the inactive X chromosome stains for Xist RNA (red) in preparation for condensation into a transcriptionally inactive Barr body. Following this random X inactivation, the same X chromosome remains inactivated in all subsequent daughter cells, and thus a mature female consists of an X chromosome mosaic, with some cells expressing the maternal X chromosome and some cells expressing the paternal X chromosome. This mosaicism can be seen in the coat colors of calico cats **(C)**, where orange coat coloration is encoded on the X chromosome. As such, calico cats are always females.

Source of images: (A) Quinlan, C. and Rheault, M.N., 2022, March. X-Linked kidney disorders in women. Sem. Nephrol. (Vol. 42, No. 2, pp. 114-121). WB Saunders. Used by permission. (B) Ng, K., et al. 2007. EMBO Rep., 8(1), pp. 34-39. Used by permission. (C) Henikoff, S. and Greally, J.M., 2016. Curr Biol., 26(14), pp. R644-R648.

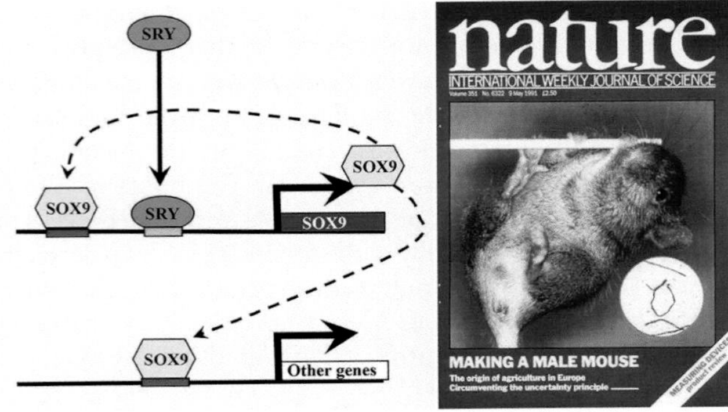

Figure 20.6 The actions of the SRY gene of the Y chromosome. The SRY protein is a transcription factor that binds to SRY DNA regulatory elements (orange line) located in one or more enhancers of its target gene, SOX9, thereby upregulating SOX9 transcription. The SOX9 protein, itself a transcription factor, binds to SOX9 DNA regulatory elements (red lines) to upregulate its own transcription (dashed line). The SOX9 protein also goes on to regulate the expression of other genes involved in the testis developmental pathway. Sex reversal in a female transgenic mouse expressing the SRY gene (*Nature* magazine cover). The external phenotype of the SRY-expressing XX transgenic mouse is identical to that of a wild-type XY male mouse.

Source of images: (B) Koopman, P., et al. 1991. Nature, 351(6322), pp. 117-121. Used by permission.

regulatory elements located in one or more enhancers of its target gene, *SOX9*,* thereby upregulating SOX9 transcription[34] (Figure 20.6). Importantly, these actions take place only in the cellular precursors in the bipotential gonad that give rise to the Sertoli cells of the testis. Following their differentiation, Sertoli cells secrete signals that induce the differentiation of steroidogenic Leydig cells that produce androgens that trigger the *virilization* (development of male features) of the embryo.

Not only does SRY function as a signal and SOX9 an effector in promoting testis development, but SOX9's downstream actions simultaneously repress gene expression associated with ovarian development.[35] Furthermore, because SOX9 also upregulates its own expression, continued SRY expression is not necessary for sustaining testis differentiation. By single-handedly causing the undifferentiated gonadal primordium to commit to a testicular fate, SRY is the testis-determining factor that promotes the elevated blood plasma levels of testosterone that are necessary for mediating the diverse aspects of male primary and secondary sex features. Importantly, although it is generally considered that SRY is the only male-determining gene on the mammalian Y chromosome, the expression of other genes located on both the Y chromosome and autosomes are required for males to be fertile.[36] The power of SRY is easy to appreciate in a 1991 photograph of what appears to be a typical male mouse on the cover of *Nature* magazine displaying, rather prominently, its normal-looking gonads (Figure 20.6). However, there was nothing normal about this particular rodent. The mouse, named Randy by the researchers who created it, was in fact genotypically female (XX) but was coaxed into developing a male phenotype by the insertion of a transgene for the mouse SRY gene.[37]

Ovary-Determining Factors

With the discovery that the presence of SRY actively switches on male sex determination, whereas its absence facilitates female sex determination, for many years it was thought that female ovarian development was a passive "default" program

* SRY-box transcription factor 9

of sorts. However, data from both human patients and animal models began to suggest that sex determination involves more than the mere presence or absence of SRY, and by the turn of the millennium this notion disappeared with the discoveries of genes that both actively promote ovarian development and suppress testicular development. For example, studies of women with an XY karyotype showed that the duplication and overexpression of the gene for the nuclear receptor, *DAX1*,† can cause male-to-female phenotypic sex reversal by acting as an anti-testis gene.[38,39] DAX1 is itself upregulated by another gene, *WNT4*,‡ which plays a key role in both the control of female development and the prevention of testes formation.[40] In addition to developing atypical genitals and gonads, XY individuals with extra copies of WNT4 can also develop a rudimentary uterus and Fallopian tubes. Other genes known to play a role in ovary determination include [Insert Omath Here 3]-catenin, *FOXL2*,§ and RSPO1.¶ Strikingly, if RSPO1 is not functioning properly, it can cause XX individuals to develop an *ovotestis*—a gonad containing regions of both ovarian and testicular development.[41] Current research suggests that sex determination is a complex process whereby the gonad's identity results from the actions of two opposing networks of self-reinforcing and cross-inhibitory gene activity, and the balance can be tipped either towards or away from the genotypic sex (Figure 20.7).

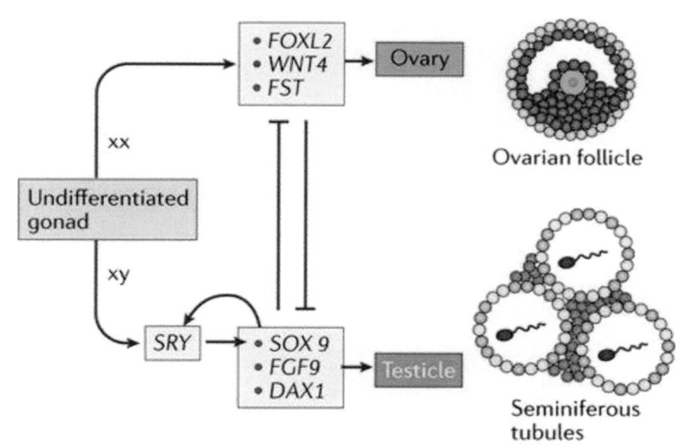

Figure 20.7 Sex determination is a complex process whereby the gonad's identity results from the actions of two opposing networks of self-reinforcing and cross-inhibitory gene activity, and the balance can be tipped either towards or away from the genotypic sex. The genes and pathways that control gonadal sex determination of the testes include sex-determining region Y protein (*SRY*), *SOX9*, fibroblast growth factor 9 (*FGF9*) and the nuclear receptor *DAX1*. Genes involved in ovary determination include forkhead box protein L2 (*FOXL2*), *WNT4*, and follistatin (*FST*).

Source of images: Karnezis, A.N., et al. 2017. Nat Rev Cancer, 17(1), pp. 65-74. Used by permission.

Sex Determination in Birds

Birds also have sex chromosomes, but in contrast to mammals the females are the heterogametic sex (ZW) and males the homogametic sex (ZZ) (see Figure 20.1). Interestingly, birds do not appear to possess an SRY gene, and the sex-chromosome pairs of most mammals and birds do not share any genes in common. However, in 2009 Smith and colleagues[42] discovered that in chickens a gene on the Z chromosome called *DMRT1*# functions as the avian sex-determining gene. Specifically, DMRT1, which codes for a vertebrate transcription factor, is specifically transcribed in the testis where it directs its development. Since males are the homogametic sex in birds, and DMRT1 is expressed in both male Z copies, it is hypothesized that a threshold amount of the gene product is required to generate a testis. However, since the female ZW gonad only has half the dosage of DMRT1, it is insufficient for the male pathway, and a female ovarian pathway ensues. Interestingly, DMRT1 also plays important roles in mammalian testis development. However, in contrast with birds, where DMRT1 is located on a sex chromosome, in humans and mice the gene is located on an autosome and appears to be expressed in the testis downstream of SOX9expression.[43]

Curiously Enough . . . Mutations in DMRT1 in XY humans has been shown to lead to male-to-female sex reversal[44] and also to impaired testis development in DMRT1 mouse mutants.[45]

Temperature-Dependent Sex Determination

As with vertebrates that utilize genotypic sex determination (GSD), the key event during environment-dependent sex determination is the commitment of the bipotential gonad to testicular or ovarian fate. Temperature-dependent sex determination (TSD) is a type of phenotypic plasticity whereby individuals of identical genotypes develop into either males or females, depending on the environmental temperature at which the eggs are incubated. In reptiles that use TSD, sex is determined midway through embryogenesis during a specific window of time when the gonads are responsive to changes in ambient temperature, referred to as the *thermosensitive period* (TSP).[46]

Roles of CYP19 Aromatase and Sex Hormone Ratio in Sex Determination

In mammals, sex steroid hormones produced by the differentiated gonads play key roles in the growth and differentiation of **secondary sexual characteristics**, or features that develop at puberty, such as facial hair, pubic hair, and breast development. However, in placental

† dosage-sensitive sex reversal, adrenal hypoplasia critical region, on chromosome X, gene 1
‡ wingless-type MMTV integration site family member 4
§ forkhead box protein L2
¶ R-spondin-1
# doublesex and mab-3-related transcription factor 1

mammals where sex is determined genotypically, these hormones do not influence primary (gonadal) sex determination, and treatment of male embryos with estradiol prior to sex determination fails to produce ovaries.[47–49] In marked contrast to placental mammals, among vertebrates that use TSD it is the androgen-to-estrogen ratio that ultimately determines whether the sexually undifferentiated gonad differentiates into a testis or ovary.[50] For example, in most reptiles treatments of embryos with exogenous estrogens induce the differentiation of ovaries, even at a male-producing temperature, whereas treatments with estrogen antagonists or inhibitors of estrogen synthesis promote testicular differentiation, even at a female-producing temperature.[11] Under natural conditions, the sex steroid ratio depends on the activity of the enzyme, *CYP19 aromatase*, which irreversibly converts androgens into estrogens in all vertebrates (Figure 20.10). Among TSD taxa, elevated expression of the aromatase gene is consistently observed in differentiating ovaries while its expression is usually suppressed during the development of testes.[51] Importantly, in reptiles, amphibians, and fish that use TSD, exposure to female-promoting temperatures during the TSP is associated with gonadal aromatase upregulation, whereas exposure to male-producing temperatures is accompanied by its suppression.[52–54] These observations suggest that transcriptional regulation of the aromatase gene by temperature during the TSP somehow biases the differentiation of bipotential gonads to a specific gonad phenotype.

Developments & Directions: Climate change is feminizing sea turtles

Sea turtles breed on the beaches of the world, incubating their eggs in sand in areas called "rookeries". Among sea turtles, warmer incubation temperatures produce more females, and cooler temperatures generate males. The **pivotal temperature** (Figure 20.8[A]) is the species and population-specific incubation temperature that produces 50% of each sex, though the transitional range of temperatures that produce 100% males or 100% females spans only a few degrees Celsius. With most sea turtle populations currently producing offspring above the pivotal temperature[55] and with average global temperature predicted to rise 2.6°C by the year 2100,[56] it is hypothesized that climate change may pose a serious threat to the survival of these populations by producing high egg mortality and generating female-only offspring.

Australia's Great Barrier Reef (GBR) is home to one of the largest green sea turtle (*Chelonia mydas*) populations in the world. Regions located in the northern GBR (Figure 20.9[A]), possesses an estimated female population size in excess of 200,000 nesting females, making it the Pacific Ocean's largest and most important green sea turtle breeding and hatching ground. In a landmark report, Jensen and colleagues[57] show that turtles originating from the relatively cool southern GBR beaches have only a moderate female sex bias (65%–69% female). However, turtles originating from the much warmer northern GBR nesting beaches were extremely female-biased (99.1% of juvenile, 99.8% of subadult, and 86.8% of adult-sized turtles) (Figure 20.9[B]).

The researchers collected turtles from foraging grounds along the GBR and determined the sex of individuals by measuring testosterone levels from blood samples, as well as by conducting microsurgical laparoscopic analysis of the gonads. They then linked individuals to their rookeries of origin using genetic analysis, enabling them to determine sex ratios produced at regional rookeries. The data represent many different cohorts of turtles hatched at different rookeries along the GBR, providing over 50 years of sex ratio records. Importantly, the researchers also estimated the sand temperatures of the rookeries corresponding with each cohort by using historical sea and air temperatures between 1960 and 2016, finding that by the 1990s the sand temperatures were consistently higher than the pivot temperature in the northern GBR. Collectively, the findings suggest that increasing sand temperatures are skewing the sex ratios of the northern GBR population such that virtually no males are currently being generated from these nesting beaches. The study highlights the urgent need for immediate management strategies focused on lowering incubation temperatures at key rookeries to generate more males and thereby avoid the collapse—or even extinction—of a population.

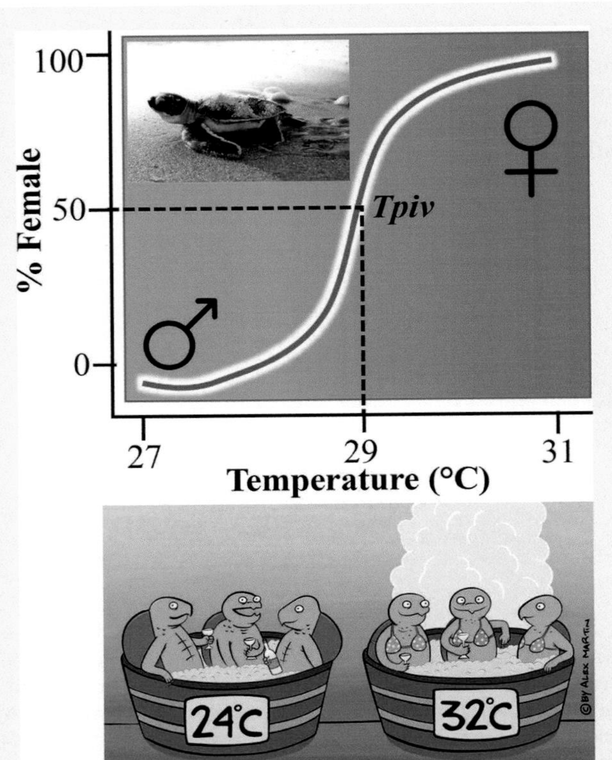

Figure 20.8 Among sea turtles, warmer egg incubation temperatures produce more females, and cooler temperatures generate males.

Source of images: Turtle inset: Courtesy of the National Park Service. Cartoon: Courtesy of Alexandra Martin.

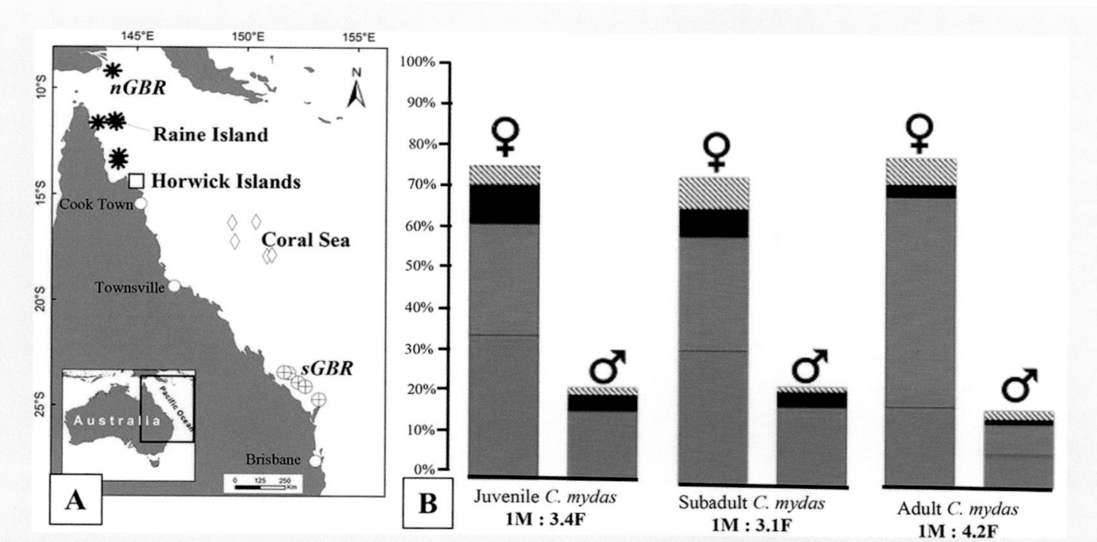

Figure 20.9 Increasing sand temperatures are skewing the sex ratios of the northern Great Barrier Reef sea turtles.

Source of images: Jensen, M.P., et al. 2018. Curr Biol, 28(1), pp. 154-159. Used by permission.

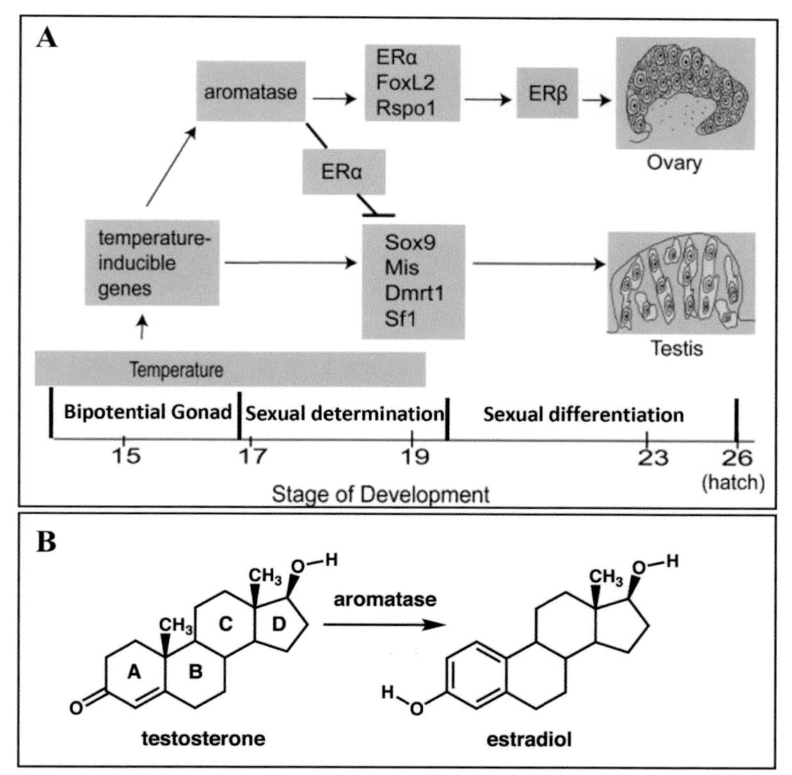

Figure 20.10 A model for temperature-dependent sex determination in the red-eared slider turtle (*Trachemys scripta*). (A) Female-producing temperatures directly or indirectly increase aromatase expression, which together with increased expression of ERα and other female-specific genes (i.e. *FoxL2*, *Rspo1*) promotes ovarian determination. Simultaneously, increased estrogen levels activate ERα expression, which inhibits male-specific gene expression during ovary development. At male-producing temperatures, aromatase expression does not increase, and testes develop in the absence of estrogen-induced inhibition. Testis determination and development is mediated by increased expression of male-specific genes such as *Sox9*, *Mis*, *DMRT1* and *Sf1*. **(B)** The enzyme aromatase induces the conversion of testosterone into estradiol.

Source of images: (A) Ramsey, M. and Crews, D., 2009. Sem. Cell Dev Biol. (Vol. 20, No. 3, pp. 283-292). Academic Press. Used by permission.

SUMMARY AND SYNTHESIS QUESTIONS

1. In mammals, SRY expression induces SOX9 expression, which goes on to promote the expression of multiple genes involved in testes development. Once SOX9 expression is initially turned on by SRY, SOX9 expression will continue even in the later absence of SRY. How?
2. Androgen insensitivity syndrome primarily affects genetically XY individuals, but rarely genetically XX individuals. Why?
3. A certain critical temperature promotes the expression of the enzyme CYP19 aromatase in the temperature-dependent sex determination of the red-eared slider turtle. Is this critical temperature a male- or female-inducing temperature? Explain your reasoning.
4. In the sea turtle study described in Developments & Directions ("Climate change is feminizing sea turtles"), why did the authors sex animals based on testosterone levels and not by genetic testing?
5. What sex phenotype would likely result in a human XY fetus with a loss of function of the SOX9 gene? Explain your reasoning.
6. In mammals, what effect, if any, will exposure of embryos to estradiol during sex determination have on the fate of gonad development?
7. In reptiles, what effect, if any, will exposure of embryos to estradiol during sex determination have on the fate of gonad development?

Sexual Differentiation of Internal and External Genitalia in Mammals

LEARNING OBJECTIVE Explain the developmental fates of the mammalian primordial internal and external genitalia in response to endocrine signaling.

KEY CONCEPTS:

- In both sexes, the sexually indifferent stage of embryogenesis is characterized by the presence of two different internal duct systems, each possessing a unipotential developmental fate.
- Following stimulation by androgens, the Wolffian ducts develop into components of the male reproductive tract, such as the vas deferens and the seminal vesicles, and in the absence of androgen stimulation the Wolffian ducts regress.
- In the absence of stimulation by anti-Müllerian hormone (AMH) from the testes, the Müllerian ducts develop into components of the female reproductive tract, such as the Fallopian tubes, uterus, and upper third of the vagina.
- When inhibited by anti-Müllerian hormone produced by the testes, the Müllerian ducts regress.
- The male and female external genitalia each arise from common, morphologically identical bipotential primordia.
- Whereas some components of the male external genitalia require exposure to testosterone for normal development, others require stimulation by the more potent androgen dihydrotestosterone.
- In the absence stimulation by androgens, the primordia of the external genitalia develop into their female forms.

In humans, sexual differentiation includes the development of internal and external genitalia, the differential development of the brain, and the development of secondary sex characteristics during puberty. In marked contrast to primary sex determination (gonadal determination), which in humans and other mammals is established genotypically by chromosomal sex and the local expression or suppression of key genes, sexual differentiation is primarily under endocrine control. The hormones of sexual differentiation include the sex steroid hormones, and in males also a glycoprotein hormone called *anti-Müllerian hormone* (AMH) is structurally related to the hormones inhibin and activin. Whereas AMH plays a key role in male sexual differentiation during fetal development, the steroid sex hormones exert their effects on this process during both fetal development and puberty. As such, the full human sexual phenotype develops sequentially, with chromosomal sex taking place at fertilization, and primary (or gonadal) sex determined during early fetal development. **Hormonal sex**, which is based on the concentrations of circulating androgens and estrogens, determines a broad array of sexual differentiation features occurring during fetal development and puberty (Figure 20.11). This array of characteristics is sometimes referred to as *phenotypic sex* or *morphological sex*. Lastly, *behavioral sex* can be ascribed based upon male and female behaviors

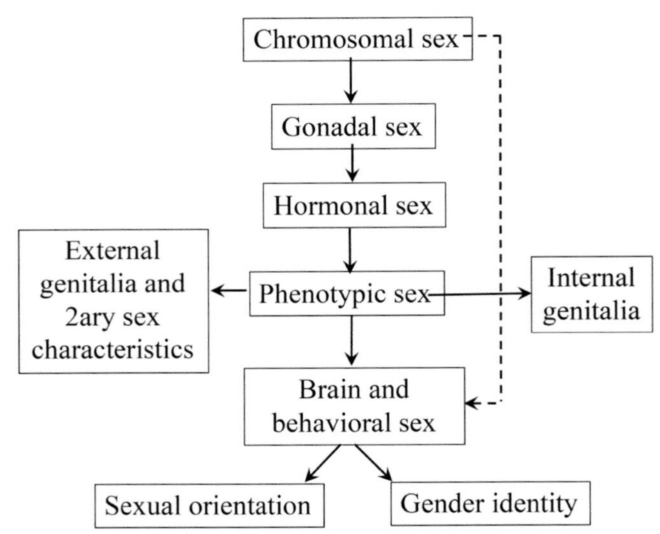

Figure 20.11 Different levels of sex determination in humans. Both chromosomal sex (dashed lines) and hormonal sex appear to influence brain and behavioral sex.

that are typical to the species. Among humans, additional categories for the classification of sex exist, such as *sexual orientation* (the development of sexual attraction to other people), *gender identity* (the psychological self-perception of being female, male, asexual, bisexual, or non-binary), and *legal sex* (the gender ascribed to an individual by a government agency).[58]

The effects of androgens and estrogens on the development of many human secondary sex characteristics during puberty will be addressed in Chapter 21: Male Reproductive System and Chapter 22: Female Reproductive System. This section will focus primarily on the effects of these hormones and AMH on sexual differentiation of the internal and external genitalia during fetal development.

Anatomy and Fate of the Indifferent Gonad and Internal Genital Ducts

Developmentally, the **primordial germ cells** (PGCs) that will give rise to spermatocytes or oocytes do not originate from the gonad, but instead migrate along the hindgut from the yolk sac to a region called the *gonadal ridge* by about the fifth week of embryogenesis in humans (Figure 20.12). Importantly, during this sexually indifferent stage, the architecture of the gonadal ridge is identical in both males and females. This region will give rise to the gonads and to the *genital ducts*. In males the genital ducts will differentiate into the epididymis, vasa deferentia, and seminal vesicles, and in females they differentiate into the Fallopian tubes, uterus, and upper third of the vagina. The gonadal ridge contains several important

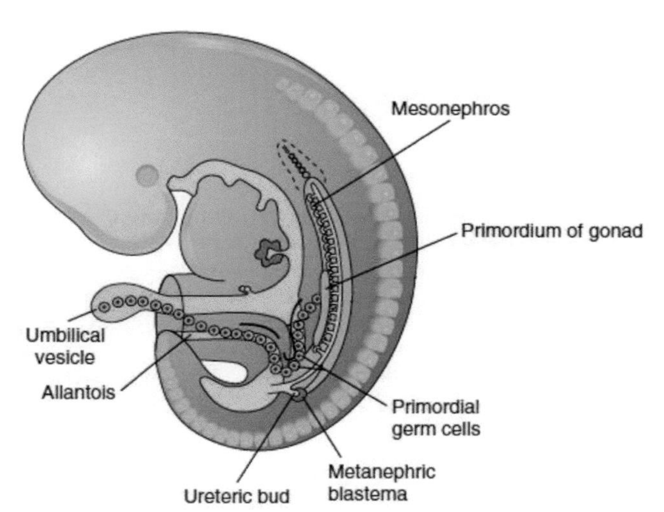

Figure 20.12 The primordial germ cells (PGCs) in both female and male human embryos originate in the yolk sac near the umbilical vesicle. The PGCs divide mitotically and migrate along the hindgut mesentery until reaching a region of the urogenital ridge called the gonadal ridge, which will form the primordium of the gonad.

Source of images: Moore, K.L., et al. 2018. The developing human-e-book: clinically oriented embryology. Elsevier Health Sciences. Used by permission.

anatomical features that will have different developmental fates, depending on the chromosomal sex of the embryo. These include:

- *Peripheral cortex of the gonad* (see Figure 20.13): In the presence of XX germ cells, the cortex thickens dramatically, forming secondary sex cords that surround the developing oogonia of the growing ovary. In the presence of XY germ cells the cortex regresses into a thin epithelial layer that surrounds the coelomic cavity of the testis.
- *Central medulla of the gonad* (see Figure 20.13): In the presence of XY germ cells, the primitive sex cords of the medulla develop into the seminiferous tubules of the testis, giving rise to Sertoli cells. In the presence of XX germ cells, the medulla regresses and is replaced by a highly vascularized loose mesenchymal tissue that forms the connective tissue covering of the ovaries.
- *Wolffian (mesonephric) ducts* (see Figure 20.13 and Figure 20.14): The mesonephros is a component of the embryonic kidney whose glomeruli and renal tubules join into a mesonephric duct. The mesonephros is a transient structure that will eventually regress or transform into other structures. In the presence of a testis, these mesonephric (Wolffian) ducts will develop into non-urethral components of the male reproductive tract, such as the vas deferens, rete testis, efferent ducts, epididymis, ductus deferens, and the seminal vesicles. In the presence of an ovary, the Wolffian ducts will regress.
- *Müllerian (paramesonephric) ducts* (see Figure 20.13 and Figure 20.14): In the presence of an ovary the Müllerian ducts develop into several regions of the female internal reproductive tract, including the Fallopian tubes, uterus, and upper third of the vagina. In the presence of a testis, the Müllerian ducts regress.

It is important to emphasize that whereas the gonad itself has a bipotential fate, the internal genital ducts develop from two different sets of *unipotential* precursors. That is, the Wolffian ducts give rise exclusively to male structures, whereas the Müllerian ducts give rise exclusively to female structures. As will be described next, the developmental fates of the Wolffian and Müllerian ducts are entirely dependent on the hormones produced by the differentiated gonad. A time course for key events in the embryonic differentiation of human gonads is described in Figure 20.15.

Endocrine Control over the Differentiation of the Internal Genitalia

Much of our understanding of the hormonal basis of sexual differentiation in mammals derives from the classic work of French scientist Alfred Jost. Beginning in 1947, Jost conducted a series of elegant experiments using rabbits to demonstrate that if the primordial testes or ovaries are surgically removed from either female (XX) or male (XY) fetuses *in utero* prior to sexual differentiation, they both developed female internal and external genitalia[47,59,60] (Figure 20.16). Specifically, in both cases the Wolffian ducts regressed and the Müllerian ducts

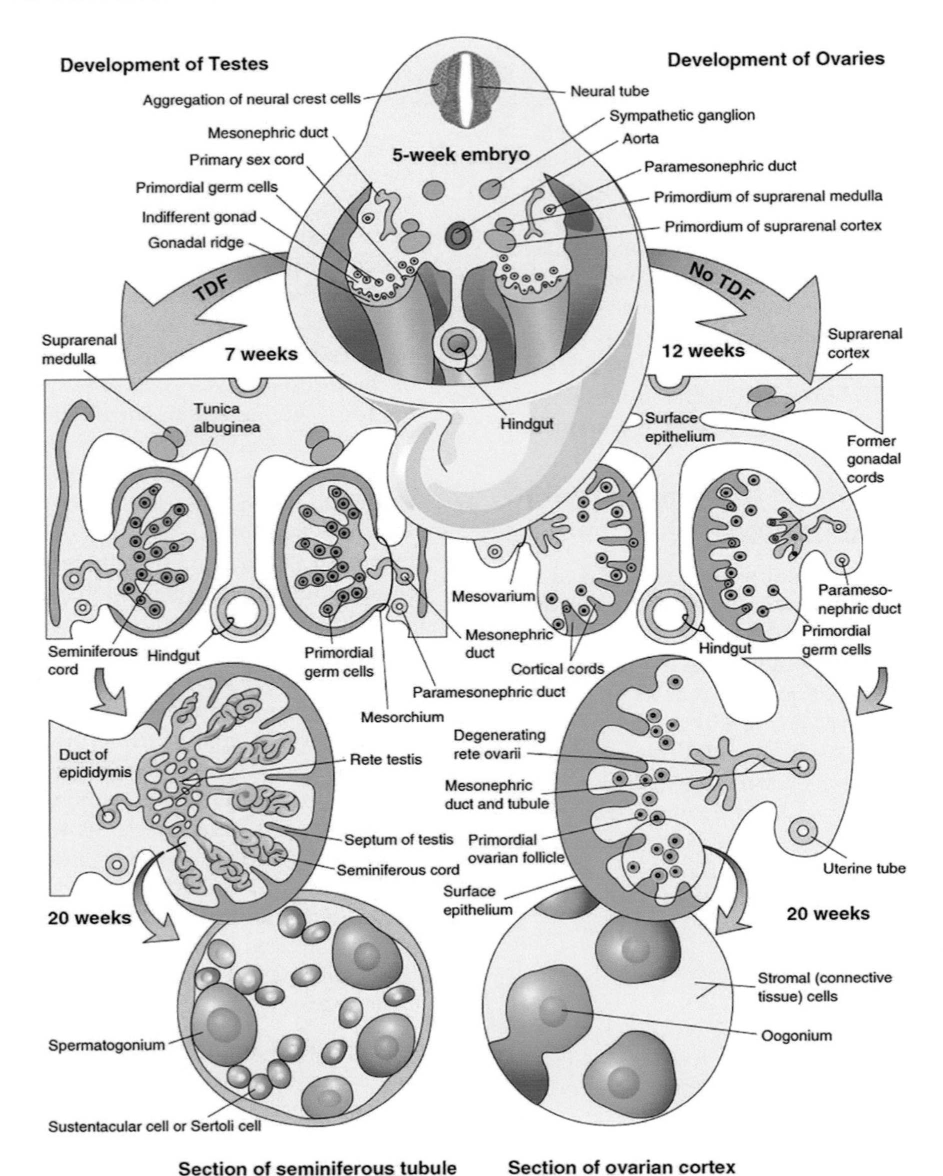

Figure 20.13 Schematic illustrations showing differentiation of the indifferent gonads in a 5-week embryo (top) into ovaries or testes. The left side of the drawing shows the development of testes resulting from the effects of the testis-determining factor (TDF) located on the Y chromosome. Note that the gonadal cords become seminiferous cords, the primordia of the seminiferous tubules. The parts of the gonadal cords that enter the medulla of the testis form the rete testis. In the section of the testis at the bottom left, observe that there are two kinds of cells: spermatogonia, derived from the primordial germ cells, and sustentacular or Sertoli cells, derived from mesenchyme. The right side of the drawing shows the development of ovaries in the absence of TDF. Cortical cords have extended from the surface epithelium of the gonad and primordial germ cells have entered them. They are the primordia of the oogonia. Follicular cells are derived from the surface epithelium of the ovary.

Source of images: Moore, K.L., et al. 2018. The developing human-e-book: clinically oriented embryology. Elsevier Health Sciences. Used by permission.

persisted, developing into the associated female phenotype. These experiments suggested that the development of phenotypic male internal features requires a factor produced specifically by differentiated testicular tissue—this factor was testosterone. However, when female (XX) fetuses were administered testosterone, the Müllerian ducts still persisted, suggesting that testosterone does not induce their regression, but that another factor produced by the testis may actively cause the Müllerian ducts to regress. This factor would turn out to be the aforementioned anti-Müllerian hormone (AMH), a glycoprotein produced by the Sertoli cells. Taken together, Jost's research suggests that in males testosterone secreted by testicular Leydig cells induces the differentiation of the Wolffian ducts, whereas AMH from the testicular Sertoli cells causes the regression of the Müllerian ducts. By

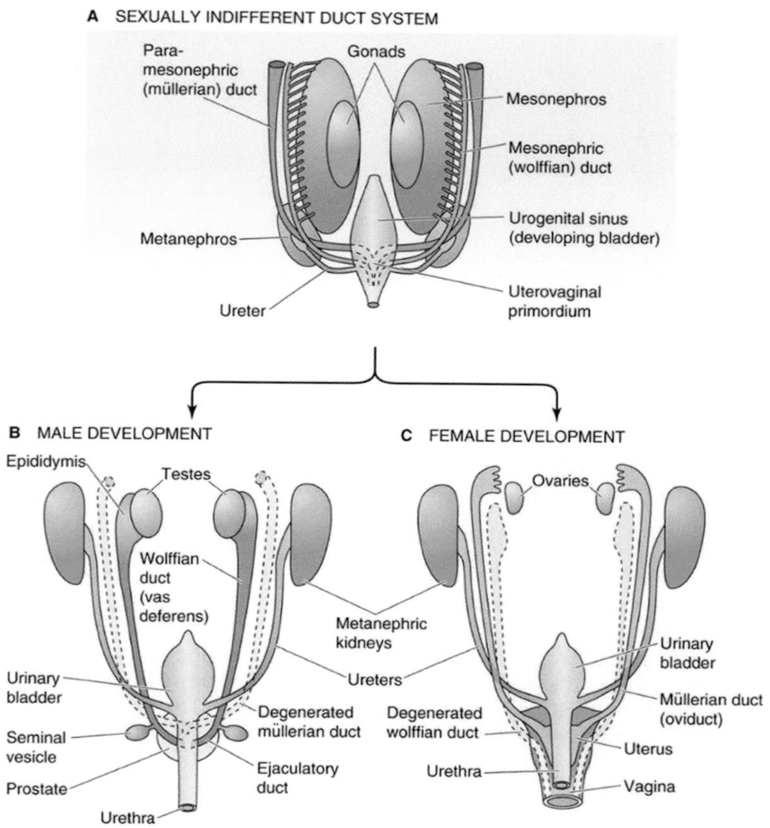

Figure 20.14 Developmental fate of the human genital ducts. (A) Prior to differentiation, the indifferent gonad is closely associated with the mesonephros (a component of the developing kidney) and the mesonephric or Wolffian duct excretory duct (a secretory duct connecting the mesonephros to the urogenital sinus). The Müllerian ducts run parallel to the Wolffian ducts and merge to form the uterovaginal primordium. **(B)** In males, the Wolffian duct develops into the vas deferens, seminal vesicles, and ejaculatory duct, and the mesonephros develops into the epididymis, whereas the Müllerian ducts degenerate. **(C)** In females, the Müllerian ducts develop into the Fallopian tubes, the uterus, the cervix, and the upper one third of the vagina, whereas the Wolffian ducts degenerate.

Source of images: Boron, W.F. and Boulpaep, E.L., 2005. Medical Physiology: A Cellular and Molecular Approach, updated 2nd ed. Elsevier. Used by permission.

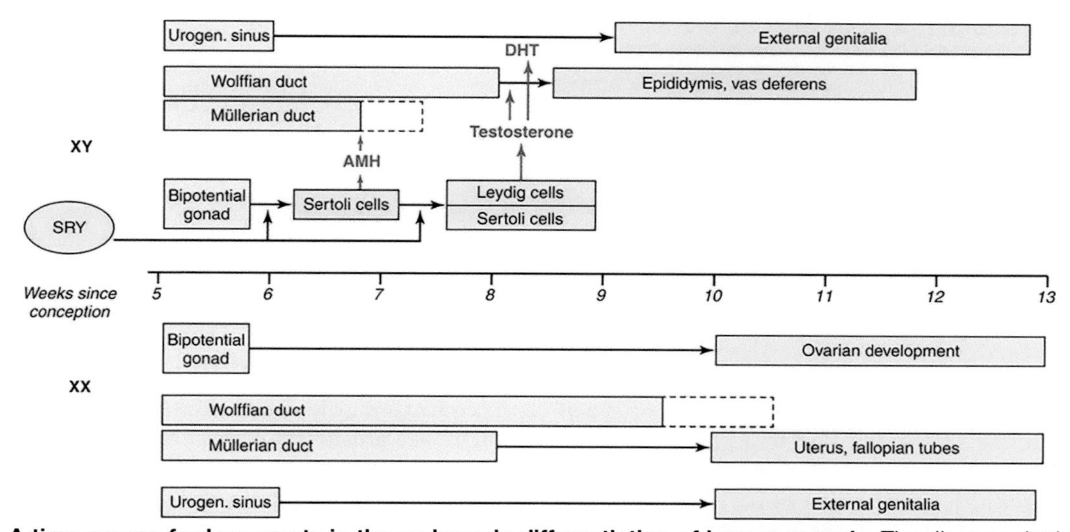

Figure 20.15 A time course for key events in the embryonic differentiation of human gonads. The diagram depicts a timeline for the differentiation of the bipotential gonad into the testis (upper half) or the ovary (lower half). AMH = anti-Müllerian hormone; DHT = dihydrotestosterone.

Source of images: Norman, A.W. and Henry, H.L., 2022. Hormones. Academic Press. Used by permission.

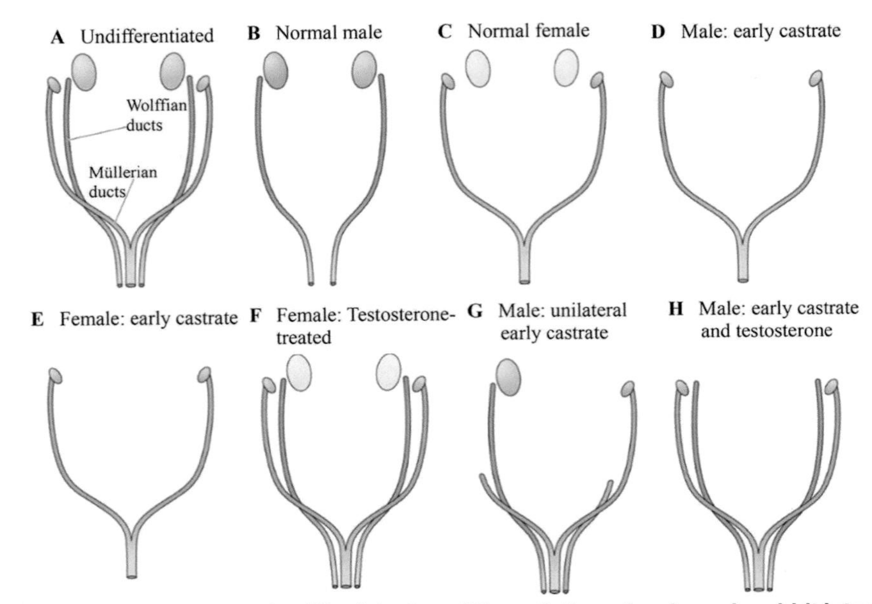

Figure 20.16 The results of some experiments by Alfred Jost on differentiation of embryonic rabbit internal genitalia. The fates of the Wolffian and Müllerian ducts are shown under normal conditions **(A–C)**, after castration **(D–E)**, following testosterone treatment **(F)**, unilateral castration **(G)**, and castration plus testosterone **(H)**. Jost's findings suggest that in males testosterone secreted by Leydig cells induces the differentiation of the Wolffian ducts, whereas AMH from the Sertoli cells causes the involution of the Müllerian ducts. By contrast, in females the Wolffian ducts spontaneously regress in the absence of testosterone and the Müllerian ducts spontaneously differentiate in the absence of AMH.

Source of images: Boron, W.F. and Boulpaep, E.L., 2005. Medical Physiology: A Cellular and Molecular Approach, updated 2nd ed. Elsevier. Used by permission.

contrast, in females the Wolffian ducts spontaneously regress in the absence of stimulation by testosterone, and the Müllerian ducts spontaneously differentiate in the absence of inhibition by AMH.

Importantly, the development of male internal genitalia requires two events: (1) *masculinization*, or retention of the Wolffian ducts and the development of male traits; and (2) *defeminization*, or regression of the Müllerian ducts and inhibition of female traits. By contrast, female internal genitalia development requires different events: (1) *feminization*, or retention of the Müllerian ducts and promotion of female traits; and (2) *demasculinization*, or regression of the Wolffian ducts and the inhibition of male traits. As such, the differentiation of internal genitalia can be considered to proceed along two dimensions: a masculinization-demasculinization continuum, and a feminization-defeminization continuum[61] (Figure 20.17). As will be seen later, the sexual differentiation of the brain and behavior also occur along similar dimensions. Since it is also possible for both masculinization and feminization events to occur in the same individual, as well as for neither to develop, these continua lead to a broad spectrum of potential disorders and differences of sexual development, many of which will be described in the last section of this chapter. Note that although estrogens do not appear to play a role in the differentiation of internal or external mammalian genitalia, they are involved in sexual differentiation of the brain, as well as the differentiation of the female breasts during puberty.

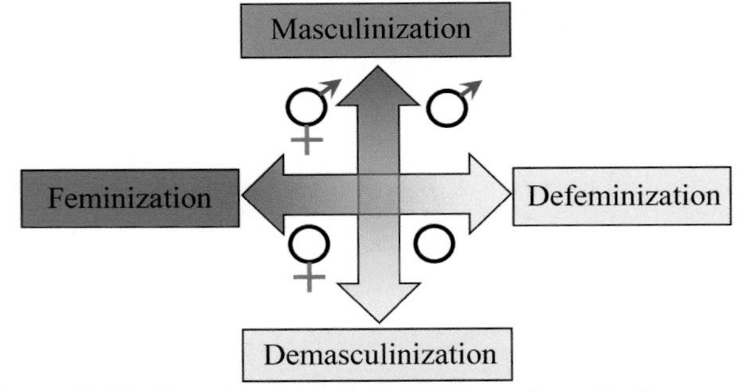

Figure 20.17 The development of internal genitalia occurs along two dimensions.

Source of images: An Introduction to Behavioral Endocrinology, 3rd edition, Fig 3.6 Randy Nelson, 2005 Sinauer Associates.

Endocrine Control over the Differentiation of the External Genitalia

In contrast to the differentiation of the internal genitalia, whose precursors consist of a set of two unipotential ducts (the Wolffian and Müllerian ducts), the male and female external genitalia each arise from common, morphologically identical bipotential primordia that include the genital tubercle, urethral folds and groove, and labioscrotal genital swellings (Figure 20.18). The male and female tissue counterparts that each develop from the same embryonic tissue (e.g. glans of the penis and the

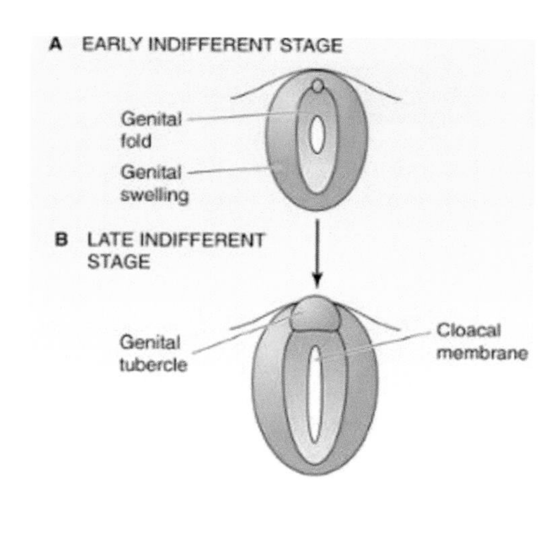

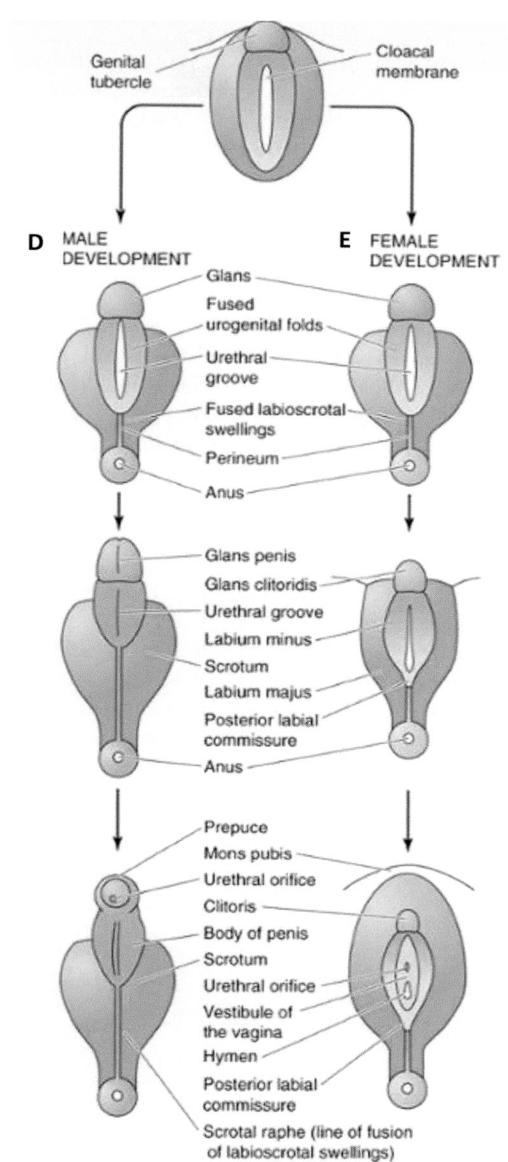

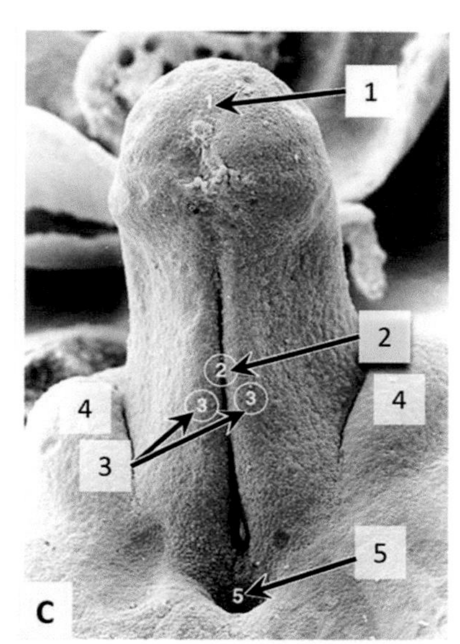

Figure 20.18 (parts a and b) Development of the external genitalia from a common bipotential primordia. Part 1: **(A)** In both sexes the external genitalia are morphologically identical during the early indifferent stage. **(B)** Also in both sexes, the genital tubercle enlarges to form the phallus (green) during the late indifferent stage. **(C)** Scanning electron micrographs of the developing external genitalia from a 7-week embryo. 1, Developing glans penis with the ectodermal cord; 2, urethral groove continuous with the urogenital sinus; 3, urethral folds; 4, labioscrotal swellings; 5, anus. Part 2: **(D)** In males, the genital tubercle develops into the glans penis (green), whereas the urogenital folds (brown) fuse to form the shaft of the penis and the labioscrotal swellings (blue) develop into the scrotum. **(E)** By contrast, in females, the genital tubercle differentiates into the clitoris (green). The urogenital folds remain separate as the labia minora (brown). The unfused portion of the labioscrotal swellings become the labia majora, whereas the ventral and dorsal labioscrotal swellings fuse to form the mons pubis and posterior labial commissure, respectively.

Source of images: (A, B, D, E) Boron, W.F. and Boulpaep, E.L., 2005. Medical Physiology: A Cellular and Molecular Approach, updated 2nd ed. Elsevier. Used by permission. (C) Moore, K.L., et al. 2018. The developing human-e-book: clinically oriented embryology. Elsevier Health Sciences. Used by permission.

clitoris; the scrotum and the outer labia) are described as developmentally *homologous* structures. Importantly, Alfred Jost's aforementioned experiments with rabbits also demonstrated that the developmental fates of these homologous tissues are based entirely upon the levels of exposure of the embryonic primordia to androgens. For example, secretion of androgens by the fetal testes induces the fusion of the genital swellings to form a scrotum and development of the penis from the bipotential genital tubercle, while absence of androgen stimulation during development leads to the female pattern of external genitalia. In particular, the normal development of the male external genitalia requires exposure to *dihydrotestosterone* (DHT), which is synthesized from testosterone via the enzyme **5α-reductase** (Figure 20.19). Growth of the penis, scrotum, and prostate gland during puberty is also mediated primarily by DHT, whereas the production of seminal fluid and other accessory secretions is promoted by testosterone (Figure 20.19). DHT not only binds the androgen receptor with greater specificity than testosterone does, but it also has a two- to ten-fold higher potency than testosterone in androgen-responsive tissues.

If female fetuses are exposed to androgens during the period of external genitalia differentiation, their external genitalia will become "masculinized", or develop into the male homolog. By contrast, if male fetuses are castrated during this period, this results in the "feminization" of the external genitalia. Since both the male and female external genitalia develop from a common precursor, their differentiation proceeds along a single masculine-feminine continuum (Figure 20.20), as the development of one form prevents the development of the other.[61] As was true for the development of the internal genitalia, estrogens do not play any role in differentiation of the external genitalia, and if the ovaries are removed from a female fetus during early development, the female external genitalia will still develop.

In male fetuses, a final endocrine-mediated event associated with development of the male phenotype is the descent of the testes from the abdominal region into the scrotum. This process, which is normally completed by the seventh month of pregnancy, requires exposure to *insulin-like peptide 3* (INSL3), a hormone produced during embryogenesis by the testes, and in its absence the testes will not descend.[62] Unlike testosterone, INSL3 is a constitutive product of the testicular Leydig cells and is not regulated by the HPG axis.[63]

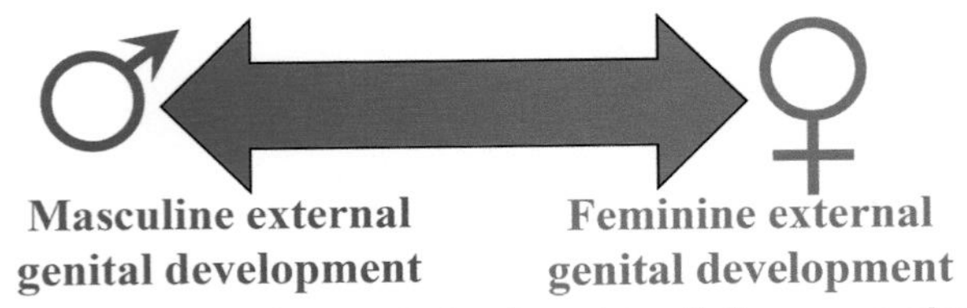

Figure 20.20 Development of external genitalia occurs along one dimension.

Source of images: An Introduction to Behavioral Endocrinology, 3rd edition, Fig 3.6 Randy Nelson, 2005 Sinauer Associates.

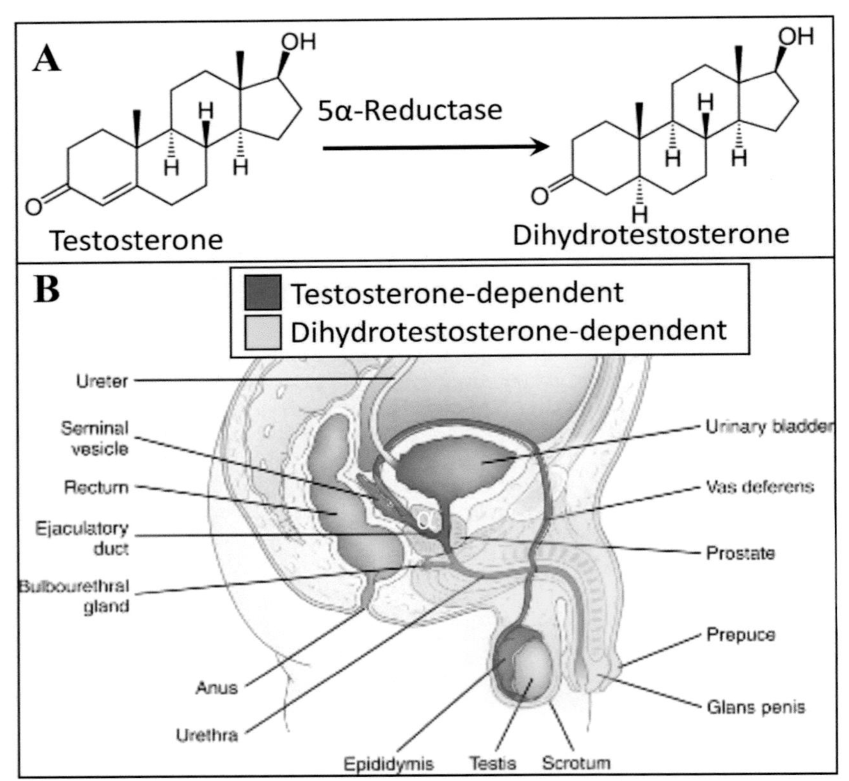

Figure 20.19 Androgen target tissues of the human male internal and external genitalia. (A) The enzyme 5α-reductase converts testosterone into the more potent androgen dihydrotestosterone (DHT). **(B)** Testosterone- and DHT-dependent regions of the male genitalia.

Source of images: (B) Norman, A.W. and Henry, H.L., 2022. Hormones. Academic Press. Used by permission.

Developments & Directions: The spotted hyena: a perplexing case of natural female virilization

Weighing up to 180 pounds with a standing shoulder height up to 30 inches, the spotted hyena (*Crocuta crocuta*) is the next largest carnivore in Africa, second only to the African lion. In marked contrast to virtually every other mammalian species studied (even compared with other species of hyena), many of the sex roles and features of the spotted hyena are reversed. For example, adult females are about 10% larger than adult males, as well as significantly more aggressive and socially dominant than the males.[64] The behavioral masculinization of female cubs has been attributed to exposure of the fetuses to elevated levels of androgens during pregnancy, in part due to a high placental level of the enzyme 17β-HSD that converts androstenedione to testosterone.[65] For example, dominant females have higher levels of androgens during the last trimester of pregnancy, when behavioral masculinization is most likely, compared with subordinate females,[66] and cubs of both sexes born to mothers with elevated androgen levels are more aggressive and exhibit higher rates of mounting behavior than cubs born to mothers with low concentrations.[67] Furthermore, even adult females who were exposed to high androgen levels *in utero* display higher aggression than those exposed to lower androgen concentrations in the womb. Thus, prenatal androgen exposure appears to influence both juvenile and adult female aggression. Importantly, although these observations may help explain why female behaviors are masculinized, they do not explain why the females are dominant to males, as male fetuses and adults are still exposed to higher levels of androgens compared to females.[65]

Perhaps the most unusual sex-reversed phenotype in the spotted hyena is the striking morphology of the female's external genitalia, whose clitoris is virtually indistinguishable from the male's penis (Figure 20.21). Specifically, the female's hypertrophied clitoris is traversed by a central urogenital canal opening at the tip of the glans, and the outer labia are fused to form a pseudo-scrotum located in the position where an external vaginal orifice would be in other mammals. These female pseudo-penises even display erections similar to those of males, with dominant hyenas inspecting the erect appendage of subordinate members of the clan during meeting ceremonies.[68] In fact, the female and male external genitalia are so similar that from the time of Aristotle through the 18th century many scholars considered these animals to be hermaphrodites. These unusual external genitalia give spotted hyenas the dubious distinction of being the only mammalian females to urinate, copulate, and give birth through a penis-like canal. Despite their highly masculinized external genitalia, the female's internal genitalia are typical for most female mammals, composed of ovaries, oviducts, and a uterus. Male internal structures, such as the Wolffian ducts and their derivatives, are absent in the female, suggesting that the presence of elevated placental androgens throughout most of fetal life is insufficient to prevent their regression.[69]

Initially, researchers hypothesized that exposure to androgens during fetal development could explain the masculinized external genitalia of the female spotted hyena, similar to their effects on behavioral masculinization. However, although the treatment of pregnant females with antiandrogens, such as finasteride (a 5α-reductase inhibitor) and flutamide (a selective antagonist of the androgen receptor), inhibited the development of some sexually dimorphic internal phallic structures, it failed to affect the gross masculinization of overall phallic size and shape of male and female fetuses.[68] This is in marked contrast to experiments with dogs and rodents treated with similar concentrations of antiandrogens, which results in development of males with small penises no longer traversed by a urethra, as well as the presence of a blind vaginal pouch. Remarkably, and in contrast to all mammals studied to date, prepubertal gonadectomy or castration also failed to significantly affect growth of the penis and only had a slight effect on growth of the clitoris. The researchers conclude that the formation of the male and female spotted hyena phalluses must occur independently of androgens, and that patterns of normal clitoral and/or penile growth are only minimally affected by exogenous androgens.

Considering how well established the linkage between androgens and masculinization of external genitalia is in other mammals, these striking findings for the spotted hyena are without precedent, and there are currently few hypotheses to explain these observations. The researchers speculate that in this species estrogens may have assumed some of the roles traditionally assigned to androgens in other species, or that the androgen receptor may be activated in a ligand-independent manner.[70]

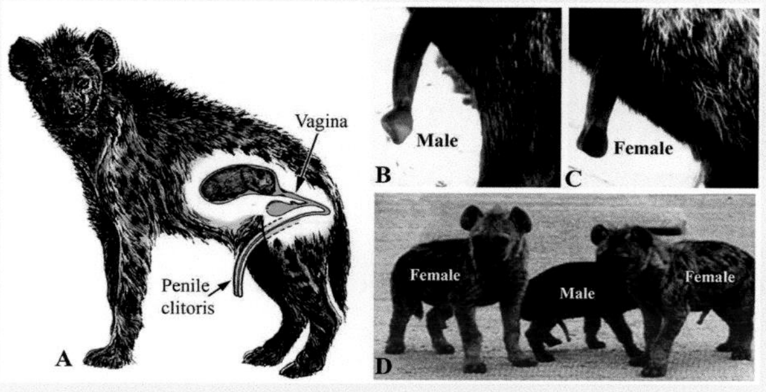

Figure 20.21 The spotted hyena (*Crocuta crocuta*). (A) Illustration of a pregnant female spotted hyena with a fetus in a uterine horn. The adult erect hyena penis **(B)** and clitoris **(C)** each have a distinct glans shape. **(D)** The sexes of the three infant spotted hyenas (3–4 months old) can be determined based upon glans shape.

Source of images: Cunha, G.R., et al. 2014. Differentiation, 87(1-2), pp. 4-22. Used by permission.

SUMMARY AND SYNTHESIS QUESTIONS

1. Compare and contrast the developmental fates of precursors to mammalian gonads, internal genitalia. and external genitalia.
2. Unlike most reptiles and fish, in placental mammals hormones do not influence primary (gonadal) sex determination, and treatment of male embryos with estradiol prior to sex determination fails to produce ovaries. Why?
3. Testosterone induces male mating behaviors in female guinea pigs, but treatment of female rodents with estradiol also induces masculinization of the brain just as effectively. How does administration of estradiol produce the same effect as testosterone?
4. Why aren't female fetuses masculinized by high levels of estradiol from their ovaries or their mother's ovaries?

5. Why is it incorrect to consider testosterone a "male" hormone and estradiol a "female" hormone?
6. In Young's guinea pig experiments, why didn't testosterone injections into adult control females induce mounting behavior?
7. Draw a flowchart using arrows and boxes to summarize the endocrine control over the normal differentiation of the internal and external genitalia in XX females and XY males.

Sexual Differentiation of the Brain

LEARNING OBJECTIVE Provide an overview of the different factors that contribute to the sexual differentiation of the mammalian brain and behavior.

KEY CONCEPTS:

- Some regions of the brain show distinct sexually dimorphic circuitry.
- Whereas testosterone has "organizational" effects during the prenatal period, it exerts "activational" effects during adulthood.
- In male mammalian fetuses, the masculinizing effects of testosterone on the brain are mediated by the estrogen receptor, which binds to estradiol following its local conversion from testosterone by the enzymeCYP19 aromatase.
- Female mammalian fetuses are protected from the masculinizing effects of high estradiol levels by the binding and sequestration of blood estradiol to alpha-fetoprotein.

Although much of the brain and its functions are indistinguishable between the sexes, some regions of the brain display a distinct sexually dimorphic circuitry. One clear example of this is the response of the female mammalian hypothalamus to both negative and positive feedback by estradiol and progesterone during different phases of the ovarian cycle, whereas in males the hypothalamus only responds via negative feedback to sex steroid hormones. Most vertebrates display certain sexually dimorphic behaviors, the most prominent of which are stereotyped displays that enhance reproductive success. Well-characterized examples include male aggression and territoriality in many species, bird song and frog mate calling, stereotypical positioning of rodents during sex ("mounting" behavior for males, and "lordosis", or arching of the back, for females), greater simulated fighting or "rough and tumble" play in males versus females among primates and many other mammals, and urinating posture and territorial marking behavior after puberty in dogs. In addition to sexually dimorphic courtship and copulatory behaviors, parental behaviors such as nest building, foraging for food, and nursing can also take different forms in females and males. The described examples of sexually dimorphic behaviors among vertebrates are instinctual and develop from a sexually differentiated nervous system, which is in turn influenced by hormonal, genetic, and epigenetic factors throughout the life of the animal.[71,72] As such, this neural circuitry is not always hard-wired, but may be plastic and influenced by environment and experience.

Sex Hormone-Dependent Effects on Brain Differentiation

William C. Young's Organizational/Activational Hypothesis

The notion that gonadal hormones play key roles in sexual differentiation of brain and behavior is as old as the field of endocrinology itself. Recall that in 1849 Arnold Berthold demonstrated that in contrast with capons (gonadectomized male chicks), castrated male chicks with subsequent testis transplants permitted them to not only develop typical adult male rooster morphology (e.g. size and shape of wattles and comb), but also develop typical male behaviors (e.g. crowing, strutting, and an interest in mating with hens). A century after Berthold's discovery, Jost reported on the profound influence of androgenic steroid hormones in mediating sexual differentiation of the genitalia of rabbits, and researchers began to ask if these hormones acted similarly on brain development. To test this idea, in 1959 William C. Young and his colleagues[73] published a classic paper that rivals Berthold's work in providing many of the fundamental principles of behavioral endocrinology that are still used today. In their experiments (summarized in Figure 20.22), the researchers began by injecting either low or high concentrations of testosterone into pregnant guinea pigs throughout their gestation. They then observed that whereas male pups developed normal male external genitalia, the female pups exposed to high testosterone levels were born with external genitalia that were indistinguishable from those of their male siblings. They went on to gonadectomize the male and female siblings, treat them with estradiol as adults, and make some striking behavioral observations:

1. Females that were administered testosterone prenatally displayed a reduced tendency for *lordosis* (arching of the back for copulation) as adults.
2. Females that were administered testosterone prenatally increased their tendency to display *mounting behavior* (normally a male copulation behavior in these animals) following testosterone treatment as adults.
3. No change in mounting behavior was observed in adult males that were administered testosterone prenatally.

Several important implications can be drawn from these observations. The first is that the prenatal effects of testosterone are distinct from those during adulthood. In their study, Young and colleagues introduced two terms that have had lasting value in describing these different effects. The prenatal actions of testosterone were referred

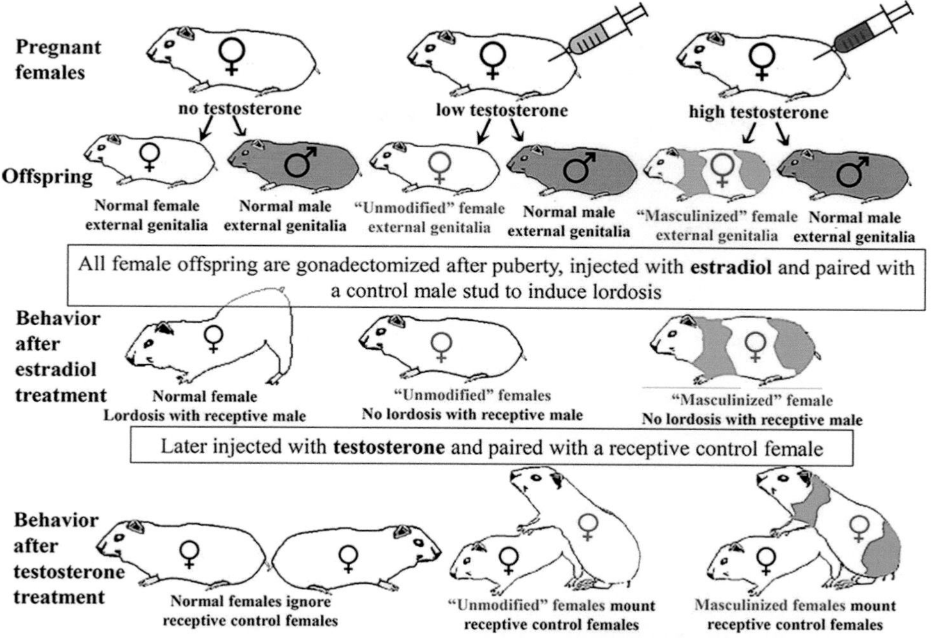

Figure 20.22 William C. Young's classic experiments with guinea pigs demonstrating the effects of testosterone concentration and timing of injection on the differentiation of external genitalia and sexual behavior. In their experiments the researchers began by injecting either low or high concentrations of testosterone into pregnant guinea pigs throughout their gestation. They then observed that whereas male pups developed normal external genitalia, the female pups exposed to high testosterone levels were born with external genitalia that were indistinguishable from those of their male siblings, and they referred to these as "masculinized females". They also observed that pups exposed to low testosterone levels had apparently normal looking external genitalia, and they referred to these as "unmodified females". Following puberty, both groups of testosterone-exposed females, as well as male and female controls, were gonadectomized. Next, all groups were injected with estradiol and paired with a male stud guinea pig to stimulate female sexual behavior (i.e. lordosis). Lastly, following a period of time, all groups were then injected with testosterone and paired with a receptive control female to stimulate male sexual mounting behavior. The observations and implications of the findings are discussed in the text.

to as having *organizational effects*—that is, they cause the neural target tissues to irreversibly "organize" (today we would say "differentiate") into components that can be induced later in development to bring about male- or female-specific behaviors. The later actions of testosterone in adults were called *activational effects*, as they activate (today we would say "induce") the expression of the differentiated neural substrate. In contrast to organizational effects, activational effects were considered reversible since the ability of testosterone to promote adult male mounting behavior lasted only as long as the testosterone was present.

A second key implication of their results was that the permanent masculinizing effects of testosterone are limited to specific *critical periods* of development. For example, in order for the external genitals to be masculinized, testosterone must be present during the middle and/or late periods of gestation, the time when genital differentiation occurs. Therefore, treatment of females with testosterone before birth masculinizes their genitals, but not if they are treated after birth. Similarly, specific behavioral circuits or brain regions experience critical periods during which testosterone acts to masculinize them. A third implication is that the differentiation of neural tissues involved in mating behavior by sex steroid hormones is in some ways analogous to the differentiation of internal genitalia observed by Jost. That is, in males the brains normally experience both masculinization and defeminization, whereas female brains are both feminized and demasculinized.

The Estrogen Paradoxes

Soon after Young's observations that testosterone could induce male mating behaviors in female guinea pigs, other researchers discovered, quite unexpectedly, that treatment of female rats with estradiol could also induce *masculinization* of the brain just as effectively as testosterone treatment.[74–77] How could the administration of a "female" hormone make females behave more like males? To explain this paradox, the "aromatization hypothesis" was developed, postulating that testosterone from the male gonads was converted locally in the brain to estradiol by the enzyme CYP19 aromatase. This hypothesis was confirmed with the discovery in rodents that brain nuclei that express sex differences also express high levels of the enzyme[78,79] and that the blockade or absence of this neuronal enzyme during the critical period inhibits differentiation of the male rodent brain.[80–82] Therefore, in rodents estradiol is in fact the active hormone in males that acts within neurons to both induce masculinization and promote defeminization of the brain. Importantly, in humans and other primates it is currently thought that the aromatization of testosterone to estradiol is not as important as it is in rodents, and that instead it is testosterone itself that differentiates the brain into a male brain during fetal development.[83] In primates, the bipotential brain appears to differentiate into a female brain in the absence of testosterone exposure, although the possibility remains for an as yet undiscovered role for estrogens in human brain development.[84]

Considering estradiol's newfound masculinizing effect, a second paradox emerged: why aren't female rodent fetuses masculinized by the estradiol produced by their ovaries or their mother's ovaries? This paradox was resolved by the discovery that a liver protein, **alpha-fetoprotein**, is secreted into the blood of female mice before and after birth.[85] This protein binds potently to blood estrogens, preventing them from entering the brain to induce masculinization. By contrast, in male rats, testosterone is not bound by alpha-fetoprotein, and it therefore can enter the brain where it is aromatized to estradiol in specific nuclei. In fact, the concentration of estradiol in the developing male brain is more than double that of females, especially in areas subject to sexual differentiation.[86,87] A summary of the rodent model for the sexually dimorphic actions of steroid hormones is presented in Figure 20.23. Importantly, since human alpha-fetoprotein does not appear to bind estrogens,[88] its function in humans remains less clear.

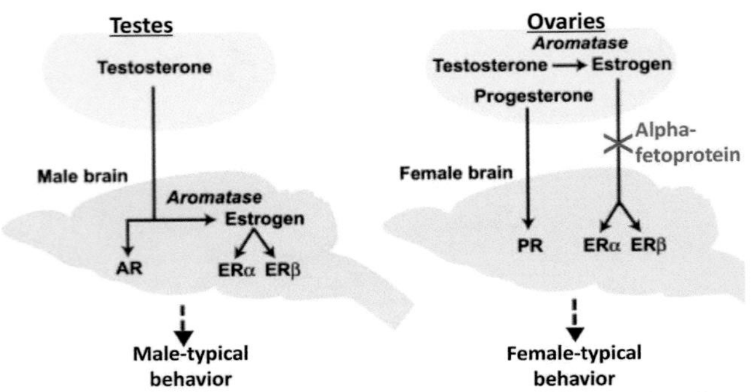

Figure 20.23 Regulation of sexually dimorphic brain development by sex steroid hormones, intracellular aromatase, and blood alpha-fetoprotein in the perinatal mouse. Sex steroid hormones produced by the gonads cross the blood-brain barrier and bind to hormone receptors in neurons to regulate sex-specific behaviors. In males, the effects of testosterone are mediated both directly by the androgen receptor (AR) and indirectly by the estrogen receptor (ER) following testosterone's conversion to estrogen by aromatase. In females, estrogen and progesterone regulate behavior via their receptors ER and PR, respectively. During the perinatal period, the liver protein alpha-fetoprotein is secreted into the blood of female rodents and prevents them from entering the brain to induce masculinization by binding tightly to blood estrogens. *Note:* Since human alpha-fetoprotein does not bind estrogen, its function in humans is less clear.

Source of images: Yang, C.F. and Shah, N.M., 2014. Neuron, 82(2), pp. 261-278. Used by permission.

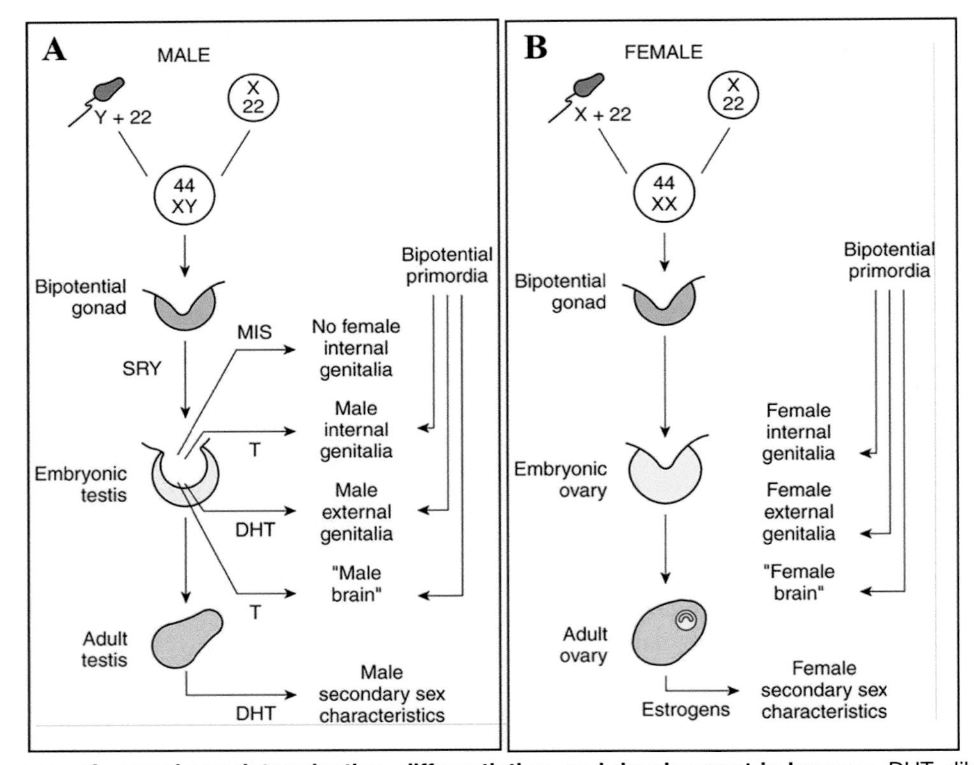

Figure 20.24 Summary of normal sex determination, differentiation, and development in humans. DHT, dihydrotestosterone, MIS, Müllerian inhibiting substance; T, testosterone.

Source of images: Barrett, K.E., et al. 2010. Ganong's review of medical physiology twenty. McGraw Hill LLC. Used by permission.

Challenges to Dogma

Interestingly, more recent studies have shown that estradiol may in fact be necessary for the development of the female brain.[89] Using aromatase knockout (ArKO) mice that cannot synthesize estrogens from androgens, Bakker and colleagues[90] showed that female mice had reduced levels of female sexual behavior in adulthood, even following ovariectomy and subsequent treatment with estradiol and progesterone. Furthermore, administration of estradiol over a specific prepubertal period almost completely restored female sexual behavior in female ArKO mice.[91] These findings challenge the classical theory of a default organization of the female brain and also question the notion that sex differences are established prior to or soon after birth, and that after the perinatal period sex hormones only have so-called activational effects on the brain. Instead, the brain may remain plastic and sensitive to organizational actions of sex hormones for a much longer period than was originally thought.

Clearly, the physiology of sex determination, differentiation, and development are particularly complex processes, which in mammals are influenced by interacting combinations of genetic and endocrine variables. A summary of these influences is presented in Figure 20.24.

SUMMARY AND SYNTHESIS QUESTIONS

1. A mouse line contains mutations in the gene coding for alpha-fetoprotein. Will the offspring of mice with this mutation exhibit primarily male or female mating behaviors? Explain your reasoning.
2. A mouse line exhibits low levels of expression of the enzyme CYP19 aromatase in the brain. Will the male offspring of mice with this mutation exhibit primarily male or female mating behaviors? Explain your reasoning.
3. Sex-reversed XY females (this condition is now known as 46,XY complete gonadal dysgenesis), likely have a mutation in what gene on the Y chromosome?

Disorders/Differences of Human Sexual Development

LEARNING OBJECTIVE Outline the various categories of human disorders and differences of sexual development (DSDs), and provide detailed examples of each.

KEY CONCEPTS:

- Disorders/differences of sex development (DSDs) *constitute diverse divergences of urogenital differentiation* arising from either chromosomal abnormalities or the disruption of the genetic networks that underlie sexual differentiation.
- Sex chromosomal DSDs are characterized by an abnormal number of X or Y chromosomes, often resulting from chromosomal nondisjunction during meiosis.
- 46,XY DSDs occur when a 46,XY individual experiences altered testicular development, androgen synthesis, or androgen action. This can produce a broad spectrum of phenotypes, including the development of ambiguous genitalia, demasculinization, and even complete sex reversal.
- 46,XX DSDs include conditions of androgen excess that cause 46,XX infants to virilize with varying degrees of masculinization.

As we have learned in this chapter, the proper determination, differentiation, and development of the gonads and internal/external genitalia relies on a tightly regulated network of transcription factors, signaling molecules, and exposure to adequate levels of specific hormones during critical periods of development. Given this complexity in sexual differentiation, it is not surprising that human neonates may be born with various genital anomalies that do not conform with their genetic sex, providing insight into the development of human sexual dimorphism. Such anomalies typically arise from either chromosomal abnormalities or the disruption of the genetic networks that underlie sexual differentiation and are referred to broadly in the medical literature as **disorders of sex development** (DSDs). While the term "disorder" is apt in the context of describing pathological aspects of a DSD, like infertility or greater susceptibility to developing illnesses like cancer, a growing number of people with DSDs and healthcare practitioners are beginning to reject the pathologization associated with the word "disorder". Instead, many now prefer to use a modification of DSD terminology by replacing "disorder" with "difference", which allows DSD to instead stand for **"differences of sex development"**.

DSDs include at least 50 different conditions of atypical urogenital differentiation (Table 20.1) and may affect between 1:100 to 1:10,000 individuals in a population, depending on the specific condition.[92,93] DSDs comprise a broad spectrum of phenotypes, ranging from relatively mild cases of **cryptorchidism** (undescended testes) and **hypospadia** (the urethral opening is located below its normal location at the head of the penis) in males to more severe cases of *gonadal dysgenesis* (the gonads are replaced with fibrous tissue), ambiguous genitalia, varying degrees of *ovotestis* (both ovary and testicular tissue are present), and even sex reversal in both males and females. Such conditions may be identified at different times of life, ranging from fetuses and newborns with clearly ambiguous internal and/or external genitalia, to later diagnoses in individuals with delayed or accelerated puberty, unanticipated virilization, **gynecomastia** (breast development in males), infertility, or gonadal tumors. In contrast with anomalous testes development, anomalies of ovarian development do not typically cause abnormalities of the internal and external genitalia since estrogens are not required for their development. However, estrogens are required for breast development and aspects of brain differentiation. Atypical

Table 20.1 Classifications of Some Disorders/Differences of Sex Development (DSD)

DSD Category	DSD Name	DSD Description
Sex Chromosomal DSDs		Abnormal number of sex chromosomes.
	47,XXY Klinefelter syndrome	Hypogonadism and infertility with a predominantly male phenotype with tall stature.
	45,X Turner syndrome	Female phenotype with ovarian dysgenesis, delayed or lack of puberty, and short stature.
	45,X/46,XY Mosaicism	Genital phenotype varies from normal female, through a continuum of ambiguous genitalia, to a normal penis.
	Chromosomal ovotesticular DSD (46,XX/46,XY" chimerism or 45,X/46,XY" mosaic	Both ovaries and testes, or an ovotestis, may be present. External genitalia are typically ambiguous, but may range from normal male to normal female phenotypes.
46,XY DSD		Disorders/differences of testicular development or disorders/differences in androgen synthesis/action
	Gonadal dysgenesis	A single gene mutation in one of several genes involved in testicular determination. Partial to complete gonadogenesis may produce a broad range of phenotypes ranging from ambiguous genitalia to hypospadias and infertility.
	17β-hydroxysteroid dehydrogenase	Reduced levels of testosterone accompanied by increased levels of androstenedione and DHEA produce a range of phenotypes from ambiguous to the female phenotype.
	5-alpha-reductase syndrome	Reduced levels of dihydrotestosterone produce female-like genitalia at birth, but individuals may partially virilize at puberty.
	Lipoid congenital adrenal hyperplasia	Disruption of adrenal and gonadal steroidogenesis, resulting in genotypic males that are phenotypic females.
	Androgen insensitivity syndrome	Partial to complete resistance of androgen to its receptor may produce a completely female phenotype in spite of normal levels of circulating testosterone.
	Persistent Müllerian duct syndrome	Retention of the Müllerian ducts and derivatives due to a mutation in the gene coding for either the AMH receptor or the AMH gene itself.
46,XX DSD		Disorders/differences of androgen excess. Phenotypes range from extreme virilization of the external genitalia to normal female genitalia accompanied by hirsutism (excess facial hair).
	Congenital adrenal hyperplasia, 21-hydroxylase deficiency	Adrenal gland disorder caused by impaired cortisol biosynthesis and increased androgen biosynthesis resulting in varying degrees of virilization.
	Fetal-placental aromatase deficiency	An inability to convert testosterone into estrogen results in excess levels of testosterone and virilization of the female fetus during pregnancy.

development of the chromosomal, gonadal, or phenotypic sex can be associated with increased risk of gonadal dysfunction, infertility, and incidence of gonadal cancer. Importantly, emerging evidence also suggests that an association may exist between *endocrine-disrupting chemicals* (EDCs, synthetic compounds, such as plastic derivatives, agricultural chemicals, and pharmacological/industrial waste that disrupt endogenous hormone signaling) and the development of certain DSDs such as cryptorchidism, hypospadias, and possibly others.[94]

Sex Chromosome DSDs

Sex chromosome aneuploidy is the condition of having less than (monosomy) or more than (polysomy) the normal number of X and Y chromosomes, often resulting from chromosomal nondisjunction during meiosis. With an estimated incidence in the general population of between 1:400 to 1:1,000, sex chromosome aneuploidy accounts for about half of all human chromosomal abnormalities.[95] Disorders of sex chromosomes can not only manifest as numbers of chromosomes, but also as structural deletions of regions within chromosomes. Compared with autosomal abnormalities, the effects of sex chromosome abnormalities are typically not as severe, are rarely lethal, and are generally compatible with normal life expectancy.[96]

47,XXY Klinefelter Syndrome

Klinefelter syndrome is the most common form of sex chromosome DSD, with a reported incidence of 1:500 to 1:1,000 live births.[97] It is also the most common form of hypogonadism and infertility in males. It is most typically associated with a 47,XXY karyotype (polysomy X) caused by nondisjunction during gametogenesis (Figure 20.25), although about 10% of individuals exhibit a mosaic (46,XY/47,XXY) form of this condition. The condition always produces testes and male external genitalia, highlighting the key role of the Y chromosome in testis determination and subsequent fetal androgen synthesis. In the most severe phenotypes, the testes are small due to seminiferous tubule dysgenesis, with low plasma testosterone levels at puberty[98] and infertility. Individuals with severe phenotypes also often possess elevated serum estradiol

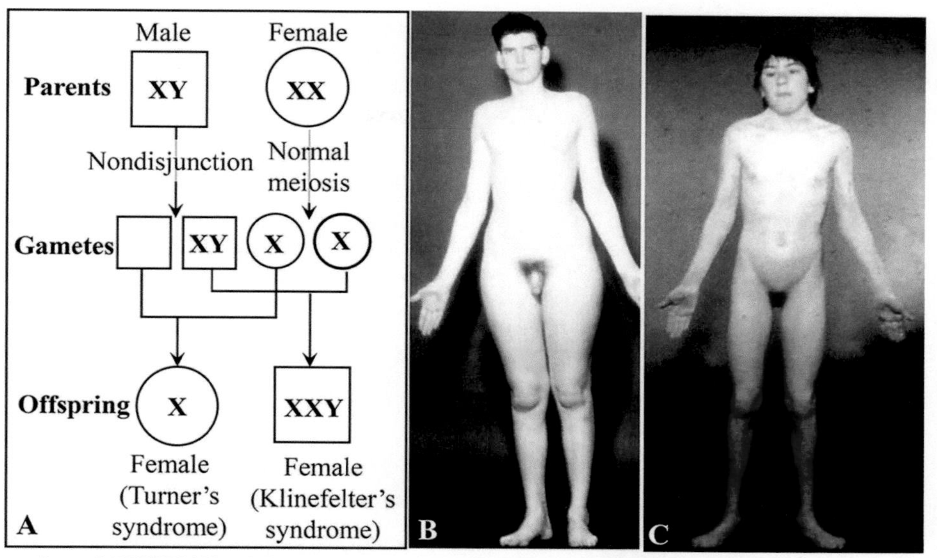

Figure 20.25 The genetics of 47,XXY Klinefelter syndrome and 45,X Turner syndrome. (A) Both syndromes occur due to a nondisjunction event during male gametogenesis. Fertilization of the normal female XX oocyte by the abnormal male gamete produces one of two possible abnormal karyotypes in the embryo. If an X oocyte is fertilized by an XY spermatozoon, a 47,XXY Klinefelter phenotype results **(B)**. If an X oocyte is fertilized by an spermatozoon lacking any sex chromosomes, a 45,XO Turner's phenotype results **(C)**.

Source of images: (B) Brookes, M. and Zietman, A., 1998. Clinical embryology: a color atlas and text. CRC Press. Used by permission.

levels that contribute to gynecomastia and the development of broad hips during adolescence (Figure 20.25) and may exhibit delayed onset of speech, learning difficulties, and abnormal motor skills. Klinefelter patients tend to have a tall stature due to the presence of three copies of *SHOX*♦ genes, one on each sex chromosome. SHOX codes for an osteogenic factor located on the X and Y chromosomes' pseudoautosomal regions, and humans expressing this gene at low levels possess a short stature (see below in Turner syndrome).[99,100]

45,X Turner Syndrome

Turner syndrome is the second most common form of sex chromosome DSD, with an occurrence of about 1:2,500.[97] Although the 45,X karyotype (monosomy X) is associated with about 50% of individuals with this condition, about 25% display a mosaic (45,X/46,XX) karyotype, with the remaining patients exhibiting structural abnormalities of the X chromosome (Figure 20.25). The absence of a Y chromosome in the karyotype promotes development of a female phenotype, but typically one with ovarian dysgenesis (accelerated germ cell apoptosis and subsequent oocyte atresia occur after the third month of gestation), underscoring the necessity for two X chromosome copies for normal ovarian development. Patients with severe ovarian dysgenesis are typically diagnosed with hypogonadism, manifesting in either a lack of sexual development after puberty or a delayed puberty. In individuals with sex chromosome mosaicism, the degree of ovarian dysgenesis may range from a typical 45,X phenotype to a normal female phenotype with sufficient estradiol synthesis for puberty and menstruation to begin in adolescence.[101] Other features commonly associated with the 45,X karyotype include multiple somatic anomalies, such as short stature, cardiac defects, a broad or web-like neck, a broad chest and short fingers and toes. The short stature results from the haploinsufficiency of SHOX genes due to a missing X chromosome. Specifically, lower levels of the SHOX gene are expressed due to the absence of one X chromosome. Since the SHOX gene is located on the X chromosome's pseudoautosomal region, it normally escapes X chromosome inactivation. Severe forms of Turner syndrome may be accompanied by cognitive deficiencies.

45,X/46,XY Mosaicism

Sex chromosome mosaicism occurs when different cells in a body possess a variable array of sex chromosomes due to a nondisjunction event in the zygote, and thus a range of phenotypes can be produced depending on where expression occurs (Figure 20.26). One of the most common sex chromosome mosaics associated with human ambiguous genitalia is 45,X/46,XY mosaicism.[102] The genital phenotype associated with a *45,X/46,XY mosaic* karyotype varies greatly, ranging from normal female external genitalia on one end of the spectrum, through a continuum of ambiguous genitalia, to a normal penis on the other end of the spectrum.[103,104] The gonadal phenotypes range from undescended testes with testicular dysgenesis to normal testes located in the inguinoscrotal region, and, rarely, the presence of scattered ovarian primary follicles embedded

♦ short stature homeobox

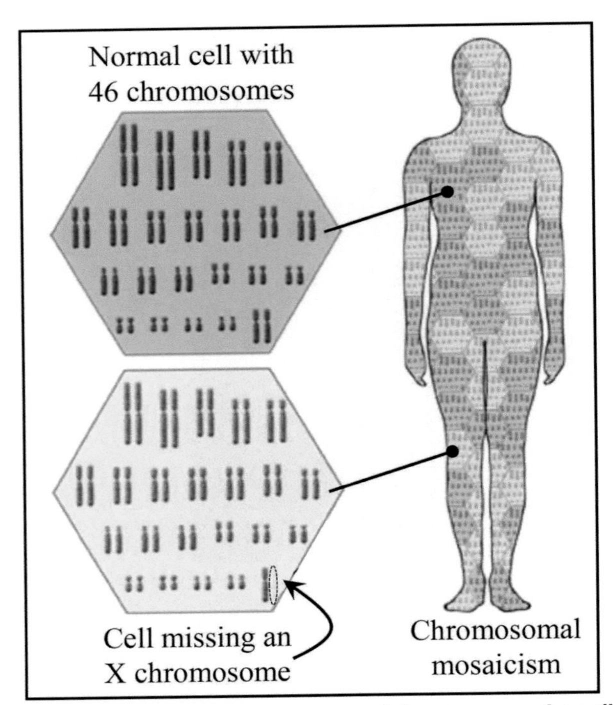

Figure 20.26 Sex chromosome mosaicism occurs when different cells in a body possess a variable array of sex chromosomes due to a nondisjunction event in the zygote, and thus a range of phenotypes can be produced.

Source of images: Avigdor, B.E., et al. 2021. Mosaicism in rare disease. In Genomics of Rare Diseases (pp. 151-184). Academic Press. Used by permission.

in the testes (mosaic-type ovotesticular DSD). In the most severe cases of gonadal dysgenesis, Müllerian structures may form due to reduced AMH synthesis by the Sertoli cells, which are malformed or absent. In some instances where there are varying degrees of dysgenesis between the left and right gonads, oviducts and a partial uterus may be present on the side that is producing less AMH.[105]

46,XY DSDs

A *46,XY DSD* manifests when a 46,XY individual experiences a disorder of testicular development, androgen synthesis, or androgen action. This can produce a broad spectrum of phenotypes ranging from ambiguous genitalia and demasculinization of secondary androgen target tissues to a completely female form virtually indistinguishable from a typical 46,XX female.

Gonadal Dysgenesis

Gonadal dysgenesis within the 46,XY karyotype may be caused by a single gene mutation in one of the genes involved in testicular determination, such as the aforementioned SRY, SOX9, WNT4, and DMRT1, as well as other genes. In the most severe cases of complete testicular dysgenesis, Müllerian-derived structures (such as oviducts and uterus) may persist due to insufficient AMH production. Partial gonadal dysgenesis, and resulting partial androgenization, may be associated with a continuum of genital phenotypes, ranging from ambiguous genitalia to hypospadias and infertility.

Disorders of Androgen Synthesis

Although the testes may be intact in a 46,XY individual, testosterone synthesis may still be impaired. Enzymatic defects anywhere along the pathway of androgen synthesis (such as mutation of a gene coding for a steroidogenic enzyme) can result in impaired androgenization in 46,XY individuals. Three examples of such mutations are described next.

17β-Hydroxysteroid Dehydrogenase and 5-Alpha-Reductase Deficiencies

17β-hydroxysteroid dehydrogenase (17β-HSD) is an enzyme located in both the adrenal cortex and testis that catalyzes the conversion of the androgen precursors androstenedione and dehydroepiandrosterone (DHEA) into testosterone. As such, a deficiency in 17β-HSD is characterized biochemically by reduced levels of testosterone accompanied by increased levels of androstenedione and DHEA. 46,XY infants born with such a testosterone deficiency typically display normal Wolffian duct-derived internal structures (e.g. seminal vesicles, and ejaculatory ducts) but have undescended testes and undervirilized external genitalia ranging from ambiguous to the female phenotype.[106]

Recall from earlier that the enzyme *5-alpha-reductase* locally converts testosterone into the more potent androgen, DHT. DHT, in turn, promotes the virilization of specific features in the fetus (e.g. growth of the penis and prostate gland, and fusion of the urogenital folds into a scrotum) and the adult (e.g. pubic and facial hair growth, and development of sebaceous glands of the skin). A deficiency in 5-alpha-reductase in 46,XY individuals is characterized by an elevated testosterone-to-DHT ratio, and such individuals are born with undescended testes, a blind vaginal pouch, and a clitoris-like, hypospadic penis resembling the female external genitalia (Figure 20.27). As such, children with this condition are often initially raised as females. However, the testes and male internal ducts develop normally, with no Müllerian structures present. Affected XY males may partially virilize at puberty due to increased testosterone production, displaying a deepening of the voice, increased muscle mass, increased penis length, enlargement of the testes, and descent of the testes into the labioscrotal folds. In a genetically isolated region of the Dominican Republic with a relatively high incidence of 5-alpha-reductase deficiency, this phenomenon is called "guevedoces" (which literally means "egg at twelve", and is slang for testes or penis), reflecting the sudden change in the individual's appearance at puberty. Although initially raised as girls, such individuals often decide to adopt a male gender identity after puberty.[107] In 2003, author Jeffrey Eugenides wrote

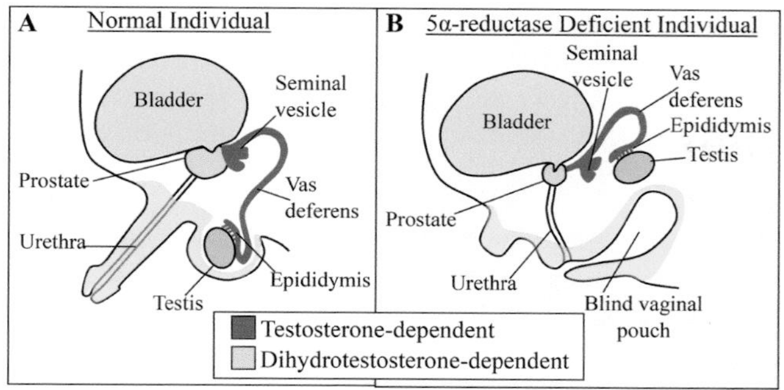

Figure 20.27 A deficiency in 5-alpha-reductase in 46,XY individuals is characterized by a deficiency in dihydrotestosterone (DHT) synthesis and the development of DHT-mediated genitalia. Compared with unaffected individuals **(A)**, affected individuals are born with undescended testes, a blind vaginal pouch, and a clitoris-like, hypospadic penis resembling the female external genitalia **(B)**.

Source of images: Cheon, C.K., 2011. Eur. J. Pediatr., 170(1), pp. 1-8.

a Pulitzer Prize-winning novel called *Middlesex* about a Greek immigrant protagonist living in Detroit who has 5-alpha-reductase deficiency[108] (refer to the chapter's opening quotation).

Disorders of Androgen and AMH Action

Considering that over 80% of individuals with 46,XY DSDs exhibit normal testosterone synthesis, the majority of male phenotypic disorders must result from varying degrees of impairment of androgen and AMH action on its target tissues.[109] In these cases insensitivity to androgens or AMH by target tissues often manifests as loss-of-function mutations in the corresponding receptors.

Complete and Partial Androgen Insensitivity Syndrome

Perhaps no other disorder illustrates the pivotal role that androgens play in the development of male physical and behavioral phenotypes than **complete androgen insensitivity syndrome** (CAIS), a condition in 46,XY individuals with normally developed testes and testosterone and AMH secretion who nonetheless display a completely female external phenotype (Figure 20.28). CAIS is an X-linked disease characterized by loss-of-function mutations in the androgen receptor gene, resulting in peripheral androgen resistance.[110] Because the fetal genitalia of the genetically male individual are insensitive to the actions of androgens, this results in the development of female external genitalia, with a clitoris, labia, and vagina. Since AMH secreted from the testes still exerts its action on the Müllerian ducts, these regress and the affected individuals lack a uterus and Fallopian tubes. Androgen resistance by the fetus typically results in the degeneration of Wolffian structures. Though hormonally functional, the testes are usually incompletely descended, remaining

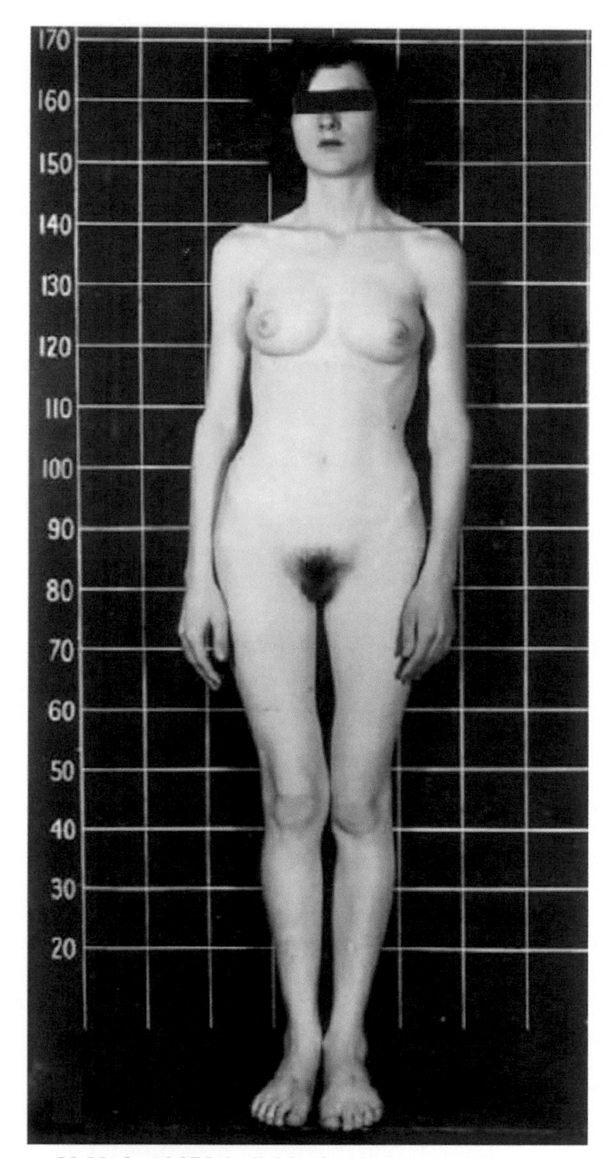

Figure 20.28 A 46,XY individual with complete androgen insensitivity syndrome (CAIS). CAIS is characterized by loss-of-function mutations in the androgen receptor gene, resulting in peripheral androgen resistance and development of a completely female external phenotype, despite developed testes and secretion of testosterone and MIH.

Source of images: Moore, K.L., et al. 2018. The developing human-e-book: clinically oriented embryology. Elsevier Health Sciences. Used by permission.

in the labial folds, inguinal canal, or abdominal cavity. Furthermore, at puberty CAIS individuals have normal development of the breast and broad hips due to the normal peripheral aromatization of testosterone into estradiol, but pubic hair is typically sparse. CAIS individuals are typically born and raised as girls, and the disorder is often identified at puberty when they fail to menstruate but have still undergone the normal growth spurt and breast development.

Depending on the specific mutation associated with the androgen receptor gene, individuals with androgen insensitivity syndrome may exhibit phenotypes ranging from the aforementioned extreme of complete AIS to *partial androgen insensitivity syndrome* (PAIS) caused by varying degrees of residual activity by the mutant androgen receptors.[110,111] For example, the retention of the Wolffian duct derivatives, such as well-developed vas deferens and epididymis, has been described in many patients with AIS who display an otherwise external female phenotype.[111] The less severe end of the PAIS spectrum may range from ambiguous genitalia at birth to hypospadias and a male phenotype with undervirilization or infertility.

Persistent Müllerian Duct Syndrome

Persistent Müllerian duct syndrome (PMDS) in 46,XY individuals is caused by a mutation in the gene coding for either the AMH receptor or the AMH gene itself.[112] In male fetuses born with PMDS, the Müllerian ducts fail to regress, and instead form rudimentary female internal genitalia as well as Wolffian duct derivatives (epididymis, vas deferens, and seminal vesicles). Newborns present with normal male external genitalia but with cryptorchidism.

46,XX DSDs

46,XX DSDs can be divided into two major categories: disorders of ovarian development (ovarian dysgenesis) and disorders of androgen excess. Since estrogen synthesis does not appear to play a role in the development of female internal and external genitalia, infants with 46,XX ovarian dysgenesis are born with normal internal and external genitalia. Such disorders do not manifest until puberty, when the failure of estrogen synthesis inhibits breast development and menstruation. This section will illustrate examples of disorders of androgen excess, which are typified by varying degrees of masculinization of the 46,XX fetus.

Congenital Adrenal Hyperplasia (21-Hydroxylase Deficiency)

Recall that whereas the fetal testis is the primary source of androgens in the male, in females it is the fetal adrenal gland that produces the androgen precursors necessary for driving adrenarche, which promotes the appearance of axillary and pubic hair independent of the gonads and puberty. **Congenital adrenal hyperplasias** (CAHs) are a category of adrenal gland disorders characterized by impaired cortisol biosynthesis, subsequently increased ACTH secretion due to feedback inhibition by cortisol, and ultimately hyperplasia of the adrenals and disordered steroidogenesis. The most common cause of CAH, and an important contributor to 46,XX DSDs, is *21-hydroxylase deficiency* (21OHD). Described in detail in Chapter 13: The Multifaceted Adrenal Gland (specifically, refer to Figures 13.2 and 13.3 of that chapter), mutations of 21OHD inhibit the biosynthesis of both glucocorticoids and mineralocorticoids, instead diverting substrates towards the synthesis of abnormally high levels of adrenal androgens. In 46,XX individuals, this form of CAH results in varying degrees of virilization, ranging from the development of ambiguous genitalia in the fetus to hirsutism (development of excess facial hair) at puberty. In particular, excess DHT synthesis by the adrenal gland during the critical window between 8 and 12 weeks of gestational age drives labioscrotal fusion and phallic growth. The severity of 46,XX DSD due to 21OHD depends on both the amount of DHT in the circulation and the gestational age when the exposure occurs.[113] As in other forms of untreated adrenal insufficiency, hyponatremia, hypoglycemia, and hypotensive crises may occur.

Fetal-Placental CYP19 Aromatase Deficiency

Recall that the enzyme CYP19 aromatase is the only cytochrome enzyme known to catalyze the conversion of androgens to estrogens. This aromatase is expressed in many tissues and plays many critical functions, including protecting the female fetus from excessive androgen exposure *in utero*, facilitating the synthesis of estrogens from the ovary during puberty and in the adult and promoting development of the brain, breast, bone, vascular endothelium, and adipose tissue. In the presence of aromatase deficiency in a 46,XX fetus, estradiol cannot be synthesized by the placenta, and large quantities of placental androgens are transferred to the fetal circulation, resulting in virilization of the female fetus during pregnancy.[114] As such, 46,XX female infants with aromatase deficiency are born with varying degrees of ambiguous or masculinized external genitalia but normal Müllerian structures. At puberty, in the absence of ovarian aromatase, the ovaries of affected individuals develop multiple, enlarged follicular cysts, and patients exhibit hypergonadotropic hypogonadism, usually fail to develop female secondary sexual characteristics, and show progressive virilization with age.

Intersex Identity

Intersex is a general term for people born with differences in sex traits or reproductive anatomy.[115] This may include variations in chromosomes, gonads, genitals, or sex hormones described earlier. Although some people with intersex traits self-identify as intersex, others do not.[116,117] The number of births where the baby is categorized as intersex is not well reported around the world, though Intersex Human Rights Australia (https://ihra.org.au/) recommends an upper-bound estimate of 1.7% of all live births.[118] Importantly, intersex traits aren't always evident at birth and may become apparent during childhood or puberty. Like the rest of the population, some intersex persons may be assigned a sex at birth and be raised as a girl or boy, but then identify with another gender later in life.

SUMMARY AND SYNTHESIS QUESTIONS

1. One of the DSDs listed in Table 20.1 is "Fetal-placental aromatase deficiency". How would this cause a male phenotype in XX females?
2. 47,XXX females and 47,XYY males (not discussed in this chapter) each display normal sexual development but usually have tall stature. What explains their tall stature?
3. Postpubertal males with 5-alpha-reductase deficiency undergo many aspects of normal male puberty, such as increased testosterone production, the deepening of the voice, and development of increased muscle mass. However, they typically do not experience the development of acne or temporal hair recession. Why not?
4. In 46,XY lipoid congenital adrenal hyperplasia (CAH), infants are born with female external genitalia, although Müllerian structures have regressed normally. What explains these observations?
5. How does CAH influence sexual development in 46,XY and 46,XX individuals?
6. A 46,XX female is born with a mutated AMH gene. What phenotype, if any, would result?
7. 49,XXXXY syndrome is an extremely rare aneuploidic sex chromosomal abnormality not discussed in this chapter. What phenotypic sex do you predict these individuals have, and why?

Summary of Chapter Learning Objectives and Key Concepts

LEARNING OBJECTIVE Provide an overview of different strategies of sex determination among animals.

- In organisms with genotypic sex determination (GSD) the sex of an individual is determined entirely by its genotype, usually by genes located on sex chromosomes.
- In the X/Y chromosomal sex determination system of mammals, males are the heterogametic sex (XY) and females the homogametic sex (XX).
- In the Z/W chromosomal sex determination system of birds, males are the homogametic sex (ZZ) and females are the heterogametic sex (ZW).
- In organisms with environment-dependent sex determination, the sex of an individual is determined during egg incubation by external variables, such as temperature.

LEARNING OBJECTIVE Compare and contrast the molecular mechanisms of genotypic and environmental sex determination in vertebrates.

- The embryonic gonadal primordium is sexually indifferent and has the potential to differentiate into either a testis or an ovary.
- In mammals, the SRY gene on the Y chromosome is the testis-determining factor that is responsible for masculinizing the embryo.
- Ovary-determining factors on the mammalian X chromosome suppress testis development.
- In birds, a double dose of DMRT1 gene expression from the ZZ karyotype promotes male development, whereas a half dose of the gene's expression from the ZW karyotype promotes female development.
- Following the divergence of birds and mammals from a common ancestor, one mammalian X chromosome experienced dramatic gene loss and shortening, retaining only a fraction of the genes from the ancestral X, and is now the Y chromosome.
- In order to compensate for imbalances in gene dosage between the sexes, mammals evolved the process of X chromosome inactivation, the epigenetic silencing of one of a female's two X chromosomes in each somatic cell.
- In vertebrates that use temperature-dependent sex determination, the development of testes or ovaries is determined by the androgen-to-estrogen ratio.
- The enzyme CYP19 aromatase converts androgens into estrogens, and its activity helps establish sex in vertebrates that use temperature-dependent sex determination.

LEARNING OBJECTIVE Explain the developmental fates of the mammalian primordial internal and external genitalia in response to endocrine signaling.

- In both sexes, the sexually indifferent stage of embryogenesis is characterized by the presence of two different internal duct systems, each possessing a unipotential developmental fate.
- Following stimulation by androgens, the Wolffian ducts develop into components of the male reproductive tract, such as the vas deferens and the seminal vesicles, and in the absence of androgen stimulation the Wolffian ducts regress.
- In the absence of stimulation by anti-Müllerian hormone (AMH) from the testes, the Müllerian ducts develop into components of the female reproductive tract, such as the Fallopian tubes, uterus, and upper third of the vagina.
- When inhibited by anti-Müllerian hormone produced by the testes, the Müllerian ducts regress.
- The male and female external genitalia each arise from common, morphologically identical bipotential primordia.
- Whereas some components of the male external genitalia require exposure to testosterone for normal development, others require stimulation by the more potent androgen dihydrotestosterone.
- In the absence stimulation by androgens, the primordia of the external genitalia develop into their female forms.

LEARNING OBJECTIVE Provide an overview of the different factors that contribute to the sexual differentiation of the mammalian brain and behavior.

- Some regions of the brain show distinct sexually dimorphic circuitry.
- Whereas testosterone has "organizational" effects during the prenatal period, it exerts "activational" effects during adulthood.

- In male mammalian fetuses, the masculinizing effects of testosterone on the brain are mediated by the estrogen receptor, which binds to estradiol following its local conversion from testosterone by the enzyme CYP19 aromatase.
- Female mammalian fetuses are protected from the masculinizing effects of high estradiol levels by the binding and sequestration of blood estradiol to alpha-fetoprotein.

LEARNING OBJECTIVE Outline the various categories of human disorders and differences of sexual development (DSDs), and provide detailed examples of each.

- Disorders/differences of sex development (DSDs) *constitute diverse divergences of urogenital differentiation* arising from either chromosomal abnormalities or the disruption of the genetic networks that underlie sexual differentiation.
- Sex chromosomal DSDs are characterized by an abnormal number of X or Y chromosomes, often resulting from chromosomal nondisjunction during meiosis.
- 46,XY DSDs occur when a 46,XY individual experiences altered testicular development, androgen synthesis, or androgen action. This can produce a broad spectrum of phenotypes, including the development of ambiguous genitalia, demasculinization, and even complete sex reversal.
- 46,XX DSDs include conditions of androgen excess that cause 46,XX infants to virilize with varying degrees of masculinization.

LITERATURE CITED

1. Barnes J. *Complete Works of Aristotle, Volume 1: The Revised Oxford Translation*, Vol 192. Princeton University Press; 1984.
2. Sarre SD, Georges A, Quinn A. The ends of a continuum: genetic and temperature-dependent sex determination in reptiles. *Bioessays*. 2004;26(6):639–645.
3. Platt A. *De Generatione Animalium*. Clarendon Press; 1910.
4. Wilson E. *The Cell in Development and Inheritance*. The Macmillan Company; 1896.
5. Schön I, Martens K, van Dijk P. *Lost Sex. The Evolutionary Biology of Parthenogenesis*. Springer; 2009.
6. Levan A. Chromosomes in cancer tissue. *Ann N Y Acad Sci*. 1956;63(5):774–792.
7. Ford C, Hamerton J. The chromosomes of man. *Human Heredity*. 1956;6(2):264–266.
8. Jacobs PA. A case of human intersexuality having a possible XXY sexdetermining mechanism. *Nature*. 1959;183:302–303.
9. Ford C, Jones K, Polani P, De Almeida J, Briggs J. A sex-chromosome anomaly in a case of gonadal dysgenesis (Turner's syndrome). *Lancet*. 1959;1:711–713.
10. Fraccaro M, Kaijser K, Lindsten J. Chromosome complement in gonadal dysgenesis (Turner's syndrome). *Lancet*. 1959;273(7078):886.
11. Pieau C, Dorizzi M, Richard-Mercier N. Temperature-dependent sex determination and gonadal differentiation in reptiles. *Cell Mol Life Sci*. 1999;55 (6–7):887–900.
12. Bull J. Sex determination in reptiles. *Q Rev Biol*. 1980;55(1):3–21.
13. Radder RS, Quinn AE, Georges A, Sarre SD, Shine R. Genetic evidence for co-occurrence of chromosomal and thermal sex-determining systems in a lizard. *Biol Lett*. 2008;4(2):176–178.
14. Yamamoto Y, Zhang Y, Sarida M, Hattori RS, Strüssmann CA. Coexistence of genotypic and temperature-dependent sex determination in pejerrey Odontesthes bonariensis. *PLoS One*. 2014;9(7):e102574.
15. Chen S, Zhang G, Shao C, et al. Whole-genome sequence of a flatfish provides insights into ZW sex chromosome evolution and adaptation to a benthic lifestyle. *Nat Genet*. 2014;46(3):253.
16. Holleley CE, O'Meally D, Sarre SD, et al. Sex reversal triggers the rapid transition from genetic to temperature-dependent sex. *Nature*. 2015;523(7558):79.
17. Crews D, Bergeron JM, Bull JJ, et al. Temperature-dependent sex determination in reptiles: Proximate mechanisms, ultimate outcomes, and practical applications. *Dev Gen*. 1994;15(3):297–312.
18. Hoffmann AA, Sgrò CM. Climate change and evolutionary adaptation. *Nature*. 2011;470(7335):479.
19. Avise J, Mank J. Evolutionary perspectives on hermaphroditism in fishes. *Sex Dev*. 2009;3(2–3):152–163.
20. Kraak SB, Pen I. Sex-determining mechanisms in vertebrates. In: *Sex Ratios: Concepts and Research Methods*. Cambridge University Press; 2002:158–177.
21. Liu H, Todd EV, Lokman PM, Lamm MS, Godwin JR, Gemmell NJ. Sexual plasticity: A fishy tale. *Mol Reprod Dev*. 2017;84(2):171–194.
22. Warner RR, Swearer SE. Social control of sex change in the bluehead wrasse, thalassoma bifasciatum (Pisces: Labridae). *Biol Bull*. 1991;181(2):199–204.
23. Nakamura M, Hourigan TF, Yamauchi K, Nagahama Y, Grau EG. Histological and ultrastructural evidence for the role of gonadal steroid hormones in sex change in the protogynous wrasse Thalassoma duperrey. *Environ Biol Fishes*. 1989;24(2):117–136.
24. Kelley JL, Yee M-C, Brown AP, et al. The genome of the self-fertilizing mangrove rivulus fish, Kryptolebias marmoratus: a model for studying phenotypic plasticity and adaptations to extreme environments. *Gen Biol Evol*. 2016;8(7):2145–2154.
25. Hughes JF, Page DC. The biology and evolution of mammalian Y chromosomes. *Ann Rev Gen*. 2015;49:507–527.
26. Warren WC, Hillier LW, Graves JAM, et al. Genome analysis of the platypus reveals unique signatures of evolution. *Nature*. 2008;453(7192):175.
27. Skaletsky H, Kuroda-Kawaguchi T, Minx PJ, et al. The male-specific region of the human Y chromosome is a mosaic of discrete sequence classes. *Nature*. 2003;423(6942):825.
28. Marshall Graves JA. Human Y chromosome, sex determination, and spermatogenesis—a feminist view. *Biol Reprod*. 2000;63(3):667–676.
29. Dewing P, Chiang CW, Sinchak K, et al. Direct regulation of adult brain function by the male-specific factor SRY. *Current Biology*. 2006;16(4):415–420.
30. Case LK, Wall EH, Dragon JA, et al. The Y chromosome as a regulatory element shaping immune cell transcriptomes and susceptibility to autoimmune disease. *Genome Res*. 2013.
31. Berletch JB, Yang F, Disteche CM. Escape from X inactivation in mice and humans. *Genome Biol*. 2010;11(6):213.
32. Jost A, Vigier B, Prépin J, Perchellet JP. Studies on sex differentiation in mammals. Paper presented at: Proceedings of the 1972 Laurentian Hormone Conference; 1973.
33. Sinclair AH, Berta P, Palmer MS, et al. A gene from the human sex-determining region encodes a protein with homology to a conserved DNA-binding motif. *Nature*. 1990;346(6281):240.
34. Sekido R, Lovell-Badge R. Sex determination involves synergistic action of SRY and SF1 on a specific Sox9 enhancer. *Nature*. 2008;453(7197):930.

35. Kashimada K, Koopman P. Sry: the master switch in mammalian sex determination. *Development*. 2010;137(23):3921–3930.
36. Mittwoch U. Sex determination: science & society series on sex and science. *EMBO Rep*. 2013;14(7):588–592.
37. Koopman P, Gubbay J, Vivian N, Goodfellow P, Lovell-Badge R. Male development of chromosomally female mice transgenic for Sry. *Nature*. 1991;351(6322):117.
38. Ludbrook LM, Harley VR. Sex determination: a "window" of DAX1 activity. *Trends Endocrinol Metab*. 2004;15(3):116–121.
39. Sekido R, Lovell-Badge R. Sex determination and SRY: down to a wink and a nudge? *Trends Genet*. 2009;25(1):19–29.
40. Jordan BK, Mohammed M, Ching ST, et al. Up-regulation of WNT-4 signaling and dosage-sensitive sex reversal in humans. *Am J Human Genet*. 2001;68(5):1102–1109.
41. Tomaselli S, Megiorni F, Lin L, et al. Human RSPO1/R-spondin1 is expressed during early ovary development and augments β-catenin signaling. *PLoS One*. 2011;6(1):e16366.
42. Smith CA, Roeszler KN, Ohnesorg T, et al. The avian Z-linked gene DMRT1 is required for male sex determination in the chicken. *Nature*. 2009;461(7261):267.
43. Bagheri-Fam S, Sinclair AH, Koopman P, Harley VR. Conserved regulatory modules in the Sox9 testis-specific enhancer predict roles for SOX, TCF/LEF, Forkhead, DMRT, and GATA proteins in vertebrate sex determination. *Int J Biochem Cell Biol*. 2010;42(3):472–477.
44. Ferguson-Smith M. The evolution of sex chromosomes and sex determination in vertebrates and the key role of DMRT1. *Sex Dev*. 2007;1(1):2–11.
45. Raymond CS, Murphy MW, O'Sullivan MG, Bardwell VJ, Zarkower D. Dmrt1, a gene related to worm and fly sexual regulators, is required for mammalian testis differentiation. *Genes Dev*. 2000;14(20):2587–2595.
46. Schroeder AL, Metzger KJ, Miller A, Rhen T. A novel candidate gene for temperature-dependent sex determination in the common snapping turtle. *Genetics*. 2016;203(1):557–571.
47. Jost A. Hormonal factors in the sex differentiation of the mammalian foetus. *Phil Trans R Soc Lond B*. 1970;259(828):119–131.
48. Greco TL, Payne AH. Ontogeny of expression of the genes for steroidogenic enzymes P450 side-chain cleavage, 3 beta-hydroxysteroid dehydrogenase, P450 17 alpha-hydroxylase/C17–20 lyase, and P450 aromatase in fetal mouse gonads. *Endocrinology*. 1994;135(1):262–268.
49. Fisher CR, Graves KH, Parlow AF, Simpson ER. Characterization of mice deficient in aromatase (ArKO) because of targeted disruption of the cyp19 gene. *Proc Nat Acad Sci*. 1998;95(12):6965–6970.
50. Pieau C, Dorizzi M. Oestrogens and temperature-dependent sex determination in reptiles: all is in the gonads. *J Endocrinol*. 2004;181(3):367–377.
51. Matsumoto Y, Buemio A, Chu R, Vafaee M, Crews D. Epigenetic control of gonadal aromatase (cyp19a1) in temperature-dependent sex determination of red-eared slider turtles. *PLoS One*. 2013;8(6):e63599.
52. Sakata N, Tamori Y, Wakahara M. P450 aromatase expression in the temperature-sensitive sexual differentiation of salamander (Hynobius retardatus) gonads. *Int J Dev Biol*. 2004;49(4):417–425.
53. Ramsey M, Shoemaker C, Crews D. Gonadal expression of Sf1 and aromatase during sex determination in the red-eared slider turtle (Trachemys scripta), a reptile with temperature-dependent sex determination. *Differentiation*. 2007;75(10):978–991.
54. D'Cotta H, Fostier A, Guiguen Y, Govoroun M, Baroiller JF. Aromatase plays a key role during normal and temperature-induced sex differentiation of tilapia Oreochromis niloticus. *Mol Reprod Dev*. 2001;59(3):265–276.
55. Hays GC, Mazaris AD, Schofield G, Laloë J-O. Population viability at extreme sex-ratio skews produced by temperature-dependent sex determination. *Proc R Soc B*. 2017;284(1848):20162576.
56. Pachauri RK, Allen MR, Barros VR, et al. *Climate Change 2014: Synthesis Report. Contribution of Working Groups I, II and III to the Fifth Assessment Report of the Intergovernmental Panel on Climate Change*. IPCC; 2014.
57. Jensen MP, Allen CD, Eguchi T, et al. Environmental warming and feminization of one of the largest sea turtle populations in the world. *Curr Biol*. 2018;28(1):154–159. e154.
58. Nelson RJ, Kriegsfeld LJ. *An Introduction to Behavioral Endocrinology*. Sinauer Associates.
59. Jost A. * Recherches sur la differenciation sexuelle de lembryon de lapin. 2. action des androgenes de synthese sur lhistogenese genitale. *Arch Anat Microsc Morphol Exp*. 1947;36(3):244–270.
60. Jost A. Problems of fetal endocrinology: the gonadal and hypophyseal hormones. *Recent Progr Hormon Res*. 1953;8:379–413.
61. Nelson RJ. *An Introduction to Behavioral Endocrinology*. Sinauer Associates; 2011.
62. Harrison SM, Bush NC, Wang Y, et al. Insulin-like peptide 3 (INSL3) serum concentration during human male fetal life. *Front Endocrinol (Lausanne)*. 2019;10:596.
63. Ivell R, Heng K, Anand-Ivell R. Insulin-like factor 3 and the HPG axis in the male. *Front Endocrinol (Lausanne)*. 2014;5:6.
64. Holekamp KE, Smith JE, Strelioff CC, Van Horn RC, Watts HE. Society, demography and genetic structure in the spotted hyena. *Mol Ecol*. 2012;21(3):613–632.
65. French JA, Mustoe AC, Cavanaugh J, Birnie AK. The influence of androgenic steroid hormones on female aggression in "atypical" mammals. *Philos Trans Royal Soc Lond B Biol Sci*. 2013;368(1631):20130084.
66. Dloniak SM, French JA, Place NJ, Weldele ML, Glickman SE, Holekamp KE. Non-invasive monitoring of fecal androgens in spotted hyenas (Crocuta crocuta). *Gen Comp Endocr*. 2004;135(1):51–61.
67. Dloniak S, French J, Holekamp K. Rank-related maternal effects of androgens on behaviour in wild spotted hyaenas. *Nature*. 2006;440(7088):1190.
68. Cunha GR, Place NJ, Baskin L, et al. The ontogeny of the urogenital system of the spotted hyena (Crocuta crocuta Erxleben). *Biol Reprod*. 2005;73(3):554–564.
69. Licht P, Hayes T, Tsai P, et al. Androgens and masculinization of genitalia in the spotted hyaena (Crocuta crocuta). 1. Urogenital morphology and placental androgen production during fetal life. *J Reprod Fert*. 1998;113(1):105–116.
70. Cunha GR, Risbridger G, Wang H, et al. Development of the external genitalia: perspectives from the spotted hyena (Crocuta crocuta). *Differentiation*. 2014;87(1–2):4–22.
71. Manoli DS, Fan P, Fraser EJ, Shah NM. Neural control of sexually dimorphic behaviors. *Curr Opin Neurobiol*. 2013;23(3):330–338.
72. Bayless DW, Shah NM. Genetic dissection of neural circuits underlying sexually dimorphic social behaviours. *Phil Trans R Soc B*. 2016;371(1688):20150109.
73. Phoenix CH, Goy RW, Gerall AA, Young WC. Organizing action of prenatally administered testosterone propionate on the tissues mediating mating behavior in the female guinea pig. *Endocrinology*. 1959;65(3):369–382.
74. BOOTH JE. Sexual behaviour of neonatally castrated rats injected during infancy with oestrogen and dihydrotestosterone. *J Endocrinol*. 1977;72(2):135–141.
75. Feder HH, Whalen RE. Feminine behavior in neonatally castrated and estrogen-treated male rats. *Science*. 1965;147(3655):306–307.
76. McEwen BS, Lieberburg I, Chaptal C, Krey LC. Aromatization: important for sexual differentiation of the neonatal rat brain. *Horm Behav*. 1977;9(3):249–263.
77. McEwen BS, Lieberburg I, Maclusky N, Plapinger L. Do estrogen receptors play a role in the sexual differentiation of the rat brain? *J Steroid Biochem*. 1977;8(5):593–598.
78. Naftolin F, MacLusky N. Aromatization hypothesis revisited. In: *Sexual Differen-*

tiation: Basic and Clinical Aspects. Raven Press; 1984:79–91.

79. Reddy V, Naftolin F, Ryan K. Conversion of androstenedione to estrone by neural tissues from fetal and neonatal rats. *Endocrinology*. 1974;94(1):117–121.
80. Lewis C, McEwen BS, Frankfurt M. Estrogen-induction of dendritic spines in ventromedial hypothalamus and hippocampus: effects of neonatal aromatase blockade and adult GDX. *Dev Brain Res*. 1995;87(1):91–95.
81. Bakker J, Brand T, van Ophemert J, Slob AK. Hormonal regulation of adult partner perference behavior in neonatally ATD-treated male rats. *Behav Neurosci*. 1993;107(3):480.
82. Bakker J, Honda S, Harada N, Balthazart J. Restoration of male sexual behavior by adult exogenous estrogens in male aromatase knockout mice. *Horm Behav*. 2004;46(1):1–10.
83. Wallen K. Hormonal influences on sexually differentiated behavior in nonhuman primates. *Front Neuroendocrinol*. 2005;26(1):7–26.
84. Arnold AP, McCarthy MM. Sexual differentiation of the brain and behavior: a primer. In: Arnold AP, McCarthy MM, Pfaff DW, Volkow ND. (eds.), *Neuroscience in the 21st Century*. Springer; 2016:1–30.
85. Andrews G, Dziadek M, Tamaoki T. Expression and methylation of the mouse alpha-fetoprotein gene in embryonic, adult, and neoplastic tissues. *J Biol Chem*. 1982;257(9):5148–5153.
86. Amateau SK, Alt JJ, Stamps CL, McCarthy MM. Brain estradiol content in newborn rats: sex differences, regional heterogeneity, and possible de novo synthesis by the female telencephalon. *Endocrinology*. 2004;145(6):2906–2917.
87. Rhoda J, Corbier P, Roffi J. Gonadal steroid concentrations in serum and hypothalamus of the rat at birth: aromatization of testosterone to 17 β-estradiol. *Endocrinology*. 1984;114(5):1754–1760.
88. De Mees C, Laes J-F, Bakker J, et al. Alpha-fetoprotein controls female fertility and prenatal development of the gonadotropin-releasing hormone pathway through an antiestrogenic action. *Mol Cell Biol*. 2006;26(5):2012–2018.
89. Bakker J. The sexual differentiation of the human brain: role of sex hormones versus sex chromosomes. In: *Neuroendocrine Regulation of Behavior*. Springer; 2018:45–67.
90. Bakker J, Honda S-I, Harada N, Balthazart J. The aromatase knock-out mouse provides new evidence that estradiol is required during development in the female for the expression of sociosexual behaviors in adulthood. *J Neurosci*. 2002;22(20):9104–9112.
91. Brock O, Baum MJ, Bakker J. The development of female sexual behavior requires prepubertal estradiol. *J Neurosci*. 2011;31(15):5574–5578.
92. Ostrer H. Disorders of sex development (DSDs): an update. *J Clin Endocrinol Metab*. 2014;99(5):1503–1509.
93. Arboleda VA, Sandberg DE, Vilain E. DSDs: genetics, underlying pathologies and psychosexual differentiation. *Nat Rev Endocrinol*. 2014;10(10):603.
94. Main KM, Skakkebæk NE, Virtanen HE, Toppari J. Genital anomalies in boys and the environment. *Best Pract Res Clin Endocrinol Metab*. 2010;24(2):279–289.
95. Samango-Sprouse C, Kırkızlar E, Hall MP, et al. Incidence of X and Y chromosomal aneuploidy in a large child bearing population. *PLoS One*. 2016;11(8):e0161045.
96. Abramsky L, Chapple J. 47, XXY (Klinefelter syndrome) and 47, XYY: estimated rates of and indication for postnatal diagnosis with implications for prenatal counselling. *Prenatal Diag*. 1997;17(4):363–368.
97. Melmed S. *Williams Textbook of Endocrinology*. Elsevier Health Sciences; 2016.
98. Wikström AM, Dunkel L. Testicular function in Klinefelter syndrome. *Horm Res Paediatr*. 2008;69(6):317–326.
99. Rao E, Weiss B, Fukami M, et al. Pseudoautosomal deletions encompassing a novel homeobox gene cause growth failure in idiopathic short stature and Turner syndrome. *Nat Genet*. 1997;16(1):54.
100. Ellison JW, Wardak Z, Young MF, Gehron Robey P, Laig-Webster M, Chiong W. PHOG, a candidate gene for involvement in the short stature of Turner syndrome. *Hum Mol Genet*. 1997;6(8):1341–1347.
101. Styne DM, Grumbach MM. Puberty: ontogeny, neuroendocrinology, physiology, and disorders. In: *Williams Textbook of Endocrinology* (12th ed.). Elsevier; 2012:1054–1201.
102. Khan MR, Bukhari I, Junxiang T, et al. A novel sex chromosome mosaicism 45, X/45, Y/46, XY/46, YY/47, XYY causing ambiguous genitalia. *Ann Clin Lab Sci*. 2017;47(6):761–764.
103. Knudtzon J, Aarskog D. 45, X/46, XY mosaicism. *Eur J Pediatr*. 1987;146(3):266–271.
104. Telvi L, Lebbar A, Del Pino O, Barbet JP, Chaussain JL. 45, X/46, XY mosaicism: report of 27 cases. *Pediatrics*. 1999;104(2):304–308.
105. Melmed S, Polonsky K, Larsen P, Kronenberg H. *Williams Textbook of Endocrinology*. (12th ed.). Saunders Elsevier. 2011.
106. Ory SJ. *Androgen Biosynthetic Defects Producing Male Pseudohermaphroditism*. Global Library of Women's Medicine; 2004.
107. Zhu, Y, Imperato-McGinley, J, *Glob. libr. women's med.*, (ISSN: 1756-2228) 2008; DOI 10.3843/GLOWM.10350. This chapter was last updated: November 2008
108. Eugenides J. *Middlesex*. Bloomsbury. 2003 [2022].
109. Kovacs WJ, Ojeda SR. *Textbook of Endocrine Physiology* (6th ed.). Oxford University Press; 2011.
110. Hiort O. The differential role of androgens in early human sex development. *BMC Med*. 2013;11(1):152.
111. Hannema SE, Scott IS, Hodapp J, et al. Residual activity of mutant androgen receptors explains wolffian duct development in the complete androgen insensitivity syndrome. *J Clin Endocrinol Metab*. 2004;89(11):5815–5822.
112. Hutson JM, Grover SR, O'Connell M, Pennell SD. Malformation syndromes associated with disorders of sex development. *Nat Rev Endocrinol*. 2014;10(8):476.
113. Auchus RJ, Chang AY. 46, XX DSD: the masculinised female. *Best Pract ResClin Endocrinol Metab*. 2010;24(2): 219–242.
114. Shozu M, Akasofu K, Harada T, Kubota Y. A new cause of female pseudohermaphroditism: placental aromatase deficiency. *J Clin Endocrinol Metab*. 1991;72(3):560–566.
115. Carpenter M. The human rights of intersex people: addressing harmful practices and rhetoric of change. *Reprod Health Mat*. 2016;24(47):74–84.
116. Preves SE. *Intersex and Identity: The Contested Self*. Rutgers University Press; 2003.
117. Harper C. *Intersex*. Berg; 2007.
118. Carpenter M. *The "Normalization" of Intersex Bodies and "Othering" of Intersex Identities in Australia*. Springer; 2018.* short stature homeobox

CHAPTER

21

Male Reproductive System

CHAPTER LEARNING OBJECTIVES:

- Describe the functional gross morphology of the male reproductive system, and discuss how testicular Leydig and Sertoli cells interact with each other to support spermatogenesis and the development of male secondary sexual characteristics.
- Describe the key genomic and cellular changes that take place during spermatogenesis and spermiogenesis.
- Outline the feedback mechanisms along the hypothalamic-pituitary-testis axis, and describe the actions of luteinizing hormone and follicle-stimulating hormone on testis function.
- Describe how testosterone and its metabolites influence fetal and adult target tissues, and list endocrine changes associated with common male reproductive pathologies and aging.
- Describe some notable endocrine-disrupting chemicals that may contribute to male infertility and other reproductive health disorders, and discuss their mechanisms of action.

OPENING QUOTATIONS:

"The rooster wears a comb, which is, so to speak, the flag that is hoisted to announce the presence of his testes to the hens. On removal of the testes the flag is lowered: the comb atrophies and the rooster has become a capon."

—M. Tausk (1978). *Organon: De geschiedenis van een bijzondere Nederlands Onderneming.* Dekker en Van de Vegt, Niimegen, the Netherlands. Cited in N. Oudshoorn. 1994. *Beyond the Natural Body: An Archeology of Sex Hormones.* Routledge, London, UK, p. 49

"The testes, unlike the ovaries, never had to be discovered. They have always been there, palpable, and the subject of experimentation and speculation throughout most of human civilization."

—Ramon Pinon, 2002[1]

Introduction and Historical Perspective

Reproductive systems consist of the organs and tissues used by members of a species to generate offspring. Among vertebrates, such organs include the brain and gonads (i.e. the hypothalamic-pituitary-gonadal axis), as well as the diverse targets of gonadal steroids and other hormones that mediate gametogenesis, the onset of puberty, the development of secondary sexual characteristics, behavioral aspects of mating, sexual differentiation, birth/hatching, nourishment of the offspring, and many other parameters. Considering how fundamental reproduction is to the success of a species, reproductive systems have been under intense selective pressure throughout evolutionary history, resulting in similar mechanisms for achieving reproductive success among vertebrate taxa.[2] The next two chapters provide a basic understanding of the endocrinology of vertebrate reproductive systems primarily by addressing humans and other placental mammals. While certainly no single class of animals can be broadly representative of vertebrate reproductive systems, the advantage of an in-depth discussion of mammalian systems is that the endocrinology of mammalian reproduction is the most thoroughly understood among vertebrates. Furthermore, the mammalian model is frequently used as a COMPARISON for studying the reproductive systems of non-mammalian taxa and their many specialized adaptations to diverse environments.

Ancient Notions of Male Reproduction

The first types of "experiments" with testes likely took place 8,000–9,000 years ago with the domestication of the first animals. **Castration**, the removal of the testes, was used for the purposes of controlling population numbers, reducing male aggression, and tenderizing the quality of the meat (the energy diverted away from reproduction can instead be stored as fat).[3] In humans, castration was an ancient form of punishment for adultery and sexual offenses prescribed by laws as old as the Babylonian Code of Hammurabi (ca. 1754 BC). The existence of this and other ancient laws suggest that the effect of the intact testes in promoting the human male sex drive was well known. In fact, the castration of prepubescent boys was used for purposes other than legal punishment, to produce *eunuchs* for guarding the women of harems

DOI: 10.1201/9781003359241-28

in the Middle East and China. In Europe, the castration of prepubertal boys was even used to maintain a supply of male soprano singers through the late 19th century, suggesting that that it was also known that the testes were responsible for producing male secondary sexual characteristics, such as the deepening of the voice.

Curiously Enough . . . The word "testis" derives from the Latin word *testis*, which had two definitions in ancient Rome. The first was a Roman legal term meaning "witness", or one who is present at a legal transaction.[4] The second signifies "testicle" (i.e. the gonad), which serves as a "witness to one's manhood". Under Roman law, people without testes (i.e. women and eunuchs) could not legally "bear witness".[4] This notion survives today in the words *testify* and *testimony*.

Dawn of a Modern Endocrine Understanding

As described in Chapter 1: The Scope and Growth of Endocrinology (Figure 1.17), the first experiment that clearly demonstrated that a blood-borne substance produced by the testes influences the development of adult male features was that by Arnold Berthold, who showed in the mid-19th century that castrated rooster chicks failed to develop male secondary sex characteristics, and that transplantation of the testes back into castrated roosters rescued these secondary sex characteristics.[5] This seminal experiment is generally considered by endocrinologists to be the first proof of the existence of hormones, founding the field of endocrinology. Progress on the isolation of the active ingredient produced by the testes picked up rapidly in the 1920s when Moore and colleagues demonstrated that injection of a lipid-soluble extract from the testes rescued the effects of castration in birds and mammals.[6] Then, in 1931, Adolf Butenandt in Germany went on to isolate a few milligrams of the first androgen steroid hormone ever discovered (androsterone) from 15,000 liters of policemen's urine, and in 1935 Ernst Laqueur and Coroli David coined the name "*testosterone*" (*testo* = testes, *ster* = sterol, *one* = ketone) after isolating crystalline testosterone from bull testes. Swiss biochemist Leopold Ruzicka soon determined the structure of testosterone, and by 1939 he and Butenandt's group had synthesized several sex hormones from cholesterol. For their work in demonstrating the structure of steroids, Ruzicka and Butenandt shared the *1939 Nobel Prize in Chemistry*. As early as 1932, McCullagh[7] had predicted that the testis in fact produced two different types of hormones: one was a lipid-soluble substance, which turned out to be testosterone, and the other was found in aqueous extracts and suppressed the enlargement of certain cells of the pituitary following damage to the seminiferous tubules. He named this latter substance *inhibin*, a factor that 50 years later would be determined to be a glycoprotein hormone that inhibits the secretion of follicle-stimulating hormone (FSH) by the pituitary.

This chapter addresses the "bottom" of the hypothalamic-pituitary-gonadal (HPG) axis, or the mechanisms by which the pituitary gonadotropins, follicle-stimulating hormone (FSH), and luteinizing hormone (LH), promote gonad maturation and function in males. The roles of gonadal steroids and other hormones in mediating spermatogenesis and maintenance of male reproductive physiology will also be discussed. The "top" of the HPG axis, or the induction and timing of pituitary LH and FSH synthesis in response to hypothalamic GnRH signaling to promote puberty and seasonal reproduction, will be discussed in Chapter 23: The Timing of Puberty and Seasonal Reproduction.

Form and Function of the Testis

LEARNING OBJECTIVE Describe the functional gross morphology of the male reproductive system, and discuss how testicular Leydig and Sertoli cells interact with each other to support spermatogenesis and the development of male secondary sexual characteristics.

KEY CONCEPTS:

- The testis is a dual function organ that generates the male germ cells, as well as synthesizes androgens.
- Each testis consists of a collection of lobules comprised of convoluted seminiferous tubules, wherein sperm cells develop.
- The lobules are divided into two general "compartments", each separated by a basal lamella.
- The interstitial compartment is comprised in part of testosterone-secreting Leydig cells.
- The intratubular compartment is composed of developing germ cells as well as Sertoli cells that support sperm development and other testicular functions.
- Sertoli cells secrete androgen-binding protein (ABP) into the seminiferous tubule lumen, where it sequesters testosterone at the elevated levels necessary to promote spermatogenesis.

The testis is a dual function organ: it generates the male germ cells, as well as synthesizes the androgens that promote male secondary sex characteristics and maintain sperm. This section describes the testis anatomy, its role in mediating spermatogenesis, its regulation of the biosynthesis of androgens and other sex hormones, as well as its maintenance of male secondary sex characteristics.

Structure and Organization

In 1668 Regnier De Graaf performed a simple but brilliant experiment on a mouse testicle (Figure 21.1):

> if the tunica albuginea is removed and the tubules thrown into a basin of water and shaken about a little, you will behold a delightful and surprising sight; the tubules will separate from one another in such a way that without any help from instruments it can be seen with absolute clarity that the substance of the testicles consists wholly of tubules.[8]

Figure 21.1 De Graaf's experiment showing that a mouse testicle consists of tubules. De Graaf's illustration depicts a mouse testicle hanging from a the cap of a bottle filled with an aqueous solution following removal of the testis outer covering, the tunica albuginea. De Graaf concludes, "if anyone asks us what is really the character of the substance of the testicles we shall say that is it is simply a collection of minute vessels or tubules which confect semen; if these same tubules were disentangled without being broken and tied to one another, they would far exceed 20 Dutch ells in length" (De Graaf, 1668, as translated by Jocelyn and Setchell, 1972). The length of one Dutch ell is 68.6 cm.

Source of Images: Lindholm, J. and Nielsen, E.H., 2009. Pituitary, 12(3), pp. 226-235. Used by permission.

Indeed, each testis is divided into multiple *lobules*, each of which consists of a collection of highly convoluted *seminiferous tubules* wherein sperm cells are produced (Figure 21.2). Each lobule contains a single seminiferous tubule that forms a loop, with both ends connecting to the rete testis. In humans the total length of these tubules is estimated to be approximately 500 m! The tubules of each testis are encapsulated by a sheath of connective tissue called the *tunica albuginea*. The tubules intersect with a network of ducts called the *rete testis*, which converges with ducts leading to the *epididymis*. Smooth muscle contraction by the epididymis and the *vas deferens* induces ejaculation, where mature spermatozoa are transported through a duct system, passing the *seminal vesicles* and the *prostate* where (in response to androgens) secretions that make up most of the volume of the ejaculate are produced. Ultimately, the ejaculate leaves the penis via the urethra. The functions of the testes and accessory male reproductive organs are summarized in Table 21.1.

Histologically, the lobules are effectively divided into two general "compartments", each separated by an

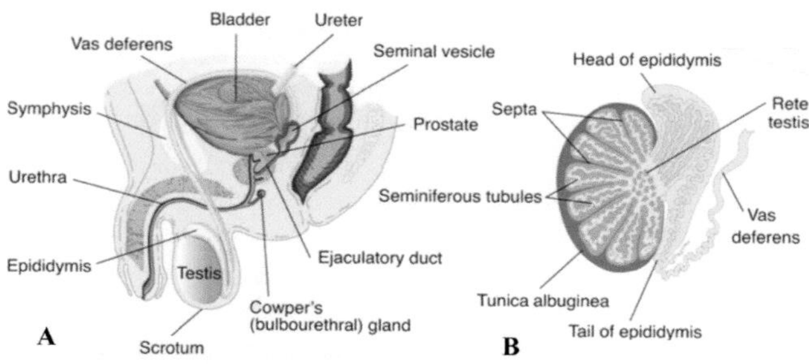

Figure 21.2 Gross anatomy of the male reproductive system (A) and sagittal section of the testis and epididymis (B).

Source of Images: Barrett, K.E., et al. 2010. Ganong's review of medical physiology twenty. McGraw Hill LLC. Used by permission.

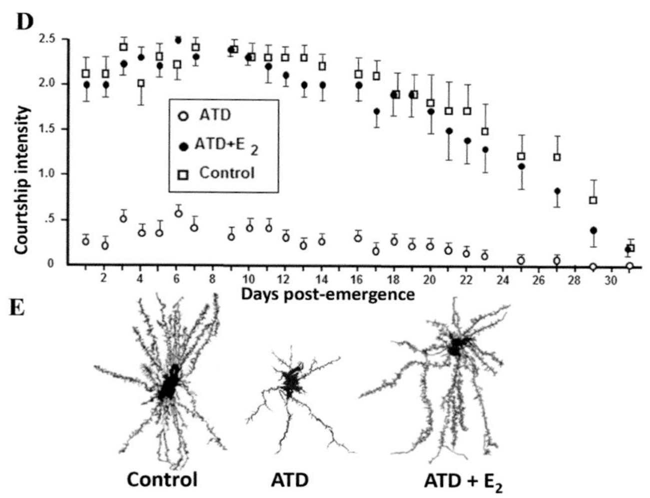

Figure 21.2a and b (parts a and b) (A) Reproductive patterns among seasonally breeding vertebrates. (B) Thamnophis sirtalis. (C) The major physiological and behavioral events in the annual reproductive cycle of the male red-sided garter snake in Canada. Animals spend most of the year underground in hibernation. In the spring, they emerge and mate with the females before dispersing to summer feeding grounds. Since all metabolic processes slow down during the cold months, androgen levels in the male will be elevated in the spring if he entered hibernation with elevated levels (dotted lines); however, androgen levels usually are basal on emergence (solid line). Sperm are produced during the summer after mating and are stored in the vas deferens (heavy squiggle line next to testis) over winter. **(D)** The aromatization of androgen into estrogen during hibernation is necessary for the development of male courtship behavior in the spring (see text for details) and for the development of dendritic spines (see panel E) on neurons in pathways critical for the initiation and control of courtship and mating. **(E)** Camera lucida tracings from photomicrographs of representative neurons from animals receiving either control, ATD, or ATD + E2 treatment.

Source of Images: (A and C) Crews, D., et al. 2009. Hormones, 2, pp. 771-816. Used by permission. (B) Castoe, et al. 2011. Stand. Genom. Sci., 4(2), pp. 257-270. (D and E) Krohmer, R.W., 2020. J. Exp. Zool. A: Ecological and Integrative Physiology, 333(5), pp. 275-283. Used by permission.

Table 21.1 Functions of the Testes and Accessory Male Reproductive Organs

Testes: This pair of organs is located in the scrotum, secured at either end by a structure called the spermatic cord. Within the testes are coiled tubes called seminiferous tubules that both generate sperm and synthesize testosterone.
Epididymis: A long, coiled tube located on the dorsal side of each testicle. It transports and stores sperm cells produced in the testes. This organ is necessary for the maturation of immature sperm derived from the testes. During sexual arousal, smooth muscle contractions force the sperm into the vas deferens.
Vas deferens: A long, muscular tube that transports mature sperm from the epididymis to the urethra, the tube that carries urine or sperm to outside of the body, in preparation for ejaculation.
Ejaculatory ducts: Formed by the fusion of the vas deferens and the seminal vesicles, these ducts empty into the urethra.
Urethra: The tube that carries urine from the bladder to outside of the body and, in males, has the additional function of ejaculating semen at orgasm.
Seminal vesicles: Sac-like pouches that attach to the vas deferens near the base of the bladder. The seminal vesicles produce a fructose-rich fluid that provides sperm with an energy source to facilitate their motility. The fluid from the seminal vesicles constitutes most of the volume of the ejaculate.
Prostate gland: A structure that contributes additional fluid to the ejaculate that nourishes the sperm.
Bulbourethral glands: Structures located on the sides of the urethra inferior to the prostate gland. The fluid produced by the glands empties directly into the urethra and serves to lubricate the urethra and to neutralize acidity due to any residual urine in the urethra.

acellular basal lamina: the *interstitial compartment*, comprised of interstitial cells, connective tissue, blood vessels, and immune cells, and the *intratubular compartment*, composed of a complex *seminiferous epithelium* (Figure 21.3). The interstitial compartment contains the primary endocrine cells of the testis, clusters of testosterone-secreting cells called *Leydig cells*. This compartment also contains an extensive capillary network that transports testosterone into systemic circulation, as well as transports nutrients to the Sertoli cells of the intratubular compartment. Another important function of the vasculature in this compartment is to help maintain a testis temperature that is about 2 degrees Celsius below body temperature, a necessity for normal sperm development. This is accomplished by the presence of the *pampiniform plexus*, a vascular network arising from the testicular venous outflow. Due to the antiparallel arrangement of juxtaposed arterial and venous blood flow, this plexus functions as a *vascular countercurrent heat exchanger* that lowers incoming arterial blood temperature upon arrival at the testis. This, in combination with the extra-abdominal positioning of the testes, ensures an optimal temperature for spermatogenesis. In fact, if not surgically corrected during the first few years of infancy, a failure of the testes to descend into the scrotum during fetal development can result in sterility due to abnormal spermatogenesis caused by an elevated temperature.

The seminiferous epithelium of the intratubular compartment is composed primarily of two categories of cells: (1) the *germ cells* whose progeny move progressively towards the lumen of the tubules as they undergo spermatogenesis and spermiogenesis, and (2) *Sertoli cells*, which are large epithelial cells that span from the basal lamella to the lumen. Sertoli cells serve several important functions that include support, endocrine, and exocrine activities.

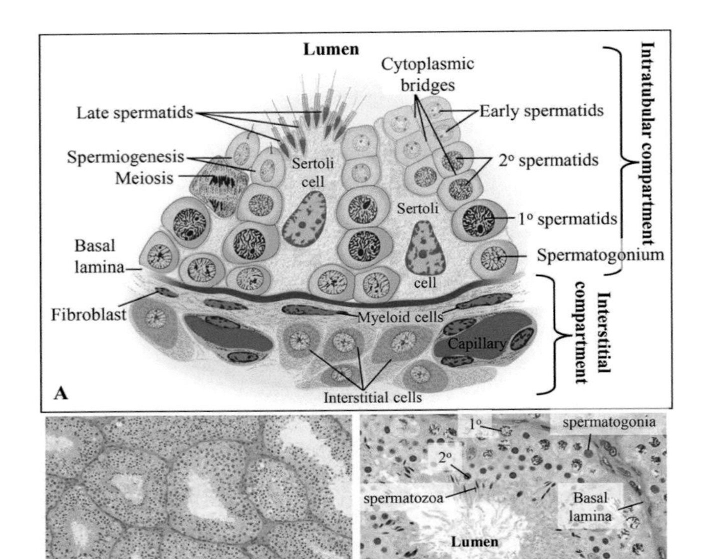

Figure 21.3 The lobules of the testes consist of two distinct compartments divided by a basal lamina (basement membrane). (A) The interstitial compartment consists of clusters of Leydig (interstitial) cells, capillaries, myoid smooth muscle cells, and fibroblasts. The intratubular compartment consists of two cell populations in close contact with one another: large Sertoli cells that span from the basal lamina to the lumen of the seminiferous tubules, and spermatogenic lineage cells in various stages of development. The spermatogonia are in contact with the basal lamina. As the germ cells undergo meiosis to develop into primary and secondary spermatocytes and spermatids, they move towards the lumen of the tubule. The mature spermatozoa eventually disassociate from the Sertoli cells on the lumen side and move from the seminiferous tubule through the ejaculatory duct system. Hematoxylin and eosin-stained testis under low **(B)** and high **(C)** magnifications.

Source of Images: (A) Barrett, K.E., et al. 2010. Ganong's review of medical physiology twenty. McGraw Hill LLC. Used by permission. (B-C) Courtesy of Dr. Peter Takizawa, Yale University.

Support Functions

- Beginning at puberty, Sertoli cells form tight junctions among adjacent Sertoli cells, creating a *blood-testis bar-*

rier that protects the exposure of more mature germ cells from antibodies and other substances generated in the peritubular compartment. Exposure of spermatozoa to the immune system can generate an immune response that may lead to infertility due to an autoimmune inflammation of the testis.
- Sertoli cells provide nutrients to the developing germ cells, such as glucose, fructose, iron, transferrin, and lactate. Like antibodies, these substances cannot cross the blood-testis barrier but are selectively and actively transported across the Sertoli cell membranes.
- Sertoli cells associate closely with sperm cells during all stages of development, forming gap junctions with them and guiding sperm cells towards the lumen as they mature via the breakdown and reformation of these connections. As such, Sertoli cells appear to be responsible for the spatial and temporal organization of spermatogenesis.

Endocrine and Paracrine Functions

- Sertoli cells synthesize *inhibin*, a glycoprotein hormone that exerts negative feedback onto the pituitary gonadotropes, ensuring that FSH levels are maintained at the optimal concentration.
- During fetal development, Sertoli cells produce *anti-Müllerian hormone*, which promotes the degradation of the embryonic Müllerian duct, a structure that would otherwise differentiate into the female reproductive tract.
- Sertoli cells express both *androgen receptors* and *FSH receptors*, whose actions (when bound to their respective hormones) are essential for promoting spermatogenesis.
- In response to FSH, Sertoli cells secrete *glial cell line-derived neurotropic factor* (GDNF), which promotes the mitotic proliferation of spermatogonia.[9,10]
- Sertoli cells express the enzyme *5α-reductase*, which converts small amounts of testosterone into the more active androgen *dihydrotestosterone* (DHT), which acts locally in the testis.
- These cells also express the enzyme *CYP19 aromatase*, which converts small amounts of testosterone into *17ß-estradiol*, which may act locally to enhance spermatogenesis.

Exocrine and Other Functions

- A key function of Sertoli cells is to secrete *androgen-binding protein* (ABP) into the seminiferous tubule lumen, where its role is to sequester testosterone and maintain the elevated levels necessary to promote normal spermatogenesis.
- Sertoli cells generate fluid that helps move the immobile spermatozoa from the seminiferous tubules to the epididymis.
- These cells phagocytose (engulf) dead sperm cells, as well as residual bodies (segments of membrane-bound cytoplasm) shed by sperm cells during spermiogenesis.

The testes continuously generate large numbers of sperm (about 120 million every day, or 1,000 with each heartbeat) throughout the reproductive life-span.[11] As a *spermatogenic cycle* (the generation of four mature spermatozoa from one spermatogonium) progresses, the germ cells move from the basal portion of the germinal epithelium adjacent to the basal lamina to the apical region where mature spermatozoa are released into the lumen.

SUMMARY AND SYNTHESIS QUESTIONS

1. Vasectomy, a common male contraceptive surgery, is accomplished by ligation of the vas deferens. Although this surgery prevents sperm from leaving the testes, it does not affect volume of the ejaculate—why not?

Male Gametogenesis: A Conceptual Overview

LEARNING OBJECTIVE Describe the key genomic and cellular changes that take place during spermatogenesis and spermiogenesis.

KEY CONCEPTS:

- In complex multicellular organisms, only the germ cells retain the capacity to transmit the materials and instructions required to initiate the assembly of a new organism.
- Diploid germ cells undergo two reductive meiotic divisions to give rise to haploid gametes, which in males are the spermatids, via a process called spermatogenesis.
- Spermatids transform morphologically into mature spermatozoa by a process called spermiogenesis.

The first sexually reproducing organisms were unicellular eukaryotes, where the division of each cell creates an entirely new organism. With the advent of more complex multicellular organisms, only a specialized subset of the organism's cells, called *germ cells*, retained the capacity to transmit the materials and instructions required to initiate the assembly of a new organism. The remainder of an organism's cells, those comprising tissues and organs that perform nonreproductive functions of the organism, are called *somatic cells*. In many animals, including vertebrates, insects, and roundworms, germ cells are distinguishable from somatic cells early in development. Germ cells give rise to two types of functionally distinct reproductive cells that together are known as *gametes* (derived from the Greek word for "marriage"). These are the *spermatozoa* (sperm cells, which in vertebrates reside in the testes) and *ova* (egg cells, which in vertebrates are located in the ovaries). Whereas all somatic cells arise from mitotic divisions and have a chromosome state that is *diploid*, or possess two copies of each chromosome, one of paternal origin and one maternal copy, the gametes result from meiotic divisions of germ cells and have a chromosome state that is *haploid*, possessing only one parental copy of each chromosome. A review of some key features of meiosis that will be needed to understand the details of spermatogenesis and, in the next chapter, oogenesis, appears in the following **Foundations & Fundamentals** box. This section of the chapter provides a conceptual overview of spermatogenesis, which will be integrated into an understanding of endocrine and physiological functions of the testes described in other chapter sections.

Foundations & Fundamentals: Meiosis: The Reductive Division of Chromosomes

The meiotic pathways of cell division are extraordinarily conserved among all organisms. Diploid germ cells, after having previously undergone several rounds of mitotic divisions, eventually undergo two meiotic *reductive divisions* that lower the total number of chromosomes from the diploid number (2N, where in humans 2N = 46) to the haploid number (N, where N = 23 in humans). Prior to the first division, or meiosis I (MI), DNA replication generates primary gametocytes where each chromosome is comprised of two *sister chromatids*. Therefore, at this stage the chromosomes are diploid, but the cell contains four copies of DNA for each chromosome, so the cell's DNA content can be described as being "4C", where "C" refers to the number of copies of DNA (Figure 21.4). By the end of MI, two haploid cells have been generated, and the DNA content has been reduced to 2C. No DNA replication takes place during meiosis II (MII), and by the completion of MII the chromosomal state for each cell remains haploid, but the DNA content has been further reduced to 1C. Prophase I, the most protracted phase of meiosis, is characterized by "crossing over" and the recombination of genetic material between *homologous chromosomes* (a matching pair of chromosomes, one from the mother and one from the father). The rest of MI involves the subsequent segregation of the homologous chromosomes from each other. In contrast to MI, MII is characterized by the segregation of the sister chromatids, and therefore mechanistically this phase is analogous to a mitotic division. Whereas in male mammals each cell undergoing meiosis generates four viable gametes (shown), in females only one viable gamete is produced, along with three nonviable cell products (not shown).

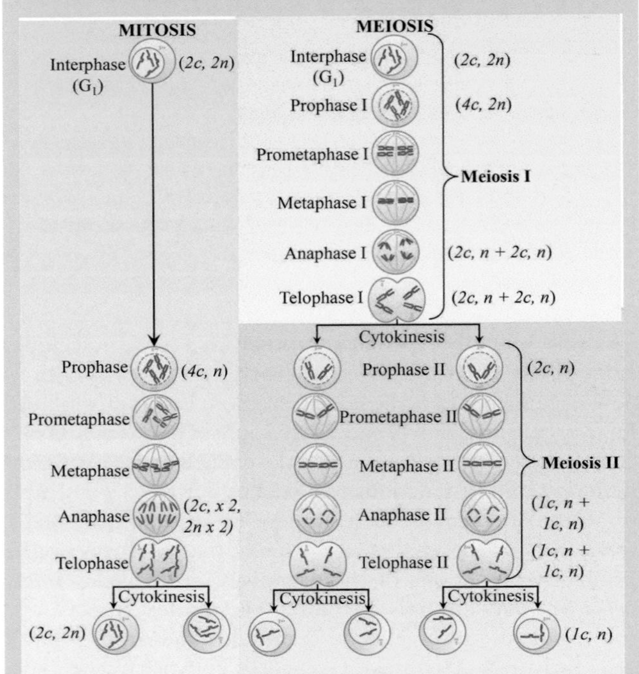

Figure 21.4 A comparison of key events of meiosis with mitosis. For simplicity, the top of the diagram depicts cells containing only two pairs of homologous chromosomes (red = maternal, blue = paternal). Progression towards the bottom of the diagram shows their respective mitotic and meiotic fates. *n* = number of copies of each chromosome (homologous chromosomes) in each cell, *c* = total number of copies of DNA for each chromosome (including sister chromatids) in each cell.

Spermatogenesis

Germ cells do not originate in the gonads, but instead arise during embryogenesis from progenitor cells called **primordial germ cells** (PGCs). The PGCs originate in the hindgut and migrate to a region called the *gonadal ridge*, the precursor to the gonads (see Figure 20.12 of Chapter 20: Sexual Determination and Differentiation, for a review). During their journey to the gonadal ridge, the PGCs undergo multiple mitotic divisions before reaching their final destination as germ cells, which will eventually divide meiotically to produce the gametes. In males, the PGCs that colonize the testes will give rise to germ cells called *spermatogonia*, undifferentiated male germ cells that will ultimately undergo **spermatogenesis** and produce mature *spermatozoa* (Figure 21.5). Upon arrival in the human embryonic testes, the spermatogonia are temporarily developmentally arrested in the G_1 phase of interphase. Following birth, the spermatogonia resume their mitotic divisions throughout childhood, but in a manner that differs from previous embryonic mitotic divisions. Specifically, these divisions are characterized by an *incomplete cytokinesis*, whereby all resulting daughter cells remain interconnected via cytoplasmic bridges called *syncytia* (Figure 21.5). Beginning at puberty, and following exposure to pituitary gonadotropins and testosterone, the diploid spermatogonia begin their first meiotic division (meiosis I), where they are now known as *primary spermatocytes*. By the end of meiosis I, the resulting haploid daughters are called *secondary spermatocytes*. The secondary spermatocytes begin their second meiotic division (meiosis II) almost immediately, yielding smaller-sized haploid cells called *spermatids*. Spermatids will go on to develop into mature spermatozoa in a process called spermiogenesis, described later. An important feature of spermatogenesis is that following birth the spermatogonia continue to divide mitotically throughout the life of the male, ensuring a continuous supply of spermatogonia from which spermatozoa can be generated sequentially through meiotic divisions. In humans, the process of spermatogenesis, from spermatogonia to spermatozoa, lasts approximately 74 days.

Spermiogenesis

The spermatids that result from the two rounds of meiotic divisions will subsequently transform morphologically into mature *spermatozoa* by a process called **spermiogenesis**. Spermiogenesis is characterized in part by cytoplasmic

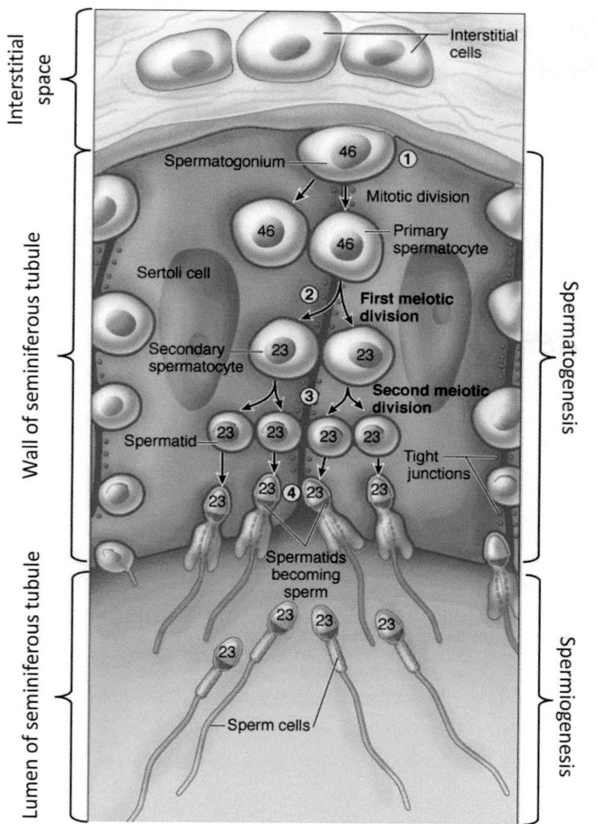

Figure 21.5 Spermatogenesis and spermiogenesis in humans. Following the migration of the primordial germ cells to the gonad during early embryogenesis, they develop into spermatogonia, diploid cells containing 23 pairs of chromosomes (1). Starting at puberty, under the influence of FSH, the spermatogonia divide mitotically for many rounds. A subset of these spermatogonia, called primary spermatocytes, will enter meiosis I and divide into haploid cells containing 23 chromosomes (2). Notice that cytokinesis is incomplete, forming syncytia (the progeny of divided cells remain attached to each other, sharing their cytoplasm). During meiosis II, the secondary spermatocytes divide, each producing two spermatids possessing a haploid number of unduplicated chromosomes (3). The spermatids will finally undergo spermiogenesis, a morphological transformation that produces the mature sperm (spermatozoa) (4). Major morphological changes associated with spermiogenesis include chromatin condensation, development of the acrosome, shedding of excess cytoplasm and organelles as residual bodies, reorganization of the mitochondria to the mid-piece region, and assembly and growth of the sperm flagellum.

Source of Images: Junqueira, L.C. and Mescher, A.L., 2013. Junqueira's basic histology: text & atlas/Anthony L. Mescher. McGraw Hill LLC. Used by permission.

reduction, the generation of a flagellum, and the formation of an *acrosome* (a digestive enzyme-filled vesicle that will assist in fertilization by breaking down the outer protein coating of the ovum) (Figure 21.5). Prior to ultimately becoming capable of fertilizing an ova, the spermatozoa must undergo a process called *capacitation*, characterized by the gain of increased motility by the flagellum and acquisition by the head of the acrosome's capacity to become chemically reactive. Capacitation typically requires the sperm to reside in the female reproductive tract (the folds of the cervix) for a period of time prior to fertilization.

Developments & Directions: Why don't the testes of many African mammals descend? Molecular forensics solves the mystery

In all mammalian embryos, the testes initially develop deep inside the abdomen at a position close to the kidneys, and in most mammals the testes descend by infancy or adulthood to a position located either externally in a scrotum or within the lower

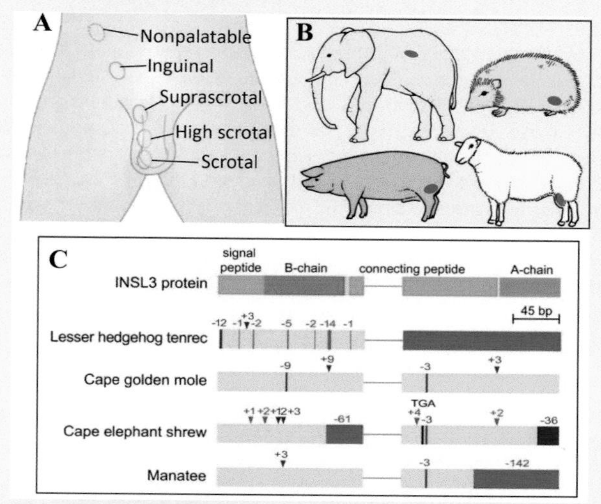

Figure 21.6 In humans, the testes are normally located within the abdominal cavity until the end of fetal life (A). In the fetus the testes are associated with a long fibrous structure called the gubernaculum, which is anchored to the scrotum. As the fetus grows the gubernaculum testis become progressively shorter, which helps facilitate the descent of the testes into the scrotum near birth or soon after birth. Depicted are a continuum of locations associated with cryptorchidism, or the pathological arrested descent of the testes. **(B)** The testes in varying degrees of descent in different mammals: in elephants they remain in their original fetal location near the kidney; in hedgehogs at the internal ring; in pigs in the superficial inguinal pouch; and in sheep and humans at the bottom of a pendulous scrotum. **(C)** Gene-inactivating mutations in the functional domains of the INSL3 protein and the exon-intron structure of the INSL3 gene. The exon-intron structure of the coding region of the gene is shown as boxes (exons, drawn to scale) and lines (introns, not drawn to scale). A vertical red line/arrowhead indicates a frameshifting deletion/insertion, with the number of deleted/inserted bases given above. Stop codon mutations are shown as a black vertical line. Splice site mutations are indicated by the mutated dinucleotide. A blue vertical line indicates a frame-preserving deletion. Red boxes are exons that either are deleted or accumulated numerous mutations that destroy any sequence similarity.

Source of Images: (A) Bay, K., et al. 2011. Nat. Rev. Urol., 8(4), pp. 187-196. Used by permission. (B) Seevagan, T., et al. 2019. Testes structure and function. Blandy's Urology, pp. 729-739. Used by permission. (C) Sharma, V., et al. 2018. PLoS Biol, 16(6), p. e2005293.

abdomen (Figure 21.6[A]). As optimal sperm development in most mammals requires a temperature lower than that found in the body core, the translocation of the testes to a more peripheral location helps to facilitate this. Indeed, in humans a failure of the testes to descend may result in sterility, if not corrected surgically. In most seasonally breeding mammals, the testes reside intra-abdominally for the majority of the year, descending into the scrotum only during mating season.

While mammalian taxa typically exhibit varying degrees of testicular descent (Figure 21.6[B]), several African species that include elephants, tenrecs, golden moles, manatees, elephant shrews, and rock hyraxes differ from other mammals by lacking any testicular descent. Instead, as adults they retain their testes at their initial position within the abdomen (see elephant example in Figure 21.6[B]). Until recently, it has remained unclear as to whether these African species lost testicular descent as part of their development, or whether other mammals gained that feature. In order to answer this puzzle, Sharma and colleagues[12] analyzed DNA sequence data of 71 mammals and made the remarkable discovery that these African mammals all possess gene-inactivating mutations of two genes, *insulin-like peptide 3* (INSL3) and *relaxin family peptide receptor 2* (RXFP2), that are required for testicular descent in other mammals Figure 21.6[C]). The findings of these "molecular vestiges" suggest that functional versions of these genes once existed in the ancestors of the African mammals that lack testicular descent today. The ancestral testicular descent process was subsequently lost in extant African mammals.

SUMMARY AND SYNTHESIS QUESTIONS

1. Primary and secondary spermatocytes undergo "incomplete cytokinesis". What is meant by this, and what is its significance?
2. Mature spermatocytes are much smaller than their precursor cells. What process accounts for this?
3. In humans, the androgen receptor is encoded by the AR gene located on the X chromosome. How many total number of copies of the AR gene are present in one human male somatic cell at each of the following stages of a cell cycle? G1, S, G2, prophase of mitosis.
4. Regarding the previous question about the AR gene, how many total number of copies of the AR gene are present in one human *female* somatic cell at each of the following stages of a cell cycle? G1, S, G2, anaphase of mitosis.
5. Regarding the AR gene, how many total number of copies of the AR gene are present in one human male *germline* cell at each of the following stages of a cell cycle? Prophase of meiosis I, prophase of meiosis II, after the completion of meiosis.
6. Regarding the AR gene, how many total number of copies of the AR gene are present in one human *female* germline cell at each of the following stages of a cell cycle? Prophase of meiosis I, prophase of meiosis II, after the completion of meiosis.

Endocrinology of Testicular Function

LEARNING OBJECTIVE Outline the feedback mechanisms along the hypothalamic-pituitary-testis axis, and describe the actions of luteinizing hormone and follicle-stimulating hormone on testis function.

KEY CONCEPTS:

- Testosterone produced by Leydig cells and inhibin produced by Sertoli cells exert negative feedback along the hypothalamic-pituitary axis.
- The pituitary gonadotropin luteinizing hormone (LH) targets testicular Leydig cells, inducing androgen steroidogenesis.
- The pituitary gonadotropin follicle-stimulating hormone (FSH) targets testicular Sertoli cells, which respond by synthesizing enzymes and other proteins that support spermatogenesis and testicular function.
- Sertoli cells and peripheral tissues can convert testosterone into either the more potent androgen dihydrotestosterone (DHT) or the estrogen estradiol via the actions of 5α-reductase and CYP19 aromatase, respectively.

Feedback along the Hypothalamic-Pituitary-Testis Axis

The influence of the hypothalamus and pituitary gland over gonad development was described in Chapter 8: The Pituitary Gland and Its Hormones, where hypophysectomy (removal of the pituitary gland) in male rodents is accompanied by reduced spermatogenesis and a testicular atrophy. The gonadotropins FSH and LH are the primary pituitary hormones that regulate both the spermatogenic and steroidogenic functions of the testes, and these gonadotropins are controlled by the secretion of hypothalamic gonadotropin-releasing hormone (GnRH) into the hypophyseal portal veins (Figure 21.7). Various hormones along the HPG axis, in turn, modulate axis activity by exerting negative feedback at various levels. Some of these hormones and their actions are described next.

Sex Steroid Hormones

The steroid (testosterone) and glycoprotein (inhibin) hormones produced by the testes each exert negative feedback along the HPG axis (Figure 21.7). In human males, negative feedback by testosterone occurs at the levels of both the hypothalamus and the pituitary primarily through activation of the estrogen receptor alpha (ERα) by estrogens derived from the local aromatization of testosterone.[13–17] The local conversion of testosterone to 17ß-estradiol is accomplished by expression of the enzyme, CYP19 aromatase (Figure 21.8). Interestingly, neither physiological levels of testosterone itself, nor

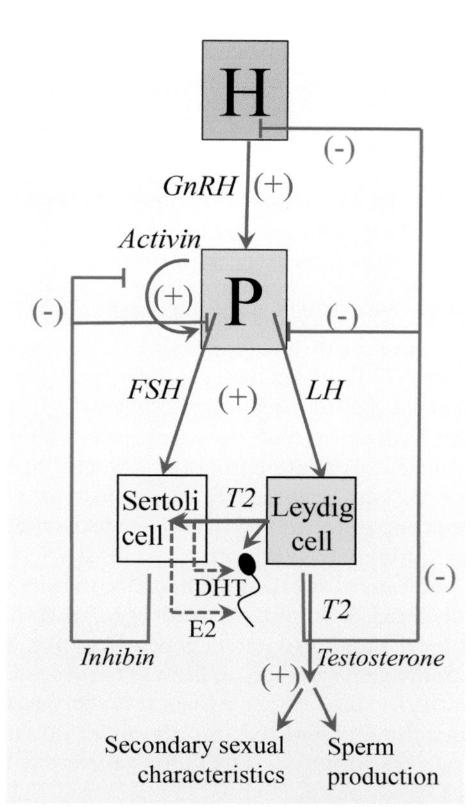

Figure 21.7 The hypothalamic-pituitary-testis (HPT) axis. Stimulatory and inhibitory actions are denoted by green arrows and red lines, respectively. T2: testosterone, DHT: dihydrotestosterone, E2: estradiol, GnRH: gonadotropin-releasing hormone, FSH: follicle-stimulating hormone, LH: luteinizing hormone.

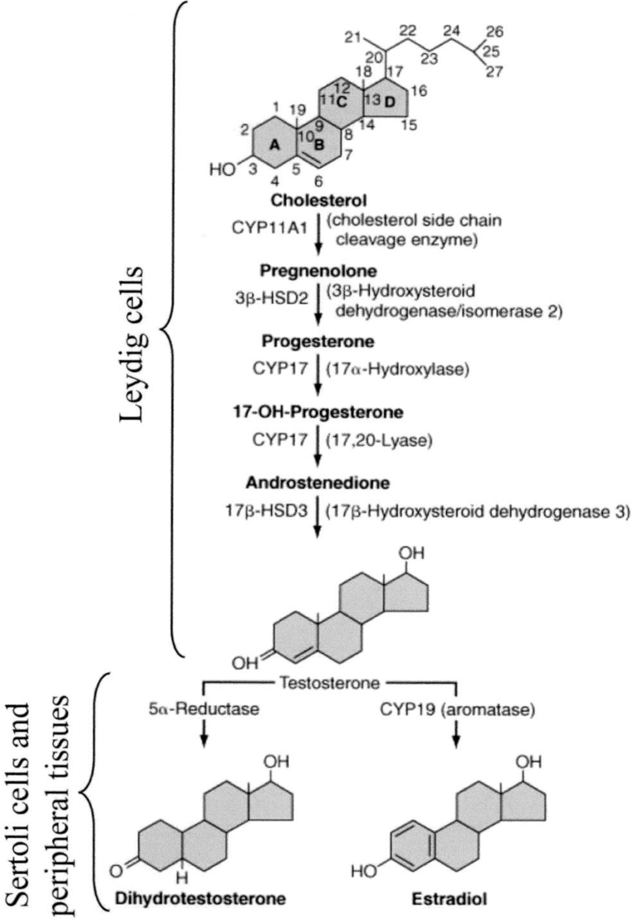

Figure 21.8 Biosynthesis of androgens and estradiol from cholesterol by the testes. Testosterone produced by the Leydig cells of the testis can be converted into the estrogen, 17β-estradiol, by aromatase (CYP19) or into the more potent androgen dihydrotestosterone by the enzyme 5α-reductase. These conversions can occur in the Sertoli cells of the testis and in diverse peripheral tissues such as the brain, skin, bone, and prostate gland.

Source of Images: Fauci AS, et al. Harrison's Principles of Internal medicine, 17th edition. McGraw Hill LLC. Used by permission.

the more potent androgen dihydrotestosterone (DHT), appear to play significant roles in negative feedback along this axis.[18]

Inhibins and Activins

Inhibins are heterodimeric glycoproteins composed of an α-subunit covalently bound via a disulfide bond to either a β_A or β_B subunit, forming inhibin A or inhibin B, respectively (Figure 21.9). In humans, *inhibin B* is the physiologically active form. Inhibin B, produced by the Sertoli cells in response to FSH signaling, feeds back negatively on the pituitary gonadotropes to inhibit FSH synthesis. In fact, in healthy adult males, inhibin B is the primary mediator of FSH negative feedback, superseding even that of sex steroids.[19]

Activins are composed of homodimers of either β_A subunits (activin A) or β_B subunits (activin B) and also form heterodimers consisting of a single β_A subunit and a single β_B subunit (activin AB) (Figure 21.9). In contrast to inhibins, activins are not produced by the testes in significant amounts but are synthesized by pituitary gonadotropes and function as autocrine regulators that promote both sensitization of gonadotropes to GnRH stimulation and induce FSHβ synthesis, which increases FSH secretion.[20] The actions of activins are counteracted by inhibin B targeting the gonadotropes, functioning as a selective antagonist of the activin receptor.

Prolactin

From rodents to humans, growing evidence suggests that the pituitary hormone prolactin may influence the development, function, and pathology of the male gonads in most mammals.[21] In 1955 it was shown that the inhibition of prolactin in rats during embryogenesis reduced the size of the prostate gland to 80% of the normal size, suggesting that prolactin is important in prostate differentiation and development.[22] Indeed, prolactin receptors are now known to be present in the prostate gland, and their overexpression can contribute to the development of prostate cancer.[23] Prolactin has also been implicated

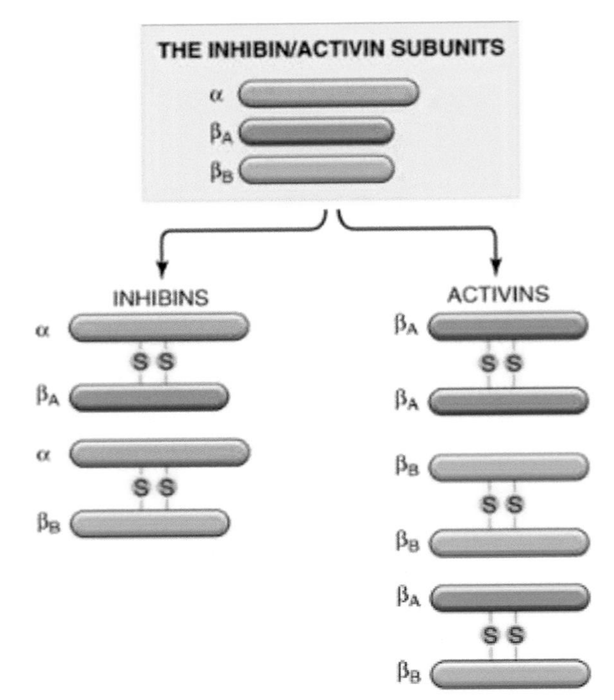

Figure 21.9 Inhibins and activins are related peptide hormones that consist of two subunits joined by a disulfide bond. Whereas inhibins are always composed of one ?? subunit dimerized to either a β_A or a β_B subunit, activins are composed only of β dimers. Activins and inhibins function as inhibitors of each other's activity.

Source of Images: Boron, W.F. and Boulpaep, E.L., 2005. Medical Physiology: A Cellular and Molecular Approach, updated 2nd ed. Elsevier. Used by permission.

in the synthesis of testosterone through upregulation of LH receptors on Leydig cells.[24–27] Additionally, acute hyperprolactinemia caused by a pituitary lactotrope adenoma suppresses testosterone synthesis and reduces male fertility by inhibiting the secretion of GnRH through prolactin receptors on hypothalamic dopaminergic neurons.[28]

The Two-Cell Theory

The roles of the two gonadotropins in testicular endocrinology are independent but cooperative in nature, with LH targeting primarily the Leydig cells and FSH targeting Sertoli cells. This interaction between two cell types mediated by two gonadotropins is known as the *two-cell theory* for testicular function.

Actions of LH on Leydig Cells

The primary function of Leydig cells is to synthesize androgens, which are required for both the development of male secondary sex features and also spermatogenesis. In rodents, androgens have been shown to be essential for the completion of both meiosis[29] and spermiogenesis,[30] and the deletion of the androgen receptor from Sertoli cells results in infertility caused by the arrest of spermatogenesis at the primary spermatocyte stage. Leydig cells are the only testicular cells to express LH receptors. The androgen biosynthetic pathway is initiated following activation of the LH receptor (a G protein-coupled receptor, GPCR), leading to the stimulation of adenylyl cyclase, generation of the cAMP second messenger, and activation of cAMP-dependent kinase protein kinase A (PKA) (summarized in Figure 21.10). The activated PKA induces the transcription of genes coding for diverse enzymes involved in steroidogenic pathways. These include the cytochrome P450 side-chain cleavage family of enzymes encoded by the CYP genes, the hydroxysteroid dehydrogenases (HSDs), and the steroidogenesis acute regulatory protein (StAR). The series of reactions mediated by these proteins occurs in the cytoplasm, mitochondria, and smooth endoplasmic reticulum, ultimately synthesizing primarily testosterone from its cholesterol substrate. This testosterone supports spermatogenesis by binding to androgen receptors inside the Sertoli cells and also enters general circulation where it exerts androgenic actions on diverse tissues and organs. Importantly, in some target tissues testosterone may first be peripherally metabolized into the more potent androgen *dihydrotestosterone* (DHT) or converted into the estrogen *17β-estradiol*.

Actions of FSH on Sertoli Cells

The initial steps that take place following the binding of FSH to its receptor located on the plasma membranes of Sertoli cells are similar to those described for the LH receptor. Namely, like the LH receptor, the FSH receptor is a GPCR that ultimately activates PKA by way of the adenylyl cyclase effector pathway. However, instead of inducing the transcription of steroidogenic enzymes, the most important genes transcribed in response to FSH include *androgen-binding protein* (ABP), CYP19 aromatase, 5α-reductase, and inhibin (summarized in Figure 21.10). The synthesis of some of these proteins, such as ABP and the androgen receptor, complement the androgen-secreting functions of the Leydig cells by sensitizing the Sertoli cells to testosterone. Paradoxically, although many genes and proteins are thought to be upregulated in response to stimulation by testosterone in Sertoli cells, very few genes have actually been characterized that respond to the androgen via the classical nuclear receptor signaling pathway.[31] This suggests that Sertoli cells likely also respond to testosterone signaling through as of yet poorly identified nongenomic pathways to regulate spermatogenesis. In addition to converting small amounts of testosterone into dihydrotestosterone and 17ß-estradiol for local use to enhance the support of spermatogenesis, the Sertoli cells also respond to FSH by synthesizing the glycoprotein hormone inhibin, which helps modulate feedback along the HPG axis.

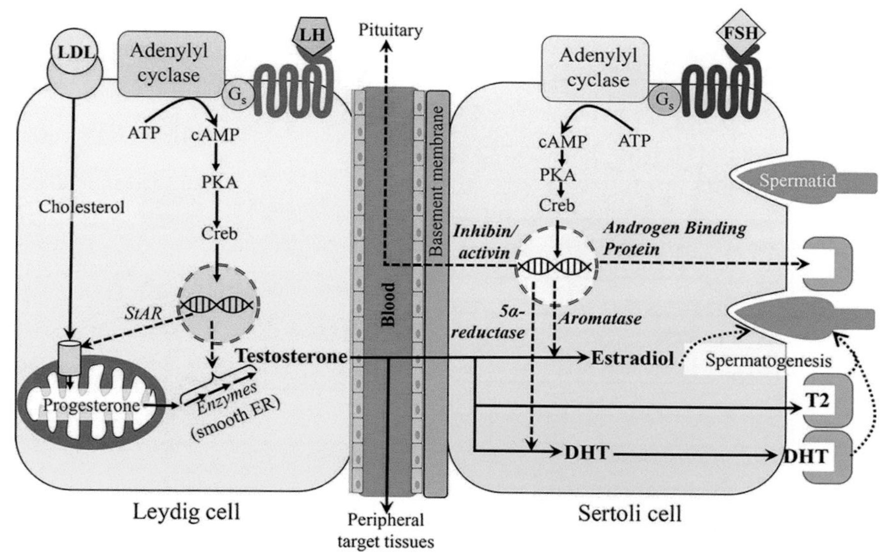

Figure 21.10 The two-cell theory of interaction between the Leydig and Sertoli cells of the testis. *Left side*: The binding of LH to its receptor on the Leydig cell plasma membrane initiates a cascade of events that promotes the biosynthesis of testosterone. The LH receptor is a G protein-coupled receptor (GPCR) whose activation of adenylate cyclase (AC) converts ATP to cAMP, the major second messenger of LH action in Leydig cells. cAMP subsequently activates the cAMP-dependent protein kinase (PKA), an event that triggers a series of cytosolic and nuclear reactions, including the activation of steroidogenesis acute regulatory protein (StAR), which is involved in cholesterol transport into mitochondria. Steroidogenic enzymes are also expressed in the mitochondria and smooth endoplasmic reticulum, which convert cholesterol into testosterone. Testosterone is released into general circulation and also moves into the adjacent Sertoli cells. *Right side*: The binding of FSH to its receptor (also a GPCR) activates PKA, ultimately inducing the expression of several proteins. These include inhibin (a hormone that feeds back on the pituitary to inhibit gonadotropin secretion), androgen-binding protein (which is secreted into the lumen of the seminiferous tubules where it sequesters testosterone for sperm development), aromatase (which converts some androgens into estradiol, which acts on the Leydig cells and is necessary for sperm development), and 5α-reductase (which converts some testosterone into the more active dihydrotestosterone). The Sertoli cells also synthesize growth factors (not shown) that act on the Leydig cells.

Developments & Directions: **Aromatase in the brain, and a dissociated reproductive pattern in male red-sided garter snakes**

Reproductive behaviors in seasonally breeding vertebrates typically manifest as one of several temporal relationships with gonadal activity[32] (Figure 21.11[A]). Individuals exhibiting a *constant reproductive pattern* (hatched line) have gonads that are maintained at nearly maximal development so that when breeding conditions arise, breeding can occur immediately. An *associated reproductive pattern* (solid line), the most common strategy, occurs when the production of gametes and/or increased sex steroid hormone secretion immediately precedes or coincides with the display of sexual behaviors. By contrast, if the display of mating behavior is temporally uncoupled from gonadal activity, this is termed a *dissociated reproductive pattern* (dashed line).

Among reptiles, the most thoroughly studied example of a dissociated reproductive pattern is the red-sided garter snake, *Thamnophis sirtalis parietalis* (Figure 21.11[B]), with much of this research conducted by the laboratory of David Crews.[33] In this species, males and females mate in the spring after emerging from a winter hibernation (Figure 21.11[C]). Remarkably, during mating the male's testes are regressed, and males use stored sperm to inseminate females. Elevated gonadal androgen production and spermatogenesis by the male occurs after mating in the late summer and fall, and the resulting sperm are stored during the winter hibernation. During spring mating, male plasma androgen levels are reported to range from basal to elevated. Importantly, male mating behavior does not appear to require a concomitant elevation of gonadal androgen levels, as the castration of males in the spring does not suppress this behavior, with males continuing to exhibit courtship behavior for up to 3 years following castration.[34,35] Furthermore, the administration testosterone does not induce male mating behavior outside of the breeding season.[34]

Interestingly, although elevated androgen levels during the spring are not required for male mating behavior, exposure of the male brain to sex steroid hormones during the winter hibernation does appear to be necessary for males to display their spring mating behavior. Specifically, Randolph Krohmer's group has found that the local aromatization of androgens into estrogens by different regions of the forebrain during winter hibernation is necessary to induce the changes in neuroanatomy and neurophysiology required to elicit male reproductive behavior in the spring.[36,37] This was demonstrated by implanting males during hibernation with either a CYP19 aromatase inhibitor containing 1,4,6-androstatriene-3,17-dione (ATD), ATD + 17β estradiol (E2), or blank (control) implants. Whereas those with blanks or ATD + 17β estradiol displayed normal mating behavior upon spring emergence, those implanted only with ATD showed

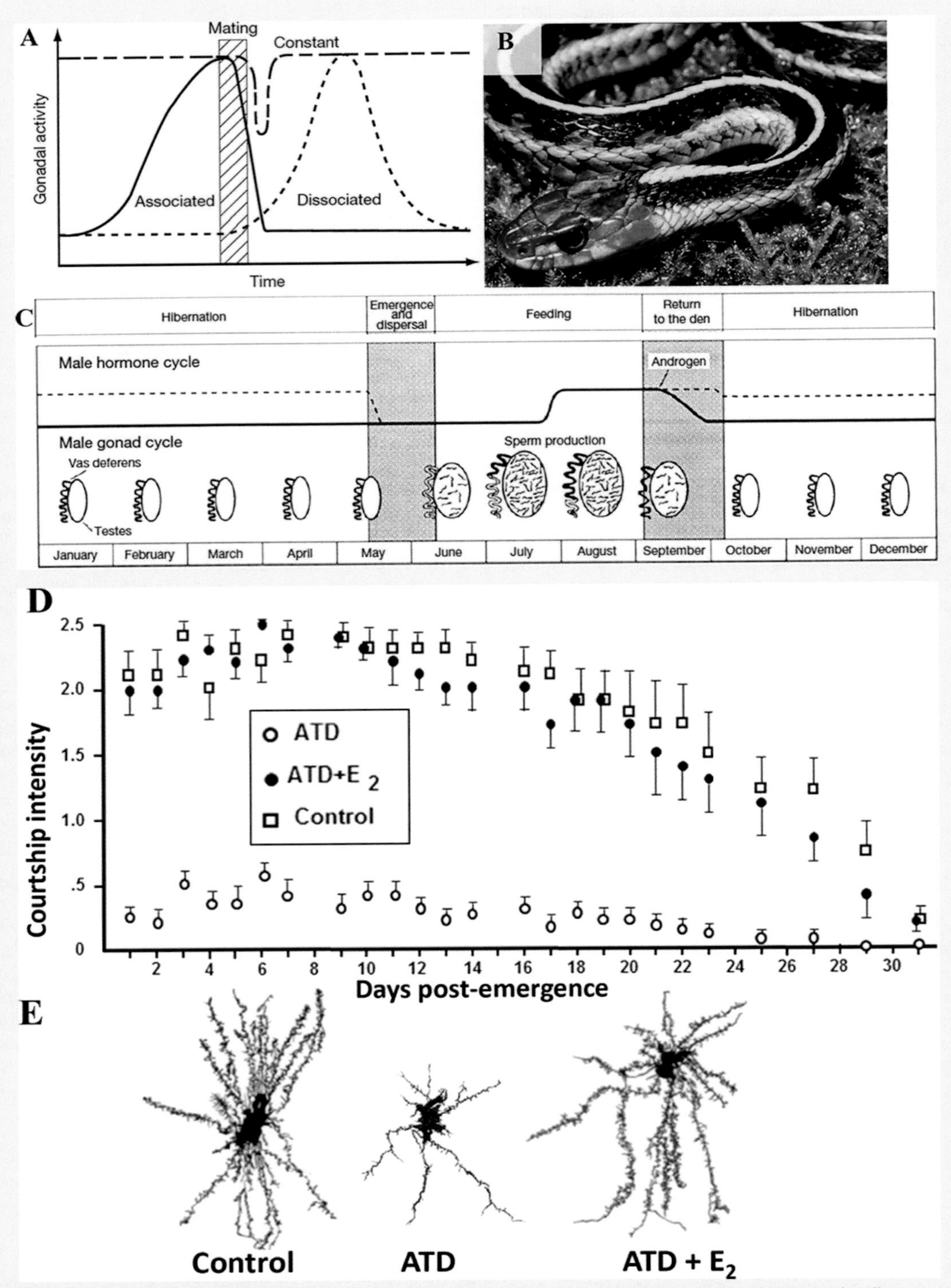

Figure 21.11 Dissociated reproductive pattern in the red-sided garter snake, *Thamnophis sirtalis parietalis.*

little or no courtship behavior (Figure 21.11[D]). Importantly, the number of dendritic spines on neurons in pathways critical for the initiation and control of courtship and mating were significantly lower in the ATD group compared with either control or ATD+E2 groups (Figure 21.11[E]), suggesting that E2 exposure during hibernation is required for their remodeling, which is necessary for the development of courtship behavior. Together, these studies show that while elevated androgens are necessary for spermatogenesis to occur in males, in these snakes male reproductive behavior ultimately requires the actions of estrogen on the brain weeks to months prior to mating for this behavior to manifest.

SUMMARY AND SYNTHESIS QUESTIONS

1. The abuse of anabolic steroids is typically accompanied by a significant reduction in the size of the testes. Why?
2. Testosterone is a hormone that is essential for normal sperm development. Considering this, how can testosterone be used as a contraceptive?
3. While testosterone exerts potent developmental effects on many tissues, this hormone can sometimes also be considered a "prohormone". Explain how this is true.
4. In the male "2 cell" model for testes function, two different pituitary hormones each exert their effects on two distinct cells in the testes. Summarize this model.
5. The Developments & Directions box describes how red-sided garter snake male reproductive behavior ultimately requires the actions of estrogen on the brain. Describe an analogy to this for the development of male sexual behaviors in rodents that you learned about in Chapter 20: Sexual Determination and Differentiation.

Roles of Androgens and Estrogens in Male Physiology

LEARNING OBJECTIVE Describe how testosterone and its metabolites influence fetal and adult target tissues, and list endocrine changes associated with common male reproductive pathologies and aging.

KEY CONCEPTS:

- Within the blood, most testosterone is bound to sex hormone-binding globulin (SHBG) and albumin.
- The physiological effects of androgens and estrogens are primarily mediated by nuclear receptors that function as ligand-activated transcription factors.
- Testosterone and dihydrotestosterone (DHT) bind to the same androgen receptor, but DHT binds with 100-fold greater affinity.
- The two androgens affect tissues differentially, with testosterone promoting masculinization of fetal internal genitalia and spermatogenesis and DHT promoting the masculinization of fetal external genitalia and secondary sexual characteristics during puberty.
- The aromatization of testosterone into estradiol by target tissues is necessary for the regulation of pituitary gonadotropin secretion and the pubertal bone growth spurt.

Steroid Hormone Transport, Peripheral Conversion, and Receptor Function

Testicular Steroid Hormones

Although testosterone is the primary sex steroid of testicular origin circulating in the blood plasma of males, the testes also secrete much lower amounts of other androgens and also estradiol (Table 21.2). Within the blood, most of the testosterone is bound to serum proteins, primarily *sex hormone-binding globulin* (SHBG) and *albumin*, with about 2% remaining unbound (Figure 21.12). Albumin has approximately a 1,000-fold lower affinity to sex steroids than SHBG, and the combination of free and weakly bound (i.e. albumin-bound) testosterone, known as the "bioavailable fraction", is that which is considered available to enter cells and bind to the androgen receptor. Importantly, small amounts of testosterone entering peripheral targets may be converted into either DHT (via 5α-reductase) or 17ß-estradiol (via CYP19 aromatase) (see Figure 21.10), which

Table 21.2 Normal Ranges for Gonadal Steroids, Pituitary Gonadotropins, and Prolactin in Men

Hormone	Ranges
Testosterone, total	260–1000 ng/dL (9.0–34.7 nmol/L)
Testosterone, free	50–210 pg/mL (173–729 pmol/L)
Dihydrotestosterone	27–75 ng/dL (0.9–2.6 nmol/L)
Androstenedione	50–250 ng/dL (1.7–8.5 nmol/L)
Estradiol	10–50 pg/mL (3.67–18.35 pmol/L)
Estrone	15–65 pg/mL (55.5–240 pmol/L)
FSH	1.6–8 mIU/mL (1.6–8 IU/L)
LH	1.5–9.3 mIU/mL (1.5–9.3 IU/L)
PRL	2–18 ng/mL (87–780 nmol/L)

Source: Gardner, D.G. and Shoback, D.M., 2017. *Greenspan's basic and clinical endocrinology*. McGraw-Hill Education.

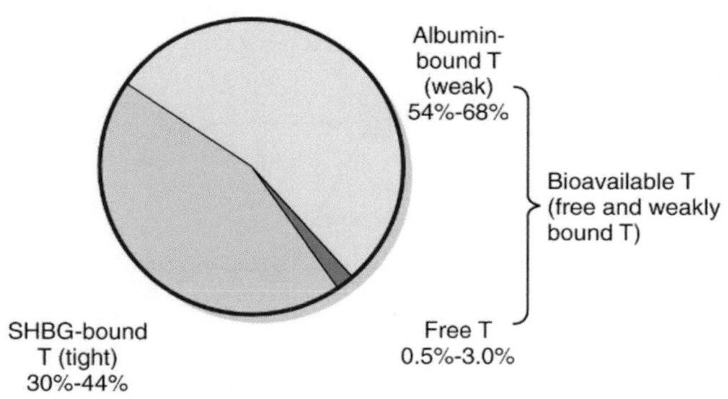

Figure 21.12 Testosterone fractions in the blood serum of human males. Blood testosterone is present in three states: bound to a high-affinity carrier protein (sex hormone-binding globulin, SHBG), bound to a low affinity carrier (albumin), or unbound (free fraction). Of the three states, the free and albumin bound partitions are considered to be bioavailable (may interact with receptors in cells).

Source of images: Melmed, S., et al. 2011. Williams Textbook of Endocrinology E-Book. Elsevier Health Sciences. Used by permission.

differentially affects target tissues. In these instances testosterone functions as a prohormone. DHT not only binds the androgen receptor with greater specificity than testosterone, but it also has a two- to ten-fold higher potency than testosterone in androgen-responsive tissues[38] (Table 21.3).

Adrenal Androgen Precursors

Recall from Chapter 13: The Multifaceted Adrenal Gland that the adrenal gland is the source of *adrenal androgen precursors*, steroids (such as androstenedione, DHEA (dehydroepiandrosterone), and its sulfated form DHEAS) that are converted by peripheral tissues into active androgens and estrogens. In fact, DHEAS is the most abundant steroid in adult human circulation, with serum concentrations 100–500 times higher than testosterone. In target tissue, DHEAS is desulfated enzymatically to produce DHEA and oxidized to make androstenedione, which is in turn converted into various active estrogenic and androgenic compounds capable of binding to the androgen or estrogen receptors. These adrenal androgen precursors, which become elevated during a component of puberty called **adrenarche**, promote pubic hair development, the production of adult body odor, and other components of puberty that occur independently of the gonads. Although adrenal androgens contribute significantly to the total androgen content of males, its effects appear to be masked by the actions of testosterone of testicular origin during **gonadarche** (increased sex hormone synthesis by the testes during puberty) and may not be essential for driving puberty in males. Females, by contrast, rely entirely on the production of androgen precursors for some features of puberty, such as pubic hair growth.

Table 21.3 Relative Androgenic Activity of Androgens

Steroid	Activity
Dihydrotestosterone	300
Testosterone	100
Androstenedione	10
DHEA, DHEAS	5

Source: From Gardner, D.G. and Shoback, D.M., 2017. *Greenspan's basic and clinical endocrinology*. McGraw-Hill Education.

Androgen and Estrogen Receptor Function

The physiological effects of androgens and estrogens are thought to be primarily mediated by nuclear receptors that function as ligand-activated transcription factors. Like other steroid hormones, sex steroids are also known to exert actions through nongenomic pathways through interactions with both classical steroid hormone receptors and plasma membrane-localized GPCRs. In the classical nuclear receptor pathway, androgen and estrogen receptors are inactive in the absence of ligand and are thought to be sequestered in the cytoplasm in association with chaperone proteins. When bound to hormone, the receptors translocate to the nucleus where they bind to specific DNA "hormone response elements" (HREs) in the promoter regions of target genes as homodimerized transcription factors that modulate gene expression. The generalized actions of androgens and estrogens on target cells are depicted in Figure 21.13. Once inside the target

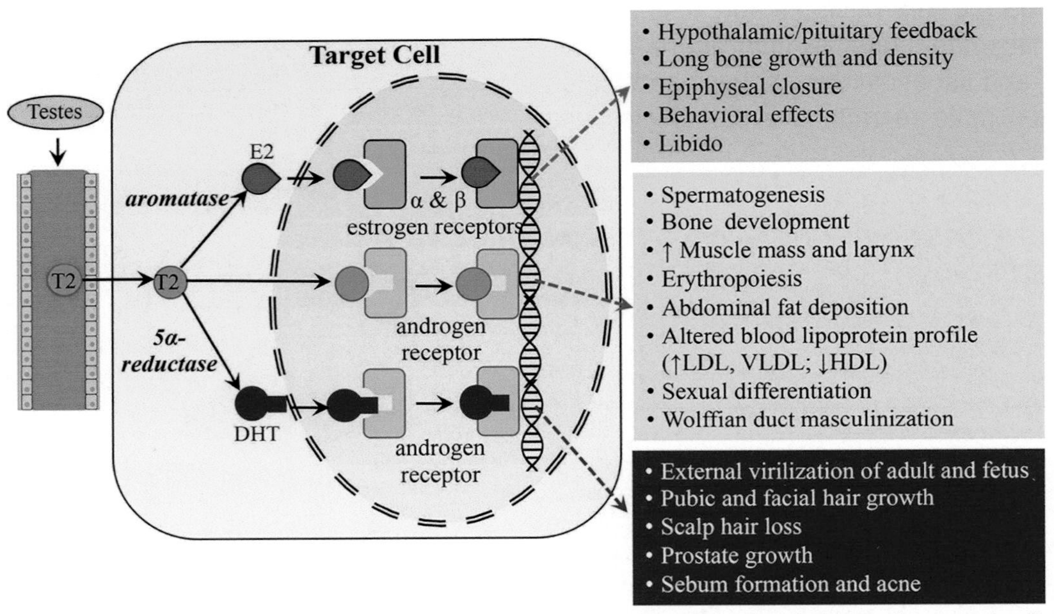

Figure 21.13 Some of the actions of androgens and estrogens on target cells. Following their entry into target cells, testosterone (T2) can exert its developmental effects by modulating gene expression using one of several different pathways, depending on the target cell: (1) it can bind directly to the androgen receptor, which functions as a ligand-dependent gene transcription factor; (2) it can first be converted by 5α-reductase into the more potent androgen dihydrotestosterone (DHT), which has a much higher affinity to the androgen receptor; or (3) it can be converted by aromatase into the estrogen estradiol (E2), which modulates gene expression by interacting with one of two different estrogen receptors (ER), ERα and ERβ. Some of the actions of these steroid hormones are shown on the right.

cell, testosterone can be converted into either DHT, via the action of 5α-reductase, or 17ß-estradiol via CYP19 aromatase. Although both testosterone and DHT can bind to the androgen receptor, DHT has a higher affinity to the androgen receptor than testosterone. Importantly, the two androgens are thought to target tissues differentially, with testosterone regulating spermatogenesis and masculinization of the Wolffian ducts during sexual differentiation and DHT promoting the masculinization of the fetal external genitalia and of secondary sexual characteristics during puberty (Figure 21.13). The actions of 17ß-estradiol, by contrast, are mediated by two different nuclear receptors, *estrogen receptors* α *and* β (ERα and ERβ), whose actions are important in regulating pituitary gonadotropin secretion and bone growth. Some of the actions of estradiol are also known to be mediated by a receptor called *GPER*, a plasma membrane-localized GPCR, though its influence in male development is only beginning to be understood.[39]

Developments & Directions: Spermageddon? Sperm counts are plummeting in Western men

In 1992, a landmark study by Carlsen and colleagues was published titled "Evidence for decreasing quality of semen during past 50 years".[40] This *meta-analysis* (a statistical analysis combining the results of multiple scientific studies) of published reports describing data on semen quality among men worldwide without a history of infertility indicated that sperm density had declined by a marked 42% from 1938 to 1990 (Figure 21.14[A]). Additionally, they found a significant decrease in mean seminal volume from 3.40 ml to 2.75 ml

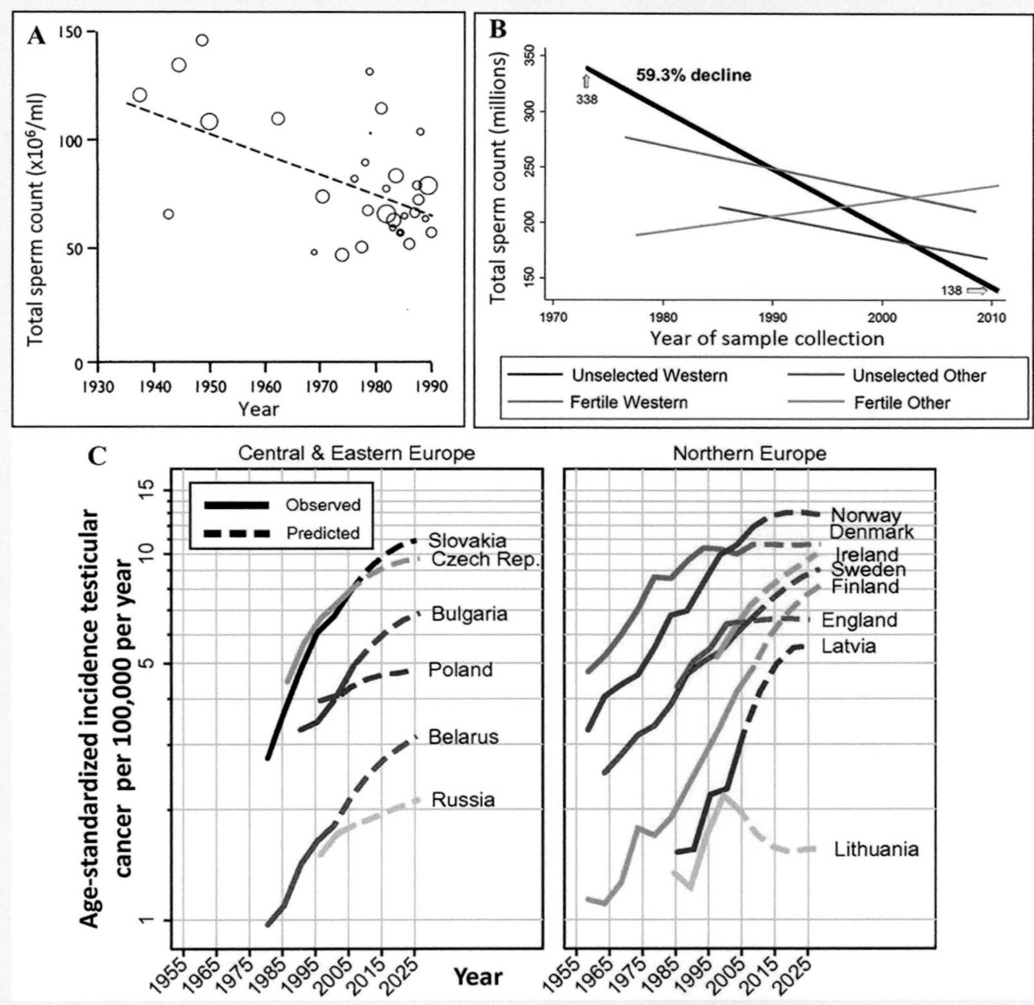

Figure 21.14 Global declines in human sperm counts correspond with increasing rates of testicular cancer. Meta-regression models of mean sperm density reported from 61 studies from 1938 to 1990 **(A)**, and for mean sperm count from 185 studies published from 1973 to 2011 controlled for fertility ("unselected by fertility" versus "fertile men") and geographic groups ("Western", including North America, Europe, Australia, and New Zealand, and "other", including Asia, Africa, and South America) **(B)**. Sperm concentrations declined significantly between 1973 and 2011. Some European trends in testicular cancer **(C)**.

Source of Images: (A) Carlsen, E., et al. 1992. Brit. Med. J., 305(6854), pp. 609-613. Used by permission. (B) Levine, H., et al. 2017. Hum. Reprod. Update, 23(6), pp. 646-659. Used by permission. (C) Skakkebaek, N.E., et al. 2016. Physiol. Rev., 96(1), pp. 55-97. Used by permission.

during the same period, indicating an even more dramatic decrease in total sperm count. Importantly, the authors of the study highlighted the concomitant increase in genitourinary abnormalities including testicular cancer, hypospadia (a congenital abnormality of the urethra where the urinary opening is not located on the head of the penis), and cryptorchidism (undescended testicles) that may be associated with declining sperm numbers. The authors hypothesized, "Such remarkable changes in semen quality and the occurrence of genitourinary abnormalities over a relatively short period is more probably due to environmental rather than genetic factors". The findings by Carlsen and colleagues, though dramatic, were criticized at the time for a relatively low sample size, a scarcity of data from the first 30 years of the analysis, and for the disproportionate amount of data drawn for Western men (74%) versus non-Westerners (26%).

In 2017, another much larger study titled "Temporal trends in sperm count: a systematic review and meta-regression analysis" was published. Working with an international team of researchers, Levine and colleagues[41] analyzed the findings of 185 sperm count studies published globally from 1973 to 2011. The results showed a 52.4% decline in sperm concentration (a –1.4% per year decline) and a 59.3% decline in total sperm count (a –1.6% per year decline) among European, North American, Australian, and New Zealand men (Figure 21.14[B]). By contrast, no statistically significant declines in sperm concentrations or numbers were observed in men from Africa, Asia, or South America. Importantly, there was no indication that the sperm decline in Western men was "leveling off" in recent years. The decline in mean sperm counts indicates that a growing proportion of Western men have sperm numbers below 40 million/ml, a threshold that is generally associated with subfertility or infertility.[42,43] This trend is no longer confined just to Western countries. In a study titled "Decline in semen quality among 30,636 young Chinese men from 2001 to 2015", Huang and colleagues[44] observed that among sperm donor applicants in Hunan Province, China, 56% qualified with sperm that met standards of healthiness in 2001, but by 2015 only 18% qualified.

In addition to reduced total sperm counts, the number of viable sperm that possess a morphology necessary to fertilize eggs plummeted from 60 to 20 million per milliliter, and testosterone levels have also been dropping over the same time period (Figure 21.15). Beyond fertility concerns, the decline in sperm count coincides with rising incidences in other reported male reproductive health indicators, such as testicular germ cell tumors (Figure 21.14[B]), cryptorchidism, delayed onset of male puberty, and reduced testosterone levels.[45] Although neither of the described studies was designed to directly address the causes of the reported sperm declines, these have been associated with several environmental and lifestyle influences that are particularly prevalent in modern industrialized societies, including obesity and endocrine disruption caused by exposure to numerous chemicals, such as cigarette smoke, fertilizers, pesticides, and other persistent organic pollutants.

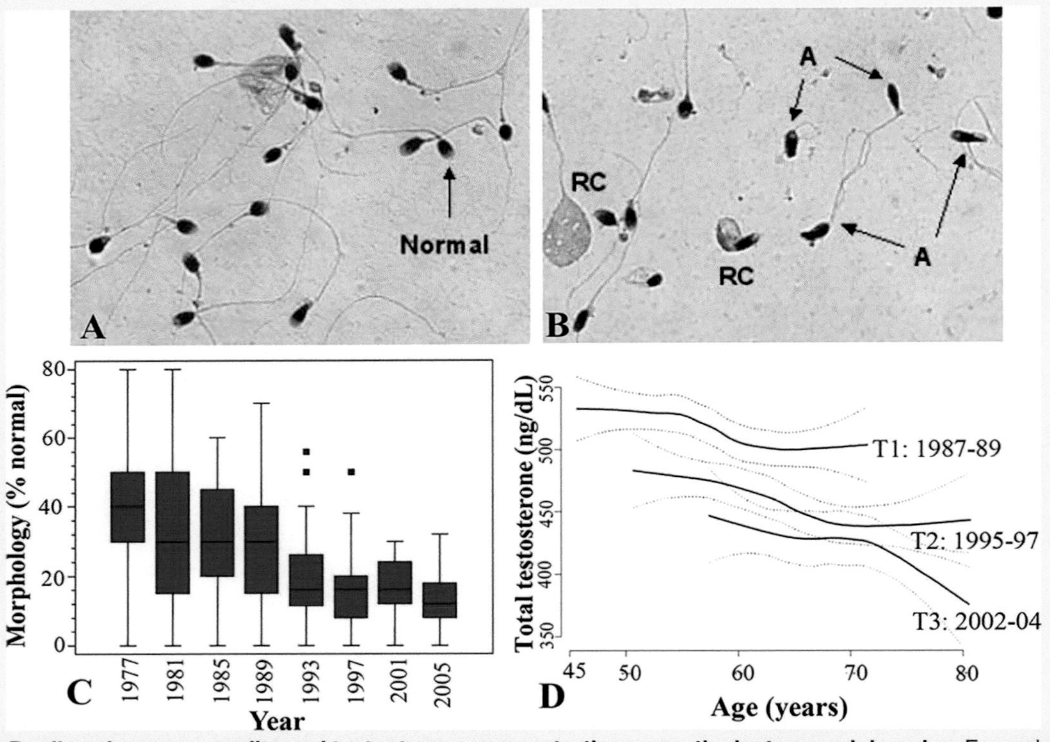

Figure 21.15 Declines in semen quality and testosterone concentrations over the last several decades. Examples of smears of semen samples with **(A)** good quality (18.5% morphologically normal spermatozoa) and **(B)** poor quality (0.05% normal spermatozoa). Men having < 9% morphologically normal spermatozoa are considered to be in a subfertile range. "RC", residual cytoplasma; "A", abnormal sperm head. **(C)** Decreasing sperm morphology and quality for candidate sperm donors in Belgium, based on the data of Comhaire et al. (2007). **(D)** Blood testosterone levels from American men declined more than 1% per year during 1987–2004. The graph shows average levels for each for men of different ages in each of the three measurement periods (T1–T3) with confidence bands (dotted lines).

Source of Images: (A-B) Skakkebæk, N.E., et al. 2006. Int. J. Androl., 29(1), pp. 2-11. Used by permission. (C) De Coster, S. and Van Larebeke, N., 2012. J Environ Public Health. (D) Travison, T.G., et al. 2007. J. Clin. Endocrinol. Metab., 92(1), pp. 196-202. Used by permission.

Target Tissues

During fetal development, testosterone peaks to levels almost as high as those in adult males. During this time, testosterone and DHT play critical roles in mediating sexual differentiation and the development of male primary sexual characteristics (e.g. differentiation of internal and external genitalia). After birth, between 3 and 6 months of age, males experience a second surge in testosterone levels that appears to influence testicular descent, growth of the penis, and (in conjunction with FSH) stimulation of Sertoli cell proliferation and spermatogonial development.[18] The increase in testosterone and its active metabolites (DHT and estradiol) to adult levels during puberty stimulates the development of male secondary sexual characteristics. These characteristics are summarized in Figure 21.13.

Sex Organs

Pubertal growth of the penis, scrotum, and prostate gland are mediated primarily by DHT, whereas the production of seminal fluid and other accessory secretions is promoted by testosterone. Although in males estrogen is present in low concentrations in blood, in rodents and bulls its concentrations can be much higher in seminal fluids, as high as 250 pg/ml in the rete testis,[46,47] which is even higher than blood serum estradiol in females.[48] Studies using estrogen receptor knockout (ERKO) mice have shown multiple defects in reproduction, including infertility, low sperm count, excess androgen biosynthesis, and reduced mating frequency.[49–51]

Hair and Skin

Locally converted DHT by the skin induces sebaceous glands to generate sebum (an oily secretion), which may contribute to the formation of acne. The growth of pubic hair and facial hair is driven exclusively by DHT. In puzzling contrast to the stimulation of pubic hair growth, for unknown reasons DHT induces the recession of the male frontal hairline and may also cause male pattern baldness (androgenic alopecia) in genetically susceptible individuals. Current research suggests that male pattern baldness results from an abnormal sensitivity of scalp hair follicles to circulating androgens, characterized by the progressive miniaturization of scalp hair and caused by increased DHT or 5α-reductase activity.[52,53] Supporting this hypothesis, clinical trials using finasteride, a 5α-reductase inhibitor, showed reductions in scalp DHT content, a trend towards reversal of scalp hair miniaturization, and improvements in scalp hair growth.[52]

Enlargement of the Larynx and Vocal Chords

Rising testosterone promotes increases in the growth and size of the laryngeal cartilages, muscles and ligaments. Together, these typically lead to a drop of about one octave in the pitch of the voice.[54]

Adipose Tissue

During puberty, elevated testosterone promotes a redistribution of body fat such that it generally accumulates more centrally in the abdominal viscera. In males, the fat-to-lean body mass ratio decreases during puberty. White adipose tissue is an endocrine organ that contains, among many other factors, CYP19 aromatase activity that converts testosterone into 17ß-estradiol. In obese males, therefore, puberty may be delayed due to reduced levels of circulating testosterone caused by excessive peripheral conversion to estradiol. Furthermore, excess adipose tissue in obese boys often promotes **gynecomastia** (proliferation of breast glandular tissue in the male).

The Pubertal Growth Spurt

In both males and females, the adolescent growth spurt is primarily induced by low levels of estrogen. This estrogen acts to increase the activity of the *growth hormone-insulin-like growth factor I* (GH–IGF-I) axis, and these hormones promote growth and elongation of the long bones. However, at higher levels, estrogen is also the key hormone that directly promotes growth plate fusion, thus terminating growth. Therefore, estrogen appears to have a biphasic dose-response relationship for epiphyseal growth, with stimulation at low levels and inhibition at high levels. Additionally, testosterone appears to stimulate growth by a direct effect on growth plate chondrocytes. Unlike estrogen, testosterone does not cause growth plate fusion. Both testosterone and estrogen are known to increase bone mineral density and help maintain skeletal mass in both females and males. Interestingly, although testosterone promotes bone growth, a protein found in bone, *osteocalcin*, has been found to promote the synthesis of testosterone in male mice by binding to its receptor on Leydig cells,[55] and thus the skeletal system may play a role in promoting male puberty (for more information on this hormone, refer to the Chapter 12: Calcium/Phosphate Homeostasis, Skeletal Remodeling, and Growth, Developments & Directions: Osteocalcin: endocrine regulation of male fertility by the skeleton).

Increased Skeletal Muscle Mass

By mass, skeletal muscle is the largest site of testosterone action. *In vitro* studies suggest that testosterone promotes the differentiation of pluripotent mesenchymal cells preferentially towards the myogenic rather than the adipogenic lineage,[56] which may account for the change in lean-to-fat ratio in males. The increase in muscle mass in males during puberty is in part due to hypertrophy (increase in individual cell size) of mature muscle fibers, but testosterone has also been shown to increase the proliferation of *satellite cells* (adult muscle stem cells), promoting their differentiation into myoblasts and their fusion with preexisting myofibers to form new muscle tissue.[57,58] In this context, testosterone and synthetic androgens are sometimes referred to as "anabolic steroids", or hormones that promote the building of muscle. Such steroids have clinical and therapeutic applications but can also be abused to the detriment of a person's health.

Blood Cholesterol and Cardiovascular Changes

In males, elevated testosterone beginning at puberty promotes increases in circulating cholesterol-laden low-density lipoprotein (LDL) and very-low-density lipoprotein (VLDL), with a concurrent decrease in high-density lipoprotein (HDL). From a clinical perspective, this altered LDL/HDL ratio makes males more susceptible to developing high blood cholesterol and associated cardiovascular complications compared with females, whose higher estrogen levels exert cardioprotective effects.[18]

Erythropoiesis

Testosterone enhances blood oxygen-carrying capacity by increasing hematocrit (percent red blood cells by volume) through stimulation of *erythropoiesis* (red blood cell synthesis). This appears to be accomplished by either direct or indirect induction of the hormone **erythropoietin** by the kidney, and also by promotion of erythroid differentiation by the bone marrow.[59,60]

Libido

Testosterone is critical for maintaining male libido, or sex drive, and its decline with aging is accompanied by a concomitant reduction of libido.[61] Interestingly, in males estrogen is also thought to promote libido, as the administration of exogenous estradiol to men with diminished testosterone has been shown to increase libido.[62] In one male patient with CYP19 aromatase deficiency and hypogonadism, concurrent treatment with exogenous estrogen and testosterone were both necessary to increase libido, suggesting a role for estrogen in this process.[63] Furthermore, studies with rats demonstrated that castrated animals given exogenous estrogen showed increased sexual activity in a dose- and time-dependent manner.[64] A strong influence of estrogen on the adult male brain is not surprising, considering that classical estrogen receptors and CYP19 aromatase enzyme are known to be widely distributed in the brains of male humans and other mammals.[65]

Reproductive Aging

Reproductive aging in males is associated with a progressive loss of sperm quality, sperm quantity, and libido. Furthermore, between the ages of 30 and 75 years, the mean percent muscle mass may decrease by 20%–50%, bone mineral density declines by 20%, and the percentage of body fat increases up to 100%.[66] Each of these changes is attributable to a progressive decline in testosterone levels (and related increases in LH and FSH), with total testosterone decreasing by about 30% between the ages of 30 and 75 and free testosterone levels declining by about 50% (Figure 21.16). Interestingly, the more dramatic reduction in free testosterone levels is due to a concurrent increase in SHBG, which begins to rise by about 1% per year after early adulthood.[67–69] Furthermore, since SHBG binds testosterone with a higher affinity compared with estradiol,[70] the reduction in the free testosterone-estradiol ratio increases the possibility of signs of feminization such as gynecomastia. The reasons for increased SHBG with male reproductive aging remain uncertain but may be related to age-related decline in GH and IGF-1, hormones known to suppress SHBG.[71] Notably, DHEA appears to exert antidepressant effects,[72] and its decline with aging may help explain age-related increases in depression.[73,74] In comparison to female menopause, or the abrupt cessation of gonadal function during middle age, a sudden arrest of gonadal function in middle-aged men does not occur, with fertility typically persisting until very old age.[75] Male reproductive aging appears distinct from menopause in that the decline of endo- and exocrine testicular function is gradual (Figure 21.17).

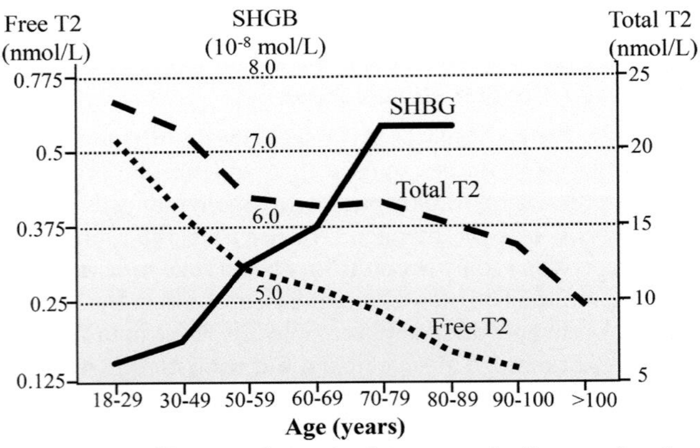

Figure 21.16 Changes in testosterone and other endocrine parameters with aging in males. Although total testosterone levels begin to decrease during middle age, free (bioavailable) testosterone declines more dramatically due to increasing levels of sex hormone-binding globulin (SHBG), which has a high affinity to testosterone.

Source of Images: Data extrapolated from Vermeulen, A., 1993. Ann. Med., 25(6), pp. 531-534.

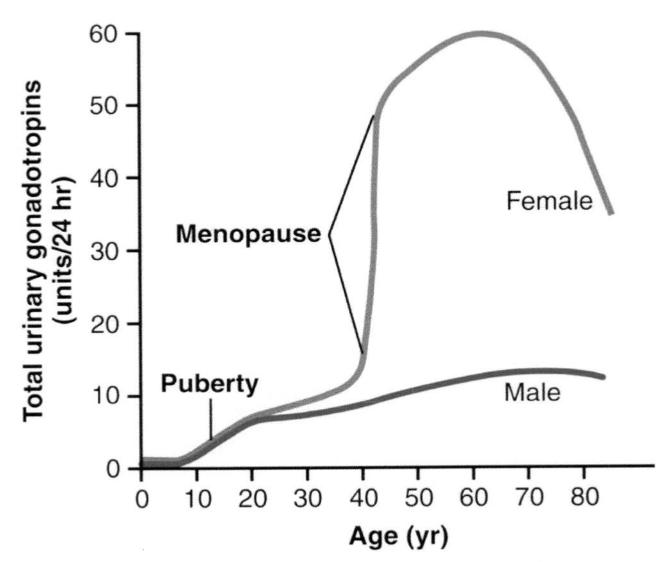

Figure 21.17 In contrast to women, who experience an abrupt increase in gonadotropin secretion at menopause, the changes in middle-aged males occur more gradually.

Source of Images: John, E., 2011. Guyton and Hall Textbook of Medical Physiology 12th Edition. Elsevier. Used by permission.

Contraception

The condom generally provides safe, cost-effective, user-controlled contraception with few side effects. Currently, the only alternative to the condom for male contraception is *vasectomy*, or the surgical interruption of the vas deferens, which is a relatively simple, safe, and effective procedure. In addition to not affecting the normal production of androgens and gonadotropins, the procedure can be successfully reversed 80%–90% of the time, although subsequent pregnancy rates are only 30%–40%.[76] In comparison to hormonal contraceptives available for women, there are as of yet no pharmacological options available for men, despite the proof of principle established for androgen-based hormonal suppression of spermatogenesis.[77] Public sector clinical research since the 1970s has generally revolved around suppressing gonadotropin synthesis (and, hence, spermatogenesis) via weekly injections of testosterone either alone or in combination with non-androgenic steroids (estrogens or progestins). One drawback to this approach is that because it takes over nine weeks to produce sperm, there is a significant delay in time before any effects of the contraceptive hormone becomes apparent. Additionally, when contraception is no longer required, there is a similar delay before sperm production resumes.

The World Health Organization commissioned a large-scale study designed to test the contraceptive efficacy and safety in men of a regimen of intramuscular injections of a long-acting progestogen administered in combination with replacement doses of a long-acting androgen.[78] Although the results of the study demonstrated a nearly complete and reversible suppression of spermatogenesis, reports of mild to moderate mood disorders and acne were relatively high, and the study was stopped after an independent review panel found that the drug had too many safety concerns and side effects. Such safety concerns may seem somewhat ironic, considering that many women on hormonal contraception exhibit similar side effects and that males never need to balance the significant health risks of pregnancy against the health risks of the contraceptive. The search for a safe and effective hormone-based male contraceptive continues but highlights the basic challenge that it is, in principle, more difficult to inhibit the production of millions of sperm per day in males than it is to prevent the generation of a single ovum per month in females.

Endocrine Pathophysiology of Male Reproductive Function

Endocrine-based pathology associated with dysfunction of the male reproductive system may develop during embryogenesis as a result of abnormal sexual differentiation or during childhood as delayed puberty. Endocrine-related dysfunctions may also develop any time during childhood through adulthood as an acquired pathology. Some of these pathologies are listed in Table 21.4.

Table 21.4 Some Acquired Endocrine Pathologies of the Male Reproductive System

	Effector Cell or Organ	Effect
Hypogonadism		
Primary	Adult Leydig cell failure or deficiency	Low testosterone
	Adult Sertoli cell dysfunction	Low sperm count
	Antibodies against sperm	Low sperm count
Secondary	Pituitary: hypogonadotropic hypogonadism	Low gonadotropins and testosterone
Tertiary	Hypothalamic hypogonadism (inhibited GnRH synthesis or pulsatility)	Low gonadotropins and testosterone
Hypergonadism		
Primary	Androgen-secreting Leydig cell tumor	High testosterone
Secondary	Pituitary: hypergonadotropic hypergonadism	High gonadotropins and testosterone
Tertiary	Hypothalamic hypergonadism (excess GnRH synthesis or pulsatility)	High gonadotropins and testosterone
Other endocrine disturbances		
Excess estrogen activity	Estrogen-secreting testicular or adrenal tumor; increased estrogen: testosterone ratio; increased SHBG; environmental estrogens	Prostate hyperplasia caused by estrogen-induced local upregulation of androgen receptor; gynecomastia
Hypothyroidism	Pituitary: hypogonadotropic hypogonadism	Low gonadotropins and testosterone
Hyperthyroidism	Pituitary: hypergonadotropic hypergonadism	Increased testosterone, estrogen, SHBG; gynecomastia
Hyperprolactemia	Inhibited GnRH secretion	Low gonadotropins and testosterone
Drugs Opioids marijuana amphetamines cocaine	Generally, exert adverse effects on the hypothalamic-pituitary-testicular axis, inhibit sperm function, and can cause infertility	

Clinical Considerations: The vicious cycle of obesity-induced hypogonadism

The growing global obesity and infertility epidemics may be interrelated, with obesity proposed as one factor that can promote infertility in males.[83] One cause of male infertility, hypogonadism (the production of low levels of testosterone), may be induced through the over-production of specific hormones and adipokines by adipocytes in the obese state.

Central hypogonadism caused by adipocyte-derived estrogen

Recall that adipose tissue is an endocrine tissue that expresses high levels of CYP19 aromatase in the obese state. Furthermore, recall from earlier that in human males, negative feedback by sex steroid hormones occurs at the levels of both the hypothalamus and the pituitary primarily through activation of the estrogen receptor by estrogens derived from the local aromatization of testosterone by expression of CYP19 aromatase. Add to this observations that male body mass index (BMI) correlates negatively with blood testosterone but positively with estradiol[84] and a plausible physiological mechanism for obesity-induced hypogonadotropic hypogonadism begins to form. Specifically, obese males are susceptible to increased feedback inhibition of the hypothalamic-pituitary axis by elevated estrogen, resulting in lowered testosterone production. Because the decrease in testosterone occurs without a compensatory increase in gonadotropin (i.e. hypogonadotropic hypogonadism), this may lead to a so-called *hypogonadal–obesity cycle*[85] whereby an increased mass of adipose tissue functions as a major peripheral source of estrogens, which, in turn, simultaneously increases adiposity and inhibits the hypothalamic-pituitary axis (Figure 21.18[A]). Importantly, estrogen is a positive regulator of adipogenesis, stimulating the growth and proliferation of adipocytes.[86] The subsequent reduction in circulating testosterone leads to a preferential deposition of more abdominal adipose tissue,[87] further increasing CYP19 aromatase activity and circulating estrogen and simultaneously reducing circulating testosterone levels. The progressive hypogonadal state leads to further deposition of abdominal fat and promotes infertility.

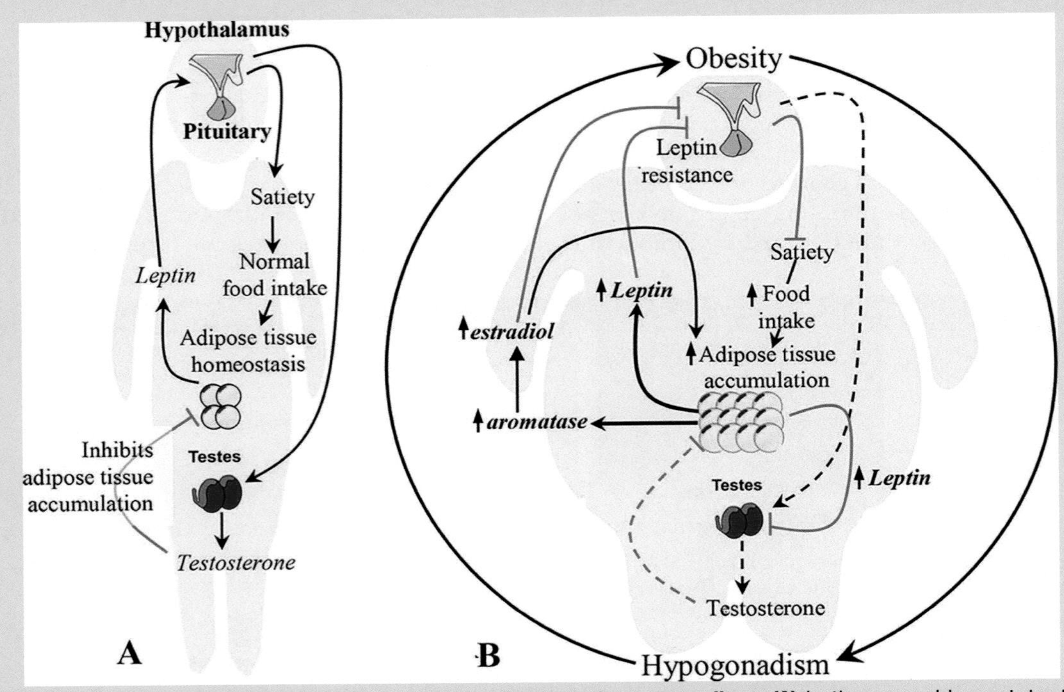

Figure 21.18 The vicious cycle of obesity and male hypothalamic hypogonadism. (A) In the normal lean state, testosterone prevents the gain of excess adipose tissue mass. Leptin synthesis by adipose tissue targets the hypothalamus and helps promote satiety, establishing adipose tissue mass homeostasis. Leptin also indirectly regulates pituitary gonadotropin secretion at the level of the hypothalamus by crossing the blood-brain barrier to modulate the activities of kisspeptinergic and GnRH neurons in the arcuate nucleus, stimulating testosterone production by the testes. **(B)** The hypogonadal–obesity cycle is induced by peripheral estrogen synthesis. Expression of the enzyme aromatase by adipocytes converts testosterone and other androgens to estrogens, like estradiol (E2). Peripheral E2 becomes elevated as a result of increased adipose tissue mass, further suppressing the hypothalamus-pituitary-testis axis. In addition, E2 promotes additional adipocyte differentiation, exacerbating a vicious cycle. As testosterone inhibits adipogenesis, the hypogonadal state even further promotes the abdominal and visceral deposition of adipose tissue distribution. Excess leptin levels inhibit testosterone production both centrally and peripherally. Elevated leptin in the obese state promotes brain leptin resistance, whereby leptin is unable to cross the BBB to stimulate the top of the HPG axis, resulting in hypogonadism. Elevated levels of leptin also appear to exert direct effects on the testes by negatively regulating testosterone synthesis by the Leydig cells.

Source of Images: Carrageta, D.F., Oliveira, P.F., Alves, M.G. and Monteiro, M.P., 2019. Obesity Rev., 20(8), pp. 1148-1158. Used by permission.

Central hypogonadism caused by adipocyte-derived leptin

Leptin is an adipocytokine that plays a key role in the neuroendocrine regulation of energy homeostasis and body weight. Importantly, mice lacking the leptin gene (ob/ob mice) are both morbidly obese and infertile,[88] and similar phenotypes are reported for mice deficient in the leptin receptor.[89] Clinical features of human patients with congenital leptin deficiency include hyperphagia (overeating), the early onset of obesity, hypogonadotropic hypogonadism, and delayed onset of puberty.[90,91] Importantly, leptin indirectly regulates pituitary gonadotropin secretion at the level of the hypothalamus by modulating the activities of kisspeptinergic and GnRH neurons in the arcuate nucleus[92] (Figure 21.18[B]). Interestingly, in contrast to the hypogonadism observed in humans with congenital leptin deficiency or in animals with deleted leptin genes or receptors, obesity-induced hypogonadism is associated with *elevated* serum leptin. It has been proposed that under conditions of elevated serum leptin, the blood-brain barrier (BBB) transport system for leptin becomes saturated, resulting in *central* (brain) *leptin resistance* and *central leptin insufficiency*.[93] Therefore, obesity-induced central leptin resistance mirrors the leptin deficiency phenotype observed in humans and animal models with congenital genetic gene deletions. Ultimately, reduced hypothalamic leptin stimulation may produce both morbid obesity (due to the absence of leptin-induced suppression of appetite) and hypogonadotropic hypogonadism (low testosterone levels due to low gonadotropin synthesis).

Peripheral hypogonadism caused by adipocyte-derived leptin

In addition to the effects of central leptin resistance, elevated serum leptin appears to impart a "double-whammy" on testosterone production by also exerting direct effects on the testes (Figure 21.18[B]). For example, in rats and humans Leydig cells express leptin receptors, and *in vitro* studies with rats have shown that leptin inhibits chorionic gonadotropin-stimulated testosterone synthesis in testicular tissue.[94] Since, unlike the BBB, leptin transport across the blood-testis barrier does not appear to be inhibited by saturation,[95] elevated leptin under obese conditions may inhibit testosterone production, as well as the development and differentiation of spermatocytes.[96,97]

Infertility

The World Health Organization has defined *infertility* as "a disease of the reproductive system defined by the failure to achieve a clinical pregnancy after 12 months or more of regular unprotected sexual intercourse".[79] Approximately 15% of couples are infertile, with male-specific infertility accounting for 30% of reported cases, female-specific infertility responsible for 45%, and couple factors for 25%.[80] The most common cause of male infertility is *idiopathic infertility*, meaning that the cause is uncertain, with the occurrence of abnormal spermatogenesis. Other common causes include Klinefelter syndrome and other types of genetic disorders, testicular cancer, testicular damage caused by chemotherapy or irradiation for treating cancer, trauma, infection, and drug abuse. Although endocrine causes have been estimated to account for only 2%–4% of male infertility cases,[80,81] identification and understanding of its causes is important since hormonal therapy is frequently successful in these cases.[82]

Erectile Dysfunction

The causes of *erectile dysfunction* (impotence) are numerous and often overlapping, and include neurovascular dysfunction, psychosocial, metabolic (e.g. obesity and diabetes), and medication-related causes. Ultimately, however, erectile dysfunction is a condition characterized by reduced blood flow to the penis, preventing erection. The drug sildenafil (also known as "Viagra") treats this condition by reducing the degradation of the gas nitric oxide (NO), which functions as a second messenger in the M3 muscarinic GPCR signaling pathway that promotes vascular smooth muscle dilation, restoring both blood flow and erectile function to the penis (reviewed in Chapter 5: Receptors, Figure 5.11 of that chapter).

SUMMARY AND SYNTHESIS QUESTIONS

1. Results of the study by Levine et al.,[41] "Temporal trends in sperm count: a systematic review and meta-regression analysis", suggest that whereas sperm counts in males from Europe, North America, Australia, and New Zealand are declining, sperm counts from Africa, Asia, and South America are not. What may explain the discrepancy between the two populations?
2. Why is administration of both testosterone plus either estradiol or progesterone more effective as a contraceptive than testosterone alone?
3. Considering that the administration of estradiol or progesterone alone is sufficient for male contraception, why should testosterone be co-administered at all?
4. Male reproductive aging is accompanied by increases in blood sex hormone-binding globulin (SHBG). How is this problematic?
5. What accounts for the significant rises in LH and FSH that accompany male aging?
6. What are some negative effects of rising DHT and prolactin levels in males undergoing reproductive aging?
7. What effect, if any, would congenital CYP19 aromatase deficiency have on a man's adult height? Why? What might be an effective treatment for this condition?
8. Obesity in males can produce a "triple-whammy" effect that promotes hypogonadism. That is, obesity appears to induce hypogonadism via three separate mechanisms. Describe these, and how they feed off of each other to produce hypogonadism.

Endocrine-Disrupting Compounds and Male Infertility

LEARNING OBJECTIVE Describe some notable endocrine-disrupting chemicals that may contribute to male infertility and other reproductive health disorders, and discuss their mechanisms of action.

KEY CONCEPTS:

- Testicular dysgenesis syndrome (TDS) comprises a constellation of interrelated male reproductive disorders resulting from a disruption of normal androgen synthesis or action during the fetal development of the testes.
- Because male mammalian sex differentiation relies heavily on sex steroid hormones for normal masculinization, male fetuses may be more susceptible than females to the disruptive effects of certain EDCs on sexual differentiation.
- Higher urinary estrogenic and antiandrogenic EDC concentrations correlate with lower sperm counts, reduced testosterone levels, and rising rates of testicular cancer and reproductive disorders.

Science fiction can sometimes have an uncanny way of foreshadowing contemporary circumstances, not all of which are particularly desirable. Take, for instance, the dystopian worlds portrayed in Margret Atwood's 1985 novel, *The Handmaid's Tale*, and in P.D. James' 1992 novel, *The Children of Men*. Both novels share a common theme: following decades of skyrocketing global population growth, humanity is suddenly on the brink of collapse due to abruptly crashing fertility rates. In Atwood's world, humanity is rendered largely sterile due to rampant chemical, biological, and nuclear pollution, and James' story portrays a world where male sperm counts have mysteriously plummeted to zero, with humanity facing imminent extinction. Although humankind has not yet reached such apocalyptic levels of reproductive dysfunction, these novels of fiction bring attention to some very real and sobering facts. For instance, recall from earlier that sperm density, viable sperm morphology, semen quality, and testosterone levels among men worldwide has been declining markedly (particularly in Western nations) (refer to the Developments & Directions box: Spermageddon? Sperm counts are plummeting in Western men). Importantly, the rapidity in the declines in male fertility and the steep rise in male reproductive disorders suggests environmental and/or lifestyle factors (including obesity and poor diet), rather than genetic, causes to these maladies. It is now widely hypothesized that exposures to EDCs make significant contributions to the etiology of infertility and reproductive health dysfunction.[98,99]

Testicular Dysgenesis Syndrome

Epidemiological data show that increasing rates of male reproductive abnormalities, including infertility and subfertility, testicular cancer, hypospadias, cryptorchidism, delayed onset of puberty, and reduced testosterone levels, often occur concomitantly.[45,100] These dysfunctions each have their origins in fetal development, leading a group of Danish researchers to propose that the disorders are interrelated, resulting from a disruption of normal androgen synthesis or action during the development of the testes, and comprise a *testicular dysgenesis syndrome* (TDS) (Figure 21.19).[101–103] Recall that the prevailing view of mammalian sex differentiation is that whereas female fetuses develop largely independently of sex steroid hormones, male sex differentiation relies heavily on sex steroid hormones for normal masculinization. Therefore, male fetuses may be more susceptible than females to the disruptive effects of certain EDCs on sexual differentiation. Androgens, such as testosterone, generated by the fetal testes are required for the normal differentiation and development of the epididymis, vas deferens, and the seminal vesicles from the Wolffian duct precursor tissue. Furthermore, the development of the prostate gland and external genitalia requires the presence of the more potent testicular androgen 5α-dihydrotestosterone (DHT), which is catalyzed both locally and in peripheral tissues from testosterone by the enzyme 5α-reductase. Some pharmacological compounds are designed to disrupt 5α-reductase, such as finasteride, a drug used to treat an enlarged prostate or hair loss in men and also used as a component of

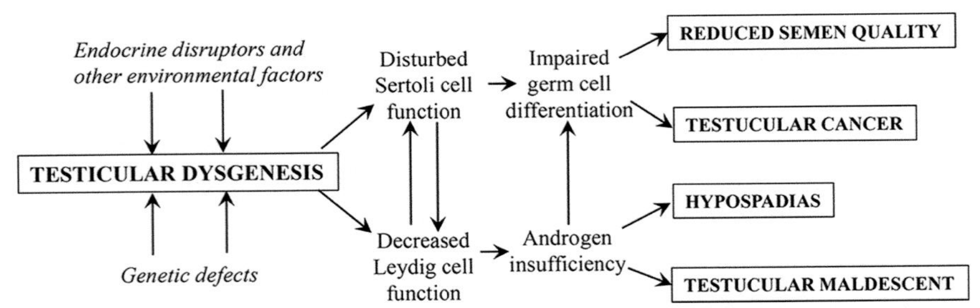

Figure 21.19 Schematic representation of pathogenetic links between testicular dysgenesis and the clinical manifestations of testicular dysgenesis syndrome.

Source of Images: Skakkebæk, N.E., et al. 2001. Hum. Reprod., 16(5), pp. 972-978.

hormone therapy for transgender women. Others interfere with the binding of DHT with the androgen receptor, such as flutamide, a medication used to treat prostate cancer. However, exposures of fetal rats and monkeys to these compounds can inhibit proper development of the

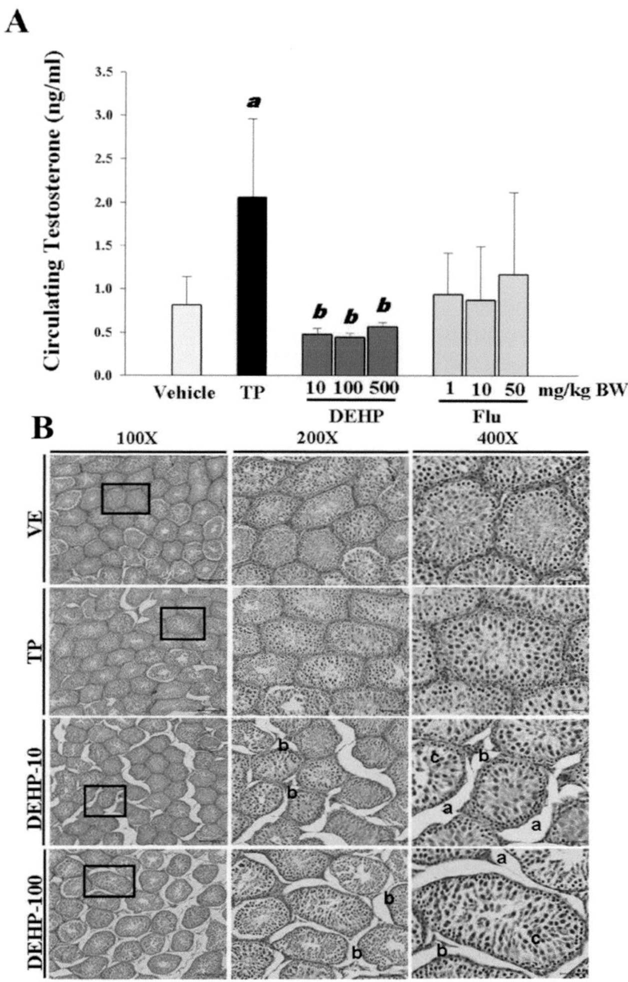

Figure 21.20 The effects of a phthalate on testosterone synthesis and testis morphology in immature rats. Immature male rats were treated daily with testosterone propionate (TP; a known androgen receptor agonist), flutamide (Flu; a known androgen receptor antagonist), or di-(2 ethylhexyl) phthalate (DEHP; an EDC with suspected antiandrogen effects) on postnatal days 21 to 35 in a dose-dependent manner. **(A)** Compared with vehicle or TP, a significant decrease in the levels of testosterone was observed when rats were exposed to all doses of DEHP. **(B)** Corresponding effects of DEHP (10, 100 mg/kg BW/day) on histopathological changes in immature male rat testes. Compared with vehicle and TP, dilatation of the tubular lumen (a: stained signals), degeneration of Leydig cells (b: stained signals), and disorder of germ cells (c: stained signals) were observed in the testes of immature male rats. Tissues were stained with hematoxylin and eosin and histopathological changes were assessed under a light microscope. Results are shown at 100×, 200× and 400× magnifications.

Source of Images: Vo, T.T., et al. 2009. Reprod. Biol. Endocrinol., 7(1), p. 104.

prostate and external genitalia, resulting in abnormalities like cryptorchidism and hypospadias.[104,105]

Phthalates and TDS

Considering that antiandrogen medications can disrupt male reproductive development in the fetuses of animal models, it should not be surprising to learn that EDCs with antiandrogenic and/or estrogenic actions, such as DDT and its breakdown product, DDE, as well as phthalates, BPA, dioxins, and PCBs, have also been implicated with similar effects on sexual development.[106] Of these EDCs, one category that is particularly ubiquitous and well studied and has been linked to TDS are the *phthalates*, a category of compounds broadly used as plastic softeners and in cosmetics and personal-care products, food packaging, pharmaceuticals, and medical tubing. Indeed, with the exception of testicular cancer (for which there is yet no effective rodent model), all aforementioned male developmental aberrations can be induced by administering phthalates to pregnant rats[107] (Figure 21.20). In contrast to the aforementioned antiandrogen medications, finasteride and flutamide, phthalates do not produce their effects by inhibiting 5α-reductase or by antagonizing the androgen receptor, but instead by interfering with the synthesis of androgens by the fetal testis by promoting PPAR-RXR heterodimerization.[108]

Strikingly, human infants housed in neonatal intensive care units possess among the highest levels of urinary phthalates that have been documented, presumably emanating from the flexible plastic tubing in IVs and other medical devices.[109,110] Among infant boys, exposure to phthalates during pregnancy has been associated with the development of a feminized (smaller) anogenital distance.[111] Furthermore, in human adults, epidemiological findings have correlated higher urinary phthalate concentrations with both lower sperm counts and a higher likelihood of sperm with damaged DNA.[112–115]

SUMMARY AND SYNTHESIS QUESTIONS

1. It has been hypothesized that males are more susceptible to the disruptive effects of EDCs on sexual differentiation compared with females. What is the basis for this hypothesis?
2. From the perspective of EDCs, how does it make sense that whereas girls are beginning puberty at earlier ages, boys are doing so at later ages?

Summary of Chapter Learning Objectives and Key Concepts

LEARNING OBJECTIVE Describe the functional gross morphology of the male reproductive system, and discuss how testicular Leydig and Sertoli cells interact with each other to support spermatogenesis and the development of male secondary sexual characteristics.

- The testis is a dual function organ that generates the male germ cells, as well as synthesizes androgens.
- Each testis consists of a collection of lobules comprised of convoluted seminiferous tubules, wherein sperm cells develop.
- The lobules are divided into two general "compartments", each separated by a basal lamella.
- The interstitial compartment is comprised in part of testosterone-secreting Leydig cells.
- The intratubular compartment is composed of developing germ cells as well as Sertoli cells that support sperm development and other testicular functions.
- Sertoli cells secrete androgen-binding protein (ABP) into the seminiferous tubule lumen, where it sequesters testosterone at the elevated levels necessary to promote spermatogenesis.

LEARNING OBJECTIVE Describe the key genomic and cellular changes that take place during spermatogenesis and spermiogenesis.

- In complex multicellular organisms, only the germ cells retain the capacity to transmit the materials and instructions required to initiate the assembly of a new organism.
- Diploid germ cells undergo two reductive meiotic divisions to give rise to haploid gametes, which in males are the spermatids, via a process called spermatogenesis.
- Spermatids transform morphologically into mature spermatozoa by a process called spermiogenesis.

LEARNING OBJECTIVE Outline the feedback mechanisms along the hypothalamic-pituitary-testis axis, and describe the actions of luteinizing hormone and follicle-stimulating hormone on testis function.

- Testosterone produced by Leydig cells and inhibin produced by Sertoli cells exert negative feedback along the hypothalamic-pituitary axis.
- The pituitary gonadotropin luteinizing hormone (LH) targets testicular Leydig cells, inducing androgen steroidogenesis.
- The pituitary gonadotropin follicle-stimulating hormone (FSH) targets testicular Sertoli cells, which respond by synthesizing enzymes and other proteins that support spermatogenesis and testicular function.
- Sertoli cells and peripheral tissues can convert testosterone into either the more potent androgen dihydrotestosterone (DHT) or the estrogen estradiol via the actions of 5α-reductase and CYP19 aromatase, respectively.

LEARNING OBJECTIVE Describe how testosterone and its metabolites influence fetal and adult target tissues, and list endocrine changes associated with common male reproductive pathologies AND aging.

- Within the blood, most testosterone is bound to sex hormone-binding globulin (SHBG) and albumin.
- The physiological effects of androgens and estrogens are primarily mediated by nuclear receptors that function as ligand-activated transcription factors.
- Testosterone and dihydrotestosterone (DHT) bind to the same androgen receptor, but DHT binds with 100-fold greater affinity.
- The two androgens affect tissues differentially, with testosterone promoting masculinization of fetal internal genitalia and spermatogenesis and DHT promoting the masculinization of fetal external genitalia and secondary sexual characteristics during puberty.
- The aromatization of testosterone into estradiol by target tissues is necessary for the regulation of pituitary gonadotropin secretion and the pubertal bone growth spurt.

LEARNING OBJECTIVE Describe some notable endocrine-disrupting chemicals that may contribute to male infertility and other reproductive health disorders, and discuss their mechanisms of action.

- Testicular dysgenesis syndrome (TDS) comprises a constellation of interrelated male reproductive disorders resulting from a disruption of normal androgen synthesis or action during the fetal development of the testes.
- Because male mammalian sex differentiation relies heavily on sex steroid hormones for normal masculinization, male fetuses may be more susceptible than females to the disruptive effects of certain EDCs on sexual differentiation.
- Higher urinary estrogenic and antiandrogenic EDC concentrations correlate with lower sperm counts, reduced testosterone levels, and rising rates of testicular cancer and reproductive disorders.

LITERATURE CITED

1. Pinon R. *Biology of Human Reproduction*. University Science Books; 2002.
2. Norris DO, Carr JA. *Vertebrate Endocrinology* (5th ed.). Academic Press; 2013.
3. Cheney VT. *A Brief History of Castration*. AuthorHouse; 2006.
4. Scarborough J. *Medical and Biological Terminologies: Classical Origins*. University of Oklahoma Press; 1992.
5. Berthold AA. Transplantation of testes. *Arch Anat Physiol Wissenschaft Med* 1849:42–46.
6. Moore CR, Gallagher TF, Koch FC. The effects of extracts of testis in correcting the castrated condition in the fowl and in the mammal. *Endocrinology*. 1929;13:367–374.
7. McCullagh DR. Dual endocrine activity of the testes. *Science*. 1932;76(1957):19–20.
8. Jocelyn HD, Setchell BP. Regnier de Graaf on the human reproductive organs: an annotated translation of Tractatus de virorum organis generationi inservientibus (1668) and De mulierum organis generationi inservientibus tractatus novus (1672). *J Reprod Fertil Suppl*. 1972;17:1–222.
9. Meng X, Lindahl M, Hyvonen ME, et al. Regulation of cell fate decision of undifferentiated spermatogonia by GDNF. *Science*. 2000;287(5457):1489–1493.
10. Hofmann MC, Braydich-Stolle L, Dym M. Isolation of male germ-line stem cells; influence of GDNF. *Dev Biol*. 2005;279(1):114–124.
11. Amann RP, Howards SS. Daily spermatozoal production and epididymal

spermatozoal reserves of the human male. *J Urol.* 1980;124(2):211–215.

12. Sharma V, Lehmann T, Stuckas H, Funke L, Hiller M. Loss of RXFP2 and INSL3 genes in Afrotheria shows that testicular descent is the ancestral condition in placental mammals. *PLoS Biol.* 2018;16(6):e2005293.
13. Hayes FJ, Seminara SB, Decruz S, Boepple PA, Crowley WF, Jr. Aromatase inhibition in the human male reveals a hypothalamic site of estrogen feedback. *J Clin Endocrinol Metab.* 2000;85(9):3027–3035.
14. Rochira V, Zirilli L, Genazzani AD, et al. Hypothalamic-pituitary-gonadal axis in two men with aromatase deficiency: evidence that circulating estrogens are required at the hypothalamic level for the integrity of gonadotropin negative feedback. *Eur J Endocrinol.* 2006;155(4):513–522.
15. Bagatell CJ, Dahl KD, Bremner WJ. The direct pituitary effect of testosterone to inhibit gonadotropin secretion in men is partially mediated by aromatization to estradiol. *J Androl.* 1994;15(1):15–21.
16. Pitteloud N, Dwyer AA, DeCruz S, et al. The relative role of gonadal sex steroids and gonadotropin-releasing hormone pulse frequency in the regulation of follicle-stimulating hormone secretion in men. *J Clin Endocrinol Metab.* 2008;93(7):2686–2692.
17. Pitteloud N, Dwyer AA, DeCruz S, et al. Inhibition of luteinizing hormone secretion by testosterone in men requires aromatization for its pituitary but not its hypothalamic effects: evidence from the tandem study of normal and gonadotropin-releasing hormone-deficient men. *J Clin Endocrinol Metab.* 2008;93(3):784–791.
18. Melmed S, Polonsky KS, Larsen PR, eds. *Williams Textbook of Endocrinology* (12th ed.). Saunders; 2011.
19. Boepple PA, Hayes FJ, Dwyer AA, et al. Relative roles of inhibin B and sex steroids in the negative feedback regulation of follicle-stimulating hormone in men across the full spectrum of seminiferous epithelium function. *J Clin Endocrinol Metab.* 2008;93(5):1809–1814.
20. Gregory SJ, Kaiser UB. Regulation of gonadotropins by inhibin and activin. *Sem Reprod Med.* 2004;22(3):253–267.
21. Bole-Feysot C, Goffin V, Edery M, Binart N, Kelly PA. Prolactin (PRL) and its receptor: actions, signal transduction pathways and phenotypes observed in PRL receptor knockout mice. *Endocr Rev.* 1998;19(3):225–268.
22. Grayhack JT, Bunce PL, Kearns JW, Scott WW. Influence of the pituitary on prostatic response to androgen in the rat. *Bull Johns Hopkins Hosp.* 1955;96(4):154–163.
23. Jacobson EM, Hugo ER, Borcherding DC, Ben-Jonathan N. Prolactin in breast and prostate cancer: molecular and genetic perspectives. *Discov Med.* 2011;11(59):315–324.
24. Purvis K, Clausen OP, Olsen A, Haug E, Hansson V. Prolactin and Leydig cell responsiveness to LH/hCG in the rat. *Arch Androl.* 1979;3(3):219–230.
25. Dombrowicz D, Sente B, Closset J, Hennen G. Dose-dependent effects of human prolactin on the immature hypophysectomized rat testis. *Endocrinology.* 1992;130(2):695–700.
26. Rubin RT, Poland RE, Tower BB. Prolactin-related testosterone secretion in normal adult men. *J Clin Endocrinol Metab.* 1976;42(1):112–116.
27. Gunasekar PG, Kumaran B, Govindarajulu P. Prolactin and Leydig cell steroidogenic enzymes in the bonnet monkey (Macaca radiata). *Int J Androl.* 1988;11(1):53–59.
28. Gill-Sharma MK. Prolactin and male fertility: the long and short feedback regulation. *Int J Endocrinol.* 2009;2009:687259.
29. Johnston DS, Russell LD, Friel PJ, Griswold MD. Murine germ cells do not require functional androgen receptors to complete spermatogenesis following spermatogonial stem cell transplantation. *Endocrinology.* 2001;142(6):2405–2408.
30. De Gendt K, Swinnen JV, Saunders PT, et al. A Sertoli cell-selective knockout of the androgen receptor causes spermatogenic arrest in meiosis. *Proc Natl Acad Sci U S A.* 2004;101(5):1327–1332.
31. Walker WH, Cheng J. FSH and testosterone signaling in Sertoli cells. *Reproduction.* 2005;130(1):15–28.
32. Crews D, Sanderson N, Dias B. Hormones, brain, and behavior in reptiles. 2009.
33. Crews D. Control of male sexual behaviour in the Canadian red-sided garter snake. In: *Hormones and Behaviour in Higher Vertebrates.* Springer; 1983:398–406.
34. Crews D, Camazine B, Diamond M, Mason R, Tokarz RR, Garstka WR. Hormonal independence of courtship behavior in the male garter snake. *Horm Behav.* 1984;18(1):29–41.
35. Crews D. Trans-seasonal action of androgen in the control of spring courtship behavior in male red-sided garter snakes. *Proc Nat Acad Sci.* 1991;88(9):3545–3548.
36. Krohmer RW, Boyle MH, Lutterschmidt DI, Mason RT. Seasonal aromatase activity in the brain of the male red-sided garter snake. *Horm Behav.* 2010;58(3):485–492.
37. Krohmer RW. Courtship in the male red-sided garter snake is dependent on neural aromatase activity during winter dormancy. *J Exp Zool Part A Ecol Integr Physiol.* 2020;333(5):275–283.
38. Gao W, Bohl CE, Dalton JT. Chemistry and structural biology of androgen receptor. *Chem Rev.* 2005;105(9):3352–3370.
39. Chimento A, De Luca A, Nocito MC, et al. Role of GPER-mediated signaling in testicular functions and tumorigenesis. *Cells.* 2020;9(9):2115.
40. Carlsen E, Giwercman A, Keiding N, Skakkebaek NE. Evidence for decreasing quality of semen during past 50 years. *BMJ.* 1992;305(6854):609–613.
41. Levine H, Jorgensen N, Martino-Andrade A, et al. Temporal trends in sperm count: a systematic review and meta-regression analysis. *Hum Reprod Update.* 2017;23(6):646–659.
42. Bonde JPE, Ernst E, Jensen TK, et al. Relation between semen quality and fertility: a population-based study of 430 first-pregnancy planners. *Lancet.* 1998;352(9135):1172–1177.
43. Skakkebæk NE, Lindahl-Jacobsen R, Levine H, et al. Environmental factors in declining human fertility. *Nat Rev Endocrinol.* 2021:1–19.
44. Huang C, Li B, Xu K, et al. Decline in semen quality among 30,636 young Chinese men from 2001 to 2015. *Fert Steril.* 2017;107(1):83–88. e82.
45. Skakkebaek NE, Rajpert-De Meyts E, Buck Louis GM, et al. Male reproductive disorders and fertility trends: influences of environment and genetic susceptibility. *Physiol Rev.* 2016;96(1):55–97.
46. Ganjam VK, Amann RP. Steroids in fluids and sperm entering and leaving the bovine epididymis, epididymal tissue, and accessory sex gland secretions. *Endocrinology.* 1976;99(6):1618–1630.
47. Free MJ, Jaffe RA. Collection of rete testis fluid from rats without previous efferent duct ligation. *Biol Reprod.* 1979;20(2):269–278.
48. Smith MS, Freeman ME, Neill JD. The control of progesterone secretion during the estrous cycle and early pseudopregnancy in the rat: prolactin, gonadotropin and steroid levels associated with rescue of the corpus luteum of pseudopregnancy. *Endocrinology.* 1975;96(1):219–226.
49. Weiss J, Bernhardt ML, Laronda MM, et al. Estrogen actions in the male reproductive system involve estrogen response element-independent pathways. *Endocrinology.* 2008;149(12):6198–6206.
50. Eddy EM, Washburn TF, Bunch DO, et al. Targeted disruption of the estrogen receptor gene in male mice causes alteration of spermatogenesis and infertility. *Endocrinology.* 1996;137(11):4796–4805.
51. Korach KS. Insights from the study of animals lacking functional estrogen receptor. *Science.* 1994;266(5190):1524–1527.
52. Kaufman KD. Androgens and alopecia. *Mol Cell Endocrinol.* 2002;198 (1–2):89–95.
53. Ustuner ET. Cause of androgenic alopecia: crux of the matter. *Plast Reconstr Surg Glob Open.* 2013;1(7):e64.
54. Rubin J, Sataloff R, Korovin G. The larynx: a hormonal target. In: Sataloff R, ed. *Diagnosis and Treatment of Voice Disorders.* Plural Publishing; 2005:392–417.

55. Oury F, Sumara G, Sumara O, et al. Endocrine regulation of male fertility by the skeleton. *Cell*. 2011;144(5): 796–809.
56. Singh R, Artaza JN, Taylor WE, Gonzalez-Cadavid NF, Bhasin S. Androgens stimulate myogenic differentiation and inhibit adipogenesis in C3H 10T1/2 pluripotent cells through an androgen receptor-mediated pathway. *Endocrinology*. 2003;144(11):5081–5088.
57. O'Connell MD, Wu FC. Androgen effects on skeletal muscle: implications for the development and management of frailty. *Asian J Androl*. 2014;16(2):203–212.
58. Dubois V, Laurent M, Boonen S, Vanderschueren D, Claessens F. Androgens and skeletal muscle: cellular and molecular action mechanisms underlying the anabolic actions. *Cell Mol Life Sci*. 2012;69(10):1651–1667.
59. Bachman E, Travison TG, Basaria S, et al. Testosterone induces erythrocytosis via increased erythropoietin and suppressed hepcidin: evidence for a new erythropoietin/hemoglobin set point. *J Gerontol A Biol Sci Med Sci*. 2014;69(6):725–735.
60. Delev D, Rangelov A, Ubenova D, Kostadinov I, Zlatanova H, Kostadinova I. Mechanism of action of androgens on erythropoiesis—a review. *Int J Pharm Clin Res*. 2016;8(11):1489–1492.
61. Travison TG, Morley JE, Araujo AB, O'Donnell AB, McKinlay JB. The relationship between libido and testosterone levels in aging men. *J Clin Endocrinol Metab*. 2006;91(7):2509–2513.
62. Wibowo E, Schellhammer P, Wassersug RJ. Role of estrogen in normal male function: clinical implications for patients with prostate cancer on androgen deprivation therapy. *J Urol*. 2011;185(1):17–23.
63. Carani C, Granata AR, Rochira V, et al. Sex steroids and sexual desire in a man with a novel mutation of aromatase gene and hypogonadism. *Psychoneuroendocrinology*. 2005;30(5):413–417.
64. Davidson JM. Effects of estrogen on the sexual behavior of male rats. *Endocrinology*. 1969;84(6):1365–1372.
65. Gillies GE, McArthur S. Estrogen actions in the brain and the basis for differential action in men and women: a case for sex-specific medicines. *Pharmacol Rev*. 2010;62(2):155–198.
66. Rudman D. Growth hormone, body composition, and aging. *J Am Geriatr Soc*. 1985;33(11):800–807.
67. Feldman HA, Longcope C, Derby CA, et al. Age trends in the level of serum testosterone and other hormones in middle-aged men: longitudinal results from the Massachusetts male aging study. *J Clin Endocrinol Metab*. 2002;87(2):589–598.
68. Vermeulen A, Kaufman JM, Giagulli VA. Influence of some biological indexes on sex hormone-binding globulin and androgen levels in aging or obese males. *J Clin Endocrinol Metab*. 1996;81(5):1821–1826.
69. Leifke E, Gorenoi V, Wichers C, Von Zur Muhlen A, Von Buren E, Brabant G. Age-related changes of serum sex hormones, insulin-like growth factor-1 and sex-hormone binding globulin levels in men: cross-sectional data from a healthy male cohort. *Clin Endocrinol (Oxf)*. 2000;53(6):689–695.
70. Knochenhauer ES, Boots LR, Potter HD, Azziz R. Differential binding of estradiol and testosterone to SHBG. Relation to circulating estradiol levels. *J Reprod Med*. 1998;43(8):665–670.
71. Gafny M, Silbergeld A, Klinger B, Wasserman M, Laron Z. Comparative effects of GH, IGF-I and insulin on serum sex hormone binding globulin. *Clin Endocrinol (Oxf)*. 1994;41(2):169–175.
72. Souza-Teodoro LH, de Oliveira C, Walters K, Carvalho LA. Higher serum dehydroepiandrosterone sulfate protects against the onset of depression in the elderly: Findings from the English Longitudinal Study of Aging (ELSA). *Psychoneuroendocrinology*. 2016;64:40–46.
73. Berr C, Lafont S, Debuire B, Dartigues JF, Baulieu EE. Relationships of dehydroepiandrosterone sulfate in the elderly with functional, psychological, and mental status, and short-term mortality: a French community-based study. *Proc Natl Acad Sci U S A*. 1996;93(23):13410–13415.
74. Kroboth PD, Salek FS, Pittenger AL, Fabian TJ, Frye RF. DHEA and DHEA-S: a review. *J Clin Pharmacol*. 1999;39(4):327–348.
75. Vermeulen A. Andropause. *Maturitas*. 2000;34(1):5–15.
76. Kovacs WJ, Ojeda SR. *Textbook of Endocrine Physiology* (6th ed.). Oxford University Press; 2011.
77. Handelsman DJ. Male contraception. In: De Groot LJ, Chrousos G, Dungan K, et al., eds. *Endotext*. Endotext.org; 2015.
78. Behre HM, Zitzmann M, Anderson RA, et al. Efficacy and safety of an injectable combination hormonal contraceptive for men. *J Clin Endocrinol Metab*. 2016;101(12):4779–4788.
79. Zegers-Hochschild F, Adamson GD, de Mouzon J, et al. International Committee for Monitoring Assisted Reproductive Technology (ICMART) and the World Health Organization (WHO) revised glossary of ART terminology, 2009. *Fertil Steril*. 2009;92(5):1520–1524.
80. Braunstein GD. Testes. In: Gardner DG, Shoback DM, eds. *Greenspan's Basic and Clinical Endocrinology* (9th ed.). McGraw Hill; 2011.
81. Baker HWG, Baker HG, deKretser DM. Relative incidence of etiological disorders in male infertility. In: Santen RJ, Swerdloff RS, eds. *Male Reproductive Dysfunction*. New York: Marcel Dekker; 1986: 341–372.
82. McClure RD. Endocrinology of male infertility. In: Patton PE, Battaglia DE, eds. *Office Andrology*. Humana Press; 2005.
83. Kasturi SS, Tannir J, Brannigan RE. The metabolic syndrome and male infertility. *J Androl*. 2008;29(3):251–259.
84. Phillips KP, Tanphaichitr N. Mechanisms of obesity-induced male infertility. *Expert Rev Endocrinol Metab*. 2010;5(2):229–251.
85. Cohen PG. The hypogonadal-obesity cycle: role of aromatase in modulating the testosterone-estradiol shunt—a major factor in the genesis of morbid obesity. *Med Hypotheses*. 1999;52(1):49–51.
86. Price TM, O'Brien SN, Welter BH, George R, Anandjiwala J, Kilgore M. Estrogen regulation of adipose tissue lipoprotein lipase--possible mechanism of body fat distribution. *Am J Obst Gynecol*. 1998;178(1 Pt 1):101–107.
87. Marin P, Oden B, Bjorntorp P. Assimilation and mobilization of triglycerides in subcutaneous abdominal and femoral adipose tissue in vivo in men: effects of androgens. *J Clin Endocrinol Metab*. 1995;80(1):239–243.
88. Tena-Sempere M, Barreiro ML. Leptin in male reproduction: the testis paradigm. *Mol Cell Endocrinol*. 2002;188 (1–2):9–13.
89. Cohen P, Zhao C, Cai X, et al. Selective deletion of leptin receptor in neurons leads to obesity. *J Clin Invest*. 2001;108(8):1113–1121.
90. Strobel A, Issad T, Camoin L, Ozata M, Strosberg AD. A leptin missense mutation associated with hypogonadism and morbid obesity. *Nat Genet*. 1998;18(3):213–215.
91. Ozata M, Ozdemir IC, Licinio J. Human leptin deficiency caused by a missense mutation: multiple endocrine defects, decreased sympathetic tone, and immune system dysfunction indicate new targets for leptin action, greater central than peripheral resistance to the effects of leptin, and spontaneous correction of leptin-mediated defects. *J Clin Endocrinol Metab*. 1999;84(10):3686–3695.
92. Gottsch ML, Cunningham MJ, Smith JT, et al. A role for kisspeptins in the regulation of gonadotropin secretion in the mouse. *Endocrinology*. 2004;145(9):4073–4077.
93. Couce ME, Green D, Brunetto A, Achim C, Lloyd RV, Burguera B. Limited brain access for leptin in obesity. *Pituitary*. 2001;4(1–2):101–110.
94. Chou SH, Mantzoros C. 20 years of leptin: role of leptin in human reproductive disorders. *J Endocrinol*. 2014;223(1):T49–62.
95. Banks WA, McLay RN, Kastin AJ, Sarmiento U, Scully S. Passage of leptin across the blood-testis barrier. *Am J Physiol*. 1999;276(6 Pt 1):E1099–E1104.
96. Herrid M, O'Shea T, McFarlane JR. Ontogeny of leptin and its receptor expression in mouse testis during the postnatal period. *Mol Reprod Dev*. 2008;75(5):874–880.

97. El-Hefnawy T, Ioffe S, Dym M. Expression of the leptin receptor during germ cell development in the mouse testis. *Endocrinology*. 2000;141(7):2624–2630.
98. Hauser R, Skakkebaek NE, Hass U, et al. Male reproductive disorders, diseases, and costs of exposure to endocrine-disrupting chemicals in the European Union. *J Clin Endocrinol Metab*. 2015;100(4):1267–1277.
99. Hunt PA, Sathyanarayana S, Fowler PA, Trasande L. Female reproductive disorders, diseases, and costs of exposure to endocrine disrupting chemicals in the European Union. *J Clin Endocrinol Metab*. 2016;101(4):1562–1570.
100. Wohlfahrt-Veje C, Main KM, Skakkebæk NE. Testicular dysgenesis syndrome: foetal origin of adult reproductive problems. *Clin Endocrinol*. 2009;71(4):459–465.
101. Boisen K, Main K, Rajpert-De Meyts E, Skakkebaek N. Are male reproductive disorders a common entity? The testicular dysgenesis syndrome. *Ann N Y Acad Sci*. 2001;948(1):90–99.
102. Skakkebæk N-E, Meyts R-D, Main K. Testicular dysgenesis syndrome: an increasingly common developmental disorder with environmental aspects: opinion. *Hum Reprod*. 2001;16(5):972–978.
103. Skakkebæk NE, Jørgensen N, Main KM, et al. Is human fecundity declining? *Int J Androl*. 2006;29(1):2–11.
104. Herman RA, Jones B, Mann DR, Wallen K. Timing of prenatal androgen exposure: anatomical and endocrine effects on juvenile male and female rhesus monkeys. *Horm Behav*. 2000;38(1): 52–66.
105. Mylchreest E, Sar M, Cattley RC, Foster PM. Disruption of androgen-regulated male reproductive development by di (n-butyl) phthalate during late gestation in rats is different from flutamide. *Toxicol Appl Pharmacol*. 1999;156(2):81–95.
106. Rehman S, Usman Z, Rehman S, et al. Endocrine disrupting chemicals and impact on male reproductive health. *Transl Androl Urol*. 2018;7(3):490.
107. Fisher JS, Macpherson S, Marchetti N, Sharpe RM. Human “testicular dysgenesis syndrome”: a possible model using in-utero exposure of the rat to dibutyl phthalate. *Human reproduction*. 2003;18(7):1383–1394.
108. David RM. Proposed mode of action for in utero effects of some phthalate esters on the developing male reproductive tract. *Toxicol Pathol*. 2006;34(3):209–219.
109. Calafat AM, Ye X, Wong L-Y, Reidy JA, Needham LL. Exposure of the US population to bisphenol A and 4-tertiary-octylphenol: 2003–2004. *Environ Heal Perspect*. 2007;116(1): 39–44.
110. Weuve J, Sánchez BN, Calafat AM, et al. Exposure to phthalates in neonatal intensive care unit infants: urinary concentrations of monoesters and oxidative metabolites. *Environ Heal Perspect*. 2006;114(9):1424–1431.
111. Swan SH, Main KM, Liu F, et al. Decrease in anogenital distance among male infants with prenatal phthalate exposure. *Environ Heal Perspect*. 2005;113(8):1056–1061.
112. Duty SM, Singh NP, Silva MJ, et al. The relationship between environmental exposures to phthalates and DNA damage in human sperm using the neutral comet assay. *Environ Heal Perspect*. 2003;111(9):1164–1169.
113. Hauser R. Urinary phthalate metabolites and semen quality: a review of a potential biomarker of susceptibility. *Int J Androl*. 2008;31(2):112–117.
114. Pant N, Shukla M, Patel DK, et al. Correlation of phthalate exposures with semen quality. *Toxicol Appl Pharmacol*. 2008;231(1):112–116.
115. Wirth JJ, Rossano MG, Potter R, et al. A pilot study associating urinary concentrations of phthalate metabolites and semen quality. *Syst Biol Reproduct Med*. 2008;54(3):143–154.
116. Bulun SE. Aromatase and estrogen receptor alpha deficiency. *Fertil Steril*. 2014;101(2):323–329.

CHAPTER

22

Female Reproductive System

CHAPTER LEARNING OBJECTIVES:

- Describe the key genomic and morphological changes that take place during oogenesis, follicle development, ovulation, and fertilization.
- Describe the structure of an ovarian follicle, and the chronology of follicular recruitment, development, and atresia.
- Describe the two-cell theory of ovarian hormone biosynthesis, the defining endocrine features of the follicular and luteal phases of the ovarian cycle, and the feedback mechanisms along the hypothalamic-pituitary-ovary axis.
- Describe how estrogens and progesterone are transported in the blood, and outline sex steroid hormone receptor function.
- Compare and contrast the human menstrual cycle with the estrous cycle in other mammals.
- Describe modern approaches to contraception and fertility control, and discuss the non-contraceptive risks and benefits of hormone-based contraceptives.
- Outline the major changes in endocrine and health parameters associated with menopause, and discuss the health benefits and potential risks of hormone replacement therapy for treating the symptoms of menopause.
- Describe some notable endocrine-disrupting chemicals that may contribute to female reproductive dysfunctions in humans and animal models, and discuss their mechanisms of action.

OPENING QUOTATION:

"If Shakespeare had been a chemist, he would have loved estrogen, a hormone fit for comedy, tragedy and a sonnet or two. Estrogen can be heroic, governing human fertility and nurturing the heart, bones, blood vessels and brain so persuasively that soon estrogen may be prescribed for middle-aged men as it is today for women. It can play Lady Macbeth, with blood on her hands: estrogen has been implicated in cancers of the breast, ovary and uterus, autoimmune diseases, asthma, fibroids, mood disorders and migraines. And just when scientists think they've got estrogen figured out, the hormone turns Puckish, mocking their assumptions and rewriting the script."

—Natalie Angier, "New Respect for Estrogen's Influence", *New York Times*, June 24, 1997

Introduction

Ancient Notions

Prior to about the middle of the 19th century, there was very little knowledge of the function of the ovaries, with the larger and more accessible uterus featuring more prominently in female reproductive theory. However, even in ancient times some understanding of a role for the ovaries in reproduction was appreciated, at least by Aristotle (384–322 BC), who recorded, "the ovaries of sows are excised with the view of quenching in them sexual appetites and of stimulating growth in size and fatness".[1] That is, the removal of the ovaries (spaying) from livestock not only eliminates their sex drive, but also diverts the energy of reproduction into fat storage. The first detailed anatomical drawings of the human ovaries and female reproductive organs were provided in 1543 by Andreas Vesalius in his monumental *De Humani Corporis Fabrica Libri Septem* ("Of the Structure of the Human Body") (Figure 22.1). However, the specific function of the ovaries still remained vague, and Vesalius referred to them as "the testes of women". The first suggestions that mammalian ovaries likely functioned as egg-generating counterparts to the bird ovary were put forth in the 17th century by the Dutch anatomist Regnier De Graaf, the same De Graaf who showed, in the previous chapter, that the mouse testicle consisted of tubules. Quite remarkably, De Graaf supported his notion that the human ovum was analogous to a bird egg by dissecting the ovaries from a cadaver and cooking them to demonstrate that their taste was similar to a hen's egg![2] Mild cannibalistic tendency aside, De Graaf made numerous contributions to the nascent field of reproductive biology, and his legacy is remembered today through the eponym, **Graafian follicle**, or the mature stage of an ovarian follicle, in his honor.

DOI: 10.1201/9781003359241-29

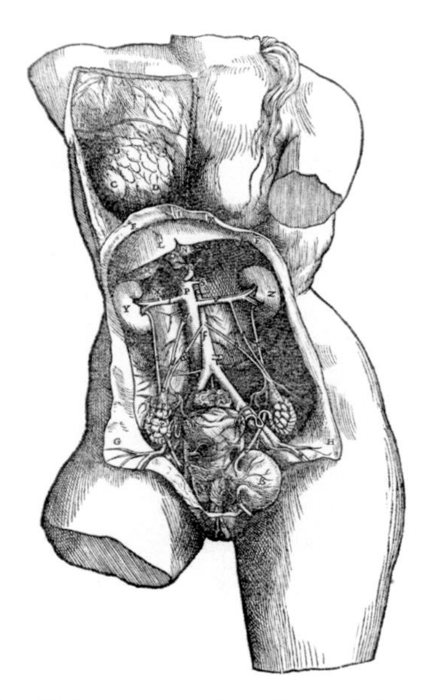

Figure 22.1 In 1543 Andreas Vesalius published *De Humani Corporis Fabrica Libri Septem*, one of the most important books in medicine. Despite the clear anatomical depiction of the female reproductive organs in this figure, the specific functions of the ovaries and uterus remained poorly described at the time.

Source of Images: Courtesy of https://www.metmuseum.org/art/collection/search/358129.

Dawn of a Modern Endocrine Understanding

Hints that the ovary also exerts endocrine functions emerged in 1856 when Carl Ludwig observed that ovariectomy in humans led to both the termination of menstruation and the shrinking of the uterus.[3,4] A critical experiment by Emil Knauer in 1896 using guinea pigs showed that ovariectomy followed by the transplantation of ovaries to different locations in the body inhibited atrophy of the uterus, suggesting that blood-borne internal secretions produced by the ovaries must affect the uterus.[3,4] New discoveries moved at an astounding pace in the first decades of the 20th century following Edgar Allen's and Edward Doisy's 1923 identification of the ovarian follicle as a major source of *estrogenic* (i.e. estrus-generating) activity in lab animals.[5] Finally, in 1929, Doisy and colleagues in the United States[6] and Adolf Butenandt in Germany[7] independently purified the first sex steroid hormone, estrone, from hundreds of gallons of urine from pregnant women. Butenandt would go on to share with Leopold Ruzicka the *1939 Nobel Prize in Chemistry* for work on sex hormones, and Doisy would share the *1943 Nobel Prize in Physiology or Medicine* with Henrik Dam for their discovery of vitamin K and its role in blood clotting. Discoveries of the more potent estrogen estradiol and of the gestation-maintenance hormone progesterone would soon follow. In addition to sex steroid hormones, diverse peptide and protein hormones of hypothalamic, pituitary, and placental origin that are critical for normal female reproductive functioning would begin to be discovered during the second half of the 20th century.

Overview of Female Reproductive Organs

As is the case for the male, the components of the female urogenital tract share similar features among all vertebrates, generally possessing paired oviducts (Müllerian ducts) with varying degrees of fusion of the uteri. A comparison of some important differences between human male and female reproductive systems is summarized in Table 22.1. This chapter primarily addresses the endocrinology of female reproduction in humans, the most thoroughly studied of the vertebrate reproductive systems. While a brief overview of female reproductive organ functions is presented in Figure 22.2 and Table 22.2, the emphasis of this chapter is on the synthesis and actions of ovarian and

Table 22.1 Comparing and Contrasting Adult Male and Female Reproductive Systems

Male	Female
The gonads (testes) are located outside the pelvic cavity in the scrotum	The gonads (ovaries) are located inside the pelvic cavity
The testes are connected directly to the rest of the reproductive tract (i.e. epididymis, vas deferens, urethra)	The ovaries are not connected directly to the reproductive tract (i.e. Fallopian tubes, uterus, vagina)
The testes are comprised of tubules	The ovaries are made up of follicles
The gametes (spermatocytes) are continuously released from the gonads	One gamete (ovum) is released episodically from the gonads once per month
The gametic reserve is replenished by spermatogonia throughout life	Oogonia are absent in adult females, and the gametic reserve is depleted by menopause
Testosterone regulates pituitary LH and FSH secretion via negative feedback	Estrogen regulate pituitary LH and FSH secretion via both negative and positive feedback
The male reproductive tract supports spermatocyte maturation and transport	The female reproductive tract supports both male and female gamete maturation and transport, as well as fertilization, placentation, gestation, and delivery
The male reproductive system shows no obvious functional rhythmicity	The female reproductive system operates on a monthly menstrual cycle (in the nonpregnant state), or (in the pregnant state) 9 months of gestation
The primary gonadal steroid is testosterone	During the first half of the menstrual cycle the primary gonadal steroid is estrogen, whereas during the second half it is progesterone
The male reproductive system plays no role in maintenance of the newborn infant	During pregnancy the female's breasts prepare for milk production to feed the newborn infant

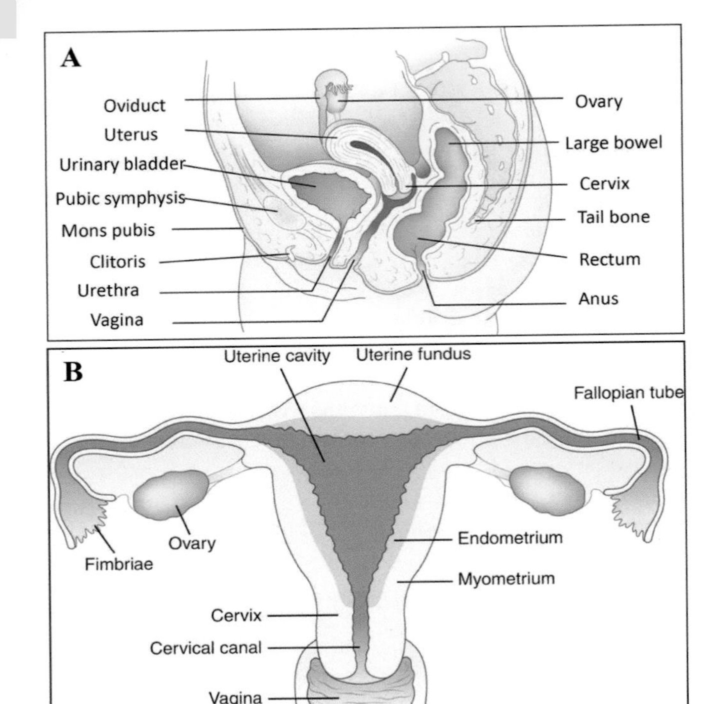

Figure 22.2 Internal anatomy of the female reproductive system. (A) The reproductive organs are shown from a mid-sagittal view. **(B)** A mid-frontal section depicting the arrangement of the ovary, Fallopian tubes, uterus, and vagina.

Source of Images: Norman, A.W. and Henry, H.L., 2022. Hormones. Academic Press. Used by permission.

non-ovarian hormones in regulating reproductive physiology of the nonpregnant adult female. The roles of the mammary glands, placenta, and uterus will be addressed in Chapter 24: Pregnancy, Birth, and Lactation.

Oogenesis and Fertilization

LEARNING OBJECTIVE Describe the key genomic and morphological changes that take place during oogenesis, follicle development, ovulation, and fertilization.

KEY CONCEPTS:

- After reaching their peak numbers in the embryonic ovary via mitotic divisions, germ cell numbers decrease abruptly by birth.
- Female germline stem cells are absent by birth, and the numbers of oocytes in the ovaries gradually decline throughout life and disappear by postmenopause.
- Oocytes undergo two periods of meiotic arrest: by birth all are arrested in prophase I, and following ovulation the oocyte is arrested in metaphase II.
- Meiotic divisions of oogonia are asymmetrical, each ultimately yielding one large viable oocyte and two much smaller and nonviable polar bodies.
- Following ovulation, the oocyte arrested in metaphase II only completes its final meiotic division following fertilization.

Oogenesis

As was the case for males, in female embryos **primordial germ cells** (PGCs) originating from the hindgut also migrate to the gonadal ridge. There they will colonize the fetal ovaries and develop into *oogonia*, undifferentiated germ cells that will ultimately either die by apoptosis or undergo **oogenesis** and produce mature *ova*. The oogonia proliferate mitotically within the fetal ovaries (in humans reaching a peak number of about 6–7 million by 5 months of gestation), and the germ cell number then decreases abruptly via programmed cell death to approximately 30% of the peak number by the time of birth[8] (Figure 22.3). In marked contrast with spermatogenesis, where meiosis of the spermatogonia begins at puberty, in females all oogonia initiate meiosis synchronously during the fetal period,[9] developing into *primary oocytes* (Figure 22.4). By birth all of the primary oocytes within the follicles have entered a prolonged state of meiotic arrest in prophase I called the *dictyotene stage*, which can last 50 years. The resumption of meiosis I and the conclusion of the first division takes place periodically in small cohorts of follicles years later, beginning as early as puberty. Although the vast majority of these follicles will be lost to *atresia* (degeneration by programmed cell death, followed by resorption), about one follicle per group ultimately completes meiosis I immediately prior to

Table 22.2 Overview of Female Reproductive Organ Functions

Ovaries: Paired structures within the pelvic cavity with the dual functions of promoting gametogenesis and hormonogenesis. Following ovulation, the egg is swept into an oviduct.

Oviducts: Also called the Fallopian tubes, the oviducts facilitate the transport of the ovulated oocyte from the ovary to the uterus. Fertilization, if it takes place, typically occurs in an oviduct, and the resulting embryo is transported to the uterus for implantation.

Uterus: A highly muscularized, thick-walled organ that functions as the site for embryonic implantation, fetal development, and growth.

Cervix: The inferior-most projection of the uterus that is continuous with and extends into the vagina. The cervix has several major functions. First, it regulates sperm viability and entry into the uterus by either facilitating or impeding (depending on the stage of the ovarian cycle) the entry of sperm. Second, the cervix prevents microbes from ascending into the upper reproductive tract, particularly during pregnancy. Third, the cervix physically supports the weight of the fetus. During childbirth, the dilation and softening of the cervix facilitates the movement of the newborn and the placenta out of the uterus and into the vagina.

Vagina: A dual function organ that serves as a copulatory structure and as the birth canal.

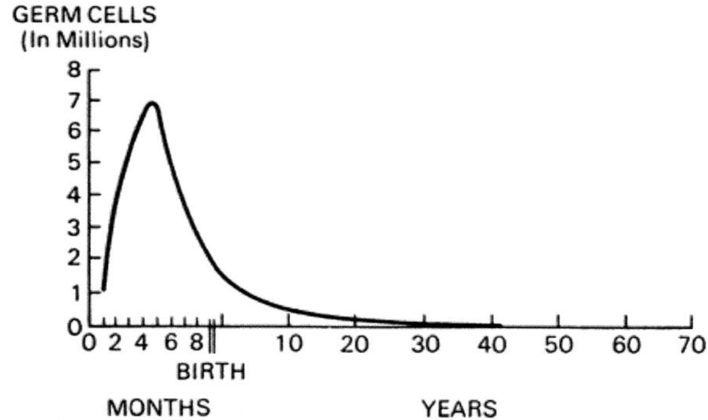

Figure 22.3 Changes in the number of oocytes with age in a typical woman.

Source of Images: Mattison, D.R., et al. 1983. Env. Health Perspect., 48, pp. 43-52.

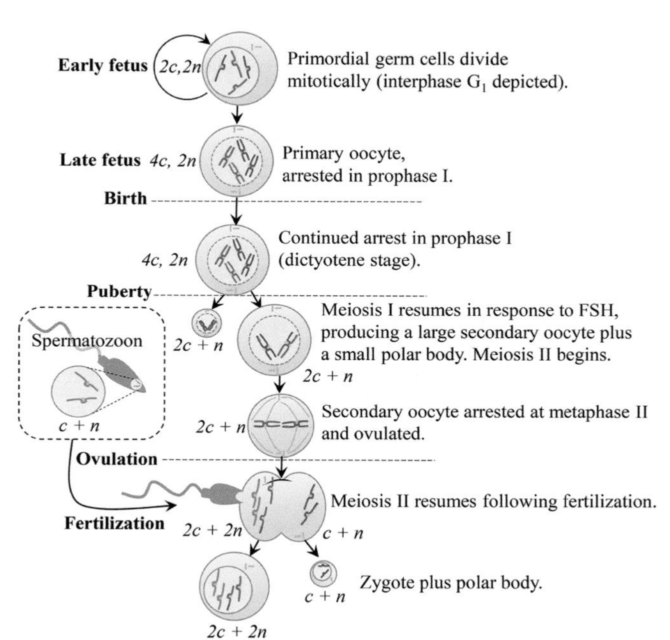

Figure 22.4 Summary of the major events in human oogenesis, with the assumption of fertilization. For simplicity, the top of the diagram depicts germ cells containing only two pairs of homologous chromosomes (red = maternal, blue = paternal). Progression towards the bottom of the diagram shows their respective meiotic fates. n = number of copies of each chromosome (homologous chromosomes) in each cell, c = total number of copies of DNA for each chromosome (including sister chromatids) in each cell.

ovulation each month. The resumption of meiosis I occurs in response to rising circulatory levels of luteinizing hormone (LH). Following the completion of the first meiotic division, the primary oocyte initiates meiosis II, generating a *secondary oocyte* that will be ovulated. The ovulated secondary oocyte then undergoes a second developmental arrest in metaphase II.[10] Typically, the second meiotic division is completed only if fertilization occurs.

In addition to the initiation of meiosis during the fetal stage and the presence of two meiotic developmental arrests, oogenesis differs from spermatogenesis in other important respects. Whereas for each spermatogonium the two rounds of meiotic cytokinesis are symmetric, generating four equal-sized and viable spermatozoa, the meiotic divisions of oogonia are asymmetric, yielding only one very large and viable egg, as well as two much smaller nonviable cells called **polar bodies** (Figure 22.4). This size asymmetry results from the shunting of most organelles and cytosol into the secondary oocyte during meiosis I, and then into the egg after fertilization during meiosis II. The functions of polar bodies, which ultimately degenerate, are to facilitate chromosomal reduction and have been described succinctly as enabling "the preovulatory oocyte to jettison redundant chromosomes while retaining sufficient cytoplasmic resources to sustain preimplantation embryo development".[11] A further contrast between oogenesis and spermatogenesis is the fate of gamete-generating stem cells. Whereas males are born with a population of spermatogonial stem cells that are constantly renewed by mitosis resulting in continuous spermatogenesis throughout the reproductive lifespan, the oogonial stem cell population is thought to be absent by birth. Thus, females are born with a finite number of oocytes in their ovaries that are not replenished over their reproductive life. Out of a total of approximately 400,000 primary oocytes present in the ovaries at the start of puberty, only 400–500 typically mature sufficiently to *ovulate*, or be expelled from the ovary on a monthly basis, over the entire course of a female's reproductive years, which in humans is generally between 13 and 46 years old. By *postmenopause*, the age of reproductive capacity culmination, most, if not all, of the remaining follicles have undergone atresia.

Fertilization and the Completion of Meiosis

The oocyte, arrested in metaphase II following ovulation, only completes its final meiotic division following fertilization, producing a second polar body (Figure 22.4). The process of *fertilization*, or the fusion of a spermatozoon with a secondary oocyte, transforms two haploid gametes into a single diploid cell called a *zygote*. The intermixing of the male and female chromosomes can be considered as the final step of fertilization and the beginning of embryogenesis.[12]

SUMMARY AND SYNTHESIS QUESTIONS

1. Create a table that compares and contrasts oogenesis with spermatogenesis (Chapter 21: Male Reproductive System) from embryogenesis through reproductive senescence.

Development of Ovarian Follicles

LEARNING OBJECTIVE Describe the structure of an ovarian follicle, and the chronology of follicular recruitment, development, and atresia.

KEY CONCEPTS:

- Ovarian follicles each consist of an oocyte surrounded by epithelial cells in various stages of growth and development.
- Subsets of follicles are continuously recruited to undergo further development and growth into a single mature follicle that will be ovulated.
- Follicles that are not ovulated eventually undergo programmed cell death and are resorbed, a process called "atresia".
- Following ovulation, the remnants of the follicle in the ovary develop into a primarily progesterone-secreting organ called the corpus luteum.
- In the absence of fertilization, the corpus luteum degenerates into a scar-like tissue called the corpus albicans.

The dual ovarian functions of oogenesis and hormonogenesis are exquisitely intertwined. This section discusses ovarian anatomy, its role in mediating oogenesis, its regulation of the biosynthesis of sex steroids hormones and other hormones, as well as its maintenance of female reproduction and secondary sex characteristics.

Structure and Organization of the Ovary

Ovaries consist primarily of germ cells, epithelial cells, and mesenchymal cells. Morphologically, the ovary can be divided into three general regions (Figure 22.5): the largest region, the *cortex*, contains the functional components of the ovary, the **ovarian follicles**. These follicles, which each consist of an oocyte surrounded by epithelial cells, appear in different states of development throughout the cortex: resting, growing, atretic (degenerating), or ready for ovulation. As the primary functional unit of the ovary, the ovarian follicle exerts both gametogenic and hormonogenic functions.

Development of the Ovarian Follicle

Recall from the prior section (22.2: Oogenesis and Fertilization) that by birth about 2 million primary oocytes

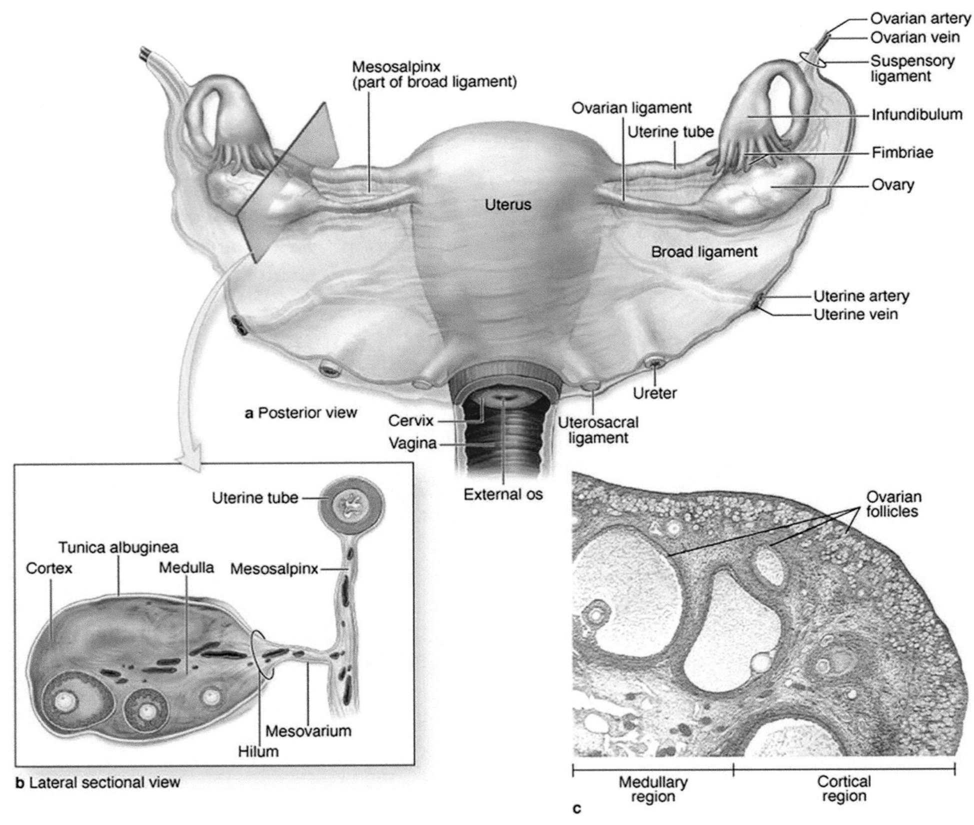

Figure 22.5 The human female reproductive system and overview of ovarian morphology. (A) The diagram shows the principal internal organs of the female reproductive system, which includes the ovaries, uterine (Fallopian) tubes, uterus, and vagina. **(B)** A lateral sectional view of an ovary shows the ovary and supporting mesenteries and ligaments. **(C)** Micrograph of a sectioned ovary stained with hematoxylin and eosin, indicating the medullary and cortical regions, with follicles of several different sizes in the cortex.

Source of Images: Junqueira, L.C. and Mescher, A.L., 2013. Junqueira's basic histology: text & atlas McGraw Hill LLC. Used by permission.

are present in the ovaries, and that number is reduced by atresia to approximately 400,000 by the start of puberty. Indeed, of the hundreds of thousands of follicles that a woman was born with, over her lifetime only about 400 will go on to ovulate, and the rest die trying. Importantly, by birth all oocytes have associated closely with surrounding epithelial cells to form structures called *follicles*. This section will address the possible fates of a follicle from birth through reproductive age.

Growth, Recruitment, and Atresia

At birth each oocyte is surrounded by a single layer of flattened epithelial cells called *pregranulosa* cells, together forming the *primordial follicle* (Figure 22.6). Although the process of follicle recruitment and development starts before birth, most follicle development occurs after puberty, as soon as a regular menstrual cycle is established. Approximately 1,000 primordial follicles are *recruited* (selected for further development) in cohorts per month, with the goal being for each cohort to ultimately produce a single dominant follicle that will release an egg during an ovulation phase. The majority of primordial follicles will not be recruited and may persist in an arrested state through puberty, or even menopause. Those that are recruited develop slowly, over the course of about six months, into *primary follicles*. These follicles are larger than their precursors, and their surrounding epithelial cells have developed into mature cuboidal *granulosa cells*. At puberty, and under stimulation by pituitary FSH, a small subset of primary follicles is recruited to develop into *secondary follicles*, a process accompanied by increased oocyte size and the proliferation of granulosa cells resulting in multiple layers. During this time, the granulosa cells produce a mucus-like substance, the *zona pellucida*, which surrounds the follicle. Simultaneously, some interstitial cells adjacent to the follicle become integrated into the follicle as a concentric shell of cells called *theca cells*. The theca cells remain separated from the granulosa cells via a thin basement membrane.

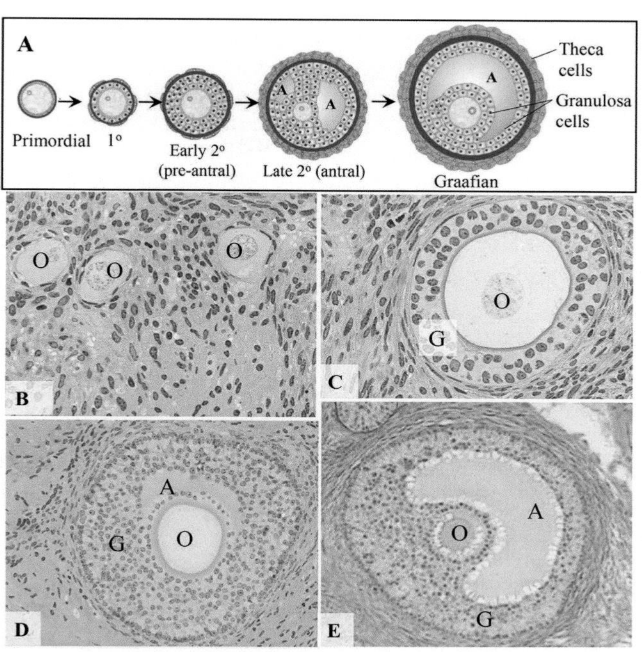

Figure 22.6 Schematic and histological stages of ovarian follicle development. **(A)** Schematic portrayal of folliculogenesis. (B–E) Hematoxylin (purple nuclei) and eosin (pink cytoplasm) histological stains. **(B)** Primordial follicle. **(C)** Primary follicle. **(D)** Secondary follicle. **(E)** Graafian follicle. "O", oocyte; "A", antrum; "G", granulosa cells.

Source of Images: (A) Araújo, V.R., et al. 2014. Reprod. Biol. Endocdrinol., 12(1), pp. 1-14. (B-D) Courtesy of Dr. Peter Takizawa, Yale University. (E) Sokol, E, Glob. libr. women's med., (ISSN: 1756-2228) 2011; DOI 10.3843/GLOWM.10001.

Within 3 weeks of the development of the secondary follicles, a further subset of about ten secondary follicles will begin to develop multiple fluid-filled spaces between granulosa cells. Each of these spaces is called an *antrum*, and these follicles are known as *antral follicles*, or "early tertiary follicles" (Figure 22.6). Among this small group

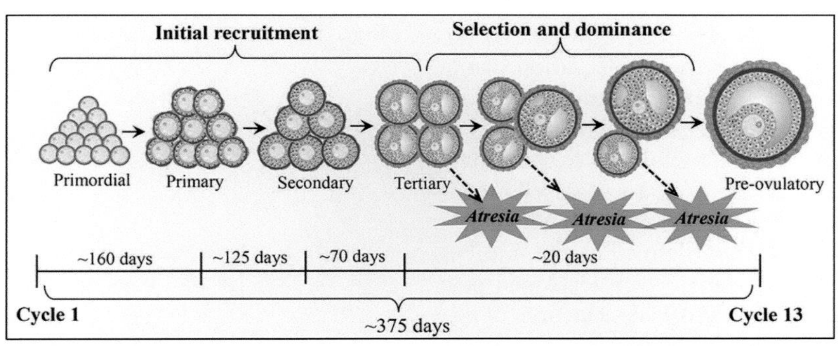

Figure 22.7 Chronology of folliculogenesis for a single cohort of follicles. Recruitment of a cohort of follicles, their percent atresia at different developmental stages, and selection of the final preovulatory follicle takes place over 13 menstrual cycles, or over a year-long period. The dominant follicle appears to be selected from a cohort of mid-tertiary follicles in the 13th cycle, and subsequently requires about 15 to 20 days to grow and develop to the preovulatory stage. During the entire process of folliculogenesis, the follicle will increase about 1,000-fold in size.

Source of Images: https://www.ovulationcalculator.com

of antral follicles, one of the leading follicles, the *dominant follicle*, typically grows faster than the others of the cohort, and its multiple antrums have converged to form a single large antrum. This dominant follicle will develop into the **Graafian follicle**, also called the "mature" or "preovulatory" follicle, which is the final stage of follicular development prior to ovulation.

Chronology of Human Folliculogenesis

It takes approximately 375 days, or 13 menstrual cycles, for a primordial follicle to be recruited and to develop to the preovulatory stage[13] (Figure 22.7). As such, it is only during the follicular phase of the 13th cycle that a follicle grows to the preovulatory stage, with the vast majority of follicular growth occurring during the preceding year as a member of a dwindling cohort.

Each cohort of recruited follicles derives from both ovaries, with follicles competing with each other within a cohort for dominance, and the dominant follicle will ultimately release an egg during ovulation. Over the next 12 cycles, most follicles will die apoptotically, with just a few (about five per ovary) surviving to the 13th cycle. Typically, a single dominant preovulatory follicle is selected in the 13th cycle and rapidly grows to over 20 mm in diameter, causing it to rupture and release the oocyte (ovulate). The mechanisms by which recruitment and atresia of follicles are determined are not well understood, but larger antral follicles appear to be selected preferentially.[14] Although FSH is critical to follicle maturation up until about mid-follicular stage, its levels decline thereafter due to inhibin feedback from the growing antral follicles. However, during late preovulatory follicular growth, the response of FSH promotes increased expression of the luteinizing hormone receptor (LHR), making granulosa cells progressively more responsive to LH and less dependent on FSH stimulus for continued development.[15] The LHR-rich follicle is now prepared for its imminent ovulatory response to LH.

A key ovarian hormone involved in regulating folliculogenesis is *anti-Müllerian hormone* (AMH). You first became familiar with AMH in Chapter 20: Sexual Determination and Differentiation in the context of male sex differentiation, where AMH is produced by embryonic testicular Sertoli cells and induces the degeneration of the Müllerian ducts, which are the developmental precursors to the internal female genital system. However, AMH is also produced postnatally by growing follicles in the female ovaries. The production of AMH by granulosa cells regulates folliculogenesis by inhibiting the recruitment of follicles from the *resting follicle pool*, defined as the number of follicles in the ovaries that have not yet developed past the primordial stage,[16] which aids in the selection of the dominant follicle.[17,18] Since AMH is produced by follicles that have made the transition from the resting to the growing follicle pool, AMH may serve as a molecular biomarker for estimating the size of the *ovarian reserve*, a clinical term that is often used to describe a woman's reproductive potential based on the number and quality of oocytes she possesses.[19]

Clinical Considerations: Polycystic ovary syndrome (PCOS)

Polycystic ovary syndrome (PCOS) is a common disorder affecting 5%–20% of women of reproductive age worldwide[20] and is a leading cause of subfertility and infertility. Its primary symptoms include:

Hyperandrogenism: Excess levels of androgen production by the ovaries, primarily testosterone and dihydrotestosterone, often leading to *hirsutism* (male pattern hair growth on the face, chest, or back), severe acne, and androgenic *alopecia* (male pattern baldness). Elevated androgens result from increased androgen synthesis by follicular theca cells, which display increased expression of several genes that code for steroidogenic enzymes.[21] While the ovaries are the primary source of excess androgens in PCOS patients, in 20% to 30% of these patients hyperandrogenism stems from adrenocortical hyperfunction.[22]

"Polycystic" ovarian morphology: Compared with normal ovaries, there are an excessive number of preantral and small antral follicles, with most antral follicles in the ovarian cortex arrested at the mid-stage of development (Figure 22.8). Fully mature Graafian follicles are rare. Although these follicles are frequently described as "cysts", this is an unfortunate misnomer as cysts are defined medically as "membranous sacs or cavities of abnormal character containing fluid".[23]

Ovulatory dysfunction: *Oligo-ovulation* (infrequent or irregular ovulation) or *anovulation* (the absence of ovulation) manifests due to the lack of selection of a dominant follicle as a consequence of insufficient FSH secretion and inhibition of local FSH action.[24,25] The precise cause of PCOS has not been established, but current studies suggest that there is a strong genetic component, with hyperandrogenism inducing the polycystic ovarian

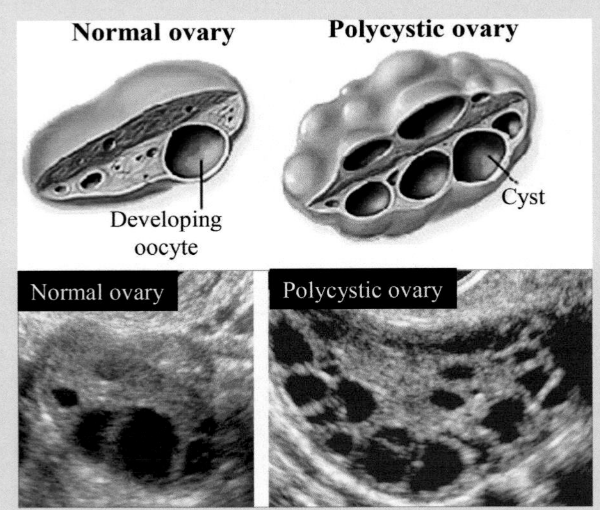

Figure 22.8 Normal and polycystic ovary morphology. Ovaries shown by transvaginal ultrasonography during the follicular phase of a menstrual cycle. The fluid-filled antrum of developing follicles appear as dark blotches. When compared with a normal ovary, the polycystic ovary is enlarged and contains increased numbers of developing follicles.

Source of Images: Norman, R.J., et al. 2007. Lancet, 370(9588), pp. 685-697. Used by permission.

morphology that perpetuates endocrine disruption.[26] One neuroendocrine abnormality in PCOS that may promote hyperandrogenism is an increased pulse frequency of GnRH secretion, which in turn elevates the LH/FSH ratio, enhancing androgen hypersecretion by the ovarian theca cells and impairing follicular development.[27] Underproduction of FSH may be further compounded by increased levels of anti-Müllerian hormone (AMH), which reduces the sensitivity of ovarian follicles to FSH.[28] Since reduced FSH activity results in lower CYP19 aromatase activity by granulosa cells, the conversion of androgens to estrogens is blocked, further contributing to hyperandrogenism.

A common pharmacological therapy for PCOS is treatment with combined estrogen-progestin oral contraceptives, which suppress gonadotropin secretion and ovarian androgen synthesis. Furthermore, the estrogen component increases the hepatic production of sex hormone-binding globulin (SHBG), reducing the bioavailability of androgen.[29]

Ovulation

Ovulation is immediately preceded by a surge in gonadotropins. The surge of LH, in particular, promotes radical changes in Graafian follicle structure that facilitate ovulation. Some of the important events mediating ovulation are summarized in Figure 22.9 and described as follows.

1. *Expansion and detachment of the cumulus-oocyte complex*. Prior to ovulation, the cells of the cumulus-oocyte complex are tightly connected to each other and to the oocyte via cell adhesion complexes and gap junctions. The LH surge stimulates cumulus cells to produce *hyaluronan*, a carbohydrate polymer that is a major constituent of the extracellular matrix.[30] As the cumulus cells synthesize the new extracellular matrix, they expand and separate from one another and move away from the oocyte.
2. *Thinning of the follicular and ovarian walls*: In response to the LH surge, the theca and granulosa

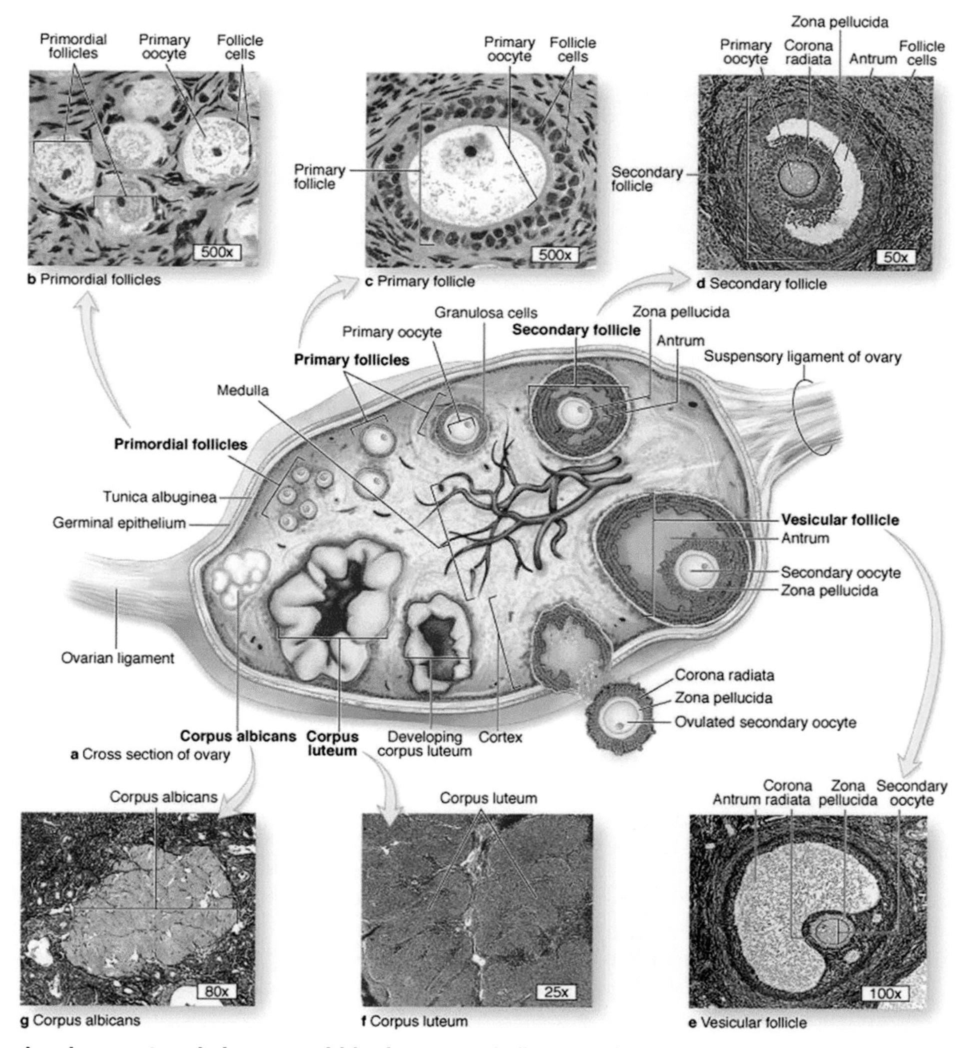

Figure 22.9 Follicle development and changes within the ovary. A diagram of a sectioned ovary (a), shows the different stages of follicle maturation (b–e), ovulation, and corpus luteum formation (f) and degeneration (g).

Source of Images: Junqueira, L.C. and Mescher, A.L., 2013. Junqueira's basic histology: text & atlas/Anthony L. Mescher. McGraw Hill LLC. Used by permission.

cells of the follicle secrete matrix metalloproteases, hydrolytic enzymes, and inflammatory cytokines that promote the degeneration of the collagenous connective tissue within the follicular and ovarian walls, thinning them.

3. *Rupture of the ovarian walls*: In combination with the thinning walls of the ovary, the growing antrum of the mature follicle causes it to progressively squeeze against the surface of the ovary, forming a swelling called a *stigma*. Ultimately, this contributes to the rupture of the ovarian wall at ovulation.[31,32]
4. *Repair of the ruptured ovarian wall.* The erosion of the ruptured ovarian wall must be repaired each month, and this is accomplished through the rapid division of neighboring epithelial cells. Immediately following ovulation, the remains of the ovulated follicle form a transient structure, the *corpus hemorrhagicum* ("bleeding body", in Latin), which fills with blood, clots, and promotes the healing of the ruptured ovarian wall. Subsequently, the corpus hemorrhagicum develops into a new endocrine structure, the corpus luteum.
5. *Generation of the corpus luteum.* Following the healing of the trauma caused by ovulation, the corpus hemorrhagicum remodels to form a new endocrine organ called the **corpus luteum** ("yellow body", in Latin). This structure consists of luteinized granulosa cells, in addition to thecal cells, fibroblasts, and extensive vascularization. The process of granulosa cell luteinization includes the development of cholesterol-rich cytoplasmic droplets that promote the yellow coloration. In the absence of fertilization, the corpus luteum will continue to synthesize progesterone for 10–14 days prior to undergoing *luteolysis*, or the degeneration of the corpus luteum and cessation of progesterone synthesis.[33] Eventually the remnants of the corpus luteum are destroyed by macrophages and replaced with fibroblasts, forming a white-colored, non-endocrine, scar-like tissue named the **corpus albicans** ("white body", in Latin). If fertilization does take place, the corpus luteum is "rescued" from its programmed senescence by *chorionic gonadotropin*, an LH-like hormone that is produced by the implanted embryo.[34] Following rescue, the progesterone-secreting structure remains present through the first 8 weeks of pregnancy, prior to degenerating.[35]

SUMMARY AND SYNTHESIS QUESTIONS

1. The process of granulosa cell luteinization includes the development of cholesterol-rich cytoplasmic droplets that promote the yellow coloration. What is the endocrine significance of these oil droplets?
2. From an endocrine perspective, which ovarian cell types are most similar to the Sertoli and Leydig cells, respectively, of the testes (justify your answers)? How do the ovarian cells differ from those of the testes?

The Ovarian Cycle: Hormone Production and Feedback along the HPG Axis

LEARNING OBJECTIVE Describe the two-cell theory of ovarian hormone biosynthesis, the defining endocrine features of the follicular and luteal phases of the ovarian cycle, and the feedback mechanisms along the hypothalamic-pituitary-ovary axis.

KEY CONCEPTS:

- Steroidogenically analogous to Leydig cells of the testis, a primary function of theca cells is to synthesize androgens in response to LH.
- Analogous to Sertoli cells of the testis, a primary function of granulosa cells is the conversion of androgens into estrogens in response to FSH.
- The first 14 days of the ovarian cycle comprise the follicular phase, characterized by the growth and maturation of the dominant follicle.
- Ovulation demarcates the start of the luteal phase of the ovarian cycle, characterized by the formation of the corpus luteum.
- During the follicular phase, a follicle's granulosa cells synthesize estrogens.
- During the luteal phase, a follicle's granulosa cells synthesize progesterone.
- Rapidly rising gonadotropin levels during the mid-late follicular phase is driven by positive feedback from peaking synthesis of estrogens.
- Falling gonadotropin levels during the luteal phase is mediated by negative feedback from rising levels of progesterone, estradiol, and inhibin.

The Two-Cell Theory for the Production of Estrogens

Although both granulosa and theca cells have steroidogenic capacity, each possess unique morphological and biochemical properties that require them to cooperate to synthesize estrogens. First, look back at Figure 22.6 and notice how the theca cells are located superficially on the follicle, whereas granulosa cells are localized more internally. The superficial positioning of theca cells places them closer to blood vessels carrying low-density lipoprotein (LDL), which is an accessible source of cholesterol for the synthesis of steroid hormones. Theca cells have the ability to efficiently generate androgen biochemical precursors to estrogens, but, importantly, they lack the *CYP19 aromatase* enzyme that is needed to convert the androgens into estrogens. Compared with theca cells, the internally located granulosa cells are not well vascularized, have less access to LDL, and lack the enzyme *17α-hydroxylase* that is required for synthesizing androgens. However, because granulosa cells do express CYP19 aromatase, they convert androgens derived from the theca into estrogens. An important biochemical difference between the two cells is that whereas granulosa cells express receptors for both FSH and LH, theca cells only express LH receptors. The mechanisms by which gonadotropins facilitate the synthesis of estrogens via interactions between these two cells, known as the *two-cell theory* of ovarian function, are summarized in Figure 22.10 and described next.

Theca Cell

Steroidogenically analogous to Leydig cells of the testis, a primary function of theca cells is to synthesize androgens. Like their male counterpart, theca cells possess LH receptors and lack FSH receptors. The binding of LH to its GPCR receptor stimulates the same cAMP-mediated intracellular signaling pathway previously discussed for Leydig cells (refer to Chapter 21: Male Reproductive System), ultimately resulting in the synthesis of androstenedione and smaller quantities of testosterone. These androgens diffuse into the neighboring granulosa cells where they are processed further.

Granulosa Cell

FSH, whose receptor also activates the cAMP-mediated intracellular signaling pathway, stimulates the synthesis of the CYP19 aromatase enzyme by activating its transcription and translation. FSH also stimulates the synthesis of the protein hormone inhibin, which is involved in negative feedback along the HPG. Androstenedione is converted into estradiol by the granulosa cell through its conversion into estrone by CYP19 aromatase, followed by estrone's processing into estradiol by 17β-HSD. A second pathway converts androstenedione into testosterone via 17β-HSD, and testosterone is subsequently processed into estradiol. In both cases, estradiol is released into general circulation where it will exert effects on diverse targets. This stimulation of CYP19 aromatase and protein sex hormones by FSH is analogous to the effects of FSH in the male Sertoli cells. Granulosa physiology differs from Sertoli cells in that granulosa cells also express LH receptors and are capable of progesterone synthesis. Preovulatory granulosa cells synthesize and upregulate the LH receptor in response to FSH- and estradiol-stimulated LH receptor gene transcription mediated via the cAMP signaling

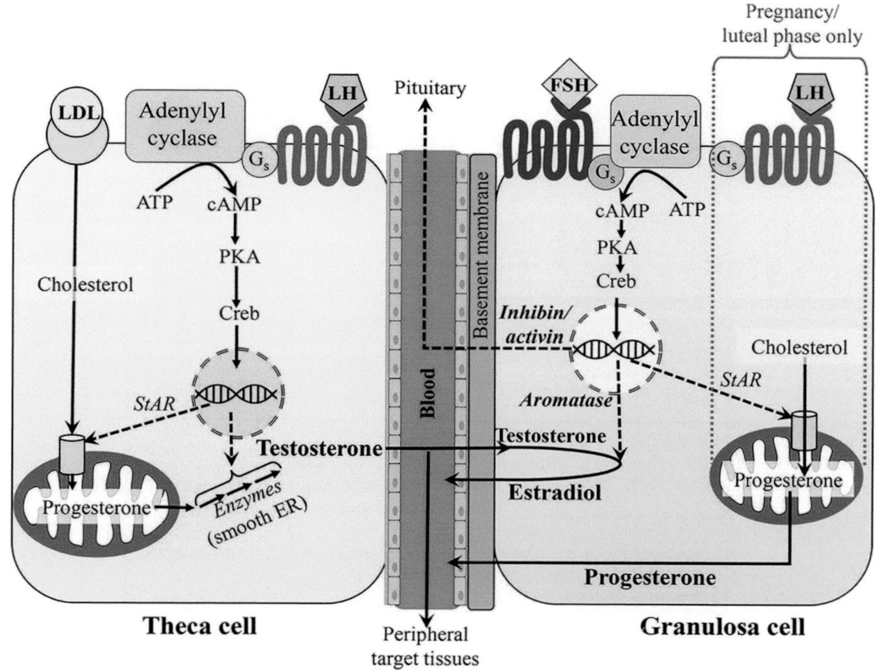

Figure 22.10 The two-cell theory of interaction between theca and granulosa cells of the ovary. *Left side*: The binding of LH to its receptor on the theca cell plasma membrane initiates a cascade of events that promotes the biosynthesis of testosterone. The LH receptor is a G protein-coupled receptor (GPCR) whose activation of adenylate cyclase (AC) converts ATP to cAMP, the major second messenger of LH action in theca cells. cAMP subsequently activates the cAMP-dependent protein kinase (PKA), an event that triggers a series of cytosolic and nuclear reactions including the activation of steroidogenesis acute regulatory protein (StAR), which is involved in cholesterol transport into mitochondria. Steroidogenic enzymes are also expressed in the mitochondria and smooth endoplasmic reticulum, which convert cholesterol into androstenedione and testosterone. The androgens diffuse into the adjacent granulosa cells. *Right side*: During the follicular phase, the primary hormone produced is estradiol. The binding of FSH to its receptor (also a GPCR) activates PKA, ultimately inducing the expression of aromatase, which converts androgens into estradiol that diffuses into general circulation. During the luteal phase as well as during pregnancy, the corpus luteum synthesizes both progesterone and some estradiol. The vascularization of the corpus luteum in the luteal phase makes blood low-density lipoprotein (LDL) accessible to the granulosa cells, and progesterone is synthesized by both the theca and granulosa cells during this time.

pathway.[36] This steroidogenic capacity of granulosa cells, however, plays a much greater role in the corpus luteum (see next) compared with the follicle.

Progesterone Synthesis by the Corpus Luteum

The corpus luteum forms from the residual post-ovulatory follicle. Specifically, the remnants of the follicle's theca cells luteinize into small *thecal-lutein* cells, and granulosa cells luteinize into large, hypertrophied *granulosal-lutein* cells. Upon the completion of corpus luteal maturation, both cells synthesize progesterone (Figure 22.10).[36] Progesterone synthesis by both cell types is thought to be mediated exclusively by the stimulation of LH receptors in response to rising levels of LH and, during pregnancy, chorionic gonadotropin.

The corpus luteum also synthesizes estrogens, but at lower levels compared with progesterone. As with the previous theca cells, thecal-lutein cells can synthesize progesterone and androgens, but not estrogens. The granulosa-lutein cells, like their follicular counterparts, possess CYP19 aromatase and use this enzyme to produce estrogens using the thecal-lutein-derived androgens as a substrate. In just a few days, the daily steroidogenic output of the ovary increases by an amazing 100-fold, from a few hundred micrograms of estrogens by the preovulatory follicle to 20 milligrams or more of progesterone by the corpus luteum.[36]

Peptide Hormone Synthesis by the Follicle and Corpus Luteum

Like the Sertoli cells of the testes, the granulosa cells of the follicle and granulosa-lutein cells of the corpus luteum synthesize the heterodimeric glycoprotein *inhibin* in response to FSH stimulation. Inhibin feeds back negatively on the pituitary gonadotropes to inhibit FSH synthesis. Similar to male physiology, in females activin is produced by extragonadal tissues, like the pituitary and hypothalamus, and promotes FSH synthesis through previously described autocrine-paracrine mechanisms.

Ovarian Hormone Interactions with the H-P Axis

In contrast to the adult testis, which maintains a relatively constant state of spermatogenesis, ovarian activity is cyclical in nature. In humans, the ovarian cycle is approximately 28 days long and is characterized by marked fluctuations in pituitary and ovarian hormone synthesis that vary according to the phase of the cycle (Figure 22.11). The first

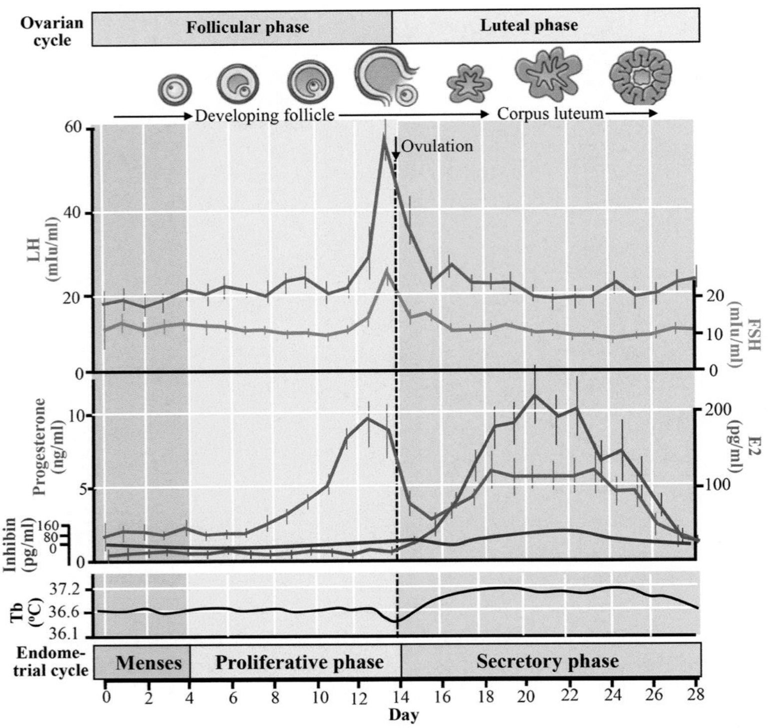

Figure 22.11 Changes in pituitary and ovarian hormone concentrations during the human menstrual cycle obtained from clinical samples. Vertical lines denote standard error, or variability of concentrations among clinical samples for any given day. The preovulatory LH surge is induced by positive feedback effects of peak concentrations of estradiol. The post-ovulatory drop in gonadotropin levels is mediated by negative feedback from rapidly rising levels of progesterone, estradiol, and inhibin. Basal body temperature may rise by 1 degree C during the luteal phase as hormones reset the hypothalamic thermostat.

Source of Images: Boron, W.F. and Boulpaep, E.L., 2005. Medical Physiology: A Cellular and Molecular Approach, updated 2nd ed., and from Thorneycroft, I.H., et al. 1971. Am. J. Obstet. Gynecol., 111(7), pp. 947-951.

14 days of the cycle comprise the *follicular phase*, characterized by the growth, development, and maturation of the dominant follicle. The follicular phase can be divided into two components: (1) an FSH-dependent follicular growth phase and (2) an LH-dependent follicular maturation phase.[37] The early follicular phase begins with FSH at relatively high levels carried over from the preceding cycle. This elevated FSH promotes the upregulation of LH receptors that will be necessary for initiating the process of dominant follicle maturation.[38] FSH gradually declines throughout most of the follicular phase in response to rising levels of inhibin and estradiol produced by the growing follicle. However, when the secretion of estrogens attains a sufficient threshold during the late follicular phase, this induces *positive* feedback along the H-P axis, ultimately initiating the preovulatory gonadotropin surge. Increased production of LH during the surge stimulates the follicle to produce increasing levels of progesterone. The rising progesterone augments the positive feedback effects of estradiol on the H-P axis, further amplifying the gonadotropin surge and ultimately promoting ovulation.

Ovulation demarcates the start of the **luteal phase of the ovarian cycle**, characterized in part by a rapid increase in progesterone synthesis by the newly formed corpus luteum (Figure 22.11), which in conjunction with an increase in inhibin production feeds back negatively along the H-P axis to inhibit the surge in gonadotropins. Note in Figure 22.11 that compared with modest peak values during the follicular phase, the ratio of the peak progesterone-to-estradiol levels during the luteal phase is dramatically higher. Progesterone powerfully inhibits GnRH secretion when bound to its receptors located in the hypothalamic anteroventral periventricular (AVPV) and arcuate (ARC) nuclei.[39] When administered before or concurrent with estradiol, progesterone inhibits the estradiol-induced positive feedback and abolishes the preovulatory GnRH and gonadotropin surge.[39,40] Indeed, when administered singly or in combination with estrogens, progesterone treatment can be used as an effective contraceptive (discussed later in Section 22.7: Hormonal Contraception and Fertility Control).

Beginning approximately 48 hours post-ovulation, the corpus luteum begins to gradually increase levels of estradiol production in parallel with the rise of serum progesterone. Throughout most of the luteal phase, gonadotropin levels are suppressed via negative feedback from increasing concentrations of three hormones: progesterone, estradiol, and inhibin. During the late luteal phase, the gradual degradation of the corpus luteum is accompanied by lower levels of progesterone, estradiol, and inhibin synthesis, allowing gonadotropin levels to begin to rise again and progress into the follicular phase of the next cycle.

Circulating ovarian steroid and peptide hormones exert feedback along the H-P axis (Figure 22.12). Importantly, whereas negative feedback by sex steroid hormones on the H-P axis predominates in females during the luteal phases of the ovarian cycle, *positive* feedback predominates during the mid- to late follicular phase. The endocrine basis of differential ovarian cycle phase-specific negative and positive

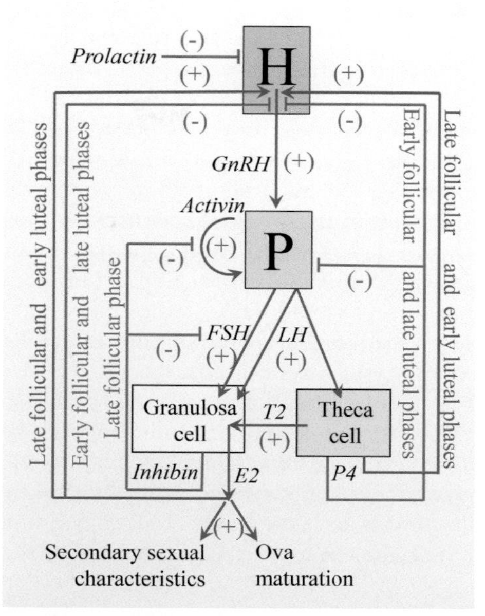

Figure 22.12 The hypothalamic-pituitary-ovarian (HPO) axis. Note the role of positive feedback in the ovarian axis during the late follicular and early luteal phases, as estradiol and progesterone promote the ovulatory GnRH/gonadotropin surge. Also worth noting here is that elevated levels of the pituitary hormone prolactin, which is normally produced by lactating women, inhibits GnRH synthesis. Stimulatory and inhibitory actions are denoted by green arrows and red lines, respectively. T2, testosterone; E2, estradiol; P4, progesterone.

feedback signaling is addressed in Chapter 23: The Timing of Puberty and Seasonal Reproduction. Briefly, in female mammals kisspeptin/neurokinin B/dynorphin (KNDy) neurons of the hypothalamic *arcuate nucleus* (ARC) respond to elevated estrogens and progesterone with negative feedback, repressing GnRH and subsequent steroid hormone synthesis. In contrast, KNDy *anteroventral periventricular nucleus* (AVPV) neurons respond to rising estrogens with positive feedback, increasing GnRH and the synthesis of estrogens.

As with the male testis, the female ovary produces inhibins that feed back onto the pituitary to negatively inhibit gonadotropin synthesis. Females differentially synthesize both inhibin A and B isoforms during different phases of the ovarian cycle. Activin functions to stimulate pituitary FSH production by increasing FSH mRNA synthesis. Its actions are modulated by inhibin, an antagonist of the activin receptor.

SUMMARY AND SYNTHESIS QUESTIONS

1. In the absence of pregnancy, the corpus luteum degenerates by the end of the luteal phase. What maintains the corpus luteum during the luteal phase, why does it degenerate, and how does pregnancy maintain the corpus luteum?

2. Clomiphene citrate is a selective estrogen receptor modulator that primarily exerts *anti*estrogenic properties. For over 50 years this drug has been the first line of treatment for reproductive-aged women with subfertility or infertility. How would the use of an antiestrogenic drug promote ovulation?
3. You are culturing granulosa and theca cells together in a petri dish. The cell culture media contains the following compounds in it: LDLs, LH, and FSH. What effects would each of the following drugs, when added singly to the culture and following incubation for several hours, have on the levels of (1) estradiol, (2) testosterone, and (3) progesterone in the culture media? Specifically, do levels of each steroid hormone remain *unchanged*, *increase*, or *decrease*? Explain your reasoning (assume that the cells contain baseline levels of all enzymes needed for steroidogenesis). A. An antagonist of the FSH receptor. B. An antagonist of the LH receptor. C. An inhibitor of adenylate cyclase. D. The CYP19 aromatase inhibitor, letrozole. E. 8-bromo-cAMP, a cAMP analog that diffuses across plasma membranes. F. An inhibitor the enzyme phosphodiesterase (phosphodiesterase converts cAMP to AMP, inactivating it). G. Removal of theca cells from the culture. H. Removal of granulosa cells from the culture. I. An inhibitor of LDL receptor function.

Ovarian Hormone Blood Transport and Receptor Interactions

LEARNING OBJECTIVE Describe how estrogens and progesterone are transported in the blood, and outline sex steroid hormone receptor function.

KEY CONCEPTS:

- In the blood, most estrogens are transported by low- or high-affinity binding proteins.
- The small unbound fraction of hormone is considered to be bioavailable.
- Most of the effects of estrogens and progesterone are mediated by nuclear-localized receptors that operate as ligand-dependent gene transcription factors.
- Some of the effects of sex steroid hormones are also driven by the actions of nongenomic, membrane-associated G protein-coupled receptors.

Blood Transport

In females, about 60% of estradiol is bound with low affinity to the blood protein *albumin*, 38% is transported bound with high affinity to *sex hormone-binding globulin* (SHBG), and the remaining 2% is unbound.[41] The *bioavailable* estradiol fraction consists of the combination of free and weakly bound estradiol. Therefore, changes in SHBG concentrations will affect the bioavailability of estradiol. Because estrogens induce SHBG synthesis by the liver, SHBG levels in women are generally twice as high as in men, and SHBG levels increase ten-fold in women during pregnancy.[42] In contrast to estrogens and androgens, which bind preferentially to SHBG, progesterone (as well as glucocorticoids) binds preferentially and with high affinity to *corticosteroid-binding globulin* (CBG). Like other steroids, progesterone also associates with the low-affinity transport protein albumin.

Receptor Function and Distribution

As is the case for other steroid hormones, the actions of estrogens and progesterone are thought to be mediated primarily via nuclear-localized receptors that operate as homodimeric, ligand-dependent gene transcription factors. Whereas in mammals the effects of androgens are mediated through a single androgen receptor (AR) gene localized to the X chromosome, the classical actions of estrogens are mediated through two estrogen receptors, ERα and ERβ, each encoded by separate genes located in humans on chromosomes 6 and 14, respectively. The two estrogen receptors are structurally and functionally distinct, each targeting specific genes in a cell- and tissue-specific manner,[43] and their actions may be exerted through homo- or heterodimerization of ERα and ERβ. The progesterone receptor (PR) is encoded by a single gene located on human chromosome 11, though this gene expresses two different progesterone receptor isoforms (PR B and a shorter, truncated version, PR A) due to alternative translational start sites.

Curiously Enough . . . Estrogen receptors appear to be the most ancient of all receptors belonging to the steroid hormone receptor family[44,45] and are thought to have arisen early in vertebrate evolution following gene duplication and mutation events.

Importantly, some of the effects of these sex steroid hormones are also known to be driven by the actions of nongenomic, membrane-associated receptors. For example, one estrogen receptor, a GPCR called *G protein-coupled estrogen receptor* (GPER), localizes to both the plasma membrane and the smooth endoplasmic reticulum membrane and has been shown to mediate both estrogen-dependent kinase activation and transcriptional responses.[46,47] Similarly, plasma membrane-localized progesterone receptors are known to contribute to the actions of progesterone.[48]

The influences of ovarian hormones over hypothalamic, pituitary, and ovarian function were described above, and their influence over uterine function will be discussed below under the sub-section, **The Uterine/Endometrial Cycle.** Some important nonreproductive effects of these hormones on some tissues is summarized in Table 22.3.

Table 22.3 Some Nonreproductive Effects of Ovarian Steroid Hormones on Target Tissues

Target	Hormone	Effect
Vascular endothelial and smooth muscle cells	Estrogens	Cardiovascular protective role Reduced blood pressure
Cholesterol profile	Estrogens	Promote healthier blood cholesterol profile Increase high-density lipoprotein (HDL) to low-density lipoprotein (LDL) ratio
Skeletal system	Estrogens	At low levels, promote adolescent growth spurt At high levels, promote growth plate fusion and terminates growth promote bone mineral density
Brain	Estrogens	Exert neuroprotective effects on the brain that reduce neuronal loss following stroke or multiple sclerosis Protect against neurodegenerative disease, depression, anxiety, and drug abuse.
Fat deposition	Estrogens	Promotes greater fat storage as subcutaneous depots
Body temperature	Estrogens	Alter the set-point of the hypothalamic thermostat The sensation of "hot flashes" associated with the perimenopausal period may result from fluctuating estrogens
Water balance Antidiuretic hormone (ADH)	Estrogens	Lowers the threshold for the secretion of ADH
Renin-angiotensin-aldosterone system (RAAS)	Estrogens and progesterone	Stimulate angiotensinogen synthesis and renin synthesis
Atrial natriuretic peptide (ANP) receptor	Estrogens and progesterone	Estrogens promote ANP's inhibitory effects on aldosterone secretion Progesterone blocks ANP's inhibitory effects on aldosterone secretion

SUMMARY AND SYNTHESIS QUESTIONS

1. Following menopause, or the cessation of the ovarian cycle, will women have lower or higher levels of SHBG? Explain your reasoning.
2. Considering your answer to the previous question, after menopause will women have higher or lower levels of bioavailable *testosterone*?

The Uterine/Endometrial Cycle

LEARNING OBJECTIVE Compare and contrast the human menstrual cycle with the estrous cycle in other mammals.

KEY CONCEPTS:

- The endometrium of the uterus is the site of implantation for an embryo.
- The thickness of the endometrium changes dramatically throughout the menstrual cycle in response to the ovarian steroid hormones.
- The shedding of the unfertilized oocyte and the uterine lining (menstruation) occurs approximately every 28 days for most women.
- The proliferative phase of the menstrual cycle parallels the follicular phase of the ovarian cycle, and the secretory phase parallels the luteal phase of the ovarian cycle.
- Whereas increasing levels of estrogens during the proliferative phase promote a thickening of the uterine endometrium, decreasing concentrations of progesterone and estrogens during the late luteal phase promotes the degeneration and shedding of the outer endometrial layers ("bleeding").
- The vast majority of mammals do not menstruate, but instead exhibit estrous cycles characterized, in part, by the resorption of the endometrium instead of its excretion.
- In contrast to menstruating species, whose females may be sexually receptive throughout the menstrual cycle, most mammals with estrous cycles are only sexually receptive at specific times of the month or year.
- In induced ovulators, ovulation is induced directly via copulatory stimulation during mating.

The Uterus

The uterus is a highly muscularized, thick-walled organ that functions as the site for embryonic implantation, fetal development, and growth. The largest component of the uterus, the *myometrium*, consists of smooth muscle (Figure 22.13). Overlying the myometrium is the *endometrium*, a highly vascularized, deeply involuted, multilayered tissue specialized for embryonic implantation and the support of pregnancy. The endometrial surface consists of simple columnar epithelial tissue with many associated exocrine glands. The uterus is an amazingly distensible organ, capable of expanding during pregnancy from the size of a fist to the size of a full term baby.

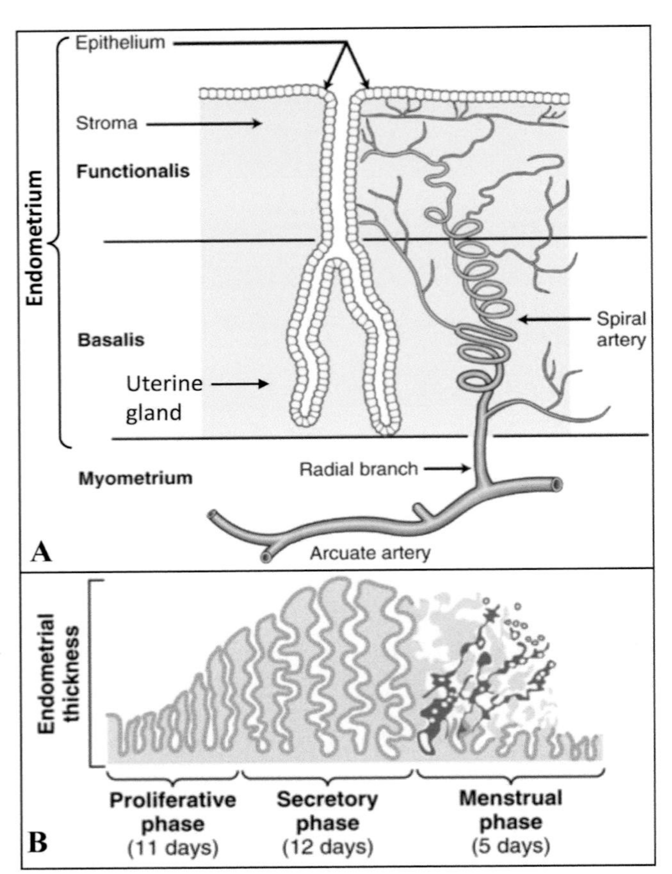

Figure 22.13 Functional anatomy of the uterine wall. (A) The endometrium is a multilayered structure penetrated by the spiral arteries and their capillaries. The surface of the endometrium is composed of a continuous layer of epithelial cells. Deep invaginations of the superficial layer almost juxtapose with the myometrium-endometrium junction, forming uterine glands. Whereas the functionalis (superficial layer) is shed during menstruation, basalis is a permanent layer that regenerates the endometrium following each menstruation. The myometrium constitutes a thick smooth muscle layer. **(B)** The phases of endometrial growth and degeneration during each monthly menstruation. The proliferative phase denotes the growth and regeneration of the functionalis layer; the secretory phase denotes active secretion by the mature uterine glands; the menstrual phase denotes degeneration of the functionalis layer and associated vasculature.

Source of Images: (A) Melmed, S., et al. 2011. Williams textbook of endocrinology, expert consult. 12th. Section IV. Elsevier. Used by permission. (B) Hall, J.E. and Hall, M.E., 2020. Guyton and Hall textbook of medical physiology e-Book. Elsevier Health Sciences. Used by permission.

The thickness of the endometrium changes dramatically throughout the menstrual cycle in response to the ovarian steroid hormones estrogens and progesterone. Therefore, the phases of the uterine cycle correspond with and are directly regulated by the phases of the ovarian cycle, and changes in the endometrium are considered one of the most sensitive indicators of hypothalamic-pituitary-ovarian hormonal signaling status. Near the time of ovulation, the uterus generates a thick layer of vascularized endometrial tissue in preparation to receive an embryo. If the oocyte has not become fertilized, resulting decreases in progesterone and estrogens from the ovaries will induce the endometrial blood vessels to atrophy and the outermost layers of the endometrium to be shed. The shedding of the uterine lining is known as **menstruation** ("bleeding"), which for most women occurs approximately every 28 days (Figure 22.13). Specifically, the *stratum functionalis*, which consists of the epithelial lining and uterine glands, is shed during menstruation. Although the underlying *stratum basalis* remains intact, it houses vasculature that contributes to the menstrual flow. Following menstruation, a new stratum functionalis is generated from the stratum basalis. If fertilization occurs, the embryo will implant itself into the endometrial lining, where it will develop into a fetus, causing changes within the endometrium that promote the formation of the placenta. Ovarian hormones also affect the size of the myometrium, particularly during pregnancy, when smooth muscle cell numbers rise and individual cell lengths increase by up to ten-fold. The myometrium plays a critical role during childbirth, with hormone-triggered waves of smooth muscle contraction ultimately pushing the fetus out of the uterus, through the cervix and vagina, and out of the mother's body.

The Human Menstrual Cycle

The lining of the uterus, or *endometrium*, is a morphologically dynamic target of ovarian sex steroid hormones and is thus strongly influenced by the phases of the ovarian cycle. These cyclic morphological transformations are generally referred to as the *uterine* or *endometrial cycle*, and in humans it is also called the *menstrual cycle*. The menstrual cycle is divided into three phases: the proliferative phase, which parallels the **follicular phase of the ovarian cycle**, the secretory phase, which parallels the luteal phase of the ovarian cycle, and menses, which occurs in response to falling estrogen and progesterone levels between the secretory and proliferative phases (Figure 22.14). Rising levels of estrogens end menses and initiate the proliferative phase. A menstrual cycle's duration consists of the number of days between the first day of menstrual bleeding and the beginning of menstrual bleeding of the next cycle, which on average lasts 28 days.

The **proliferative phase**, which begins from the first day of menses and lasts until ovulation, is characterized by the proliferation of endometrial cells, enlargement of endometrial glands, and enrichment of the endometrium with blood vessels, all in response to rising levels of estradiol. Following ovulation the **secretory phase** begins, characterized by the secretion of progesterone and estradiol from the corpus luteum. This promotes an increased thickening of the endometrium, as well as secretion of fluids by the endometrial glands that are essential for survival and

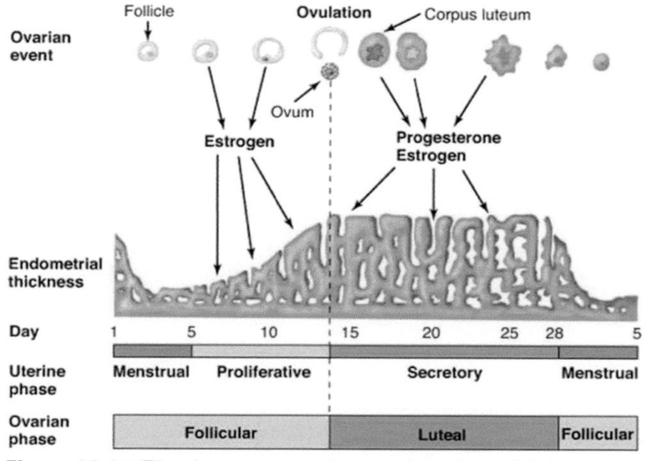

Figure 22.14 The human ovarian and endometrial cycles. The menstrual cycle is comprised of parallel ovarian and endometrial cycles. Day 0 is defined as the start of the menses and follicular phase. Ovulation occurs on approximately day 14, and the total cycle lasts approximately 28 days. During the proliferative phase, the endometrium is stimulated by the estrogen estradiol. After ovulation, the secretion of estradiol and progesterone during the ovarian cycle's luteal phase corresponds with the maintenance of the endometrium's vascularity and the secretion by uterine exocrine glands during the secretory phase. Following the degeneration of the corpus luteum, the endometrium degenerates and the uterus reenters menses.

Source of Images: Barrett, K.E., et al. 2010. Ganong's review of medical physiology twenty. McGraw Hill LLC. Used by permission.

development of a prospective embryo. In the absence of a pregnancy, declining progesterone levels are accompanied by decreased blood flow to the superficial endometrial layers (the compacta and spongiosa), starving them of their blood supply and causing them to degenerate. Furthermore, the degenerating tissue detaches from the surrounding extracellular matrix and connective tissue via the actions of *matrix metalloproteinases* (MMPs),[49] leading to *menses* (menstruation). MMPs also play key roles in remodeling the remaining tissue, preparing it for the next cycle. Finally, prostaglandins are released from the endometrium that induce uterine smooth muscle contractions and the sloughing off of the degraded endometrial tissue. In women who suffer from excessive menstrual bleeding (menorrhagia), the amount of menstrual bleeding can be reduced by treatment with prostaglandin synthetase inhibitors.[50]

Menstrual fluid is composed primarily of endometrial tissue, blood, and proteolytic enzymes. Following menstruation, the inner lining of most of the uterus lacks epithelial cells, consisting primarily of remnant glands and nonepithelial stromal cells. Notable exceptions to this are the areas close to the Fallopian tubes and the lower regions of the uterus, which retain epithelial cells. Upon the initiation of the proliferative phase of the next cycle, and stimulated by rising estradiol levels, epithelial cells from the aforementioned regions of the uterus proliferate and repopulate the uterine lining. Estradiol also induces the stromal component of the endometrium to proliferate, differentiate, and thicken. The proliferative effects of estradiol are thought to be regulated by the upregulation

Clinical Considerations: The enigma of endometriosis

Endometriosis is a condition defined as the presence of endometrial tissue outside of the uterus, primarily, but not solely, in the pelvic peritoneum and ovaries.[53] It is an estrogen-dependent condition most often affecting women between 25 years and 35 years of age. The malady is characterized in part by pelvic pain, chronic bleeding, and infertility caused by a persistent state of inflammation, and it is also associated with a 50% increase in the risk of developing ovarian cancer.[54] Endometriosis occurs exclusively in menstruating mammals, such as humans and other closely related primates.[55]

The ultimate cause of endometriosis remains enigmatic. A prevailing theory is known as the *retrograde menstruation hypothesis*, which postulates that during menstruation desynchronized uterine contractions drive endometrial fragments through the Fallopian tubes, and these subsequently may implant, grow, and invade into the pelvic peritoneum or the ovary.[56] As such, the likelihood of developing endometriosis increases with any factor that could augment pelvic contamination, such as long duration of menstrual flows or large quantities of backwashed menstrual tissue caused by vaginal obstruction of outflow.[57] Indeed, the risk of developing endometriosis increases in women with regular and abundant menstrual flows.[58]

The growth of ectopic endometrium is thought to be fueled by estrogens, and ectopic tissue has also been shown to express higher levels of estrogen receptor β (ERβ) and altered expression of estrogen receptor α (ERα) compared with normal tissue.[57] Furthermore, the incidence of endometriosis increases in women who were exposed to estrogen-mimicking chemicals such as diethylstilbestrol (DES) *in utero*.[59] Interestingly, under conditions of endometriosis the promoter region of the gene encoding ERβ has been suggested to be hypomethylated, resulting in the pathological overexpression of ERβ[57] (Figure 22.15). Subsequently, ERβ overexpression suppresses both ERα and progesterone receptors. This leads to progesterone resistance and a compromised ability to inactivate estradiol synthesis in endometriotic tissues, enhancing their survival and inflammation. Therefore, in addition to a genetic predisposition to developing endometriosis, an epigenetic basis (a nongenetic, possibly environmentally induced influence on gene expression) may also be present.

Treatments for endometriosis are designed to temporarily mitigate its pain and other accompanying symptoms, though no current treatment exists that can cure the disease. Treatments include the surgical removal of endometriotic tissue, oral hormone contraceptives, gonadotropin-releasing hormone agonists and antagonists, selective estrogen receptor modulators (SERMS), and CYP19 aromatase inhibitors. Low estrogen-dose oral contraceptives are often used for chronic treatment to induce a pseudopregnant state that causes the differentiation and subsequent atrophy of the ectopic, as well as normal, endometrium.[54]

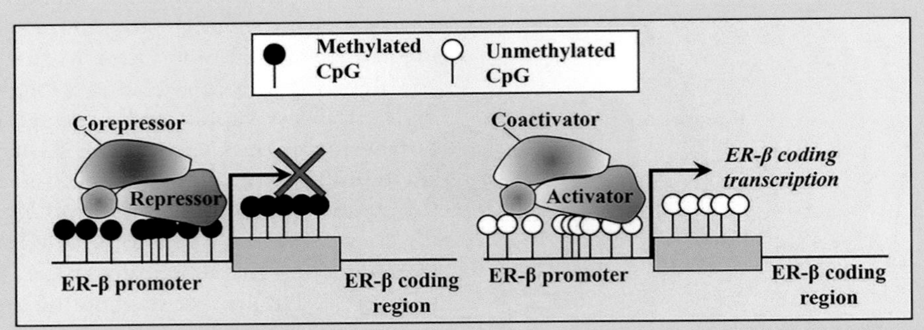

Figure 22.15 Differential methylation of CpG-rich regions of the estrogen receptor β (ERβ) promoter modulate its expression in the endometrium. A lack of promoter methylation may be associated with promoter activation and pathologic overexpression of these nuclear receptors in endometriotic stromal cells.

of *proto-oncogenes*, genes that are normally involved in mediating cell proliferation, but when mutated or overexpressed may promote the development of cancer.[51] In contrast to estradiol, progesterone exerts anti-proliferative effects on the epithelial lining while inducing its differentiation,[52] but promotes proliferation of the underlying stroma. Should fertilization occur, the secretion of *chorionic gonadotropin*, a luteinizing hormone receptor agonist, by the developing embryo and placenta ensures continued progesterone secretion by the corpus luteum, maintaining the endometrium.

Estrous Cycles

In the vast majority of mammals, reproductive cyclicity is described as an **estrous cycle**, whose timing and frequency is discussed in Chapter 23: The Timing of Puberty and Seasonal Reproduction. While the profile of ovarian hormone secretion is similar in menstrual and estrous cycles, the two strategies differ in several important ways (summarized in Table 22.4 and Figure 22.16). Physiologically, the most obvious difference is that whereas in the absence of fertilization the uterine lining is shed in menstruating animals, it is resorbed in animals with estrous cycles. Although during the estrous cycle some animals may display a bloody discharge, unlike true menstruation this fluid does not contain endometrial stromal or epithelial tissue. From an endocrine perspective, in animals with estrous cycles the corpus luteum typically produces relatively low levels of estradiol compared with menstruating animals. Therefore, whereas the menstrual cycle is characterized by a primary surge of estradiol at the end of the follicular phase followed by a secondary surge during the luteal phase, the estrous cycle generally only has one pronounced estradiol surge prior to ovulation. The relative lengths of the follicular and luteal phases also differ between these reproductive cycle strategies, with the two phases being relatively equal in the menstrual cycle, but in the estrous cycle the follicular phase is generally short, with most follicular growth taking place during the preceding luteal phase.

Behaviorally, animals with estrous cycles exhibit a distinct period of sexual receptivity called **behavioral estrus** (note that the spelling of behavioral *estrus* is different from *estrous* cycle). Also known as "heat", behavioral estrus occurs during or immediately following the proestrus stage of the estrous cycle, a period of time near ovulation when the females are most sexually attractive, proceptive, and receptive to males.[60] As such, behavioral estrus is an outward manifestation of the peri-ovulatory phase of the ovarian cycle. In contrast, menstruating animals may be sexually active throughout their menstrual cycle. Finally, whereas animals with estrous cycles often (but not always) breed seasonally, menstruating animals typically breed throughout the year.

Table 22.4 Comparison of a Typical Estrous Cycle with That of the Human Menstrual Cycle

Typical Estrous Cycle	Human Menstrual Cycle
Exhibit *behavioral estrus* ("heat"): recurrent periods corresponding to ovulation when the female is receptive to the male.	Females are receptive to males throughout the menstrual cycle. Receptivity is dissociated from the timing of ovulation.
Follicular phase is short, with most follicular growth taking place during the preceding luteal phase.	Follicular phase is approximately the same length as the luteal phase.
In the absence of fertilization, declining progesterone levels during the late luteal phase correspond with resorption of the endometrial lining.	In the absence of fertilization, declining progesterone levels during the late luteal phase correspond with breakdown and shedding of endometrial lining and bleeding.
One surge of estradiol during the follicular phase, but no secondary surge in the luteal phase.	Two surges in estradiol production: one at the end of the follicular phase, and one secondary surge during the luteal phase.
Cycle may be seasonal.	No seasonality in cycling.

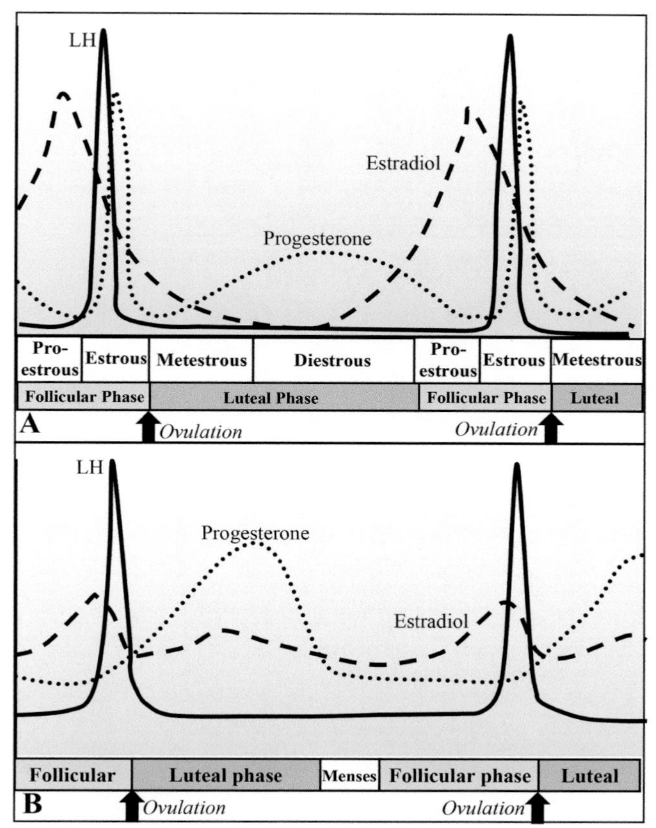

Figure 22.16 Relative reproductive hormone concentrations from the estrous cycle of an idealized mammal (A) compared with that of the human menstrual cycle (B). Proestrus corresponds with the regression of the corpus luteum (progesterone declines) and start of the follicular phase during which the follicles are rapidly growing, differentiating into secondary and Graafian follicles, and producing increasing amounts of estrogens and gonadotropins. Ovulation usually occurs during estrus, soon after the peak in gonadotropins is reached, though some mammals, like cows, ovulate during metestrus. Proestrus and estrus comprise the follicular phase of the ovarian cycle. Corpora lutea develop during metestrus and function at optimal capacity during diestrus. Metestrus and diestrus comprise the luteal phase.

Source of Images: Julian Lombardi. 1998. Comparative Vertebrate Reproduction. Springer.

If conception does not occur during a particular estrous cycle, animals reabsorb the endometrial lining of the uterus, and the cycle begins anew. In the case of humans, apes, Old World monkeys, elephant shrews, and some species of bats, the ovarian cycle is more accurately called the *menstrual cycle*, which primarily differs from the estrous cycle such that in nonpregnant females the uterine lining is sloughed off (i.e. "bleeding") instead of being resorbed. Interestingly, among most primates (with the notable exception of humans) the menstrual cycle is characterized by the display of a behavioral estrus. Humans, in contrast, are one of the only species whose menstrual cycle occurs in the absence of an obvious behavioral estrus and has been termed a "hidden estrus" or a "concealed ovulation".[60,61]

Curiously Enough . . . The term "estrus" derives from *oistros*, the Greek word for gadfly. Estrus literally means "in a frenzied state", referring to the similarity in cattle between the hyperactive behavior exhibited by females at estrus and the frenzied response that occurs when gadflies swarm around cattle.

The Pap Test

Not only does the uterus change with the ovarian cycle, but vaginal cytology is also altered throughout ovarian follicle development (Figure 22.17). An important advance in the study of female reproduction and behavior was the development of a relatively simple, rapid, and non-surgical technique by Stockard and Papanicolaou in 1917[62,63] of swabbing cells acquired from the vagina of guinea pigs and observations that changes in their histology were closely correlated with the ovarian cycle. This powerful technique is now used to study behavioral and physiological changes associated with different stages of the estrous cycle in many animals (Figure 22.18). A modified form of this test that is part of a typical gynecological exam in humans, called the "Pap" test (named after Dr. Papanicolaou), obtains cells from the cervix instead of the vagina.

Spontaneous and Induced Ovulators

Female mammals may be placed into one of two categories based on their mechanism of initiating ovulation. **Spontaneous ovulators** (e.g. humans, horses, cows, sheep, and most rodents) ovulate at regular intervals, and their ovulation is dependent upon the positive-feedback induction of gonadotropins by estradiol during the late follicular phase. The generation of the surge in LH occurs independently of any mating cues, and therefore occurs "spontaneously". By contrast, following the completion of follicular growth, **induced ovulators** (e.g. domestic cats, rabbits, ferrets, alpacas, llamas, and camels) remain in a state of sexual receptivity during estrus until ovulation is induced directly via copulatory stimulation during mating. Ovulation may occur even in the absence of positive feedback from estradiol.[64] Instead, in induced (also called *reflex*) ovulators, genital-somatosensory stimulation by the penis during copulation appears to activate noradrenergic pathways in the brainstem and midbrain that promote the immediate release of GnRH from the hypothalamus, which in turn induces a rapid surge in pituitary LH, driving ovulation[65] (Figure 22.19). Interestingly, since subminimal stimulation is insufficient for ovulation, the male genitals of some species of induced ovulators, like domestic cats, possess morphological attributes such as penile spines that may augment stimulation in the female, increasing the probability of inducing ovulation.[66] In addition to somatosensory stimulation of ovulation, other stimuli may include chemical signaling compounds present in the semen,[64] as well as visual, olfactory, pheromonal, auditory, tactile, and emotional signals.[67,68]

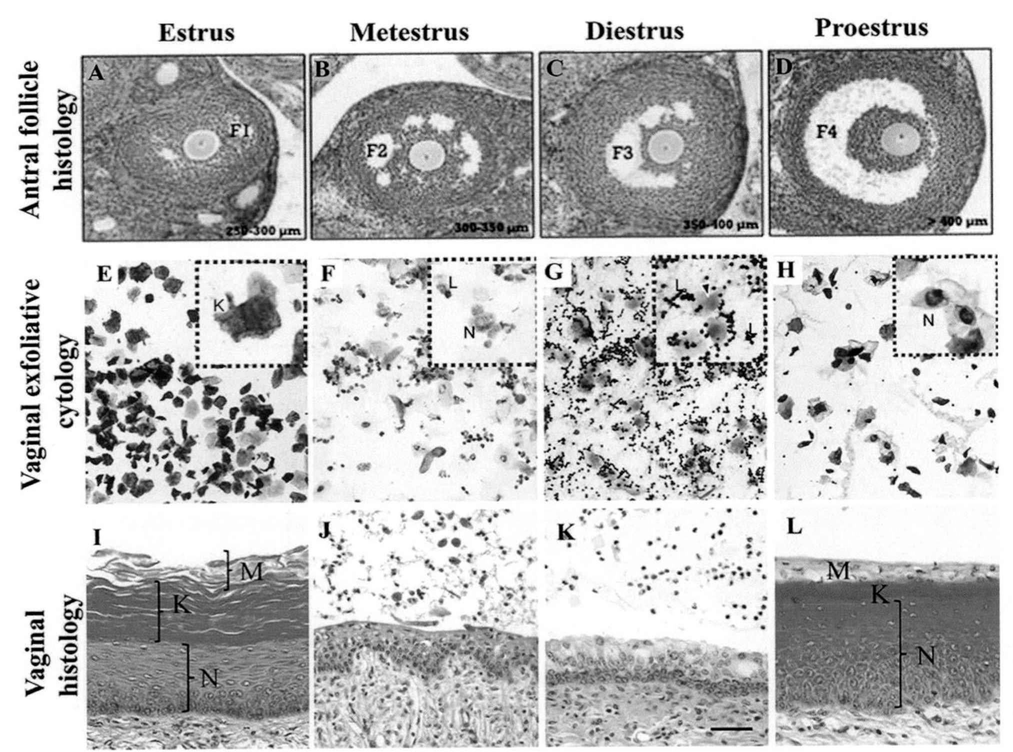

Figure 22.17 Changes in vaginal cytology correspond with ovarian follicle development in the mouse estrous cycle. (A–D) Ovarian antral follicle progression based on antrum size from the F1 to preovulatory F4 stage. (E–H) Progression of vaginal exfoliative cytology in each phase of the estrous cycle. In estrus (at receptivity), only keratinized cells (*K*) are present. At metestrus, nucleated epithelial cells (*N*, which are primarily mucoid cells at this stage) and leukocytes (*L*) are present. At diestrus, only leukocytes can be observed, accompanied by few nucleated epithelial cells. At proestrus, mostly nucleated epithelial and few keratinized cells are present along with some leukocytes. (I–L) Progression of vaginal histology in each phase of the estrous cycle. In estrus, there is gradual shedding of superficial mucoid epithelial (*M*) and keratinized layers. At metestrus, there has been a complete detachment of the keratinized layer as well as some underlying nucleated epithelial cells. Leukocyte infiltration is also evident. By diestrus, the proliferation of new epithelial cells has begun, and there is little shedding of remaining epithelial cells. By proestrus, the mucoid and keratinized layers, as well as underlying epithelial cell layers, have regenerated to their maximum thicknesses.

Source of Images: (A-D) Gaytan, F., et al. 2017. Sci. Rep., 7(1), pp. 1-11. (E-L) Sugiyama, M., et al. 2020. Cell Tiss. Res., pp. 1-8. Used by permission.

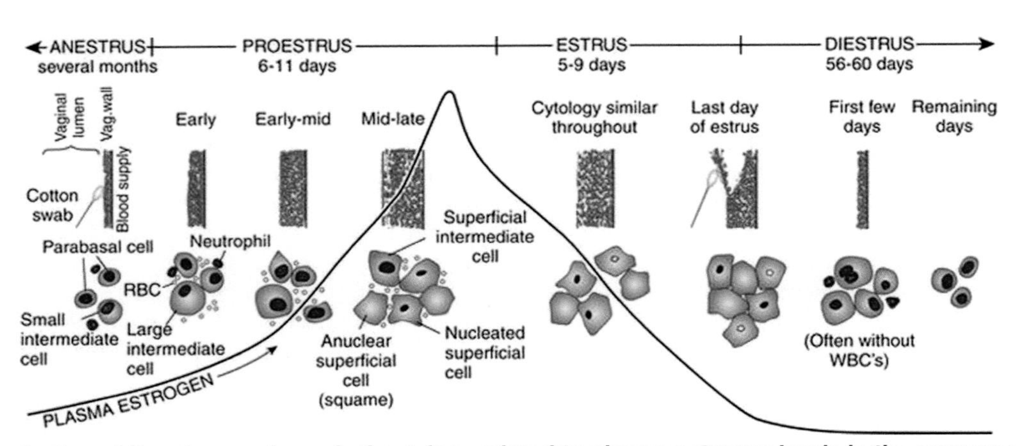

Figure 22.18 Illustration of the changes in vaginal cytology related to plasma estrogen levels in the average canine estrous cycle.

Source of Images: Feldman EC, Nelson RW: Ovarian Cycle and Vaginal Cytology. In: Feldman EC, Nelson RW, editors: Canine and feline endocrinology and reproduction, ed 3, Philadelphia, PA, 2004, Saunders, p. 755.) Used by permission.

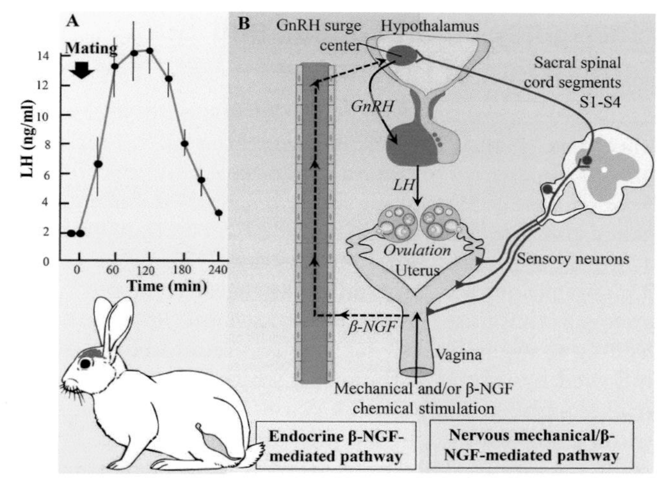

Figure 22.19 Copulation induces a surge of LH secretion and ovulation in rabbits. (A) The LH surge increases over seven-fold almost immediately after copulation. Data depict the average LH concentration for nine rabbits. **(B)** Depiction of the copulation-induced neuroendocrine reflex pathway that ends with stimulation of ovulation by the LH surge. Semen induces ovulation in rabbits via an endocrine- and a nervous-mediated pathway by which β-NGF, mainly synthesized in the uterus, acts on uterine/cervix afferent neurons projecting to the GnRH hypothalamic surge centers. Mechanical stimulation during copulation also stimulates the neural reflex arc.

Source of Images: Tsou, R. C., et al. (1977). Endocrinology 101, 534-5.

SUMMARY AND SYNTHESIS QUESTIONS

1. From the perspective of maximizing fecundity, what is a theoretical advantage of induced versus spontaneous ovulation?
2. How could each of the following drugs be used to treat endometriosis? A. Oral hormone contraceptives. B. Gonadotropin-releasing hormone (GnRH) antagonists. C. GnRH agonists. D. Certain selective estrogen receptor modulators (SERMs), like raloxifene. E. CYP19 aromatase inhibitors.

Hormonal Contraception and Fertility Control

LEARNING OBJECTIVE Describe modern approaches to contraception and fertility control, and discuss the non-contraceptive risks and benefits of hormone-based contraceptives.

KEY CONCEPTS:

- The contraceptive effects of progesterone are significantly enhanced in the presence of estrogens.
- The combined oral contraceptive pill (COCP) prevents conception by inhibiting both follicle growth and ovulation through its negative feedback actions on the hypothalamus and pituitary.
- Progestin-only oral contraceptives are estrogen-free tablets that increase the level of progestogen to mimic pregnancy and ensure that ovulation and implantation cannot occur.
- COCPs confer health benefits that extend beyond the prevention of pregnancy, such as reduced symptoms associated with painful or irregular menstruation, reductions in conditions caused by elevated androgen hormones, and reduction in the risk of developing epithelial ovarian and endometrial cancer.
- Estrogen-based contraceptives may slightly increase the risk of developing breast or cervix cancers, as well as thromboembolic and cardiovascular diseases.

Contraception, which literally means "against conception", refers to any one of several methods used to prevent fertilization. While not technically correct, the term is often extended to also include strategies of fertility control that either inhibit implantation of the embryo soon following fertilization ("emergency contraception") or remove a newly implanted embryo. Methods of contraception and fertility control include surgical sterilization (e.g. ligation of the Fallopian tubes), prescription of oral, transdermal, subdermal, or intrauterine applied steroid hormones (i.e. *hormonal contraception*), use of physical barriers to insemination (condoms, diaphragms, sponges), and intrauterine devices that inhibit implantation by inducing a chronic state of endometrial inflammation. Methods of behavioral contraception include the natural rhythm method (i.e. avoiding sex near the time of ovulation) and withdrawal (removal of the penis from the vagina before ejaculation). Of these, the most effective preventatives are hormonal and surgical contraception, with the least effective being the rhythm and withdrawal techniques (a list of failures in contraceptive function during the first year of use is listed in Appendix 10). This section will focus on the endocrine basis of various types of hormonal contraception.

Combined Oral Contraceptive and Progestin-Only Pills

By the early 20th century, it was known that during pregnancy the corpus luteum somehow inhibits further ovulation. In 1921, the Austrian physiologist Ludwig Haberlandt demonstrated that transplanting the ovaries of pregnant rabbits into nonpregnant females rendered them infertile for several months.[69] He also suggested that extracts from the ovaries might be developed as a method to sterilize women, and thus may be considered the first scientist to seriously consider the possibility of hormonal sterilization. Following the isolation and synthesis of progesterone, Makepeace and colleagues[70] reported that injection of the steroid hormone prevented rabbits from ovulating. In the early 1950s, Margaret Sanger, a nurse and reproductive rights activist, met with researcher Gregory Pincus to begin a decade-long journey to transform hormonal contraception in the United States. Gregory Pincus and others investigated the clinical effectiveness of the synthetic progestogen *norethynodrel* as an oral contraceptive pill for human use. During these clinical trials, it was discovered that the initial norethynodrel preparations contained trace amounts of a contaminant, *mestranol*, a synthetic estrogen. Importantly, this serendipitous discovery led to the finding that the contraceptive effects of progesterone were significantly enhanced in the presence of estrogens. It is now known that estrogens potentiate progesterone action in part by stimulating the synthesis of progesterone receptors.[71] The first major report on the clinical efficacy of a contraceptive pill in a large number of women, using a combination of 10 mg of norethynodrel and 0.15 mg of ethinyl estradiol (*very* high concentrations of hormone, by today's standards), was published by Pincus and colleagues in 1959.[72] In 1960 the U.S. Food and Drug Administration approved the use and development of the first *combined oral contraceptive pill* (COCP), so named because of the presence of two active ingredients, a progestin and an estrogen. The development of "the pill" helped usher in the sexual revolution of the 1960s, and in industrialized countries is credited today as an important tool helping to empower women's reproductive, economic, and educational freedom.[73,74]

The COCP prevents conception by inhibiting both follicle growth and ovulation through its negative feedback actions on the hypothalamus and pituitary. Whereas the estrogenic component primarily suppresses FSH synthesis, the progestin primarily inhibits LH. In essence, the pill mimics the negative feedback actions of estrogens and progesterone, which are normally produced concurrently at high levels during the luteal phase of the ovarian cycle and during pregnancy. The COCPs are generally taken daily for 21 days, followed by 7 days of placebo pills that allow menstrual bleeding to occur. The hormone-free period can be eliminated or modified to improve contraceptive success, reduce the effects of hormone withdrawal, and reduce problems associated with pelvic pain, dysmenorrhea, and anemia.[75] The efficacy of COCPs requires that the pills are administered at approximately the same time each day, as ovulation may resume if a day is skipped or a pill is taken at an irregular time of the day. When the hormone concentrations in each pill are identical throughout the 21 days of treatment, the regimen of COCP administration is known as *monophasic*, or fixed combination. Another regimen of COCP administration is known as *multiphasic*, where hormone concentrations vary over the course of administration, complementing natural levels to ensure that the critical balance required for ovulation is never reached. A third category of pills known as *progestin-only* oral contraceptives consist of estrogen-free tablets that increase the level of progestogen to mimic pregnancy, ensuring that ovulation and implantation cannot occur. High levels of progestogen also inhibit conception by thickening the cervical mucus, inhibiting the entry of sperm into the uterus.

Emergency Contraception and Early Termination of Pregnancy

Emergency contraception, such as levonorgestrel (also known as "Plan B" or "the morning after pill") is used to prevent pregnancy for women who have had unprotected sex or whose contraception method has failed. Such emergency contraception is intended for backup only and not as a primary method of contraception. The pill contains a combination of a large dose of progesterone and some estrogens that raise the sex steroid hormone levels quickly, leading to negative feedback suppression of LH and FSH, followed by steroid withdrawal, which triggers endometrial sloughing. Because this causes loss of the endometrium prior to or about the time any fertilized egg might reach the uterus, it has to be taken within the first 5 days after unprotected sex. "Morning-after pills" do not end a pregnancy that has implanted.

Whereas "morning-after pills" prevent the establishment of pregnancy, or implantation of the embryo into the uterus, *mifepristone* (RU486) is an antagonist of the progesterone receptor (and at higher concentrations, also the glucocorticoid receptor) and is used as an early medical abortion option following implantation. It is safest for use during the first semester of pregnancy. By blocking the action of progesterone, mifepristone induces bleeding and causes the endometrium to shed. Mifepristone is typically used in conjunction with misoprostol (a pharmacological analog of prostaglandin E1), which induces the cervix to soften and the uterus to contract, promoting the expulsion of the uterine lining.

Non-Contraceptive Health Benefits of Hormonal Contraception

Not only is the COCP an extremely effective method of contraception, but its use also confers some health benefits that extend beyond the prevention of pregnancy (summarized in Table 22.5). These include reduced symptoms associated with menstruation (e.g. heavy, painful, or irregular periods), reductions in conditions caused by elevated androgen hormones (e.g. acne, hirsutism), and treatment of endometriosis and menorrhagia (menstrual periods with abnormally heavy or prolonged bleeding).[76] Furthermore, COCP use is associated with a long-lasting reduction in the risk of developing epithelial ovarian and endometrial cancer as well as benign breast diseases. Potential beneficial effects on COCP administration on bone health and colon cancer prevention may exist, but require further investigation.

Risks of Hormonal Contraception

Having been prescribed to millions of women, COCPs are amongst the most thoroughly tested drugs ever produced. Despite the aforementioned non-contraceptive health benefits conveyed by COCPs, health risks may also be associated with their use (summarized in Table 22.5). Though the scientific consensus is that the overall risks to most

Table 22.5 Some Potential Benefits and Risks of Hormonal Contraceptives

Benefits	Risks
Reduced risk of ovarian, endometrial, and colorectal cancers	Coronary heart disease (especially in smokers)
Increased bone mineral density	Venous thrombosis
Increased menstrual cycle regularity	Cerebral vein thrombosis
Reduced migraines during menstruation	Stroke
Treatment of pelvic pain from endometriosis	Hypertension
Treatment of premenstrual syndrome	Breast cancer
Treatment of acne	Cervical cancer
Treatment of hirsutism	

Source: Adapted from: ACOG practice bulletin no. 110: noncontraceptive uses of hormonal contraceptives. *Obstet Gynecol*. 2010;115(1):206.

women are very small and are outweighed by the contraceptive benefits, contraindications to contraceptive use include women who are heavy smokers and over the age of 35, or who have a history of thromboembolic disease, abnormal liver function, cardiovascular disease, or breast or cervical cancers.[77] For example, estrogens, particularly in combination with progesterone, appear to activate the renin-angiotensin-aldosterone system (RAAS) via stimulation of angiotensinogen synthesis by the liver and increasing plasma levels of renin. As such, in some women the use of COCPs may increase blood pressure. COCPs have been found to increase the risk of arterial thrombosis (heart attack or stroke), possibly in relation to high blood pressure and increased risk of blood clot formation associated with the oral contraceptives. Oral contraceptive use may slightly increase the risk of developing breast or cervix cancer.[78,79] For breast cancer, these risks may be even greater for women with a family history of cancer, or if they carry the *BRCA** gene mutation.

For women with contraindications to the estrogenic effects of COCP, the progestin-only contraceptive may be an alternative. Progestin-only pills act to impair implantation and thicken the cervical mucus, but they are not as effective at inhibiting ovulation as COCPs. As such, the progestin-only contraceptives have a higher contraceptive failure rate than COCPs. Users of progestin-only pills also report a high incidence of *breakthrough bleeding*, or midcycle bleeding, due to an insufficient amount of circulating estrogens. Other side effects of progestin-only pills, like acne and hirsutism, are caused by the androgenic actions of the synthetic progestins that are not counteracted by estrogens. Compared with the relatively high concentrations of hormones used in first generation of contraceptive pills (10 mg and 150 μg of progestin and estrogens, respectively), the hormone levels in today's pills are generally much lower (1–3 mg of progestin and 20–30 μg of estrogens).

* BRCA1 (*br*east *ca*ncer gene 1) and BRCA2 (*br*east *ca*ncer gene 2) are tumor-suppressor genes that produce proteins that help repair damaged DNA.

SUMMARY AND SYNTHESIS QUESTIONS

1. Women who take oral contraceptive pills are thought to have lower rates of ovarian cancer.[80] Propose a hypothesis that explains this.
2. The drug RU486 (mifepristone) is a powerful progesterone receptor antagonist and can be used to induce abortions within the first week following conception. Specifically, how would use of this drug terminate a pregnancy?
3. Drugs like *letrozole* are aromatase inhibitors. Would treatment with such a drug be most effective for treating infertility or for preventing pregnancy? Explain your reasoning.

Menopause and Reproductive Senescence

LEARNING OBJECTIVE Outline the major changes in endocrine and health parameters associated with menopause, and discuss the health benefits and potential risks of hormone replacement therapy for treating the symptoms of menopause.

KEY CONCEPTS:

- The onset of the perimenopausal period is associated with a dramatic acceleration of follicular loss and lowered estradiol synthesis.
- Reduced estradiol synthesis makes women susceptible to developing osteoporosis, as well as various androgenic side effects such as weight gain, hirsutism, and thinning hair.
- Selective estrogen receptor modulators (SERMs) are compounds that when bound to the estrogen receptor produce a spectrum of responses in a tissue-specific manner, ranging from estrogenic agonism to an antiestrogenic activity.
- Some SERMs may have estrogen-like actions in bone but antiestrogenic effects in the breast, introducing the possibility that both osteoporosis and breast cancer risk can be simultaneously reduced.

Menopause technically refers to the final menstrual period in a woman's life, which on average is at about 50 years of age and therefore can only be determined retrospectively.

A more useful term is **perimenopause**, which describes the endocrine and physiological changes that indicate the approach of menopause, as well as the transition period between menopause and the start of *postmenopause*. On average, perimenopause lasts 4–5 years but can be as short as a few months or last up to 8 years.[81] Many changes in a woman's physiology begin to take place beginning in her mid-30s, and these include decreased fertility, decreased bone mass, and increased probability of conceiving a child with chromosome abnormalities. The perimenopausal period itself typically begins about 5 years before menopause, and is associated with decreasing levels of ovarian estradiol synthesis leading to increasingly irregular menstrual periods. In addition, due to the changing sex steroid hormone profile, many women experience a range of symptoms that include "hot flashes", mood changes, sleep disturbances, and other symptoms (summarized in Table 22.6). Beginning with perimenopause and continuing through postmenopause, women are at higher risk for developing cardiovascular disease, osteoporosis, dementia, and other maladies.

The Cause of Menopause

Although a progressive loss of ovarian follicles occurs throughout life due to atresia, the perimenopause period is associated with a dramatic acceleration of follicular loss (Figure 22.20). The human ovary contains approximately 400,000 follicles at puberty, but by menopause very few primary follicles remain. The loss of follicles during perimenopause coincides with dramatically decreasing levels of estradiol and estrone synthesis and a concurrent increase in gonadotropin production (Figure 22.21). The loss of follicles and accompanying decline in estrogens during perimenopause is thought to be the primary factor leading to menopause and its accompanying symptoms. Following menopause, a weaker estrogen, estrone, continues to be produced at low levels and becomes the primary estrogen during the postmenopausal period. The primary source of estrone is thought to be adipose tissue, caused by peripheral aromatization of androgen, and estrone levels are higher in obese postmenopausal women than in lean.[82] Furthermore, adrenal androgen synthesis continues throughout life, increasing the androgen/estrogen ratio, which leads to various androgenic side effects such as weight gain, hirsutism, and thinning hair.

Table 22.6 Some Symptoms Associated with Menopause

Menopausal Syndrome	Physical Changes
Vasomotor instability	Atrophy of vaginal epithelium
Hot flashes	Changes in vaginal pH
Night sweats	Decreases in vaginal secretions
Mood changes	Reduced circulation to vagina and uterus
Short-term memory loss	Pelvic relaxation
Sleep disturbances	Loss of vaginal tone
Headaches	Cardiovascular disease
Loss of libido	Osteoporosis
	Alzheimer's disease

Source: Adapted from *Medical Physiology*, 2e, Boron, W.F. and Boulpaep, E.L., 2012. Medical physiology, 2e updated edition e-book: with student consult online access. *Elsevier health sciences*.

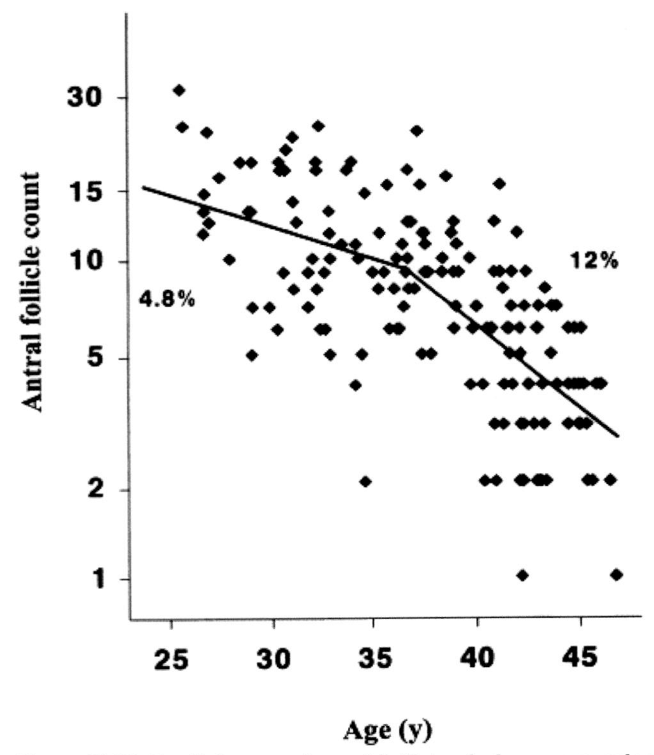

Figure 22.20 Declining number of follicles in human ovaries during adulthood. Whereas the number of follicles decreases progressively from approximately 1,000,000 at birth to about 24,000 by 37 years of age, at 37, the rate of decrease doubles from age 37 through 51 (or menopause).

Source of Images: Scheffer, G.J., et al. 1999. Fertil. Steril., 72(5), pp. 845-851. Used by permission.

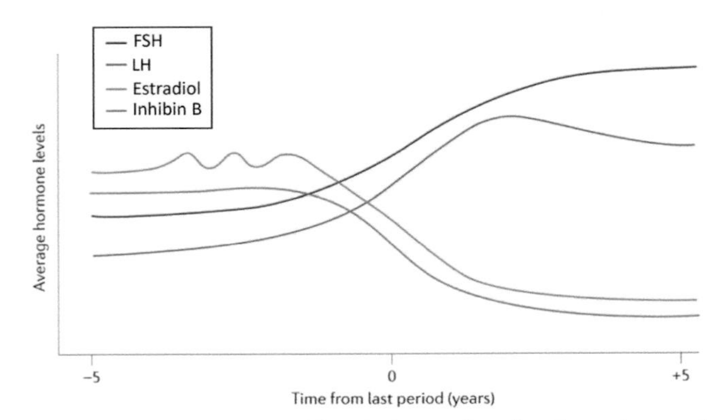

Figure 22.21 Changes in blood plasma levels of ovarian and pituitary hormones during the transition through menopause.

Source of images: Davis, S., et al. Menopause. Nat Rev Dis Primers 1, 15004 (2015). Used by permission.

Hormone Replacement Therapy for Menopause

Considering the diverse roles played by estradiol in both reproductive and nonreproductive tissues, the notion of replacing the estrogens lost during perimenopause for the purposes of alleviating perimenopausal symptoms and to offset dangerous nonreproductive maladies, osteoporosis via *hormone replacement therapy* (HRT) is appealing. Although estrogen replacement alone improves the symptoms of menopause, it may also increase the risk of endometrial cancer.[83] Therefore, most women who still have a uterus (those who have not had a hysterectomy) taking HRT use a combination of estrogens and progesterone together, known as *combined hormone therapy*, which significantly reduces the risk of endometrial cancer.

Nonetheless, as described earlier under the "risks of hormonal contraception" section, even combined hormone therapy still has its drawbacks. Initiated in 1991, the Women's Health Initiative (WHI) was a landmark study conducted by the U.S. National Institutes of Health (NIH) as the first large randomized study testing the effects of combined hormone replacement therapy on women.[84] Eleven years after the start of the study, WHI was terminated early when combined hormone replacement therapy was observed to be associated with increased rates of breast cancer and cardiovascular disease. Although the use of combined hormone replacement therapy dropped significantly after this finding in 2002, millions of women still take hormones for the control of menopausal symptoms. Importantly, the mean age of women in the WHI study was 63–64, significantly older than the average age of women who currently take hormone therapy for menopausal symptoms (aged 50–59).[85] Therefore, the findings of the WHI study, which may be biased toward postmenopausal women, may not accurately reflect the situation for women in peri- and early postmenopause. Indeed, new analyses of the WHI data suggest that short-term hormone replacement therapy given prior to age 60 had no association with increased risk for breast cancer, stroke, or thromboembolic occurences.[86–88]

Importantly, there is no current medical consensus on the right path forward for the treatment of menopause and perimenopause symptoms, and more research is clearly required to address the critically important question of the efficacy of hormone supplementation for these symptoms. An emerging technology that aims to retain the beneficial effects of estrogens while avoiding its adverse effects is the use of **selective estrogen receptor modulators** (SERMs). This exciting and promising frontier was discussed in the Chapter 5: Receptors Clinical Considerations box titled "Selective estrogen receptor modulators (SERMs)".

Interestingly, in addition to the decline in estrogen synthesis that accompanies menopause, another important though less well-studied variable also changes with aging.

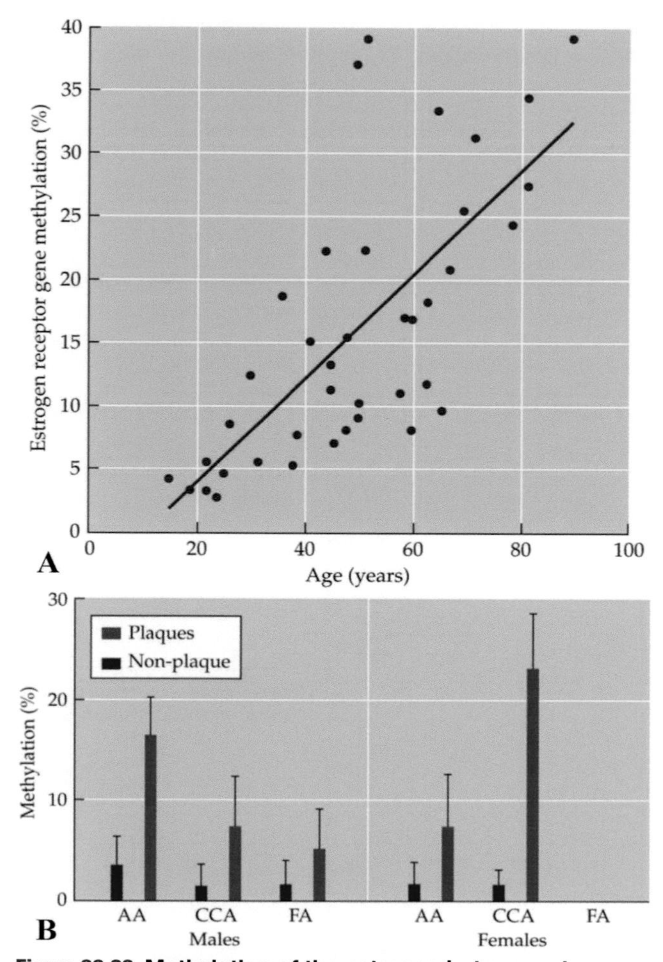

Figure 22.22 Methylation of the estrogen beta receptor gene occurs as a function of normal aging (A) and also in association with atherosclerotic plaques in diseases of the vascular system (B). Ascending aorta (AA), common carotid artery (CCA), femoral artery (FA).

Source of Images: Gilbert, S.F. and Epel, D., 2015. Ecological developmental biology: the environmental regulation of development, health, and evolution. Sinauer Associates, Incorporated Publishers. Used by permission.

Methylation of the promoter regions for both the α and β estrogen receptor genes has been shown to increase linearly with age[97] (Figure 22.22 A), which is thought to suppress their expressions in the smooth muscle cells of blood vessels. This decline in estrogen receptor expression could inhibit the ability of estrogens to maintain the elasticity of these muscles, contributing to the "hardening" of the arteries. This hypothesis is supported by findings that compared with surrounding tissues, increased methylations of estrogen receptor genes is evident in atherosclerotic plaques that occlude the vasculature[98,99] (Figure 22.22 B). This epigenetic, methylation-associated inactivation of estrogen receptor genes in these cells may, therefore, contribute to the age-related deterioration of the vascular system

Clinical Considerations: Breast cancer and the menopause–obesity connection

Despite low levels of circulating estrogens after menopause, most estrogen-dependent breast cancers occur during postmenopause,[89,90] and the risk of developing breast cancer increases with obesity.[91] Considering that the local production of estrogens by the breast itself is an important driver of breast cancer growth,[92–95] Brown and colleagues[96] have hypothesized that local CYP19 aromatase expression by breast tissue may increase during postmenopause, elevating the susceptibility of women developing breast cancer during this time. Also, taking into account that aromatase is expressed by adipose cells, the researchers further postulated that obese women may have even higher breast aromatase activity, accounting for their increased risk for developing breast cancer.

To test these hypotheses, the researchers conducted a cross-sectional study of 102 premenopausal (age 27 to 56) and 59 postmenopausal (age 45 to 74) women who underwent mastectomy for breast cancer treatment or prevention. Breast tissue was assessed for aromatase expression and activity in relation to body mass index (BMI). The researchers' hypotheses were supported by findings that aromatase levels were higher in breast tissue of postmenopausal women, and BMI was positively correlated with aromatase mRNA in both pre- and postmenopausal women (Figure 22.23). The study's authors concluded that elevated aromatase activity in the setting of obesity may provide a mechanism for the higher incidence of estrogen-dependent breast cancer in obese women after menopause.

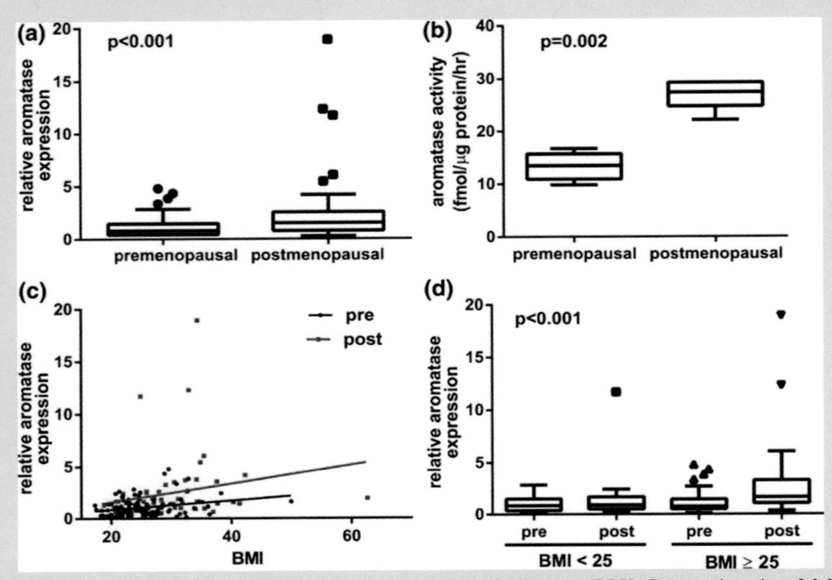

Figure 22.23 Effect of menopausal status on breast aromatase in relation to BMI. Breast levels of **(a)** aromatase mRNA and **(b)** activity are higher in postmenopausal compared with premenopausal women. **(c)** Aromatase transcript expression is positively correlated with BMI in pre- and postmenopausal (red) women. **(d)** Menopause and overweight/obesity are associated with elevated levels of aromatase mRNA. Transcript expression was assessed in n = 161 (102 premenopausal, 59 postmenopausal), whereas aromatase activity was measured in n = 6/group.

Source of Images: Brown, K.A., et al. 2017. J. Clin. Endocrinol. Metab., 102(5), pp. 1692-1701. Used by permission.

SUMMARY AND SYNTHESIS QUESTIONS

1. If the ovaries no longer produce estrogens after menopause, how is the estrogen estrone still present in circulation?
2. Following menopause, the androgen/estrogen ratio increases, potentially leading to various androgenic side effects such as weight gain, hirsutism, and thinning hair. What factors are responsible for the change in androgen/estrogen ratio?
3. If you wanted to purify LH from the urine of women, would you find higher levels of LH in cycling or postmenopausal women? Explain.
4. Many cancers, including estrogen-dependent breast cancer, occur more frequently after menopause, despite low levels of circulating estrogens. A study by Brown and colleagues (2017)[96] reported that breast tissue expresses higher levels of aromatase activity in postmenopausal women than in premenopausal women. Use this finding to formulate a hypothesis that helps explain why breast cancer is more prevalent following menopause.
5. In addition to reduced estrogen synthesis, what other changes occur to the estrogen system with aging?

Endocrine-Disrupting Compounds (EDCs) and Female Reproductive Disorders

LEARNING OBJECTIVE Describe some notable endocrine-disrupting chemicals that may contribute to female reproductive dysfunctions in humans and animal models, and discuss their mechanisms of action.

KEY CONCEPTS:

- Disrupted ovarian development during fetal life by EDCs is suspected to promote the development of polycystic ovary syndrome (PCOS), ovarian cancers, precocious puberty, infertility, and early menopause in the human adult.
- Compared with the limited human data that exists, the negative effects of EDCs on reproductive health in female animal models, particularly rodents, has been demonstrated very clearly.
- Some EDCs, like diethylstilbestrol (DES) and bisphenol A (BPA), are estrogen mimics that exert their effects by binding to estrogen receptors.

Compared with men, biomarkers for assessing infertility in women are more difficult to obtain on a large scale, as it is not possible to easily observe these in females without invasive procedures. Nonetheless, according to the Centers for Disease Control and Prevention (CDC), about 10% of women in the United States between ages 15–44 have difficulty becoming pregnant or staying pregnant.[100] Alarming reproductive health trends are also being reported for women and girls, such as earlier ages for the start of breast development and menarche (refer to Chapter 23: Timing of Puberty and Seasonal Reproduction, Developments & Directions box: The changing timing of puberty in girls throughout the ages), as well as rising rates of breast cancer, ovarian cancer, and endometriosis.

In contrast to male mammals, in females sex steroid hormones are not thought to play critical roles in fetal sexual differentiation. However, mounting evidence suggests that several reproductive disorders that can manifest in girls and women, such as polycystic ovary syndrome (PCOS), ovarian cancers, precocious puberty, infertility, and early menopause, may often have shared etiologies during fetal life. Analogous to *testis dysgenesis syndrome* (TDS) in males, a corresponding *ovarian dysgenesis syndrome* (ODS) has been hypothesized.[101] ODS can be defined as fetal or neonatal alterations in ovarian structure/function caused by genetic or environmental factors that promote the disruption of reproductive health in the adult. Such perturbations may lead to impaired primordial germ cell proliferation or migration, altered meiosis and follicle development, or abnormal gonadal development that increase susceptibility to developing the aforementioned disorders later in life (Figure 22.24). Considering that the damage to the ovary in

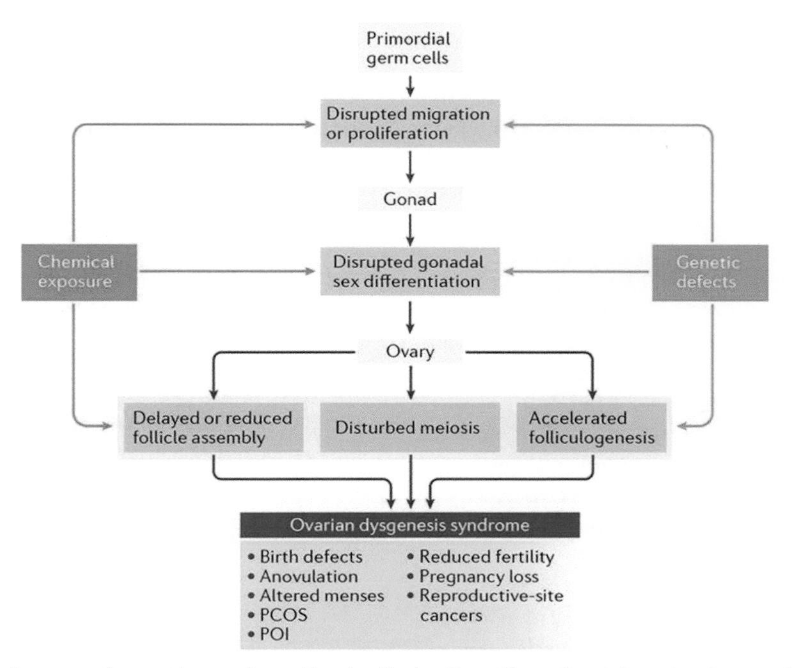

Figure 22.24 The ovarian dysgenesis syndrome hypothesis. Early disruption of ovarian structure or function, caused by either genetic or environmental factors, leads to the impairment of reproductive function later in life. Perturbations of early developmental processes can result in disrupted migration or proliferation of primordial germ cells, disrupted gonadal sex differentiation, delayed or reduced follicle assembly, disturbed meiosis, and accelerated folliculogenesis. Ultimately, the syndrome may manifest in the form of various pathologies. Abbreviations: PCOS, polycystic ovary syndrome; POI, premature ovarian insufficiency.

Source of Images: Johansson, H.K.L., et al. 2017. Nat Rev Endocrinol, 13(7), p. 400. Used by permission.

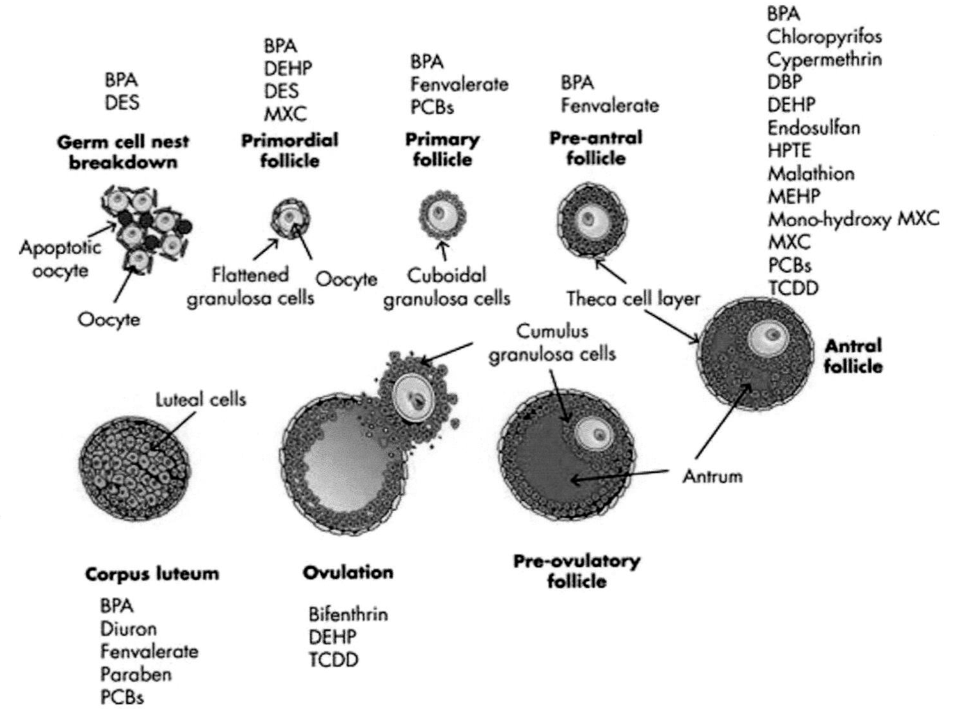

Figure 22.25 Some targets of EDCs on ovarian development using rodent models. This schematic shows the normal developmental stages of ovarian follicles beginning with germ cell nest breakdown around birth, formation of primordial follicles, and their growth to primary follicles, preantral follicles, antral follicles, and finally, preovulatory follicles. This schematic also shows ovulation and the formation of the corpus luteum. Examples of EDCs that adversely affect the ovary are listed above or below their likely site of action.

Source of Images: Gore, A.C., et al. 2015. Endocr. Rev., 36(6), pp. E1-E150. Used by permission.

ODS occurs *in utero*, providing evidence that directly links ODS to reproductive disorders in human adult patients is challenging. However, one well-documented example of an EDC whose exposure during the fetal period has profound health effects in adulthood is **diethylstilbestrol** (DES), an estrogenic compound administered to pregnant women in the 1940s–1970s as a "pregnancy enhancer". Tragically, *in utero* exposure to DES was found to not only promote a rare form of vaginal cancer in girls and young women, but also led to increased rates of endometriosis, uterine fibroids, reproductive-site cancers, and infertility.

Beyond DES, a drug whose use by individual patients could be precisely traced and documented with prescriptions, a linkage between *in utero* exposure to EDCs of environmental origin on adult female reproductive disorders is much more difficult to demonstrate. Nonetheless, epidemiological studies examining the estrogenic EDC *bisphenol A* (BPA) suggest that women with infertility possessed higher blood BPA levels than did fertile women, and that in women undergoing in vitro fertilization (IVF) treatments BPA concentrations were inversely proportional to peak estradiol concentrations, number of oocytes retrieved, degree of oocyte maturation, fertilization rates, and embryo quality.[102] These findings suggest that the success rates of IVF treatment decreases with increasing BPA levels. Additionally, higher blood BPA concentrations have been found to correspond with PCOS, infertility, endometriosis, and recurrent miscarriages.[103–106]

In contrast to studies with human females, the negative effects of EDCs on reproductive health in female animal models, particularly rodents, has been demonstrated very clearly. For example, prenatal and perinatal exposures of mice and rats to estrogenic or antiandrogenic EDCs, such as phthalates and BPA, can reduce the numbers of follicle reserves and accelerate reproductive aging[107–110] and also disrupt the functions of female reproductive tissues and organs in offspring later in life.[111–114] Importantly, the phenotypes observed in animal models exposed to EDCs are similar to reproductive disorders seen in humans. A summary of the known effects of some EDCs on ovarian development and other reproductive target tissues in animal models is provided in Figure 22.25.

SUMMARY AND SYNTHESIS QUESTIONS

1. Like testis dysgenesis syndrome (TDS) in males, damage to the ovary that causes ovarian dysgenesis syndrome (ODS) in females is thought to take place at what time point during the life cycle?
2. By what mechanism are EDCs like DES and BPA thought to exert their biological effects?

Summary of Chapter Learning Objectives and Key Concepts

LEARNING OBJECTIVE Describe the key genomic and morphological changes that take place during oogenesis, follicle development, ovulation, and fertilization.

- After reaching their peak numbers in the embryonic ovary via mitotic divisions, germ cell numbers decrease abruptly by birth.
- Female germline stem cells are absent by birth, and the numbers of oocytes in the ovaries gradually decline throughout life and disappear by postmenopause.
- Oocytes undergo two periods of meiotic arrest: by birth all are arrested in prophase I, and following ovulation the oocyte is arrested in metaphase II.
- Meiotic divisions of oogonia are asymmetrical, each ultimately yielding one large viable oocyte and two much smaller and nonviable polar bodies.
- Following ovulation, the oocyte arrested in metaphase II only completes its final meiotic division following fertilization.

LEARNING OBJECTIVE Describe the structure of an ovarian follicle, and the chronology of follicular recruitment, development, and atresia.

- Ovarian follicles each consist of an oocyte surrounded by epithelial cells in various stages of growth and development.
- Subsets of follicles are continuously recruited to undergo further development and growth into a single mature follicle that will be ovulated.
- Follicles that are not ovulated eventually undergo programmed cell death and are resorbed, a process called "atresia".
- Following ovulation, the remnants of the follicle in the ovary develop into a primarily progesterone-secreting organ called the corpus luteum.
- In the absence of fertilization, the corpus luteum degenerates into a scar-like tissue called the corpus albicans.

LEARNING OBJECTIVE Describe the two-cell theory of ovarian hormone biosynthesis, the defining endocrine features of the follicular and luteal phases of the ovarian cycle, and the feedback mechanisms along the hypothalamic-pituitary-ovary axis.

- Steroidogenically analogous to Leydig cells of the testis, a primary function of theca cells is to synthesize androgens in response to LH.
- Analogous to Sertoli cells of the testis, a primary function of granulosa cells is the conversion of androgens into estrogens in response to FSH.
- The first 14 days of the ovarian cycle comprise the follicular phase, characterized by the growth and maturation of the dominant follicle.
- Ovulation demarcates the start of the luteal phase of the ovarian cycle, characterized by the formation of the corpus luteum.
- During the follicular phase, a follicle's granulosa cells synthesize estrogens.
- During the luteal phase, a follicle's granulosa cells synthesize progesterone.
- Rapidly rising gonadotropin levels during the mid-late follicular phase is driven by positive feedback from peaking synthesis of estrogens.
- Falling gonadotropin levels during the luteal phase is mediated by negative feedback from rising levels of progesterone, estradiol, and inhibin.

LEARNING OBJECTIVE Describe how estrogens and progesterone are transported in the blood, and outline sex steroid hormone receptor function.

- In the blood, most estrogens are transported by low- or high-affinity binding proteins.
- The small unbound fraction of hormone is considered to be bioavailable.
- Most of the effects of estrogens and progesterone are mediated by nuclear-localized receptors that operate as ligand-dependent gene transcription factors.
- Some of the effects of sex steroid hormones are also driven by the actions of nongenomic, membrane-associated G protein-coupled receptors.

LEARNING OBJECTIVE Compare and contrast the human menstrual cycle with the estrous cycle in other mammals.

- The endometrium of the uterus is the site of implantation for an embryo.
- The thickness of the endometrium changes dramatically throughout the menstrual cycle in response to the ovarian steroid hormones.
- The shedding of the unfertilized oocyte and the uterine lining (menstruation) occurs approximately every 28 days for most women.
- The proliferative phase of the menstrual cycle parallels the follicular phase of the ovarian cycle, and the secretory phase parallels the luteal phase of the ovarian cycle.
- Whereas increasing levels of estrogens during the proliferative phase promote a thickening of the uterine endometrium, decreasing concentrations of progesterone and estrogens during the late luteal phase promotes the degeneration and shedding of the outer endometrial layers ("bleeding").
- The vast majority of mammals do not menstruate, but instead exhibit estrous cycles characterized, in part, by the resorption of the endometrium instead of its excretion.
- In contrast to menstruating species, whose females may be sexually receptive throughout the menstrual cycle, most mammals with estrous cycles are only sexually receptive at specific times of the month or year.
- In induced ovulators, ovulation is induced directly via copulatory stimulation during mating.

LEARNING OBJECTIVE Describe modern approaches to contraception and fertility control, and discuss the non-contraceptive risks and benefits of hormone-based contraceptives.

- The contraceptive effects of progesterone are significantly enhanced in the presence of estrogens.
- The combined oral contraceptive pill (COCP) prevents conception by inhibiting both follicle growth and ovulation through its negative feedback actions on the hypothalamus and pituitary.
- Progestin-only oral contraceptives are estrogen-free tablets that increase the level of progestogen to mimic pregnancy and ensure that ovulation and implantation cannot occur.
- COCPs confer health benefits that extend beyond the prevention of pregnancy, such as reduced symptoms associated with painful or irregular menstruation, reductions in conditions caused by elevated androgen hormones, and reduction in the risk of developing epithelial ovarian and endometrial cancer.
- Estrogen-based contraceptives may slightly increase the risk of developing breast or cervix cancers, as well as thromboembolic and cardiovascular diseases.

LEARNING OBJECTIVE Outline the major changes in endocrine and health parameters associated with menopause, and discuss the health benefits and potential risks of hormone replacement therapy for treating the symptoms of menopause.

- The onset of the perimenopausal period is associated with a dramatic acceleration of follicular loss and lowered estradiol synthesis.
- Reduced estradiol synthesis makes women susceptible to developing osteoporosis, as well as various androgenic side effects such as weight gain, hirsutism, and thinning hair.
- Selective estrogen receptor modulators (SERMs) are compounds that when bound to the estrogen receptor produce a spectrum of responses in a tissue-specific manner, ranging from estrogenic agonism to an antiestrogenic activity.
- Some SERMs may have estrogen-like actions in bone but antiestrogenic effects in the breast, introducing the possibility that both osteoporosis and breast cancer risk can be simultaneously reduced.

LEARNING OBJECTIVE Describe some notable endocrine-disrupting chemicals that may contribute to female reproductive dysfunctions in humans and animal models, and discuss their mechanisms of action.

- Disrupted ovarian development during fetal life by EDCs is suspected to promote the development of polycystic ovary syndrome (PCOS), ovarian cancers, precocious puberty, infertility, and early menopause in the human adult.
- Compared with the limited human data that exists, the negative effects of EDCs on reproductive health in female animal models, particularly rodents, has been demonstrated very clearly.
- Some EDCs, like diethylstilbestrol (DES) and bisphenol A (BPA), are estrogen mimics that exert their effects by binding to estrogen receptors.

LITERATURE CITED

1. Aristotle. The Works of Aristotle. Translated by Thompson, D'Arcy W. In: Smith JA, Ross WD, eds., Thompson DA, trans. *Historia Animalium, liber IX, 50.* Clarendon Press; 1910.
2. Jocelyn HD, Setchell BP. Regnier de Graaf on the human reproductive organs: an annotated translation of Tractatus de virorum organis generationi inservientibus (1668) and De mulierum organis generationi inservientibus tractatus novus (1672). *J Reprod Fertil Suppl.* 1972;17:1–222.
3. Short RV. The discovery of the ovaries. In: Zuckerman S, Weir BJ, eds. *The Ovary.* Academic Press; 1979:1–39.
4. Corner GW. *The Hormones in Human Reproduction.* Princeton University Press; 1943.
5. Allen E, Doisy EA. Landmark article Sept 8, 1923. An ovarian hormone. Preliminary report on its localization, extraction and partial purification, and action in test animals. By Edgar Allen and Edward A. Doisy. *JAMA.* 1983;250(19):2681–2683.
6. Doisy EA, Veler CD, Thayer S. Folliculin from urine of pregnant women. *Am J Physiol.* 1929;90:329–330.
7. Butenandt A. U ?ber "Progynon" ein krystallisiertes weibliches Sexualhormon. *Naturwissenschaften.* 1929;17:879.
8. Fulton N, Martins da Silva SJ, Bayne RA, Anderson RA. Germ cell proliferation and apoptosis in the developing human ovary. *J Clin Endocrinol Metab.* 2005;90(8):4664–4670.
9. Handel MA, Schimenti JC. Genetics of mammalian meiosis: regulation, dynamics and impact on fertility. *Nat Rev Genet.* 2010;11(2):124–136.
10. Sen A, Caiazza F. Oocyte maturation: a story of arrest and release. *Front Biosci (Schol Ed).* 2013;5(5):451–477.
11. Wells D, Hillier SG. Polar bodies: their biological mystery and clinical meaning. *Mol Hum Reprod.* 2011;17(5):273–274.
12. Boron WF, Boulpaep EL. *Medical Physiology* (3rd ed.). Elsevier Saunders; 2016.
13. Gougeon A. Regulation of ovarian follicular development in primates: facts and hypotheses. *Endocr Rev.* 1996;17(2):121–155.
14. Abbara A, Vuong LN, Ho VNA, et al. Follicle size on day of trigger most likely to yield a mature oocyte. *Front Endocrinol.* 2018;9(193).
15. Hillier S. Gonadotropic control of ovarian follicular growth and development. *Mol Cell Endocrinol.* 2001;179(1–2):39–46.
16. de Bruin JP, Dorland M, Spek ER, et al. Ultrastructure of the resting ovarian follicle pool in healthy young women1. *Biol Reprod.* 2002;66(4):1151–1160.
17. Pellatt L, Rice S, Mason HD. Anti-Müllerian hormone and polycystic ovary syndrome: a mountain too high? *Reproduction.* 2010;139(5):825–833.
18. Kollmann Z, Bersinger NA, McKinnon BD, Schneider S, Mueller MD, von Wolff M. Anti-Müllerian hormone and progesterone levels produced by granulosa cells are higher when derived from natural cycle IVF than from conventional gonadotropin-stimulated IVF. *Reprod Biol Endocrinol.* 2015;13:21.
19. Tal R, Seifer DB. Ovarian reserve testing: a user's guide. *Am J Obst Gynecol.* 2017;217(2):129–140.
20. Azziz R, Carmina E, Chen Z, et al. Polycystic ovary syndrome. *Nat Rev Dis Primers.* 2016;2:16057.
21. McAllister JM, Legro RS, Modi BP, Strauss JF, 3rd. Functional genomics of PCOS: from GWAS to molecular mechanisms. *Trends Endocrinol Metab.* 2015;26(3):118–124.

22. Yildiz BO, Azziz R. The adrenal and polycystic ovary syndrome. *Rev End Metabo Dis*. 2007;8(4):331–342.
23. Escobar-Morreale HF. Polycystic ovary syndrome: definition, aetiology, diagnosis and treatment. *Nat Rev Endocrinol.* 2018;14(5):270–284.
24. Dumesic DA, Oberfield SE, Stener-Victorin E, Marshall JC, Laven JS, Legro RS. Scientific statement on the diagnostic criteria, epidemiology, pathophysiology, and molecular genetics of polycystic ovary syndrome. *Endocr Rev*. 2015;36(5):487–525.
25. Fauser BC, Van Heusden AM. Manipulation of human ovarian function: physiological concepts and clinical consequences. *Endocr Rev*. 1997;18(1):71–106.
26. Chang, R, Kazer, R, *Glob. libr. women's med*., (ISSN: 1756-2228) 2014; DOI 10.3843/GLOWM.10301
27. Burt Solorzano CM, McCartney CR, Blank SK, Knudsen KL, Marshall JC. Hyperandrogenaemia in adolescent girls: origins of abnormal gonadotropin-releasing hormone secretion. *BJOG*. 2010;117(2):143–149.
28. Broekmans FJ, Visser JA, Laven JS, Broer SL, Themmen AP, Fauser BC. Anti-Müllerian hormone and ovarian dysfunction. *Trends Endocrinol Metab*. 2008;19(9):340–347.
29. McCartney CR, Marshall JC. Polycystic ovary syndrome. *N Engl J Med*. 2016;375(14):1398–1399.
30. Kawashima I, Liu Z, Mullany LK, Mihara T, Richards JS, Shimada M. EGF-like factors induce expansion of the cumulus cell-oocyte complexes by activating calpain-mediated cell movement. *Endocrinology*. 2012;153(8):3949–3959.
31. Gaytan F, Tarradas E, Morales C, Bellido C, Sanchez-Criado JE. Morphological evidence for uncontrolled proteolytic activity during the ovulatory process in indomethacin-treated rats. *Reproduction*. 2002;123(5):639–649.
32. Matousek M, Carati C, Gannon B, Brannstrom M. Novel method for intrafollicular pressure measurements in the rat ovary: increased intrafollicular pressure after hCG stimulation. *Reproduction*. 2001;121(2):307–314.
33. Kovacs WJ, Ojeda SR. *Textbook of Endocrine Physiology* (6th ed.). Oxford University Press; 2011.
34. Baird DD, Weinberg CR, McConnaughey DR, Wilcox AJ. Rescue of the corpus luteum in human pregnancy. *Biol Reprod*. 2003;68(2):448–456.
35. Glanc P, Salem S, Farine D. Adnexal masses in the pregnant patient: a diagnostic and management challenge. *Ultrasound Quarterly*. 2008;24(4):225–240.
36. Gibson M. Corpus Luteum. *Glob libr women's med*. 2008.
37. Hattori K, Orisaka M, Fukuda S, et al. Luteinizing hormone facilitates antral follicular maturation and survival via thecal paracrine signaling in cattle. *Endocrinology*. 2018;159(6):2337–2347.
38. Erickson G, Wang C, Hsueh A. FSH induction of functional LH receptors in granulosa cells cultured in a chemically defined medium. *Nature*. 1979;279(5711):336–338.
39. He W, Li X, Adekunbi D, et al. Hypothalamic effects of progesterone on regulation of the pulsatile and surge release of luteinising hormone in female rats. *Sci Rep*. 2017;7(1):1–11.
40. McCartney CR, Gingrich MB, Hu Y, Evans WS, Marshall JC. Hypothalamic regulation of cyclic ovulation: evidence that the increase in gonadotropin-releasing hormone pulse frequency during the follicular phase reflects the gradual loss of the restraining effects of progesterone. *J Clin Endocrinol Metab*. 2002;87(5):2194–2200.
41. Wu CH, Motohashi T, Abdel-Rahman HA, Flickinger GL, Mikhail G. Free and protein-bound plasma estradiol-17 beta during the menstrual cycle. *J Clin Endocrinol Metab*. 1976;43(2):436–445.
42. Hammond GL. Diverse roles for sex hormone-binding globulin in reproduction. *Biol Reprod*. 2011;85(3):431–441.
43. Tee MK, Rogatsky I, Tzagarakis-Foster C, et al. Estradiol and selective estrogen receptor modulators differentially regulate target genes with estrogen receptors alpha and beta. *Mol Biol Cell*. 2004;15(3):1262–1272.
44. Thornton JW. Evolution of vertebrate steroid receptors from an ancestral estrogen receptor by ligand exploitation and serial genome expansions. *Proc Natl Acad Sci U S A*. 2001;98(10):5671–5676.
45. Saez PJ, Lange S, Perez-Acle T, Owen GI. Nuclear receptor genes: evolution. In: *Encyclopedia of Life Sciences (ELS)*. John Wiley & Sons, Ltd; 2010.
46. Revankar CM, Cimino DF, Sklar LA, Arterburn JB, Prossnitz ER. A transmembrane intracellular estrogen receptor mediates rapid cell signaling. *Science*. 2005;307(5715):1625–1630.
47. Prossnitz ER, Arterburn JB, Smith HO, Oprea TI, Sklar LA, Hathaway HJ. Estrogen signaling through the transmembrane G protein-coupled receptor GPR30. *Annu Rev Physiol*. 2008;70:165–190.
48. Thomas P, Pang Y. Membrane progesterone receptors: evidence for neuroprotective, neurosteroid signaling and neuroendocrine functions in neuronal cells. *Neuroendocrinology*. 2012;96(2):162–171.
49. Curry TE, Jr., Osteen KG. Cyclic changes in the matrix metalloproteinase system in the ovary and uterus. *Biol Reprod*. 2001;64(5):1285–1296.
50. Reed BG, Carr BR. The normal menstrual cycle and the control of ovulation. In: De Groot LJ, Chrousos G, Dungan K, et al., eds. *Endotext*. endotext.org; 2015.
51. Weinstein IB, Joe AK. Mechanisms of disease: Oncogene addiction—a rationale for molecular targeting in cancer therapy. *Nat Clin Pract Oncol*. 2006;3(8):448–457.
52. Li Q, Kannan A, DeMayo FJ, et al. The antiproliferative action of progesterone in uterine epithelium is mediated by Hand2. *Science*. 2011;331(6019):912–916.
53. Giudice LC. Clinical practice. Endometriosis. *N Engl J Med*. 2010;362(25):2389–2398.
54. Vercellini P, Vigano P, Somigliana E, Fedele L. Endometriosis: pathogenesis and treatment. *Nat Rev Endocrinol*. 2014;10(5):261–275.
55. D'Hooghe TM, Debrock S. Endometriosis, retrograde menstruation and peritoneal inflammation in women and in baboons. *Hum Reprod Update*. 2002;8(1):84–88.
56. Burney RO, Giudice LC. Pathogenesis and pathophysiology of endometriosis. *Fertil Steril*. 2012;98(3):511–519.
57. Bulun SE. Endometriosis. *N Engl J Med*. 2009;360(3):268–279.
58. Darrow SL, Vena JE, Batt RE, Zielezny MA, Michalek AM, Selman S. Menstrual cycle characteristics and the risk of endometriosis. *Epidemiology*. 1993;4(2):135–142.
59. Missmer SA, Hankinson SE, Spiegelman D, Barbieri RL, Michels KB, Hunter DJ. In utero exposures and the incidence of endometriosis. *Fertil Steril*. 2004;82(6):1501–1508.
60. Tarin JJ, Gomez-Piquer V. Do women have a hidden heat period? *Hum Reprod*. 2002;17(9):2243–2248.
61. Burt A. "Concealed ovulation" and sexual signals in primates. *Folia Primatol (Basel)*. 1992;58(1):1–6.
62. Stockard CR, Papanicolaou GN. The existence of a typical oestrous cycle in the guinea-pig-with a study of its histological and physiological changes. *Am J Anat*. 1917;22(2):225–283.
63. Stockard CR, Papanicolaou GN. A rhythmical "heat period" in the guinea-pig. *Science*. 1917;46(1176):42–44.
64. El Allali K, El Bousmaki N, Ainani H, Simonneaux V. Effect of the Camelid's seminal plasma ovulation-inducing factor/beta-NGF: A Kisspeptin target hypothesis. *Front Vet Sci*. 2017;4:99.
65. Bakker J, Baum MJ. Neuroendocrine regulation of GnRH release in induced ovulators. *Front Neuroendocrinol*. 2000;21(3):220–262.
66. Larivière S, Ferguson SH. Evolution of induced ovulation in North American Carnivores. *J Mammol*. 2003;84(3):937–947.
67. Fernandez-Baca S, Madden DH, Novoa C. Effect of different mating stimuli on induction of ovulation in the alpaca. *J Reprod Fertil*. 1970;22(2):261–267.
68. Jochle W. Current research in coitus-induced ovulation: a review. *J Reprod Fertil Suppl*. 1975(22):165–207.
69. Haberlandt L. Ueber hormonale Sterilisierung des weiblichen Tierkorpers. *Munch Med Wochenschr*. 1921;68:1577–1578.

70. Makepeace AW, Weinstein GL, Friedman MH. The effect of progestin and progesterone on ovulation in the rabbit. *Am J Physiol.* 1937;119:512–516.
71. Ing NH, Tornesi MB. Estradiol up-regulates estrogen receptor and progesterone receptor gene expression in specific ovine uterine cells. *Biol Reprod.* 1997;56(5):1205–1215.
72. Pincus G, Garcia CR, Rock J, et al. Effectiveness of an oral contraceptive; effects of a progestin-estrogen combination upon fertility, menstrual phenomena, and health. *Science.* 1959;130(3367):81–83.
73. Goldin C, Katz LF. The power of the pill: Oral contraceptives and women's career and marriage decisions. *J Political Econ.* 2002;110(4):730–770.
74. Bailey MJ. Fifty years of family planning: new evidence on the long-run effects of increasing access to contraception. *Brookings Pap Econ Act.* 2013;2013:341–409.
75. Melmed S, Polonsky KS, Larsen PR, eds. *Williams Textbook of Endocrinology* (12th ed.). Saunders; 2011.
76. Eshre-Capri-Workshop-Group. Noncontraceptive health benefits of combined oral contraception. *Hum Reprod Update.* 2005;11(5): 513–525.
77. Rosen MP, Cedars MI. Female reproductive endocrinology and fertility. In: Gardner DG, Shoback DM, eds. *Greenspan's Basic and Clinical Endocrinology* (9th ed.). McGraw Hill; 2011.
78. Morch LS, Skovlund CW, Hannaford PC, Iversen L, Fielding S, Lidegaard O. Contemporary hormonal contraception and the risk of breast cancer. *N Engl J Med.* 2017;377(23):2228–2239.
79. Gierisch JM, Coeytaux RR, Urrutia RP, et al. Oral contraceptive use and risk of breast, cervical, colorectal, and endometrial cancers: a systematic review. *Cancer Epidemiol Biomarkers Prev.* 2013;22(11):1931–1943.
80. Berchuck A, Schildkraut J. Oral contraceptive pills. Prevention of ovarian cancer and other benefits. *N C Med J.* 1997;58(6):404–407; discussion 408.
81. Harlow SD, Paramsothy P. Menstruation and the menopausal transition. *Obst Gynecol Clin.* 2011;38(3):595–607.
82. Freeman EW, Sammel MD, Lin H, Gracia CR. Obesity and reproductive hormone levels in the transition to menopause. *Menopause.* 2010;17(4):718–726.
83. Weiderpass E, Adami H-O, Baron JA, et al. Risk of endometrial cancer following estrogen replacement with and without progestins. *J Nat Cancer Inst.* 1999;91(13):1131–1137.
84. Rossouw JE, Anderson GL, Prentice RL, et al. Risks and benefits of estrogen plus progestin in healthy postmenopausal women: principal results From the Women's Health Initiative randomized controlled trial. *JAMA.* 2002;288(3):321–333.
85. Chen WY. Postmenopausal hormone therapy and breast cancer risk: current status and unanswered questions. *Endocrinol Metab Clin North Am.* 2011;40(3):509–518, viii.
86. Roehm E. A reappraisal of women's health initiative estrogen-alone trial: long-term outcomes in women 50–59 years of age. *Obstet Gynecol Int.* 2015;2015:713295.
87. Lobo RA. Where are we 10 years after the Women's Health Initiative? *J Clin Endocrinol Metab.* 2013;98(5): 1771–1780.
88. Flores VA, Pal L, Manson JE. Hormone therapy in menopause: concepts, controversies, and approach to treatment. *Endocr Rev.* 2021;42(6):720–752.
89. Jatoi I, Chen BE, Anderson WF, Rosenberg PS. Breast cancer mortality trends in the United States according to estrogen receptor status and age at diagnosis. *J Clin Oncol.* 2007;25(13):1683–1690.
90. Henderson VW, John JAS, Hodis HN, et al. Cognition, mood, and physiological concentrations of sex hormones in the early and late postmenopause. *Proc Nat Acad Sci.* 2013;110(50): 20290–20295.
91. Ligibel JA, Alfano CM, Courneya KS, et al. American Society of Clinical Oncology position statement on obesity and cancer. *J Clin Oncol.* 2014;32(31):3568.
92. Wang X, Simpson ER, Brown KA. Aromatase overexpression in dysfunctional adipose tissue links obesity to postmenopausal breast cancer. *J Steroid Biochem Mol Biol.* 2015;153:35–44.
93. O'Neill JS, Elton RA, Miller WR. Aromatase activity in adipose tissue from breast quadrants: a link with tumour site. *Br Med J (Clin Res Ed).* 1988;296(6624):741–743.
94. Bulun SE, Price TM, Aitken J, Mahendroo MS, Simpson ER. A link between breast cancer and local estrogen biosynthesis suggested by quantification of breast adipose tissue aromatase cytochrome P450 transcripts using competitive polymerase chain reaction after reverse transcription. *J Clin Endocrinol Metab.* 1993;77(6):1622–1628.
95. Sasano H, Nagura H, Harada N, Goukon Y, Kimura M. Immunolocalization of aromatase and other steroidogenic enzymes in human breast disorders. *Hum Pathol.* 1994;25(5):530–535.
96. Brown KA, Iyengar NM, Zhou XK, et al. Menopause is a determinant of breast aromatase expression and its associations with BMI, inflammation, and systemic markers. *J Clin Endocrinol Metab.* 2017;102(5):1692–1701.
97. Issa J-PJ, Ottaviano YL, Celano P, Hamilton SR, Davidson NE, Baylin SB. Methylation of the oestrogen receptor CpG island links ageing and neoplasia in human colon. *Nat Genet.* 1994;7(4):536–540.
98. Post WS, Goldschmidt-Clermont PJ, Wilhide CC, et al. Methylation of the estrogen receptor gene is associated with aging and atherosclerosis in the cardiovascular system. *Cardiovasc Res.* 1999;43(4):985–991.
99. Kim J, Kim JY, Song KS, et al. Epigenetic changes in estrogen receptor β gene in atherosclerotic cardiovascular tissues and in-vitro vascular senescence. *Biochim Biophys Acta Mol Basis Dis.* 2007;1772(1):72–80.
100. CDC. National Center for Health Statistics. Key statistics from NSFG 2015–2017. U.S. Department of Health and Human Services Centers for Disease Control and Prevention National Center for Health Statistics Hyattsville, Maryland; December 2018.
101. Johansson HKL, Svingen T, Fowler PA, Vinggaard AM, Boberg J. Environmental influences on ovarian dysgenesis—developmental windows sensitive to chemical exposures. *Nat Rev Endocrinol.* 2017;13(7):400.
102. Ziv-Gal A, Flaws JA. Evidence for bisphenol A-induced female infertility: a review (2007–2016). *Fert Steril.* 2016;106(4):827–856.
103. Vom Saal FS, Akingbemi BT, Belcher SM, et al. Chapel Hill bisphenol A expert panel consensus statement: integration of mechanisms, effects in animals and potential to impact human health at current levels of exposure. *Reprod Toxicol (Elmsford, NY).* 2007;24(2):131.
104. Vandenberg LN, Maffini MV, Sonnenschein C, Rubin BS, Soto AM. Bisphenol-A and the great divide: a review of controversies in the field of endocrine disruption. *Endocr Rev.* 2009;30(1):75–95.
105. Chapin RE, Adams J, Boekelheide K, et al. NTP-CERHR expert panel report on the reproductive and developmental toxicity of bisphenol A. *Birth Def Res Part B Dev Reprod Toxicol.* 2008;83(3):157–395.
106. Lauretta R, Sansone A, Sansone M, Romanelli F, Appetecchia M. Endocrine disrupting chemicals: effects on endocrine glands. *Front Endocrinol.* 2019;10.
107. Chao H-H, Zhang X-F, Chen B, et al. Bisphenol A exposure modifies methylation of imprinted genes in mouse oocytes via the estrogen receptor signaling pathway. *Histochem Cell Biol.* 2012;137(2):249–259.
108. Johansson HKL, Jacobsen PR, Hass U, et al. Perinatal exposure to mixtures of endocrine disrupting chemicals reduces female rat follicle reserves and accelerates reproductive aging. *Reprod Toxicol.* 2016;61:186–194.
109. Rodríguez HA, Santambrosio N, Santamaría CG, Muñoz-de-Toro M, Luque EH. Neonatal exposure to bisphenol A reduces the pool of primordial fol-

licles in the rat ovary. *Reprod Toxicol.* 2010;30(4):550–557.

110. Zhang XF, Zhang LJ, Li L, et al. Diethylhexyl phthalate exposure impairs follicular development and affects oocyte maturation in the mouse. *Environ Mol Mutagen.* 2013;54(5):354–361.
111. Fernández M, Bourguignon N, Lux-Lantos V, Libertun C. Neonatal exposure to bisphenol a and reproductive and endocrine alterations resembling the polycystic ovarian syndrome in adult rats. *Environ Heal Perspect.* 2010;118(9):1217–1222.
112. Gao H, Yang B-J, Li N, et al. Bisphenol A and hormone-associated cancers: current progress and perspectives. *Medicine.* 2015;94(1).
113. Susiarjo M, Hassold TJ, Freeman E, Hunt PA. Bisphenol A exposure in utero disrupts early oogenesis in the mouse. *PLoS Genet* 3: e5.2007.
114. Wang W, Hafner KS, Flaws JA. In utero bisphenol A exposure disrupts germ cell nest breakdown and reduces fertility with age in the mouse. *Toxicol Appl Pharmacol.* 2014;276(2):157–164.

CHAPTER 23

The Timing of Puberty and Seasonal Reproduction

CHAPTER LEARNING OBJECTIVES:

- Discuss the importance of GnRH pulsatility on the onset of puberty.
- Discuss the neuroendocrine basis of negative and positive feedback signaling along the HPG axis in response to hormones throughout the ovarian cycle.
- Describe the major endocrine benchmarks of puberty in human males and females and developmental changes in the HPG axis from birth through adulthood.
- Discuss the influence of metabolic status, obesity, and stress on the timing of puberty.
- Define "estrous cycle", and describe how endocrine cues are interpreted differently by long-day and short-day seasonal breeding animals.

OPENING QUOTATION:

"Foremost among the unsolved mysteries of modern biology are the precise neurophysiological mechanisms that drive the onset of adolescence, and direct the individual's progress through normal puberty into the transitional state of young adulthood."

—J.D. Veldhuis (1996). Neuroendocrine mechanisms mediating awakening of the human gonadotropic axis in puberty. *Pediatric Nephrology*. 10, 304–317

Introduction

In 1955 Geoffrey Harris, a professor of anatomy at Oxford University, suggested that "a major factor responsible for puberty is an increased rate of release of pituitary gonadotrophin" and proposed that "a neural (hypothalamic) stimulus, via the hypophyseal portal vessels, may be involved".[1] With the discoveries of the two pituitary gonadotropins, *luteinizing hormone* (LH) and *follicle-stimulating hormone* (FSH), and the hypothalamic decapeptide *gonadotropin-releasing hormone* (GnRH), Harris was proven correct on both accounts.

A simplified view of the regulatory components of the hypothalamic-pituitary-gonad (HPG) axis is depicted in Figure 23.1. In both males and females, GnRH functions as the hypothalamic-releasing factor that induces the simultaneous synthesis and release of the two adenohypophysis gonadotropins, FSH and LH. The HPG axis of males is under negative feedback control mediated through synthesis and release of the testis end organ's hormones, which include the steroid hormone testosterone, as well as the peptide hormone inhibin. The female HPG axis also implements a similar negative feedback control during the luteal phase of the ovarian cycle, mediated by the ovarian hormones inhibin, estradiol, and progesterone. However, the female HPG axis varies from the male as *positive* feedback in response to ovarian steroid hormones and also plays an essential role in triggering the ovulatory gonadotropin surge during the ovarian cycle's follicular phase. This chapter will focus

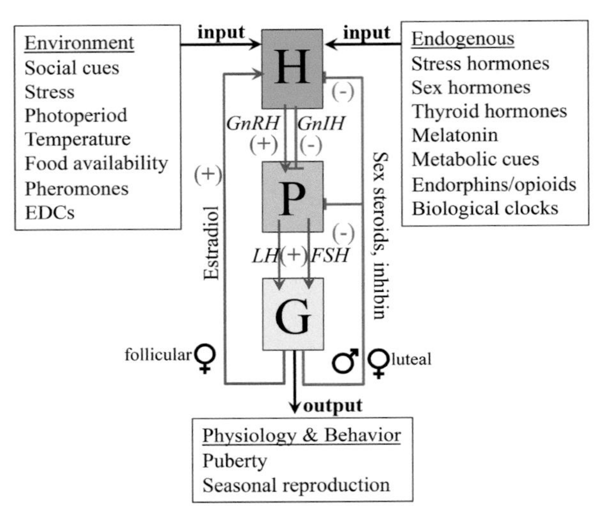

Figure 23.1 Diverse environmentally and endogenously derived inputs are processed by the HPG axis to modulate the timing of puberty and seasonal reproduction. Green arrows denote stimulatory input, red arrows denote inhibitory input. The presence of positive versus negative feedback loops varies between the sexes. In males, negative feedback mediated via inhibin and sex steroids prevails, whereas in females negative feedback prevails in the luteal phase of the ovarian cycle and positive feedback mediated by sex steroids in the follicular phase.

DOI: 10.1201/9781003359241-30

Developments & Directions: An African cichlid fish model for studying the social regulation of the hypothalamic-pituitary-gonadal (HPG) axis

Astatotilapia burtoni is an African cichlid fish that lives in East Africa's Lake Tanganyika river and estuary system. The males of the species exhibit a remarkable example of *phenotypic plasticity*, or the ability of one genotype to produce multiple phenotypes following exposure to different environmental conditions. Specifically, the males live as two distinct and reversible phenotypes: (1) *dominant* males, which are brightly colored, possess distinct stripes and spots, and are highly territorial against rival males, actively courting and spawning with females; and (2) *subordinate* males, which are dull in coloration, nonterritorial, nonreproductive, and school with females and other subordinate males[2] (Figure 23.2). Importantly, their HPG axis and reproductive activity are tightly coupled to social status,[3] and this species has thus become an excellent model for examining how genomic changes along the reproductive axis are regulated by social information. Along the HPG axis, dominant males have larger GnRH1 neurons in the preoptic area of the brain, higher GnRH-receptor 1 levels in the pituitary, and larger testes compared with subordinate males (Figure 23.2). Dominant males also have higher steroid and kisspeptin receptor mRNA levels in the brain,[4–6] higher FSHβ and LHβ mRNA levels in the pituitary,[7,8] higher levels of circulating steroid hormones and gonadotropins,[8–11] and higher sperm density and spermatogenic potential[12,13] compared with subordinate males. *Social ascent* in this species is the perception of a social opportunity causing the subordinate to transition to a dominant male. Interestingly, social ascent is associated with a rapid (20–30 min) induction of changes in GnRH1 neurons,[2,3,14] followed by ensuing downstream physiological changes in the pituitary gland and testes, as well as behavioral changes, over the course of the next several days.

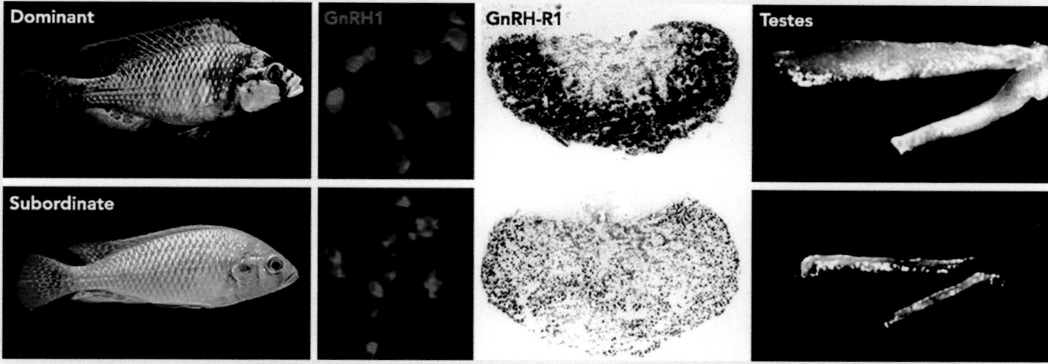

Figure 23.2 Some phenotypic features of reproductively active dominant males (top) and socially suppressed subordinate males (bottom) of the African cichlid fish, *Astatotilapia burtoni*. Dominant males have larger GnRH1 neurons (red; immunohistochemical staining) in the preoptic area of the brain, higher GnRH-R1 levels (black, GnRH-R1 *in situ* hybridization; purple, cresyl violet counterstain) in the pituitary gland, and larger testes compared with subordinate males.

Source of Images: Maruska, K.P. and Fernald, R.D., 2011. Physiology, 26(6), pp. 412-423. Used by permission.

primarily on regulation of the top of this axis, or the induction and timing of pituitary LH and FSH synthesis in response to hypothalamic GnRH signaling to promote puberty and seasonal reproduction.

The Hypothalamic-Pituitary-Gonadal (HPG) axis

LEARNING OBJECTIVE Discuss the importance of GnRH pulsatility on the onset of puberty.

KEY CONCEPTS:

- The onset of puberty is promoted by the maturation of the GnRH pulse generator, whose pulse frequency and amplitude modulate the release of the pituitary gonadotropins.
- A cyclic up/downregulation of GnRH receptors in response to fluctuating GnRH concentrations promotes the pulsatile release of gonadotropins.
- Changes in GnRH pulse frequency alter the FSH/LH ratio, with a high pulse frequency preferentially stimulating LH and low pulse frequency stimulating FSH.

GnRH: the Central Regulator of Reproduction

The central regulation of sexual maturity is influenced by diverse endogenous and environmental variables (Figure 23.1) whose disturbances can lead to either precocious or delayed puberty, as well as disruption of seasonal breeding behaviors that are present in many animals. Puberty is ultimately promoted by the maturation of the hypothalamic-pituitary-gonadal (HPG) axis. The most critical final common mediator of exogenous and endogenous signals regulating the timing of puberty and seasonal reproduction by the central nervous system is GnRH, which in humans is synthesized by about 1,000–1,500 neurons that are dispersed among the medial preoptic area

(POA) and the arcuate nucleus of the hypothalamus.[15] These nuclei form a diffuse neuronal network with projections to the median eminence, where GnRH is secreted into portal blood vessels and is thereby transported to the anterior pituitary.

In mammals, GnRH neurons are developmentally unique as they originate from outside of the brain in the nasal placode and migrate along the olfactory bulb to their final destination in the hypothalamus during mid-gestation.[16] A consequence of this GnRH neuron migration is that GnRH cell bodies are scattered throughout the basal forebrain, with each cell stopping somewhere along the migratory route.[17] This scattered topography raises many questions about how GnRH neurons can operate as synchronous, functional units. Importantly, a defect in GnRH neuron migration leads to **Kallmann syndrome**, characterized by hypogonadotropic hypogonadism due to a deficiency of GnRH synthesis that manifests as reduced LH and FSH synthesis. The importance of hypothalamic GnRH in modulating reproduction is clear, as a dysfunction of its synthesis or action inhibits gonadal development and function. This is evident not just in human patients with Kallmann syndrome, but also in mice lacking either the GnRH receptor (Figure 23.3) or the GnRH gene, each condition producing a similar gonadal atrophy phenotype (compare Figure 23.3 with Figure 7.1 in Chapter 7: Neurosecretion and Hypothalamic Control of the Pituitary). Some hypothalamic GnRH axons do not

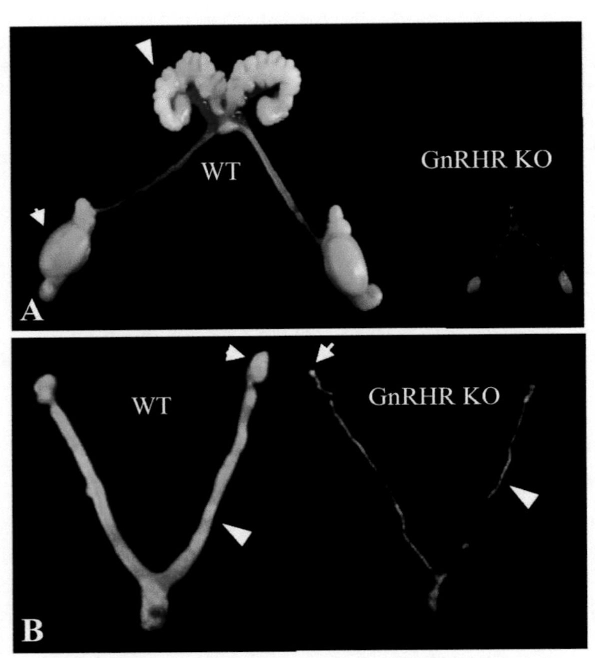

Figure 23.3 Effects of GnRH receptor (GnRHR) gene knockout (KO) on reproductive organ morphology of male and female mice. (A) Testes (*white arrows*) and seminal vesicles (*white arrowheads*) showing organs from male wild-type (WT) GnRHR KO mice. **(B)** Ovaries (*arrows*) and uterine horns (*arrowheads*) from female WT and GnRHR KO mice.

Source of Images: Wu, S., et al. 2010. Endocrinology, 151(3), pp. 1142-1152. Used by permission.

Clinical Considerations: Kallmann syndrome patients don't go through puberty due to an inability to produce GnRH

In the late 19th century, there was a medical report of a man displaying two unusual symptoms: he had both *anosmia*, or lacking a sense of smell due to the absence of olfactory nerves, and hypogonadism, where his genitalia were much smaller than normal.[18] Patients with these symptoms are now known to have the X-linked disease called Kallmann syndrome. How are these seemingly disparate symptoms related? The olfactory receptor neurons of the nose are known to originate from outside the brain in the *vomeronasal organ*, the olfactory epithelium in the nose rudiment. Whereas the cell bodies of these neurons remain in the developing nose, the axons and termini normally migrate to the brain and synapse with the olfactory bulb. However, patients with Kallmann syndrome lack an olfactory bulb in the brain due to a failure of these axonal projections to migrate to the brain.[19] Therefore, the anosmia is due to the lack of neurons in the brain that receive input from the axons coming from nasal neurons.

In 1989 two laboratories[20,21] made the surprising discovery that the GnRH-secreting neurons do not originate in the hypothalamus. Instead, they also derive from the same epithelium as olfactory receptor neurons and normally migrate into the hypothalamus during fetal development (Figure 23.4). The

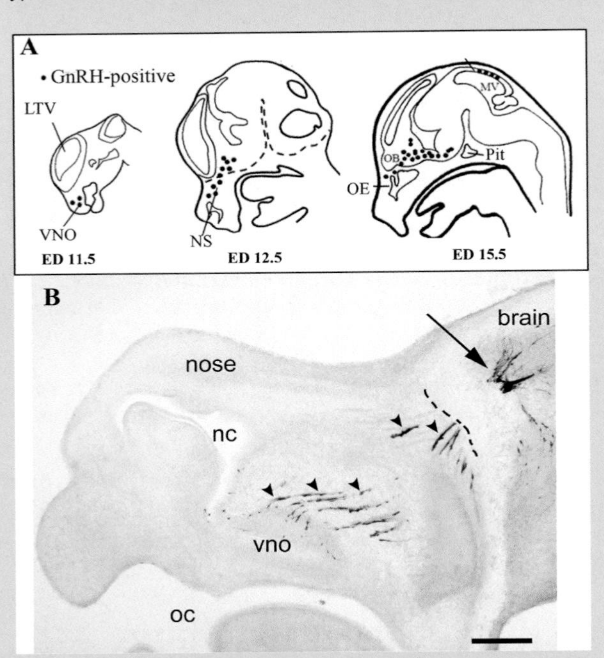

Figure 23.4 Model for the etiology of Kallmann syndrome. (A) The series of sagittal head sections from embryonic mice shows the migration of GnRH-secreting neurons from the nose anlage near the vomeronasal organ (VMO), past the nasal septum (NS), and into the hypothalamic portion of the brain. This migration does not occur in Kallmann syndrome. **(B)** Immunostaining of GnRH sagittal section of the head on embryo (ED) day 13.5. Arrow indicates GnRH neurons already located in the brain, whereas the arrow heads designate migrating GnRH neurons still in the nasal compartment. LTV, left telencephalic vesicle; OB, olfactory bulb; OE, olfactory epithelium; Pit, pituitary; MV, mesencephalic vesicle. (Photograph courtesy of John Gill and Pei-San Tsai, Univ. of Colorado.)

Source of Images: (A) Larco, D.O., et al. 2013. Front. Endocrinol., 4, p. 83. (B) Norris, D.O. and Carr, J.A., 2020. Vertebrate endocrinology. Academic Press. Used by permission.

hypogonadal defect in Kallmann syndrome is due to the failure of the GnRH-secreting neurons to migrate into the brain from the olfactory placode.[20] Failure of this migration results in an absence of these hypothalamic neurons, with downstream effects on the hypothalamic-pituitary-gonadal axis, mediated by the adenohypophysis. The hypogonadism and sterility result from of a lack of stimulation by GnRH and subsequent low levels of the gonadotropins FSH and LH that are necessary for genital maturation. The gene, KAL-1, codes for the protein, *anosmin-1*, whose expression is normally detected in the basement membranes of several organs, including the olfactory system. Anosmin is critical for the branching and extension of the olfactory bulb output neurons, and its absence or inhibition causes the syndrome by preventing their migration to the hypothalamus.[22–25]

connect with the median eminence, but instead project to other areas of the brain and may mediate sexual behavior. Considering its long reach, it is not surprising that the pharmacological stimulation or suppression of normal GnRH pulsatility and/or GnRH receptor activation/deactivation is therapeutically advantageous for numerous clinical purposes, including contraception, the treatment of infertility, precocious and delayed puberty, endometriosis, polycystic ovary syndrome, breast and prostate cancer, and other conditions.

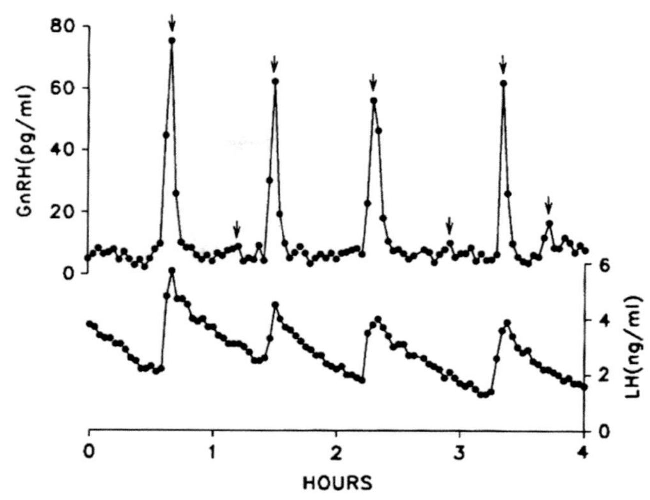

Figure 23.5 The close temporal relationship between the secretion of hypothalamic GnRH and LH from the anterior pituitary gland. GnRH secretion was measured by serial sampling of hypophyseal portal blood of an ovariectomized ewe, and LH levels were measured in jugular venous blood. Though strong pulses of GnRH drive the secretion of corresponding LH pulses, weak pulses do not.

Source of Images: Clarke (1993) Endocrinology. 133:1624-32. Used by permission.

GnRH Pulsatility Modulates Gonadotropin Release

Two distinct patterns of GnRH secretion have been described in mammals: "pulsatile" and "surge" profiles.[26] The *pulsatile* profile, which is present in both male and female adults, describes an episodic release of GnRH, with clear peaks of GnRH secretion into the portal circulation and undetectable GnRH concentrations between pulses. The GnRH *surge* profile occurs only in females during the preovulatory phase, when either GnRH appears to be persistently elevated in the portal circulation or the frequency of pulses is too high to differentiate GnRH peaks from troughs.[15] Experiments in mammals sampling blood concurrently from hypothalamic portal blood vessels to measure GnRH and from systemic blood to measure gonadotropins have established that each is secreted in a pulsatile manner over time. In such an experiment performed with the ewe, LH pulses occur in a *circhorial* manner, or about one pulse per hour, preceded by a strong GnRH pulse (Figure 23.5). The pulse frequency of GnRH, therefore, sets the rhythm for gonadotropin pulsatility,[27] which varies according to reproductive status. It is well established that the pulsatile release of GnRH into the median eminence is responsible for driving pulsatile gonadotropin secretion in virtually all mammals,[28] with species-specific variability in the pulse frequency and amplitude. This GnRH pulsatility is not merely an intriguing characteristic of these hypothalamic neurons, but its episodic nature is critical for the release of pituitary gonadotropins.

One of the first studies demonstrating the significance of GnRH pulsatility was performed using rhesus monkeys, where endogenous GnRH production was first eliminated by lesioning GnRH-producing nuclei in the mediobasal hypothalamus, which also inhibited the synthesis of pituitary gonadotropins.[29] Subsequently, elevated LH and FSH production was restored using an intravenous infusion pump programmed to supply pulses of exogenous GnRH at a frequency similar to that occurring naturally (Figure 23.6). Intriguingly, administration of a continuous, nonpulsatile supply of GnRH delivered at the same concentration as the previous pulsatile treatment failed to restore FSH and LH production by the pituitary, and these gonadotropins fell to almost undetectable levels. However, a return to pulsatile GnRH treatment restored gonadotropins to pre-lesion levels.

Interestingly, changes in GnRH pulse frequency in monkeys has been shown to alter the FSH/LH ratio, with a high pulse frequency preferentially stimulating LH beta mRNA and low pulse frequency stimulating FSH

Curiously Enough . . . Many human reproductive pathologies stem from GnRH pulse frequencies that are either abnormally high, such as precocious puberty and polycystic ovary syndrome (PCOS), or low, for example delayed puberty, hypothalamic amenorrhea, obesity-related male hypogonadism, and opioid-induced hypogonadism.[15]

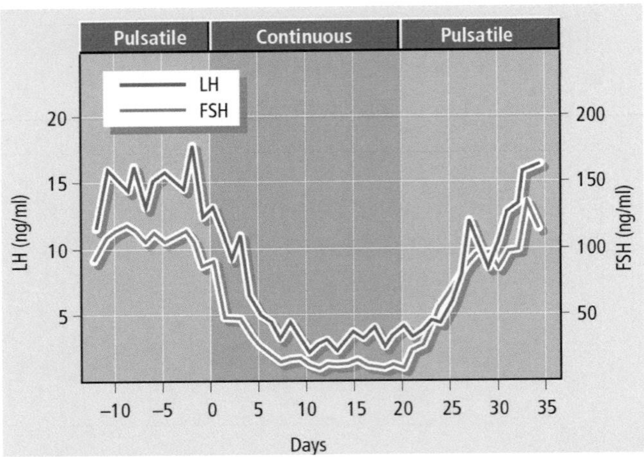

Figure 23.6 The effects of continuous versus pulsatile GnRH on FSH and LH levels. Endogenous GnRH secretion in an ovariectomized rhesus monkey was ablated by lesioning the GnRH-secreting nuclei of the hypothalamus. Normal gonadotropin secretion was reinstituted by the pulsatile infusion of exogenous GnRH during the first 10 days of the experiment (1 μg/min for 6 minutes once per hour). Gonadotropin secretion was suppressed under continuous GnRH infusion (1 μg/min; days 0–20) and restored with the pulsatile GnRH regimen (days 20–35).

Source of Images: M.H. Johnson (2007) Essential Reproduction, 6th edition. Wiley-Blackwell. Used by permission.

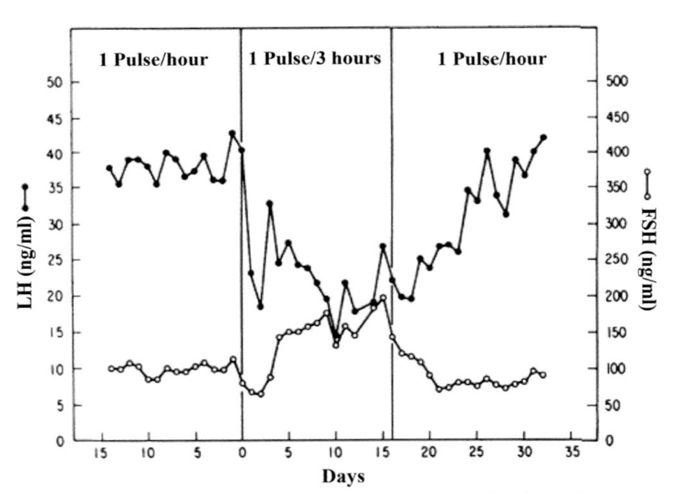

Figure 23.7 Modulation of the FSH:LH ratio by changes in GnRH pulse frequency in a monkey. Note the dramatic increase in the ratio with the decrease in GnRH pulse frequency.

Source of Images: Wildt L, et al. Endocrinology 109:376, 1981. Used by permission.

beta mRNA[30–33] (Figure 23.7). In adult humans, there are important differences in GnRH pulsatility between females and males. In males both gonadotropins, and presumably GnRH, are secreted in pulses at a constant frequency of about once every 2 hours. By contrast, in females the frequency of the pulses varies throughout

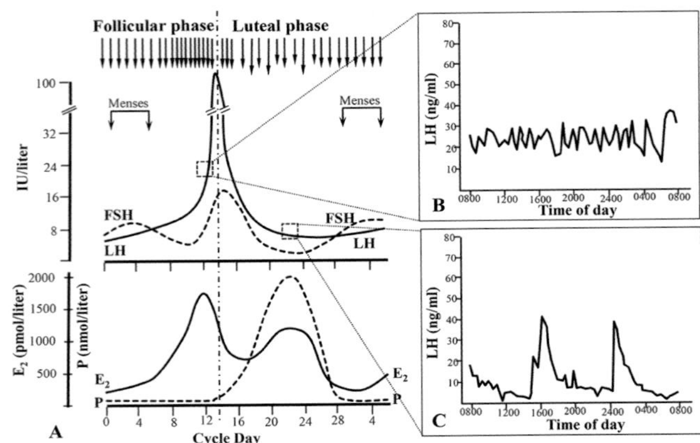

Figure 23.8 Schematic showing how the GnRH pulse generator promotes changes in gonadotrope pulse frequency and amplitude throughout the human ovarian cycle. (A) GnRH is secreted at a high frequency/low amplitude pulse rate during the follicular phase, leading to a preferential secretion of LH and the LH surge (which induces ovulation) shown on Day 14. During the second half of the ovarian cycle, the luteal phase, GnRH is secreted at a lower pulse frequency and a higher amplitude that reduces LH secretion and promotes the preferential secretion of FSH, particularly towards the end of the luteal phase. **(B)** Examples of the pulsatile pattern of LH secretion in a woman during the late follicular phase and **(C)** midluteal phase of the menstrual cycle. Note the dramatic slowing of pulsatile LH secretion during the luteal phase.

Source of Images: (A) Marshall JC, et al. Recent Prog Horm Res. 1991; 47:155-187. Used by permission. (B-C) Modified from Soules, M.R., et al. 1984. J. Clin. Endocrin. Metab. 58:378-383.

the menstrual cycle, with high frequency/low amplitude pulses about once every 60–90 minutes favoring LH secretion during the follicular phase promoting ovulation, and low frequency/high amplitude pulses of about once every 200 minutes favoring FSH secretion in the late luteal phase[34] (Figure 23.8).

SUMMARY AND SYNTHESIS QUESTIONS

1. From a public health perspective, how is an earlier start to puberty concerning?
2. Note on Figure 23.8 that during the end of the luteal phase in the second half of the ovarian cycle, LH secretion is dramatically reduced and FSH is preferentially secreted. What, specifically, induces the change in FSH and LH secretory profiles?

Regulation of GnRH Pulsatility

LEARNING OBJECTIVE Discuss the neuroendocrine basis of negative and positive feedback signaling along the HPG axis in response to hormones throughout the ovarian cycle.

KEY CONCEPTS:

- The GnRH pulse generator includes hypothalamic kisspeptinergic neurons that synapse with GnRH neurons to regulate their secretion.
- GnRH secretion is influenced by innervation from two major kisspeptin-secreting neuronal populations in the hypothalamus: the arcuate (ARC) and the anteroventral periventricular (AVPV) nuclei.
- In male and female mammals, kisspeptinergic neurons of the ARC promote GnRH pulsatility and respond to elevated sex steroid hormones with negative feedback.
- In female mammals, kisspeptinergic AVPV neurons respond to rising estrogens during the follicular phase of the ovarian cycle with positive feedback, which facilitates the ovulatory surge in GnRH.
- A higher amplitude and increased cycling rate of gonadotropins is observed in castrated males and in women following menopause.

The GnRH Pulse Generator

The pulsatility of GnRH suggests the need for synchronization among at least a subpopulation of the dispersed GnRH neurons. However, the precise nature of the *GnRH pulse generator*, or the subset of neurons that modulate GnRH pulsatility, remains unclear. The GnRH pulse generator appears to be localized to the hypothalamus, as the pulsatility of GnRH neuronal firing occurs in hypothalamic explant cultures, where the GnRH neurons are completely separated from all other afferent inputs.[35] Current debate revolves around whether the GnRH pulse generator is *intrinsic*, or mediated entirely by a scattered population of GnRH neurons, or *extrinsic*, where GnRH pulsatility is driven by other cells that connect to GnRH neurons.[17,36] Growing support for extrinsic GnRH pulse generation via a population of kisspeptinergic hypothalamic neurons in the *arcuate nucleus* (ARC) and *anteroventral periventricular nucleus* (AVPV) of the hypothalamus stems from several types of experimental observations, including the presence of shared synapses on GnRH neuron dendrites[37,38] and the absence of pulsatile LH secretion in mutations in the kisspeptin and kisspeptin receptor genes in humans[39,40] and rodents.[41,42]

GnRH Regulation by Kisspeptin

One of the most powerful known inducers of GnRH is *kisspeptin*, a neuropeptide encoded by the Kiss1 gene[43,44] whose administration increases circulating LH levels in both animal and human studies.[45–48] Axonal projections from both hypothalamic kisspeptin neuron populations are in close juxtaposition with GnRH neurons in a broad range of species,[49,50] and approximately 50%–75% of GnRH neurons express the kisspeptin receptor.[51–53] Kisspeptin actually constitutes a family of peptides all derived from the Kiss1 gene, each formed from the differential proteolysis of a common prepro-kisspeptin precursor. In human circulation, the most abundant neuroactive kisspeptin is kisspeptin-54, which can be further cleaved into several shorter peptides.[54] In rodents, hypothalamic expression of kisspeptin and its receptor (a G protein-coupled receptor known as *GPR54*) have been localized in AVPV and ARC hypothalamic neuron populations[45] (Figure 23.9). In the hypothalamus of humans and other primates, kisspeptin is expressed primarily in a region equivalent to the ARC in these animals, called the infundibular nucleus.[55] Importantly, ARC kisspeptin neurons are distinguishable from AVPV neurons in that they coexpress *neurokinin B* (NKB) and the opioid *dynorphin A* (Dyn) and are thus often denoted as *kisspeptin-NKB-Dyn* (KNDy) neurons. KNDy neurons have been strongly implicated with a role in GnRH pulse generation, with NKB and Dyn thought to respectively stimulate and inhibit KNDy neuron activity, promoting the episodic release of kisspeptin, which in turn drives pulses of GnRH release.[56,57] By contrast, kisspeptin AVPV neurons, which are virtually exclusive to the female brain, mediate the positive feedback of sex steroids that leads to the preovulatory LH surge. The current model for hypothalamic kisspeptinergic neuron function in the rodent presumes that whereas KNDy neurons in the ARC mediate the negative feedback effects of sex steroids in both sexes, kisspeptinergic neurons of the AVPV elicit the preovulatory gonadotropin surge through positive feedback mechanisms.[45,58–60]

Curiously Enough . . . Discovered in 1996 by a lab in Hershey, Pennsylvania, kisspeptin was named after the famous "Hershey's Kisses" chocolate candies.

Clear roles of kisspeptin as a major regulator of reproduction come from studies using kisspeptin receptor knockout mice showing inhibited gonadal growth and development (Figure 23.10). Note how this phenotype is strikingly similar to that of the aforementioned GnRH receptor knockout (compare to Figure 23.3). Also, landmark studies using human patients with congenital hypogonadotropic hypogonadism[40,61] discovered a number of inactivating mutations in the kisspeptin receptor gene, paving the way for many other studies showing similar results for mutations in both the kisspeptin receptor gene and in mutations of the Kiss1 gene itself.[62] Furthermore, Kiss1 and Kiss1 receptor-activating mutations have also been described in children with central precocious puberty.[63,64] Finally, the exogenous administration of kisspeptin antagonists inhibited GnRH release in pubertal monkeys[65] and delayed puberty in rats,[66] whereas treatment with kisspeptin induced earlier puberty in rats[67] and monkeys.[68] Taken together, these findings strongly implicate kisspeptin with stimulating hypothalamic GnRH neurons to release GnRH into the hypothalamic-pituitary portal circulation, promoting the release of gonadotropes from the anterior pituitary into general circulation and inducing gonad growth and maturation.

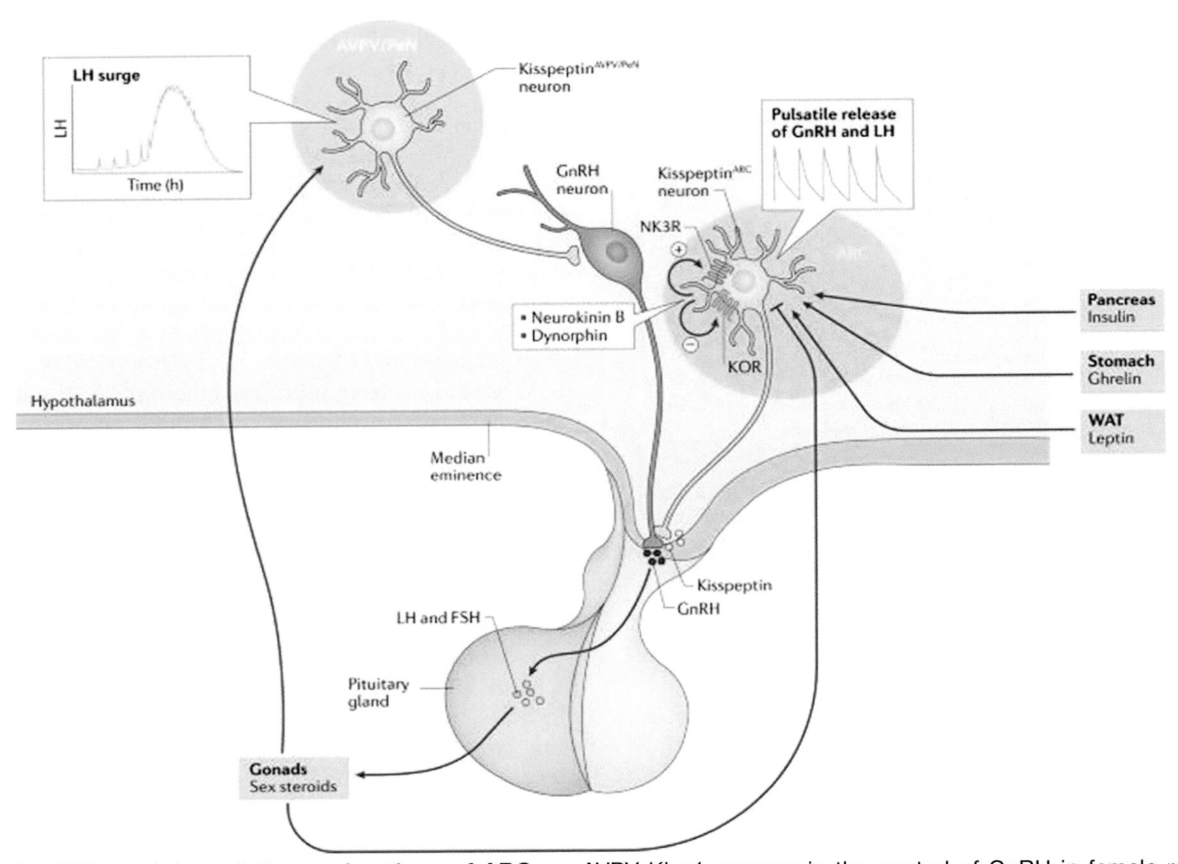

Figure 23.9 Differential regulation and actions of ARC vs. AVPV Kiss1 neurons in the control of GnRH in female rodents. Kisspeptinergic neuronal bodies, located in both the anteroventral periventricular (AVPV) and arcuate (ARC) nuclei, project to the cell bodies and axonal termini of GnRH neurons located in the hypothalamic preoptic area (POA). Arcuate kisspeptin (kisspeptin^{ARC}) neuron together with neurokinin B and dynorphin are involved in the tonic (pulsatile) release of kisspeptin (positive and negative symbols indicate the effect on kisspeptin release) and, therefore, gonadotropin-releasing hormone (GnRH). By contrast, anteroventral periventricular/periventricular nucleus kisspeptin (kisspeptin$^{AVPV/PeN}$) neurons are involved in the control of the luteinizing hormone (LH) surge and are almost absent in the male brain. Major peripheral metabolic factors are depicted (leptin, insulin, and ghrelin) acting at the level of the brain to regulate kisspeptin output. ARC, arcuate nucleus; FSH, follicle-stimulating hormone; HPG axis, hypothalamic-pituitary-gonadal axis; KOR, κ-opioid receptor; NK3R, neurokinin B receptor; WAT, white adipose tissue.

Source of Images: Navarro, V.M., 2020. Nat Rev Endocrinol, 16(8), pp. 407-420. Used by permission.

Sexually Dimorphic Kisspeptin Signaling

Male and female humans and other mammals exhibit conserved differences in the hypothalamic architecture of kisspeptinergic neurons and also in gonadotropin responses to kisspeptin. For example, compared with male hypothalami, those of females have more kisspeptin cell bodies and fibers,[15,69,70] a feature that is particularly prominent in the AVPV nuclei of mice[71] (Figure 23.11). Responsivity to kisspeptin also differs between the sexes, as the peptide is a potent stimulator of LH in human males, but in females its effect varies with the phase of the menstrual cycle with minimal responsivity during the early follicular phase and greater responsivity during the preovulatory phase.[72] These anatomical and functional differences likely reflect sexually dimorphic functions of kisspeptin that influence differences in the feedback of sex steroids on GnRH and gonadotropin secretion. Additionally, in rats the estrogen receptor β (ERβ) is expressed in kisspeptinergic AVPV neurons, and there is a dramatic sexually dimorphic expression of ERβ in the AVPV, with females expressing more than males[73] (Figure 23.11). This observation may explain why administration of estrogens to males is unable to generate a gonadotropic surge.[74]

Lessons from Castration and Menopause

The influence of negative feedback by circulating gonadal sex steroids along the HPG axis is clearly evident in male mammals following castration (removal of the testes), causing reduced testosterone synthesis (Figure 23.12). Note that in this example the loss of feedback inhibition

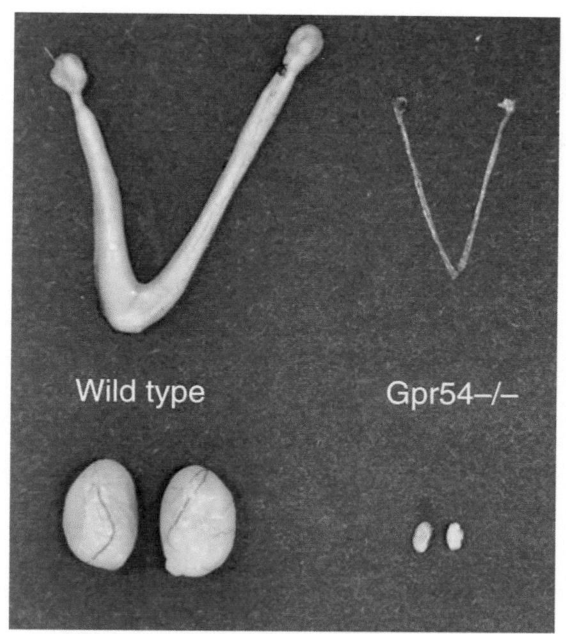

Figure 23.10 Kisspeptin receptor knockout (Gp54-/-) mice display inhibited gonadal growth and development. Uteri and ovaries (top); testes (bottom).

Source of Images: Kirilov, M., et al. 2013. Nat Comm, 4(1), pp. 1-11. Used by permission.

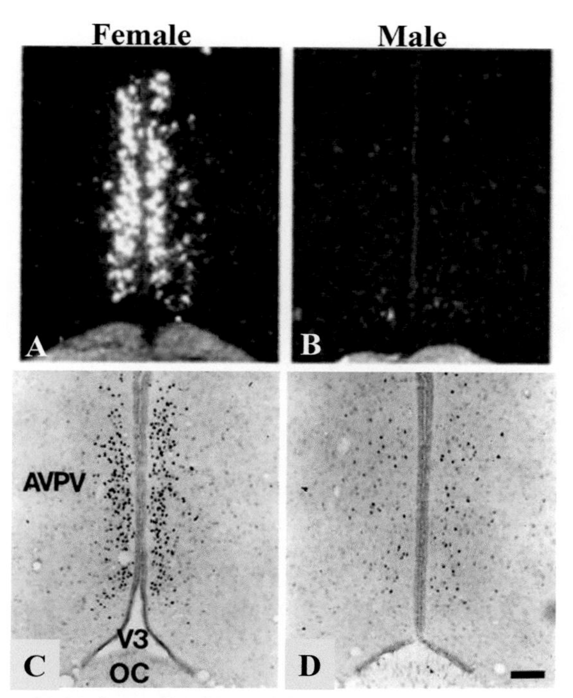

Figure 23.11 Sexually dimorphic distribution of *Kiss1* mRNA expression and ERβ protein in the AVPV of wild-type adult female and male rodents. (A–B) *Kiss1* mRNA in mice visualized by *in situ* hybridization. **(C–D)** ERβ protein in rats visualized by immunohistochemistry.

Source of Images: Gill, J.C., et al. 2010. PLoS One, 5(7), p. e11911.

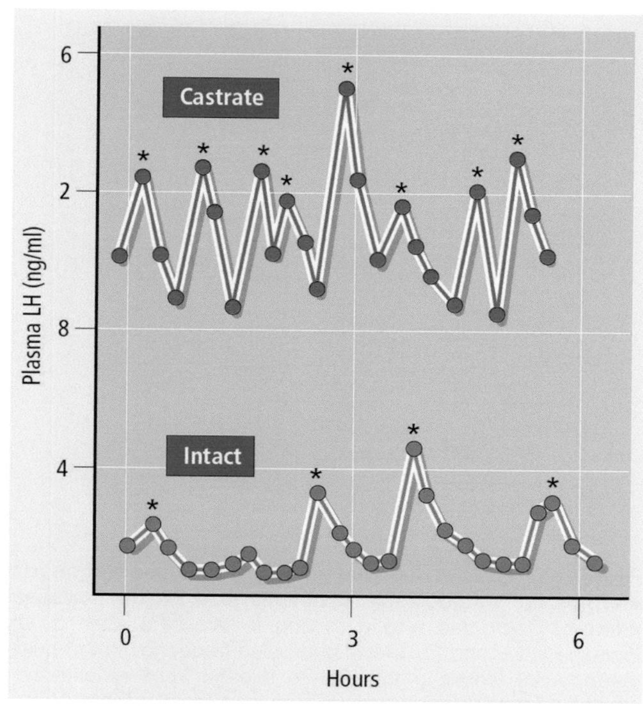

Figure 23.12 Loss of negative feedback inhibition by gonadal steroids on gonadotropin synthesis is evident following castration in a deer. Compared with the intact deer, in the castrated deer LH amplitude and pulse frequency increase as a consequence of the loss of negative feedback inhibition by the testis.

Source of Images: M.H. Johnson (2007) Essential Reproduction, 6th edition. Wiley-Blackwell. Used by permission.

in the red deer promotes increases in both LH pulse frequency and amplitude, suggesting that testosterone influences the hypothalamic pulse regulator, altering pulse frequency and the magnitude of the pituitary gonadotrope response. In human males, negative feedback by testosterone occurs at the levels of both the hypothalamus and the pituitary through activation of the androgen receptor (AR) and/or estrogen receptor alpha (ERα) by estrogens derived from the local aromatization of testosterone into estrogens.[75–77] Interestingly, at the level of the pituitary most, if not all, of testosterone-induced negative feedback appears to be mediated by the local aromatization of testosterone to estradiol rather than by direct androgen action.[76] Furthermore, testosterone appears to upregulate brain CYP19 aromatase, thereby increasing its own conversion into estradiol.[78,79] Since few GnRH neurons express either AR or ERα,[80–82] a key role for kisspeptinergic neurons that express these receptors[72,83,84] has been suggested for mediating feedback in both males and females. Similar to the effects of castration in males, in women following menopause (the cessation of steroid hormone synthesis by the ovaries), a high amplitude and increased cycling rate of gonadotropins is also observed (Figure 23.13).

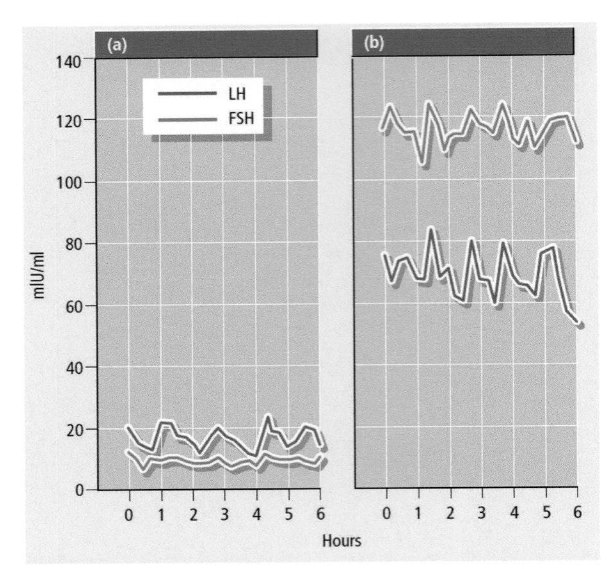

Figure 23.13 Loss of negative feedback inhibition by gonadal steroids on gonadotropin synthesis in a postmenopausal woman. Compared with a cycling female (left), there is an increase in LH and FSH level and pulse frequency in the postmenopausal female (right) due to release from inhibition by ovarian hormones.

Source of Images: M.H. Johnson (2007) Essential Reproduction, 6th edition. Wiley-Blackwell. Used by permission.

SUMMARY AND SYNTHESIS QUESTIONS

1. Whereas administration of estrogens to adult female mammals can sometimes induce LH and sometimes repress it, in males administration of estrogens always suppresses LH. Explain how estrogens have variable effects on LH synthesis in females, but always have suppressive effects on males.
2. You have learned that continuous administration of GnRH at high concentration and/or pulse frequency results in the suppression of gonadotropin release, resulting in chemical gonadectomy. However, if continuous GnRH treatment is used to treat steroid hormone-sensitive cancers, the initiation of treatment is often accompanied by the so-called flare phenomenon characterized by an initial acceleration in cancer activity prior to its decrease. What likely causes this flare phenomenon, and how does continuous GnRH eventually suppress steroid hormone synthesis?
3. The drug Buserelin is a GnRH receptor agonist. Interestingly, this drug can be used as either a pro-fertility drug or as an anti-fertility drug, depending on how it is administered. You are a doctor and have three different patients to whom you prescribe Buserelin treatment via a portable, computer-controlled pump. You have the option of programming the pump to administer either (a) a low pulse frequency, (b) a high pulse frequency, or (c) a continuous flow (i.e. no pulsing). Your first patient is a female with endometriosis, a condition that is treated by inhibiting both FSH and LH. Which setting would you use to treat endometriosis? Your second patient is a male who is producing low levels of testosterone due to low levels of LH. Which setting would you administer to increase testosterone synthesis? Your third patient is a female who is unable to synthesize estrogens from its precursor due to low levels of CYP19 aromatase synthesis caused by low circulating FSH. Which setting would you administer to increase aromatase synthesis? Explain your reasoning for each answer.
4. Think back to what you learned about Kallmann syndrome, which manifests as hypogonadism due to a failure of GnRH neurons to migrate to the hypothalamus. Assume, for the moment, that the neurons still possess the ability to synthesize GnRH. Think of two reasons why a failure of these neurons to migrate properly would cause hypogonadism.

Regulating Puberty in Humans

LEARNING OBJECTIVE Describe the major endocrine benchmarks of puberty in human males and females and developmental changes in the HPG axis from birth through adulthood.

KEY CONCEPTS:

- Adrenarche, the earliest endocrine change during puberty, denotes the increased synthesis of androgen precursor hormones by the adrenal gland in both males and females.
- Mediated by ovarian hormones, thelarche represents the first morphological changes in puberty in females and is characterized by the beginning of breast development, growth acceleration, and redistribution of body fat.
- Menarche denotes the time of the first menstruation in females in response to increasing maturation of the GnRH pulse generator and follicular development.
- Gonadarche describes testicular enlargement and increased testosterone synthesis in males, generally the first signs of puberty in boys.
- The HPG axis is established and active in fetal development where a mid-gestation increase in gonadotropins occurs.
- Although the HPG axis is established *in utero*, children experience a "juvenile hiatus" in gonadotropin secretion, a period of time when GnRH and gonadotropin secretion is nonpulsatile and highly sensitive to negative feedback suppression.
- The start of puberty is characterized by a reactivation of GnRH pulsatility and reduced sensitivity to negative feedback by sex steroid hormones.

What Is Puberty?

The attainment of full gametogenic and hormonal capacity by the gonads is termed *puberty*. The interval between hatch/birth and onset of sexual maturity varies dramatically among species, ranging from 6 weeks in mice, to 15 years in elephants, to over 20 years in the European eel.[85,86] Importantly, the timing of puberty can also vary remarkably within a species. For example, although in the United States the mean age of first menstruation in human females is 12.3 years, the normal range spans from 9.7 to 16.7 years, resulting in significant variability in pubertal change within the same chronological age.[85] Diverse behavioral, morphological, and physiological changes accompany puberty. Some changes are associated with reproductive tissues present since birth, such as growth and development of the ovaries, uterus, vagina, testes, and penis in humans. Others, called **secondary sexual characteristics**, involve changes in somatic tissues that emerge only at puberty and include the development of axillary hair and the deepening of the voice (some secondary sexual characteristics in females and males associated with puberty are listed in Appendix 11). Morphological changes in external genitalia in human males and changes in breast development and growth of pubic hair in females during puberty are categorized into *Tanner stages* that range from prepubertal through the completion of puberty (Appendix 12 describes Tanner stages of development for males and females). These Tanner stages generally correspond with the reaching of identifiable endocrine benchmarks, described next.

Adrenarche denotes the increased synthesis of the androgen precursor steroid hormones DHEA and DHEA-S by the adrenal gland in both male and female humans and some other closely related primates. Indeed, these occur independently of the gonads and are the earliest hormonal changes to take place during puberty. Adrenarche promotes the development of *apocrine sweat glands* in the skin and an accompanying change of sweat composition that produces adult body odor and increases oiliness of the skin (which can lead to acne) and hair in both males and females. In girls, the maturation of the GnRH pulse regulator and gonadotropin secretion promotes ovarian follicular development, synthesis of estrogens, and, eventually, ovulation. *Thelarche* is a pubertal benchmark characterized morphologically by the beginning of breast development and growth acceleration in girls, as well as changes in body fat distribution and composition. Thelarche typically represents the first morphological signs of the start of puberty in females induced by the production of gonadotropin-driven ovarian estrogens. In contrast to adrenarche, thelarche appears to be mediated independently of the adrenal glands, and instead via production of estrogens by the ovaries.[87,88] *Menarche* denotes the time of the first menstruation in females in response to increasing maturation of the GnRH pulse generator and follicular development and is typically the last obvious marker of female sexual maturation. A regular pattern of ovulatory cycling generally develops within two years of menarche. **Gonadarche** describes testicular enlargement and increased testosterone synthesis in males, generally the first signs of puberty in boys. Gonadarche promotes masculinization, a developmental progression that includes increasing penile and scrotal size, musculoskeletal changes, and male pattern hair development.

The Gonadotropin Profile Changes throughout Life

Although GnRH plays a central role in initiating puberty during adolescence, in humans and some other mammals the HPG axis is already established and active much earlier in life, extending even back into fetal development where a mid-gestation increase in gonadotropins occurs[89,90] (Figure 23.14). This mid-gestational gonadotropin peak subsequently declines in response to rising placental steroids that exert negative feedback on both the hypothalamus and pituitary.[34] Following the disappearance of placental steroids at birth, gonadotropin levels transiently rise again, persisting for several months in male infants and about 2 years in female infants.[17] For the next 10–14 years, human children experience a **juvenile hiatus** in gonadotropin secretion, or a period of time during which GnRH and gonadotropin secretion is suppressed by both gonad-dependent and gonad-independent factors. This juvenile hiatus has been conceptualized as a *prepubertal brake* imposed upon the GnRH pulse generator mediated by either the presence of poorly understood inhibitory signals to the pulse generator, the loss of stimulatory signals, or a combination of both.[91] Importantly, during the juvenile hiatus period, the HPG axis can be artificially reactivated by treatment with neurotransmitters such as kisspeptin[68,92] and glutamate,[93,94] suggesting that the prepubertal brake normally inhibits these and other signals.

Release from the prepubertal brake, in combination with a gain in stimulatory signals, terminates the juvenile period of primate development, leading to a reactivation of GnRH pulsatility.[91] At puberty, the resumption of pulsatile gonadotropin secretion is initially circadian in profile, with greater prominence at night (or during sleep) before achieving the normal adult reproductive pattern with pulsatility that persists throughout the 24-hour period[95] (Figure 23.14). As with other primates, in humans the onset of pulsatile secretion typically begins earlier in females than males.[17] In addition, towards the end of puberty, females begin to exhibit the aforementioned surges in LH secretion that will ultimately initiate ovulation. Following menopause, GnRH pulse characteristics change once again with decreasing levels of the secretion of estrogens (Figure 23.14). Specifically, the subsequent loss of negative feedback by estrogens promotes increased GnRH secretion, with a gradually decreasing pulse frequency as women age from the fifth to the eighth decade.[96]

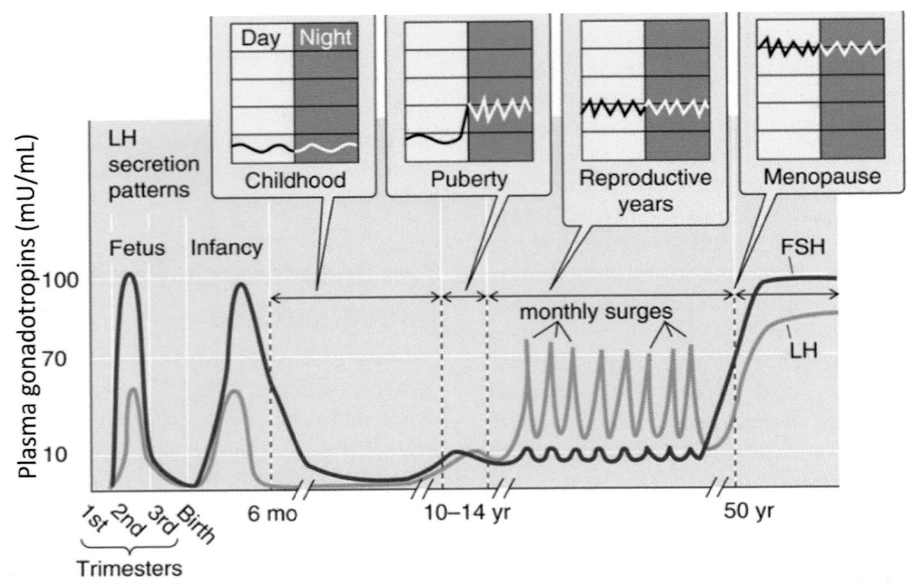

Figure 23.14 Gonadotropin levels throughout female life. The levels of both LH and FSH peak during fetal life and again during early infancy, before falling to low levels throughout the rest of childhood. At the onset of puberty, LH and FSH levels slowly rise and then begin to oscillate at regular monthly intervals. At menopause, gonadotropin levels rise to very high levels. The four insets show circadian changes in gonadotropin levels.

Source of Images: Boron, W.F. and Boulpaep, E.L., 2005. Medical Physiology: A Cellular and Molecular Approach, updated 2nd ed. Elsevier. Used by permission.

SUMMARY AND SYNTHESIS QUESTIONS

1. Imagine that a girl is born with primary ovarian insufficiency and her ovaries do not synthesize sex steroid hormones. Which of the following components of puberty, if any, will she experience: (a) pubic hair growth, (b) breast development, (c) menstruation?
2. Explain why the following statement is incorrect: "The HPG axis first becomes functional during puberty".

Environment and the Timing of Human Puberty

LEARNING OBJECTIVE Discuss the influence of metabolic status, obesity, and stress on the timing of puberty.

KEY CONCEPTS:

- A younger age of puberty carries important public health consequences caused, in part, by a variable exposure time to sex hormones associated with a higher risk of developing cancers and cardiovascular and metabolic diseases.
- Variability in the timing of puberty is influenced by both genetic and environmental factors.
- Puberty begins when a threshold of energy stores, optimal metabolic status, and critical body mass are attained.
- Leptin acts as an essential permissive factor for the initiation of puberty.
- Obesity in girls tends to promote an earlier onset of puberty, whereas in boys it can either accelerate or delay puberty, depending on its severity.
- Stress hormones, such as adrenal glucocorticoids and hypothalamic corticotropin-releasing hormone (CRH), can delay puberty and inhibit fertility.

How does the body "know" when to begin puberty? What information prompts the pubertal reactivation of the GnRH pulse generator? The timing of puberty is remarkably variable in humans and carries important public health consequences caused, in part, by a variable lifetime exposure to sex hormones. Earlier puberty, in particular, is associated with a higher risk of developing cancers that are sensitive to sex hormones in later life, such as ovarian, breast, and endometrial cancers in women and prostate cancer in men.[97] Earlier puberty has also been linked to a higher risk of cardiovascular and metabolic disease, including type 2 diabetes and hypertension in both women and men.[98] Variability in the timing of puberty is influenced by both genetic and environmental factors. Variance in the age at menarche of 50%–80% has been attributed to genetics,[99] and hundreds of genetic signals associated with puberty timing have been identified.[100]

The remaining 20%–50% of variability result from diverse environmental effects, including nutrition, body weight, pre- and postnatal growth rates, stress (both physical and psychosocial), and possibly exposure to *environmental endocrine disruptors*, which are cocktails of synthetic and naturally occurring chemical compounds that mimic hormones and/or interfere with their synthesis, action, and metabolism. This section will focus on environmentally derived metabolic and stress parameters that affect the timing of puberty.

Metabolic Cues Influence the Timing of Puberty

Reproduction is an energetically demanding process, as evident by the enormous resources diverted to pregnancy and lactation in female mammals and partner selection and territoriality in the males of many species. Therefore, nutritional status has long been considered a critical factor for pubertal development. Classic studies using rats have shown that the start of puberty corresponds with body size, but not chronological age,[101] and in humans Rose E. Frisch proposed that a critical fat/lean ratio, rather than a critical age, had to be attained for menarche to occur in girls.[102,103] Thus, puberty is suggested to occur only when a threshold of energy stores, optimal metabolic status, and adult somatic size are attained.[104–106] Supporting this "critical weight" hypothesis is the observation that although the age at which girls reach menarche has dropped since the 19th century, the average body weight of girls reaching puberty has remained constant[107] (Figure 23.15). Furthermore, growth restriction during the juvenile period in rodents,[101] sheep,[108,109] and human female athletes with markedly diminished body fat[110,111] have been shown to delay the onset of puberty, whereas increased adiposity accelerates puberty in in rats,[112] cattle,[113,114] and human girls.[110,115,116]

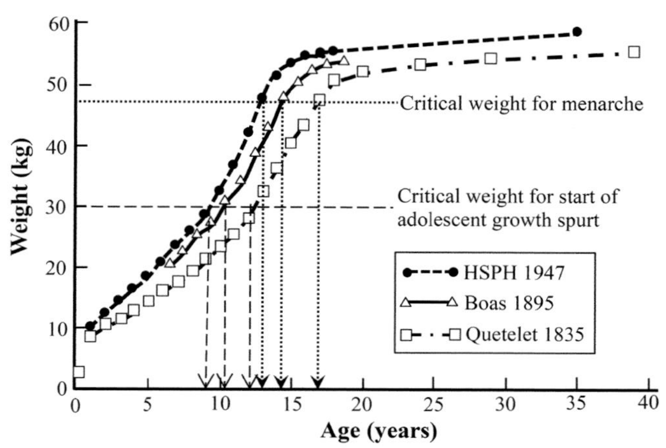

Figure 23.15 Age plotted against body weight in three populations of girls in 1835, 1895, and 1947. Note the constant weights at the initiation of the growth spurt (30 kg) and at menarche (47 kg). Populations: Belgian girls in 1835 from data of Adolphe Quetelet, 1869; American girls in 1895 adapted from data of Franz Boas; and American girls in 1947 adapted from data of Reed and Stuart, 1959, Harvard School of Public Health (HSPH).

Source of Images: Frisch, R.E., 1972. Pediatrics, 50(3), pp. 445-450.

Developments & Directions: The changing timing of puberty in girls throughout the ages

Significant changes in the timing of female puberty throughout human history indicate that the onset of puberty is responsive to changes in the environment. Paleontological evidence suggests that in Neolithic times (early agricultural and village communities dating from 9,000–3,000 years ago) females reached menarche at an early age, approximately 7–13 years old, likely driven by the

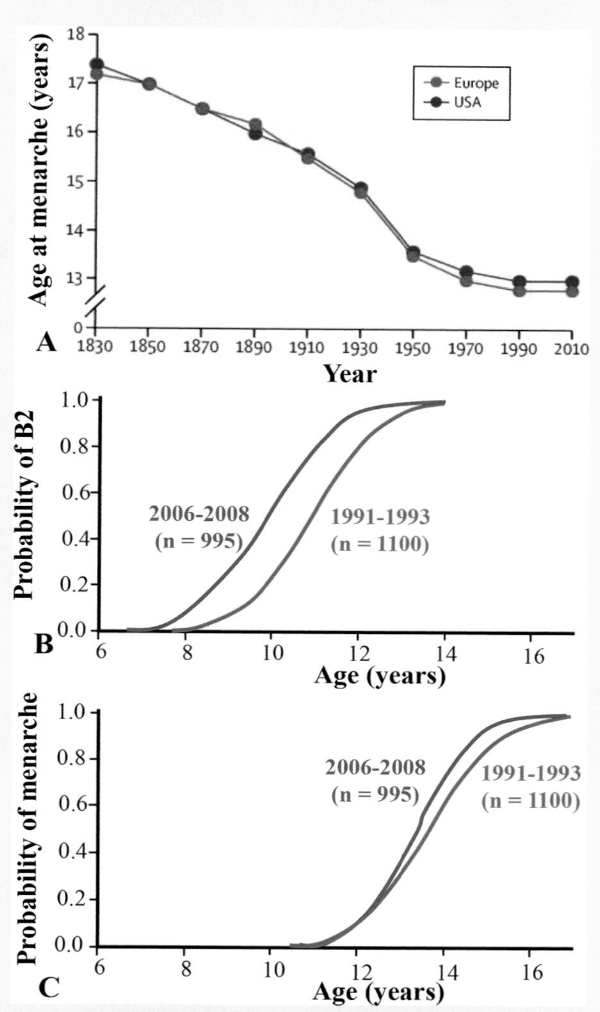

Figure 23.16 (A) Decreasing age at menarche in Europe and the United States from 1830 to 2010. This trend has been slowing since 1965, and has leveled off to about 13 years. **(B)** Recent decline in age at breast development. Girls aged 5.6 to 20.0 years were studied in 1991–1993 (1991 cohort; n = 1,100) and 2006–2008 (2006 cohort; n = 995). The onset of puberty (B2), defined as mean estimated age at attainment of glandular breast tissue (Tanner breast stage 2+), occurred significantly earlier in the 2006 cohort (estimated mean age: 9.86 years) when compared with the 1991 cohort (estimated mean age: 10.88 years). **(C)** The ages at menarche (13.42 and 13.13 years in the 1991 and 2006 cohorts, respectively) did not differ significantly.

Source of Images: Reinehr T, Roth CL. Lancet Child Adolesc Health. 2019 Jan;3(1):44-54. Used by permission.

selective advantage of reaching reproductive age quickly in the face of a relatively short lifespan.[117] Post-Neolithic periods were characterized by dramatically increased population density, malnutrition, deteriorating hygiene, and childhood disease.[118,119] By the advent of the Industrial Revolution (about 1760–1840), the age of menarche was delayed to about 17 years of age.[120] Interestingly, there has been a well-documented subsequent trend of *decreasing* age of menarche from 1850 through 1960 in many European countries and the USA,[121–124] stabilizing to about 13 years of age in the 1940s (Figure 23.16[A]). This decline in menarche age is thought to result from improved nutrition, hygiene, disease control, and overall living conditions over the past two centuries, and in countries like China and India where the standard of living has improved only relatively recently menarcheal age has shown a very rapid contemporary decrease.[121] Altogether, these data are consistent with the "critical weight" hypothesis originally proposed by Frisch and Revelle,[102,103] and subsequent research[104–106] has emphasized the role of nutrition as measured by adiposity in the timing of puberty. Indeed, many studies have noted the association of elevated body mass index (BMI) and obesity with earlier puberty in modern girls.[125,126]

Strikingly, beginning in the 1990s a renewed downward trend in the age of puberty has been recorded in girls in both Europe and the USA, specifically with the age at breast development occurring 1–2 years earlier compared to data collected 10–15 years prior[127] (Figure 23.16[B]). In contrast to breast development, the age at menarche has only changed minimally, suggesting that the total duration of the pubertal transition has recently become more prolonged.[128] Toppari and Juul[129] have suggested that the current obesity epidemic cannot fully explain the worrisome trend for earlier glandular breast tissue development over such a short period of time and have proposed that this recent progression may instead result from exposure to endocrine-disrupting chemicals such as pharmaceuticals, polychlorinated biphenyls, dioxin, and dioxin-like compounds, components of plastics such as bisphenol A (BPA) and phthalates, and DDT and other pesticides. These and other compounds have been implicated with roles in disrupting or exacerbating the signaling actions of sex hormones (as well as other hormones), a topic addressed further in Unit VIII: Endocrine-Disrupting Chemicals.

Leptin as a Regulator of Puberty

In 1994 the discovery of *leptin*, a hormone produced by white adipose tissue that conveys energy state information to the brain, introduced an endocrine mechanism by which a critical energy balance threshold optimal for reproduction could be measured by the brain. Although other endocrine signals reflective of metabolic status, such as growth hormone, insulin-like growth factor-1, and circulating metabolites, may also contribute to the timing of puberty, leptin has received the most attention due to its ability to communicate levels of adipose tissue reserves. Importantly, in addition to obesity, infertility is a key characteristic of mouse leptin knockout (*ob⁻/ob⁻*) models, and this is reversible with leptin treatment.[130] Similarly, rare cases of humans with homozygous leptin gene mutations display both obesity and hypogonadotropic hypogonadism, symptoms that are each reversible with leptin administration.[131–137] Furthermore, humans with leptin receptor deficiency also present varying degrees of hypogonadotropic hypogonadism, further emphasizing the importance of leptin signaling for proper HPG function.[138]

Rather than functioning as a trigger for the onset of puberty in humans, leptin is considered to act as an essential permissive factor for its initiation.[139,140] This view is supported by studies in leptin-deficient human children, where the initiation of recombinant leptin treatment fails to induce puberty immediately but instead puberty begins only following prolonged treatment with the hormone and after a normal pubertal bone age has been reached.[135,136,141] Furthermore, although obese humans produce higher levels of circulating leptin, they also exhibit leptin resistance at the level of the brain, and elevated leptin in obese individuals is therefore unlikely to directly stimulate the HPG axis and promote an earlier onset of puberty. However, in girls, specifically, other indirect effects of obesity, such as increased conversion of androgens into estrogens by adipose tissue, may contribute to precocious puberty.

Hypothalamic Amenorrhea

Since GnRH neurons do not appear to have receptors for leptin,[142] leptin's puberty-gating effects are likely mediated by other leptin-sensitive neurons that influence GnRH neuron activity. For example, in the mouse approximately 40% of kisspeptinergic neurons in the hypothalamic ARC nucleus express mRNA for the leptin receptor, and leptin deficient *ob⁻/ob⁻* mice display reduced ARC kisspeptin mRNA levels compared with wild-type controls.[143] Interestingly, in extreme cases of fasting and negative energy balance, hypothalamic kisspeptin and neurokinin B (NKB) receptor expression in the ARC is inhibited in several species.[140,144] Therefore, in times of energy insufficiency, metabolic cues may be used to inhibit not just puberty, but also fertility, even in sexually mature adults. For example, **hypothalamic amenorrhea** is a condition characterized by a cessation of menstrual cycling in women due to a dysfunction of the hypothalamic-pituitary-gonadal axis in the absence of disease or ovarian failure. This disorder is associated with chronic energy deficiency, typically caused by long-term strenuous exercise (Figure 23.17) and/or reduced food intake, such as starvation or extreme dieting[145] (Figure 23.18). Women with hypothalamic amenorrhea generally have low serum leptin levels that reflect their energy deficits, and leptin administration can improve their LH pulsatility, ovulation, and reproductive function.[146–148]

Stress and the Timing of Puberty

Life history theory argues that the distribution of an organism's energetic resources is a trade-off between survival and reproduction.[168,169] *Stress* can be broadly defined as any state of threatened or actual disruption to homeostasis and the resulting increased energy consumption by an organism that is required to restore or recalibrate the homeostatic state. Therefore, under stressful conditions the physiological stress response reallocates energetic resources in part by prioritizing an individual's survival over reproduction. As such, environmental conditions that affect health will also influence an organism's age at

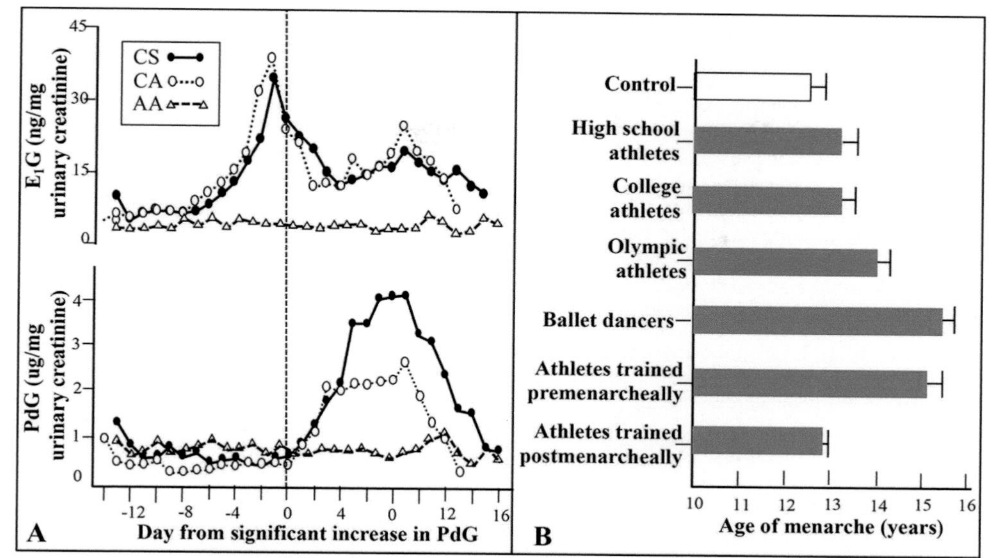

Figure 23.17 The influence of athletic activity on the ovarian cycle and the age at menarche. (A) Steroid hormone patterns during the ovarian cycle of amenorrheic athletes (AA) compared with cycling athletes (CA) and cycling sedentary (CS) controls. Note the low estrogen and progesterone levels in the AA group. Though there are no differences in estrogen levels between the CA and CS groups, the CA group displays significantly lower progesterone during the luteal phase compared with the CS group. E_1G, estrone-glucuronide; PdG, pregnanediol-glucuronide. **(B)** Ballet dancers and athletes trained premenarcheally consistently exhibit a longer delay in the start of menarche than do other athletes.

Source of Images: (A) Modified from Loucks, A.B., 1990. Med Sci Sports Exerc., 22(3), pp. 275-280. (B) Modified from Cumming D.C. (1990) Sem. Reprod. Med. 8(01):15-24.

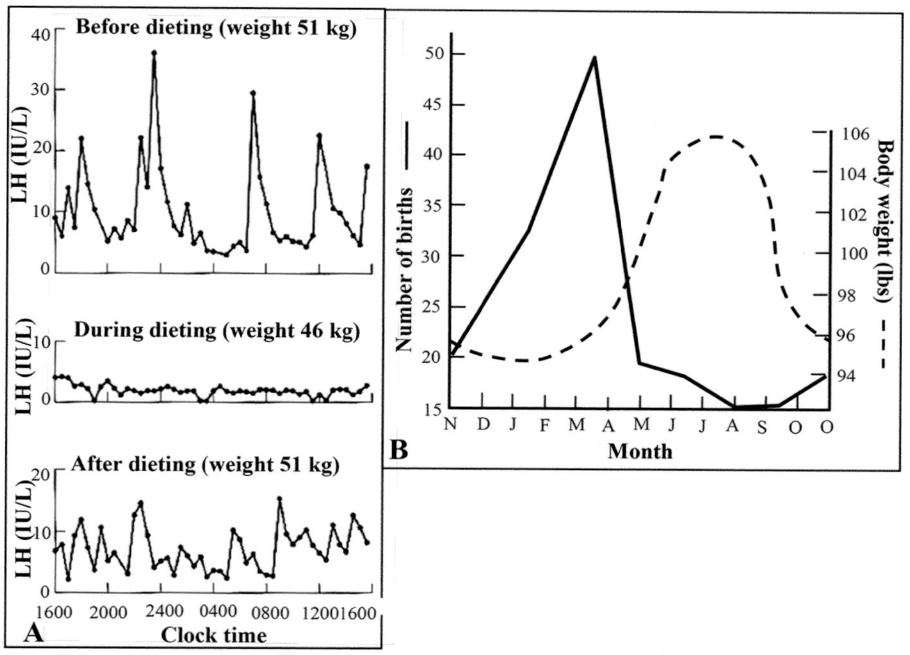

Figure 23.18 Effects of voluntary and involuntary food restriction on female gonadotropin cycling and fertility. (A) Luteal phase LH profile in a healthy young woman before, during, and after two and a half weeks of strict dieting. **(B)** Relationship between the month of birth, body weight, and numbers of births among females of the hunter/gatherer !Kung-San people of the Kalahari Desert. Note that peak weights (106 lb) are attained during the rainy season in July and August, and births peak about 9 months after during March and April.

Source of Images: (A) M.M. Fichter and K.M. Pirke (1984). Hypothalamic-pituitary function in starving healthy subjects. In: The Psychology of Anorexia Nervosa. K.M. Pirke and D. Ploog, Eds. Springer-Verlag GmbH Ltd. Used by permission. (B) Van der Walt, L. A., et al. (1978). J. Clin. Endocrinol. Metab. 46: 658-663. Used by permission.

Developments & Directions: The co-rise of obesity and pubertal perturbance—coincidence or causality?

During the last several decades, childhood obesity has become a significant public health concern due to its association with diverse metabolic disorders such as hypertension, atherosclerosis, and type 2 diabetes.[149] Recent findings suggest that excess adiposity may also influence pubertal development, in particular by playing a role in advancing the age at the start of puberty in girls.[126,150] For boys, the relationship between childhood adiposity and the timing of puberty is more mysterious and controversial, with some studies finding that excess adiposity delays puberty while others report the opposite.[151] More recent studies suggest that these discrepancies among boys might be explained by a nonlinear relationship between BMI and puberty, whereby relatively mild excess adiposity (i.e. being "overweight") relates to early puberty, whereas severe excess adiposity (i.e. being "obese") associates with delayed puberty.[152,153] Although it remains unclear exactly how excess adiposity disrupts the timing of puberty in girls and boys, the following potential physiological and endocrine mechanisms are currently under considerable scrutiny:

1. **Elevated levels of circulating androgens (hyperandrogenemia) in girls**. Increased adrenal androgen synthesis during adrenarch typically represents the first hormonal change at the start of puberty. It has been hypothesized that androgens normally play a critical role in initiating puberty, with increasing androgens promoting a decrease in GnRH pulse generator sensitivity to negative feedback suppression by other sex steroids, such as progesterone and estrogens.[154] Importantly, obesity in peripubertal girls has been shown to be accompanied by elevated blood testosterone levels[155] (Figure 23.19), possibly resulting from abnormal adrenal steroidogenesis stimulated by hyperleptinemia and hyperinsulinemia.[126,156,157] In this scenario, elevated androgen levels during obesity may promote an earlier pubertal increase of pulsatile GnRH secretion, possibly leading to an earlier onset of puberty.[126] Compounding the effects of increased testosterone production, hyperinsulemia can also decrease circulating levels of hepatic *sex hormone-binding globulin* (SHBG), which increases the bioavailable fraction of sex steroids to target tissues (Figure 23.19).
2. **Increased peripheral conversion of androgens to estrogens**. Adipose tissue is a highly active metabolic and endocrine organ, secreting numerous hormones including sex and stress steroids. In particular, adipocytes in the obese state express high levels of *cytochrome P450 (CYP19) aromatase*, an enzyme that converts circulating androgens into estrogens.[158,159] Indeed, increased levels of estrogens have been documented in obese prepubertal girls[160,161] and may play a role in promoting an earlier onset of puberty in girls. In adult men, obesity can be associated with hypogonadotropic hypogonadism related to an increase in the aromatization of androgens to estrogens, resulting in a decreased ratio of testosterone to estrogens,[162,163] followed by feedback inhibition of gonadotropin secretion by the estrogens.[164] It is possible that a similar mechanism occurring in obese boys plays a role in delaying puberty. However, such a mechanism would only apply to obese boys, given that overweight (but not obese) boys appear to have an earlier onset of puberty.[161] Other potential mechanisms contributing to increased estrogens in obesity include the aforementioned insulin-induced reduction of SHBG that increases the bioavailability of estradiol and a decrease in the metabolic clearance of estrogens by the liver.[159]
3. **Obesogens**. "Obesogens" have been defined as "dietary, pharmaceutical, and industrial compounds that may alter metabolic processes and predispose some people to gain weight".[165] Many of these adipogenic compounds are both lipophilic (stored in fat tissue) and estrogenic (mimic the effects of estrogens),[166] and thus are considered endocrine-disrupting compounds (EDCs). Examples of such estrogen-disrupting obesogens in laboratory models include the pharmaceutical compound diethylstilbestrol (DES), the herbicide atrazine, and the plastic softener bisphenol A (BPA). Exposure to such "environmental estrogens" may produce early thelarche in girls, thereby promoting secondary sexual characteristics in the absence of central H-P axis-mediated puberty.[126] Since EDCs are lipophilic, as the amount of body fat increases the EDC body burden proportionally, and obese people are therefore capable of storing large amounts of EDCs compared with healthy weight people. To compound matters, the adipogenic effects of EDCs function to further increase adiposity, permitting even greater amounts of EDC storage in an upwards spiral.[167] The extent to which high EDC loads in obese children play a role in perturbing the onset of puberty remains a particularly active area of investigation. The topic of obesogens is discussed in Chapter 19: Metabolic Dysregulation and Disruption.

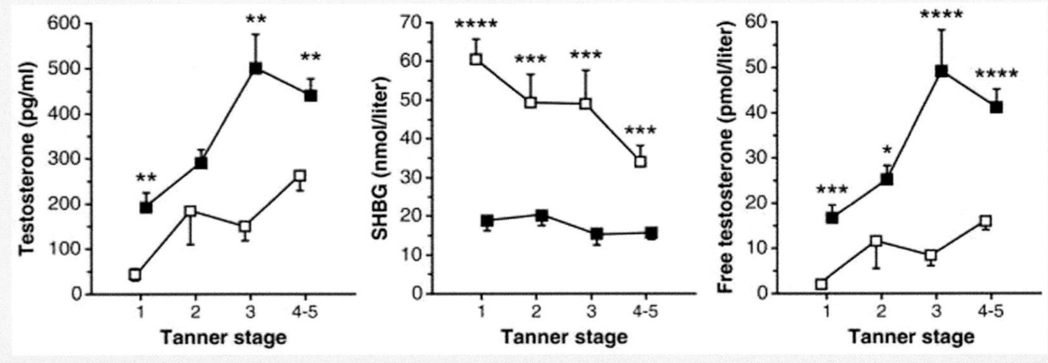

Figure 23.19 Total testosterone, sex hormone-binding globulin (SHBG), and free testosterone concentrations in obese (closed squares) and normal-weight (open squares) girls grouped by Tanner stage. Data are shown as mean ± SEM. Differences were assessed with Wilcoxon rank sum tests: *, $P < 0.05$; **, $P \leq 0.01$; ***, $P \leq 0.001$; ****, $P \leq 0.0001$ before Bonferroni correction.

Source of Images: McCartney, C.R., et al. 2007. J. Clin. Endocrinol. Metab. 92(2), pp. 430-436. Used by permission.

puberty and first reproduction, the size and number of its offspring, and the duration of its reproductive lifespan.[170] In humans, physical, psychological, or psychosocial stress immediately prior to or during puberty may delay the start or progression of puberty,[171–173] whereas pubertal advancement has been reported in children who experienced such stress as neonates or infants.[174,175] The response to stress in humans can not only alter the timing of puberty, but also is detrimental to reproductive function throughout life, with high perceived stress during pregnancy increasing the risk for preterm labor,[176,177] physical and emotional stressors reported to alter female menstrual cycling,[178,179] and higher levels of stress associated with a longer time to pregnancy and an increased risk of infertility.[180,181] Indeed, stress-induced trade-offs to reproductive output are evident across species, with other mammals, birds, reptiles, and even plants all responding to stress by decreasing reproductive function in both sexes.[170]

The notion that long-term stress can impact mammalian reproduction was first established in 1939, when Hans Selye proposed that an activated stress axis was capable of inhibiting the reproductive axis. The stress response is typically associated with the activation of the hypothalamic-pituitary-adrenal axis, resulting in an increased secretion of *corticotropin-releasing hormone* (CRH) from the hypothalamus, which in turn stimulates *adrenocorticotropic hormone* (ACTH) by the pituitary, which ultimately promotes glucocorticoid secretion by the adrenal cortex. Glucocorticoids are the primary mediators of stress-induced reproductive dysfunctions, and they act on all levels of the HPG, as well as directly on the reproductive organs themselves,[182–196] observations that are summarized in Figure 23.20. In addition to glucocorticoids, CRH can exert glucocorticoid-independent central effects on the hypothalamic-pituitary-gonadal axis (HPG).[197–203]

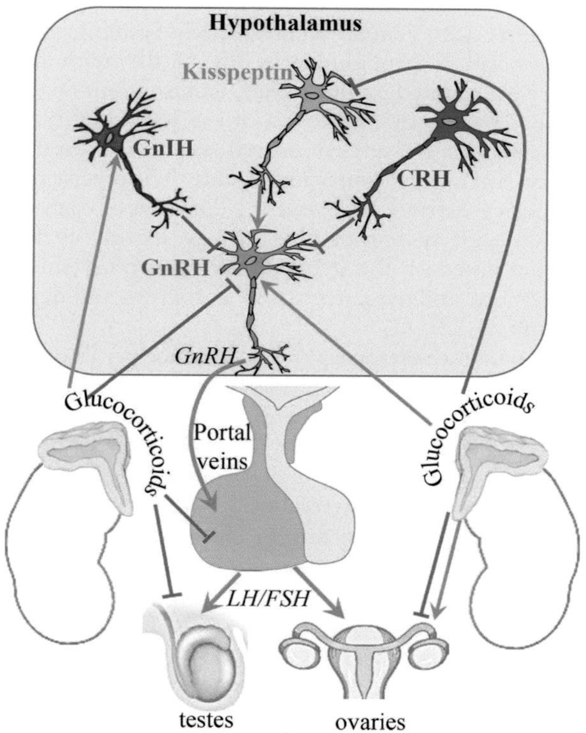

Figure 23.20 A summary of some effects of CRH and glucocorticoid stress hormones along the HPG axis in birds and mammals. Glucocorticoids emanate from the adrenal cortex, and CRH is produced by neuroendocrine cells in the hypothalamus. Green arrows denote a stimulatory effect; blunt-ended red lines denote an inhibitory effect. In some instances, glucocorticoids exert context-dependent stimulatory or inhibitory effects on the same tissue.

SUMMARY AND SYNTHESIS QUESTIONS

1. The data in Figure 23.3B shows that whereas the age of menarche in girls from 1991 to 2008 has only changed minimally, the age at thelarche has advanced significantly. What are the implications of these observations?
2. Although many hormones influence GnRH synthesis (leptin, sex steroids), there are few receptors for these hormones in GnRH nuclei. How, then, do these hormones influence GnRH synthesis?
3. There are two panels in Figure 23.17: *Effects of voluntary and involuntary food restriction on female gonadotropin cycling and fertility.* How does the information in the study in panel A explain the observations of a completely different study in panel B?
4. When young, prepubertal female mice are raised on a low-calorie diet, they do not undergo puberty. From a mouse population perspective, what is the adaptive value of the inhibition of puberty?
5. Figure 23.19 shows that in females, obesity is accompanied by elevated levels of total and free testosterone, but lower levels of sex hormone-binding protein (SHBP). What is the significance of reduced SHBP levels?
6. From the perspective of an organism's survival, how does it make sense that stress inhibits reproduction?
7. Abuse of opioid drugs (e.g. heroin, morphine) can cause hypogonadism, which inhibits fertility. What is the endocrine mechanism by which this occurs?
8. Whereas very obese male children have a high probability of delayed sexual maturity and hypogonadism, obese female children tend to begin puberty at much earlier ages than average. Explain why the effects of obesity on sexual maturity differ in males vs. females.
9. Adipocytes contain the enzyme 11β-hydroxysteroid dehydrogenase, which deactivates cortisol into cortisone. How could this knowledge help explain why obese pubertal girls have elevated levels of circulating adrenal *androgens*? Lower levels of circulating cortisol would result in increased ACTH synthesis by the pituitary gland.

Periodicity in Seasonally Breeding Animals: Photoperiod-HPG Interactions

LEARNING OBJECTIVE Define "estrous cycle", and describe how endocrine cues are interpreted differently by long-day and short-day seasonal breeding animals.

KEY CONCEPTS:

- The estrous cycle is the ovarian cycle in most mammals and is characterized by changes in sexual behavior and fertility.
- The menstrual cycle, present in humans and some closely related primates, differs from the estrous cycle such that in nonpregnant females the uterine lining is sloughed off instead of being resorbed.
- Seasonal polyestrous cycles can be divided into long-day breeders (estrous cycles occur in spring and summer) and short-day breeders (estrous cycles occur during the fall and winter).
- The most reliable measure of seasonality among species is photoperiod, which is transduced into a nocturnally produced endocrine signal, melatonin.
- Melatonin regulates seasonal breeding by modulating the actions of thyroid hormone and the synthesis of gonadotropin-inhibiting hormone (GnIH).

Periodicity, whether it be circadian, circalunar, or circannual, is a common feature of endocrine processes. One of the clearest examples of periodicity in most animals is seasonal breeding. Irrespective of where they live, most species do not breed year-round, but are **seasonal breeders** that display seasonal variability in their frequency of ovulation, spermatogenic activity, gamete quality, and also sexual behavior.[204] Even humans, who are considered behaviorally to be *non-seasonal breeders* (also referred to as "continuous breeders"), show seasonal variability in births due mostly to seasonal variation in the frequency of successful conception.[205–207] Other examples of animals that exhibit non-seasonal, year-round breeding behaviors are the common mouse (*Mus musculus*), as well as cattle and pigs, the latter two of which have been artificially selected during domestication for the higher yield of offspring and animal products that accompanies non-seasonal breeding.

The most reliable measure of seasonality among species is photoperiod. Exceptions to this occur at equatorial latitudes where photoperiod does not change dramatically over the course of the year, so these residents must instead rely more on other cues for timing their breeding, such as visual, auditory, temperature, nutritional, and rainfall cues.[208] Compared with low equatorial latitudes, mid- to high latitudes have significant seasonal variations in both photoperiod and ambient temperature. Animals living in these regions often develop their gonads and display reproductive behavior only during specific times of the year in order to ensure that their offspring are born between spring and early summer, the time of year when climate and food availability are at their most favorable levels for maximizing the probability of survival of their offspring. In contrast, equatorial and tropical latitudes are inhabited by more animals that display long and year-round breeding seasons, since photoperiod and temperature vary little throughout the year in these regions.[209] The relationship between latitude and seasonal vs. non-seasonal breeding is strikingly evident in white-footed mice, deer mice, and other related mice of the genus *Peromyscus* that display very high variability in the annual patterns of reproduction. Whereas at high latitudes *Peromyscus* mouse populations breed once per year, at low latitudes they breed continuously throughout the year[210] (Figure 23.21). Other mammals, such as deer within the genus *Odocoileus*, display a similar latitudinal shift in breeding strategy.

Types of Reproductive Cyclicity among Mammals

Puberty in females can be defined as the first estrous cycle accompanied by ovulation. The **estrous cycle**, which refers to the ovarian cycle in most mammals, is divided into several stages that reflect changing behavior, ovarian and pituitary hormone concentrations, and fertility (described in Chapter 22: Female Reproductive System). Estrous cycles continue throughout the life of the adult female and are interrupted by pregnancy, lactation/nursing, and, in many species, by the season of the year. Additionally, cycling may halt if environmental conditions are unusually stressful or nutrition is inadequate. The precise timing of a species' estrous cycle and the duration of each phase will vary with its reproductive ecology. Estrous cycles are classified based on the frequency of occurrence throughout the year, and are categorized as follows and depicted in Figure 23.22:

- *Monoestrous*: estrous cycles occur once per year; representative species include bears, wolves, dogs, and foxes.
- *Non-seasonal polyestrous*: display a uniform distribution of cycles that occur regularly throughout the year, irrespective of season; representative species include cattle, pigs, and many rodents.
- *Seasonal polyestrous*: display clusters of estrous cycles that occur only during a specific season of the year. Seasonal polyestrous cycles can be further subdivided as follows:
- *Long-day breeders*: estrous cycles occur in spring and summer when the days are longer; representative species include horses, hamsters, groundhogs, and mink.
- *Short-day breeders*: estrous cycles occur in the autumn or winter when the days are shorter; representative species include sheep, goats, red deer, elk, and moose.

The timing of the breeding period typically relates to the length of the gestation (in mammals) or incubation (in birds) period. For example, small birds and rodents typically have incubation/gestation periods lasting only several

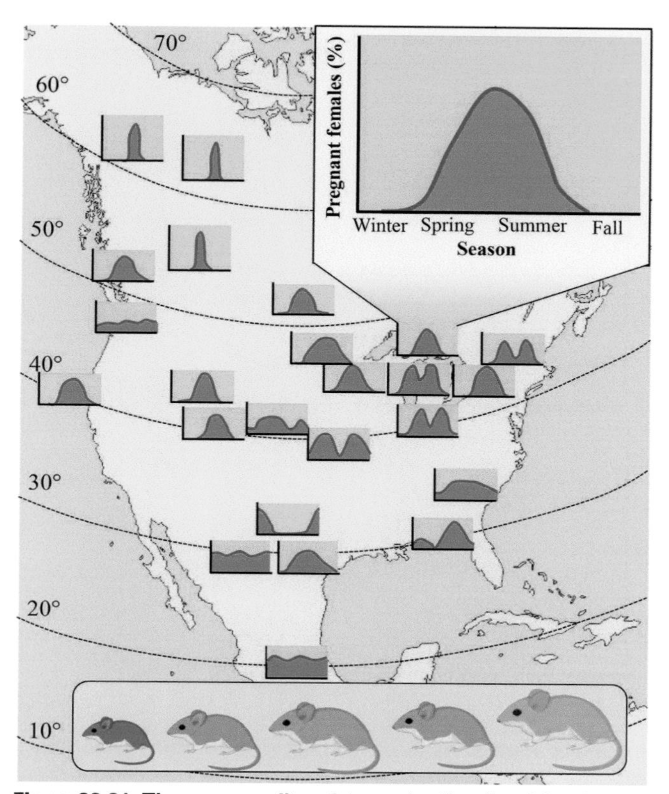

Figure 23.21 The seasonality of reproduction in white-footed mice, deer mice, and other related mice within the genus *Peromyscus* varies with latitude. Populations of the many species of this genus live in diverse habitats ranging from tropical latitudes (about 15° of latitude) to near the arctic circle (about 60° of latitude), but they generally breed year-round below 30° latitude. Around 60° of latitude these mice typically have a short two- to three-month breeding season during which they produce only a single litter. At 45° of latitude they typically show five- to seven-month breeding seasons during which several litters are produced. From 30° to 45° of latitude, *Peromyscus* show a mixture of seasonal and non-seasonal patterns, with large year to year variation within a population. Below 30° they tend to breed year-round, though habitats with seasonal rainfall patterns may promote seasonal breeding. Some examples of *Peromyscus* species and their northernmost ranges in latitude, from left to right: *P. polionotus* (35°), *P. maniculatus* (55°), *P. leucopus* (50°), *P. eremicus* (35°), and *P. californicus* (35°).

Source of Images: Map and line graphs: Modified from F. H. Bronson, Mammalian Reproduction: An Ecological Perspective, Biology of Reproduction, Volume 32, Issue 1, 1 February 1985, Pages 1–26. Peromyscus spp. Drawings: Yawitz, T.A., Barts, N. and Kohl, K.D., 2022. Comp. Biochem. Physiol. Part A: Molecular & Integrative Physiology, 271, p. 111265. Used by permission.

weeks, and they therefore breed from spring to early summer (long-day breeders) to ensure that their offspring are hatched/born under optimal conditions in the late spring or summer. Horses are also long-day breeders but have much longer gestation periods (about a year long) than smaller animals, so mate in the spring in order to give birth the following spring. Large animals with gestation periods lasting about six months must, by contrast, mate during the fall

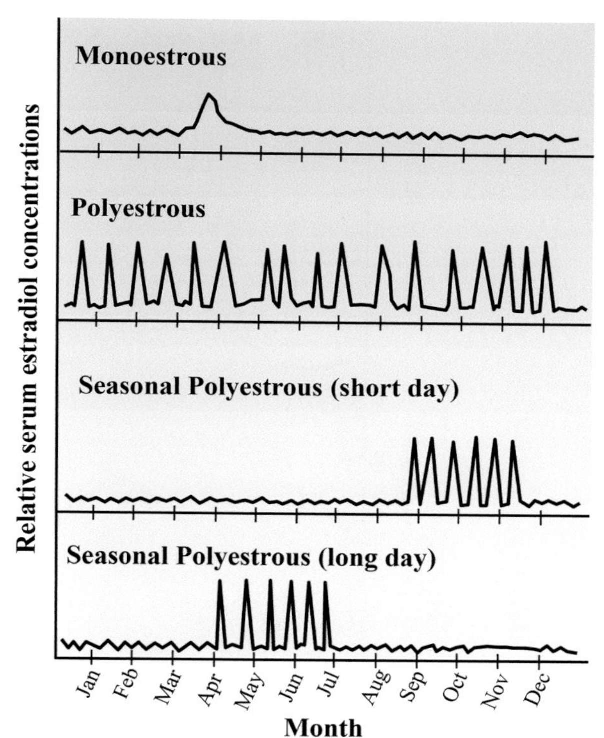

Figure 23.22 Categories of estrous cycles as determined by annual estradiol profiles.

and early winter (short-day breeders) to ensure that their offspring are born in the spring or early summer. The powerful influence of photoperiod in driving seasonal breeding is clearly illustrated in an experiment by Lincoln and colleagues[211] using Soay rams under conditions where light-dark cycling was artificially manipulated (Figure 23.23). That is, when the rams, which are short-day breeders, were switched abruptly from 4 months of long days (16 hours of light and 8 hours of darkness, or 16L:8D) to short days (8D:16L), this induced increases in both gonadotropins, testosterone, and testicular size in 2–4 weeks. When rams were returned to the long-day photoperiod, each of these parameters fell to non-breeding levels.

Endocrine Responses to Seasonal Changes in Photoperiod

Melatonin and Seasonality

In mammals, photoperiodic information is received by the eyes, transmitted via a nonvisual neuronal pathway directly to the suprachiasmatic nuclei (SCN) of the hypothalamus, and relayed to the *pineal gland*, which releases the hormone *melatonin* into general circulation. Melatonin is secreted in a circadian manner, with basal secretion during the day and peak secretion occurring at night. The importance of melatonin as an endocrine conveyor of photoperiod information is made clear through observations that pinealectomy blocks the behavioral and physiological responsiveness to photoperiod in every mammalian species studied.[212] Melatonin appears to be

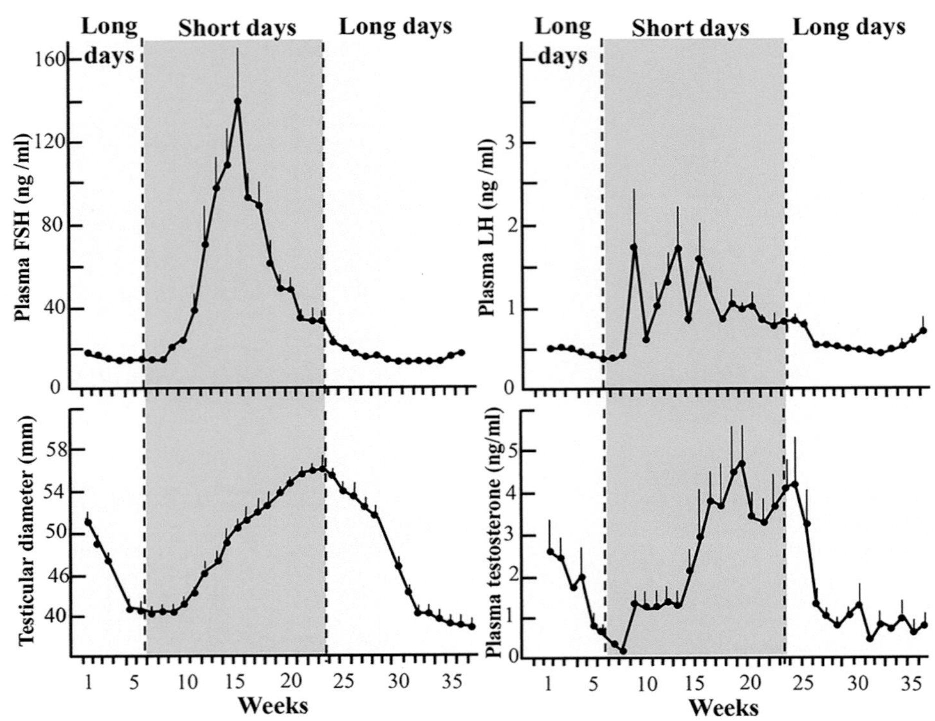

Figure 23.23 Changes in levels of plasma FSH, LH, and testosterone in six adult Soay rams (a short-day breeder) under long-day photoperiod (16L:8D) and short-day photoperiod (8L:16D). Testicular diameter follows increases with rising gonadotropins.

Source of Images: Lincoln, G.A., et al. 1977. J. of Endocrinol., 72(3), pp. 337-349.

involved in avian regulation of both gonadotropin secretion and gonadal activity, as its administration inhibits reproductive activities in quail and chicken.[213–217]

In mammals, the seasonal changes in duration of elevated nocturnal melatonin levels serve to transduce the photoperiodic signal of day length into an endocrine one that is interpreted differently by long- and short-day breeders. Whereas in long-day breeders, such as hamsters, the extended duration of melatonin during the short fall and winter days inhibits reproduction, in short-day breeders, like sheep, the longer circadian presence of melatonin promotes breeding. Strikingly, in the long-day breeding hamster pinealectomy results in continuous reproduction throughout the year[218] (Figure 23.24), whereas in short-day breeding sheep melatonin treatment out of the breeding season (e.g. during the summer) promotes reproductive development.[219–221]

Thyroid Hormone and Seasonality

Considering that reproduction is a high energy expenditure phenomenon and that thyroid hormones (TH) are central players in the maintenance of energy balance, it is not surprising that TH is critical in the regulation of seasonal reproduction in diverse vertebrates including fish, birds, and mammals.[222–224] For example, early studies of birds, ducks,[225] and starlings[226] demonstrated that removal of the thyroid gland dramatically altered the seasonal response of gonads. Furthermore, not only does thyroidectomy block many of the seasonal responses to photoperiod in the Japanese quail (a long-day breeder), but a single injection of thyroxine (T_4) can restore the seasonal response.[227] Thyroidectomy studies in sheep (a short-day breeder) showed that the normal photo-stimulated transition to sexual inactivity (called "anestrous", when the gonads start to regress) at the end of the winter is blocked and can be restored by administration of T_4.[228,229] Therefore, in quail and sheep, T_4 treatment appears to mimic the effects of a long-day photoperiod.

GnIH and Seasonality

Birds

The majority of birds living in temperate zones are long-day breeders that can show over 100-fold seasonal changes in gonad size.[230,231] With such dramatic changes in reproduction, bird research has contributed significantly to our understanding of endocrine mechanisms of photoperiod transduction. Using quail as a model, Ubuka and colleagues[217] showed that the removal of the eyes and pineal gland (the two main sites of melatonin synthesis in birds) reduced GnIH mRNA and peptide expression levels in the hypothalamus. In contrast, treatment of these birds with melatonin increased GnIH mRNA and peptide expression, at least in part, via the direct binding of melatonin to receptors on GnIH cells. Therefore, during short-day

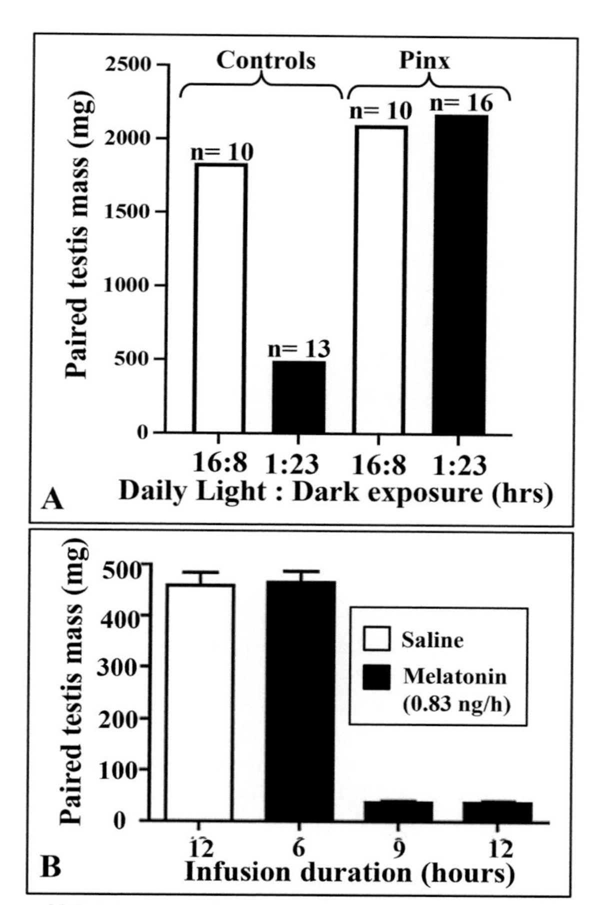

Figure 23.24 Prolonged exposure to melatonin promotes testicular atrophy in hamsters, a long-day breeder. (A) Exposure of male Siberian hamsters to "short"-day cycles of 1 hour of light and 23 hours of darkness causes atrophy of the gonads. Pinealectomy prevents this atrophy, but has no effect on animals exposed to "long"-day light-dark cycles of 16:8. **(B)** Daily melatonin infusions of 9 h or 12 h duration inhibited testicular development in pinealectomized juvenile Djungarian hamsters, while daily infusions of 6 h duration did not prevent gonadal growth.

Source of Images: (A) Modified from Hoffman, R.A. and Reiter, R.J., 1965. Science, 148(3677), pp. 1609-1611. (B) Modified from Goldman, B.D., et al. 1984. Endocrinology, 114(6), pp. 2074-2083.

photoperiods when the birds are not reproductively active, melatonin appears to induce GnIH expression and release, which in turn inhibits GnRH release, ensuring that gonad development remains suppressed during the fall and winter (Figure 23.25). Under long-day photoperiods, the melatonin-induced GnIH secretion attenuates, allowing GnRH to promote gonad growth and breeding.

Mammals

Among mammals, sheep (which are short-day breeders) and hamsters (long-day breeders) are often used to study the effects of photoperiod on reproduction, though their seasonal changes in gonad mass are less dramatic compared with those of birds. In short-day breeding sheep, melatonin exerts an opposite effect on GnIH compared with long-day breeding birds—in the winter, when melatonin levels are higher, GnIH mRNA and protein expression in the sheep hypothalamus are inhibited (Figure 23.25), permitting GnRH levels to rise and facilitate reproduction.[232,233] In sheep, GnIH expression increases during long days when melatonin levels are lower and the animals are not reproductive. GnIH mRNA and protein expression in the hypothalamus are also increased during long days, but at a time when these animals are not reproductive.[232,233]

Though the opposing effects of melatonin on GnIH production in long-day breeding quail and short-day breeding sheep seem logical, the dynamics of melatonin-GnIH interactions on well-studied long-day breeding mammals, hamsters, are less intuitively obvious. Similar to short-day breeding sheep, expression of GnIH *decreases* in long-day breeding Siberian and Syrian hamsters exposed to short-day photoperiod[217,234–236] (Figure 23.25). Furthermore, treatment with melatonin inhibited the expressions of GnIH, and the inhibitory effect of short-day photoperiod was not seen if the hamsters were pinealectomized. It is possible that in hamsters melatonin-GnIH interactions function to

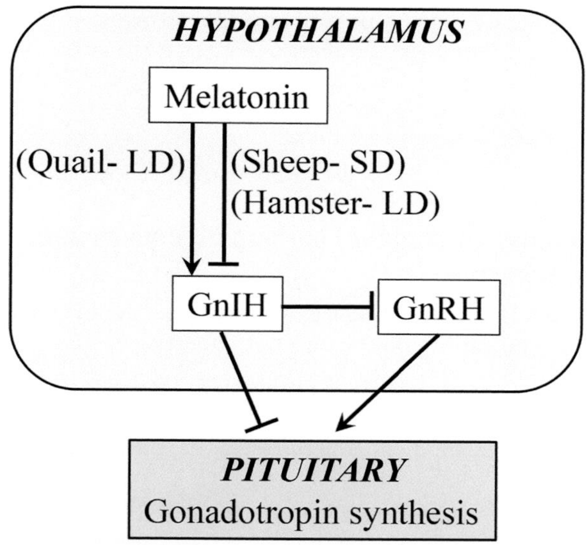

Figure 23.25 Influence of melatonin on gonadotropin synthesis in long- and short-day breeding animals. In quail, a long-day (LD) breeder, increased melatonin synthesis during short days promotes GnIH synthesis, which inhibits both GnRH and gonadotropin synthesis, ensuring that breeding does not take place during short days. In contrast, in short-day (SD) breeding sheep, increased melatonin during the fall and winter inhibits GnIH, permitting breeding during this time. Although hamsters are LD breeders, like birds, elevated melatonin levels inhibit GnIH in a manner similar to sheep. The adaptive function of melatonin suppressing GnIH outside the breeding season in hamsters remains unclear, though it is possible that in hamsters melatonin-GnIH interactions function to more precisely modulate gonadotropin concentrations under long-day photoperiod, when gonadotropin levels are very high.

Source of Images: Tsutsui, K., et al., 2012. Gen Comp. Endocrinol., 177(3), pp. 305-314.

fine-tune gonadotropin expression such that under long-day photoperiod, when gonadotropin levels are very high, short durations of melatonin stimulate GnIH to more precisely modulate gonadotropin concentrations.[236] Thus, despite the differences in breeding strategies, the photoperiodic regulation of GnIH is conserved in both long- and short-day breeding mammals, with GnIH expression being induced under long-day photoperiod when the duration of melatonin exposure is relatively low. The downstream targets of the GnIH system, however, diverged between long- and short-day breeding mammals to accommodate their different reproductive strategies.[237]

SUMMARY AND SYNTHESIS QUESTIONS

1. A team of Italian researchers[238] found that children who watched TV regularly at night produced lower levels of urinary melatonin compared with children who did not watch TV at night. Though controversial, the researchers propose that this observation could, in theory, help explain why children are entering puberty at earlier ages than in the past. Propose a viable endocrine mechanism that explains how watching TV at night accelerates the onset of puberty. Note that the researchers were not interested in the content of the TV shows, but simply the fact that TV is a powerful light source.

Summary of Chapter Learning Objectives and Key Concepts

LEARNING OBJECTIVE Discuss the importance of GnRH pulsatility on the onset of puberty.

- The onset of puberty is promoted by the maturation of the GnRH pulse generator, whose pulse frequency and amplitude modulate the release of the pituitary gonadotropins.
- A cyclic up/downregulation of GnRH receptors in response to fluctuating GnRH concentrations promotes the pulsatile release of gonadotropins.
- Changes in GnRH pulse frequency alter the FSH/LH ratio, with a high pulse frequency preferentially stimulating LH and low pulse frequency stimulating FSH.

LEARNING OBJECTIVE Discuss the neuroendocrine basis of negative and positive feedback signaling along the HPG axis in response to hormones throughout the ovarian cycle.

- The GnRH pulse generator includes hypothalamic kisspeptinergic neurons that synapse with GnRH neurons to regulate their secretion.
- GnRH secretion is influenced by innervation from two major kisspeptin-secreting neuronal populations in the hypothalamus: the arcuate (ARC) and the anteroventral periventricular (AVPV) nuclei.
- In male and female mammals, kisspeptinergic neurons of the ARC promote GnRH pulsatility and respond to elevated sex steroid hormones with negative feedback.
- In female mammals, kisspeptinergic AVPV neurons respond to rising estrogens during the follicular phase of the ovarian cycle with positive feedback, which facilitates the ovulatory surge in GnRH.
- A higher amplitude and increased cycling rate of gonadotropins is observed in castrated males and in women following menopause.

LEARNING OBJECTIVE Describe the major endocrine benchmarks of puberty in human males and females and developmental changes in the HPG axis from birth through adulthood.

- Adrenarche, the earliest endocrine change during puberty, denotes the increased synthesis of androgen precursor hormones by the adrenal gland in both males and females.
- Mediated by ovarian hormones, thelarche represents the first morphological changes in puberty in females and is characterized by the beginning of breast development, growth acceleration, and redistribution of body fat.
- Menarche denotes the time of the first menstruation in females in response to increasing maturation of the GnRH pulse generator and follicular development.
- Gonadarche describes testicular enlargement and increased testosterone synthesis in males, generally the first signs of puberty in boys.
- The HPG axis is established and active in fetal development where a mid-gestation increase in gonadotropins occurs.
- Although the HPG axis is established *in utero*, children experience a "juvenile hiatus" in gonadotropin secretion, a period of time when GnRH and gonadotropin secretion is nonpulsatile and highly sensitive to negative feedback suppression.
- The start of puberty is characterized by a reactivation of GnRH pulsatility and reduced sensitivity to negative feedback by sex steroid hormones.

LEARNING OBJECTIVE Discuss the influence of metabolic status, obesity, and stress on the timing of puberty.

- A younger age of puberty carries important public health consequences caused, in part, by a variable exposure time to sex hormones associated with a higher risk of developing cancers and cardiovascular and metabolic diseases.
- Variability in the timing of puberty is influenced by both genetic and environmental factors.
- Puberty begins when a threshold of energy stores, optimal metabolic status, and critical body mass are attained.
- Leptin acts as an essential permissive factor for the initiation of puberty.
- Obesity in girls tends to promote an earlier onset of puberty, whereas in boys it can either accelerate or delay puberty, depending on its severity.
- Stress hormones, such as adrenal glucocorticoids and hypothalamic corticotropin-releasing hormone (CRH), can delay puberty and inhibit fertility.

LEARNING OBJECTIVE Define "estrous cycle", and describe how endocrine cues are interpreted differently by long-day and short-day seasonal breeding animals.

- The estrous cycle is the ovarian cycle in most mammals and is characterized by changes in sexual behavior and fertility.
- The menstrual cycle, present in humans and some closely related primates, differs from the estrous cycle such that in nonpregnant females the uterine lining is sloughed off instead of being resorbed.
- Seasonal polyestrous cycles can be divided into long-day breeders (estrous cycles occur in spring and summer) and short-day breeders (estrous cycles occur during the fall and winter).
- The most reliable measure of seasonality among species is photoperiod, which is transduced into a nocturnally produced endocrine signal, melatonin.
- Melatonin regulates seasonal breeding by modulating the actions of thyroid hormone and the synthesis of gonadotropin-inhibiting hormone (GnIH).

LITERATURE CITED

1. Harris GW. *Neural Control of the Pituitary Gland*. Edward Arnold; 1955.
2. Maruska KP, Fernald RD. Social regulation of gene expression in the hypothalamic-pituitary-gonadal axis. *Physiology*. 2011;26(6):412–423.
3. Maruska KP. Social regulation of reproduction in male cichlid fishes. *Gen Comp Endocrinol*. 2014;207:2–12.
4. Burmeister SS, Kailasanath V, Fernald RD. Social dominance regulates androgen and estrogen receptor gene expression. *Horm Behav*. 2007;51(1):164–170.
5. Grone BP, Maruska KP, Korzan WJ, Fernald RD. Social status regulates kisspeptin receptor mRNA in the brain of Astatotilapia burtoni. *Gen Comp Endocrinol*. 2010;169(1):98–107.
6. White SA, Nguyen T, Fernald RD. Social regulation of gonadotropin-releasing hormone. *J Exp Biol*. 2002;205 (Pt 17):2567–2581.
7. Au TM, Greenwood AK, Fernald RD. Differential social regulation of two pituitary gonadotropin-releasing hormone receptors. *Behav Brain Res*. 2006;170(2):342–346.
8. Maruska KP, Levavi-Sivan B, Biran J, Fernald RD. Plasticity of the reproductive axis caused by social status change in an african cichlid fish: I. Pituitary gonadotropins. *Endocrinology*. 2011;152(1):281–290.
9. Maruska KP, Fernald RD. Behavioral and physiological plasticity: rapid changes during social ascent in an African cichlid fish. *Horm Behav*. 2010;58(2):230–240.
10. Parikh VN, Clement TS, Fernald RD. Androgen level and male social status in the African cichlid, Astatotilapia burtoni. *Behav Brain Res*. 2006;166(2):291–295.
11. Maruska KP, Fernald RD. Steroid receptor expression in the fish inner ear varies with sex, social status, and reproductive state. *BMC Neurosci*. 2010;11:58.
12. Kustan JM, Maruska KP, Fernald RD. Subordinate male cichlids retain reproductive competence during social suppression. *Proc Biol Sci*. 2012;279(1728):434–443.
13. Maruska KP, Fernald RD. Plasticity of the reproductive axis caused by social status change in an african cichlid fish: II. testicular gene expression and spermatogenesis. *Endocrinology*. 2011;152(1):291–302.
14. Burmeister SS, Jarvis ED, Fernald RD. Rapid behavioral and genomic responses to social opportunity. *PLoS Biol*. 2005;3(11):e363.
15. Marques P, Skorupskaite K, Rozario KS, Anderson RA, George JT. Physiology of GnRH and Gonadotropin Secretion. 2022 Jan 5. In: Feingold KR, Anawalt B, Blackman MR, et al., eds. *Endotext* [Internet]. South Dartmouth (MA): MDText.com, Inc.; 2000–. PMID: 25905297.
16. Forni PE, Wray S. GnRH, anosmia and hypogonadotropic hypogonadism--where are we? *Front Neuroendocrinol*. 2015;36:165–177.
17. Herbison AE. Control of puberty onset and fertility by gonadotropin-releasing hormone neurons. *Nat Rev Endocrinol*. 2016;12(8):452–466.
18. Gilbert S. *Developmental Biology*. 10th ed. Sinauer Associates, Inc.; 2014.
19. Stout RP, Graziadei PP. Influence of the olfactory placode on the development of the brain in Xenopus laevis (Daudin). I. Axonal growth and connections of the transplanted olfactory placode. *Neuroscience*. 1980;5(12):2175–2186.
20. Schwanzel-Fukuda M, Bick D, Pfaff DW. Luteinizing hormone-releasing hormone (LHRH)-expressing cells do not migrate normally in an inherited hypogonadal (Kallmann) syndrome. *Brain Res Mol Brain Res*. 1989;6(4):311–326.
21. Wray S, Grant P, Gainer H. Evidence that cells expressing luteinizing hormone-releasing hormone mRNA in the mouse are derived from progenitor cells in the olfactory placode. *Proc Natl Acad Sci U S A*. 1989;86(20):8132–8136.
22. Franco B, Guioli S, Pragliola A, et al. A gene deleted in Kallmann's syndrome shares homology with neural cell adhesion and axonal path-finding molecules. *Nature*. 1991;353(6344):529–536.
23. Legouis R, Hardelin JP, Levilliers J, et al. The candidate gene for the X-linked Kallmann syndrome encodes a protein related to adhesion molecules. *Cell*. 1991;67(2):423–435.
24. Hardelin JP, Julliard AK, Moniot B, et al. Anosmin-1 is a regionally restricted component of basement membranes and interstitial matrices during organogenesis: implications for the developmental anomalies of X chromosome-linked Kallmann syndrome. *Dev Dyn*. 1999;215(1):26–44.
25. Soussi-Yanicostas N, de Castro F, Julliard AK, Perfettini I, Chedotal A, Petit C. Anosmin-1, defective in the X-linked form of Kallmann syndrome, promotes axonal branch formation from olfactory bulb output neurons. *Cell*. 2002;109(2):217–228.
26. Maeda K, Ohkura S, Uenoyama Y, et al. Neurobiological mechanisms underlying GnRH pulse generation by the hypothalamus. *Brain Res*. 2010;1364:103–115.
27. Clarke IJ, Cummins JT. The temporal relationship between gonadotropin releasing hormone (GnRH) and luteinizing hormone (LH) secretion in ovariectomized ewes. *Endocrinology*. 1982;111(5):1737–1739.
28. Pohl CR, Knobil E. The role of the central nervous system in the control of ovarian function in higher primates. *Annu Rev Physiol*. 1982;44:583–593.
29. Belchetz PE, Plant TM, Nakai Y, Keogh EJ, Knobil E. Hypophysial responses to continuous and intermittent delivery of hypopthalamic gonadotropin-releasing hormone. *Science*. 1978;202(4368):631–633.
30. Wildt L, Hausler A, Marshall G, et al. Frequency and amplitude of gonadotropin-releasing hormone stimulation and gonadotropin secretion in the rhesus monkey. *Endocrinology*. 1981;109(2):376–385.
31. Dalkin AC, Haisenleder DJ, Ortolano GA, Ellis TR, Marshall JC. The frequency of gonadotropin-releasing-hormone stimulation differentially regulates gonadotropin subunit messenger ribonucleic acid expression. *Endocrinology*. 1989;125(2):917–924.
32. Haisenleder DJ, Dalkin AC, Ortolano GA, Marshall JC, Shupnik MA. A pulsatile gonadotropin-releasing hormone

stimulus is required to increase transcription of the gonadotropin subunit genes: evidence for differential regulation of transcription by pulse frequency in vivo. *Endocrinology*. 1991;128(1):509–517.
33. Kaiser UB, Jakubowiak A, Steinberger A, Chin WW. Differential effects of gonadotropin-releasing hormone (GnRH) pulse frequency on gonadotropin subunit and GnRH receptor messenger ribonucleic acid levels in vitro. *Endocrinology*. 1997;138(3):1224–1231.
34. Ehlers K, Halvorson LM. Gonadotropin-releasing Hormone (GnRH) and the GnRH Receptor (GnRHR). In: *Global Library of Women's Medicine*. The Foundation for the Global Library of Women's Medicine; 2013.
35. Bourguignon JP, Franchimont P. Puberty-related increase in episodic LHRH release from rat hypothalamus in vitro. *Endocrinology*. 1984;114(5):1941–1943.
36. Constantin S. Progress and challenges in the search for the mechanisms of pulsatile gonadotropin-releasing hormone secretion. *Front Endocrinol (Lausanne)*. 2017;8:180.
37. Campbell RE, Gaidamaka G, Han SK, Herbison AE. Dendro-dendritic bundling and shared synapses between gonadotropin-releasing hormone neurons. *Proc Natl Acad Sci U S A*. 2009;106(26):10835–10840.
38. Herde MK, Iremonger KJ, Constantin S, Herbison AE. GnRH neurons elaborate a long-range projection with shared axonal and dendritic functions. *J Neurosci*. 2013;33(31):12689–12697.
39. Tenenbaum-Rakover Y, Commenges-Ducos M, Iovane A, Aumas C, Admoni O, de Roux N. Neuroendocrine phenotype analysis in five patients with isolated hypogonadotropic hypogonadism due to a L102P inactivating mutation of GPR54. *J Clin Endocrinol Metab*. 2007;92(3):1137–1144.
40. Seminara SB, Messager S, Chatzidaki EE, et al. The GPR54 gene as a regulator of puberty. *N Engl J Med*. 2003;349(17):1614–1627.
41. Steyn FJ, Wan Y, Clarkson J, Veldhuis JD, Herbison AE, Chen C. Development of a methodology for and assessment of pulsatile luteinizing hormone secretion in juvenile and adult male mice. *Endocrinology*. 2013;154(12):4939–4945.
42. Uenoyama Y, Nakamura S, Hayakawa Y, et al. Lack of pulse and surge modes and glutamatergic stimulation of luteinising hormone release in Kiss1 knockout rats. *J Neuroendocrinol*. 2015;27(3):187–197.
43. Oakley AE, Clifton DK, Steiner RA. Kisspeptin signaling in the brain. *Endocr Rev*. 2009;30(6):713–743.
44. Messager S, Chatzidaki EE, Ma D, et al. Kisspeptin directly stimulates gonadotropin-releasing hormone release via G protein-coupled receptor 54. *Proc Natl Acad Sci U S A*. 2005;102(5):1761–1766.
45. Gottsch ML, Cunningham MJ, Smith JT, et al. A role for kisspeptins in the regulation of gonadotropin secretion in the mouse. *Endocrinology*. 2004;145(9):4073–4077.
46. Thompson EL, Patterson M, Murphy KG, et al. Central and peripheral administration of kisspeptin-10 stimulates the hypothalamic-pituitary-gonadal axis. *J Neuroendocrinol*. 2004;16(10):850–858.
47. Dhillo WS, Chaudhri OB, Patterson M, et al. Kisspeptin-54 stimulates the hypothalamic-pituitary gonadal axis in human males. *J Clin Endocrinol Metab*. 2005;90(12):6609–6615.
48. Dhillo WS, Chaudhri OB, Thompson EL, et al. Kisspeptin-54 stimulates gonadotropin release most potently during the preovulatory phase of the menstrual cycle in women. *J Clin Endocrinol Metab*. 2007;92(10):3958–3966.
49. Rance NE, Young WS 3rd, McMullen NT. Topography of neurons expressing luteinizing hormone-releasing hormone gene transcripts in the human hypothalamus and basal forebrain. *J Comp Neurol*. 1994;339(4):573–586.
50. Clarkson J, Herbison AE. Postnatal development of kisspeptin neurons in mouse hypothalamus; sexual dimorphism and projections to gonadotropin-releasing hormone neurons. *Endocrinology*. 2006;147(12):5817–5825.
51. d'Anglemont de Tassigny X, Fagg LA, Carlton MB, Colledge WH. Kisspeptin can stimulate gonadotropin-releasing hormone (GnRH) release by a direct action at GnRH nerve terminals. *Endocrinology*. 2008;149(8):3926–3932.
52. Herbison AE, de Tassigny X, Doran J, Colledge WH. Distribution and postnatal development of Gpr54 gene expression in mouse brain and gonadotropin-releasing hormone neurons. *Endocrinology*. 2010;151(1):312–321.
53. Tena-Sempere M. Kisspeptin signaling in the brain: recent developments and future challenges. *Mol Cell Endocrinol*. 2010;314(2):164–169.
54. Kotani M, Detheux M, Vandenbogaerde A, et al. The metastasis suppressor gene KiSS-1 encodes kisspeptins, the natural ligands of the orphan G protein-coupled receptor GPR54. *J Biol Chem*. 2001;276(37):34631–34636.
55. Rometo AM, Krajewski SJ, Voytko ML, Rance NE. Hypertrophy and increased kisspeptin gene expression in the hypothalamic infundibular nucleus of postmenopausal women and ovariectomized monkeys. *J Clin Endocrinol Metab*. 2007;92(7):2744–2750.
56. Wakabayashi Y, Nakada T, Murata K, et al. Neurokinin B and dynorphin A in kisspeptin neurons of the arcuate nucleus participate in generation of periodic oscillation of neural activity driving pulsatile gonadotropin-releasing hormone secretion in the goat. *J Neurosci*. 2010;30(8):3124–3132.
57. Navarro VM. Metabolic regulation of kisspeptin—the link between energy balance and reproduction. *Nat Rev Endocrinol*. 2020:1–14.
58. Navarro VM. New insights into the control of pulsatile GnRH release: the role of Kiss1/neurokinin B neurons. *Front Endocrinol (Lausanne)*. 2012;3:48.
59. Smith JT, Cunningham MJ, Rissman EF, Clifton DK, Steiner RA. Regulation of Kiss1 gene expression in the brain of the female mouse. *Endocrinology*. 2005;146(9):3686–3692.
60. Navarro VM, Gottsch ML, Chavkin C, Okamura H, Clifton DK, Steiner RA. Regulation of gonadotropin-releasing hormone secretion by kisspeptin/dynorphin/neurokinin B neurons in the arcuate nucleus of the mouse. *J Neurosci*. 2009;29(38):11859–11866.
61. de Roux N, Genin E, Carel JC, Matsuda F, Chaussain JL, Milgrom E. Hypogonadotropic hypogonadism due to loss of function of the KiSS1-derived peptide receptor GPR54. *Proc Natl Acad Sci U S A*. 2003;100(19):10972–10976.
62. Clarke H, Dhillo WS, Jayasena CN. Comprehensive review on kisspeptin and its role in reproductive disorders. *Endocrinol Metab (Seoul)*. 2015;30(2):124–141.
63. Teles MG, Bianco SD, Brito VN, et al. A GPR54-activating mutation in a patient with central precocious puberty. *N Engl J Med*. 2008;358(7):709–715.
64. Silveira LG, Noel SD, Silveira-Neto AP, et al. Mutations of the KISS1 gene in disorders of puberty. *J Clin Endocrinol Metab*. 2010;95(5):2276–2280.
65. Pineda R, Garcia-Galiano D, Roseweir A, et al. Critical roles of kisspeptins in female puberty and preovulatory gonadotropin surges as revealed by a novel antagonist. *Endocrinology*. 2010;151(2):722–730.
66. Roseweir AK, Kauffman AS, Smith JT, et al. Discovery of potent kisspeptin antagonists delineate physiological mechanisms of gonadotropin regulation. *J Neurosci*. 2009;29(12):3920–3929.
67. Navarro VM, Castellano JM, Fernandez-Fernandez R, et al. Developmental and hormonally regulated messenger ribonucleic acid expression of KiSS-1 and its putative receptor, GPR54, in rat hypothalamus and potent luteinizing hormone-releasing activity of KiSS-1 peptide. *Endocrinology*. 2004;145(10):4565–4574.
68. Plant TM, Ramaswamy S, Dipietro MJ. Repetitive activation of hypothalamic G protein-coupled receptor 54 with intravenous pulses of kisspeptin in the juvenile monkey (Macaca mulatta) elicits a sustained train of gonadotropin-releasing hormone discharges. *Endocrinology*. 2006;147(2):1007–1013.
69. Lehman MN, Hileman SM, Goodman RL. Neuroanatomy of the kisspeptin signaling system in mammals: comparative

and developmental aspects. *Adv Exp Med Biol*. 2013;784:27–62.
70. Hrabovszky E, Ciofi P, Vida B, et al. The kisspeptin system of the human hypothalamus: sexual dimorphism and relationship with gonadotropin-releasing hormone and neurokinin B neurons. *Eur J Neurosci*. 2010;31(11):1984–1998.
71. Gill JC, Wang O, Kakar S, Martinelli E, Carroll RS, Kaiser UB. Reproductive hormone-dependent and -independent contributions to developmental changes in kisspeptin in GnRH-deficient hypogonadal mice. *PLoS One*. 2010;5(7):e11911.
72. Skorupskaite K, George JT, Anderson RA. The kisspeptin-GnRH pathway in human reproductive health and disease. *Hum Reprod Update*. 2014;20(4):485–500.
73. Orikasa C, Kondo Y, Hayashi S, McEwen BS, Sakuma Y. Sexually dimorphic expression of estrogen receptor beta in the anteroventral periventricular nucleus of the rat preoptic area: implication in luteinizing hormone surge. *Proc Natl Acad Sci U S A*. 2002;99(5):3306–3311.
74. Dumalska I, Wu M, Morozova E, Liu R, van den Pol A, Alreja M. Excitatory effects of the puberty-initiating peptide kisspeptin and group I metabotropic glutamate receptor agonists differentiate two distinct subpopulations of gonadotropin-releasing hormone neurons. *J Neurosci*. 2008;28(32):8003–8013.
75. Hayes FJ, Seminara SB, Decruz S, Boepple PA, Crowley WF Jr. Aromatase inhibition in the human male reveals a hypothalamic site of estrogen feedback. *J Clin Endocrinol Metab*. 2000;85(9):3027–3035.
76. Rochira V, Zirilli L, Genazzani AD, et al. Hypothalamic-pituitary-gonadal axis in two men with aromatase deficiency: evidence that circulating estrogens are required at the hypothalamic level for the integrity of gonadotropin negative feedback. *Eur J Endocrinol*. 2006;155(4):513–522.
77. Bagatell CJ, Dahl KD, Bremner WJ. The direct pituitary effect of testosterone to inhibit gonadotropin secretion in men is partially mediated by aromatization to estradiol. *J Androl*. 1994;15(1):15–21.
78. Roselli CE, Resko JA. Sex differences in androgen-regulated expression of cytochrome P450 aromatase in the rat brain. *J Steroid Biochem Mol Biol*. 1997;61(3–6):365–374.
79. Balthazart J, Baillien M, Charlier TD, Cornil CA, Ball GF. Multiple mechanisms control brain aromatase activity at the genomic and non-genomic level. *J Steroid Biochem Mol Biol*. 2003;86(3–5):367–379.
80. Herbison AE, Skinner DC, Robinson JE, King IS. Androgen receptor-immunoreactive cells in ram hypothalamus: distribution and co-localization patterns with gonadotropin-releasing hormone, somatostatin and tyrosine hydroxylase. *Neuroendocrinology*. 1996;63(2):120–131.
81. McDevitt MA, Glidewell-Kenney C, Jimenez MA, et al. New insights into the classical and non-classical actions of estrogen: evidence from estrogen receptor knock-out and knock-in mice. *Mol Cell Endocrinol*. 2008;290(1–2):24–30.
82. Herbison AE. Multimodal influence of estrogen upon gonadotropin-releasing hormone neurons. *Endocr Rev*. 1998;19(3):302–330.
83. Smith JT, Dungan HM, Stoll EA, et al. Differential regulation of KiSS-1 mRNA expression by sex steroids in the brain of the male mouse. *Endocrinology*. 2005;146(7):2976–2984.
84. Pinilla L, Aguilar E, Dieguez C, Millar RP, Tena-Sempere M. Kisspeptins and reproduction: physiological roles and regulatory mechanisms. *Physiol Rev*. 2012;92(3):1235–1316.
85. Pinon R. *Biology of Human Reproduction*. Sausalito, CA: University Science Books; 2002.
86. Vidal B, Pasqualini C, Le Belle N, et al. Dopamine inhibits luteinizing hormone synthesis and release in the juvenile European eel: a neuroendocrine lock for the onset of puberty. *Biol Reprod*. 2004;71(5):1491–1500.
87. Ducharme JR, Forest MG, De Peretti E, Sempe M, Collu R, Bertrand J. Plasma adrenal and gonadal sex steroids in human pubertal development. *J Clin Endocrinol Metab*. 1976;42(3):468–476.
88. Sizonenko PC, Paunier L. Hormonal changes in puberty III: Correlation of plasma dehydroepiandrosterone, testosterone, FSH, and LH with stages of puberty and bone age in normal boys and girls and in patients with Addison's disease or hypogonadism or with premature or late adrenarche. *J Clin Endocrinol Metab*. 1975;41(5):894–904.
89. Kaplan SL, Grumbach MM, Aubert ML. The ontogenesis of pituitary hormones and hypothalamic factors in the human fetus: maturation of central nervous system regulation of anterior pituitary function. *Recent Prog Horm Res*. 1976;32:161–243.
90. Clements JA, Reyes FI, Winter JS, Faiman C. Studies on human sexual development. III. Fetal pituitary and serum, and amniotic fluid concentrations of LH, CG, and FSH. *J Clin Endocrinol Metab*. 1976;42(1):9–19.
91. Plant TM. Neuroendocrine control of the onset of puberty. *Front Neuroendocrinol*. 2015;38:73–88.
92. Navarro VM, Fernandez-Fernandez R, Castellano JM, et al. Advanced vaginal opening and precocious activation of the reproductive axis by KiSS-1 peptide, the endogenous ligand of GPR54. *J Physiol*. 2004;561(Pt 2):379–386.
93. Terasawa E, Fernandez DL. Neurobiological mechanisms of the onset of puberty in primates. *Endocr Rev*. 2001;22(1):111–151.
94. Brann DW. Glutamate: a major excitatory transmitter in neuroendocrine regulation. *Neuroendocrinology*. 1995;61(3):213–225.
95. McCartney CR. Maturation of sleep-wake gonadotrophin-releasing hormone secretion across puberty in girls: potential mechanisms and relevance to the pathogenesis of polycystic ovary syndrome. *J Neuroendocrinol*. 2010;22(7):701–709.
96. Hall JE, Lavoie HB, Marsh EE, Martin KA. Decrease in gonadotropin-releasing hormone (GnRH) pulse frequency with aging in postmenopausal women. *J Clin Endocrinol Metab*. 2000;85(5):1794–1800.
97. Perry JR, Murray A, Day FR, Ong KK. Molecular insights into the aetiology of female reproductive ageing. *Nat Rev Endocrinol*. 2015;11(12):725–734.
98. Day FR, Elks CE, Murray A, Ong KK, Perry JR. Puberty timing associated with diabetes, cardiovascular disease and also diverse health outcomes in men and women: the UK Biobank study. *Sci Rep*. 2015;5:11208.
99. Wehkalampi K, Silventoinen K, Kaprio J, et al. Genetic and environmental influences on pubertal timing assessed by height growth. *Am J Hum Biol*. 2008;20(4):417–423.
100. Day FR, Thompson DJ, Helgason H, et al. Genomic analyses identify hundreds of variants associated with age at menarche and support a role for puberty timing in cancer risk. *Nat Genet*. 2017;49(6):834–841.
101. Kennedy GC, Mitra J. Body weight and food intake as initiating factors for puberty in the rat. *J Physiol*. 1963;166:408–418.
102. Frisch RE, Revelle R. Height and weight at menarche and a hypothesis of critical body weights and adolescent events. *Science*. 1970;169(3943):397–399.
103. Frisch RE, Revelle R, Cook S. Components of weight at menarche and the initiation of the adolescent growth spurt in girls: estimated total water, lean body weight and fat. *Hum Biol*. 1973;45(3):469–483.
104. Biro FM, Khoury P, Morrison JA. Influence of obesity on timing of puberty. *Int J Androl*. 2006;29(1):272–277; discussion 286–290.
105. Cheng G, Buyken AE, Shi L, et al. Beyond overweight: nutrition as an important lifestyle factor influencing timing of puberty. *Nutr Rev*. 2012;70(3):133–152.
106. Roa J, Tena-Sempere M. Energy balance and puberty onset: emerging role of central mTOR signaling. *Trends Endocrinol Metab*. 2010;21(9):519–528.
107. Frisch RE. Weight at menarche: similarity for well-nourished and undernourished girls at differing ages, and evidence for historical constancy. *Pediatrics*. 1972;50(3):445–450.
108. Kile JP, Alexander BM, Moss GE, Hallford DM, Nett TM. Gonadotropin-releasing hormone overrides the negative effect of reduced dietary energy on gonadotropin

synthesis and secretion in ewes. *Endocrinology*. 1991;128(2):843–849.
109. Foster DL, Olster DH. Effect of restricted nutrition on puberty in the lamb: patterns of tonic luteinizing hormone (LH) secretion and competency of the LH surge system. *Endocrinology*. 1985;116(1):375–381.
110. Frisch RE, McArthur JW. Menstrual cycles: fatness as a determinant of minimum weight for height necessary for their maintenance or onset. *Science*. 1974;185(4155):949–951.
111. Warren MP. Effects of undernutrition on reproductive function in the human. *Endocr Rev*. 1983;4(4):363–377.
112. Castellano JM, Bentsen AH, Sanchez-Garrido MA, et al. Early metabolic programming of puberty onset: impact of changes in postnatal feeding and rearing conditions on the timing of puberty and development of the hypothalamic kisspeptin system. *Endocrinology*. 2011;152(9):3396–3408.
113. Gasser CL, Grum DE, Mussard ML, Fluharty FL, Kinder JE, Day ML. Induction of precocious puberty in heifers I: enhanced secretion of luteinizing hormone. *J Anim Sci*. 2006;84(8):2035–2041.
114. Cardoso RC, Alves BR, Prezotto LD, et al. Use of a stair-step compensatory gain nutritional regimen to program the onset of puberty in beef heifers. *J Anim Sci*. 2014;92(7):2942–2949.
115. Zacharias L, Wurtman RJ, Schatzoff M. Sexual maturation in contemporary American girls. *Am J Obstet Gynecol*. 1970;108(5):833–846.
116. Lee JM, Appugliese D, Kaciroti N, Corwyn RF, Bradley RH, Lumeng JC. Weight status in young girls and the onset of puberty. *Pediatrics*. 2007;119(3):e624–e630.
117. Gluckman PD, Hanson MA. Evolution, development and timing of puberty. *Trends Endocrinol Metab*. 2006;17(1):7–12.
118. Diamond J. Evolution, consequences and future of plant and animal domestication. *Nature*. 2002;418(6898):700–707.
119. Goodman AH, Martin DL. Reconstructing health profiles from skeletal remains. In: Steckel RH, Rose JC, eds. *The Backbone of History: Health and Nutrition in the Western Hemisphere*. Cambridge University Press; 2002:94–124.
120. Gluckman PD, Bergstrom CT. Evolutionary biology within medicine: a perspective of growing value. *BMJ*. 2011;343:d7671.
121. Parent AS, Teilmann G, Juul A, Skakkebaek NE, Toppari J, Bourguignon JP. The timing of normal puberty and the age limits of sexual precocity: variations around the world, secular trends, and changes after migration. *Endocr Rev*. 2003;24(5):668–693.
122. Karlberg J. Secular trends in pubertal development. *Horm Res*. 2002;57(Suppl 2):19–30.
123. Garn SM. The secular trend in size and maturational timing and its implications for nutritional assessment. *J Nutr*. 1987;117(5):817–823.
124. Ellison PT. Morbidity, morality, and menarche. *Hum Biol*. 1981;53(4):635–643.
125. Biro FM, Kiess W. Contemporary trends in onset and completion of puberty, gain in height and adiposity. *Endocr Dev*. 2016;29:122–133.
126. Burt Solorzano CM, McCartney CR. Obesity and the pubertal transition in girls and boys. *Reproduction*. 2010;140(3):399–410.
127. Toppari J, Juul A. Trends in puberty timing in humans and environmental modifiers. *Mol Cell Endocrinol*. 2010;324(1–2):39–44.
128. Aksglaede L, Sorensen K, Petersen JH, Skakkebaek NE, Juul A. Recent decline in age at breast development: the Copenhagen Puberty Study. *Pediatrics*. 2009;123(5):e932–e939.
129. Toppari J, Juul A. Trends in puberty timing in humans and environmental modifiers. *Mol Cell Endocrinol*. 2010;324(1–2):39–44.
130. Chehab FF, Lim ME, Lu R. Correction of the sterility defect in homozygous obese female mice by treatment with the human recombinant leptin. *Nat Genet*. 1996;12(3):318–320.
131. Montague CT, Farooqi IS, Whitehead JP, et al. Congenital leptin deficiency is associated with severe early-onset obesity in humans. *Nature*. 1997;387(6636):903–908.
132. Strobel A, Issad T, Camoin L, Ozata M, Strosberg AD. A leptin missense mutation associated with hypogonadism and morbid obesity. *Nat Genet*. 1998;18(3):213–215.
133. Ozata M, Ozdemir IC, Licinio J. Human leptin deficiency caused by a missense mutation: multiple endocrine defects, decreased sympathetic tone, and immune system dysfunction indicate new targets for leptin action, greater central than peripheral resistance to the effects of leptin, and spontaneous correction of leptin-mediated defects. *J Clin Endocrinol Metab*. 1999;84(10):3686–3695.
134. Clement K, Vaisse C, Lahlou N, et al. A mutation in the human leptin receptor gene causes obesity and pituitary dysfunction. *Nature*. 1998;392(6674):398–401.
135. Farooqi IS, Jebb SA, Langmack G, et al. Effects of recombinant leptin therapy in a child with congenital leptin deficiency. *N Engl J Med*. 1999;341(12):879–884.
136. Farooqi IS, Matarese G, Lord GM, et al. Beneficial effects of leptin on obesity, T cell hyporesponsiveness, and neuroendocrine/metabolic dysfunction of human congenital leptin deficiency. *J Clin Invest*. 2002;110(8):1093–1103.
137. Licinio J, Caglayan S, Ozata M, et al. Phenotypic effects of leptin replacement on morbid obesity, diabetes mellitus, hypogonadism, and behavior in leptin-deficient adults. *Proc Natl Acad Sci U S A*. 2004;101(13):4531–4536.
138. Farooqi IS, Wangensteen T, Collins S, et al. Clinical and molecular genetic spectrum of congenital deficiency of the leptin receptor. *N Engl J Med*. 2007;356(3):237–247.
139. Bianco SD. A potential mechanism for the sexual dimorphism in the onset of puberty and incidence of idiopathic central precocious puberty in children: sex-specific kisspeptin as an integrator of puberty signals. *Front Endocrinol (Lausanne)*. 2012;3:149.
140. Elias CF. Leptin action in pubertal development: recent advances and unanswered questions. *Trends Endocrinol Metab*. 2012;23(1):9–15.
141. Plant TM. Hypothalamic control of the pituitary-gonadal axis in higher primates: key advances over the last two decades. *J Neuroendocrinol*. 2008;20(6):719–726.
142. Quennell JH, Mulligan AC, Tups A, et al. Leptin indirectly regulates gonadotropin-releasing hormone neuronal function. *Endocrinology*. 2009;150(6):2805–2812.
143. Smith JT, Acohido BV, Clifton DK, Steiner RA. KiSS-1 neurones are direct targets for leptin in the ob/ob mouse. *J Neuroendocrinol*. 2006;18(4):298–303.
144. Navarro VM, Ruiz-Pino F, Sanchez-Garrido MA, et al. Role of neurokinin B in the control of female puberty and its modulation by metabolic status. *J Neurosci*. 2012;32(7):2388–2397.
145. Chou SH, Chamberland JP, Liu X, et al. Leptin is an effective treatment for hypothalamic amenorrhea. *Proc Natl Acad Sci U S A*. 2011;108(16):6585–6590.
146. Welt CK, Chan JL, Bullen J, et al. Recombinant human leptin in women with hypothalamic amenorrhea. *N Engl J Med*. 2004;351(10):987–997.
147. Chan JL, Mantzoros CS. Role of leptin in energy-deprivation states: normal human physiology and clinical implications for hypothalamic amenorrhoea and anorexia nervosa. *Lancet*. 2005;366(9479):74–85.
148. Musso C, Cochran E, Javor E, Young J, Depaoli AM, Gorden P. The long-term effect of recombinant methionyl human leptin therapy on hyperandrogenism and menstrual function in female and pituitary function in male and female hypoleptinemic lipodystrophic patients. *Metabolism*. 2005;54(2):255–263.
149. Cali AM, Caprio S. Obesity in children and adolescents. *J Clin Endocrinol Metab*. 2008;93(11 Suppl 1):S31–S36.
150. Li D, Zhang B, Cheng J, et al. Obesity-related genetic polymorphisms are associated with the risk of early puberty in Han Chinese girls. *Clin Endocrinol*. 2022;96:319–327.
151. Villamor E, Jansen EC. Nutritional determinants of the timing of puberty. *Annu Rev Public Health*. 2016;37:33–46.
152. Tinggaard J, Mieritz MG, Sorensen K, et al. The physiology and timing of male

puberty. *Curr Opin Endocrinol Diabetes Obes*. 2012;19(3):197–203.
153. Lee JM, Wasserman R, Kaciroti N, et al. Timing of puberty in overweight versus obese boys. *Pediatrics*. 2016;137(2):e20150164.
154. Blank SK, McCartney CR, Chhabra S, et al. Modulation of gonadotropin-releasing hormone pulse generator sensitivity to progesterone inhibition in hyperandrogenic adolescent girls--implications for regulation of pubertal maturation. *J Clin Endocrinol Metab*. 2009;94(7):2360–2366.
155. McCartney CR, Blank SK, Prendergast KA, et al. Obesity and sex steroid changes across puberty: evidence for marked hyperandrogenemia in pre- and early pubertal obese girls. *J Clin Endocrinol Metab*. 2007;92(2):430–436.
156. Shalitin S, Phillip M. Role of obesity and leptin in the pubertal process and pubertal growth--a review. *Int J Obes Relat Metab Disord*. 2003;27(8):869–874.
157. Kaplowitz PB. Link between body fat and the timing of puberty. *Pediatrics*. 2008;121(Suppl 3):S208–S217.
158. Dunger DB, Ahmed ML, Ong KK. Effects of obesity on growth and puberty. *Best Pract Res Clin Endocrinol Metab*. 2005;19(3):375–390.
159. Jasik CB, Lustig RH. Adolescent obesity and puberty: the "perfect storm". *Ann N Y Acad Sci*. 2008;1135:265–279.
160. Zhai L, Liu J, Zhao J, et al. Association of obesity with onset of puberty and sex hormones in Chinese girls: A 4-year longitudinal study. *PLoS One*. 2015;10(8):e0134656.
161. Mauras N, Santen RJ, Colon-Otero G, et al. Estrogens and their genotoxic metabolites are increased in obese prepubertal girls. *J Clin Endocrinol Metab*. 2015;100(6):2322–2328.
162. Tsai EC, Matsumoto AM, Fujimoto WY, Boyko EJ. Association of bioavailable, free, and total testosterone with insulin resistance: influence of sex hormone-binding globulin and body fat. *Diabetes Care*. 2004;27(4):861–868.
163. Strain GW, Zumoff B, Kream J, et al. Mild hypogonadotropic hypogonadism in obese men. *Metabolism*. 1982;31(9):871–875.
164. Hammoud AO, Gibson M, Peterson CM, Hamilton BD, Carrell DT. Obesity and male reproductive potential. *J Androl*. 2006;27(5):619–626.
165. Holtcamp W. Obesogens: an environmental link to obesity. *Environ Health Perspect*. 2012;120(2):a62–a68.
166. Phillips KP, Tanphaichitr N. Mechanisms of obesity-induced male infertility. *Expert Rev Endocrinol Metab*. 2010;5(2):229–251.
167. Darbre PD. Endocrine disruptors and obesity. *Curr Obes Rep*. 2017;6(1):18–27.
168. Stearns SC. Life-history tactics: a review of the ideas. *Q Rev Biol*. 1976;51(1):3–47.
169. Wingfield JC, Sapolsky RM. Reproduction and resistance to stress: when and how. *J Neuroendocrinol*. 2003;15(8):711–724.
170. Whirledge S, Cidlowski JA. A role for glucocorticoids in stress-impaired reproduction: beyond the hypothalamus and pituitary. *Endocrinology*. 2013;154(12):4450–4468.
171. van Noord PA, Kaaks R. The effect of wartime conditions and the 1944–45 "Dutch famine" on recalled menarcheal age in participants of the DOM breast cancer screening project. *Ann Hum Biol*. 1991;18(1):57–70.
172. Tahirovic HF. Menarchal age and the stress of war: an example from Bosnia. *Eur J Pediatr*. 1998;157(12):978–980.
173. Magner JA, Rogol AD, Gorden P. Reversible growth hormone deficiency and delayed puberty triggered by a stressful experience in a young adult. *Am J Med*. 1984;76(4):737–742.
174. Moffitt TE, Caspi A, Belsky J, Silva PA. Childhood experience and the onset of menarche: a test of a sociobiological model. *Child Dev*. 1992;63(1):47–58.
175. Wierson M, Long PJ, Forehand RL. Toward a new understanding of early menarche: the role of environmental stress in pubertal timing. *Adolescence*. 1993;28(112):913–924.
176. Mancuso RA, Schetter CD, Rini CM, Roesch SC, Hobel CJ. Maternal prenatal anxiety and corticotropin-releasing hormone associated with timing of delivery. *Psychosom Med*. 2004;66(5):762–769.
177. Rice F, Jones I, Thapar A. The impact of gestational stress and prenatal growth on emotional problems in offspring: a review. *Acta Psychiatr Scand*. 2007;115(3):171–183.
178. Allsworth JE, Clarke J, Peipert JF, Hebert MR, Cooper A, Boardman LA. The influence of stress on the menstrual cycle among newly incarcerated women. *Womens Health Iss*. 2007;17(4):202–209.
179. Luo E, Stephens SB, Chaing S, Munaganuru N, Kauffman AS, Breen KM. Corticosterone blocks ovarian cyclicity and the LH surge via decreased kisspeptin neuron activation in female mice. *Endocrinology*. 2016;157(3):1187–1199.
180. Lynch CD, Sundaram R, Maisog JM, Sweeney AM, Buck Louis GM. Preconception stress increases the risk of infertility: results from a couple-based prospective cohort study--the LIFE study. *Hum Reprod*. 2014;29(5):1067–1075.
181. Whirledge S, Cidlowski JA. Glucocorticoids, stress, and fertility. *Minerva Endocrinol*. 2010;35(2):109–125.
182. Dubey AK, Plant TM. A suppression of gonadotropin secretion by cortisol in castrated male rhesus monkeys (Macaca mulatta) mediated by the interruption of hypothalamic gonadotropin-releasing hormone release. *Biol Reprod*. 1985;33(2):423–431.
183. Kamel F, Kubajak CL. Modulation of gonadotropin secretion by corticosterone: interaction with gonadal steroids and mechanism of action. *Endocrinology*. 1987;121(2):561–568.
184. Oakley AE, Breen KM, Clarke IJ, Karsch FJ, Wagenmaker ER, Tilbrook AJ. Cortisol reduces gonadotropin-releasing hormone pulse frequency in follicular phase ewes: influence of ovarian steroids. *Endocrinology*. 2009;150(1):341–349.
185. Plant TM. A study of the role of the postnatal testes in determining the ontogeny of gonadotropin secretion in the male rhesus monkey (Macaca mulatta). *Endocrinology*. 1985;116(4):1341–1350.
186. Hayashi KT, Moberg GP. Influence of the hypothalamic-pituitary-adrenal axis on the menstrual cycle and the pituitary responsiveness to estradiol in the female rhesus monkey (Macaca mulatta). *Biol Reprod*. 1990;42(2):260–265.
187. Poisson M, Pertuiset BF, Moguilewsky M, Magdelenat H, Martin PM. [Steroid receptors in the central nervous system. Implications in neurology]. *Rev Neurol (Paris)*. 1984;140(4):233–248.
188. Takumi K, Iijima N, Higo S, Ozawa H. Immunohistochemical analysis of the colocalization of corticotropin-releasing hormone receptor and glucocorticoid receptor in kisspeptin neurons in the hypothalamus of female rats. *Neurosci Lett*. 2012;531(1):40–45.
189. Kinsey-Jones JS, Li XF, Knox AM, et al. Down-regulation of hypothalamic kisspeptin and its receptor, Kiss1r, mRNA expression is associated with stress-induced suppression of luteinising hormone secretion in the female rat. *J Neuroendocrinol*. 2009;21(1):20–29.
190. Grachev P, Li XF, O'Byrne K. Stress regulation of kisspeptin in the modulation of reproductive function. *Adv Exp Med Biol*. 2013;784:431–454.
191. University of California – Berkeley. Stress puts double whammy on reproductive system, fertility. *ScienceDaily*; 29 June 2009. www.sciencedaily.com/releases/2009/06/090615171618.htm.
192. Hyde CL, Childs G, Wahl LM, Naor Z, Catt KJ. Preparation of gonadotroph-enriched cell populations from adult rat anterior pituitary cells by centrifugal elutriation. *Endocrinology*. 1982;111(4):1421–1423.
193. Cheng KW, Leung PC. The expression, regulation and signal transduction pathways of the mammalian gonadotropin-releasing hormone receptor. *Can J Physiol Pharmacol*. 2000;78(12):1029–1052.
194. Hardy MP, Gao HB, Dong Q, et al. Stress hormone and male reproductive function. *Cell Tissue Res*. 2005;322(1):147–153.
195. Schreiber JR, Nakamura K, Erickson GF. Rat ovary glucocorticoid receptor: identification and characterization. *Steroids*. 1982;39(5):569–584.

196. Hirst MA, Northrop JP, Danielsen M, Ringold GM. High level expression of wild type and variant mouse glucocorticoid receptors in Chinese hamster ovary cells. *Mol Endocrinol*. 1990;4(1):162–170.
197. Petraglia F, Sutton S, Vale W, Plotsky P. Corticotropin-releasing factor decreases plasma luteinizing hormone levels in female rats by inhibiting gonadotropin-releasing hormone release into hypophysial-portal circulation. *Endocrinology*. 1987;120(3):1083–1088.
198. Li XF, Knox AM, O'Byrne KT. Corticotrophin-releasing factor and stress-induced inhibition of the gonadotrophin-releasing hormone pulse generator in the female. *Brain Res*. 2010;1364:153–163.
199. D'Agata R, Cavagnini F, Invitti C, et al. Effect of CRF on the release of anterior pituitary hormones in normal subjects and patients with Cushing's disease. *Pharmacol Res Commun*. 1984;16(3):303–311.
200. Rivier C, Vale W. Influence of corticotropin-releasing factor on reproductive functions in the rat. *Endocrinology*. 1984;114(3):914–921.
201. Rivier C, Rivier J, Vale W. Stress-induced inhibition of reproductive functions: role of endogenous corticotropin-releasing factor. *Science*. 1986;231(4738):607–609.
202. Ono N, Lumpkin MD, Samson WK, McDonald JK, McCann SM. Intrahypothalamic action of corticotrophin-releasing factor (CRF) to inhibit growth hormone and LH release in the rat. *Life Sci*. 1984;35(10):1117–1123.
203. Williams CL, Nishihara M, Thalabard JC, Grosser PM, Hotchkiss J, Knobil E. Corticotropin-releasing factor and gonadotropin-releasing hormone pulse generator activity in the rhesus monkey. Electrophysiological studies. *Neuroendocrinology*. 1990;52(2):133–137.
204. Chemineau P, Guillaume D, Migaud M, Thiery JC, Pellicer-Rubio MT, Malpaux B. Seasonality of reproduction in mammals: intimate regulatory mechanisms and practical implications. *Reprod Domest Anim*. 2008;43(Suppl 2):40–47.
205. Bronson FH. Seasonal variation in human reproduction: environmental factors. *Q Rev Biol*. 1995;70(2):141–164.
206. Rojansky N, Brzezinski A, Schenker JG. Seasonality in human reproduction: an update. *Hum Reprod*. 1992;7(6):735–745.
207. Plant TM, Zeleznik AJ, eds. *Knobil and Neill's Physiology of Reproduction*. Academic Press; 2014.
208. Brown GP, Shine R. Why do most tropical animals reproduce seasonally? Testing hypotheses on an Australian snake. *Ecology*. 2006;87(1):133–143.
209. Bronson FH. Mammalian reproductive strategies: genes, photoperiod and latitude. *Reprod Nutr Dev*. 1988;28(2B):335–347.
210. Bronson FH. Mammalian reproduction: an ecological perspective. *Biol Reprod*. 1985;32(1):1–26.
211. Lincoln GA, Peet MJ, Cunningham RA. Seasonal and circadian changes in the episodic release of follicle-stimulating hormone, luteinizing hormone and testosterone in rams exposed to artificial photoperiods. *J Endocrinol*. 1977;72(3):337–349.
212. Hazlerigg DG, Wagner GC. Seasonal photoperiodism in vertebrates: from coincidence to amplitude. *Trends Endocrinol Metab*. 2006;17(3):83–91.
213. Guyomarc'h C, Lumineau S, Vivien-Roels B, Richard J, Deregnaucourt S. Effect of melatonin supplementation on the sexual development in European quail (Coturnix coturnix). *Behav Processes*. 2001;53(1–2):121–130.
214. Ohta M, Kadota C, Konishi H. A role of melatonin in the initial stage of photoperiodism in the Japanese quail. *Biol Reprod*. 1989;40(5):935–941.
215. Rozenboim I, Aharony T, Yahav S. The effect of melatonin administration on circulating plasma luteinizing hormone concentration in castrated White Leghorn roosters. *Poult Sci*. 2002;81(9):1354–1359.
216. Chowdhury VS, Yamamoto K, Ubuka T, Bentley GE, Hattori A, Tsutsui K. Melatonin stimulates the release of gonadotropin-inhibitory hormone by the avian hypothalamus. *Endocrinology*. 2010;151(1):271–280.
217. Ubuka T, Bentley GE, Ukena K, Wingfield JC, Tsutsui K. Melatonin induces the expression of gonadotropin-inhibitory hormone in the avian brain. *Proc Natl Acad Sci U S A*. 2005;102(8):3052–3057.
218. Reiter RJ. The pineal and its hormones in the control of reproduction in mammals. *Endocr Rev*. 1980;1(2):109–131.
219. Fitzgerald JA, Stellflug JN. Effects of melatonin on seasonal changes in reproduction of rams. *J Anim Sci*. 1991;69(1):264–275.
220. Rekik M, Taboubi R, Ben Salem I, et al. Melatonin administration enhances the reproductive capacity of young rams under a southern Mediterranean environment. *Anim Sci J*. 2015;86(7):666–672.
221. DeNicolo G, Morris ST, Kenyon PR, Morel PC, Parkinson TJ. Melatonin-improved reproductive performance in sheep bred out of season. *Anim Reprod Sci*. 2008;109(1–4):124–133.
222. Shinomiya A, Shimmura T, Nishiwaki-Ohkawa T, Yoshimura T. Regulation of seasonal reproduction by hypothalamic activation of thyroid hormone. *Front Endocrinol (Lausanne)*. 2014;5:12.
223. Dardente H, Hazlerigg DG, Ebling FJ. Thyroid hormone and seasonal rhythmicity. *Front Endocrinol (Lausanne)*. 2014;5:19.
224. Wood S, Loudon A. Clocks for all seasons: unwinding the roles and mechanisms of circadian and interval timers in the hypothalamus and pituitary. *J Endocrinol*. 2014;222(2):R39–R59.
225. Benoit J. Role de la thyroide dans la gonado-stimulation par lumiere artificielle chez le canard domestique. *Comptes Rendus Societe de Biologie Paris*. 1936;123:243–246.
226. Woitkewitsch A. Dependence of seasonal periodicity in gonadal changes on the thyroid gland. *Doklady Akademii Nauk SSSR*. 1940;27:741–745.
227. Follett BK, Nicholls TJ. Influences of thyroidectomy and thyroxine replacement on photoperiodically controlled reproduction in quail. *J Endocrinol*. 1985;107(2):211–221.
228. Nicholls TJ, Follett BK, Goldsmith AR, Pearson H. Possible homologies between photorefractoriness in sheep and birds: the effect of thyroidectomy on the length of the ewe's breeding season. *Reprod Nutr Dev*. 1988;28(2B):375–385.
229. Webster JR, Moenter SM, Woodfill CJ, Karsch FJ. Role of the thyroid gland in seasonal reproduction. II. Thyroxine allows a season-specific suppression of gonadotropin secretion in sheep. *Endocrinology*. 1991;129(1):176–183.
230. Kriegsfeld LJ, Mei DF, Bentley GE, et al. Identification and characterization of a gonadotropin-inhibitory system in the brains of mammals. *Proc Natl Acad Sci U S A*. 2006;103(7):2410–2415.
231. Ikegami K, Yoshimura T. Seasonal time measurement during reproduction. *J Reprod Dev*. 2013;59(4):327–333.
232. Dardente H, Birnie M, Lincoln GA, Hazlerigg DG. RFamide-related peptide and its cognate receptor in the sheep: cDNA cloning, mRNA distribution in the hypothalamus and the effect of photoperiod. *J Neuroendocrinol*. 2008;20(11):1252–1259.
233. Smith JT, Coolen LM, Kriegsfeld LJ, et al. Variation in kisspeptin and RFamide-related peptide (RFRP) expression and terminal connections to gonadotropin-releasing hormone neurons in the brain: a novel medium for seasonal breeding in the sheep. *Endocrinology*. 2008;149(11):5770–5782.
234. Revel FG, Saboureau M, Pevet P, Simonneaux V, Mikkelsen JD. RFamide-related peptide gene is a melatonin-driven photoperiodic gene. *Endocrinology*. 2008;149(3):902–912.
235. Mason AO, Duffy S, Zhao S, et al. Photoperiod and reproductive condition are associated with changes in RFamide-related peptide (RFRP) expression in Syrian hamsters (Mesocricetus auratus). *J Biol Rhythms*. 2010;25(3):176–185.
236. Ubuka T, Inoue K, Fukuda Y, et al. Identification, expression, and physiological functions of Siberian hamster gonadotropin-inhibitory hormone. *Endocrinology*. 2012;153(1):373–385.
237. Kriegsfeld LJ, Ubuka T, Bentley GE, Tsutsui K. Seasonal control of gonadotropin-inhibitory hormone (GnIH) in birds and mammals. *Front Neuroendocrinol*. 2015;37:65–75.
238. Salti R, Tarquini R, Stagi S, et al. Age-dependent association of exposure to television screen with children's urinary melatonin excretion? *Neuro Endocrinol Lett*. 2006;27(1–2):73–80.

CHAPTER

24

Pregnancy, Birth, and Lactation

CHAPTER LEARNING OBJECTIVES:

- Compare and contrast different strategies for the generation of offspring, and describe some fundamental changes in placental gene expression that accompanied the evolution of vivipary in mammals.
- Outline the major fetal-placental-maternal interactions in the production of sex steroids and other hormones.
- Describe the neuroendocrine mechanisms responsible for the timing of human parturition and the major hormones involved in the promotion of labor.
- Describe the suckling-induced positive feedback reflex, and discuss how hormones of the adenohypophysis and neurohypophysis interact to regulate lactation.

OPENING QUOTATION:

"As a successful parasite, the fetal-placental unit manipulates the maternal host for its own gain but normally avoids imposing excessive stress that would jeopardize the pregnancy."

—Taylor RN, Lebovic DI, Martin-Cadieux MC. The endocrinology of pregnancy. *Basic and Clinical Endocrinology* (Ed by FS Greenspan & DG Baxter). 2001:575–602[1]

Introduction: The Evolution of Pregnancy

LEARNING OBJECTIVE Compare and contrast different strategies for the generation of offspring, and describe some fundamental changes in placental gene expression that accompanied the evolution of vivipary in mammals.

KEY CONCEPTS:

- Pregnancy describes the gestational period of all viviparous animals, from live-bearing species of sharks and other fishes to some amphibians, reptiles, and most mammals.
- One of the first steps in the transition from oviparity to viviparity among vertebrates was the evolution of the amniotic egg that permitted a shift from yolk-sac nutrition to embryonic nourishment delivered by the mother.
- Among mammals, the evolution of pregnancy was accompanied by radical changes in the expression of a vast array of genes in the uterus.
- During mammalian uterine evolution, ancient transposable elements deposited DNA binding sites for master transcriptional regulators of endometrial development (such as the progesterone receptor), imparting a new hormone responsivity to numerous genes.

Strategies for the Generation of Offspring

In order to understand the evolution of endocrine features associated with pregnancy, it is first necessary to appreciate how strategies for generating offspring differ among various vertebrates. Among all sexually reproducing vertebrates, offspring are generated from the fusion of two haploid gametes, the egg and the sperm, to form a diploid embryo. In many (but not all) fishes and amphibians, this fertilization takes place *externally* (Figure 24.1 A), and from the moment of conception the growing embryo develops with only the eggshell membranes separating it from the outside environment. In other vertebrates, such as many reptiles, all birds, and all egg-laying mammals, fertilization takes place *internally*, but the eggshell-encapsulated embryo is eventually deposited into the outside environment where the rest of development prior to hatching occurs external to the parent (Figure 24.1 B). **Ovipary** (from the Latin *ova*, or egg, and *parous*, meaning giving birth to) describes these strategies of reproduction whereby some or all of embryogenesis takes place outside of the parent in an egg containing all the nutrients necessary for development through hatching. With the exception of mammals, oviparity is the most widely used mode of reproduction among vertebrates, occurring in over 97% of fish, 90% of amphibians, 85% of reptiles, and in 100% of birds.[2,3] Because most animal species use oviparity as a strategy for generating offspring, it is assumed that this is the ancestral condition.[3]

Ovoviviparity refers to a strategy by which embryos in shelled eggs are not laid externally, but are instead

DOI: 10.1201/9781003359241-31

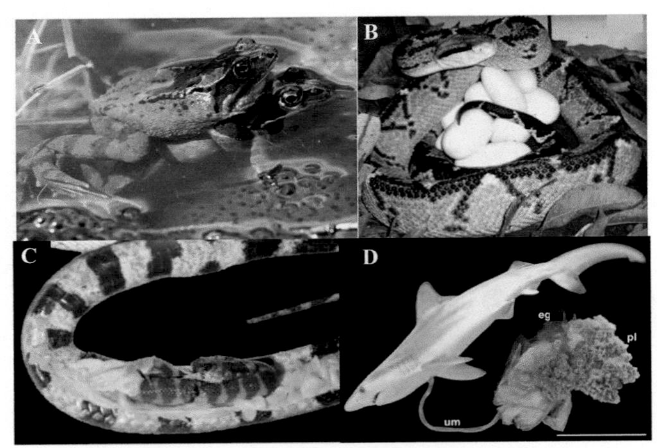

Figure 24.1 **Oviparous, ovoviparous, and viviparous reproductive strategies in vertebrates.** **(A)** External fertilization and oviparous external hatching in an amphibian. **(B)** Internal fertilization and oviparous external hatching in a reptile (python). **(C)** Internal fertilization and ovoviparous internal hatching in a reptile (South American water snake). The image shows the abdomen of a pregnant female dissected open to reveal a developing snakelet free of its eggshell and about to undergo parturition. **(D)** Internal fertilization and viviparous parturition. Note the placenta (pl) and umbilical cord (um) in a shark (dogfish).

retained inside the parent where they "hatch" prior to parturition. Although the offspring emerge shell-less, free-moving, and alive ("vivi"), besides the yolk sac there is little or no direct transfer of nutrients from the parent to the offspring due to the presence of an impermeable eggshell during most of gestation. Examples of animals using this strategy include many cartilaginous fishes and some snakes (Figure 24.1 C).

Distinct from both oviparity and ovoviparity is **viviparity** ("live birth"). This strategy refers to the delivery of "live" young that lack an eggshell and are able to interact directly with the external environment following a gestation period within the body of a parent. In contrast with ovoviparity, during this gestation period the embryo typically receives some significant form of direct nutrient transfer from the parent that supplements the yolk, often via a placenta or placenta-like structure (Figure 24.1 D). In other cases, nutrients are derived from uterine secretions, via nutrient absorption through the skin or gills, or even through cannibalization of their own developing siblings.[4,5] With the notable exception of birds, which are all oviparous, viviparity is found scattered among all other vertebrate groups.

Whereas viviparity describes a life strategy for generating offspring, the term **pregnancy** (from the Latin *pre*, meaning before, and *(g)natus*, meaning birth) refers to the period of time between fertilization and parturition during which a viviparous animal is in a gestational state. As such, pregnancy describes the gestational period of all viviparous animals, from live-bearing species of sharks and other fishes[6] to some amphibians,[7] reptiles[8] and most (but not all) mammals.

Curiously Enough . . . Pregnancy is not confined to maternal parents, but in some species can even extend to paternal parents, such as seahorses in which females deposit unfertilized eggs into a pouch in the male's abdomen where they are fertilized and gestated.[9]

Transition from Oviparity to Viviparity in Mammals

One of the first steps in the transition from oviparity to viviparity was the evolution of the amniotic egg that permitted a shift from yolk-sac nutrition to embryonic nourishment delivered by the mother. The eggs of fishes and amphibians are *anamniotic*, and, with the exception of the yolk sac,[10] possess no true extraembryonic membranes.* By contrast, the eggs of *amniotes*, which constitute mammals, birds, and reptiles, develop four extraembryonic *membranes* consisting of the *chorion*, *yolk sac*, *allantois*, and the *amnion* (Figure 24.2). The *amnion*, which

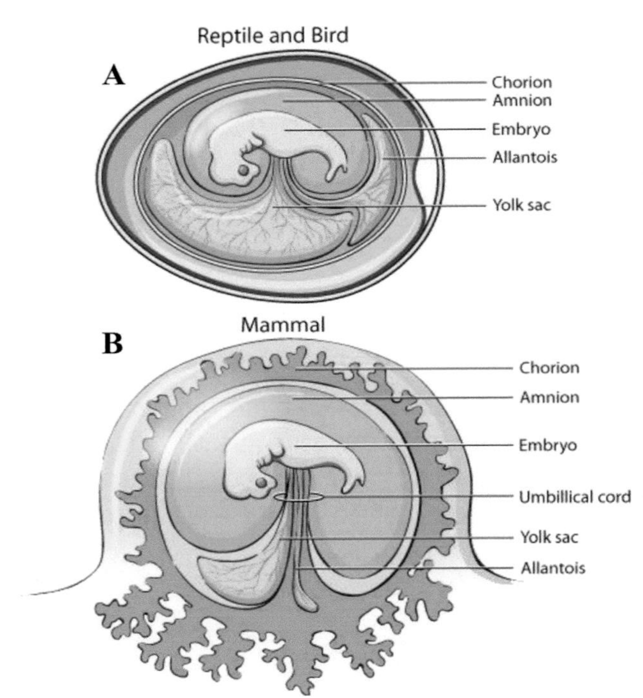

Figure 24.2 Extraembryonic membranes of amniotes. (A) The amniotic eggs of reptiles, birds, and egg-laying mammals possess four extraembryonic membranes (including the desiccation-resistant amnion) and a protective outer shell. A chicken embryo is shown in this diagram. **(B)** In placental mammals, the developing embryo has become internalized, and some of the extraembryonic membranes are integrated into the umbilical cord, which is juxtaposed to the placenta.

* Although fish and amphibians do contain a structure that is sometimes termed the "chorion", in these animals the structure is acellular and thus differs from that of amniotes.

Table 24.1 Some Features Associated with Different Reproductive Strategies

	Reproductive Strategy	Placenta Type	Eggshell	Post-Hatching/Birth Nutrition
Monotremes	Oviparity	Simple and transient, little nutritional contribution	Thin, poorly mineralized eggshell	Nipple-less mammary glands
Metatheria	Short pregnancy	Simple and transient, little nutritional contribution	Prolonged separation of maternal and embryonic tissues by an eggshell	Internal (in a pouch) mammary glands with nipples
Eutherians	Extended pregnancy	Elaborate, significant nutritional contribution	Complete loss of eggshell	External mammary glands with nipples

surrounds the embryo, provides an aqueous environment for the embryo, is desiccation resistant, and its appearance was likely essential for the transition of vertebrates to land.[11] In oviparous amniotes, the *yolk sac* functions to regulate embryonic nourishment, and in many viviparous species also contributes to maternal-fetal exchange.[12] The primary function of the *chorion* is gas exchange, whereas the *allantois* functions to sequester nitrogenous metabolic wastes and also facilitates gas exchange. Although the mineralized eggshell was ultimately lost during the evolution of viviparity, these extraembryonic membranes are retained by all amniote embryos. Importantly, in placental mammals, some of these membranes become juxtaposed to the uterine wall and have been modified to form the placenta and umbilical cord, which facilitates the exchange of gases, nutrients, water, waste products, and hormonal crosstalk between the embryo and the parent (Figure 24.2).

Phylogenetic reconstructions suggest that in mammals viviparity arose 191–124 million years ago from oviparous ancestors via the successive steps of egg retention, eggshell reduction, decreased reliance on the egg yolk for nourishing embryos, and, ultimately, placentation.[13,14] This transition (summarized in Table 24.1) is evident in the three groups that comprise extant mammals: **monotremes**, the most ancestral living mammals, which include the duck-billed platypus and several species of echidna, or spiny anteaters; **metatheria**, marsupials, such as opossums, kangaroos, and koalas; and **eutheria**, so-called placental mammals. Although eutheria are often referred to as "placental mammals", this is somewhat inaccurate, as marsupials and monotremes also have simple placentae, though compared with that of eutherian mammals those structures are very transient and rudimentary and do not make as significant a contribution to fetal nourishment.

Viviparity Altered Fetal-Maternal Endocrine Signaling

Therian mammals (metatherians and eutherians) express many uterine and placental genes that facilitate endocrine signaling between the fetus and mother in order to ensure that nutrients are directed to the fetus to promote its growth and development. Importantly, while this endocrine signaling in therians is complex, these processes arose through modifications to signaling processes that were already present in oviparous non-mammalian amniotes. The chorioallantoic membrane of birds and reptiles, for example, is an endocrine organ that produces a large diversity of hormones and other signaling factors, including progesterone.[15] Therefore, once the shell membrane was sufficiently reduced, hormones produced by the embryo would have the potential to impact the mother, and vice versa.[16,17] The synthesis of progesterone, a key hormonal regulator of reproduction in both mammals and reptiles, by the chorioallantoic placenta may have played a significant role in the evolution of viviparity, as egg retention in combination with a reduction in eggshell thickness could increase the transfer of progesterone from embryo to mother.[15]

Remarkably, most of the genes that became newly expressed by the uterus and placenta, particularly those responsive to progesterone signaling, did not evolve *de novo*, but appear to have been "repurposed" from their original roles in other organs such as the brain, blood, and gastrointestinal system. In essence, the fetal component of the placenta has "appropriated" genes expressed by these and other tissues in order to alter maternal physiology in favor of the fetus.[18] Strikingly, the expansion of the expressions of these genes into new organs appears to have been mediated by the actions of ancient transposable elements (transposons) that imparted novel responsivity to progesterone signaling.

Developments & Directions: Ancient "genomic parasites" imparted novel hormone responsivity and uterine cell functionality during the evolution of mammalian pregnancy

The transition from oviparity to viviparity and the development of associated novel structures, such as the uterus and placenta, was a giant physiological leap in mammalian evolution. A key component of establishing and maintaining pregnancy in many eutherian mammals is endometrial *decidualization*, or the differentiation of endometrial stromal fibroblasts into decidual stromal cells in response to progesterone or, in some species, to fetal signals[19] (refer to Chapter 22: Female Reproductive System for a review). The process of decidualization induces large-scale changes in endometrial gene expression leading to dramatic changes in endometrial physiology, such as vascular remodeling, the influx of immune cells, and transformation of the secretory uterine glands.[19–21] How could such broad and complex changes in gene expression that are required to mediate decidualization have evolved in a comparatively short period of time? In order to understand how mammals made this leap, Vincent Lynch and colleagues[18] used high-throughput sequencing to compare gene expression by the endometrium during pregnancy from 14 species, including eutherian mammals (dog, cow, horse, pig, armadillo, mouse, rhesus monkey, and human), a marsupial (the short-tailed opossum), a monotreme (platypus), as well as from the uteri/oviducts of several non-mammalian species including chicken, lizard, and frog. The researchers established a dataset of 19,641 protein-coding genes from the uteri of these species and used ancestral transcriptome reconstruction and functional genomics to determine the lineage in which each gene evolved endometrial expression. They showed that the evolution of pregnancy coincided with the recruitment of thousands of genes, including those involved in fetal-maternal communication and fetal immune tolerance. Interestingly, the expression of recruited genes was enriched in distinct anatomical systems, with most genes likely recruited from neural systems such as the brain (1,083 genes), the gut (717 genes), and the hemolymphoid system (661 genes) (Figure 24.3(A)).

Importantly, the researchers discovered that the DNA regulatory elements active in human decidual stromal cells are enriched with many distinct families of ancient mammalian *transposable elements* (TEs). Discovered by Barbara McClintock in the 1930s,[22,23] TEs (also known as "transposons" or "jumping genes") are DNA sequences that translocate from one location of the genome to another, leaving behind DNA "footprints" that can alter the expressions of affected or nearby genes. The researchers determined that many of these TEs inserted DNA binding sites for diverse transcription factors that mediate hormone responsiveness and endometrial cell identity (Figure 24.3(B)). Intriguingly, the TE sites are particularly enriched for DNA elements that associate with the progesterone receptor, and thus establish progesterone responsiveness to decidual stromal cells. According to the researchers, "TEs that amplified prior to the divergence of Eutherian mammals played a central role in recruiting these genes into endometrial expression and thereby in the origin of decidualization because they apparently deposited binding sites for master transcriptional regulators of endometrial stromal cell-type identity and progesterone responsiveness to numerous genes across the genome".[18] Therefore, this scenario enabled a single hormone, progesterone, to gain control over the expression of a large number of uterine genes during pregnancy. In an interview with *Science Life*, Vincent Lynch states, "Most remarkably, we found the genetic changes that likely underlie the evolution of pregnancy are linked to domesticated transposable elements that invaded the genome in early mammals. So I guess we owe the evolution of pregnancy to what are effectively genomic parasites".[24]

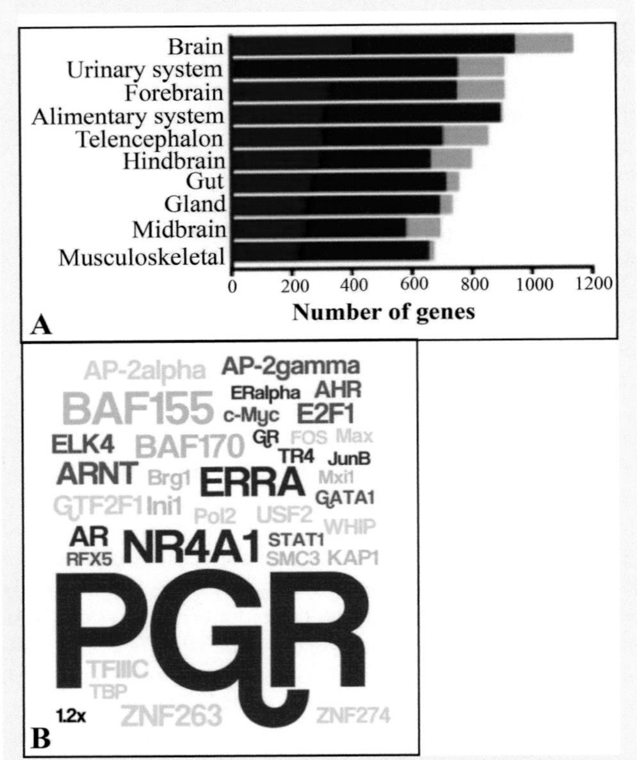

Figure 24.3 The evolution of pregnancy coincided with the recruitment of thousands of genes from distinct tissues, facilitated by transposable elements enriched for transcription factors that mediate hormone responsiveness. (A) The top ten organ systems from which the expression of recruited genes is enriched. The stacked bar chart shows the number of genes recruited into endometrial expression in the mammalian (light blue), therian (blue), and eutherian (red) stem lineages. **(B)** A "word cloud" portrayal of transcription factor binding sites enriched in ancient mammalian transposable elements. Colors indicate transcription factors that mediate hormone responses (purple), remodel chromatin (light purple), have known functions in endometrial cells or that mediate immune responses (green), or with general regulatory functions (light green). Data depict transcription factors enriched for ≥ 1.2-fold. Note that transcription factor binding sites for the progesterone receptor (PGR) are enriched over ten-fold.

Not all genes of pregnancy were repurposed from other functions, and some novel peptide hormones with roles in mammalian pregnancy arose singly via the processes of gene duplication followed by mutation. For example, in primates, *chorionic gonadotropin* (which is expressed by the embryo and placenta) arose from the duplication of the gene for *luteinizing hormone* (LH), which is expressed by the anterior pituitary gland.[25] Similarly in primates, *placental lactogen* (a hormone that allows fetal control over the development of lactation) arose from duplication and mutation of the *growth hormone* (GH) gene.[26] Another example, this time in ruminants (e.g. cattle, sheep, antelopes, and deer), is *interferon-τ*. This gene, which arose from the duplication of a type I interferon gene (originally with an antiviral function), has become the pregnancy recognition factor in ruminants.[27]

Foundations & Fundamentals: The Early Stages of Human Pregnancy

Fertilization and early embryogenesis

Fertilization typically occurs in the Fallopian tube within about 24 hours following ovulation (Figure 24.4). The initial stages of development, from *zygote* (fertilized ovum) to *morula* (Latin for "berry"; a mass of 12–16 cells), take place as the embryo is transported along the

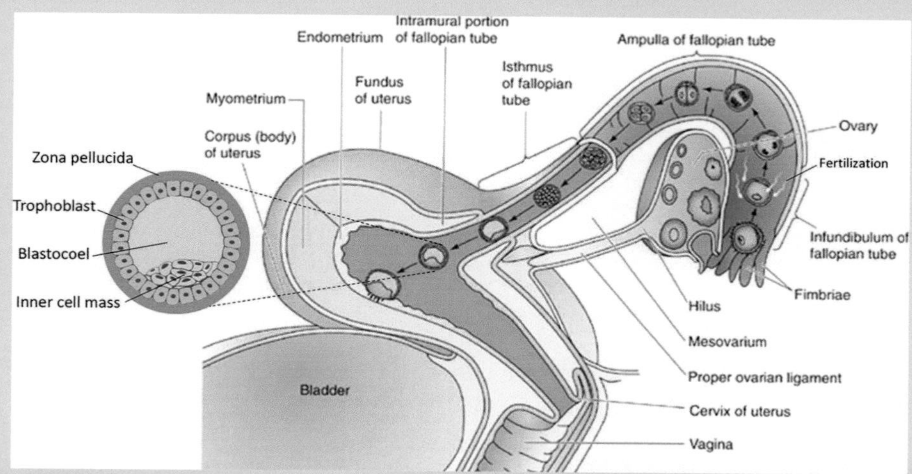

Figure 24.4 Human embryogenesis from fertilization to implantation. After ovulation, eggs are fertilized in the Fallopian tube to form the zygote. Over the first several days post-fertilization, the embryo divides mitotically to form a compact structure called the morula. Upon reaching the uterus, a fluid-filled cavity develops inside the embryo forming the blastocyst, which eventually implants into the uterine wall.

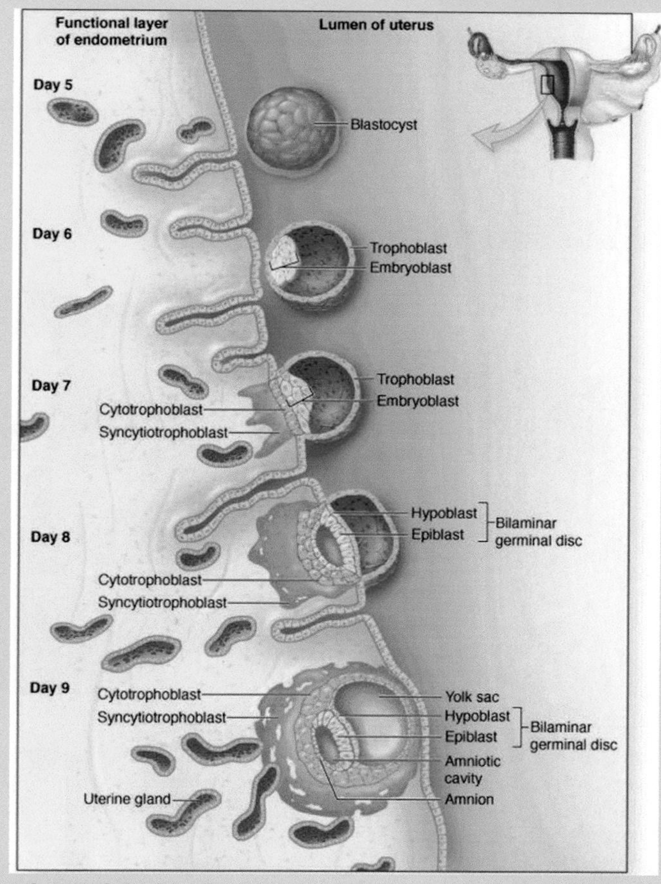

Figure 24.5 Embryonic implantation and decidualization.

Fallopian tube. During its journey to the uterus, which lasts several days, the embryo is enclosed in a non-adhesive protective coating called the *zona pellucida* (Figure 24.1). The transition from morula to *blastocyst* is characterized by the appearance of a fluid-filled inner cavity within the cell mass. Blastocyst development coincides with the differentiation of the *trophoblast*, which consists of the surface cells that will give rise to extraembryonic structures, including the placenta, and the *inner cell mass*, which gives rise to the embryo itself. Within three days of entering the uterus, the trophoblast secretes proteases that digest the zona pellucida, enabling the "hatched blastocyst" to both adhere to and implant into the receptive uterine endometrium.

Implantation

Implantation is the process by which the embryo first attaches to the endometrial surface of the uterus and then invades through the epithelium, ultimately accessing maternal circulation to form the placenta (Figure 24.5). Uterine stromal cells surrounding the implanting blastocyst differentiate into a specialized cell type that is rich in glycogen and lipids, called *decidual cells*, via a process known as *decidualization*. Decidualization is promoted by *progesterone*, whose primary source during the first 6–8 weeks of pregnancy is the corpus luteum, after which the placenta becomes the most important source of the steroid hormone.

Placentation

The *placenta* is an unusual organ in that it consists of tissues derived from two organisms: a fetal extraembryonic membrane (chorion) and the maternal decidua. When fully formed, the placenta (Latin for "flat cake") is a disc-shaped organ that averages a 22 cm diameter and a weight of about 500 grams (Figure 24.6). The first new organ to form during embryogenesis, the placenta functions as a vascular interface between the maternal and fetal circulatory systems and is required for the exchange of hormones, nutrients, respiratory gases, and waste products between the fetus and mother. In addition, this organ is involved in fetal immune protection and is a critical source of pregnancy-associated hormones.

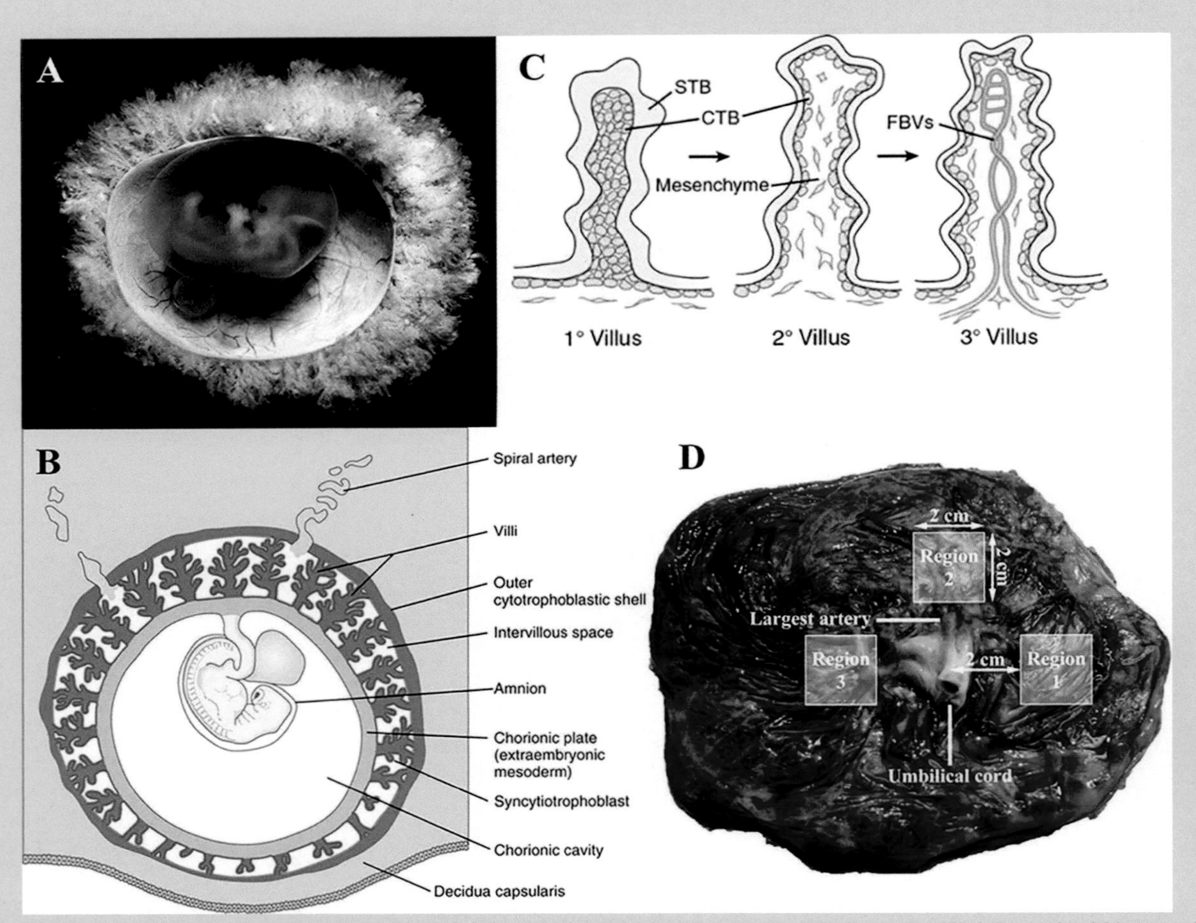

Figure 24.6 View of human fetuses, surrounding extraembryonic membranes, and placenta. (A) A 7-week-old human fetus surrounded by its amnion membrane. **(B)** A 5-week-old fetus is attached to the placenta. On the fetal side of the placenta, the placental villi converge in the center to form the umbilical artery and veins. **(C)** Development of primary (1°), secondary (2°), and tertiary (3°) villi. CTB, cytotrophoblast; FBVs, fetal blood vessels; STB, syncytiotrophoblast. **(D)** A human placenta minutes after birth. The side shown faces the baby with the umbilical cord top right. The unseen side connects to the uterine wall. The white fringe surrounding the bottom is the remnants of the amniotic sac.

SUMMARY AND SYNTHESIS QUESTIONS

1. Whereas the embryos of anamniotic vertebrates survive well in an aqueous environment, they cannot survive incubation on land. Why not?
2. In mammals viviparity arose from oviparous ancestors via the successive steps of egg retention, eggshell reduction, decreased reliance on the egg yolk for nourishing embryos, and, ultimately, placentation. From an endocrine perspective, what is the significance of eggshell reduction?
3. In mammals, viviparity arose from oviparous ancestors 191–124 million years ago. This is a very brief period of time for the evolution of the uterus, a particularly complex organ that expresses hundreds of genes in response to circulating hormones. How are these complex uterine gene expression pathways thought to have evolved in such a short period of time?
4. One of the earliest events to take place during human embryogenesis is called "hatching", a process by which proteolytic factors in the free-floating blastocyst cause the zona pellucida surrounding it to degenerate. Why is a successful completion of this event so crucial?

Hormones of the Human Fetal-Placental-Maternal Unit

LEARNING OBJECTIVE Outline the major fetal-placental-maternal interactions in the production of sex steroids and other hormones.

KEY CONCEPTS:

- Physiologically, the pregnant woman consists of three functionally distinct but interacting compartments that together constitute the fetal-placental-maternal unit.
- The placenta, which secretes a greater diversity and quantity of hormones than any other single endocrine tissue and becomes the most important source of estrogens and progesterone during pregnancy, relies on assistance from both the fetus and the mother for hormone synthesis.
- Much of pregnancy is characterized by the gradual development of maternal insulin resistance, promoting a "diabetogenic state" that facilitates the transfer of metabolic fuel to the fetus for its growth and development.

From an endocrine perspective, the pregnant human consists of three functionally distinct but interacting compartments that together constitute the **fetal-placental-maternal unit.** The concept was initially described in the 1960s by Egon Diczfalusy and his colleagues[28] as a way to illustrate how the placenta, which lacks the complete metabolic pathways to synthesize progesterone and estrogen sex steroids, must rely on assistance from both the fetus and the mother to compensate for the deficiencies of the placental enzymes. The placenta, which is supplied with precursor substrates from the fetal and maternal compartments, generates high levels of steroid hormones and in turn releases these products into maternal and fetal circulation. In addition to sex steroid hormones, the placenta also synthesizes other steroid hormones and diverse peptide hormones (Table 24.2) that act upon the fetal and maternal compartments. By the end of pregnancy, each compartment of the fetal-placental-maternal unit is exposed to a broad array of prostaglandins, steroid and peptide hormones secreted from one or more compartments.

Table 24.2 A Subset of Endocrine and Paracrine Hormones and Related Factors Expressed in Human Placenta

Steroid Hormones	Pituitary-Like Hormones	Hypothalamic-Like Hormones	Neuropeptides	Placental Cytokines	Eicosanoids
Estriol	hCG	CRH	Serotonin	TNF-α	Prostaglandins
Estradiol	hCS	Urocortins	Dynorphin	LIF	Leukotrienes
Estrone Esterol Progesterone Alloprogesterone Pregnenolone5 α-DHP Cortisone	hGH-V IGF-I IGF-II Activin Inhibin Follistatin β-Endorphin Oxytocin ACTH MSH Relaxin	GnRH-I, GnRH-II GnRH Somatostatin TRH PRH	Met-enkephalin ANP Leptin Ghrelin Neurotensin Substance P Melatonin Cholecystokinin Galanin Neuropeptide Y Endothelin VIP	Interferon-α Interferon-β Interferon-γ IL-1 IL-2 IL-6 IL-8 IL-10	Prostacyclin Thromboxane

Note: ACTH, adrenocorticotropic hormone; ANP, atrial natriuretic peptide; CRH, corticotropin-releasing hormone; 5α-DHP, 5α-dihydroprogesterone; GHRH, growth hormone-releasing hormone; GnRH, gonadotropin-releasing hormone; hCG, human chorionic gonadotropin; hCS, human chorionic somatomammotropin; hGH-V, human growth hormone variant; IGF, insulin-like growth factor; LIF, leukemia inhibitory factor; MSH, melanocyte-stimulating hormone; PRH, prolactin-releasing hormone; TNFα, tumor necrosis factor α; TRH, thyrotropin-releasing hormone; VIP, vasoactive intestinal peptide.

Source: After Polin, R.A. and Abman, S.H., 2011. *Fetal and neonatal physiology E-book.* Elsevier Health Sciences

Fetal-Placental-Maternal Interactions in the Production of Sex Steroids

During pregnancy, the levels of maternal progesterone and estrogens increase to concentrations significantly higher than those during a typical menstrual cycle. For the first 6–8 weeks of gestation, the corpus luteum is the primary source of progesterone and estrogens. However, by the eighth week of pregnancy, the placenta supplements sex steroid synthesis and becomes the most important generator of these hormones. This transition is called the *luteo-placental shift* (Figure 24.7).

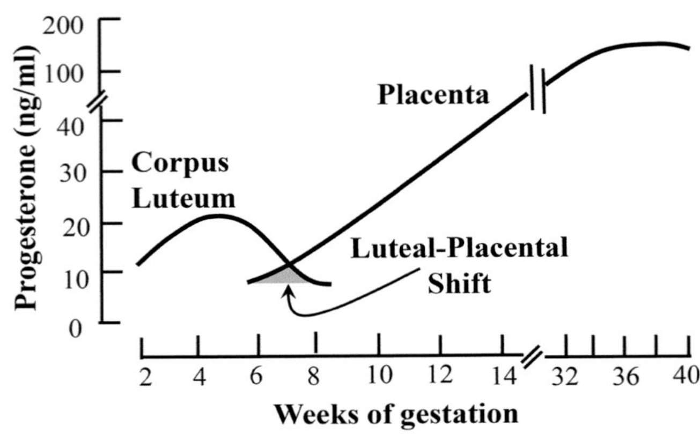

Figure 24.7 The luteo-placental shift. A shift in progesterone production from the corpus luteum to the placenta occurs at approximately the seventh to ninth week of gestation. The small, shaded area represents the estimated duration of this functional transition.

Progesterone is essential for the maintenance of pregnancy, with functions that include the inhibition of uterine smooth muscle contractility, the suppression of prostaglandin synthesis, and suppression of the maternal hypothalamic-pituitary-ovary axis, all processes that mediate parturition. Progesterone also upregulates genes whose products protect the fetus from the mother's immune system. Despite the large amount of progesterone secretion by the placenta, this organ has a limited ability to synthesize cholesterol, the principal substrate for progesterone biosynthesis. As such, the primary source of cholesterol is maternal in the form of **low-density lipoprotein** (LDL). Thus, the mother provides cholesterol to the placenta, which in turn provides maternal circulation with progesterone (summarized in Figure 24.8).

The estrogens estradiol, estriol, and estrone also play critical roles in the maintenance of pregnancy, stimulating growth of the mammary glands to prepare them for milk production, promoting relaxation of the pelvic ligaments, inducing uterine growth and blood flow, and promoting progesterone synthesis. However, the placenta lacks three key enzymes required for the synthesis of estrogens. These are (1) *17α-hydroxylase* and (2) *17,20-desmolase*, which are both are needed to convert progesterone into the androgen precursor necessary to synthesize estradiol and estrone, and (3) *16α-hydroxylase*, which is needed to synthesize the androgenic precursor to estriol. Therefore, the placenta must rely upon the acquisition of the preformed androgen *dehydroepiandrosterone sulfate* (DHEA-S) from the fetal and maternal adrenal gland

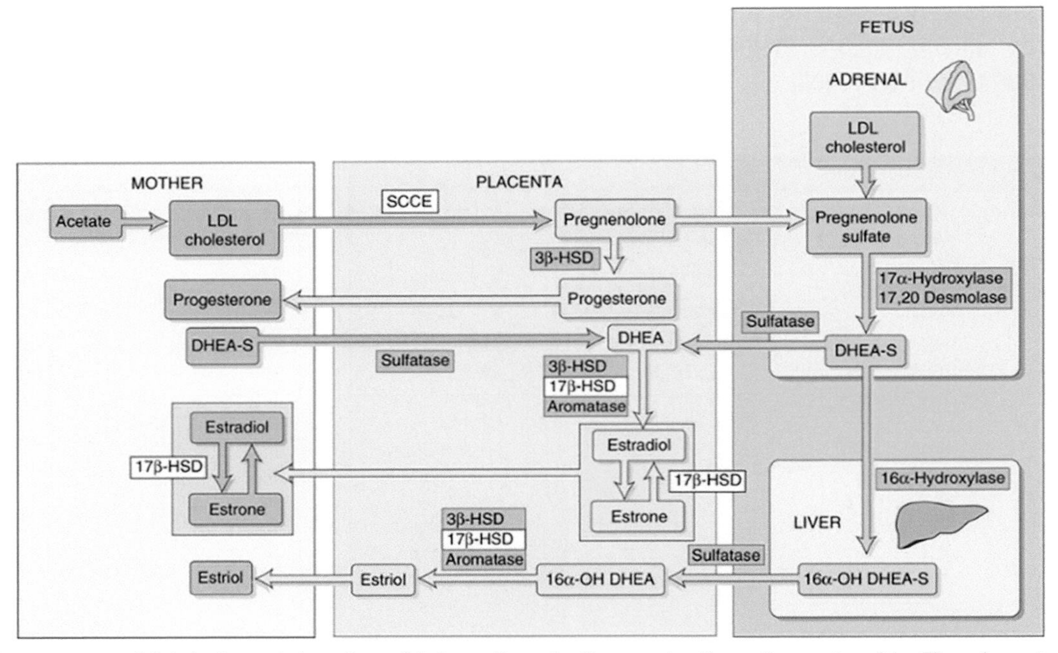

Figure 24.8 A summary of fetal-placental-maternal interactions in the production of sex steroids. The placenta, which lacks the complete metabolic pathways to synthesize progesterone and estrogen sex steroids, must rely on assistance from both the fetus and the mother to compensate for the deficiencies of the placental enzymes. The placenta, which is supplied with precursor substrates from the fetal and maternal compartments, generates high levels of steroid hormones and in turn releases these products into maternal and fetal circulation. See text for details.

compartments to synthesize estradiol and estrone, and the androgen *16α-OH DHEA-S* from the fetal liver for estriol synthesis (Figure 24.8). The conjugation of these androgenic intermediates to sulfate by the fetus appears to reduce their androgenic activity, inhibiting the masculinization of female fetuses. The removal of the sulfates by *sulfatases* present in the placenta permits the conversion of these steroids by *17β-HSD, 3β-HSD*, and *CYP19 aromatase* into estrogens that are exported to the maternal compartment. The fetal and neonatal adrenal cortex of humans contains a large transient fourth zone (the **fetal zone**) that produces high quantities of DHEA-S (described in Chapter 13: The Multifaceted Adrenal Gland; refer to Figure 13.6). At birth the neonatal adrenal cortex is as large as that from an adult, but the organ shrinks as the fetal zone recedes.

Hormones of the Placental Compartment

Remarkably, the placenta secretes a greater diversity and quantity of hormones than any other single endocrine organ, with steroid hormones secreted at the massive rate of 0.5 g/day and peptide hormones generated at over 1 g/day at term.[29] In general, the role of placental hormones is to modulate maternal metabolism in order to increase the concentrations of maternal blood glucose and other nutrients, maximize their transfer to promote fetal growth and development, and balance fetal growth and development with maternal homeostasis.[30] Importantly, because the placenta prevents most maternal hormones from entering either the placental or the fetal compartment in their active forms, the fetal-placental endocrine system in effect develops and functions independently from that of the mother.[31] In particular, placental cells express *11ß-hydroxysteroid dehydrogenase* (11ß-HSD), which inactivates most maternal cortisol into cortisone, as well as *17ß-hydroxysteroid dehydrogenase* (17ß-HSD), which converts most maternal estradiol into the less active estrone *type III deiodinase* (DIII), which inactivates excess thyroxine and triiodothyronine, and *monoamine oxidase* (MAO) and other enzymes that catabolize catecholamines.

Placental hormones can be placed into two categories: steroid hormones and peptide hormones. The biosynthesis and functions of the sex steroids progesterone and estrogens was described earlier. Intriguingly, the synthesis of diverse placental peptide hormones appears to take place via interactions between the placental *cytotrophoblast* and *syncytiotrophoblast* cells (see their locations on Figure 24.7) in a manner that is reminiscent of the hypothalamic-pituitary system. Specifically, the cytotrophoblasts secrete analogs of many hypothalamic-releasing and inhibiting hormones, and the syncytiotrophoblasts respond to these tropic hormones by releasing placental analogs of diverse pituitary adenohypophysial hormones into the maternal compartment[29,31] (Table 24.2). A subset of these peptide hormones are described next.

Human Chorionic Gonadotropin

Human chorionic gonadotropin (hCG), a glycoprotein hormone closely related to the two-subunit pituitary glycoprotein hormones (LH, FSH, and TSH), is one of the earliest hormones secreted in pregnancy and can be detected in maternal serum as early as 6–8 days following conception where it is produced by the embryonic trophoblast. As such, hCG is used frequently as a test for pregnancy detection. Plasma levels of hCG increase rapidly during pregnancy, peaking around the eighth week of gestation (Figure 24.9). By week 13, the level drops dramatically, reaching a low steady state. By this time, the placenta now produces sufficient progesterone to support pregnancy. hCG production appears to be regulated locally by a placental GnRH analog produced by the cytotrophoblasts, which stimulates hCG release from the syncytiotrophoblast.[32] The primary role of hCG, which is similar in structure to LH, is to maintain progesterone synthesis by the corpus luteum until the luteo-placental shift takes place. This is accomplished by the binding of hCG to LH receptors on the corpus luteum.

Human Placental Lactogen and Placental Growth Hormone

Human placental lactogen (hPL; also called *human chorionic sommatomammotropin*, hCS) and *human placental growth hormone* (hGH) are members of a family of closely related peptides. In contrast to hCG, hPL levels increase with advancing gestational age and plateau at term (Figure 24.9). The primary function of hPL is to augment the supply of glucose and other nutrients to the fetus by increasing maternal *insulin-like growth factor I* (IGF-I) levels and altering the maternal secretion of insulin as the pregnancy approaches term.[31] hPL essentially opposes maternal insulin action (promotes insulin resistance), induces maternal glucose intolerance, and promotes maternal lipolysis and proteolysis, favoring the transport

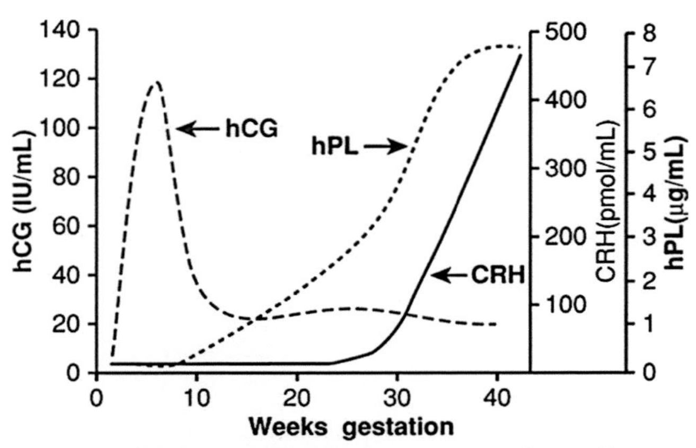

Figure 24.9 Distinct profiles for the concentrations of human chorionic gonadotropin (hCG), human placental lactogen (hPL), and corticotropin-releasing hormone (CRH) in serum of women throughout normal pregnancy.

of nutrients to the fetus.[33] In addition to its metabolic effects, hPL also has a poorly understood lactogenic activity. Similar to hPL, placental hGH also regulates maternal IGF-I levels, stimulating gluconeogenesis and lipolysis in the maternal compartment.[31]

Corticotropin-Releasing Hormone and Adrenocorticotropic Hormone

The placental *corticotropin-releasing hormone* (CRH) analog is structurally similar to the hypothalamic peptide,[34,35] and its mRNA is produced by both cytotrophoblasts and syncytiotrophoblasts.[36] Placental CRH has been shown to stimulate *adrenocorticotropic hormone* (ACTH) production.[37] Since fetal CRH can stimulate the production of estrogen precursor by the fetal adrenal gland, as well as the release of prostaglandins involved in parturition and the late third trimester surge of fetal glucocorticoids that help promote fetal maturation, CRH is often considered a "placental clock" that contributes to the timing of parturition (discussed further below in Section 24.3.1 The Timing of Birth).

Hormones of the Fetal Compartment

The fetal endocrine system is the first physiological system to begin to develop, and it functions throughout pregnancy.[42] As described previously, its function is partially dependent on the secretion of chemical precursor substrates from the placental and/or maternal compartments transported across the fetal-maternal interface. However, as the fetus develops, its endocrine system gradually matures, becoming more independent, preparing the fetus for life outside the uterus.

Hypothalamus and Pituitary

The anterior pituitary cells that develop from Rathke's pouch are capable of secreting luteinizing hormone (LH), follicle-stimulating hormone (FSH), adrenocorticotropic hormone (ACTH), and growth hormone (GH) as early as 7 weeks of gestation. However, none of the pituitary hormones are released into fetal circulation at high levels until week 20 during the second trimester (Figure 24.10), coinciding with the maturation and development of the hypophyseal portal system.

Adrenal Gland

Described in detail in Chapter 13: The Multifaceted Adrenal Gland (refer to Figures 13.5 and 13.6), the fetal adrenal gland contains a unique zone, the fetal zone, which

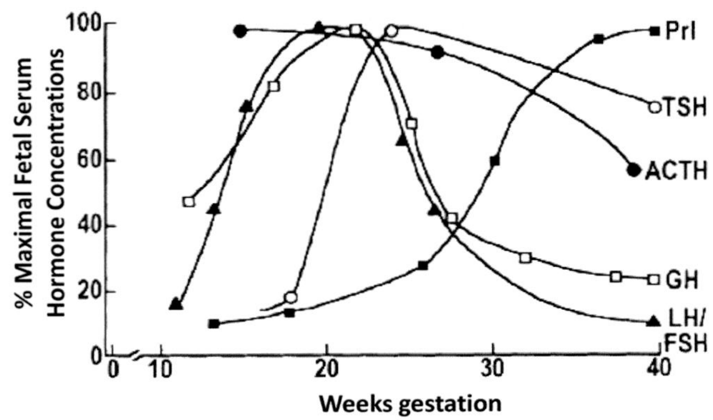

Figure 24.10 Fetal blood serum pituitary hormone levels throughout pregnancy. Prl, prolactin; TSH, thyroid-stimulating hormone; ACTH, adrenocorticotropic hormone; GH, growth hormone; LH, luteinizing hormone; FSH, follicle-stimulating hormone.

Developments & Directions: Placental genes in marsupial milk

Therian mammals are divided into two groups: (1) the "eutheria" constitute the majority of living mammals and possess complex placentas composed of many types of cells and tissues, and (2) the "metatheria" (marsupials, such as kangaroos, koalas, opossums), which possess a simpler placenta composed of only a few layers of cells. Due to their simpler placenta, marsupials have much shorter gestational periods and give birth to relatively underdeveloped young compared with eutherians. Because marsupial young must continue their development inside of their mother's pouch for an extended time following birth, compared with eutherians they are fed a more complex range of milk whose carbohydrate, lipid, and protein components change dynamically over time to support the different stages of their fetal development. Considering that marsupial milk supports fetal growth and development, a role that in eutherians is primarily fulfilled by the placenta, it has been hypothesized that the elaborate and extended lactation periods of marsupials evolved as an alternative strategy to the complex placentation used by eutherians.[38–40]

To better understand the evolution of pregnancy and lactation in marsupials, Guernsey and colleagues[41] studied the tammar wallaby (*Macropus eugenii*), a small member of the kangaroo family. Compared with the mouse, which has a 20-day pregnancy followed by a 20- to 24-day period of lactation, the tammar wallaby has a 26.5-day pregnancy followed by a 300- to 350-day period of lactation, relying much more on lactation for offspring development. The researchers analyzed transcriptome data from placental and mammary gland tissues of the tammar wallaby and compared them to those of the eutherian mouse, looking for the expression of shared or missing transcripts. In addition to showing that many genes expressed by the eutherian placenta are also expressed in the less complex marsupial placenta, they also made the remarkable observation that genes critical for eutherian placental function are also expressed by the tammar mammary gland (Figure 24.11). The genes shared between the tammar mammary gland and mouse placenta are enriched for processes involved in embryonic morphogenesis, immune function, and nutrition. These observations support the fascinating notion that reproductive genes that allow for efficient exchange between mother and offspring have been respectively co-opted in both eutherian placentation and marsupial lactation to facilitate alternative strategies for proper offspring growth and development.

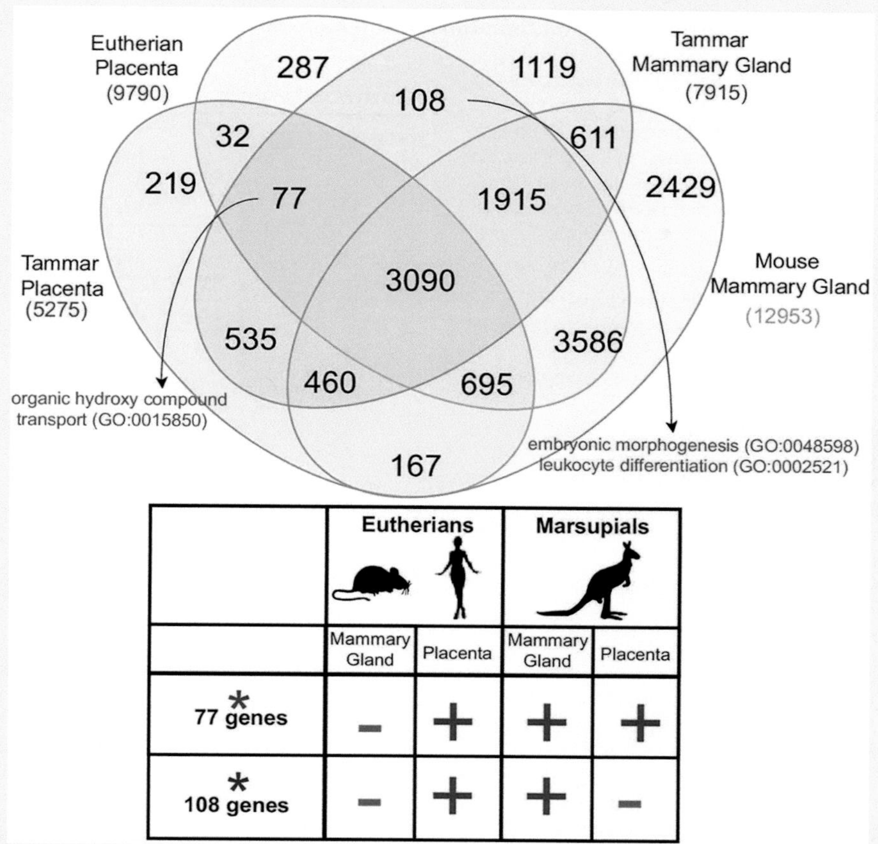

Figure 24.11 Shuffling of reproduction genes between lactation and placentation. Venn diagram (left) comparing the number of genes expressed in the lactating tammar and mouse mammary glands with the tammar and eutherian placenta. Key ontology of genes in overlapping categories is highlighted. The table depicts key categories of genes from the Venn diagram and whether they are expressed in the placenta and mammary gland tissues of the tammar and eutherians. If the gene class is present in the tissue of a given lineage, it is given a green "+", if it is absent, it is given a red "–".

by mid-pregnancy exceeds the size of the fetal kidneys themselves, and at term the adrenals are as large as those of adults. This zone regresses rapidly over the first several weeks following birth, and by one year it has completely regressed. The fetal zone secretes primarily large quantities of DHEA-S, which function as substrates for placental estrogen biosynthesis. Pituitary ACTH appears to be the primary tropic hormone of the fetal adrenal gland, promoting the steroidogenesis of DHEA-S by the fetal zone of the adrenal cortex and inducing glucocorticoid synthesis by the zona fasciculata tissue of the adrenal cortex.[43]

Glucocorticoids, whose circulating levels rise progressively throughout pregnancy and peak at term, are essential for the transition from fetal to neonatal life. They not only induce the production of pulmonary surfactant,[44] but also are critical for modulating development of the mammalian central nervous system, retina, skin, gastrointestinal tract, kidney, heart, and lungs.[45–47] As such, in cases of preterm delivery, synthetic glucocorticoids are routinely used to accelerate organ maturation and prevent suffocation and respiratory distress.

Thyroid Gland

Prior to weeks 18–20 of gestation when the thyroid gland is capable of thyroid hormone synthesis, the fetus relies on maternally derived thyroid hormones.[48] Like glucocorticoids, thyroid hormones are critical for the normal development of the brain, gut, lungs, heart, and musculoskeletal and other organ systems. As such, combined maternal and fetal hypothyroidism, typically caused by a deficiency of iodine in the diet, may lead to profound developmental disorders.

Hormones and Physiology of the Maternal Compartment

Consider for a moment some of the profound physiological and morphological changes that take place during human pregnancy in almost every organ system. These include a 30%–50% increase in blood volume, up to a 50% increase in cardiac output, a 33% increase in oxygen consumption, and a weight gain of 10–16 kg (22–36 pounds) consisting mostly of water, fetal-placental tissue, uterine

Developments & Directions: Human birth and frog metamorphosis are more similar than you might think

Consider some of the remarkable similarities that exist between the human *perinatal period,* or the time ranging from days to weeks before and after birth, and frog metamorphosis, the developmental transition from a tadpole to a frog. In each case the organism abruptly switches from an aquatic environment (amniotic fluid or water) to an air-breathing terrestrial one. Shared physiological changes that accompany this transition include (refer to Figure 24.12):

1. Lung growth and development to facilitate air breathing
2. A switch from the fetal or larval type of hemoglobin to the adult type
3. Changes in osmotic environment that necessitate remodeling of the kidney
4. Upregulation of liver urea cycle enzymes
5. Increased plasma protein synthesis by the liver
6. Changes in feeding and nutrition that require intestinal and pancreatic remodeling to the adult form
7. Transition of the single layer of fetal/larval skin cells to the stratified/keratinized adult form
8. Significant development and restructuring of the nervous system
9. Limb elongation and bone ossification

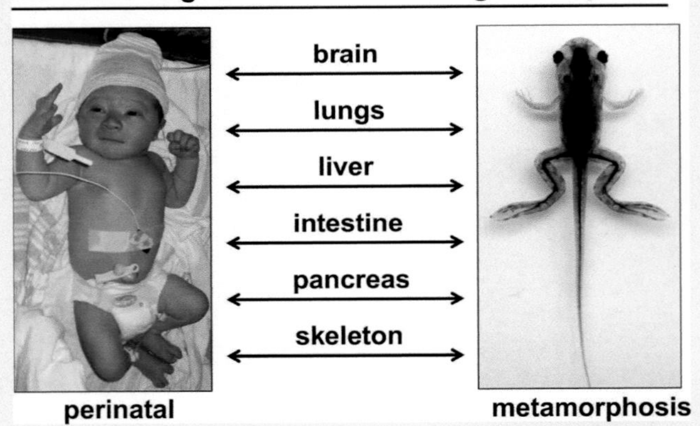

Figure 24.12 Target organs in common requiring glucocorticoid and thyroid hormone signaling for normal ontogeny in human perinatal development and frog metamorphosis.

Even more intriguing is that all of these post-embryonic developmental events are regulated by the same hormones in both humans and amphibians, with thyroid hormone (TH) and glucocorticoids peaking at both birth[49–51] and metamorphosis[52,53] (Figure 24.13), with each playing particularly prominent roles in organ development. Furthermore, frogs and humans have conserved TH and glucocorticoid receptors and pathways for regulating gene expression and comparable roles in many of the same target tissues.[54] Indeed, the study of the endocrine control of frog metamorphosis has provided key insights into the regulation of TH and glucocorticoid secretion by the hypothalamus and pituitary gland,[55,56] the influence of TH on gene expression,[57,58] and identification and characterization of TH and glucocorticoid response genes.[59–62] Considering the conservation of TH and glucocorticoid functions between frogs and humans, information gained from the frog model will continue to provide fundamental insights that further our understanding of endocrine diseases and disruptions that influence the perinatal period in humans.[54]

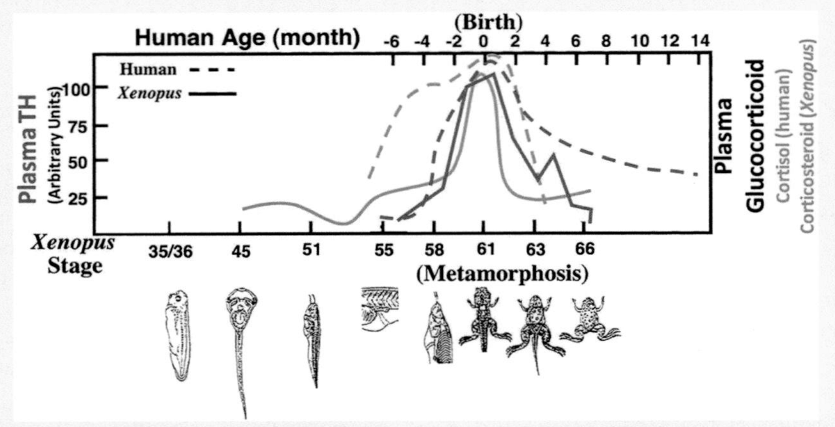

Figure 24.13 Regulation of plasma TH levels during human and ***Xenopus laevis*** **development.** Note that the peak levels of TH in both human and frog correspond to a period of dramatic tissue remodeling (birth in human and metamorphosis in frog), as well as organogenesis. The changes occurring in human and frog at this time bear considerable similarities, including (1) the initiation of air breathing through developing lungs and (2) the switch from an obligate aquatic habit (i.e. intrauterus life for human and free-aquatic life for tadpoles) to an at least optional terrestrial habit, etc.

and breast tissue, and adipose tissue.[63] These adaptations to the pregnant state, which begin just after conception and develop through delivery, are initiated principally by the production of large quantities of the aforementioned hormones from the fetal and placental compartments. These hormones in turn promote alterations in not only the production of several key maternal hormones, but also their transportation through altered levels of blood hormone-binding protein synthesis by the liver, and their clearance due to increased rates of catabolism by the placenta, increased glomerular filtration, and reduced breakdown by hepatic enzymes.[1] The general purpose of these maternal alterations is, of course, to mobilize nutrients and other resources that promote the growth and development of the fetus, to prepare for parturition and lactation, and also to accommodate the changing physiological needs of the mother. Remarkably, over a period of just weeks following parturition, most of these dramatic physiological changes almost completely revert back to the nonpregnant state. This section describes some changes in maternal hormone systems during pregnancy.

Pituitary Gland

During pregnancy, the maternal pituitary gland increases in size by approximately three-fold,[64] primarily due to hyperplasia of the lactotropes in the adenohypophysis. This occurs in response to rising estradiol levels, with lactotropes occupying 20% of the adenohypophysis in nonpregnant women and increasing to 60% by the third trimester. This increase in lactotropes gradually disappears within months after delivery due to a decrease in circulating estradiol levels. The decline in serum prolactin levels is slower in nursing women due to phases of intermittent hyperprolactinemia caused by suckling. Interestingly, the process of reduction is often incomplete, and the pituitaries of multiparous (multiple pregnancies) women are typically larger than those of nulliparous (no pregnancy) women.[65] Maternal FSH and LH decrease to undetectable levels during pregnancy due to feedback inhibition caused by elevated levels of progesterone, estrogens, and inhibin produced by the placenta.[66] The production of oxytocin by the neurohypophysis rises continuously across gestation, peaking around the onset of labor.[67] As will be discussed in detail later, *oxytocin* promotes smooth muscle contraction and plays key roles in both parturition and lactation.

Osmoregulatory Hormones

A key physiological feature of pregnancy is a decrease in mean arterial pressure caused, in part, by increased levels of estrogens, progesterone, and the circulating corpus luteal and placental hormone *relaxin*, which promotes vasodilation.[68–71] This reduced blood pressure occurs in spite of an increased induction of all components of the *renin-angiotensin-aldosterone system* (RAAS)[72,73] that promotes the reabsorption of salts and water by the kidney in an attempt to preserve intravascular volume, ultimately doubling circulating aldosterone levels by the third trimester.[74] Low mean arterial pressure during pregnancy also takes place despite a resetting of hypothalamic osmoreceptors for AVP/ADH sensitivity that lowers the threshold for thirst and promotes drinking and increased blood volume[75] (Figure 24.14 A). Together with increased RAAS activity, these changes lead to a hypervolemic state, where maternal blood plasma volume increases by 50%–60% by the end of gestation, or an increase in total body water of 6.5 to 8.5 L. The increased plasma volume during pregnancy plays a key role in maintaining circulating blood volume, blood pressure, and utero-placental perfusion.[74] This increased water retention also contributes to maternal weight gain, elevated

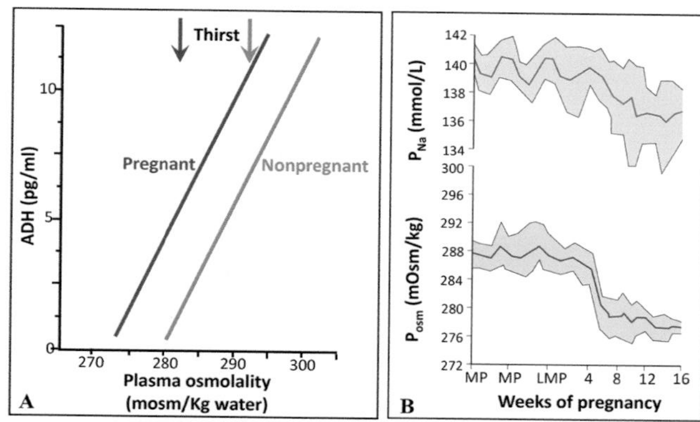

Figure 24.14 Changes in osmoregulatory parameters during human gestation. (A) Relationship between plasma ADH and osmolality in eight women before pregnancy and at the end of the third month of gestation. Arrows denote plasma osmolality at which a desire to drink (thirst) was experienced. **(B)** Plasma sodium (P_{Na}) and plasma osmolality (P_{osm}) during gestation (n = 9). LMP, last menstrual period; MP, menstrual period.

cardiac output, and physiological anemia of pregnancy caused by hemodilution (reduced hematocrit due to red blood cell dilution). Importantly, because the rate of water retention exceeds that of salt retention, pregnancy is also accompanied by low blood plasma osmotic pressure. Thus, hemodynamically pregnancy is characterized by a state of hyponatremic hypervolemia (Figure 24.14 B).

Thyroid Hormones

The maternal thyroid gland increases in size by about 18% during pregnancy, likely in response to the thyrotropic effects of hCG, which promotes increased follicular size, colloid content, and glandular blood volume.[76] As a member of the closely related family of glycoprotein hormones, at elevated levels hCG is able to stimulate the TSH receptor. Although total circulating levels of thyroxine (T_4) and triiodothyronine (T_3) also increase and may even double during pregnancy, this does not result in elevated free thyroid hormone or hyperthyroidism since serum *thyroid hormone-binding globulin* (TBG) increases simultaneously in response to exposure of the liver to estrogens (Figure 24.15). Similar increases in TBG and altered blood thyroid hormone profiles are also seen in women who use estrogen-based contraceptives.[77] As mentioned previously, the maintenance of normal levels of circulating maternal thyroid hormone is critical for normal fetal development, as the fetal brain and other organs requires thyroid hormone for development even prior to the onset of fetal thyroid hormone synthesis that takes place around 20 weeks of gestation.

Pancreas and Glucose Metabolism

Early pregnancy is characterized by the proliferation of maternal insulin-secreting pancreatic beta cells, resulting in increased insulin secretion and sensitivity.[78] Thus, early pregnancy is accompanied by an anabolic state during which maternal nutrients are stored. However, this early period is

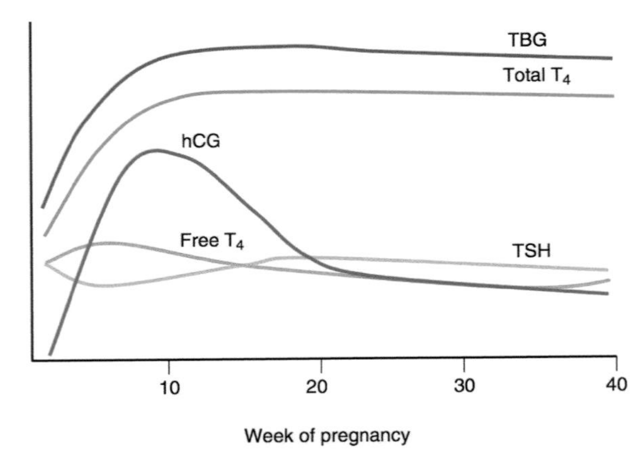

Figure 24.15 Changes in hCG and maternal thyroid parameters throughout pregnancy. hCG (human chorionic gonadotropin), T_4 (thyroxine), TBG (thyroxine-binding globulin, TSH (thyroid-stimulating hormone).

followed by the gradual development of **insulin resistance** throughout the rest of pregnancy. As such, pregnancy has been described as a progressively "diabetogenic state" manifested by maternal postprandial (following consumption of a meal) hyperglycemia and hyperinsulinemia that facilitates the transfer of metabolic fuel to the fetus while simultaneously maintaining necessary maternal nutrition.[79,80] By the third trimester of pregnancy, insulin sensitivity may decline by 50% in response to several endocrine factors, including increases in circulating placental lactogen, estrogens, and progesterone, as well as increases in maternal adiposity.[81,82] In contrast to the postprandial state, the maternal fasting state is characterized by hypoglycemia caused by increased storage of glycogen, decreased glucose production by the liver, increased peripheral glucose consumption, and uptake of glucose by the fetus.[83] Importantly, the diabetogenic effects of pregnancy are increased by maternal obesity, and if a woman's endocrine pancreatic function is impaired and she cannot overcome the insulin resistance of pregnancy, then a condition called *gestational diabetes* can develop.

Clinical Considerations: Gestational diabetes mellitus

Recall from Chapter 15: Regulation and Dysregulation of Energy Homeostasis that Type 2 diabetes mellitus (T2DM), which accounts for 90%–95% of those with diabetes, is a heterogeneous collection of progressive metabolic disorders characterized by hyperglycemia, insulin resistance in the main target organs (liver, skeletal muscle, adipose tissue), excessive hepatic glucose synthesis, abnormal fat metabolism, and impaired insulin secretion.[84] *Gestational diabetes mellitus* (GDM) occurs when pregnancy aggravates a preexisting maternal T2DM condition that manifests for the first time during pregnancy.[85] As previously described, in normal pregnancy, maternal tissues become progressively insensitive to insulin. However, in women with GDM this physiological insulin resistance takes place on a preexisting background of chronic insulin resistance, resulting in an even greater insulin resistance than in normal pregnant women (Figure 24.16). GDM develops when the pregnant woman is unable to produce enough of an insulin response to compensate for the additional normal insulin resistance of pregnancy. This culminates in maternal hyperglycemia, and fetal hyperinsulinemia occurs to counter the transfer of excess maternal glucose to the fetus. High fetal insulin levels subsequently stimulate fetal growth, resulting in *fetal macrosomia* (birth weight over 4,000 g) that increases the probability of a cesarean delivery and other complications.[86] GDM occurs in approximately 5% of pregnancies but varies considerably with population demographics. As obesity is a major cause of T2DM, it also makes pregnant women vulnerable to developing GDM, and the prevalence of GDM is projected to increase as the epidemic of obesity continues.[87] Furthermore, women with GDM have a high risk of developing T2DM after pregnancy, and their children also are at greater risk of developing obesity and T2DM early in life[88].

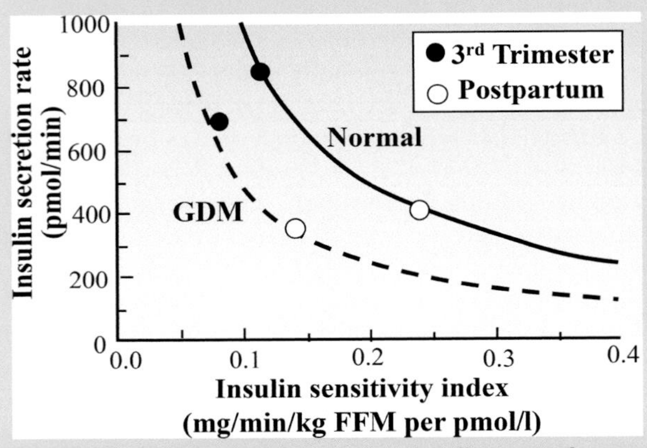

Figure 24.16 Relationship between insulin secretion and insulin sensitivity in women with gestational diabetes mellitus (GDM) and normal women during the third trimester and postpartum period.

SUMMARY AND SYNTHESIS QUESTIONS

1. Despite the large amount of progesterone secretion by the placenta, this organ has a limited ability to synthesize cholesterol (the principal substrate for progesterone biosynthesis) from acetate. How, then, can the placenta produce progesterone?
2. Some of the interactions between cytotrophoblast and syncytiotrophoblast cells of the placenta resemble those of the hypothalamic-pituitary system. Describe this.
3. Acromegaly is a condition where excess levels of GH and IGF-I are produced in the adult. You are a doctor and have a pregnant patient with acromegaly. Your patient is concerned that if left untreated her high levels of circulating GH and IGF-I will be transferred to the fetus and affect its growth rate. What is your response to this concern?
4. Most pregnancy tests measure the presence or absence of hCG in the urine. Why is hCG more

informative than other hormones of pregnancy, such as estrogens or progesterone?

5. Pregnancy is characterized by increased RAAS activity and sensitivity to ADH, resulting in a 50%–60% increase in blood volume by the end of gestation. Remarkably, pregnancy is also accompanied by a reduced blood pressure. What facilitates the reduced blood pressure?
6. What is meant by the following statement: "hemodynamically pregnancy is characterized by a state of hyponatremic hypervolemia"? What is the cause of this odd physiological status?
7. Elevated levels of placental-derived hCG during pregnancy doubles the total thyroid hormone synthesis by the maternal thyroid gland. How does this occur?
8. Although total maternal thyroid hormone levels can double with pregnancy, the bioavailable free fraction of thyroid hormone remains constant. What makes this possible?
9. Some pregnant women whose placenta overproduce the enzyme vasopressinase suffer from a form of transient diabetes insipidus. Why?
10. Pregnancy has been described as a progressively "diabetogenic state". What causes this, and how is this beneficial to the fetus?
11. During pregnancy, why do maternal FSH and LH decrease to undetectable levels?

Parturition

LEARNING OBJECTIVE Describe the neuroendocrine mechanisms responsible for the timing of human parturition and the major hormones involved in the promotion of labor.

KEY CONCEPTS:

- In humans, increasing levels of CRH of fetal and placental origin trigger parturition indirectly by promoting placental estrogen synthesis and directly by inducing myometrial contraction.
- Elevated levels of estrogens enhance myometrial contractility by increasing intermuscular gap junctions and prostaglandins, as well as enhancing receptor expression for prostaglandins and oxytocin.
- Progesterone opposes the effects of estrogens, promoting uterine quiescence.
- A shift from progesterone to estrogen dominance late in pregnancy plays an important role in promoting human parturition.
- The fetal ejection reflex is a positive neuroendocrine feedback loop between oxytocin and fetal pressure on the cervix that generates increasingly powerful contractions that facilitate labor.

The Timing of Birth

What signals the initiation of birth, and from where in the fetal-placental-maternal unit do these signals emanate? Over 2,000 years ago, Greek philosopher and physician Hippocrates suggested that the timing of labor is decided by the baby itself in response to a decreasing ability for the mother and placenta to maintain the nutritional demands of the fetus.[89,90] In the 1930s, Sir Joseph Barcroft was the first scientist to perform *in utero* experiments on pregnant sheep measuring both fetal and maternal blood oxygen levels. Barcroft proposed that as the fetus grows, its oxygen requirements gradually exceed the mother's ability to provide enough oxygen via the placenta, and that birth is initiated by dropping oxygen levels. Hippocrates and Barcroft both hypothesized that decreasing life-support provision from the mother triggers the fetus to initiate parturition. A modern interpretation of their thinking might posit that the fetal perception of physiological stress induces birth. However, the location of the fetal "stress sensor" and specific mechanisms by which the fetus initiates labor and birth remained unknown.

Clues from Sheep Gestation

Clues that the fetal brain, and more specifically, the hypothalamus and pituitary, are involved in the timing of mammalian birth arose from the field of *teratology*, the study of congenital abnormalities. In the late 1950s and 1960s, some sheep farmers in Idaho, U.S.A., began to report that when sheep grazed during early pregnancy on a particular plant, the corn lily (*Veratrum californicum*), this produced abnormally long gestation periods up to 100 days beyond the normal 150 days of pregnancy. Even more striking, these lambs possessed the fatal deformity of *holoprosencephaly*, whereby the fetal brain fails to separate into two lobes, also resulting in *cyclopia*, or development of a single eye in the center of the head (Figure 24.17), as well as impaired hypothalamus and pituitary development. It was eventually determined that this plant produces a natural steroidal alkaloid, today called "cyclopamine", and that this compound inhibits the *sonic hedgehog* signaling pathway that plays a critical role in the development of the hypothalamus and pituitary gland, among other organs.[91] Importantly, these cyclopic lambs also displayed prominent adrenal hypoplasia,[92,93] suggesting that the prolonged pregnancy may be caused by reduced adrenal glucocorticoid synthesis resulting from impairment of pituitary ACTH secretion. In short, normal parturition in sheep appeared to be controlled by activation of the fetal hypothalamic-pituitary-adrenal (HPA) stress axis.

The first direct evidence that parturition in sheep was indeed modulated by the fetal HPA axis was provided in the 1960s and 1970s. Sir Graham Liggins' research team at the National Women's Hospital in Auckland, New Zealand, showed that pregnancy in ewes could be extended dramatically following either the surgical removal of the fetal pituitary alone (hypophysectomy) or the ablation of

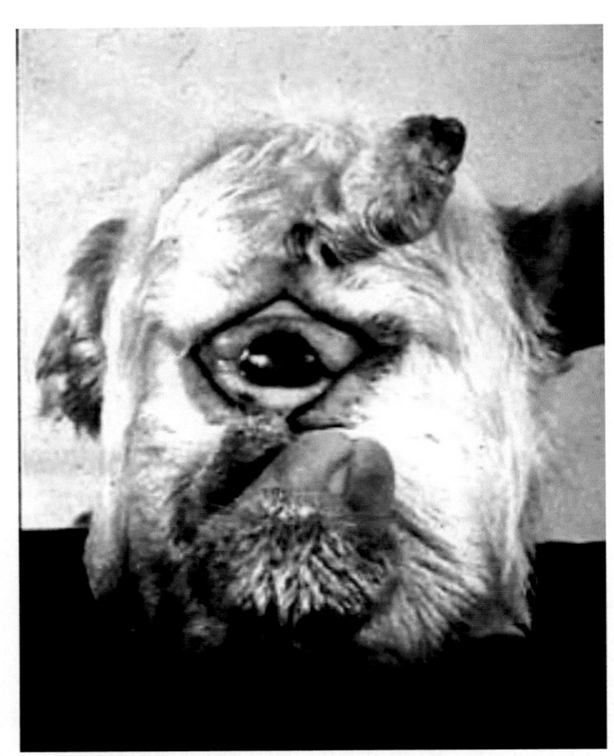

Figure 24.17 Cyclopic head of a lamb born from a sheep that ate leaves of the corn lily plant. The cyclopia is caused by the presence of cyclopamine (an inhibitor of the sonic hedgehog signaling pathway) in the plant.

both adrenal glands.[47,94,95] Later it would be shown that lesioning of the fetal paraventricular nucleus (PVN, a hypothalamic region that secretes *corticotropin-releasing hormone*, CRH) also prolongs gestation.[89] In contrast to hypothalamic-pituitary inhibition, if cortisol or ACTH was injected into the fetus following hypophysectomy, this induced the ewes to give birth within a few days. Through these elegant studies, Liggins inferred that parturition in sheep is in large part initiated by increased fetal HPA axis activity, rather than by hormones produced by the mother (Figure 24.18). It would later be discovered that in sheep, elevated cortisol of fetal origin promotes the conversion of placental progesterone into estrogen, stimulating a rise in circulating estrogen that ultimately stimulates uterine smooth muscle contractility and labor ensues.[96] Liggins' observations then begged the question, is the timing of human birth also regulated by the stress axis?

CRH: A Trigger for Human Birth

In 1933 it was observed that human fetuses were occasionally born *anencephalic*, or missing some or most of the brain and head. If these babies survived to birth, they were typically born several weeks post-term,[97] lending some credence to the possibility that, as with sheep, the human fetal HPA axis regulates the timing of birth. Surprisingly, and in marked contrast to sheep and most other non-primate mammals, treatment of pregnant women with glucocorticoids was not found to induce placental estrogen

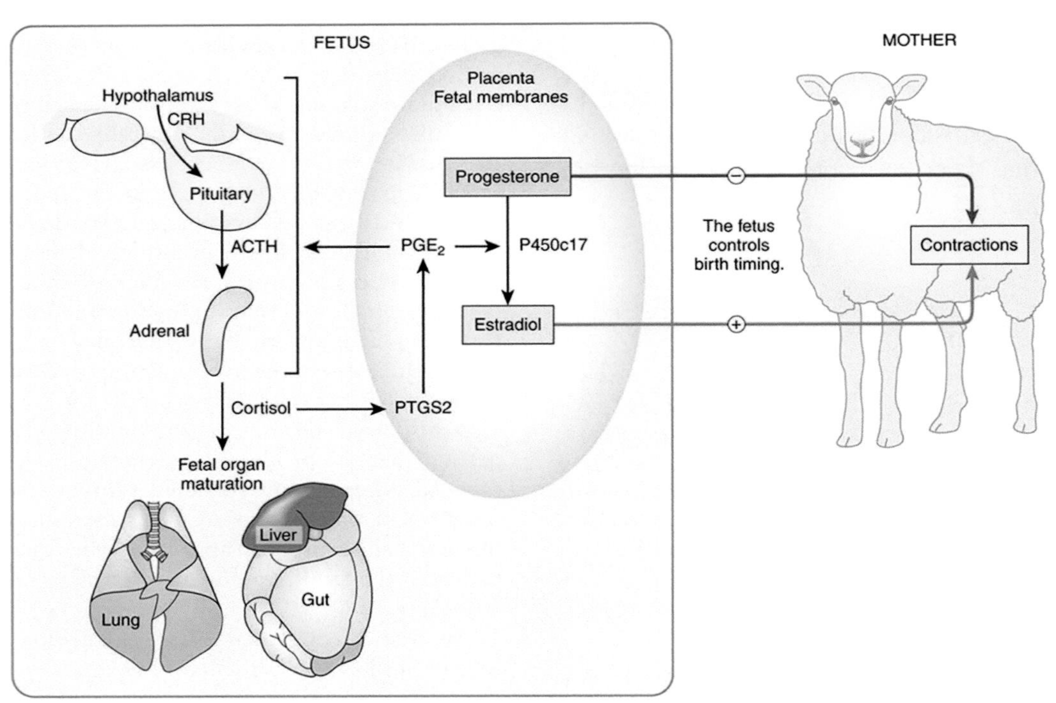

Figure 24.18 The fetal HPA axis mediates the timing of birth in sheep. Close to the middle of gestation, the fetal hypothalamus starts to secrete CRH, causing the pituitary to release ACTH, which subsequently induces the fetal adrenal cortex to produce cortisol. By inducing the synthesis of prostaglandin E2, cortisol promotes the conversion of progesterone to estrogens by the placenta. At elevated levels, estrogen prepares the uterus and cervix for labor, initiating parturition. Abbreviations: P450c17, 17α-hydroxylase/17,20 lyase; PGE$_2$, prostaglandin E$_2$; PTGS2, prostaglandin synthase 2.

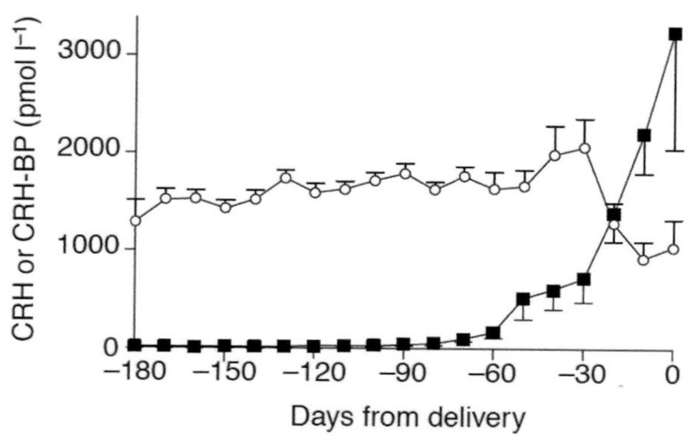

Figure 24.19 Changes in serum levels of corticotropin-releasing hormone (CRH) and its binding protein (CRH-BP) during gestation and parturition. Closed squares: CRH; open circles: CRH-BP.

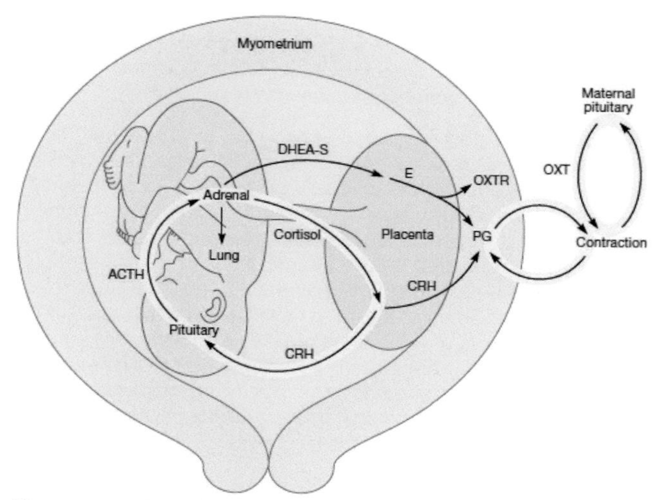

Figure 24.20 Control of human parturition by fetal CRH. CRH directly stimulates the fetal pituitary to produce ACTH, which stimulates the adrenal cortex to produce dehydroepiandrosterone sulfate (DHEA-S), which is converted by the placenta to estrogens. Estrogens, in turn, induce the expression of oxytocin receptors (OXTR) and prostaglandins (PG) in the myometrium, events that ultimately promote parturition.

secretion or labor.[98] However, McClean and colleagues found that another component of the stress axis, CRH, does induce human labor.[99] Even more striking were findings that in humans most of this CRH is derived not from the fetal hypothalamus, but from the *placenta*.[100,101] In anthropoid primates (humans and great apes), the placenta appears to be unique in secreting high levels of CRH into the bloodstream during the second and third trimesters of pregnancy, peaking exponentially during parturition[102] (Figure 24.19). Furthermore, this steep rise in maternal plasma CRH coincides with a fall in serum *CRH binding protein* (CRH-BP), resulting in a rise in circulating levels of bioavailable CRH at a time concomitant with parturition. Therefore, in humans, rising levels of placenta-derived CRH appear to function as a trigger that initiates the onset of labor. In response to threats to survival, such as reduced umbilical blood flow or low nutrition levels, the fetus appears to deploy placental CRH to adjust the fetus's developmental trajectory. Specifically, elevated CRH accelerates the rates of maturation of critical organs (e.g. lungs and central nervous system) to increase the probability of survival in a potentially hostile environment.[103]

Interestingly, in contrast to CRH of hypothalamic origin, placental CRH secretion is enhanced by adrenal glucocorticoids, providing a positive feedback mechanism by which high levels of CRH secretion can be maintained throughout pregnancy (Figure 24.20). Importantly, placental CRH also stimulates the synthesis of ACTH by both the fetal and the maternal pituitary glands, which in turn promotes the synthesis of large quantities of *dehydroepiandrosterone sulfate* (DHEA-S) by the maternal adrenal cortex and by the fetal zone of the fetal adrenal cortex. This DHEA-S is the primary substrate used by the placenta to generate high levels of circulating estrogens, which will ultimately help promote uterine smooth muscle contraction and labor. In addition, the uterine myometrium possesses several forms of CRH receptors whose ligand-activated intracellular signaling pathways either culminate in smooth muscle relaxation or, closer to term, contraction.[104–106]

Clinical Considerations: Pre- and post-term birth and the ticking of the "placental clock"

In humans, parturition typically occurs at term, which ranges from 37 to 40 weeks of gestational age. Since for the majority of gestation the fetus is physiologically unable to survive outside of the womb, the rates of fetal development must be precisely coordinated with the timing of birth. Asynchrony between these parameters may increase the risk of adverse outcomes for the newborn. For example, if the rate of fetal development is normal but the timing of birth occurs prior to 37 weeks of gestation before maturation of the lung has occurred (known as *preterm birth*), this can result in respiratory distress at birth. Preterm birth is also associated with complications to other organs, including the brain, gut, and eyes.[107] *Post-term pregnancy*, or a pregnancy that extends to 42 weeks of gestation or beyond, is associated with increased risk of fetal and neonatal mortality and morbidity as well as with increased maternal morbidity.[108]

Until relatively recently, it was thought that the length of human gestation was determined exclusively by events that occur during late pregnancy, such as functional changes in the cervix, myometrium, and fetal membranes.[109] In a remarkable discovery using a longitudinal cohort study of 485 pregnant women, McClean and colleagues[99] found that not only does placental-derived CRH act as a trigger that helps initiate normal human birth, but they also showed that patterns of plasma CRH established early during pregnancy (within the first trimester) persist throughout gestation and are associated with the aberrant timing of both preterm and post-term delivery (Figure 24.21). The authors conclude:

> These data demonstrate the existence of a longitudinal process in the placenta that is established early in pregnancy and which influences the subsequent timing of delivery. We suggest that this process is analogous to a "placental clock", which triggers the onset of parturition after a predetermined length of gestation, and that the

maternal plasma CRH level is an indicator of the rate of progress towards this event. . . . This suggests a new model for the control of parturition in humans.

Preterm birth is documented to occur in 5% to 15% of pregnancies, varying with regional demographics, and neonatal mortality following preterm birth is inversely proportional to gestational length.[110] Furthermore, survivors of preterm birth have a higher risk of developing cerebral palsy and intellectual handicaps.[111] Some pathologies that induce preterm birth, such as infection and premature rupture of the fetal membranes, are known to occur in the absence of elevated CRH, and a single measurement of low levels of CRH, therefore, has a relatively low sensitivity for ruling out preterm birth.[112] However, high CRH measurements in pregnant women do correspond with increased risk of preterm birth,[106] providing obstetricians with a new tool to alert them to and prepare in advance for a preterm pregnancy.[113,114]

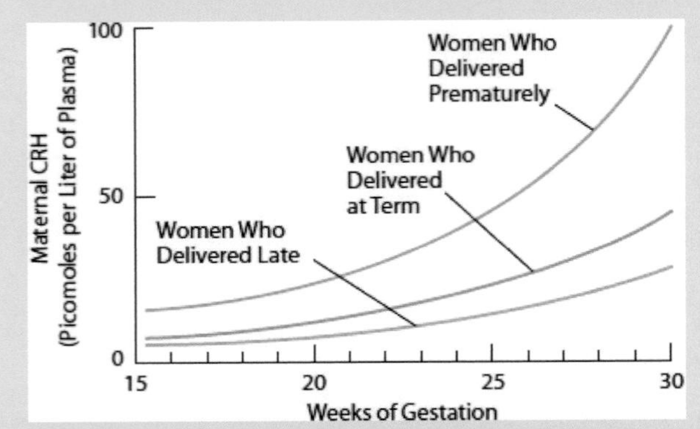

Figure 24.21 Mean plasma corticotropin-releasing hormone (CRH) in women who delivered at term (n = 308) compared with women who delivered preterm (n = 24) or post-term (n = 29).

The Hormones of Labor

Pregnancy is a period of time when a dynamic balance exists between the physiological forces that maintain uterine quiescence and those that promote contractility. Throughout most of pregnancy the uterus is in a state of muscular dormancy and the cervix remains closed, preventing the fetus from being expelled from the womb prematurely. However, at term during labor the balance between these forces tips in favor of uterine activity. The cervix softens, dilates, and thins, a process called *effacement* that occurs partly in response to increased local matrixmetalloprotease activity that decreases the aggregation of collagen fibers. Uterine myocytes transition to a highly contractile state, producing powerful, coordinated, phasic contractions that promote delivery. Labor at term is generally viewed as a hormonal release from the inhibitory effects of pregnancy on the myometrium.[115] The transition from a quiescent to a contractile myometrium requires changes in the number and activity of ion channels and pumps, as well as the formation of *gap junctions*, intercellular channels that permit direct transfer of ions and small molecules between myometrial cells that facilitate transmission of the contractile signal. Together, these changes enhance both myometrial connectivity and excitability, augmenting smooth muscle cell depolarization.[116] Ultimately, the activation of myometrial contractility and labor is promoted by local and circulating hormones.

The Antagonistic Actions of Estrogens and Progesterone

In most mammals, the onset of labor is ultimately determined by an increase in maternal estrogen levels prompted by the fetus, a fall in progesterone concentrations, or both.[116,117] Estrogens significantly enhance myometrial excitability, contractility, and coordination by increasing the synthesis of connexins (proteins that form gap junctions),[118,119] augmenting production of calmodulin and other proteins that facilitate smooth muscle contraction,[120,121] increasing the synthesis of the prostaglandins E2 and F2α (potent local stimulators of myometrial contraction), and enhancing receptor expression for prostaglandins and oxytocin.[122] Throughout most of pregnancy, estradiol (E2) is the most biologically active estrogen. However, late in pregnancy due to increased CRH-induced fetal adrenal DHEA synthesis, the levels of estriol (E3) increase more rapidly than E2, reaching an almost ten-fold higher concentration than E2.[123] Therefore, during labor it is the high E3 concentrations that appear to promote the activation of the estrogen-responsive genes that will induce labor. Interestingly, the mechanisms of action of estrogens in promoting myometrial contractility during labor have been expanded beyond the classical nuclear receptor signaling pathway to include signaling via the plasma membrane-associated *G protein-coupled estrogen receptor 30* (GPR30) in myometrial tissues.[124]

In contrast to the stimulatory effects by estrogens on myometrial contractility, progesterone appears to exert opposing effects, promoting uterine quiescence by stabilizing smooth muscle membrane potential. For example, whereas estrogens stimulate the expression of myometrial connexin-43, progesterone inhibits its expression.[125] Indeed, in many pregnant mammals treatment with the progesterone receptor antagonist RU486 promotes the onset of labor.[117] Although, in contrast to most mammals, humans exhibit no clear rise in estrogens or fall in progesterone prior to the onset of labor, growing evidence suggests that the onset of human labor is preceded by a "functional withdrawal of progesterone activity", characterized in part by changes in local progesterone metabolism,[126] a shift in the ratio of progesterone receptor isoform (PR-A/PR-B) expression,[127] and changes in progesterone receptor cofactor expression.[128] Therefore, an increasing estrogen-to-progesterone activity ratio late in pregnancy may play an important role in heralding human parturition. The balance between the effects of estrogens and progesterone in maintaining pregnancy or promoting parturition can be visualized as a hormonal "tug of war", and is summarized in Figure 24.22.

Prostaglandins

Prostaglandins are **eicosanoids**, bioactive lipids derived from arachidonic acid. These compounds typically act via paracrine or autocrine signaling by binding to specific G

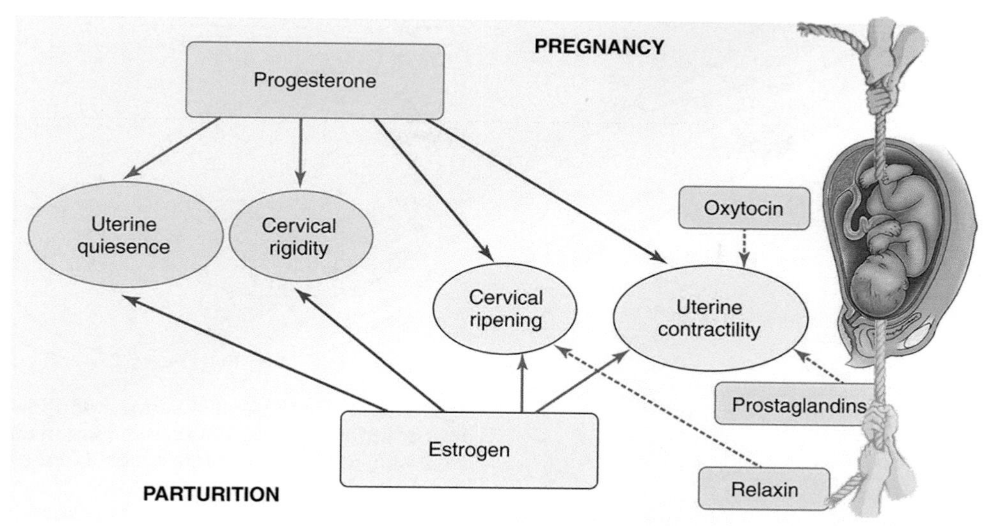

Figure 24.22 Antagonistic actions of progesterone and estrogen in pregnancy and parturition. During pregnancy progesterone maintains uterine quiescence and cervical rigidity, actions that promote the retention of the fetus in the uterus. In contrast, estrogen has opposing effects, and at high levels promotes cervical dilation and uterine contractility, actions required to induce delivery. Other hormones involved in promoting parturition include relaxin from the ovaries (which induces cervical dilation), as well as oxytocin from the neurohypophysis and prostaglandins from the uterus (which promote myometrial contraction).

protein-coupled receptors, activating intracellular signaling and gene transcription. Prostaglandins, particularly PGE2 and PGF2α, also play key roles in changing the state of the uterus during pregnancy from a quiescent to a contractile state, a process mediated in part by the differential expression of prostaglandin receptors within the myometrium and fetal membranes.[129] At term, the fetal membranes produce high levels of arachidonic acid, the precursor to prostaglandin synthesis, and the placenta and uterus also synthesize and release high levels of prostaglandins. These prostaglandins act on the myometrium and cervix via paracrine signaling to facilitate labor. Whereas estrogens appear to activate *phospholipase* A_2 (PLA2), a rate-limiting enzyme in the prostaglandin synthesis pathway, progesterone inhibits it.[130] In addition to the promotion of prostaglandin synthesis by estrogens and oxytocin, uterine mechanical stretch promoted by the growing fetus is also known to induce prostaglandin synthesis, as is the production of oxytocin.[129] Blocking of prostaglandin synthesis with cyclooxygenase inhibitors (e.g. aspirin) inhibits uterine contractility, prolonging the duration of labor in women.[129,131]

Oxytocin

Oxytocin (OT) is a member of the neurohypophyseal nonapeptide family of hormones synthesized by the hypothalamus and released by the neurohypophysis. OT plays a central role in the regulation of labor and, as will be discussed in the next section, lactation. The stimulation of uterine myometrial contraction during parturition is one of OT's most well-studied functions, and both the activation and the inhibition of its receptor are common targets in the management of dysfunctional and preterm labors, respectively. For example, synthetic OTs (e.g. Syntocinon and Pitocin) are used clinically in labor induction and augmentation, and Carbetocin is used to prevent excess uterine hemorrhage by inducing forceful uterine contractions that clamp uterine blood vessels and reduce bleeding.[132] Importantly, the resulting uterine contractions impart mechanical pressure on the cervix by the head of the baby. This induces further OT synthesis by hypothalamic magnocellular neurons via a brainstem relay involving noradrenergic cells.[133] This positive neuroendocrine feedback loop, known as the *Ferguson reflex* (also called the fetal ejection reflex) helps to generate a self-sustaining cycle of increasingly powerful contractions that facilitate labor (Figure 24.23).

Corticotropin-Releasing Hormone (CRH)

As mentioned previously, CRH indirectly stimulates uterine contraction by promoting placental estrogen synthesis. However, CRH also acts directly on the myometrium by binding to several forms of CRH receptors, all members of the G protein-coupled receptor superfamily.[104] Somewhat counterintuitively, the binding of CRH to CRHR1α, the most common form of the receptor, promotes myometrial quiescence and muscle relaxation. However, at term and under the influence of oxytocin, CRH receptors are hypothesized to change to a form that is less efficient at inducing relaxation, and instead more favorably promotes contractile pathways.[105]

Relaxin

Relaxin belongs to the insulin superfamily of structurally related hormones that includes insulin and the insulin-like growth factors (IGF-I and IGF-II).[134] The main source of relaxin in the rat, mouse, and pig is the corpus luteum. In humans, the corpus luteum also produces relaxin, but the placenta becomes the primary source of relaxin during

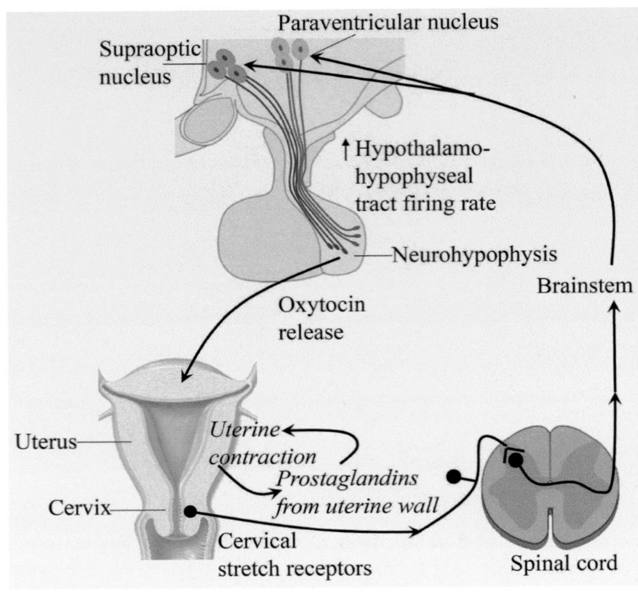

Figure 24.23 The Ferguson reflex (fetal ejection reflex) is a positive feedback neuroendocrine mechanism that regulates the synthesis and secretion of oxytocin to promote parturition. Two prominent positive feedback circuits initiate as the fetus's head exerts pressure on the cervix. First, in a neuroendocrine reflex loop, cervical stretch receptors promote neural signaling via the spinal cord to hypothalamic nuclei that ultimately cause the release of oxytocin (OT) by the neurohypophysis into circulation, and OT then directly stimulates myometrial contraction. In a second positive feedback loop, OT and myometrial contractions each indirectly stimulate the release of prostaglandins from the myometrium, which act in a paracrine manner to further stimulate myometrial contraction. These progressive uterine contractions push the baby onto the cervix with increasing force, driving increased OT and prostaglandin secretion and further uterine contractions in a positive feedback loop that ends with parturition.

the last two-thirds of human pregnancy.[135] One of the first reproductive hormones to be discovered, relaxin's injection into virgin guinea pigs shortly after estrus was found by Frederick Hisaw to relax the pubic ligament.[136] Thus, the hormone is thought to be responsible for both the generalized ligamentous relaxation and the remodeling and softening of collagenous tissues (e.g. the cervix) to facilitate birth.[137] In humans, the function and importance of relaxin in pregnancy and parturition are not as clear, since hyporelaxinemic women (the corpus luteum is absent) are still able to give birth without serious difficulty.

Integrated Endocrine Control of Parturition

As can be seen from earlier, there are multiple parallel interacting neuroendocrine and paracrine pathways involved in the onset of labor that eventually tip the myometrial balance in favor of coordinated uterine contractility and cervical dilation. These pathways are summarized in Figure 24.24. Importantly, once myometrial contractions become powerful enough to push the baby forward and irritate the cervix and the uterus, labor contractions

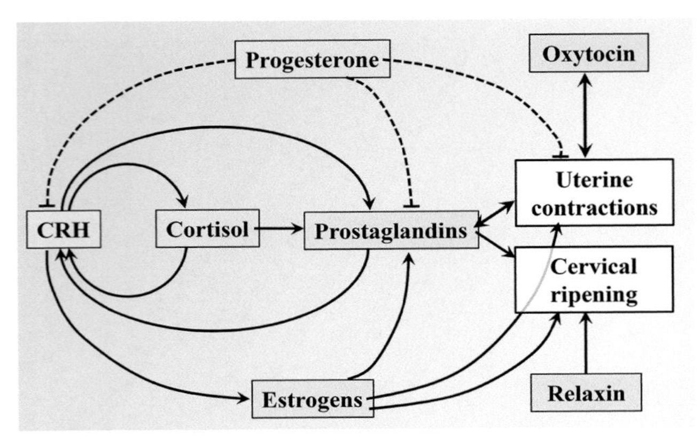

Figure 24.24 Some parallel neuroendocrine feedback cycles that contribute to the initiation of human parturition. Straight lines with arrows denote stimulation, blunt-ended dashed lines denote inhibition.

that are facilitated by positive feedback ensue, whereby subsequent contractions become more and more powerful. The contractions repeat until the baby is delivered, ending the positive feedback stimulus.

SUMMARY AND SYNTHESIS QUESTIONS

1. Pharmacological analogs of oxytocin, such as Pitocin, are often administered at term to induce labor. If Pitocin was administered prior to term, would you expect it to have similar effects in promoting labor? Explain your reasoning.
2. During childbirth, the consumption of pain killers such as aspirin, ibuprofen, and others that utilize the same mechanism of action can *prolong* the period of labor. How? (Answering this question correctly requires remembering information learned in Chapter 4: Hormone Classes and Biosynthesis.)
3. Compare and contrast the control of the timing of sheep and human parturition.
4. Considering that in humans cortisol does not stimulate parturition, explain why women carrying anencephalic fetuses (a very rare condition where these deformed fetuses have no head or pituitary gland) have prolonged pregnancy.
5. Despite oxytocin's prominent role in inducing muscle contraction during parturition, mice that lack oxytocin receptors are still able to give birth.[138] Propose a mechanism that explains how this can happen.
6. The estrogen estriol (E3) is much weaker than estradiol (E2), which is considered the most biologically active estrogen. However, during late pregnancy and labor E3 is thought to be the most important estrogen. What makes this possible?
7. Explain why in many pregnant mammals that treatment with the progesterone receptor antagonist RU486 promotes the onset of labor.

Neuroendocrine Control of Lactation

LEARNING OBJECTIVE Describe the suckling-induced positive feedback reflex, and discuss how hormones of the adenohypophysis and neurohypophysis interact to regulate lactation.

KEY CONCEPTS:

- In women, the greatest level of breast development takes place during pregnancy, largely in response to elevated levels of circulating estrogens, progesterone, and prolactin.
- During nursing, a suckling-induced positive feedback reflex promotes the release of both oxytocin from the pituitary neurohypophysis and prolactin from the adenohypophysis. These two hormones function synergistically to generate and eject milk during lactation.
- In the early stages of lactation, the suckling stimulus temporarily suppresses fertility by inhibiting GnRH and gonadotropin secretion, a period referred to as "breastfeeding-induced amenorrhea".

Hypothalamic-Pituitary-Mammary Axis Overview

In mammals, PRL secretion by the adenohypophysis is predominantly under negative control from dopaminergic neurons in the hypothalamus (Figure 24.25). Although TRH and other peptides have been postulated to function as PRL release factors (PRFs), there is little evidence that they play physiological roles in regulating normal human prolactin secretion. Interestingly, abnormally high levels of TRH, such as those associated with primary hypothyroidism, can induce hyperprolactemia,[139] so the notion that TRH can function as a PRF may have value at the pathophysiological level. Prolactin targets the mammary glands, preparing them for the synthesis of milk. Interestingly, PRL is the only major pituitary hormone that is not subject to feedback inhibition from its target tissue. That is, the mammary glands do not appear to secrete hormones that target the hypothalamus or pituitary. PRL primarily regulates its own release via a short feedback loop through autoreceptors located in the hypothalamic tuberoinfundibular dopaminergic neurons (TIDA), which release dopamine into the portal vessels.[140] Interestingly, these dopaminergic neurons of the arcuate nucleus also inhibit the release of gonadotropin-releasing hormone, explaining in part why lactating women experience oligomenorrhea or amenorrhea (infrequency or absence of menses). Additionally, PRL secretion by pituitary lactotropes is positively regulated by several hormones that lie outside the HP-PRL axis, including GnRH and estrogens. Estrogens stimulate lactotrope *hypertrophy* (increased cell size) and *hyperplasia* (increased cell number), and the number of lactotropes is consequently greater in women than in men. Furthermore, during pregnancy estrogen sensitizes the pituitary to release PRL.

During breastfeeding the HP-PRL axis experiences a form of positive feedback whereby nipple-stimulation caused by the nursing infant (the "suckling stimulus") results in a spinal reflex transmitted to the hypothalamus that leads to PRL release (Figure 24.25 and Figure 24.26). This increased PRL release during the first bout of feeding leads to more milk synthesis for the next bout of feeding. The positive feedback cycle ends when the infant is satiated. Importantly, this suckling-HP-PRL positive feedback loop operates in tandem with a separate, but parallel, positive feedback loop called the *milk letdown reflex*. This feedback loop (discussed further later) is also propelled by the mechanostimulation of nipple suckling, but induces the *neurohypophysis* to release oxytocin (OT), a hormone that induces smooth muscle contraction in (among other places) the mammary glands to expel milk into the mouth of the feeding infant. The suckling-HP-PRL and suckling-HP-OT positive feedback loops are an interesting

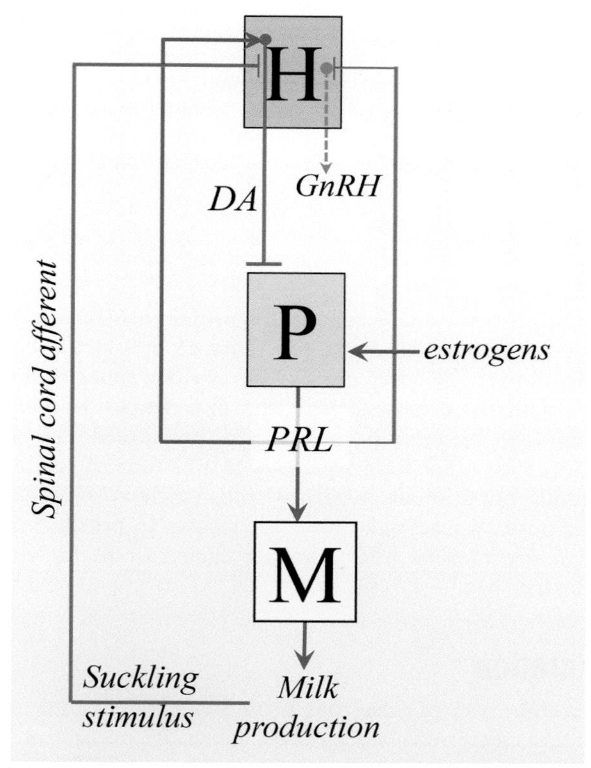

Figure 24.25 The mammalian hypothalamic-pituitary-mammary (HPM) axis. Note the roles of estrogen and neural positive feedback (suckling stimulus) in this axis. Also, note that prolactin feeds back to inhibit GnRH synthesis by the hypothalamus. Stimulatory and inhibitory actions are denoted by green arrows and dashed red blunt-ended lines, respectively.

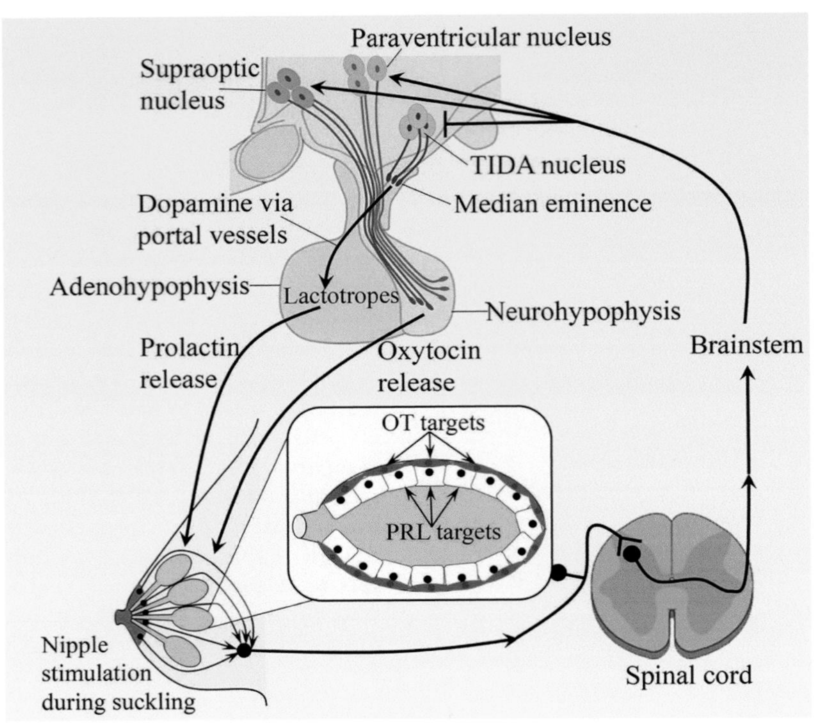

Figure 24.26 Somatosensory pathways in the suckling-induced reflex release of oxytocin (OT) and prolactin (PRL). Both suckling-induced reflexes are transmitted by sensory receptors in the nipple that transmit impulses through thoracic nerves to the spinothalamic tracts in the spinal cord, terminating in neurons in the brainstem. For the prolactin pathway, inhibitory brainstem impulses are then transmitted to the tuberoinfundibular dopamine (TIDA) neurons located in the dorsomedial arcuate nucleus of the hypothalamus, resulting in decreased dopamine (the prolactin-inhibiting factor) production, promoting the synthesis of prolactin by the lactotropes of the adenohypophysis. Prolactin targets the alveolar epithelial cells of mammary tissue (white cuboidal epithelial cells in the inset) to synthesize milk proteins. For the OT pathway, stimulatory impulses are transmitted from the brainstem to the paraventricular and supraoptic nuclei, where they stimulate the synthesis and release of OT. OT induces the myoepithelial cells (red elongated cells in the inset) of the mammary alveoli to contract, resulting in the release of milk into the lactiferous ducts and sinuses where it can then be removed by the suckling infant.

example of cooperativity between the adenohypophysis and neurohypophysis.

The positive feedback loop for prolactin makes it possible for women to nurse for years after a single birth. This allowed for "wet nurses" in previous times to nourish infants of other mothers and also allows women to pump their breast milk for others or breastfeed their own children for much longer periods of time than the children actually nutritionally need the milk (such as 4–6 years). This positive feedback also allows cows to produce milk for many months after birthing each calf in the dairy industry.

Lactation

Lactation, the process that promotes efficient milk production, secretion, and ejection, generally occurs by 4–5 days postpartum stimulated by prolactin, which acts on the alveolar epithelium and transforms it to a secretory state. Over the course of the first 3 weeks postpartum, the consistency of the milk produced will change. Initially, milk comes in the form of from *colostrum*, a low-volume yellow milk that is rich in protein, minerals, fat-soluble vitamins, and immunoglobulins, but contains less lactose and water-soluble vitamins than mature milk. The milk then transitions from *transitional milk* to *mature milk*, which is a higher volume milk with increased lactose content. Both mature milk and colostrum are enriched with immunoglobulins (maternal antibodies) and other immune factors that protect the infant against bacterial and viral infections. Breast milk also contains factors that appear to stimulate the development of the infant's own immune system, conferring long-lasting benefits. After delivery, prolactin binds to its receptor on the alveolar epithelium plasma membrane. Prolactin acts synergistically with cortisol to ultimately stimulate the transcription and translation of milk proteins (e.g. casein, γ-lactalbumin, and β-lactoglobulin) and other enzymes needed for milk synthesis (e.g. lactose synthetase and galactosyltransferase).[141–143]

Although after delivery plasma prolactin levels decrease to normal, nonpregnant levels within about 7 weeks postpartum in both lactating and non-lactating mothers, surges of prolactin are induced during nursing

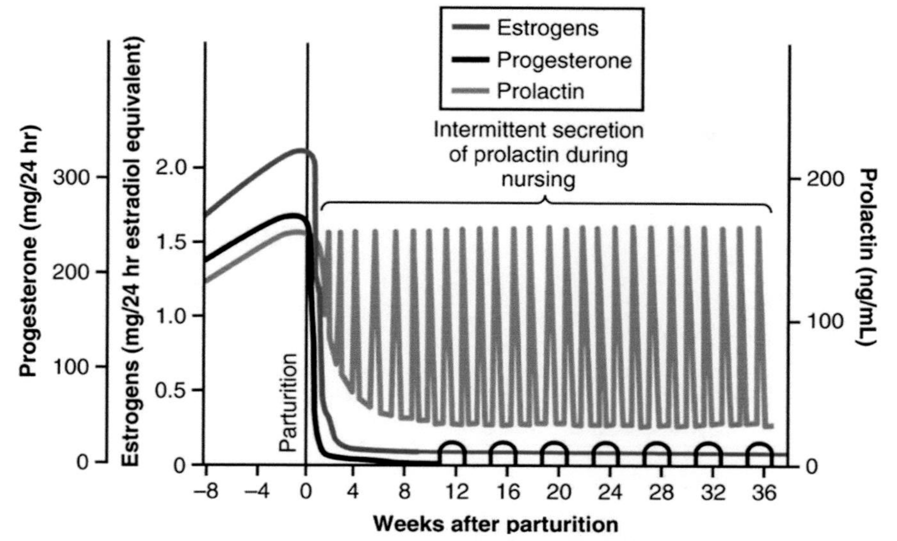

Figure 24.27 Changes in rates of secretion of estrogens, progesterone, and prolactin for before and after parturition. Although prolactin secretion returns to basal levels within several weeks following parturition, prolactin levels spike intermittently during periods of breastfeeding.

within 15 minutes of nipple stimulation (Figure 24.27). The resulting 1-hour spikes of prolactin during each feeding function to stimulate the production of milk for the next round of feeding. In effect, the suckling infant orders its next feed during the current one. With each prolactin surge, estrogen and progesterone also increase slightly. This suckling-induced reflex is transmitted by sensory receptors in the nipple that transmit impulses through thoracic nerves to the spinothalamic tracts in the spinal cord, terminating in neurons in the brainstem (Figure 24.26). Brainstem impulses are then transmitted to the hypothalamus, resulting in a decrease in tuberoinfundibular dopamine production, releasing the lactotropes of the adenohypophysis from inhibition. The suckling-induced reflex initiates a powerful positive feedback loop, as the more prolactin is produced, the greater the amount of milk is synthesized, prompting the infant to suckle more. The positive feedback loop terminates only when the infant is full and stops suckling.

A hormone critical for milk ejection is oxytocin (OT). OT stored in the neurohypophysis is also released in response to suckling via the same neural pathways mediating prolactin release up to the level of the brainstem (Figure 24.26). Beyond that point, the pathways diverge and are transmitted to the paraventricular and supraoptic nuclei, where they stimulate the synthesis and release of OT. OT induces the myoepithelial cells of the mammary alveoli to contract, resulting in the release of milk into the lactiferous ducts and sinuses where it can then be removed by the suckling infant. Eventually, in lactating women the release of OT becomes a conditioned response, requiring only audiovisual stimulation (e.g. hearing a crying baby) or conscious thought. In contrast, prolactin release does not appear to respond to such conditioning.

Importantly, milk synthesis and secretion are interdependent processes, and if synthesis is inhibited, milk secretion will terminate. This is evident in mice genetically deficient in OT that exhibit a severe lactation deficiency.[144] If the suckling stimulus is terminated for a long period, the prolonged absence of prolactin stimulates mammary gland involution, and the secretory epithelium returns to a state resembling the pre-pregnant state.

Clinical Considerations: Protective effects of pregnancy and breastfeeding from breast cancer

Breast cancer is now the most commonly diagnosed cancer and the fifth largest cause of cancer deaths in the world.[150] *Nulliparity*, or the state of never having given birth to a child, is a well-established risk factor for breast cancer, with nulliparous women having a 20%–40% higher risk of postmenopausal breast cancer than *parous* women who first gave birth before age 25[151–156] (Figure 24.28[A]). Furthermore, breastfeeding has also been shown to provide protection against developing breast cancer, with the risk of this cancer inversely associated with the duration of breastfeeding, and especially if breastfeeding lasts longer than 12 months[157–160] (Figure 24.28[B]). Developmentally, the breast is a highly plastic organ that undergoes complex developmental changes throughout a woman's life in response to hormones like estrogens and prolactin. Although the physiological basis for the protections of pregnancy and breastfeeding against cancer remain unclear, they are hypothesized to reflect a hormonally induced reduction of damage to mammary epithelial DNA via the endocrine promotion of cell differentiation, enhanced DNA repair capacity, and reduced cell proliferation.[161–163] These "oncoprotective" effects of pregnancy extend to rodent models as well, where pregnancy or treatments with pregnancy-mimicking hormones resulted in reduced mammary carcinogenesis and persistent structural and molecular changes when compared with nulliparous rodents.[164,165]

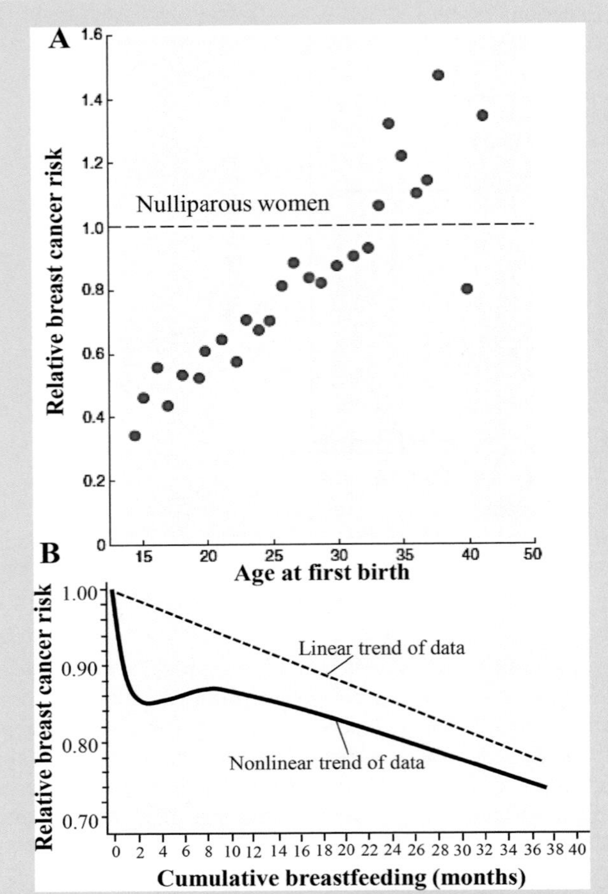

Figure 24.28 Influence of nulliparity, the timing of first birth, and the frequency of breastfeeding on incidence of breast cancer. (A) Women who gave birth prior to the age of 30 have a lower incidence of breast cancer than nulliparous women. **(B)** Dose-response relation between breastfeeding duration and breast cancer risk in parous women.

Breastfeeding-Induced Amenorrhea

Not only does the suckling stimulus promote the release of prolactin and oxytocin, but it is also well known to suppresses fertility for a variable time after birth.[145–147] This period, referred to as **breastfeeding-induced amenorrhea**, begins with an almost complete inhibition of GnRH and gonadotropin secretion in the early stages of lactation, then transitions through a period where the frequency and amplitude of GnRH and gonadotropin pulsatility is too low and erratic to promote normal follicular growth, and the first several ovulations and menses are associated with a corpus luteum function that is usually insufficient to support a pregnancy.[148] Eventually, normal menstrual cycles will resume when the rate of suckling declines below a certain threshold. It is not precisely known how the suckling stimulus reduces the pulsatile secretion of GnRH. However, it is interesting to note that hyperprolactinemia (excess prolactin synthesis, usually due to a pituitary lactotrope tumor) is a well-known cause of hypogonadotropic hypogonadism and anovulatory infertility, and that this inhibition appears to be due to reduced kisspeptin expression by a subset neurons that also express the prolactin receptor.[149] Recall that kisspeptin plays a critical role in modulating GnRH pulsatility.

SUMMARY AND SYNTHESIS QUESTIONS

1. Why are the pituitary glands of women who have had multiple pregnancies typically larger than those of women who have had no pregnancy?
2. It is a well-known fact that the abdomens of women who breastfeed their children typically return to their pre-pregnancy size much faster than do women who bottle-feed their children. Propose an endocrine basis for this observation. (*Note:* This has nothing to do with fat or weight loss, but rather the return of the distended uterus to normal size.)
3. The suckling-induced increase in prolactin secretion usually subsides by 12 weeks after parturition. What explains the decrease in serum prolactin following parturition?
4. From an evolutionary fitness perspective, what is the selective advantage of breastfeeding-induced amenorrhea? How is this a desirable trait?

Summary of Chapter Learning Objectives and Key Concepts

LEARNING OBJECTIVE Compare and contrast different strategies for the generation of offspring, and describe some fundamental changes in placental gene expression that accompanied the evolution of vivipary in mammals.

- Pregnancy describes the gestational period of all viviparous animals, from live-bearing species of sharks and other fishes to some amphibians, reptiles, and most mammals.
- One of the first steps in the transition from oviparity to viviparity among vertebrates was the evolution of the amniotic egg that permitted a shift from yolk-sac nutrition to embryonic nourishment delivered by the mother.
- Among mammals, the evolution of pregnancy was accompanied by radical changes in the expression of a vast array of genes in the uterus.
- During mammalian uterine evolution, ancient transposable elements deposited DNA binding sites for master transcriptional regulators of endometrial development (such as the progesterone receptor), imparting a new hormone responsivity to numerous genes.

LEARNING OBJECTIVE Outline the major fetal-placental-maternal interactions in the production of sex steroids and other hormones.

- Physiologically, the pregnant woman consists of three functionally distinct but interacting compartments that together constitute the fetal-placental-maternal unit.

- The placenta, which secretes a greater diversity and quantity of hormones than any other single endocrine tissue and becomes the most important source of estrogens and progesterone during pregnancy, relies on assistance from both the fetus and the mother for hormone synthesis.
- Much of pregnancy is characterized by the gradual development of maternal insulin resistance, promoting a "diabetogenic state" that facilitates the transfer of metabolic fuel to the fetus for its growth and development.

LEARNING OBJECTIVE Describe the neuroendocrine mechanisms responsible for the timing of human parturition and the major hormones involved in the promotion of labor.

- In humans, increasing levels of CRH of fetal and placental origin trigger parturition indirectly by promoting placental estrogen synthesis and directly by inducing myometrial contraction.
- Elevated levels of estrogens enhance myometrial contractility by increasing intermuscular gap junctions and prostaglandins, as well as enhancing receptor expression for prostaglandins and oxytocin.
- Progesterone opposes the effects of estrogens, promoting uterine quiescence.
- A shift from progesterone to estrogen dominance late in pregnancy plays an important role in promoting human parturition.
- The fetal ejection reflex is a positive neuroendocrine feedback loop between oxytocin and fetal pressure on the cervix that generates increasingly powerful contractions that facilitate labor.

LEARNING OBJECTIVE Describe the suckling-induced positive feedback reflex, and discuss how hormones of the adenohypophysis and neurohypophysis interact to regulate lactation.

- In women, the greatest level of breast development takes place during pregnancy, largely in response to elevated levels of circulating estrogens, progesterone, and prolactin.
- During nursing, a suckling-induced positive feedback reflex promotes the release of both oxytocin from the pituitary neurohypophysis and prolactin from the adenohypophysis. These two hormones function synergistically to generate and eject milk during lactation.
- In the early stages of lactation, the suckling stimulus temporarily suppresses fertility by inhibiting GnRH and gonadotropin secretion, a period referred to as "breastfeeding-induced amenorrhea".

LITERATURE CITED

1. Taylor RN, Lebovic DI, Martin-Cadieux MC. The endocrinology of pregnancy. In: Greenspan FS, Baxter DG, eds. *Basic and Clinical Endocrinology*. McGraw Hill Medical; 2001:575–602.
2. Dulvy NK, Reynolds JD. Evolutionary transitions among egg-laying, live-bearing and maternal inputs in sharks and rays. *Proc Royal Soc B Biol Sci*. 1997;264(1386):1309–1315.
3. Pough FH, Janis CM, Heiser JB. *Vertebrate Life*. Simon & Schuster; 1999.
4. Avise JC. *Evolutionary Perspectives on Pregnancy*. Columbia University Press; 2013.
5. Blackburn DG. Evolution of vertebrate viviparity and specializations for fetal nutrition: a quantitative and qualitative analysis. *J Morphol*. 2015;276(8):961–990.
6. Wourms JP, Grove BD, Lombardi J. The maternal-embryonic relationship in viviparous fishes. In: Hoar WS, Randall DJ, eds. *Fish Physiology*. Vol 11. Academic Press; 1988:1–134.
7. Iskandar DT, Evans BJ, McGuire JA. A novel reproductive mode in frogs: a new species of fanged frog with internal fertilization and birth of tadpoles. *PLoS One*. 2014;9(12):e115884.
8. Van Dyke JU, Brandley MC, Thompson MB. The evolution of viviparity: molecular and genomic data from squamate reptiles advance understanding of live birth in amniotes. *Reproduction*. 2014;147(1):R15–R26.
9. Whittington CM, Griffith OW, Qi W, Thompson MB, Wilson AB. Seahorse brood pouch transcriptome reveals common genes associated with vertebrate pregnancy. *Mol Biol Evol*. 2015;32(12):3114–3131.
10. Ross C, Boroviak TE. Origin and function of the yolk sac in primate embryogenesis. *Nat Commun*. 2020;11(1):1–14.
11. Power ML, Schulkin J. *The Evolution of the Human Placenta*. JHU Press; 2012.
12. Ferner K, Mess A. Evolution and development of fetal membranes and placentation in amniote vertebrates. *Respir Physiol Neurobiol*. 2011;178(1):39–50.
13. Blackburn DG. Squamate reptiles as model organisms for the evolution of viviparity. *Herpetol Monogr*. 2006;20(1):131–146.
14. dos Reis M, Inoue J, Hasegawa M, Asher RJ, Donoghue PCJ, Yang Z. Phylogenomic datasets provide both precision and accuracy in estimating the timescale of placental mammal phylogeny. *Proc Royal Soc B Biol Sci*. 2012;279(1742):3491–3500.
15. Griffith OW, Brandley MC, Whittington CM, Belov K, Thompson MB. Comparative genomics of hormonal signaling in the chorioallantoic membrane of oviparous and viviparous amniotes. *Gen Comp Endocr*. 2017;244:19–29.
16. Griffith OW, Wagner GP. The placenta as a model for understanding the origin and evolution of vertebrate organs. *Nat Ecol Evol*. 2017;1:0072.
17. Albergotti LC, Hamlin HJ, McCoy MW, Guillette LJ Jr. Endocrine activity of extraembryonic membranes extends beyond placental amniotes. *PLoS One*. 2009;4(5):e5452.
18. Lynch VJ, Nnamani MC, Kapusta A, et al. Ancient transposable elements transformed the uterine regulatory landscape and transcriptome during the evolution of mammalian pregnancy. *Cell Rep*. 2015;10(4):551–561.
19. Gellersen B, Brosens IA, Brosens JJ. Decidualization of the human endometrium: mechanisms, functions, and clinical perspectives. *Semin Reprod Med*. 2007;25(6):445–453.
20. Aghajanova L, Tatsumi K, Horcajadas JA, et al. Unique transcriptome, pathways, and networks in the human endometrial fibroblast response to progesterone in endometriosis. *Biol Reprod*. 2011;84(4):801–815.
21. Giudice LC. Elucidating endometrial function in the post-genomic era. *Hum Reprod Update*. 2003;9(3):223–235.
22. McClintock B. The origin and behavior of mutable loci in maize. *Proc Natl Acad Sci U S A*. 1950;36(6):344–355.
23. Ravindran S. Barbara McClintock and the discovery of jumping

genes. *Proc Natl Acad Sci U S A*. 2012;109(50):20198–20199.
24. Jiang K. Ancient "genomic parasites" spurred evolution of pregnancy in mammals. *Sci Life*. University of Chicago News; 2015.
25. Henke A, Gromoll J. New insights into the evolution of chorionic gonadotrophin. *Mol Cell Endocrinol*. 2008;291(1):11–19.
26. Carter AM. Evolution of placental function in mammals: the molecular basis of gas and nutrient transfer, hormone secretion, and immune responses. *Physiol Rev*. 2012;92(4):1543–1576.
27. Roberts RM. Interferon-tau, a Type 1 interferon involved in maternal recognition of pregnancy. *Cytokine Growth Factor Rev*. 2007;18(5–6):403–408.
28. Diczfalusy E. Endocrine functions of the human fetus and placenta. *Am J Obstet Gynecol*. 1974;119(3):419–433.
29. Penn AA. Endocrine and paracrine function of the human placenta. In: Abman SH, Rowitch DH, Benitz WE, Fox WW, eds. *Fetal and Neonatal Physiology (Fifth Edition)*. Elsevier; 2017.
30. Burton GJ, Fowden AL. The placenta: a multifaceted, transient organ. *Philos Trans R Soc Lond B Biol Sci*. 2015;370(1663):20140066.
31. Tal R, Taylor HS. Endocrinology of Pregnancy. 2021 Mar 18. In: Feingold KR, Anawalt B, Blackman MR, et al., eds. *Endotext* [Internet]. South Dartmouth (MA): MDText.com, Inc.; 2000–. PMID: 25905197.
32. Khodr GS, Siler-Khodr TM. Placental luteinizing hormone-releasing factor and its synthesis. *Science*. 1980;207(4428):315–317.
33. Handwerger S, Freemark M. The roles of placental growth hormone and placental lactogen in the regulation of human fetal growth and development. *J Pediatr Endocrinol Metab*. 2000;13(4):343–356.
34. Chrousos GP, Calabrese JR, Avgerinos P, et al. Corticotropin releasing factor: basic studies and clinical applications. *Prog Neuropsychopharmacol Biol Psychiatry*. 1985;9(4):349–359.
35. Stalla G, Hartwimmer J, Von Werder K, Müller O. Ovine (o) and human (h) corticotrophin releasing factor (CRF) in man: CRF-stimulation and CRF-immunoreactivity. *Acta Endocrinol*. 1984;106(3):289–297.
36. Shibahara S, Morimoto Y, Furutani Y, et al. Isolation and sequence analysis of the human corticotropin-releasing factor precursor gene. *EMBO J*. 1983;2(5):775–779.
37. Florio P, Vale W, Petraglia F. Urocortins in human reproduction. *Peptides*. 2004;25(10):1751–1757.
38. Renfree M. Marsupial reproduction: the choice between placentation and lactation. *Oxf Rev Reprod Biol*. 1983;5:1–29.
39. Tyndale-Biscoe C, Tyndale-Biscoe H, Renfree M. *Reproductive Physiology of Marsupials*. Cambridge University Press; 1987.
40. Renfree MB. Marsupials: placental mammals with a difference. *Placenta*. 2010;31:S21–S26.
41. Guernsey MW, Chuong EB, Cornelis G, Renfree MB, Baker JC. Molecular conservation of marsupial and eutherian placentation and lactation. *Elife*. 2017;6:e27450.
42. Pasqualini JR, Chetrite GS. The formation and transformation of hormones in maternal, placental and fetal compartments: biological implications. *Horm Mol Biol Clin Investig*. 2016;27(1):11–28.
43. Simpson ER, Carr BR, Parker CR Jr, Milewich L, Porter JC, MacDonald PC. The role of serum lipoproteins in steroidogenesis by the human fetal adrenal cortex. *J Clin Endocrinol Metab*. 1979 Jul;49(1):146–148. doi: 10.1210/jcem-49-1-146. PMID: 221527.
44. Grier DG, Halliday HL. Effects of glucocorticoids on fetal and neonatal lung development. *Treat Respir Med*. 2004;3(5):295–306.
45. Fowden AL. Endocrine regulation of fetal growth. *Reprod Fertil Dev*. 1995;7(3):351–363.
46. Fowden AL, Forhead AJ. Endocrine interactions in the control of fetal growth. In: *Maternal and Child Nutrition: The First 1,000 Days*. Vol 74. Karger Publishers; 2013:91–102.
47. Liggins G. The role of cortisol in preparing the fetus for birth. *Reprod Fertil Dev*. 1994;6(2):141–150.
48. Smallridge RC, Ladenson PW. Hypothyroidism in pregnancy: consequences to neonatal health. *J Clin Endocrinol Metab*. 2001;86(6):2349–2353.
49. Carr BR, Parker CR, Madden JD, MacDonald PC, Porter JC. Maternal plasma adrenocorticotropin and cortisol relationships throughout human pregnancy. *Am J Obstet Gynecol*. 1981;139(4):416–422.
50. Hume R, Simpson J, Delahunty C, et al. Human fetal and cord serum thyroid hormones: developmental trends and interrelationships. *J Clin Endocrinol Metab*. 2004;89(8):4097–4103.
51. Kawahara K, Yokoya S. Establishment of reference intervals of thyrotropin and free thyroid hormones during the first week of life. *Clin Pediatr Endocrinol*. 2002;11(1):1–9.
52. Jaudet GJ, Hatey JL. Variations in aldosterone and corticosterone plasma levels during metamorphosis in Xenopus laevis tadpoles. *Gen Comp Endocr*. 1984;56(1):59–65.
53. Leloup J. La triiodothyronine, hormone de la metamorphose des amphibiens. *CR Acad Sci Paris, D*. 1977;284:2261–2263.
54. Buchholz DR. More similar than you think: Frog metamorphosis as a model of human perinatal endocrinology. *Dev Biol*. 2015;408(2):188–195.
55. Allen BM. The endocrine control of amphibian metamorphosis. *Biol Rev*. 1938;13(1):1–19.
56. Dodd, M.H.I. and Dodd, J.M. The biology of metamorphosis. In: Lofts, B, ed. *Physiology of the Amphibia*, Vol. 3, Academic Press; 1976:467–599.
57. Tata J. Turnover of nuclear and cytoplasmic ribonucleic acid at the onset of induced amphibian metamorphosis. *Nature*. 1965;207(4995):378.
58. Tata J. Requirement for RNA and protein synthesis for induced regression of the tadpole tail in organ culture. *Dev Biol*. 1966;13(1):77–94.
59. Brown DD, Wang Z, Kanamori A, Eliceiri B, Furlow JD, Schwartzman R. Amphibian metamorphosis: a complex program of gene expression changes controlled by the thyroid hormone. *Recent Prog Horm Res*. 1995;50:309–315. doi: 10.1016/b978-0-12-571150-0.50018-4. PMID: 7740163.
60. Fu L, Das B, Matsuura K, Fujimoto K, Heimeier RA, Shi Y-B. Genome-wide identification of thyroid hormone receptor targets in the remodeling intestine during Xenopus tropicalis metamorphosis. *Sci Rep*. 2017;7(1):6414.
61. Das B, Heimeier RA, Buchholz DR, Shi Y-B. Identification of direct thyroid hormone response genes reveals the earliest gene regulation programs during frog metamorphosis. *J Biol Chem*. 2009;284(49):34167–34178.
62. Kulkarni SS, Buchholz DR. Beyond synergy: corticosterone and thyroid hormone have numerous interaction effects on gene regulation in Xenopus tropicalis tadpoles. *Endocrinology*. 2012;153(11):5309–5324.
63. Thornburg KL, Bagby SP, Giraud GD. Maternal adaptations to pregnancy. In: *Knobil and Neill's Physiology of Reproduction: Two-Volume Set*. Elsevier Inc.; 2014.
64. Elizondo G, Saldivar D, Nanez H, Todd LE, Villarreal JZ. Pituitary gland growth during normal pregnancy: an in vivo study using magnetic resonance imaging. *Am J Med*. 1988;85(2):217–220.
65. Castrique E, Fernandez-Fuente M, Le Tissier P, Herman A, Levy A. Use of a prolactin-Cre/ROSA-YFP transgenic mouse provides no evidence for lactotroph transdifferentiation after weaning, or increase in lactotroph/somatotroph proportion in lactation. *J Endocrinol*. 2010;205(1):49–60.
66. Karaca Z, Tanriverdi F, Unluhizarci K, Kelestimur F. Pregnancy and pituitary disorders. *Eur J Endocrinol*. 2010;162(3):453–475.
67. Kuwabara Y, Takeda S, Mizuno M, Sakamoto S. Oxytocin levels in maternal and fetal plasma, amniotic fluid, and neonatal plasma and urine. *Arch Gynecol Obstet*. 1987;241(1):13–23.
68. Conrad KP. Maternal vasodilation in pregnancy: the emerging role of relaxin. *Am J Physiol*. 2011;301(2):R267–R275.
69. Cheung KL, Lafayette RA. Renal physiology of pregnancy. *Adv Chronic Kidney Dis*. 2013;20(3):209–214.

70. Davison JM, Gilmore EA, Durr J, Robertson GL, Lindheimer MD. Altered osmotic thresholds for vasopressin secretion and thirst in human pregnancy. *Am J Physiol Renal Physiol.* 1984;246(1):F105–F109.
71. Walters W, Lim Y. Cardiovascular dynamics in women receiving oral contraceptive therapy. *Lancet.* 1969;294(7626):879–881.
72. Morganti AA, Zervoudakis I, Letcher R, et al. Blood pressure, the renin-aldosterone system and sex steroids throughout normal pregnancy. *Am J Med.* 1980;68(1):97–104.
73. Sealey JE, Itskovitz-Eldor J, Rubattu S, et al. Estradiol-and progesterone-related increases in the renin-aldosterone system: studies during ovarian stimulation and early pregnancy. *J Clin Endocrinol Metab.* 1994;79(1):258–264.
74. Lumbers ER, Pringle KG. Roles of the circulating renin-angiotensin-aldosterone system in human pregnancy. *Am J Physiol.* 2013;306(2):R91–R101.
75. Lindheimer MD, Barron WM, Davison JM. Osmotic and volume control of vasopressin release in pregnancy. *Am J Kidney Dis.* 1991;17(2):105–111.
76. Glinoer D. The regulation of thyroid function in pregnancy: pathways of endocrine adaptation from physiology to pathology. *Endocr Rev.* 1997;18(3):404–433.
77. Torre F, Calogero A, Condorelli R, Cannarella R, Aversa A, La Vignera S. Effects of oral contraceptives on thyroid function and vice versa. *J Endocrinol Invest.* 2020:1–8.
78. Butte NF. Carbohydrate and lipid metabolism in pregnancy: normal compared with gestational diabetes mellitus—. *Am J Clinic Nutri.* 2000;71(5):1256S–1261S.
79. Angueira AR, Ludvik AE, Reddy TE, Wicksteed B, Lowe WL, Layden BT. New insights into gestational glucose metabolism: lessons learned from 21st century approaches. *Diabetes.* 2015;64(2):327–334.
80. Catalano P. The diabetogenic state of maternal metabolism in pregnancy. *NeoReviews.* 2002;3(9):e165–e172.
81. McLachlan KA, O'Neal D, Jenkins A, Alford FP. Do adiponectin, TNFα, leptin and CRP relate to insulin resistance in pregnancy? Studies in women with and without gestational diabetes, during and after pregnancy. *Diabetes Metab Res Rev.* 2006;22(2):131–138.
82. Ryan EA, Enns L. Role of gestational hormones in the induction of insulin resistance. *J Clin Endocrinol Metab.* 1988;67(2):341–347.
83. Brizzi P, Tonolo G, Esposito F, et al. Lipoprotein metabolism during normal pregnancy. *Am J Obstet Gynecol.* 1999;181(2):430–434.
84. Powers AC. *Harrison's Endocrinology.* 3rd ed. McGraw Hill; 2013.
85. Kampmann U, Madsen LR, Skajaa GO, Iversen DS, Moeller N, Ovesen P. Gestational diabetes: a clinical update. *World J Diabetes.* 2015;6(8):1065.
86. Pedersen J. *The pregnant diabetic and her newborn: problems and management.* Munksgaard; 1977.
87. Ben-Haroush A, Yogev Y, Hod M. Epidemiology of gestational diabetes mellitus and its association with Type 2 diabetes. *Diabet Med.* 2004;21(2):103–113.
88. Group HSCR. The hyperglycemia and adverse pregnancy outcome (HAPO) study. *Int J Gynecol Obstet.* 2002;78(1):69–77.
89. Nathanielsz PW. The timing of birth. *Am Sci.* 1996;84(6):562–569.
90. Pinon R. *Biology of Human Reproduction.* University Science Books; 2002.
91. Chiang C, Litingtung Y, Lee E, et al. Cyclopia and defective axial patterning in mice lacking Sonic hedgehog gene function. *Nature.* 1996;383(6599):407.
92. Hintz R, Menking M, Sotos JF. Familial holoprosencephaly with endocrine dysgenesis. *J Pediatr.* 1968;72(1):81–87.
93. Van Kampen K, Ellis L. Prolonged gestation in ewes ingesting Veratrum californicum: morphological changes and steroid biosynthesis in the endocrine organs of cyclopic lambs. *J Endocrinol.* 1972;52(3):549–560.
94. Liggins GC, Fairclough RJ, Grieves SA, Kendall JZ, Knox BS. The mechanism of initiation of parturition in the ewe. *Recent Prog Horm Res.* 1973;29:111–159. doi: 10.1016/b978-0-12-571129-6.50007-5. PMID: 4356273.
95. Liggins GC. Premature delivery of foetal lambs infused with glucocorticoids. *J Endocrinol.* 1969;45(4):515–523.
96. Whittle W, Holloway A, Lye S, Gibb W, Challis J. Prostaglandin production at the onset of ovine parturition is regulated by both estrogen-independent and estrogen-dependent pathways. *Endocrinology.* 2000;141(10):3783–3791.
97. Malpas P. Postmaturity and malformations of the foetus. *BJOG Int J Obstet Gy.* 1933;40(6):1046–1053.
98. Smith R. The timing of birth. *Sci Am.* 1999;280(3):68–75.
99. McLean M, Bisits A, Davies J, Woods R, Lowry P, Smith R. A placental clock controlling the length of human pregnancy. *Nat Med.* 1995;1(5):460.
100. Shibasaki T, Odagiri E, Shizume K, Ling N. Corticotropin-releasing factor-like activity in human placental extracts. *J Clin Endocrinol Metab.* 1982;55(2):384–386.
101. Atsushi S, Anthony SL, Marjorie ML, Andrew NM, Toshihiro S, Dorothy TK. Immunoreactive corticotropin-releasing factor is present in human maternal plasma during the third trimester of pregnancy. *J Clin Endocrinol Metab.* 1984;59(4):812–814.
102. Alcántara-Alonso V, Panetta P, de Gortari P, Grammatopoulos DK. Corticotropin-releasing hormone as the homeostatic rheostat of feto-maternal symbiosis and developmental programming in utero and neonatal life. *Front Endocrinol.* 2017;8:161.
103. Alcantara-Alonso V, Panetta P, de Gortari P, Grammatopoulos DK. Corticotropin-releasing hormone as the homeostatic rheostat of feto-maternal symbiosis and developmental programming in utero and neonatal life. *Front Endocrinol (Lausanne).* 2017;8:161.
104. Grammatopoulos D, Thompson S, Hillhouse E. The human myometrium expresses multiple isoforms of the corticotropin-releasing hormone receptor. *J Clin Endocrinol Metab.* 1995;80(8):2388–2393.
105. Grammatopoulos DK, Hillhouse EW. Role of corticotropin-releasing hormone in onset of labour. *Lancet.* 1999;354(9189):1546–1549.
106. Smith R. Parturition. *N Engl J Med.* 2007;356(3):271–283.
107. Chaudhari B, Plunkett J, Ratajczak C, Shen T, DeFranco E, Muglia L. The genetics of birth timing: insights into a fundamental component of human development. *Clin Genet.* 2008;74(6):493–501.
108. Galal M, Symonds I, Murray H, Petraglia F, Smith R. Postterm pregnancy. *Facts Views Vis ObGyn.* 2012;4(3):175.
109. Cunningham F, Leveno K, Bloom S, Spong CY, Dashe J. Parturition. In: Cunningham F, Leveno K, Bloom S, Spong CY, Dashe J, eds. *Williams Obstetrics, 19e.* Prentice-Hall International; 1993:297–361.
110. Slattery MM, Morrison JJ. Preterm delivery. *Lancet.* 2002;360(9344):1489–1497.
111. Drummond P, Colver A. Analysis by gestational age of cerebral palsy in singleton births in north-east England 1970–94. *Paediatr Perinat Epidemiol.* 2002;16(2):172–180.
112. Smith R, Nicholson RC. Corticotrophin releasing hormone and the timing of birth. *Front Biosci.* 2007;12:912–918.
113. McGrath S, McLean M, Smith D, Bisits A, Giles W, Smith R. Maternal plasma corticotropin-releasing hormone trajectories vary depending on the cause of preterm delivery. *Am J Obstet Gynecol.* 2002;186(2):257–260.
114. Leung T, Chung T, Madsen G, Lam PK, Sahota D, Smith R. Rate of rise in maternal plasma corticotrophin-releasing hormone and its relation to gestational length. *BJOG Int J Obstet Gy.* 2001;108(5):527–532.
115. Rivera J, Europe-Finner G, Phaneuf S, Asbóth G. Parturition: activation of stimulatory pathways or loss of uterine quiescence? *Adv Exp Med Biol.* 1995;395:435–451.
116. Smith R, Paul J, Maiti K, Tolosa J, Madsen G. Recent advances in understanding the endocrinology of human birth. *Trends Endocrinol Metab.* 2012;23(10):516–523.
117. Young IR, Renfree MB, Mesiano S, Shaw G, Jenkin G, Smith R. The comparative physiology of parturition in mammals: hormones and parturition in mammals.

In: *Hormones and Reproduction of Vertebrates: Mammals*. Elsevier; 2011:95–116.
118. Petrocelli T, Lye S. Regulation of transcripts encoding the myometrial gap junction protein, connexin-43, by estrogen and progesterone. *Endocrinology*. 1993;133(1):284–290.
119. Lye S, Nicholson B, Mascarenhas M, MacKenzie L, Petrocelli T. Increased expression of connexin-43 in the rat myometrium during labor is associated with an increase in the plasma estrogen: progesterone ratio. *Endocrinology*. 1993;132(6):2380–2386.
120. Matsui K, Higashi K, Fukunaga K, Miyazaki K, Maeyama M, Miyamoto E. Hormone treatments and pregnancy alter myosin light chain kinase and calmodulin levels in rabbit myometrium. *J Endocrinol*. 1983;97(1):11–19.
121. Windmoller R, Lye S, Challis J. Estradiol modulation of ovine uterine activity. *Can J Physiol Pharmacol*. 1983;61(7):722–728.
122. Bale TL, Dorsa DM. Cloning, novel promoter sequence, and estrogen regulation of a rat oxytocin receptor gene. *Endocrinology*. 1997;138(3):1151–1158.
123. Smith R, Smith JI, Shen X, et al. Patterns of plasma corticotropin-releasing hormone, progesterone, estradiol, and estriol change and the onset of human labor. *J Clin Endocrinol Metab*. 2009;94(6):2066–2074.
124. Maiti K, Paul J, Read M, et al. G-1-activated membrane estrogen receptors mediate increased contractility of the human myometrium. *Endocrinology*. 2011;152(6):2448–2455.
125. Renthal NE, Chen C-C, Koriand'r CW, Gerard RD, Prange-Kiel J, Mendelson CR. miR-200 family and targets, ZEB1 and ZEB2, modulate uterine quiescence and contractility during pregnancy and labor. *Proc Natl Acad Sci U S A*. 2010;107(48):20828–20833.
126. Mitchell BF, Wong S. Changes in 17β, 20α-hydroxysteroid dehydrogenase activity supporting an increase in the estrogen/progesterone ratio of human fetal membranes at parturition. *Am J Obstet Gynecol*. 1993;168(5):1377–1385.
127. Mesiano S, Chan E-C, Fitter JT, Kwek K, Yeo G, Smith R. Progesterone withdrawal and estrogen activation in human parturition are coordinated by progesterone receptor A expression in the myometrium. *J Clin Endocrinol Metab*. 2002;87(6):2924–2930.
128. Condon JC, Jeyasuria P, Faust JM, Wilson JW, Mendelson CR. A decline in the levels of progesterone receptor coactivators in the pregnant uterus at term may antagonize progesterone receptor function and contribute to the initiation of parturition. *Proc Natl Acad Sci U S A*. 2003;100(16):9518–9523.
129. Khan AH, Carson RJ, Nelson SM. Prostaglandins in labor--a translational approach. *Front Biosci*. 2008;13:5794–5809.
130. Periwal SB, Farooq A, Bhargava V, Bhatla N, Vij U, Murugesan K. Effect of hormones and antihormones on phospholipase A2 activity in human endometrial stromal cells. *Prostaglandins*. 1996;51(3):191–201.
131. Walsh, S, *Glob. libr. women's med*., (ISSN: 1756-2228) 2011; DOI 10.3843/GLOWM.10315.
132. Arrowsmith S, Wray S. Oxytocin: its mechanism of action and receptor signalling in the myometrium. *J Neuroendocrinol*. 2014;26(6):356–369.
133. Russell JA, Leng G, Douglas AJ. The magnocellular oxytocin system, the fount of maternity: adaptations in pregnancy. *Front Neuroendocrinol*. 2003;24(1):27–61.
134. Sherwood OD. Relaxin's physiological roles and other diverse actions. *Endocr Rev*. 2004;25(2):205–234.
135. Klein C. The role of relaxin in mare reproductive physiology: a comparative review with other species. *Theriogenology*. 2016;86(1):451–456.
136. Hisaw FL. Experimental relaxation of the pubic ligament of the guinea pig. *Proc Soc Exp Biol Med*. 1926;23(8):661–663.
137. Vannuccini S, Bocchi C, Severi FM, Challis JR, Petraglia F. Endocrinology of human parturition. *Ann Endocrinol (Paris)*. 2016 Jun;77(2):105–113. doi: 10.1016/j.ando.2016.04.025. Epub 2016 May 5. PMID: 27155774.
138. Nishimori K, Young LJ, Guo Q, Wang Z, Insel TR, Matzuk MM. Oxytocin is required for nursing but is not essential for parturition or reproductive behavior. *Proc Natl Acad Sci*. 1996;93(21):11699–11704.
139. Ansari MS, Almalki MH. Primary hypothyroidism with markedly high prolactin. *Front Endocrinol*. 2016;7:35.
140. Moore KE, Demarest KT. Tuberoinfundibular and tuberohypophysial dopaminergic neurons. In: Ganong WF, Martini L, eds. *Front Neuroendocrinol*. Raven Press; 1982:161–190.
141. Shiu R. The prolactin target cell and receptor. *Prog Reprod Biol*. 1980;6:97–121.
142. Topper YJ. Multiple hormone interactions in the development of mammary gland in vitro. *Recent Prog Horm Res*. 1970;26:287–308. doi: 10.1016/b978-0-12-571126-5.50011-x. PMID: 4919093.
143. Bole-Feysot C, Goffin V, Edery M, Binart N, Kelly PA. Prolactin (PRL) and its receptor: actions, signal transduction pathways and phenotypes observed in PRL receptor knockout mice. *Endocr Rev*. 1998;19(3):225–268.
144. Young WS III, Shepard E, Amico J, et al. Deficiency in mouse oxytocin prevents milk ejection, but not fertility or parturition. *J Neuroendocrinol*. 1996;8(11):847–853.
145. McNeilly AS. Neuroendocrine changes and fertility in breast-feeding women. In: *Progress in Brain Research*. Vol 133. Elsevier; 2001:207–214.
146. Kennedy KI, Visness CM. Contraceptive efficacy of lactational amenorrhoea. *Lancet*. 1992;339(8787):227–230.
147. Kennedy KI, Rivera R, McNeilly AS. Consensus statement on the use of breastfeeding as a family planning method. *Contraception*. 1989;39(5):477–496.
148. McNeilly AS. Lactational control of reproduction. *Reprod Fertil Dev*. 2001;13(8):583–590.
149. Bernard V, Young J, Chanson P, Binart N. New insights in prolactin: pathological implications. *Nat Rev Endocrinol*. 2015;11(5):265.
150. Lei S, Zheng R, Zhang S, et al. Global patterns of breast cancer incidence and mortality: a population-based cancer registry data analysis from 2000 to 2020. *Cancer Commun*. 2021;41(11):1183–1194.
151. MacMahon B, Cole P, Lin T, et al. Age at first birth and breast cancer risk. *Bull World Health Organ*. 1970;43(2):209.
152. MacMahon B, Cole P, Brown J. Etiology of human breast cancer: a review. *J Natl Cancer Inst*. 1973;50(1):21–42.
153. Rosner B, Colditz GA. Nurses' health study: log-incidence mathematical model of breast cancer incidence. *JNCI: J Natl Cancer Inst*. 1996;88(6):359–364.
154. Pike MC, Kolonel LN, Henderson BE, et al. Breast cancer in a multiethnic cohort in Hawaii and Los Angeles: risk factor-adjusted incidence in Japanese equals and in Hawaiians exceeds that in whites. *Cancer Epidemiol Biomarkers Prev*. 2002;11(9):795–800.
155. Sweeney C, Blair CK, Anderson KE, Lazovich D, Folsom AR. Risk factors for breast cancer in elderly women. *Am J Epidemiol*. 2004;160(9):868–875.
156. Butt S, Borgquist S, Anagnostaki L, Landberg G, Manjer J. Parity and age at first childbirth in relation to the risk of different breast cancer subgroups. *Int J Cancer*. 2009;125(8):1926–1934.
157. Unar-Munguía M, Torres-Mejía G, Colchero MA, Gonzalez de Cosio T. Breastfeeding mode and risk of breast cancer: a dose—response meta-analysis. *J Hum Lact*. 2017;33(2):422–434.
158. Anothaisintawee T, Wiratkapun C, Lerdsitthichai P, et al. Risk factors of breast cancer: a systematic review and meta-analysis. *Asia Pac J Public Health*. 2013;25(5):368–387.
159. Chowdhary R, Sinha B, Sankar M, Taneja S, Bhandari N. Breastfeedin g and maternal health: a systematic review and met-analysis. *Acta Pediatr*. 2015;104(467):96–113.
160. Zhou Y, Chen J, Li Q, Huang W, Lan H, Jiang H. Association between breastfeeding and breast cancer risk: evidence from a meta-analysis. *Breastfeed Med*. 2015;10(3):175–182.
161. Russo J, Moral R, Balogh GA, Mailo D, Russo IH. The protective role of pregnancy in breast cancer. *Breast Cancer Res*. 2005;7(3):1–12.

162. Russo J, Rivera R, Russo I. Influence of age and parity on the development of the human breast. *Breast Cancer Res Treat.* 1992;23(3):211–218.
163. Russo J, Balogh GA, Russo IH, Participants FCCCHN. Full-term pregnancy induces a specific genomic signature in the human breast. *Cancer Epidemiol Biomarkers Prev.* 2008;17(1):51–66.
164. Blakely CM, Stoddard AJ, Belka GK, et al. Hormone-induced protection against mammary tumorigenesis is conserved in multiple rat strains and identifies a core gene expression signature induced by pregnancy. *Cancer Res.* 2006;66(12):6421–6431.
165. Ginger MR, Gonzalez-Rimbau MF, Gay JP, Rosen JM. Persistent changes in gene expression induced by estrogen and progesterone in the rat mammary gland. *Mol Endocrinol.* 2001;15(11):1993–2009.
166. Caron P, Broussaud S, Bertherat J, et al. Acromegaly and pregnancy: a retrospective multicenter study of 59 pregnancies in 46 women. *J Clin Endocrinol Metab.* 2010;95(10):4680–4687.

Unit Overview
Unit VIII

Endocrine-Disrupting Chemicals

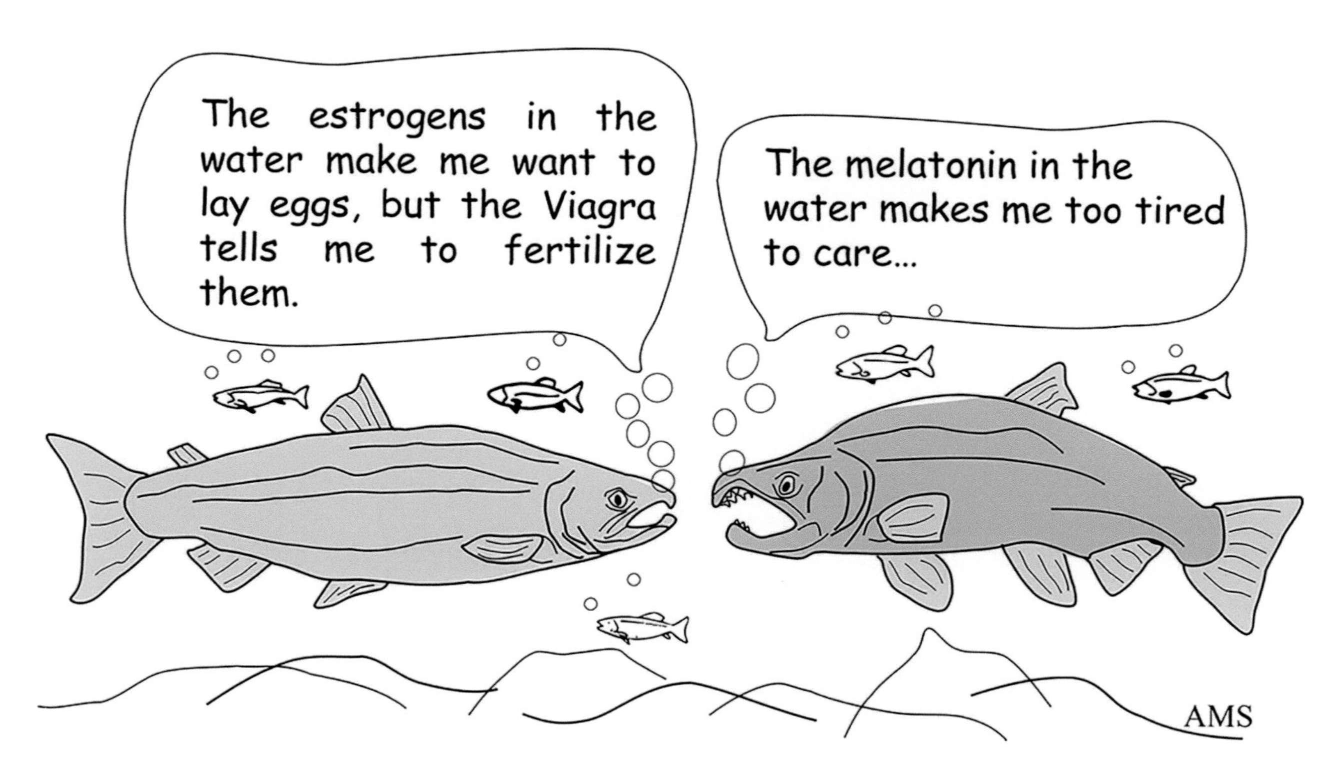

Salmon downstream of municipal effluent, by A.M. Schreiber.

DOI: 10.1201/9781003359241-32

OPENING QUOTATIONS:

> "We may have been born in a 'primeval soup,' but today, we are swimming in a 'synthetic soup' of toxins, carcinogens, and endocrine-disrupting chemicals (EDCs), all vying to confuse or derail our bodies' attempts to interpret nutritional and environmental cues."
>
> —Grün F. The obesogen tributyltin. In: *Vitamins & Hormones*. Vol 94. Elsevier; 2014:277–325[1]

The pseudonymous author, Lemony Snicket, opens his darkly comic children's book series, *A Series of Unfortunate Events*[2], with the following cautionary statement:

> "If you are interested in stories with happy endings, you would be better off reading some other book. In this book, not only is there no happy ending, there is no happy beginning and very few happy things in the middle. . . . I'm sorry to tell you this, but that is how the story goes."

Perhaps this final unit of this textbook, Unit VIII, should have its own advisory, as it addresses the dark, though decidedly non-comedic, topic of "endocrine-disrupting chemicals" (EDCs). Indeed, the topic is of growing importance to the field of public health and, after climate change, it has been described as the second greatest environmental threat of our time. Chapter 25, the unit's single chapter, begins by defining EDCs, describing their history, and stating the scope of the problem. The chapter then provides an overview of some of the most important characteristics of EDCs and their well-established pathological effects on the health of wildlife and of laboratory model organisms. The chapter poses the question, "are EDCs affecting human health?" and ends by restating the importance of integrating clinical, basic, comparative, and translational endocrine research to address this question in a meaningful way.

***Chapter 25*: EDCs: Assessing the Risk**

1. Grün F. The obesogen tributyltin. In: *Vitamins & Hormones*. Vol 94. Elsevier; 2014:277–325.
2. Snicket L, Curry T. *A Series of Unfortunate Events*. HarperCollins Egmont; 2004.

CHAPTER

25

EDCs

Assessing the Risk

CHAPTER LEARNING OBJECTIVES:

- Describe key events that led to the birth of the modern environmental movement in the 1960s and 1970s.
- Describe the mechanisms by which the pesticide DDT exerts both estrogenic and antiandrogenic effects on developing vertebrates.
- Define "environmental estrogen", and explain how wildlife species, such as fish, birds, and alligators, can function as "sentinel" species.
- Discuss the importance of diethylstilbestrol (DES) in founding the field of endocrine disruption studies.
- Describe the concept of the "fragile fetus", and explain why embryos, fetuses, and newborns are so susceptible to the effects of endocrine disruption compared with adults.
- Describe factors that influenced the coining of the term "endocrine disruptor".
- Describe the primary sources of endocrine-disrupting chemicals (EDCs).
- Describe the biochemical features and mechanisms of action of EDCs, and how the study of endocrine disruption differs from that of classical toxicology.

OPENING QUOTATIONS:

"It's incredibly obvious, isn't it? A foreign substance is introduced into our precious bodily fluids, without the knowledge of the individual, certainly without any choice."

—Brig. Gen. Jack D. Ripper, a character in Director Stanley Kubrick's 1964 movie masterpiece, *Dr. Strangelove or: How I Learned to Stop Worrying and Love the Bomb*

"Every man sitting in this room today is half the man his grandfather was."

—Professor Lou Guillette (explaining to a congressional committee in 1993 that human sperm counts have been decreasing for decades)

"So let me remind you/don't put this behind you/Atrazine ain't a good thing/it causes male frogs to grow eggs/contributes to extra legs/and exposed males don't want to sing/If that ain't enough/when you combine the stuff/with a few other pesticides/it causes greater than additive effects/unpredictable defects/exposed larvae don't grow/and they develop slow/and they contract diseases that otherwise could be beaten/you see this exposure affects their composure and determines who gonna eat and who gonna be eaten."

—Professor Tyrone Hayes, *The Atrazine Rap*[1]

Pesticides and the Birth of the Modern Environmental Movement

LEARNING OBJECTIVE Describe key events that led to the birth of the modern environmental movement in the 1960s and 1970s.

KEY CONCEPTS:

- We are living in the Anthropocene, a proposed epoch of geological time characterized by the presence of distinct human impacts on the geology and ecology of the earth, caused in part by chemical pollution, climate disruption, overpopulation, habitat destruction, and species extinction.
- With the publication of *Silent Spring* in 1962, a book describing the detrimental effects of the pesticide DDT on wildlife health, Rachel Carson ushered in the modern environmental movement and the notion that synthetic chemicals can disrupt hormone function.
- Carson's book provided impetus for establishing the U.S. National Institute of Environmental Health Sciences (NIEHS), the founding of the U.S. Environmental Protection Agency, the U.S. Clean Air and Water Acts, and international agreements to ban or restrict several key synthetic chemicals.

DOI: 10.1201/9781003359241-33

The Age of Synthetic Compounds

The earth is now experiencing its sixth mass species extinction, the largest and fastest-occurring since the dinosaurs disappeared, and the first to be caused by a single species, humans.[2,3] The anthropogenic causes of this extinction include climate disruption, chemical pollution, light pollution, human population overgrowth, and the global loss of non-human-dominated ecosystems.[4] Indeed, it has been suggested that we are currently living in the **Anthropocene**, an epoch of geological time characterized by the presence of distinct human impacts on the geology and ecology of the earth caused by the aforementioned processes.[4–6] **Metazoans** (multicellular organisms) and their endocrine systems have been evolving since their first appearance approximately 750–800 million years ago, driven by forces of natural selection that include changes in their physical and chemical environments. However, since the start of the Industrial Revolution in the 18th and 19th centuries, humanity is estimated to have introduced over 80,000 new synthetic chemicals and pollutants into the global environment, and that number is increasing by the astounding rate of about 2,000 new compounds every year.[7] Other than those chemicals specifically created for warfare, the vast majority were designed to improve human health and quality of life by enhancing agriculture and manufacturing technologies, to protect us from pathogens, and to provide us with textiles, cleansers, personal care products, and countless other conveniences of modern living. However, the introduction of massive quantities of many of these novel chemicals into the environment began so recently in geological time that living organisms have simply not had the evolutionary time to effectively adapt to them, introducing unintended consequences to human and wildlife health on an unprecedented scale. Using an analogy, if we imagine the timeline of metazoan life on earth to be 1 year, this means that animals have been exposed to these novel chemicals during just the last 8.6 seconds of that year. This chapter introduces you to the impacts of a specific category of synthetic pollutants, termed *endocrine-disrupting chemicals* (EDCs), on the health and reproduction of wildlife and, potentially, humans.

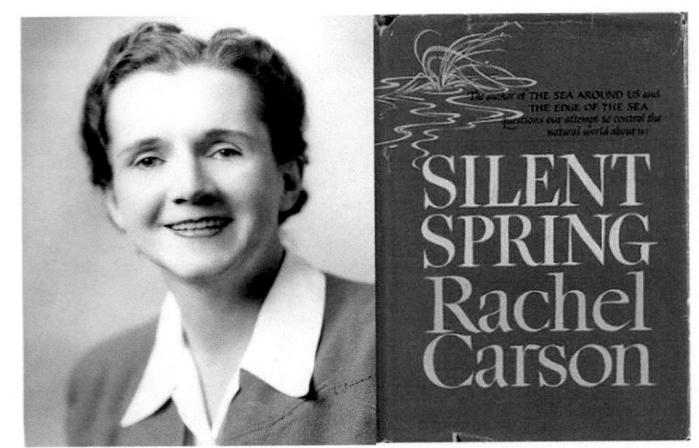

Figure 25.1 Rachel Carson (1907–1964), and the first edition of *Silent Spring*.

Source of Images: Courtesy of U.S. Fish and Wildlife Service.

"Silent Spring": A Call to Arms

On June 4, 1963, a United States Senate Government Operations subcommittee studying pesticide spraying held hearings. Speaking before the subcommittee was a 56-year-old fisheries biologist and author who, unbeknownst to most, was in the terminal stages of breast cancer. To hide her baldness, she wore a brown wig. She could barely walk down the aisle to take her seat at the large table before the Congressional panel, as the cancer had metastasized to her pelvis, perforating it with fractures. The speaker was Rachel Carson (Figure 25.1), the author of several best-selling books about the sea, *Under the Sea-Wind*,[8] *The Sea Around Us*,[9] and *The Edge of the Sea*.[10] Whereas these books were written to communicate her sense of love and wonder about nature, her more recent book, *Silent Spring*,[11] published a year before the hearings, had a much more ominous tone. In elegant prose that resonated with the general public, Carson described years of findings collected by the scientific community detailing how industrial pesticides, such as *DDT* (dichloro-diphenyl-trichloroethane) and other organochlorine compounds, not only killed insects but were also bioaccumulating and exterminating wildlife, especially of some coastal bird populations whose numbers were plummeting. She argued that the data not only showed that these pesticides were a menace to wildlife, but that they may also affect human health, especially of developing children. The book drew widespread popularity, making it impossible to be ignored by policymakers. At the time Senator Ernest Gruening, a Democrat from Alaska, told Carson, "Every once in a while in the history of mankind, a book has appeared which has substantially altered the course of history".[12] Her assessments were not received as well by the agrochemical industry, by whom she was labeled a "fanatic", a "communist", and far worse.[13–15] Importantly, Carson never called for a ban on pesticides, testifying, "I think chemicals do have a place". In *Silent Spring* Carson writes, "If we are going to live so intimately with these chemicals eating and drinking them, taking them into the very marrow of our bones—we had better know something about their nature and their power".

Carson's book functioned as a catalyst to the modern environmental movement, ultimately helping to provide impetus for establishing the U.S. National Institute of Environmental Health Sciences (NIEHS) in 1966 to study environmental influences on human health, the founding of the U.S. Environmental Protection Agency in 1970, the U.S. Clean Air and Water Acts in 1972, and international agreements to ban or restrict several key synthetic chemicals. In 1972, eight years after Carson's death, the U.S. banned all domestic sales of DDT, although U.S. companies were permitted to export the compound until the mid-1980s. China stopped manufacturing DDT in 2007, and as of 2014 the only country still manufacturing DDT was India,[16] which has no affordable alternative to combating the public health threat of mosquito-borne malaria. Ultimately, the development of new environmental standards for pollutants led to declines in the gross toxic effects and mortality rates of wildlife and the subsequent recovery of many bird population numbers,[17,18] Critically, despite this recovery, adverse health effects in the populations persisted and a new and unexpected threat emerged: the offspring of previously exposed parents still retained these compounds and their metabolites in their bodies, albeit at extremely low concentrations, and began to show numerous reproductive and developmental malformations indicative of abnormal hormone function.

SUMMARY AND SYNTHESIS QUESTIONS

1. Rachel Carson's book *Silent Spring* primarily addressed the toxic effects of pesticides on wildlife. Why did this resonate with the public?

DDT: The First Endocrine Disruptor

LEARNING OBJECTIVE Describe the mechanisms by which the pesticide DDT exerts both estrogenic and antiandrogenic effects on developing vertebrates.

KEY CONCEPTS:

- Whereas DDT exerts estrogenic effects by functioning as an estrogen receptor agonist, its primary breakdown product, DDE, exerts antiandrogenic effects by antagonizing the androgen receptor.
- The males of many wild vertebrate populations exposed to DDT become simultaneously both "feminized" and "demasculinized".
- Although DDT was ultimately banned in the United States in 1972, the chemical was so widely used and is so resistant to degradation that it remains present to this day in virtually all animals, including humans.

DDT (dichloro-diphenyl-trichloroethane) (Figure 25.2) is an organochlorine chemical that was first synthesized in 1874 but began to be widely used in the 1930s when Paul Müller, a Swiss chemist, found it to be an effective insecticide. DDT was broadly applied in the military as a delousing agent, as a public health tool to combat mosquito-borne malaria, and was also widely applied to crops to kill insect pests (Figure 25.2). Increased use of DDT following World War II marked the beginning of the modern age for widespread use of synthetic organic pesticides. Indeed, DDT was perceived to be so beneficial to humanity that Müller was awarded the *1948 Nobel Prize in Physiology or Medicine* "for his discovery of the high efficiency of DDT as a contact poison against several arthropods".[19]

However, as early as the 1950s wildlife biologists began to suspect that the pesticide affected far more than just insects. For example, Florida naturalist Charles Broley described abnormal courtship and nesting behaviors and reduced hatch rates among a diversity of bird species in the U.S.A.

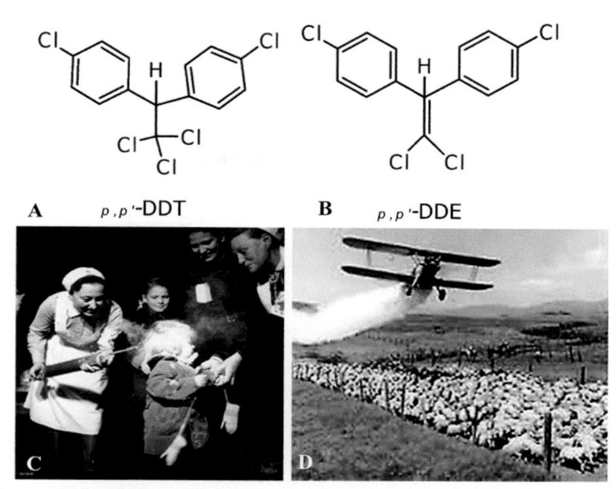

Figure 25.2 DDT was used as a delousing and anti-malarial agent during WWII and as an agricultural pesticide. The chemical structures of **(A)** DDT (dichloro-diphenyl-trichloroethane) and **(B)** its metabolic byproduct, DDE. Whereas DDT is an estrogen receptor agonist, DDE is an androgen receptor antagonist. DDT application on humans **(C)** and cattle **(D)** in the 1940s.

Source of Images: (A-B) Hongsibsong, S., et al. J. Agric. Food Chem. 60, no. 1 (2012): 16-22. Used by permission. (C-D) Carvalho, F.P., 2017. Food Energ. Secur., 6(2), pp. 48-60.

and Canada, particularly in top predators such as peregrine falcons, pelicans, osprey, and bald eagles.[20] On a stretch of the western coast of Florida, Broley documented the decimation of nesting pairs of bald eagles from about 125 in 1944, prior to DDT's introduction in 1945, to less than 10 in 1958 (Figure 25.3 A), a trend that reflected the rest of the U.S.A.[11] Broley hypothesized that consumption of DDT-contaminated fish was responsible for altering reproduction in these birds, and over the next several decades mounting evidence supporting this hypothesis began to accumulate.[21]

Evidence that DDT and its metabolites function as vertebrate endocrine disruptors arose as early as 1950, when Burlington and Lindeman[22] demonstrated that treatment of juvenile roosters with DDT decreased their testicular growth and inhibited the development of male secondary sex characteristics, such as wattles and combs. At the time, the researchers concluded, "These findings suggest that DDT may exert an estrogen-like action". Burlington and Lindeman's notion that DDT functioned as an estrogen mimic was supported by studies in the 1960s. These showed that when rats and birds were treated with a DDT isomer, o,p'-DDT, their uteri and oviducts exhibited the same responses as when they were treated with estrogen, namely increased weight, glycogen, and RNA content.[23,24] o,p'-DDT was later shown to interact with the mammalian estrogen receptor-α (ERα).[25,26] Importantly, DDT not only exerts estrogenic actions via o,p'-DDT, but much of its antiandrogenic activity has since been traced to its chief and most environmentally persistent metabolic product, *dichloro-diphenyl-dichloroethylene* (p,p'-DDE) (Figure 25.2), which inhibits androgens, such as testosterone, from binding to the androgen receptor, depressing androgen-induced transcriptional activity *in vitro* and during embryonic development in male rats.[27,28] Thus

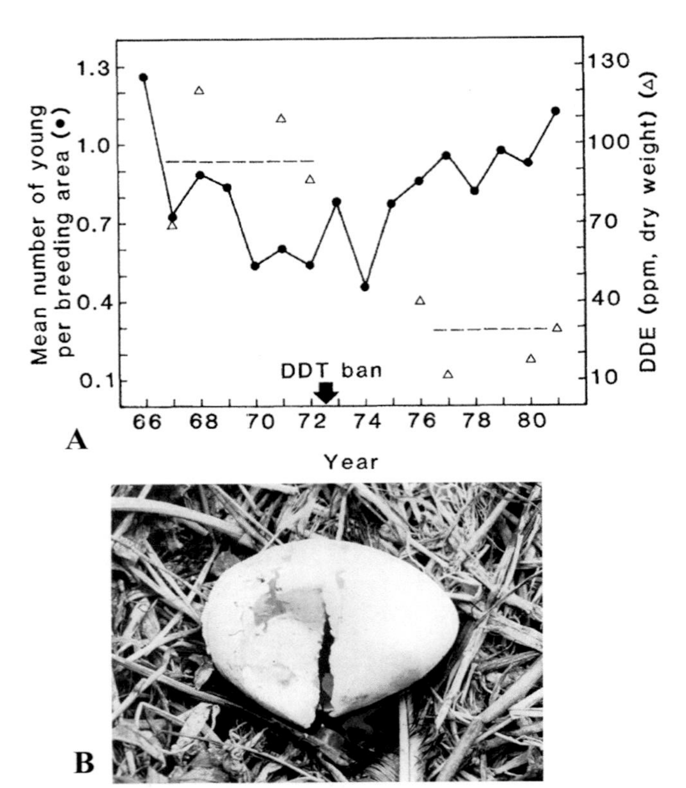

Figure 25.3 The influence of DDT on bald eagle nesting pairs and pelican eggshell thickness. (A) Average rates of bald eagle reproduction and DDE concentrations in eggs in northwestern Ontario from 1966 to 1981. Dashed lines denote mean concentrations of DDE in clutches before (1967–1972) and after (1976–1981) the ban of DDT. **(B)** A crushed egg in the nest of a brown pelican off the California coast had such a thin shell that the weight of the nesting parent's body destroyed it long before the embryo inside was ready to hatch. The concentration of DDE in the eggs of this 300-pair colony reached 2,500 parts per million; no eggs hatched.

Source of Images: (A) Modified from Grier, J.W., 1982. Science, 218(4578), pp. 1232-1235. (B) Jehl, J.R., 1973. Condor, 75(1), pp. 69-79. Used by permission.

DDT and its metabolites have the potential to act as both hormone mimics and receptor antagonists.[29]

Beyond DDT's demasculinizing effects on male birds, perhaps its most devastating effect on reproduction was that the chemical also caused the eggshells of affected birds to become extremely thin and fragile, making them more prone to predation and being accidentally killed by their parents when the thinned eggshells broke under the weight of the bird during incubation[30,31] (Figure 25.3 B). Eggshell formation is under endocrine control, and its disruption is now known to be affected by high levels of DDE that downregulate the synthesis of carbonic anhydrase and prostaglandins that are essential for transporting calcium ions by shell glands that promote calcium carbonate deposition.[32–34] Importantly, the pathological effects of DDT are now known to extend well beyond reproduction, also affecting thyroid, adrenal, and hepatic functions.[34]

Although DDT was ultimately banned in the United States in 1972, the chemical was so widely used and is so resistant to degradation that it remains present to this day in virtually all vertebrates, including humans. In addition, because DDT has a half-life of approximately 15 years, it can take over 100 years for concentrations of the compound to fall below active levels in the environment. Furthermore, the primary metabolite of DDT, DDE, is even more persistent than its parent compound and is still found in the environment. Strikingly, in humans, *in utero*, neonatal, and prepubertal exposures to DDT have been linked to increased risk of breast cancer.[35–37] Reports from regions of the world where DDT is still in use suggest that its exposure may also be associated with other pathologies, including reduced semen quality, menstrual disruption, preterm birth, and lactation defects.[38–40]

Despite DDT's ban, its legacy persists in the form of its chemical cousins, the organophosphate pesticides (e.g. monocrotophos, methamidophos, and carbofuran), which are also highly toxic to birds and other vertebrates.[41] Although these agricultural chemicals are rated as Class I toxins by the World Health Organization and are either restricted or banned in the United States and Europe, they are widely used in Central and South America, as well as the Caribbean. Not only have these pesticides been linked to a recent large decline in migratory songbird populations that use these locations as winter nesting grounds,[42,43] but these chemicals are entering the global food market via imported fruits and vegetables.

SUMMARY AND SYNTHESIS QUESTIONS

1. Although DDT use in the U.S. was banned in 1972, leading to declines in gross toxic effects and mortality rate in wildlife, its detrimental effects continue to persist decades later. Why?
2. DDT exerts a "double-whammy" of both estrogenic and antiandrogenic effects. What is its proposed mode of action?

Rise of the Environmental Estrogens

LEARNING OBJECTIVE Define "environmental estrogen", and explain how wildlife species, such as fish, birds, and alligators, can function as "sentinel" species.

KEY CONCEPTS:

- Many persistent organic pollutants, like DDT, act as "environmental estrogens", or synthetic compounds with estrogenic activity at extremely low, ecologically relevant concentrations.
- Vitellogenin, a gene naturally induced by estrogen in female fish and other vertebrates, is also induced by estrogen-mimicking chemicals in males.

- "Sentinel species" are wildlife that can be physiologically monitored for long-term health effects of environmental chemical exposures on an ecosystem.

Additional hints that the destructive actions of DDT and other *persistent organic pollutants* (POPs) on vertebrate physiology were linked to their disruptions of endocrine function began to emerge during the 1970s when ecologists started observing unusual reproductive dysfunctions in animals. For example, domesticated mink fed on a diet of fish from the North American Great Lakes virtually stopped giving birth to pups,[44,45] and the embryos of herring gulls were dying in their eggs.[46] In the 1980s, researchers began to document disturbing examples of feminization of male gull embryos by DDT, such as the development of oviducts and ovotestes.[47] Unusual sexual behaviors in gulls were also observed, including a marked disinterest in mating or pairing by male gulls and the same-sex pairing and nesting of female herring gulls.[48] Altered sex development by synthetic chemicals was not confined to vertebrates, as male-to-female sex changes in the dog whelk (a mollusk) were found to be induced by tributyltin, an antifouling agent from ship-hull paints.[49,50]

In the 1990s in England, Peter Sumpter and colleagues began to document the estrogenic effects of detergent components in sewage effluent on male fish using vitellogenesis as a biomarker.[51,52] The group would later report male fish with severe reproductive abnormalities, including some expressing vitellogenin and others with testes containing oocytes.[53,54] Around the same time, McLachlan and colleagues reported that *polychlorinated biphenyl* (PCBs) elicited estrogenic activity, affecting turtle sex determination.[55]

Developments & Directions: Vitellogenesis in male fish as a biomarker for estrogenic contamination of aquatic environments

Vitellogenin is a yolk protein that is produced by the liver following hormonal stimulation, and *vitellogenesis* is the process of yolk synthesis and deposition in growing oocytes of a female. Although expression of the vitellogenin gene is under multihormonal control, estrogens play a dominant role in vitellogenesis[56] (Figure 25.4[A]). Therefore, in female fish plasma vitellogenin concentrations rise dramatically during sexual maturation, concomitant with increasing estrogen levels, to become the major blood protein. By contrast, very little, if any, vitellogenin is normally detectable in male fish,[57] presumably due to low levels of circulating estrogen. In a series of influential studies, Peter J. Sumpter

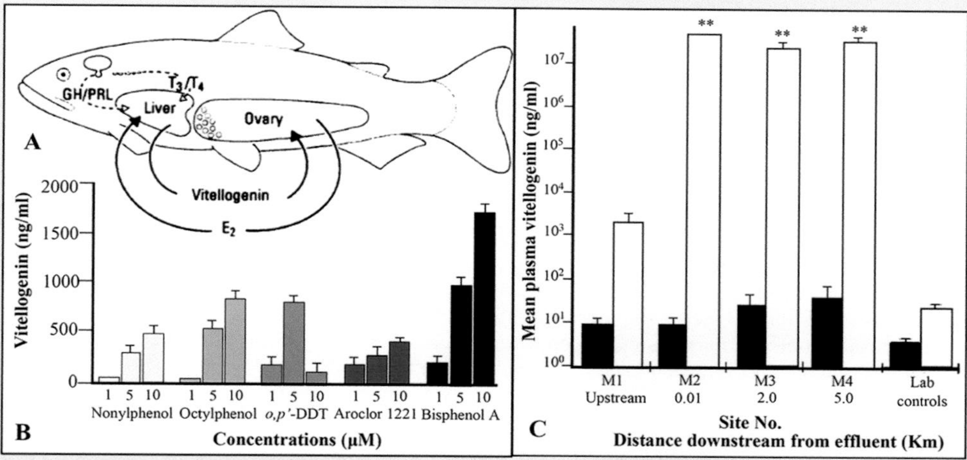

Figure 25.4 Vitellogenesis is induced by low levels of estrogenic compounds and by wastewater effluent in male trout. **(A)** Control of vitellogenin synthesis is under multihormonal control, though estradiol (E2) plays a prominent role. **(B)** Estrogenic activity of some environmentally persistent chemicals, measured by their ability to stimulate vitellogenin synthesis in cultured hepatocytes obtained from male rainbow trout. Nonylphenol and octylphenol are degradation products of widely used surfactants; o,p'-DDT is a pesticide; Aroclor is used primarily in electrical capacities and transformers; and bisphenol A is a plasticizer. All five chemicals are aquatic pollutants. Note that in each case the stimulatory effect was dose related, with the exception of o,p'-DDT in which the highest concentration tested (10 pM) was toxic to the hepatocytes. Results (mean ± SEM; n = 6) are expressed as the vitellogenin concentration in the culture medium after a 2-day exposure to the chemicals. **(C)** Mean plasma vitellogenin concentration (ng/ml) in caged male trout before and after being held in the River Aire (northern England) for 3 weeks during the summer of 1994. Sites labeled MI through M5 represent caged male rainbow trout held at various distances from the Marley Sewage Treatment Facility. "Lab controls" are control fish held in dechlorinated tap water at the laboratory. ** denotes significant ($p < 0.001$) differences between values at the sites and laboratory control values.

Source of Images: (A-B) Sumpter, J.P. and Jobling, S., 1995. Env. Health Perspec., 103(suppl 7), pp. 173-178. (C) Harries, J.E., et al. 1997. Env. Toxicol. Chem., 16(3), pp. 534-542. Used by permission.

and colleagues[51,52,58,59] used an *in vitro* system of cultured hepatocytes to demonstrate that estrogens and diverse estrogen-mimicking compounds elicit vitellogenesis in male trout tissues (Figure 25.4[B]). Though all compounds tested are weakly estrogenic, the chemicals nonetheless stimulated vitellogenin synthesis at concentrations reported to be present in the aquatic environment. Critically, the researchers also performed studies with caged male trout held downstream of sewage treatment effluent from several different cities on rivers in the United Kingdom, and found that the water significantly elevated blood vitellogenin levels, irrespective of the distance downstream[58] (Figure 25.4[C]). The group would later go on to show altered sexual development in males from wild populations of another fish, the roach (*Rutilus rutilus*), living downstream of waste treatment effluent. In addition to possessing elevated vitellogenin levels, many males displayed pathological phenotypes ranging from malformation of the germ cells and/or reproductive ducts to altered gamete production.[53] Together, these laboratory and field study findings make a strong case that very low concentrations of estrogenic compounds in these rivers are having a measurable impact on the health of fish populations.

In Lake Apopka, Florida, Louis Guillette and colleagues reported that in waters contaminated with PCBs and residues of organochlorine pesticides, alligator populations had stopped reproducing, with many males possessing small, demasculinized, dysfunctional genitalia.[60,61] The males of another Floridian top predator, the Florida panther (*Puma concolor coryi*), also demonstrated alarmingly high rates of malformed reproductive organs. Already threatened by plummeting population numbers caused by human encroachment and reduced habitat availability, researchers began to report rapidly increasing rates of **cryptorchidism**, or undescended testicles, in newborn males, a condition associated with a high incidence of damaged sperm and infertility.[62,63] Findings by Facemire and colleagues (1995) indicated that the presence of pollutants that include p,p′-DDE, PCBs, and mercury were likely contributors to this reproductive impairment. The researchers concluded, "evidence presented in this paper, including the fact that there appears to be no significant difference between serum estradiol levels in males and females, suggests that many male panthers may have been demasculinized and feminized as a result of either prenatal or postnatal exposure". In addition to these diverse observations in the field, many controlled laboratory-based studies also demonstrated that a variety of these persistent organic pollutants indeed act as "*environmental estrogens*", or synthetic compounds with estrogenic activity at extremely low, ecologically relevant concentrations.[64,65]

Developments & Directions: Of pesticides and penises: The Lake Apopka alligator apocalypse

Professor Louis (Lou) J. Guillette, Jr. (1954–2015), a comparative endocrinologist and an expert on American alligator reproduction at the University of Florida, Gainesville, was one of the most influential researchers in the field of environmental health (Figure 25.5). Early in his career he partnered with the Florida Game and Fresh Water Fish Commission to evaluate the general health of Florida alligators in large lakes. It didn't take long for Guillette to realize that something was fundamentally wrong with the alligators of one lake in particular, Lake Apopka, Florida's fourth largest freshwater body. Compared with alligator eggs in nests in other lakes that displayed hatching rates of between 70% and 80%, only 5%–20% of Lake Apopka eggs hatched, with neonates displaying very high mortality rates.[34] Of particular concern was the fact that Lake Apopka contained very high levels of organochlorine compounds from municipal waste, agricultural runoff, and a 1980 pesticide spill from the Tower Chemical Company that contained the compounds dicofol (chemically similar to DDT), as well as DDT and its metabolites.[66] Lake Apopka alligators and their eggs also contained high concentrations of these same organochlorine compounds.[67,68] Strikingly, compared to alligators from Lake Woodruff, a relatively pristine lake located within a national wildlife refuge in central Florida with no known point source of organochlorine contamination, Guillette and his colleagues documented numerous endocrine and reproductive anomalies in alligators from Lake Apopka, as well as from other lakes in Florida with varying degrees of organochlorine contaminants. These findings[61,69] include:

1. Reduced penis sizes (one-half to one-third the normal size) in male Lake Apopka alligators (Figure 25.5[A]). Since penis size is androgen-dependent in crocodilians and the potent androgen receptor antagonist, p,p′-DDE, was elevated in the serum collected from Lake Apopka juvenile alligators, Guillette and colleagues hypothesized that DDE acts to block the growth of penis tissue, as well as other androgen-dependent structures, in developing alligators.[34] Indeed, when researchers painted DDE onto the outer surface of alligator eggshells derived from pristine waters, a similar induction of reduced penis size was observed in neonates.[70]
2. Abnormally developed gonads in both sexes of juvenile Lake Apopka alligators. Compared with Lake Woodruff, male Lake Apopka alligators had testes with poorly organized seminiferous tubules and an absence of clearly defined Leydig cells and spermatogonia.[71] Strikingly, these testicular abnormalities in alligators from Lake Apopka are similar to those in human workers exposed to dibromochloropropane (DBCP, a brominated organochlorine pesticide) in that the seminiferous tubules are the affected target tissues.[60,72] Female Apopka alligators possessed abnormal ovaries with an increased number of polyovular follicles (containing more than one oocyte) and polynuclear follicles (possess more than one nucleus). It is notable that similar ovarian abnormalities were also observed in mice injected with the potent synthetic estrogen diethylstilbestrol (DES, a compound discussed further later).[73]
3. Abnormal sex steroid hormone levels in both sexes of juvenile Lake Apopka alligators. Whereas Lake Apopka males possessed plasma testosterone concentrations that were over three-fold lower than normal Lake Woodruff males and also comparable to those of normal Lake Woodruff females, plasma estradiol levels in both Apopka males and females were significantly elevated compared to those of Lake Woodruff males and females, respectively. These findings are summarized as differences in estrogen/testosterone ratios in Figure 25.5(B). The researchers concluded from these findings that "the gonads of juveniles from Lake Apopka have been permanently modified *in ovo*, so that normal steroidogenesis is not possible, and thus normal sexual maturation is unlikely".[61]

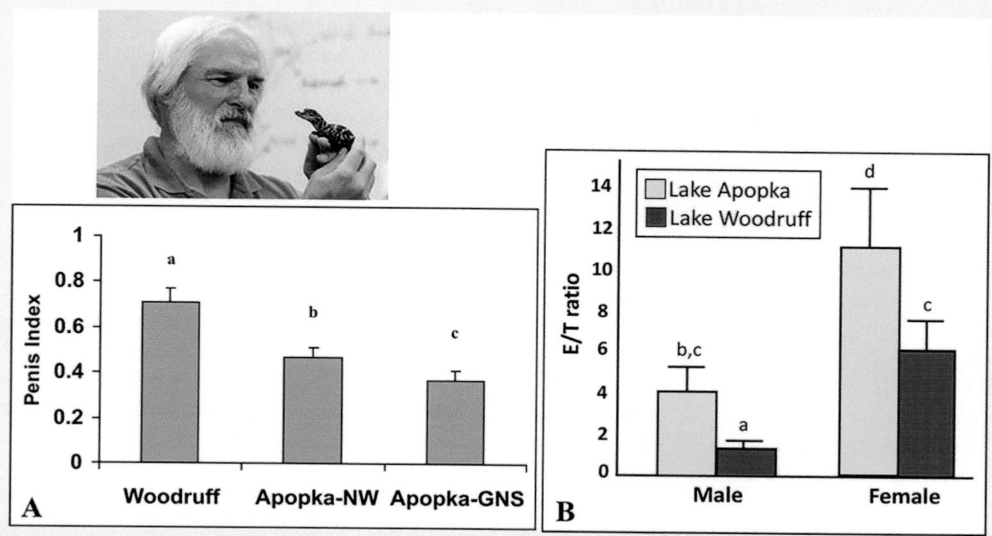

Figure 25.5 Professor Lou Gillette studied the effects of estrogen-disrupting compounds on reproduction in alligator populations. (A) Mean phallus size in alligators (*Alligator mississippiensis*) from a control lake (Lake Woodruff) and a contaminated lake (Lake Apopka). Penis size is represented as an index [(penis tip length × penis base width)/snout vent length]. The letters a, b, c indicate significant differences ($p > .05$) among lake regions. Lake Apopka samples are separated into two localities, the Gourd Neck Spring area (Apopka-GNS), where contaminants from the Tower Chemical spill entered the lake, and the northwestern part of the lake (Apopka-NW), which is farther from the spill. Redrawn from Guillette et al. (1996). **(B)** Mean (± SE) estrogen/testosterone ratios for male and female juvenile alligators from Lakes Apopka and Woodruff. Bars with different letters are significantly different.

Source of Images: Lou Guillette photograph: Courtesy of National Institute of Environmental Health Sciences. (A) UNEP, W., 2013. State of the science of endocrine disrupting chemicals-2012. WHO-UNEP, Geneva. Used by permission. (B) Guillette Jr, L.J., et al. 1994. Env. Health Perspect., 102(8), pp. 680-688. Used by permission.

Professor Guillette's research with alligators developed into one of the most thorough and convincing demonstrations that endocrine-disrupting chemicals promote reproductive dysfunction in wildlife. He and his colleagues showed that alligators may act as a powerful *sentinel species*, or animals used to measure risks to long-term wildlife health by providing advance warning of a danger. Beyond that, his research stimulated another generation of scientists to study parallel influences of endocrine-disrupting chemical exposure on human reproductive pathology.

SUMMARY AND SYNTHESIS QUESTIONS

1. Lake Apopka, FL, suffered a chemical spill in 1980. However, by the turn of the century the chemicals are entirely absent from the water and the lake is rich in alligators and turtles. Why is there still a big problem, and what is the nature of the problem?
2. What is the significance of the observation that male fish living in a certain lake have high levels of the protein vitellogenin in their blood?

Diethylstilbestrol: The First Designer Estrogen

LEARNING OBJECTIVE Discuss the importance of diethylstilbestrol (DES) in founding the field of endocrine disruption studies.

KEY CONCEPTS:

- Diethylstilbestrol (DES), the first synthetic compound to be specifically designed and marketed as a medically prescribed estrogen, was also the first known example of a human "transplacental" carcinogen.
- Fetal exposure to DES resulted in the occurrence of rare cancers and other pathologies in women and men later in life.

As alarm over the harmful effects of xenoestrogens on the health of wildlife spread, concern about how such compounds might affect human health also increased. Unfortunately, the clearest evidence that exposure to a synthetic estrogenic compound during fetal development can have profoundly detrimental effects on human health had already started to emerge by the early 1970s. This became apparent in a succession of medical tragedies involving the first synthetic compound to be purposefully designed and marketed as a medically prescribed estrogen,

diethylstilbestrol (DES). DES is a potent estrogen agonist, and has a binding affinity for ERα and ERβ that is approximately equivalent to estradiol itself.[74] Despite findings that DES induced miscarriages in rabbits and rodents,[75] DES was initially approved as a drug for *preventing* human miscarriages in the U.S. and Europe beginning in 1938. Ultimately, the drug was advertised and used as a pregnancy supplement for producing "stronger babies", and also given to newborns to augment their weight gain[21] (Figure 25.6 A). It is estimated that DES was taken by as many as 10 million pregnant women in the U.S. alone prior to its ultimate suspension in 1971. DES was also used for cosmetic purposes as an ingredient in shampoos and lotions. DES implants in poultry and livestock were used in the U.S. to stimulate animal growth by increasing fat deposition, leading to its release into the environment through feed lots and cattle waste until it was phased out in 1979 (Figure 25.6 B).[21]

Tragically, not only was DES ineffective at preventing miscarriage (indeed, it was actually found to increase the chance of miscarriage), but the compound was found to cause extensive reproductive harm to many of the developing male and female fetuses, damage that in most cases would not manifest until years to decades after birth. One of the first consequences of fetal exposure to DES was identified by physicians in 1971. They observed that girls born to mothers who were prescribed DES during pregnancy (referred to at the time as "DES daughters") were much more likely to develop an extremely rare form of cervical cancer called *cervicovaginal clear-cell adenocarcinoma* (CCAC).[76,77] As such, DES was the first known example of a human "transplacental" carcinogen, or a chemical that when administered to the mother causes cancer in her daughter.[78] In addition to increased incidences of cancer and reproductive abnormalities, DES daughters were also at higher risk for developing infertility and more complicated and unsuccessful pregnancies,[79] as well as increased rates of diverse psychiatric disorders and learning disabilities.[80] Male offspring were also not immune to damage from DES and were born with unusually high incidences of cryptorchidism and other urogenital malformations.[81,82] Additionally, as adults these DES sons were more prone to infertility caused by low sperm density and mobility,[83,84] as well as to developing testicular cancer.[85] Adverse effects of DES have even been reported in the second generation, with the sons of DES daughters being born with increased

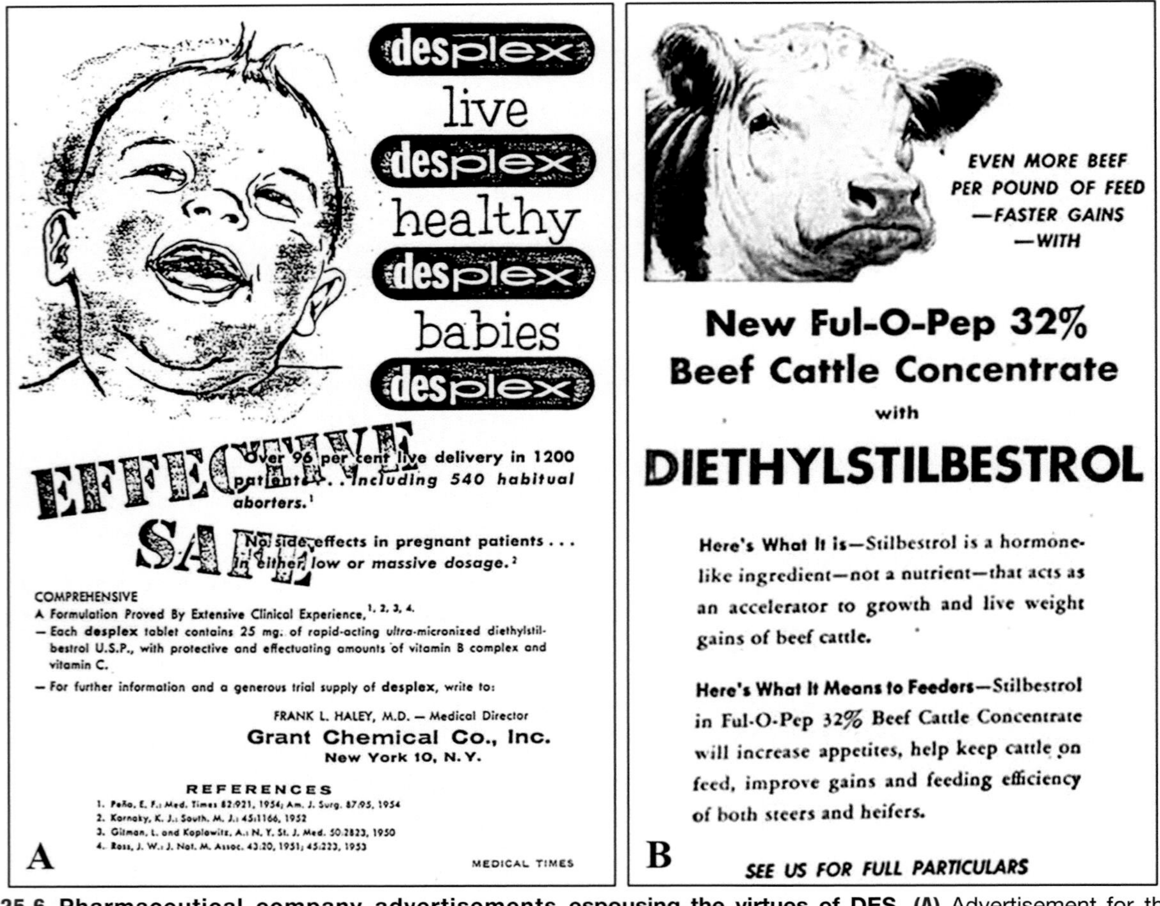

Figure 25.6 Pharmaceutical company advertisements espousing the virtues of DES. (A) Advertisement for the Grant Chemical Company, Brooklyn, NY, printed in *Medical Times* in 1957. **(B)** Advertisement in *The Western Producer*, May 1956.

Source of Images: Figures courtesy of DES Daughter.

risk of hypospadias, the condition where the opening of the urethra is on the ventral side of the penis instead of at the tip.[86] The full impact of DES's effects on human health still remains undiscovered, as the youngest DES daughters and sons entered middle age in approximately 2015, with many still unaware of their exposure to DES *in utero* and/or as infants.

SUMMARY AND SYNTHESIS QUESTIONS

1. Why do women whose mothers took DES only after the 20th week of pregnancy usually not suffer from reproductive tract deformities compared to those exposed to DES before the 10th week?

The "Fragile Fetus"

LEARNING OBJECTIVE Describe the concept of the "fragile fetus", and explain why embryos, fetuses, and newborns are so susceptible to the effects of endocrine disruption compared with adults.

KEY CONCEPTS:

- Fetuses possess several attributes that make them more susceptible than adults to the effects of endocrine-disrupting compounds, including a high metabolic rate, rapid cell proliferation and differentiation, a nonfunctional blood-brain barrier, and an underdeveloped liver and immune system.
- Subtle functional changes in specific tissues during sensitive developmental windows of early life can result in increased susceptibility to disease or dysfunction later in life.

Studies with laboratory animals confirmed the transplacental effects of DES, and in addition to CCAC, mice exposed prenatally to DES also displayed oviduct malformations and ovarian cysts.[87] Importantly, the effects of this potent estrogen on laboratory animals set the standard for evaluating the activities of other estrogenic compounds, demonstrating the critical importance of animal models in human and wildlife risk assessment. DES's tragic legacy introduced physicians and scientists for the first time to the potential of certain chemicals to be transmitted transplacentally and not just cause physical deformities at birth, but also far more subtle effects that may emerge decades later.[64] Physicians could no longer assume that a baby who was apparently healthy at birth remained unharmed by chemicals that it was exposed to during gestation. In 1992, Professor Howard Bern at the University of California, Berkeley, coined the term "the fragile fetus" to describe the extreme vulnerability of a developing embryo to disruption by environmental chemicals, especially by those with hormone-disrupting activity, at concentrations far below those that would be considered harmful in the adult.[88] Bern suggested that several fetal attributes contribute to this high vulnerability to chemical insult, including rapid cell proliferation and differentiation, a nonfunctional blood-brain barrier, and an underdeveloped liver and immune system. Furthermore, the developing fetus has an elevated metabolic rate compared to an adult, which may result in increased toxicity.

The notion that subtle functional changes in specific tissues during sensitive developmental windows of early life can result in increased susceptibility to disease or dysfunction later in life would lay the foundation for a new paradigm for understanding non-communicable disease known as the **developmental origins of health and disease** (DOHaD)[89] (described in Chapter 19: Metabolic Dysregulation and Disruption). The theoretical underpinnings of the "fragile fetus" concept were also refined in a field called *ecological developmental biology*, a discipline founded on the notion that the environment, along with genetics, codetermines an organism's phenotype.[90] Beyond the fetus, young children are generally also at greater risk of exposure to hormone- and development-disrupting chemicals due to several reasons that include (1) exposure to fat-soluble contaminants in breast milk; (2) children more frequently place their hands and objects in their mouth compared with adults; (3) children live and play close to the ground where harmful chemicals may be encountered; and (4) since they are small, they possess a higher surface area to volume ratio (skin area relative to their body weight) than adults, which permits a greater absorption of chemicals.[91]

SUMMARY AND SYNTHESIS QUESTIONS

1. Prior to the 1970s, it was assumed by doctors that the placenta protected the fetus from environmental toxins. What changed that notion?
2. Relatively low levels of EDCs may have no measurable effect on adults, but can exert profound effects on embryos and fetuses. Why?

The Founding of a New Field of Study: Endocrine Disruption

LEARNING OBJECTIVE Describe factors that influenced the coining of the term "endocrine disruptor".

KEY CONCEPTS:

- The term "endocrine disruptor" was coined in 1992, resulting from the convergence of the human DES tragedy and the growing awareness of the effects of environmental hormone-disrupting chemicals on wildlife.

Although research into the hormone-mimicking and hormone-disrupting effects of certain pharmacological and industrial compounds initially focused on their estrogenic properties and their roles in decreased wildlife fertility and increased birth anomalies, studies soon expanded to address the disruption of many other hormones, including androgens,[27] thyroid hormones,[92] and adrenal hormones,[93] as well as metabolic disruptions linked to obesity and type 2 diabetes,[94] the development of hormone-related cancers,[95] and neurodevelopmental disorders.[96] Throughout the 1970s and 1980s, the growing awareness of the effects of environmental hormone-disrupting chemicals on wildlife converged with the human DES tragedy, and in 1991 scientists from a variety of disciplines convened at the Wingspread Conference Center in Racine, Wisconsin, organized by Theo Colborn and colleagues.[97] Wingspread was a key turning point in the development of the emerging field of endocrine disruption, and it was there that the term "endocrine disruptor" was coined. The meeting's participants wrote a powerful consensus statement that started with an unambiguous declaration: "We are certain of the following: A large number of man-made chemicals that have been released into the environment, as well as a few natural ones, have the potential to disrupt the endocrine system of animals, including humans".[97] Several years after the seminal conference, in 1996, the U.S. EPA officially recognized and defined an *endocrine disruptor* as

> an exogenous agent that interferes with the production, release, transport, metabolism, binding, action, or elimination of natural hormones in the body responsible for the maintenance of homeostasis and the regulation of developmental processes.[98]

The Food Quality Protection Act was subsequently passed and the Safe Drinking Water Act was amended, mandating that the U.S. Environmental Protection Agency develop methods to evaluate the potential of suspected chemicals to disrupt human endocrine function. Today, the Endocrine Society, the world's oldest, largest, and most active organization devoted to research on hormones and the clinical practice of endocrinology, has simplified the EPA's definition to read: "An endocrine disruptor is an exogenous chemical, or mixture of chemicals, that interferes with any aspect of hormone action".[99]

In 1996, Theo Colborn and colleagues published a landmark book titled *Our Stolen Future: Are We Threatening Our Fertility, Intelligence, and Survival?: A Scientific Detective Story*.[100] As did *Silent Spring* before it, *Our Stolen Future* also stimulated public awareness of the broad dangers associated with *endocrine-disrupting chemicals* (EDCs), this time by weaving together examples of EDC-induced reproductive anomalies in wildlife, livestock, laboratory animals, and humans to make a compelling case that EDCs constitute a global threat. The book also makes a powerful argument for how lessons learned from wildlife not only provide basic knowledge of the mechanisms of action of EDCs, but also directly inform how EDCs influence human health as well. In this regard, the use of wildlife sentinel species in biomonitoring programs to detect EDCs in contaminated regions is particularly relevant to both wildlife and human welfare.[101–104]

SUMMARY AND SYNTHESIS QUESTIONS

1. Describe how observations by wildlife researchers and human clinical research converged to create the new field of endocrine disruption research in the 1990s.

Sources of EDCs

LEARNING OBJECTIVE Describe the primary sources of endocrine-disrupting chemicals (EDCs).

KEY CONCEPTS:

- The primary origins of EDCs are industrial, agricultural, residential, pharmaceutical, and heavy metal.
- Exposure to EDCs may occur via eating, drinking, breathing, or absorption via skin contact.

Today, of the approximately 80,000 known synthetic chemicals in our environment, about 1,500 have been classified as EDCs,[7] a small fraction that will certainly increase as thousands of new chemicals (most developed with little to no toxicological testing) are manufactured every year. Alarmingly, EDCs have been found to be present in virtually every human and animal tested, from the Arctic to the Antarctic, and everywhere in between.[105,106] Furthermore, one study in the U.S. showed that when pregnant women were screened for 163 synthetic compounds, over 90% had at least 62 of these chemicals in their bodies.[107] The sources of EDCs are diverse, and some examples, according to their origin, can be grouped as follows:

- Industrial: lubricants/solvents and their metabolites, including polychlorinated biphenyls (PCBs), polybrominated biphenyls (PBBs), and, dioxins
- Agricultural: pesticides (e.g. methoxychlor, chlorpyrifos, and DDT), herbicides (e.g. atrazine), fungicides (e.g. vinclozolin), and phytoestrogens (e.g. genistein)
- Residential: plastics (e.g. bisphenol A, BPA), cosmetics and fragrances (phthalates parabens)
- Pharmaceutical: e.g. DES and birth control pills
- Heavy metals: e.g. cadmium, lead, mercury, and arsenic

Some naturally occurring plant derivatives are also classified as EDCs. These include *phytoestrogens*, which are plant-derived estrogen receptor agonists, such as genistein and coumestrol. Claude Hughes, a neuroendocrinologist at Cedars-Sinai Medical Center, has proposed the interesting hypothesis that some plants, such as soy, evolved phytoestrogens specifically to reduce the fertility of herbivores, essentially producing oral contraceptives to reduce their probability of being eaten.[108] Exposure to EDCs may occur via eating contaminated food or medications, drinking affected water, breathing contaminated air, or via skin contact with agro-industrial chemicals directly or indirectly through contaminated soil. A list of some relatively well-described EDCs and their sources and physiological effects is shown in Appendix 13. As can be seen in the table, EDCs originate from different chemical groups with highly variable characteristics and mechanisms of action. This section will explore some of the characteristics of EDCs that make them so potentially threatening to the health of humans and wildlife alike.

EDC in the Spotlight: BPA

Of all the known EDCs that are in use worldwide today, perhaps none has been synthesized in as vast a quantity or has drawn more public scrutiny than *bisphenol A* (BPA).[64] BPA (Spotlight Figure 25.7(A)) is currently used as a monomer in the manufacture of common polycarbonate, polysulfone, and other plastic products found in children's toys, the lining of food cans, epoxy resins, food containers, medical tubing, dental sealants, thermal receipt paper, and many others. Indeed, due to the ubiquitous use of BPA-containing products, BPA is detectable in the blood and urine of most humans, including 93% of all Americans.[109,110] Exposure to the monomer occurs as a result of the depolymerization and migration of BPA out of these products. Importantly, BPA is a well-described estrogen mimic in humans and in rodent models,

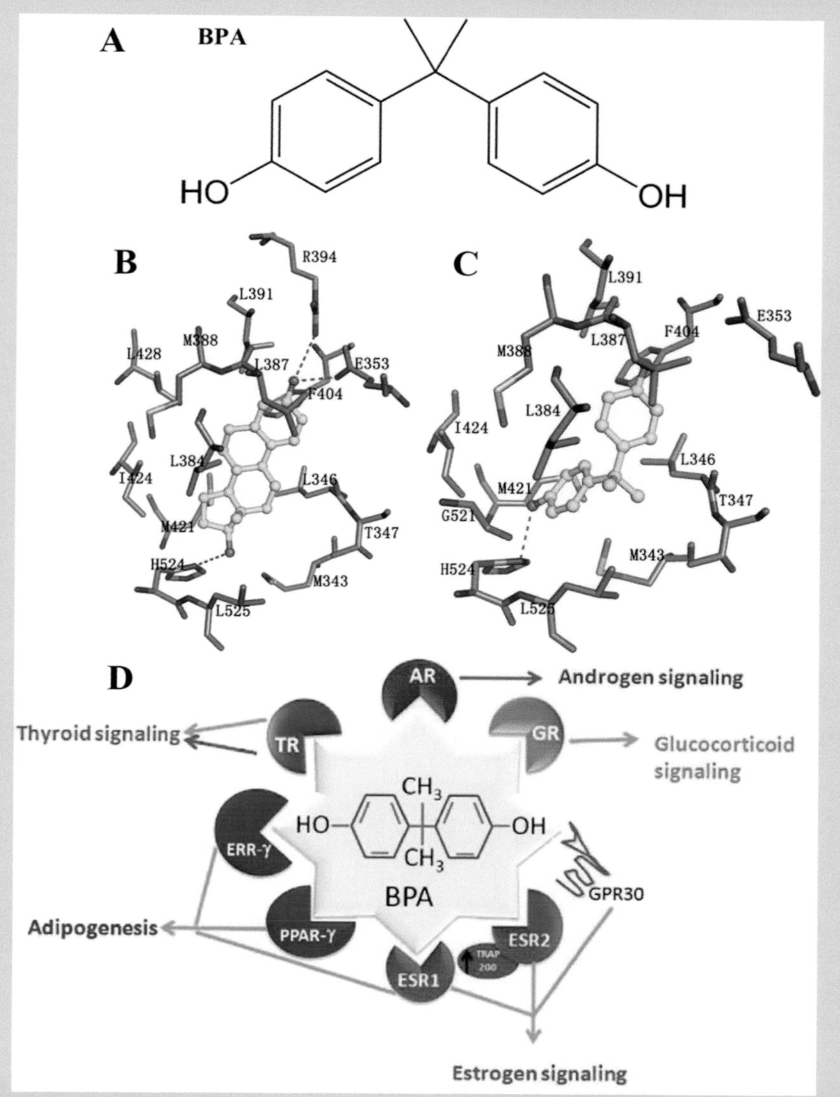

Figure 25.7 The structure of BPA (A). The interaction of the estrogen receptor at the ligand-binding domain with estradiol **(B)** and BPA **(C)**. The ligands are shown in yellow balls and sticks. hERα is shown in purple. Hydrogen bonds formed between the ligand and receptors are indicated as red dashed lines. BPA alters multiple endocrine signaling mechanisms **(D)**. Green and red arrows indicate agonistic and antagonistic effects, respectively, on various hormonal pathways. Abbreviations: estrogen receptors alpha and beta (ER alpha and ER beta); thyroid receptor (TR), androgen receptor (AR), glucocorticoid receptor (GR), estrogen membrane receptor (GPR30), peroxisome proliferator-activated receptor gamma (PPAR-γ), estrogen-related receptor-gamma (ERR-γ).

Source of Images: (A-C) Li, L., et al. 2015. PloS one, 10(3), p. e0120330. (D) Cooke, P.S., et al. 2013. Endocrine disruptors. In Haschek and Rousseaux's Handbook of Toxicologic Pathology (pp. 1123-1154). Academic Press. Used by permission.

widely reported to bind to the nuclear estrogen receptor (Spotlight Figure 25.7[B]) and impact fertility, behavior, cancer susceptibility, the development of obesity, and the development of many other maladies.[111] Indeed, BPA's estrogen-mimicking nature was well known as early as the mid-1930s when British medical researcher Edward Charles Dodds identified it in his search for synthetic compounds with estrogenic activity that could eventually be used to study the endocrinology of estrogen function.[112] Eventually, Dodds found a far more potent synthetic estrogen to work with instead, which he identified as diethylstilbestrol (DES).[113]

Critically, BPA not only binds to the 17β-estradiol receptor to exert both agonistic and antagonistic actions, but also to other nuclear receptors such as the orphan nuclear estrogen-related receptor gamma (ERR-gamma), the androgen receptor (AR), the peroxisome proliferator-activated receptor gamma (PPAR-gamma), and the thyroid hormone receptor,[114–116] as well as to the membrane-localized G protein-coupled estrogen receptor 30 (GPR30)[117,118] (Spotlight Figure 25.7[C]). Based on such observations, exposure to BPA is increasingly suspected to exert significant reproductive disruption in both males and females.[119,120] Although BPA's use in the manufacture of baby bottles and plastic nipples has been banned by the European Union, BPA has not been officially restricted by U.S. regulatory agencies. However, despite a lack of federal restrictions of BPA in the U.S., public attention to BPA's estrogenic activity has led to a remarkable shift in consumer demands, forcing manufacturers to phase the compound out of children's products and plastic water bottles.[64]

SUMMARY AND SYNTHESIS QUESTIONS

1. Why is it not surprising that BPA is an estrogen mimic?

Some Key Characteristics of EDCs

LEARNING OBJECTIVE Describe the biochemical features and mechanisms of action of EDCs, and how the study of endocrine disruption differs from that of classical toxicology.

KEY CONCEPTS:

- EDCs are chemically heterogeneous, making it difficult to predict based on structure alone whether a compound will exert endocrine-disrupting actions.
- Because many EDCs are chemically stable and lipophilic, they tend to bioaccumulate in the fats of animals, with the highest concentrations occurring in predators at the tops of food chains.
- Compared with other environmental toxicants, EDCs are active at extremely low concentrations.
- Multiple EDCs can interact to generate unpredictable interactive effects not only among themselves, but also with endogenous hormones, even at low doses that individually do not produce observable effects.
- The effects of some EDCs can be transmitted to the offspring of a parent, or even the children and grandchildren of their offspring, via multigenerational and transgenerational mechanisms.

EDCs Are Chemically Diverse

Certain classes of EDCs, such as PCBs, PBBs, dioxins, and pesticides, often contain halogen group substitutions by chlorine and bromine and may also have a *phenolic moiety*, or a class of chemical compounds consisting of a hydroxyl group bonded directly to an aromatic hydrocarbon group (Figure 25.8). Importantly, these phenolic moieties are thought to mimic steroid hormones and enable EDCs to interact with steroid hormone receptors. However, in general EDCs are highly heterogeneous (many do not share any common structure), making it is difficult to predict in advance and based on its structure alone whether a compound will exert endocrine-disrupting actions.[121] For example, even some heavy metals, such as cadmium, are thought to exert estrogenic activity (see Appendix 13).

Many EDCs Are Persistent, Are Lipophilic, and Bioaccumulate in Animals

Unlike hormones, which have relatively short half-lives, many EDCs are purposefully designed to have long half-lives, and their chemical stability makes them resistant to biodegradation. The half-life for DDT, for example, is 10–15 years. In some cases where an EDC, such as DDT, is metabolized, its metabolites (DDE and DDD) are also persistent and may be equally or more harmful than the parent compound itself.

Many but not all EDCs are lipophilic and can be stored and concentrated in lipid-rich tissues such as adipose tissue, the brain, liver, kidneys, and mammary glands for long periods of time. Maternal exposure to EDCs can result not only in the transmission of the chemicals to the fetus via the placenta during pregnancy, but also to the newborn infant through lipids in the breast milk[122] (Figure 25.9). Since EDCs tend to accumulate and persist in the lipids of tissues, once they enter the food chain they bioaccumulate in animals higher up the food chain. As such, animals at the tops of food chains, such as humans, birds of prey, alligators, polar bears, and other predatory animals, tend to have the highest concentrations of these chemicals in their bodies. Importantly, the storage of lipophilic EDCs in fatty tissues may result in not just chronic exposures to these contaminants, but also acute exposure when these compounds are mobilized during times of high fat store utilization, such as during migration, hibernation, or lactation.[34]

EDCs Are Active at Extremely Low Concentrations

A central tenet of toxicology that is attributed to the 16th-century Swiss physician and alchemist Paracelsus is

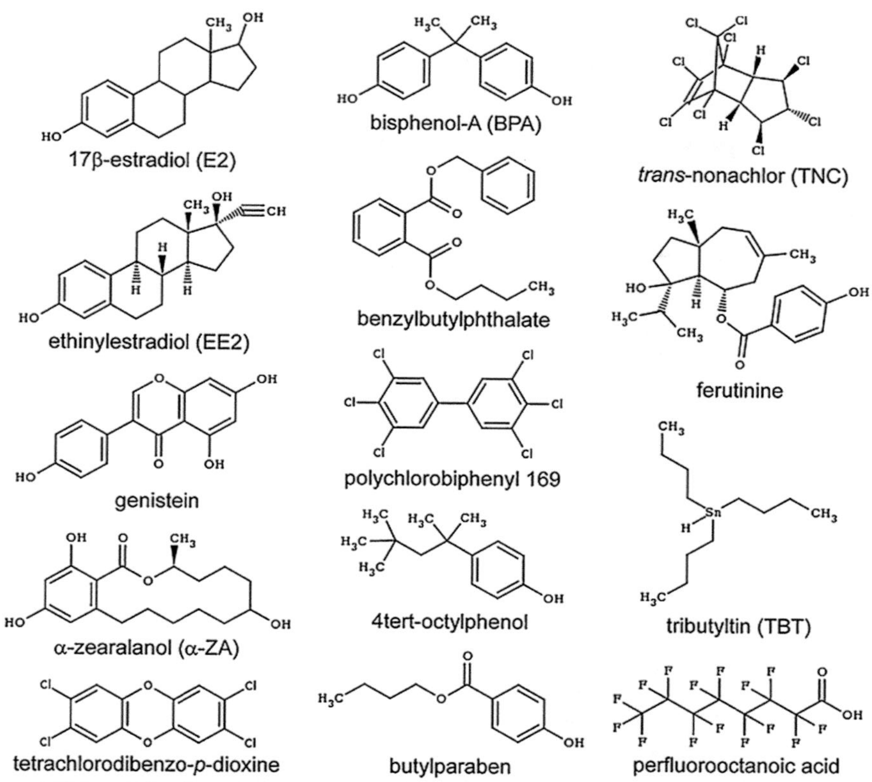

17β-estradiol (E2)
ethinylestradiol (EE2)
genistein
α-zearalanol (α-ZA)
tetrachlorodibenzo-p-dioxine
bisphenol-A (BPA)
benzylbutylphthalate
polychlorobiphenyl 169
4tert-octylphenol
butylparaben
trans-nonachlor (TNC)
ferutinine
tributyltin (TBT)
perfluorooctanoic acid

Figure 25.8 Chemical structures of some common endocrine disruptors.

Source of Images: Balaguer, P., Delfosse, V., Grimaldi, M. and Bourguet, W., 2017. Comptes rendus biologies, 340(9-10), pp. 414-420.

900
800
700
600
500
400
300
200
100
0
ng/g fat
Mothers' milk
Cord serum
Maternal serum
BDE-47
BB-153
HCB
β-HCH
Oxychlordane
Transnonachlor
4,4'-DDE
Mirex
CB-153

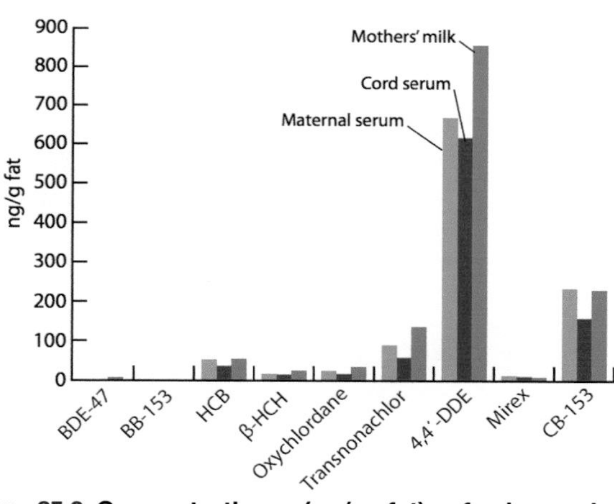

Figure 25.9 Concentrations (ng/g fat) of nine selected persistent organic pollutants (POPs) in maternal serum, cord serum, and mothers' milk from the same individual from the Faroe Islands (diagram prepared on basis of data from Needham et al., 2011). The abbreviated compounds are BDE-47 = 2,2',4,4'-tetrabromodiphenyl ether; BB-153 = 2,2',4,4',5,5'-hexabromobiphenyl; HCB = hexachlorobenzene; β-HCH = β-hexachlorocyclohe xane; 4,4'-DDE = 1,1-bis(4-chlorophenyl)-2-dichloroethene; CB-153 = 2,2',4,4',5,5'-hexachlorobiphenyl.

Source of Images: UNEP, W., 2013. State of the science of endocrine disrupting chemicals-2012. WHO-UNEP, Geneva. Used by permission.

"the dose makes the poison". This common sense notion implies that whereas a chemical can be harmless or even beneficial at lower concentrations, at higher concentrations it can be poisonous. Vitamin D supplements, for example, can be healthy at low doses, but at high concentrations can cause serious health problems, including hypercalcemia, kidney stones, high blood pressure, and hearing loss. The assumption that every chemical has a "safe exposure" has created the dogma that every compound has a distinct threshold, and that exposures to levels below that threshold are acceptable.[123] Unfortunately, many EDCs do not conform to this toxicological paradigm and have been found to be harmful at far lower concentrations than doses once considered to be safe. For example, in the 1980s, the U.S. EPA determined that a BPA dose below 50 mg/kg of body weight per day is safe, but evidence from multiple studies challenges this assumption. One of the first studies to challenge this was conducted by vom Saal and colleagues,[124] who showed that when pregnant mice were fed BPA at 2 αg/kg/day, male mice exposed *in utero* to the BPA had significantly enlarged and hypersensitized prostate glands in adulthood (Figure 25.10), altering their reproductive systems. Strikingly, the dose administered was 25,000 times lower than the EPA's "safe" dose and an environmentally relevant dose in the range consumed by humans. Other studies with mice treated with low doses of BPA have revealed alterations in maternal behaviors, earlier age at puberty, increased insulin resistance, and other developmental disturbances.[125–127]

Low Dose EDC Mixtures May Exert Interactive Effects

Perhaps one of the most worrying and least understood aspects of EDC nontraditional dose-response dynamics manifests in the ability of multiple EDCs to interact to generate sometimes unpredictable additive, synergistic, or antagonistic interactions not only among themselves, but also with endogenous hormones, even at low doses that individually do not produce observable effects.[128–132] Although wildlife and human populations are typically exposed to dozens if not hundreds of chemicals simultaneously, regulatory agencies typically test compounds individually for EDC activity, and not as mixtures or "cocktails".[130] As such, existing procedures for environmental risk assessment likely underestimate the dangers posed by EDC mixtures, ultimately generating erroneous conclusions of absence of risk.

The concept of synergistic interactions among EDCs is well exemplified in a study by Silva and colleagues,[133] who showed that estrogen-responsive cells treated with a mixture of weakly estrogenic EDCs, which individually induce only a weak reporter gene response, in mixture induce a response far greater than the predicted sum of the individual components (Figure 25.11). In another study using a whole-organism fish model, John Sumpter and colleagues[134] evaluated the combined estrogenic effects of five natural and synthetic chemicals on vitellogenin (VTG) induction in male fathead minnows (*Pimephales promelas*). VTG is a liver protein normally produced by female fish in response to elevated estrogen levels (it is deposited in the yolk of eggs), and the protein is not normally found at high levels in male fish. Two of the compounds studied, the natural steroidal estrogen 17β-estradiol (E2) and the synthetic estrogen 17α-ethynylestradiol (EE2), are waste products from human birth control pills that have been detected in water effluents that discharge into rivers at concentrations that are individually capable of inducing significant biological effects in fish.[135,136] The other

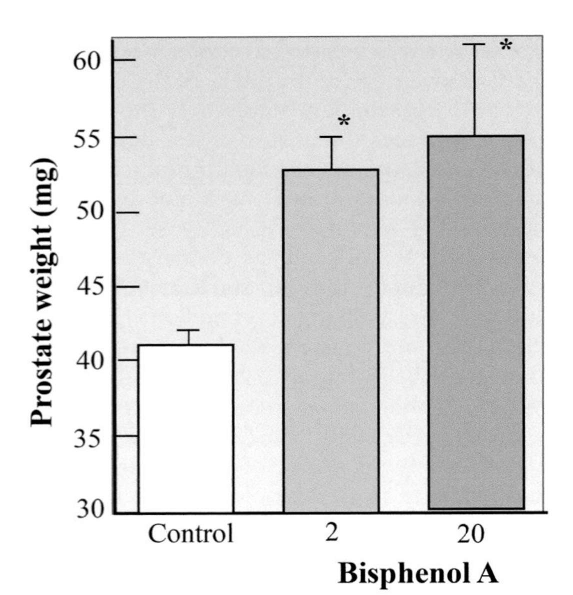

Figure 25.10 Prostate gland weight in adult male mice (6 months old) exposed as fetuses to bisphenol A. Mothers were fed 2 pg/kg body weight/day bisphenol A from days 1 to 17 of pregnancy. Error bars are the standard error; n= 7 for bisphenol A, and n= 11 for unexposed controls. *p<0.05.

Source of Images: Nagel, S.C., et al. 1997. Env. Health Perpect., 105(1), pp. 70-76.

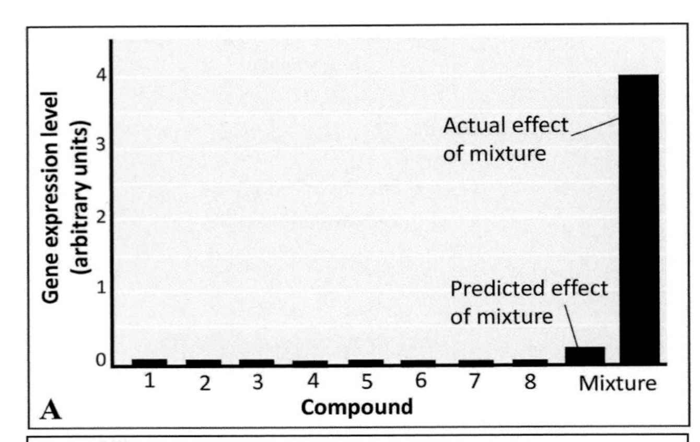

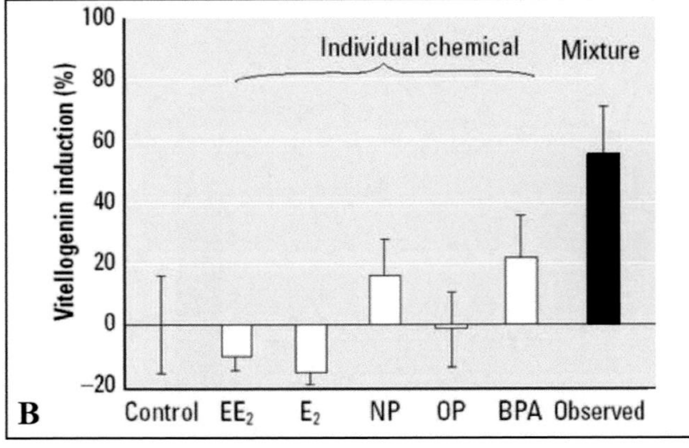

Figure 25.11 The combined estrogenic effects of multiple natural and synthetic chemicals on estrogen-responsive gene and protein expression. (A) Using a recombinant yeast estrogen screen, a mixture of eight estrogenic compounds at low concentrations activates an estrogen-responsive reporter gene at levels far greater than the predicted additive values of the individual responses. **(B)** The combined estrogenic effects of five natural and synthetic chemicals on vitellogenin (VTG) induction in male fathead minnows. At the low concentrations studied, each of the five chemicals failed to individually induce a VTG response that was significantly different from that of the controls. However, when the male fish were exposed to the same low doses in combination, VTG was significantly induced. EE2, synthetic steroidal estrogen; E2, estradiol; NP, 4-tert-nonylphenol; OP, 4-tert-octylphenol; BPA, bisphenol A.

Source of Images: Silva, E., et al. 2002. Env, Sci. Tech., 36(8), pp. 1751-1756.

EDC in the Spotlight: Atrazine

Atrazine (Spotlight Figure 25.12) is a triazine class herbicide that inhibits photosynthesis in plants and is used primarily as a weed killer in crops including sorghum, maize, and sugarcane.[137] Atrazine is one of the most widely used herbicides in the USA, second only to glyphosate, the active ingredient in the popular weed killer Roundup®. In 2001, atrazine was identified as the most abundantly detected pesticide contaminating U.S. drinking water,[138] and in 2003 it was banned by the European Union (EU) when groundwater levels were found to exceed the limits set by EU regulations.[139,140] Days after atrazine was removed from the EU market, the U.S. Environmental Protection Agency (EPA) approved the continued use of atrazine in the USA.[141]

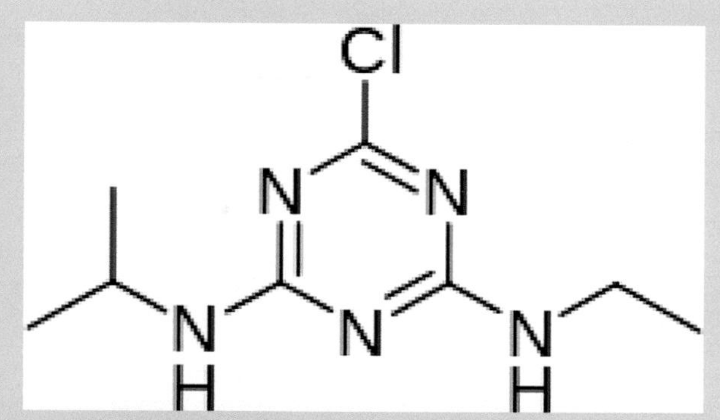

Figure 25.12 The chemical structure of atrazine.

Source of Images: Hou, X., et al. 2022. J. Mater. Sci: Materials in Electronics, 33(29), pp. 22710-22717. Used by permission.

Due to atrazine's broad use and presence in freshwater habitats of the USA, wildlife biologists have shown substantial interest in atrazine's effects on the development and health of freshwater vertebrates, and especially on amphibians that are particularly susceptible to aquatic toxicants due to their permeable skin.[142] The laboratory of Tyrone Hayes, a comparative endocrinologist and professor of biology at the University of California at Berkeley, has been especially active in studying the effects of low, environmentally relevant concentrations of atrazine on the development of frogs both in the lab and in the wild. In 2002, Hayes published his lab's first study using a model amphibian, the African clawed frog (*Xenopus laevis*), showing that exposure of tadpoles to very low concentrations of atrazine in their rearing water (0.1 ppb, a concentration 30-times below the EPA's "safe" drinking water standard of 3 ppb) dramatically reduced the size of the larynx, which in males requires exposure to high testosterone levels for normal development. Strikingly, it also caused male frogs to develop female gonads (hermaphroditism)[143] (Spotlight Figure 25.13). In the same study, the researchers also showed

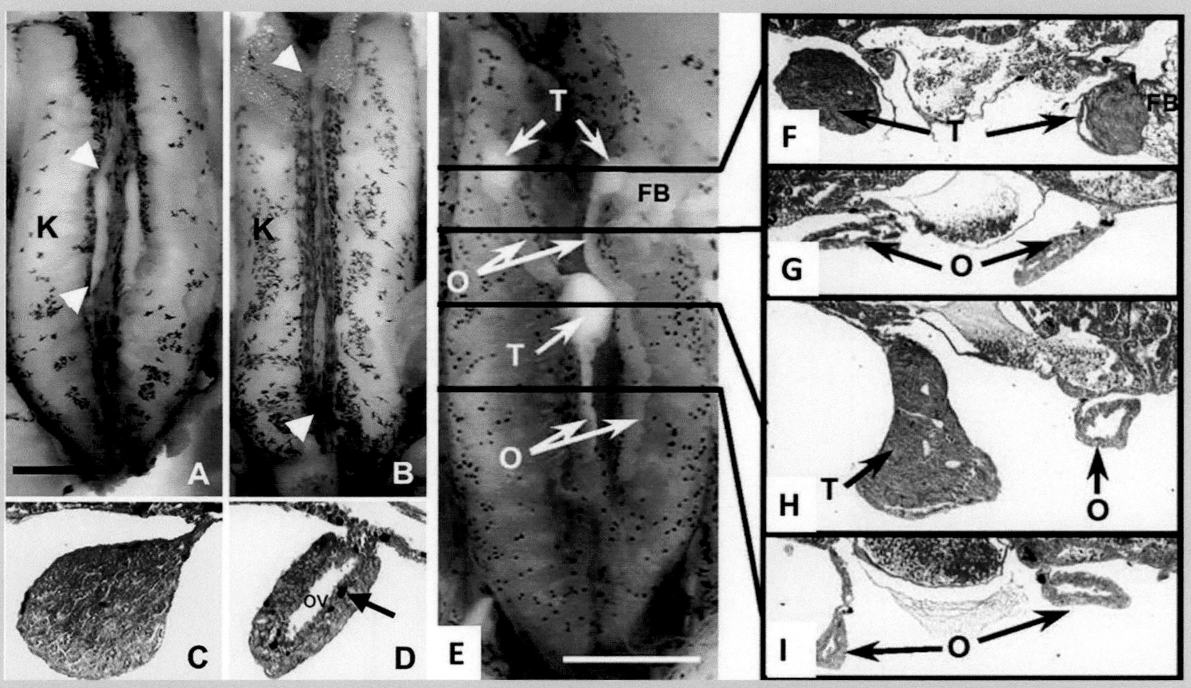

Figure 25.13 Hermaphroditic gonads in atrazine-treated African clawed frogs (*Xenopus laevis*). (A–D): Gonads of a control postmetamorphic male **(A, C)** and female **(B, D)**. Abbreviations: OV, ovarian vesicle; K, kidney. **(A, B)** Arrowheads show the rostral and caudal ends of the animal's right gonad. **(C, D)** Transverse cross sections (8 μm) through the geometric center of each animal's right gonad. Arrow indicates melanophore in the ovary. Scale bar: **(A, B)** 0.1 mm; **(C, D)** 10 μm. Figure adapted from Hayes et al. (2002a). **(E–I)**: A postmetamorphic mixed hermaphrodite treated with 0.1 ppb atrazine. Abbreviations: FB, fatbody; K, kidney; O, ovary; T, testis. **(E)** The entire dissected, Bouin's-fixed kidney-interrenal-gonadal complex. **(F–I)** Transverse cross sections (8 μm) stained in Mallory's trichrome stain. Sections were taken through areas indicated by the black lines. Note the absence of pigment in the ovaries, a conditional typical in hermaphrodites. Scale bar: **(E)** 0.1 mm; **(F–I)** 25 μm. Figure adapted from Hayes et al. (2002a).

Source of Images: Hayes, T.B., et al. 2002. Proc Nat Acad Sci USA, 99(8), pp. 5476-5480. Used by permission.

that a concentration of 25 ppb atrazine was sufficient to chemically castrate exposed males, reducing their testosterone levels to those found in females (Spotlight Figure 25.14(A)).

Soon afterwards, Hayes and colleagues published a follow-up study investigating the effects of exposure to water-borne atrazine contamination to both laboratory-raised and wild leopard frogs (*Rana pipiens*).[144] The researchers found in wild populations in different regions of the United States that 10%–92% of males possessed gonadal abnormalities, such as retarded development and hermaphroditism, of similar morphology to those induced by atrazine in the same species in the laboratory. In other experiments, Hayes and colleagues not only showed that genetically male frogs exposed to very low levels of atrazine (levels deemed "safe" by the EPA) not only became demasculinized, but that 10% of these males developed into functional females that actually copulated with unexposed males and produced viable eggs.[145]

The mechanism by which atrazine exerts the double-whammy effect of simultaneously reducing testosterone and increasing estradiol production is thought to be via its activation of the enzyme *aromatase* (Cyp19A1)[146] (Spotlight Figure 25.14[B]). Recall that elevated expression of the CYP19A1 aromatase gene is consistently observed in differentiating ovaries while its expression is usually suppressed during development of testes, particularly in taxa that experience temperature-sensitive sexual differentiation.[147] By inhibiting a specific phosphodiesterase enzyme, atrazine increases intracellular cAMP, which in turn increases CYP19A1 aromatase gene expression, and results in excess estrogen synthesis from testosterone.

Importantly, the demasculinizing (reduced androgen) and feminizing (increased estradiol) effects of atrazine on male gonads is consistent across vertebrate classes.[148] For example, in addition to its effects on amphibians, atrazine has been reported to cause declines in sperm production in fish,[149] reptiles,[150] birds,[151] and laboratory rodents.[152,153] Furthermore, atrazine has also been associated with low sperm counts in humans exposed to atrazine at levels 24,000 times lower than those experienced by farm workers who are in closer contact with the chemical.[154,155] These effects all likely stem from atrazine's inhibitory effect on androgen production and action, described earlier.[156]

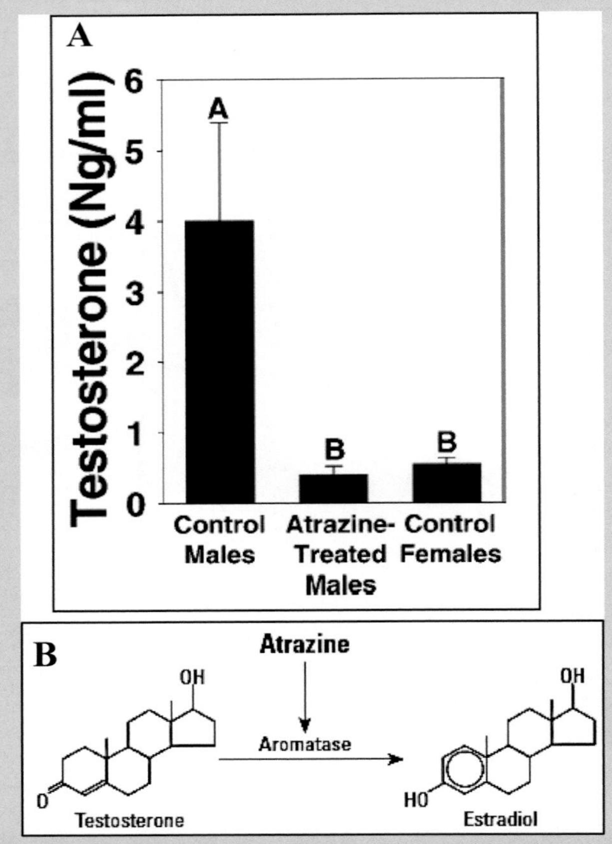

Figure 25.14 Effects of atrazine exposure on plasma testosterone levels in sexually mature male *X. laevis*. (A) Experimental animals were treated every third day with 25 ppb atrazine for 46 days. Control females are shown for comparative purposes. Letters above bars show statistical groupings (ANOVA, P 0.05). **(B)** Proposed mechanism of atrazine action on amphibians. The induction of aromatase by atrazine results in a decrease in androgens and a subsequent increase in estrogens. Therefore, atrazine both demasculinizes (chemically castrates) and feminizes amphibians.

Source of Images: (A) Hayes, T.B., et al. 2002. Proc Nat Acad Sci USA,, 99(8), pp. 5476-5480. Used by permission. (B) Hayes, T.B., et al. 2006. Env. Health Perspec., 114(Suppl 1), pp. 134-141.

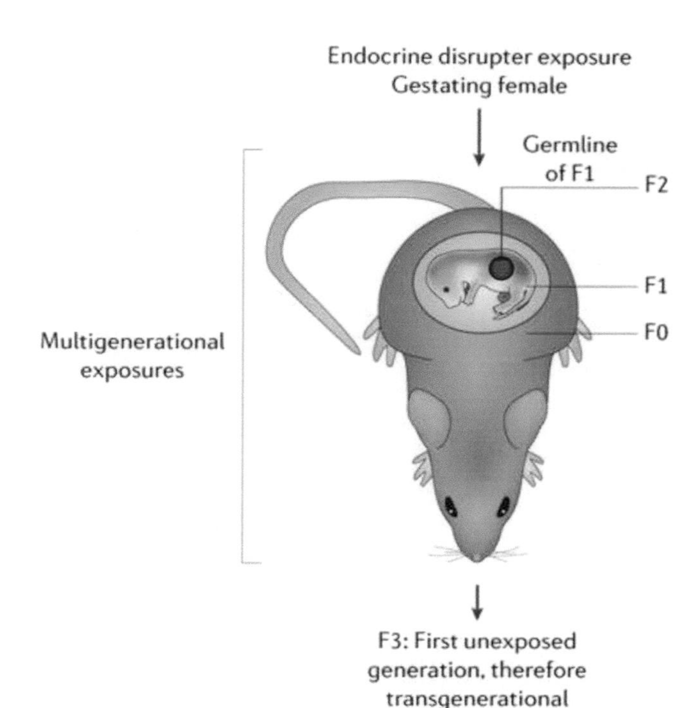

Figure 25.15 Multigenerational exposure and transgenerational inheritance of environmental insults of a pregnant female and affected generations. The simultaneously and directly exposed generations are the F0–F2. Because the F3 generation is the first without any direct exposure to the environmental insult, any phenotype induced in this generation by an ancestral environmental insult must arise from transgenerational inheritance.

Source of Images: Skinner, M.K., 2015. Nat Rev Endocrinol, 12(2), p. 68. Used by permission.

Table 25. 1 Some Examples of Mechanisms of Endocrine Disruption

Stimulation of Endocrine Pathway	Inhibition of Endocrine Pathway
Hormone receptor agonism	Hormone receptor antagonism
Increase in receptor coactivator/intracellular signaling components Stimulation of hormone biosynthetic enzymes Decreased rates of receptor degradation Reduced rates of hormone clearance	Decrease in receptor coactivators/intracellular signaling components Suppression of hormone biosynthetic enzymes Increased rates of receptor degradation Increased rates of hormone clearance Binding to hormone transport proteins

compounds, 4-tert-nonylphenol (NP), 4-tert-octylphenol (OP), and bisphenol A (BPA), are weak estrogen mimics that are present in the environment but rarely occur at concentrations that are individually considered to be biologically active. The researchers found that at the low concentrations studied, each of the five chemicals failed to individually induce a VTG response that was significantly different from that of the controls; however, when the male fish were exposed to the same low doses of all five estrogenic compounds in combination, VTG was significantly induced (Figure 25.11). The findings are striking, as they imply that low-effect concentrations of individual components may give rise to significant mixture effects.

EDC Exposure Can Affect Multiple Generations of Offspring

Recall that DES, which was given to women in the 1940s–1970s during early pregnancy to prevent miscarriages, not only increased the risk of a rare vaginal clear-cell adenocarcinoma and infertility in their daughters, as well as other effects in their sons, but second-generation effects have also been reported in grandsons and granddaughters.[86] These multigenerational effects of DES are supported by extensive research in experimental animal models.[157] When a pregnant mother is exposed to DES (or any other transplacental toxic insult), three different generations are being simultaneously and directly exposed to the compound, a phenomenon known as **multigenerational exposure**. The simultaneously and directly exposed generations are the *F0 generation* (the mother), the *F1 generation* (the fetus), and the germ cells present in the developing fetus that will go on to help generate the *F2 generation* (the mother's grand-offspring) (Figure 25.15). The *F3 generation* (the mother's great-grand-offspring) is the first generation without any direct environmental exposure to the EDC, and in the case of human exposures to DES virtually no incidences of abnormalities have thus far been reported to this generation of offspring.[157] However, the worrisome phenomenon of *transgenerational inheritance*, or transmission of a chemical insult to the F3 generation and, possibly, beyond has been documented for other EDCs.[158–162]

Broad Mechanisms of EDC Action

EDCs can disrupt endocrine signaling in diverse ways that go well beyond simply mimicking a hormones structure. These include disruption at the levels of hormone biosynthesis, hormone receptor turnover, hormone coactivator/co-suppressor and second messenger intracellular signaling pathways, and by altering rates of hormone clearance, to list just a few.[163] Generally, the more complex a hormone's biosynthetic and signaling pathway is, the more opportunities there are for EDCs to disrupt that hormone system. Some broad mechanisms of EDC action are listed in Table 25.1, and some key characteristics of EDCs and their actions are summarized in Figure 25.16.

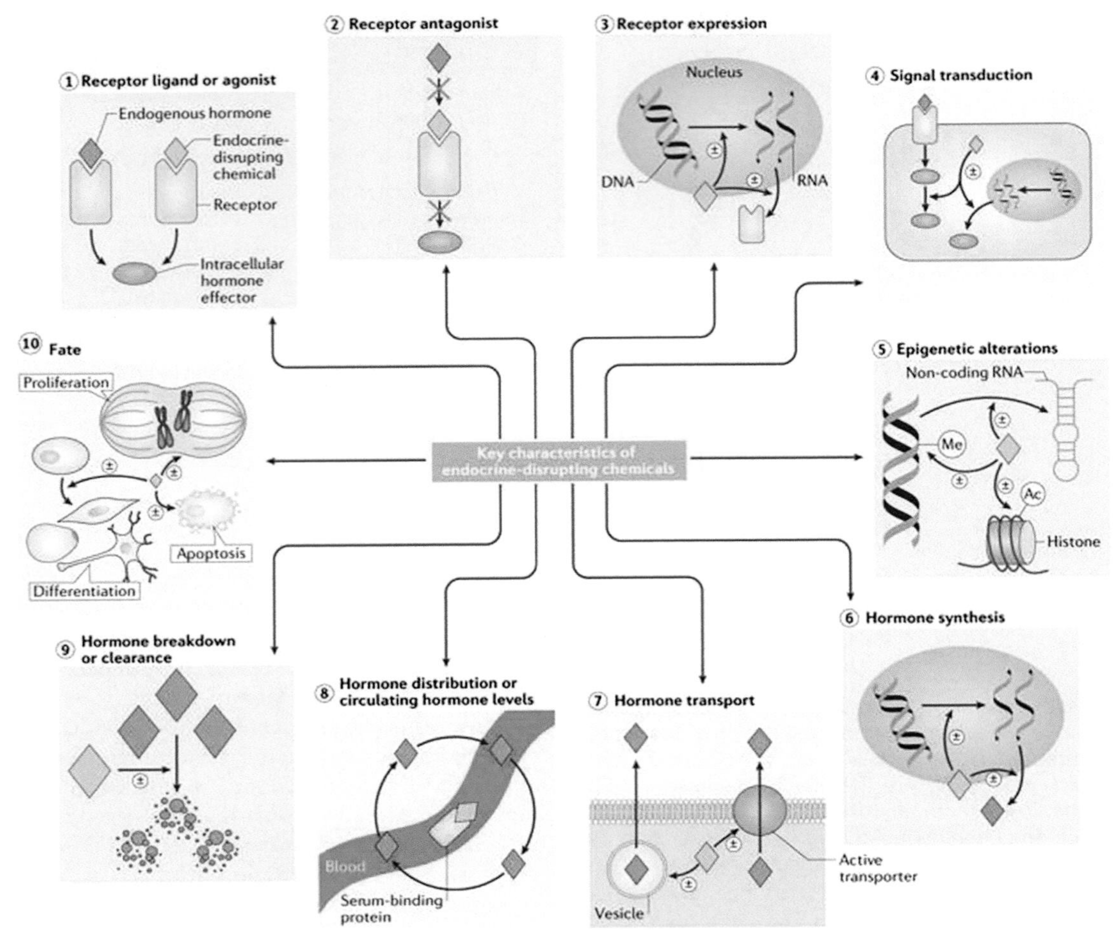

Figure 25.16 Some key characteristics of endocrine-disrupting chemicals (EDCs). Arrows identify ten key characteristics of endocrine-disrupting chemicals. (1) An EDC can interact with or activate hormone receptors. (2) An EDC can antagonize hormone receptors. (3) An EDC can alter hormone receptor expression. (4) An EDC can alter signal transduction (including changes in protein or RNA expression, posttranslational modifications and/or ion flux) in hormone-responsive cells. (5) An EDC can induce epigenetic modifications in hormone-producing or hormone-responsive cells. (6) An EDC can alter hormone synthesis. (7) An EDC can alter hormone transport across cell membranes. (8) An EDC can alter hormone distribution or circulating hormone levels. (9) An EDC can alter hormone metabolism or clearance. (10) An EDC can alter the fate of hormone-producing or hormone-responsive cells. Depicted EDC actions include amplification and attenuation of effects. Ac, acetyl group; Me, methyl group. The ± symbol indicates that an EDC can increase or decrease processes and effects.

Source of Images: La Merrill, M.A., et al. 2019. Nat Rev Endocrinol, pp. 1-13.

SUMMARY AND SYNTHESIS QUESTIONS

1. Which animal, a turkey or a bald eagle, is more prone to being affected by DDT and why?
2. Why might marine mammals, such as dolphins, orcas, and polar bears, be particularly valuable sentinel species for studying EDCs?
3. Although it may take a Beluga whale mother several decades to accumulate high levels of EDCs, her baby will accumulate even higher levels by the time it is only 2 years old. Why?
4. Periods of an animal's life cycle when they may experience acute exposure to EDCs include periods of high energy utilization, such as during migration, hibernation, or lactation. Why?
5. Regulatory agencies that evaluate chemicals for endocrine-disrupting activity often test chemicals individually. Why is the examination of one chemical at a time likely to underestimate that chemical's potential degree of toxicity in a real-world setting? What is an alternative way to test chemicals for toxicity?
6. Like DDT, atrazine also exerts both estrogenic and antiandrogenic effects, but via a different mechanism. What is its proposed mode of action?
7. What is a weakness of the traditional toxicological paradigm, as it pertains to EDCs?

Are EDCs Affecting Human Health?

LEARNING OBJECTIVE Discuss what challenges exist to linking specific EDCs with human chronic disease, and describe how researchers can study this topic most effectively.

KEY CONCEPTS:

- Although there is strong correlational epidemiological evidence connecting the presence of certain EDCs with chronic human diseases, correlational evidence alone is insufficient to demonstrate causation.
- In contrast to the controlled experiments researchers conduct to directly test the effects of EDCs on laboratory model organisms, confounding ethical and logistical barriers thwart the ability to gather similar data for humans.
- Today, researchers rely primarily on the collective assessment of findings derived from clinical, basic, comparative, and translational endocrine research to reasonably infer causality from certain EDCs on some human chronic diseases.

There is broad scientific consensus that EDCs exert adverse effects on the health of wildlife and of laboratory model organisms at environmentally relevant concentrations, and some examples of the data supporting this have been presented throughout this textbook. Indeed, the global threat that EDCs pose to the health of wildlife alone is a more than sufficient reason to continue to study and mitigate the detrimental effects of these compounds around the globe. However, do EDCs contribute specifically to the development of human chronic diseases in the general population? This is a challenging question to answer with precision on the basis of current human data alone.

Epidemiological information collected over the last 50 years shows clearly that the incidence of chronic human health problems associated directly or indirectly with endocrine dysfunction has been increasing steadily.[164–166] These include rising rates of obesity, type 2 diabetes, neurodevelopmental disorders, breast cancer, prostate cancer, testicular cancer, cardiovascular disorders, plummeting fertility rates (Figure 25.17), and others. Furthermore, a trend towards an earlier age of female puberty has also been widely reported, as is increased incidence of male infants born with hypospadias, ambiguous genitalia, and other genital malformations. Indeed, strong correlational epidemiological evidence does exist connecting the presence of certain EDCs in human tissues with child and adult obesity, impaired glucose tolerance, gestational diabetes, reduced semen quality, polycystic ovarian syndrome, endometriosis, breast cancer, prostate cancer, and other chronic conditions.[167] In a meta-analysis study published in the medical journal *The Lancet*, Leonardo Trasande's group estimated the strength of the human evidence for the causation of certain chronic diseases with

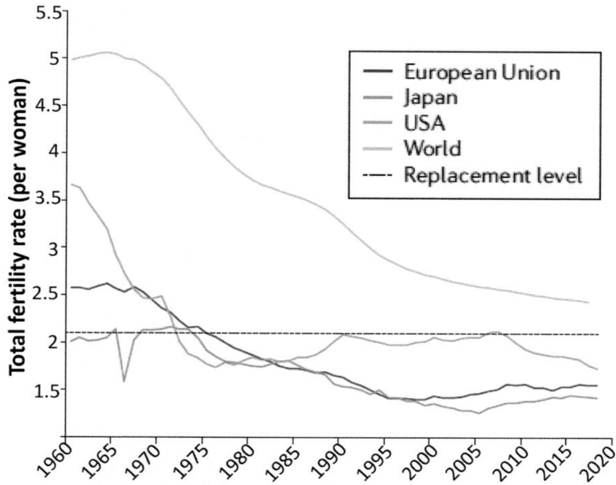

Figure 25.17 Total fertility rates in the European Union, Japan, and the U.S.A., 1960–2018. The dashed line represents a fertility rate of 2.1, below which a population cannot be sustained (total fertility rate is the average number of children per woman). Despite higher birth rates in non-industrialized parts of the world, even the total fertility rate of the total world population seems to be declining towards 2.1. Additional information on trends in fertility rates in 43 countries across North America, South America, Europe, Asia, and Africa, 1960–2019, is shown in Supplementary Figure 1. Data from Databank, World Development Indicators; Country: European Union, Japan, USA; Series: Fertility rate, total (births per woman); time: 1960–2018.

Source of Images: Skakkebæk NE, et al. Nat Rev Endocrinol 2021:1-19. Used by permission.

exposure to specific EDCs[167] (Table 25.2) and concluded, "Although systematic evaluation is needed of the probability and strength of these exposure–outcome relations, the growing evidence supports urgent action to reduce exposure to EDCs".

However, while these reports are clearly concerning, human correlational evidence alone is insufficient to demonstrate causation.[168] Showing definitively that general population-level exposures to EDCs contributes to the development of a human disease would require deliberately exposing large numbers of people to a specific EDC or mixture of EDCs in a controlled manner, and then observing the resulting disease. Although this testing strategy is conducted on humans when designing new pharmaceuticals for disease prevention and treatment, it would, of course, be unethical to directly test the impact of potentially toxic substances on humans, which is why such studies are instead conducted using proxy model organisms. Other complications to addressing this question definitively include:

- Humans are exposed to complex chemical mixtures across their lifespans, making it challenging to determine if a disease results from exposure to one compound or arises from collective interactive effects of multiple chemicals.
- The outcomes of exposure to many environmental pollutants may only manifest years or decades later. In par-

Table 25.2 Exposure-Outcome Associations with a Probability of Evidence for Causation Identified up to 2015

	Outcome	Strength of Human Evidence	Probability of Causation, %
Prenatal PBDEs	IQ loss and intellectual disability	Moderate to high	70%–100%
Prenatal organophosphate pesticides	IQ loss and intellectual disability	Moderate to high	70%–100%
Multiple prenatal exposures	Attention-deficit disorder	Low to moderate	20%–69%
Multiple prenatal exposures	Autism spectrum disorder	Low	20%–39%
Prenatal DDE	Childhood obesity	Moderate	40%–69%
Prenatal BPA	Childhood obesity	Very low to low	20%–69%
Adult DEHP	Adult obesity	Low	40%–69%
Adult DEHP	Adult diabetes	Low	40%–69%
Prenatal DDE	Adult diabetes	Low	20%–39%
Prenatal PBDEs	Cryptorchidism	Low	40%–69%
Prenatal PBDEs	Testicular cancer	Very low to low	0%–19%
Adult phthalates	Low testosterone, resulting in increased early mortality	Low	40%–69%
Adult benzyl and butyl phthalates	Male infertility, resulting in increased use of assisted reproductive technology	Low	40%–69%
Adult DEHP	Endometriosis	Low	20%–39%
Lifetime DDE	Fibroids	Low	20%–39%

Note: BPA, bisphenol A; DDE, dichlorodiphenyldichloroethylene; DEHP, di-2-ethylhexyl phthalate; PBDE, polybrominated diphenyl ether; IQ, intelligence quotient.

Source: Kahn, L.G., Philippat, C., Nakayama, S.F., Slama, R. and Trasande, L., 2020. Endocrine-disrupting chemicals: Implications for human health. *The Lancet Diabetes & Endocrinology*, 8(8), pp. 703–718.

ticular, there is often a lack of exposure data during critical periods of development (such as fetal and childhood periods) that influence later functioning in adult life.

- It is difficult to compare and integrate results from different human studies, as data are often collected at different time periods, under different exposure conditions, and using different experimental designs.

Therefore, we might appear to currently lack the capacity to answer the question posed, "do EDCs contribute to the development of human chronic diseases?" with direct experimental certainty. However, researchers still possess a powerful arsenal of tools that allow us to use the next best approach: to reasonably infer causality from the collective assessment of findings derived from clinical, basic, comparative, and translational endocrine research. Decades of findings compiled from wildlife studies, laboratory animal models, human cell- and molecular-based assays, human clinical findings from patients with known acute exposures to a particular EDC, and human epidemiological observations correlating disease propensity with EDC body burdens are now beginning to converge.[121,165,169,170] Based upon the analysis of data from such a diversity of sources, the Endocrine Society, the world's oldest and largest organization dedicated to research on hormones and the clinical treatment of patients with endocrine diseases, analyzed 1,800 studies on EDCs and published an evidence-based scientific statement in 2015 concluding:

> As we move forward, we believe that the evidence is sufficient to recommend greater regulation, more precaution, better communication between healthcare professionals and patients, and efforts to avoid introducing new EDCs in a misguided effort to replace previous chemicals in the absence of proper testing.[171]

Are EDCs affecting human health? The preponderance of the evidence suggests that EDCs will continue to be scrutinized as potentially serious threats to public health into the foreseeable future. A resolution to this question can only be addressed using an integration of knowledge and tools from both general and comparative endocrinology.

SUMMARY AND SYNTHESIS QUESTIONS

1. This chapter has described numerous examples of how EDCs have been linked to disease causation in mammalian wildlife and lab animals. Why are these findings likely translatable to humans?

Summary of Chapter Learning Objectives and Key Concepts

LEARNING OBJECTIVE Describe key events that led to the birth of the modern environmental movement in the 1960s and 1970s.

- We are living in the Anthropocene, a proposed epoch of geological time characterized by the presence of

distinct human impacts on the geology and ecology of the earth, caused in part by chemical pollution, climate disruption, overpopulation, habitat destruction, and species extinction.
- With the publication of *Silent Spring* in 1962, a book describing the detrimental effects of the pesticide DDT on wildlife health, Rachel Carson ushered in the modern environmental movement and the notion that synthetic chemicals can disrupt hormone function.
- Carson's book provided impetus for establishing the U.S. National Institute of Environmental Health Sciences (NIEHS), the founding of the U.S. Environmental Protection Agency, the U.S. Clean Air and Water Acts, and international agreements to ban or restrict several key synthetic chemicals.

LEARNING OBJECTIVE Describe the mechanisms by which the pesticide DDT exerts both estrogenic and antiandrogenic effects on developing vertebrates.

- Whereas DDT exerts estrogenic effects by functioning as an estrogen receptor agonist, its primary breakdown product, DDE, exerts antiandrogenic effects by antagonizing the androgen receptor.
- The males of many wild vertebrate populations exposed to DDT become simultaneously both "feminized" and "demasculinized".
- Although DDT was ultimately banned in the United States in 1972, the chemical was so widely used and is so resistant to degradation that it remains present to this day in virtually all animals, including humans.

LEARNING OBJECTIVE Define "environmental estrogen", and explain how wildlife species, such as fish, birds, and alligators, can function as "sentinel" species.

- Many persistent organic pollutants, like DDT, act as "environmental estrogens", or synthetic compounds with estrogenic activity at extremely low, ecologically relevant concentrations.
- Vitellogenin, a gene naturally induced by estrogen in female fish and other vertebrates, is also induced by estrogen-mimicking chemicals in males.
- "Sentinel species" are wildlife that can be physiologically monitored for long-term health effects of environmental chemical exposures on an ecosystem.

LEARNING OBJECTIVE Discuss the importance of diethylstilbestrol (DES) in founding the field of endocrine disruption studies.

- Diethylstilbestrol (DES), the first synthetic compound to be specifically designed and marketed as a medically prescribed estrogen, was also the first known example of a human "transplacental" carcinogen.
- Fetal exposure to DES resulted in the occurrence of rare cancers and other pathologies in women and men later in life.

LEARNING OBJECTIVE Describe the concept of the "fragile fetus", and explain why embryos, fetuses, and newborns are so susceptible to the effects of endocrine disruption compared with adults.

- Fetuses possess several attributes that make them more susceptible than adults to the effects of endocrine-disrupting compounds, including a high metabolic rate, rapid cell proliferation and differentiation, a nonfunctional blood-brain barrier, and an underdeveloped liver and immune system.
- Subtle functional changes in specific tissues during sensitive developmental windows of early life can result in increased susceptibility to disease or dysfunction later in life.

LEARNING OBJECTIVE Describe factors that influenced the coining of the term "endocrine disruptor".

- The term "endocrine disruptor" was coined in 1992, resulting from the convergence of the human DES tragedy and the growing awareness of the effects of environmental hormone-disrupting chemicals on wildlife.

LEARNING OBJECTIVE Describe the primary sources of endocrine-disrupting chemicals (EDCs).

- The primary origins of EDCs are industrial, agricultural, residential, pharmaceutical, and heavy metal.
- Exposure to EDCs may occur via eating, drinking, breathing, or absorption via skin contact.

LEARNING OBJECTIVE Describe the biochemical features and mechanisms of action of EDCs, and how the study of endocrine disruption differs from that of classical toxicology.

- EDCs are chemically heterogeneous, making it difficult to predict based on structure alone whether a compound will exert endocrine-disrupting actions.
- Because many EDCs are chemically stable and lipophilic, they tend to bioaccumulate in the fats of animals, with the highest concentrations occurring in predators at the tops of food chains.
- Compared with other environmental toxicants, EDCs are active at extremely low concentrations.
- Multiple EDCs can interact to generate unpredictable interactive effects not only among themselves, but also with endogenous hormones, even at low doses that individually do not produce observable effects.
- The effects of some EDCs can be transmitted to the offspring of a parent, or even the children and grandchildren of their offspring, via multigenerational and transgenerational mechanisms.

LEARNING OBJECTIVE Discuss what challenges exist to linking specific EDCs with human chronic disease, and describe how researchers can study this topic most effectively.

- Although there is strong correlational epidemiological evidence connecting the presence of certain EDCs with chronic human diseases, correlational evidence alone is insufficient to demonstrate causation.

- In contrast to the controlled experiments researchers conduct to directly test the effects of EDCs on laboratory model organisms, confounding ethical and logistical barriers thwart the ability to gather similar data for humans.
- Today, researchers rely primarily on the collective assessment of findings derived from clinical, basic, comparative, and translational endocrine research to reasonably infer causality from certain EDCs on some human chronic diseases.

LITERATURE CITED

1. Singer M. *Ecosystem Crises Interactions: Human Health and the Changing Environment*. John Wiley & Sons; 2021.
2. Ceballos G, Ehrlich PR, Dirzo R. Biological annihilation via the ongoing sixth mass extinction signaled by vertebrate population losses and declines. *Proc Natl Acad Sci U S A*. 2017;114(30): E6089–E6096.
3. IPBES. Media Release: Nature's Dangerous Decline "Unprecedented" Species Extinction Rates "Accelerating". Intergovernmental Science-Policy Platform on Biodiversity and Ecosystem Services (IPBES). www.ipbes.net/news/Media-Release-Global-Assessment. Published 2019. Accessed 2019.
4. Barnosky AD, Brown JH, Daily GC, et al. Introducing the scientific consensus on maintaining humanity's life support systems in the 21st century: Information for policy makers. *Anthr Rev*. 2014;1(1):78–109.
5. Steffen W, Grinevald J, Crutzen P, McNeill J. The Anthropocene: conceptual and historical perspectives. *Philos Trans R Soc A: Math Phy Eng Sci*. 2011;369(1938):842–867.
6. Steffen W, Rockström J, Richardson K, et al. Trajectories of the Earth System in the Anthropocene. *Proc Natl Acad Sci U S A*. 2018;115(33):8252–8259.
7. TEDX. The Endocrine Disruptor Exchange. TEDX List of Potential Endocrine Disruptors. www.endocrinedisruption.org/. Published 2018. Accessed December 2018, 2018.
8. Carson R. *Under the Sea-Wind: A Naturalist's Picture of Ocean Life*. Simon & Schuster; 1941.
9. Carson R. *The Sea Around Is*. Vol Republished 2003: Oxford University Press; 1951.
10. Carson R. *The Edge of the Sea*. Houghton Mifflin Harcourt; 1955.
11. Carson R. *Silent Spring*. Houghton Mifflin Company; 1962.
12. Griswold E. How "Silent Spring" ignited the environmental movement. *The New York Times*. 2012;21.
13. Lear L. *Rachel Carson: Witness for Nature*. Macmillan; 1998.
14. Souder W. *On a Farther Shore: The Life and Legacy of Rachel Carson*. Crown; 2012.
15. Orlando L. *Industry Attacks on Dissent: From Rachel Carson to Oprah Forty Years After the Publication of Silent Spring, Corporations Are Still Trying to Silence Critics*. Dollars and Sense; 2002:26–30.
16. Van Den Berg H, Manuweera G, Konradsen F. Global trends in the production and use of DDT for control of malaria and other vector-borne diseases. *Malar J*. 2017;16(1):401.
17. Grier JW. Ban of DDT and subsequent recovery of reproduction in bald eagles. *Science*. 1982;218(4578):1232–1235.
18. Newton I, Wyllie I. Recovery of a sparrowhawk population in relation to declining pesticide contamination. *J Appl Ecol*. 1992:476–484.
19. Sourkes TL, Stevenson LG. *Nobel Prize Winners in Medicine and Physiology, 1901–1965*. Abelard-Schuman; 1967.
20. Broley CL. The plight of the American bald eagle. *Audubon Magazine*. 1958;60:162–163, 171.
21. Patisaul HB, Adewale HB. Long-term effects of environmental endocrine disruptors on reproductive physiology and behavior. *Front Behav Neurosci*. 2009;3:10.
22. Burlington H, Lindeman V. Effect of DDT on testes and secondary sex characters of white leghorn cockerels. *Proc Soc Exp Biol Med*. 1950;74(1):48–51.
23. Bitman J, Cecil HC, Harris SJ, Fries GF. Estrogenic activity of o, p'-DDT in the mammalian uterus and avian oviduct. *Science*. 1968;162(3851):371–372.
24. Welch R, Levin W, Conney A. Estrogenic action of DDT and its analogs. *Toxicol Appl Pharmacol*. 1969;14(2):358–367.
25. Shelby MD, Newbold RR, Tully DB, Chae K, Davis VL. Assessing environmental chemicals for estrogenicity using a combination of in vitro and in vivo assays. *Environ Heal Perspect*. 1996;104(12):1296.
26. Klotz DM, Ladlie BL, Vonier PM, McLachlan JA, Arnold SF. o, p'-DDT and its metabolites inhibit progesterone-dependent responses in yeast and human cells. *Mol Cell Endocrinol*. 1997;129(1):63–71.
27. Kelce WR, Stone CR, Laws SC, Gray LE, Kemppainen JA, Wilson EM. Persistent DDT metabolite p, p'—DDE is a potent androgen receptor antagonist. *Nature*. 1995;375(6532):581.
28. Xu L-C, Sun H, Chen J-F, Bian Q, Song L, Wang X-R. Androgen receptor activities of p, p'-DDE, fenvalerate and phoxim detected by androgen receptor reporter gene assay. *Toxicol Lett*. 2006;160(2):151–157.
29. Rooney AA, Guillette LJ Jr. *Contaminant Interactions with Steroid Receptors: Evidence for Receptor Binding*. Taylor & Francis; 2000.
30. Cooke A. Shell thinning in avian eggs by environmental pollutants. *Environ Pollut (1970)*. 1973;4(2):85–152.
31. Ratcliffe DA. Decrease in eggshell weight in certain birds of prey. *Nature*. 1967;215(5097):208.
32. Holm L, Blomqvist A, Brandt I, Brunström B, Ridderstråle Y, Berg C. Embryonic exposure to o, p'-DDT causes eggshell thinning and altered shell gland carbonic anhydrase expression in the domestic hen. *Environ Toxicol Chem*. 2006;25(10):2787–2793.
33. Lundholm C. DDE-induced eggshell thinning in birds: effects of p, p'-DDE on the calcium and prostaglandin metabolism of the eggshell gland. *Comp Biochem Physiol Part-C: Toxicol Pharmacol*. 1997;118(2):113–128.
34. Guillette LJ Jr, Kools S, Gunderson MP, Bermudez DS. DDT and its analogues: new insights into their endocrine disruptive effects on wildlife. In: *Endocrine Disruption: Biological Bases for Health Effects in Wildlife and Humans*. Oxford University Press; 2006:332–355.
35. Cohn BA. Developmental and environmental origins of breast cancer: DDT as a case study. *Reprod Toxicol*. 2011;31(3):302–311.
36. Cohn BA, La Merrill M, Krigbaum NY, et al. DDT exposure in utero and breast cancer. *J Clin Endocrinol Metab*. 2015;100(8):2865–2872.
37. Clapp RW, Jacobs MM, Loechler EL. Environmental and occupational causes of cancer: new evidence 2005–2007. *Revi Environ Health*. 2008;23(1):1–38.
38. Beard J, Collaboration ARHR. DDT and human health. *Sci Total Environ*. 2006;355(1–3):78–89.
39. Rogan WJ, Chen A. Health risks and benefits of bis (4-chlorophenyl)-1, 1, 1-trichloroethane (DDT). *Lancet*. 2005;366(9487):763–773.
40. Venners SA, Korrick S, Xu X, et al. Preconception serum DDT and pregnancy loss: a prospective study using a biomarker of pregnancy. *Am J Epidemiol*. 2005;162(8):709–716.
41. Roberts JR, Reigart JR. *Recognition and Management of Pesticide Poisonings*. U.S. Environmental Protection Agency; 2013.
42. Stutchbury B. *Silence of the Songbirds*. Bloomsbury Publishing; 2009.

43. Stutchbury B. Did your shopping list kill a songbird? *The New York Times*. 2008;30(3):2008.
44. Aulerich R, Ringer R, Iwamoto S. Reproductive failure and mortality in mink fed on Great Lakes fish. *J Reprod Fertil, Suppl*. 1973;19:365–376.
45. Aulerich RJ, Ringer RK. Current status of PCB toxicity to mink, and effect on their reproduction. *Arch Environ Contam Toxicol*. 1977;6(1):279–292.
46. Gilbertson M, Reynolds L. Hexachlorobenzene (HCB) in the eggs of common terns in Hamilton Harbour, Ontario. *Bull Environ Contam Toxicol*. 1972;7(6):371–373.
47. Fry DM, Toone CK. DDT-induced feminization of gull embryos. *Science*. 1981;213(4510):922–924.
48. Peakall DB, Fox GA. Toxicological investigations of pollutant-related effects in Great Lakes gulls. *Environ Heal Perspect*. 1987;71:187.
49. Gibbs P, Pascoe P, Burt G. Sex change in the female dog-whelk, Nucella lapillus, induced by tributyltin from antifouling paints. *J Mar Biol Assoc UK*. 1988;68(4):715–731.
50. Mensink BP, Kralt H, Vethaak AD, et al. Imposex induction in laboratory reared juvenile Buccinum undatum by tributyltin (TBT). *Environ Toxicol Pharmacol*. 2002;11(1):49–65.
51. Jobling S, Sumpter J. Detergent components in sewage effluent are weakly oestrogenic to fish: an in vitro study using rainbow trout (Oncorhynchus mykiss) hepatocytes. *Aquat Toxicol*. 1993;27(3–4):361–372.
52. Jobling S, Sumpter JP, Sheahan D, Osborne JA, Matthiessen P. Inhibition of testicular growth in rainbow trout (Oncorhynchus mykiss) exposed to estrogenic alkylphenolic chemicals. *Environ Toxicol Chem*. 1996;15(2):194–202.
53. Jobling S, Beresford N, Nolan M, et al. Altered sexual maturation and gamete production in wild roach (Rutilus rutilus) living in rivers that receive treated sewage effluents. *Biol Reprod*. 2002;66(2):272–281.
54. Tyler C, Jobling S, Sumpter J. Endocrine disruption in wildlife: a critical review of the evidence. *Crit Rev Toxicol*. 1998;28(4):319–361.
55. Bergeron JM, Crews D, McLachlan JA. PCBs as environmental estrogens: turtle sex determination as a biomarker of environmental contamination. *Environ Heal Perspect*. 1994;102(9):780.
56. Specker J. Vitellogenesis in fishes: status and perspectives. *Perspect Comparat Endocrinol*. 1994:304–315.
57. Copeland P, Sumpter J, Walker T, Croft M. Vitellogenin levels in male and female rainbow trout (Salmo gairdneri Richardson) at various stages of the reproductive cycle. *Comp Biochem Physiol B Biochem Mol Biol*. 1986;83(2):487–493.
58. Harries JE, Sheahan DA, Jobling S, et al. Estrogenic activity in five United Kingdom rivers detected by measurement of vitellogenesis in caged male trout. *Environ Toxicol Chem*. 1997;16(3):534–542.
59. Sumpter JP, Jobling S. Vitellogenesis as a biomarker for estrogenic contamination of the aquatic environment. *Environ Heal Perspect*. 1995;103(suppl 7):173–178.
60. Semenza JC, Tolbert PE, Rubin CH, Guillette LJ Jr, Jackson RJ. Reproductive toxins and alligator abnormalities at Lake Apopka, Florida. *Environ Heal Perspect*. 1997;105(10):1030.
61. Guillette LJ Jr, Gross TS, Masson GR, Matter JM, Percival HF, Woodward AR. Developmental abnormalities of the gonad and abnormal sex hormone concentrations in juvenile alligators from contaminated and control lakes in Florida. *Environ Heal Perspect*. 1994;102(8):680.
62. Facemire CF, Gross TS, Guillette LJ Jr. Reproductive impairment in the Florida panther: nature or nurture? *Environ Heal Perspect*. 1995;103(Suppl 4):79.
63. Raloff J. The gender benders. *Sci News*. 1994;145(2):24–27.
64. Schug TT, Johnson AF, Birnbaum LS, et al. Minireview: endocrine disruptors: past lessons and future directions. *Mol Endocrinol*. 2016;30(8):833–847.
65. Fry DM. Reproductive effects in birds exposed to pesticides and industrial chemicals. *Environ Heal Perspect*. 1995;103(Suppl 7):165.
66. EPA US. *Tower Chemical Company Superfund Site Biological Assessment*. U.S. Environmental Protection Agency; 1994.
67. Heinz GH, Percival HF, Jennings ML. Contaminants in American alligator eggs from lake Apopka, lake Griffin, and lake Okeechobee, Florida. *Environ Monit Assess*. 1991;16(3):277–285.
68. Guillette L Jr, Brock J, Rooney A, Woodward A. Serum concentrations of various environmental contaminants and their relationship to sex steroid concentrations and phallus size in juvenile American alligators. *Arch Environ Contam Toxicol*. 1999;36(4):447–455.
69. Guillette LJ Jr, Woodward AR, Crain DA, Pickford DB, Rooney AA, Percival HF. Plasma steroid concentrations and male phallus size in juvenile alligators from seven Florida lakes. *Gen Comp Endocr*. 1999;116(3):356–372.
70. Matter JM, Crain DA, Sills-McMurry C, et al. Effects of endocrine-disrupting contaminants in reptiles: alligators. In: *Principles and Processes for Evaluating Endocrine Disruption in Wildlife*. Oxford University Press; 1998:267–289.
71. Guillette LJ Jr, Gross TS, Masson GR, Matter JM, Percival HF, Woodward AR. Developmental abnormalities of the gonad and abnormal sex hormone concentrations in juvenile alligators from contaminated and control lakes in Florida. *Environ Heal Perspect*. 1994;102(8):680–688.
72. Biava CG, Smuckler EA, Whorton D. The testicular morphology of individuals exposed to dibromochloropropane. *Exp Mol Pathol*. 1978;29(3):448–458.
73. Iguchi T, Fukazawa Y, Uesugi Y, Takasugi N. Polyovular follicles in mouse ovaries exposed neonatally to diethylstilbestrol in vivo and in vitro. *Bio Reprod*. 1990;43(3):478–484.
74. Korach KS, Metzler M, McLachlan JA. Estrogenic activity in vivo and in vitro of some diethylstilbestrol metabolites and analogs. *Proc Natl Acad Sci U S A*. 1978;75(1):468–471.
75. Dodds EC, Goldberg L, Lawson W, Robinson R. Oestrogenic activity of certain synthetic compounds. *Nature*. 1938;141(3562):247.
76. Herbst AL, Green TH, Ulfelder H. Primary carcinoma of the vagina: an analysis of 68 cases. *Am J Obstet Gynecol*. 1970;106(2):210–218.
77. Herbst AL, Ulfelder H, Poskanzer DC. Adenocarcinoma of the vagina: association of maternal stilbestrol therapy with tumor appearance in young women. *N Engl J Med*. 1971;284(16):878–881.
78. McLachlan JA, Arnold SF. Environmental estrogens. *Am Sci*. 1996;84(5):452–461.
79. Palmlund I. Exposure to a xenoestrogen before birth: the diethylstilbestrol experience. *J Psychosom Obstet Gynecol*. 1996;17(2):71–84.
80. Vessey M, Fairweather D, Norman-Smith B, Buckley J. A randomized double-blind controlled trial of the value of stilboestrol therapy in pregnancy: long-term follow-up of mothers and their offspring. *BJOG Int J Obstet Gy*. 1983;90(11):1007–1017.
81. Gill W, Schumacher G, Bibbo M. Structural and functional abnormalities in the sex organs of male offspring of mothers treated with diethylstilbestrol (DES). *J Reprod Med*. 1976;16(4):147–153.
82. Palmer JR, Wise LA, Robboy SJ, et al. Hypospadias in sons of women exposed to diethylstilbestrol in utero. *Epidemiology*. 2005:583–586.
83. Stenchever M, Williamson R, Leonard J, et al. Possible relationship between in utero diethylstilbestrol exposure and male fertility. *Am J Obstet Gynecol*. 1981;140(2):186–193.
84. Wilcox A, Baird D, Weinberg C, Hornsby P, Herbst A. Fertility in men exposed prenatally to diethylstilbestrol. *Int J Gynecol Obstet*. 1996;52(2):225–225.
85. Strohsnitter WC, Noller KL, Hoover RN, et al. Cancer risk in men exposed in utero to diethylstilbestrol. *J Natl Cancer Inst*. 2001;93(7):545–551.
86. Brouwers M, Feitz W, Roelofs L, Kiemeney L, De Gier R, Roeleveld N. Hypospadias: a transgenerational effect of diethylstilbestrol? *Hum Reprod*. 2005;21(3):666–669.

87. McLachlan JA, Newbold RR, Bullock BC. Long-term effects on the female mouse genital tract associated with prenatal exposure to diethylstilbestrol. *Cancer Res*. 1980;40(11):3988–3999.
88. Bern H. The fragile fetus. In: Theo Colborn and Coralie Clement, (eds.), *Chemically Induced Alterations in Sexual and Functional Development the Wild Life/Human Connection*. Princeton Scientific Pub. Co.; 1992:9–16.
89. Barker DJ. The origins of the developmental origins theory. *J Intern Med*. 2007;261(5):412–417.
90. Gilbert SF, Epel D. *Ecological Developmental Biology: Integrating Epigenetics, Medicine, and Evolution*. Sinauer Associates; 2009.
91. Landrigan PJ, Etzel RA. *Textbook of Children's Environmental Health*. Oxford University Press; 2013.
92. Zoeller TR. Environmental chemicals targeting thyroid. *Hormones (Athens)*. 2010;9(1):28–40.
93. Harvey PW. Adrenocortical endocrine disruption. *J Steroid Biochem Mol Biol*. 2016;155:199–206.
94. Heindel JJ, Blumberg B, Cave M, et al. Metabolism disrupting chemicals and metabolic disorders. *Reprod Toxicol*. 2017;68:3–33.
95. Scsukova S, Rollerova E, Mlynarcikova AB. Impact of endocrine disrupting chemicals on onset and development of female reproductive disorders and hormone-related cancer. *Reprod Biol*. 2016;16(4):243–254.
96. Parent A-S, Naveau E, Gerard A, Bourguignon J-P, Westbrook GL. Early developmental actions of endocrine disruptors on the hypothalamus, hippocampus, and cerebral cortex. *J Toxicol Env Heal, B*. 2011;14(5–7):328–345.
97. Colborn T, Clement C, Mehlman M. The statement of consensus. In: Colborn T, Clement C, eds. *Chemically-Induced Alterations in Sexual and Functional Development: The Wildlife/Human Connection*. Princeton Scientific Publishing; 1992:1–8.
98. Kavlock RJ, Daston GP, DeRosa C, et al. Research needs for the risk assessment of health and environmental effects of endocrine disruptors: a report of the US EPA-sponsored workshop. *Environ Heal Perspect*. 1996;104(Suppl 4):715.
99. Zoeller RT, Brown TR, Doan LL, et al. Endocrine-disrupting chemicals and public health protection: a statement of principles from the Endocrine Society. *Endocrinology*. 2012;153(9):4097–4110.
100. Colborn T, Dumanoski D, Myers JP, Murden M. *Our Stolen Future: Are We Threatening Our Fertility, Intelligence, and Survival? A Scientific Detective Story*. Dutton; 1996.
101. Carere C, Costantini D, Sorace A, Santucci D, Alleva E. Bird populations as sentinels of endocrine disrupting chemicals. *Ann Ist Super Sanita*. 2010;46:81–88.
102. Ortiz-Zarragoitia M, Bizarro C, Rojo-Bartolomé I, de Cerio O, Cajaraville M, Cancio I. Mugilid fish are sentinels of exposure to endocrine disrupting compounds in coastal and estuarine environments. *Mar Drugs*. 2014;12(9):4756–4782.
103. Gouveia D, Bonneton F, Almunia C, et al. Identification, expression, and endocrine-disruption of three ecdysone-responsive genes in the sentinel species Gammarus fossarum. *Sci Rep*. 2018;8(1):3793.
104. Bossart G. Marine mammals as sentinel species for oceans and human health. *Vet Pathol*. 2011;48(3):676–690.
105. Bergman Å, Heindel JJ, Kasten T, et al. The impact of endocrine disruption: a consensus statement on the state of the science. *Environ Heal Perspect*. 2013;121(4):a104.
106. Kortenkamp A, Martin O, Faust M, et al. State of the art assessment of endocrine disrupters. *Final Rep*. 2011;23.
107. Woodruff TJ, Zota AR, Schwartz JM. Environmental chemicals in pregnant women in the United States: NHANES 2003–2004. *Environ Heal Perspect*. 2011;119(6):878–885.
108. Hughes CL Jr. Phytochemical mimicry of reproductive hormones and modulation of herbivore fertility by phytoestrogens. *Environ Heal Perspect*. 1988;78:171–174.
109. Shahidehnia M. Epigenetic effects of endocrine disrupting chemicals. *J Environ Anal Toxicol*. 2016;6(4):381.
110. Calafat AM, Ye X, Wong L-Y, Reidy JA, Needham LL. Exposure of the US population to bisphenol A and 4-tertiary-octylphenol: 2003–2004. *Environ Heal Perspect*. 2007;116(1):39–44.
111. Mileva G, Baker SL, Konkle A, Bielajew C. Bisphenol-A: epigenetic reprogramming and effects on reproduction and behavior. *Int J Environ Res Public Health*. 2014;11(7):7537–7561.
112. Dodds EC, Lawson W. Synthetic strogenic agents without the phenanthrene nucleus. *Nature*. 1936;137(3476):996.
113. Dickens F. Edward Charles Dodds. 13 October 1899–16 December 1973. *Biogr Mem Fellows R Soc*. 1975;21:227–267.
114. Matsushima A, Kakuta Y, Teramoto T, et al. Structural evidence for endocrine disruptor bisphenol A binding to human nuclear receptor ERRγ. *J Biochem*. 2007;142(4):517–524.
115. Okada H, Tokunaga T, Liu X, Takayanagi S, Matsushima A, Shimohigashi Y. Direct evidence revealing structural elements essential for the high binding ability of bisphenol A to human estrogen-related receptor-γ. *Environ Heal Perspect*. 2008;116(1):32–38.
116. Richter CA, Birnbaum LS, Farabollini F, et al. In vivo effects of bisphenol A in laboratory rodent studies. *Reprod Toxicol*. 2007;24(2):199–224.
117. Dong S, Terasaka S, Kiyama R. Bisphenol A induces a rapid activation of Erk1/2 through GPR30 in human breast cancer cells. *Environ Pollut*. 2011;159(1):212–218.
118. Wozniak AL, Bulayeva NN, Watson CS. Xenoestrogens at picomolar to nanomolar concentrations trigger membrane estrogen receptor-α—mediated Ca2+ fluxes and prolactin release in GH3/B6 pituitary tumor cells. *Environ Heal Perspect*. 2005;113(4):431–439.
119. Erler C, Novak J. Bisphenol a exposure: human risk and health policy. *J Pediatr Nurs*. 2010;25(5):400–407.
120. De Toni L, De Rocco Ponce M, Petre GC, Rtibi K, Di Nisio A, Foresta C. Bisphenols and male reproductive health: from toxicological models to therapeutic hypotheses. *Front Endocrinol*. 2020;11:301.
121. Diamanti-Kandarakis E, Bourguignon J-P, Giudice LC, et al. Endocrine-disrupting chemicals: an Endocrine Society scientific statement. *Endocr Rev*. 2009;30(4):293–342.
122. Stefanidou M, Maravelias C, Spiliopoulou C. Human exposure to endocrine disruptors and breast milk. *Endocr Metab Immune Disord-Drug Targets*. 2009;9(3):269–276.
123. Gore AC, Crews D, Doan LL, La Merrill M, Patisaul H, Zota A. *Introduction to Endocrine Disrupting Chemicals (EDCs). A Guide for Public Interest Organizations and Policy-Makers*. The Endocrine Society, 2014:21–22.
124. Nagel SC, vom Saal FS, Thayer KA, Dhar MG, Boechler M, Welshons WV. Relative binding affinity-serum modified access (RBA-SMA) assay predicts the relative in vivo bioactivity of the xenoestrogens bisphenol A and octylphenol. *Environ Heal Perspect*. 1997;105(1):70.
125. Palanza P, Gioiosa L, vom Saal FS, Parmigiani S. Effects of developmental exposure to bisphenol A on brain and behavior in mice. *Environ Res*. 2008;108(2):150–157.
126. Alonso-Magdalena P, Quesada I, Nadal A. Prenatal exposure to BPA and offspring outcomes: the diabesogenic behavior of BPA. *Dose-Response*. 2015;13(2).
127. Nah WH, Park MJ, Gye MC. Effects of early prepubertal exposure to bisphenol A on the onset of puberty, ovarian weights, and estrous cycle in female mice. *Clin Exp Reprod Med*. 2011;38(2):75–81.
128. Crews D, Putz O, Thomas P, Hayes T, Howdeshell K. Wildlife as models for the study of how mixtures, low doses, and the embryonic environment modulate the action of endocrine-disrupting chemicals. *Pure Appl Chem*. 2003;75(11–12):2305–2320.
129. Kortenkamp A. Ten years of mixing cocktails: a review of combination effects of endocrine-disrupting chemicals. *Environ Heal Perspect*. 2007;115(Suppl 1):98.

130. Ribeiro E, Ladeira C, Viegas S. EDCs mixtures: a stealthy hazard for human health? *Toxics*. 2017;5(1):5.
131. Rajapakse N, Ong D, Kortenkamp A. Defining the impact of weakly estrogenic chemicals on the action of steroidal estrogens. *Toxicol Sci*. 2001;60(2):296–304.
132. Rajapakse N, Silva E, Kortenkamp A. Combining xenoestrogens at levels below individual no-observed-effect concentrations dramatically enhances steroid hormone action. *Environ Heal Perspect*. 2002;110(9):917.
133. Silva E, Rajapakse N, Kortenkamp A. Something from "nothing"– eight weak estrogenic chemicals combined at concentrations below NOECs produce significant mixture effects. *Environ Sci Technol*. 2002;36(8):1751–1756.
134. Brian JV, Harris CA, Scholze M, et al. Accurate prediction of the response of freshwater fish to a mixture of estrogenic chemicals. *Environ Heal Perspect*. 2005;113(6):721.
135. Desbrow C, Routledge E, Brighty G, Sumpter J, Waldock M. Identification of estrogenic chemicals in STW effluent. 1. Chemical fractionation and in vitro biological screening. *Environ Sci Technol*. 1998;32(11):1549–1558.
136. Thorpe KL, Cummings RI, Hutchinson TH, et al. Relative potencies and combination effects of steroidal estrogens in fish. *Environ Sci Technol*. 2003;37(6):1142–1149.
137. Needham LL, Özkaynak H, Whyatt RM, et al. Exposure assessment in the National Children's Study: introduction. *Environ Heal Perspect*. 2005;113(8):1076–1082.
138. Gilliom RJ, Barbash JE, Crawford CG, et al. *Pesticides in the Nation's Streams and Ground Water, 1992–2001*. US Geological Survey; 2006.
139. Commission E. Commission decision of 10 March 2004 concerning the non-inclusion of atrazine in Annex I to Council Directive 91/414/EEC and the withdrawal of authorisations for plant protection products containing this active substance. 2004/248/EC. *Off J Eur Union*. 2004;78:53–55.
140. Ackerman F. The economics of atrazine. *Int J Occup Environ Health*. 2007;13(4):437–445.
141. Aviv R. A valuable reputation: After Tyrone Hayes said that a chemical was harmful, its maker pursued him. *The New Yorker*; 2014. https://www.newyorker.com/magazine/2014/02/10/a-valuable-reputation
142. Rohr JR. Atrazine and Amphibians: A story of profits, controversy, and animus. In: *The Encyclopedia of the Anthropocene*, Vol. 5. 2018:141–148.
143. Hayes TB, Collins A, Lee M, et al. Hermaphroditic, demasculinized frogs after exposure to the herbicide atrazine at low ecologically relevant doses. *Proc Natl Acad Sci U S A*. 2002;99(8):5476–5480.
144. Hayes T, Haston K, Tsui M, Hoang A, Haeffele C, Vonk A. Herbicides: feminization of male frogs in the wild. *Nature*. 2002;419(6910):895.
145. Hayes TB, Khoury V, Narayan A, et al. Atrazine induces complete feminization and chemical castration in male African clawed frogs (Xenopus laevis). *Proc Natl Acad Sci U S A*. 2010;107(10): 4612–4617.
146. Sanderson JT, Seinen W, Giesy JP, van den Berg M. 2-Chloro-s-triazine herbicides induce aromatase (CYP19) activity in H295R human adrenocortical carcinoma cells: a novel mechanism for estrogenicity? *Toxicol Sci*. 2000;54(1):121–127.
147. Matsumoto Y, Buemio A, Chu R, Vafaee M, Crews D. Epigenetic control of gonadal aromatase (cyp19a1) in temperature-dependent sex determination of red-eared slider turtles. *PLoS One*. 2013;8(6):e63599.
148. Hayes TB, Anderson LL, Beasley VR, et al. Demasculinization and feminization of male gonads by atrazine: consistent effects across vertebrate classes. *J Steroid Biochem Mol Biol*. 2011;127(1–2):64–73.
149. Moore A, Waring CP. Mechanistic Effects of a triazine pesticide on reproductive endocrine function in mature male atlantic salmon (salmo salarL.) parr. *Pestic Biochem Physiol*. 1998;62(1):41–50.
150. Rey F, González M, Zayas MA, et al. Prenatal exposure to pesticides disrupts testicular histoarchitecture and alters testosterone levels in male Caiman latirostris. *Gen Comp Endocr*. 2009;162(3):286–292.
151. Hussain R, Mahmood F, Khan MZ, Khan A, Muhammad F. Pathological and genotoxic effects of atrazine in male Japanese quail (Coturnix japonica). *Ecotoxicology*. 2011;20(1):1–8.
152. Kniewald J, Jakominić M, Tomljenović A, et al. Disorders of male rat reproductive tract under the influence of atrazine. *J Appl Toxicol Intern J*. 2000;20(1):61–68.
153. Victor-Costa AB, Bandeira SMC, Oliveira AG, Mahecha GAB, Oliveira CA. Changes in testicular morphology and steroidogenesis in adult rats exposed to Atrazine. *Reprod Toxicol*. 2010;29(3):323–331.
154. Swan SH, Kruse RL, Liu F, et al. Semen quality in relation to biomarkers of pesticide exposure. *Environ Heal Perspect*. 2003;111(12):1478–1484.
155. Lucas AD, Jones AD, Goodrow MH, et al. Determination of atrazine metabolites in human urine: development of a biomarker of exposure. *Chem Res Toxicol*. 1993;6(1):107–116.
156. Hayes TB, Hansen M. From silent spring to silent night: Agrochemicals and the anthropocene. *Elem Sci Anth*. Elem Sci Anth, 2017;5.
157. Newbold RR. Lessons learned from perinatal exposure to diethylstilbestrol. *Toxicol Appl Pharmacol*. 2004;199(2):142–150.
158. Nilsson EE, Skinner MK. Environmentally induced epigenetic transgenerational inheritance of disease susceptibility. *Transl Res*. 2015;165(1):12–17.
159. Crews D, Gore AC, Hsu TS, et al. Transgenerational epigenetic imprints on mate preference. *Proc Natl Acad Sci U S A*. 2007;104(14):5942–5946.
160. Skinner MK, Anway MD, Savenkova MI, Gore AC, Crews D. Transgenerational epigenetic programming of the brain transcriptome and anxiety behavior. *PloS One*. 2008;3(11):e3745.
161. Anway MD, Leathers C, Skinner MK. Endocrine disruptor vinclozolin induced epigenetic transgenerational adult-onset disease. *Endocrinology*. 2006;147(12):5515–5523.
162. Manikkam M, Tracey R, Guerrero-Bosagna C, Skinner MK. Plastics derived endocrine disruptors (BPA, DEHP and DBP) induce epigenetic transgenerational inheritance of obesity, reproductive disease and sperm epimutations. *PloS One*. 2013;8(1):e55387.
163. La Merrill MA, Vandenberg LN, Smith MT, et al. Consensus on the key characteristics of endocrine-disrupting chemicals as a basis for hazard identification. *Nat Rev Endocrinol*. 2019:1–13.
164. De Coster S, Van Larebeke N. Endocrine-disrupting chemicals: associated disorders and mechanisms of action. *J Environ Public Health*. 2012;2012.
165. Bergman Å, Heindel JJ, Jobling S, Kidd K, Zoeller TR, World Health Organization. *State of the Science of Endocrine Disrupting Chemicals 2012*. World Health Organization; 2013.
166. Skakkebæk NE, Lindahl-Jacobsen R, Levine H, et al. Environmental factors in declining human fertility. *Nat Rev Endocrinol*. 2021:1–19.
167. Kahn LG, Philippat C, Nakayama SF, Slama R, Trasande L. Endocrine-disrupting chemicals: Implications for human health. *Lancet Diabetes Endocrinol*. 2020;8(8):703–718.
168. Gore A, Chappell V, Fenton S, et al. Executive summary to EDC-2: the Endocrine Society's second scientific statement on endocrine-disrupting chemicals. *Endocr Rev*. 2015;36(6):593–602.
169. Gore A, Chappell V, Fenton S, et al. Executive summary to EDC-2: the endocrine society's second scientific statement on endocrine-disrupting chemicals. *Endocr Rev*. 2015; 36(6):593.
170. Kumar M, Sarma DK, Shubham S, et al. Environmental endocrine-disrupting chemical exposure: role in non-communicable diseases. *Front Public Health*. 2020;8:549.
171. Gore AC, Chappell VA, Fenton SE, et al. EDC-2: the Endocrine Society's second scientific statement on endocrine-disrupting chemicals. *Endocr Rev*. 2015;36(6): E1–E150.

Appendices

Appendix 1

Endocrine-Related Nobel Prizes Awarded in the Last Century

Year	Name	Nobel Committee Rationale for Research Addressing:	The Nobel Prize in:
1903	Niels Ryberg Finsen	"the treatment of diseases, especially lupus vulgaris, with concentrated light radiation"	Physiology or Medicine
1909	Emil Theodor Kocher	"the physiology, pathology and surgery of the thyroid gland"	Physiology or Medicine
1923	Frederick Grant Banting and John James Rickard Macleod	"for the discovery of insulin"	Physiology or Medicine
1927	Heinrich Otto Wieland	"for his investigations of the constitution of the bile acids and related substances"	Chemistry
1928	Adolf Otto Reinhold Windaus	"for the services rendered through his research into the constitution of the sterols and their connection with the vitamins"	Chemistry
1936	Sir Henry Hallett Dale and Otto Loewi	"for their discoveries relating to chemical transmission of nerve impulses"	Physiology or Medicine
1939	Adolf Friedrich Johann Butenandt	"for his work on sex hormones"	Chemistry
	Leopold Ruzicka	"for his work on polymethylenes and higher terpenes"	
1943	George de Hevesy	"for his work on the use of isotopes as tracers in the study of chemical processes"	Chemistry
1947	Gerty Cori, Carl Ferdinand Cori, and Bernardo Alberto Houssay	"for their discovery of the course of the catalytic conversion of glycogen"	Physiology or Medicine
1950	Edward Calvin Kendall, Tadeus Reich-stein, and Philip Showalter Hench	"for their discoveries relating to the hormones of the adrenal cortex, their structure and biological effects"	Physiology or Medicine
1953	Hans Adolf Krebs	"for his discovery of the citric acid cycle"	Physiology or Medicine
	Fritz Albert Lipmann	"for his discovery of co-enzyme A and its importance for intermediary metabolism"	
1955	Axel Hugo Theodor Theorell	"for his discoveries concerning the nature and mode of action of oxidation enzymes"	Physiology or Medicine
1955	Vincent du Vigneaud	"for his work on biochemically important sulphur compounds, especially for the first synthesis of a polypeptide hormone"	Chemistry
1958	Frederick Sanger	"for his work on the structure of proteins, especially that of insulin"	Chemistry
1964	Dorothy Crowfoot Hodgkin	"for her determinations by X-ray techniques of the structures of important biochemical substances"	Chemistry
1964	Konrad Bloch and Feodor Lynen	"for their discoveries concerning the mechanism and regulation of the cholesterol and fatty acid metabolism"	Physiology or Medicine
1966	Peyton Rous	"for his discovery of tumor-inducing viruses"	Physiology or Medicine
	Charles Brenton Huggins	"for his discoveries concerning hormonal treatment of prostatic cancer"	
1970	Sir Bernard Katz, Ulf von Euler, and Julius Axelrod	"for their discoveries concerning the humoral transmitters in the nerve terminals and the mechanism for their storage, release and inactivation"	Physiology or Medicine
1971	Earl W. Sutherland Jr.	"for his discoveries concerning the mechanisms of the action of hormones"	Physiology or Medicine

Year	Name	Nobel Committee Rationale for Research Addressing:	The Nobel Prize in:
1977	Rosalyn Yalow	"for the development of radioimmunoassays of peptide hormones"	Physiology or Medicine
	Roger Guillemin and Andrew V. Schally	"for their discoveries concerning the peptide hormone production of the brain"	
1982	Sune K. Bergström, Bengt I. Samuelsson, and John R. Vane	"for their discoveries concerning prostaglandins and related biologically active substances"	Physiology or Medicine
1985	Michael S. Brown and Joseph L. Goldstein	"for their discoveries concerning the regulation of cholesterol metabolism"	Physiology or Medicine
1985	Herbert A. Hauptman and Jerome Karle	"for their outstanding achievements in the development of direct methods for the determination of crystal structures"	Chemistry
1986	Stanley Cohen and Rita Levi-Montalcini	"for their discoveries of growth factors"	Physiology or Medicine
1992	Edmond H. Fischer and Edwin G. Krebs	"for their discoveries concerning reversible protein phosphorylation as a biological regulatory mechanism"	Physiology or Medicine
1994	Alfred G. Gilman and Martin Rodbell	"for their discovery of G-proteins and the role of these proteins in signal transduction in cells"	Physiology or Medicine
1997	Paul D. Boyer and John E. Walker	"for their elucidation of the enzymatic mechanism underlying the synthesis of adenosine triphosphate (ATP)"	Chemistry
	Jens C. Skou	"for the first discovery of an ion-transporting enzyme, Na+, K+ -ATPase"	
1998	Robert F. Furchgott, Louis J. Ignarro, and Ferid Murad	"for their discoveries concerning nitric oxide as a signaling molecule in the cardiovascular system"	Physiology or Medicine
2000	Arvid Carlsson, Paul Greengard, and Eric R. Kandel	"for their discoveries concerning signal transduction in the nervous system"	Physiology or Medicine
2004	Richard Axel and Linda B. Buck	"for their discoveries of odorant receptors and the organization of the olfactory system"	Physiology or Medicine
2010	Robert G. Edwards	"for the development of in vitro fertilization"	Physiology or Medicine
2012	Robert J. Lefkowitz and Brian K. Kobilka	"for studies of G-protein-coupled receptors"	Chemistry
2013	James E. Rothman, Randy W. Schekman, and Thomas C. Südhof	"for their discoveries of machinery regulating vesicle traffic, a major transport system in our cells"	Physiology or Medicine
2017	Jeffrey C. Hall, Michael Rosbash, and Michael W. Young	"for their discoveries of molecular mechanisms controlling the circadian rhythm"	Physiology or Medicine
2019	William G. Kaelin Jr., Sir Peter J. Ratcliffe, and Gregg L. Semenza	"for their discoveries of how cells sense and adapt to oxygen availability"	Physiology or Medicine

Appendix 2

Some Examples of Endocrine Breakthroughs Using Non-Mammalian Model Organisms

Discovery	Contribution
First endocrine experiment conducted with chickens	Arnold Adolph Berthold described the first sophisticated endocrine experiment in which he castrated cockerels and found that this caused regression of secondary sex characters, such as the wattles and comb, and the loss of male-typical sexual behavior (Berthold, 1849).
Discovery of neuroendocrine signaling in invertebrates and fish	Using caterpillars, Stephan Kopeć demonstrated that a signal secreted from the brain was required for metamorphosis (Kopeć, 1917, 1922). Ernst Scharrer developed the concept of vertebrate neurosecretion based on work with the minnow, *Phoxinus laevis*, postulating that specific hypothalamic neurons possessed endocrine activity related to pituitary function (Scharrer, 1928). Berta Sharrer described neuroendocrine activity in various invertebrates, including the sea slug, *Aplysia*, and insects (Scharrer, 1935, 1937). Wolfgang Bargmann and Ernst Scharrer subsequently postulated, correctly, that in mammals hypothalamic nuclei (clusters of neuronal cell bodies with similar functions) and are transported via axons to the posterior pituitary where they are stored until released (Bargmann and Scharrer, 1951).
Hypothalamic-to-pituitary blood flow discovered in toads	Bernardo Houssay was the first to show that blood flowed from the hypothalamus to the pituitary gland (Houssay, 1936).
Hypophysectomy (pituitary removal) in toads leads to breakthroughs in studying diabetes	While studying pancreatic regulation of blood sugar in a South American toad, *Rhinella arenarum*, Bernardo Houssay discovered that hypophysectomy alleviated the diabetic symptoms that normally followed removal of the pancreas (Houssay et al., 1942). This led to the development of hypophysectomized and pancreatectomized dogs (called Houssay animals) for use in the clinical study of diabetes mellitus.
Pigeon prolactin was the first pituitary hormone to be crystallized and purified in 1937, leading to the subsequent purification of mammalian prolactin	Oscar Riddle (Riddle et al., 1933) demonstrated that an avian pituitary factor that promoted growth of the pigeon crop sac was identical to a mammalian pituitary factor that earlier had been found to initiate and maintain milk secretion in mammals. Riddle called this avian factor prolactin and the response of the crop sac provided a sensitive assay for the detection of human prolactin in pituitary extracts.
The toad urinary bladder is a useful system for studying sodium transport	The toad urinary bladder proved to be a simple substitute for studying the role of the mineralocorticoid aldosterone in the regulation of sodium transport across an epithelium (Bentley, 1966).
Study of a newt showed that hypothalamic gonadotropin-releasing hormone (GnRH) neurons originate in the nasal epithelium	This study demonstrated that in all vertebrates the hypothalamic GnRH neurons migrate to the hypothalamus from the nasal epithelium, explaining the connection between symptoms of anosmia (inability to perceive smells) and reproductive failure in humans with Kallmann syndrome (Muske and Moore, 1988).
Flies used to discover that steroid hormones bind to nuclear receptors to regulate gene expression	This concept first came from studies of the insect steroid hormone ecdysone that was found to induce "puffing" of the giant polytene chromosomes in the salivary glands of midges and flies (Clever and Karlson, 1960). This phenomenon was later expanded into a theory of a transcriptional cascade of hormone action by Ashburner (Ashburner et al., 1974).
Rapid, nongenomic actions of steroid hormones were first discovered in amphibians	Godeau (Godeau et al., 1978) showed rapid, membrane-mediated effects of progesterone on frog oocyte maturation. The first discovery and pharmacological characterization of a membrane steroid receptor located in neuronal membranes was carried out in the male rough-skinned newt in which the stress hormone corticosterone causes rapid inhibition of males' clasping behavior (Orchinik et al., 1991).
Frogs used to demonstrate that thyroid hormone receptors (TR) exert dual gene repressor/inducer functions	The dual function model for TR function proposed by Yun-Bo Shi and Laurent Sachs and colleagues (Sachs et al., 2000; Shi et al., 1996) has since been adopted to explain TR function and dysfunction in all vertebrates, including humans.
Circadian clocks first discovered in birds and insects	In the 1950s Karl von Frisch (using bees), Gustav Kramer and Klaus Hoffman (using birds), and Colin Pittendrigh (using fruit flies) each independently discovered compelling evidence for internal circadian clocks in animals, and their investigations marked the beginning of the modern field of circadian rhythms research (von Frisch, 1967).

LITERATURE CITED

Ashburner, M., Chihara, C., Meltzer, P., Richards, G., 1974. Temporal control of puffing activity in polytene chromosomes. In *Cold Spring Harbor Symposia on Quantitative Biology*, Vol. 38. Cold Spring Harbor Laboratory Press, 1974, 655–662.

Bargmann, W., Scharrer, E., 1951. The site of origin of the hormones of the posterior pituitary. *Am Sci.* 39, 255–259.

Bentley, P., 1966. The physiology of the urinary bladder of amphibia. *Biol Rev.* 41, 275–314.

Berthold, A. A., 1849. Transplantation der hoden. *Arch Anat Physiol.* Berlin: Veit, 1849, 16, 42–46.

Clever, U., Karlson, P., 1960. Induktion von Puff-Veränderungen in den Speicheldrüsenchromosomen von Chironomus tentans durch Ecdyson. *Exp Cell Res.* 20, 623–626.

Godeau, J. F., Schorderet-Slatkine, S., Hubert, P., Baulieu, E.-E., 1978. Induction of maturation in Xenopus laevis oocytes by a steroid linked to a polymer. *Proc Natl Acad Sci U S A.* 75, 2353–2357.

Houssay, B. A., 1936. What we have learned from the toad concerning hypophyseal functions. *N Engl J Med.* 214, 913–926.

Houssay, B. A., Foglia, V., Smyth, F., Rietti, C., Houssay, A., 1942. The hypophysis and secretion of insulin. *J Exp Med.* 75, 547–566.

Kopeć, S., 1917. Experiments on metamorphosis of insects. *Bull Int Acad Sci Cracovie Classe Sci Math Nat Ser B.*, 57–60.

Kopeć, S., 1922. Studies on the necessity of the brain for the inception of insect metamorphosis. *Biol Bull.* 42, 323–342.

Muske, L. E., Moore, F. L., 1988. The nervus terminalis in amphibians: anatomy, chemistry and relationship with the hypothalamic gonadotropin-releasing hormone system. *Brain Behav Evol.* 32, 141–150.

Orchinik, M., Murray, T. F., Moore, F. L., 1991. A corticosteroid receptor in neuronal membranes. *Science.* 252, 1848–1851.

Riddle, O., Bates, R. W., Dykshorn, S. W., 1933. The preparation, identification and assay of prolactin—a hormone of the anterior pituitary. *Am J Physiol Legacy Cont.* 105, 191–216.

Sachs, L. M., Damjanovski, S., Jones, P. L., Li, Q., Amano, T., Ueda, S., Shi, Y.-B., Ishizuya-Oka, A., 2000. Dual functions of thyroid hormone receptors during Xenopus development. *Comp Biochem Physiol Part B Biochem Mol Biol.* 126, 199–211.

Scharrer, B., 1935. Uber das Hanstromsche Organ X bei Opisthobranchiern. *Pubbl Stn Zool Napoli.* 15, 132–142.

Scharrer, B., 1937. Uber sekretorisch tatige Nervenzellen bei wirbellosen Tieren. *Naturwissenschaften.* 25, 131–138.

Scharrer, E., 1928. Die Lichtempfindlichkeit blinder Elritzen (Untersuchungen fiber das Zwischenhirn der Fische). *Z Vergl Physiol.* 7, 1–38.

Shi, Y. B., Wong, J., Puzianowska-Kuznicka, M., Stolow, M., 1996. Tadpole competence and tissue-specific temporal regulation of amphibian metamorphosis: roles of thyroid hormone and its receptors. *Bioessays.* 18, 391–399.

von Frisch, K., 1967. *The Dance Language and Orientation of Honeybees.* Belknap.

Appendix 3

Classical and Modern Concepts in Endocrinology

	Classical	Modern
Endocrine tissues	Cells that secrete a particular hormone form clusters or tight aggregates with each other	Many hormones are also produced by endocrine cells scattered loosely throughout organs and tissues (e.g. enteroendocrine cells)
	All hormones are secreted by distinct glands devoted to their synthesis (e.g. thyroid, adrenal, testes/ovaries)	Many/most organs and tissues with other functions secrete hormones or their precursors (e.g. bone, heart, kidney, skin, gut, brain)
	A particular hormone is generated from only one tissue	The same hormone can be produced by many different tissues and organs (e.g. ghrelin is made by both the stomach and the hypothalamus)
Hormones and receptors	One gene codes for one peptide hormone or hormone receptor	• The protein coded by one gene can generate several different hormones via alternative posttranslational processing (e.g. proopiomelanocortin (POMC) gives rise to adrenocorticotropic hormone (ACTH), α/β-melanocyte-stimulating hormone, lipocortin, and endorphins) • The RNAs of hormone receptors can be alternatively spliced to express different isoforms of the receptor with varying functions
	One cell type synthesizes only one hormone	One cell can generate more than one hormone (e.g. pancreatic β cells produces not only insulin but also amylin and chromogranin, and many pituitary cells synthesize several hormones)
	One hormone binds to a single receptor	• One hormone can bind to multiple receptors (e.g. estrogen binding to estrogen receptor (ER)α, ERβ, and even to a nongenomic receptors) • Conversely, many receptors behave promiscuously and bind several hormones
	Each hormone is associated with a single function	• Pleiotropy: individual hormones can exert variable, even opposing functions on different targets • At the physiological level, e.g. AVP functions as both an osmoregulatory hormone and a mediator of the stress response pathway • At the cellular level, epinephrine induces contraction of smooth muscle associated with blood vessels, but dilation of smooth muscle associated with bronchioles
Signaling	Hormones are always transported via the bloodstream to act on distant targets	Hormones produced by cells can also exert local effects on adjacent cells or even on the cells where the hormone originated, effects that are independent of blood transport (e.g. paracrine, autocrine, intracrine signaling)
	Endocrine signaling (blood-borne chemicals) and neuronal signaling (chemical communication across a synapse) are distinct processes	Many cells communicate via "neuroendocrine" signaling, whereby classical neurotransmitters become blood-borne chemical signals • Many classical neurotransmitters can also function as hormones, and vice versa
	Hormones are endogenously produced signals	Hormone signals can also originate exogenously (e.g. nutrient-like signals, such as free fatty acids stimulating the peroxisome-proliferator-activated receptor, and even exogenous ligands such as light and pheromones can behave as hormonal signals)

Appendix 4

Some Vertebrate and Arthropod Hormones and Their Characteristics

Source	Hormone Name	Chemical Class	Target Tissue	Solubility	Action by Target
Vertebrate hormones					
Hypothalamus	Thyrotropin-releasing hormone (TRH)	Polypeptide	Adenohypophysis	Water	Releases TSH
	Gonadotropin-releasing hormone (GnRH)	Polypeptide	Adenohypophysis	Water	Releases LH/FSH
	Corticotropin-releasing hormone (CRH)	Polypeptide	Adenohypophysis	Water	Releases ACTH
	Somatostatin (SST or GH-RIH)	Polypeptide	Adenohypophysis	Water	*Inhibits* GH release
	Growth hormone-releasing hormone, or somatocrinin (GHRH)	Polypeptide	Adenohypophysis	Water	Releases GH
	Prolactin-inhibiting hormone (PIH or Dopamine)	Amine-derived	Adenohypophysis	Water	*Inhibits* PRL release
	Prolactin-releasing hormone (PRH)	Polypeptide	Adenohypophysis	Water	Releases PRL
	Melanotropin-inhibiting hormone (MIH)	Polypeptide	Adenohypophysis	Water	*Inhibits* MSH release
	Melanotropin-releasing hormone (MRH)	Polypeptide	Adenohypophysis	Water	Releases MSH
	Orexin	Polypeptide	-	Water	Stimulates hunger
Hypothalamus/ Neurohypophy-sis	Oxytocin (OT)	Polypeptide	Uterus, vas deferens	Water	Smooth muscle contraction
	Antidiuretic hormone or Arginine vasopressin (ADH or AVP)	Polypeptide	Kidney, brain	Water	Water reabsorption, drinking behavior
Adenohypophysis	Thyroid-stimulating hormone (TSH)	Glycoprotein	Thyroid gland	Water	Synthesis and release of TH
	Luteinizing hormone (LH)	Glycoprotein	Gonads	Water	Androgen synthesis; progester-one synthesis; gamete release
	Follicle-stimulating hormone (FSH)	Glycoprotein	Gonads	Water	Gamete formation; estrogen synthesis
	Somatotropin or growth hormone (GH)	Polypeptide	Liver, muscle	Water	Synthesis of IGF and other proteins
	Prolactin (PRL)	Polypeptide	Mammary gland	Water	Protein synthesis
	Corticotropin, or adrenocortico-tropic hormone (ACTH)	Polypeptide	Adrenal cortex	Water	Corticosteroid synthesis
	Melanotropin, or melanocyte-stimulating hormone (MSH)	Polypeptide	Melanin-produc-ing cells	Water	Melanin synthesis
Thyroid gland	Thyroxine (T_4), triiodothyronine (T_3)	Amine-derived	Most tissues	Lipid	Increases metabolism, controls development
	Calcitonin (CT)	Polypeptide	Bone	Water	Reduce blood calcium (store in bone)
Parathyroid gland	Parathyroid hormone (PTH)	Polypeptide	Bone, kidney	Water	Increase blood calcium

Source	Hormone Name	Chemical Class	Target Tissue	Solubility	Action by Target
Ovary	Estriol (E3)	Steroid	Primary & secondary sexual structures; brain	Lipid	Stimulates development; reproductive behavior
	Estradiol (E2)	Steroid	Primary & secondary sexual structures; brain	Lipid	Stimulates development; reproductive behavior
	Estrone (E1)	Steroid	Primary & secondary sexual structures; brain	Lipid	Stimulates development; reproductive behavior
	Progesterone (P4)	Steroid	Uterus	Lipid	Prepares uterus for egg implantation; maintains pregnancy
	Inhibin	Glycoprotein	Adenohypophysis	Water	Blocks FSH release
	Activin	Glycoprotein	Adenohypophysis	Water	Enhances FSH biosynthesis
	Relaxin	Polypeptide	Uterus	Water	Relaxes smooth muscle
Testis	Testosterone (T2)	Steroid	Primary & secondary sexual structures; brain	Lipid	Stimulates development; reproductive behavior
	Dihydrotestosterone (DHT)	Steroid	Primary & secondary sexual structures; brain	Lipid	Stimulates development; reproductive behavior
	Inhibin	Glycoprotein	Adenohypophysis	Water	Blocks FSH release
	Activin	Glycoprotein	Adenohypophysis	Water	Enhances FSH biosynthesis
Adrenal cortex	Aldosterone (A)	Steroid	Kidney	Lipid	Sodium reabsorption
	Corticosterone/Cortisol (B/F)	Steroid	Liver, muscle	Lipid	Convert protein into glucose
	Androstenedione (A4)	Steroid	Primary & secondary sexual structures; brain	Lipid	Stimulates development; reproductive behavior
	Dehydroepiandrosterone (DHEA)	Steroid	Primary & secondary sexual structures; brain	Lipid	Stimulates development; reproductive behavior
Adrenal medulla	Epinephrine, norepinephrine	Modified amine	Liver, muscle	Water	Glycogen breakdown
Pancreas	Insulin	Polypeptide	Liver/muscle	Water	Glycogen storage/glucose uptake
	Glucagon	Polypeptide	Liver/adipose tissue	Water	Anti-insulin actions
	Somatostatin (SST)	Polypeptide	Pancreatic islet cells	Water	Suppress insulin secretion
Stomach	Gastrin	Polypeptide	Stomach glands	Water	Stimulates acid secretion
	Ghrelin	Polypeptide	Brain	Water	Stimulates hunger
Small intestine	Secretin	Polypeptide	Exocrine pancreas	Water	Release of HCO_3 into duodenum
	Cholecystokinin (CCK)	Polypeptide	Exocrine pancreas/gall bladder	Water	Release of proteases into duodenum/eject bile into duodenum
	Gastrin-releasing peptide (GRP)	Polypeptide	Stomach gastrin cells	Water	Release gastrin
	Gastric inhibitory peptide (GIP)	Polypeptide	Endocrine pancreas	Water	Release of insulin
	Motilin	Polypeptide	Stomach	Water	Stimulates pepsinogen and gastric motility
	Vasoactive intestinal peptide (VIP)	Polypeptide	Enteric blood vessels	Water	Increases blood flow to intestines

Source	Hormone Name	Chemical Class	Target Tissue	Solubility	Action by Target
Liver	Insulin-like growth factors (IGFs)	Polypeptide	Many tissues	Water	Stimulates proliferation
	Angiotensinogen (AGT)	Polypeptide	Blood vessels and adrenal cortex	Water	Stimulates vasoconstriction and aldosterone release
Adipose tissue	Leptin (LEP)	Polypeptide	Brain	Water	Inhibits hunger
	Adiponectin (AdipoQ)	Polypeptide	Liver, muscle, brain	Water	Enhances insulin sensitivity
	Interleukin-6 (IL6)	Polypeptide	Immune system	Water	Proinflammatory cytokine
Kidney	Erythropoietin (EPO)	Glycoprotein	Bone marrow	Water	Stimulates RBC proliferation
	Renin (a protease)	Polypeptide	Substrate in blood	Water	Catalyzes angiotensin synthesis
Cardiovascular	Atrial natriuretic peptide (ANP)	Polypeptide	Kidney, blood vessels	Water	Reduces blood pressure, induces vasodilation
	Endothelin	Polypeptide	Blood vessels	Water	Induce vasoconstriction
Placenta	Chorionic gonadotropin (CG)	Glycoprotein	Corpus luteum	Water	Maintains corpus luteum during beginning of pregnancy
Pineal gland	Melatonin (MT)	Modified amine	Brain	Lipid	Control biological clock
Thymus	Thymosins	Polypeptide	Lymphocyte-producing tissue	Water	Production of lymphocytes
Bone	Osteocalcin	Polypeptide	Bone, testis	Water	Causes bone calcification; increases testosterone production
Bone marrow	Erythroferrone		Red blood cells	Water	Stimulates iron uptake by red blood cells
Skin	Calciferol (provitamin D3)	Secosteroid		Lipid	Stimulates calcium uptake by intestine
Produced by many sites	Eicosanoids (prostaglandins, thromboxanes, leukotrienes)	Arachidonic acid-derived		Lipid	Diverse functions: smooth muscle contraction (i.e. childbirth), alter thermoregulation (produce fever), mediate cell growth
Insect Hormones					
Brain	Prothoracicotropic hormone (PTTH)	Polypeptide	Prothoracic glands	Water	Initiates molting by stimulating release of ecdysone from prothoracic glands
	Allotropin	Polypeptide	Axon terminals extending to corpora allata	Water	Induces JH release from corpora allata
	Allostatin	Polypeptide	Axon terminals extending to corpora allata	Water	Inhibits JH release from corpora allata
	Corazonin	Polypeptide	Inka cells	Water	Promotes PETH and ETH secretion
	Eclosion hormone	Polypeptide	Inka cells	Water	Promotes PETH and ETH secretion
	Buriscon	Polypeptide	Cuticle and epidermis	Water	Tans and hardens new cuticle
Prothoracic glands	Ecdysone	Steroid	Epidermis in larva/nymph; fat body in adult	Lipid	Following conversion to 20-hydroxyecdysone, promotes destruction of old cuticle and synthesis of a new one; stimulates yolk protein synthesis on adult

Source	Hormone Name	Chemical Class	Target Tissue	Solubility	Action by Target
Corpora allata	Juvenile hormone (JH)	Terpene (fatty acid derivative)	Epidermis in larva/nymph; ovary in adult	Lipid	Opposes formation of adult structures and promotes formation of larval/nymph ones; acts like a gonadotropin in adults
Inka cells of tracheae	Pre-ecdysis triggering hormone (PETH)	Polypeptide	Brain	Water	Prepares motor programs for cuticle shedding
	Ecdysis triggering hormone (ETH)	Polypeptide	Brain	Water	Prepares motor programs for escaping from old cuticle
Crustacean hormones					
Y organ homolog of the insect prothoracic gland)	Ecdysone	Steroid	Epidermis	Lipid	Stimulates molting and metamorphosis
Mandibular organ (MO)	Methylfarnesoate (MF, a hormone similar to juvenile hormone in insects)	Terpene	Epidermis	Lipid	Opposes formation of adult structures and promotes formation of larval/nymph ones
X organ-sinus gland complex	Molt-inhibiting hormone (MIH)	Polypeptide	Y organ	Water	Inhibits ecdysone synthesis by the Y organ
	Mandibular organ-inhibiting hormone (MOIH)	Polypeptide	MO	Water	Inhibits MF secretion by the MO
	Crustacean hyperglycemic hormone (CHH)	Polypeptide	Brain and other tissues	Water	Regulates energy balance during molting

Appendix 5

Some Diseases Associated with GPCR Dysfunction

Disease	Affected Receptor	Comments
Blomstrand chondrodysplasia	Parathyroid hormone receptor 1, PTHR1	Advanced skeletal maturation at birth; loss-of-function mutation
Central hypogonadism	Gonadotropin-releasing hormone receptor, GNRHR	Impairment of pubertal maturation and reproductive function; loss-of-function mutation
Central hypothyroidism	Thyrotropin-releasing hormone receptor, TRHR	Insufficient TSH secretion resulting in low levels of thyroid hormones; loss-of-function mutation
Color blindness	Cone opsins	X-linked, loss-of-function mutation
Congenital hypothyroidism	Thyroid-stimulating hormone receptor, TSHR	Increased plasma TSH and low levels of thyroid hormone; loss-of-function mutation
Congenital night blindness	Rhodopsin	Impaired night vision, decreased visual acuity, nystagmus, myopia, and strabismus; congenital loss-of-function mutation
Familial ACTH resistance	Adrenocorticotropic hormone, ACTH	Loss-of-function mutation
Familial hypocalcemia	Ca^{2+} sensing receptor; CASR	Hypocalcemia and hyperphosphatemia; neuromuscular irritability, calcification of the basal ganglia, brittle hair, mental retardation; gain-of-function mutation
Familial male precocious puberty	Luteinizing hormone receptor, LHR	Gain-of-function mutation
Familial non-autoimmune hyperthyroidism	Thyroid-stimulating hormone receptor, TSHR	Gain-of-function mutation
Growth hormone deficiency	Growth hormone-releasing hormone receptor, GHRH	Loss-of-function mutation
Ovarian dysgenesis 1 (ODG1), also called hypergonadotropic ovarian failure	Follicle-stimulating hormone receptor, FSHR	Lack of menstruation accompanied by severe osteoporosis, gonadal dysgenesis, often with somatic abnormalities; loss-of-function mutation
Jansen metaphyseal chondrodysplasia	Parathyroid hormone receptor 1, PTHR1	Extreme disorganization of the metaphysis of the long bones and of the metacarpal and metatarsal bones; gain-of-function mutation
Male pseudohermaphroditism	Luteinizing hormone/choriogonadotropin receptor, LHCGR	Loss-of-function mutation
Morbid obesity	Melanocortin 4 receptor, MC4R (binds α-melanocyte-stimulating hormone, α-MSH)	Mutations in this gene are the most frequent genetic cause of severe obesity; loss-of-function mutation
Nephrogenic diabetes insipidus	Vasopressin V2 receptor, AVPR2	Inability of the renal collecting ducts to absorb water in response to antidiuretic hormone (ADH)/arginine vasopressin (AVP); loss-of-function mutation
Retinitis pigmentosa	Rhodopsin	Loss-of-function mutation
Sporadic hyperfunctional thyroid adenomas	Thyroid-stimulating hormone receptor, TSHR	Gain-of-function mutation
Sporadic Leydig cell tumors	Luteinizing hormone/choriogonadotropin receptor, LHCGR	Gain-of-function mutation

Appendix 6

Some Diseases Associated with Nuclear Receptor Dysfunction and Their Treatments

Name	Disease/Function	Natural Ligand	Therapeutic Ligands (Trade Name)	Therapeutic Relevance
Thyroid hormone receptor (TR)	Hypothyroidism, obesity	Thyroid hormone	Levothyroxine (Synthroid)	Thyroid deficiency
Retinoic acid receptor (RAR)	Inflammatory skin disorders, leukemia	Retinoic acid	Isotretinoin (Accutane)	Acne
Peroxisome prolif-erator-activated receptor (PPAR)	Diabetes, coronary heart disease, obesity	Fatty acids, eicosanoids	Fibrates, GW501516, thiazolidinediones (TNZ)	Dyslipidemia, diabetes and insulin sensitization
Liver X receptor (LXR)	Atherosclerosis	24-Hydroxycholesterol	-	Lipid and cholesterol metabolism, atherosclerosis
Vitamin D receptor (VDR)	Osteoporosis, calcium homeostasis, cancer prevention	Vitamin D, bile acids	Calcitriol (Rocaltrol)	Hypocalcemia, osteoporosis, renal failure
Pregnane X receptor (PXR)	Xenobiotic metabolism	Xenobiotics	St. John's wort, rifampicin	Protection from toxic metabolites
Retinoid X receptor (RXR)	Leukemia, coronary heart disease	All-trans retinoic acid	LG1069 (Targretin)	Skin cancer
Estrogen receptor (ER)	Breast cancer, osteopo-rosis, atherosclerosis, CNS	Estradiol, estrogens	Tamoxifen, raloxifene	Menopausal symptoms, osteoporosis prevention, breast cancer
Glucocorticoid receptor (GR)	Immunological disorders, metabolic disorders	Cortisol, glucocorticoids	Prednisone, dexamethasone	Inflammatory and immuno-logical diseases, asthma, arthritis, allergic rhinitis, cancer, immune suppressant for transplant
Mineralcorticoid receptor (MR)	Hypertension, myocardial hypertrophy	Aldosterone, deoxycorticosterone	Spironolactone (Aldactone), epleronone (Inspra)	Hypertension, heart failure
Progesterone receptor (PR)	Breast cancer, infertility, pregnancy maintenance	Progesterone, progestins	RU486 (Mifepristone)	Abortifacient, menstrual control
Androgen receptor (AR)	Prostate cancer, X-linked androgen insensitivity, spinal/muscular atrophy	Testosterone, androgens	Flutamide, bicalutamide (Casodex)	Prostate cancer

Appendix 7

The 48 Known Members of the Human Nuclear Receptor Family Categorized According to Sequence Homology

Subfamily	Group	#	Ligand(s)
Steroid hormone receptors	Estrogen receptor (ER)	2	Estrogens
	Estrogen-related receptor (ERR)	3	*Orphan*
	3-ketosteroid receptors Glucocorticoid receptor (GC) Mineralcorticoid receptor (MR) Progesterone receptor (PR) Androgen receptor (AR)	4	 Cortisol Aldosterone Progesterone Testosterone
Thyroid Hormone Receptor-like	Thyroid hormone receptor (TR)	2	Thyroid hormones
	Retinoic acid receptor (RAR)	3	Vitamin A and related compounds
	Peroxisome proliferator-activated receptor (PPAR)	3	Fatty acids, prostaglandins
	Rev-ERBa	2	Heme
	RAR-related orphan receptor (ROR)	3	Cholesterol derivatives
	Liver X receptor-like (LXR) Farnesoid X receptor (FXR)	2 2	Oxysterols, bile acids Oxysterols
	Vitamin D receptor (VDR) Pregnane X receptor (PXR) Constitutive androstane (CAR)	1 1 1	Vitamin D Xenobiotics Androstane
	Nuclear receptors with two DNA-binding domains (2DBD-NR)	3	*Orphan*
Retinoid X Receptor-like	Retinoid X receptor (RXR)	3	Retinoids
	Hepatocyte nuclear factor-4 (HFN4)	2	Fatty acids
	Testicular receptor (TR)	2	*Orphan*
	TLX/PNR	2	*Orphan*
	COUP/EAR	3	*Orphan*
Other	Steroidogenic factor-like	2	Phosphatidylinositols
	Nerve growth factor IB-like	3	*Orphan*
	Germ cell nuclear factor-like	1	*Orphan*
	DAX/SHP	2	*Orphan*

Source: Assembled from Nuclear Receptors Nomenclature Committee (April 1999). "A unified nomenclature system for the nuclear receptor superfamily". *Cell*. 97 (2): 161–3, and from Laudet V (December 1997). "Evolution of the nuclear receptor superfamily: early diversification from an ancestral orphan receptor". *Journal of Molecular Endocrinology*. 19 (3): 207–26.

Appendix 8

Primary Enzymes of Steroid Hormone Biosynthesis

Common Name	Gene Name	Activity	Primary Tissue of Expression	Subcellular Location
Steroidogenic acute regulatory protein	StAR	Mediates mitochondrial import of cholesterol	All steroidogenic tissues except placenta and brain	Mitochondria
Desmolase, P450ssc	CYP11A1	Cholesterol-20, 23-desmolase	Steroidogenic tissues	Mitochondria
3β-hydroxysteroid dehydrogenase type 1	HSD3B2	3β-hydroxysteroid dehydrogenase	Steroidogenic tissues	Smooth endoplasmic reticulum
P450c11	CYP11B1	11 β-hydroxylase	Zona fasciculata and zona reticularis of adrenal cortex	Mitochondria
P450c17	CYP17A1	17α-hydroxylase and 17,20-lyase	Steroidogenic tissues	Smooth endoplasmic reticulum
P450c21	CYP21A2	21-hydroxylase	Not expressed in zona reticularis	Smooth endoplasmic reticulum
Aldosterone synthase	CYP1B2	18α-hydroxylase	Exclusive to zona glomerulosa of adrenal cortex	Mitochondria
Estrogen synthase	CYP19A1	Aromatase	Gonads, brain, adrenals. Adipose, bone	Smooth endoplasmic reticulum
17β-hydroxysteroid dehydrogenase type 3	HSD17B3	17-ketoreductase	Steroidogenic tissues	Smooth endoplasmic reticulum
Sulfotransferase	SULT2A1	Sulfotransferase	Liver, adrenals	Smooth endoplasmic reticulum
5α-reductase type 2	SRD5A2	5α-reductase	Steroidogenic tissues	Smooth endoplasmic reticulum

Appendix 9

A Summary of Steps in Human Digestion

Location	General Digestive Functions	Exocrine Secretions
Mouth	Mechanical breakdown of food (chewing), lubrication of food, initiate digestion of starches	Some contents of saliva produced by salivary glands: —Salivary amylase: starches → dextrin and maltose —Mucus: lubricates food
Tongue	Voluntary swallowing of food	
Esophagus	Involuntary peristaltic contractions transport food to stomach	
Stomach	Mechanical breakdown of food (churning) and enzymatic conversion of proteins into polypeptides	—Hydrochloric acid —Pepsinogen (active form is pepsin)
Small Intestine —Duodenum	—Receives partially digested food from stomach —Receives various exocrine secretions produced by accessory organs (gall bladder and pancreas): amylases, proteases, lipases, nucleases, bicarbonate, bile —Has its own endogenous digestive enzymes associated with brush border	Small intestinal brush border enzymes: —Maltase: maltose→ 2 glucose —Isomaltase: isomaltase → 2 glucose —Sucrase: sucrose → glucose + fructose —Lactase: lactose → glucose + galactose —Aminopeptidase: hydrolyzes terminal peptide bond at amino end —Dipeptidase: hydrolyses pairs of amino acids —Enteropeptidase: trypsinogen → trypsin; procarboxypeptidases → carboxypeptidases
Liver/Gall Bladder	Liver synthesizes bile (a lipid emulsifier), which is stored in the gall bladder and transported to the duodenum via the common bile duct	Bile emulsifies fats to facilitate digestion of lipids by lipases in the intestine
Pancreas	Pancreatic exocrine cells produces diverse digestive enzymes and bicarbonate that are transported to duodenum by the common bile duct	—Bicarbonate: neutralizes HCl from stomach, creating a favorable duodenal pH (6–7) optimal for pancreatic enzyme function —Pancreatic amylase: starch → maltose + dextrin —Trypsinogen: peptidase —Chymotrypsinogen: peptidase (activated by trypsin) —Procarboxypeptidases A & B: hydrolyze terminal peptide bond at carboxy end —Ribonuclease: RNA digestion —Deoxyribonuclease: DNA digestion
Small Intestine —Ileum and Jejunum	Continued digestion, absorption, and propulsion of food received from the duodenum	Small intestinal brush border enzymes: same as duodenum, listed above
Large Intestine	—Compaction of intestinal contents —Reabsorption of water	Bacterial action: —Fermentation of some undigested carbohydrate and amino acid residues —Bile degradation —Biosynthesis of vitamin K —Produces diverse metabolites and neuroactive substances involved in brain-gut communication

Appendix 10

Failures in Contraceptive Function during the First Year of Use

Method	% of Women Experiencing an Unintended Pregnancy within the First Year with Typical Use	Category of Contraceptive
No method	85	-
Withdrawal	20	Behavioral
Fertility awareness-based methods*	2–34	Behavioral
Copper IUD	0.8	Long-acting and reversible
Implant	0.1	Long-acting and reversible
NuvaRing	7	Long-acting and reversible
Patch	7	Long-acting and reversible
Combination pill	7	Short-acting
Progestin-only pill	7	Short-acting
Diaphragm (with spermicidal cream or jelly)	17	Short-acting
Sponge (when used by women who have given birth)	27	Short-acting
Spermicides	21	Short-acting
Male condom	13	Short-acting
Female condom	21	Short-acting
Female sterilization (tubal surgery)	0.5	Permanent
Male sterilization (vasectomy)	0.15	Permanent

Notes: Typical-use failure rates express effectiveness among all women who use the method, including those who use it inconsistently and incorrectly. IUD=intrauterine device.

*Range of estimates comes from a small number of moderate-quality studies and may not apply to all populations.

Sources: Statistics compiled from Hatcher RA et al., *Contraceptive Technology*, 21st ed., New York: Managing Contraception, 2018; Sundaram A et al., Contraceptive failure in the United States: estimates from the 2006–2010 National Survey of Family Growth, Perspectives on Sexual and Reproductive Health, 2017, 49(1):7 – 16; Peragallo Urrutia R et al., Effectiveness of fertility awareness-based methods for pregnancy prevention: a systematic review, *Obstetrics & Gynecology*, 2018, 132(3):591 – 604; and Urrutia RP and Polis CB, Fertility awareness based methods for pregnancy prevention, *BMJ*, 2019, 366:l4245.

Appendix 11

Secondary Sexual Characteristics in Females and Males Associated with Puberty

Change	Mean age
Females	
Initiation of growth spurt and deposition of fat	10
Widening of the pelvis	11
Growth and maturation of the internal genitalia and the vagina	12
Axillary hair begins to appear	13
Skeletal growth decreases; sweat and sebaceous gland development, sometimes accompanied with acne; first ovulation	14
Voice deepens, slightly	15
Adult height reached	16
Males	
Initiation of androgen production after childhood quiescence	10
Fat deposition begins	11
Skeletal growth begins; spontaneous erections; growth of seminal vesicles and prostate gland	12
Spontaneous nocturnal ejaculations begin	13
Growth of vocal cords and deepening of voice; appearance of axillary hair and hair on upper lip	14
First fertile ejaculation	15
Appearance of chest, body, and facial hair; sweat and sebaceous glands develop, often with acne; loss of body fat	17
Muscle growth and increase in muscle strength; broadening of shoulders	17
Adult height may be reached, although often growth may continue into the early 20s	18

Source: From Piñón, R. and Piñón, R., 2002. *Biology of human reproduction*. Sterling Publishing Company.

Appendix 12A

Tanner Stages of Breast and Pubic Hair Development in Females

Stage	Mean age
Breast	
I. Prepubertal; slight elevation of papilla	
II. Elevation of breast and papilla; areola diameter increases	11.2
III. Enlargement of breast tissue; no separation of breast and areola	11.2
IV. Areola and papilla form a secondary mound above the level of the breast	13.1
V. Mature stage; erect papilla projecting above areola	15.3
Pubic hair	
I. Prepubertal; no pubic hair	
II. Sparse, curly pigmented hair appearing along the lower labia	11.7
III. Spread of darker, coarser hair across the lower pubis	12.4
IV. Abundant adult type hair, but limited to labia area	13.0
V. Spread of pubic hair to form an inverted triangle; spread of hair along the upper inner thigh	14.4

Appendix 12B

Tanner Stages of External Genitalia and Pubic Hair Development in Males

Stage	Mean age	Range (years)
External genitalia		
I. Prepubertal		
II. Enlargement of scrotum and testes; scrotum becomes pigmented	11.6	9.0–14.7
III. Growth of penis; continued growth of scrotum and testes	12.9	10.3–15.5
IV. Increase in length and breadth of penis; growth of glans of penis	13.8	11.2–16.3
V. Adult size and shape	14.9	12.2–17.7
Pubic hair		
I. Prepubertal; no pubic hair		
II. Sparse growth of slightly curled hair along the base of the penis	13.4	10.8–16.0
III. Spread of darker, coarser hair above penis	13.9	11.4–16.5
IV. Abundant adult type of hair, but limited to genitalia	14.4	11.7–11.71
V. Adult hair in type and quantity; spread of hair along the inner thigh and above the penis	15.2	12.5–17.9

Source: From Piñón, R. and Piñón, R., 2002. *Biology of human reproduction*. Sterling Publishing Company.

Appendix 13

Classifications of Some Common Endocrine-Disrupting Chemicals

Category	Example Compounds	Source/Use	Associated Diseases
Plasticizers (plastic softeners)	Phthalates (DEHP, BBP, DBP, DiNP)	Plastics, children's toys	Low sperm count, obesity, birth defects, asthma, neurobehavioral disorders, endometriosis
Plastics	Bisphenol A, Bisphenol F, Bisphenol S	Polycarbonate plastics, food and beverage containers, thermal receipts, children's toys	Breast and prostate cancers, obesity, puberty, neurobehavioral
Cosmetics and detergent additives	Alkylphenols and p-Nonyl-phenol, Butylated hydroxyanisole (BHA), Phthalates, Parabens, Resorcinol, Octylphenol	Fragrances, lotions, hair dyes, nail polish	Breast cancer, thyroid disorders
Pesticides	DDT, DDE, Carbaryl, Malathion, Chlorpyrifos	Insecticides	Cancers, developmental toxicity
Herbicides and fungicides	Atrazine, Glyphosate, Mancozeb, Vindozolin	Weed killers and anti-fungal agents	Alterations in pubertal development
Metals and organo-metallic chemicals	Lead	Drinking water, paint, gasoline	Neurological disorders, premature birth, kidney disorders
	Mercury and methylmercury	Burning coal, seafood	Neurological disorders, diabetes
	Cadmium	Tobacco smoke, fertilizers	Cancers, reproductive disorders
	Arsenic	Drinking water, animal feed, herbicides, fertilizers	Cancers, diabetes, immune suppression, neurodevelopment, cardiovascular disease
Phytoestrogens	Isoflavones (e.g. Genistein, Daidzein), Coumestans (e.g. Coumestrol), Mycotoxins (e.g. Zearalenone), Prenylflavonoids (e.g. 8-prenylnaringeriin)	Present in plant-derived foods, such as soy	ER agonists
Natural hormones	Estradiol, Estrone, Testosterone	ER and AR agonists	Breast cancer, prostate cancer, and other cancers
Pharmaceuticals and synthetic hormones	Diethylstilbestrol, Ethinylestradiol, Tamoxifen, Levonorgestrel, Flutamide	Sex steroid hormone receptor agonists, antagonists, or biosynthesis inhibitors	Breast and other cancers
Industrial wastes	PCB Dioxin	Electrical coolant Industrial byproduct	Cancers, developmental disorders sperm quality, fertility, neurobehavioral disorders, anti-thyroid
Flame retardants	PBDEs	Furniture foam cushions, clothes, building materials, and electronics	Thyroid disruption, neurological disorders

Sources: Schug, T.T., Johnson, A.F., Birnbaum, L.S., Colborn, T., Guillette Jr, L.J., Crews, D.P., Collins, T., Soto, A.M., Vom Saal, F.S., McLachlan, J.A. and Sonnenschein, C., 2016. Minireview: endocrine disruptors: past lessons and future directions. *Molecular Endocrinology*, 30(8), pp. 833–847; Shahidehnia, M., 2016. Epigenetic effects of endocrine disrupting chemicals. *J Environ Anal Toxicol*, 6(381), pp. 2161–0525.

Glossary

Note: A list of common hormones and their actions is found in Appendix 4: Some Vertebrate and Arthropod Hormones and Their Characteristics.

5α-reductase: The enzyme that converts testosterone into the more active dihydrotestosterone.

Achondroplasia: The most common and recognizable form of human dwarfism. It is caused by a glycine mutation in the transmembrane domain of the fibroblast growth factor receptor 3 (FGFR3), leading to premature termination of chondrocyte cell division.

Acinar cells: Cells that comprise the exocrine portion of the pancreas.

Acromegaly: Excessive GH secretion that begins after puberty and after epiphyseal fusion of the long bones has taken place. Although long bone growth has ceased, appositional, cartilage, and membranous bone growth continues. Therefore, acromegaly is typically accompanied by growth of membranous bones of the skull, forehead, and jaw, enlargement of the hands and feet, as well as growth of cartilaginous structures, such as the nose and ears.

Adaptive developmental plasticity: The capacity of the same genotype to produce variable phenotypic outcomes depending upon inputs received from the internal or external environment during earlier development.

Addison's disease: Primary adrenal insufficiency due to the destruction of all adrenocortical tissue.

Additive response: The combined effect of two or more hormones equals the sum of their separate actions.

Adenohypophysis: One of the pituitary's two lobes, it is made up of epithelial tissue-derived glandular endocrine cells and is thus a classical endocrine gland. Often called the "anterior" lobe of the pituitary in humans and primates.

Adenylate cyclase: An enzyme found in all eukaryotic cells and a common effector used by multiple hormone signaling pathways. The activated enzyme converts ATP into the potent second messenger, cAMP, which binds to and activates protein kinase A, initiating a phosphorylation cascade that ultimately induces a change in cell physiology.

Adrenarche: A period of increased rate of DHEA and DHEA-S synthesis by the zona reticularis that in humans typically begins at around 10 to 11 years, with increasing rates of synthesis continuing through puberty. The primary effects of adrenarche in humans are the promotion of pubic hair development, the development of apocrine glands in the skin and an accompanying change of sweat composition that produces adult body odor, and increased oiliness of the hair and skin, which can lead to acne.

Adrenergic receptors: Receptors that bind to epinephrine and/or norepinephrine.

Adrenocortical component of the stress response: The "general adaptation" response mediated by the hypothalamus-pituitary-adrenal cortex and, in particular, adrenal glucocorticoids.

Agonist: An endogenously produced or exogenously derived natural or synthetic hormone analog that occupies the receptor and mimics the effects of the endogenous hormone.

Agouti and agouti-related peptide: Endogenous antagonists of melanocortin receptors.

Allostasis: The process of achieving homeostasis through physiological change that actively promotes adaptation.

Allostatic load: The cumulative energetic costs and wear and tear associated with maintenance of the allostatic state.

Allostatic overload: A pathological state that imparts damage to the body and predisposes the organism to disease.

Alpha-fetoprotein: Secreted into the blood of female mice before and after birth, this protein binds potently to blood estrogens, preventing them from entering the brain to induce masculinization.

Alternative posttranslational processing: Alternative processing of the prohormone protein into variable products depending upon cell type and what prohormone convertases the cell's vesicles contain.

Alternative splicing: The generation of multiple mRNA variants through the selective exclusion of specific exons is a contributor to the sequence diversity of many types of proteins.

Ametabolous development: A strategy of "no metamorphosis" in insects, whereby a genitalia-lacking miniature version of the adult emerges from the egg and simply increases in size throughout the succeeding immature nymphal stages.

Anorexigenic signals: Signals that suppress appetite and feeding.

Antagonist: Hormone analogs that occupy the receptor but inhibit activation by the hormone.

Anthropocene: An epoch of geological time characterized by the presence of distinct human impacts on the geology and ecology of the earth caused by the aforementioned processes.

Antidiuresis: A state of reduced urine volume production.

Aporeceptor: A receptor in its unbound form.

Autocrine signaling: A compound released by one cell that affects acts upon the same cell that released it.

Barr body: A transcriptionally inactivated X chromosome in a female mammal.

Behavioral estrus: Animals with estrous cycles exhibit this distinct period of sexual receptivity, also known as "heat". Behavioral estrus occurs during a period of time near ovulation when the females are most sexually attractive, proceptive, and receptive to males. As such, behavioral estrus is an outward manifestation of the peri-ovulatory phase of the ovarian cycle.

Beige adipose tissue (BeAT): Thermogenic BAT-like adipocytes that are found interspersed within WAT.

Bilaterian: Metazoan animal with bilateral symmetry.

Bioregulators: Refers to the three main interacting chemical signaling classes: neurotransmitters, hormones, and cytokines.

Bone modeling: Refers to bone lengthening and shape changes taking place from birth to puberty.

Bone remodeling: The continual process of bone turnover, as well as the strengthening or repairing of damaged bone.

Bound fraction: The hormone fraction in blood that is bound to transport proteins.

Brain-gut-microbiome axis: The notion that the gut's microbial makeup is not only influenced by neural and endocrine signaling from the brain, but the microbes themselves also influence diverse neurodevelopmental processes, energy homeostasis, metabolic health, and associated diseases.

Breastfeeding-induced amenorrhea: The suckling stimulus suppresses fertility for a variable time after birth with an almost complete inhibition of GnRH and gonadotropin secretion in the early stages of lactation, then transitions through a period where the frequency and amplitude of GnRH and gonadotropin pulsatility is too low and erratic to promote normal follicular growth, and the first several ovulations and menses are associated with a corpus luteum function that is usually insufficient to support a pregnancy.

Brown adipose tissue (BAT): Type of adipose tissue involved primarily in energy consumption and thermogenesis.

Cardiotoxicosis: Excess thyroid hormone synthesis may manifest itself as atrial fibrillation and congestive heart failure.

Castration: Removal of the testes.

Catecholamines: Tyrosine-derived neurotransmitters and hormones, such as epinephrine, norepinephrine and dopamine.

Caudal neurosecretory system: Neurosecretory cells located in the tails of fish, and specifically consists of neuroendocrine cells located in the terminal segments of the spinal cord that project to a neurohemal organ, the urophysis, from which neuropeptides are released.

Cephalic phase of digestion: Refers to a set of food intake-associated autonomic and endocrine responses to the stimulation of sight, smell, and taste sensory systems located in the head's eyes and oropharyngeal cavity.

Chordate: A phylum of animals that possess a rigid internal structure called a notochord during development.

Chromaffin cells: Modified postganglionic sympathetic neurons located in the adrenal medulla of adrenal glands and interrenal glands.

Circadian clocks: Biological clocks that operate on a daily cycle and maintain a periodicity of about (but never exactly) 24 hours to follow the earth's rotation around its axis.

Class I nuclear receptors: Bind all vertebrate steroid hormones. Current dogma states that in the absence of ligand the receptors are inactive and are thought to be sequestered as monomers in the cytoplasm, bound to heat shock proteins.

Class II nuclear receptors: In the absence of ligand the receptors form heterodimers with RXR, bind to hormone response elements, and repress gene transcription. In the presence of ligand they activate transcription.

Comparative endocrinology: A discipline that studies the differences and similarities of hormone systems among diverse vertebrate and invertebrate species, including humans.

Competitive binding assay: Increasing amounts of a test compound (hormone or drug) are incubated with cells expressing the receptor in question, in the presence of a fixed amount of radiolabeled hormone already known to bind to the receptor. If the test compound displays any specific binding affinity to the receptor, then increasing concentrations of that test compound should competitively displace the radioligand.

Complete androgen insensitivity syndrome (CAIS): A condition in 46,XY individuals with normally developed testes and testosterone and AMH secretion who nonetheless display a completely female external phenotype characterized by loss-of-function mutations in the androgen receptor gene, resulting in peripheral androgen resistance.

Congenital adrenal hyperplasia (CAH): A 21-hydroxylase deficiency prevents the conversion of 17-hydroxyprogesterone precursors common to both glucocorticoid and androgen synthesis. This substrate is instead diverted into the synthesis of abnormally high levels of active adrenal androgens, like testosterone, and their precursors DHEA and DHEAS, giving rise to varying degrees of virilization, or the acquisition of male sexual features.

Conjugation: The inactivation of hydrophobic hormones, such as steroid hormones and thyroid hormones, via their conversion into more hydrophilic compounds to ensure their solubility at high concentrations in biological fluids, prior to their excretion in the urine.

Corpus albicans: The remnants of the corpus luteum are destroyed by macrophages and replaced with fibroblasts, forming a white-colored, non-endocrine, scar-like tissue.

Corpus luteum: Ovarian remnants of the ovulated follicle that include luteinized granulosa cells, in addition to thecal cells, fibroblasts, and extensive vascularization. The structure produces primarily progesterone, as well as some estrogens.

Corticotropes: ACTH-producing cells of the adenohypophysis.

Cretinism: The most severe manifestation of a spectrum of hypothyroid disorders that is, in part, characterized by a severely stunted physical stature and mental development caused by a deficiency of thyroid hormone during fetal and early postnatal development.

Cryptorchidism: Undescended testes.

Cushing's disease: Typically caused by a tumor in the pituitary gland that produces large amounts of adrenocorticotropic hormone (ACTH), causing the adrenal glands to produce elevated levels of cortisol.

Cyanoketone: A drug that inhibits the activity of 3b-HSD and is therefore an effective inhibitor of both adrenal and gonadal steroid hormone synthesis.

Cyclooxygenases (COX): COX 1 and 2 convert arachidonic acid to prostaglandin H2, products that can be further processed into additional varieties of leukotrienes and prostaglandins.

CYP19 aromatase: An enzyme that irreversibly converts androgens into estrogens in all vertebrates.

Cytochrome P-450 monooxidases (CYPs): Enzymes localized to the inner mitochondria matrix or smooth endoplasmic reticulum where they facilitate electron transfer, functioning as oxidases, aromatases, hydroxylases, and lyases.

Cytokines: Signaling chemicals typically released by cells of the immune system.

Deiodinases: Intracellular transmembrane selenoproteins that convert T_4 to the more biologically active T_3 (in the case of D1 and D2), or catalyze the direct degradation of T_4 to the inactive reverse T_3 (rT_3), as well as T_3 to 3,3' T_2 (in the case of D3).

Deuterostomes: Early embryonic structure of a metazoan where the blastopore develops into the anus.

Developmental origins of health and disease (DOHaD): The notion that subtle functional changes in specific tissues during sensitive developmental windows of early life can result in increased susceptibility to disease or dysfunction later in life.

Diabesogen: An endocrine-disrupting chemical that can induce both diabetes and obesity.

Diabetes: The term denotes a high rate of urine production.

Diabetes insipidus: High rates of production of a "tasteless" (glucose-free) urine caused by the inability of the neurohypophysis to secrete antidiuretic hormone.

Diabetes mellitus: Refers to the production of a high volume of sweet-tasting urine due to the presence of glucose in the urine. Describes a group of metabolic diseases characterized by hyperglycemia resulting from defects in insulin secretion, insulin action, or both.

Diabetogen: An endocrine-disrupting chemical that promotes the development of diabetes.

Diethylstilbestrol (DES): The first synthetic compound to be specifically designed and marketed as a medically prescribed estrogen mimic for human use.

Disorders/differences of sex development (DSDs): Human neonates born with various genital anomalies that do not conform with their genetic sex.

Diuresis: A state of increased urine volume production.

DNA methyltransferases (DNMTs): Enzymes that transfer a methyl group to the 5-position of cytosine rings, usually in the context of GC-rich regions. In general, patterns of DNA methylation tend to correlate with chromatin structure, with active regions of the chromatin associated with hypomethylated DNA, whereas hypermethylated DNA is associated with inactive chromatin.

Downregulating: A decrease of receptor numbers on the cell surface, making the cell less responsive to subsequent exposure to the hormone.

Dwarfism: Short stature resulting from a genetic or medical condition, is generally defined as an adult height of 4 feet 10 inches (147 centimeters) or less.

Ecdysis: Molting, or the stripping off of an old cuticle in an arthropod to facilitate growth.

Eicosanoids: A family of hydrophobic hormones derived from the C20 fatty acid, arachidonic acid. Includes prostaglandins, thromboxanes, and leukotrienes.

Endocrine-disrupting compounds (EDCs): Chemicals that disturb endocrine signaling by mimicking hormones, blocking hormone–receptor interactions, or altering signaling pathways, rates of hormone synthesis, or degradation.

Endocrine signaling: A compound released by one cell is transported via the bloodstream to targets located some distance away.

Endocrinology: A branch of physiology that studies a specific class of blood-borne chemical signals called hormones, as well as the cells that generate hormones, and the resulting hormone actions on target cells and tissues.

Endostyle: A grooved, mucus-secreting pharyngeal organ that traps food particles and facilitates filter feeding in adult protochordates. It is also an exocrine organ that secretes thyroid hormones into the gut for absorption, and is likely represents the ancestral form of the thyroid gland.

Enteroendocrine cells: Cells of the gut that comprise a diffuse endocrine system that is scattered along the gut lining from the stomach to the colon.

Environmental endocrinology: The study of how animals transduce signals received from their natural biotic and abiotic environment into endocrine responses that mediate diverse adaptive behavioral and physiological changes.

Epigenetic: Heritable changes in gene expression that do not result from changes in the underlying DNA sequence. Examples of epigenetic modifications to histone tails include acetylation, methylation, phosphorylation, and ubiquitination.

Epiphyseal growth plates: Mediate the ordered process of growth plate chondrocyte proliferation, hypertrophic differentiation, apoptosis, and subsequent new bone formation at these plates, which promote linear growth until adulthood.

Equilibrium dissociation constant (K_d): The affinity of a hormone for its receptor, where K_d is the concentration of hormone at equilibrium that is required for binding of 50% of its receptor sites.

Estrous cycle: Reproductive cyclicity is described as an estrous cycle in the vast majority of mammals. While the profile of ovarian hormone secretion is similar in menstrual and estrous cycles, the most obvious difference is that whereas in the absence of fertilization the uterine lining is shed in menstruating animals, it is resorbed in animals with estrous cycles. Although during the estrous cycle some animals may display a bloody discharge, unlike true menstruation this fluid does not contain endometrial stromal or epithelial tissue.

Eutheria: So-called placental mammals, possess elaborate placentae.

Exons: Nucleotide sequences *ex*pressed in the mature mRNA.

External secretion: A product secreted external to the blood such as onto an epithelial surface associated with either a lumen of an organ or the skin.

Fetal-placental-maternal unit: From an endocrine perspective, the pregnant human consists of these three functionally distinct but interacting compartments.

Fetal zone: A transient fourth zone of the adrenal cortex that produces large quantities of DHEA and DHEA-S.

Follicular phase of the ovarian cycle: The first 14 days of the cycle comprise the follicular phase, characterized by the growth, development, and maturation of the dominant follicle under the influence of FSH and LH.

Free fraction: Considered the blood's bioavailable fraction, this hormone is not bound to any protein.

Free-running rhythm: In the absence of external time cues, circadian activity will drift because the periodicity of the endogenous circadian clock is not exactly 24 hours.

G protein-coupled receptors: Cell-surface receptors that interact with proteins possessing GTPase activity to induce an intracellular signaling cascade.

Gastric phase of digestion: The entry of food into the stomach from the esophagus provides both mechanical and chemical stimulation to the gastric wall. Gastric phase mechanical and chemical stimulations build upon events already put into motion during the cephalic phase, promoting additional exocrine, endocrine, and neural responses not only by the stomach itself, but also by accessory organs and the intestine.

Gene divergence: Following a gene duplication event, the asymmetric accumulation of mutations among the daughters of the duplicated genes over evolutionary time promotes functional disparity in the duplicated gene lineages.

Gene family: Genes that derive from a common ancestral gene that underwent one or more rounds of gene duplication and divergence.

Gene loss: The loss of redundant gene duplicates from an organism's genome.

General endocrinology: Originally established as a field of internal medicine, general endocrinology typically refers to clinical human and veterinary applications, as well as basic research on fundamental principles of endocrine signaling without the bias of application.

Genotypic sex determination (GSD): Denotes that the sex of an individual is determined entirely by its genotype, usually by genes located on sex chromosomes, and is fixed at fertilization.

Gigantism: A condition characterized by excessive growth and height that is not just in the upper 1% of the population in question, but several standard deviations above the mean. The condition is most typically caused by hypersecretion of GH before puberty due to a tumor on the adenohypophysis.

Gluconeogenesis: The synthesis of glucose from non-carbohydrate precursors, like amino acids and lipids.

Glycogenesis: Storage of glucose monomers as glycogen polymers in liver and skeletal muscle.

Glycogenolysis: Breakdown of stored glycogen polymers in liver and skeletal muscle into glucose monomers.

Glycoproteins: Polypeptides that have carbohydrate moieties covalently attached to them.

Goiter: An enlarged thyroid gland most commonly caused by increased TSH production due to a hypothyroid state. Increased TSH induces the thyroid gland to synthesize and store more thyroglobulin.

Gonadarche: Describes testicular enlargement and increased testosterone synthesis in males, generally the first signs of puberty in boys.

Gonadotropes: LH- and FSH-producing cells of the adenohypophysis.

Graafian follicle: The "mature" or "preovulatory" follicle, which is the final stage of follicular development prior to ovulation.

Graves' disease: An autoimmune disorder that results in the production of antibodies that bind to and activate the TSH receptor, stimulating follicular hypertrophy and hyperplasia, thyroid gland enlargement, and increased TH synthesis.

Gynecomastia: Breast development in males.

Half-life: Amount of time required for a hormone to fall to half of its initial concentration in the blood.

Hashimoto's thyroiditis: A disease characterized by the development of several cell- and antibody-mediated immune processes that destroy the thyroid gland, with antibodies most frequently targeting thyroid peroxidase and thyroglobulin proteins.

Hemimetabolous development: A strategy of "incomplete metamorphosis" in insects where the external wing primordia grow sequentially larger throughout the nymphal stages and are visible on the outside of the body in the later nymphal stages.

High-density lipoproteins: Blood lipoproteins that return excess unused cholesterol from cell back to the liver for storage.

Histone acetyltransferases (HATs): Enzymes that neutralize the positively charged lysine residues on histone tails, promoting dissociation of the tails from the negatively charged DNA and an overall decondensation of the chromatin, which promotes transcription.

Histone deacetylase (HDAC): Enzymes that promote nucleosome condensation and inhibit transcription.

Histone methyltransferases: Enzymes that can either activate or further repress gene transcription depending on the histone tail and specific residue being methylated and the presence of other histone tail modifications in the vicinity.

Holometabolous development: A strategy of "complete metamorphosis" in insects. The immature stages of these insects, called larval stages, are typically widely divergent from the adult form. Holometabolous larvae undergo a radical transformation by first passing from the larval to the pupal stage, an outwardly quiescent but developmentally active stage when larval structures are replaced with adult legs, wings, features of the head, genitalia, and a distinct adult cuticle.

Holoreceptor: A receptor bound to a ligand.

Homeostasis: The physiological constancy of the internal environment that is actively maintained by organisms.

Homolog: Genes with shared ancestry.

Hormonal sex: Based on the concentrations of circulating androgens and estrogens, determines a broad array of sexual differentiation features occurring during fetal development and puberty.

Hormone: A blood-bourne chemical secreted by cells or tissues.

Hormone response elements (HREs): Short DNA sequences localized predominantly to the regulatory 5′-flanking regions of target genes, typically within several hundred base pairs upstream of the transcription start site. Nuclear receptors bind to HREs to regulate gene transcription.

Hydroxysteroid dehydrogenase (HSD): Localized to the smooth endoplasmic reticulum, these enzymes can activate or deactivate steroid hormones by catalyzing the dehydrogenation of hydroxysteroids.

Hyperglycemia: Elevated blood sugar, often associated with diabetes mellitus.

Hypogonadism: Reduced sex steroid synthesis by the testes or ovaries.

Hypophyseal portal vessels: Blood vessels that drain blood from capillary loops originating in the median eminence down to the adenohypophysis.

Hypophysectomy: Surgical removal of the pituitary.

Hypospadia: A condition where the urethral opening is located below its normal position at the head of the penis.

Hypothalamic amenorrhea: A condition characterized by a cessation of menstrual cycling in women due to a dysfunction of the hypothalamic-pituitary-gonadal axis in the absence of disease or ovarian failure. This disorder is associated with chronic energy deficiency, typically caused by long-term strenuous exercise and/or reduced food intake, such as starvation or extreme dieting.

Hypothalamus: That portion of the vertebrate "diencephalon", or the posterior part of the forebrain that sits on top of the brainstem, that lies inferior to the thalamus.

Hypovolemic thirst: Experienced following a reduction of blood volume due to bleeding, diarrhea, or excessive perspiration.

Imaginal discs: Internally developing adult structures in the larva formed from special epidermal cells whose growth is suppressed during larval life and have no functions in the survival of the larvae.

Induced ovulators: Animals that remain in a state of sexual receptivity during estrus until ovulation is induced directly via copulatory stimulation during mating. Includes domestic cats, rabbits, ferrets, alpacas, llamas, and camels.

Insulin resistance: A defect in the ability of tissues to respond to insulin, and an important medical condition that predisposes individuals to developing type 2 diabetes mellitus and other metabolic disorders.

Internal secretion: A product secreted directly into the blood, like a hormone or glucose.

Interrenal cells: The homolog of the mammalian adrenal cortex, in non-mammalian vertebrates these clusters of cells are often found between or within the kidneys.

Intersex: A general term for people born with differences in sex traits or reproductive anatomy.

Intestinal phase of digestion: The delivery of chyme from the stomach to the duodenum of the small intestine triggers the inhibition of gastric acid secretion, acceleration of gastric emptying, release of digestive fluids from accessory organs, and deceleration of the rate of food transit in the intestine to enhance the absorption of food substrates.

Introns: Intervening sequences that are not expressed in the mature mRNA.

Iodothyronines: Iodinated derivatives of tyrosine residues, such as the thyroid hormones thyroxine and triiodothyronine.

Ionotropic receptors: Cell-surface receptors that function as ligand-gated ion channels.

Islets of Langerhans: Comprise the endocrine portion of the pancreas.

Jet lag: Also known as circadian desynchrony, refers to a number of pathologies resulting from rapid changes in the day–night cycle incurred following jet travel across multiple time zones.

Juvenile hiatus: A period of time in human children during which GnRH and gonadotropin secretion is suppressed by both gonad-dependent and gonad-independent factors.

Kallmann syndrome: Characterized by hypogonadotropic hypogonadism due to a deficiency of GnRH synthesis that manifests as reduced LH and FSH synthesis. Caused by a failure of GnRH cells to migrate to the hypothalamus from the embryonic vomeronasal organ.

Ketoacidosis: The severe reduction in blood pH, a condition that can lead to cardiac arrhythmia, coma, and death. Often results from the over-reliance of ketones as a metabolic fuel in type 1 diabetes.

Ketone bodies: Derived from increased fatty acid catabolism, ketones are additional fuels that can be used by the brain and other organs in the absence of glucose.

Kinases: Enzymes that catalyze the transfer of the terminal phosphate group from ATP molecules onto specific amino acids within another protein.

Klinefelter syndrome: The most common form of sex chromosome DSD, typically associated with a 47,XXY karyotype (polysomy X) caused by nondisjunction during gametogenesis.

Lactotropes: PRL-producing cells of the adenohypophysis.

Light exposure at night (LEN): Light pollution caused by indoor and outdoor lighting.

Lipoxygenases (LOX): Converts the substrate arachidonic acid to leukotriene B_4.

Long-term adiposity signals: Signals that circulate in proportion to body adiposity and participate in the negative feedback control of fat stores for the maintenance of optimal energy balance and body weight.

Low-density lipoproteins (LDL): Export cholesterol and other lipids from the liver and transport them to target cells that may convert the cholesterol into steroid hormone.

Luteal phase of the ovarian cycle: Characterized by a rapid increase in progesterone synthesis by the newly formed corpus luteum, which in conjunction with an increase in inhibin production feeds back negatively along the H-P axis to inhibit the surge in gonadotropins.

Median eminence: A region of the hypothalamus where nerve endings of hypothalamic nuclei that communicate with the adenohypophysis terminate and are juxtaposed to the hypophyseal portal vessels.

Melanotropes: MSH-producing cells of the adenohypophysis.

Menarche: Denotes the time of the first menstruation in females in response to increasing maturation of the GnRH pulse generator and follicular development, and is typically the last obvious marker of female sexual maturation.

Menopause: Technically refers to the final menstrual period in a woman's life, which on average is about 50 years of age, and therefore can only be determined retrospectively.

Menstruation: The shedding of the uterine lining ("bleeding") that for most women occurs approximately every 28 days. Specifically, the stratum functionalis, which consists of the epithelial lining and uterine glands, is shed during menstruation. Although the underlying stratum basalis remains intact, it houses vasculature that contributes to the menstrual flow. Following menstruation, a new stratum functionalis is generated from the stratum basalis.

Metabolic syndrome: Describes the presence of three or more risk factors in a patient, which include abdominal obesity, high triglycerides, low- and high-density lipoprotein cholesterol, high blood pressure, and elevated fasting blood glucose. Individuals with metabolic syndrome are at increased risk for developing type 2 diabetes mellitus and cardiovascular disease.

Metabolism-disrupting chemical (MDC): Any endocrine-disrupting chemical that alters susceptibility to developing metabolic disorders.

Metamorphosis: In the most common biological usage of the term, metamorphosis refers to a period of spectacular development that defines the transition from a morphologically distinct larval form to a new juvenile (adult-like) form.

Metatheria: Marsupials, such as opossums, kangaroos, and koalas. Possess simple placentae.

Metazoans: Multicellular organisms with differentiated tissues.

Metyrapone: A selective inhibitor of 11b-hydroxylase, and thus effectively blocks cortisol synthesis by adrenocortical cells.

Model organisms: Animals that have been very broadly studied, usually because they are relatively easy to breed in a laboratory environment, have a rapid generation time, are cost-effective to raise, and can be genetically manipulated.

Monotremes: The most ancestral living egg-laying mammals, which include the duck-billed platypus and several species of echidna, or spiny anteaters.

Multigenerational exposure: The simultaneously and directly exposed generations to any endocrine-disrupting compound are the F0 generation (the mother), the F1 generation (the fetus), and the germ cells present in the developing fetus that will go on to help generate the F2 generation (the mother's grand-offspring).

Myxedema: Severe adult-onset hypothyroidism, typically associated with late stages of Hashimoto's thyroiditis.

Negative feedback: The detection of a regulated variable away from its optimal value, followed by corrective responses that return the perturbed variable back to pre-perturbation levels.

Neurohemal organ: The juxtaposition of multiple neurosecretory neuron axonal termini with a capillary bed.

Neurohypophyseal component: (of the stress response) AVP of magnocellular nuclear origin is activated specifically in response to hypovolemic and/or osmoregulatory stress.

Neurohypophysial neurons: Magnocellular neurons that originate from the paraventricular (PVN) and supraoptic (SON) nuclei whose axons traverse the hypothalamic-pituitary stalk. They release the neuropeptide hormones arginine vasopressin and oxytocin from nerve endings in the neurohypophysis.

Neurosecretion: The secretion of hormones into the bloodstream by neurons.

Neurosecretory neurons: A category of neurons that specialize in the synthesis, storage, and release of hormones into the circulatory system.

Neurotransmitters: Chemicals released by neurons that typically enter the synaptic space in between the neuron and an adjacent cell.

Non-bilaterian: Metazoan animal without bilateral symmetry.

Non-steroidal anti-inflammatory drug (NSAID): Drugs, like aspirin, that suppress prostaglandin synthesis by inhibiting both COX 1 and 2, and are hence potent inhibitors of the inflammatory response.

Nonadditive response: The combined effect of multiple hormones is less than the sum of their actions.

Nuclear receptors: A "superfamily" of ligand-activated transcription factors that regulate genetic networks controlling diverse aspects of physiology, such as development, homeostasis, reproduction, immunity, and metabolism.

Obesity: Broadly defined as excessive fat accumulation that presents a risk to health, and is a harbinger to metabolic diseases like type 2 diabetes and cardiovascular disease.

Obesogen: An endocrine-disrupting chemical that alters the regulation of energy balance to favor weight gain and obesity.

Oligopeptides: Polypeptides containing fewer than about 20 amino acids in length.

Oogenesis: Production of a mature haploid ova from diploid oogonia via two rounds of meiotic division.

Orexigenic: Signals that stimulate appetite and feeding.

Organotherapy: The practice of eating a specific organ from an animal for medicinal purposes in order to treat a disease or deficiency in a person's corresponding malfunctioning organ.

Ortholog: Related genes in different genomes (i.e. different species) that arose via gene duplication and divergence.

Osmoregulation: The maintenance of a constant intracellular and extracellular ionic and osmotic condition that creates the necessary environment for diverse metabolic processes and is essential for the normal functioning of all cells and the organisms they constitute.

Osmotic thirst: A potent thirst that occurs in response to drops in *intracellular* fluid volume caused by water being drawn out of cells due to an increased blood plasma osmotic pressure.

Osteoblasts: Cells that synthesize protein components of the extracellular matrix (osteoid).

Osteoclasts: Cells that mediate bone resorption by attaching to bone, forming a sealed space between the osteoclast and the bone.

Osteoporosis: A progressive bone disease that is characterized by a decrease in bone mass and density which can lead to an increased risk of fracture.

Ovarian follicle: An oocyte surrounded by theca and granulosa epithelial cells.

Ovipary: Describes a strategy of reproduction whereby some or all of embryogenesis takes place outside of the parent in an egg containing all the nutrients necessary for development through hatching.

Ovoviviparity: Refers to a strategy by which embryos in shelled eggs are not laid externally, but are instead retained inside the parent where they "hatch" prior to parturition. Although the offspring emerge shell-less, free-moving, and alive ("vivi"), besides the yolk sac there is little or no direct transfer of nutrients from the parent to the offspring due to the presence of an impermeable eggshell during most of gestation.

Paedomorphosis: In salamanders, this describes a life history strategy of retaining of larval morphological features (e.g. presence of gills and tail fins, and the absence of lungs) in the adult. Paedomorphic salamanders neither undergo metamorphosis nor transition to a terrestrial existence but mature into breeding adults and maintain a completely aquatic life.

Paracrine signaling: A compound released by one cell affects an adjacent cell without entering the bloodstream.

Paralog: Related genes within the same genome that arose via gene duplication and divergence.

Pars distalis: Distal part of the adenohypophysis, this region produces at least six major hormones, the greatest diversity of hormones of any pituitary region.

Pars intermedia: Intermediate part of the adenohypophysis, in many mammals these cells form a boundary layer between the adenohypophysis and neurohypophysis.

Pars tuberalis: A collar of tissue that forms the outer portion of the pituitary stalk connecting the pituitary to the hypothalamus.

Perimenopause: Describes the endocrine and physiological changes that indicate the approach of menopause, as well as the transition period between menopause and the start of postmenopause. On average, perimenopause lasts 4–5 years but can be as short as a few months or last up to 8 years.

Permissive action: A hormone elicits no effect alone, but its presence is required for another hormone to elicit an effect.

Pheochromocytoma: The hypersecretion of adrenal medullary catecholamines, a potentially life-threatening cause of hypertension.

Phosphodiesterase: An enzyme that inactivates the second messenger, cAMP, by converting it into AMP by breaking its phosphodiester bond.

Physiology: The discipline in the life sciences that studies the forms, functions, and mechanisms of action of living systems, ranging from biomolecules and cells to tissues, organs, and organisms.

Pinealocytes: A group of neurosecretory neurons located in the brain found in the pineal gland and whose major function is the synthesis of melatonin, a hormone that modulates circadian rhythms.

Pituitary gland: Located beneath the middle region of the hypothalamus, the pituitary is a bi-lobed endocrine structure, consisting of the adenohypophysis and the neurohypophysis.

Pivotal temperature: The species and population-specific incubation temperature that produces 50% of each sex.

Plasticity: Adaptability of an organism to changes in its environment.

Pleiotropy: The production of different effects on different tissues by a single hormone.

Polar bodies: The meiotic divisions of oogonia are asymmetric, yielding only one very large and viable egg, as well as two much smaller nonviable cells called polar bodies.

Polycystic ovary syndrome (PCOS): A common disorder affecting 5%–20% of women of reproductive age worldwide, and is a leading cause of subfertility and infertility. Its primary symptoms include (1) excess levels of androgen production by the ovaries, (2) an absence of fully mature Graafian follicles, and (3) anovulation (the absence of ovulation).

Polyphenism: A type of phenotypic plasticity prominent in insects with cast systems of development where larvae molt into one of two or more alternative morphologies, depending on nutritional and other environmental signals they receive.

Positive feedback: A condition where a variable is temporarily driven away from its set points in a manner that transiently exacerbates the magnitude and direction of an initially small perturbation.

Post-traumatic stress disorder (PTSD): A psychological and physiological condition resulting from exposure to a significantly traumatic event or stressor to which the person responded with fear, helplessness, or horror. PTSD is associated with a uniquely paradoxical profile whereby corticotropin-releasing hormone (CRH) levels are increased, while urinary and plasma levels of cortisol are lower or at least not elevated, compared to non-exposed persons without PTSD.

Pregnancy: Refers to the period of time between fertilization and parturition during which a viviparous animal is in a gestational state.

Prehormone: Inactive nascent polypeptide that requires the cleavage of the signal sequence by a signal peptidase in the RER for direct conversion into the mature hormone.

Preprohormone: Inactive nascent polypeptide that requires the cleavage of both the signal sequence and further excision of "pro" sequences within the prohormone to reach full maturation.

Primordial germ cells (PGCs): Cells that will give rise to spermatocytes or oocytes that do not originate from the gonad, but instead migrate along the hindgut from the yolk sac to a region called the gonadal ridge by about the fifth week of embryogenesis in humans.

Prohormone: An inactive or less active form of a hormone that requires activation by enzymes to become functional.

Prohormone convertases: Enzymes that convert prohormones into their mature forms.

Proliferative phase: (of the uterine cycle) Begins from the first day of menses and lasts until ovulation, is characterized by the proliferation of endometrial cells, enlargement of endometrial glands, and enrichment of the endometrium with blood vessels, all in response to rising levels of estradiol.

Proopiomelanocortin (POMC): A large pituitary prohormone that is cleaved to produce several different hormones, including ACTH, β-endorphin, β-lipotropin, melanocyte-stimulating hormone, and other peptides.

Protandry: A male transforms into a reproductively viable female.

Protochordates: The most ancestral living non-vertebrate chordates, the cephalochordates (lancelets, such as amphioxus) and ascidians (sea squirts, such as *Ciona*).

Protogyny: A female transforms into a reproductively viable male.

Protostome: Early embryonic structure of a metazoan where the blastopore develops into the mouth.

Pseudogene: Genes that remain in the genome, but have lost some or all functionality.

Radioimmunoassay: A method that uses radioactively labeled antibodies to specifically recognize and bind to miniscule levels of a hormone of interest in the blood or another body fluid and accurately quantify its concentration.

Rathke's pouch: Early developmental anlage of the adenohypophysis.

Receptor tyrosine kinase (RTK): The most common category of enzyme-linked cell-surface receptor, RTKs possess an extracellular domain containing the ligand-binding site, and an intracellular domain containing intracellular domain containing tyrosine kinase activity.

Recruiter receptors: A cell-surface receptor that, upon binding a ligand, recruits separate enzymes that associate with the receptor to mediate an intracellular signaling pathway.

Regional duplication (of a gene): The generation of a second copy of a portion of a gene, a single gene, a portion of a chromosome, or an entire chromosome.

Retinohypothalamic tract (RHT): Nonvisual pathway of intrinsically photosensitive retinal ganglion cells (ipRGC) that communicates light information to the SCN.

Rickets: A condition characterized by the defective mineralization of bones and the development of weak and toneless muscles due to impaired metabolism of vitamin D, phosphorus, or calcium.

Seasonal breeders: Species that display seasonal variability in their frequency of ovulation, spermatogenic activity, gamete quality, and sexual behavior.

Secondary sexual characteristics: Features that develop at puberty, such as facial hair, pubic hair, and breast development.

Secosteroids: Cholesterol-derived relatives of steroids, such as vitamin D, that possess a "broken" aromatic ring.

Secretory phase: (of the uterine cycle) Characterized by the secretion of progesterone and estradiol from the corpus luteum. This promotes an increased thickening of the endometrium, as well as secretion of fluids by the endometrial glands that are essential for survival and development of a prospective embryo.

Selective estrogen receptor modulators (SERMs): Compounds that selectively block the negative actions of estrogen on one tissue (like the breast) breast tissue, while simultaneously promoting estrogen's beneficial effects on another tissue (like bones).

Sequential hermaphroditism: A condition where an individual changes its sex during its lifetime.

Sex chromosome aneuploidy: The condition of having less than (monosomy) or more than (polysomy) the normal number of X and Y chromosomes, often resulting from chromosomal nondisjunction during meiosis.

Sex chromosome mosaicism: Occurs when different cells in a body possess a variable array of sex chromosomes due to a nondisjunction event in the zygote, and thus a range of phenotypes can be produced depending on where expression occurs.

Sex determination: The process that establishes whether the gonads of a sexually reproducing organism will develop into either an ovary or a testis, a course of development that begins early during embryogenesis.

Sexual differentiation: The process that follows gonadal determination whereby internal structures, external genitalia, the brain, and secondary sex characteristics that arise at puberty acquire their female or male phenotypes. In vertebrates, the process of sexual differentiation begins during fetal development and is completed at puberty.

Short-term hunger and satiety signals: Situational- and meal-related signals that are proportional to the size of the meal being consumed and are crucial to regulating the size and timing of individual meals.

Signal transduction pathway: Interpretation of extracellular signals by the intracellular machinery via complex, highly integrated and ordered branching networks of consecutive protein–protein interactions.

Somatotropes: GH-producing cells of the adenohypophysis.

Spare receptors: The number of receptors that exist in excess of the minimum number required for a full biological response. Spare receptors increase the sensitivity of a cell to low levels of hormone and to small changes in hormone concentration and may prolong the duration of the biological response in the presence of slowly declining hormone concentrations.

Spermatogenesis: Production of haploid spermatids from diploid spermatogonia via two rounds of meiotic division.

Spermiogenesis: Morphological transformation of spermatids into mature spermatozoa.

Spontaneous ovulators: Animals that ovulate at regular intervals, and their ovulation is dependent upon the positive-feedback induction of gonadotropins by estradiol during the late follicular phase. The generation of the surge in LH occurs independently of any mating cues, and therefore occurs "spontaneously". Includes humans, horses, cows, sheep, and most rodents.

SRY: "Sex-determining region on the Y chromosome" encodes a transcription factor that induces a testis-forming pathway beginning at about 7 weeks post-fertilization.

Steroidogenic acute regulatory (StAR): Rate-limiting protein in steroid hormone synthesis that imports cholesterol into mitochondria for the initial steps of steroidogenesis.

Stress: Can be broadly defined as any state of threatened or actual disruption to homeostasis and the resulting increased energy consumption by an organism that is required to restore or recalibrate the homeostatic state.

Suprachiasmatic nucleus (SCN): The hypothalamic nucleus that functions as the master clock in mammals by responding indirectly to light information. SCN temporal output in the form of both endocrine and neural signals in turn synchronizes the timing of diverse peripheral oscillators that ultimately regulate local circadian physiological outputs.

Sympathetic component: (of the stress response) The "flight or fight response" that is mediated by the adrenal medulla and the sympathetic nervous system.

Temperature-dependent sex determination (TSD): The temperature of incubation that exists during the time of embryonic gonad formation biases the percentage of male or female offspring.

"Thrifty phenotype" hypothesis: Hypothesis for the development of type 2 diabetes mellitus, whereby fetal malnutrition programs develop, at least in part, by promoting hypoplasia (reduced cell division) of the endocrine pancreas, resulting in insulin hyposecretion by the pancreatic beta cells.

Thyroglobulin: A large extracellular thyroid hormone precursor protein to thyroid hormone that is sequestered in the colloid of thyroid follicles.

Thyrotoxicosis: Describes a generalized presence of elevated serum thyroid hormones independent of the source.

Thyrotropes: TSH-producing cells of the adenohypophysis.

Turner syndrome: The second most common form of sex chromosome DSD. Although the 45,X karyotype (monosomy X) is associated with about 50% of individuals with this condition, about 25% display a mosaic (45,X/46,XX) karyotype, with the remaining patients exhibiting structural abnormalities of the X chromosome.

Type 1 diabetes mellitus (T1DM): A failure of the pancreas to produce insulin, results from the destruction of the pancreatic β cells, often by autoimmune mechanisms.

Type 2 diabetes mellitus (T2DM): A heterogeneous collection of progressive metabolic disorders characterized by hyperglycemia, insulin resistance in the main target organs (liver, skeletal muscle, adipose tissue), excessive hepatic glucose synthesis, abnormal fat metabolism, and impaired insulin secretion.

Upregulating: An increase of receptor numbers on the cell surface, making the cell more responsive to subsequent exposure to the hormone.

Viviparity: This reproductive strategy refers to the delivery of "live" young that lack an eggshell and are able to interact directly with the external environment following a gestation period within the body of a parent. In contrast with ovoviparity, during this gestation period the embryo typically receives some significant form of direct nutrient transfer from the parent that supplements the yolk, often via a placenta or placenta-like structure.

White adipose tissue (WAT): Type of adipose tissue involved primarily in energy storage and release.

Zeitgeber: An environmental cue such as light or temperature that provides information about the external time.

Zona fasciculata: The middle and largest zone of the adrenal cortex that produces glucocorticoids.

Zona glomerulosa: The most superficial layer of the adrenal cortex that secretes primarily the mineralocorticoid aldosterone.

Zona reticularis: The innermost zone of the cortex that synthesize the androgen precursors, androstenedione, dehydroepiandrosterone (DHEA), and its sulfated derivative DHEA-S.

Index

Note: Page numbers in *italics* indicate a figure and page numbers in **bold** indicate a table on the corresponding page.

B

O

P

R

S